IRRIGATION ENGINEERING

[INCLUDING HYDROLOGY]

[For B.E./B.Tech. Students of Civil Engineering]

INCLUDING OBJECTIVE TYPE QUESTIONS

R.K. SHARMA
Consultant, Water Resources

T.K. SHARMA
B.E.

S Chand And Company Limited
(ISO 9001 Certified Company)

S Chand And Company Limited

(ISO 9001 Certified Company)

Head Office: Block B-1, House No. D-1, Ground Floor, Mohan Co-operative Industrial Estate, New Delhi – 110 044 | Phone: 011-66672000

Registered Office: A-27, 2nd Floor, Mohan Co-operative Industrial Estate, New Delhi – 110 044 Phone: 011-49731800

www.**schandpublishing.com**; e-mail: **info@schandpublishing.com**

Branches

Chennai	:	Ph: 23632120; chennai@schandpublishing.com
Guwahati	:	Ph: 2738811, 2735640; guwahati@schandpublishing.com
Hyderabad	:	Ph: 40186018; hyderabad@schandpublishing.com
Jalandhar	:	Ph: 4645630; jalandhar@schandpublishing.com
Kolkata	:	Ph: 23357458, 23353914; kolkata@schandpublishing.com
Lucknow	:	Ph: 4003633; lucknow@schandpublishing.com
Mumbai	:	Ph: 25000297; mumbai@schandpublishing.com
Patna	:	Ph: 2260011; patna@schandpublishing.com

First Edition 2002
Second Edition 2004; Third Edition 2007; Reprints 2008 (Twice), 2009, 2010, 2012, 2013, 2014
Reprint 2020

ISBN : 978-81-219-2128-2 **Product Code :** H3IRR46CIVE10ENAC07O

PRINTED IN INDIA

By Vikas Publishing House Private Limited, Plot 20/4, Site-IV, Industrial Area Sahibabad, Ghaziabad – 201 010 and Published by S Chand And Company Limited, A-27, 2nd Floor, Mohan Co-operative Industrial Estate, New Delhi – 110 044.

PREFACE

The first edition of this treatise on Irrigation Engineering duly subsidised by National Book Trust, Government of India, published in 1984, was highly acclaimed by the engineering teachers and taughts and its revised edition appeared in 1990 (through M/S Oxford & IBH Publishing Co. Pvt. Ltd., New Delhi). The dynamism inherent in the subject necessitated drastic changes in the text, prompted by the overwhelming response of irrigation and agriculture engineering students and practising engineers in the country and abroad duly patronised by the doyen of publications, Shri Ravindra Kumar Gupta, Managing Director, S. Chand & Company Ltd., New Delhi.

This new millenium edition is a thoroughly revised, updated version with a new orientation keeping in view latest developments on the subject and the changing scenario of developing Indian nation of over one billion beset with water scarcity vis-a-vis ever increasing demand for irrigation and water power, droughts and recurring floods. Accordingly the development of new means and methods of advanced irrigation techniques, better and economical water carrier channels, efficient hydraulic structure and storage dams to meet increasing water and power demands as also to avert recurring droughts and floods, enhancing quality of life of the nation, besides. The presently increased environmental sensitivity to these edifices of prosperity to the posterity have been given added importance to ensure right amount of water availability for particular use at the right time and of right quality with emphasis to foster the motivation for conservation of water and economy in its use for sustained development.

The special features of the book are :

- The only book of national character on the subject embodying the practices being followed in the various states of the Indian Union, and Central Water Commission. Standardisation by the Bureau of Indian Standards, unlike the contemporary books which are not up-to-date besides having only regional application.
- The Theoretical principles have been blended with field practices for better appreciation of the principles and practices of irrigation engineering by the engineering students.
- The text has been profusely illustrated with diagrams, tables and solved examples to help in comprehension of the principles involved.
- The maximum available details and latest advances on the subject have been incorporated.
- Incorporates essence of planning design and construction of one of world's highest Bhakra dam, Pong dam and world's largest Sardar Sarovar canal systems.

The book covers completely the syllabus of undergraduate and postgraduate students of the various engineering colleges, I.I.Ts, section B examination of Institution of Engineers (India). It is an indispensable reference book to the practising engineers and students of agriculture engineering as well.

In the preparation of this work, we owe a great deal to authors and publishers of various books, journals and publications of Central Board of Irrigation and Power and Director General, Bureau of Indian Standards, New Delhi as referred to in the text. For further details, references are invited to the standards which are available for sale from Bureau of Indian Standards, New Delhi and its regional and branch offices at Ahmedabad, Bangalore, Bhubaneswar, Mumbai, Chandigarh, Chennai, Hyderabad, Jaipur, Kolkata, Kanpur, Patna and Thiruvananthapuram.

We are indebted to Dr. P.V. Rao former Prof. and Head of Postgraduate Hydraulics and Irrigation Department, Punjab Engineering College, Chandigarh, Er. A.M. Singal Ex-Chief Engineer, Irrigation (Haryana), Er. G.D. Gupta, Chief Engineer, Bhakra dam, Dr. C.D. Thatte, Secretary General International Commission on Irrigation and Drainage, Er. P.L. Dewan, CMD, WAPCOS, Er. R.S. Goel, Director Environment, Central Water Commission, Er. K.V.S. Thakur, Director, Vigilance, BBMB, Chandigarh, and Er. R.K. Sharma, Consultant, Chandigarh for their inspiration and guidance in the finalisation of this work.

A word of praise for forebearance is also due to my wife Mrs. Janak Dulari Sharma and Siblings Ashish, Amit and Savita.

I must also place on record the deep sense of appreciation for the writing efforts put in by Shri R.S. Saxena, Shri P.J. Singh and computer section of S.Chand & Co. for the excellent layout of the text, diagrams and get up of the book.

Suggestions for the improvement of the book and pointing out mistakes and errors in the text will be gratefully acknowledged and incorporated in the next edition of the book .

We are grateful to **Mr. N.S. Elangovan, M.E., Asst Prof. in Civil Engg. Department Jerusalem Engg. College Chennai** for preparing the new Chapter on **Irrigation Water Management** which is added as Chapter 22 in this book.

Authors

Anna University, Tamil Nadu

B.E. Syllabus Civil Engineering Course- EB 130

OBJECTIVE

At the end of the semester, the student shall understand the need and mode of irrigation. The student also shall know the irrigation management practices of the past, present and future. The structures involved, the elementary hydraulic design of different structures and the concepts of maintenance shall also form part. Finally, the student shall be in a position to conceive and plan any type of irrigation project.

1. INTRODUCTION 12

Irrigation – Need and mode of irrigation – Merits and demerits of irrigation – Crop and crop seasons – consumptive use of water – Duty – Factors affecting duty – Irrigation efficiencies – Planning and Development of irrigation projects.

2. IRRIGATION METHODS 10

Canal Irrigation – Left irrigation – Tank irrigation – Flooding methods – Merits and demerits – *Sprinkler* irrigation – Drip irrigation.

3. DIVERSION AND IMPOUNDING STRUCTURES 14

Weirs – elementary profile of a weir – weirs on pervious foundations – Types of impounding structures – Tanks, Sluices and Weirs – Gravity dams – Earth dams – Arch dams – Spillways – Factors affecting location and type of dams – Forces on a dam – Hydraulic design of dams.

4. CANAL IRRIGATION 14

Alignment of canals – Classification of canals –Canal drops –Hydraulic design of drops – Cross drainage works – Hydraulic design of cross drainage works – Canal Head works – Canal regulators – River Training works.

5. IRRIGATION WATER MANAGEMENT 10

Needs of optimum use of water – Minimising irrigation water losses – On farm development works – Percolation ponds – Participatory irrigation management – Water users associations – Changing paradigms in water management – Performance evaluation.

TEXT BOOKS

1. **Sharma R.K. and Sharma T.K.– Irrigation Engineering (Including Hydrology) S. Chand & Co. Ltd.**
2. Asawa, G.L. "Irrigation Engineering", New Age International Publishers
3. Gupta, B.L. & Amir Gupta, "Irrigation Engineering", Satya Prakashan, New Delhi

REFERENCES

1. Dilip Kumar Majumdar, "Irrigation Water Management (Principles & Practices)", Prentice Hall *of India (P)*, Ltd.
2. Basak, N.N., "Irrigation Engineering", Tata McGraw-Hill Publishing Co.
3. Garg, S.K., "Irrigation Engineering"

CONTENTS

PART I : IRRIGATION PRINCIPLES AND PRACTICES

PART II : RIVER ENGINEERING

PART III : CANAL STRUCTURES

PART IV : HYDROLOGY

PART V : DAM ENGINEERING

PART - I

IRRIGATION PRINCIPLES AND PRACTICES

- Introduction
- Irrigation Water Requirement and Crops
- Irrigation Systems
- Methods of Irrigation
- Unlined Canals
- Lined Canals
- Drainage Engineering

1

INTRODUCTION

1.1 INTRODUCTION

Irrigation is defined as the artificial application of water to arid land for growing crops. It is a profession as well as a science. A crop requires certain amount of water at certain fixed intervals throughout its period of growth. Irrigation is not required if this requirement is met with from sufficient rainfall as in England. In tropical countries like India the first two of three essential requirements of plant growth, that is, heat and light are available in abundance, but the third, that is, moisture needs to be supplemented frequently by artificial application of water. Thus, irrigation is supplementary to rainfall when it is either deficient or comes irregularly or at unreasonable times.

The objective of an irrigation engineer is to make water available to the cultivators with respect to location, time and quantity as per the crop requirements. He has essentially to be associated with investigation, planning, design, construction, operation and maintenance of dams and their appurtenant works; canal system and hydraulic structures connected therewith, river training including flood control, drains, tubewells, watershed management, sedimentation measurement, soil conservation, etc. The irrigation engineering is thus a multi-disciplinary science encompassing hydrology, agriculture, geology, climatology, river engineering, agronomy, forestry, social science, hydraulics, river soil mechanics, snow hydrology and ground water hydrology.

With the ever increasing demand for irrigation, drinking water and power generation etc, the development of new means and methods of irrigation, advanced irrigation techniques, better and economical water carrier channels and efficient hydraulic structures, cutting down transmission losses and flow breakdowns have attained more relevance as the country has entered into a new millennium which essentially is an era of conservation, eco-friendly and sustainable water management and development.

1.2 NEED FOR IRRIGATION

As already mentioned, unless rainfall is sufficient to meet the water requirement of crops throughout their period of growth, need for irrigation water is unavoidable. The necessity for irrigation may thus be summarised as under:

1. *Deficient rainfall.* When rainfall is less than 100 cm, irrigation water is essentially required. Rainfall deficiency vis-a-vis irrigation requirement by crops can be briefed as follows:

Rainfall (cm)	*Irrigation requirement*
100	Rainfall needs to be supplemented by irrigation.
100-50	Rainfall is insufficient. Irrigation is essential.
50-25	Irrigation is essentially required.
Less than 25	No crop can be grown without irrigation.

2. *Non-uniformity of rainfall.* Where rainfall is sufficient but is not uniform, concentrated as it usually is in monsoon months, there is acute requirement of irrigation in other periods.

3. *Augmentation of crop yields.* New high yielding varieties of crops have higher water requirement for giving higher yields. Sugarcane and rice have higher requirement of water.

4. *Exacting water requirement.* The high yielding varieties of crops have more exacting requirement of water.

5. *Cash crops cultivation.* Cash crops require higher and assured supply of water with frequent waterings for maturity.

6. *Assured water supply.* For successful farming, availability of water in needed quantum and at right times is very essential.

7. *Orchards and gardens.* Fruit trees grown in orchards and gardens have higher requirement of water.

1.3 ADVANTAGES AND DISADVANTAGES OF IRRIGATION

Advantages

I. Direct Benefits. These include (*i*) Increase in food output through higher yield to attain self-sufficiency in food, (*ii*) Cultivation of cash crops, (*iii*) With the introduction of irrigation land value appreciates manifold which makes wealthy the land holders, the State and the country, (*iv*) Growing of fruit trees and development of gardens, (*v*) Protection from famine-irrigation makes agriculture and economy drought proof, (*vi*) Prevention of damage through floods, (*vii*) Domestic water supply to towns and villages. Important cities like Delhi, Jaipur, Bikaner and Chandigarh depend on canal water for public water supply, (*viii*) Hydel power generation at dam site and canal falls, (*ix*) The rise in subsoil water level in dry areas assists in meeting demands of domestic water supply by pumping the ground water, (*x*) Means of communication where navigation is possible in the canal. Network of canals with inspection bank/boundary roads along them improve the communication in remote command areas, (*xi*) Revenue from recreation facilities such as boating, fishing and swimming, (*xii*) Fish and wild life preservation and development of pisciculture. (*xiii*) Afforestation - plantation is raised along the banks of canals and field boundaries, (*xiv*) Irrigation substantially lowers production risks and farmers are greatly encouraged to raise productivity through input intensification, (*xv*) Makes agriculture competitive and profitable, (*xvi*) Industrial and thermal plant requirements are met with from canal water, (*xvii*) The reduced risk of crop failures and increased food production, (*xviii*) Improve the nutrition of the people and cattle considerably which leads to increased resistance to diseases and hence improved health.

II. Indirect Benefits. These are (*i*) Increase in gross domestic product of the country, (*ii*) Increase in revenue from sales tax on foodgrains, (*iii*) Increase in employment. Retards migration to cities for livelihood, (*iv*) Improvement in groundwater storage, (*v*) Increased revenue to government from other departments such as custom, excise, posts and telegraph, railways, taxes on vehicles etc., (*vi*) Increase in value of land property, (*vii*) General development of country, (*viii*) Farm labourers are benefited who get higher wages, (*ix*) Creation of more jobs and incomes, and (*x*) Rise to whole array of agro-based industries.

Disadvantages. (*i*) Climate becomes damp and cold, causing malarial diseases, (*ii*) Over-irrigation coupled with poor drainage in an areas where water-table is high leads to waterlogging of the area; and causes efflorescence. Crop yield is drastically reduced as a result, (*iii*) Low land revenue in certain cases where irrigation is extended as a protective measure, and (*iv*) Excessive seepage from unlined canals leads to waterlogging of lands adjacent to canals.

1.4 ENVIRONMENTAL EFFECTS OF IRRIGATION

Irrigation is to agriculture as blood is to the human body. Introduction of irrigation results in changes in vegetation, the fauna and the flora thereby altering the ecology of the command area These improvements have added advantage of a chain reaction in many spheres which lead to a more prosperous life for the people of the area. Lands which were once barren, infertile and sandy hum with activity with green pastures, verdant forests and teeming population. The human environment too changes as a result thereof. Development of irrigation is alone effective in combating the natural disasters like droughts and the single most promising short and long term means of reducing poverty and generating employment. It helps removing disparity in living standard of the population in rainfed and irrigated areas; the per capita income in the latter being three to four times that of the former. Irrigation has brought about stability in agricultural production and has made significant contribution in the attainment of self sufficiency in food and hence agricultural independence. All reforms mooted in agriculture will not bear fruit if irrigation facilities are not there to fructify them. Irrigation is an instrument of national integration by developing backward and barren areas and thereby bringing them into national main stream. The long distance mass transfer of water stored in man-made lakes virtually amount to transportation of snow and rainfall from Himalayas to water starved areas like Thar desert of Rajasthan thereby removing regional imbalance. The development of irrigation to meet the ever increasing food requirements of the country has, however, certain adverse environmental effects. With the greater emphasis on multiple cropping, the intensity of irrigation in most of the irrigation systems is increasing substantially which can lead to waterlogging and salinity. However, irrigation development by itself is not responsible for drainage related problems. Lack of adequate drainage provision, improper water management, seepage from canals, obstruction to natural drainage on account of various developmental activities have gradually resulted in waterlogging and soil salinity/alkalinity in some command areas. Proper drainage of the command areas is a *sine qua non* to mitigate the adverse environmental effects of irrigation.

1.5 HISTORY OF IRRIGATION DEVELOPMENT IN INDIA

Irrigation is a very ancient science. Irrigation has been practised in India from time immemorial and so has been the construction of canals. The Vedas are replete with references to wells, tanks, canals and dams. Samritis too contain evidence of early irrigation works. Construction of tanks and wells were considered pious and holy acts. Irrigation from shallow wells was, as it is now, traditionally at the initiative of the individual effort, while that from the rivers and streams was a community or State enterprise. The ancient rulers of India took keen interest in the provision of irrigation facilities. The early irrigation works were not primitive but had scientific basis. The most outstanding example of engineering talent in ancient time is manifest in the bold conception and construction of grand anicut across river Kaveri in the second century A.D. The medieval rulers too continued the tradition of construction of irrigation canals. Ferozeshah Tughlaq, in the fourteenth century, took off a number of canals from the Satluj and Yamuna rivers. Mughals too constructed a few canals here and there during their reign. Akbar in the sixteenth century renovated the abandoned Satluj canal. Shahjahan remodelled the Hissar canal, besides he constructed some new canals. When the British conquered the country, there were numerous canals in the upper areas in the North and scores of inundation canals along the Indus system of rivers, numerous small and large tanks in Central and Southern India. However, the total area under irrigation was relatively small. The British too interested themselves in irrigation development. They started irrigation development in the nineteenth century. To begin with, they remodelled the existing canals. Later they constructed a number of diversion canals on various rivers such as Satluj, Ravi, Ganga Krishna, Godavari and Kaveri. They also constructed dams such as Periyar and Mettur; the Nizamsagar and Krishnaraja-sagar were constructed by the princes in their native States.

The British introduced a definite irrigation policy in 1854 with the setting up of Public Works Department and instituting a separate fund for irrigation works. Prior to 1854, whatever irrigation works had been constructed were financed from the general revenues and no separate account for the revenue and expenditure was kept. After 1854, two categories of irrigation works, namely Minor works and Major works came into existence. Minor works which were undertaken in principle more for the sake of protecting the existing cultivation and revenue from retrogression than as revenue producing works continued to be financed out of the general revenues. Later minor irrigation works generally included private works (particularly their renovation) and private irrigation works formed a major constituent of Grow More Food Campaign launched. Public tubewells were also included in the category of minor irrigation works when the department of agriculture was constituted. Major works were henceforth financed by raising public loans. Each major work was required to satisfy the productivity criterion. Later the Famine Commission (1880) and the First Irrigation Commission (1928) laid great stress on encouraging private works (wells, tanks, etc.) to overcome recurrent famines. During the last 150 years, eight severe famines have occurred, the last one was in Bengal as late as in 1943. In tropical and subtropical countries like India, famines occurred due to drought conditions. Agriculture was a gamble in most parts of the country due to uncertainty of rains which even in normal times were inadequate. Table 1.1 indicates the progressive increase in the irrigation in the country in the pre-independence period:

Table 1.1 Pre-independence development of irrigation

Year	*Area (million ha)*	
1800	0.8	Undivided India
1900	13.3	
1920-21	19.3	
1945	23.5	
1947	19.4	Independent India

Planned Irrigation Development

The partition of India (in 1947) in the wake of independence, dealt a severe blow to the agricultural economy of that part of the country which is now called India. India was left with 80 percent of the pre-partition population, but most of the irrigated area on which the country was largely dependent for cereals, fibres and oil seeds went to the share of Pakistan. Drought and war torn economy of India caused a massive deficit in foodgrains. At independence attaining self sufficiency in food was given paramount importance and accordingly development of irrigation received a great fillip to meet the needs of the growing population. Thus the planned development of irrigation was set in motion in 1950-51. Irrigation schemes were divided into three categories viz., Major schemes, Medium schemes and Minor schemes. The total irrigated area stood at 22.6 million ha in 1950-51, 9.7 million ha from major and medium irrigation projects and 12.9 million ha from minor irrigation projects. As a result of massive investment made in the successive plan periods, the total irrigation potential created has been increased four-fold. Total irrigation potential at the end of the Eighth Plan stands at 92 million ha. The net irrigated area is 37 per cent of net sown area and 29 per cent of total culturable area. The ultimate irrigation potential is 113.5 million ha. Table 1.2 shows the development of irrigation potential during the successive plan periods since 1950-51.

Table 1.2. Development of irrigation potential since 1950-51

Period	*Irrigation potential created-cumulative (million ha)*					*Potential utilised*
	Major and medium	*Minor*			*Total*	
		Ground	*Surface*	*Total*		
Pre-plan (1950-51)	9.70	6.50	6.40	12.90	22.60	22.60
End of First Plan (1955-56)	12.20	7.63	6.43	14.06	26.26	25.04
End of Second Plan (1960-61)	14.33	8.30	6.45	14.75	29.08	27.80
End of Third Plan (1965-66)	16.57	10.52	6.48	17.00	33.57	32.17
End of Annual Plan (1968-69)	18.10	12.50	6.50	19.00	37.10	35.75
End of Fourth Plan (1973-74)	20.70	16.50	7.00	23.50	44.20	42.19
End of Fifth Plan (1977-78)	24.72	19.80	7.50	27.30	52.02	48.46
Annual Plan (1978-80)	26.61	22.00	8.00	30.00	56.61	52.64
Sixth Plan (1980-85)	27.70	27.82	9.70	37.52	65.22	58.82
Seventh Plan (1985-90)	29.92	35.62	10.99	46.61	16.53	68.59
Annual Plan (1990-92)	30.74	38.89	11.46	50.35	81.09	72.86
End of Eighth Plan (1996-97)	33.82	45.73	12.20	57.97	91.79	82.02
Ultimate feasible (Provisional)	58.50	40.00	15.00	55.00	113.50	

There has been progressive development of groundwater structures since the planned era. In the country as a whole, about one-third of the ultimate potential of groundwater development has so far been utilised. Special efforts are required to speed up groundwater development. Latest estimate of annual replenishable groundwater is 43.2 million ha m, out of which 36 M ha m is available for irrigation, corresponding to an irrigation potential of 80 million ha, double of the earlier estimated groundwater potential of 40 million ha. Presently about 34 million ha are irrigated from groundwater resources.

The significant and important projects undertaken and completed since the start of the planning era are indicated in Table 1.3:

Table 1.3 Important irrigation projects completed since 1950-51

River system	*Projects completed*
Indus System	Bhakra Nangal project, Beas project, Indira Gandhi canal project (distribution system under construction), Ranjit Sagar project.
Ganga River System	Ramganga project, Sarda sahayak project, Kosi project, Matatila, Gandhisagar, Ranapratap Sagar, Jawahar Sagar, Kosi and Kangsabati projects.
Krishna River System	Kadana, Ukai, Hirakud, Jayakwadi and Nagarjunasagar, Tungabhadra, Malaprabha, Ghataprabha and Upper Krishna project.

From less than 300 large dams existing at the beginning of planned development, the number of dams constructed (including about 700 dams under construction) has gone to over 4,300.

1.6 PERSPECTIVE IRRIGATION DEVELOPMENT

1. National Water Policy

The country has adopted a National Water Policy in 1987, which recognised water as one of the crucial elements in development planning. The policy document, inter alia, lays down that planning and development of water need to be governed by the national perspective.

The need for national water policy is that water as a resource is one and indivisible. Rainfall, river waters, surface ponds and lakes and groundwater are all parts of one system; water is also a part of a larger ecological system. Common policies and strategies need to be adopted for planning and implementation of individual irrigation or multipurpose projects, though done at the State level, involve a number of aspects and issues such as environmental protection, rehabilitation of project-affected people and livestocks, public health consequences of water impoundment, dam safety etc. Hence the need to develop, conserve, utilise and manage this important resource by national perspectives on an integrated and environmentally sound basis keeping in view the needs of the State concerned. The essence of national water policy are:

(*i*) *Maximising availability*: Water resources availability be brought within the category of utilisable resources to the maximum possible extent. The resource should be conserved and availability augmented by suitable measures for maximising retention and minimising losses.

(*ii*) *Resource planning*: The resource planning should be done for a hydrological unit such as a drainage basin as a whole, or for a sub-basin.

(*iii*) *Appropriate organisation*: Appropriate organisation of special multidisciplinary units should be established in the States for the planned development and management of a river basin as a whole taking into account not only needs of irrigation but also harmonising various other water uses.

(*iv*) *National perspective*: Water should be made available to water deficit areas by transfer from other areas including transfers from one river basin to another based on a national perspective.

(*v*) *Recycling and re-use*: Recycling and re-use of water should be an integral part of water resources development.

(*vi*) *Project planning*: Water resources development projects should be planned and developed as multi-purpose projects with provision of drinking water as a primary consideration. The impact of a project during construction and later on human lives, settlements, occupations, economics and other aspects should form an essential component of project planning such as quality of environment and ecological balance, catchment treatment and management, rehabilitation of affected people and command area development. Time and cost overruns and deficient realisation of benefits should be overcome by upgrading the quality of project preparation and management.

(*vii*) *Maintenance and modernisation*: Structures and systems created should be properly maintained, monitored and modernisation programmes undertaken.

(*viii*) *Safety of structures*: Organisational arrangements at the National and State levels for ensuring the safety of storage dams and other water related structures be made.

(*ix*) *Groundwater development*: There should be a periodical reassessment on a scientific basis of the groundwater potential taking into consideration the quality of the water available and economic viability. Exploitation should not exceed the recharging possibilities as also to ensure social equity.

(*x*) *Conjunctive water use*: Integrated and co-ordinated development of surface water and groundwater and their conjunctive use should be envisaged right from the project planning stage and should form an integral part of the project.

(*xi*) *Water allocation priorities*: In planning and operation of systems, water allocation priorities should broadly be drinking water, irrigation, hydro-power, navigation, industrial and other uses.

(*xii*) *Water rates*: Water rates should be such as to convey the scarcity value of the resource to the users and to foster motivation to economy in water use.

(*xiii*) *Public participation*: Participation of farmers and voluntary organisations in various aspects of management of irrigation system particularly in water distribution and collection of water rates be encouraged.

(*xiv*) *Water quality*: Both surface water and groundwater should be regularly monitored for quality.

(*xv*) *Water zoning*: There should be water zoning of the country and the economic activities should be guided and regulated in accordance with such planning.

The Water Policy is being reviewed and updated. It includes inter-alia adequate safeguards against over exploitation of groundwater.

2. National Perspective Plan

The National Water Plan has underlined the need for making water available to water short areas by transfer from other areas including transfer from one basin to another, based on national perspective. The Ministry of Water Resources has prepared a National Perspective Plan for developing water resources, one of the greatest water development projects of the world, envisages 18 Mha m of storage which along with the interlinks will facilitate additional utilisation of about 24 M ha m of water for beneficial use in the country. This would give additional benefits of 35 M ha of irrigation and generation of 34,000 MW of hydropower, apart from the benefits of flood control, navigation, etc. The plan comprises two major components, namely Himalayan and Peninsular rivers development :

1. *Himalayan rivers development.* It envisages construction of storage reservoirs on the main Ganga and Brahmaputra and their principal tributaries in India and Nepal along with interlinking canal system to transfer surplus flows of the eastern tributaries of the Ganga to the west, apart from linking of the main Brahmaputra with the Ganga. It shall provide irrigation to an additional area of about 22 million ha and generate about 30,000 MW of hydropower, besides substantial flood control in the Ganga-Brahmaputra basin and benefit Uttar Pradesh, Bihar, West Bengal, and Madhya Pradesh besides helping Nepal and Bangladesh. It will also provide 1130 cumecs of water to Calcutta port.

2. *Peninsular rivers development.* It envisages providing additional benefit of irrigation to about 10 million ha by surface water, 5 million ha by groundwater and generation of 4000 MW of hydropower. It comprises four major parts described below.

(*a*) Interlinking of Mahanadi-Godavari-Krishna-Pennar-Kaveri rivers. It envisages the diversion of the surplus water of Mahanadi to Godavari and transfer of water from the Godavari and its tributaries to the Krishna basin. This will enable irrigation in the drought-prone areas of Maharashtra,

Karnataka, Andhra pradesh and Tamil Nadu, besides providing benefits of power generation to Orissa and Madhya Pradesh.

(*b*) Interlinking of west flowing rivers, north of Mumbai and south of Tapi. It envisages construction of storage dam in Kerala and diversion of a part of the waters of west flowing rivers of Kerala to the east for irrigation of drought-prone areas of Tamil Nadu apart from providing additional irrigation and water supply to Kerala.

(*c*) Interlinking of southern tributaries of Yamuna like the Ken with Chambal. It envisages interlinking of southern tributaries of the Yamuna like the Ken and Chambal in addition to construction of dams on small intermediate tributaries to provide irrigation benefits in Ujjain-Indore region of Madhya Pradesh as well as upper areas in Rajasthan, besides benefiting the areas in Ken, Dhasan, Betwa and other basins in Uttar Pradesh and Madhya Pradesh.

(*d*) Diversion of west flowing rivers. It envisages construction of storage and interlinking small rivers flowing along the west coast north of Mumbai and south of Tapi. It will not only help in meeting the drinking and industrial needs of Metropolitan Mumbai but also in providing irrigation facilities to the coastal areas in Maharashtra and drought-prone areas in Saurashtra and Kutch regions of Gujarat.

1.7 CLASSIFICATION OF IRRIGATION PROJECTS

The irrigation projects are classified as under:

1. Classification Based on Source of Water
   (*i*) *Direct Irrigation Project*: In this project, water is directly diverted from the river into the canal by the construction of a diversion weir or barrage across the river with some pondage to take care of diurnal variations as also to raise water level to feed the offtaking channel.
   (*ii*) *Indirect Irrigation Project*. In this project, water is stored in a reservoir with the construction of a dam across the river or stream to provide dependable supply into the offtaking canals.
2. Classification Based on Purpose Served
   (*i*) *Single Purpose Irrigation Project*: A single purpose irrigation project, as the name implies, is meant for a particular purpose *i.e.*, irrigation. The pre-independence irrigation projects were mostly single purpose, but the necessity to derive the maximum benefits from the surface water resources inevitably led to the concept of multipurpose projects.
   (*ii*) *Multipurpose Irrigation Project*. It is a project which is constructed for two or more purposes such as irrigation, flood control, power generation, navigation, domestic and industrial water supply etc. etc.
3. Classification Based on Administrative Convenience

Class of project	*Description of works*	*C.C.A. (ha)*
Major project	Multipurpose river valley projects, barrages, weirs etc.	Over 10,000
Medium project	Low earthen dams and canal irrigation schemes	10,000–2,000
Minor project	Extension of irrigation, tubewell and well irrigation, lift irrigation	Less than 2,000

4. Classification Based on Financial Return
   (*i*) *Productive Irrigation Projects*: Productive irrigation projects are those the benefit-cost ratio in respect of which is not less than 1.5.
   (*ii*) *Unproductive Irrigation Projects*: Unproductive irrigation projects are those the benefit-cost ratio in respect of which is less than 1.5.
   (*iii*) *Protective Irrigation Projects*: Protective irrigation projects are unproductive but taken up as for protection against famine and are financed from famine relief fund.

5. Classification Based on Physical Criterion

   (*i*) *Minor Irrigation Schemes*: All groundwater schemes, whether they consist of open wells, deep or shallow tubewells. However, where a battery of tubewells is set up to supplement water of an existing canal, or to feed a new canal, the scheme is classified as Major, or Medium, depending on its size.

   (*ii*) *Surface Water Schemes*: These envisage a net irrigation not more than 1600 ha, whether they consist of small irrigation, tanks, small canals or lift schemes.

1.8 PLANNING FOR IRRIGATION PROEJCTS

The planning for the development of water resources is made to meet the ever increasing, demand for irrigation, hydropower, drinking water etc. besides other objectives such as (*i*) national economic development, (*ii*) national integration, (*iii*) quality of environments improvement, (*iv*) regional welfare, (*v*) priority and quick return, (*vi*) optimum utilisation, (*vii*) better water management, (*viii*) integrated water development, and (*ix*) employment generation. National water policy has underlined that planning for water resources project has to be done for hydrological unit such as the drainage basin as a whole or for sub-basin by the States (Fig. 1.1).

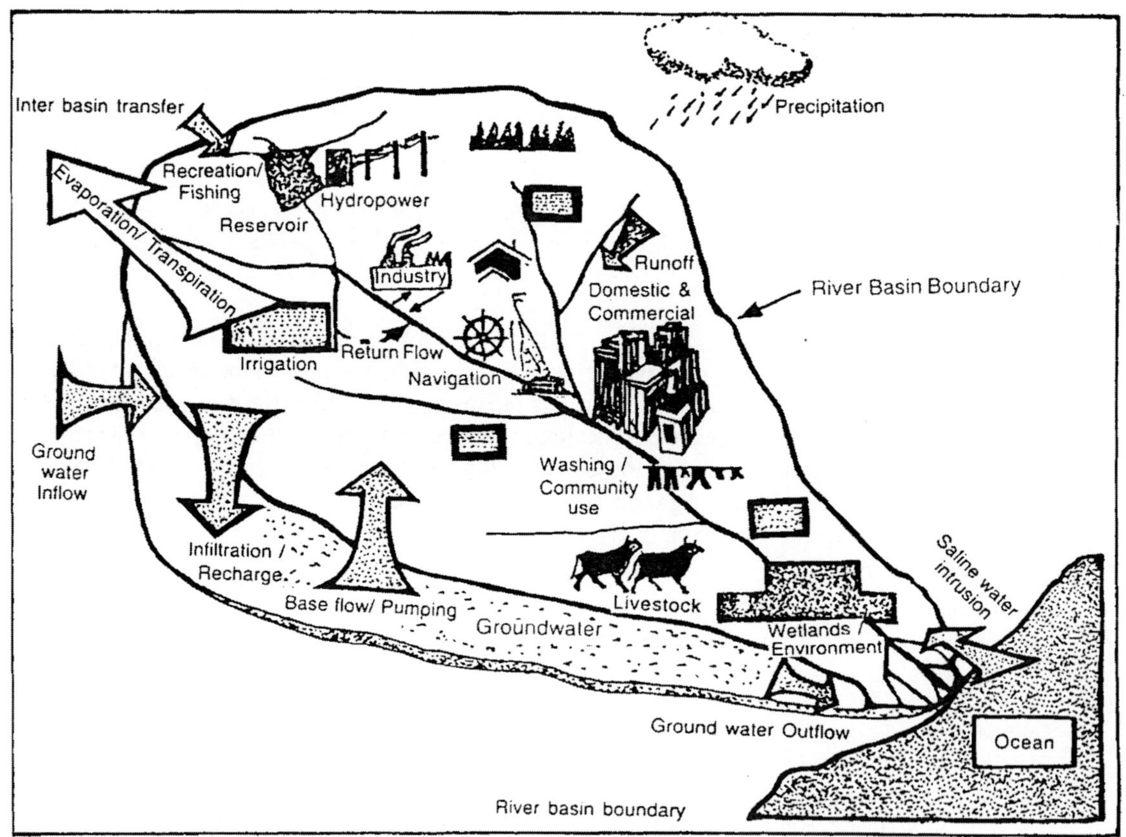

Fig. 1.1 River basin schematic
(*Source*: World Bank publication)

The tools which are available to policy makers include a number of *hardware* (*e.g.* construction-related) and *software* (*e.g.* policy-related) options, both on the supply-side and demand-side as shown in Table 1.4.

Table 1.4 Tools for water resources planning and management

	Supply-side	*Demand-side*
"Hardware" Options	• Dams/canals • Inter-basin transfers • Desalination plants • Pumps	• Conveyance and use efficiency • Conservation • Recycling • Process changes
"Software" Options	• Water (Reallocation/negotiations/agreements/water markets) • Subsidies • Reservoir releases/pumping patterns • Institutions	• Pricing/subsidies • Cropping patterns • Re-allocation/water markets • Institutions/public awareness

Planning for the development of water resources is very complex in view of the competing demands for utilisation. It is thus imperative that latest modern techniques of study and evaluation by adopting systems approach to the development of water resources are resorted to. In planning, development and management of water resources it is essential that realistic appraisal of current and anticipated demands for water by various uses is made. The demand for irrigation is expected to vary considerably is space and time depending on type of crops that may develop, needs of the country, responses of the cultivators to the present and of the future and also the future technological developments in agriculture. The water requirement for irrigation is better determined scientifically. Most of the project reports are prepared on the basis of standard duty/delta concept. Of the various formulae available, modified Penman method gives the best results for determining net irrigation requirement. The percolation and pre-sowing requirements are added. Cutoff period of each crop is determined as no irrigation water is required to be given towards the maturity of crops. On the basis of these parameters, net irrigation requirement is determined for different months and 10 daily periods during lean seasons. The gross irrigation requirement is determined by taking into account field application (85% for paddy and 70% for non-paddy) and distribution efficiencies (unlined canal 60%, lined canal 75%) of the system. The overall efficiency is taken as 63.75% and 52.5% for ponding crops and non-ponding crops respectively. The gross irrigation requirement at canal head is net irrigation requirement as modified/increased by the overall efficiency as indicated above. Generally, groundwater should be preferred to be used for Rabi, and hot weather crops, as the river supplies as well as regulated releases from storage will not be adequate during that period. It is imperative that integrated operation of water resources development systems including conjunctive use of groundwater is made. The groundwater can be utilised to supplement surface irrigation water in tail areas of canal system, when canal supply is short and when there is peak demand for palewa. A proper balance between the surface supplies and pumping from the groundwater is maintained at economical depth.

Water resources development covers a large field encompassing many scientific and engineering disciplines such as meteorology, hydrology, hydraulic engineering including foundation engineering, dam engineering, agricultural engineering and agronomy. Besides such a wide interaction of disciplines and vivid purposes, its two other important aspects are environmental and ecological protection. Ecological impact of large storage reservoirs as also the rehabilitation of the population uprooted due to submergence of land in the reservoirs, forest submergence are necessarily considered. Planning for suitable soil conservation structures and measures upstream of reservoir should form an integral part of any irrigation system; and they should be taken up as integrated one on watershed basis. Likewise, development of waterlogging, soil salinity and alkalinity, for which although irrigation is not responsible being drainage related problems, and broadly drainage should also form integral part of planning for an irrigation project. Command area development be essentially associated to ensure proper water management, land levelling, field channel development etc. Proper maintenance of the irrigation system has to be planned to ensure equitable distribution of water in head and tail reaches.

Surface irrigation systems should be designed to irrigate compact blocks, the blocks being dispersed over a large area to a large number of farmers. In areas with abundance of both surface and groundwater, the latter should be exploited for local use and, if feasible, surface water transferred to low rainfall zones.

Apart from the determination of source, availability, quality and requirement of water, the other essential components of planning an irrigation project are land classification, legal title to water, economic aspects, environmental aspects, canal distribution network design, and that of the dam and appurtenant works, organisational set up including operation and maintenance. Typical organisational set up is indicated in Fig. 1.2.

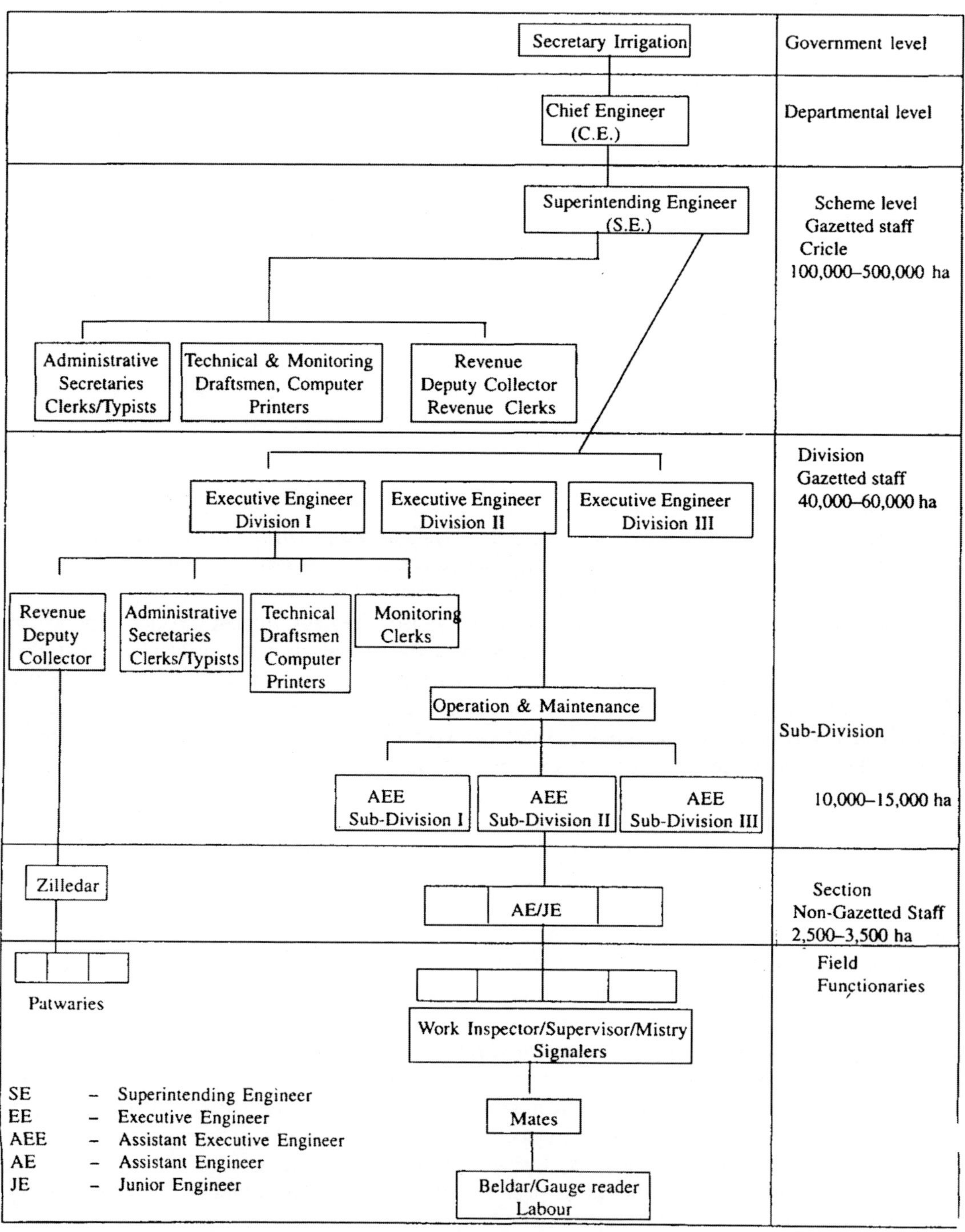

Fig. 1.2 General organisational pattern of an irrigation department
(*Source*: World Bank Publication)

The project formulation is based on the collected basic data and projections for the future water availability (at least 75% dependability) and demands, various alternatives and selecting the optimum one taking into consideration overall aspects of cost, benefits, time value of money represented through an appropriate discount rate and environmental aspects.

1.9 ECONOMIC EVALUATION OF IRRIGATION PROJECTS

The economic viability of a project stipulates that the discounted benefits, accruing as a result of project development should exceed the discounted cost. It is done by either estimating net present value (NPV) of the project *i.e.,* present value of the benefit minus present value of cost or benefit cost ratio (BCR) or by determining the internal rate of return (IRR):

1. *Cost per Hectare Criterion*: The economic feasibility of a new irrigation project is judged with reference to norms of capital outlay per hectare of irrigation. The major limitation of this criterion is that it is based purely on the income derived from water rate levied and does not consider the role of project in improving the socio-economic conditions in the command area. The actual benefits accruing to the command area in monetary terms are thus not reflected by this criterion.

2. *Benefit-Cost Ratio Criterion*: It is the widely used criterion for assessing the economic feasibility of irrigation projects. Benefit-cost ratio means that the benefits which may accrue from the project are in excess of the estimated cost of construction. The Irrigation Commission, India (1972) adopted the benefit-cost ratio as a standard for the evaluation and ranking of irrigation projects. The commission has laid down benefit-cost ratio criterion of 1.5 for the irrigation projects located in normal rainfall areas and 1.0 for drought prone areas.

3. *Net Present Value (NPV)*: In the method, future benefits as well as costs are discounted at the appropriate social discount rate to determine their value. The overall benefit cost ratio (BCR) is then determined by dividing the present value of discounted benefits by present value of discounted cost.

4. *Internal Rate of Return Method*: The method envisages the basic technique of discounting costs and benefits accruing in different periods and expressing them all in a common value at any one point of time. If the net present value of project works out negative *i.e.,* if the discounted value of the benefits is less than the discounted value of the cost then the project is not fit for sanction. To take note of inflation and increase due to economic change over the long construction period, an appropriate indicator of price rise (*i.e.* an adjustment factor) is constructed and the increase so obtained is added to the estimate as a supplementary provision for adjusting cost estimates of plan projects.

1.10 WATER RESOURCES DEVELOPMENT

India is endowed with vast water and land resources. Average annual flow available in the rivers is about 4 per cent of global water supply. It is the largest country of the world with 2 per cent of the world's area and second most populous. The country is blessed with a large network of rivers comprising 12 major and 48 medium river basins having combined catchment area of about 277 M ha within the country. The average annual precipitation, including snowfall, as per estimate by Central Water Commission is about 400 million ha m (actual volume in 1993 rainfall 1% above normal was 393 M ha m), of which the monsoon rainfall (June-September) is around 300 M ha m. The average natural runoff available in the rivers is about 188 M ha m as per Central Water Commission estimation. Owing to topographic, hydrological and other constraints, the utilisable water resource is assessed by National Commission (1976) as 70 M ha m. The total groundwater, a dynamic resource, (an important source of irrigation and caters to more than 45 per cent of total irrigation in the country) is assessed by Central Water Commission as 85 M ha m, of which utilisable potential is 43.2 M ha m. The other sources of water are relatively insignificant quantities of lake waters, snow and ice in Himalayan mountains which constitute an important source for contribution to annual surface flows particularly in the form of snowmelt in hot weather season to enable round the year flows in the Indo-Gangetic rivers. Thus the average yearly utilisable water resources presently assessed are 113.2 M ha m.

4010 large dams in the country (1998) account for total reservoir capacity of 27 M ha m and live storage of about 17.7 M ham which is a milestone in the water resources development. Besides, dams to create additional storage capacity of 7.54 M ha m are under construction and 13.2 M ha m is expected through dams under planning stage. The ultimate capacity of surface storage in the context of National perspective plan implementation has been assessed as 37.5 M ha m.

The present utilisation of surface water and groundwater is about 70 per cent and 32 per cent respectively of the utilisable resources. The present pattern of utilisation and the projected utilisation (demand) by 2025 AD are as given in Table 1.5. Irrigation is the largest user of water accounting for 93 per cent of gross water and its use is increasing rapidly.

Table 1.5 Present and projected water utilisation (M ham)

Use	*Present Utilisation (M ham)*		*Projected utilisation*
	1997	*2000 AD (Projected)*	*2025 AD*
Irrigation	50.1	63	77
Domestic	3	3.3	5.2
Industrial	2	3	12
Hydropower	2	2.7	7.1
Others	3.4	3	3.7
Total	60.5	75	105
	Surface water	50	70
	Groundwater	25	35

Per Capita Water Availability

The average water availability in the country remains more or less fixed according to the natural hydrologic cycle but the per capita water available is reducing progressively due to increasing population. The per capita water availability in 1955 was 5300 cu m which has come down to 2200 cu m. However, the average availability in Brahmaputra basin is as high as 18,400 cu m while it is as low as 380 cu m in some of the east flowing rivers of Tamil Nadu as against the world average and Asian average of 7,420 cu m and 3,240 cu m respectively. With the projected population growth, urbanisation and industrialisation, the availability will dwindle to only 1500 cu m by 2025 AD as national average, (Fig. 1.3) while it will be much less in some of the regions. As per the United Nations criteria any situation of water availability of less than 1000 cu m/ per capita is considered as critical.

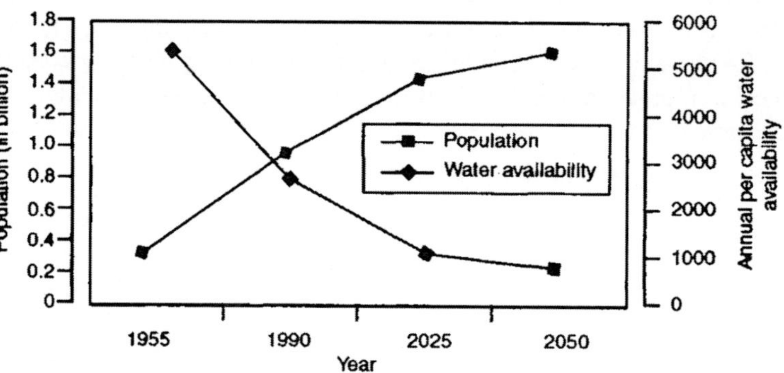

Fig. 1.3 Population and water availability trends in India (*Source*: Population Action International, 1995; World Bank, 1997)

The water use efficiency in Indian irrigation projects is about 45 percent. In case the water use efficiency is increased to 60 percent, in Israel it is 80 percent, the irrigated land could be doubled without developing any new water sources. FAO report indicates that our land and water resources are capable of sustaining a population of 2.6 billion people provided they are properly managed.

Land Resources

India is the seventh largest country in the world and covers a geographical area of 329 million hectares, of which the cultivable area is about 188 M ha which is the second highest in the world. The present cultivated area is 145 M ha, also called net sown area, and the corresponding gross* cropped area is 188 M ha, of which about 92 M ha (49%) is irrigated. Area sown more than once is 45 M ha. Out of 145 M ha, about 90 M ha is under rainfed agriculture. The gross sown area expected by 2025 AD is 210 M ha, taking into consideration the areas sown more than once. The national perspective plan envisages an additional irrigation benefit of 25 M ha from surface water and 10 M ha from groundwater. Thus when the available water resources are harnessed fully, it may be that the entire culturable area of 188 M ha may not be adequately served with irrigation facilities. Even in ultimate stage of irrigation development, rainfed areas may be covering two-third cropped area in the country.

Nearly 70 per cent of the cultivated area lies in the region of moderate to low rainfall wherein water deficiencies lead to drought conditions. One of the important and somewhat paradoxical feature of the land resources is that about 33 percent is drought prone and about 12.5 percent especially in the Indo-Gangetic and Brahmaputra basins are flood prone. Thus the spectra of drought and flood occur almost in the same year and even at the same period of the year.

The present population of the country is 1.0 billion which may increase to 1.4 billion by 2025AD and stabilise at 1.5 to 1.8 billion by 2050 AD. The total food supplies at the present level of consumption will need to be about 310 and 500 M tonnes by the year 2025 AD and 2050 AD respectively as against the present production of 200 M tonnes. The per capita availability of cultivated area which has declined from 0.33 ha in 1950 to 0.2 ha in 1980 and has reduced further to 0.15 ha by the year 2000 AD, against the world average of 0.27 ha.

Land degradation to the extent of 174 M ha, of which nearly two-third requires special treatment to restore to productive and protective use, is due to (*i*) deforestation and on account of loss of vegetative cover resulting in a low water table and loss of top soil, and (*ii*) environmental degradation due to misapplication of yield increasing inputs like water, chemical fertilisers and pesticides causing waterlogging and salinity.

Irrigation Potential

Irrigation potential is the aggregate area that can be irrigated annually by the amount of water that is made available at the field. Area under irrigation in 1947 was 19.5 M ha. The irrigation potential created by 1997 is 92 M ha, irrigation potential utilised is 82 M ha, the potential to be achieved by 2010 AD is 113.5 M ha; 58.5 M ha through major and medium schemes, 15 M ha and 40 M ha through minor surface water and minor groundwater schemes. Total irrigation potential is 115.6 M ha. On implementation of National perspective plan, the irrigation potential will increase to 150 M ha.

1.11 IRRIGATION WATER MANAGEMENT

Agriculture with two-third of production based on irrigation and accounting for 93 per cent of consumptive water use is critically dependent on water and evidently greater scope for economy in water use is possible in this field through effective management of water. India has one of the finest network of irrigation canal system. Even though India can be proud of its achievements in water resources development, we are still backward in efficient utilisation of water particularly at the farm level. The traditional irrigation system is mainly designed for protective irrigation and not adapted to the modern intensive agriculture based on high yielding varieties of crops and multiple cropping with increased use of fertilizers. There is considerable wastage of water through the existing irrigation systems. The old systems need modernisation by way of modifications, remodelling and addition of engineering works. The operation of the canal systems calls for radical changes so that effective control is exercised, co-operation of the cultivators sought and equitable supplies of water

*The term "Gross" accounts for cropping intensity over the same year which is 132 per cent

are resumed. The entire operation programmes need to be geared to the actual need of the crops taking into consideration the rainfall and the soil characteristics. Communication system needs to be improved. Linking of the channels should be provided where necessary. In short as irrigation potential as well as our water resources are limited, the applied technology need to strategically shift from irrigation development to better and efficient irrigation management for sustained agricultural development.

Water is a manageable input in agriculture and as such needs to be managed on scientific lines to conserve and to efficiently use country's water resources and avoid damage created by waterlogging and accumulation of harmful salts so that water, land and labour are used more efficiently for sustained benefits. Water management involves a large number of disciplines, including various facets of engineering; hydrology, soils, plant physiology, agronomy, water economics, water institutions etc. The need for better management of the available water resources is well recognised by planners and policy makers in the country and accordingly National Water Policy (1987) is being updated.

Objectives of Water Management

Scientific management of land and water use is basic to economic advance and survival. Rational use of water not only helps in higher production but also makes the benefits available to as many as possible, thus meeting the ends of social use. The principal objectives of water management are (*i*) Allocation and supply of irrigation water to the irrigators according to their entitlement at the right time for optimum yield of crops with the allocated water, (*ii*) Equitable distribution of supplies to irrigators especially at the tail, (*iii*) Achievement of optimum irrigation efficiencies commensurate with agricultural production, (*iv*) Ensured increased yield for a prolonged period without loss of fertility of land with time due to soil erosion and nutrients, waterlogging, salinity and/or deterioration of soil structure, that is protection of environment, (*v*) Economical and efficient operation and maintenance of irrigation system, and (*vi*) Realisation of sense of justice and equity among the irrigators by proper implementation of water management rules.

An efficient irrigation water management thus envisages integrated development of water resources, efficient methods of conveyance and distribution of water on the farm, judicious methods of water application, proper soil management practices and cropping patterns for high water use efficiency, scientific timing of irrigation according to the developmental rhythm of the plant and the removal of excess water. Efficient soil and water management can lessen or at times avoid the need for artificial drainage.

Components of Water Management

The various components of water management are as under:

1. *Watershed management.* It includes all such measures as are effective in preventing or delaying the movement of soil from the point of origin in the watershed to the reservoir. Afforestation in the catchment area is very effective to reduce the silt inflow into the reservoir besides other erosion control and soil conservation measures such as regrassing and grass land cultivation processes, gulley control, reservoir shore line protection, highway and railway line erosion control etc. Improving the erosion resistance through polymers may be considered when much technological advancement is made in this direction.

2. *Reservoir management.* The reservoir management is designed and operated to cater for the crop requirements during different parts of the year, in addition to flood control. The reservoir management encompasses reduction of losses due to (*i*) evaporation, (*ii*) seepage, and (*iii*) sedimentation.

3. *Conveyance system management.* In irrigation projects, in general only about 40 per cent of the water delivered at the canal head reaches the field, mainly due to excessive losses in canal

distribution systems. Water in conveyance from storage reservoir through canal system down to the farm suffers losses due to evaporation and seepage. The lining of canal system and water courses thus achieves the effective water management in conveyance system. Lining of water courses minimises weed infestation, helps in maintaining designed flow of water as also reduces maintenance cost.

4. *Land management.* The intensity of cropping that is raising productivity per unit area per unit time needs to be increased which amounts to increasing gross cropped area. The present intensity of cropping is projected to be raised to 150 per cent. In Ludhiana (Punjab) cropping intensity is 185 per cent.

5. *Rain water management.* The management of rain water through appropriate land shaping can optimise in situ moisture conservation as also help conserve excess water by storage for irrigation use.

6. *Water Management in Farms.* Water management in the farm underlines the importance of supply of water to the fields in required quantity and at the required time so that with the limited water resources it is used more efficiently. There is need to educate the irrigators in this regard together with the adoption of the following water management steps in the farm:

(*i*) *Land shaping and levelling.* Land shaping and levelling and bunding etc. of the fields is imperative to obtain uniform crop growth, higher irrigation efficiency and full benefits from irrigation.

(*ii*) *Right amount of irrigation.* Soil holds only limited amount of water in the root zone. Sandy soils hold less water than clayey soils. The amount of water to be applied shall be just sufficient to meet the water deficit created due to extraction by crop. Water applied in excess of soil water deficit is likely to be lost through deep percolation. The deep percolation losses can be reduced by the application of right amount of water at each irrigation.

(*iii*) *Rotational use of water.* Rotational use of water not only helps in higher production but also make the benefits available to as many as possible, thus meeting the ends of social justice.

(*iv*) *Irrigation methods.* Irrigation is an age old practice in India but the methods in vogue are generally inefficient. The surface irrigation methods suitable for various crops are as under:

Methods of irrigation	*Suitable for crops*
Border strip method	Wheat, leafy vegetables, barseem and other fodders etc.
Furrow method	Cotton, sugarcane, potatoes etc.
Basin method	Orchard trees.

Amongst the various methods to conserve the water, sprinkler and drip irrigation systems though little costly are commonly adopted all over the world. Almost all advanced countries have their major irrigation through these methods only.

(*v*) *Crop planning.* The cropping pattern need to be developed on scientific basis keeping in view the availability of water, soil types, and regional agro-climatic conditions. Crop varieties as give equivalent yield with less moisture and short duration be popularised. Scientific contribution in the form of double or multiple cropping can be achieved if the sowing of crops such as paddy, groundnut, arhar etc. is advanced, if necessary, by raising the nurseries with the help of groundwater. Crop rotation is so selected that optimum yields are obtained per unit of water applied. The average crop productivity is lower than average world productivity and needs to be raised through proper management.

(*vi*) *Irrigation scheduling.* The agriculture need to be based on climate-crop-soil relationship. The application of irrigation should be such that the available water in the soil above the permanent wilting point is fully utilised by the crop, before requiring recouping through irrigation for such depletion of soil moisture. Warabandi (*i.e.* turn schedule of water according to which water is made available to each farmer in the command area of an outlet for a specific period in proportion to the size of the holding and according to a turn schedule prepared in advance) is prepared and enforced for equitable distribution and efficient utilisation of irrigation water, particularly for the farmers at the tail end of the canal. Since the quantity of water allocated to each irrigator is thus fixed, the warabandi induces the farmer for making more efficient use of water to bring maximum area of his holding under irrigation with the allocated amount of water. Strict enforcement of schedules and vigilance against infringement is desirable and at the same time it is to be ensured that designated amount of water shall be made available at each outlet according to the specified schedule.

(*vii*) *Agronomic practices.* Agronomic practices such as plant density, seed rate, sowing time, planting method, weed control, insect and pest control etc. shall be at optimum levels for maximum production and efficient use of water.

(*viii*) *Irrigation practices.* The irrigation practices should ensure (*i*) optimum availability of moisture in root zone, (*ii*) minimise salinity built up in root zone, and (*iii*) avoid waterlogging and aeration problems. With a view to effect improvement in irrigation efficiency, presently as low as 45 per cent, there should be regular and systematic observation of irrigation efficiencies on almost every irrigation system of size and their regular evaluation done.

7. *Conjunctive use of water.* Conjunctive use of surface water and groundwater is an important approach for water management in irrigation systems. The conjunctive use deserves maximum emphasis not only to increase intensity and stability of irrigation but also to control waterlogging and salinisation and to reduce the requirements of drainage. Public tubewells and private tubewells and dug wells need to be encouraged. Conjunctive use system when developed with other water management practices will be the ultimate in the discipline of water management.

8. *User participation.* A management system with active user participation and control is the desirable improvement in running irrigation system. The irrigators need to be actively associated with the distribution and release of water at the outlet. It shall ensure willingness on the part of irrigators to effect the improvement in the use of water in the most efficient way. Sound water management methods should be demonstrated and introduced right from the time the irrigation commences. This will not only lead to economy in the use of water but will avoid the danger of waterlogging.

9. *Soil management.* Soil management practices aim at obtaining uniform distribution of irrigation water on the farm, storing large quantities of rain water within the crop zone and improving the soil structure for increased water availability. The most important soil management practices in relation to irrigation are land grading, land preparation and cropping practices. Fine levelling is considered to save as much as 15-20 per cent of water and with proper grade drainage can be effectively achieved.

10. *Command Area Development.* The CAD is another concept in water management to ensure efficient water utilisation and integrated area developments in the irrigation command including modernisation of the distribution system, provision of drainage and the maintenance and operation of both the distribution and drainage systems.

The CAD authority has functions and responsibilities such as (*i*) Modernisation, maintenance and efficient operation of the irrigation system upto the farm outlet, (*ii*) Development and maintenance of the main and intermediate drainage systems, (*iii*) Planning and executing programmes of land development within a catchment including realignment and lining of watercourses, land levelling, soil reclamation, farm drainage, field channels etc. through institutional finance and otherwise, (*iv*) Redrawing of field boundaries, consolidation of holdings, exchange of portion of holdings to organise the farms in the command for the best use of irrigation, (*v*) Enforcement of proper system of 'Warabandi' and fair distribution of water to individual fields, (*vi*) Selection and introduction of suitable cropping pattern, (*vii*) Development of groundwater to supplement surface irrigation, (*viii*) Supply of all inputs and services, including credit, (*ix*) Development of marketing and processing facilities and communications, (*x*) Preparing individual programmes for agricultural labour, (*xi*) Diversification of agriculture through livestock development, horticulture, farm forestry etc., (*xii*) Soil conservation, ravine control and forestry programmes, and (*xiii*) Town planning, village development, rural housing etc.

11. *Canal Automation*: Canal automation is an arrangement of electronic, electrical and mechanical components that commands or directs the regulation of canal system. It includes collection of data regarding canal operation, like water depth, gate position, flow, trasmitting it to the control element at the headquarter or control supervisor and trasmitting back signals for canal operation at the sensing location in order to adjust canal system. Elements of a simple feedback control system are illustrated in Fig. 1.4. In India, sardar Sarovar Project (Gujarat) will be covered under canal automation system.

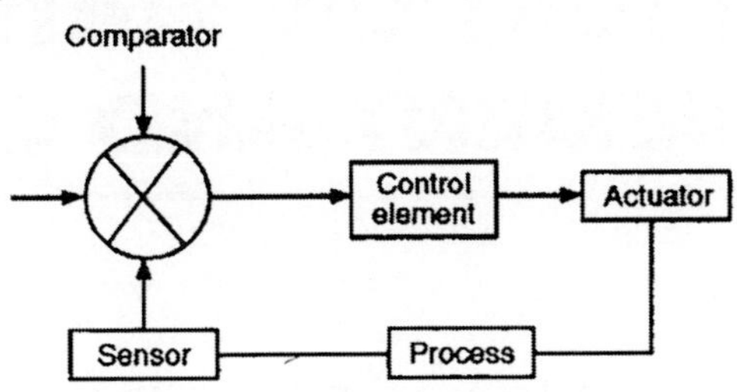

Fig. 1.4. Control system

QUESTIONS

1. What do you understand by the term irrigation? Explain the necessity of irrigation in a tropical country like India. Discuss the advantages and ill effects of irrigation.
2. Discuss briefly the history of irrigation development in India. What is the scope of development of irrigation in the country.
3. Irrigation is principal user of water, explain.
4. Independence has set in an era of irrigation development, discuss.
5. Discuss National water plan.
6. How do you classify irrigation projects.
7. Describe various criteria for financial soundness of an irrigation project. What are the factors involved in planning an irrigation project? Describe.
8. Discuss various environmental effects of irrigation.
9. Describe National water policy, briefly mention its salient features. Why is it being revised?
10. What do you understand by National perspective plan? How is it going to benefit the nation?
11. Discuss classification of irrigation projects. What is the measure to classify a scheme as productive or unproductive.
12. How will you describe planning for irrigation projects, its objectives and various parameters.
13. Water resources development encompasses many scientific and engineering disciplines, discuss.
14. Give an outline of typical irrigation operation and maintanance set up.
15. What do you understand by per capita water availability? Why is it dwindling in India? What is the norm for considering a situation as cirtreal.
16. Define irrigation water management and its objectives. Why is it said that a stage has reached for the country to shift from irrigation development to irrigation management.
17. What do you understand by canal automation.

2

IRRIGATION WATER REQUIREMENT AND CROPS

2.1 INTRODUCTION

The soil, water and plant are important natural resources very vital for the survival of the men and animals. These three resources need to be managed scientifically and efficiently for maximising food production to meet the requirements of ever increasing population.

Soil is the natural medium for the growth of plants. It is a natural mass of unconsolidated material on earth's surface, formed by physical and chemical disintegration of surface rock formation, *i.e.*, residual soils or by the action of water, wind or glacier. From the standpoint of agriculture, soil is simply the upper layer of surface which supports growth of crops under favourable conditions of temperature and moisture. Soil has three essential constituents namely solid particles, water and air in varying proportions briefed as under :

1. *Solid matter.* It mainly consists of inorganic particles derived from the parent rock, mineral material and a small proportion of organic matter.

2. *Air and water.* The solid matter is seldom closely packed to occupy the entire volume of solid mass. There are some open spaces, called voids, which are filled by either air or water or both in different proportions. The presence of air and moisture in the soil imparts many important properties to soil. Change in the proportion of moisture and air affects almost all properties of the soil. Besides these three basic components, soil contains living organisms like bacteria, fungi, algae, insects and animals which too affect soil structure and plant growth. They aerate the soil thereby creating favourable conditions for the growth of plant roots.

2.2 SOIL PROFILE

The soil consists of various layers different in colour and composition, known as horizons. The succession of horizons is called profile of the soil.

The development of soil as end product of the process of decay and decomposition of rocks proceeds through three recognizable stages. These stages appear as a series of zones in any section of the soil. These zones are layers approximately parallel to the soil surface. Such zones, arranged in descending order, indicate different stages in the development of soil which collectively constitute soil profile. The three zones, beginning from the surface downward, are discussed as below:

1. *Zone A.* It is a zone of leaching and decomposition in which the minerals and organic matter are decomposed and converted into colloids. It is characterized by finely divided particles extending from a few centimetres to 60 cm below the surface. It is the zone in which the grass-roots, etc., are embedded. It contains humus in moist region. The zone has maximum biological activity. The plough layer is generally 20-30 cm thick.

2. *Zone B.* It lies immediately below zone A. It is a zone of deposition. It extends from a few centimetres to 100 cm thickness. It varies from zone A in colour, structure, clay content and organic matter. Organic matter is low. The main size is rather coarse.

3. *Zone C.* It is layer of unconsolidated material like decomposed rock or freshly deposited alluvium. It is characterized by a coarse grained, rather pebbly texture, marking a transition between actual solid rock and the finally decomposed rock. Roots penetrate into this zone. It extends usually up to 2 m in thickness.

Individual soils may contain one or more of these zones but not necessarily all of them. Recently deposited alluvium may be of uniform texture and structure up to substantial depth.

2.3 PHYSICAL PROPERTIES OF SOILS

1. *Colour.* Colour is one of the most useful and important characteristics for soil identification. A great deal about the soil can be inferred from its colour. Colour of the soil depends upon its composition, drainage condition, porosity and age. Concentration of organic matter imparts a grey, black or dark-brown colour to the soil. Diffusion of iron oxides imparts red and yellow colour to the soil. Manganese dioxide and hydrated iron oxides may also contribute red colour. A light grey colour may indicate a very low content of organic matter and iron. Effective drainage tends to change these colours appreciably. Soil colour changes with moisture content.

2. *Texture.* The size of particles comprising the soil determines its texture. The texture of a soil is perhaps its most nearly permanent characteristic, although soil structure can be quickly modified by management. The texture commonly refers to the fineness or coarseness of soil as a whole as to the proportion of the particle groups of different sizes. It is controlled by climate and mineralogical composition of the parent rock. The texture of soils is commonly described by size limits discussed under "Particle size" in soil classification 2.4.

The three textural types of soils generally recognized are sandy soils ($\geq$ 50% sand particles), silty soils ($\geq$ 50% silt particles) and clayey soils ($\geq$ 50% clay particles). Loamy soil has mixed particles of sand, silt and clay.

In nature soil is seldom made up of one group. Depending on the clay content, soils are classified (Table 2.1) as (*i*) light, (*ii*) medium, and (*iii*) heavy.

Table 2.1 Soil classification vis-a-vis water requirement.

Soil type	*Percentage of clay*	*Water requirement for irrigation*	*Crops grown*
Light	2 to 10	Suitable for crops with less water requirements	Jowar, bajra, pulses, sesame, mustard, groundnut, fodder etc.
Normal or medium	10 to 20	Ordinary water requirement	Cotton, wheat, gram, maize, oil seeds, peas, vegetables etc.
Heavy	40 and above	Large quantity of water at regular intervals	Rice, sugarcane etc.

3. *Structure.* It signifies the arrangement of soil particles, sand, silt, clay, organic matter etc. in soil mass into porous compounds called aggregates as also to the arrangement of these aggregates separated as they are by pores and cracks. However, unlike texture, soil structure is not permanent but with cultivation practices the top soil becomes granular. The soil structure is important to soil productivity, soil permeability, root growth and soil genesis. The capability of any soil for growth of plants and its response to management depends as much on its structure as on its fertility. The common types of soil structure are granular, prismatic, blocky and massive. Fine grained soils have granular structure in which water entry is rapid with better seed germination and as such are desirable for irrigation as large voids provide space for circulation of air. In prismatic structure, the movement of water is mainly vertical and as a result the water supply to the roots is poor. In top soil, massive structure blocks the entry of water, seed germination is

difficult and aeration is poor. In blocky top soil, water entry is moderate. From the standpoint of agriculture, permeability of soil to water, air and roots provided by favourable soil structure is important. Favourable soil structure is the key to soil fertility.

4. *Porosity.* Porosity and void ratio are defined as under:

$$\text{Porosity} = \frac{\text{Volume of voids}}{\text{Total volume of soil mass including that of water and air}}$$

It is expressed as a percentage

$$\text{Void ratio} = \frac{\text{Volume of voids}}{\text{Volume of soil solids}}, \text{ expressed as a ratio.}$$

The porosity of soil varies with (*i*) structure of soil, (*ii*) size, shape and gradation of soil particles, and (*iii*) presence of organic matter.

Voids in the soil influence production through their effect on water holding capacity of the soil, movement of air, water and roots.

5. *Permeability.* It is an important quality of the soil that enables it to transmit water or air. It is defined as the velocity of flow under a unit hydraulic gradient. It depends on presence of minute openings in the soil as also on the size, shape and arrangement of such openings. Sandy soils of coarse texture have large pore spaces and hence high permeability. Clayey soils of fine texture have relatively small pore spaces and hence low permeability. Permeability is affected by water temperature, texture and structure of the soil.

Infiltration rate of soil is the velocity at which water seeps into it, measured by depth (mm) of the water layer that the soil absorbs in an hour. Infiltration rate is low if it is less than 15 mm/hr, medium if between 15-50 mm/hr and high in case it is more than 50 mm/hr.

6. *pH value of soils.* pH value or hydrogen ion concentration is a measure of the intensity of acidity or alkalinity of a soil. Its value ranges from 0 to 14, of which 7 is neutral in the sense of chemical reaction. Below 7 the soil is acidic, above 7 it is alkaline. Soil productivity increases as pH approaches neutrality. Most irrigation soils have pH ranging between 6.0 and 8.5. pH value vis-a-vis soil fertility is given in Table 2.2.

Table 2.2. pH value vis-a-vis soil fertility

pH value	Below 6	6.5-7.5	7.5-8.5	8.5-9.0	11 and above
Yield	Special treatment needed	Best	Normal	Declines	Infertile soil

7. *Depth of soil.* Sufficient depth of soil for the storage of irrigation water and penetration of roots is essential. Shallow soils require frequent irrigation for optimum growth of crops. Deep soils of medium texture and granular structure provide for (*i*) adequate storage of water, and (*ii*) root penetration and sustain relatively long periods between irrigation. Soils of arid regions are relatively deep compared to humid regions.

8. *Infiltration capacity.* Infiltration capacity of a soil is the maximum rate at which soil is capable of absorbing water in a given condition. Alternatively, infiltration is the downward flow of water from the surface into the soil. Infiltration capacity is expressed as cm per hour over an area. Larger the voids greater is the infiltration capacity. Vegetated soils have more infiltration capacity.

2.4 SOIL CLASSIFICATION

Soils are classified from the standpoint of agriculture to ascertain their suitability or otherwise for crops and their fertility (Fig. 2.1):

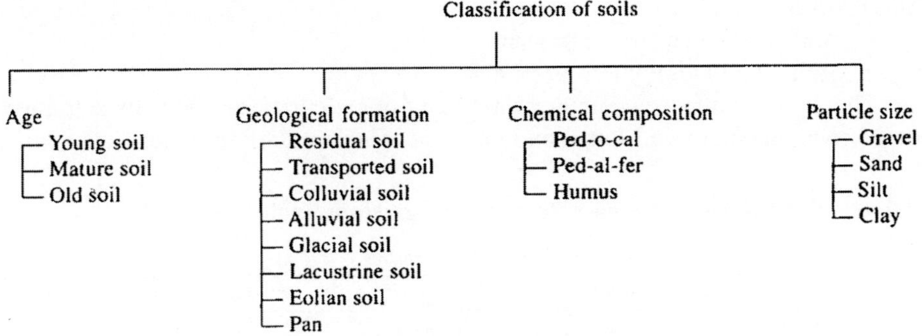

Fig. 2.1 Soil classification

Age of Soil

1. *Young soils.* Also termed as immature soils and are characterized by incomplete weathering and little leaching. They are fully pervious.

2. *Mature soils.* These are the soils generally met with. They show influence of climate and vegetation with pronounced effects of leaching and weathering. They have low permeability.

3. *Old soils.* Highly differentiated soils with fully developed soil profiles.

Geological Formation

1. *Residual soils.* Also termed as soils of degradation. These soils have suffered very little or no transport during or after their formation and are found more or less covering the parent rock. The thickness of such soils vary from place to place depending on climate, lithology and the extent to which soil forming processes have advanced.

2. *Transported soils.* Also termed as soils of aggradations. They comprise materials which have been moved from the place of their origin and redeposited during the weathering processes or during some phase of those processes, and consist of or are weathered from unconsolidated formations. These soils are of heterogeneous nature.

Transported soils are further classified as indicated in Fig. 2.1.

(*a*) *Colluvial soils.* They are poorly sorted material near the base of strong slopes that has been moved by gravity, frost action, soil creep and local wash.

(*b*) *Alluvial soils.* They consist of sediments deposited by streams. They are made by river floods and are found in flood plains. Such soils may occur in terraces above the present streams or in the normally flooded bottoms of the streams. They may be nearly homogeneous rock and soil material at the base of slopes and along small streams or alluvium of mixed origin along major stream courses. The flood plains are composed of alluvial soils, such as Indo-Gangetic alluvium.

(*c*) *Glacial soils.* These consist of material picked up, mixed, disintegrated, transported and deposited through the action of glacial ice or water resulting primarily from the melting of glaciers.

(*d*) *Lacustrine soils.* These are the deposits consisting of materials that have settled out of quiet waters of lakes. They show stratification. These soils are very fertile.

(*e*) *Eolian soils.* These are the soils deposited by wind and are composed of fine grained silt and clay fractions.

3. *Pan.* In soils rich in humus but devoid of free drainage, layer after layer accumulation of salts of calcium, magnesium, iron and other bases is called Pan.

(*a*) *Clay Pan.* It is the layer of formation which is relatively impervious and compact clay but is not cemented.

(*b*) *Hard Pan.* It is a hard formation layer and is cementitious. Water and roots cannot penetrate into this formation.

Chemical Composition

1. *Ped-o-Cal.* Soils rich in calcium carbonate.
2. *Ped-al-fer.* Soils rich in aluminium or iron salts.
3. *Humus.* The well decomposed complex mixture of organic materials in soils is termed as humus. It is dark brown in colour and forms an important constituent of agricultural top soil.

Particle Size

The classification of soil based on particle size is as given in Table 2.3.

Table 2.3 Soil classification

Soil type	*Average dia. of particles (mm)*	*Soil type*	*Average dia. of particles (mm)*
Gravel	60–2	Silt coarse	0.06–0.02
Sand coarse	2–0.60	Silt medium	0.02–0.006
Sand medium	0.6–0.20	Silt fine	0.006–0.002
Sand fine	0.2–0.06	Clay	less than 0.002

2.5 INDIAN SOILS

Soils of India have been formed over different geological periods from parent rocks of varied origin. They are derived from a wide variety of minerals.

Alluvial Soils

They are the transported soils carried away and deposited by the tributaries of Ganges and Brahmaputra system of rivers. These soils vary in character according to the nature of soils that occur in the region of their transportation. Generally they are deep soils. The Indo-Gangetic alluvium is the most productive soil and the largest of all soil groups in India and occupies thickly populated areas of the country. It covers Assam, West Bengal, Bihar, Maharashtra, Tamil Nadu, Orissa, Punjab, Haryana and Uttar Pradesh. Suitable crops in alluvial soils are wheat, maize, gram, rice, oil seeds, sugarcane, spiked millet, pulses, jute, tobacco, etc.

Black Soils

These are clay soils varying from clay to loam with clay content 40 to 50 per cent. These are formed by decomposition of rocks by long continued weathering. Their thickness vary from 1 to 4 m or more. These soils vary in colour from dark grey, black or blue black. Black soils are fine grained containing high proportions of carbonates of calcium and magnesium. They possess a high water holding capacity but suffer from poor drainage and consequent waterlogging. They are sticky when wet and undergo considerable contraction on drying. Thus swelling and shrinkage with increase or decrease is moisture content is characteristic of black soils. Black soils are distinguished into shallow (30 cm or less), medium (30–100 cm) and deep (over 100 cm).

Black soils are found in Andhra Pradesh, Maharashtra, western parts of Madhya Pradesh, Tamil Nadu, Gujarat and Karnataka.

The principal crops grown in such soils are cotton, sugarcane, groundnut, great millet, spiked millet, wheat, maize, tobacco, etc.

Red Soils

These are red in colour due to the presence of iron oxide. The colour varies from red to black with intermediate shade of dark brown. These are formed due to decomposition of granites and gneisses and rocks rich in iron and magnesium bearing minerals. Red soils are mostly sandy loam or sandy clay in texture. They are deficient in nitrogen, phosphorus, lime and organic matter. They are responsive to farming practices, manures, fertilizers and irrigation.

Red soils are found in Tamil Nadu, Karnataka, Maharashtra, Madhya Pradesh, Orissa and Andhra Pradesh.

The principal crops are rice, maize, groundnut, tobacco, pulses, millets, barley, oats, etc.

Laterite Soils

These are formed as a result of decay of many types of rocks. They are red in colour. They are deficient in colour in nitrogen, phosphorus, potassium and lime. These soils are developed on the summits of hills of Karnataka, Kerala, Madhya Pradesh, Orissa and Assam. They are poor in crop production.

The principal crops are barley, rice, great millets, coffee, tea, potato, fruits, etc.

Forest and Hill Soils

These soils, typical of forests and mountains, occur along the slopes or in depressions and valleys in forested areas. They are very shallow, steep and stoney. They show high content of organic matter and nitrogen and are developed in Himachal Pradesh, Punjab and Tamil Nadu.

Terai Soils

These soils are deposited as a result of their movement through water erosion of the Himalayan ranges. They lie at the foot of the Himalayan range. These soils are developed all along the foot hills in northern parts of Uttar Pradesh, Bihar and West Bengal. They are deep and moderately fertile soils but become highly productive with the provision of drainage.

Desert Soils

These are sandy soils occurring in low rainfall areas of Gujarat, Rajasthan and parts of Punjab. They are abundant in soluble salts but low in organic matter. They are productive when irrigated.

The principal crops are barley, coconut, millets, oils seeds, etc.

Saline and Alkaline Soils

Saline soils contain toxic concentration of salts in root zone. Electrical conductivity is taken as a measure of salts. If the solution extracted from the saturated paste of a soil has an electrical conductivity of more than 4 mmhos/cm, the soil is termed as saline. High salt content in soil results in reduction in yield. In alkali soils exchangeable sodium constitutes more than 15 per cent of total exchangeable cations. pH value of these soils is usually more than 8.5.

2.6 FUNCTIONS OF IRRIGATION SOILS

The essential chemical elements to crops for growth are provided through the medium of soil. They move from soil to crops as ions and molecules in a water solution. The soil ideal for optimum crop production is a rich loam soil high in organic matter, having a gentle slope, maintained constantly moist by quality irrigation water or rainfall with favourable temperature for the crop growth.

The functions of an irrigation soil for crop growth are thus: (*i*) Adequate moisture holding capacity to meet water requirements of crops, (*ii*) Provide proper circulation of air to a suitable depth for the development of root system of plants, (*iii*) Free from harmful concentration of soluble salts and parasites, (*iv*) Promote growth of bacteria which make the soil richer in organic matter for crop growth, (*v*) Supply nutrients for crop yield, (*vi*) Resistant to soil erosion or soil depletion under the cropping system, (*vii*) Suitable for agricultural implements, and (*viii*) Suitable for providing anchor for plant roots.

2.7 MAINTAINING SOIL FERTILITY

Soil is a store of nutrients for crop growth. It supplies 13 out of 16 elements needed continuously in balanced proportions for the nutrition of higher plants, in addition to mechanical support, sunshine, adequate oxygen around plant roots, optimum amount of soil water and non-existence of toxic substances to plants such as soluble aluminium, excess of soluble salts etc. The five factors of soil formation *i.e.*, parent material, climate, topography, bio-sphere and time determine the kind of soil existing at any one time and place. All plants depend on food produced by them through photosynthesis, as such balanced nutrition through fertilizer addition to plants is essential. The nutrients in the soil get consumed if the same type of crop is grown repeatedly. Soil fertility can be maintained by the methods as follow :

1. *Rest to the land.* Sufficient rest to the soil is necessary for recuperation after harvesting the crop and before the next crop is grown.

2. *Additions of manures.* Manures and fertilizers supply deficient plant nutrient in the soil.

3. *Crop rotation.* Crop pattern or crop rotation is so selected that optimum yields are obtained per unit of water applied. Crop rotation implies changing the crops to be grown in the land every year for the three reasons, (*a*) Since different crops require different nutrient and in different proportions, by changing the crop every year the same type of nutrient is not used up from the soil every year, (*b*) Crop rotation keeps the soil free from certain soil diseases. If the same crop is grown every year, the crop may impart some disease associated with it to the soil. Thus crop rotation controls insect pests, weed growth and plant diseases, and (*c*) Different crops have different depths of root zone. By growing different crops, some of which may have shallow roots while the other deep roots, optimum utilization of the available moisture of the soil is made. Proper crop rotation helps maintain soil structure, improve tilth, increase moisture holding capacity of the soil, maintain adequate supply of organic matter and nitrogen, and improve quality of crop products.

The usual crop rotation is (*i*) Wheat-great millet-gram, (*ii*) Rice-gram, (*iii*) Cotton-wheat-gram, (*iv*) Cotton-wheat-sugarcane, and (*v*) Cotton-great millet-gram.

Gram appears in each rotation because gram being a leguminous crop gives nitrogen to the soil to enrich it.

2.8 SOIL-WATER-PLANT RELATIONSHIP

Water is essential to crop plants for their growth and development. It is a basic input influencing crop production. Water plays several vital roles in plant life. The amount of water required for a given crop depends on (*i*) State of development of soil, (*ii*) Quantity and type of fertilizer given, (*iii*) Quality of water used, and (*iv*) The climatic conditions. Soil-water-plant relationship is thus a process that requires to be regulated for maximization of yields with a given unit of water. An understanding of soil-water-plant relationship is essential in order that water management principles are applied to various climatic, soil and cropping regions of both rainfed and irrigated lands. Several soil and plant factors influence the storage and movement of water in the soil-plant system.

Soil Factors

Infiltration. The downward flow of water from the surface into the soil is termed as infiltration. Infiltration rates influence selection of irrigation methods, slope needed for the land, length of run, irrigation application time in relation to depth of water required to replenish root zone of systems with slope, etc. Infiltration rates also influence runoff and thus water storage in the soil for rainfed conditions. Infiltration rates range from excessive to negligible, reflecting physical characteristics of the soil profile from natural processes and soil, crop and water management practices.

The soil and water parameters affecting infiltration rates are (*i*) *Texture.* Per cent sand, silt and clay. Deep sands or soils low in clay and colloid may have rather high infiltration rate even when near saturation because swelling is not appreciable, (*ii*) *Structure.* Degree of soil particle aggregation. Deep soil silt loams that are highly aggregated and have large pores show the highest rates of infiltration, (*iii*) *Bulk density.* Weight of soil per unit volume of soil. Higher the bulk density of soil lower is the infiltration rate, (*iv*) *Sodium salts.* High concentration of sodium salts in soil or irrigation water have a harmful effect on physical conditions of soil and reduces infiltration, (*v*) *Crop grown.* Inter-tilled orchards have a lower infiltration compared to orchards in sod. Beneath inter-tilled crops, viz. corn, cotton, potatoes, infiltration is considerably less than beneath grass, trees. The gain in infiltration due to change in vegetation is more quantitative on deep permeable soils than on shallow tight soils, (*vi*) *Irrigation water.* Fine silt and clay particles in the irrigation water also influence infiltration rate. Fines in the flowing surface water fill the pores in the strata below thereby forming a relatively impervious seal resulting in low infiltration, (*vii*) *Temperature.* Soil or water temperature and initial soil water content may change the rate of infiltration, (*viii*) *Tillage.* Tillage and incorporation of crop residues into the soil increase infiltration, and (*ix*) *Water in soil.* The amount of available water in the soil is dependent mainly on the percentage of sand, silt, and clay of the soil. Other factors influencing the amount of available water are bulk density and structure of the soil, soil temperature and salts in soil water. Generally, the more clay or silt, the greater is the available water. Depths of available water may range from 6 to 25 cm of water per metre depth of soil in sand loam and clay respectively.

Permeability. Permeability is the property of soil to transmit water. Permeability is also termed as hydraulic conductivity which is defined as the effective flow velocity at unit hydraulic gradient and has the dimensions of velocity. It is defined as velocity of flow under a unit hydraulic gradient in which driving force is one kg/kg of water. Permeability depends on (*i*) soil texture and structure, (*ii*) presence of plant roots, reactions of base exchange depending on the salts present in the soil, and (*iii*) changes in temperature of water. The coefficient of permeability K is given by the relation, $K = QL/A(H_1-H_2)$, where, Q = discharge/unit time, A = cross sectional area through which water flows, H_1-H_2 = hydraulic head, and L = percolation path length. Values of K (cm/s) are as under:

Soil	Gravel (clean)	Coarse sand (clean)	Sand (mixture)	Fine sand	Silty sand	Silt	Clay
K	1.0 or more	1.0–0.01	0.01–0.005	0.05–0.001	0.002–0.0001	0.0005–0.00001	0.00001 or <

A good irrigation soil should be permeable. Sandy soils of coarse texture have a high permeability. Soils having medium texture are sufficiently permeable. Clayey soils of fine texture are less permeable. On permeability depends the available water holding capacity of the soil, depth to which roots can penetrate and drainage of excess water through underground drainage system. Permeability, infiltration, and other characteristics such as depth, texture, structure, nature of subsoil, etc., determine the suitability or otherwise of a soil for irrigation. Permeability is measured by permeameters, as under:

1. *Constant Head Permeameter.* It is used for determining permeability of coarse and medium textured soils. Soil of cross sectional area A is filled in a metallic cylinder (Fig. 2.2) to a depth L. Water is stored over the soil to a constant level with the help of a feeder and an overflow pipe. Under the constant head h a steady flow exists through the soil. Permeability is calculated from the relation: $K = QL/A\,h$ (m/hr)
where, Q = discharge (m^3/hr), h = loss of head (m).

2. *Variable Head Permeameter.* It is used for determining permeability of fine textured compact soils of low permeability. The permeameter shown in Fig. 2.3 consists of a cylinder with conical top through which a glass tube (of small bore) of diameter at least 10 times less than the diameter of the cylinder passes. A small drop of water level in the cylinder is represented by a substantial drop in the glass tube. Soil is filled in the permeameter to a thickness L cm. Water is

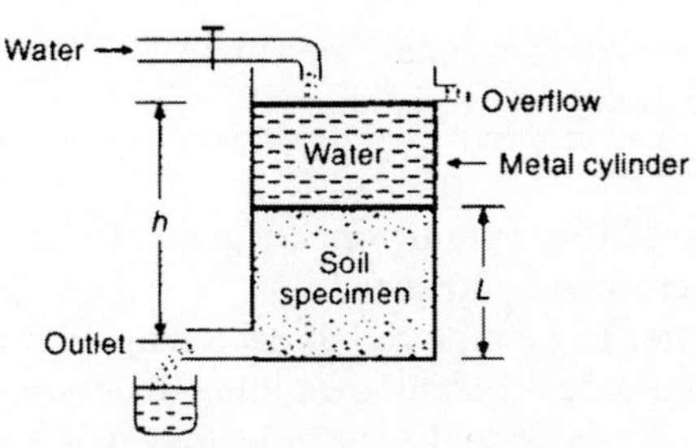

Fig. 2.2 Constant head permeameter.

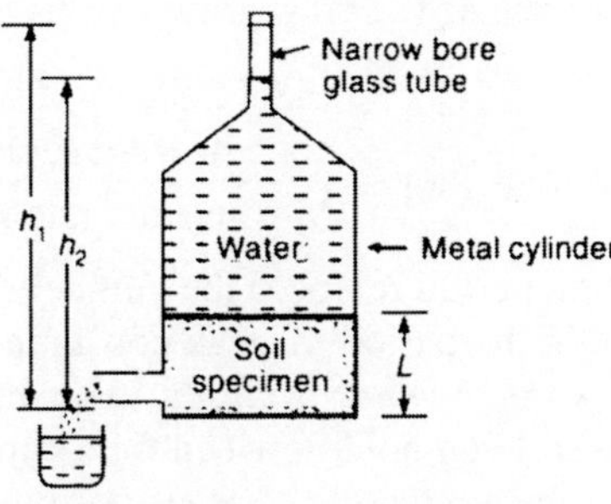

Fig. 2.3 Variable head permeameter

stored over the soil till it reaches a height h_1 cm above the bottom of the cylinder in the glass tube. The water level falls to h_2 cm after an interval of t hours. Permeability is calculated from the formula:

$$K = \frac{2.303\,aL}{AT} \log_{10} \frac{h_1}{h_2}$$

where, K = permeability (cm/*hr*), a = cross sectional area of glass tube (cm^2), A = cross sectional area of cylinder (cm^2), L = flow length (cm), t = time interval (hours), h_1 = initial water level (cm), and h_2 = final water level (cm).

Drainability and Leachability: It is the property of soil to drain off excess water easily without raising the watertable within the root zone reach. The principal factors in predicting drainability of a soil is its permeability (hydraulic conductivity) and hydraulic gradient. Leachability is the removal of soluble salts by passing good quality water through the soil. For most soils it is directly related to drainability.

Erodibility. Irrigation is associated with erosion of soil if stream flow is excessive.

Plant Factors

1. *Rooting characteristics.* The distribution and activity of roots determine how much water is extracted from various depths of the soil profile. Plants may have (*i*) tap roots that penetrate deeply into the soil under favourable conditions, (*ii*) shallow primary roots, and (*iii*) lateral roots. Root extent and distribution is altered by the physical, chemical and biological properties of the soil. A high watertable limits the root growth due to lack of sufficient aeration. Farming practices also influence rooting characteristics.

2. *Evapo-transpiration.* The water removed from the soil profile by evapo-transpiration depends on the distribution and activity of roots within the soil profile. A general rule is that 40, 30, 20 and 10 per cent of the total evapo-transpiration is removed respectively from each successively deeper one-quarter of the rooting depth. (Fig. 2.7) Seasonal evapo-transpiration values of crops are needed to evaluate and determine seasonal irrigation water supplies. Peak-period evapo-transpiration rates are needed in designing irrigation systems for peak needs.

3. *Effect of soil water level on crop growth and yield.* Water between field capacity and permanent wilting point is not equally available for crop growth. Crop growth and transpiration generally decreases as the wilting point approaches. The point at which growth or transpiration of a plant is retarded for want of soil water is influenced by distribution and depth of root system, conductivity of soil for water, crop characteristics, low or high evaporative demand, etc.

Water Factors

1. *When to irrigate.* Irrigation frequency means the time, in days, between irrigations during periods of no rainfall. Generally irrigation shall start when 50 per cent, but not over 60 per cent of the available moisture is used from the root zone. Irrigation frequency used for the design of irrigation system refers to the number of days between two irrigations in the period of highest consumptive use of the crop grown.

$$\text{Design frequency} = \frac{\text{Field capacity of soil in effective crop root zone} - \text{moisture content of the same zone at the starting of irrigation}}{\text{Moisture use rate of crop in peak period}}$$

Irrigation period refers to the number of days allowed for applying one irrigation to the given design area in the period of peak consumptive use of crop being irrigated.

2. *How much water to apply.* The amount of water to be replaced at each irrigation is the amount the soil can hold between field capacity and the water content level immediately prior to irrigation or the amount that has evapo-transpired by the crop since the last irrigation. The amount

to be replaced is usually 40 to 50 per cent of the available water in the root zone of soils having a uniform available water capacity with depth.

3. *Water application method.* The method of applying irrigation water is influenced by the quantity of available water supply, type of soil, topography and crops to be grown. Irrigation water may be applied by controlled surface flooding method, sprinkler method and drip method. Whatever be the method of irrigation, the essential requirement in water use is the application of the right amount of water and its uniform distribution in the field so as to wet the root zone to its storage capacity.

2.9 SOIL WATER

Soil water is defined as suspended water in the uppermost belt of soil of zone of aeration lying near enough to surface to be discharged into atmosphere by transpiration of plants or by evaporation from the soil. It includes gravitational water, capillary water and hygroscopic water (Fig. 2.4).

1. *Gravitational water.* It is the water in the unsaturated zone in excess of hygroscopic and capillary water which moves out of the soil under favourable drainage conditions.

2. *Capillary water.* It is the water held by surface tension in the capillary spaces and as a continuous film around the particles, free to move under the influence of capillary forces and available to plants.

3. *Hygroscopic water.* It is the wa*er held in static state with the atmospheric water vapour. Hygroscopic coefficient is the percentage of moisture the dry soil will take up from the saturated atmosphere in order to be in equilibrium with atmospheric saturated water vapour at that particular temperature. Hygroscopic water is not capable of any movement by gravity or capillary forces.

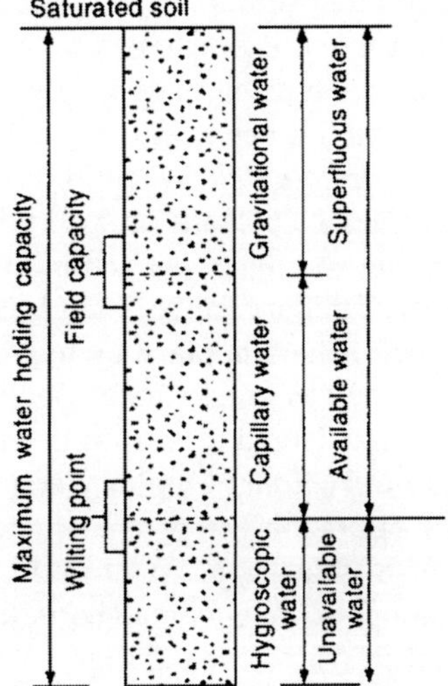

Fig. 2.4 Classification of soil moisture.

Water is also classified as unavailable, available and superfluous water (Fig. 2.4) on the basis of availability of soil water to crops.

4. *Unavailable water.* It refers to hygroscopic water which is not available to crops due to its inability to move by gravity or capillary forces. It is the soil moisture held so firmly by molecular forces that it cannot ordinarily be absorbed by plant roots with sufficient rapidity to produce growth.

5. *Available water.* It refers to the capillary water which readily contributes to the plant roots. Plants wilt if the capillary water is used up. The available water in the capillary water zone is limited up to a permanent wilting point and accordingly it forms the line of demarcation between the capillary water and hygroscopic water.

6. *Superfluous water.* It refers to the gravitational water which drains down so deeper that plant roots cannot draw it.

2.10 SUBSURFACE WATER

The water that enters soil is termed as subsurface water. The subsurface water occurs between the ground surface and the lower limit of water bearing rock formations, called groundwater. The subsurface water is divided into two zones, (*i*) Zone of aeration and (*ii*) zone of saturation.

1. Zone of aeration

It is very important from agriculture point of view. It comprises three belts (Fig. 2.5) discussed as under :

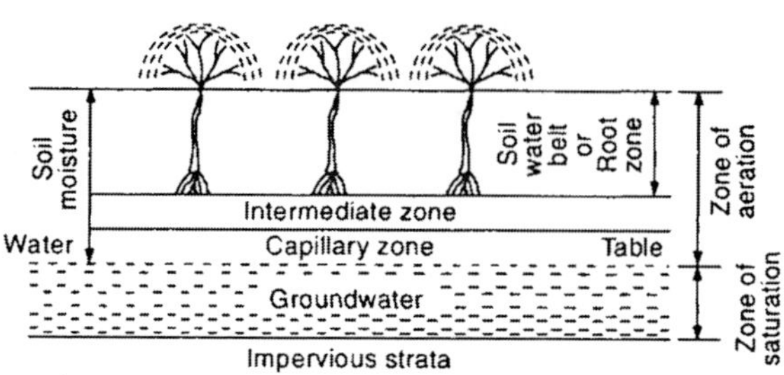

Fig. 2.5 Soil moisture.

Soil Water Belt: It refers to the zone of water adjacent to the ground surface consisting of top soil and the subsoil. The depth of soil water, depending on the type of soil and vegetation, varies from a few to several metres. Water in the zone moves downward by infiltration and upward by evaporation and transpiration by the plants. Water from the belt is evaporated to the atmosphere and is also used by plants through evapo-transpiration. This belt is a source of supply to all vegetative growth and is of importance to agriculture. Water passed down this belt goes beyond the reach of root zone and is not available for plant growth.

Intermediate Belt: Water in its continued downward movement enters from the soil water into the intermediate belt and as such this belt serves as passage for the downward movement of water through the soil under the action of gravity. The thickness of the belt may vary from almost zero (when the capillary fringe extends into the soil) to several metres. Non-moving or pellicular water in this belt is held in place by hygroscopic and capillary forces. Temporarily water moves downward this belt as gravitational water. The water in this belt, known as suspended water or vadose water, is considered dead storage and is not available for use.

Capillary Zone: Capillary zone is sandwiched between the intermediate belt at the top and zone of saturation below and provides passage for the water moving downward under gravity to the zone of saturation. Water in this zone is held above the zone of saturation by capillary force. The amount of water contained in the zone and the thickness of the fringe depends on the type of material in which the capillary fringe is located. Thickness of the zone may vary from a few to 60 cm or more depending on the type of soil. The capillary rise is 20-50 cm in coarse sand, 50-80 cm in medium sand and over 80 cm to several metres in fine clay.

2. Zone of saturation

It extends from water table at the top to the lower limit where rock formation is impermeable. Water table is defined as the locus of point in the soil water system at which the water has atmospheric pressure. At certain locations, the zone of saturation is overlain by unsaturated material that contains an impervious formation above which a local zone of saturation called perched watertable exists. All pores and spaces in the zone of saturation are filled with water thereby forming a huge natural reservoir from which springs, streams and wells draw relatively uniform supply and therein lies the importance of the zone of saturation or groundwater.

2.11 SOIL-MOISTURE-IRRIGATION RELATIONSHIP

Certain definitions in the context of soil-moisture-irrigation relationship are as follows:

Saturation capacity. The soil whose all soil pores are filled with water and there is no air left in it is termed as saturated soil. Most crops except rice cannot withstand saturated soil condition for more than 2-5 days. Saturation capacity, termed as saturation point or maximum moisture holding capacity, is the amount of water required to fill all the pore spaces between the soil particles by replacing the entire air held in pore spaces. It is the upper limit of possible moisture content. It is expressed as equivalent cm of water per metre of soil depth. Thus if the porosity of a soil is 60 per cent by volume, the moistuie in each metre of saturated soil is equivalent to depth of 60 cm on the field surface.

Field capacity: After the irrigation is stopped, part of water held in the larger pores moves downward called drainage or percolation. As the drainage stops the large soil pores are filled with both air and water whereas smaller zones are still full of water. The soil, at this stage, is said to be at field capacity. The water and air contents of the soil at field capacity are highly conducive for the crop growth. Field capacity is thus defined as the amount of water held in the soil after excess gravitational water has been drained, *i.e.*, the moisture percentage of the soil expressed on dry weight basis in the field 2 or 3 days after the soil profile is thoroughly wetted by rain or irrigation water, provided there is no watertable within capillary reach of the root zone:

$$\text{Field capacity} = \frac{\text{Weight of water retained in a certain volume of soil}}{\text{Weight of same volume of soil}} \times 100$$

Soil moisture content: The amount of water present in a soil is termed as soil moisture content, expressed as mm of water depth present in 1 m depth of soil. It is not constant but varies with time.

Permanent wilting point. Soil water is not replenished by irrigation or rainfall over long period; extending over weeks and months. The soil acts as a moisture reservoir for plant, the capacity of which in root zone of a crop is limited by the field capacity as the upper limit and permanent wilting point of the soil as the lower limit. Plant roots which penetrate into capillary zone draw continuous supply but those which remain up to the soil zone can draw a limited amount of moisture. As water stored in a soil is taken up by the roots of plants or evaporated from the top soil into the atmosphere, in case irrigation water is not added to the soil, the soil becomes drier and tightly retain the remaining water and plant roots are not in a position to extract it easily. When the water is rendered insufficient to meet the requirement of the plant, its green leaves turn yellow and the plant wilts and finally dies out. The soil water content at the stage when the plant dries is called permanent wilting point. Alternatively moisture content in the soil at which permanent wilting of plants occurs is termed as permanent wilting point. In any given soil, all forms of vegetation wilt when the moisture content is reduced to a certain percentage known as *wilting coefficient percentage* or *wilting coefficient.* It is defined as the moisture content of soil expressed as a percentage of the dry weight, at the time when the leaves of a plant growing in the soil first undergo a permanent reduction in their moisture content as a result of the deficiency in the soil moisture supply. Alternatively it is defined as the percentage moisture in a soil at which the plants wilt and fail to recover the turgidity when placed in an atmosphere saturated with water vapour. The permanent wilting percentage of soils varies greatly for different soils. It may vary from 3 to 23 per cent for a coarse sandy soil to fine clay respectively, corresponding to field capacity range of 2 to 40 per cent.

Available moisture. It is the amount of water in the soil at any time in excess of the wilting coefficient, expressed as percentage by weight of dry soil or as equivalent of water per unit depth of soil. Alternatively it is the difference in soil moisture from field capacity to wilting point (Fig. 2.6). It is a measure of availability of maximum quantity of water for use by plant without replenishment. Available moisture usually considered is the amount retained between field capacity or the moisture content 2 or 3 days after a soil has been wetted to a depth of 1 m, and the permanent wilting percentage or that moisture content at which plants cease to grow.

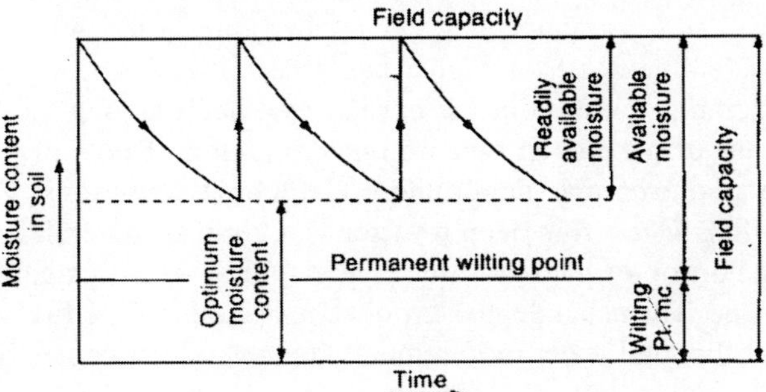

Fig. 2.6 Optimum soil moisture level

The available moisture content depends on soil texturé and structure (Table 2.4):

Table 2.4 Available moisture

Soil	*Available moisture (mm/m)*
Sand	25–100
Loam	100–175
Clay	175–250

Readily available moisture: It is the portion of available moisture that is most easily extracted by plants. It varies from 75 to 80 per cent of the available moisture.

Soil moisture deficiency: Also called field moisture deficiency. It is the amount of water that must be applied to a soil to bring moisture content of a soil to its field capacity.

Moisture equivalent: It is the ratio of water which a soil after saturation will retain against a centrifugal force of 1000 times the force of gravity to weight of the soil when dry. Moisture equivalent is slightly less or roughly equal to field capacity for medium textured soil.

Moisture equivalent field capacity = 1.8 to 2 permanent wilting point.

Depth of water stored in root zone: The depth of water stored in root zone of a soil containing water up to field capacity is estimated as under:

Assume d = depth of root zone (m), F = field capacity (expressed as ratio), γ = density of soil, and w = unit weight of water.

Considering unit area of 1 m^2 of soil area,

$$F = \frac{\text{Weight of water retained in unit area of soil}}{\text{weight of soil of unit area}}$$

$$= \frac{\text{Weight of water retained in unit area}}{\gamma \times 1 \times d}$$

Therefore, Weight of water retained in unit area = $F\gamma d$.

Depth of water stored (which is available for evapo-transpiration) = $F\gamma d/w$ (m)

Available moisture depth = $\gamma d/w$ (Field capacity-wilting point moisture content).

Optimum moisture content: For satisfactory growth, plants require readily available water in the soil, *i.e.*, water which can be easily extracted by them. Excessive flooding and deficient soil moisture are not conducive to plant growth. Thus, optimum moisture content in the soil is required for optimum plant growth. The optimum moisture percentage is the moisture content in the soil corresponding to which optimum growth of plants takes place.

Root zone depth: It is the depth below the ground surface in which the crops develop roots system to derive water for growth. The plants may have (*i*) tap roots that penetrate deeply into the soil under favourable conditions, (*ii*) shallow primary roots, and (*iii*) lateral roots. The depths of root zone in irrigated lands depend on (*i*) soil type, (*ii*) subsoil formation, (*iii*) depth of watertable, (*iv*) kind of crops grown, and (*v*) amount of water applied during irrigation. In clay soils, roots cannot penetrate to considerable depth. A high watertable limits the root growth due to lack of aeration. There is considerable variation in the normal root habit among the various species and varieties of plants, because of genetic and environmental factors. Roots of perennial field crops and fruit trees extend to a considerable depth under favourable conditions. Vegetables and grass roots do not normally utilize water from deep root zones. Crops, in general, develop most of their root system in the upper portion of the root zone. The depth to which irrigation water percolates limits the depth of root zone. The usual depth of root zone is 1.25 m. The first quarter of this depth contributes 40 per cent of the total water requirement, the next 30 per cent and the further next 20 per cent and the last quarter only 10 per cent (Fig. 2.7). The size of the wet bulb formation more or less tallies with this moisture absorption pattern.

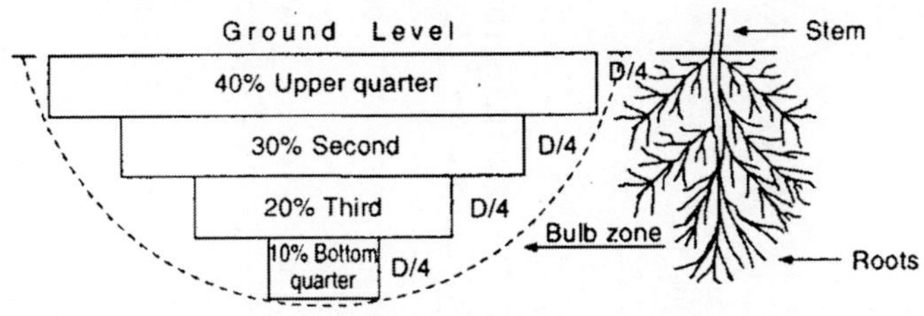

Fig. 2.7 Pattern of moisture absorption

2.12 WATER REQUIREMENT OF CROPS

Water is essentially a basic input influencing assured crop production. Water dissolves mineral nutrients which move in the plant along with its stem. At the end of the life cycle of a plant, water is also a constituent of the economic product, which may be seed, stem, leaf, flower or fruit. In crop production water is also needed sometimes for certain special purposes. The functions of irrigation water in crop production are: (*i*) Triggers activity in a seed, setting a chain of biochemical reactions, (*ii*) Dissolves mineral nutrients for their rise from the soil to the plant, (*iii*) Promotes chemical action within the plant for its growth, (*iv*) Promotes and supports life of bacteria beneficial to plant growth, (*v*) Helps temperature control of the soil as also minimizes the effect of frost and (*vi*) At the end of life cycle of the plant, water is still a constituent of the product.

Water requirement of a crop is defined as the quantity of water required by a crop in a given period of time for normal growth under field conditions. It includes evaporation and other economically unavoidable waste. It may also be expressed as the seasonal consumptive use plus unavoidable percolation. Usually water requirement of a crop is expressed in water depth per unit area. The total water requirement may be defined as the quantity of water needed for potential production per unit of land for sustained production and is sum of three entities, viz. consumptive use (A), application and conveyance losses (B), and other special needs (C). Numerically water requirement, $WR = A + B + C$. It is also equal to water requirement — effective rainfall + contribution of groundwater. Unit water requirement is defined as weight of water actually used by plants in producing unit weight of dry matter. Table 2.5 indicates the principal crops and their water requirements. The quantity of water needed for net irrigation on various soil types per metre depth of soil profile at 50 per cent of soil moisture availability is as under:

Soil type	*Irrigation depth (cm)*	*Efficiency (%)*
Sandy	25–50	60
Sandy loam	45–80	65
Loam	70–110	70
Clay loam	80–120	75
Heavy clay	100–140	80

Table 2.5 Principal crops and their water requirements

S. No.	Crop	Suitable soil	Optimum time of sowing	Seed (kg/ha)	Irrigations: Number of waterings	Irrigations: Delta (cm)	Harvesting time	Yield (kg/ha)
1	2	3	4	5	6	7	8	9
				RABI CROPS				
1.	Wheat	Well drained clay loams, loams and sandy loams.	Hills-4th week Oct-1st half Nov. Plains-2nd fortnight Oct.- 1st week Nov. for rainfed areas.	60-75, 90-100 for rainfed areas. 100-130 (high yielding varieties)	Ist irrigation 4 weeks after sowing, last irrigation around 15th March, 5-6 waterings of 7-10 cm depth for high yielding wheat and 3-4 waterings for ordinary wheat. Total 4 to 5.	37.5	April	5000-high yielding. 1000-1500 ordinary wheat
2.	Barley	Well suited to dry farming and saline areas, light soil	15 Oct.-15 Nov.	85-130	1 to 2 irrigations post-sowing, one at jointing stage and another at booting stage. Total 6-7	45	Mar.-Apr.	1000-2000
3.	Winter Maize Pulse	Loam to clay loam	25 Oct. - 10th Nov.	30	Ist watering 3-4 weeks after sowing, subsequent at 4-5 weeks interval up to mid March, thereafter 1-2 weeks interval.			
4.	Gram	Alluvial sandy loam to loam soils.	Desi gram, 10th-25th Oct. Kabli gram end Oct.	35-45 90	Heavy pre-sowing rain followed by post-sowing irrigation between mid-Dec. and end Jan. Total 4.	30	End Mar. mid.-Apr.	1100-2500
5.	Peas	Alluvial sandy loam to heavy black clay soils.	End Oct.-mid. Nov.	40-50	Irrigated once or twice. Total 7.	50	Mar.-Apr.	1200
6.	Rape seed and mustard (Sarson)	Loam to sandy loam.	10 Oct.-30 Oct.	4	Heavy pre-sowing irrigation, 1st irrigation 3 to 4 weeks after sowing, another irrigation at intervals of 7 to 10 days. Total 6 to 7.	45	Mar.	700-1500
7.	Potatoes	All types of soils	Oct., In high hills up to early Apr., 2nd crop some times taken in Feb.-Apr.	500-1000	Usually irrigated 8 to 10 times	60-90	Feb.	6,000-20,000
8.	Toria	Loam to clay loam	1st fortnight Sept.	5	1 to 2 irrigations.	40-50	End Dec.	
9.	Linseed (Alsi)	Light soils	Oct.-Nov.	10-15	Irrigations at intervals of 15 days. Total 2-3.	40-50	Mar.	400-1000
10.	Massar	All soils except alkaline and waterlogged soils.	End Oct.-15 Nov. Last sowing up to 15th Dec. Perennial crop	30-35	2 to 3 irrigations. Generally grown under barani condition.		Apr.	

1	2	3	4	5	6	7	8	9
					KHARIF CROPS			
1.	Rice	Clay loam to clay.	Nursery 10-20 May-Basmati 15 Jun. to 30 Jun. Transplanting Mid Jul. to 1st week Aug.	20-24	Continuous ponding for first two, thereafter at two days intervals.	125-150	End Sept.-Nov.	3000-5000 (Dhan)
2.	Maize	Loam to clay loam well-drained soil.	20 Jun. to 7 Jul. in submountaine region and 1-15 Jul. in other regions.	20-22	4 to 6 irrigations depending on rain.	45	Sep.-Oct.	2500-5000
3.	Spiked millets (Bajra)	Light loam.	July	5 for grain, 6 for hybrid and 10 for fodder	2 to 3 irrigations, water should not stand.	30	Oct.-Nov.	2000-4000
4.	Great millet (Jowar)	Light loam	Jun.-Jul.	12.5	2 to 3 irrigations	30	Sep.-Oct.	1200-3000
5.	Cotton-American	All types except too sandy, saline or waterlogged soils.	Mid Apr.-mid May	20	3 to 5 irrigations	25-40	Sep.-Oct.	350-1500
	Desi	Soils less fertile than those required for American cotton.	Apr.-May	12-15	3 to 5 irrigations	25-40	Sep.-Nov.	500
6.	Groundnut	Sandy to sandy loam.	Mid Jun.-early Jul.	44 kg of kernels at 35 × 15 cm spacing	1 to 5 irrigations depending on rains. Generally sown under rainfed condition.	45	End Oct.-mid. Nov.	1500
7.	Pulses (Arhar)	Loam	Jul.-Aug.	12.5	Irrigated when leaves get drained. Total 2-3	30	Mar.-Apr.	700-1000
8.	Til	Loam	Jul.-Aug.	1.25	Generally not to be irrigated, but better to irrigate once.	30	Oct.-Nov.	350
9.	Sugarcane	Well drained loam.	Mid Feb.-end Mar.	2.75 tonnes of cut pieces	10-12	90	Dec.	33,000-108,000

2.13 CROPPING PATTERN

The suitability of crops in a place and its economic return largely determine the choice of crops to be grown. The crops may be selected depending upon the water availability and the profit desired. When water resources are limited first preference is given to oilseed crops, followed by pulses and the cereals. The cropping pattern suitable for an area depends on (*i*) Availability of water, (*ii*) Size of farms, (*iii*) Resources and (*iv*) Specific situation or need. The cropping pattern is never fixed and is dynamic with reference to the various factors mentioned above.

In the formulation of any cropping pattern, various aspects considered are (*i*) A fixed cropping pattern may accentuate certain diseases, insect and weed problems; so continuous fixed rotations on the same field are avoided, (*ii*) A cropping sequence aims at building the soil fertility and improving the soil conditions. Legumes should find place in rotation with cereals, (*iii*) A cropping pattern should not form a drain on some specific plant nutrients from a particular depth in the soil, and (*iv*) It should provide employment and income to the farmers all the year round and ensure best utilisation of his resources particularly irrigation water, fertilizers, pesticides, labour, machinery and power.

2.14 FREQUENCY OF IRRIGATION

Soil moisture in the root zone varies between field capacity (upper limit) and wilting point (lower limit). The soil moisture is not allowed to deplete up to wilting point. For satisfactory growth the moisture content in the soil should be between field capacity and optimum moisture content, which is the readily available moisture. Thus irrigation is aimed at supplying the amount of water such that water content is equal to field capacity. After irrigation, moisture content starts falling as water (soil moisture) is consumed by plants through their roots and when the soil moisture in the soil reaches optimum value, next irrigation is required to raise moisture content again to field capacity of the soil. The moisture will again start falling necessitating fresh irrigation application when the soil moisture content reaches optimum moisture content level. Thus water contained in the root and the consumptive use of water by plants dictates the frequency of irrigation. The frequency of irrigation and quantity of water to be supplied are affected by the soil fertility. The most desirable frequency of irrigation to give optimum crop yield is only when a supply of readily available nitrogen is present in the soil. Phosphorus is taken up easily by the plants from moist soil and as such light frequent irrigations may produce optimum growth. The water application should be for a long enough period to get sufficient water in the soil to moisten the dry soil without appreciable amount penetrating into the moist soil below. The application of water should be slow enough to avoid erosion and wastage through runoff. The crops draw water from the root zone in the soil with a suction between one-third bar (at field capacity) and 15 bars (at wilting percentage). Irrigation may be started when soil moisture is about halfway between the field capacity and wilting point. Adequate amount of water is applied to bring soil moisture to field capacity (Fig. 2.8). The depth of water to be stored in root zone during each irrigation is given by the relation:

$$D = \frac{\gamma d}{w}(F - M_0)$$

where, D = depth of water (m), γ = unit weight of soil (g), d = depth of root zone (m), w = unit weight of water (g), F = field capacity (expressed as ratio), and M_0 = optimum moisture content (expressed as ratio).

Frequency of irrigation, $f = D/C_u$

where, f = frequency of irrigation, D = depth of water (cm), and C_u = rate of daily consumptive use (cm).

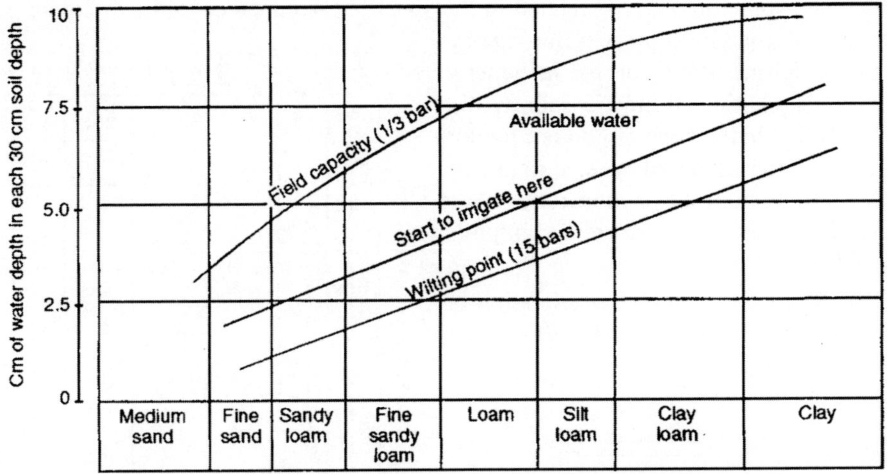

Fig. 2.8 Amount and frequency of Irrigation
(*Source*: Water-the Yearbook of Agriculture USDA)

2.15 QUALITY OF WATER FOR IRRIGATION

The term quality, as applied to water, embraces its combined physical, chemical and biological characteristics. The quality of water for irrigation is as important as nature of soil. Good quality water improves the soil because of its calcium content. Conversely, if water applied for irrigation is not of suitable quality soil deteriorates and crop yield decreases. The suitability of water for irrigating a particular crop grown on a particular soil requires consideration of (*i*) Sodium and calcium salts dissolved in it, (*ii*) Its pH, (*iii*) Texture of soil and salts present in it, and (*iv*) Sensitivity of crop and drainage conditions of soil. Water in streams in humid areas is generally suitable for irrigation. Streams in industrial locations are, however, polluted with industrial waste rendering it unsuitable for irrigation. Irrigation water is generally obtained from rivers, canals, reservoirs, groundwater, and tanks. In India most of the river flow is contributed by rainfall during June to September. During this period river water becomes highly turbid on account of soil erosion in the catchment area. In dry months local pollution develops due to discharge of sewerage and industrial effluents. Reservoirs yield better quality water than rivers because of beneficial effects of impoundment. However, water quality characteristics of reservoirs are highly variable and depend on many factors, most of which are too complex. Storage ponds or tanks, an important source of irrigation water, are of good quality for irrigation except in Andhra Pradesh which show low to medium salinity.

1. Quality of surface water for irrigation.

Irrigation water is classified into three classes as given in Table 2.6.

Table 2.6 Standards for irrigation waters

Class of water	*Total dissolved salts (mg/l)*	*Electrical conductivity (micromhos/cm)*	*Exchangeable sodium percentage (ESP)*	*Sulphates (mg/l)*	*Chlorides (mg/l)*	*Boron (mg/l)*	*Suitability for irrigation*
I	0–700	0–1000	0–60	0–192	0–142	0–0.5	Excellent to good
II	700–2000	1000–3000	60–75	192–480	142–355	0.5–2.0	Good to injurious
III	over 2000	over 3000	over 75	over 480	over 355	over 2	Unfit

The salient characteristics of irrigation water determining its quality are discussed as under:

1. Total Concentration of Solids

Total dissolved solids in water is related to the specific conductance. Salts of calcium, magnesium, sodium and potassium present in irrigation water may prove detrimental to crops. In general, water is good if its position in the US salinity diagram (Fig. 2.8) is within the zone of good or moderate waters. The diagram is based on electrical conductivity (EC) and sodium adsorption ratio where concentration of constituents are expressed in milli-equivalent per litre.

IS: 11624-1986 has specified classification on the basis of hazardous effects of total salt concentration into four major groups, as under:

Sl.No.	*Class*	*Range of EC (Micromhos/cm)*
1.	Low	Below 1500
2.	Medium	1500 – 3000
3.	High	3000 – 6000
4.	Very high	Above 6000

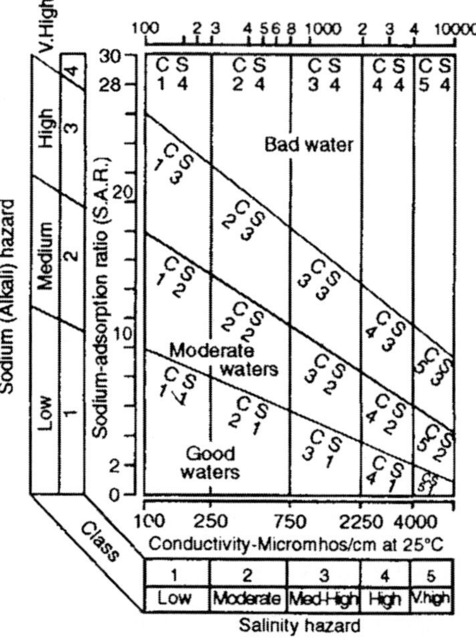

Fig. 2.9 U.S. salinity diagram for classifying irrigation waters

2. Electrical Conductivity

Electrical conductance is the ability of water solutions to conduct an electric current and is measured in mhos. It is a function of temperature, type of ions present and concentration of various ions. The classification of irrigation water based on electrical conductivity is as per Table 2.7:

Table 2.7 Classification of irrigation water based on electrical conductivity

Type of water	*Classification*	*Electrical conductance (micromhos / cm at 25° C)*	*Suitability for irrigation*
Fresh water		0 – 100	Excellent to good
Low saline (C_1)	Excellent	100–250	All crops and all soils except extremely low permeable soils.
Medium saline (C_2)	Good	250–750	Normal salt tolerant plants with moderate leaching.
Saline (C_3)	Permissible	750–2000	Only high-salt tolerant plants, drainage is required.
Highly saline (C_4)	Doubtful	2000–3000	Bad water for irrigation.
Very highly saline (C_5)	Unsuitable	over 3000	Unsuitable.

3. Sodium

Sodium salts are generally present in irrigation waters but if in a high proportion they may be absorbed on the soil particles resulting in impeding the movement of water and air when the soil is wet and formation of hard clods when the soil is dry. Classification of irrigation water on sodium content, reported as sodium adsorption ratio (SAR) = $Na^+ / \sqrt{\frac{Ca^{++} + Mg^{++}}{2}}$ is given in Fig. 2.8 and its suitability for irrigation is, as follows:

Type of water	*(SAR value)*	*Suitability for irrigation*
Low sodium water (S_1)	Below 10	All soils and all crops except those which are highly sensitive to sodium.
Medium sodium water (S_2)	10–18	May be used on coarse textured or organic permeable soils. Addition of gypsum either to water or soil is required for use on fine textured soils. The soils tend towards alkaline because of increase in pH value.
High sodium water (S_3)	18–26	May be used provided gypsum is added, and good drainage and high leaching is provided.
Very high sodium water (S_4)	above 26	Generally not suitable.

Na, Ca, Mg represent the concentration in milliequivalent per litre. Exchangeable sodium percentage is determined by the relation:

$$ESR = \frac{100(0.01475SAR - 0.0126)}{1 + (0.01475SAR - 0.0126)}$$

4. Boron

Boron is present as a minor constituent in all natural waters. It is nutrient but is toxic if present slightly in excess of that required for optimum growth. Beet and sugarcane may tolerate large concentration of boron in irrigation water but groundnut, beans, etc., are very sensitive to the presence of excessive boron. The permissible limits for different crops are as given in Table 2.8:

Table 2.8 Classification based on boron content (IS: 11624-1986)

Class	*Boron content (mg/l)*	*Suitability for irrigation*
Low	Below 1.0	Excellent
Medium	1.0 – 2.0	Most crops
High	2.0 – 4.0	For more boron-tolerant crops
Very high	Above 4.0	Doubtful for irrigation

5. Soluble Salts

Salts of calcium, magnesium, sodium and potassium present in irrigation water may prove detrimental to crops. Some crops are more salt tolerant whereas others are easily damaged by excess salt concentration.

6. Bicarbonate Concentration

Irrigation water may be high in bicarbonate. The use of such water as is low in total salt content but high in bicarbonates result in the increase of relative proportion of sodium ion, thus aggravating the sodium problem. Water relatively low in total salts and sodium percentage (Na × 100/total cation) within safe limits is questionable if the amount of bicarbonate is appreciably in excess of the calcium and magnesium present. This excess bicarbonate over calcium plus magnesium is referred as residual sodium carbonate as indicated in Table 2.9:

Table 2.9 RSC vis-a-vis water suitability for irrigation

Class	*RSC (mg/l)*	*Suitability for irrigation*
Low	Below 1.5	Probably safe
Medium	1.5 – 3.0	Marginally safe
High	3.0 – 6.0	Not suitable
Very high	Above 6.0	

7. Soil Texture

Texture of soil to be irrigated also determines the suitability of water for irrigation. Water with total dissolved solids as high as 1500 mg/l but a low amount of bicarbonates though found unsuitable for irrigating heavy soils has shown better yields when applied to soils of very light texture.

8. Crop Grown

Suitability of irrigation water also depends on the crop grown. Some crops are more salt tolerant whereas others are easily damaged by excess salt concentration. Crops are divided into three categories according to the salt tolerance as indicated in Table 2.10:

Table 2.10 Salt Tolerance of crops

Sensitive	*Semi-tolerant*	*Tolerant*
Gram	Wheat, rice, millets, maize	Barley
Arhar	Tomato, cabbage, potato	Beat, tobacco
Moong	Onion, karela, guava, mango	Mustard, cotton
Peas	Banana, pear, apples, orange, lemon, papaio	Sugarcane

9. Chloride and Sulphate Content

Chloride-ion accumulation in plant tissues is frequently toxic. Many species are not sensitive to chloride but to equal concentration of sulphate salts.

10. Sediment Content

Sediment brought by canal water has manurial value. Groundwater being sediment free has lesser manurial value. However, silt may reduce permeability of the soil.

11. pH Value

Irrigation water with pH more than 8.5 will cause sodium hazard.

12. Soil type and Crop Tolerance to salts

IS: 11624-1986 specified upper limit of EC, CAR, RSC and boron content for the semi-tolerant and tolerant crops vis-a-vis soil types and quality of ground water are given in Table 2.11.

Table 2.11 Suitability of irrigation water for semi-tolerant and tolerant crops in different soil types

Soil textural group	*Upper limit of*							
	Salinity EC (micromhos/cm)		*Sodicity SAR (millimole/l)*		*Boron RSC (me/l)*		*B (mg/l)*	
	*S.T**	*T**	*ST**	*T**	*ST**	*T**	*ST**	*T**
Above 30% clay Sandy clay, clay loam, silty clay loam, silty clay, clay	1500	2000	10	15	2.	3	2	3
20–30% clay, sandy clay loam, Loam, silty loam	4000	6000	15	20	3	4	2	3
10–20% clay, sandy Loam, loam, silty loam	6000	8000	20	25	4	5	2	3
Below 10% clay, sand, Loamy sand, sandy loam, silty loam, silt	8000	1000	25	30	5	6	1	2

Use of water of 4000 micromhos/cm EC and above be confined to winter season crops only.

2. Quality of Ground Water for Irrigation

Suitability of ground water for irrigation depends upon the effect of mineral constituents of water on both plants and soils as also on the piping system of the tubewell. Quality of ground water varies from place to place, from stratum to stratum and from season to season. The suitability of ground water for irrigation is determined on the basis of chemical, physical and bacterial characteristics. Chemical analysis requires determination of the concentrations of inorganic constituents, measurement of pH and specific electrical conductance. Physical analysis requires determination of the colour, odour, taste, temperature, turbidity etc. Bacterial analysis is done to determine the presence of coliform organisms. Dissolved gases and radiomidides may also be present. The criteria for suitability of ground water for irrigation is briefly discussed as under :

1. Total Dissolved Solids (TDS). The total concentration of dissolved solids indicates the overall suitability of water. Generally a water is considered unsuitable for irrigation if TDS is above 1000 mg/l. The limit of TDS is simply a guide and suitability of water is finally judged from U.S. salinity diagram.

Most ground waters carry some calcium and magnesium bicarbonates in solution and in waters of many regions this may be the principal salt present so that from these waters calcium carbonate is deposited on the well screen or in the soil when water is applied for irrigation.

Gases dissolved in ground water are oxygen, nitrogen, carbon dioxide, hydrogen sulphide, methane, oxides of nitrogen, sulphur dioxide, and ammonia. Dissolved gases occur in concentrations of 1 to 100 mg/l in most ground waters.

The mineral content of ground water causes corrosion and incrustation of well screen leading to its failure. Corrosion is caused by dissolved oxygen in water. Chemical corrosion generally takes place in water low in total dissolved solids and pH value (acidic). Electrochemical corrosion takes place in water with high total dissolved solids (> 1000 mg/l) and high pH value (alkaline).

The rating limits fixed for well irrigation water analysis are given as under :

Texture of soil to be irrigated	*Conductivity (micromhos/cm)*		*Residual sodium carbonate*	*Boron (mg/l)*
	Tolerant Crops	*Semi-tolerant Crops*		
Sandy	Upto 8000	Upto 6000	2.5	< 2.0
Sandy loam	Upto 6000	Upto 4000	2.5	< 2.0
Loan	Upto 4000	Upto 3000	2.5	< 2.0
Clay loan	Upto 3000	Upto 2000	2.5	< 2.0
Clay	Upto 2000	Upto 1500	2.5	< 2.0

Residual sodium carbonate = (carbonate + bicarbonate) – (Ca + Mg) all in milli equivalent per litre.

Under high water table condition, half of the above conductivity values in each category is taken for judging the quality of irrigation water.

(Source : Punjab Agriculture University, Ludhiana Handbook - 1993).

2.16 CONSUMPTIVE USE OF WATER

Considerable part of water applied for irrigation is lost by evaporation and transpiration. The two processes being difficult to separate are taken as one and called evapo-transpiration or consumptive use. Consumptive use is thus the sum of two terms (*i*) Transpiration which is the water entering the plant roots and used to build plant tissue or being passed through leaves of the plant into the atmosphere, and (*ii*) Evaporation which is the water evaporating from the adjacent soil, water surface or from the surfaces of leaves of the plant. Water deposited by dew, rainfall or sprinkler irrigation and subsequently evaporated without entering the plant system is part of consumptive use. Consumptive use is expressed in water depth or depth area units per unit area for specified periods such as days, months or seasons. Consumptive use of water by a crop is the depth of water consumed by the plant in transpiration and evaporation during the crop growth. The values of consumptive use vary not only with the kind of crop but also for the same crop at different times and places. The consumptive use for a given crop at a given place varies throughout the day, throughout the month and as such throughout the crop period. For a given crop and at a given place, values of daily consumptive use or monthly consumptive use are determined to ascertain values of monthly consumptive use over the entire crop period in order to compute the irrigation requirement of the crop.

Consumptive use varies with temperature, humidity, wind velocity, soil topography, sunlight hours, available moisture, viz. precipitation, irrigation water or natural groundwater, method of irrigation, depth of water applied for irrigation, cropping pattern, seasons and mean monthly temperature.

Optimum consumptive use: It is the consumptive use which produces a maximum crop yield.

Potential evapotranspiration: It is the amount of water transpired by a green crop of about the same colour as green grass, which completely covers the ground, and which has an adequate water supply.

Seasonal consumptive use: Depths of water consumed by evapotranspiration during crop growth till maturity including water used by accompanying weed growth.

Consumptive Irrigation Requirement (CIR)

It is the amount of irrigation water required to meet the consumptive use of crop during the growth period. It is the same as consumptive use exclusive of effective precipitation, stored soil moisture or groundwater.

$$\text{CIR} = C_u - R_e$$

where, C_u = consumptive use, and R_e = rainfall effective *i.e.* the rainfall falling during the growth period of a crop and is available to meet evapotranspiration needs of the crop exclusive of the rainfall lost through deep percolation below the root zone or the water lost as surface runoff.

Net Irrigation Requirement (NIR)

It is the amount of irrigation water required to meet the evapotranspiration need of the crop as also other needs such as leaching.

$$\text{NIR} = C_u - R_e + L_e$$

where, L_e = water lost as percolation in satisfying other needs like leaching.

Determination of Consumptive Use

Consumptive use can be measured or estimated by following methods:

1. Direct Measurement

Vapour transfer method. It is the only method for the measurement of evapotranspiration from a field or any other natural surface without disturbing vegetation cover. When evaporation takes place, the amount of moisture is maximum in the air near the land surface and decreases with height above it. Water vapour when it enters the atmosphere from the ground or from plants is carried upwards by the moving air in small eddies or bodies of air that are replaced by drier eddies from above. The rate and amount of evapotranspiration can be computed by determining (*a*) the rate at which the air near the land surface mixes with the air above it, and (*b*) the difference in water vapour content at the two levels.

The method has limitation in use because it requires precise physical measurements than usually made, and the coefficient of turbulent transfer of air not only varies with time and place but also with the height at a given time and place.

Climatic conditions: The rate of evapotranspiration principally depends on (*a*) climate, (*b*) supply of soil moisture, (*c*) plant cover, (*d*) soil type and structure, and (*e*) land management. When the soil moisture is maintained at the optimum, land management and soil type or structure have little effect on the rate of evapotranspiration. When the root zone of the soil is well supplied with water, the amount used by vegetation depends more on the amount of solar energy received by the surface and the resultant temperature than on the kind of vegetation grown in the area. The water loss under optimum conditions of soil moisture, the potential evapotranspiration, thus appears to be determined principally by climatic conditions.

Field plot method: It is a more reliable method and is based on natural conditions. Irrigation water is applied to experimental field plots such that there is no runoff or deep percolation. In irrigated land, water input, *i.e.*, water applied for irrigation, rainfall water and outflow are measured and the fraction of water applied that does not runoff, adjusted for changes in soil moisture, is taken as evapotranspiration, usually in terms of cm of depth. Deep percolation, being not susceptible to precise measurement, introduces an error in the result and in order to limit it irrigation water is applied in small quantities.

Tanks and lysimeters: It is a reliable method provided the tanks and lysimeters are properly designed and the soil within the tank simulates approximately field conditions and plant density and exposure are representative. Tanks are watertight containers 10 m square and 3 m deep installed in ground with their rims nearly flush with the ground surface. Large size tanks are preferred as they permit more nearly natural root development. The consumptive use is computed by measuring the quantity of water required for maintaining constant moisture conditions in the tank.

Lysimeters are the containers with pervious bottom and simulate more closely natural growing conditions. The difference between the water applied at the surface and that draining out through the pervious bottom and collected in a pan, adjusted for changes in moisture content, gives the consumptive use. Satisfactory results are obtained if arrangement is made to apply a suction force at the bottom comparable to that in the natural soil profile.

Irrigation method. It is the summation of the products of (*a*) unit consumptive use for each crop times its area, plus (*b*) the unit consumptive use of native vegetation times its area, plus (*c*) water surface evaporation times water surface area, plus (*d*) evaporation from bare land times its area. For successful application of this method, it is necessary to know unit consumptive use of water and the areas of various classes of agricultural crops, native vegetation, bare land and water surface. Thus the annual consumptive use for the whole area is determined in hectare-metres.

2. Empirical Formulae

(*i*) *Lowry-Johnson method*

Consumptive use, $U = 0.00015\,H + 0.9$

where, H = accumulated degree days during the growing season computed from the maximum temperature above 32°C.

(*ii*) *Blaney-Criddle method.* Monthly consumptive use factor F is obtained by multiplying the mean monthly temperature t, by the monthly percentage of day light hours of the year:

$$C_u = \frac{kP}{40}(1.8\,t + 32) = KF$$

where, C_u = monthly consumptive use (cm), K = empirical coefficient (annual or irrigation season or growing season), P = monthly percent of annual day light hours, t = mean monthly temperature (°C), F = sum of monthly consumptive use factor for the period, *i.e.*, sum of the products of mean temperature and percent of annual day light hours = $t \times p/100$.

(*iii*) *Hargreaves, class A pan evaporation method.*

$$C_u = KE_p$$

where, C_u = consumptive use coefficient, which is different for different crops, different for the same crop at different places, and different at different stages of the same crop, E_p = Class A pan evaporation, value determined from Christiansen formula:

$$E_p = 0.459\ RC_tC_wC_hC_sC_e \qquad \text{(cm)}$$

where, R = extra-terrestrial radiation (cm or mm as of E_p), C_t = coefficient for temperature $= 0.393 + 0.02796\,T_c + 0.0001189\,T_c^2$,

(T_c = mean temperature °C), C_w = wind velocity coefficient = $0.708 + 0.0034\,W - 0.0000038\,W^2$, W = mean wind velocity in km/day at 0.5 m above ground surface, C_h = coefficient for relative humidity $= 1.25 - 0.0087\,H + 0.75 \times 10^4\,H^2 - 0.83 \times 10^{-8}\,H^4$

where, H = mean percentage relative humidity at noon or average relative humidity for 11 and 18 hours, C_s = coefficient for percent of possible sunshine

$= 0.542 + 0.008\,S - 0.78 \times 10^{-4}\,S^2 + 0.62 \times 10^{-6}\,S^3$

where, S = mean sunshine percentage, C_e = coefficient of elevation

$= 0.97 + 0.00984E$, E being elevation in100 m.

Table 2.12 gives the average consumptive use coefficients for selected crop groups. These coefficients multiplied by class A evaporation rates give the consumptive use of crops.

Table 2.12 Average consumptive use (evapotranspiration) coefficients for selected crop groups

Group No.	Crops	Range of monthly CUC's	Peak period CUC	Weighted average CUC
1. High water-intensity crops	Rice	0.80–1.30	1.30	1.01
	Tropical pastures, orchards, cover crops, plantation, etc.	0.90	0.90	0.90
	Sugarcane, alfalfa	0.5 – 1.0	1.0	0.75
2. Medium water-intensity crops	Maize, jowar, cotton, beans, peas, potatoes, sugar-beet	0.20–1.0	1.0	0.64
	Citrus crops, grapes, etc.	0.60	0.60	0.60
3. Low water-intensity crops	Wheat, barley, flax, etc.	0.08–0.90	0.90	0.50
	Hybrid jowar, dates, olives, plums, walnuts, tomatoes, etc.	0.15–0.75	0.75	0.60
	Melons, onions, carrots, hops, grapes, etc.	0.12–0.60	0.60	0.38

2.17 DEFINITIONS

Gross commanded area (GCA): It is defined as the gross irrigable area which can be commanded by irrigation. The boundary of GCA. is usually defined by the drainage on either side across which irrigation cannot be extended economically.

Gross irrigable area (GIA): It is defined as the gross commanded area less such areas within irrigation limits as may be excluded, for any reason, for irrigation by the canal, project or scheme.

Culturable commanded area: It is that portion of the culturable irrigable area which is commanded by irrigation, *i.e.,* gross commanded area less the area of unculturable land included in the gross area such as areas under small drainage, ponds, abadies, reserve forest, alkali soils, roads, railway lines and watercourses, etc. In the formulation of projects and schemes, CCA is roughly taken as 90 to 80 per cent of GCA depending on the configuration of land.

Intensity of irrigation: It is defined as the percentage CCA proposed to be irrigated annually or seasonly. It may depend on the objective of the irrigation project. In intensive irrigation system, the intensity may be 100 per cent or even more depending on multiple cropping once or twice under different crops in the same area. In the protective irrigation scheme, the intensity may be low to extend the benefits to wider area. The various factors considered in fixing the intensity of irrigation are climate, soil, topography, cropping pattern, drainage, groundwater condition, amount and nature of available water, socio-economic needs of the area, water application method, dispersal of benefits. Rainfall and its distribution has a marked relation to intensity of irrigation. In high rainfall areas intensity is high because water needs are less and in semi-arid areas intensity is low due to frequent irrigation needs. With the adoption of water saving methods such as spinkler irrigation and drip irrigation, more area can be brought under irrigation with the available water and hence higher intensity of irrigation. High intensity has to be accompanied by efficient drainage otherwise fraught with risk of waterlogging.

Area to be irrigated: It is the area to which irrigation is to be done, *i.e.,* the product of CCA and intensity of irrigation. It is normally worked out separately for each crop season.

Crop ratio: Crop ratio or Kharif-Rabi ratio is defined as the ratio between the areas anticipated to be irrigated in these two crops. Crop ratio adopted in Punjab and Haryana is 1 : 1.5, Uttar Pradesh 1 : 1.2, Maharashtra, Madhya Pradesh, Orissa, West Bengal, Andhra Pradesh 1 : 1.

Overlap allowance: Crop of one season may extend into the other season. In such a period of overlapping, both the crops require irrigation simultaneously. Thus there is extra demand during this period. To cater for this, usually 5 percent of canal discharge is provided as overlap allowance which implies that the canal discharge is increased by 5 per cent for this period of overlap.

Time factor: It is expressed as the number of days the canal has actually run to the number of days the canal was designed to run for a particular period of watering. Alternatively, it is defined as the number of days the canal is in flow to the base days. For example, time factor of a canal run for 12 days, out of 20 days it was supposed to run, is 12/20.

Capacity factor: It is defined as the rate of mean supply to authorized full supply or capacity of the canal at a certain point. Thus if it were possible to run a canal system at full supply discharge or closed, then the capacity factor and the time factor would be the same. Owing to rotational working, branches and distributaries do not run for the same number of days as either the main canal or each other. The volume of discharge of a canal is given in cumec days by the sum of daily discharges for the period in question. It is equal to (*i*) average discharge multiplied by the number of days the canal is in flow, and (*ii*) mean discharge multiplied by the number of days in the crop.

Crop period: It is the time that a crop takes from the instant of its sowing to that of its harvesting.

Base Period: It is the time in days between first watering of a crop at the time of sowing to its last watering before harvesting, *i.e.*, the number of the days in a crop or more precisely the number of days over which duty is measured. Base period of a crop is thus slightly less than its crop period, but for practical purposes it is generally taken equal to the crop period. It numbers 183 days for Kharif and 182 days for Rabi.

Delta: It is an expression used in irrigation practice to mean the depth of water that would result over a given area from a given discharge for a certain length of time. Alternatively, it may be defined as the total volume of water delivered divided by the area over which it has been spread. It is a term equivalent to duty of irrigation water when the latter is expressed in water-depth units and refers to irrigation projects under operation.

Delta is the total depth of water in cm required by a crop to come to maturity. It depends on the amount of each watering and the interval between successive waterings during the base period.

Table 2.13 indicates the average value of delta for certain important crops. Delta is stated with reference to the place at which it is measured, that is, delta at farm, delta at outlet, delta at distributary head, delta at the head of main canal.

Table 2.13 Average delta

Crop	*Delta (cm)*	*Crop*	*Delta (cm)*
Rice	125–150	Gram	30
Sugarcane	90	Cotton	25–40
Tabacco	60	Wheat	37.5
Groundnut, Maize	45	Fodder	22

Duty: Duty or duty of water is the relation between the area irrigated, or to be irrigated, and the quantity of water used, or required to irrigate it for the purpose of maturing its crop. When applied to a channel it is the area irrigated during a base period divided by the mean supply utilized in cumecs. It may be defined as the area irrigated by unit discharge which means the number of hectares under a particular crop brought to maturity by a constant supply of 1 cubic metre of water per second flowing continuously for the base period. Duty for a channel is usually calculated on the head discharge. Duty based on discharge passed through the outlet and thus excluding all losses in the canal system is called the outlet discharge factor. Duty is expressed in (*i*) water depth units, (*ii*) depth-area units per unit area, (*iii*) area per unit rate of flow or per unit volume of water, and (*iv*) volume of water or rate of flow per unit area. In practice duty is expressed in hectares per cubic metre per second and is represented by D. If 1 cubic metre/sec water flowing continuously for a base period B, matures 175 hectares, then the duty of water for that crop is defined as 175 hectares per cumec to the base of B days. It is a measure of degree of efficient utilization of irrigation water from a canal.

Cumec day: A unit of volume used in irrigation practice and means the volume of water resulting from a discharge of 1 cumec for one day (24 hours). It amounts to 8.64 hectare metres.

Relation between Duty and Delta:

Let D = duty of water (hectares/cumec), B = base period (days), and Δ = delta of water (m).

(*i*) Volume of one cubic metre flowing for one day

$$= 1 \times 24 \times 60 \times 60 = 86{,}400 \text{ m}^3 = 8.64 \text{ ha m.}$$

(*ii*) Volume of one cubic metre flowing for B days = 8.64 B (ha m).

(*iii*) By the definition of duty (D), one cubic metre = 10^4 D m^2 of area supplied for B days matures D hectares of land.

Total depth of water applied on the land = Volume/area = $86400B/10^4 D$.

$$= 8.64 \; B/D \text{ metre}$$

(*iv*) Now, delta is total depth of water, $\Delta = 8.64 \; B/D$ m

or $D = 8.64 \; B/\Delta$ m

$D = 864 \; B/\Delta$ cm

Hence, delta (Δ) = $864 \; B/D$ cm

where, Δ = delta (cm), B = base period (days), and D = duty (hectares/cumec).

Gross duty: It is the duty of water measured at the source of diversion of irrigation supplies.

Nominal duty: It is the duty sanctioned as per schedule of an irrigation department.

Economic water duty: It is the duty of water which results in the maximum yield, (*i*) per unit area when land is the limiting factor, and (*ii*) per unit of irrigation water when water is the limiting factor.

Designed duty: It is the duty of water assumed in an irrigation project for designing capacities of the channels.

Paleo: It is the first watering before the crop is sown in order to add sufficient water to unsaturated zone of the soil for initial growth of the crop.

Kor watering: It is the first watering after the crop has grown to a few centimetres or the second watering from the beginning, the first (before crop is grown) being paleo. The depth of water applied is the maximum and is designated as kor depth. All subsequent waterings are of lesser depth, the least being at the time when the crop is mature. The portion of the base period in which kor watering is needed, is termed as *kor period.* While designing capacity of an irrigation canal, kor watering is taken into account in view of maximum water requirement during this period.

Total number of waterings required depend on (*i*) condition of soil at the time of waterings, (*ii*) kind of crop grown, (*iii*) length of base period, (*iv*) water requirement of plants at various stages of growth, and (*v*) pattern and amount of rainfall. Table 2.5 indicates the number of waterings required for principal crops.

Demand: It is defined as the amount of water needed for irrigation based on elements of time and quantity. It is related to a particular point along the irrigation system such as demand at distributary head, demand at head of canal.

2.18 DUTY IN IRRIGATION SYSTEM

By definition duty is the number of hectares per cumec. Duty is stated with reference to a base period and the place where the discharge is measured. Duty at distributary head is the duty of water measured at the head of a distributary. Duty measured at the farm is termed as *Farm duty* or *Net duty.* At the field, duty is maximum because all the supply reaching is utilised for irrigation. Duty is obviously lesser at outlet head due to the losses enroute the watercourse up to the field. Likwise, duty of water at canal head is lesser than that at outlet head, and so on. Duty at the outlet head is important because control on water thereafter transfers to cultivators. This is called outlet discharge factor.

Flow Duty: Unless specified otherwise, duty is determined at head discharge of a channel. It is called flow duty.

Quantity Duty: In storage reservoirs duty may be expressed in terms of volume of water stored, *i.e.*, so many hectare metres or million cubic metres of water available.

Duty Inclusive of Rainfall: During the base period of the crop, rainfall also meets part of requirement of the crop. Thus duty inclusive of rainfall is high.

Duty Exclusive of Rainfall: It is the duty calculated without taking rainfall into consideration. Thus duty exclusive of rainfall is low.

Factors Affecting Duty

Various factors affecting duty of water are (*i*) *Kind of Crop Grown:* Different crops require different amounts of water and hence have different duties. Duty of water for a crop requiring more water is less and vice versa. Obviously duty of water for rice is less than that for wheat, (*ii*) *Nature of Soil:* Evidently porous soil under a crop has less duty due to high absorption loss. Duty of water is more in heavy (clay) soils than light (sandy) soils, (*iii*) *Cultivation Methods:* Properly ploughed and tilled field before irrigation retains higher amount of water in its saturated zone, thereby reducing the number of waterings and increasing the duty. Better tillage envisages less evaporation losses from soil surface, besides soil is properly aerated and results in better yield. More efficient the cultivation method, higher is the duty, (*iv*) *Method of Water Application:* The more efficient the mode of application of water, higher is the duty. Flooding irrigation system has lesser duty than furrow irrigation. Basin irrigation too has less duty. Sub-irrigation system has higher duty. Sprinkler and drip irrigation have the highest duty, (*v*) *Irrigation System:* Perennial irrigation system has higher duty than inundation canal system, because in the former case not only the soil is continuously moist requiring lesser amount of water for initial saturation of soil but also there is lesser wasteful use of water than in the latter system of irrigation. Gravity flow irrigation has lesser duty due to transmission losses in the canal water. Tubewell irrigation and tank irrigation have higher duty of water due to low transmission losses and more efficient use of water, (*vi*) *Water Quality:* Better the quality of water, higher is the duty of water and vice versa. Liberal quantity of water is required to be applied to leach off salts in case the water is of inferior quality, (*vii*) *Climate and Season:* Since water duty includes losses of water which, in turn, vary with climate and season, more precisely temperature, humidity, wind velocity, etc., the duty, therefore, varies from season to season and from time to time, during the same season. The values of duties for various crops generally referred to are the average values considered over whole crop period, (*viii*) *Rainfall:* Useful rainfall in the crop period meets the part requirement of water and hence duty of water is higher. Duty exclusive of rainfall is low, (*ix*) *Base Period:* Different crops have different base periods. Smaller the base period lesser is the amount of water required and higher is the duty and vice versa, (*x*) *Method of Assessment:* Volumetric system of assessment implies more efficient and economical use of water and hence higher duty of water because the cultivator has the tendency to cultivate more area with the same amount of water. In the area under conventional method of assessment, the duty is low, because the farmer has a tendency to use more water in the same area, (*xi*) *Canal Section:* Earthen section of the canal entails more transmission losses and hence low duty of water. Lined section results in low transmission losses and hence higher duty of water. Canal Network: Canal with small network, *i.e.*, having concentrated command area means higher duty of water and vice versa, and (*xii*) *Topography of Land:* A level field means uniform application of water, but if it is not level, low portions receive more depth of water than higher level portions which implies wasteful use of water and hence less duty. Properly levelled field means more economical use of water and high duty.

Methods of Improving Duty

Various methods of improving duty are (*i*) Land should be properly ploughed, and tilled, (*ii*) Correct quantity and timing of water application mean higher duty. Drip irrigation is the most modern and efficient water application method, (*iii*) Canals be lined to cut down transmission losses in the canal system. Higher velocity in lined section also results in reduced evaporation losses, (*iv*) The canal should be nearest to the command area so that idle length of the canal is minimum and hence reduced transmission losses, (*v*) Good quality of water should be used for irrigation. Pollution enroute the canal should be avoided. Canal alignment through the soils which may dissolve harmful salts in the canal water should be avoided, (*vi*) Rotation of crops be practised, and (*vii*) Volumetric assessment of water is enforced so that efficient and economical use of the canal water is made by the cultivators.

2.19 IRRIGATION SYSTEM EFFICIENCY

Irrigation system efficiency, implies minimum loss from the time water is taken from the canal outlet till it becomes available in the root zone for plant utilization. The objective of the efficiency concept is to show where improvements can be made so as to have more efficient irrigation. Method of measurement and evaluation of performance in terms of efficiency are prerequisites for proper use of irrigation water. Generally speaking water use in an irrigation project under development has lower efficiency. However, as water becomes more and more scarce and as the need for it becomes more pressing, its utilization improves. Various concepts of irrigation efficiencies are as under:

1. *Irrigation Efficiency:* It is defined as the ratio of water output to the water input, *i.e.*, the ratio or percentage of the irrigation water consumed by the crops of an irrigated farm, field or project to the water delivered from the source of supply and is usually expressed as percentage. In view of water losses occurring at various stages before the plants ultimately utilize it, several partial efficiencies have been defined. For maximum overall efficiency the various partial efficiencies should individually be maximum:

$$E_i = \frac{W_c}{W_r} \times 100$$

where, E_i = irrigation efficiency (%), W_c = irrigation water consumed by crop during its growth period in an irrigation project, and W_r = water delivered from canals during the growth period of crops.

In most irrigation projects, the irrigation efficiency ranges between 12 to 34 per cent.

2. *Water Conveyance Efficiency:* It is a measure of efficiency of water conveyance system from canal network to watercourses and field channels. It is the ratio of water delivered in fields at the outlet head to that diverted into the canal system from the river or reservoir. Water losses occur in conveyance from the point of diversion till it reaches the farmer's fields which can be evaluated by water conveyance efficiency, as under:

$$E_c = (W_f/W_r) \times 100$$

where, E_c = water conveyance efficiency, per cent, W_f = water delivered to the farm by conveyance system (at field supply channel), and W_r = water introduced into the conveyance system from the point of diversion.

Water conveyance efficiency is generally low; about 21% losses occur in earthen watercourses only.

3. *Water Application Efficiency:* It is a measure of efficiency of water application in the field. It is the ratio of volume of water that is stored in the root zone of crops and ultimately consumed by transpiration or evaporation or both to the volume of water actually delivered at the field. Alternatively, it may be defined as the percentage of water applied that can be accounted for as increase in soil moisture in soil as occupied by principal rooting system of the crop. It is also termed as *farm efficiency* as it takes into account water lost in application at the farm.

$$E_a = (W_s/W_f) \times 100$$

where, E_a = water application efficiency, percent, W_s = irrigation water stored in the root zone of farm soil, and W_f = irrigation water delivered to the farm (at field supply channel).

In general, water application efficiency decreases as the amount of water during each irrigation increases. Water losses due to inefficient application of water in the field vary from 28 to 50 percent.

Common sources of loss of irrigation water during application are represented as R_f = surface runoff from the farm, and D_f = deep percolation below the farm root zone soil. Neglecting evaporation losses during application:

$$W_f = W_s + D_f + R_f$$

$$E_a = \left[\frac{W_f - (D_f + R_f)}{W_f}\right] \times 100$$

4. *Water Use Efficiency:* Having conveyed water to the point of use and having applied it, the next efficiency concept of concern is the efficiency of water use. It is expressed in kg/ha cm. The proportion of water delivered and beneficially used on the project can be calculated using the following formula:

$$E_u = (W_u/W_d) \times 100$$

where, E_u = water use efficiency, percent, W_u = water beneficially used, and W_d = water delivered.

Water use efficiency is also defined as (*i*) crop water use efficiency and (*ii*) field water efficiency.

(*a*) *Crop Water Use Efficiency:* It is the ratio of yield of crop (*Y*) to the amount of water depleted by crop in evapotranspiration (*ET*) *i.e., Y/ET.*

(*b*) *Field Water Use Efficiency:* It is the ratio of yield of crop (*Y*) to the total amount of water used in the field *i.e., Y/WR.*

5. *Water Storage Efficiency:* Also termed as water storage factor. It is defined as the ratio of the water stored in the root depth by irrigation to the water needed in the root depth to bring it to the field capacity *i.e.,* $E_s = (W_s/W_w) \times 100$

where, E_s = water storage efficiency, percent, W_s = water stored in the root zone during the irrigation, and W_w = water needed in the root zone prior to irrigation, *i.e.,* field capacity–available moisture.

6. *Water Distribution Efficiency:* It is the expression for distribution efficiency to evaluate the extent to which the water is uniformly distributed *i.e.,* $E_d = (1 - d/D) \times 100$

$$= \left(1 - \frac{\text{average deviation}}{\text{average depth applied}}\right) \times 100$$

where, E_d = water distribution efficiency, per cent, d = average numerical deviation in depth of water stored from average depth stored during irrigation, and D = average depth of water stored along the run during irrigation.

A water distribution efficiency of 80% means that 10% of water was applied in excess and consequently 10% was deficient in comparison to the average depth of application.

7. *Consumptive Use Efficiency:* It is defined as the ratio of consumptive water use by the crop of irrigated farm or project and the irrigation water stored in the root zone of the soil on the farm or the project area. After irrigation water is stored in the soil, it may not be available for use by the crop because water may evaporate from the ground surface or continuously move downward beyond the root zone as it may happen in a wide furrow spacing. The loss of water by deep penetration and by surface evaporation following an irrigation is evaluated from the following expression:

$$E_{cu} = (W_{cu} / W_d) \times 100$$

where, E_{cu} = consumptive use efficiency, percent, W_{cu} = normal consumptive use of water, W_d = net amount of water depleted from root zone soil.

Consumptive use efficiency is useful in explaining the difference in crop response from different methods of irrigation.

2.20 ASSESSMENT OF IRRIGATION WATER

Agriculture accounts for substantial use of water resources. Irrigation projects are planned, executed, owned and operated by State Governments. Irrigation projects are financed by State Governments out of their own resources, market borrowing and loans and grants made available by the Central Government. The States have to repay the loan and interest, besides they incur expenditure on operation and maintenance of the projects for supply of water to cultivators year after year. The beneficiaries in commanded areas derive considerable benefits by way of (*i*) increased production with insurance against crop failure, (*ii*) use of fertilizers, improved seeds, pesticides, etc., with the availability of irrigation facilities, (*iii*) increased cost of land due to advent of irrigation. It is, therefore, imperative that the beneficiaries pay for it. The fixation of such charges is termed as assessment of irrigation water. The basis of recovery of water rates is (*i*) Adequate to cover the annual running and maintenance charges and a part of fixed costs *i.e.*, interest on capital and depreciation, (*ii*) Levied on 'crop basis' except in tubewell irrigation and to 'Water users' Associations, (*iii*) For fixing rates, irrigation should be divided into categories A, B and C depending on the quality of service, that is, on the basis of quantity and timeliness of supply. Lower rates may be fixed where on account of good rainfall, the demand for irrigation water is less or where the supply is inadequate and uncertain, (*iv*) Rates should be uniform for same class of supply over a State or region and the disparity in rates for same class of supply between the regions, if any, should be minimum, and (*v*) Rates should be revised after every five years.

Irrigation water rates vary from State to State. They comprise one or more of the following:

(*i*) The kind of crop as water rate is levied on crop area basis.

(*ii*) Betterment levy representing Government's share in the increase in value of land accruing with the provision of irrigation facility.

(*iii*) Irrigation cess being the annual charges on the irrigable area whether water is actually taken or not to be adjusted in the water rate when the facility is made available.

Irrigation Commission, India (1972) has recommended that the charge to cultivator should be 5 to 12 per cent of gross income, the upper limit being applicable to cash crops and lower limit to food and fodder crops. IS : 14519 – 1998 has stipulated that average cost per unit of water, that is, the total amount of the variable and fixed costs, per unit of water in respect of each class of irrigation projects in a State or region for 3 years immediately preceding the last year or the current block period of 5 years should be worked out. Quantum of water to be considered would be the annual utilisation on each project. Interest on capital cost is to be taken at 10 per cent per annum. The life of major and medium irrigation projects is generally assumed as 100 years and of minor irrigation projects as 50 years. Accordingly depreciation is 1 percent and 2 percent of capital cost respectively. For lift irrigation schemes, life of civil works is considered as 25 years and that of pumping machinery and ancellaries as 12 years and the depreciation is accordingly 4 and 8.33 percent respectively.

The cost per unit of water for the first year and the fifth year of the next block period of 5 years should then be worked out by applying escalation at the rate of 5 per cent per annum to the variable cost. The average of these two values give the cost per unit of water for the next block period of 5 years.

Methods of Assessment

1. *Assessment on Area Basis or Crop Rate Basis.* In this method the rates are related neither to the cost of supplying water nor to the quantity of water used. These are fixed charges for the different types of crops on area basis. These rates are quite low, having been fixed long back. The water rate should be within 5 and 12 percent of gross income from that crop.

This method has disadvantages: (*i*) Cultivators make wasteful use of water as the charges are not levied on the basis of actual quantity of water used, (*ii*) Cultivators near the head reaches of the canal use more water than those near the tail resulting in inequitable distribution of supply, and (*iii*) Tendency amongst cultivators for over-irrigation of the limited area, rather than bringing more area under irrigation by judicious application of water.

2. *Volumetric Assessment:* In this method, water rates are charged on the basis of actual volume of water supplied at outlet head. Essentially it requires installation of numerous water meters at all irrigation outlets in the canal system to measure the quantity of supply delivered at a high initial and subsequent maintenance costs. It is an ideal method and the irrigator has the incentive for economical use of water but is not practicable due to certain limitations. The method has, however, found use in tubewell irrigation where water is sold out on volumetric basis, *i.e.*, litres per rupee or unit rate of electric power consumed on motor horse power basis. The method can be enforced, with advantage, in case water is sold to cultivators through their co-operative societies or panchayats who in turn are made responsible for payment of water charges and internal water distribution.

Limitations. (*i*) Individual holdings are small and numerous and it is not possible to provide a meter to each and every small holding, (*ii*) No suitable cheap and foolproof water measuring device has been developed as yet which can be used in the existing water distribution system, (*iii*) High initial cost on account of large number of meters required, (*iv*) Internal distribution of volumetric supply and apportionment of the charges amongst land holders may be difficult, (*v*) Large strength of staff required to keep records of meters, and (*vi*) Meters susceptible to damage by silt and debris.

3. *Assessment on Seasonal Basis:* This method is not in vogue in India. In this method assessment is made on the basis of kind of crops grown in an area during certain crop season.

4. *Composite Rate Basis:* In this method composite rate which is a combination of water charges and land revenue are levied.

5. *Permanent Assessment or Betterment Levy Basis:* It pertains to a particular area where irrigation is introduced as protection against drought and normally the cultivators use their own means of irrigation. A fixed rate, termed as betterment levy, is charged every year. In a drought year, the cultivators are allowed to use canal water without paying extra charges above betterment levy.

To irrigate one crop hectare with canal water it costs the nation Rs. 4000 in fixed and variable expenses (in 2000 AD) but barely 5 per cent of this cost was recovered through irrigation fees levied on farmers.

SOLVED EXAMPLES

Example 2.1 *After how many days you will supply water to soil in order to ensure sufficient irrigation of the given crop, if field capacity of the soil = 30%, permanent wilting point = 12%, density of soil = 1.25 gm/cc, effective depth of root zone = 60 cm, and daily consumptive use of water for the given crop = 12.5 mm.*

Assume that readily available moisture is 80% of the available moisture.

Solution. Available moisture = Field capacity-permanent wilting point

= 30 – 12 = 18%.

Readily available moisture @ 80% = 18 × 80/100 = 14.4%.

Optimum moisture = Field capacity—readily available moisture

= 30–14.4 = 15.6%.

It follows that moisture will be filled by irrigation between optimum moisture and field capacity.

$$\text{Depth of water} = \frac{\gamma d}{w}(F - M_0) = \frac{1.25 \times 0.60}{1.0}(0.30 - 0.156) \times 100 = 10.80 \text{ cm.}$$

No. of days after which water should be supplied = D/C_u = 10.80/1.25 ≈ 9 days.

Example 2.2 *Compute the depth and frequency of irrigation required for a certain crop if root zone depth = 100 cm, field capacity = 22%, wilting point = 12%, apparent specific gravity of soil = 1.5 gm/cc, consumptive use = 25 mm/day, and efficiency of irrigation = 50%. Assume 50% depletion of available moisture before application of irrigation water at field capacity.*

Solution: Available moisture = Field capacity—permanent wilting point.

= 22 – 12 = 10%

Readily available moisture at 50% depletion of available moisture

= 10 × 50/100 = 5%

Optimum moisture = Field capacity—readily available moisture

= 22 – 5 = 17%

Depth of water stored in root zone

$$= \frac{\gamma d}{w}(F - M_0)$$

$$= \frac{1.5}{1.0} \times 1.0\,(0.22 - 0.17) \times 100 = 7.5 \text{ cm}$$

Field irrigation required at 50% efficiency = 7.5/0.50 = 15 cm

Water application frequency = D/C_u = 15/2.5 = 6 days.

Example 2.3 *Water is released at the rate of 10 cumecs at the head sluice. If duty at the field is 125 hectares/cumec and loss of water in transit is 30%, find the area of the land that can be irrigated.*

Solution. Area irrigated with 10 cumecs = 125 × 10 = 1250 ha

Actual area irrigated with 30% wastage = 1250 × 0.70 = 875 ha.

Example 2.4 *Determine seasonal consumptive use for wheat when mean monthly temperature and monthly percentage of day-time hours in the months of Nov., Dec., Jan., Feb. are 23, 20 16, 18 and 7.40, 7.42, 7.53, 7.14 respectively. Consumptive coefficient = 0.70.*

Solution.

$$C_u = K \Sigma\, p/40\,(1.8\,T + 32)$$

$$= 0.70/40\,[7.4\,(1.8 \times 23 + 32) + 7.42\,(1.8 \times 20 + 32) + 7.53\,(1.8 \times 16 + 32) + 7.14\,(1.8 \times 18 + 32)]$$

$$= 0.70/40\,(1965.36) = 34.24 \text{ cm.}$$

Example 2.5 (*a*) *Determine pan evaporation for the month of March at a place (latitude 25°N, mean elevation + 500 m), monthly temperature 25°C, mean wind velocity at 0.5 m above ground is 100 km/day, mean relative humidity 40%, mean sunshine percent = 80%, R = 42.647.* (*b*) *What is the consumptive use for a crop having consumptive use coefficient 0.70.*

Solution.

$$C_t = 0.393 + 0.02796 \times 25 + 0.0001189\,(25)^2 = 1.166$$

$$C_w = 0.708 - 0.0034 \times 100 - 0.0000038\,(100)^2 = 1.086$$

$$C_h = 1.25 - 0.0087 \times 25 + 0.75 \times 10^{-4}\,(25)^2 - 0.85 \times 10^{-8}\,(25)^4 = 1.076$$

$$C_s = 0.542 + 0.008 \times 80 - 0.78 \times 10^{-4}\,(80)^2 + 0.62 \times 10^{-6}\,(80)^3 = 1.000$$

$$C_e = 0.97 + 0.00984 \times 5.0 = 1.019.$$

Now, pan evaporation,

$$E_p = 0.459\,R\,C_t\,C_w\,C_h\,C_s\,C_e$$

$$= 0.459 \times 42.647 \times 1.166 \times 1.086 \times 1.076 \times 1.0 \times 1.019 = 23.31 \text{ cm}$$

(*b*) C_u = 0.70 × 23.31 = 16.32 cm.

Example 2.6 *Determine consumptive use and net irrigation requirement for a crop from the data as under:*

Period	*Oct. 10–31*	*Nov. 1–30*	*Dec. 1–31*	*Jan. 1–31*	*Feb. 1–15*
Pan evaporation, E_p (cm)	5	10	9	8	4
C_u Coefficient, K	0.5	0.55	1.0	0.9	0.7
Effective rainfall, R_e (cm)	2	1.5	1	0.5	0.2

Assume $N_a = N_c$ = 70%. Also determine field irrigation requirement and gross irrigation requirements.

Solution.

Period	Oct. 10–31	Nov. 1–30	Dec. 1–31	Jan. 1–31	Feb. 1–15	Total
$C_u = K E_p$	2.5	5.5	9	7.2	2.8	27cm
$CIR = C_u - R_e$	0.5	4.0	8.0	6.7	2.6	21.8 cm

Total consumptive use = 27 cm
Consumptive irrigation requirement = 21.8 cm
Net irrigation requirement = CIR as Le is zero
Field irrigation requirement = 21.8/0.70 = 31.43 cm
Gross irrigation requirement = 31.43/0.70 = 44.90 cm.

Example 2.7 *Determine delta for a crop having base period 120 days, water depth for one irrigation 12 cm and successive irrigation interval of 12 days.*

Solution. Number of waterings required = Base period/interval = 120/12 = 10
Delta = 10 × 12 = 120 cm

Example 2.8 *The gross area of an irrigation project is 50,000 ha. Out of this, about 5,000 ha have been utilized for construction of dwellings, roads, bridges, etc. The area to be cultivated during rabi (i.e., winter season) is 25,000 ha and during kharif (i.e., summer season) is 24,000 ha. The duty of canal water rabi crops is 5,000 ha per cumec and for kharif crops is 3,000 ha per cumec. Find the design discharge for the canal after giving 10% allowance for peak discharge and loss of water in transit. What would be the annual intensity of irrigation?*

Solution.

CCA (ha)	Season	Area to be irrigated (ha)	Duty (ha/cumec)	Peak discharge @10% allowance (cumec)
50,000 – 5000 = 45,000	Rabi	25000	5000	25000/5000 = $\frac{5 \times 100}{90}$ = 5.56
	Kharif	24000	3000	24000/3000 = $\frac{8 \times 100}{90}$ = 8.89
Intensity of irrigation	Rabi	(25000/45000) × 100 = 55.5%		
	Kharif	(24000/45000) × 100 = 53.3%		

Annual intensity (combined) = [(25000 + 24000)/45000] × 100 = 109%

Example 2.9 *The culturable commanded area for a distributary is 15,000 ha. The intensity of irrigation is 40% for rabi and 10% for rice. If kor period is 4 weeks for rabi and 2.5 weeks for rice, determine the outlet discharge. Outlet factor for rabi and rice may be assumed as 1800 ha/m³/sec and 775 ha/m³/sec. What is design discharge at distributary head at 10% conveyance loss.*

Solution.

Season	Area to be irrigated (ha)	Outlet factor (ha/cumec)	Outlet discharge (cumec)
Rabi	15000 × 0.40 = 6000	1800	6000/1800 = 3.33
Kharif	15000 × 0.10 = 1500	775	1500/775 = 1.94

Design discharge of canal is 3.70 cumec (3.33/0.90) being higher than kharif season discharge.

Example 2.10 *A watercourse has a culturable commanded area of 1500 hectares. The intensity of irrigation for crop A is 45% and for B is 40%, both the crops being rabi crops. Crop A has a kor period of 20 days and crop B has kor period of 15 days. Calculate the discharge of watercourse if the kor depth for crop A is 10 cm and for B it is 16 cm.*

Solution.

Crop	Area under irrigation (ha)	Kor period (days)	Kor depth (cm)	Outlet factor (ha/cumec)	Discharge (cumec)
A	1500 × 0.45 = 675	20	10	(8.64 × 20)/0.10 = 1728	675/1728 = 0.39
B	1500 × 0.40 = 600	15	16	(8.64 × 15)/0.16 = 810	600/810 = 0.74

∴ Design discharge of watercourse in rabi = 0.39 + 0.74 = 1.13 cumecs.

Example 2.11 *Calculate the design discharge of canal at 0, 1, 2 and 3 km from headworks. The gross commanded area at the head of the canal is 60,000 hectares and after each km it is reduced by 10,000 hectares. Of this command, the culturable area is only 75%. The intensity of irrigation for rabi and kharif seasons is 35% and 18% respectively. Assume total losses below km 3 = 0.5 cumec, channel losses per km = 3% of discharge at beginning of each km, kor period rabi (wheat) and kharif (rice) is 4 and 2.5 weeks respectively. Kor depth for rabi and kharif is 14 and 20 cm respectively.*

Solution. Given data is tabulated as under:

Below km	Gross commanded area (ha)	CCA @ 75% of GCA (ha)
At head	60,000	45,000
1	50,000	37,500
2	40,000	30,000
3	30,000	22,500

Outlet discharge

Rabi	Outlet discharge	Kharif
Outlet factor = (8.64 × 28)/0.14	= 1728 ha/cumec	(8.64 × 17.5)/0.20 = 756 ha/cumec
Discharge required = 0.35 CCA/1728	= CCA/4937	(0.18 CCA)/756 = CCA/4200

Canal discharge at various points during Kharif being more than Rabi.

Below km	CA (ha)	Discharge = CA/4200	Discharge below reach	Losses in reach	Total Q in reach (cumecs)
0	45,000	10.71	13.58 + 8.93 = 22.51	(22.51 × 3/100) = 0.68	23.19
1	37,500	8.93	6.04 + 7.14 = 13.18	13.18 × 3/100 = 0.40	13.58
2	30,000	7.14	5.86	(5.86 × 3/100) = 0.18	6.04
3	22,500	5.36	0.50	Nil = 0.50	5.86

Total discharge at the head (R.D.O.) is 33.90 cumecs (23.19 + 10.71).

Example 2.12 *Calculate the design discharge with the data given in example 2.11 if water losses in the canal are of the order of 2 cumec/million sq m of wetted perimeter.*

Solution. The design discharges at various points:

RD (km)	GA	CCA @75% (ha)	Discharge CCA/4200 (cumecs)	Discharge d/s of reach	Loss in reach (at 2 cumecs/10^6 m² of P_w) $P_w = 4.75\,Q^{1/2}$	Actual discharge through the reach (cumecs)	Discharge at reach head (cumecs)
0	60,000	45,000	10.71	22.51	$\frac{4.75(22.51)^{1/2}\times 2\times 10^4}{10^6} = 0.45$	22.96	33.67
1	50,000	37,500	8.93	13.23	$\frac{4.75(13.23)^{1/2}\times 2\times 10^4}{10^6} = 0.35$	13.58	22.51
2	40,000	30,000	7.14	5.86	$\frac{4.75(5.86)^{1/2}\times 2\times 10^4}{10^6} = 0.23$	6.09	13.23
3	30,000	22,500	5.36	0.50	—	5.86	5.86

Example 2.13 *Determine reservoir capacity for:*
Command area of 40,000 ha, canal losses = 10% and reservoir losses = 10%
Base period, outlet factor and intensity of irrigation are as under:

Crop	*Base period (days)*	*Outlet factor (ha/ cumec)*	*Irrigation intensity (%)*	*Crop*	*Base period (days)*	*Outlet factor*	*Irrigation intensity*
Sugarcane	360	1700	20	Rice	120	700	15
Cotton	180	1500	10				
Wheat	120	1800	20	Vegetables	120	700	15

Solution.

Crop	*Area under irrigation (C.A. × intensity) (ha)*	*Outlet factor*	*Discharge (Area/outlet factor) (cumecs)*	*Base period days*	*Volume of water (cumec days = cumecs × base period)*
Sugarcane	8000	1700	4.71	360	1695.60
Cotton	4000	1500	2.67	180	480.60
Wheat	8000	1800	4.44	120	532.80
Rice	6000	700	8.57	120	1028.40
Vegetables	6000	700	8.57	120	1028.40
				Total	4765.80

Volume of water required for these crops
(a) In the field = 4765.80 cumec days
(b) At canal head with = 5295.33 cumec days
10% losses, (4765.80 × 100/90)
(c) In reservoir, with 10% losses = (5295.33 × 100/90) = 5,883.70 cumec days,
or 5883.70 × 8.64 = 50,835 ha m

Example 2.14 *Determine uniformity coefficient for depths of penetration at five points in a field 1.0, 0.75, 0.60, 0.50 and 0.40 m.*

Solution.

Mean water depth, $D = (1.0 + 0.75 + 0.60 + 0.50 + 0.4)/5 = 0.65$ m

Values of deviation from mean = (1.0 – 0.65 = 0.35), (0.75 – 0.65 = 0.10), (0.65 – 0.60 = 0.05), (0.50 – 0.65 = – 0.15), (0.40 – 0.65 = – 0.25)

Absolute values of deviation from the mean = 0.35, 0.10, 0.05, 0.15, 0.25 respectively.

Average of absolute values of deviation from the mean, $d = (0.35 + 0.10 + 0.05 + 0.15 + 0.25)/5 = 0.18$ m.

Uniformity coefficient $= (1 - d/D) = (1 - 0.18/0.65) = 0.723$.

Example 2.15 *Work out irrigation efficiency if water conveyance and delivery loss = 40%, deep percolation and surface runoff in farm = 30%, and water stored in soil lost by evaporation = 20%.*

Solution. Water delivered to farm in given period of time,

$W_f = 0.6\ W_r$ at 40% water conveyance and delivery loss.

Water stored in the root zone of soil, $W_s = 0.7\ W_f = 0.7 \times 0.6\ W_r = 0.42\ W_r$.

Water consumed, $W_c = 0.8\ W_s = 0.8 \times 0.42\ W_r = 0.34\ W_r$.

Hence, irrigation efficiency, $E_i = (W_c/W_r) \times 100 = [(0.34 \times W_r) / W_r] \times 100 = 34\%$.

Example 2.16 *A stream of 125 lit/s was diverted from a canal and 100 lit/s were delivered to the field. An area of 1.6 hectares was irrigated in 8 hours. Effective depth of root zone was 1.7 m. The runoff loss in the field was 420 cum. The depth of water penetration varied linearly from 1.7 m at the head end of the field to 1.3 m at the tail end. Available moisture holding capacity of the soil is 20 cm per metre depth of soil. Determine water conveyance, water application, water storage and water distribution efficiencies. The irrigation was started at a moisture extraction level of 50% of the available moisture.*

Solution.

(*i*) Water conveyance efficiency (η_c) = (100/125) × 100 = 80%.

(*ii*) Water supplied to field = 100 × 8 × 60 × 60 = 2880 cu m.

Water stored in the root zone = 2880 – 420 = 2460 cu m.

Water application efficiency (η_a) = (2460/2880) × 100 = 85.4%.

(*iii*) Moisture holding capacity of soil = 20 × 1.7 m = 34 cm.

Moisture already available = 34 × 0.5 = 17 cm.

Additional water depth required = 34 – 17 = 17 cm.

Additional water required in root zone in 1.6 ha = 0.17 × 1.6 × 10000 = 2720 cu m.

Water storage efficiency (η_s) = (2460/2720) × 100 = 90%

(*iv*) Water distribution efficiency (η_d)

Mean depth of water stored in root zone = (1.7 + 1.3) / 2 = 1.5 m

Deviation from mean value at upper end (absolute values) = 1.7 – 1.5 = 0.2 m

Deviation from mean value at lower end = 1.5 – 1.3 = 0.2 m.

Average absolute values of deviation from mean (*d*) = (0.2 + 0.2) / 2 = 0.20

Water distribution efficiency (η_d) = (1 – 0.20/1.5) × 100 = 86.7%.

Example 2.17 *An area of 300 ha is to be irrigated from a minor channel with one outlet, C.C.A. is 80% of total area. The intensity of irrigation is 50% for rabi crop and 30% for kharif crop. Taking loss in conveyance system as 5% of outlet discharge, determine the head discharge of the channel. Take outlet discharge factor for wheat season as 1,500 ha/m³/sec and for rice season 1,000 ha/m³/sec.*

Solution.

Crop	*C.C.A (ha)*	*Area under irrigation (ha)*	*Outlet discharge (cumec)*	*Conveyance loss @ 5%*	*Head discharge (cumec)*
Rabi	300 × 0.8 = 240	240 × 0.5 = 120	120/1500 = 0.08	0.004	0.084
Kharif	240	240 × 0.3 = 72	72/1000 = 0.072	0.004	0.076

∴ Design discharge of channel at head = 0.084 cumec.

QUESTIONS

1. Define the following terms as used in irrigation practice:
   (*i*) available soil moisture, (*ii*) effective rainfall, (*iii*) application efficiency of irrigation, and (*iv*) base period.
2. Define the following terms as used in irrigation practice:
   (*i*) field capacity, (*ii*) available moisture, and (*iii*) consumptive use.
3. Explain depth and frequency of irrigation.
4. Discuss rotation of crops.
5. Distinguish between:
   (*a*) Alkaline soil and saline soils (*b*) Crop rotation and mixed cropping
   (*c*) Base period and kor period
6. (*a*) Define duty, delta and base period. Derive a relation between them.
   (*b*) Find field capacity of a soil if root zone depth = 2.5 m, existing water content = 7.5%, dry density of soil = 1.5 gm/cm³, water applied to the soil = 600 m³, water losses due to evaporation = 15% and area of plot = 1000 sq m.
   Solution
   Water actually used = 0.85 × 600 = 510 m³ = 510 × 10^6 gm
   Total dry weight of the soil = (2.5 × 1.5 × 1000) × 10^6 gm

Percentage of water added $= \dfrac{510 \times 10^6}{2.5 \times 1.5 \times 1000 \times 10^6} \times 100 = 13.6\%$

New water content $= 13.6 + 7.5 = 21.1\%$

$\therefore$ Field capacity $= 21.1\%$.

7. (*a*) Explain field capacity, moisture equivalent and wilting point. How are these terms related to the quantity of irrigation water to be supplied and frequency of application? Explain (*b*) After how many days the farmer should apply water to his field to ensure efficient use of irrigation water, if the field capacity is 27%, permanent wilting point 14%, density of soil 1500 kg/m^3, effective root zone depth 0.75 m and daily consumptive use of water is 11 mm.
8. Define: (*i*) base period, (*ii*) intensity of irrigation, (*iii*) duty, (*iv*) culturable commanded area, and (*v*) capillary water.
9. Name the principal rabi crops of your region and give an account of the agricultural and climatic requirements for sowing, growth and harvesting of one of them. Also indicate the normal requirement of seed per hectare and average yield per hectare for the crop.
10. What is meant by the term duty of canal water? Obtain the relationship between duty and delta for a given base period.
11. Define the terms as used in irrigation practice: (*i*) field capacity, (*ii*) available moisture, (*iii*) consumptive use, and (*iv*) duty.
12. Define the following terms as used in irrigation practice: (*i*) available soil moisture, (*ii*) effective rainfall, (*iii*) application efficiency of irrigation, (*iv*) base period.
13. Explain how the following factors effect duty of a crop: (*i*) soil and subsoil condition, (*ii*) stage of growth, (*iii*) temperature, (*iv*) rainfall.
14. Define consumptive use. What are the factors which influence consumptive use of water for any crop and how consumptive use is related to the total requirement of water for the crop?
15. What is meant by evapotranspiration?
16. (*a*) Explain the terms: (*i*) delta, (*ii*) duty, (*iii*) culturable commanded area, and (*iv*) inten sity of irrigation.

    (*b*) The gross area of an irrigation project is 50,000 hectares. Out of this, about 5,000 hectares have been utilized for construction of dwellings, roads, bridges, etc. The area to be cultivated during rabi (*i.e.*, winter season) is 35,000 hectares and during kharif (*i.e.*, summer season) 25,000 hectares. The duty of canal water for rabi crop is 5,000 hectares per cumec and for the kharif crop is 3,000 hectares per cumec. Find the design discharge for the canal after giving a 10% allowance for peak discharge and loss of water in transit. What would be the annual intensity of irrigation?
17. (*a*) Explain what you understand by the term irrigation efficiency?

    (*b*) The culturable commanded area for a reservoir is 60,000 hectares. Find out the reservoir capacity if canal losses are 7.5%, reservoir losses 10.5% and other data are given as follows:

Crop	*Base period (days)*	*Duty (ha/cumec)*	*Irrigation intensity (%)*
Rice	130	1000	10
Wheat	120	2400	15
Cotton	180	1800	12
Sugarcane	360	3000	18

18. Why is it necessary to collect the charges for irrigation water supplied to the cultivators?
19. Distinguish between saline and alkaline soils. Explain the process of reclamation of such soils.

20. (a) What are the methods of maintaining soil fertility?
    (*b*) How will you judge the suitability of water for an irrigation project?
21. Discuss physical properties of soils in the context of irrigation. What are the functions of irrigation soils?
22. Classify soils from the stand point of agriculture.
23. Mention the method for maintaining soil fertility.
24. Discuss various factors which determine the soil-water-plant relationship.
25. Define permeability. Why is it necessary that a good irrigation soil be permeable? How is permeability determined?
26. Write brief notes on: (*i*) Weather is relation to crops, (*ii*) Soil texture and soil structure, (*iii*) Field capacity, (*iv*) Wilting coefficient, (*v*) Crop rotation. [P.U.]
27. (a) Define : (*i*) Base period, (*ii*) Intensity of irrigation, (*iii*) Culturable commanded area, (*iv*) Capiliary water and (*v*) Field capacity.
    (*b*) A watercourse has a culturable commanded are of 1200 hectares. The intensity of irrigation for wheat crop is 45% and for rice 20%. Wheat has a kor period of 20 days and rice has a kor period of 35 days. Calculate the head discharge of watercourse if kor depths for wheat and rice are 10 cm and 16 cm respectively and losses in the conveyance system are 5% of outlet discharge. [P.U.]
28. (a) Define GCA, CCA and intensity of irrigation. How are these estimated?
    (*b*) A watercourse has a CCA of 1000 hectares. The intensity of irrigation for wheat crop is 45% and for rice crop is 30%. Wheat has a kor period of 28 days and rice has a kor period of 20 days. Calculate the outlet discharge. Kor depths for wheat and rice are respectively 12 cm and 20 cm. [P.U. Nov., 1980]
29. Discuss volumetric sale of water versus sale on basis of area. [P.U.]
30. (a) Explain the necessity of irrigation in India.
    (*b*) Define duty of a canal. State briefly the measures you would recommend for improving duty of the canal. [Kurukshetra University]

3

IRRIGATION SYSTEMS

3.1 INTRODUCTION

Precipitation is the principal source of water, part of which percolates underground and a part flows over the surface as runoff. Thus there are two main sources of water for irrigation viz., surface water and groundwater. The rivers carrying surface water may be snowfed or perennial capable of feeding the offtaking canals throughout the year or they may carry seasonal supply and the canals taking off from them may carry water for seasonal irrigation only. Furthermore, the source of water may be at such a level that it can command the area by gravity flow or alternatively the water may have to be lifted to the desired level to irrigate the fields. The design, equipment and technique of replenishing the soil water deficit by applying irrigation water is referred to as irrigation system.

3.2 IRRIGATION SYSTEMS

The various systems of irrigation have been developed depending on the nature of source of water and location of command area such as gravity irrigation, lift irrigation (which includes water lifted from canals, wells, tubewells etc.), tank irrigation, etc. as illustrated in Fig. 3.1.

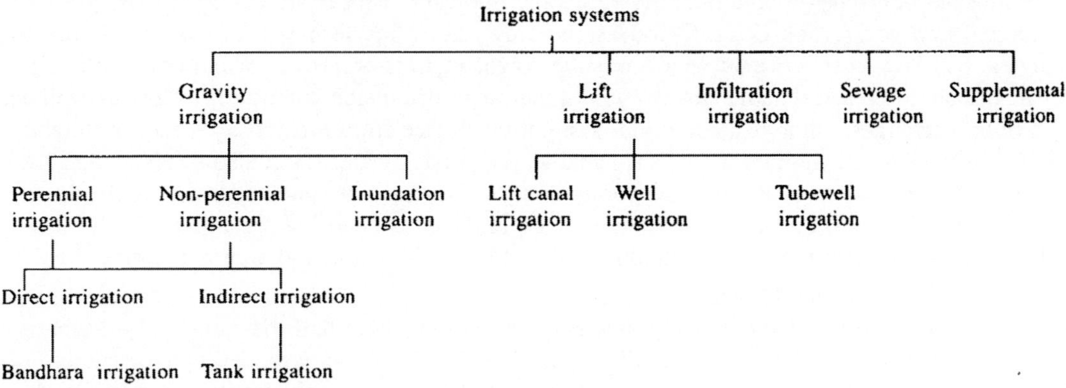

Fig. 3.1 Types of irrigation systems

3.2.1 Gravity Irrigation

Gravity or flow irrigation is the type of irrigation in which water is available at a higher level so as to enable supply to the land by gravity flow. In flow irrigation water is supplied to the fields through the canals offtaking from headworks. Gravity flow irrigation is cheaper compared to lift irrigation. The gravity irrigation is further classified as under:

Perennial Irrigation: In this system assured supply of water throughout the crop period to meet irrigation requirement of the crops is made available to the command area through storage of water done at the dam or diversion of supply made by means of headworks at the offtake point of the canal. Perennial irrigation may be either direct or indirect, as follows:

Direct irrigation. In direct irrigation system water is directly diverted from the river into the canal by construction of diversion weir or barrage across the river without attempting to store water. This method is practised where the river has adequate perennial supply to feed the canal system at all times of crops periods.

Indirect irrigation. It is also termed as storage *irrigation.* Here water is stored in a reservoir during monsoon period by construction of a dam across the river for supply into the offtaking canals. Evidently indirect irrigation is adopted where the river is non-perennial or flow in the river is inadequate during lean period. Storage irrigation has greater irrigation potential than direct irrigation but is costly due to the cost of construction of dam.

Non-Perennial irrigation. Also called restricted irrigation. Canal supply is generally made available in non-monsoon period from the storage in small dams as in Kandi areas which is inadequate to feed all the year round, and/or canal water is not required during monsoon due to adequate rainfall in the command area.

3.2.2 Inundation Irrigation

Inundation irrigation is done by a canal taking off from a river in flood without any diversion work. It depends on the periodical rise in water level of the river and the supply is drawn through open cuts in the river bank or creeks which are called *heads.* Owing to changes in the river course, the heads have often to be changed. A regulator is, however, provided at the canal about 5 km downstream from the offtake, where the discharge passing below in the canal is controlled and the surplus supply is escaped back into the river. Inundation canals usually flow only during the summer months and bring in large quantity of silt beneficial to crops.

Selection of Offtake Point: The various factors which govern the selection of offtake point are (*i*) Site which is least likely to silt, (*ii*) Site on a creek of suitable size is preferred, since it is more constant in discharge than main river and also subject to milder attacks by floods, (*iii*) Site at the tail end of a creek is preferred as the flow there has less silt and is less liable to changes, (*iv*) Site on concave curve towards the downstream is preferred for minimum silt entry. A part of concave reach remains always within the range of outside curvature for low as well as high river discharges, otherwise a site in a straight reach free from erosion upstream is suitable, (*v*) Site with stable and high river banks is suitable, (*vi*) A back water reach also offers a good site, (*vii*) Wider river section at the offtake point which embodies two benefits, low velocity in the river and as such low silt charge in the water drawn in the canal, and low variation in water level, and (*viii*) If other conditions are favourable, a site close to the command area is preferred.

Design of Inundation Canal

The bed level in the channel is kept low enough to draw about half the full supply discharge of the canal in low river, *i.e.*, in the middle of April and again in October, and narrow enough to limit the excessive high floods to the minimum possible. Full supply level is fixed with due consideration to the steady water level in the river during about 1½ months, usually fair irrigation season. Beyond the head regulator, design considerations of unlined canals apply equally to the inundation canals. However, higher values of silt factor f and water surface slope are adopted than those allowed for canals offtaking from permanent headworks. Generally the slope ranges between 0.20 and 0.25 m/km. Manning's, Chezy or Lacey formulae may be used for the design of inundation canal.

Advantages. (*i*) Economical in cost being the simplest system of irrigation as permanent headworks is not to be constructed, (*ii*) Silty water carried by the canal has manurial value, and (*iii*) Minimum length of the canal, the area to be irrigated being close to the river.

Disadvantages. (*i*) High maintenance cost, and (*ii*) Seasonal irrigation.

Perennial Canal vs. Inundation Canal

Perennial canal	*Inundation canal*
Irrigation all the year round.	Irrigation during rainy season or winter floods only.
Supply in the canal is immune from fluctuations in river within working limits.	Supply in the canal depends on periodical rise of water level due to floods in the river.
Provided with a permanent headworks at offtake point.	Not provided with any headwork but a regulator at about 5 km downstream from offtake.
Escape channel is not normally required.	Escape channel is essentially provided.
Offtaking point is permanently fixed.	Provided with a number of heads or open cuts in the river bank.

Bandhara Irrigation

It envisages construction of a series of small dams or bandharas 1.25 to 4.5 m high across a river or stream to raise water level locally to feed offtaking canal and passing downstream excess supply over the top of bandhara which forms overflow crest. The supply downstream can be further arrested by constructing other bandharas in series depending on the availability of command area at each bandhara site. Usually one canal takes off from a bandhara and the supply into the canal is regulated by a head regulator. The length of canal is usually less than 7 km. Once the bandhara system is developed, the cultivators can maintain it. It is practised in Maharashtra state.

It is an economical type of irrigation. The area served by a bandhara may vary from a few hectares to 400 hectares. The area irrigated is termed as Thal which is divided into a number of parts called phad each of size 8 to 40 ha. A phad may contain a number of fields owned by various cultivators. In a particular year same kind of crop is grown in a particular phad. The other phad may have other type of crop. The area near the outlet is irrigated first and then the next until tail-end field is reached.

Site selection: The various considerations which govern selection of a site for a bandhara are: (*i*) Adequate supply, preferably perennial, (*ii*) Below the confluence of a tributary, (*iii*) Good foundation at reasonable depth at site to facilitate construction of bandhara, (*iv*) River section straight, narrow and well-defined, (*v*) High river banks at site, (*vi*) Command nearest to site, (*vii*) Accessibility of site, (*viii*) Good command of the offtaking canal, and (*ix*) Economical.

Design of Bandhara

A bandhara is a small diversion weir generally constructed in masonry on a stream or rivulet for utilisation of the perennial flow. The upstream face is usually vertical and the downstream face has a batter of 1 : 2 to 1 : 5. The crest width is kept equal to $\sqrt{h}$ or 1.25 m, where h is the height of bandhara. The section of the bandhara is checked for two conditions (*a*) When there is overflow over the bandhara, there is afflux and differential head, and (*b*) When there is no flow in the stream but water is headed up to the crest level on the upstream side and the downstream is dry.

The length of bandhara is determined from the formula:

$$Q = 1.7\, L.H^{3/2}$$

where, Q = discharge (cumecs), L = length of bandhara (m), and H = depth of water over the crest (m).

Sluices or suitable openings 1.0 × 1.25 m to serve as undersluices, are provided only in the part length of the bandhara near the head regulator. The sluices remove silt and sand thereby pass silt-free water in the canal (Fig. 3.2).

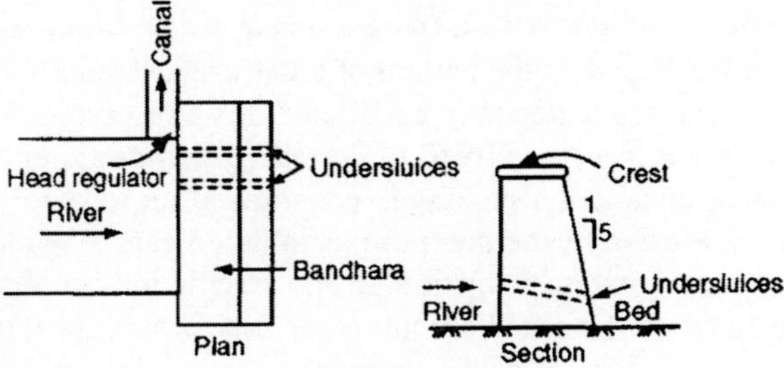

Fig. 3.2 Bandhara

Advantages. (*i*) Economical and eco-friendly type of irrigation, (*ii*) Once developed, the cultivators can maintain it, (*iii*) Utilisation of the small quantity of water available in the river or stream which would have otherwise run to waste, (*iv*) Intensive irrigation with high duty, and (*v*) Low seepage and evaporation losses as the canal length is very small.

Disadvantages. (*i*) Irrigable area being fixed, the supply if available in excess cannot be used, (*ii*) Supply is uncertain if the river is non-perennial, and (*iii*) By damming up whatever small supply is available in the river or stream, the areas lying down are deprived of this water.

Tank Irrigation

Tanks on local streams form a significant source of irrigation especially in the peninsular areas in the States of Karnataka, Maharashtra and Tamil Nadu. Tank irrigation belongs to the category of storage irrigation. Tanks are small sized reservoirs formed by small earthen embankments to store runoff for irrigation. The site is selected within a watershed protected by vegetation and containing minimum of cultivated land so as to ensure minimum rate of sedimentation which lowers its storage capacity. Adequate soil conservation measures are essentially adopted to ensure quantity and quality of water inflow into the tank. The essential components of irrigation tanks are tank embankment, surplus or escape weir, and outlet sluice. A suitable breaching section is also sometimes provided to ensure that the tank embankment is not overtopped in the event of excessive discharge from the catchment. The breaching section is a low level embankment of a certain length designed to have a localised breach to escape excessive inflow.

Irrigation tanks are classified, as under, according to the nature of supply of water:

1. *System tanks*: The system tanks get assured supply from nearby rivers or canal system and as such they may not have their own catchment.

2. *Non-system tanks:* Also called 'isolated' tanks. The non-system tanks depend on the runoff from their own catchment. They are not connected to any other tank.

3. *Grouped tank:* The grouped tanks, as the name implies, consist of a series of tanks connected together such that outflow from the upper tank is stored in the lower one for irrigation.

They usually have limited depth of 5 to 10 m and fill up two or three times in the rainy season and redistribute to some extent the available supplies and tide over breaks in the monsoon. The maximum flood discharge from the catchment is usually determined from Ryve's formula. The length of escape weir is worked out from the formula, $Q = CLH^{3/2}$, where C is a coefficient of discharge with value 1.66 for broad crested weir and 1.84 for short crested weir. Water losses from tanks are enormous as they usually have more wetted area for the given storage capacity. The water losses due to evaporation and absorption are taken as 1.75 m per year in Southern India and about 1.25 m in Mumbai.

3.2.2 Lift Irrigation

In lift irrigation water is lifted from a river or a canal to the bank to irrigate the lands which are not commanded by gravity flow. Lift irrigation is being increasingly practised in India. In every State such areas exist where irrigation can be extended only by lift canals. Lift irrigation also includes tubewell irrigation but the latter is not feasible in areas where scarcity of water exists, climate is dry and groundwater is low, *i.e.*, groundwater is in insufficient quantity and of unsuitable quality. Lift canal then constitutes the only means of extension of irrigation to such perched lands. A lift canal can cater for much larger areas than a tubewell and is suitable where supplies either from a river or a canal are available for lifting to higher elevation.

Lift area is defined as the area the level of which is too high to permit irrigation by gravity flow from the source, but which can be irrigated by lifting water to the necessary level by means of pumps. *Gross lift area* is the portion of gross irrigable area which can be irrigated only by pumping.

Surface water may be lifted from a reservoir, a river, or from a gravity flow canal. From a river, water may be lifted by installing pumps either on the ground or on floating barges. Floating barge installation envisages economical and expeditious execution of the project for lift irrigation. Moreover, the pumping installation can be made mobile by rowing to the next pumping site after completing irrigation at one installation. In a canal, water may be lifted at the point of the offtake of the lift distributary or inbetween the length of a canal wherefrom high tract to be irrigated starts. Fig. 3.3 shows the longitudinal section of a lift canal.

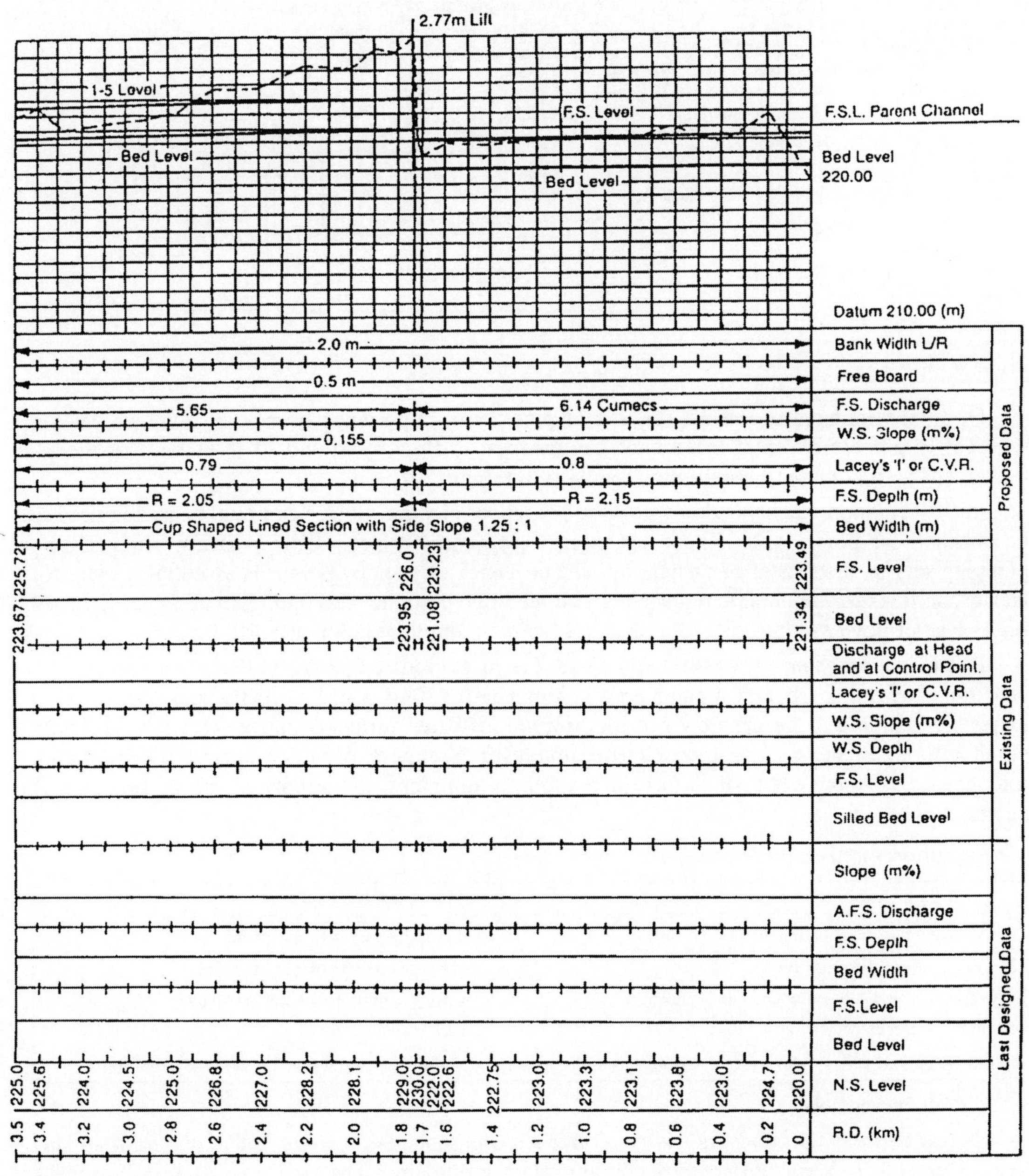

Fig. 3.3 Longitudinal section of lift distributary

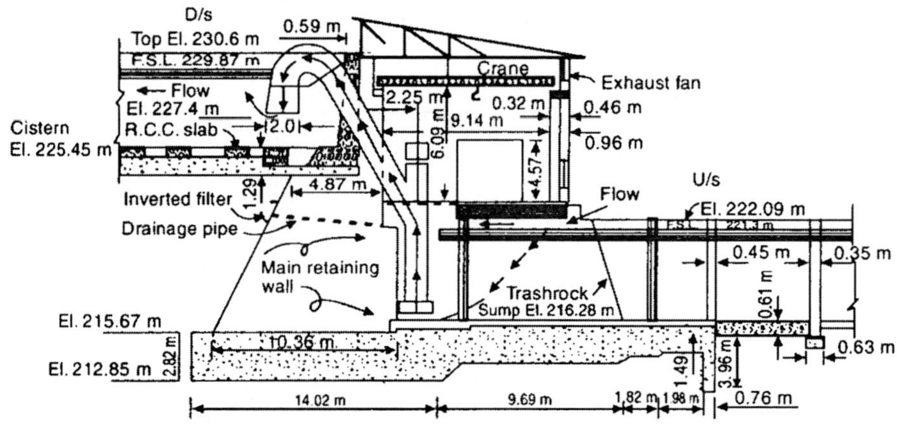

Fig. 3.4 Typical pumping station

Optimal alignment of the lift canal is of paramount importance because it involves dealing with costly process of water lifted and then diverted to flow by gravity. The essence of the arrangement is to command maximum area with minimum lift.

In lift irrigation mechanical devices like pumps, or electric motors and pumps are required to be installed for lifting water. Electrical pumps are generally provided for lifting water. Diesel pumping sets are also installed as standby. The provision of standby pumps is essential to cater for lack of facilities for immediate replacement of defective pumping sets and to switch over to diesel set in case of power failure. Fig. 3.4 shows a typical pumping station at a lift canal.

Lift irrigation canals are essentially lined due to obvious advantages of a lined section and hence low pumping lift and greater area commanded. Lifted water is costly and transmission losses are required to be cut down to the minimum. Water surface slope is kept the minimum possible to command greater area with minimum lift. Escape is provided upstream of pumping station so that in the event of failure of pumping equipment, the supply reaching upstream is disposed off.

Lift Irrigation vs. Gravity Irrigation

Lift irrigation	*Gravity flow irrigation*
1. Costly means of irrigation.	Cheapest means of irrigation.
2. Less manurial silt in water.	Silt in water has manurial value.
3. Working dependent on operation of machinery.	Lifting equipment is not involved.
4. Higher water rates.	Lowest water rates.
5. Lift irrigation is a complex system and by and large costly.	Simple and economical system of irrigation.

3.2.3 Well Irrigation

Groundwater: Groundwater is generally a more dependable source of irrigation than surface water and is free from weed seeds and plant organisms. The first cost of installation is, however, high. The best water bearing stratum or aquifer is coarse gravel free from sand but such formations are rare to find. An aquifer is a saturated formation which creates groundwater reservoir and yields sufficient quantity of water to wells or springs. These are made of unconsolidated formations like sand, gravel, fractured rocks. The various types of aquifers (Fig. 3.5) are as under:

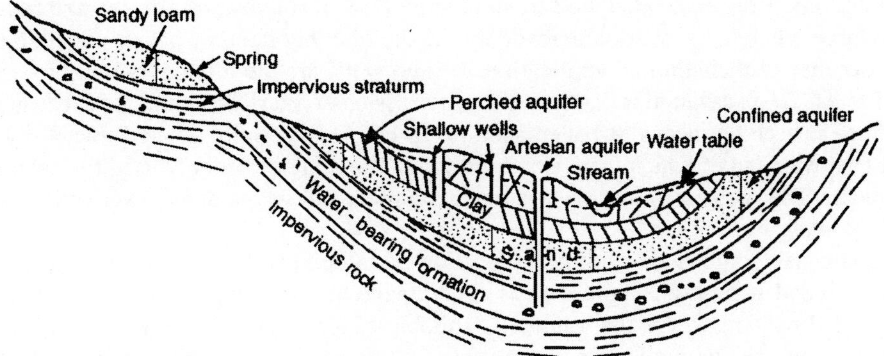

Fig. 3.5 Aquifers

Rich aquifer: It is the aquifer which has massive groundwater reservoir at a reasonable depth such as sand and gravel of fairly uniform size and moderately compacted.

Confined aquifer: It is the aquifer in which the water is confined *i.e.*, held under pressure greater than atmosphere between two impervious layers. Also called pressure aquifer.

Artesian aquifer: It is an aquifer which is fed from some source laterally at a higher elevation than natural ground surface and is subjected to hydrostatic pressure. A well in this aquifer flows without pumping.

Unconfined Aquifer: It is the aquifer which is exposed to atmosphere or water table is the same as piezometric level.

Perched aquifer: It is a special case of an unconfined aquifer that occurs whenever an impervious layer of limited horizontal area is located between the water table of a phreatic aquifer and ground surface.

Aquitard: It is a saturated formation which contains water and allows movements of water but yields inappreciable quantity of water. Also called semi-confining layer. An aquifer underlain by an aquitard is termed as leaky or semi-confined aquifer.

Aquiclude. It is a geological saturated formation which contains water but does not transmit appreciable quantity of water. There is no yield of water.

Aquifuge. It is a geological formation which neither transmits nor stores water and yields no water.

Well irrigation

Well irrigation is essentially a type of lift irrigation extracting water from the groundwater reservoir. An open well is essentially a vertical hole of 2 to 6 m diameter and of shallow depth. It has small discharge capacity, usually 0.004 cumec. It draws water from one pervious stratum only. The contribution of water is from the bottom and sides in the case of an unlined well, and from bottom only (not sides) in a lined well. It is preferable that the well penetrates to such a depth below water table that there is a water depth of 3 to 4 m during a dry year. From economic standpoint, the depth of open well is limited to 30 m.

Yield of an open well can be increased to a considerable extent by boring 8 to 10 cm diameter hole in the centre of the well penetrating into the lower rich water bearing formation. A suitable arrangement consists of digging open well up to a clayey or kankar layer (mota formation) which shall support the well steining and a bore hole made in the centre of well to tap the lower sandy strata.

The type of well depends on local ground conditions and availability of equipment. Open wells may be either dug, bored, drilled or driven. Large-sized wells which have to be lowered to shallow depths through comparatively soft soils are usually dug manually or by semi-mechanical means. Radial wells are a variation of the dug wells. The collecting pipes are driven horizontally from the bottom of the excavated well to yield large flow in shallow aquifer but are costly. Bored wells are those wherein excavation is made by the use of either hand or power augers but are now obsolete because of their limited application. Drilled wells are the most common type of modern irrigation wells. The excavation is made by either percussion or rotary drills and the excavated material is brought to surface by means of bailer, sand pump etc. Driven wells are constructed by driving a casing at the lower end of which there is a drive point. These are commonly used for obtaining small flows of water. They consist of a drive pipe that has a point and screen at its lower end.

Location of Open Well

Various considerations governing the location of an open well are (*i*) In the heart of the area to be irrigated and at the highest point so as to command the area by gravity flow, and economy in pumping, (*ii*) Location where the aquifer is suitable to give dependable yield, (*iii*) Sufficiently removed from an existing well so as not to lead to interference between the wells and consequent low yield, (*iv*) Location consistent with the property lines, *i.e.*, the circle of influence under maximum drawdown does not interfere with the adjoining boundary lines. Well operation may undermine property if the circle of influence extends into the structure, base and (*v*) Removed from sources of pollution.

Advantages of Well Irrigation

(*i*) Owned by the cultivator and as such sole control on the source of irrigation, (*ii*) Located nearest to the place where it is intended to provide irrigation and on the same property, (*iii*) Particularly suitable for intensive irrigation to raise cash crops. Two or more crops can be grown, (*iv*) Irrigation can be introduced in stages, (*v*) Reliable means of irrigation as yield from wells generally exhibit less fluctuations, (*vi*) It can supplement canal irrigation, (*vii*) Particularly suitable for isolated area, where other means of irrigation are not available, (*viii*) Operation at will as and when required for crops. Benefit of rainfall can be had, (*ix*) Supply is more uniform in quality, temperature and chemical composition and as such more conducive to crop yield, (*x*) Supply is free from weed seeds, plant organism, turbidity and bacterial pollution, (*xi*) Centrally located well entails low transit losses which occur in small length of water course only, and (*xii*) Economical water use as irrigator has to put in labour for lifting water and hence higher duty and low delta.

Disadvantages: (*i*) Costly means of irrigation as it is essentially a kind of lift irrigation, (*ii*) Low discharge and consequently small area irrigated by a well, and (*iii*) Water is silt free and devoid of manurial value. More fertilisers are required to be added on agricultural lands to raise crop yields.

Well Irrigation vs. Canal Irrigation

Well irrigation	*Canal irrigation*
1. Instant source of supply.	Dependent on water availability in canal system.
2. Supply can be drawn as per requirements. Benefit of rainfall can be had.	Supply received from the canal has to be used. Benefit of rainfall may or may not be possible.
3. Lesser fluctuations in supply.	Higher fluctuations in supply.
4. Supply more uniform in temperature and soluble mineral contents. Well water warmer in winter and cooler in summer is suitable for crops.	Canal supply is not uniform in temperature. It is colder in winter and warmer in summer. It may pick up minerals enroute.
5. Phased investment as irrigation develops.	High investment all at once on canal system, canal structures, etc.
6. Volumetric sale of water possible.	Volumetric sale of water not in use as yet.
7. Low transit loss in small watercourse length only.	High water loss in transit in long canal system and watercourses.
8. Effective anti-waterlogging measure.	Leads to waterlogging, if drainage is not provided.
9. High working expenses, usually to owner's account.	Maintenance of canal system is done by canal department.
10. Lifting equipment difficult to maintain.	No such equipment is involved in gravity flow system.
11. High cost of irrigation.	Low cost of irrigation.
12. Limited life.	Permanent source.

Classification of Wells

Wells are classified as illustrated in Fig. 3.6.

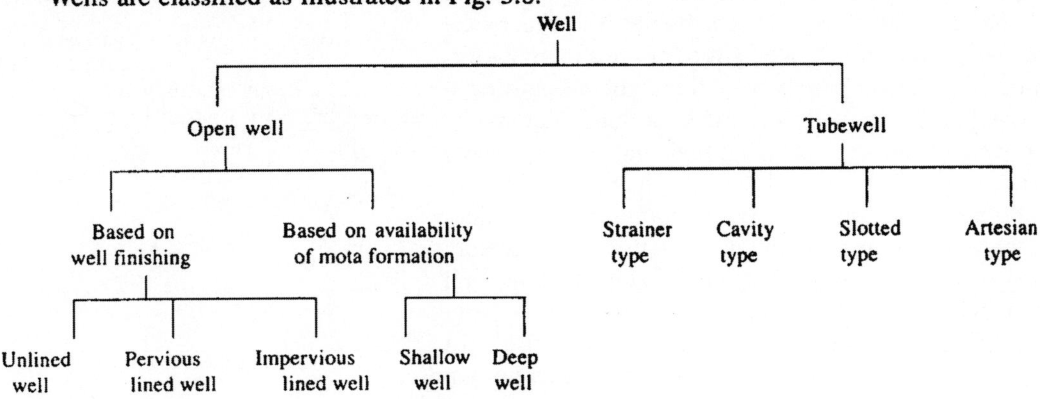

Fig. 3.6 Classification of wells.

1. Classification Based on Well Finishing

(*i*) *Unlined well:* Also known as *Kacha* well *i.e.,* without any lining on the sides and bottom (Fig. 3.7). Such wells are shallow pits, 3 to 5.5 m deep, dug in stable pervious soils or hard formation as can stand vertical and where the groundwater table can be tapped within this depth. The water contribution is from the bottom as well as sides. These wells are very cheap but temporary as they collapse after sometime.

(*ii*) *Pervious lined well:* The wells with dry brick or stone lining or steining with open joints, resting on a well curb are termed as pervious lined wells (Fig. 3.8). The space behind the lining, in sandy formation , is packed with brick ballast of size 20 to 25 cm with a view to avoid flow of sand into the well. These wells are of comparatively small depth and are suitable for shallow water table and in gravel or coarse sand formation. The contribution of water is from the bottom and sides and as such flow is a combination of radial and spherical pattern. Generally pervious wells are plugged at the bottom and the flow pattern is radial.

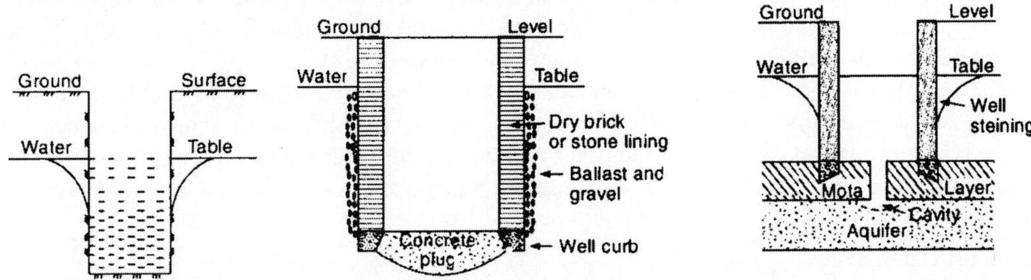

Fig. 3.7 Unlined well **Fig. 3.8** Pervious lined well **Fig. 3.9** Impervious lined well

(*iii*) *Impervious lined well.* Impervious lined wells have a comparatively impervious lining of brick masonry or stone masonry in cement or lime mortar, plain or reinforced concrete or precast concrete blocks or sections. Well steining 30 to 60 cm thick is supported on well curb constructed of reinforced concrete, wood or iron. Impervious lined wells are sunk to depths ranging between 5 to 60 m specially where water bearing strata is available in this range. The flow is spherical as water seeps into the well through the open bottom where a cavity is formed (Fig. 3.9). A better type of impervious well draws water from deeper sandy formation through a pipe sunk down from the centre of the well (Fig. 3.10). It is a very common type of well and forms a permanent source of irrigation. It is suitable for all types of formations and has greater yield.

2. Classification Based on Availability of Mota Formation

(*i*) *Shallow well.* A shallow well, contrary to what the name may suggest, has nothing to do with the actual depth of the well. Rather it is a technical name assigned to a well which does not rest on mota formation (Fig. 3.10). A shallow well may be deeper than a deep well. A shallow well penetrates pervious stratum only from which it draws supply but does not rest on a mota layer. The supply of a shallow well, drawn as it is from the top pervious formation, is susceptible to contamination by rain water percolating in the vicinity and possibly taking with it decomposed organic matter. Shallow well has limited yield.

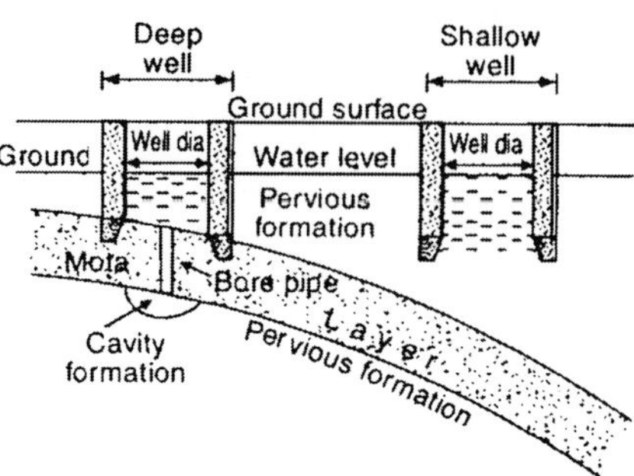

Fig. 3.10· Shallow and deep wells

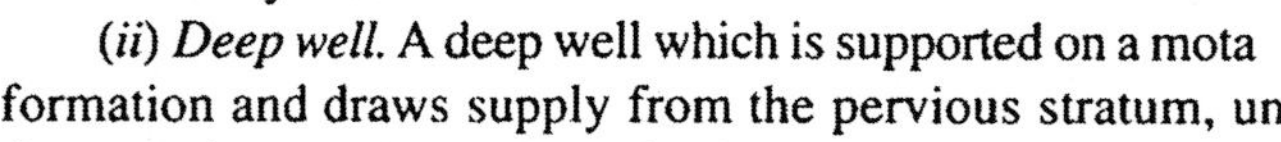

(*ii*) *Deep well.* A deep well which is supported on a mota formation and draws supply from the pervious stratum, underneath the mota layer, through a hole bored across this layer is termed as a deep well (Fig. 3.10). The depth of a deep well depends on the depth below the ground surface where mota formation is met with. The mota formation provides support to the well steining as also due to it a large cross sectional area in the form of cavity is developed below the mota layer which gives higher yield.

Shallow vs. Deep Wells

Shallow well	*Deep well*
1. Economical in first cost.	Initial cost may be high depending on the depth of mota layer.
2. Supply is proportional to the cross sectional area of well bottom.	Large cross sectional area due to wider cavity developed.
3. Suitable where water table is high.	Only indicated where a mota layer is met with.
4. Low yield.	High yield.
5. Water liable to contamination by impurities seeped from the surface.	Better quality water. Not liable to contamination.

3. Classification Based on Conditions of Flow

(*i*) *Gravity well:* A well in which surface of water in the water bearing formation surrounding the well is at atmospheric pressure is called gravity well. The water flows under gravity into the well and rises to the height of saturated material surrounding it.

(*ii*) *Pressure well*: Also called artesian flowing well. It is a well in which water in the aquifer is at a pressure higher than atmospheric because the aquifer is confined or sandwiched between two impervious strata, one above and the other below, so that water flows under pressure into the well. In some cases the pressure is so high that the water from the well rises above the ground surface.

Yield of an Open Well

Safe yield of an open well expressed in either cubic metre/hour or litre/minute is defined as the rate at which water percolates into the well under safe maximum working head or critical depression head. Discharge formula is expressed as:

$$Q = KH = \frac{K}{A} \times \frac{A}{H}$$

where, Q = discharge is cumec or m^3/hour, A = cross sectional area (m^2) of the cavity formed at the well bottom; the area is taken as 4/3 of actual cross sectional area of the well bottom, V = mean velocity of percolating water into the well (m/sec), H = depression head (m), and K/A = specific yield of the well (m^3/hr per m^2 of the area through which water percolates under 1 m of depression head). Usual values of K/A are 0.25 for clay, 0.50 per fine sand and 1.0 for coarse sand.

3.2.4 Tubewell Irrigation

A tubewell is a smaller diameter but large discharging well and draws water from underground reservoir. The usual capacity of a tubewell is 0.042 cumec but the optimum capacity is 0.07 cumec. The selection of site for a tubewell is governed by the same considerations as described for an open well. However, certain additional factors are (1) Extensive water bearing formation of adequate depth and of transmission coefficient more than 1×10^{-2} cm/sec, (2) Economical pumping with hole depth not exceeding 125 m, (3) Availability of electric supply to run the tubewell, and (4) Located nearest to a motorable road and a town so that mechanical defects, as and when they develop during running, can be attended to promptly and economically.

After a well is bored the design of a tubewell involves selecting appropriate dimensions of various components such as depth and diameter of housing pipe, optimum diameter, length and material of screen and choosing proper materials to be used in its construction which includes mechanical analysis of samples collected from various depths and preparation of a bore log, selection of strata to be screened and fixing length of screen, selection of type of well and size of gravel for shrouding. A good design of tubewell aims at efficient utilization of aquifer, long useful life, low initial, maintenance and operation costs.

Parts of Tubewell

Various parts of a tubewell (Fig. 3.11) are discussed below:

Pump. To lift water by creating a strong vacuum head under which it rushes at a high velocity through the strainer into the well pipe and further into the delivery pipe.

Motor. To run the pump.

Impeller. The rotating element producing head. It receives water and impels it to bowel passage. It may be of enclosed or semi-enclosed type.

Well casing. To support the hole temporarily during drilling (in cable-tool method) or installed afterwards to complete the well assembly.

Strainer. To admit sand-free water into the well and to support loose formation materials in unconsolidated formation. It consists of very fine mesh surrounding a slotted or perforated pipe.

Blind pipe. To serve as a conduit to carry further the water that has entered through the strainer. It is put in the impermeable stratum.

Delivery pipe. It is a water conductor from pump to the watercourse which carries water to the field.

Plug. It is to close the bottom-most end of the blind pipe so that no water enters from the bottom as also to house any sand which has found entry into the well during installation of well assembly or development for ultimate removal through the water to the surface.

Types of Tubewells

1. **Strainer Type Tubewell:** It is the most common and widely used type of tubewell. It draws water from an unconfined aquifer of unlimited extent or from one or more confined aquifers (Fig. 3.11). It consists of a strainer or screen placed against the water bearing strata through which water enters into the well, while in the rest of bore hole length plain pipe, called *blind pipe*, is provided. At the bottom end a short length of blind pipe is provided which is plugged at the bottom with cement concrete.

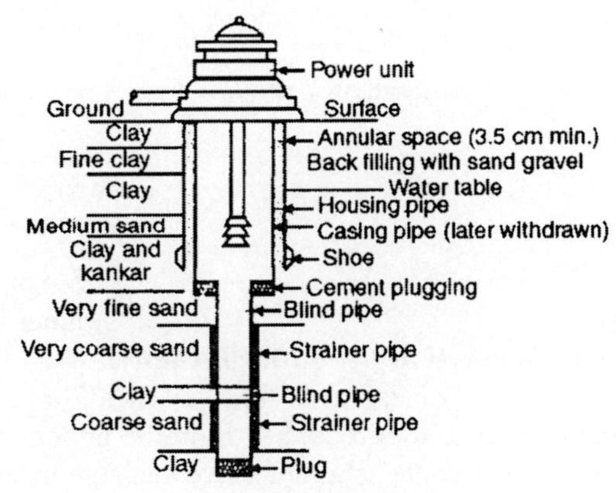

Fig. 3.11 Strainer tubewell

The tubewell is bored by a casing pipe of diameter 5 to 10 cm larger than the well pipe. The casing pipe is later withdrawn and well assembly consisting of strainer pipe and blind pipe is lowered into the bore hole. The space left in the bore hole by the casing pipe is packed with gravel of suitable size around the screen. Gravel packing improves the discharging capacity of tubewell. Strainer type tubewell is not considered suitable for very fine sandy strata.

Discharge Formulae

(1) *Fully Penetrating Tubewell in Confined Aquifer*

Notations: r = radius of tubewell (m), R = radius of influence of tubewell (m), H = depth of original watertable above the bottom impermeable layer (m), h = depth of water in the tubewell above the impermeable laye: (m), b = thickness of confined aquifer (m), K = coefficient of permeability (m/s), and Q = discharge of tubewell (cumec). Fig. 3.12 shows a fully penetrating tubewell in confined aquifer. Let P be any point on the drawdown curve whose coordinates are x, y taking origin at point O, the centre of the well at the bottom.

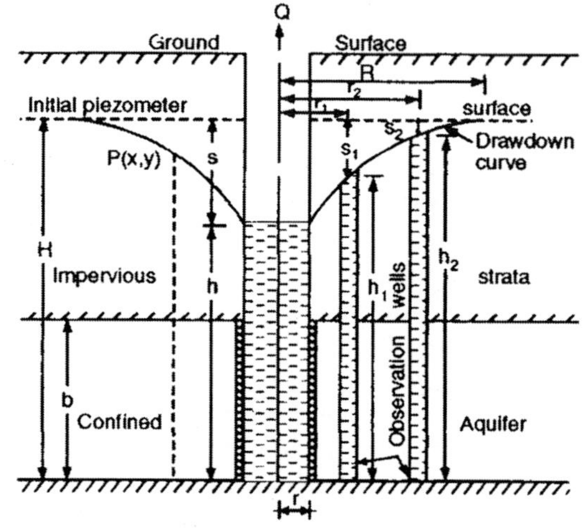

Fig. 3.12 Confined aquifer tubewell

From Darcy's law $Q = KA_xI_x$

where, A_x = cross sectional area of flow at $P = 2\pi xb$, and I_x = hydraulic gradient at $P = dy/dx$

Putting values in equation, $Q = K.\ 2\pi xb.\ dy/dx$

or $$Q\frac{dx}{x} = 2\pi Kb\, dy$$

Integrating, $$Q\int_r^R \frac{dx}{x} = 2\pi Kb\int_h^H dy$$

or $$Q\Big|\log_e x\Big|_r^R = 2\pi Kb\Big|y\Big|_h^H$$

$$Q\log_e R/r = 2\pi Kb\,(H-h)$$

$$Q = \frac{2\pi Kb(H-h)}{\log_e R/r}$$

or $$Q = \frac{2.72\,Kb(H-h)}{\log_{10} R/r}$$

Taking drawdown at well $(H - h)$ = S and coefficient of transmissibility T = bk, we get:

$$Q = \frac{2.72\,TS}{\log_{10} R/r}$$

Also $$Q = \frac{2\pi KLS}{2.3\log_{10} R/r}$$

where, L = strainer length (m), and S = depression head (m)

(*ii*) **Tubewell in Unconfined Aquifer**

Notations: r = radius of tubewell (m), R = radius of influence of tubewell (m), H = thickness of unconfined aquifer reckoned from the impermeable layer to the initial level of watertable (m), h = water depth in the tubewell above the impermeable layer (m), S = drawdown at tubewell (m), and

K = coefficient of permeability (m/sec). Fig. 3.13 shows a tubewell in unconfined aquifer.

Let P be any point on the drawdown curve whose coordinates are x, y taking origin at point 0, centre of the tubewell at the bottom. From Darcy's law

$$Q = K.A_x I_x$$

where, Q = discharge in cubic metre per day if value of K is m³/day/m² of the area of sub-soil formation, A_x = cross sectional area of saturated part of the aquifer at $P = 2\pi xy$, and I_x = hydraulic gradient at $P = dy/dx$.

Putting the values in equation

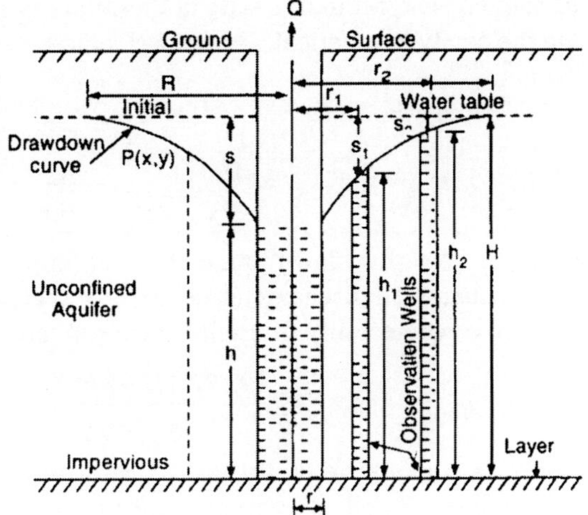

Fig. 3.13 Unconfined aquifer tubewell

$$Q = K\, 2\pi\, xy\, dy/dx$$

or $$Q \frac{dx}{x} = 2\pi\, Ky\, dy$$

Integrating, $$Q \int_r^R \frac{dx}{x} = 2\pi K \int_h^H y dy$$

$$Q \left| \log_e x \right|_r^R = 2\pi K \left| \frac{y^2}{2} \right|_h^H$$

or $$Q \log_e R/r = \pi K (H^2 - h^2)$$

$$Q = \frac{\pi K (H^2 - h^2)}{\log_e R/r}$$

or $$Q = \frac{\pi k (h_2^2 - h_1^2)}{2.3 \log_{10} r_2/r_1}$$

where h_1, h_2 are depths of water in observation wells and r_1, r_2 are radii.

$$Q = \frac{1.36 K (H^2 - h^2)}{\log_{10} R/r}$$

and $$K = \frac{Q \log_{10} R/r}{1.36 (H^2 - h^2)}$$

Also $$Q = \frac{2\pi KS (L + S/2)}{2.3 \log_{10} R/r}$$

where L is the strainer length (m).

2. **Cavity Type Tubewell:** Cavity type tubewell is without any strainer. It draws water from only one aquifer located at the bottom where cavity is formed (Fig. 3.14). It requires a very hard clayey stratum to form a strong and dependable roof over the cavity. It consists of a pipe holed through the formation and resting on the bottom of the clayey roof. In the initial stage of pumping, fine sand comes up with water resulting in the formation of a cavity. With further pumping, ultimately equilibrium condition is reached when clean water is

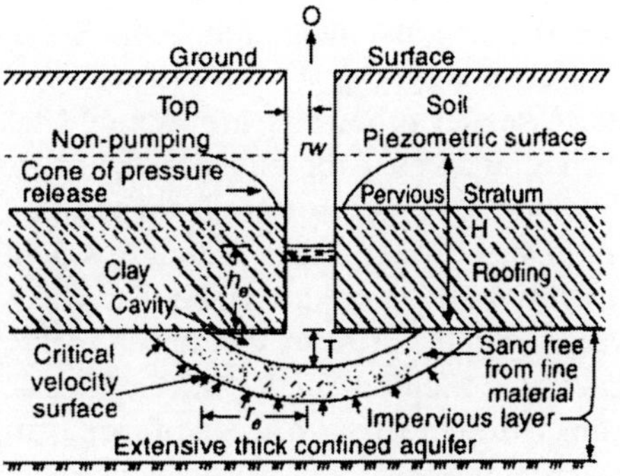

Fig. 3.14 Cavity tubewell

discharged. The depth of cavity at the centre is 15 to 30 cm and radius 6 to 10 m. The water flow into the cavity is spherical. Cavity wells have shorter life. Failure is caused due to collapse of clay roofing.

Discharge Formula

(*i*) *Unsteady flow*: $S = \dfrac{Q\sqrt{S_s}\, t^{-\frac{1}{2}}}{\sqrt{4\pi K^{3/2}}} + \dfrac{Q}{2\pi K}$

where, S = drawdown in the observation pipe placed at a distance r from the cavity well (m), Q = constant well discharge (cumec), S_s = specific storage coefficient (*i.e.*, for unit aquifer thickness), K = permeability of aquifer material, and t = time since pumping began.

(*ii*) *Steady flow*: $Q = \dfrac{2\pi TK(H - he)}{(1 - rw/R)}$ and $re = \sqrt{(2rw - T)T}$

where, T = depth of cavity (m), H = piezometric head (m), h_e = depth of water in well (m), rw = radius of well, K = coefficient of permeability, R = radius of influence measured from the centre of spherical cavity (m) and re = radius of cavity (m).

3. **Slotted Type Tubewell:** It consists of a slotted wrought iron pipe having slots 25 × 3 mm at 10–12 mm spacing shrouded with gravel consisting of graded mixture of gravel and coarse sand against the slots to prevent entry of fine sand into the well. The well penetrates to some depth (say 5 m) into the confined aquifer of rich water bearing stratum located at the bottom of the pipe (Fig. 3.15). The drilling process is the same as for the strainer tubewell. After the casing of 40 cm diameter is lowered to penetrate about 5 m into the aquifer, the well pipe, commonly called education pipe, of 15 cm diameter consisting of 5 m slotted pipe and the rest of plain length is lowered. Gravel is constantly poured into the annular space between the casing and education pipe, as the former is withdrawn 5 cm at a time to a depth about 3 to 4 m higher than the top level of the slotted portion of the pipe.

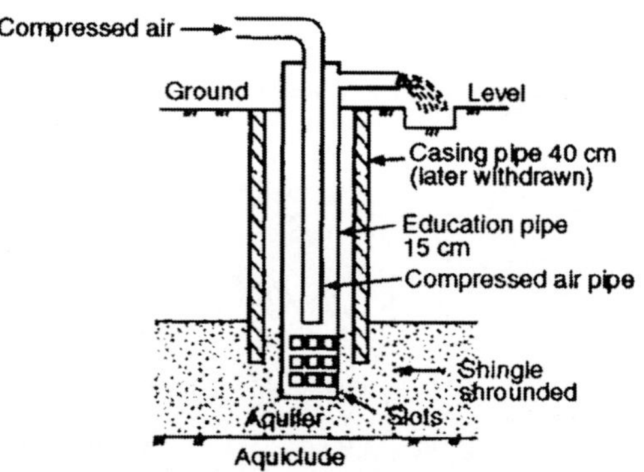

Fig. 3.15 Slotted tubewell

4. **Artesian Tubewell:** In terai regions of Uttar Pradesh and some parts of Tamil Nadu, Himachal Pradesh and Punjab peculiar geological formations consisting of an aquifer sandwiched between two impervious strata and bent like a trough outcropping at both the ends exist. Artesian tubewells are adapted to such terai areas. A borehole is drilled at the bottom of such a trough shaped basin to yield water under pressure which rises above the ground surface. The yield varies with the pressure developed in the artesian conditions under different seasons depending on recharge at outcropping ends. Fig. 3.16 shows such a free flowing artesian tubewell.

Fig. 3.16 A free flowing artesian well

Ideal Tubewell

A tubewell is said to be ideal if it meets the requirements such as (*i*) The chak is suitable for intensive irrigation, (*ii*) Designed discharge is available in all seasons, (*iii*) Installed in an extensive water bearing formation of adequate depth, (*iv*) Located centrally or at the highest point in the chak to command all the fields with minimum length of watercourse, (*v*) Yields silt-free water, and (*vi*) Gives trouble-free working, preferably run by electric power.

Suitability of an Area for Tubewell Irrigation

The tubewell irrigation is very suitable for the area where subsoil formation is suitable for storing water such as alluvial and there are massive groundwater reservoirs. The source of water supply is such that the yield of the tubewell does not deteriorate in the course of time and is of optimum quality to raise crops. Suitable location is available at the highest place in the area to give good command; preferably near the centre of the area to be irrigated so that radiating water-courses are not more than 1.5 km in length to ensure irrigation of the entire area at the least expanse of distribution system.

With irrigation intensity 100 per cent, a 0.056 cumec tubewell shall irrigate 200 hectares of land after providing due margin for rainfall. Deep tubewell can cater for up to 300 ha.

Optimum Capacity of a Tubewell

The optimum capacity of a tubewell is defined as the maximum yield at the least unit cost. The capacity of a conventional tubewell is 0.056 cumec based on the quantum of water that a single cultivator can handle easily. It is too small a capacity keeping in view the advance in tubewell engineering and mechanised farming. It is one-third of the capacity adopted in USA. Simulation model of Indus river basin for conjunctive use of surface and groundwater has indicated that the discharge of a tubewell should be 0.1 cumec and the wells spaced 2 km apart. However, the optimum capacity of a tubewell considered most suitable and compatible with the conditions and irrigation practices in Northern India is 0.07 cumec.

Yield of a Tubewell

Theoretically discharge of a tubewell can be increased by increasing the length of strainer, depression head and diameter of strainer. But for a given transmissibility coefficient of aquifer, if the discharge is increased by increasing strainer length, the well becomes deep. Depression head cannot be exceeded to increase discharge. The diameter of strainer too is limited, because increase in diameter is accompanied by nominal increase in discharge. Yield of a well is a function of its diameter, though the two are not directly proportional. Large diameter well has more yield but percentage increase in discharge vis-a-vis increase in diameter may be relatively small. Doubling the diameter may increase yield by 5 to 20 per cent due to improvement of screening conditions. Table 3.1 shows the approximate increase in the yield of a well in unconfined aquifer by increasing its diameter, other conditions remaining the same. In confined well, increase in yield is still less because for the same discharge rates, the radius of influence is greater in confined aquifer.

Table 3.1 Effect of increase in well diameter on well yield

Well diameter (cm)	10	15	20	30	45	60	90
Increase in yield (per cent)	0	5	10	15	23	28	38
		0	5	10	18	23	33
			0	5	13	18	28
				0	8	13	23
					0	5	15
						0	10
							0

Safe yield is defined as the amount of water that can be withdrawn from an aquifer without producing adverse result. The safe yield of a basin is governed by water supply, economics, water quality and water rights.

Strainer

In tubewells installed in unconsolidated water bearing formations, a screen or strainer is provided to allow entry of sand-free water into the well. An important aspect of design of a well screen is the determination of its diameter, length, percentage open area, size and shape of each slot, thickness and material of the strainer. The length of strainer and its placement depends on the thickness and location of the aquifers. In case adequate thickness of water bearing formations are available, the length of the screen, its diameter and percent open area are governed by the head loss through the screen and its effect on the losses in the aquifer, in addition to the initial cost of the screen. Table 3.2 indicates the optimum diameter of the strainer corresponding to various values of discharge and permeability of the aquifer.

Table 3.2 Optimum diameter of strainer

Tubewell discharge		*Optimum diameter (cm) for permeability coefficient K in cm/s*		
litre/minute	*cumec*	0.04	0.09	0.16
2400	0.04	15	18	22
4800	0.08	20	25	30
7200	0.12	23	28	33
9600	0.16	25	30	35

The size of slot openings is governed by the size of the gravel or aquifer material which it has to retain. A strainer with continuous slot openings is least liable to get clogged. The most suitable shape for the screen openings is the slots sloping and widening inwards with a sharp edge offering least chance of clogging. An annular space is allowed between the perforated or slotted pipe and the wire mesh lest the open area of the pipe perforations is covered by the wires of the screen. The strainer is placed opposite the water bearing formations in the hole. Blind pipe or well pipe is provided against the non-water bearing formations. Thus well assembly consists of strainer(s) and well pipe(s).

Types of Strainers

Following types of strainers are commonly used:

1. Agricultural strainer: It is a strainer fabricated by wrapping a copper mesh on perforated steel pipe with threaded ends to provide screwed or socketted joints. A spiral wire of 3.2 mm diameter is wound round the perforated pipe or 9.6 × 3.2 mm hoops are welded to the pipe to provide an annular space between the pipe and the mesh. The mesh openings are kept small to prevent entry of sand grains since these strainers are placed directly in contact with the aquifer.

2. Cook strainer: It is a superior type of strainer of American manufacture but is costly. It consists of a solid drawn brass pipe slotted with wedge-shaped horizontal slots. The slots of gauge 0.15 to 0.40 mm, depending on the coarseness of the sand particles, are wider inside and narrower outside which provide large surface to inlet ratios without sacrificing strength.

3. Phoenix strainer: It is a strainer similar to cook type strainer wherein slots of desired width are made from inside the pipe. It is chromium-plated to ensure against choking and corrosion caused by chemical action.

4. Tej strainer: It is a cheap and widely used type of strainer. It is indigenously manufactured and is similar to cook strainer with the essential difference that brass sheet is first slotted and then bent into the form of a tube with brazed vertical joint. The V-cut is kept inside and the outer surface represents the sharp edge slits to eliminate the chance for silt deposition. The slot size varies between 0.35 to 0.45 mm giving percentage open area of 4 per cent. It is generally marketed in diameter of 7.5 cm and above, in multiples of 2.5 m length. The joints between individual lengths are made by means of screwed brass collars.

5. Ashford strainer: It is a very delicate type of strainer. It consists of a perforated pipe with a wire wound around it to provide annular space between the perforated pipe and the wire mesh which is soldered to the wire. The wire mesh is further protected and strengthened by a wire net over the mesh.

6. Brownlie strainer: It is considered the best type of strainer. It consists of a polygonal convoluted perforated steel pipe, surrounded by a wire mesh consisting of heavy parallel copper wire/brass wire, woven with copper ribbons with a little space between the two. The use of convolute pipe by Brownlie was to have a least contact with the iron pipe.

7. Coconut coir cord strainer: It is a comparatively inexpensive strainer and is commonly used in the cultivator owned tubewells. Generally it is of 10 cm internal diameter. It consists of steel frame, fabricated by welding the ends of eight number 6 mm M.S. square bars to the periphery of 13 cm long pieces of steel pipe, around which a coconut coir cord of 6 mm diameter is wound (Fig. 3.17). The steel pieces are threaded to enable jointing through the sockets.

8. Porous concrete pipe strainer: It is a strainer prepared from 1:6 cement gravel mixture reinforced with six equally spaced longitudinal M.S. bars to the ends of which flanges are welded for making joints of the pipes. This type of strainer is not found satisfactory in hydraulic performance.

9. Prepacked filter strainer: It consists of a slotted pipe to the outer surface of which pea gravel is attached by a special adhesive. Ashim Filters market this type of strainer. (Fig. 3.17).

10. M.S. slotted strianer: Mild steel slotted strainers are conventionally used in tubewells. These strainers are made of medium or heavy duty M.S. pipes. The size of slot opening is such that the slots are wide enough to suit the varying characteristics of aquifers. Usually the slot size is such that it retains about 90 per cent of the gravel. The size may be reduced for corrosive waters depending on the extent of corrosion. The usual slot size is 3.2 × 75 mm or 3.2 × 100 mm, about 9 cm long with a minimum spacing of 3 mm between slots. The slots are made in groups of 2 to 4 slots such that the slots of one row are not in the line with that of the adjacent row in order to ensure that the strainer section is of requisite strength. The recommended open area is 20 per cent. M.S. strainers are subject to corrosion and incrustation especially in the areas where the groundwater is saline and alkaline. They cause greater head loss to flow of water due to their rough surface which also contributes to incrustation.

Fig. 3.17 Prepack strainer, coir strainer and polythene strainer.

11. Fibreglass strainer: It is a variety of slotted type strainer. It is made by cutting slots 1.25 mm wide and slot length inside 5.75 cm and on the outside 6 cm. The solts are rectangular and are cut in groups. The open area varies from 12 to 15 per cent. It is resistant to corrosion. Advantages of easy to handle, lower cost of slotting, and simplicity of coupling adjoining pipe lengths offset the increase in first cost.

12. Cement asbestos strainer: It is also a variety of slotted type strainer and is suitable for shallow tubewells. It is made by cutting slits in the body of cement asbestos pipe which has been recently spun but has not yet completely hardened. The slits have winding of coir string. The strainer is usually available in 3 m lengths and made to any diameter.

13. Polythene strainer: It is similar to coir strainer in construction. It consists of high density polythene rings mounted closely on a steel framework which is made of steel bars welded over steel rings and provided with sockets on either side of the framework. The rings vary in width from 6 to 10 mm. The space between adjacent rings allows percolation of water from aquifer into the tube (Fig. 3.17). It is resistant to corrosion and deterioration.

Gravel Packing

The strainers are shrouded by gravel. The size of gravel to be shrouded, depending on the size of aquifers tapped, is determined by pack aquifer (P/A) ratio *i.e.,* D50 of gravel/D50 of aquifer. Central Board of Irrigation and Power has recommended P/A ratio of 9-12.5 for uniform aquifer with uniform gravel pack and 11-15.5 for non-uniform aquifer with uniform gravel pack based on minimum head loss through gravel pack and minimum sand movement.

For a tubewell tapping more than one formation, the gravel pack designed for the finest formation is provided for all the formations provided the average grain size of the material in the coarsest aquifer is less than 4 times the 50 per cent size of the materials in the finest aquifer. The thickness of gravel for sand-free discharge is 15-20 cm.

Well Pipes

Various types of well pipes commonly used are as under:

1. *Mild steel pipes.* These are commonly used in tubewells M.S. pipes are prone to corrosion and entail high head loss due to flow of water, their internal surface being rough. Galvanized M.S. pipes are also widely used in tubewells. They are prone to incrustation.

2. *Stainless steel pipes.* Stainless steel pipes have longer life but are costly. Stainless steel continuous slot screens are widely used in USA as an anti-corrosion measure.

3. *Rigid PVC pipes.* PVC pipes are resistant to corrosion and are light. These are at the same time rigid and strong. However, these pipes require careful handling and storage to prevent damage. These pipes are used in corrosive groundwater environment. Maximum installation depth is limited to about 100 m due to potential vertical stress failure.

4. *Fibreglass reinforced pipes.* These are widely used in tubewells in Pakistan, Nepal and Bangladesh. These pipes are rigid, light in weight, have high tensile strength, and are not susceptible to corrosion. The smooth internal surface of the pipes entails smaller head loss to the flow of water. These are found more useful than PVC pipes.

Casing Pipes

Casing pipes are used to support the hole during drilling in cable tool method and are extracted after lowering of the well assembly. The size of the casing pipes depends on the size of the hole which in turn depends on the capacity of the well and the depth of aquifer tapped. The upper portion of the casing, if used to house the pump, is conventionally known as housing pipe. The diameter of the casing is required to be sufficiently large (usually at least 5 cm bigger than the nominal diameter of the pump) when used as a housing pipe to accommodate the pump with sufficient clearance between the pump and housing pipe for installation and operation. Various considerations in the selection of casing pipe are (*i*) It should have a smooth exterior to minimise resistance offered by the formation due to friction, (*ii*) Inside of the casing is smooth and of adequate size to permit passage of drilling tools, installation of pump and operation of development equipment, (*iii*) Wall thickness is sufficient, (*iv*) Capable of withstanding potentially corrosive groundwater environment, and (*v*) Field connections are leak-proof and equal to casing strength.

Development of Tubewells

Development or education of a tubewell is one of the essential operations of tubewell construction. Soon after construction of a tubewell, it is required to develop it to its maximum sand-free discharge before it is put into service especially in formations containing sand, silt or clay. The operation is also done in sick and incrusted tubewells to revitalise them to their maximum yield. The operation involves flushing testing and equipping the tubewell. The principle of development is to agitate water in the formation with a view to remove the finer materials near the casing and to prevent bridging of the sand particles across the screen openings and in the formation around the tubewell. The development of a tubewell thus consists in removal of silt, sand and very fine gravel from around the well screen for increasing porosity and permeability of the surrounding aquifer material. This facilitates steady flow of water into the well, thereby resulting in increased effective radius of the well and consequent high yield.

Functions of Tubewell Development. The functions of the development of a tubewell are (*i*) To bring tubewell to its maximum yield of sand-free water at minimum head loss, (*ii*) To increase useful life of a tubewell, (*iii*) To stabilize the sand formation (aquifer) and gravel pack around the well screen so that the well will yield sand-free water. The removal of fines from the formation in contact with the gravel pack also reduces the chances of fines getting deposited in the gravel after being dislodged at some later stage, (*iv*) To increase the effective radius of the tubewell by increasing porosity and permeability of the natural formation in the vicinity of the well screen, (*v*) To clear the water bearing formation clogged by mud in drilling operation, and (*vi*) To create a natural filter around the well screen by keeping coarsest material nearest to the tubewell.

Methods of Tubewell Development: Different methods for the development of tubewells are as follows:

1. Overpumping method: It is the simplest and most common method of developing a tubewell. In this method the well is pumped at a higher rate, than it would pump if it is put in service, by means of a large capacity variable speed pump. The high rate of pumping implies that sufficiently high velocities, than those considered desirable during the regular service of the well, are induced in the immediate vicinity of the well to remove the fines. Starting with slow speed, the rate of withdrawal is increased in stages. The discharge of water during development shall correspond to depression head 50 per cent higher than the normal depression obtained at the well and shall be free from sand during the operating test run. The development is continued until the stabilization of sand and gravel pack is completely ensured.

Limitations: (*i*) There is a limit to the degree of development in cases where the aquifer is either too coarse or too fine, (*ii*) Due to limited removal of fines, the well is generally not developed to its maximum efficiency, (*iii*) Pump of a larger capacity than is normally used in tubewells is required. The pumping set is likely to get damaged due to abrasion and sandlocking, and (*iv*) The method is not effective for developing large capacity tubewells and its use is restricted to small wells.

2. Backwashing method: Backwashing method envisages agitation of the formation in the vicinity of the tubewell to cause breakdown of bridging of sand particles thereby permitting them to be effectively washed out. The surging effect or reversal of flow required to develop the formation can be produced by three backwashing methods, viz., backwashing with bailer, backwashing under pressure, and backwashing by rawhiding. In one method the pump is started and stopped intermittently, *i.e.*, by first lifting the water to the ground surface by pumping and then letting water to fall down suddenly into the well and outflow of water through the screen openings into the aquifer. The methods of backwashing, though simple to operate, are seldom satisfactory unless the surging action so produced is very strong.

3. Compressed air method: It is a very rapid and effective method and is widely used to develop new tubewells as also to revitalize the sick wells. The method is a combination of surging and pumping carried out alternately. In this method, compressed air is introduced into the well through a small diameter air pipe placed inside the discharge pipe of the well. The compressed air forces out water back through the well screen into the surrounding formation. A large quantity of compressed air is continuously supplied through the air pipe into the well, while the air pipe is moved up and down. During downward motion of the air pipe, well water is forced out into the aquifer whereas water enters the well from the surrounding formation when the air pipe is raised bringing with it fine sand dislodged by the process. The process is repeated for the full depth of the aquifer till the sand flow is stopped. The method has the limitation that it does not suit the weak well screens which may get ruptured.

4. High velocity jetting method: It is generally the most effective method of well development. The principal items of the equipment are a relatively simple jetting tool, a high pressure pump and necessary hose and piping. In this method, a horizontal high velocity jet is operated inside the well which projecting through the screen openings loosen the drilling mud, clay and fine sand from the formation in the vicinity of the well screen. The force due to the jet agitates and rearranges the sand and gravel particles of the formation and also causes the loose fines to flow in the well casing from where they are removed by pumping or bailing.

Advantages: (*i*) Simple to apply and is not likely to give trouble in case of over application, (*ii*) The energy is concentrated over a small area with greater effectiveness, (*iii*) Every part of the screen is covered separately by slowly rotating the jetting tool and gradually raising or lowering it, as need be, thereby ensuring complete development of the well, (*iv*) Little chance of over-development, although judgement has to be exercised to determine the extent to which it should be carried out, and (*v*) Most effective where the well screen is strong and has a high percentage of open area.

5. Mechanical surging: It is an effective method of tubewell development. The development is done by operating a plunger in a casing like a piston in a cylinder. The tool normally used is called a surge plunger or surge block. A heavy bailer is used to produce the surging action, but is not as effective as the close-fitting surge block. During the downward stroke the plunger forces water outward through the screen to break the bridging of finer particles. After lowering, the plunger is started slowly and the speed is gradually increased until it reaches the maximum limit the system can work. The repeated applications of surging forces draw the fine particles into the well, leaving coarser particles intact in the aquifer. The surging and bailing out is carried out till little or no sand is drawn into the well. The surge plunger should not be operated unless a sufficiently free flow of water has been established so that the tool runs smoothly and freely. It is used continuously if the overlying formation consists principally of fine sand, silt or soft clay.

Commonly Adopted Methods

The commonly used methods for developing a new tubewell are compressed air and high velocity jetting, while backwashing and surging are often employed to treat sick or incrusted tubewells. A comparison of the two methods is as under:

(*i*) Compressed air method is most suited to gravel packed wells provided with conventional screens. It is rapid and most effective.

(*ii*) Overpumping method is the second best method.

(*iii*) Method of jetting is not considered suitable for developing tubewells provided with conventional slotted screens.

Interference Between Tubewells

The symmetrical cone of depression of a tubewell is considered on the assumption that there exists an homogeneous aquifer of great extent. The circle of influence of a heavy pumped tubewell may extend to a kilometre or so in diameter. As such the tubewells are located sufficiently far apart. When two or more tubewells are so closely located to each other that their areas of influence overlap, *i.e.*, their cones of depression intersect, the wells are considered to be interfering. The draft of one well effects the draft of the other well or wells (Fig. 3.18). The interference in closely spaced tubewells may be so great that a given group of tubewells may behave like a single well witn a very large cone of depression and the drawdown at a point is the sum total of the drawdowns corresponding to the individual wells. The magnitude of interference in any well is represented by the ratio of reduction in its discharge to its discharge without interference. The actual reduction in pumping rate from the wells, however, depends on the formation characteristics, distance between the wells and the pumping time. The drawdown at any point in the area of influence caused by simultaneous operation of a number of tubewells is worked out by superposing the drawdown caused by each well individually. Thus,

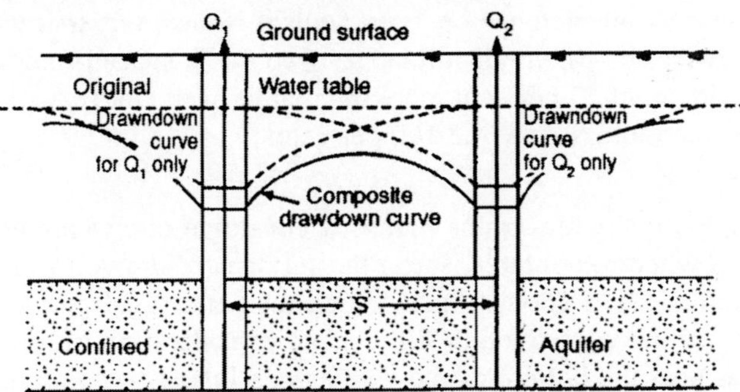

Fig. 3.18 Interference between tubewells

$$S_t = S_1 + S_2 + \dots + S_n$$

where, S_t = total drawdown at a given point and S_1, S_2, ..., S_n denote the drawdowns at the point caused by each well individually. The discharge through each well is given by the following simplified relation:

1. For two identical wells in confined aquifer at a distance s apart.

$$Q_1 = Q_2 = Q_i = \frac{2\pi T\,(h_0 - h)}{2.3\log_{10} R^2 / rS}$$

where $(h_0 - h) = S_t$ or the combined drawdown at one of the wells, *i.e.*, the difference between the initial level of the water table (h_0) and water level in the well (h).

2. For three identical wells distance s apart, in the pattern of an equilateral triangle:

$$Q_1 = Q_2 = Q_3 = Q_i = \frac{2\pi T\,(h_0 - h)}{2.3\ \log_{10} R^3 / rS^2}$$

Similar expressions can be derived for other combinations of wells by superposing drawdowns at the point of intersection due to various wells.

3. For two identical wells in unconfined aquifer.

$$Q_1 = Q_2 = Q_i = \frac{\pi k\,(h_0^2 - h^2)}{2.3\log_{10} R^2 / rS} = \frac{1.365\,KS(L + S/2)}{\log_{10} R^2 / rS}$$

where K = coefficient of permeability of the aquifer.

The interference effect can be expressed by the ratio of discharge with interference to the discharge without interference for two tubewells by, $\dfrac{Q_i}{Q} = \dfrac{\log R}{\log R^2 / S}$.

The tubewells are so located that they do not interfere with each other *i.e.*, their radii of influence do not overlap. Table 3.3 indicates the radius of influence of wells in different soils.

Table 3.3 Radius of influence of wells in different soils

Strata	*Radius of influence (m)*	
	Shallow well	*Deep well*
Fine sand layers with some silt and clay	30–60	250–300
Fine to medium sand layers fairly clean, free from silt and clay	60–90	300–350
Coarse sand and fine gravel layers, free from silt and clay	90–120	350–400

In order to avoid mutual interference between wells it is desirable that wells are spaced at a distance not less than twice the radius of influence. Two wells spaced one-fourth the radius of influence apart may yield about 75 per cent more than either well singly, with a spacing of twice the radius of influence, the increase may be 100 per cent.

Pumping

Pumping is generally done with turbine pumps. Centrifugal pumps are economical than turbine pumps and are used where water table is near the surface and drawdown is not excessive. The practical maximum suction lift is about 5 m. As such the centrifugal pumps are sometimes installed in pits at the top of the well. Suction lift is of no consideration in turbine pumps because the pump is installed as near the bottom of the well as possible. The pump may be driven by an engine-powered or electric powered vertical driven shaft, or by a submersible electric motor close-coupled to the pump. The well diameter has to be bigger to allow installation of a pump of requisite size.

Constraints of Tubewell Irrigation

The principal constraints in accelerated development of tubewell irrigation are (*i*) High initial investment for the installation of a tubewell, (*ii*) Susceptibility of groundwater conditions and the speed at which these can vary in specific areas, (*iii*) Limited capacity of the farmers to pay for first cost, fuel and other operation and maintenance costs, (*iv*) Lack of institutional capacity with respect to supervision, maintenance and proper irrigation distribution, (*v*) Slow rate of formation of water-user groups, though this might be rapidly overcome with the new emphasis on tubewell development, (*vi*) Involves a lot of mechanical hardware like strainers, electric motor and pump which need proper maintenance and availability of facilities for quick repairs, and (*vii*) Life of pumping equipment is limited.

Duty and Delta of Tubewell Irrigation

Duty. The duty of a tubewell is defined as the area (in hectares) irrigated by it anually to bring the crops grown on the land to maturity. The duty of different tubewells is different depending on the lift involved, cropping pattern and water management practices adopted for efficient and economical use of the well water, rainfall pattern in the region etc. However, the duty of well irrigation is higher than that of canal irrigation due to the tendency on the part of the owner to make efficient and economical use of water.

Delta. The amount of water used to mature a crop by a well is less than that required from a canal and as such delta of well irrigation is about two-third of that from canal irrigation due to (*i*) Economical use of water owing to costly pumped water, (*ii*) Shorter length of watercourse and usually lined means less transmission losses, (*iii*) Better crop cultivation practices, and (*iv*) Reduced wastage of water use in the fields.

3.2.5 Infiltration Galleries

Infiltration galleries are horizontal or nearly horizontal small tunnels circular in section or consist of pipes constructed through water-bearing strata parallel to a river bank for the collection of inflow of groundwater to the river. The groundwater, seeping from the bed and sides of the river, can enter the tunnel through weep holes provided in the sides. With this end in view the tunnels are covered with graded filter all around (Fig. 3.19). The water collected in the gallery is led to a sump wherefrom it is pumped for water supply. The infiltration galleries provide an effective source of supply at places where groundwater is available in sufficient quantity just below the ground level. The galleries are placed at the full depth of the aquifer for maximum yield. The yield may be of the order of 1.5×10^4 litres/day/metre length of infiltration gallery.

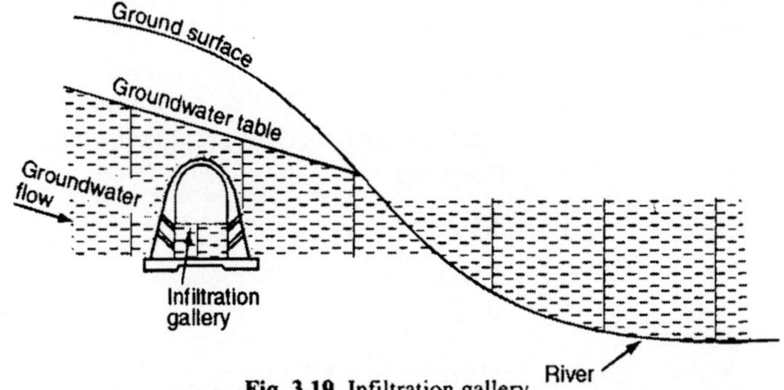

Fig. 3.19 Infiltration gallery

Yield of an infiltration gallery is given by the formula: $Q = k(H^2 - h^2) / 2L$.
where, Q = discharge (m^3/m length), K = coefficient of permeability, L = distance through which seepage takes place (m), H = depth of permeable stratum above the bottom of the infiltration gallery (m), and h = depth of water in infiltration gallery (m).

3.2.6 Sewage Irrigation

Sewage water can be recycled for agriculture. The treated sewage affluent is a usable supplemental water resource and can be considered in integrated water development and utilisation. However, use of sewage effluent has to be managed in a sanitary manner without polluting the soil or contaminating crops raised on the sewage farm. It is also a valuable source for artificial recharging of aquifers in areas of rapid depletion of underground water sources. Sewage is a water borne waste from community residences, institutions, offices, commercial and industrial establishments. Its characteristics and composition depends upon the source and vary from time to time and at the same place. IS: 3307-1977 recommended standard of treated sewage water for irrigation is as under:

Total concentration of solids (inorganic)	$\leq$200 mg/l
Percentage of sodium with respect to total content of sodium, calcium, magnesium and potassium	$\leq$60 mg/l
Boron	$\leq$2 mg/l
P^H of sewage	$\leq$ 5.5 to 9.0
BOD	$\leq$500

In India, raw sewage, settled sewage as well as treated sewage is used for sewage farming both by gravity and lifting. The command area for effluent irrigation shall not be located within 300 m from the residential area and 1000 m from the city waterworks or in close proximity of water bearing aquifers. The groundwater table in the area should be more than 2 m below the surface.

3.2.7 Supplemental Irrigation

In humid areas with adequate rainfall to meet the irrigation water requirements of crops, normally artifical means of irrigation are not required. But to take care of water requirement of crops during seed germination or other critical periods in plant growth in the serious drought periods extending to several days, supplemental irrigation is needed. Sprinkler irrigation or furrow irrigation with perforated pipes instead of ditches is usually adopted.

SOLVED EXAMPLES

Example 3.1 *Design an open well in a fine sandy subsoil formation to yield 0.004 cumec discharge under a depression head of 5 m.*

Solution. Substituting the values in the discharge formula.

$$Q = (K/A)\,AS$$

$$Q = 0.004 \text{ cumec or } (0.004 \times 60 \times 60) = 14.4 \text{ m}^3\text{/hour}$$

K/A = 0.5 m^3/hour per m^2 of area under unit depression head for fine sand.

$$\therefore \quad 14.4 = 0.5\, A.S.$$

or

$$A = \frac{14.4}{0.5 \times 5} = 5.76 \text{ m}^2$$

Well diameter,

$$d = \sqrt{\frac{4A}{\pi}} = \sqrt{\frac{4 \times 5.76}{22/7}} \simeq 2.75 \text{ m.}$$

Example 3.2 (*a*) *A well penetrates fully in a confined aquifer 10 m thick (saturated thickness) having coefficient of permeability of 0.0005 m/sec. The radius of well is 10 cm. There is a drawdown of 4 m at the well face and its radius of influence is 300 m. Calculate the steady state discharge which can be withdrawn from this well.*

(*b*) *What will be the percentage increase in the discharge if the radius of the well is doubled?*

[P.U. April, 1980]

Solution. (*a*)

Discharge formula,

$$Q = \frac{2.72\, KbS}{\log R/r}$$

Substituting the given values,

$$Q = \frac{2.72 \times 0.0005 \times 10 \times 4}{\log 300/0.10}$$

$$Q = 0.0156 \text{ cumec}$$

(*b*) Now Q with $r = 0.2$ m,

$$Q = \frac{2.72 \times 0.0005 \times 10 \times 4}{\log 300/0.20} = 0.0171 \text{ cumec}$$

Increase in discharge $= 0.0171 - 0.0156 = 0.0015$ cumec

$$\text{Percentage increase in discharge} = \frac{0.0015 \times 100}{0.0156} = 9.6\%$$

Thus by doubling diameter, the increase in discharge is hardly 10%.

Example 3.3 *In an unconfined aquifer (coefficient of permeability $k = 10^{-3}$ m/sec) resting above an impervious base a fully penetrating well of diameter 0.3 m draws a steady state discharge of 120 m^3/hr. Due to pumping the water table is lowered by 1.5 m at a point 12 m from the centre of the well. If the water table was 15 m above the impervious base before the pumping started, compute the drawdown at 30 m from the well.*

Solution. Discharge, $(Q) = 120/60 \times 60 = 0.03$ cumec

Water depth in observation well at 12 m from the well (h_1) $= 15.0 - 1.5 = 13.5$ m

Now,

$$Q = \frac{1.36\, K(h_2^2 - h_1^2)}{\log r_2/r_1}$$

$$\text{or} \qquad 0.03 = \frac{1.36\times10^{-3}(h_2^2 - 13.5^2)}{\log 30/12}$$

$$\text{or} \qquad h_2 = 13.82 \text{ m}$$

$\therefore$ Drawdown at 30 m $= 15.0 - 13.82 = 1.18$ m.

Example 3.4 *A fully penetrating well of diameter 0.4 m is abstracting water from a 2.5 m thick confined aquifer. The steady state drawdowns at 8 m and 50 m were observed to be 2.5 m and 0.6 m respectively. Compute the steady state discharge from the well. The coefficient of permeability of aquifer is 10^{-3} m/sec.*

Solution. It is a case of confined aquifer. If h_1 and h_2 are the depths of water measured in two observation wells located at radial distances r_1 and r_2 respectively from the tubewell (Fig. 3.12).

$$Q = K I_x A_x$$

$$Q = K\left(\frac{dy}{dx}\right)(2\pi xb)$$

$$\text{or} \qquad Q\frac{dx}{x} = 2\pi kb\, dy$$

$$\text{Integrating} \qquad Q = \int_{r_1}^{r_2}\frac{dx}{x} = 2\pi kb\int_{h_1}^{h_2} dy$$

$$\text{or} \qquad Q\Big|\log x\Big|_{r_1}^{r_2} = 2\pi kb\Big|y\Big|_{h_1}^{h_2}$$

$$\text{or} \qquad Q = \frac{2\pi kb(h_2 - h_1)}{\log_e r_2/r_1} \qquad ...(i)$$

$$Q = \frac{2.72kb(h_2 - h_1)}{\log_{10} r_2/r_1}$$

Now S_1, the drawdown in well No. 1 $= H - h_1$

or $h_1 = H - S_1$

and S_2, the drawdown in well No. 2 $= H - h_2$

or $h_2 = H - S_2$

$$\therefore \qquad h_2 - h_1 = (H - S_2) - (H - S_1)$$

$$= S_1 - S_2$$

Substituting in equation (*i*),

$$Q = \frac{2.72\, kb(S_1 - S_2)}{\log_{10} r_2/r_1}$$

Putting given values,

$$Q = \frac{2.72\times10^{-3}\times2.5(2.5-0.6)}{\log_{10} 50/8} = 0.016 \text{ cumec}$$

Example 3.5 *A tubewell is driven in a confined aquifer of 24 m thickness. The aquifer is met at 25 m below ground level. The static water table is 15 m below ground level. The discharge of the well is found to be 6000 m³/day, when the depression head is 12 m. Permeability of the stratum is 24.5 m/day. Find the diameter of the tubewell when the radius of influence is 300 m.*

Solution. Drawdown, $H - h = 12$ m

Using relation,

$$Q = \frac{2.72bk(H-h)}{\log_{10} R/r}$$

$$6000 = \frac{2.72\times24\times24.5\times12}{\log_{10} 300/r}$$

$$\log_{10}\frac{300}{r} = \frac{2.72\times24\times24.5\times12}{6000}$$

$$2.4771 - \log_{10} r = 3.1987$$

$$\log_{10} r = -0.7216$$

$$\log_{10}(r^{-1}) = 0.7216$$

$$r^{-1} = 5.27$$

$$r = 0.19 \text{ m}$$

Example 3.6 *A fully penetrating well (diameter 0.3 m) draws water from a confined aquifer of permeability 10^{-3} m/sec and thickness 15 m. If steady state discharge is found to be 1/30 m³/sec, compute the drawdown at 10 m from the centre of the well. Radius of influence 1000 m.*

Solution. $$Q = \frac{2.72kb(H-h)}{\log_{10} R/r}$$

where h is water depth in the well at steady condition or $H - h$ is drawdown at well face.

$$\therefore \quad \frac{1}{30} = \frac{2.72\times10^{-3}\times15\times S}{\log_{10}1000/0.15}$$

or $$S = 3.12 \text{ m}$$

1. Drawdown at 10 m from the well.

$$Q = \frac{2.72kb(S-S_1)}{\log_{10} r_1/r}$$

$$\frac{1}{30} = \frac{2.72\times10^{-3}\times15(3.12-S_1)}{\log_{10}10/0.15}$$

or $$S_1 = 1.63 \text{ m}$$

Example 3.7 *Two tubewells penetrating fully a 10 m thick aquifer are located 200 m apart. The salient data of tubewells are tube diameter 30 cm, depression head 5 m, radius of influence 300 m, and coefficient of permeability 10^{-3} m/sec.*

(a) Are the tubewells interfering? (b) Discharge of tubewell when one well is working, and (c) Percentage decrease in discharge of the well if both wells are working under a depression head of 5 m.

Solution. (*a*) Since radius of influence of each tubewell is 300 m and tubewells are located 200 m apart, they are interfering because they lie within the radius of influence of each other.

(*b*) When one tubewell is working, $Q = \dfrac{2.72bkS}{\log_{10} R/r}$

$$= \frac{2.72\times10\times10^{-3}\times5}{\log_{10}300/0.15} = 0.041 \text{ cumec}$$

(*c*) When both tubewells are working

$$Q_1 = Q_2 = \frac{2.72\,KbS}{\log_{10} R^2/rs}$$

Now, $$Q_1 = Q_2 = \frac{2.72\times10\times10^{-3}\times5.0}{\log\dfrac{300\times300}{0.15\times200}} = 0.039 \text{ cumec}$$

Percentage decrease in discharge = $\dfrac{0.041-0.039}{0.041} \times 100 \simeq 5\%$

Example 3.8 *Find the percentage decrease in discharge in Example 3.7 if a third tubewell is also located at 200 m spacing.*

Solution.

$$Q_1 = Q_2 = Q_3 = 2\pi T \frac{(h_0 - h)}{2.3 \log_{10} R^3 / rs^2}$$

$$= \frac{2\pi \times 10^{-3} \times 10 \times 5.0}{2.3 \log_{10} \dfrac{300 \times 300 \times 300}{0.15 \times 200 \times 200}} = 0.037 \text{ cumec}$$

Percentage decrease in discharge = $\dfrac{0.041-0.037}{0.041} \times 100 \simeq 10\%$

Example 3.9 *Determine discharge from 0.3 m diameter fully penetrating artesian tubewell. Drawdown 5 m, radius of influence 250 m, average coefficient of permeability 7000 cm/day, effective strainer length is 30 m.*

Solution.

$$Q = \frac{2\pi KS(L + S/2)}{2.3 \log_{10} R/r}$$

$$= \frac{[2 \times 3.14 (7000 / 100 \times 24 \times 60 \times 60) \times 5 \times (30 + 5/2)]}{2.3 \log 250 / 0.15}$$

$= 0.11$ cumec.

Example 3.10 *A well penetrating an unconfined aquifer 20 m thick below water table is pumped at a uniform rate of 600 litre/minute, till the water level in the well becomes steady. Two observation wells drilled radially at a distance of 20 and 80 m from the centre of the well show depression of 3.2 m and 1.10 m. Determine the coefficient of permeability.*

Solution. Discharge = 0.01 cumec

$$Q = \frac{\pi K (h_2^2 - h_1^2)}{2.3 \log r_2 / r_1}$$

or

$$K = \frac{Q}{\pi (h_2^2 - h_1^2)} 2.3 \log \frac{r_2}{r_1}$$

$$h_2 = 20 - 1.10 = 18.9 \text{ m}$$

$$h_1 = 20 - 3.2 = 16.8 \text{ m}$$

$\therefore$

$$K = \frac{0.01}{-\pi[(20-1.1)^2 - (20-3.2)^2]} 2.3 \log 80 / 20$$

$$= 5.87 \times 10^{-5} \text{ cm/s.}$$

QUESTIONS

1. Write an essay on the types of irrigation system. Name the most economical type of irrigation system.
2. What are the various types of irrigation? What are the merits and demerits of each?
3. Discuss the inundation irrigation and compare it with perennial irrigation. What are the advantages and disadvantages of inundation irrigation? How does design of inundation canal differs from that of a perennial canal?
4. What is meant by Bandhara irrigation? Give briefly the advantages and disadvantages of bandhara irrigation.
5. Lift irrigation is a costly system of irrigation, explain. Discuss the salient features of lift irrigation. Compare lift irrigation and gravity irrigation.

6. How does design of lift canals differ from gravity canals?
7. Discuss well irrigation. What are the advantages and disadvantages of well irrigation. Compare well irrigation and canal irrigation.
8. Well irrigation is an anti-waterlogging measure, explain.
9. (*a*) Write an essay on various types of irrigation wells. Which one you consider most suitable?
   (*b*) Compare tubewell irrigation with canael irrigation. Explain demerits of tubewell irrigation.
10. What do you understand by the term shallow well and deep well? Compare shallow well and deep well.
    Design an open well in a fine sandy subsoil formation of 0.005 cumec discharge under a depression head of 3 m.
11. Define tubewell irrigation. What are the various types of tubewell? Discuss the merits and demerits of tubewell irrigation.
12. Discuss various parts of tubewell and their functions.
13. (*a*) What is the difference between an open well and a tubewell? How does the flow into a well occur in the two cases?
    (*b*) How will you determine safe discharge from a tubewell?
    (*c*) Compare tubewell irrigation with canal irrigation.
14. Derive discharge formula of a fully penetrating tubewell in confined aquifer.
    A tubewell yielding 0.1 cumec is supplied water from a confined aquifer 7 m thick. Two observation wells completely penetrating through the aquifer and located at the distances of 40 m and 100 m indicate drawdowns of 0.71 m and 0.29 m respectively. Determine the coefficient of permeability of the aquifer.
15. Define cavity tubewells. Give a sketch of such a tubewell and state when such tubewells can be constructed.
16. Describe the various types of strainers used for tubewells in your area.
17. Derive discharge formula for a tubewell in unconfined aquifer.
    Design a tubewell to deliver 7.20 cum water per hour with a drawdown of 6 m. The average water level in the well is 20 m below the ground level in monsoon and 30 m in summer. Assume the radius of influence to be 200 m. Geological investigations have yielded the following results:

Depth below ground (m)	*Type of strata*	*Depth below ground (m)*	*Type of strata*
0 – 5 m	Soft earth	35 – 50	Coarse sand
5 – 15 m	Hard murrum	Below 50 m	Hard rock (impermeable)
15 – 35 m	Fine clay		

18. (*a*) What are different types of strainers used in tubewells? Discuss the factors which govern the selection of a strainer.
    (*b*) Derive an expression for discharge from a confined aquifer, stating clearly the assumptions made. (*P.U.*)
19. (*a*) What is the function served by a strainer in a tubewell? What are various types of strainers in common use?
    (*b*) Differentiate between an unconfined aquifer and a confined aquifer.
    (*c*) Which type of aquifer exists in Haryana?
    (*d*) Discuss the relative advantages and disadvantages of open well and tubewell.
    (*Kurukshetra University*)
20. (*a*) What are the advantages and disadvantages of well irrigation as compared to flow irrigation?

(*b*) Calculate the discharge in cum/day from a tubewell of diameter = 30 cm, drawdown = 1.5 m, length of strainer = 35 m, radius of influence = 300 m, and coefficient of permeability = 10×10^{-4} m/sec. The aquifer is unconfined.

Solution.

$$Q = \frac{2.72ks(L+s/2)}{\log R/r}$$

$$= \frac{2.72 \times 0.1 \times 10^{-2} \times 1.5(35+0.75)}{\log(300/0.15)} = 3818 \text{ cum/day.}$$

21. (*a*) Define:

    (*i*) Confined aquifer.

    (*ii*) Unconfined aquifer.

    (*b*) Derive the equation for discharge from an unconfined aquifer.

    (*c*) A well penetrates fully a 15 m thick water bearing stratum of medium sand having coefficient of permeability 50 m/day. The well radius is 15 cm and is to be worked under a drawdown of 5 m at the well face. Calculate the discharge from the well.

22. Explain the term duty and delta. Derive a relationship between the two. The discharge available from a tubewell is 120 m^3/hr. Assuming 3200 hours of working for a tubewell in a year, estimate culturable area that it can command. The intensity of irrigation is 50% and the average depth of Rabi and Kharif crop is 48 cm.

23. An unconfined aquifer has permeability of 0.48 mm/day. A tubewell with 0.3 m diameter has strainer length of 10 m. Calculate yield under the drawdown of 2 m. Assume radius of influence to be 300 m.

4

METHODS OF IRRIGATION

4.1 INTRODUCTION

The design, equipment and technique of replenishing the soil water deficit by applying irrigation water is referred to as irrigation method. Five basic methods of the application of irrigation water to the fields to suit different crops, soil type, topography, water availability and its quality, climatic conditions, individual holdings and costs are (*i*) flooding irrigation, (*ii*) furrow irrigation, (*iii*) sprinkler irrigation, (*iv*) drip irrigation, and (*v*) subirrigation, besides various subclasses as illustrated in Fig. 4.1:

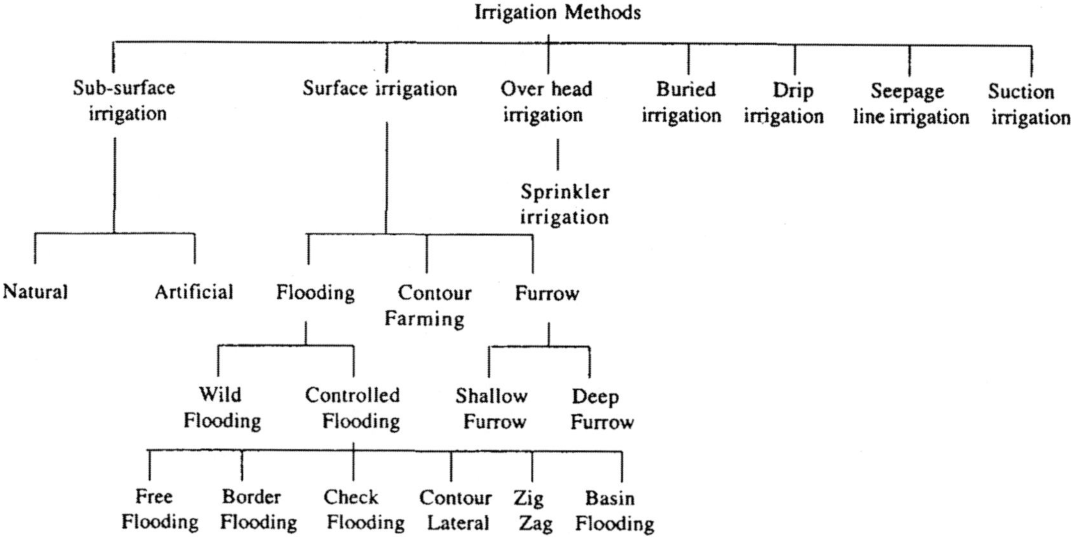

Fig. 4.1 Irrigation methods

The primary objective of any irrigation method is to supply water to soil so that moisture will be readily available at all times for crop growth but without indiscriminately adding to the water table, as well as avoiding influence of soil salinity. An efficient irrigation method is that which best suits local conditions such as (*i*) soil characteristics, (*ii*) kind of crop and its age, (*iii*) crop rotation, (*iv*) topographic condition, (*v*) available water flow, (*vi*) underground water table condition (high or low), and (*vii*) state of soil salinity.

4.2 REQUIREMENTS OF AN IRRIGATION METHOD

The essential requirements for adoption of any irrigation method are (1) Application is within desirable limits. Stream flow is adequate so that the quality of irrigation is such that the depth of wetting and hence the stand of crop are approximately uniform. The irrigator achieves high productivity so that during a day's work he can irrigate a large hectarage, (2) Afford a uniform water distribution in root zone of a crop with as small as 6 cm applications for light irrigation, (3) Allow heavy uniform application of 15 to 20 cm of water depth and under some conditions as much as 25 cm per irrigation for salt leaching where such a problem exists, (4) Allow use of large concentrated water flows for reduction of conveyance losses, field channel network and labour cost, (5) Suitable for use with economical conveyance structure, (6) Facilitate mechanized farming, (7) Occupy minimum land under bunds, etc, (8) Inexpensive and economically justifiable, (9) High efficiency of water application *i.e.*, the ratio of water stored in the root zone to that delivered to the field should be maximum, (10) Minimum wastage of water either through surface runoff or through deep percolation below the root zone of a crop.

A suitable irrigation method shall ensure maximum yield at optimum water utilization, *i.e.*, conservation of water resource. The choice of an irrigation method suitable to a specific situation is an important aspect of water management of crops.

4.3 SUB-SURFACE IRRIGATION

In sub-surface or sub-irrigation water is applied beneath the ground by creating and maintaining an artificial water table at some depth, usually 30 to 75 cm, below the ground surface. Moisture moves upwards towards the land surface through capillary action to meet requirements of the crops in plant roots. Water is applied through underground distribution system consisting of a properly designed main field ditches, laterals, laid 15 to 30 m apart. Water may be obtained from wells, streams, lakes etc. Water is introduced into soil profile through open ditches, mole drains or tile drains. Open ditches are preferred because they are relatively inexpensive and suitable for all types of soils. Tiles and mole drains are suitable only for organic soils. Sub-surface irrigation requires little field preparation and labour. It entails minimum evaporation loss and surface waste. The irrigation water is essentially required to be of good quality to prevent excessive soil salinity. The flow rate in supply ditches is required to be low to prevent waterlogging of the field. The use of sub-irrigation is limited because it requires certain soil condition that is the soil is permeable in root zone, underlain by an impervious horizon or high water table.

Essential Requirements

The essential requirements for a successful sub-surface irrigation are (*i*) Availability of adequate supply of good quality water throughout growth period of the crop, (*ii*) Fields must be nearly level and smooth. Ground slope is moderate. Land is approximately parallel to water table, (*iii*) Availability of a layer of permeable soil such as sandy loam or loam immediately below the surface soil to permit free and rapid movement of water laterally and vertically, (*iv*) Availability of a relatively impervious layer at 2 to 3 m in the substratum to prevent deep percolation of water or a permanently high natural water table on which an artificial water table can be built, (*v*) A well planned distribution system of main ditches, field laterals, etc., which raises the water table to a uniform depth below the ground surface over the entire area, (*vi*) Availability of adequate outlet for drainage of the area so irrigated particularly in humid areas, (*vii*) Subsoil water table is within 2 to 3 m below the ground surface, (*viii*) Topographical conditions are uniform, and (*ix*) Soil is capable of lifting moisture from the water table to the root zone. Also the soil permits lateral and downward movement of water. The efficiency of water use depends on soil characteristics, topography and operation and maintenance management. In good system, the efficiency is 70-75 per cent.

Classification of Sub-surface Irrigation

The sub-surface irrigation may be classified into (*i*) natural sub-irrigation, and (*ii*) artificial sub-irrigation.

1. *Natural sub-irrigation.* This method is applicable to low lying lands where the water table is high and within the capillary reach of root zone of crops. The watertable is charged by seepage from irrigation canals. It is better controlled lest it may develop into waterlogging conditions. The method, applicable where above conditions exist, offers most economical means of raising crops.

2. *Artificial sub-irrigation.* It is a very expensive method. Its use is indicated only under favourable water supply and subsoil conditions for high yielding crops. The water is provided to crops by capillarity through a network of buried perforated pipes which carry water under pressure to percolate into soil.

Advantages of Sub-surface Irrigation. (*i*) Minimum water requirement for raising crops, (*ii*) Suitable for most crops, (*iii*) Minimum evaporation and deep percolation losses, (*iv*) High crop yield, (*v*) Most economical method of irrigation, (*vi*) Involves no wastage of land, (*vii*) No interference in free movement of farm machinery, (*viii*) Cultivation operations can be carried out without concern for the irrigation period, and (*ix*) Little field preparation and labour.

Disadvantages. (*i*) Requires a special combination of natural conditions so that only limited areas are suitable for irrigation by this method, (*ii*) There is danger of development of waterlogging, (*iii*) Possibility of choking of the pipes laid underground, and (*iv*) High cost.

4.4 SURFACE IRRIGATION

It is by far the most common type of irrigation. In surface irrigation method, since water is applied to the field in varied quantities and at different times, the flow remains unsteady. The method involves diverting a stream of water from the head of a field into furrows or borders, and allowing it to flow downward. Water infiltrates into the soil while traversing the furrow. By subsequent ponding and lateral movement, the soil is restored to its full water holding capacity to a depth that depends on the quantity of water introduced, the duration and rate of stream flow, the gradient and the soil structure and texture. Generally under open-ditch conveyance and surface irrigation methods, less than one-half of the water released reaches plants. Highly efficient irrigation can be achieved by an appropriate combination of size of the irrigation stream, uniform application of water, minimum soil erosion, minimum labour requirement, maximum land use, size and shape of field and use of machinery. The surface irrigation is essentially supplemented with efficient water disposal system.

Advantages: (*i*) Allows use of machinery for land preparation, cultivation and harvesting, and (*ii*) Helps to store the required amount of water in the capillary zone of the soil for supply to the root zone of plants.

Disadvantages: (*i*) Greater loss of water by surface runoff and deep percolation, larger requirement of water per unit area, (*ii*) Water is lost in infiltration and deep percolation, (*iii*) Low efficiency due to imperfect control over the water flow, (*iv*) Inferior quality crops with a low yield, (*v*) Wasteful use of water compared to better irrigation methods, and (*vi*) Costly and time consuming preparation of land.

Classification of Surface Irrigation

Surface irrigation may be classified as flooding method, furrow method and contour farming method, with further divisions as illustrated in Fig. 4.1.

1. **Flooding Method:** In flooding irrigation, water is allowed to cover the surface of land in a continuous sheet, the water standing just long enough in the field for the soil to absorb the water applied to refill the root zone. A properly designed size of irrigation stream aims at proper balance against the intake rate of soil, the total depth of water to be stored in the root zone and the area to be covered so as to give reasonably uniform coverage of water over the entire field. The flooding may be (*i*) wild, and (*ii*) controlled.

(*a*) **Wild Flooding:** It is also called *uncontrolled flooding*. It is the primitive and most inefficient method of irrigation. In this method water is spread over the smooth or flat field without much control over the flow or prior preparation. The water is spread into the field from the ditch excavated either on the contour or up and down the slope. This method is applicable to inundation irrigation system or for pastures or forage crops where water is available in abundance at the highest elevation and is inexpensive or the crop values do not justify adoption of better method. The water distribution is quite uneven. The method is suitable for all medium to fine texture soils. It has low cost and does not interfere with tillage. The disadvantages of the method are (*i*) wasteful use of water, (*ii*) non-uniform distribution of water, (*iii*) excessive soil erosion on steeper slopes, (*iv*) require drainage arrangement to reduce ponding.

(*b*) **Controlled Flooding:** (*i*) *Free flooding*. Also called *ordinary flooding*. It is the commonly adopted method where irrigation water is in abundance and cheap. The land is divided into plots or kiaries of suitable size depending on porosity of soil. Water is spread over the field from watercourse. The irrigation operation begins at the higher area and proceeds towards the lower levels. The flow is stopped when the lower end of the field has received the desired depth of water. The field watercourse is properly spaced, the spacing depends on the topography, soil texture, depth of soil and size of stream. The spreading may vary from less than 15 m to more than 60 m. Porous soil requires close spacing than tight soil. The method is most suitable for soils of medium texture and with moderate slopes.

(*ii*) **Border flooding:** In this method, the field is divided into narrow strips by low parallel ridges on the sides. The border strip is formed by levelling and grading the land between the ridges (Fig. 4.2). The size of the strip depends on the type of soil, slope of the land and size of irrigation stream. The width of the strip varies from 5 to 15 m and length varies from 60–100 m for sandy loam soils, 100–120 m for medicum loam and 150–300 for clay loam and clay soils. A longitudinal gradient of 0.02% to 0.05% for clay to clay loam, 0.20% to 0.40% for medium loam and 0.25% to 0.60% for sandy laom to sandy soil is provided in each strip without any cross slope.

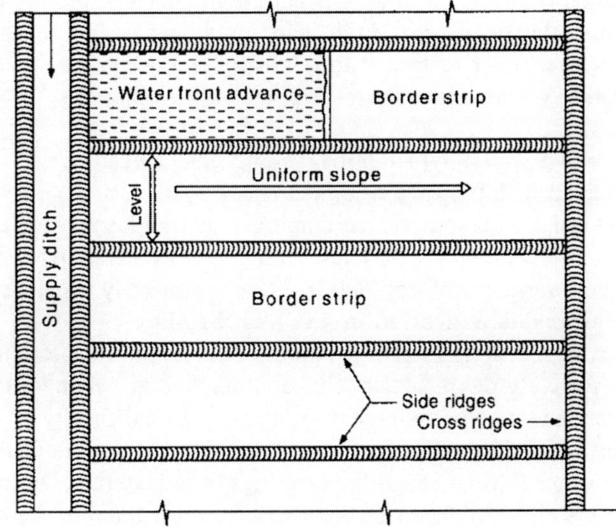

Fig. 4.2 Border irrigation system

The entire land including the rounded border ridges is utilized for cultivation. Thus no land is lost to production. Water is let into each strip in a thin sheet through a set of outlets or siphon pipes and underground concrete pipes through risers placed along the upper end of the border. Water flows down the slope along the length of the strip at non-erosive velocity covering the entire width of the strip. Thus sufficient moisture is provided in the soil profile to the entire length of the border. Even distribution of water is ensured if the strip is levelled at a more or less uniform longitudinal slope.

The method is applicable to most soil textures but is best adapted to soils having moderately low to moderately high infiltration rates. It is suitable for high value crops and all close growing crops and is used for some row crops like cotton. In general, it is suitable for hay, pastures and gram crops on lands having slopes up to 3 per cent.

Advantages: (*i*) Utilizes large water streams safely, (*ii*) Requires less labour and time; low maintenance cost, (*iii*) Provides uniform wetting and efficient use of water.

Disadvantages: (*i*) Requires proper levelling, (*ii*) High initial cost, and (*iii*) A large supply of water is needed.

(*i*) *Check flooding*. It is a widely practised method of irrigation in India. In consists of applying water to relatively level check basins enclosed by small bunds (Fig. 4.3). The checks are square, rectangular or irregular plots. If the ground has some initial slope, levees may follow the contours. The size of check basin may range from 3 × 2 m to 30 × 30 m or even larger. In orchards.

checks are used to irrigate each tree or a group of few trees in a check basin. The levees are 2 to 3 m wide at the base and 25 to 30 cm high in order to avoid obstruction to the movement of farm machinery as also to permit growth of crops over them.

In highly permeable soils, large size stream is required to quickly impound the check basin with the required quantity of water without allowing deep percolation losses on the side of the irrigation ditch. In highly impermeable soil with low infiltration rate, the water is impounded in the basin and is slowly allowed to infiltrate in the soil. The method is specially suited to fine textured soils which require holding of water in order to secure the desired penetration. This method is suitable for growing grain and forage crops. The use is generally restricted to relatively smooth lands because of the expenditure involved in levelling the plots.

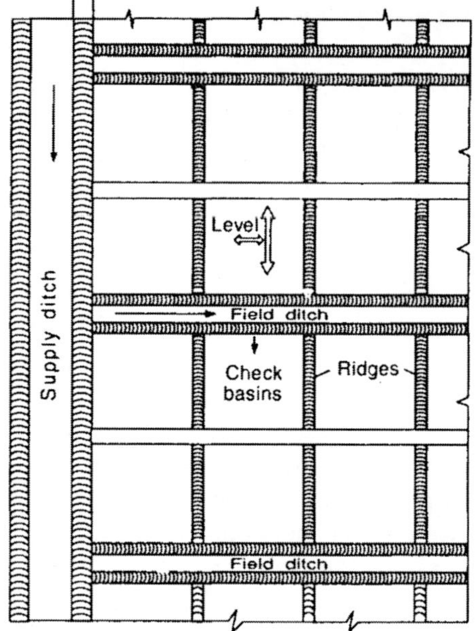

Fig. 4.3 Check basin irrigation system

Advantages: (*i*) High irrigation efficiency can be achieved with properly designed check system, and (*ii*) Unskilled labour can be employed as there is no danger of erosion.

Disadvantages: (*i*) High labour requirement, (*ii*) Levees impose restrictions in the use of modern farm machinery, and (*iii*) Use is generally restricted to relatively smooth lands because of the expenditure involved in levelling the plots.

(*d*) **Contour lateral method:** This method is best suited to steeper terrain. The field is cut by relatively dense network of contour laterals; their spacing depends on the (*i*) grade of land between two adjacent ditches or laterals, (*ii*) uniformity of slope, and (*iii*) type of soil. It may vary from 15 to 50 m. The contour ditches may have practically little grade if they are not too long. The water flows through the openings in the laterals. On the other side of contour lateral, a small bund is provided to divert the water from the main to the lateral. The stream size per unit area is the same as in border irrigation.

Contour lateral flooding is adopted on close growing crops on sloping or rolling lands not subject to any degree of levelling necessary for other methods of irrigation. Water flows down slope between closely spaced ditches running along the contour of the land.

Disadvantages: (*i*) High labour cost of irrigation, and (*ii*) Not suitable if other methods of irrigation can be adopted.

(*e*) **Zig zag method:** It is a special method of flooding suitable for relatively level fields. But the method is unsuitable for mechanical farming operation. In this method, land is divided into square or rectangular plots; each plot is further subdivided with the help of low bunds or levees. Water enters at the upper end of the plot and follows circuitous route to reach the lower end of the plot when the supply is cut off.

(*f*) **Basin flooding:** It is essentially a check method of flooding adapted for irrigation of orchards. In orchards basins are made around one or more trees depending on the soil condition and topography. For efficient irrigation, each basin has to be level, grade less than 0.1 per cent with no cross slope. It is adapted especially to flat lands. Water is filled quickly in the basin and is allowed to percolate into the soil. It is desirable for close-growing crops and orchards on medium to coarse textured soils as also rice grown on fine-textured, slowly permeable soils. Water is conveyed to the basin or preferably through small ditches feeding individual basins. The merits and demerits of the method are broadly the same as indicated for border irrigation method.

2. **Contour Farming:** It is adapted to hilly areas with steep slopes and quickly falling contours. The land is divided into longitudinal curved plots, the bunds of the plots following the contours. The irrigation water stored higher up in some depressions flows between the bunds. The cultivation of crops is along the contour lines instead of the usual down the slope. Contour cultivation reduces runoff and soil loss.

3. **Furrow Method:** It is the common method of irrigating row crops. A furrow consists of a narrow ditch between the rows of crops. Water is applied in small streams between rows of crops, grown on ridges or in furrows (Fig. 4.4). The characteristic feature of furrow irrigation is that, unlike flooding method, only half to one-fifth of the irrigated land surface is wetted, thereby cutting down evaporation losses. Water flowing in the furrow seeps into the soil and spreads laterally to irrigate areas between the furrows. For distribution of water to the furrows, the main ditch is placed on the higher side of the field.

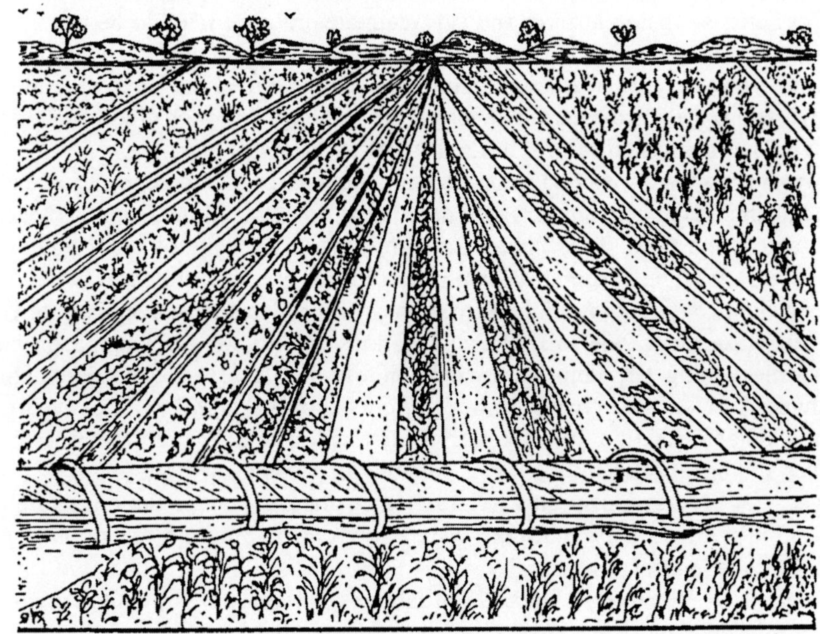

Fig. 4.4 Furrow irrigation

The size and shape of the furrows depend on the soil, crop spacing, equipment used for the formation of furrow, and intercultivation. Long furrows are advantageous in intercultivation. However, too long furrow results in deep percolation losses and soil erosion near the upper end of the field, and too little water near the tail. It may range from 70 m on light soils to 200 m on heavy soils. The usual size of furrow for row crops is about 25 cm wide and 8 to 10 cm deep. Furrow grade may vary from 0.1 to 0.5 per cent for different soils and stream sizes. The maximum non-erosive stream size for a furrow is determined from the following relation:

$$Q = \frac{0.60}{S}$$

where, Q = maximum furrow stream size (litre/sec), and S = slope of furrow expressed as a per cent.

The average depth of water applied during an irrigation is calculated from the following relation:

$$D = \frac{Q \times 3600}{S \times L}$$

where, D = average depth of water (cm) applied in one hour, and Q = stream size (litre/sec), S = furrow spacing (m), and L = furrow length (m) irrigated in one hour.

Furrow irrigation is adaptable to all row crops orchards etc. but not suitable for very light soils having high infiltration capacity, as water is wasted on the upper end of the furrow due to deep percolation.

Advantages: (*i*) Low evaporation losses as only one-half to one-fifth of irrigated land surface is wetted, (*ii*) It is possible to cultivate earlier, especially in heavy soils, (*iii*) Furrows serve as drainage ways for surface runoff in areas of heavy rainfall, (*iv*) Especially suitable for crops like maize that are injured by contact with water, (*v*) No wastage of land in field ditches, (*vi*) Fairly efficient in use of water, (*vii*) Relatively cheap to construct and maintain, and (*viii*) Efficient in its use of water.

Disadvantages: (*i*) Not recommended for very light soils having high infiltration capacity, (*ii*) Ditches may interfere with tillage, (*iii*) Usually expensive from the consideration of time and labour cost, (*iv*) Serious erosion hazard, and (*v*) Adequate drainage need to be provided.

Furrow irrigation may be contour furrow, shallow furrow and deep furrow.

(*a*) **Contour furrow:** It is an adaptation of furrow irrigation. If the slope of land is too steep for effective irrigation, the furrows are laid across the slope on a proper grade and are termed as contour furrows. The furrows are useful in the land of uneven topography.

(*b*) **Shallow furrow:** Also termed as corrugation method. In this method, the stream of water is guided through small channels or corrugations running down the slope from the head ditch and evenly spaced across the field. The corrugations are made by means of a corrugator generally after the crops are sown. The corrugations are primarily for guiding stream of water, rather than conveying water. Some overtopping of the furrow may occur. The size and spacing of corrugation depend on water intake rate of the soil. The corrugations are more closely spaced in more pervious soils to permit wetting between the corrugation without excessive deep percolation losses. These are V-shaped or U-shaped channels about 10 cm deep, spaced 40 to 75 cm apart. Corrugation irrigation is suitable for fine to moderately coarse textured soils that take water slowly and tend to seal over and bake when flooded. It is suitable for close growing crops.

Advantages: (*i*) Low evaporation losses as water does not cover the entire field surface, (*ii*) Useful for fine textured soils that take water slowly, (*iii*) Suitable for sloping and rolling lands, and (*iv*) Provides uniform wetting.

Disadvantages: (*i*) Relatively expensive in labour cost, and (*ii*) Not suitable where a large flow capacity is available and where irrigation can be done almost round the clock.

(*c*) **Deep furrows:** Deep furrows 15 to 30 cm deep, spaced 1 to 2 m apart and of length less than 30 m are more suitable for orchards and sugarcane. The longitudinal slope may vary from 0.2 to 0.5 per cent. This method is adapted to soils of low permeability.

4.5 SPRINKLER IRRIGATION

Sprinkler irrigation is an improvement over conventional surface irrigation. It simulates natural rainfall to spread water in the form of rain uniformly over the land surface just when needed and as much needed at a uniform pattern and at a rate less than the infiltration rate of the soil so as to avoid surface runoff from irrigation. It is generally indicated for undulating lands, poor water availability, sandy or shallow soils and where uniform application of water is required. No land levelling is required for sprinklers thereby saving considerable cost. The soil is not disturbed to expose the less fertile subsoil. Even on sand dunes crops can be grown under sprinkler irrigation. It enables economical use of water through control over the application rate and uniform application of water. The fertilizer is applied through the sprinkler system to ensure even distribution that avoids waste, saves cost and gives immediate plant response. The sprinkler irrigation is suitable for all types of soils, more particularly coarse, sandy and gravelly soils and for almost all crops like wheat, jowar, cotton, potatoes, tobacoo, groundnut, vegetables, ragi, etc. However, it is not recommended for crops having high water requirement such as rice, jute and plantation crops like coffee, tea, cardamom. It is particularly suitable for production of high yielding crops or for having continuous and quick growth of valuable crops. With a dependable supply of water, such as ground or surface water and electric power supply, sprinkler irrigation serves as a crop insurance in wet or semiwet regions, while in dry areas or tracts unsuitable for conventional irrigation, it provides economical means for production of crops. The sprinkler irrigation helps to conserve precious water up to 50 per cent and can irrigate 2 to 3 times the area compared to surface irrigation. This system is suitable for any size of farm.

Sprinkler irrigation is widely used in the U.S.A., Canada, England, France, Germany and Israel. In India, sprinkler irrigation was initially confined to commercial crops like tea gardens in Assam and a few experimental plots. Presently it is being tried on a large scale in Tamil Nadu, Karnataka, Haryana and Punjab and lift irrigation areas where land is sandy and undulating.

Factors Governing Selection

The conditions favourable for the adoption of this system or selection and use of sprinkler method is governed by various factors, such as (*i*) Land of undulating topography (sandy dunes) where land levelling is either impracticable or uneconomical, (*ii*) Land of steep slopes and easily erodible soils by surface irrigation, (*iii*) Land with shallow soil cover, sandy soils or soils with high infiltration rate, (*iv*) Highly porous or relatively impermeable soils unsuitable for proper water distribution by surface methods of irrigation, (*v*) Limited water supply and high cost of water as in lift irrigation, (*vi*) Need for light and frequent irrigation, (*vii*) Costly and unreliable farm irrigation, and (*viii*) Lands need to be brought into production quickly without waiting for land development and construction of channels etc.

Sprinkler Irrigation System

In sprinkler irrigation system, water is pumped under pressure, carried through high pressure main line and let out through sprinkler nozzles placed at regular intervals on lateral lines to form a gentle rain.

The system is designed considering agro-climatic conditions, general land condition, maximum difference in elevation, cropping pattern as different crops require different amounts of water and different irrigation schedules, cover-crop requirements and their effect on peak water use, requirement and rates, availability of labour, matching pump and power unit, pipe sizes economical to instal and operate, water supply source, quantity and quality as essentially clean and debris-free water is required. Considering these factors, the system is so designed that it delivers the right amount of water with optimum rate of application, depth of application and period of application, as under:

Layout: The layout of main and laterals is dependent on the slope and size of the farm and location of water source. The source of water near the centre of the area ensures most economical combination of pipe sizes. The main lines run up and down the slope. The laterals are at right angle to the main line and run across the slope, as nearly on the level as possible, to minimise variation of pressure. Usually the distance between laterals is kept 12 m as well as the distance between two sprinklers is also maintained at 12 m. Fig. 4.5 shows the set up found useful at

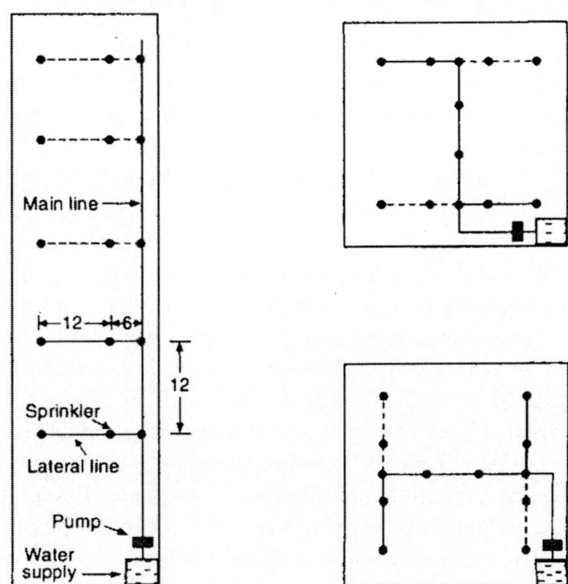

Fig. 4.5 Sprinkler irrigation set up

least in initial stages. Fig. 4.6 indicates principal parts of the piping system.

Water application rate: The water application rate has to be less than infiltration capacity of the soil to be irrigated. The maximum rate at which water can be applied without causing pudding, caping or surface runoff is controlled by the nature of soil, crop grown and the topography of the area to be irrigated. Normally it varies from 0.25 cm/hr for clay to 5.5 cm/hr for very light soils. The application rate (cm/hr) is given by the relation:

$$\frac{\text{Discharge (lt/ s)} \times 282.60}{\text{Spacing of sprinkler (m)} \times \text{spacing of laterals (m)}}$$

Also, $R = qA/360$, where R = application rate (cm/hr) q = discharge of sprinkler (lt/s), and A = wetted area of sprinkler (m^2).

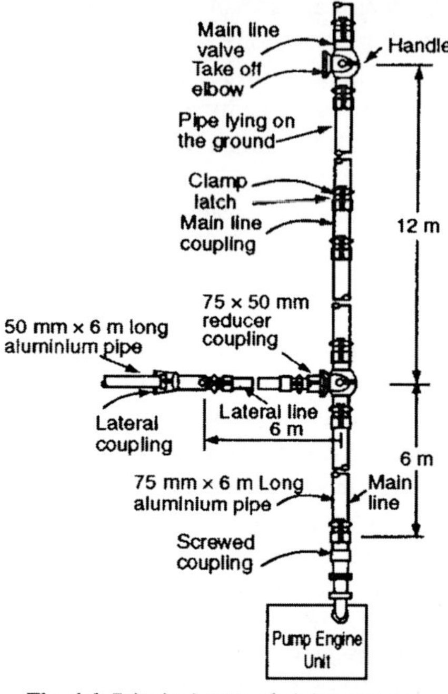

Fig. 4.6 Principal parts of piping system

Water application in each irrigation. The amount of water to be applied in each irrigation is determined by using not more than 50 per cent of the available moisture in the soil. The area to be irrigated in one setting depends on the type and pattern of sprinkler and operating pressure adopted.

Sprinkler losses. The sprinkler losses depend on wind velocity, temperature, fineness of spray, humidity, soil texture and vegetation cover.

Capacity of system. The capacity of efficient sprinkler system is required to be adequate to meet the peak demand of the area under crops during the hottest and driest periods keeping in view the crop rotation likely to be developed in future with due allowance for water losses. The capacity of the pipe system depends on the rate of application of water and the area to be irrigated in one setting.

The system capacity is given by the relation

$$Q = \frac{A \times D \times 27.8}{I \times H \times E}$$

where, Q = discharge (lt/s), A = area (ha), D = depth of water application (cm), I = interval between successive irrigation (days), H = operating hours (hr/day) and E = field application efficiency.

Operating pressure. The size of droplets comprising spray is limited in order to avoid damage to the soil. Small nozzles operated at high pressure produce relatively finer spray with small droplets. Roughly a pressure of 2.75 kg/cm^2 is adequate for 3 mm nozzle size with additional 0.35 kg/cm^2 pressure for each 0.75 mm increase in nozzle size. Usually the sprinklers operate with a water pressure of 0.5 to 10 kg/cm^2.

Sprinkler spacing and distribution pattern. The variation in pressure between any two sprinklers or sprinkler locations in the field is not to be more than 20 per cent of the higher pressure irrespective of elevation and friction losses. This will limit the variation in sprinkler discharge to be about 10 per cent. The sprinkler spacing, nozzle sizes, nozzle pressures and lateral layouts may need to be adjusted to provide adequate water distribution. Fig. 4.7 shows the distribution pattern at too low, too high and satisfactory pressure. Essentially the recommended pressure for the sprinkler is adopted. Overlap of the sprinkler throw has to be adequate.

The maximum move interval for typical intermediate pressure sprinkler is 18 m.

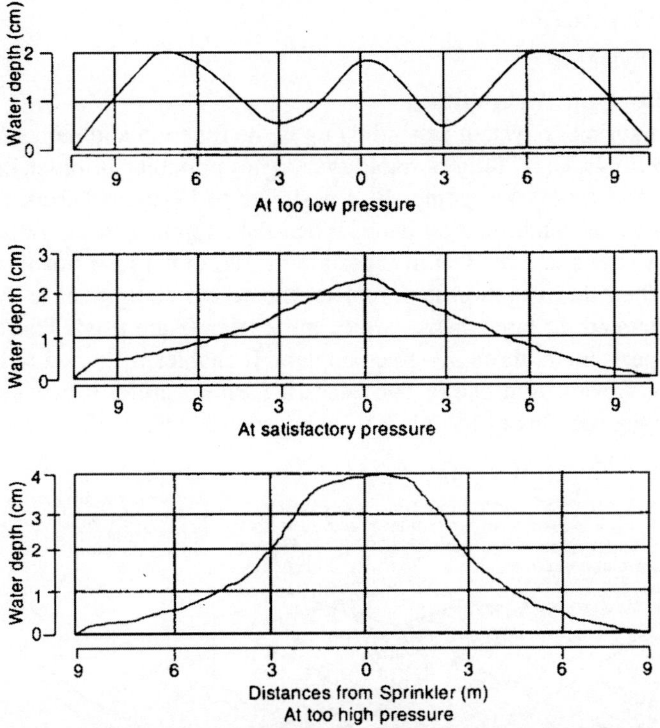

Fig. 4.7 Water distribution under different pressures

Main and lateral pipe sizes: The size of the pipelines for mains and laterals are determined by the maximum rate of flow and nature and length of pipes involved. The variation in depth of penetration attained at various points in the irrigated area is not to exceed 10 per cent and accordingly the variation of pressure in the lateral, due to friction loss etc. is usually restricted to within 20 per cent of the operating head. The pressure loss in main line is generally restricted to about 0.7 kg/cm^2.

Pumping unit. The peak consumptive use for the crops grown, area to be irrigated, cropping pattern, water availability in well, operating period (hrs/day) are determined to compute discharge of the pump required. The efficiency of the pump and motor and that of conveyance and distribution of water are also considered. The selection of the pumping set is made from the characteristics curve of the pumpset supplied by the manufacturer.

Discharge of sprinkler. The discharge of a sprinkler is given by the relation:

$$Q = \frac{Sl \times Sm \times R}{360}$$

where, Q = discharge (lt/s) of a sprinkler, Sl = spacing (m) of sprinkler along laterals, Sm = spacing (m) of laterals along the mains, and R = optimum application rate (cm/hr).

The discharge of sprinkler nozzle is given by the formula, $q = Ca\sqrt{2gh}$. Where, q = nozzle discharge (cumec), a = sectional area of the nozzle (m^2), H = pressure head of nozzle (m), g = aceleration due to gravity (m/s^2), C = coefficient of discharge.

Water spread area of sprinkler:

$$R = 1.35\sqrt{dh}$$

where, R = radius of wetted area covered by sprinkler (m), d = diameter of nozzle (m), h = pressure head at nozzle (m).

Types of Sprinkler Systems

1. Classification Based on Portability

Various types of sprinkler system depending upon the function and requirement are as under:

1. *Fully portable system.* A fully portable system has portable sprinkler lines, portable main lines and submains and a portable pump. It is designed to be moved from field to field or to different pump sites in the same field by hand. When the irrigation in the first position has been completed pump is stopped and the lateral pipes are moved to the next position on the main line and reassembled. When the irrigation is complete, the whole equipment is manually moved to another field and operated. In large areas, two or three laterals are used. These are connected to main line, which is permanent, through valve couplers. It enables irrigation to continue while the other lateral is being moved from one or two laterals. Farmers prefer it because it is economical and also allows for greater flexibility but labour is required. Fig. 4.8 shows the working of a portable sprinkler system.

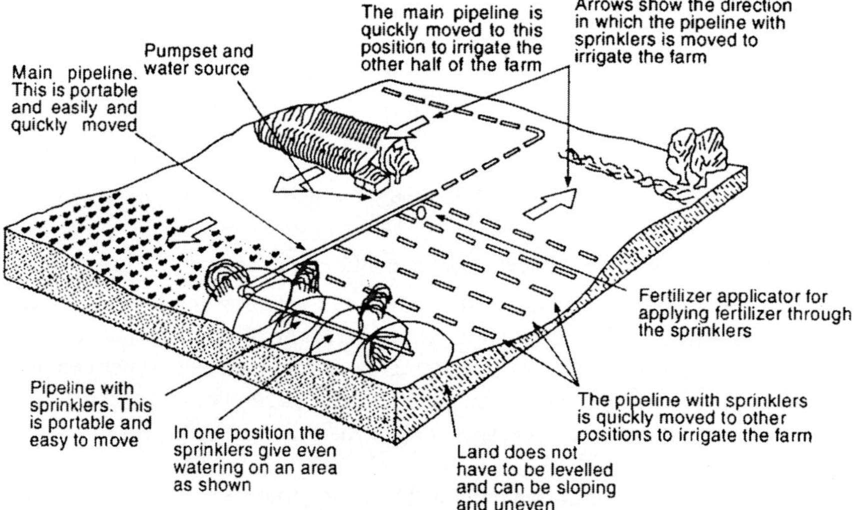

Fig. 4.8 Working of portable sprinkler system
Courtesy: Premier Irrigation. (Calcutta).

2. *Semi-portable system*: It is a variation of the fully portable system in that the water source and pumping plant are fixed. With an extended main line, the system is used on more than one field.

3. *Permanent system:* It consists of permanently buried main and laterals, either of galvanized iron pipes, aluminium, fibre glass or generally plastic pipes, fixed pumping plant, and permanent sprinklers located on each riser. A variation of the system which reduces the number of sprinklers has sprinklers mounted on quick coupling risers capable of moving along lateral lines. The flow is directed from one part of the system to another by hydrants and valves. The purpose of the system is to reduce labour cost by eliminating movement of lateral lines, especially where tree foliage is very heavy. It is used on permanently irrigated areas and for relatively high value crops such as orchards, grapes, vineyards and nurseries and where labour is expensive.

4. *Semi-permanent sprinkler system:* In this system main lines and submains are permanent and usually buried, but lateral lines are portable, fed by a stationary water source and pumping plant. The nozzles are mounted on risers at suitable intervals. It is of three types:

(*i*) *Sprinkler hop system*: It is similar to portable system but sprinklers are placed at alternate points along lateral. When irrigation is complete, sprinklers are disconnected and moved or hopped along the next point to irrigate but flow is not stopped in the lateral. Each sprinkler connection is fitted with a valve to automatically stop the flow when the sprinkler is removed. The lateral is moved to the next position and irrigation process resumed.

(*ii*) *Pipe grid system:* It is similar to permanent system. The pipes are laid over the area to stay for the irrigation season with a view to avoid movement of pipes. Two sprinklers are connected to each lateral, one near the top, the other half way down. When irrigation is complete, each sprinkler is disconnected and moved along the lateral to next point and so on.

(*iii*) *Hose pull system*: In this system main line and laterals are permanently laid on or buried underground. Plastic hose pipe of small diameter supply water from the lateral to one or more rotary sprinklers. A sprinkler is moved easily to new position. It is used for irrigation of under tree orchard crops and other row crops. It is cheap.

2. Classification Based on Spraying Pattern

1. *Rotary head or revolving system*: In this system, the nozzles are mounted on riser pipes at uniform intervals along the lateral pipe which may be laid on the ground or installed on posts above the height of crop. The nozzle is rotated through 90° to irrigate a rectangular strip of the field; the rotating head is rotated with a small hammer actuated by the water thrust striking against a vane attached to it. The variations of the system are (*i*) Boom type or self-propelled sprinkler, in which a boom sprinkler is connected to each lateral. The boom is a nozzled, slowly rotating pipeline suspended from a portable post. After completing irrigation, the tower and laterals are moved with a tractor to next position and (*ii*) Self-propelled sprinkler which comprises a radial pipe of 1.75 to 2.5 m height, at about 30 m interval on towers suitably mounted on a trailer. At irrigation location in the centre of the field, the radial pipe is slowly rotated around by water pressure actuators or electric motors at each tower. The rotating sprinkler fitted to the pipeline sprays water as the pipeline is moving.

2. *Perforated sprinkler system:* Also termed as perfo-spray system. It is a simple, low cost sprinkler. In this system water is carried through the main pipeline and let out through lateral pipelines which have small perforations of 5 mm or smaller diameter. Water is applied at a relatively high rate than rotating sprinklers. Water is sprayed over the crop by hundreds of small gentle sprays directed to both sides of the pipe to irrigate a strip of land 6 to 15 m in width. This system is suitable for sandy soils having moderately high infiltration rates but is not recommended for use on bare heavy clay soils. It is adapted to lawns, gardens and small vegetable fields and other short plants of 45-60 cm height. Plants taller than this block the spray.

3. Classification Based on Arrangement of Spraying

1. *Fixed head type:* Various types of fixed head sprinklers are used for irrigation of turf, ornamental gardens and irrigating young orchards before trees have established an extensive root system. Parallel pipes with fixed nozzles are installed at about 15 m apart, supported on rows of posts. Water under pressure passing through the pipes discharges perpendicularly from the pipelines. In order to irrigate entire width between the pipelines, the pipes are turned to about 135°.

Disadvantages: (*i*) The diameter of coverage is small, and (*ii*) Too high proportion of the spray droplets land near the sprinkler and too low a proportion near the periphery.

2. *Rotary type:* Most agricultural sprinklers are of this type. They range from single nozzle sprinklers to multiple nozzle sprinklers working under high pressures. Usually they can be adopted for almost any desired angle but are not useful at edges and corners of fields. Water is thrown to a considerable distance and sprinklers are spaced farther apart to ensure that actual application rates with an overlapping pattern are low enough to match infiltration capacity of the soil. Large orifice minimizes clogging by debris in water.

Sprinkler Losses

Losses during sprinkling occur in the form of evaporation, some deep percolation, drift of spray due to wind depending on fineness of spray, rate of application, wind temperature, humidity, soil texture and cover, and conveyance loss which is negligible. Evaporation loss is minimum at night and maximum in afternoon. Windy and hot weather conditions increase sprinkler losses. During sprinkling of standing crops, a part of water is intercepted and evaporated without reaching the soil. The rate of evaporation, as such, may appear to be high during sprinkling.

Efficiency

Sprinkler irrigation has high efficiency. It, however, varies according to climatic conditions; 60% on hot day, 70% in moderate climate and 80% in humid or cool climate.

Advantages of Sprinkler Irrigation

(*i*) *Saves water, irrigates more land.* Compared to surface irrigation only 30 to 50 per cent of the water is required depending on the type of soil. Thus with same quantity of water 2 to 3 times area can be irrigated, (*ii*) *Low water loss*: Water loss is estimated as 18 per cent against 54 per cent in a surface irrigation system with lined channels and 71 per cent in a system without any lining, (*iii*) *Effective water management*: Since the water application is controlled, over or under irrigation is avoided. Just the right quantity of water at the right time can be applied according to the needs of the crops, slope and soil texture, (*iv*) *Saving in land:* Saving in land which otherwise is occupied by the canal section, banks and watercourses in surface irrigation, (*v*) *Saving in fertilizers*: Application of soluble fertilizers is efficient and automatic. Nearly half of fertilizers which are lost due to excessive leaching below the root zone and drainage in surface irrigation are saved, (*vi*) *Land levelling not necessary*: Even undulated lands can be irrigated effectively without land levelling. Time and money to be spent in land levelling are saved. Moreover, when the land is levelled, top fertile soil gets disturbed and it takes years to develop new soil suitable for cultivation, (*vii*) *Soil conserved*: Due to controlled rate of water application, water is not allowed to flow but is absorbed by the soil fully. Hence there is no soil erosion, (*viii*) *Soil condition is maintained*: The soil is kept open and its structure unbroken whilst soluble fertilizers and soil amendments are easily applied with irrigation water, (*ix*) *Soil is stabilized*: Light and frequent waterings are observed to have stabilized sand dunes, (*x*) *Better seed germination*: More seeds germinate, often earlier and growth is vigorous. Fine droplets absorb more oxygen and supply to root zone for growth of crops, (*xi*) *Frost control*: Crops are protected against frost damage, (*xii*) *Instant irrigation*: Irrigation at the turn of a valve, (*xiii*) *Use of limited source*: Even a limited source of water can make use of sprinkler irrigation to irrigate more area, (*xiv*) *Uniform application*: Uniform application of water is possible on all kinds of soils, (*xv*) *Controls climate:* Light sprinkling reduces temperatures to safe level, (*xvi*) *Sport grounds*: Best for watering lawns and sport grounds, (*xvii*) *Free aeration of root zone*: Top soil becomes porous and there is free aeration of the root zone which helps better growth of plants, (*xviii*) *Poor soils irrigated*: Poor soils such as sandy soil and black cotton soil can be irrigated with equal ease and efficiency, (*xix*) *Drainage problems eliminated*: Surface runoff of irrigation water is eliminated with the controlled water application, (*xx*) *Improved soil fertility organism*: The better aeration also enables micro-organism to work to their fullest extent to improve soil fertility, (*xxi*) *Weeds and pests controlled*: Harmful ditch weeds do not appear, cannot establish themselves because sprinkled water washes plants regularly, (*xxii*) *High crop yield and quality*: Crops yield is appreciably higher and quality is very often improved. Three or even four crops per year can be grown, (*xxiii*) *Reduced labour requirement:* It means low running cost comparable to lift irrigation. Being automated manpower needed is less, (*xxiv*) *People's participation*: People's participation in the development is achieved as the individual farmer who adopts sprinkler irrigation helps the Nation in the efficient and economical use of water and fertilizers, otherwise, the State would have been called upon to make public investment on storage dams and canal system to irrigate these areas.

According to a study made by the Central Water Commission, sprinkler irrigation reduces wastage of water in the field from 42 to 6 per cent and increases utilisation of water by crops from 29 to 82 per cent. There is a fairly good control over the rate of water application. The farmer can mix water soluble farm chemicals for plant protection and weed control with the irrigation water at no extra cost.

Limitations: (*i*) High initial cost, (*ii*) Poor application efficiency in windy weather and high temperature, (*iii*) Higher evaporation losses in spraying water. It is on this account that night time watering is beneficial, (*iv*) Not suitable for jute and rice, (*v*) Cannot be used with rotational supply systems of water distribution in canal irrigated areas, (*vi*) Water supply free of soil and debris is needed for economical use of equipment, (*vii*) Assured source of surface or groundwater supply is needed, (*vii*) It cannot be used on fine textured and heavy clay soils having low infiltration rates, (*viii*) Higher power requirement, (*ix*) Uniformity coefficient is low compared to drip irrigation system, (*x*) Poor distribution efficiency, (*xi*) Some of the irrigation water is lost by infiltration into the underground, (*xii*) Excessive use of saline water on foliage is inadvisible, (*xiii*) Use of marginal (recycled sewage) water is restricted for health reasons, (*xiv*) Equipment need careful handling, and (*xv*) Nozzles need screened water supply, or otherwise likely to be plugged.

4.6 BURIED IRRIGATION

Buried irrigation system is a substitute of canals by pipelines. In this system water is delivered by a canal from the source of water supply to the irrigation area, and inside area it is distributed by gravity pressure pipelines. The distributing pipelines receive water from the farm pipe line (or open canal) laid along the contour lines or at a small angle to them. When water supply to field pipelines is by pumping, they are laid straight and parallel to each other at 400 m spacing which is the optimum distance between them. From distribution pipelines water runs to the field pipelines which in turn discharge into the flexible irrigation pipeline. Water is lifted mechanically to the buried network when the level in the source of water supply is below the elevation of the area to be irrigated or when the head created by the gradient is insufficient. Field pipelines are provided with hydrants, sprinklers, or watering machines. High density polyethylene pipes 10-30 cm in diameter are placed at a depth of 0.75-1.25 m. Depending on irrigation methods, underground irrigation pipelines mainly operate under a pressure up to 6 atmosphere.

Water flows in the pipelines due to the head created by the natural gradient. The flow of water is possible when the gradient is more than 0.003. The design gradient shall also provide for the velocity sufficient to prevent silting in the pipelines. For clear water *i.e.*, without suspended sediment, the diameter of pipeline is calculated for the maximum velocity subject to the natural gradient. This velocity is not to exceed 3.5 m/s to prevent water hammering when using the valves. For water containing suspended sediment, the design velocity shall exceed the non-silting velocity. The non-silting velocity is determined, in USSR, according to the formula of the transporting flow capacity derived for the silt content up to 10 kg/m^3 and the size of particles up to 0.5 mm.

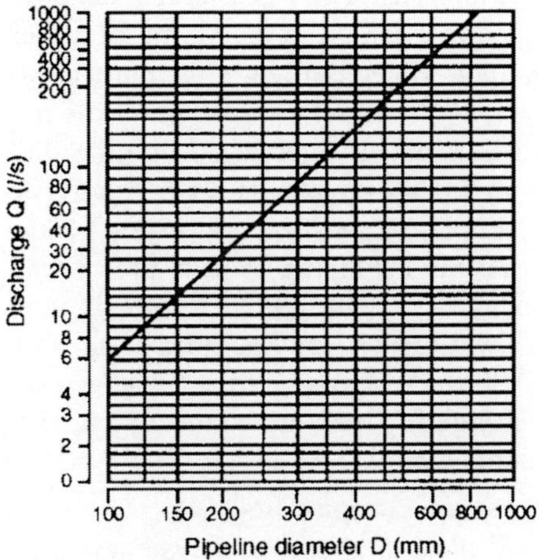

Fig. 4.9 Lobachev graph

$$V = \frac{v^2 f}{0.0000232u^{0.25} \times 8g} \text{ kg/ m}^3$$

where, V = current velocity of flow (m/s), u = fall velocity of sediment (mm/s), and g = acceleration due to gravity (9.81 m/s/s).

The Lobachev graph (Fig. 4.9) can be used for preliminary determination of most economical and efficient diameter for a pipeline of the buried irrigation system. The diameter of the pipe so determined is checked to verify that pipeline will not silt and, if necessary, the diameter is changed.

Advantage: It is a reliable method of irrigation for control of (*i*) water losses from irrigation canal, (*ii*) waterlogging, and (*iii*) salinization of lands.

4.7 DRIP IRRIGATION

Also called *Trickle irrigation.* It is the latest developed method of irrigation. It was first introduced in Israel. In India, it is being increasingly practised in Gurjarat, Maharashtra, Kerala and Karnataka. In this system water is applied in the form of drops directly to the plants through drip nozzles from which it drops into the soil slowly and frequently to keep the soil moisture within the desired range for healthy plant growth so that the plants do not experience any moisture stress throughout their life cycle. It is particularly suited to soils with very low and very high infiltration rates under the conditions of water scarcity and in areas where drainage of excess water is difficult. It is more economical for orchard crops than for other crops and vegetables since in the orchards plants as well as the rows are widely spaced. It is most successful for high income crops because of the relatively high first cost of the installation. Coconut is the major crop, followed by grapes, vegetables, citrus and sugarcane. Drip irrigation limits the water supplied for consumptive use of the plant by maintaining a minimum soil moisture in the root zone, thereby maximizing the water saving. A dripper discharges water (and nutrients) directly into the area of the root system. It permits fine control of application of moisture and nutrients at frequencies prescribed for the individual plants. Drip irrigation involves the lateral spread of water on the surface to be irrigated by conducting the water under pressure to a relatively closely spaced grid of outlets, and discharge the water (together with fertilizer solution) through these outlets at virtually zero pressure. Irrigation is done through drippers fitted on small diameter lateral lines. The dripper located at suitable spacing delivers the water to the soil surface near the base of plant.

The basis of this potential water saving method is that frequent but just sufficient amount of irrigation guards the plant free of water stress by maintaining soil moisture level at the field capacity. Water is filtered to remove the suspended impurities so as not to block the fine drip nozzles. The effective working of the drippler lines lies in efficient functioning of the filter. The fertilizer mixed water is led into the filter tank from which dirt free solution gets into the main supply lines with laterals spread on the field in rows. The laterals hold the drippers through which the fertilizer-cum-irrigation water trickles.

At the beginning of each plot there is a secondary line which distributes the water into the trickle lines. To the secondary line is connected a discharge regulator whose capacity is in accordance with the size and number of nozzles being used. The automatic valve at the head is adjusted to deliver desired quantity of water and irrigation terminates automatically after this amount is discharged. Pipes are made of cheaper materials like polyethylene, polyvinyl chloride or alkathyelene.

Irrigation Components

Typical drip irrigation system is depicted in Fig. 4.10, and discussed as under:

1. *Head:* The water lifted is stored in a head tank, usually 3 × 3 × 3 m, resting on a raised platform to maintain pressure head of 3 to 5 m. The head is connected to the central water supply. Its functions are to (*i*) regulate the pressure and amount of water applied, (*ii*) filter the water, and (*iii*) add nutrient material. It is so located that the main lateral is conveniently taken in the middle of the field.

2. *Main lines:* Plastic main lines deliver water to submains. These are usually of 20 to 40 mm diameter, suitable for desired discharge and distance over which water is to be carried. A water meter is installed in it close to the tank for the measurement of discharge passed.

3. *Laterals:* Submains deliver water to laterals which in turn convey it to emitters. Small diameter laterals, usually 10 to 20 mm in diameter, fitted to the submains are generally placed along plant row. There is one lateral for each row of crops. The laterals are perforated at a distance equal to the spacing of the crop, usually 50 cm for vegetables and cotton.

4. *Emitter:* The application of water to the root zone is through the emitter, a precise metering device, which has a very low rate of flow, such as a few litres, usually 1 to 9 litres per hour depending on the crop requirement.

5. *Chemical injection unit*: Chemical tank and injection equipment are an integral part of drip system. Control of organic growth requires injection of chemicals such as chlorine. Also required is an efficient means of applying fertilizers (except phosphorus) through water.

6. *Monitoring and control equipment*: It includes controller, pressure gauges, tensiometers and a flow meter. The control components include pressure and flow regulators, control valves and flow control valves.

7. *Fertilizer tank:* A tank filled with a concentrated nutrient solution and connected by means of a tubing to the head directly in front of filter unit and the nutrient solution is introduced into the water flowing to the trickle network. The fertilizer dispenser injects fertilizers into the system at a predetermined rate; thus water drips from the nozzle fully enriched with the amount of fertilizers required to serve the crops. Filters are either of a screen type or media (such as sand bed) type and may be manually or automatically cleaned. A system often has more than one filter.

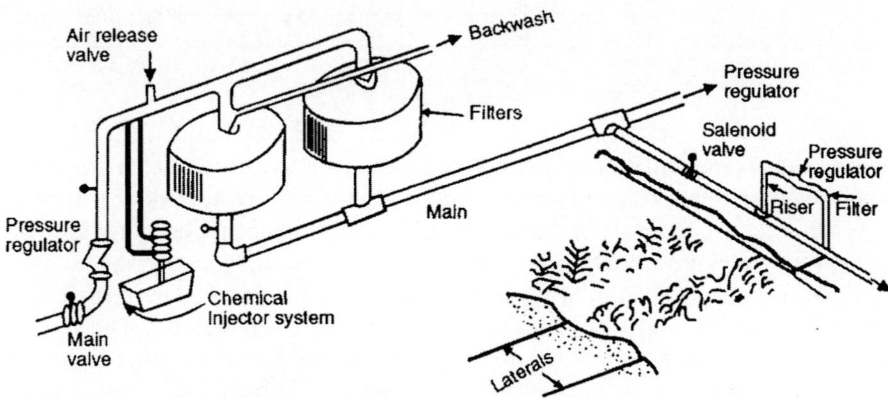

Fig. 4.10 Typical drip irrigation system

Design of Drip Irrigation System

In plantation, trickle lines are laid as far as 6 m apart (Fig. 4.11) to which high discharge nozzles, three beside each tree, one adjacent to the trunk, one 0.5 m before and the third 0.5 m beyond the tree are fixed. The irrigation requirement is more at the time of flowering and fruiting than in the beginning or at the time of maturity or harvesting of the crop. Accordingly, additional nozzles are added along the trickle lines to provide a spacing of 1 to 1.5 m between the nozzles as the plant matures. The flow pressure available at a hole in the pipeline of the drip system is calculated as excess of the head (pressure) loss that has occurred due to friction.

The irrigation system has to have design capacity adequate to satisfy the peak irrigation water demand of each and all crops to be irrigated within the design area, as also to apply a stated amount of water to the design area in a specified net operation period. The minimum design capacity is sufficient to deliver the peak daily irrigation water requirement in about 90 per cent of the time available or not more than about 22 hours of operation per day with due regard to the soil water-holding capacity and climate conditions.

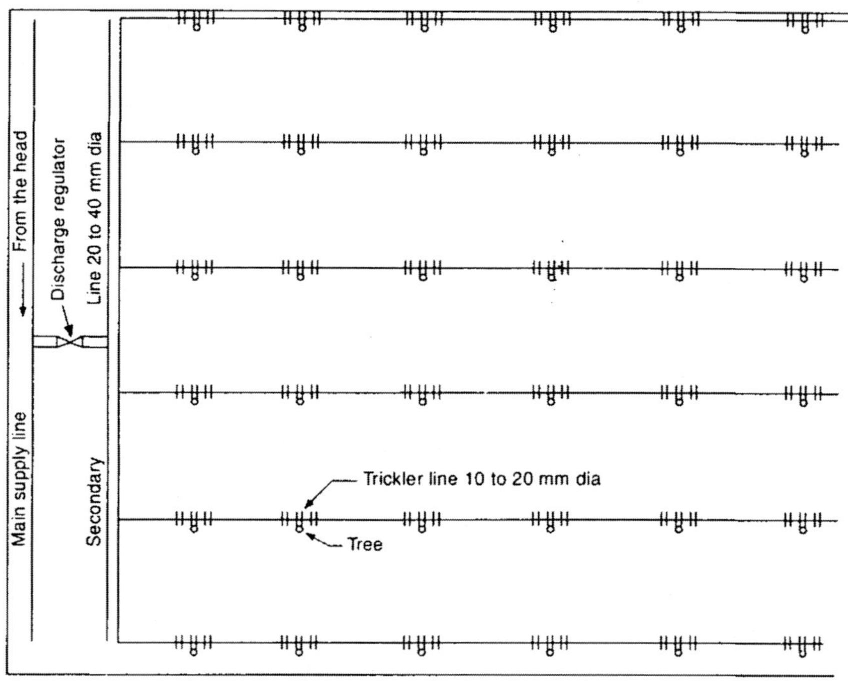

Fig. 4.11 Arrangement of drip irrigation in field.

1. *Irrigation water requirement*: The irrigation water requirement (or equivalent depth of water application) per irrigation is mainly influenced by the crop evapotranspiration rate, the interval between irrigations, and the water application uniformity. It shall not allow accumulation of salts or other toxic chemicals in the root zone of crops but ensure leaching salts that may have entered the root zone by infiltrating rainfall.

2. *Emitter discharge ratio*: Proper emitter discharge rate is determined on the basis of experience with similar soil types and prescribed by the manufacturer. The discharge rate for point source emitters are generally less than 12 l/hr for single-outlet emitters, and for line-source emitters these are usually less than 12 l/h/m of the lateral. Allowing the flow to pass through the pipes the emitters are adjusted by trial and error to get uniform discharge through all the outlet openings.

3. *Number and spacing of emitters*: The number and spacing of emitters along the lateral line depend upon the emitter discharge rate, system capacity, soil water-holding capacity, depth of irrigated crop root zone, lateral spread of water from the emission points, and desired water application uniformity. In arid areas a minimum 33 per cent of the potential crop root zone is wetted and up to 75 per cent may be aimed at. The wetting between 25 and 50 per cent is sufficient in humid areas.

4. *Operating pressure:* It has to be in accordance with the manufacturer's recommendation. The water is usually fed at a reduced pressure of 0.5 to 0.7 kg/cm^2 using pressure reducer in the feeding lines or using an elevated tank. Laterals operate under pressure as low as 0.15-0.20 kg/cm^2 and as high as 1.0-1.75 kg/cm^2.

5. *Water application uniformity*: The water application uniformity is affected by the operating pressure, emitter spacing, land slope, pipeline size, emitter discharge rate, and emitter variability. The recommended classification of manufacturers' coefficient of variation (*Cv*) is as given in Table. 4.1.

Table 4.1 Classification of coefficient of variation

Emitter type	*Cv range*	*Classification*
	< 0.05	Good
Point source	0.05 – 0.10	Average
	0.10 – 0.15	Marginal
	> 0.15	Unacceptable
	< 0.10	Good
Line-source	0.10 – 0.20	Average
	> 0.20	Marginal to unacceptable

The design emission uniformity: The design emission uniformity in terms of *Cv* and pressure variation at the emitter is estimated from the equation

$$Eu = 100\left(1.0 - \frac{1.27}{n} Cv\right)\frac{q_m}{q_a}$$

where, *Eu* is design emission uniformity in percent, *n* for a point-source emitter on a permanent crop, the number of emitters per plant, for a line-source emitter on an annual row crop, either the spacing between plants divided by the same unit length of lateral line used to calculate *Cv* or 1, whichever is greater, *Cv* = manufacturer's coefficient of variation for point or line source emitters, q_m is minimum emitter discharge rate for the minimum pressure in the system (l/h) and q_a is average or design emitter discharge rate (l/h). The recommended range of *Eu* is 95-70.

6. *Laterals*: Laterals are usually 10-20 mm in diameter and 300 m long. The laterals are laid down slope for slopes of less than 5 per cent; for steeper terrain these are laid along the field contour, pressure compensating emitters specificed, or pressure control devices used along down slope laterals.

7. *Varying size of outlet openings*: As the flow advances more and more downstream in a pipe, the pressure suffers in its value an increasing loss due to the friction offered by the pipe wall as well as the viscosity of the fluid. The head loss occurring in a segmental pipe length *l* is given by:

$hf = (fl\, v^2)/2\, gd$, with usual notations. Thus, the pressure available at any downstream section goes on falling, which means that the size of the openings spaced at equal intervals should be monotomically increased for getting a constant discharge.

The filtration system, chemical treatment, fertilization system, injection system, flow monitoring by flow meter shall be as per manufacturer's recommendation.

Drip Irrigation System

Drip irrigation is adaptable to (*i*) all vegetables and a vast assortment of crops, except paddy which requires standing water, (*ii*) all climatic variations, and (*iii*) highly permeable soils, like sandy soils, sandy loam, gravelly soils and less permeable soils like clay, clay loam, silty clay or shallow soils underlain with impermeable strata. However, it is most suited to coarse sandy formation. Under the adverse condition of high salinity of both soil and water, high temperatures and low relative humidity, crops of record yields and superb quality are obtained. Fig. 4.12 shows cheap drip irrigation system.

A unique feature of drip irrigation is its excellent adaptability to saline water. Since the frequency of irrigation is quite high, the plant base always remains wet which keeps the salt concentration in the plant root zone below the critical.

Efficiency: Irrigation efficiency is more than 90 per cent as compared to about 65 to 70 per cent in sprinkler and about 50 per cent in lined network of conventional surface irrigation.

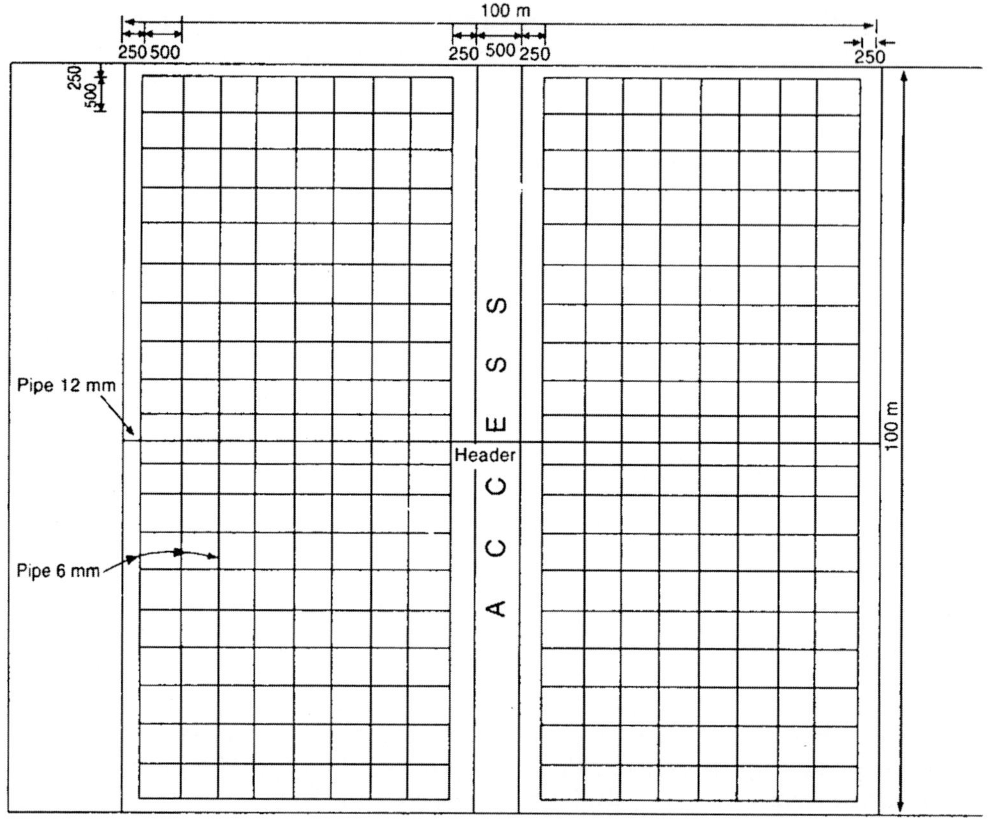

Fig. 4.12 Cheap drip irrigation system.

Advantages:

(*i*) *Water saving:* Saves 40-70% water when compared with the conventional methods, (*ii*) *Reduced losses:* Since only small areas around the plants are wetted with precise water amounts, hence water is saved through reduced evaporation, runoff and deep percolation. Quantity of water delivered is reduced to 20-15 per cent of what is normally given. More accurate water applications ensure no waterlogging and consequently no need for drainage facilities, (*iii*) *Uniform water distribution:* Water distribution is highly uniform, over 90 per cent, and also fully controlled, (*iv*) *Over-irrigation avoided:* Over-irrigation required in surface methods to provide sufficient moisture even at the deficient areas is avoided, (*v*) *No land levelling:* It does not require levelling of fields. It is suitable even for highly sloping terrains, (*vi*) *Application rate to suit need*: Application rates can be adjusted by using a different size nozzle for a higher or lower discharge; distances between nozzles along the lateral and between the laterals can be varied to suit soil and crop conditions, (*vii*) *No soil erosion:* There is no surface flow of water hence no soil erosion or tail water loss. Erosion problems encountered in the rolling topography as well as seepage problem inherent under coarse structured soil condition are overcome, (*viii*) *Weed control:* Weed growth in the land is minimum as only the root zone of plants are irrigated and the other portions of the land are left dry, (*ix*) *Land saving*: There is saving in land which otherwise is occupied by canals. As such more agricultural land is brought under cultivation, (*x*) *Nutrient preservation:* Water enriched with fertilizers (nutrients), pest and insect combating chemicals are conducted directly into the root, (*xi*) *Salinity control*: Frequent irrigations move salts to the perimeter of the wetted pattern and salinity problems are reduced, (*xii*) *Marginal soil*: Can work on marginal, rocky or very sandy soils because it delivers precise amount of water and nutrients directly to the plant zone thereby preserving nutrients, (*xiii*) *Use of unsuitable supply:* Highly saline soil or saline or brackish water source, normally considered unsuitable for agriculture, can be utilized, (*xiv*) *Less labour cost:* Labour cost is less as no field channels, furrows are required to be constructed. Less operational cost due to ease of application and simplicity of operation, (*xv*) *High surface temperature:* By not wetting the entire soil surface, the surface temperature is kept high thereby hastening the ripening or maturity of the produce, (*xvi*) *Leaf shedding minimised*: By keeping the root zones moist and cool, the tendency of leaf shedding is minimized, (*xvii*) *Soil in most congenial form:* It continuously maintains the physical conditions of soil in the most congenial form by keeping the soil-water-air-proportion in the optimum range, thereby promotes better return from crops, (*xviii*) *Uninterrupted cultivation operation*: Many times irrigation can continue while harvesting the crop, (*xix*) *Higher yield:* It results in improvement in quality and quantity of the yield. Daily irrigation or even more than one irrigation per day produces the highest yields. There is early maturation and fruiting, (*xx*) *Improved water penetration*: Application of water at very slow rates to limited areas around the plants improves water penetration, (*xxi*) *Energy saving*: Lower operating pressures and lower flow rates require less energy for pumping. The pump and pipe network is smaller and cheaper.

Disadvantages: (*i*) High initial cost, (*ii*) The drippers are susceptible to blockade, (*iii*) High skill is required for design, installation, operation and maintenance, (*iv*) Not all crops respond well to localised wetting only, (*v*) Trees grown may develop shallow confined root zones limiting the area of extraction of soil nutrients resulting in poor anchorage of the roots and consequent danger of the toppling of the trees under strong wind, (*vi*) Use in saline conditions has to be managed properly otherwise reverse pressure gradients in the soil cause flow of salts towards plant roots with resulting detrimental effect, (*vii*) Application of insoluble or slightly soluble fertilizers, like superphosphate or sulphate of potash, through the flow is not readily possible, (*viii*) On soils of high levels of salts or with highly saline water, the salts may accumulate at the edge of wetted zone.

Surface Irrigation vs. Drip Irrigation

Surface irrigation	*Drip irrigation*
Economical means of irrigation.	Costly means of irrigation.
Brackish or saline source of water is unsuitable.	Highly saline soil or saline or brackish water source, can be utilized.
Wasteful method of irrigation.	Highest water saving method.
Low yield.	Highest yield.
Land is lost under canal section and field channels.	Maximum use of land.
Suitable for all crops.	Suitable for close growing crops ,vegetables and orchards.
Suitable for all soils except very light soils.	Suitable for all soils, especially coarse textured soils.
Localized application of fertilizer is not possible.	Localized application of fertilizer minimizes the volatilization as well.
High percolation loss and leaching down of fertilizers.	Minimum percolation loss; reduces leaching down of fertilizers.
Less effective in water distribution efficiency.	More effective in water distribution efficiency.
Requires levelling.	Does not require levelling.
All the surface soil is saturated, 20-50% wetted even in furrow irrigation.	Only root zone is saturated during irrigation.
Soil moisture depletes even by 50 per cent before the next irrigation and salt concentration is more if salts are present in the soil or water.	Moisture is always present in the root zone more or less at the field capacity and the relative concentration of salts is less.
Soil factor plays an important role in establishing frequency of irrigation.	The soil factor plays a less important role in establishing the frequency of irrigation.
Moisture movement is normal to the furrow shape.	Ultimate moisture distribution is roughly two-dimensional.
Variation in supply difficult to make.	Supply into field can be regulated with ease by regulating the valve.

Surface irrigation	Drip irrigation
Surface flow is associated with soil erosion.	No surface flow and no soil erosion.
Water distribution is not uniform, it is as low as 25 to 50 per cent in furrow irrigation.	Water distribution is highly uniform and controlled as each nozzle distributes with close proximity the same quantity of water.
High labour cost.	Low labour cost.
Lowest irrigation efficiency (50%).	Highest irrigation efficiency (90%).

Drip vs. Sprinkler Irrigation

Drip irrigation	Sprinkler irrigation
Localized application of fertilizer.	Localized application of fertilizer is absent.
Not affected by windy condition.	Seriously affected by windy condition.
Suitable for all including less permeable soils.	Suitable for all including highly permeable soils.
Surface temperature is kept high, reduced crop period.	Surface temperature is reduced, longer crop period.
Limited wetting of soils surface, water drops in roots.	Water sprinkled over entire area and wetting it including foliage.
Free from need for risers.	Requires risers of requisite height and spacing.
Soil erosion is controlled.	No soil erosion.
Minimum evaporation loss.	Higher evaporation loss.
Highest yield, that too irrespective of water quality.	High yield depending on water quality.
Higher rate of crop growth. Quicker germination of seeds.	Comparatively slower rate of crop growth and seed germination.
Pipes used are of low pressure and cheaper.	Pipes used are of high pressure and are expensive.
Free from any movable components.	Has movable parts.
Operates on low pressure hence least energy requirement, small capacity pump.	Operates at high pressure with higher energy requirement and large capacity pump.
Highest water application efficiency (90 per cent).	Lower water application efficiency (65-70%).
Lowest water requirement.	Higher water requirement.
Water distribution efficiency is poor when a low pressure system is installed on undulating land.	Comparatively higher water distribution efficiency.
Low initial cost and operation (labour) costs.	High initial cost and operation charges.
Highest uniformity in water application.	Poor application uniformity under windy and high temperature conditions.
Weed growth is less.	Weed growth is least.
Highly automated	Complete automation is rare.

4.8 SEEPAGE LINE IRRIGATION

A relatively new cost-effective concept in irrigation is seepage line irrigation method as has been tried out in Kandi area of Hoshiarpur district in Punjab, known as Makowal watershed project (seepage line) which provides limited but assured irrigation facilities to crops during critical times, one or two irrigations at the critical juncture of crop growth, besides water for bathing purposes. Four wells have been developed at points in and around where big water seepage was observed. These wells were connected by pipelines to a central pond in the village which has been dug up to act as a water tank. Water is supplied to individual farms from this tank through pipelines at nominal cost by an irrigation water consumer society formed by the villagers.

4.9 SUCTION IRRIGATION

It is a variant of drip irrigation. In this method of irrigation, uniform and controlled water is supplied to the root zone of a plant automatically. It was initiated by Dr. R.K. Yadav of S.K.N. College of Agriculture, Rajasthan as also in North-East Brazil. In the system, spindle shaped burnt clay emitters with neck on either side are used (Fig. 4.13). Water seeps out of the wall of porous, emitters and the plant receives water directly at the root zone. The water is absorbed by the soil due to soil water tension and by the plant due to water absorption capacity of roots. The absorption is governed by the prevailing moisture status of the soil.

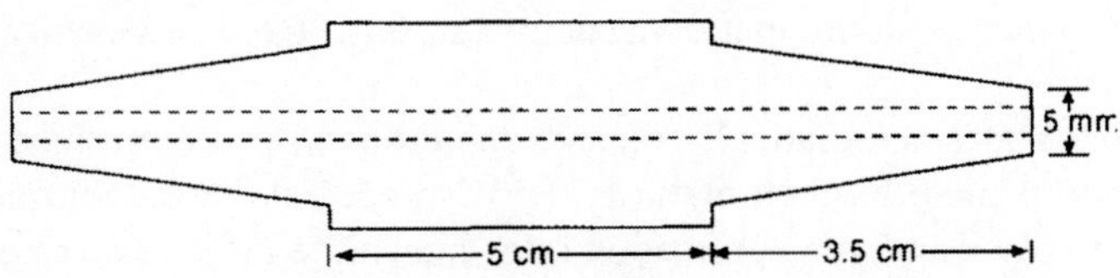

Fig. 4.13 Gilly Emitter

The layout of the system (Fig. 4.14) similar to drip system, consists of pipe network connected to a source of water. The main line is connected to submains and submains to the laterals with intermittent emitters. The laterals are kept 0.6 m apart and in each lateral the spacing between two emitters is also 0.6 m. The laterals are buried 10 cm below the ground surface. The main line is connected to a tank of 1 m^3 capacity with a regulating valve. The entire network of pipes is filled with water and entrapped air is removed through the ends of laterals but subsequently these ends are closed by plugs. For distributing water in the field as per plants requirements in the lateral line, the submains act as a control system which adjusts the water pressure in order to deliver the required amount of flow in each lateral. The mainline serves as a conveyance system for delivering the total amount of water for the irrigation system. The flow in the lateral has to be controlled so as to be laminar under all conditions and, therefore, limiting length of the lateral is to be determined. The submains, the laterals have changing flow capacities with respect to the length. Both have more discharge in the upstream sections than in the downstream sections. The energy drop for a given length of pipe gives the energy slope or energy gradient. With the energy slope determined, the length upto which water could reach with the necessary head can be found out. For simplicity a straight energy line is assumed for the design of laterals. The Hazen Poiseuille's equation, as under, is used for the design of laterals.

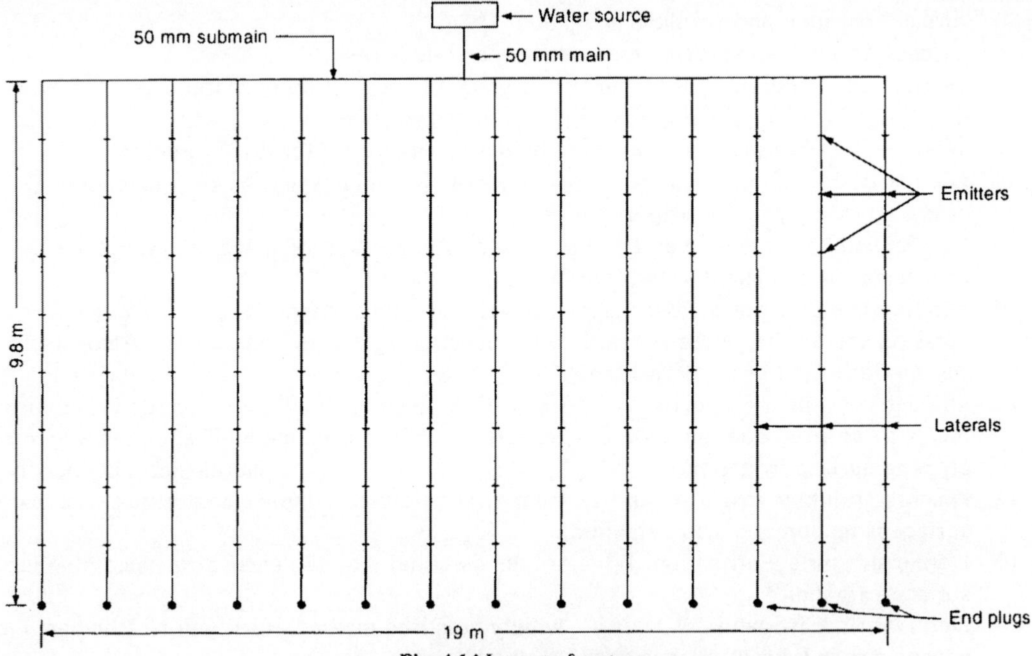

Fig. 4.14 Layout of system
(*Source*: Suction Irrigation and conservation of water wealth)
Deshmukh M.B. Pujari R.M. and Paseband S.P.

Head loss $H_L = (32\ \mu\ VL)/wd^2$

where, μ = coefficient of viscocity, L = pipe length, W = specific weight of water, Reynold's

Number, $RN = Vd/v$, where v = kinematic viscocity *i.e.*, 0.01009, V = velocity of water, d = pipe diameter.

Advantages: (*i*) Water automatically reaches the crops in predetermined quantity, (*ii*) Continuous water supply to plant hastens maturity, (*iii*) No special site conditions are necessary to adopt the system, (*iv*) Slow and low application rates match the infiltration rates thereby reducing the problems of ponding and erosion, (*v*) Low head is required than drip irrigation system, (*vi*) Upper soil layer remains dry hence interculture weeds control activity is obviated, (*vii*) Deep percolation losses are minimised, restricted to root zone, (*viii*) Evaporation of water from soil surface is restricted, (*ix*) Higher yield to the extent of 1.5 times, (*x*) More land can be brought under irrigation and hence considerable increase in agricultural production.

Limitations: (*i*) High initial cost, (*ii*) Quality of water used is important particularly the turbidity count, (*ii*) Emitters made of burnt clay cannot be reused after using for one crop, (*iii*) Emitters of different porosity are required for various types of crops, (*iv*) Everyday maintenance is required for non-breakage of emitters in the system, and (*v*) Not suitable for crops like rice wherein ponding of water is needed.

QUESTIONS

1. What are the various methods of irrigation? What are the merits and demerits of each?
2. Draw neat sketches to illustrate the methods of irrigation, (*a*) Border irrigation, (*b*) Basin irrigation, and (*c*) Furrow irrigation, State under what circumstances you will recommend their use.
3. What are the situations in which one would recommend sprinkler irrigation. Discuss its advantages and disadvantages (limitations). Describe drip method of irrigation explaining its component parts.
4. Explain why sprinkler irrigation is preferred to other methods of irrigation. Compare surface irrigation and sprinkler irrigation.
5. Discuss drip irrigation. What are its merits and demerits.
6. Discuss the conditions that favour use of graded furrow method of application of irrigation water. What are the merits and demerits of the same.
7. With the help of neat sketch, explain the design procedure for drip irrigation.
8. Discuss design of sprinkler irrigation. What do you understand by drip irrigation. Discuss component parts of drip irrigation.
9. (*a*) Explain the furrow method and the sprinkler method of irrigation. What are their advantages as compared to the free flooding methods?
10. Differentiate between a cavity type tubewell and strainer type tubewell. Derive an expression for discharge from a well fully penetrating a confined aquifer. Also discuss assumptions used in the derivation.
11. Discuss the primary objectives of an irrigation method. What are the essential requirements of an irrigation method? Discuss briefly different methods of applying water to crops in surface irrigation. [Kurukshetra University]
12. Discuss sprinkler irrigation and how far it is suitable in Indian conditions. Compare surface irrigation and drip irrigation.
13. Define sub-surface irrigation. What are the essential requirements for a successful sub-surface irrigation?
14. Define surface irrigation. Why is it a widely practised method of irrigation? What are the advantages and disadvantages of the method?
15. Give a comparison of drip irrigation and sprinkler irrigation.
16. Discuss buried irrigation and suction irrigation.

5

UNLINED CANALS

5.1 INTRODUCTION

Irrigation is an ancient art. Irrigation has been practised from time immemorial and so has been the construction of irrigation canals. There are constructions dating back to many centuries in India, Egypt and other countries. In earlier periods irrigation was restricted to comparatively smaller areas close to the rivers by means of inundation canals. Later with the taking up of large irrigation schemes, perennial irrigation came into being and with that the subject of design of irrigation canals, economical in first cost and maintenance with minimum liability to silting and scouring engaged the attention of engineers. However, formulation of laws governing the flow of water and sediment transport has been done in recent times.

It is only in the nineteenth century that canals have been designed on a more or less scientific basis. The formula of Chezy with coefficients derived from Kutter's and Bazin's formulae is currently in use in India and many other countries. Manning's formula is equally widely used. Kennedy's formulae and the rational empirical Lacey's regime theory, although originated in India, are used in many countries too. The tractive force theory, though not developed for general and practical adoption on a wider scale, is currently mainly used as a check over the parameters calculated by one or other empirical formula. Thus both empirical and analytical concepts are increasingly being used.

In any irrigation project, the network of canals is the key component and is dominant in the financial outlay of the project. Proper design of the canals is thus of great importance. The efficiency of canal design has a bearing also on its working. The canal efficiency suffers from silting and scouring of the canal due to faulty design necessitating heavy maintenance and at times even remodelling if the original design is past redemption. The other serious problems whose origin lie in a faulty design of canals are weed growth infection, heavy seepage losses entailing development of waterlogging alongside the canals. Proper design of canal has thus wide implications.

Canal design practices also depend on the conditions, particularly the soil formations, sediment transport characteristics, operational needs and desired standards of maintenance. Usually irrigation canals are constructed in alluvial soil and they essentially take supplies from a river and carry silt charged water. The canals taking their supplies from a reservoir carry comparatively lesser amount of silt charge. The sediment passed on into the offtaking canal of low velocity flow deposits it in the canal bed resulting in silting it up and consequent loss of carrying capacity necessitating frequent silt clearance. A high velocity, on the other hand, entails scouring or erosion of the bed thereby lowering the full supply level resulting in loss of command. Thus both silting and scouring are not desirable in an irrigation channel. With this end in view, the design of irrigation canals implies a section which is stable, *i.e.*, neither silts nor scours with the given discharge, water surface slope and silt charge.

5.2 DESIGN APPROACH OF IRRIGATION CHANNELS

Empirical

Chezy Formula (1775); $V = C\sqrt{RS}$

where, V = mean velocity (m/sec), C = Chezy's constant, it is only a factor of flow resistance, R = mean hydraulic radius or hydraulic mean depth (m), *i.e.*, cross sectional area/wetted perimeter, S = sine of water surface slope.

This is a practical and fundamental formula for uniform flow in open channels and is the basis of all modern velocity formulae. It is based on the assumptions that (*i*) force resisting flow per unit of area of channel bed is proportional to the square of mean velocity, and (*ii*) the effective component of gravity force causing the flow must be balanced by total force of resistance to flow.

Advantages: (*i*) Recognized as a classical equation of flow, and (*ii*) Applicable in all cases where uniform flow is assumed or is in existence.

Limitations: (*i*) It is not dimensionally correct, since for a given channel, C and S having numerical values, V varies as $\sqrt{R}$, and the dimensional equation is m/sec = $\sqrt{m}$, (*ii*) It does not take into account the physical character of the channel, (*iii*) It is derived from a limited number of observations, but aim at application to all types of rugosities, (*iv*) Its assumption that hydraulic gradient runs parallel to energy gradient is a condition which is realized in fairly straight regular channels in steady or nearly steady discharge rates, and (*v*) It is valid only for channel cross section resembling wide rectangles with roughly level bottom and parallel flow at right angle thereto.

***Kutter's Formula* (1869):**

$$C = \left[\frac{\dfrac{1}{N} + \left(23 + \dfrac{0.00155}{S}\right)}{1 + \left(23 + \dfrac{0.00155}{S}\right)\dfrac{N}{\sqrt{R}}}\right]$$

or

$$V = \left[\frac{\dfrac{1}{N} + \left(23 + \dfrac{0.00155}{S}\right)}{1 + \left(23 + \dfrac{0.00155}{S}\right)\dfrac{N}{\sqrt{R}}}\right]\sqrt{RS}$$

where, N = Kutter's roughness factor (coefficient of rugosity) having different values for various grades of roughness of channels. Value of N (Table 5.1) depend on the roughness of channel, depth is considered not to effect N to any appreciable extent in alluvial soils, S = water surface slope, and R = mean hydraulic radius or depth (m).

Table 5.1 Values of Kutter's roughness factor N

Nature of channel surface	*Value of N*			
	Perfect	*Good*	*Fair*	*Bad*
Alluvium regime	0.017	0.020	0.0225	0.025
Loose gravel	0.0275	0.030	0.033	0.035
Rock cut, smooth and regular	0.025	0.030	0.033	0.035
Rock cut and irregular	0.035	0.040	0.045	0.050
With cobble bed and earthen banks	0.025	0.030	0.035	0.040
With earthen bed and rubble sides	0.025	0.030	0.033	0.035

Central Water and Power Commission recommended values of Kutter's N applicable for the canals taking off from reservoirs carrying silt free water flowing through rocky cuts range between 0.035–0.05 and for the soils other than rock varies from 0.020–0.030 corresponding to discharge between 0.14 and over 14 cumecs.

Limitations: Kutter's formula is widely used but it suffers from limitations (*i*) It is unnecessarily complicated, (*ii*) Coefficient *C* increases with slope for values of hydraulic radius (*R*) less than 1 m but drops when the slope rises for *R* greater than 1 m, (*iii*) It is not more accurate than Manning's formula, and (*iv*) Coefficient of *N* is assumed to depend only on the nature of the material.

Manning's Formula (1889)

Manning, an Irish engineer, evolved the formula: $V = \frac{1}{N} R^{\frac{2}{3}} S^{\frac{1}{2}}$

It gives better results than Chezy formula. According to this formula, the value of Chezy coefficient is given by the relation: $C = R^{1/6}/N$

where, C = Chezy coefficient. It is considered dependent only on the roughness and hydraulic radius, thereby simplifying their computations, and N = coefficient of rugosity. Its values in normal ranges of slopes and hydraulic mean radius generally seem to be numerically very close to Kutter's N for practical purposes.

Manning's formula, in terms of discharge, for trapezoidal cross section can be written as

$$Q = \frac{PR^{5/3} S^{1/2}}{N}$$

A simplified form of Manning's formula is $Q = KD^{5/3}$, where D is the depth of water in the channel. It is commonly used for the formulation of discharge tables.

Limitations: (*i*) Exponent of R (in relation $C = R^{1/6}/N$) is not a constant but varies depending mainly upon the channel shape and roughness, and (*ii*) Proper selection of value of N to be used is a complex problem. Great difficulty lies in judicious selection of value of N.

Bazin Formula (1897)

Bazin considered Chezy C to be a function of R but not of S. He evolved the following formula for the determination of Chezy coefficient C. The formula gives fairly comparable results.

$$C = \frac{87}{1 + M/\sqrt{R}}$$

where, M = coefficient of roughness which depends on the roughness of channel surface; values for different materials in which the channel is constructed are as given in Table 5.2.

The values of M are more nearly constant than N in Kutter's formula for fairly smooth channels under variations of R and V under identical conditions but reverse is true in the case of erodible channels.

Table 5.2 Value of Bazin's M

Type of channel section	*M*	*Type of channel section*	*M*
Earthen channel — in bad condition	1.75	Smooth rubble masonry	0.46
— in usual condition	1.30	Timber, ashlar, neat brickwork	0.16
Plaster, very smooth earth	0.85	Smooth plaster, planed timber	0.06

The formula is derived mainly from data collected from small experimental channels and is less satisfactory than Kutter's formula.

Pavlovsky's Formula (1925)

Pavlovsky proposed the following uniform flow equation as an improvement over Manning's formula. The formula is widely used in U.S.S.R.

$$C = \frac{1}{n} R^{y}$$

where, $y = 2.5\sqrt{n} - 0.13 - 0.75\sqrt{R}(\sqrt{n} - 0.10)$. It depends on the roughness coefficient and hydraulic radius. The formula is valid for values of R between 0.1 m and 3.0 m and n between 0.011 and 0.040.

These formulas have been extensively used in the design of artificial channels but they are deficient as they have no reference to silt which the water carries in the earthen irrigation channels. The result was that channels designed with these formulas regularly silted badly requiring frequent silt clearance at huge annual expenditure.

Regime Approach

Kennedy Formula (1895)

Kennedy, a Punjab engineer, gave a classic empirical equation correlating mean velocity V in a regime canal with the vertical depth D, as measured on the approximately horizontal silted bed. According to this relation: $V_0 = C.m \, . \, D^n$

or $V_0 = 0.55D^{0.64}$

where, V_0 = critical (non-silting, non-scouring) velocity (m/sec). Fine silt has lower critical velocity than coarse silt, D = full supply depth (m) of water over the bed portion of the canal, Kennedy's modified equation: $V_0 = 0.55 \text{ m } D^{0.64}$.

Kennedy gave hydraulic diagrams which are extensively used for designing irrigation channels in India. From these diagrams it is possible to design any number of channel sections for a given slope with different bed width and depth. It was considered doubtful if all these would be equally efficient regarding stability.

Garret's Diagrams (1913)

Er. Garret in India presented a set of hydraulic diagrams for the design of non-silting channels. These diagrams provide a graphical solution of Kennedy's and Kutter's equations. The diagrams are available for any discharge to over 335 cumecs, for any bed slopes from 1/100 to 1/10000 and for any value of Kutter's N from 0.018 to 0.03. These diagrams are used for the design of irrigation channels.

Wood Normal Data

From Kennedy diagrams a number of channel designs could be worked out for a given discharge and for the same V_0 and evidently all these designs could not be suitable. Er. Woods, Chief Engineer, Punjab, gave a table of the ratios of bed width to depth (Table 5.3).

Table 5.3 Wood's normal design table (N = 0.0225)

Discharge (cumecs)	$\frac{B}{D}$ *ratio*	*B (m)*	*D (m)*	*Slope (1 in)*	$\frac{V}{V_0}$	*Mean V (m/sec)*
0.28	2.9	1.45	0.5	3333	0.92	0.34
0.7	3.4	2.2	0.66	3636	1.01	0.42
1.4	3.7	3.15	0.84	4000	1.0	0.48
2.8	4.2	4.4	1.04	4444	1.0	0.55
7.0	4.5	6.7	1.43	4444	1.01	0.70
14	5.7	9.75	1.72	5000	1.0	0.77
28	7.6	15.28	1.98	5000	1.03	0.88
56	11.3	25.4	2.26	5714	1.03	0.945
141	22.5	56.5	2.5	6666	0.98	0.975
283	41.0	105	2.59	6666	1.02	1.03

Lindley Formula (1919)

$V = 0.567 \, D^{0.57}$, connecting mean velocity with vertical depth D,

$V = 0.274 \, B^{0.355}$, connecting mean velocity with bed width,

$B = 7.86 \, D^{1.61}$, connecting bed width and depth.

Lindley is credited as the first to postulate the regime concept. His definition of regime channel is that, (*i*) When an artificial channel is constructed in alluvium to carry silty water, its bed and banks would silt or scour until the depth, slope and width attain a state of balance, to which he designated as regime, (*ii*) Regime dimensions depend on discharge, nature of bed and quantity of silt, (*iii*) For a channel to carry a certain discharge and sand charge will ultimately develop a cross section and a slope so that equilibrium is established between the forces acting; it is presumed that the channel can freely develop its regime.

Limitations: (*i*) The Lindley observations did not relate to channels in regime. He simply selected straight and regular reaches, (*ii*) No attempt was made to correlate rugosity and silt grade, (*iii*) A formal cross section consisting of a horizontal bed and ½ : 1 side slope was assumed, and it is not practicable to apply his equations to natural channels.

Lacey's Formula (1929)

Lacey, a U.P. Engineer, developed relationships for determining regime slope and channel dimensions. Lacey's set of formulae uniquely fix the size, shape and slope of stable alluvial channels, as under:

Silt factor (f). Lacey introduced silt factor f which is very important in the use of his formulae, to denote silt grade by linear ratio rather than by velocity ratio and dependent on the following factors:

(*a*) Average size of boundary material in the channel, $f = 1.76\sqrt{m_r}$

where, m_r is the average particle size of boundary material (mm). The value of 'f' is estimated from this relation after determining m_r size of the boundary material. Suggested values of 'f' for various values of boundary material are given in Table 5.4.

When related to Kennedy's critical velocity ratio, $f = (V/V_0)^2$

Table 5.4 Values of 'f' for different boundary materials

Type of boundary material		*Grain size (mm)*	f	*Type of boundary material*		*Grain size (mm)*	f
Silt	Very fine	0.05	0.40	Bajri mixed	Medium	1.29	2.0
	Fine	0.12	0.60	with sand	Coarse	2.4	2.75
	Medium	0.23	0.85	Gravel	Medium	7	4.75
	Standard (Kennedy)	0.32	1.0		Heavy	26	9.0
Sand	Medium	0.51	1.25	Boulders	Small	50	12
	Coarse	0.73	1.5		Medium	72.5	15
	Fine	0.99	1.75		Large	188.8	24

(*b*) f is proportional to V^2/R or $f = 2.5\ V^2/R$. (2)

where, V = average velocity (m/sec), and R = hydraulic mean depth (m).

Coefficient of Rugosity N_a–f Relation: Lacey defined N_a as the coefficient of absolute rugosity which is dependent only on grade and density of boundary material. The standard grade of silt is that which connotes a rugosity coefficient of 0.0225 corresponding to hydraulic mean depth of 1 m. It is an overall rugosity coefficient, which in addition to boundary friction allows for shock losses in a channel due to irregularities and bends.

$$N_a = 0.0225\, f^{1/4} \quad (3)$$

$$N/N_a = 1.104\, R^{1/12} \quad (4)$$

When related to Manning's N, values of N_a for silt factor f 1.0 is 0.0225, for f 0.9, N_a = 0.0223 and corresponding to f 0.08, N_a = 0.020.

Suggested values of N_a for different earthen channels are as under:

Channel condition	*Na*	*Channel condition*	*Na*
Excellent order	0.020	Bad order	0.0275
Moderate order	0.0225	Very bad order	0.030
Poor order	0.025		

Wetted Perimeter-discharge Relationship. Lacey gave a very important relationship in which wetted perimeter (P) of a stable channel is constant for a given discharge and is independent of the sediment size, $P = 4.75\ Q^{0.5}$. The coefficient has been found to have some variation but the average value is very nearly as given by Lacey. The formula is important for the calculation of waterway and scour depth for the design of bridges and drainage crossings on rivers and streams, as river in peak flood condition is considered to be in temporary regime.

Regime Scour Depth, R: Lacey gave another important relation for computing depth of maximum scour for the bridge foundation in large rivers viz, $R = 0.47\ (Q/f)^{1/3}$, and also $R = 1.35\ (q^2/f)^{1/3}$, where, q is discharge per metre width $= 0.207\ Q^{1/2}$.

Regime Equation. Lacey's regime equations, known as fundamental equations, also known as V-Q-f relation are:

$$V = \sqrt{(2/5)\, fR} \quad \text{or} \quad = 0.639\ \sqrt{fR} \text{ or } R = 2.5\ V^2/f$$

and $$Af^2 = 140V^5 \quad \text{or } V = (Qf^2/140)^{1/6},$$ where A is area of channel section (m^2), besides regime flow equation described hereinafter.

Regime Flow Equation: $V = 10.8\ R^{2/3}\ S^{1/3}$.

The value of rugosity coefficient cannot be accurately fixed but since it effects the velocity and discharge in inverse proportion, it goes to the credit of Lacey that he gave the above regime formula in which rugosity coefficient and silt factor has been eliminated. These two parameters were considered to be implicit in the actual depth and slope attained by the channel in loose alluvium.

Non-regime formula: For the channels between initial and final regime, Lacey derived the relation, $V = \frac{1}{N} R^{3/4}\ S^{1/2}$ which is in the form of Manning's formula, and gives the same velocity when R is one metre.

Regime Slope. Lacey postulated that the required slope and channel dimensions are dependent on the characteristics of the boundary material which he quantified in terms of silt factor f:

$$S = \frac{1}{3316}\frac{f^{5/3}}{Q^{1/6}}, \text{ further modified to } \frac{1}{3340}\frac{f^{5/3}}{Q^{1/6}},$$

Also, $$S = 0.0003\, f^{5/3}/Q^{1/6}.$$

A number of other important relations are:

$$V = 0.4382\ (Qf^2)^{1/6} \text{ and } A = 2.28\ q^{5/6}/f^{1/2}.$$

Lacey's diagrams for design of canal sections are shown in Fig. 5.1, 5.2 and 5.3.

Bose Formula: Bose, on statistical analysis of Punjab channels, derived the following relationships:

$$P = 5.07\ Q^{0.5}$$
$$S \times 10^4 = 4.73\ d^{0.86}/Q^{0.21}$$
$$R = 0.47\ Q^{1/3}$$

where d is weighted mean diameter (mm) of the bed material which in his analysis varied between 0.16 to 0.41 mm.

5. Regime Type Fitted Equations

Statistical approach was adopted for evolving canal equations by Bose besides Lacey. The data used were limited. The fitted equations, on the other hand, are based on much more extensive data and are preferable. The Central Water and Power Research Station, Pune, has evolved the equations (Table 5.5) for the design of channels in alluvial soil and are considered more accurate than Lacey's.

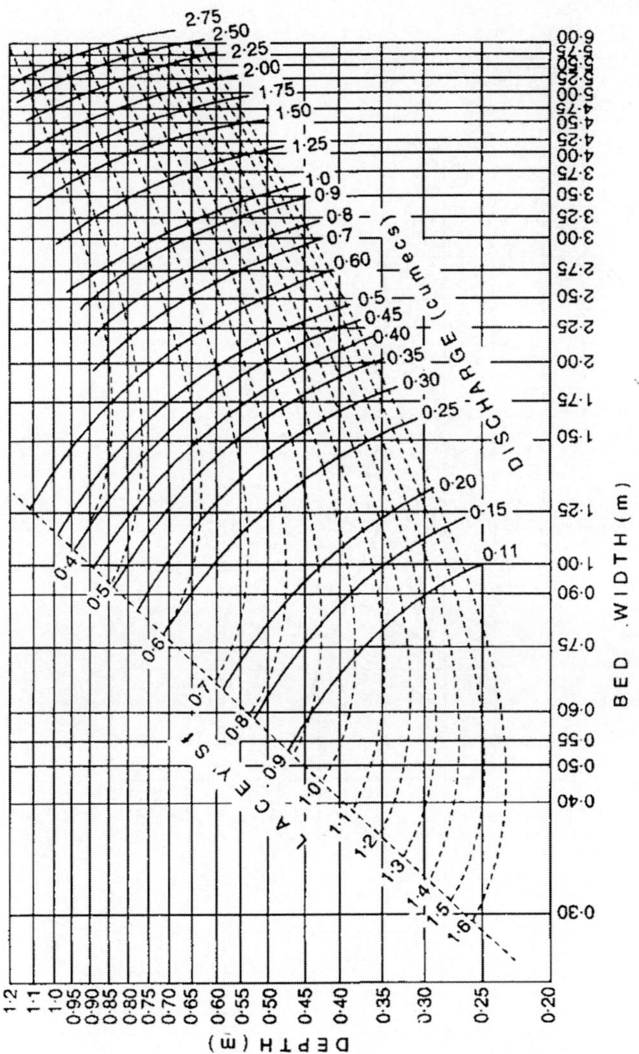

Fig. 5.1 Lacey diagram for canal discharges 0.11 to 2.75 cumecs. Silt factor f 0.4 to 1.6.

Table 5.5 Regime type fitted equations

Hydraulic parameter	*All India canals*	*Punjab canals*	*U.P. canals*	*Bengal canals*
Slope, S	$\frac{0.000315}{Q^{0.1651}}$	$\frac{0.000251}{Q^{0.0961}}$	$\frac{0.00036}{Q^{0.1450}}$	$\frac{0.0001346}{Q^{0.0375}}$
Wetted perimeter, P	$4.3\ Q^{0.5231}$	$7.0\ Q^{0.6019}$	$3.98\ Q^{0.502}$	$5.52\ Q^{0.419}$
Hydraulic mean depth, R	$0.515\ Q^{0.3406}$	$0.446\ Q^{0.3389}$	$0.448\ Q^{0.3649}$	$0.438\ Q^{0.4454}$

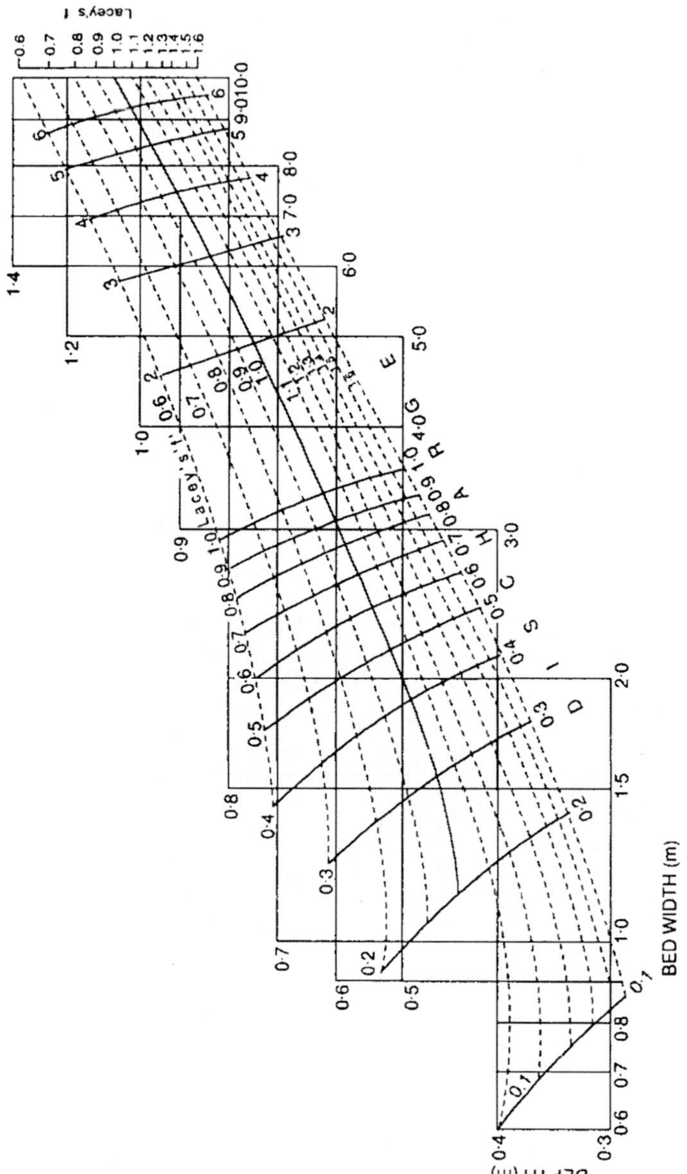

Fig. 5.2 Lacey's diagram for canal discharges up to 6 cumecs.

Limitations: (*i*) The equations are valid only in the areas covered by respective canal systems or in areas having analogous conditions, (*ii*) Observations in respect of Bengal canals on which the equation is based are considered to be insufficient, and (*iii*) More research is needed to fully establish the relations for adoption with confidence.

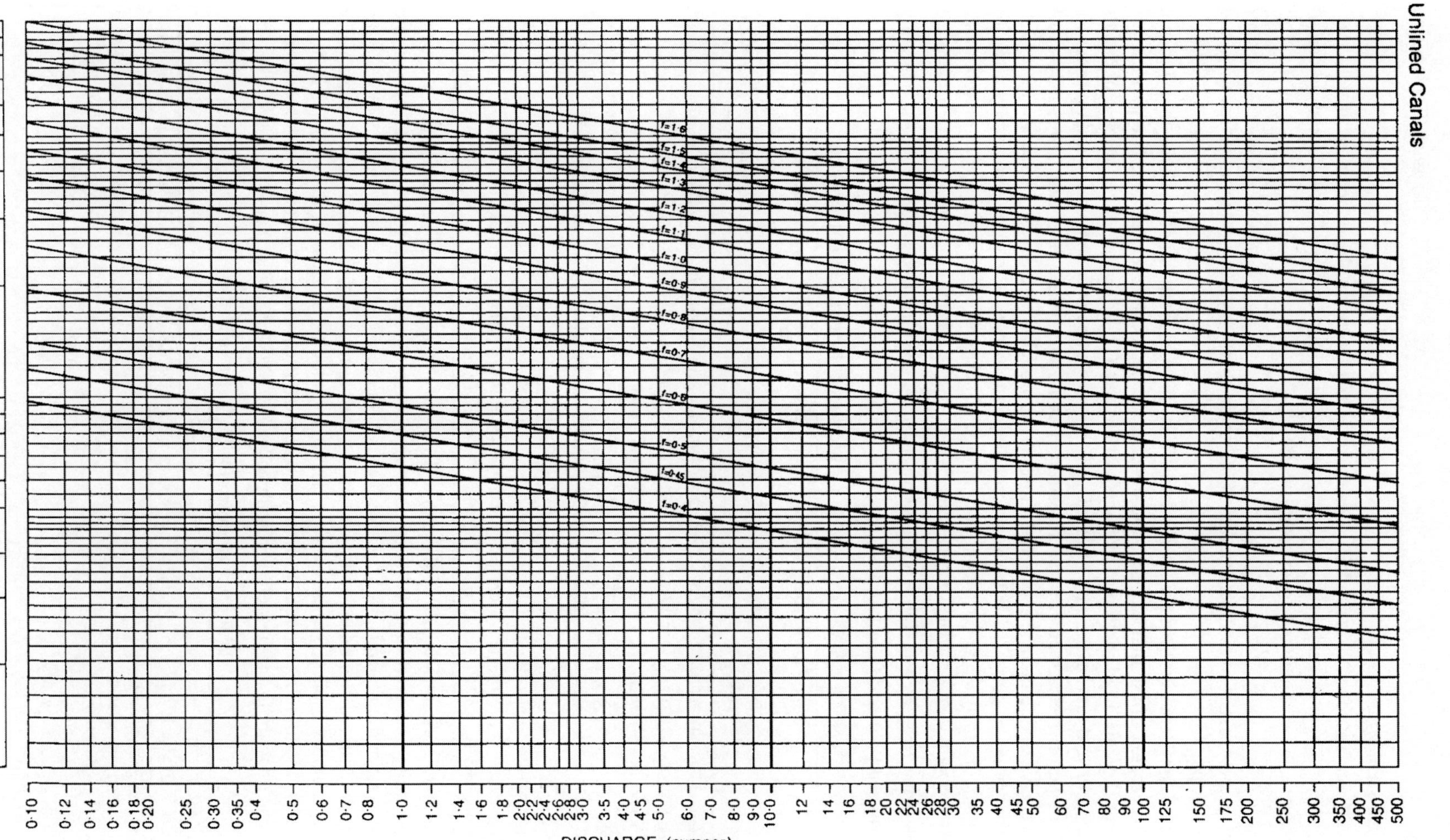

Fig. 5.3 Lacey's diagram to determine regime slope.

3. Tractive Force Approach

Tractive force approach is essentially Lane's approach based on the concept of critical tractive force for the incipient conditions of bed movement. The unit tractive force exerted on bed of a running canal can be calculated from the formula:

$$\tau = WRS \qquad (i)$$

where, W = unit weight of water (usually 1000 kg/m^3), R = hydraulic mean radius (m), and S = channel slope.

The permissible tractive force may be defined as the maximum tractive force that will not cause serious erosion of the material forming the channel bed on a level surface. The permissible tractive forces (kg/cm^2) is a function of average particle size (D_{50}) of channel bed in case of canals in sandy soils and void ratio in the case of canals in clayey soils and sediment concentration (gm/m^3). The values of permissible tractive force for straight canal are better determined by analysis of observed data on existing canals to provide a rational approach to the design of section of regime channels. The values of permissible tractive force for sinuous canals may be reduced by 10% for slightly sinuous ones, by 25% for moderately sinuous ones and by 40% for very sinuous ones. Lane's tentative values of critical tractive force, velocity and Manning's rugosity coefficient for different materials are given in Table 5.6. Allowable maximum tractive force (τ_{max}) is generally assumed between 0.24 and 0.34 kg/m^2.

Table 5.6 Critical tractive force and bed velocity for different bed materials

Material	*Critical tractive force (kg/m²)*	*Approximate bed velocity (m/sec)*	*Manning's n*	*Material*	*Critical tractive force (kg/m²)*	*Approximate bed velocity (m/sec)*	*Manning's n*
Sandy loam (non-colloidal)	0.20	0.50	0.020	Medium sand (non-colloidal)	0.17	0.50	0.020
Silt loam (non-colloidal)	0.25	0.60	0.020	Coarse sand (non-colloidal)	0.25	0.60	0.020
Alluvial silt (non-colloidal)	0.25	0.60	0.020	Fine gravel	0.37	0.75	0.020
Ordinary firm loam	0.37	0.75	0.020	Coarse gravel	1.47	1.25	0.025
Volcanic ash	0.37	0.75	0.020	Cobbles and shingles	4.40	1.55	0.035
Stiff clay (very colloidal)	1.22	1.15	0.025	Graded loam to cobbles	1.96	1.15	0.030
Alluvial silt (colloidal)	1.22	1.15	0.025	when non-colloidal			
Shales and hard pans	3.18	1.85	0.025	Graded silts to cobbles	2.20	1.25	0.030
Fine sand (non-colloidal)	0.12	0.45	0.020	when colloidal			

The sediment concentration X of the canal flow and the D_{50} size of bed material in case of non-cohesive soils and void ratio of bed material in case of cohesive soils is determined and thence the corresponding permissible tractive force. A suitable bed slope is then selected and the value of R is determined from equation (i), knowing the value of R and assuming a suitable value of n for the channel (Table 5.7), the average desirable velocity of flow in the channel is determined by using Manning's formula. In India, for normal alluvial soils, value of n is assumed as 0.0225 for canals with discharge below 15 cumecs and 0.020 for higher capacity canals.

Table 5.7 Rugosity coefficient (n) for unlined canals

Type of canal	*Rugosity coefficient n*		
	Minimum	*Normal*	*Maximum*
1. *Earthen straight and uniform*			
Clean, recently completed	0.016	0.018	0.020
Clean, after weathering	0.018	0.022	0.025
Gravel, uniform section, clean	0.022	0.025	0.030
With short grass, few weeds	0.022	0.027	0.033

Type of canal	Rugosity coefficient n		
	Minimum	Normal	Maximum
2. *Earthen winding and sluggish*			
No vegetation	0.023	0.025	0.030
Grass, some weeds	0.025	0.030	0.033
Dense weeds or aquatic plants in deep channels	0.030	0.035	0.035
Earth bottom and rubble sides	0.030	0.035	0.040
Stoney bottom and weedy banks	0.025	0.035	0.040
Cobble bottom and clean sides	0.030	0.040	0.050
3. *Dragline excavated or dredged*			
No vegetation	0.025	0.028	0.033
Light bush on banks	0.035	0.050	0.060
4. *Rock cuts*			
Smooth and uniform	0.025	0.035	0.040
Jagged and irregular	0.035	0.040	0.050
5. *Channels not maintained: weeds and bush uncut*			
Dense weeds, high as flow depth	0.050	0.080	0.120
Clean bottom, bush on sides	0.040	0.050	0.080
Same, highest stage of flow	0.045	0.070	0.110
Dense bush, high stage	0.080	0.100	0.140

Thus the area of cross section required may be determined and knowing R and A, the desirable channel bed width (B) or depth (D) is calculated.

4. Rational Approach

The empirical approach to arrive at the design of regime channels suffers from certain drawbacks in that estimation of sediment charge to be carried by a canal is essential. Observed data are scent to permit the use of formula in which sediment charge is a parameter. The rational approach aims at to develop the equations such as resistance law, and sediment transport law in addition to the equation of continuity and the equations for the channel width and bank resistivity. The important contributors to this approach in India are as follows:

1. *Chitale Formula* (1966)

$P = 4.298Q^{0.533}$, $R = 0.498Q^{0.341}$, $S = 0.00028Q^{-0.165}$, $V = 7.335R^{0.5}(R^{0.5}S)^{0.293}$.

2. *Bharat Singh Formula*

$B = 17\,S^{0.182}\,Q^{0.545}$, or $B = 4.25\,Q^{0.545}$, when the slope is expressed in terms of discharge by Lacey's relationship and silt factor unity.

$A = 2.08\,Q^{0.89}$.

5.3 DESIGN OF CHANNELS ON ALLUVIAL SOILS

5.3.1 Kennedy Theory

Kennedy, as described earlier, did pioneering research for obtaining a stable non-silting, non-scouring irrigation canal system. On the basis of observations made in certain reaches of Upper Bari Doab Canal system in Punjab, which were found to be fairly stable, *i.e.*, required no silt clearance for about three decades, Kennedy concluded that, (*i*) The flowing water is to counteract friction against the bed of the channel resulting in generation of vertical eddies rising up gently to the water surface and work up against the depth of channel. These eddies can always be observed over the full width of the channel. These eddies keep most of the silt in suspension. The vertical component of the eddies tend to move the sediment up, and the weight of sediment tends to bring it down, thereby keeping the sediment in suspension, (*ii*) The silt supporting power of a channel cross section is mainly dependent on these eddies. Some of the eddies may start from the sides of the channel but these are for most of their part horizontal and as such of no silt supporting power. The silt supporting power of the channel is, therefore, proportional to the bed width, (*iii*) A velocity sufficient to generate these eddies keeps the sediment in suspension thereby avoiding silting up of the channel. He designated it as critical velocity (V_0) defined as the mean velocity which just keeps the channel free from silting or scouring. The silt supporting power of a channel is proportional tc $V_0^{5/2}$, (*iv*) Safe velocity against erosion for canals in Punjab soil is 1 m/sec corresponding to a depth of not more than 3 m, (*v*) The amount of silt held in suspension is proportional to the upward acting force of vertical eddies and varies as bed width and some power of the velocity of flow in the channel, (*vi*) A regime channel is that which neither silts nor scours, (*vii*) Coefficient of roughness n average for all irrigation channels is 0.0225. Later, he found out that this value is not constant and suggested value of 0.02 for large canals and 0.025 for small channels.

Kennedy concluded that the channels of various sizes with velocities based on this formula did not silt or scour their beds. However, the formula gives correct results only when correct shape of the channel is chosen.

Kennedy later, as more data became available, recognized that the sediment size played a significant role in this relationship and regarded the silt of Upper Bari Doab Canal as the standard silt. Since critical velocity in a channel depends upon the grade of silt, he introduced another factor, Kennedy *critical velocity ratio,* CVR, denoted by symbol 'm' in the equation. It is the ratio of critical velocity in a channel to the critical velocity obtained at Upper Bari Doab Canal only. The factor was utilized to define the grade of silt and silt charge carried in the canals whose regime differed from that given by the Kennedy formula. The modified equation is as under:

$$V = 0.55\, m\, D^{0.64}$$

where, V = velocity (m/s), m = critical velocity ratio (V/V_0), with values varying from 1.1 to 1.2 for sand coarser than the standard and 0.8 to 0.9 for finer sands.

The non-silting velocity by the above formula is relevant to channels which have silt of the same character as that in the observed channels of UBDC by Kennedy but different for silt charge of different size and grade. The suggested values of constant C and m for different silt charge in the general form of the formula $V_0 = Cm\, D^n$ are as under:

Type of silt grade	*C*	*m*	*Type of silt grade*	*C*	*m*
Coarse silt	0.70	1.3	Coarse light sandy silt	0.59	1.1
Sandy loam silt	0.65	1.2	Light sandy silt	0.53	1.0

Kennedy equations applicable in different regimes are as under:

$V_0 = 0.391\, D^{0.55}$, Godavari delta

$V_0 = 0.530\, D^{0.52}$, Krishna Western delta

$V_0 = 0.567\, D^{0.57}$, Lower Chenab canal

$V_0 = 0.283\, D^{0.73}$, Egyptian canals

A general feature of these formulae was that the value of C is proportional to the size of the bed material.

Kennedy later (1904) recognized that the ratio of bed width to depth varied with the size of the channel. He stated that the proportion of bed to depth will, of course, vary with the size of the channel, as per Table 5.8.

Table 5.8 Bed width-depth ratio

Channel capacity (cumecs)	*Bed width depth ratio*	*Channel capacity (cumecs)*	*Bed width depth ratio*
Up to 0.28	3.5	5.6	5.0
0.7	4.0	14	6.0
2.8	4.5	28	7.0
			and so on, these figures being rather indefinite

Practical Application

1. *Critical Velocity Ratio.* It is defined as the ratio of actual mean velocity (V) to critical velocity (V_0) given by Kennedy formula. It does not reflect departure from regime but (*i*) is indicative of variation in silt in the channel from the standard silt observed by Kennedy in UBDC system, and (*ii*) is a factor which takes into account the scale effect of the channel dimensions.

The criteria of V/V_0 is not a dominant factor to canals taking off from a reservoir and as such not carrying much silt. V/V_0 value of 1.1 at the head of a main canal taking off from a river depending on location of the head and quantity and grade of silt entering the canal and 0.85 towards the tail end of a canal system is a general practice in India, the lowest value being 0.8. A value below the critical value implies silting tendency and higher value means scouring. CVR adopted in various States of India vary from 1.1 to 0.8.

CVR values vis-a-vis various soil formations are given in Table 5.9.

Table 5.9 CVR values vis-a-vis soil formation

Soil type	V/V_0	*Soil type*	V/V_0
Earth	0.6 to 0.3	Boulders	1.75 to 1.5
Ordinary murum	0.9 to 0.6	Soft rock	2.5 to 1.75
Hard murum	1.25	Hard rock	3

2. *Bed Width-depth Ratio.* The Kennedy depth formula gives the maximum depth that can be allowed in an earthen channel but has no relationship with the shape of the channel, *i.e.*, bed width-depth ratio. The losses due to absorption and evaporation as also the minimum cross-sectional area required for given capacity of a canal are governed by bed width-depth ratio. Experience of the working of the irrigation canals in alluvial soils has dictated that certain bed width-depth ratio exists depending upon the discharge or in other words the size of channels. Thus whenever there is free choice for selecting dimensions of channel width or depth, Central Water Commission (India) recommended bed width-depth ratio (Fig. 5.4.) is adopted. Ratio of 1.25 to 5 is beneficial for canals in cutting in very hard and very porous strata.

Limitations: (*i*) Kennedy did not investigate to find out the correct slope formula applicable directly to the design of canals, and (*ii*) He simply took Kutter's formula and adopted value of N equal to 0.0225 as the average value for all channels. No attempt was made to correlate Kutter's N with CVR, and (*iii*) Kennedy made no correlation between water surface slope of regime channels and the mean velocity or the depth. He relied on Kutter's equation to give slopes for the channels.

Drawbacks in Kennedy's Theory

Kennedy's theory suffers from the drawbacks (*i*) It does not take cognisance of the width or the shape of the channel which have to be assumed, *i.e.*, importance of bed width and depth ratio is ignored, (*ii*) It involves use of Chezy's formula and Kutter's N for working out mean velocity and as such incorporates limitations of those relations in this theory, (*iii*) Adoption of arbitrary value of N as 0.0225 is not correct, (*iv*) Design of only average regime channel was aimed at, (*v*) Did not specify regime water surface slope relation. His diagrams, however, show that steeper slopes are required for small channels and flatter ones for large channels, (*vi*) Silt concentration and bed load were not considered, (*vii*) Silt grade and silt charge were not defined, (*viii*) Design of channel involves trial and error as the velocity worked out with assumed depth should give the required discharge for the section and at the same time satisfy Kennedy's equation, and (*ix*) Simply stated that CVR varies according to the silt charge and silt grade but gave no method to measure value of m or CVR applicable for channels of different silt grades.

Design of Canals by Kennedy Theory

Given: Discharge (Q), water surface slope (S), coefficient of rugosity (N), and critical velocity ratio (m).

Procedure

Assume reasonable depth (m), and calculate velocity from formula, $V_0 = 0.55\ m\ D^{0.64}$ m/sec and $A = Q/V\ (m^2)$.

Bed width is calculated from the relation:

$$A = BD + SD^2 = BD + 0.5\ D^2 \text{ (with ½ : 1 side slope)}$$

Wetted perimeter P and hydraulic mean depth R are worked out for the calculated cross section:

$$P = B + \sqrt{5}\,D \text{ (m) and } R = \frac{BD + 0.5\,D^2}{B + \sqrt{5}\,D}, \text{ (m) for ½ : 1 side slope}$$

Actual mean velocity is calculated from Kutter's formula by substituting value of R calculated above; if it agrees with that calculated from Kennedy's relation ($V_0 = 0.55$ m $D^{0.64}$), the depth, and hence bed width assumed is correct, otherwise trial values of D are adopted till the two values of velocities work out to be approximately the same. The problem is simplified if B/D ratio is assumed or read from Central Water Power Commission curve (Fig. 5.4).

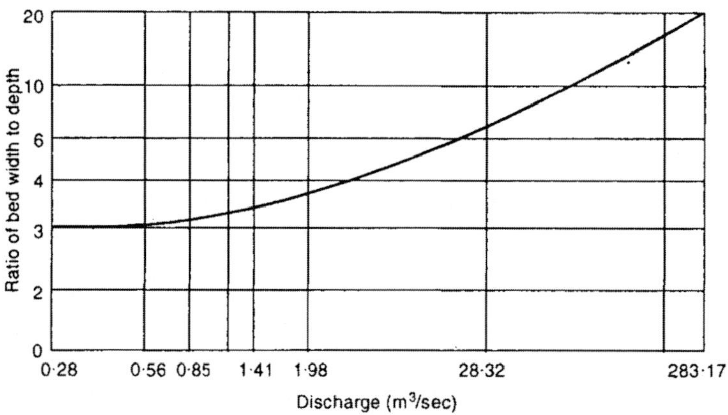

Fig. 5.4 Bed width depth ratio.

5.3.2 Lacey Theory

Lacey attempted to remove the deficiencies in various formulae developed earlier by Kennedy and Lindley which did not take into account the physical dimensions and properties of the channel such as slope, size and bed material. Lacey on the analysis of data from a large number of natural drainages and canals running for long put forth his theory, the salient features of which are (*i*) If a channel runs indefinitely with constant discharge and sediment charge rates, it will attain a definite stable section having a definite slope, (*ii*) He used Kennedy data and substituted silt factor f for m and stated that required slope and channel dimensions are dependent on the silt factor f, (*iii*) Lacey considered that the silt is kept in suspension by the vertical components of eddies generated at all points by forces normal to the wetted perimeter of the channel and accordingly adopted wetted perimeter and hydraulic mean depth as the flow parameters instead of the surface width and depth of flow assumed by Kennedy, (*iv*) There is only one channel section and one longitudinal slope for a given discharge and particular silt factor, (*v*) Lacey stated that rugosity coefficient of the channel solely depends on the size of the boundary material forming the bed and sides and is independent of channel condition. Regime channel has constant rugosity coefficient for given size of silt grade, (*vi*) Lacey's standard silt is that which connotes a coefficient of rugosity of 0.0225 when hydraulic mean depth is 1.0 m, (*vii*) Lacey assumed channel section as semi-elliptical, (*viii*) Ratio of bed width to depth effects channel capacity, *i.e.*, shape of channel for a given discharge is a function of silt grade, channel in finer material being narrow and deeper (Fig. 5.5), (*ix*) Lacey considered hydraulic mean depth as more relevant variable. In elliptical or semi-circular channel section assumed by Lacey there is no side in true sense and hence hydraulic mean depth is the correct basis. According to him concept of 'R' in place of 'D' is because eddies which keep the silt in suspension are generated from bed and sides both normal to the surface of generation and not from bed only as stated by Kennedy, (*x*) Channels designed with Lacey's theory for $f = 0.8$ were smaller in section than those designed with Kennedy's $N = 0.0225$ for the same discharge and slope, (*xi*) Lacey considered that his formulae are applicable to channels of all sizes, and (*xii*) Lacey enunciated that a channel flowing in its own silt, if not interfered, reaches final stability and in case the discharge and silt remain constant final regime is ultimately attained. He thus propounded regime theory of channels.

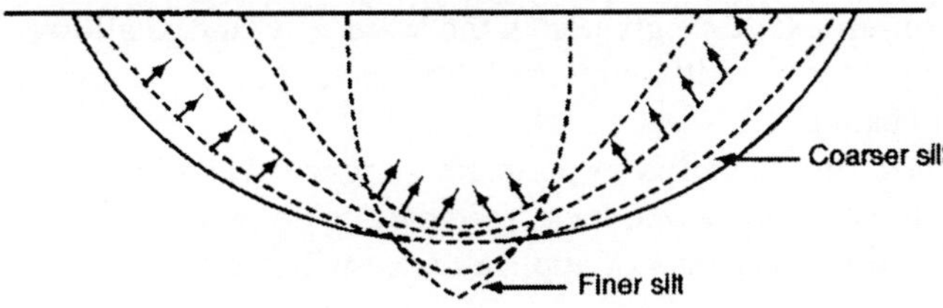

Fig. 5.5 Channel section according to Lacey theory.

Regime: Lacey was not satisfied with the definition of regime channel, *i.e.*, which neither silts nor scours as given by Kennedy. He stated that even a channel in non-silting and non-scouring condition is not in regime.

Lacey's concept of regime channel is (*i*) A non-silting, non-scouring equilibrium channel carries its normal that is a certain kind and quantity of suspended charge, (*ii*) A channel is said to have attained a regime condition when a balance between the silting and scouring and dynamic equilibrium in the forces generating and maintaining the canal cross section and gradient are obtained, (*iii*) A channel running indefinitely with constant discharge and sediment charge rates will attain a definite stable section having a definite slope, (*iv*) A channel designed with too small a section for the given discharge and steeper slope than required develops scouring till final regime is attained, and (*v*) A channel with large section for the given discharge and flatter slope than required develops silting till true regime is attained.

Fundamental Conditions: Fundamental conditions to be satisfied for the establishment of regime conditions in a stable channel, as per Lacey's conception, are (*i*) Channel is flowing uniformly in unlimited incoherent alluvium of the same character as that transported. Loose granular material which can be scoured out as easily as deposited was considered incoherent alluvium, (*ii*) Discharge is constant, and (*iii*) Silt grade and silt charge are constant.

True Regime: The essential characteristics for true regime conditions according to Lacey are (*i*) The channel is flowing in an unlimited incoherent alluvium of the same grade as that of the material transported, (*ii*) Discharge in the channel is constant, (*iii*) Silt grade or silt charge is constant, (*iv*) There is complete freedom for lateral movement which can be achieved in sandy rivers in incoherent alluvium by meandering to adjust their length and slope, and (*v*) A constant discharge transporting silt of given grade, *i.e.*, of a fixed regime silt grade and flowing in self transported alluvium plane of the same grade, eventually tends to assume a gradient solely determined by the discharge and silt grade, mean velocity, hydraulic mean radius and wetted perimeter also tends to unique determination.

Evidently irrigation canals cannot be in true regime but can generally be in initial regime or in final regime under favourable conditions. Accordingly, Lacey classified the regime for unlined irrigation channels into initial regime and final regime, described below:

Initial Regime: Lacey's conception of initial regime is (*i*) Channels excavated initially with defective slopes and somewhat narrow dimensions are free by immediately throwing down incoherent silt on the bed to increase their slope and by generation of increased velocity assume a non-silting regime and are said to be in initial regime, (*ii*) Channels whose bed slope vary while their cross section or wetted perimeter remaining unaffected can also be in non-silting, non-scouring condition to be called in initial regime, and (*iii*) Channels with rigid banks are not free to erode the sides and as a consequence their slope and velocity are higher and cross section narrower. Such channels are not in fact in alluvium and the working stability attained by them is initial regime. Lacey's regime theory is not applicable to the channels in initial regime.

Final Regime: Final regime envisages the eventual condition set up in theory when all variables *i.e.*, wetted perimeter, depth, slope, etc., which are equally free to vary get adjusted for the given discharge and silt grade. Such a channel assumes a semi-elliptical section. The coarser the silt, flatter is the semi-ellipse, *i.e.*, greater is the waterway and shallower is the depth of the channel (Fig. 5.5). The finer the silt, more nearly the section attains a semi-circle. Regime theory is applicable to such channels.

Permanent Regime: A channel is in permanent regime when its section is provided with a lining to protect against scouring action thereby imparting permanency to the channel section and longitudinal slope. Regime theory is not applicable to such channels.

Lacey's Shock Theory

Lacey's absolute coefficient of rugosity N_a is dependent on the grade and density of boundary material and that it is constant for channels in similar material and irrespective of their condition is not true. Taking into consideration the channel condition, Lacey later introduced his shock theory according to which slope may include shock due to bends or irregularities and also due to channel conditions. Provided that the grade of silt transported is perfectly incoherent or of constant coherence, the slope should be a function of the bed-silt grade. Shock in such channels is indicated by departure from the regime slope, and a corresponding departure from normal bed silt grade. He divided the slope into two parts, one to overcome friction and the other to meet shock losses. He modified non-regime flow formula in the context of shock theory, as under:

$$V = \frac{1}{N_a} R^{3/4} (S - s)^{1/2} \text{ m/sec}$$

where, V = velocity (m/s), S = slope required to overcome friction, s = slope required to withstand shock losses, and N_a = Lacey's absolute rugosity coefficient independent of channel condition.

For a channel in good condition, $N_a = 0.025$ and for that in very good condition $N_a = 0.0225$,

thus
$$\frac{R^{3/4}(S-s)^{1/2}}{0.0225} = \frac{R^{3/4}\,S^{1/2}}{0.025}$$

or
$$(S-s)^{1/2} = 0.9\,S^{1/2}$$

or
$$s = 0.19\,S$$

that is for a channel in good condition 19% of the gross slope is accounted for by shock losses and remaining 81% is available for frictional resistance. Similarly in a channel in fair and bad condition, shock losses account for 33% and 43.7% respectively. Lacey's shock theory of separating form resistance from frictional resistance is unique, but is not used in practice, there being no reliable method for calculating form resistance (shock losses) and as such different values of N_a are used for different channel conditions. Buckley's suggested values of shock factor s for different channel conditions are as given in Table 5.10.

Table 5.10 Values of shock factors

Channel condition	N	s
Very good	0.0225	0.050 S
Good	0.0250	0.1905 S
Indifferent	0.0275	0.3315 S
Bad	0.0300	0.4375 S

Limitations of Lacey's Theory

Lacey's regime equations, though in common use in India and other countries, have certain shortcomings, (*i*) True regime conditions are seldom met with in practice, (*ii*) Choice of appropriate value of silt factor f occurring in Lacey's formula for depth and slope present major difficulty in the application of Lacey's equations, (*iii*) Lacey channels carry 12 to 20 mg/l of bed load in the discharge range of 2.83 to 283 cumecs but outside this range of bed load content, Lacey's equations are not applicable according to Bharat Singh, (*iv*) Significant divergence has been observed in dimensions given by Lacey's equation in existing stable channels. Lacey put forth shock theory to explain these divergences but this was not generally accepted, (*v*) Lacey's assumption that rugosity coefficient N_a solely depends on boundary material forming bed and sides and is independent of channel condition is not accepted in practice, (*vi*) Wetted perimeter depends not only on discharge, but on silt charge as well. $P = 4.75\ Q^{1/2}$ is the minimum wetted perimeter for channels in incoherent alluvium. This value is not rigid but varies between 3.7 and 5.7 for Punjab soils, (*vii*) Even for a constant silt grade, the ratio of wetted perimeter to hydraulic mean depth steadily diminishes with reduction in discharge, (*viii*) Non-silting, non-scouring velocity depends not only on the grade of silt but also on the quantity of the silt load transported, (*ix*) Lacey's final regime slope formula $\left(S \propto f^{5/3}/Q^{1/6}\right)$ shows that there is no rigidity about the modified constant 1/3340 in formula, $S = \dfrac{f^{5/3}}{3340\ Q^{1/6}}$, (*x*) Lacey's equations in terms of P and R involved a limiting semicircular cross section and a limiting velocity of 0.3 m/sec for any type of material. Lacey later considered that as far as cross section was concerned, the correct parameters to employ were not P and R but the horizontal width and vertical depth, approximately measured, (*xi*) Bed material carried by a canal system is not same, main and large branches carry coarser material than smaller branches and distributaries, (*xii*) Lacey's formulae are applicable up to certain values of discharge beyond which they give indeterminate results. Kennedy's formula may be used for small discharges, and (*xiii*) Kennedy's 'D' has been replaced by 'R' and it is assumed variable. However, it is observed that there is hardly any difference in values of R and D in wide channels.

Design of Canals by Lacey's Theory

Discharge and mean diameter of silt (*mm*) are given.

Design Procedure: (*i*) Fix the value of silt factor f, *i.e.*, $f = 1.76\ \text{m}^{0.5}$, (*ii*) Calculate velocity, $V = (Qf^2/140)^{1/6}$, (*iii*) Calculate hydraulic mean depth, $R = 2.5\ V^2/f$, (*iv*) Calculate area, $A = Q/V$, (*v*) Calculate wetted perimeter, $P = 4.75\ Q^{1/2}$, (*vi*) Assuming side slope ½ : 1, calculate bed width B and depth D, A and P being known:

$$A = BD + 0.5\ D^2,\ P = B + \sqrt{5}\ D, \text{ and } A = R \times P$$

or, alternatively,

$$D = \frac{P - \sqrt{P^2 - 6.944\ A}}{3.472}$$

and,

$$B = P - \sqrt{5}\ D$$

(*vii*) Calculate

$$R = \frac{BD + 0.5\ D^2}{B + \sqrt{5}\ D}$$

It should tally with the value obtained in step (*iii*) above.

(*viii*) Calculate slope, $S = 0.0003\ f^{5/3}/Q^{1/6}$.

Alternatively discharge and water surface slope (within the range of Lacey's $f = 1.0$ to 0.8) are given. The Lacey's f is first determined from the known discharge and slope by the relation, $S = 0.0003\ f^{5/3}/Q^{1/6}$. The remaining procedure is as above. Lacey's diagrams for design of canal sections are used as shown in Fig. 5.1 to 5.3.

5.4 DRAWBACKS IN KENNEDY AND LACEY THEORIES

Both the well known theories are deficient in the respects (*i*) Both aimed to fined out average regime conditions but none took cognisance of the effect of varied silt conductive powers of outlets and offtakes on the regime of the channel, (*ii*) Both ignored the effect of silt attrition, (*iii*) Both did not consider effect of silt left in channel by the water lost in absorption, (*iv*) Both did not consider scale effect. River and a minor or a distributary may have Lacey's silt factor unity or Kennedy's CVR unity but in fact they carry silt charge and silt grade many times different, and (*v*) Both did not precisely define the silt charge and the channel size for unity silt factor or unity CVR.

5.5 COMPARISON OF LACEY AND KENNEDY THEORIES

	Lacey theory	*Kennedy theory*
1.	Only one regime section for given discharge and silt factor.	There can be as many sections for the given discharge.
2.	Did not leave silt grade as guess work but correlated silt factor f to diameter of silt and rugosity coefficient.	Simply stated CVR varied according to the silt grade but did not give any method to measure CVR.
3.	Discarded arbitrary factors and gave general regime equation independent of rugosity coefficient, viz. $V = 10.8\ R^{2/3}\ S^{1/3}$	Gave no equation for regime velocity. Adoption of Kutter's N and Chezy equation to work out mean velocity incorporated the limitations of those relations.
4.	Considered hydraulic mean depth R as relevant variable and derived velocity formula in terms of R: $V = 0.639\ (fR)^{1/2}$	Considered depth as variable and gave critical velocity formula in terms of depth D, $V_0 = 0.55\ m\ D^{0.64}$.
5.	Specified regime slope for the given discharge and silt factor.	Did not specify regime slope.
6.	Silt is kept in suspension by vertical components of eddies caused by friction against wetted perimeter.	Eddies rise on account of roughness of bed and work up against depth. The eddies generated from sides are ignored.
7.	Considered channel section semi-elliptical.	Considered channel section trapezoidal.
8.	Section is wider and shallower.	Section is tighter and deeper.
9.	Considered that channels cannot be in true regime and classified them into initial and final regime.	Considered all channels in regime when they neither silt nor scour.
10.	Gave formula for definite wetted perimeter for the given discharge. $P = 4.75\ \sqrt{Q}$.	Did not consider importance of bed width-depth ratio. Wood's B-D ratios are not rigid.
11.	Applicable to irrigation channels and rivers as well.	Applicable to irrigation channels only.

5.6 SELECTION OF METHOD OF DESIGN

For the design of irrigation channels, India can be broadly divided into two regions, the northern States comprising the Punjab, Haryana, Himachal Pradesh, Uttar Pradesh, Rajasthan, Bihar, West Bengal and some parts of Madhya Pradesh where canals draw their supplies from rivers carrying heavy sediment laden water. These canals carry sediment similar to that forming their bed and banks. In Southern India, the canals carry much lesser sediment and the boundary material of canal section is dissimilar to the sediment transported. On the basis of above dissimilarities, Central Water Commission (India) has recommended the following practices for the design of irrigation channels:

1. *Channels Carrying Low Silt Content and Flowing through Hard Soils*

These are the channels taking off from reservoirs. The channels passing through hard soils or disintegrated or fresh rock are designed with as high a velocity as the soil can withstand without erosion consistent with ground slope. The channel section is designed from Chezy's formula with Kutter's values of C and N.

2. *Channels Carrying heavy Sediment Flowing in Alluvium*

These are the channels taking off from barrages or weirs carrying heavily sediment laden water and generally pass through alluvial soils. Lacey formulae are usually adopted for the design of unlined irrigation channels with suitable value of f. If necessary value of f may be correlated to Kennedy's CVR. For small capacity silt-bearing channels as also the channels with low silt factor in cohesive soils, the slope may be determined from Lacey's theory and channel section worked out from Kennedy formula.

Which theory preferred

Lacey's theory has earned recognition for the design of canals. It is preferred on account of its merits (*i*) Lacey's theory is more rational as it gives quantitative relation between the discharge, average grain size of silt and coefficient of rugosity, (*ii*) There is only one canal section, wetted perimeter and hydraulic mean depth for given discharge, slope and silt factor and in due consideration to other features which determine the canal section, (*iii*) Ease in calculation work; no trial and error is involved as is the case with Kennedy theory, and (*iv*) Specified separate design equations for perfect regime and those postulating between initial and final regime.

5.7 DESIGN OF CHANNELS ON NON-ALLUVIAL SOILS

Non-alluvial soils are hard soils or disintegrated or fresh rock, compact clay, murum, boulders etc. These are considered stable for the design of irrigation channels. Water is comparatively silt free and there is no silt problem in those channels. Such channels are designed for high velocities close to the critical velocity for the soil without erosion tendency consistent with ground slope. The canals called incised unlined channels in peninsular plateau of Southern India carry much less sediment; the boundary material of the canal section is dissimilar to that transported and is more scour resistant.

Chezy, Kutter's, Kennedy and Manning's formulae are used in the design of these channels. Values of n vary between 0.035 and 0.040 for rocky surface and between 0.022 and 0.03 for murum irrespective of discharge velocity. The velocity adopted is 1 m/s, 1.25 m/s, 1.5 m/s, 2.5 to 3.5 m/s for murum, gravel, disintegrated rock and hard rock respectively as per CBIP practice. Critical velocity ratio of 1.1 at head and decreasing to 0.8 in tail reach is adopted. The criterion of V/V_0 is needed to prevent deposition or scour of the canal bed; applicable only to the canal carrying bed material load.

Side slope: The usually adopted side slopes in hard rock vary from 0.125 : 1 to 0.25 : 1, in soft rock, 0.25 : 1 to 0.5 : 1, 0.75 : 1 in hard clay or gravel and 0.75 : 1 to 1.5 : 1 in murum.

Bed width-Depth Ratio: Comparatively lesser B/D ratio than that adopted in alluvial channels is adopted; between 1.25 to 5 is recommended for minimising losses due to absorption and evaporation and economy in cutting. For wider channels, bed width-depth ratio curve indicated in Fig. 5.4 is adopted.

Free Board: The recommended free board is 0.5, 0.6, 0.75, 1.0 for discharge < 0.7, 0.7–1.4, 1.4–8.5 and above 8.5 cumecs respectively.

Longitudinal slope: It varies from 1/1000 to 1/10000 and also adopted as given by Lacey's formula. In the design of channels, cross section is assumed and corresponding cross sectional area, hydraulic mean depth, velocity is determined using the indicated formulae and check is made for CVR.

5.8 LAYOUT OF UNLINED CANALS

Canals are artificial channels constructed for carrying water from a reservoir or river to the fields for irrigation. Fig. 5.6 show the typical layout of a canal system.

The planning and layout of a canal system is governed by the command area to be irrigated and source of water supply. The layout of main canals and branches is generally fixed on the consideration of economy as normally no irrigation is done directly from them but in the case of distributaries and minors the points of offtake are suitably selected but their layout is broadly governed by the tracts of areas to be irrigated with due regard to ridges, drainages as outlets for irrigation are provided on them.

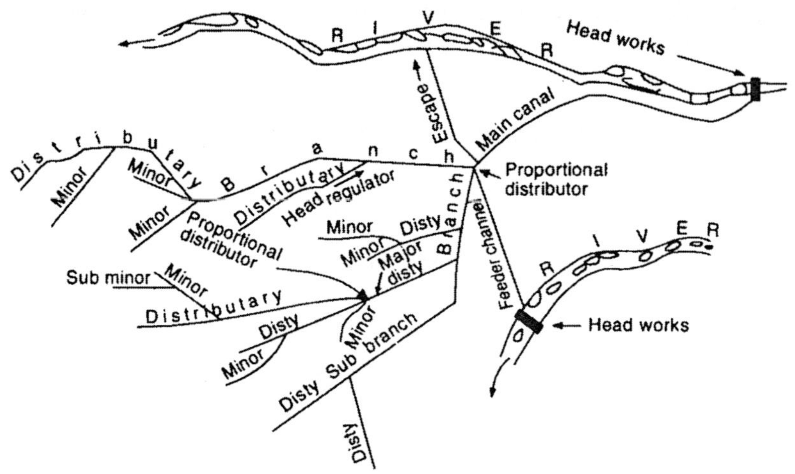

Fig. 5.6 Typical layout of canal system.

The canal taking off from a reservoir gets regulated and dependable water supply. The canal taking off from a diversion work is essentially designed for 75 percent dependability determined by flow duration curves. On the topographical maps of the command area on a scale of 1:15000, the layout of the canal system is marked in consideration of ridges, drainages, habitations, roads, railway lines, soil types etc. The areas that are at higher contours and not commandable are excluded from the culturable commanded area. In consideration of intensity of irrigation, cropping pattern and their water requirement, and duty for various crops, the head discharge of the canal is determined.

5.9 CLASSIFICATION OF CANALS

Classification Based On Soil Through Which Constructed

1. *Alluvial Canal.* The canals constructed through alluvial soils are termed as alluvial canals. The bed and banks of the canal comprise the same material as is transported by them. The alluvial bed is readily silted or scoured. Their design is based on Lacey's regime theory, *i.e.*, neither silting nor scouring. The canals usually take supply from a reservoir, barrage or weir.

2. *Non-alluvial Canals.* The canals constructed through hard soils or disintegrated or fresh rock are termed as non-alluvial canals. Such canals are stable and are designed for higher velocities which the soil can withstand without erosion consistent with ground slope. Manning's, Chezy's and Kutter's formulae are used in their design. These canals usually take their supplies from reservoirs and barrages.

Classification Based on Source of Supply

1. *Permanent Canals.* Also termed as perennial canals. These canals have permanent source of supply, viz, a dam or a barrage to take off and carry assured supply for perennial irrigation. Such canals have permanent works for regulation and distribution of supplies.

2. *Inundation Canals.* These are non-perennial canals which flow only in summer months when the river is high. They have no permanent headworks or barrages across the river. The offtakes of the canals have often to be changed due to the changing river course as discussed in detail in Chapter 2.

Classification Based on Function Served

1. *Feeder Canal.* It is a canal located outside the command area meant primarily to convey water from one source of supply or system to another or within the same system. It is also called link canal.

2. *Conveyance Canal.* It is a canal which carries water either from the headworks or from the feeder canal up to the distribution canal network and also does irrigation.

3. *Distribution Canal.* It is a canal composed of distributaries and minors which have direct outlets to the fields.

4. *Hydel Canal.* It is a canal on which power houses are set up for hydel power generation. A hydel channel can serve as irrigation canal as well.

5. *Navigation Canal.* It is a canal which is primarily used for transportation by water. It may or may not serve as irrigation canal as well.

6. *Multipurpose Canal.* A canal meant for two or more purposes such as irrigation, navigation, power generation, domestic and industrial water supply.

Classification Based on Capacity

1. *Main Canal.* It is the principal channel of a canal system taking off from the headworks or a reservoir or tail reach of a feeder. Also called main line. It is a large capacity channel and usually there is no direct irrigation from it. Small capacity ditch distributaries running parallel to the canal are taken off from the main canal to irrigate the adjoining areas. Main canals deliver supply to branch canal and main distributaries.

2. *Branch Canal.* Branch canals of capacity over 10 cumecs take their supply from the main canal and convey to the distributaries. Very little direct irrigation is done from the branch canals. 'Sub-branch' is a canal which takes off from branch canal but has capacity higher than a distributary.

3. *Major Distributary.* It is a distributing channel designated by virtue of having been taken off from a main canal, branch canal or sub-branch and has discharge capacity lesser than that of a branch canal. It supplies water to distributaries and minors. Irrigation is done through outlets fixed along it.

4. *Distributary.* It is a channel receiving supply from branch canal or major distributary and has discharge less than that of major distributary. Minors take off from it, besides irrigation is done from it through outlets.

5. *Minor.* It is a small capacity channel usually taking its supply from a distributary. It supplies water to watercourses.

6. *Sub Minor.* It is a small capacity channel taking its supply from a minor.

Classification Based on Alignment

1. *Contour Canal.* A contour canal has its alignment conforming generally to the contours of the country traversed by giving such slope along its length as is necessary to produce the required velocity of flow (Fig. 5.7). As the line of flow of surface drainage is at right angles to the ground contours such a channel cuts across the natural drainage lines of the country traversed. The command area is on one side of the canal and as such it irrigates only on one side. As the ground level on the other side is high sometimes bank is not constructed on the high side which lends it the name 'single bank canal'. It has to cross valleys and natural drainages.

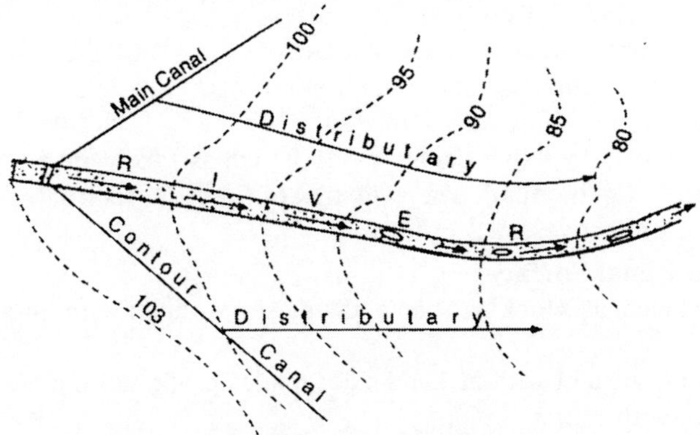

Fig. 5.7 Contour canal.

2. *Watershed (ridge) Canal.* As the name implies it follows the ridge line for the most of its length. A ridge canal has command area on both the sides and as such irrigates on both sides. It does not cross drainage line since drainage is away from the ridge line. Main canal has essentially to be on ridge line except when the ridge line takes a very sharp loop (Fig. 5.8), the canal is aligned straight, off the ridge line, for a short distance and then again mounts on the ridge line. It results in saving in the length of canal. Likewise, the ridge line may have to be abandoned to bypass habitations situated on the watershed.

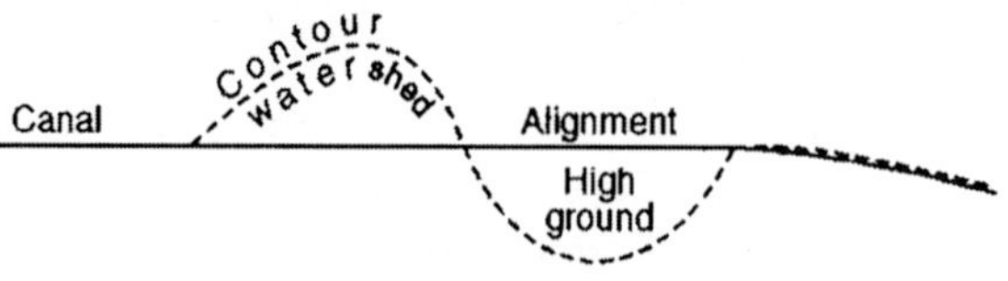

Fig. 5.8 Watershed canal.

3. *Side Slope Canal.* It is a canal which is aligned roughly at right angle to contours of the country (Fig. 5.9) but not on watershed or valley. The canal thus runs roughly parallel to the natural drainage of the country and as such cross drainage works are avoided. The side slope channel has the advantage of not intercepting cross drainage works but its course must follow the shortest route to the nearest valley and such channel shall be along a line of steepest possible slope except in very flat areas. Only the smaller of the distributary should be so aligned. Just like contour canal it irrigates only on one side.

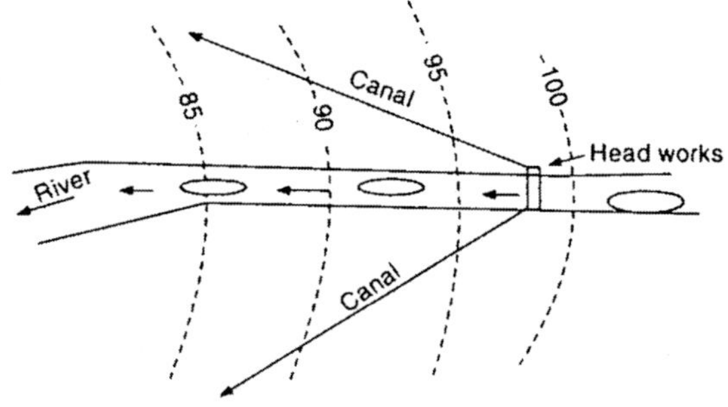

Fig. 5.9 Side slope channel.

4. *Detour Canal.* Detour canal envisages adoption of a lengthy alignment by-passing ridges and valleys and resulting in deep cutting or tunnels in crossing ridges and high filling in crossing valleys so as to have minimum cost of construction of the canal and related hydraulic structures. The lengthy alignment of detour canal (Fig. 5.10) may result in loss of command and increased number of cross drainage works.

Classification Based On Financial Output

1. *Productive Canal.* A productive canal is one the net revenue (gross receipts minus working expenses) derived from which within 10 years after the date of its completion is equivalent to or more than the prevailing rate of interest on the total capital outlay or the prescribed productive rate. The revenue considered for the productivity test accrues mainly from water rates and productivity rates vary from time to time with reference to prevailing rate of interest.

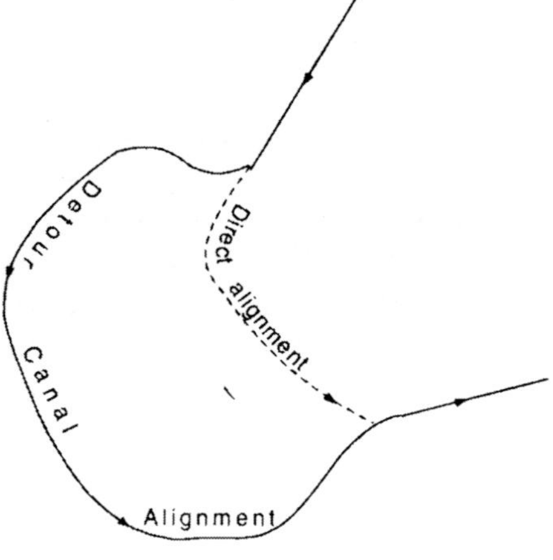

Fig. 5.10 Detour canal.

2. *Protective Canal.* A protective canal is one which is taken up for protection against famine and is financed from famine fund. The yardstick of financial return does not form the sole criterion in sanctioning a protective canal project.

Classification Based on Canal Surface

1. *Lined canal:* It is a canal which has bed and side slopes lined with impervious material to cut down seepage losses.

2. *Unlined canal:* It is a canal with bed and sides made up of natural material through which it traverses; not provided with lining.

5.10 DESIGN PARAMETERS

Working Heads

The full supply level in the canal system, from outlet head to barrage or weir site are fixed in consideration of the minimum working heads as per Table 5.11 and illustrated in Fig. 5.11.

Table 5.11 Minimum working heads

Location in canal system	*Working head (m)*
At outlet head from minor or distributary	0.15
At minor head (*a*) from a distributary	0.25
(*b*) from a major distributary or sub-branch	0.30
At distributary or major distributary head	
(*a*) from sub-branch, or branch canal	0.50
(*b*) from main canal	0.60
At branch head (*a*) from main branch	0.50
(*b*) from main canal	0.60
At main branch head from main canal	0.60
At main canal head from barrage or weir	1.0

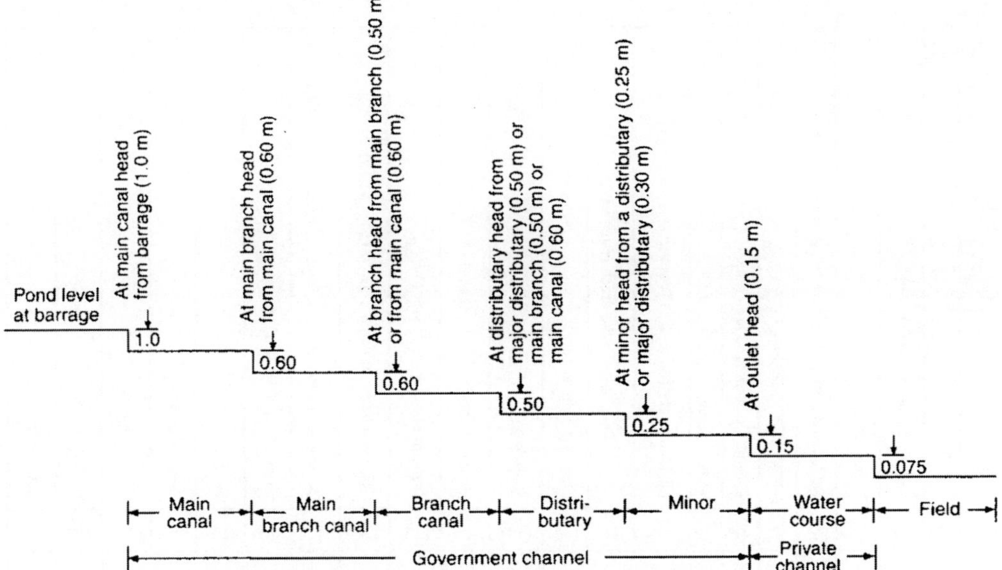

Fig. 5.11 Working head at relevant points in a canal system.

Alignment

Channels are normally aligned along ridge lines and as a general rule all the ridges in the command area form the alignment of distributaries and minors. The distribution system is so aligned that maximum area is served with the least length of channels. The principal factors influencing the alignment of a channel are (*i*) It shall follow ridge line to ensure better command as also few drainages to be intercepted, thereby avoiding cross drainage works and irrigation is done on both sides of the canal, (*ii*) It shall mount ridge line in minimum possible length from the point of offtake from the river, the lowest point, (*iii*) It shall take direct route, as far as possible, to the area to be irrigated to ensure economy, minimum loss of head and transmission losses, (*iv*) When passing near a village or town, it shall pass on lower side even at additional cost, (*v*) It shall run through the heart of the commanded area, as far as possible, to keep the cost of distributing channels minimum, (*vi*) It shall have optimum alignment in the head reach where it is generally in deep cutting and intercepted by large drainages making it expensive, (*vii*) It shall run straight if watershed line makes a sharp loop, skirts abadies, religious places and other valuable properties, (*viii*) The number of cross drainage works shall be the minimum, (*ix*) It shall avoid the country where soil is sandy, alkaline, waterlogged, perched watertable, and hard to excavate. It is expeditious and economical to excavate in the best soil formation, (*x*) The curves and bends are minimum and of optimum radius, (*xi*) Minimum the length, or shorter the alignment, the lesser are the absorption losses and lower the maintenance charges, (*xii*) It shall normally be so aligned that the angle of crossing with the metalled roads and railway lines is 90° or not sharper than 60°; the sharpest angle of crossing is limited to 45° in exceptional cases, (*xiii*) With respect to drainage, it shall cross where the stream is straight, with minimum catchment area upstream and minimum waterway, (*xiv*) Aligned at balancing depth. Heavy filling not only increases cost and seepage losses but also is prone to breaches. Deep cutting is also avoided, (*xv*) It shall not unduly upset the well irrigation or canal irrigation system in the area if any irrigation exists, (*xvi*) Length of water courses is not more than 3 km, and (*xvii*) Blind length of canal that is where it does not have outlets to irrigate adjoining area should be the minimum. These ideal conditions may not occur in practice and the optimum alignment may mean a compromise between various considerations.

Typical Capacity Statement

The capacity of a canal is worked out on the basis of culturable commanded area, water allowance and transmission losses. Typical capacity statement is given as under:

Sl. No.	Reach		Length of reach (m) (3)-(2)	Name of offtake	Reduced distance of offtake (m) and side L or R	Gross commanded area (ha)	Culturable command area (ha)	Water allowance (per ha of CCA)	Discharge at offtake head (cumecs) (8) × (9)	Discharge in the reach (cumec) Σ (10)	Wetted perimeter (m)	Wetted area (m^2)	Rate of transmission losses ($m^3/s/10^6\ m^2$)	Transmission losses (13) × (14)	Total discharge in reach (m^3/s) (11) + (15)	Discharge at head of reach (cumec) [discharge at tail + (16)]	Slope of reach m/1000 m	N	Bed width (m)	F.S. depth (m)	Velocity (m/s)	Discharge (cumecs)	Remarks
	From	To																					
(1)	(2)	(3)	(4)	(5)	(6)	(7)	(8)	(9)	(10)	(11)	(12)	(13)	(14)	(15)	(16)	(17)	(18)	(19)	(20)	(21)	(22)	(23)	(24)

Curves in Canals

The canals are preferably aligned straight to run on the shortest possible route from the source of supply to the command area but curves are inevitable. The radius of curvature suitable for unlined canals is influenced by various factors such as size or capacity of the canal, velocity of flow, characteristics of the soil and dimensional elements of the canal section. In a canal the curve provided is of the largest possible radius to ensure streamlined flow. It is fixed in considerations of water depth, water surface width, bed width and discharge. The minimum radius is taken as 150 times the water depth. In Gurjarat, Maharashtra and Karnataka States the radius is taken as 20 times the bed width and in Tamil Nadu 10 times and 15 times the bed width for canal with discharge less than 0.7 cumec and above 0.7 cumec respectively. IS : 5968–1987 recommended radii of curvature, related to discharge capacity, are as per Table 5.12.

Table 5.12 Radii of curves for canals

Discharge (cumecs)	*Minimum radius (m)*	*Discharge (cumecs)*	*Minimum radius (m)*
Up to 0.3	90	15–30	600
0.3–3	150	30–80	1000
3–15	300	80 & above	1500

Longitudinal Slope

The longitudinal water surface slope or slopes is defined as the inclination of the water surface expressed as change of elevation per unit of horizontal length of the canal. The longitudinal slope of a canal is fixed primarily from the considerations of the country slope consistent with economy and command such as (*i*) It may generally approximate the ground slope unless command considerations demand flatter slope, (*ii*) From the standpoint of command, the slope is determined mainly by the command levels required at the head of the offtaking channels and also by the levels required at the head of outlets, (*iii*) A steeper slope is required for a smaller discharge at the tail end and is gradually made flatter as the discharge increases towards the head of the canal, (*iv*) The requirement of providing maximum possible irrigation indicates flatter slope guided by the minimum permissible velocity, (*v*) It is also controlled to some extent by critical velocity ratio, *i.e.*, 1.1 near the head and 0.8 near tail end of the canal as also Lacey's f 1.0 at the head and 0.8 at the tail as adopted in Bhakra Nangal canal system, and (*vi*) In general, the slope required to obtain maximum permissible velocity makes the construction economical.

Coefficient of Roughness

It may be defined as a factor in Kutter, Manning, Bazin and other formulae expressing the character of a channel as effecting the friction slope of water flowing therein. The value of n is not constant but varies within wide ranges depending on several factors, such as surface roughness, vegetation, channel irregularity, suspended material and bed load, stage and discharge, silting and scouring and channel alignment, etc. The recommended values of n are as per Table 5.13.

Table 5.13 Values of n

Channel condition	*Recommended value*	*CWPC practice*	
		Discharge (cumecs)	*n*
Good condition	0.02		
Moderate condition	0.025	up to 0.15	0.030
Bad condition	0.027	0.15-1.5	0.025
Very bad condition	0.0295	1.5-14	0.0225
Normal section	0.025	above 14	0.020
Hard murum and boulders	0.03		
Soft rock and rugged hard rock	0.035-0.04		

Maximum and Minimum Permissible Velocities

Permissible velocity is defined as the highest velocity at which water may be carried safely in a channel. Maximum permissible velocity is specified so that the discharge is passed within the smallest section without causing erosion. The maximum permissible velocities adopted in various States of India, depending on the soil characteristics, are as per Table 5.14.

Table 5.14 Maximum permissible velocity

Soil type	*Velocity (m/s)*	*CBIP Practice Velocity (m/s)*
Ordinary soil	0.6-1.0	0.6-0.9
Very light loose sand to average sand	3-0.6	0.3-0.6
Sandy loam, black cotton and similar	0.5-1.0	0.6-0.8
Murum, hard	0.75-1.25	0.9-1.0
Gravel	1.25	1.25
Rock	1.25-2.0	1.5

Stipulation of minimum velocity is considered desirable. Minimum velocity is taken as a fraction of maximum permissible velocity, say 0.8 V_{max}. The minimum non-silting and non-weed velocity is considered as 0.3 to 0.5 m/sec. Objections to the slack velocity are (*i*) It tends to encourage weed growth, (*ii*) It increases transmission losses through the canal, and (*iii*) It results in increased cross sectional area and hence uneconomical section.

Side Slopes

Side slope is an important parameter in the design and construction of an irrigation channel. Economy dictates provision of steeper side slopes. The side slopes depend upon (*a*) nature of soil in which the channel is excavated, (*b*) depth of cutting, and (*c*) height of filling of the embankment. In alluvial soils these are generally fixed in consideration of the factors (*i*) sudden drawdown in the canal for inner slopes and (*ii*) canal running full and there is plenty of rainfall for outer slopes.

The designed inner side slopes and those adopted at the time of construction are different in alluvial canals. It is generally assumed that the channel shall attain an inner slope of ½ :1, on the consideration that the sides get silted up to attain this slope, although during construction the channel section has a slope of 1:1 in cutting (excavated slope) or 1.5:1 in filling (Fig. 5.12(*a*) and 5.12. (*b*).

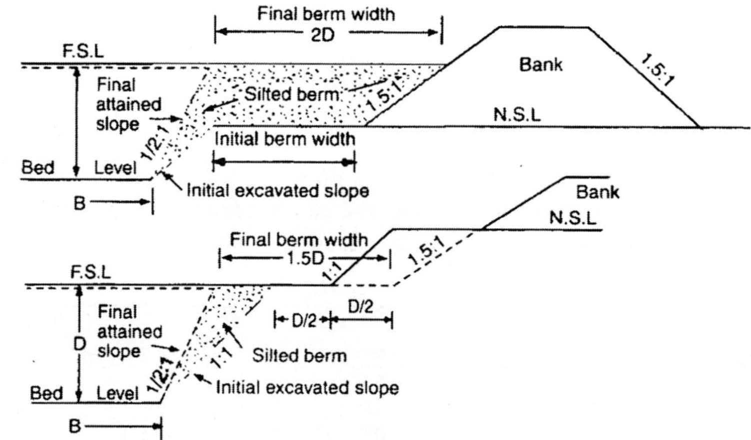

Fig. 5.12 (*a*) Initial excavated slope and final attained slope.

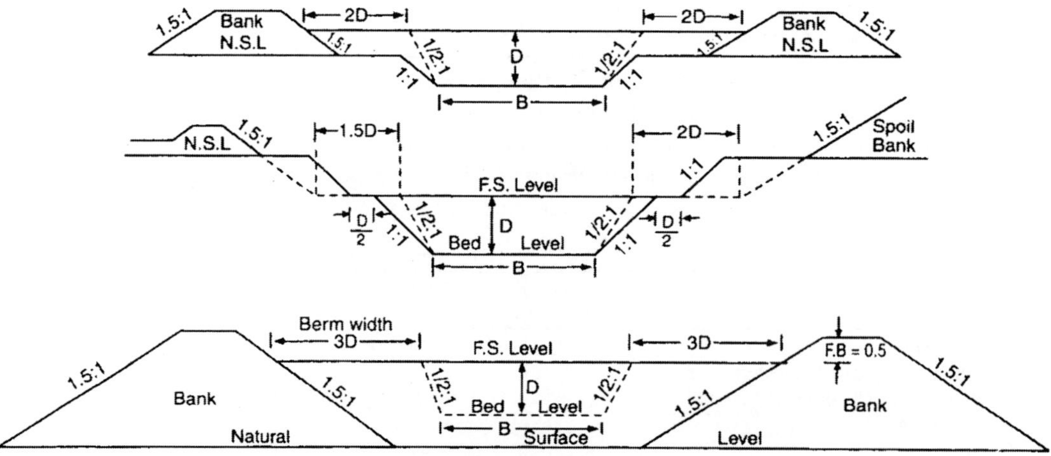

Fig. 5.12 (*b*) Berm

Side Slope of Banks

The side slope of the banks is flatter when in embankment and steeper when in cutting. It is also made flatter when soil is non-cohesive than when it is cohesive. Generally, the banks have a side slope of 1.5:1 in filling. For filling over 3 m, a 1.5 m wide berm is provided at intervals of 3m instead of flattening the slope, 1.5:1 in cutting through sand, and 1:1 in cutting through loamy soil. Minimum bank side slopes are as Table 5.15.

Table 5.15 Recommanded minimum sides slope of channels

Nature of soil	*Side slope*	
	Filling	*Cutting*
Very light loose sand to average sandy soil	2:1 to 3:1	1.5:1 to 2:1
Sandy loam, loamy soil and similar soil	2:1	1:1 to 1.5:1
Sandy soil or gravel	2:1	1:1 to 2:1
Soft clay and alluvial soils	1.5:1 to 2:1 (in very high filling)	1:1
Murum, hard soil, etc.	1.5:1	0.75:1 to 1.5:1
Rock	1.5:1	0.25:1 to 0.5:1

Berm Width

Berm is defined as the space initially left at the FS level between the upper edge of water section and the inner toe of bank along the canal. The full berm width is, however, reckoned from silted berm edge to the inner toe of the bank at designed full supply level of the canal. The width of berm, otherwise variable depending on the position of the canal bed level and natural ground level, is kept in multiples of channel depth so that the bed line and bank line remains parallel. The purposes served by berm are (*i*) They reduce bank loads which may otherwise cause sloughing of earth into the canal section, (*ii*) Berms developed from fine silt deposition (being impervious) provide a fairly impervious lining which reduces seepage losses, (*iii*) Make channel to flow in incoherent unlimited alluvium of the material similar to that transported by water, (*iv*) Provide additional strength to the bank, besides shifting the flow away from the bank on development of berm. They also protect the bank from erosion due to wave action, (*v*) Origin of the hydraulic gradient line is shifted advantageously, on berm formation, by a distance equal to the berm width, as shown in Fig. 5.13, (*vi*) To provide wider waterway above F.S.L. to limit the height of rise of water at the time of positive fluctuations, (*vii*) Break the flow of rain water down the bank slope and thus prevent guttering, (*viii*) Provide scope for widening of the canal section within the designed bank width, (*ix*) Provide earth for borrowing to close a breach, (*x*) Make driving along the canal bank more safe, (*xi*) Help attain regime conditions, and (*xii*) Reduce maintenance cost of banks.

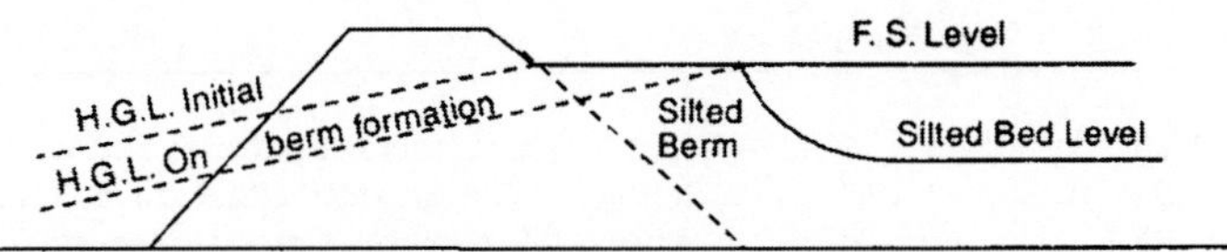

Fig. 5.13 H.G.L. through silted berm.

The channel section is usually excavated to 1:1 slope, but it is assumed that the final silted slope in the regime section will be ½ : 1. The berm width is reckoned from the edge of ½ : 1 final slope of the section. Generally the berm width is kept as 0.5 m + 0.25 of width of combined side slopes of cutting and embankment. In practice, berm width of 3D is provided when a canal is in filling, 2D when in cutting, completely or partly. In distributaries partly in cutting and partly in filling 1.5D berm is provided on left side and 2D on right side (spoil bank side).

Free Board

Free board is defined as the vertical distance between the designed full supply level in the canal and the top level of canal banks. It is provided to allow for wave action, floating debris, or any other condition or emergency without over-topping the banks of the channel. The height of the dowel portion is not considered for free board purposes. Free board is fixed in consideration of (*a*) canal size or capacity, (*b*) water surface fluctuations caused by wind action, regulation, etc. (*c*) soil characteristics, (*d*) hydraulic gradient, (*e*) location of the canal, (*f*) velocity of water flow in the canal, (*g*) storm water inflow, (*h*) road traffic requirement, and (*i*) availability of excavated material, etc. A minimum free board of $(0.3 + 0.1D)$ metres, where D is FS depth (m), is provided.

IS:7112–1973 recommended free board is 0.75 m in main lines and branches, capacity above 10 cumecs, and 0.5 m in distributaries and minors.

Dowel

Dowel is a mini-guard bank constructed on canal side of the inspection bank (Figs. 5.14(*a*) and 5.14 (*b*)). It is provided (*i*) To act as a wheel guard for the safety of the vehicular traffic plying on the bank, (*ii*) To prevent cutting up of the slope of the bank due to rain, and (*iii*) To keep the limits of the inspection bank and the inside slope of the bank discernible. The usual section of dowel is 0.5 m as top width and height above bank level and side slopes 1.5:1. Dowels are provided on both sides of the banks of lined main canal and branches.

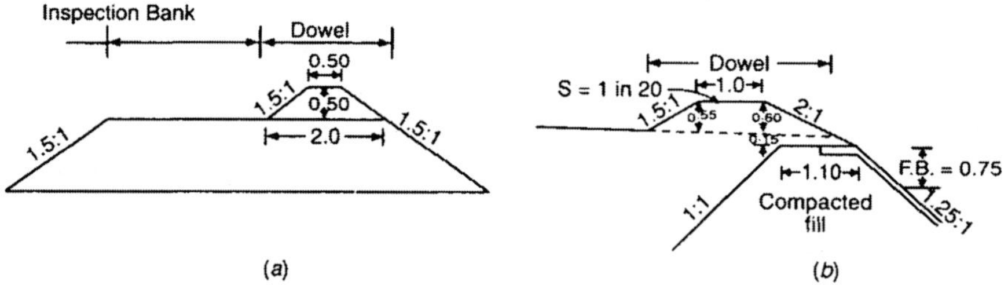

Fig. 5.14 (*a*) Dowel in unlined canal, (*b*) in lined canal.

Bank Width

Banks retain the canal supply till the formation of berms and as such they withstand water pressure above ground level. For filling up to 8 m design for bank width is not made but is fixed arbitrarily. The width of bank is fixed in consideration of (*a*) canal size or capacity, (*b*) soil characteristics, (*c*) inspection road requirement, (*d*) availability of excavated material and (*e*) location of the canal. The suggested formula for bank width is $B_w = D + 0.3\,(B/D)^{1/2}$, where, B_w = bank width (m), D = FS depth (m), and B = bed width (m).

Bank widths generally adopted are given in Table 5.16.

Table 5.16 Bank widths

Canal capacity (cumecs)	*Bank width (m)*		*Remarks*
	Left bank	*Right bank*	
Distributary			
Up to 1.4	1.25	1.25	(*i*) In minors and distributaries 5 m wide boundary road (Fig. 5.17) is provided at the ground level on left side for inspection purposes. Combining together of canal bank and boundary road is considered where land is costly.
1.4 – 2.8	1.5	1.5	
2.8 – 5.6	1.75	1.75	
5.6 – 9.9	2.5	2.5	
Branch			
10 – 15	6.0	2.5	
15 – 30	7.0	3.5	
30 and above	8.0	5.0	(*ii*) In main canal and branches, above 10 cumecs capacity, left bank is made motorable for inspection purposes. Boundary road at ground level is provided on right side. (Fig. 5.18).
C.W.C. Practice and IS : 7112–1973 Recommendation			
0.15 – 7.5	5.0	1.5	
7.5 – 10.0	5.0	2.5	
10.0 – 15.0	6.0	2.5	
15.0 – 30.0	7.0	3.5	
30 and above	8.0	5.0	

Hydraulic Gradient

Also called *saturation line.* It is defined as the line across the banks on either side of a canal in filling or partly in filling up to which the banks get saturated after the canal has been running for sometime. When water runs against fill banks the lines of saturation bend downwards from the water surface through the embankment material. The rate of bend or the hydraulic gradient depends mainly on the characteristics and relative placement of the different types of material in the embankment. For embankments more than 5 m in height, the true position of the saturation line is ascertained by laboratory tests and stability of the slope is checked. In practice, the HGL is arbitrarily considered as a straight line and the empirical hydraulic gradients (horizontal to vertical) as adopted for banks of less than 5 m height are 4:1 for good clayey soils, silty clay and clayey gravels, 5:1 for average clayey soils, clayey sand, silty gravel, and 6:1 for bad loamy soil, average loamy soil, very fine sand, or silt fine to medium, sand poorly graded·silty.

Counter Berm or Pushta

The hydraulic gradient line may fall out of the bank section in the case of banks in filling or partly in filling. The extra layer of earth (sloping, horizontal or broken horizontal steps) provided near the toe of outer slope of banks to cover the saturation line when it cuts the slope above the ground level is defined as counter berm or pushta. The hydraulic gradient line is provided a cover of 0.3 m. In very high filling, a second pushta is provided so as to cover HGL (Fig. 5.15). Usually in the case of distibutaries and minors, sloping pushta parallel to HGL is provided (Fig. 5.17). For embankments more than 5 m in height, the true position of saturation line shall be worked out and stability of slope checked as in earthen dams.

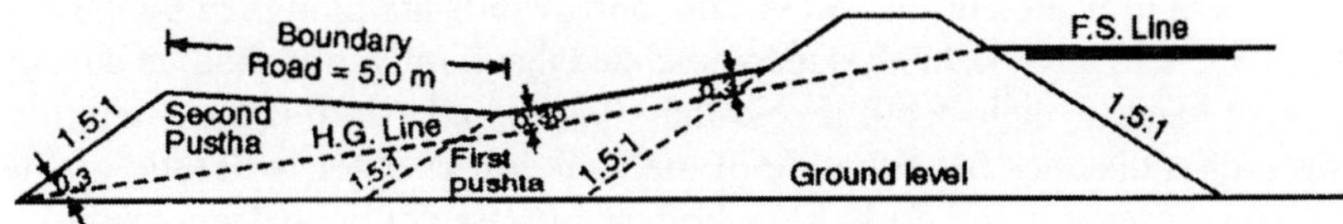

Fig. 5.15 Second pushta.

Spoil Bank

Also called *waste bank.* It is an auxiliary bank composed of waste earth, *i.e.,* disposal of the surplus excavated material from the canal section. When the quantity of surplus earth is small, it is better disposed off in raising the bank. However, in deep cutting entailing lot of surplus excavated earth, it is disposed off, behind the canal bank, in the form of spoil bank in the sequence shown in Fig. 5.16. The section of the spoil bank is top width 1.5 m minimum, and height maximum 4 m by manual labour, 6 m by machinery and side slope 1.5:1.

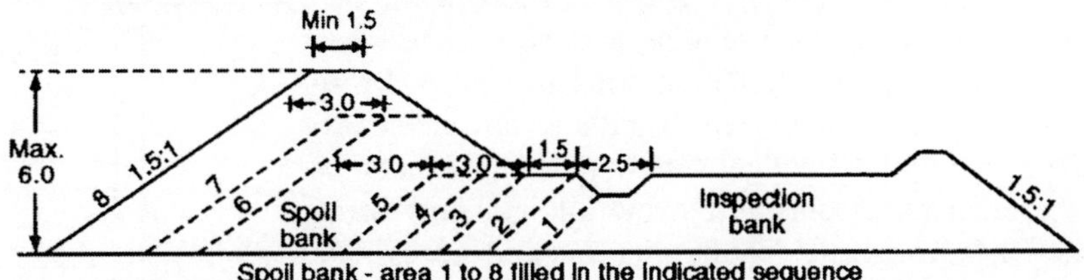

Fig. 5.16 Sequence of filling spoil bank.

The top of the spoil is dressed with a slight outward slope. A space of 4 m in the case of branch canals and 3 m in the case of distributaries is left between the toe of the spoil bank and the outer edge of the inspection bank, out of which 1.5 m is taken up by the line of trees and 1.5 m by side drain (see details in Fig. 5.18). 3 m wide gaps between the spoils are left at 75 m interval for the disposal of the storm water coming through the drain. In the case of small distributaries and minors, it is suggested to provide spoil bank on right side only in order to economize on land width and lead (Fig. 5.17).

Boundary Road

Boundary road is defined as a road on the ground surface adjacent to the canal bank. It serves for inspection of the distributary system (Fig. 5.17). The boundary road is 5 m wide. It is usually provided along the left side in the case of distributaries and minors and at right side (the left bank being already an inspection bank) on main canals and branches (Fig. 5.18).

Land Width

It is defined as the width of land occupied by or acquired for a canal section including its inside berms, bank, spoil bank, if any (Figs. 5.17. and 5.18). It is the permanent land width the ownership of which is transferred to the State Irrigation department. The land width constitutes the distance between the outer edge of the drain at one end and additional land up to 1.5 m beyond the toe of the bank at other end (Figs. 5.17 and 5.18) so that the land width is in full multiples of a local measure called *karam* (1.68 m). Stones installed on either side of the canal mark the boundary of land permanently acquired for the canal.

Borrow Pits

Borrow pit is defined as a pit excavated for obtaining embankment material. When the quantity of earth filling required for building banks to the designed section exceed the excavated material from the canal section, the extra earth is obtained from borrow pits. Borrow pits 0.3 m deep and 3 m away from the canal boundary are generally laid to borrow earth from outside for making good the earth filling. Where deeper borrow pits are insisted upon by the land owners they may follow the pattern indicated in Fig. 5.19. The borrow pits are ploughed by the department to make the land cultivable IS:4701–1968 has recommended that borrow pits shall be dug within 5 m of the toe of the canal bank, if their depth is less than 0.5 m, and 10 m, if their depth is more than 0.5 m or within such a distance from the toe of the bank where a 4:1 hydraulic gradient cuts the ground surface, whichever is more. Further the borrow pits are not to be deeper than 1 m and larger than 25 m in length. A clear distance of 1 m is left between the pits.

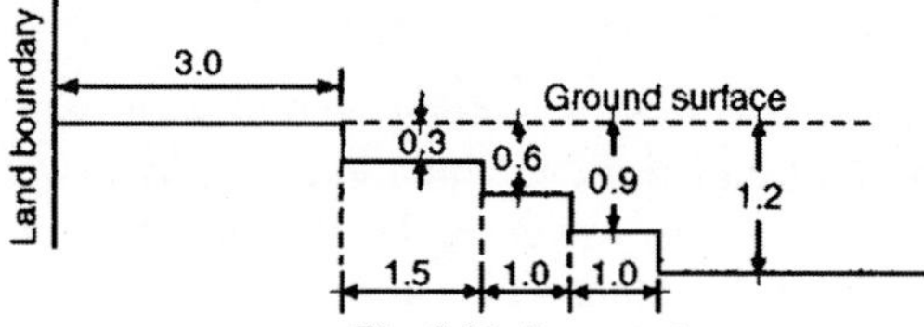

Fig. 5.19 Borrow pits.

The borrow pits are also laid sometimes in the bed of the canal up to a maximum depth of 1 m and 1 m short of the bed width of the canal and 10 m long leaving a ridge 3 m between two adjacent borrows (Fig. 5.20). The internal borrow pits provide economical way for borrowing earth and also obviate necessity for outside borrow pits. They get silted up during the course of time. On the other hand, outside borrow pits not only bring into disuse the land brought under the pits but also necessitate payment of compensation for the borrowed earth besides ploughing off them by the department to make the land cultivable.

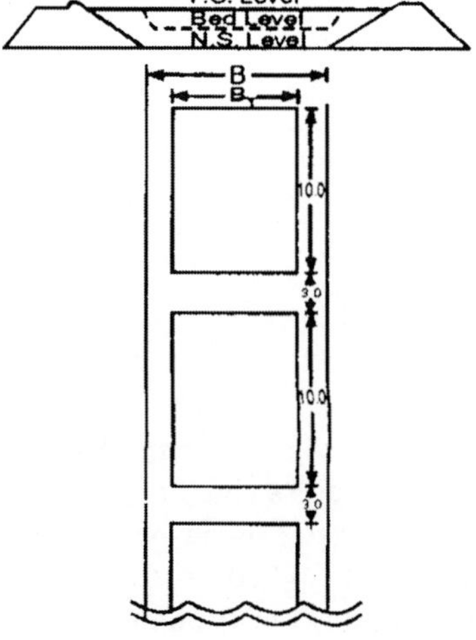

Fig. 5.20 Internal borrow pits.

Typical Cross Sections

Figs. 5.17 and 5.18 show the typical cross sections of unlined distributaries and unlined branch canals respectively.

5.11 TRANSMISSION LOSSES IN CANALS

While transporting water in the irrigation canal part of it is lost as seepage to subsoil and some part as evaporation to the

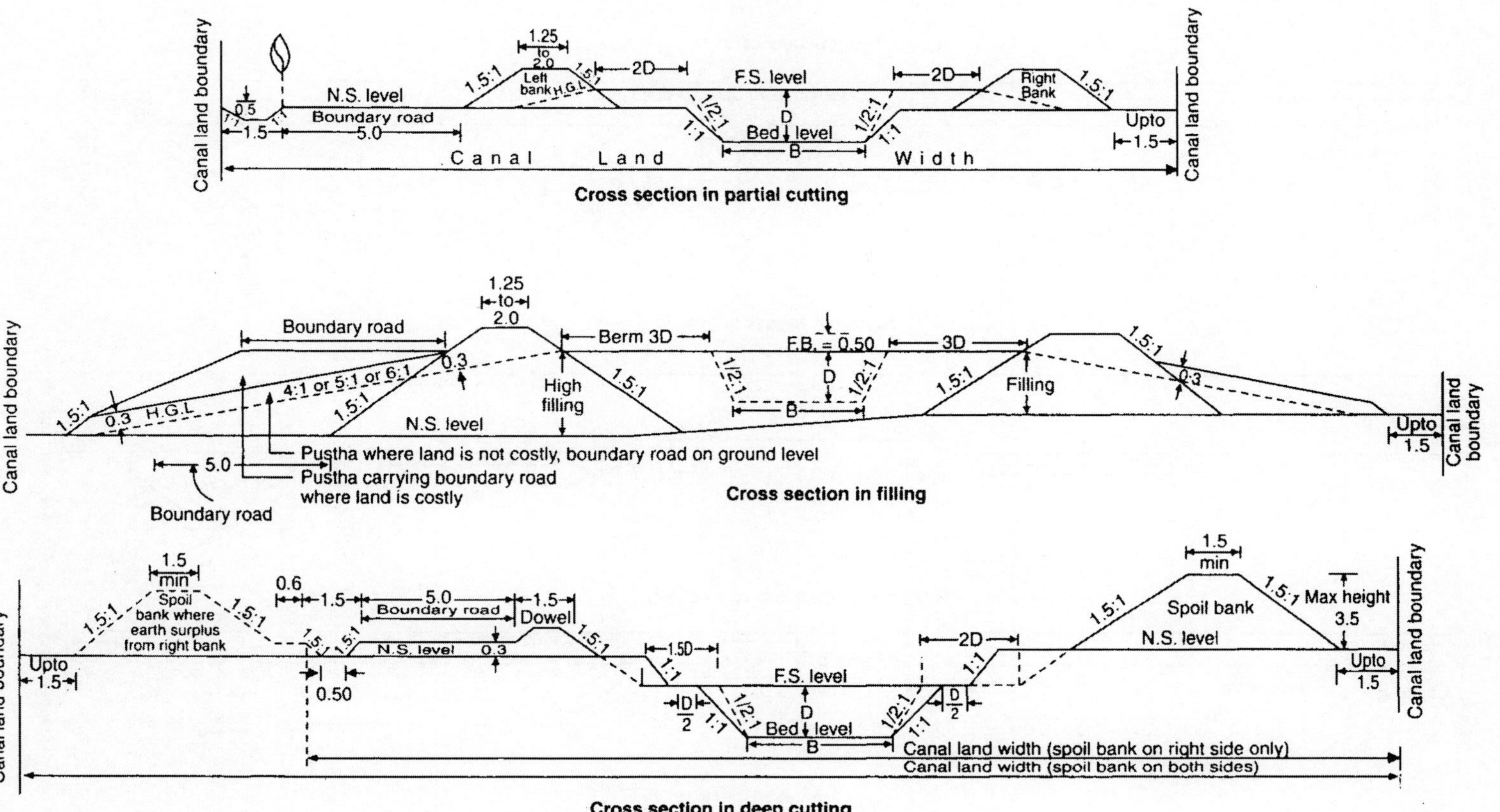

Fig. 5.17 Typical cross sections of unlined distributaries and minors.

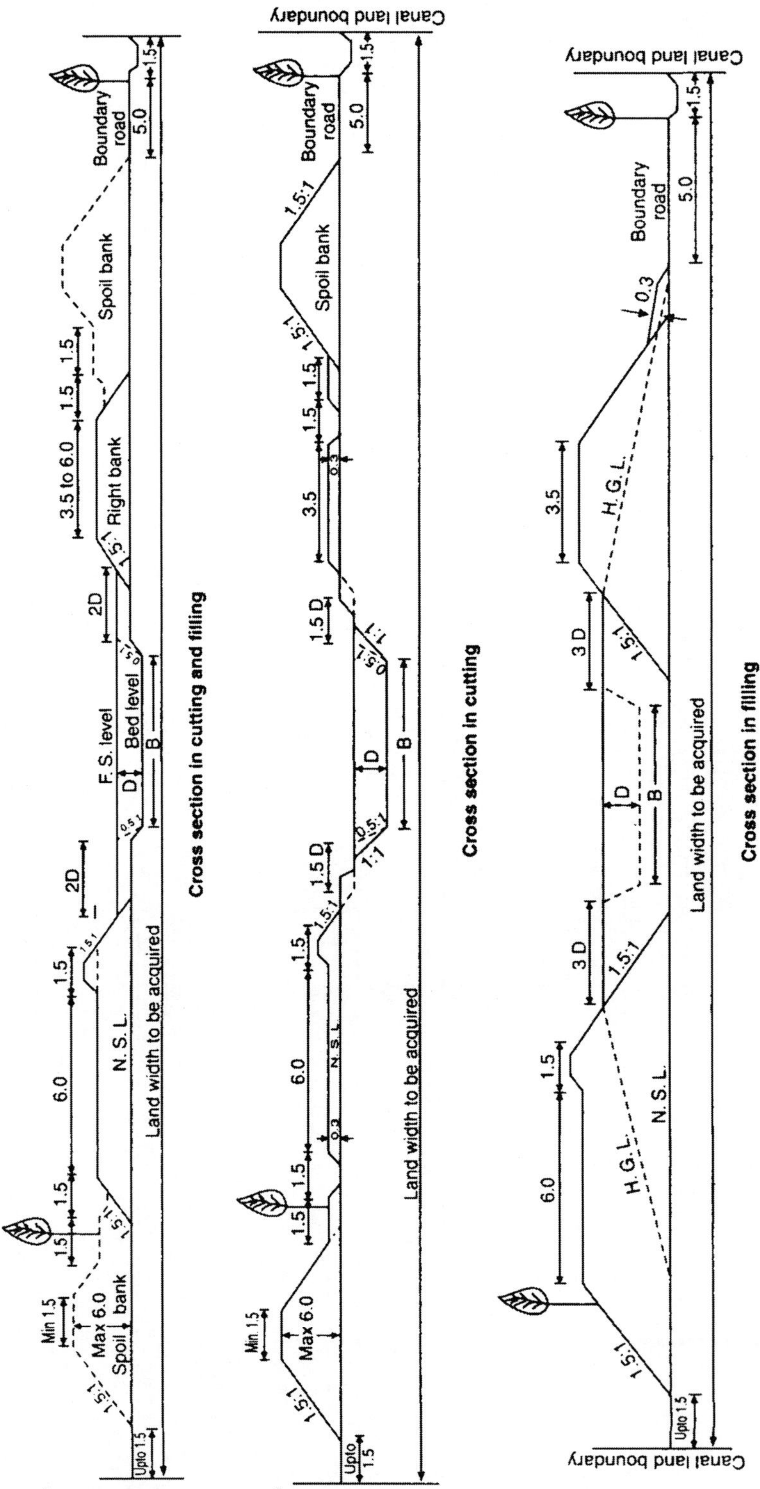

Fig. 5.18 Typical cross sections of unlined branch canals.

atmosphere. The loss of water by seepage from unlined canals in India generally varies from 0.3 to 7 cumecs/$10^6 m^2$ of wetted perimeter. Besides, large quantities of water are lost by seepage and evaporation from distributaries, watercourses and field channels. It is considered that in Indo-Gangetic Plains about 45% of the water diverted at the canal headworks is lost by seepage in the distribution system before it reaches the farmer's field. Losses in the watercourse and in the field are substantial, as shown below:

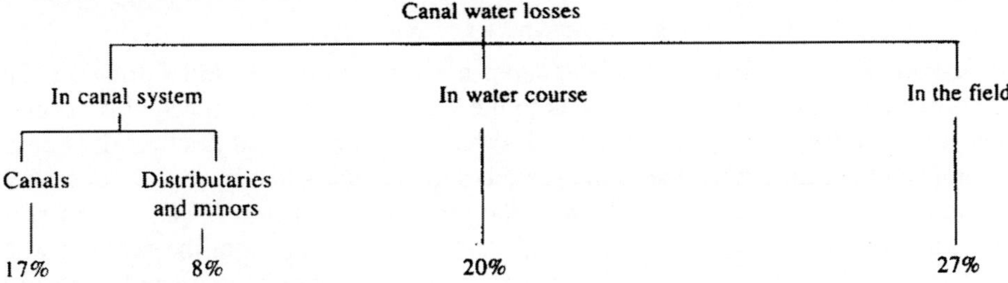

Causes of water Losses in Canals

Water losses in the canals take place due to evaporation, seepage and transpiration.

1. *Evaporation Loss.* Loss of irrigation water in a canal system depends upon:

(*a*) *Climatic factors* — Temperature, humidity and wind velocity. Evaporation varies directly with temperature and wind velocity and inversely with humidity. Maximum loss on this account is in summer months of May and June. Evaporation loss is more in dry and desert areas having low rainfall, (*b*) *Canal factors* — Evaporation varies directly as water surface area and inversely as water depth and velocity of flow in the canal. Losses are maximum in unlined canals due to wider water surface area, shallower water depth and low velocity but are the minimum in lined canals. Evaporation losses due to climatic factors cannot be helped but those due to canal factors are minimized by lining the canals which imply minimum water surface area, maximum full supply depth and maximum permissible velocity. The evaporation losses in a canal are insignificant, compared to seepage loss, being 1.5 to 2 percent of the total losses. Evaporation losses in canals on an average are less than 0.25 to 1.0 percent of the flow.

2. *Seepage Loss.* Seepage loss constitutes the major loss of irrigation water from a canal system and as such usually the seepage losses are considered to be the transmission losses. The two different conditions of seepage loss, depending on the position of watertable, are due to absorption, and percolation.

(*a*) *Absorption loss:* There does not exist a continuous and constant flow of water from the canal to the ground water reservoir, where the watertable is at a considerable depth below the ground surface, because an unsaturated zone intervenes. In the soil formation under a canal there exist two saturated soil zones separated by a layer of soil having decreasing moisture content from the two ends towards the centre and through which no water flows (Fig. 5.21). (*i*) *Upper zone*—It is a small but completely saturated zone which exists around the canal section saturated by the seepage from the canal, and (*ii*) *Lower zone*— It is the zone which exists above the watertable saturated by capillary action.

The seepage water from the canal first fills the pores of the soil around it but subsequently the loss is balanced partly by the evaporation and transpiration of seepage water which re-emerges at the ground surface by capillary action. The rate of loss, in this case, is independent of seepage head but is dependent upon the depth of water in the canal as the saturation line in a deeper canal has large area close to the surface resulting in increased evaporation. In due course of time, seepage loss is considerably reduced as a comparatively impervious lining is formed by the deposition of fine silt carried by the canal water. The absorption loss is more when a canal is in heavy cutting reach. Other conditions being constant, absorption loss varies directly as the wetted perimeter and depth of channel.

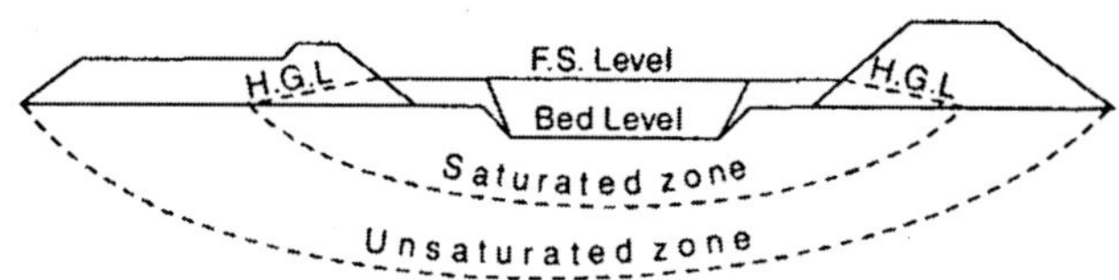

Fig. 5.21 Absorption loss from a canal.

(*b*) *Percolation loss:* Water percolates from the canal to the subsoil formation. There is a zone of continuous saturation from the canal to the watertable and a direct flow is set up. Most of the water lost from the canal joins the ground water reservoir, a small fraction rises up by capillary action from the saturation zone to the ground surface and is lost by evaporation and transpiration. The loss of water varies directly with the total seepage head, *H i.e.*, the difference of a water surface level in the canal and watertable level but is independent of the depth of water in the canal (Fig. 5.22). The seepage loss through percolation is very high compared to absorption

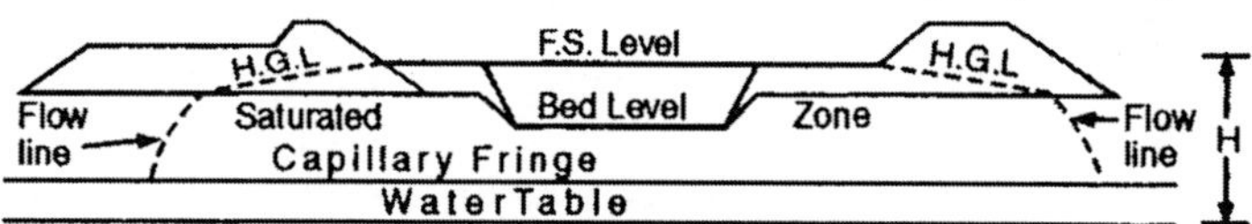

Fig. 5.22 Percolation loss from a canal.

loss and it depends on various factors such as (*i*) Permeability of the soil formation through which the canal passes-greater the permeability of the soil comprising the bed and banks, higher are the losses. Silt in canal water tend to make the wetted perimeter impermeable, (*ii*) Position of subsoil watertable and drainage condition of subsoil–the higher the water table, more are the losses, (*iii*) Condition of canal-seepage loss reduces with age of canal, (*iv*) Lined section-seepage losses are minimum from a lined canal, and (*v*) Water depth-greater the water depth in canal, higher are the losses. In practice, evaporation and seepage losses from canals are considered together.

3. *Transpiration Loss:* It occurs through vegetation and weeds along canal bed/berm but is of negligible amount. Canal section is required to be clear of vegetation and weeds.

Methods of Determination of Water Losses

The two methods for determination of water losses from a canal are ponding method, and inflow and outflow method.

Ponding Method

In this method, as the name implies, seepage loss is determined by the loss of water from a ponded reach of a canal of capacity of about 15 cumecs. The reach of the canal selected shall be straight for a length of about 30 to 40 times of bed width/water depth ratio of 20 to 5 as per IS:9452 (Part I): 1993 and in which there is no outflow of water. Two earthen embankments are constructed across the canal at the two ends of the straight reach to impound water for determining seepage losses. Evidently such a test can be performed only during a long closure. The test is better made during dry weather so that there is no element of addition of water due to rainfall. Two procedures to compute the rate of seepage losses by this method are:

1. *Falling Level Pond:* The pond is filled and gauges are installed to measure the rate of drop in the water surface in the pond. From the rate of loss of water (duly adjusted for evaporation loss) from the ponded reach and the dimension of the wetted perimeter of the pond, seepage losses are determined.

2. *Constant Level Pond:* Water surface level in the pond is maintained at constant level by adding water to the pond. From the volume of water added and wetted perimeter of the pond, seepage losses are determined.

Merits: (*i*) Suitable even for large capacity canal, (*ii*) Particularly suitable for measuring small seepage losses, and (*iii*) Reasonably accurate results can be obtained which may be used as a standard of comparison for the seepage losses determined by other methods.

Limitations: (*i*) Normal running of the canal is interrupted, cross bunds constructed have to be removed, (*ii*) Actual running conditions of the canal are not simulated; losses determined from standing pool with tendency of suspended material to settle down and seal the wetted perimeter of the ponded reach may be different from the losses taking place from the flowing water in the canal, (*iii*) There may be variation in nature of material of the bed of canal in this reach and the rest of the canal, (*iv*) There may be variation in ground watertable throughout the length of canal from that in the ponded reach, and (*v*) Arrangement for filling the pond and maintaining at constant level (by constant level method) may be difficult, (*vi*) Can be applied to smaller reaches of a canal, and (*vii*) The losses worked out pertain to water table, temperature and humidity conditions in the reach at the time of test.

Inflow and Outflow Method

1. *Steady State Method*: The method consists in measuring the water that flows into and out of the selected section of canal. The difference of water quantities flowing into and out of the reach is the seepage. Evaporation and precipitation is also taken into consideration. In this method a long reach, say about 100 kilometres long, is selected for observation. The outlets in the reach are preferably closed or if essential some of them may be kept open to constant opening. Known discharge, which is kept constant during the test, is let into the test reach. At steady state condition, indicated by steady gauges fixed at various sites, discharge observations are made to measure the quantity of water entering at the head and leaving at the tail of the reach. Discharges of the outlets as are kept open are also measured. Water level at various sites in the reach for which gauge-discharge relationship is known are recorded at an interval of three to six hours. The difference between inflow and outflow, adjusted for evaporation from the water surface of the canal, and precipitation, if any, constitutes the water loss in the reach. The average loss in m^3/sec per unit area of wetted perimeter or water loss as a percentage of the steady discharge in the reach is determined.

Advantages: (*i*) The observations are made in actual running condition of the canal, and (*ii*) Closure of the canal is not required to be made and as such irrigation and power generation, if any, in the canal system downstream is not affected.

Limitations: (*i*) Not an accurate method, (*ii*) Steady state condition may be attained after a long time, (*iii*) Outflow and leakages through the offtakes and the outlets kept open have to be determined, (*iv*) Losses determined represent those occurring in the test reach for that temperature, humidity, water table and bed material conditions, and (*vi*) Most of the outlets in the canal have to be closed.

2. *Dynamic Method*: The World Bank dynamic method is more precise as it envisages measurement of losses in running conditions of canal. Fig. 5.23 depicts a typical layout of the canal system covered by a distributary alongwith its minors and sub-minors selected for the purpose. At all the head regulators measuring devices are installed. Automatic gauge recorders are installed near all the head regulators and outlet points to record fluctuations of water level in the whole canal system. Measurements of canal discharges is made in the distributary and minors with the help of currents meters. The observations are made over a period of 7 days or so. The canal losses are determined from such actual measurements of the discharges in various canals in the whole system thus selected.

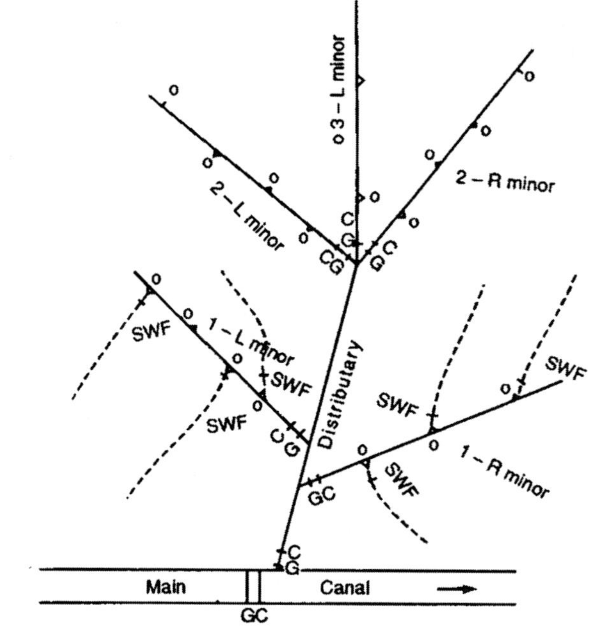

D – Distributary
O – Outlet
G – Automatic gauge recorder
C – Current meter site
SWF – Standing wave flume

Fig. 5.23 Measurement of canal losses by dynamic method.

3. *Special Method*: These are tracer technique, electrical logging or resistivity measurement, piezometric surveys, remote sensing and seepage meter. The seepage meter cannot be used where the canal has rocky bottom or heavy weed growth. The tracer technique is best suited to canals in homogeneous and isotropic formation. Seepage meter is a convenient tool for measuring seepage rates in a canal or ponds. The accurate measurements of rate of water loss from an isolated surface area of a canal requires that the water head acting on that area is identical to that acting simultaneously on the adjacent surface outside. The seepage meter (Fig. 5.24), duplicates the outside head conditions inside the meter and determines the flow from the seepage meter as a direct measure of the seepage from the area enclosed by a cyclinder. IS: 9452 (Part I) : 1993 contains details about measurement of seepage in a canal.

Advantages: (*i*) Normal working conditions of canal are not disturbed, (*ii*) Can measure seepage from bed as well as sides, (*iii*) Prompt, and (*iv*) Cheap.

Limitations: (*i*) Meter is difficult to use in deep canals, particularly if water is turbid and/or velocity is high, (*ii*) Can be used in small canals of less than 2 cumecs capacity, (*iii*) Not used in canals, having heavy weed growth, rocky or gravelly bottom or lined canals.

Estimation of Transmission Losses

The losses are expressed in different ways discussed as under. However, generally water losses are expressed in terms of cumec per million square metre of wetted perimeter.

1. *Cumec Per Million Square Metre of Wetted Perimeter.* Davis and Wilson Formula

$$S_L = 0.45\, C\, \frac{P_w L}{4\times10^6 + 3650V^{1/2}}\, D^{1/3}$$

where, S_L = seepage losses in a canal of length L unit long, wetted perimeter P_w, depth D and velocity V, C = constant depending on type of lining material; value 1 for concrete lining, 4 for mass clay lining 150 mm thick, 5 for light bitumen lining, 8 for cleaned clay lining 75 mm thick, and 10 for bitumen or cement mortar lining.

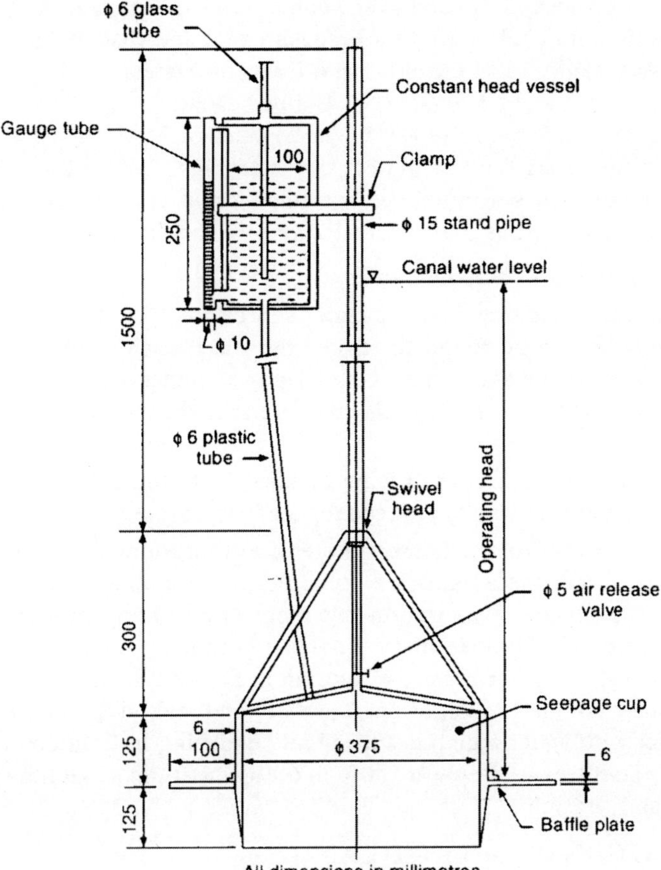

Fig. 5.24 Seepage meter.

2. *Percentage of Canal Discharge.* The usually adopted percentage losses in a canal system are Main canal and branches = 17%, vary between 15% to 20%, Distributary and minors = 8%; vary between 6% to 8%, and Watercourses = 20%; vary between 17.5% to 22%.

3. Percentage loss per km length of canal.

(*a*) *U.P. practice.* $\Delta Q = (1/200)(B + D)^{2/3}$

Percentage loss per km length = $(\Delta Q/Q) \times 100$

where, ΔQ = discharge loss in canal (cumecs/km length of canal), B = bed width of canal (m), D = depth of water in canal (m) and Q is canal discharge (cumecs)

(*b*) *Pavlovsky's formula*

$$S_l = 0.0116\, B_{ws} + 2yC_1$$

where, S_l = seepage loss (cumec/km length of canal), B_{ws} = canal width at water edge (m), y = depth of flow (m), and C_1 = coefficient of soil infiltration in m/day; the values are 0.05 for heavy loam, 0.05-0.10 for medium and light loam, 0.10-0.50 for sandy loam, 0.25-0.50 for loess, 0.50-1.0 for dusty sand, 1.0-5.0 for fine sand, and 5.0-20.0 for medium grained sand.

4. *Central Water Commission and States Practices*: The transmission loss (cumecs/10^6 m^2 of wetted perimeter) as per CWC is-rock (0.91), black cotton soil (1.83), alluvial soil (2.74), decayed rock or gravel (3.0). Tamil Nadu-rock (0.91), black cotton soil (1.52), alluvial or red soil (2.44), decayed rock or gravel (3.0), Orissa-canals (2.44), distributaries and minors (10%),

Maharashtra-upto 7 cumecs (3.42) and over 7 cumecs (3.0), Bihar and Karnataka-canals (2.44), distributaries and minors (3.96), Andhra Pradesh-canal and distributaries (1.83), Rajasthan-ordinary soil (2.44), sandy (3.05), Punjab and Haryana States:

$$K = 1.6\ Q^{0.0625} \quad \text{Unlined canals}$$

$$K = 0.349\ Q^{0.056} \quad \text{Lined canals}$$

where, K is absorption loss/10^6 m^2 of wetted perimeter. Figs. 5.25 and 5.26 depict the curves for determining transmission losses in unlined and lined canals respectively. As per IS: 4745-1968 seepage losses are considered as 0.6 cumecs/10^6 m^2 of wetted perimeter of lined section.

Minimizing Transmission Losses in Canals

Unlined main canal and branches are major sources of loss of water in a canal system and in case some methods can be found to reduce the transmission losses, it will mean a large saving in quantity of water and units of hydel power generation lost on this account. The reduction in transmission losses in canals can be effected by minimizing seepage losses and evaporation losses.

1. *Reduction in Seepage Losses.* Lining of canals with impermeable lining of cement concrete or tile lining is the best way to reduce seepage losses from earthen canals.

2. *Reduction in Evaporation Losses.* Rate of evaporation is directly proportional to the water surface area, larger the area, more is the evaporation and vice versa. Water depth has little to do with the surface evaporation but there is relatively higher surface evaporation from shallow water depth. Unlined main canals and branches entail more loss of water since they expose large water surface area to evaporation. So far it is not possible to reduce evaporation loss in canals. Alcohols such as cetyl alcohol and stearyl alcohol form a monomolecular film on contact with water which is sufficiently enduring in field conditions. The invisible film is nontoxic, retards evaporation by 50% to 60% and at the same time allows free passage of rain, oxygen and sunlight.

5.12 CONSTRUCTION OF UNLINED CANALS

Various steps involved in the construction of an unlined canal are (*i*) *Layout*: The centre line of the canal is marked by stones or pegs fixed at an interval of 30 m. The curves are laid out, all half breadths, top and bottom edges of the excavation and toe of the embankments are suitably marked by pegs with reference to an established bench mark, (*ii*) *Site clearance*: The entire land width between the outer toes of the canal is cleared of all trees, stumps, roots, bushes, rubbish and other objectionable matter, (*iii*) *Stripping loose materials*: All loose materials close to the area to be excavated as is liable to fall in or may otherwise endanger the working area is removed, (*iv*) *Excavation:* The excavation is carried out by manual labour or by means of earthmoving equipment in accordance with the drawings. The excavation is done to the full depth and width and finished to prescribed lines and grades. Where there is both cutting and filling, spoil from the cutting, as far as is economical, is utilised for the filling, (*v*) *Bank construction*: The entire ground surface under the canal banks is stripped to a depth of 15 cm minimum. The toe of the slope on each side of the banks is dagbelled and marked at every 20 m by pegs firmly driven into the ground. Profiles, 3 m long, are constructed at about 50 m intervals in straight reaches and about 25 m apart at curves. The banks are constructed to the full width including the side slopes as per approved cross sections of the canal in layers not exceeding 30 cm in thickness. The settlement allowance of 10 percent is allowed.

Fig. 5.25 Absorption losses in unlined canal, $K = 1.6\ Q^{0.0625}$

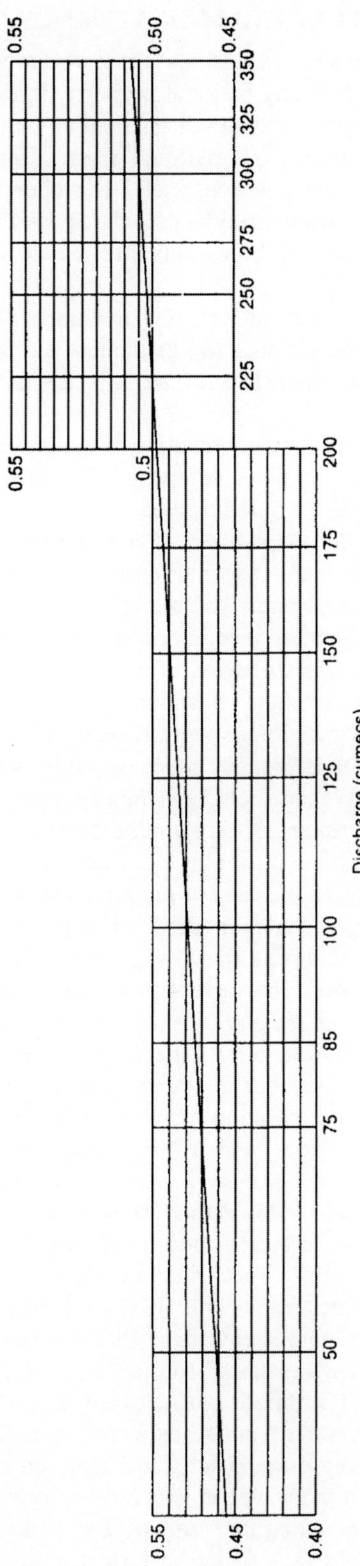

Fig. 5.26 Absorption losses in lined canal, $K = 0.349\ Q^{0.056}$

5.13 OPERATION AND MAINTENANCE OF UNLINED CANALS

Irrigation water is a national asset developed at a very high cost. The canal distribution system too is a very costly national property. It is essential that the irrigation water is economically and efficiently conveyed and distributed among the cultivators. Each cumec of irrigation water, if not properly utilised, is a national waste. Herein lies the importance of proper operation and maintenance of irrigation channels. The operation and maintenance of canal system calls for a particular multi-disciplinary approach. Special training is called for operational personnel so that correct use is made of the various information that reach them with improved communications in operation of canals.

The unlined canals suffer from (*i*) Silting of channels, (*ii*) Erosion of bed and banks, (*iii*) Weed and plant growth in the channel, (*iv*) Settlement and sloughing, (*v*) Seepage, (*vi*) Breaches in the channels, (*vii*) Cuts in the channels, (*viii*)Weak banks failure, and (*ix*) Surface deterioration of patrol bank/boundary road.

The main problem confronting the engineers responsible for the maintenance and operation of such channels involves complex relationships governing the flow of water and the sediment which it transports, briefed as under:

1. *Hydraulic functioning of a channel*: The irrigation channels which have run for long periods may have attained final regime and there may be little difficulty in their operation. There may be channels wherein the regime is yet to be attained. It is essential to carry out hydraulic survey of the channels at regular intervals to diagnose their proper hydraulic functioning.

2. *Equitable distribution of supplies*: The irrigation supplies should not only be properly maintained and distributed but should appear to be so to the satisfaction of the cultivators. For this purpose, the distribution network should be well maintained and flow of water closely controlled so that more area is brought under irrigation with the same supply and there is less wastage. The tail end of the system should essentially have adequate supplies.

3. *Silt clearance*: Silting up of the channel in the bed and sides reduces its carrying capacity and impair its efficient running necessitating removal of the deposited silt periodically to restore it to its capacity. In the case of main canals and branches it is virtually impracticable to silt clear the canal. The practicable method is to remodel them. In the case of distributaries and minors, silt clearance is done during the available closures. The channel section is cleared up to theoretical bed level and sides. The various methods adopted for silt clearance are (*i*) *Desilting*: The deposited silt is removed manually to restore the theoretical cross section of the channel, (*ii*) *Silt traps*: These are the pits dug in the canal bed to arrest the silt. The pits are laid scattered in the bed. The pits get filled up with silt in course of time necessitating digging up of fresh pits, (*iii*) *Dredging*: The under-water excavation of silt by dredgers is a very costly method of silt clearance and is rarely used in practice, and (*iv*) *Flushing*: It is a rarely used method. The deposited silt is disturbed and carried down by generating scouring velocity with the passing of higher discharge through the canal section rather the section is scoured beyond the theoretical section so that the operation has not to be repeated frequently.

4. *Weed control*: The irrigation channels are infested with aquatic weed and plant growth in the canal bed, sides, banks and the channel water. The various control measures are (*i*) *Regulation measures*: This method envisages alternate wetting and drying. The supplies are so regulated that the canal is run with authorised discharge for the required period and is closed for 2-3 days in a 10-day period. The duration of closure is such that the canal bed is completely dried. In Bhakra canals drawing water from storage reservoir the minimum drying period adopted was of 5 days duration. The exposure to sun for 5 days and repeated for 6 years the channels were rendered completely free from weeds, (*ii*) *Higher velocity*: Higher velocity of flow (over 0.6 m/sec) is generated by regrading the slope of the channel to keep the silt in suspension. This turbid water cuts off sun light rays which in turn prevents weed growth, (*iii*) *Manual control*. The manual

removal of weeds is done at the time of usual annual clearance, (*iv*) *Mechanical control*: It consists of weed removal by using scythes, sackles etc. by power operated movers, cutters, under-water saws etc. mounted on power launch by chaining, dredging, ploughing etc. (*v*) *Chemical control*: The chemical application for killing weeds is done with selective herbicides, non-selective translocated sprays on perennial weeds primarily, (*vi*) *Biological control:* Some fish species consume aquatic weeds.

5. *Jungle clearnance:* The canal banks and berms are cleared of all grass and bush wood as their presence induces abnormal and irregular growth resulting in obstruction in canal section.

6. *Closing of canal breach*: Irrigation channels are sometimes cut by cultivators to unauthorisedly draw supply. Breach in a canal is quite distinct from a cut because the former is not man made but is caused due to a variety of reasons, such as (*i*) *Weak bank*: A weak bank of insufficient top width and free board and constructed of poor soil is prone to breach especially in high filling reach, more particularly in a newly constructed unlined channel, (*ii*) *Leakage*: Holes made in the canal bank by burrowing animals or insects and rats may cause development of leakage through the bank and eventual failure, (*iii*) *Overflowing*: Canal is overflowed and hence breached when there is excessive supply, and (*iv*) *Piping*: As a consequence of raised water level (Fig. 5.27) the hydraulic gradient line goes out of the toe of the banks which develops into a breach by piping action, *i.e.*, (*i*) seepage from the canal and progressive dislodgement of soil particles from the outer slope of the bank, and (*ii*) soil in the portion of the bank (between LM and L'M' in Fig. 5.27) gets wet for the first time and settles down (the dry earth of the bank above which stays on by arching action) forming virtually a pipe (L M and L' M').

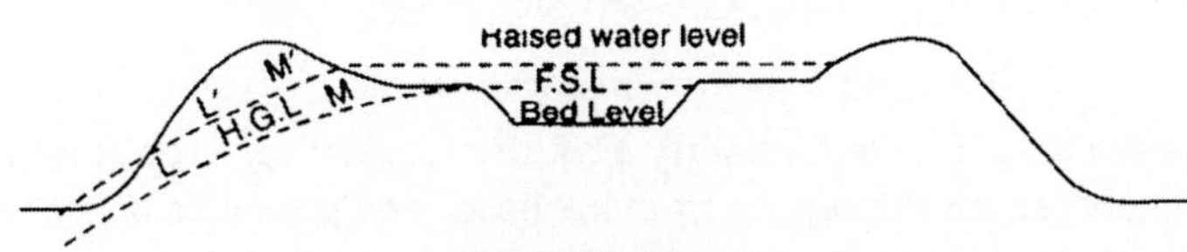

Fig. 5.27 Piping

7. *Strengthening of canal banks*: The canal banks are essentially maintained to the designed width, level, stable slopes etc. The top of bank is at such a level as to have minimum designed free board at designed discharge for the safe passage of supplies through the channel. The existing weak sections are essentially strengthened by encouraging silting and berm formation.

8. *Proper maintenance of patrol banks and boundary roads*: The patrol bank of main/branch canal and boundary road along distributaries and minors is maintained smooth to regular longitudinal grade.

Work Plan for Maintenance of Canals

The work plan for the maintenance of canal include monitoring, silt removal, vegetation control (chemical or mechanical) and seepage control. Likewise, the canal structures utilised for conveyance, regulation and control of water are modernised. The general maintenance problems are due to age, natural attrition and design or construction inadequacies.

5.14 CLOSING BREACH IN A CANAL

Closing of canal breach broadly involves two operations (*i*) Reduction of supply in the canal from the head or through an escape, if any, upstream of the breach site so as to make the closing operation easier, and (*ii*) Closing operation which involves stockpiling of earth on existing bank on both sides of the breach and closing of the breach by quick dumping of the earth simultaneously from both ends of the breached section. The actual closing operation of breach depends primarily on the capacity of canal, as under:

1. Closing Breach in Small Distributary or Minor: The closing operation is comparatively easier, the flow in the distributary or minor being manageable. The closing process is started from both ends by slipping the earth from the heaps and protecting the canal side by grassy clods usually available from the berms. No grassy clod is allowed to be washed down into the breach site. With a rush of earthwork at the end, the breach may be closed straightaway, progressing from the bank.

2. Closing Breach in a Major Distributary or a Small Branch Canal: The closing operation is very difficult, the flow in the canal being large. The flow in the canal is preferably reduced from the head. Straight closure is not possible. A double line of stakes is driven as indicated in Fig. 5.28 and planks or mattresses are put against them, if available, and if not then jungle is filled in between the stakes pressing it down with bags filled with sand and then men walking over them. When the flow through the breach has been so controlled to some extent, the closing is started from both sides by slipping earth from the head in the form of a ring as shown in Fig. 5.28. The earth is slipped from a heap and in no case thrown in water. The last gap of about 3 m is closed with a rush when enough earth has been collected on both sides.

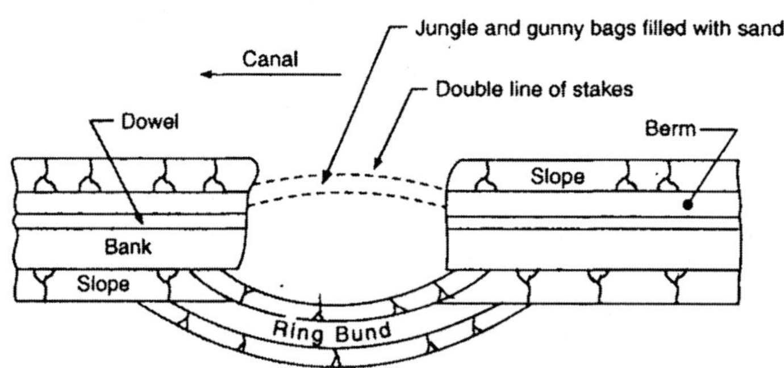

Fig. 5.28 Closing breach of canals.

3. Closing Breach in a Large Capacity Canal: In view of large flow in the canal, it is desirable to immediately reduce the supply from the head. The sequence of closure is the same as indicated for major distributary except that the space between the double lines of stakes shall be packed with gunny bags filled with sand. The inter-spaces are plugged with earth borrowed from berm. A temporary bank of gunny bags is raised in the position of stakes and bushing is done as indicated in Fig. 5.28. The closing of breach is then done by constructing a ring bund behind the temporary bank.

5.15 CANAL DISCHARGE OBSERVATION

In canal discharge observation, a higher degree of accuracy and precision is aimed at to enable regulation of supplies for proportionate distribution amongst various offtakes and outlets. A lined section of short length in a straight reach of the canal provides ideal section of definite dimensions for permanent discharge site. In unlined canal, a part of the canal is chosen conforming to the requirements (*i*) A straight reach of uniform section with trimmed berms and of minimum length 10 times the canal width, (*ii*) A straight run of 15, 30 or 60 m in the middle of the reach depending upon the velocity in the canal, (*iii*) The straight reach is sufficiently removed from outlets, bridges or other structures, (*iv*) A site above a fall usually has a constant relation between gauge height and discharge, (*v*) The water current is divergent from the prevailing wind direction, and (*vi*) The site is easily accessible at all times of the year.

The site after selection is demarcated. The direction of water flow is noted. Permanent graduated enamelled vertical staff gauges are erected, two at the side of the canal, in gauge wells 150 or 300 m both upstream and downstream of the central rope and the third very close to the *central* rope where current meter observations are made. The verticals for depth and velocity measurements are kept the same (Table 5.17). In general, the cross-section is divided into five main segments-central, sides and slopes as follows (Fig. 5.29):

Table 5.17 **Segmentation for canals**

Canal capacity (cumecs)	*Surface width (m)*	*No. of verticals*
Upto 15	Upto 15	5
15–85	15–35	9
above 85	above 35	11

Central Segment: Central segment ($6W_3$) equals the surface width minus 3 to 4 times the water depth, and made a multiple of 6, so that W_3 is a whole number. If this is not possible it shall be a multiple of 3.

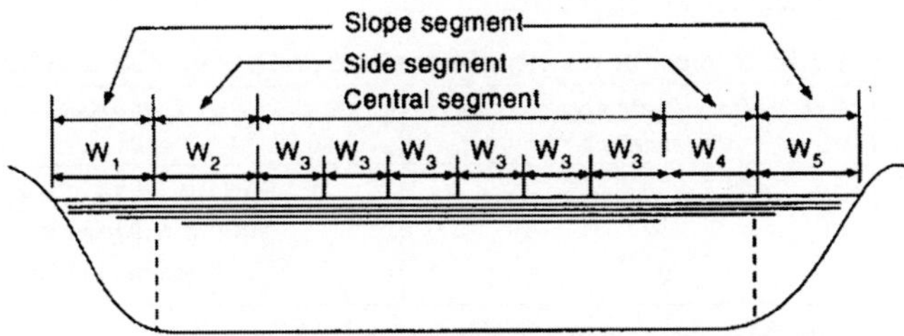

Fig. 5.29 Segmentation of canal section

Side Segments: 2 No. side segments (W_2 and W_4) shall be equal to each other and shall extend as nearly to the toe of the slope as possible.

Slope Segments: 2 No. slope segments (W_1 and W_5) obtained by dividing the balance of full surface width equally.

Classes of Observation

There are three classes of canal discharge observation, depending upon the object of discharge measurement or the nature of the section throughout the run.

1. First Class Discharge Measurement

In this method maximum accuracy is aimed at. Three ropes are stretched, 15-30 m apart across the canal, one at each end and the third at middle of the run. Tags are attached to the section rope at all points of the principal soundings which are also velocity stations. Depth is measured at the upper, lower and central float runs and mean depth adopted. Soundings (d_1 and d_{11}) are taken in the centre of the slope segments and velocities (V_1 and V_{11}) observed when practicable. When velocities V_1 and V_{11} cannot be observed, these are assumed as ½ V_2 and ½ V_{10} respectively. Intermediate soundings shall be taken only in the central segment at intervals of $^1/_3$ W_3 in the case of first class observation. The intermediate soundings may be omitted when the surface width does not exceed 21 times the general depth. In case velocity observation is made with the help a current meter, it shall be observed at the central section only but the depths are measured at all the three sections. Typical section, given below, shows the general arrangement :

Width of Segments	W_1	W_2	← $6W_3$ →						W_4	W_5
Spaces between Velocity stations	W_1	W_2	W_3	W_3	W_3	W_3	W_3	W_3	W_4	W_5
Principal soundings	d_1	d_2	d_3	d_4	d_5	d_6	d_7	d_8	d_9	d_{10} d_{11}
Intermediate soundings	—	—	d_3'' d_4'	d_4'' d_5'	d_5'' d_6'	d_6'' d_7'	d_7'' d_8'	d_8'' d_9'	—	—
Mean depths		D_2	D_3	D_4	D_5	D_6	D_7	D_8	D_9	D_{10}
Mean velocities	V_1	V_2	V_3	V_4	V_5	V_6	V_7	V_8	V_9	V_{10} V_{11}

Note — $D_3 = \frac{1}{3}(d_3 + 2d_3'')$ and $D_9 = \frac{1}{3}(2d_9' + d_9)$

D_4 to $D_8 = \frac{1}{3}(d' + d + d'')$

Float Observations

In this case, soundings are taken at upper, central and end sections. The rod is delivered into the canal about 15 m above the run to enable the float to attain a constant velocity by the time it reaches the cross section. It is collected by a man downstream of the end section. In the case of large canal, the rod is released from a boat moving along a rope stretched across the canal about 150 m upstream of the upper section and is collected at downstream end with the help of another boat, if available. The number of velocity observations made at a segment and the permissible deviation of the float from the pendants at the upper rope are given in Table 5.18. For greater deviations the velocities observed are rejected.

Table 5.18 Number of velocity observations and permissible deviation

Segment	*No. of velocity observations*	*Deviation*
Central	5 for V_4, V_6 and V_8	Half the width of each subsegment $W_9/2$
Side slope	3	Not more than ½ W_2
Slope	3	Not more than ½ W_1

2. Second Class Discharge Measurement

In second class observation, no central rope is required and soundings are taken only at the upper and lower ends of the float run.

3. Third Class Discharge Measurement

In the third class of observation, a single rope is used in the middle of the run, the soundings taken at this section only. Poles erected opposite each other at the end of the float run serve to show when the floats pass.

Discharge Table

The discharge table is a very useful tool in regulation of canal supply. Discharge tables are maintained for each distributary head and for the indenting sites at the beginning and end of canal sub-division where generally a fall-cum-meter flume is available or a regular pitched discharge site 45 m in length is provided. A series of discharge measurements, gauge readings and area of waterway at fully supply, ¾ full supply ½ full supply and ¼ full supply are made against a permanently fixed gauge. Mean of these four observations is determined. The mean depth 'D' is then calculated by dividing the area by the mean width. From these four values, the relation between discharge (Q) and depth (D) is established from the equation:

$Q = KD^n$, usually value of of n is taken as 5/3. Finally the values of 'D' are determined corresponding to the values of 'G' and discharge for intermediate values worked out as illustrated in the example 5.7.

Discharge Measurement of Distributaries and Minors

The sectional area of the small channels with sloping sides is usually obtained by measuring the surface width, taking soundings at 0.25 or 0.5 m interval and multiplying each interval by the corresponding mean depth. Sum of the result gives the sectional area.

An approximate method for computing the sectional area is by deducting 1/5th from the product of the surface width and the greatest depth.

Velocity Measurement

1. *Central Velocity Method.* It envisages measurement of the velocity at the centre of the surface width of a channel by a float. The mean velocity is derived from the observed central surface velocity by the application of a reduction factor of 0.70 in the case of a distributary, 0.65 for minors and 0.60 for watercourses. Product of mean velocity and the sectional area of the channel gives the discharge.

2. *With Pigmi Current Meter.* It is the most accurate method and is always preferred. Pigmi current meter, especially suitable for velocity measurement of small canals, is capable of measuring velocities upto 1 m/sec. The velocity is generally observed by lowering current meter to 0.6 depth or just below the water surface, clamped to the wading rod.

Discharge Measurement of Watercourses

A sufficient length above and below the selected straight reach is cleared for observation.

By Float. The central surface velocity (V) is observed with the help of a stop watch by floating a cow-dung or float for a distance of about 10 m in a straight reach. An average of five readings is taken as surface velocity. Water surface width (W) and central water depth (D) is measured (Fig. 5.29). Area is considered as 2/3 WD and accordingly discharge is computed by the formula $Q = \frac{2}{3} \times 0.6$ W.D.V. = 0.4 W.D.V.

By Portable Flume. A portable flume gives more accurate measurement of water course discharge. It consists of a steel structure comprising two parts, viz. (*i*) upstream approach and crest, and (*ii*) glacis (Fig. 5.30). One part can be folded in the other and carried by two men. The discharge is determined for the observed depth of water (H) over the crest of the flume. Discharge is given by the relation, $Q = 1.67\ BH^{3/2}$ cumec.

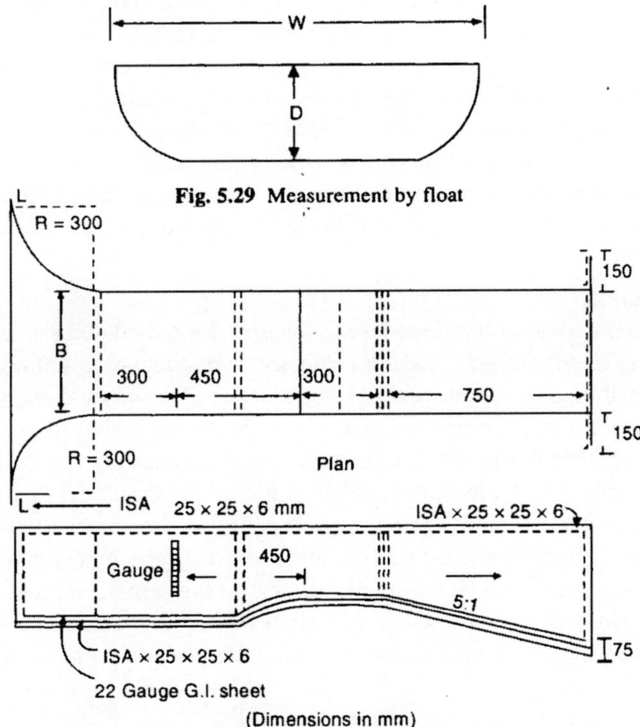

Fig. 5.29 Measurement by float

Fig. 5.30 Portable flume

Various precautions in the use of discharge observation by the flume are (*i*) Essentially the crest is kept absolutely level, (*ii*) Entry of water into the flume is smooth, and (*iii*) No leakage of water below and along the sides of the flume.

5.16 REMODELLING OF CHANNELS

Remodelling of a channel is necessitated due to certain reasons such as (*i*) Non-attainment of regime section, (*ii*) Defective slope, (*iii*) Heavy silting up, (*iv*) Drowned bridges, (*v*) Chronic shortage at tail, (*vi*) Defective outlets, (*vii*) Poor command, (*viii*) Increased design capacity, (*ix*) Defective hydraulic structures, and (*x*) Inadequate cross regulators.

Formulation of Remodelling Scheme

Formulation of the remodelling scheme involves the following steps:

(*i*) *Survey and investigation.* Having diagnosed the causes of trouble from which the channel is suffering and the necessity of remodelling decided upon, the irrigators are notified that the channel is being remodelled and they can put forth their requirement, if any, for the adjustment of outlet chaks and/or site of outlets. Remodelling of the channel is based on accurate data and for that land survey and hydraulic survey of the channel is carried out.

1. *Land survey.* Reliable benchmarks by double levelling along closed circuits are set up along the channel to be remodelled, to observe cross sections at every 30 m indicating all the information contained therein.

2. *Hydraulic survey.* A detailed hydraulic survey of the channel is carried out and longitudinal section and cross sections are observed and plotted. The hydraulic survey comprises all the information relating to the canal, outlet and hydraulic structures.

3. *Revision of longitudinal section.* The capacity statement and command statement of the channel is prepared on the basis of corrected chakbandi incorporating the adjustments in transfer of areas or shifting of outlet sites sought by the irrigators and considered technically advisable or extension or curtailment of canal found essential for proper command of the area. In the formulation of revised longitudinal section of the channel, various points considered are (*i*) Control points are essentially required to help the channel attain designed regime conditions. A control point at every 10 km length of the channel with a minimum head of 75 cm, besides one at about 3 km downstream of the head regulator and the other about 3 km upstream of the tail are essentially required, (*ii*) Distributors are essential at the heads of offtakes for proportionate distribution of supplies, and (*iii*) All outlets shall be provided with a minimum working head of 0.25 m.

In the case of the channels not originally designed with Lacey's equations, it is not generally convenient and economical to adopt Lacey's design. Lacey-cum-Kutter or Kennedy-cum-Kutter formulae may be used.

Remodelling Operation

Consequent upon approval of the remodelling scheme, the remodelling of a channel is taken up.

1. *Remodelling of distributaries.* The silt clearance of distributaries and minors, remodelling of outlets and bridges shall be done in one and the same closure. Raising and strengthening of banks is continued. The remodelling of outlets is essentially done in the beginning of kharif or rabi crops.

2. *Remodelling of Branch Canals.* Main canals and branches have to be kept running almost throughout the year and very little period is available when the canal is dry. Silt clearance is usually not practicable. The earthen section is preferably lined to cater for increased capacity at the same water surface slope. The lined section of the existing channel is generally determined in such a manner that the depth is maintained more or less at the same level and the bed is tightened on one side or both sides usually or wider and shallow section instead or an ideal cup shaped section (Fig. 5.31). The hydraulic structures are remodelled in one and the same closure in so far

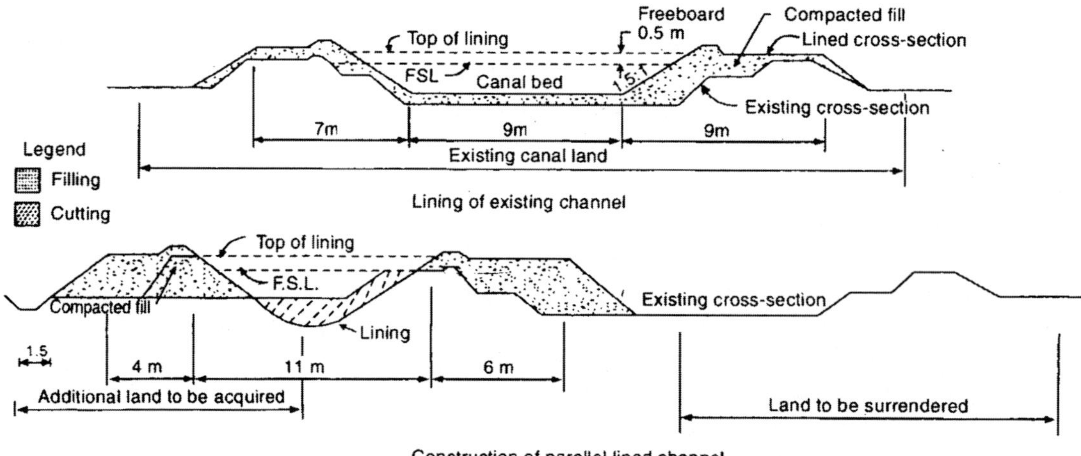

Fig. 5.31 Typical x-section of a distributary (After Er. A.M. Singal)

as the work in the canal bed is concerned. The remodelling of the canal section is taken up in reaches 5 to 10 km long defined by control points at the ends. The channel section is designed with Lacey-cum-Kutter's or Kennedy-cum Kutter's formulae keeping economy in view.

SOLVED EXAMPLES

Example 5.1 *Design a channel by Kennedy formula, discharge = 15 cumecs, slope = 0.0002, n = 0.0225 and m = 1.*

Solution. Assume FS depth = 1.75 m

Kennedy $V_0 = 0.55\, mD^{0.64}$

$= 0.55 \times 1 \times (1.75)^{0.64} = 0.786$ m/s.

$A = Q/V = 15/0.786 = 19.06\text{m}^2$

Also $A = (B + 0.5D)D$, assuming ½ : 1 side slope

$19.06 = (B + 0.5 \times 1.75)\, 1.75$

or $B = 10$ m

Now $P = B + \sqrt{5}\,D = 10 + \sqrt{5} \times 1.75 = 13.91$ m

$R = A/P = 19.06/13.91 = 1.37$ m

Kutter's
$$C = \frac{23+\left(\dfrac{0.00155}{0.0002}+\dfrac{1}{0.0225}\right)}{1+\left(23+\dfrac{0.00155}{0.0002}\right)\dfrac{0.0225}{\sqrt{1.37}}} = 47.3$$

Also $V = C\sqrt{RS} = 47.3\sqrt{137 \times 0.0002} = 0.783$ m/s

which is almost the same as worked out earlier, hence acceptable. Therefore, Bed width = 10 m, FS depth = 1.75 m and Side slope = ½ : 1.

Alternative method:

For 15 cumecs discharge, b/d ratio read from Fig. 5.4 = 5.75

∴ Bed width $= 5.75 \times 1.75 = 10$ m

Area $= (10 + 0.5 \times 1.75)\, 1.75 = 19.03\ \text{m}^2$

$P = (10 + 2.236 \times 1.75) = 13.9$ m

$R = A/P = 19.03/13.9 = 1.37$

$$C = \frac{23+\left(\dfrac{1}{0.0225}+\dfrac{0.00155}{0.0002}\right)}{1+\left(23+\dfrac{0.00155}{0.0002}\right)\dfrac{0.0225}{\sqrt{1.37}}} = 47.26$$

$V = C\sqrt{RS} = 47.26\sqrt{1 \cdot 37 \times 0.0002} = 0.78$ m/s

$Q = \text{A.V} = 19.03 \times 0.78 \approx 15$ cumecs

C.V.R. $= V/V_0 = 0.78/0.786 = 0.99.$

Example 5.2 *An irrigation canal having side slope 1:1 has bottom width of 3 m. It runs at a depth of 1 m with a bed slope of 1 in 2500. Manning's value of n = 0.028. Determine whether the canal will be silting or scouring or remain stable. Use Manning's and Kennedy's equations.*

Solution. $A = (3 + 1)1 = 4\ \text{m}^2$

$P = 3 + 2\sqrt{2} = 5.83$ m

$R = A/P = 4/5.83 = 0.69$ m

By Manning's equation $V = \dfrac{1}{n} R^{2/3} S^{1/2}$

$$V = \frac{1}{0.028}(0.69)^{2/3}(1/2500)^{1/2} = 0.558 \text{ m/s.}$$

Kennedy's non-silting, non scouring velocity, assuming m = 1.0

$$V_0 = 0.55\,D^{0.64}, = 0.55\,(1.0)^{0.64} = 0.55 \text{ m/s.}$$

Since the velocity as per Manning's formula is approximately the same as Kennedy's non-silting, non-scouring velocity, the channel is stable.

Example 5.3 *Design an irrigation canal in clayey alluvial soil for full supply discharge = 35 cumecs, coefficient of roughness 0.025, canal side slope 1:1, longitudinal slope 1 in 5000. Also check for critical velocity ratio, allowable CVR is 0.9 to 1.1.*

Solution. Assume depth = 2.0 m

As per Kennedy's formula,

$$V_0 = 0.55\text{ m } D^{0.64}, \text{ assuming m} = 1.0$$

$$V_0 = 0.55\,(2.0)^{0.64} = 0.86 \text{ m/s}$$

Now, $A = Q/V = 35/0.86 = 40.70 \text{ m}^2$

Also $A = 40.70 = (B + 2)2$

or $B = 18.35$, say 18.5 m

$$P = B + 2.828\,D$$

$$P = 18.5 + 2.828 \times 2 = 24.16 \text{ m}$$

$$R = A/P = 40.70/24.16 = 1.68$$

As per Kutter's formula,

$$C = \frac{23 + \dfrac{1}{0.025} + \dfrac{0.00155}{0.0002}}{1 + \left(23 + \dfrac{0.00155}{0.0002}\right)\dfrac{0.025}{\sqrt{1.68}}} = 44.50$$

Therefore, $V = C\sqrt{RS} = 44.50\sqrt{1.68 \times 0.0002} = 0.82 \text{ m/s}$

$$CVR = V/V_0 = 0.82/0.86 = 0.95$$

which is within the allowable value of 0.9 to 1.1, hence O.K.

Example 5.4 *Design the canal in example 5.3 by Lacey theory.*

Solution. $S = f^{5/3}/3340\,Q^{1/6}$

$$1/5000 = f^{5/3}/3340\,(35)^{1/6}$$

or $f = 1.12$

Also $N = 0.0225\,f^{1/4}$

$$0.025 = 0.0225\,f^{1/4}$$

or $f = 1.52$

Thus, we have got two different values of f. Further it is stipulated that CVR allowable is 0.9 to 1.1.

But $f = (V/V_0)^2$ or $(0.9)^2$ to $(1.1)^2$, which implies that the value of f as 1.12 is acceptable.

Now $V = \left(\dfrac{Qf^2}{140}\right)^{1/6} = \left[\dfrac{35(1.12)^2}{140}\right]^{1/6} = 0.82 \text{ m/s}$

$$A = Q/V = 35/0.82 = 42.68 \text{ m}^2$$

$$R = 2.5\,V^2/f = 2.5\,(0.82)^2/1.12 = 1.5 \text{ m}$$

Also $A = PR$

or $42.68 = P \times 1.5$ or $P = 28.45$ m

Also $42.68 = BD + D^2$ with 1:1 side slope (*i*)

And $28.45 = B + 2.828\,D$ (*ii*)

From (*i*) and (*ii*) $D = 1.68$ m

From (*i*) with $D = 1.68$, $B = 23.75$ m

$$N = 0.0225 f^{1/4} = 0.0225\,(1.12)^{1/4} = 0.02315$$

Channel by Lacey is shallow, and wide 1.68 × 23·75 m as against 2.0 × 18.5 m by Kennedy formula in Example 5.3.

Example 5.5 *Design a canal by Lacey's theory for 40 cumecs discharge, and $f = 0.9$.*

Solution.

$$R = 0.47\,(40/0.9)^{1/3} = 1.66 \text{ m}$$

$$P = 4.75\,(40)^{0.5} = 30 \text{ m}$$

$$D = \frac{P - \sqrt{P^2 - (6.944 \times R \times P)}}{3.472}$$

$$D = \frac{30.0 - \sqrt{(30.0)^2 - 6.944 \times 1.66 \times 30)}}{3.472}$$

$$D = 1.86 \text{ m}$$

$$\begin{aligned} B &= P - 2.236\,D \text{ with } \tfrac{1}{2} : 1 \text{ side slope} \\ &= 30.0 - 2.236 \times 1.86 \\ &= 25.84 \text{ Say } 26 \text{ m} \\ V &= [40\,(0.9)^2/140]^{1/6} \\ &= 0.78 \; m/s \\ S &= (0.9)^{5/3}/3340\,(40)^{1/6} = 1/7356 \end{aligned}$$

Example 5.6 *Design an irrigation channel in non-alluvial soil by Chezy-Kutter and Kennedy formula for a discharge of 30 cumecs, $n = 0.02$, slope 0.000125 and side slope ½ : 1.*

Solution. Assume bed width 16 m and depth 2.15 m (as may be read from Fig. 5.4).

Cross sectional area, $A = (16 + 1.075)\,2.15 = 36.71 \text{ m}^2$

Wetted perimeter, $P = (16 + 1.118 \times 2.15 \times 2) = 20.81$ m

Hydraulic mean depth, $R = A/P = 36.71/20.81 = 1.764$ m

Kutter's

$$C = \frac{23 + \dfrac{1}{N} + \dfrac{0.00155}{S}}{1 + \left(23 + \dfrac{0.00155}{S}\right)\dfrac{N}{\sqrt{R}}}$$

$$C = \frac{23 + \dfrac{1}{0.02} + \dfrac{0.00155}{0.000125}}{1 + \left(23 + \dfrac{0.00155}{0.000125}\right)\dfrac{0.02}{\sqrt{1.764}}} = 55.78$$

Chezy, $V = C\sqrt{RS} = 55.78\,\sqrt{1.764 \times 0.000125}$

$V = 0.83$ m/s.

Now, $Q = A.V = 36.71 \times 0.83 = 30$ cumecs

Critical velocity $V_0 = 0.55\,D^{0.64} = 0.55\,(2.15)^{0.64} = 0.90$ m/s

Critical velocity ratio, $CVR = 0.83/0.90 = 0.92$.

Example 5.7 *Prepare discharge table of a canal, the observed discharge at gauge 1.42 m is 8.8 cumecs and 5.7 cumecs at 1.17 m gauge. Is the canal silted or scoured?*

Solution: Suppose actual depth of water with 1.17 gauge = D

∴ Actual depth of water with 1.42 gauge = $(D + 0.25)$

$$Q = KD^{5/3}$$

$$8.8 = K(D + 0.25)^{5/3} \qquad (i)$$

and

$$5.7 = KD^{5/3} \qquad (ii)$$

Dividing (*i*) and (*ii*)

$$\frac{8.8}{5.7} = \frac{K(D+0.25)^{5/3}}{KD^{5/3}}$$

or $$D = 0.85 \text{ m.}$$

The actual water depth against the gauge of 1.17 is 0.85 m. Thus the canal is silted up by 0.32m, *i.e.*, (1.17–0.85).

Now from equation (*ii*)

$$5.7 = K(0.85)^{5/3}$$

or $$K = 7.5$$

By putting the values of actual water depths and K equal to 7.5 in the relation $Q = KD^{5/3}$ discharge corresponding to various gauges is worked out and discharge table formulated as given in Table 5.19.

Table 5.19 Discharge Table

Gauge (m)	*Water depth (m)*	*Discharge (cumec)*	*Gauge (m)*	*Water depth (m)*	*Discharge (cumec)*
0.32	—	—	1.0	0.68	3.94
0.50	0.18	0.45	1.1	0.78	4.95
0.6	0.28	0.90	1.2	0.88	6.06
0.7	0.38	1.50	1.3	0.98	7.26
0.8	0.48	2.18	1.42	1.10	8.8
0.9	0.58	3.0			

Example 5.8 *Work out balancing depth for the canal section given in Fig. 5.32.*

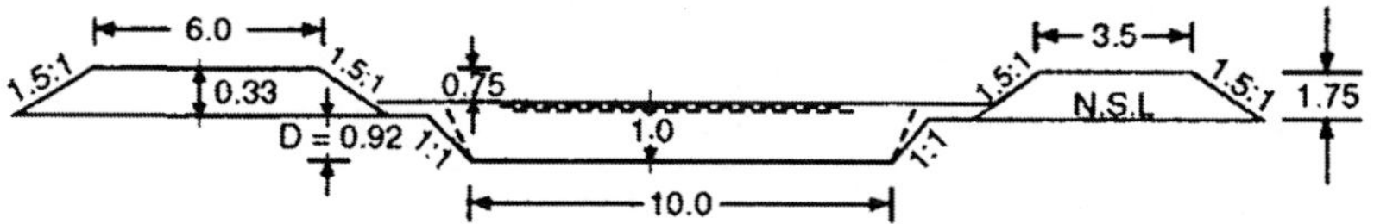

Fig. 5.32 Balancing depth.

Solution. Let NSL be at a depth D above the designed bed level of the canal.

Cutting $= (10 + D)\,D = 10D + D^2$

Filling $= [6 + (1.75–D)\,1.5]\,(1.75–D) + [3.5 + (1.75–D)\,1.5]\,(1.75–D)$

$= 3D^2–19.99D + 25.80$

Equating cutting and filling, we get

$$10D + D^2 = 3D^2–19.99D + 25.80$$

or $$D = 0.92 \text{ m}$$

Check, considering 1m length of canal

Cutting $= 1 \times (10 + 0.92)\,0.92 = 10.05 \text{ m}^3$

Filling $= 1 \times (6 + 0.83 \times 1.5)\,0.83 + (3.5 + 0.83 \times 1.5)\,0.83$

$= 9.96 \text{ m}^3$

which is almost equal to cutting. Thus the balancing depth is when NSL is 0.92 m above bed level of the canal. Fig. 5.32 shows the section of canal in balancing depth.

QUESTIONS

1. Explain Lacey's regime theory.
2. Explain what is meant by unlimited incoherent alluvium in the context of Lacey's theory and discuss the concept of regime embodied in this theory for the design of earth channels in alluvium.
3. Describe the salient features of the empirically derived silt theories of Kennedy and Lacey for design of canals. Discuss the limitations in their applicability.

4. Discuss the salient features of Kennedy's theory for the design of earthen channels based on the critical velocity concept and mention its limitations.
5. Design an earthen canal section to carry 50 cumecs discharge at a slope of 0.25 m/km, having been given that N = 0.0225 and m = 1.0, where the symbols have their usual meanings. Comment on the designed channel section.
6. (*a*) What is meant by regime? Differentiate between regime in natural rivers and in artificial channels.
   (*b*) Design an irrigation channel section for discharge = 40 cumecs, silt factor = 1.0, side slope = 1:1, Determine the longitudinal slope also.
7. State Kennedy's silt theory and Lacey's silt theory and distinguish between the two.
8. (*a*) How would you design a canal using Kennedy's theory? Write various steps.
   (*b*) Design a channel section using Lacey's theory for discharge = 15 m^3/s, mean diameter of silt particles = 0.33 mm, and side slopes = ½ : 1,
   Also find the longitudinal slope. **Ans.** B = 15.5 m, D = 1.35 m, S = 1/5246.
9. A canal has bed width = 6 m, depth = 1.2 m, discharge = 4.60 cumecs, slope = 1 in 5000, Manning's N = 0.0225, and critical velocity ratio = 0.95. Check whether the chosen section satisfies Kennedy's theory. (*P.U.*)
10. (*a*) Draw a neat sketch showing various components of a flow irrigation system.
    (*b*) Name a few important canals of Northern India.
11. Diffferentiate between (*i*) Alluvial canal and non-alluvial canal, (*ii*) Permanent canal and inundation canal, (*iii*) Branch canal and distributary canal, (*iv*) Contour canal and watershed canal, and (*v*) Protective canal and productive canal.
12. Enumerate the various considerations which govern the selection of the optimum alignment of a canal.
13. Give an account of procedure for the construction of an unlined canal.
14. What are the various causes of seepage and evaporation losses from a canal? Can these be reduced?
15. How will you determine seepage loss from a canal?

6

LINED CANALS

6.1 INTRODUCTION

Conservation of water supplies is important as the demand continues to increase and new sources of supply are becoming increasingly scarce. The principle of conservation requires that full use be made of available water supplies, the most promising being reduction in the amount of water lost through seepage during conveyance in the canals. Avoidable seepage losses from unlined canal system not only entail sizeable wastage of investment made on the construction of dam to impound water, and canal system to distribute it, but also calls for further investment to control the menace of waterlogging created in the canal irrigated areas. Lining offers a solution because it helps conserve the costly impounded water otherwise lost during conveyance due to high seepage losses in unlined section. Thus, the principal objects of lining irrigation channels are to (*i*) Minimize losses due to seepage, (*ii*) Protect the area prone to waterlogging due to rise in watertable, and (*iii*) Pass increased discharge through the canal due to the adoption of higher velocity, (*iv*) Rigid lining also results in improvement of command and larger working head for power generation. Lining of water courses in general and in the areas irrigated by tubewells assumes significance as pumped water is costly.

6.2 NECESSITY OF LINING CANALS

Lining of irrigation channels is necessitated to achieve all or some of the objectives, keeping in view economy as well. These are (*i*) Reduction of seepage losses through unlined canal section thereby making available additional supplies for extension of irrigation to new areas, (*ii*) Prevention of waterlogging in the area adjacent to the canal by reducing seepage of canal water to water table, (*iii*) Improvement of discharging capacity of existing channel, (*iv*) Improvement of operational efficiency, (*v*) Elimination of silting problems inherent in unlined canals due to higher velocities possible in lined channels, (*vi*) Reduction in maintenance cost, (*vii*) Reduction in cost of excavation due to reduced cross section (and thereby saving in land) with increased velocity permissible in lined section, (*viii*) Increase in available head for power generation as also to command more area as a flatter gradient can be provided in lined channels, and (*ix*) Prevention of weed growth. The necessity of lining has to be established by quantifying the contribution of each of the above factors and assessing the overall economy to be achieved by lining a canal.

The benefits that accrue from lining of the channels generally justify the initial capital cost and as a consequence there is now better appreciation of the need for lining of channels.

6.3 ADVANTAGES AND DISADVANTAGES OF CANAL LINING

The lining of an irrigation canal has the advantages (*i*) Reduction in seepage losses *from* canals reaching water table and raising it resulting in waterlogging and reduction in yield, (*ii*) Reduced seepage losses by as much as 75 per cent means saving of water which otherwise would have required construction of bigger reservoir and dam for the same amount of actual water delivered to the field which implies more capital expenditure without much gain, (*iii*) Reduction in losses and thereby making available more water for extension of irrigation to new areas and improvement of irrigation facilities in the areas already under irrigation, (*iv*) Flatter slope in lined canal system results in low height of dam and consequent saving in cost of dam construction, (*v*) Stable channel section, (*vi*) Brings more area under command due to very flat slope possible, (*vii*) Steeper side slopes and bed slope possible as the lined section is immune from erosion, (*viii*) Higher velocity permissible, resulting in proportionate saving in cross sectional area, land width, quantum of earthwork excavation and construction of bridges and cross drainage works which in certain cases may offset completely the extra cost of lining, (*ix*) Possibility of breaches are remote owing to sound structural stability of lined section, hence improvement in operational efficiency, (*x*) More hydel power generation possible with saving in water from losses and conservation of head losses due to flatter bed slope possible, (*xi*) Low coefficient of rugosity resulting in high velocity with the same slope and hence reduced cross sectional area compared with unlined section, (*xii*) Lined canal water does not pick up harmful salts from the soil through which it passes because canal water does not come in contact with the subsoil, (*xiii*) Permits more winding alignment resulting in saving in embankment and cutting costs, (*xiv*) Prevents weed growth thereby resulting in saving of expenditure incurred on weed removal in the case of earthen channels, (*xv*) Stable section which in the case of distributaries and minors reduces remodelling and alteration of outlets, (*xvi*) Theft of water by cultivators is stopped, (*xvii*) High velocity provided in lined section shall carry forward the blown in sand in sandy tracts, (*xviii*) Recurring charges on silt clearance inherent in unlined canals are avoided in the case of a silting canal, with the same slope. Existing silting channel when lined at the same slope generates higher velocity to carry forward the sediment, (*xix*) Considerable economy in the acquisition of cultivable land due to relatively narrow section of the canal, (*xx*) Improvement in cropping pattern, (*xxi*) Environmental betterment, (*xxii*) Greatly reduced maintenance and operational charges of the canal, (*xxiii*) Reduction in evaporation and transmission losses due to reduced exposed area, (*xxiv*) Reduction in erosion which occurs in unlined channels constructed in steep lands.

Disadvantages of Lining

The widely publicized lining of channels is not free from disadvantages, some of which are (*i*) Higher initial investment; a lined canal is 3 to 4 times costlier than an unlined one of the same capacity, (*ii*) Costly repairs, (*iii*) Shifting of outlets is very costly because it involves dismantling and relaying of lining, (*iv*) Longer construction period, and (*v*) More sophisticated construction equipment and skilled labour are required.

6.4 SELECTION OF SUITABLE TYPE OF LINING

The essential requirements of a satisfactory type of lining are (*i*) low cost, (*ii*) impermeability, (*iii*) hydraulic efficiency (*i.e.*, reduction in rugosity coefficient, (*iv*) durability, (*v*) resistance to erosion, (*vi*) repairability, and (*vii*) structural stability. The selection of particular type of lining is governed by the following considerations, the important factors being imperviousness and smoothness, position of water table, climatic conditions, size of canal, etc.

Imperviousness. To save seepage losses and as an important antiwaterlogging measure, it should ensure maximum degree of water-tightness. Seepage is proportional to the number and size of cracks. The type of lining selected should, therefore, ensure minimum cracks to limit seepage losses to the minimum. Cement concrete lining is more impervious than tile lining.

Hydraulic Efficiency. The carrying capacity of a channel varies inversely with the value of coefficient of rugosity of the lined surface. The coefficient of rugosity increases with the deterioration of lined surface with the passage of time. A reasonable smooth surface, *i.e.*, low coefficient of rugosity is desirable to ensure maximum velocity with corresponding economy in section. Concrete and tile linings are hydraulically most efficient.

Durability: The lining should be strong and durable. It should be resistant to wearing, weathering, chemical action of salts present in the soil, thermal and moisture changes.

Structural Stability: The lining should be reasonably stable to withstand the differential subsoil water pressures due to subsoil water and backfill getting saturated through seepage or due to sudden drawdown, though the lining to have economical thickness is not intended to serve as a retaining wall.

Economy: Lining is justified in case the benefits accruing from it offset the first cost and subsequent maintenance and give a reasonable return on the capital investment. The lining should be economical in initial cost as also in repair and maintenance. The most economical lining is that the investment on which is within the extra revenue accruing from the water saved from losses, other advantages apart.

High Velocity: Lining is intended to withstand maximum velocity of flow so that the section is the minimum possible.

Life: The life of the lining should be as intended. Cement concrete lining has the longest proved life (over 60 years) and one that needs least maintenance. The material used for lining should not deteriorate in the course of time, compressive strength being not of much importance.

Weed Growth: The lining should be inpenetratable to roots of plants, entirely eliminate possibility of undesirable weed growth and keep the flow smooth, clear and perfect.

Availability of Construction Materials: The economical lining is that which makes maximum use of the available construction materials at or near the site.

Labour Strength Available. The optimum utilization of the kind and strength of labour available for the work should be possible with the type of lining selected.

Operational and Maintenance Charges. The type of lining selected shall require least operational and maintenance charges; easy repairability and at economical cost as in the case of tile, brick, boulder and precast slab lining compared to *in situ* concrete lining.

Sub grade. On account of interdependence of the satisfactory performance of lining on the nature of the subgrade which supports it, adaptability of the type of lining selected to the given subgrade is of paramount importance.

IS : 10430-1982 recommendations for the selection of type of lining on the basis of the capacity of a channel are as under:

Channel bed width (m)	*Preference to type of lining*		
	1st	*2nd*	*3rd*
Up to 3.0	Single tile or brick lining.	Precast cement concrete if tiles are not available	Flexible membrane lining with adequate earth/tile cover.
3.0 – 8.0	Single tile lining	Precast concrete lining	Composite lining (flexible membrane in bed and rigid lining on sides) for existing stable channels with no danger of scouring.
Over 8.0	Cement concrete lining.	Cement concrete lining in bed and P.C.C. slab lining on sides.	Single tile lining in bed and double tile on sides.

The capacity-wise recommended lining as per research studies are

Capacity (cumecs)	*Subgrade*	*Channel condition*	*Type of lining*
Up to 3	Sandy or clayey	In bank	Polythene film laid in a rectangle with earth cover.
Up to 15	Clayey, sand wiched	In cutting	Tiles on sides and polythene film with 45 cm soil cover or single tile in bed.
15 .	Sandy	In cutting	Polythene film with single tile in sides and polythene film with 45 cm soil cover, slightly compacted in bed.

Resistance to Erosion. Sediments carried by canal water damage the lining by abrasion. Lining should be resistant to abrasion caused by the sediment carried by the canal water. Concrete and boulder lining is more resistant to abrasion compared to tile lining.

6.5 FINANCIAL JUSTIFICATION FOR LINING

Lining of a canal is economically justified only in case the annual benefits accruing therefrom exceed the initial cost plus interest on the capital investment and maintenance charges, *i.e.*, when its cost inclusive of interest on capital and maintenance charges can be repaid in monetary benefits accrued during the life of the lining.

The maximum rate of expenditure on lining that is economically justifiable as per IS : 10430–1982 is as under:

Annual value of water lost by seepage from unlined canal $= pLsdW$ Rs.

Annual saving in value of water otherwise lost by seepage $= \{(pLsdW-PLSdW)$ Rs.

$= [Ldw(ps-PS)]$ Rs.

Total annual benefits resulting from lining of canal, 'a' $= \{LdW(ps-PS)+B+M]$ Rs...(*i*)

Additional capital expenditure on construction of lined canal = TLC Rs.

Net present worth (NPW) of the total annual benefits 'a',

over the life of the canal, y,

$$NPW = a\frac{(1+x)^y-1}{x(1+x)^y} \qquad (ii)$$

For lining to be economically feasible, the additional initial cost of the lined canal (TLC) should be equal to or less than net present worth of savings, *i.e.*, TLC $\leq$ NPW, *i.e.* value of 'a' worked out in (*ii*) is equal to or less than (*i*). Where, C = cost of lining (Rs m^2) including the additional cost of dressing the banks for lining and accounting for the saving, if any, resulting from the smaller cross sections and hence smaller area of land, quantity of earth work and structures required for the lined sections. This saving will be available on new canal excavated to lined cross sections from the beginning, but not on lining of existing unlined canals, s and S = seepage losses in unlined and lined canals respectively (m^3/m^2 of wetted perimeter/day of 24 hr), p and P = wetted perimeter (m) of unlined and lined sections respectively, T = total perimeter of lining (m), d = number of running days of the canal per year, W = value of water saved (Rs/m^3), L = length of canal (m), y = life of canal (*yr*) assumed as 60 years for concrete, brick tile and boulder lining (IS: 10430–1982), M = annual saving (Rs) in operation and maintenance due to lining, B = annual estimated value (Rs) and of other benefits for the length of canal which include prevention of waterlogging, reduced cost of drainage for adjoining lands, reduced risk of breaching, etc., x = per cent rate of interest per year. Transmission loss of 0.60 m^3/s per 10^6 m^2 of wetted perimeter of lined canal is generally assumed.

6.6 DESIGN OF LINED CHANNELS

The section of lined canal is governed by the following considerations:

1. Economical section, *i.e.*, maximum cross sectional area for the minimum wetted perimeter.

2. Maximum velocity through the section so that the cross sectional area is the minimum. Higher velocity can be allowed in lined section carrying water free from sand and silt. However, velocity depends as it is on the slope which is limited from the consideration of command.

3. Higher silt carrying capacity so that the capacity of lined section is not reduced by silting.

Lined Section

The lined section generally adopted is circular with sloping sides. The best and the most economical section has the bed a curved one with its centre at the full supply level of the canal and sides tangential to the curve at the same slope as the angle of repose of the soil (Fig. 6.1). Two standardized lined sections are Circular section and Trapezoidal section.

1. *Circular Section.* Also called cup shaped section. The circular section with curved bed (Fig. 6.1) is suitable for discharge up to 85 cumecs and depth 5.5 m and velocity 1.75 m/sec. This section has higher silt carrying capacity than a trapezoidal one. The velocities during falling supply are higher in a circular section than in a trapezoidal section as illustrated in Fig. 6.2. Heading up of supplies at the offtake sites lead to silt deposition but it effects the coefficient of rugosity on a smaller portion of the perimeter and hence cannot affect the velocities. The effect on f is also correspondingly less. Side slope adopted is 1 : 1 for depths up to 3.5 m and 1.25:1 for depths over 3.5 m provided the angle of repose of the soil is not flatter than 1.5:1.

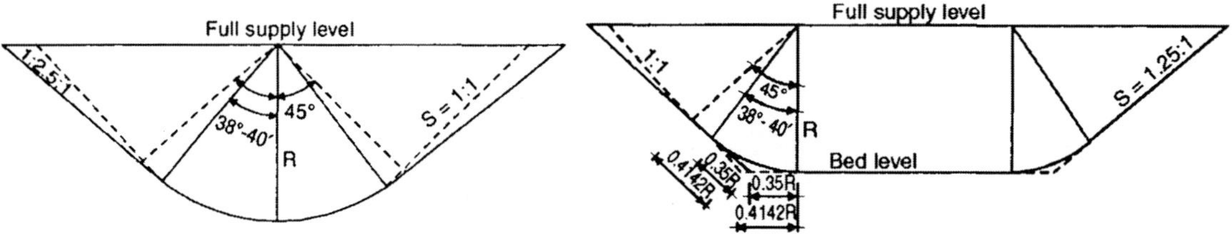

Fig. 6.1 Circular section. **Fig. 6.3** Trapezoidal section.

2. *Trapezoidal Section.* For discharges up to 85 cumecs but velocities higher than 1.75 m/sec and depths over 5.5 m trapezoidal or flat bottom section (Fig. 6.3) is adopted. Also for discharges exceeding 85 cumecs, it becomes necessary to provide horizontal bed to keep the depth within reasonable limits. The bed is, however, joined to side slopes in a curve of radius equal to the full supply depth. The side slopes are preferably kept nearly the same as the angle of repose of the natural soil, as under, so that the slopes are such that no earth pressure is exerted on the back of lining. The slopes steeper than these have to be designed as gravity sections and require greater thickness and are consequently uneconomical.

Material	*Angle of repose*	*Side slope*
Clay, well-drained	45°	1:1
Clay, loam, average sandy loam	36° to 33°	1.25 to 1.5:1
Sandy or gravelly soil	33° to 26°	1.5 to 2:1

Fig. 6.4. shows the typical cross sections of lined canal in filling, cutting and deep cutting. Fig. 6.5 shows the cross sections of Narmada main canal, the largest capacity canal in the world. The side slopes recommend by IS : 10430 – 1982 for very light loose sand to average sandy soil is 2:1 to 3:1, sandy soil or gravel 1:1 to 2:1, murum, gravel mixed soil 1:1 to 1.5:1 black cotton soil 2:1 to 3.5:1, clayey soils 1.5:1 to 2.5:1 in embankment and in rock 0.25:1 to 0.5:1.

The side slopes usually adopted are 1:1 to 1.25:1. Present practice is to adopt 1.5:1 and 2:1 side slopes for reasons of stability of the section.

6.7 DESIGN PARAMETERS

Full Supply Depth

Various considerations in fixing full supply depth of the lined section are (*i*) To achieve balancing depth as far as possible, (*ii*) Velocity is within permissible limit for the type of lining selected, (*iii*) The bed of the channel is higher than the subsoil water level to the extent possible so as to avoid pumping during construction and uplift pressures after closure of the canal.

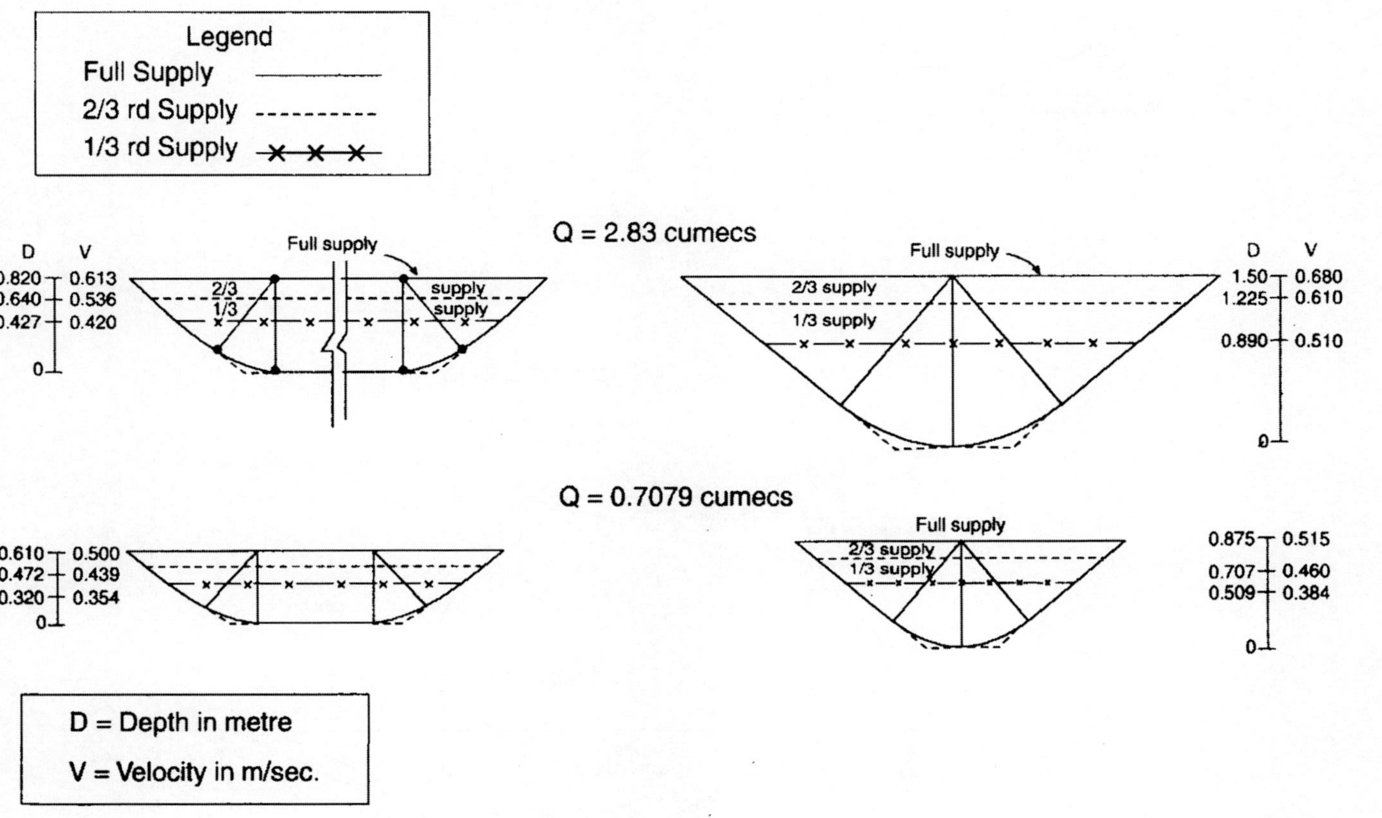

Fig. 6.2 Velocities in cup shaped and trapezoidal sections

(After Er. A.M. Singal)

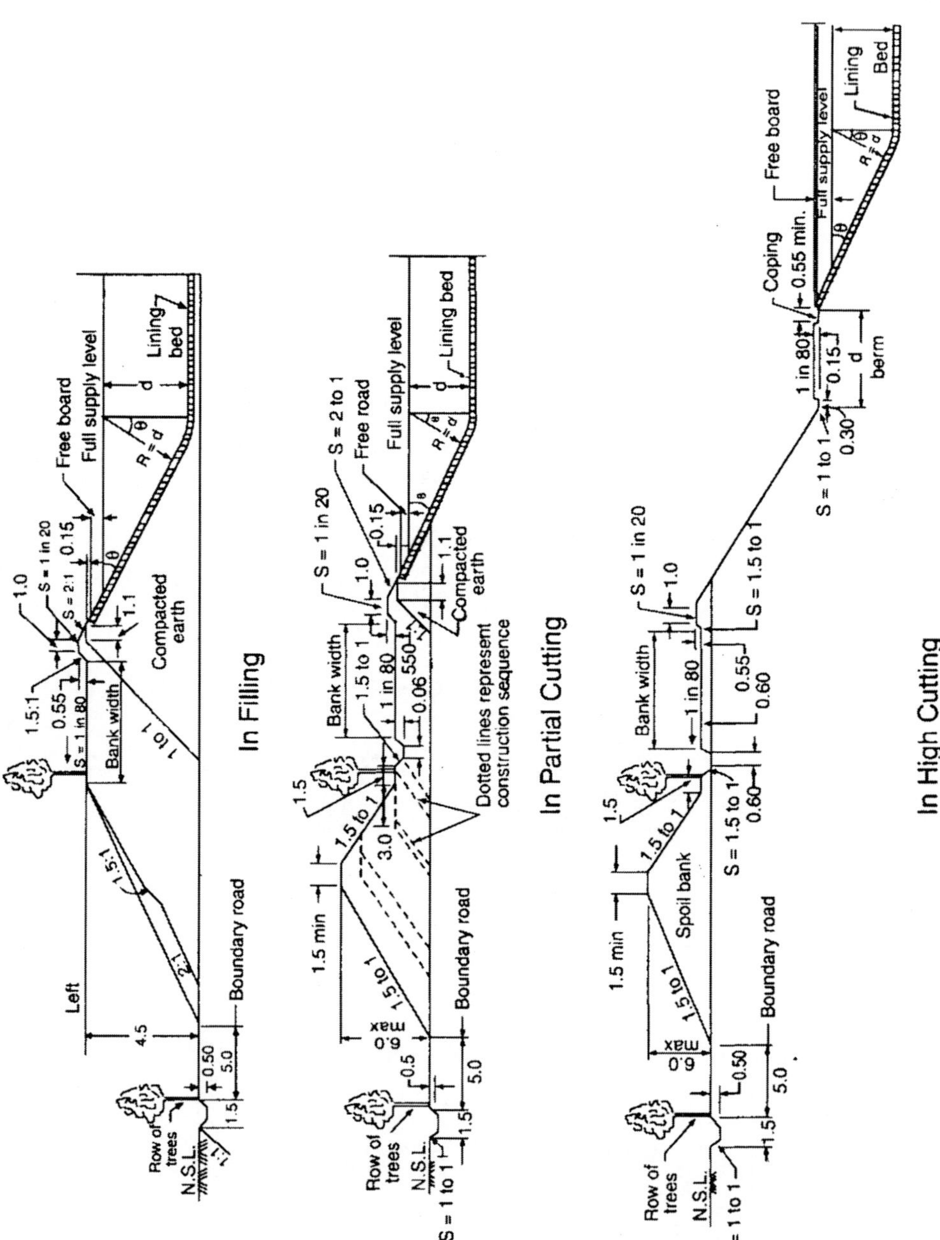

Fig. 6.4 Typical cross sections of lined canals

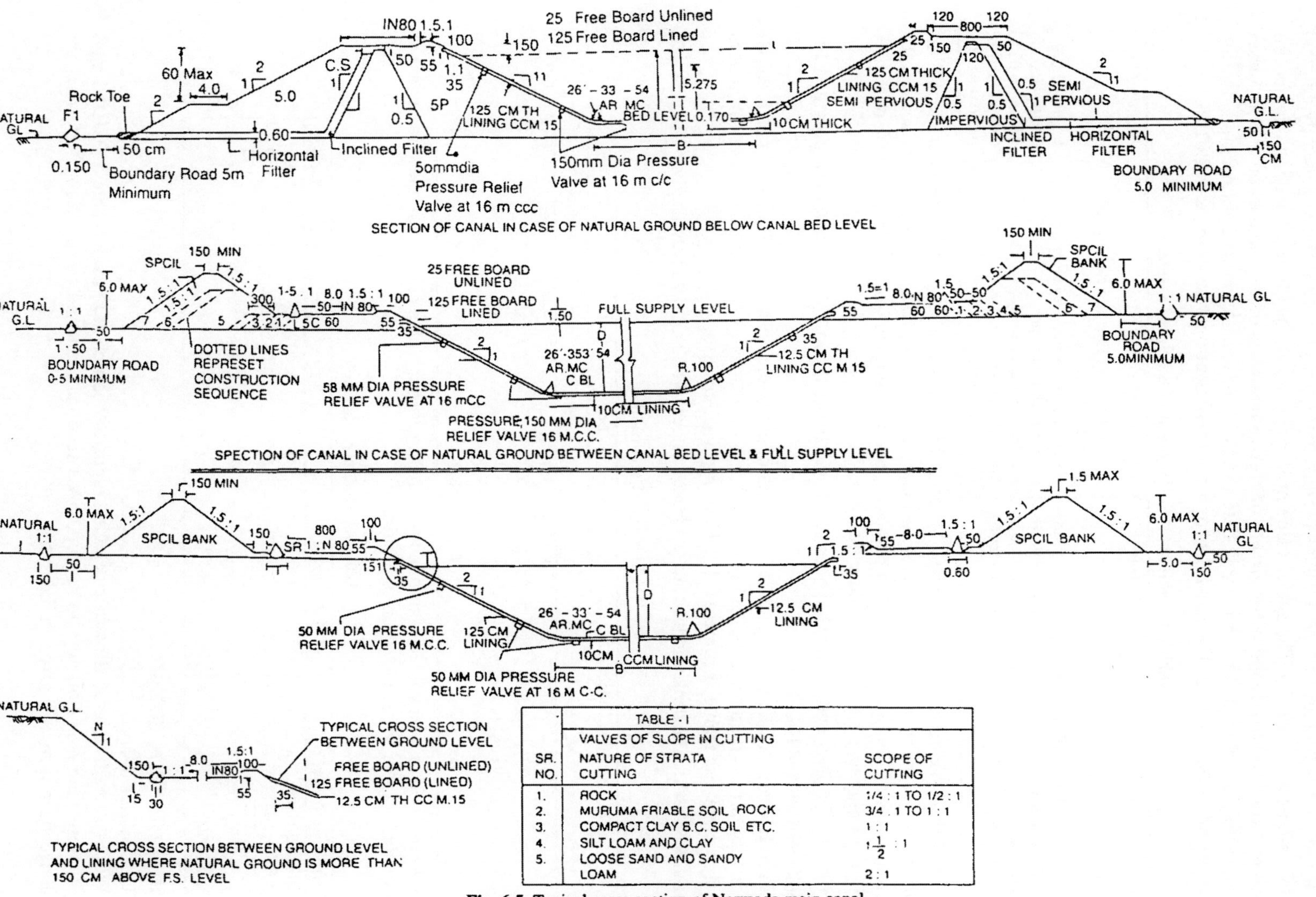

TABLE - I

VALVES OF SLOPE IN CUTTING

SR. NO.	NATURE OF STRATA CUTTING	SCOPE OF CUTTING
1.	ROCK	1/4 : 1 TO 1/2 : 1
2.	MURUMA FRIABLE SOIL ROCK	3/4 : 1 TO 1 : 1
3.	COMPACT CLAY B.C. SOIL ETC.	1 : 1
4.	SILT LOAM AND CLAY	$1\frac{1}{2}$: 1
5.	LOOSE SAND AND SANDY LOAM	2 : 1

Fig. 6.5 Typical cross section of Narmada main canal.

Rugosity Coefficient *n*: The *n* values for different types of lined surfaces generally adopted in India are as per Table 6.1 The actual value of *n* in a concrete or tile lined section adopted for design purposes is 0.018 to allow for growth of algae, sinuosities of the canal as no energy losses are separately accounted for the curves, deterioration of concrete surface with aging. During initial period of operation of the canal the rugosity coefficient may be lower, say 0.0166.

Table 6.1 Value of '*n*' for lined channels (IS: 10430–1982)

Surface characteristics	*Value of 'n'*	*Surface characteristics*	*Value of 'n'*
Concrete with surface as indicated below:		Concrete bottom float finished and sides as indicated below:	
Formed, no finish/PCC tiles or slabs	0.018 – 0.020	Hammer dressed stone masonry	0.019 – 0.021
Trowel/float finish	0.015 – 0.018	Coursed rubble masonry	0.018 – 0.020
Gunited finish	0.018 – 0.022	Random rubble masonry	0.020 – 0.025
Brick tile lining	0.018 – 0.020	Masonry plastered	0.015 – 0.017
		Dry boulder lining	0.020 – 0.030

Maximum Velocity: Water surface slope is kept as steep as practicable consistent with command to get the maximum permissible velocity. The safe velocity through a lined section for different types of lining is 2.75 m/s in cement concrete lining but usually limited to 2 m/s, 1.75 m/s in tile lining and 1.5 m/s in boulder lining.

Berm: In deep cutting reaches, berms of width equal to depth of canal are provided for facility of maintenance.

Free Board: A free board of 0.75 m for main and branch canals, 0.60 m for branch canals and distributaries with discharge 10-5 cumecs, 0.5 m for discharge 5-1 cumecs, 0.3 m for minors with discharge less than 1.0 cumec, 0.1-0.15 m for watercourses.

Bank Width: The bank widths usually adopted as per IS : 10430-1982 are as given in Table 6.2.

Table 6.2 Bank width

Canal capacity (cumecs)	*Bank width (m)*	
	Inspection Bank	*Non-inspection bank*
0.15 – 7.5	5.0	1.5
7.5 – 10	5.0	2.5
10 – 15	6.0	2.5
15 – 30	7.0	3.5
30 and above	8.0	5.0

Curves: ISI recommended radii of curves are as given in Table 6.3. Where these radii cannot be provided, proper super-elevation is provided.

Table 6.3 Radius of curves

Discharge (cumecs)	*Radius (m)*	*Discharge*	*Radius*
280 and above	900	< 40 – 10	200
< 280 – 200	750	< 10 – 3	150
< 200 – 140	600	< 3 – 0.3	100
< 140 – 70	450	< 0.3	50
< 70 – 40	300		

6.8 DESIGN PROCEDURE

1. *Circular Section*

With assumed value of allowable velocity, rugosity coefficient *n;* usual value is 0.018, longitudinal slope, and side slope of the section, cross sectional area and radius (depth) of canal are calculated from formulae given in Table 6.4.

Table 6.4 Design parameters for circular section

Design parameter	*Side slope*			
	1:1	1.25:1	1.5:1	2:1
Sectional area (A)	$1.785r^2$	$1.925r^2$	$2.088r^2$	$2.463r^2$
Wetted perimeter (P)	$3.57r$	$3.85r$	$4.176r$	$4.926r$
Hydraulic mean depth (R)	$0.5r$	$0.5r$	$0.5r$	$0.5r$
Central angle	90°	77°–20′	67°	53°
Velocity, V	$\frac{1}{n}R^{2/3}S^{1/2}$	Discharge, Q	$= \frac{1}{n}AR^{2/3}S^{1/2}$	

2. *Trapezoidal Section*

(*i*) With assumed velocity, rugosity coefficient n, and longitudinal slope, hydraulic mean depth R is worked out from the Manning's formula: $V = \frac{1}{n} R^{2/3} S^{1/2}$. Also R in terms of B and D for the given slope is taken from Table 6.5, (*ii*) With the assumed velocity, work out cross sectional area, $A = Q/V$. Also A in terms of B and D is taken from Table 6.5 for the same side slope, (*iii*) The two equations in terms of B and D are solved simultaneously to get bed width and depth, and (*iv*) Determine Lacey's f, from the relation, $f = 2.5\ V^2/R$, its minimum value is 1.2.

Table 6.5 Design parameters for trapezoidal section

Side slope	*1:1*	*1.25:1*	*1.5:1*	*2:1*
Sectional area	$BD + 1.785D^2$	$BD + 1.925D^2$	$BD + 2.088D^2$	$BD + 2.463D^2$
Wetted perimeter (P)	$B + 3.570D$	$B + 3.85D$	$B + 4.176D$	$B + 4.926D$
Hydraulic mean depth (R)	Sectional area/wetted perimeter			
Angle	45°	38°– 40′	33°– 42′	26°– 30′
Velocity (V)	$\frac{1}{n}R^{2/3}\ S^{1/2}$	(m/s)		
Tangent length	$0.4142D$	$0.35D$	$0.30D$	$0.235D$

Alternatively, (*i*) Assume channel section, (*ii*) Calculate V *i.e.* Q/A in terms of B, (*iii*) Velocity by Manning's formula is also expressed in terms of B, (*iv*) Value of B is determined after equating steps, (*ii*) and (*iii*) above, (*vii*) Design is checked for permissible velocity of > 1.75 m/s and minimum silt factor 1.2.

Figure 6.4 gives the curves for the design of lined channels.

6.9 TYPES OF LINING

1. Cement Concrete Lining

The ability of concrete to be formed into a structural material having a wide variety of sizes and shapes at relatively low cost has resulted in its wide usage in various types of canal lining. Cement concrete lining is economical where natural aggregate is locally available. Cement concrete lining is more preferable than any other type of lining where channel is to carry high velocity water.

(*i*) *In Situ Cement Concrete Lining*

It is the widely used type of lining for hydel channels and irrigation canals owing to the superior durability, impermeability, hydraulic efficiency, longer proved life, weed-free surface but has high initial cost. It has better performance in high water table reaches than tile lining. The use of mechanized equipment for laying concrete reduces construction cost. Slip-form machines supported on rails placed along both berms for laying of concrete lining offer great advantages.

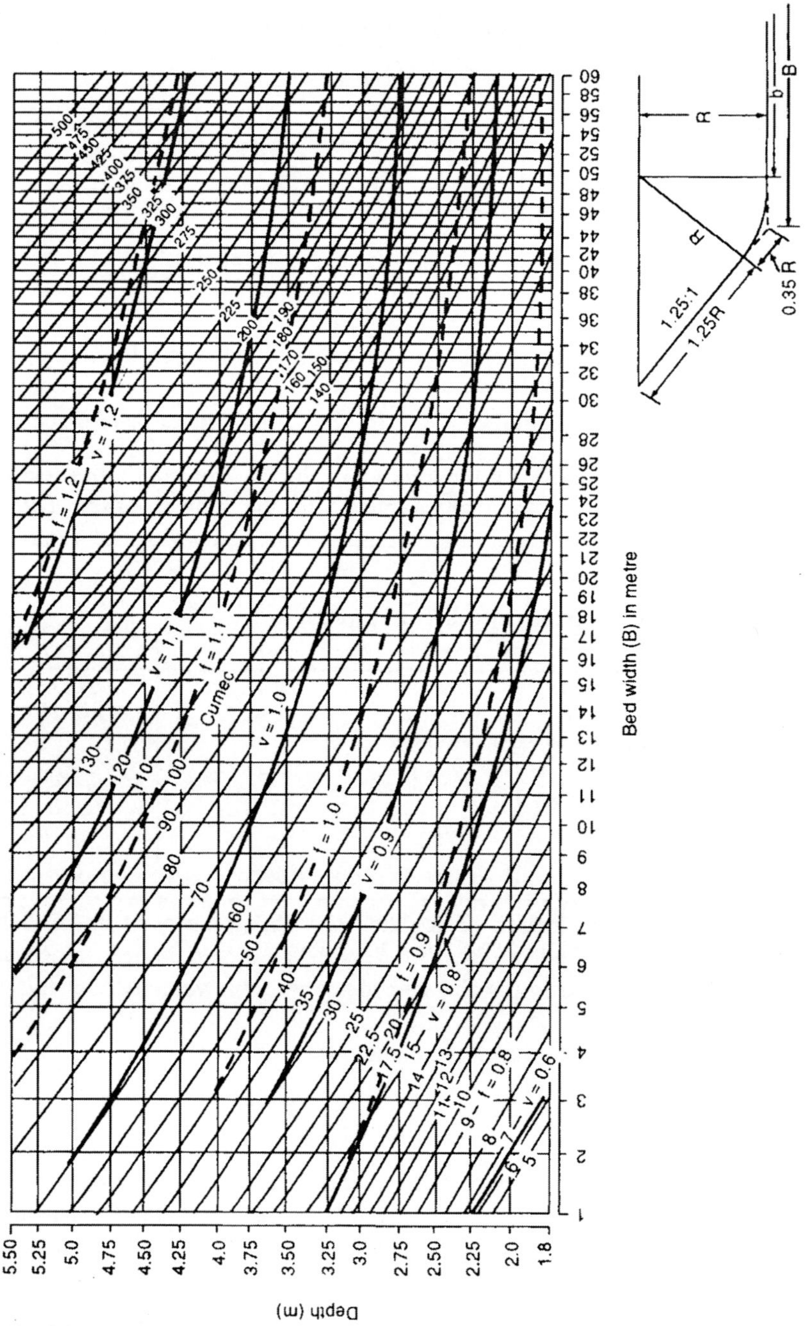

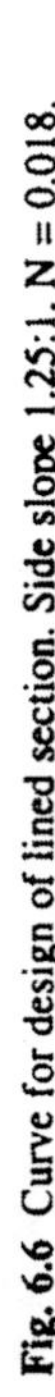

Fig. 6.6 Curve for design of lined section. Side slope 1.25:1, N = 0.018.

(A) For Other Values of N

1. To Determine 'R'

Read bed width and depth against discharge multiplied by N_1/N ($N = 0.018$)

$N_1 = 0.015$ 0.016 0.020 0.022

$N_1/N = 0.833$ 0.8889 1.111 1.222

2. To Determine 'f'

Read for discharge calculated above & multiply by $(N/N_1)^2$

$N_1 = 0.015$ 0.016 0.020 0.022

$(N/N_1)^2 = 1.44$ 1.266 0.810 0.669

3. To Determine V.

Read for discharge calculated above & multiply by N/N_1 ($N = 0.018$)

$N_1 = 0.015$ 0.016 0.020 0.022 $(N/N_1) = 1.200$ 1.125 0.900 0.818

(B) For Other Value of Slopes

1. Read bed width & depth against discharge divided by $(S_1/S)^{1/2}$

where S_1 is the new slope

1. To Determine 'f'

Read for discharge calculated above & multiply by $(S_1/S)^{1/2}$

3. To Determine V.

Read for the Q calculated above & multiply by $(S_1/S)^{1/2}$

Thickness of lining depends on full supply depth, carrying capacity, nature of channel *i.e.* irrigation or hydel; the latter may have greater thickness because of drawdown effect. Deeper channels should have more thickness than shallow channels. The usual thickness of cement concrete lining is 5 to 15 cm. The recommended thickness of lining as per IS: 3873-1992 is given in Table 6.6. On Nangal hydel channel (354 cumecs capacity) 12 cm thick cement concrete lining of 1:3:6 mix in bed and 15 cm on sides has been provided.

Table 6.6 Thickness of cement concrete lining

Canal discharge (cumecs)	*Water depth (m)*	*Lining thickness (cm)*	
up to 5	up to 1.0	5	Tolerance in thickness is ± 10 mm
5 – 50	1.0 – 2.5	6 – 7.5	
50 – 200	2.5 – 4.5	7.5 – 10	
200 – 300	4.5 – 6.5	9 – 10	
300 – 700	6.5 – 9.0	12 – 15	

Expansion and Contraction joints: Cement concrete lining is susceptible to cracking. To cater for initial shrinkage and cracks, concreting is done in alternate panels or bays of 2 to 3 m width. The construction joints should be either parallel or perpendicular to direction of flow. In case of high depth, construction joints are provided along the direction of flow to divide the length of the panel. In case lining is laid by mechanical mover, rubber water stops are provided at a spacing not more than 4 m centre to centre.

Air Entraining : Entrainment of air from 4.5 to 8% by volume corresponding to maximum aggregate size of 10 to 40 mm improves finishing and workability, makes the concrete lining impervious, durable, free from honeycombs and bleeding.

Advantages: (*i*) Most commonly used type of lining, (*ii*) Higher durability, impermeability, hydraulic efficiency and longer proved life, (*iii*) Permits high reduction in seepage losses, (*iv*) Can be used to any thickness, (*v*) Low coefficient of rugosity resulting in high velocity, (*vi*) High velocity prevents silting tendency in the canal and also evaporation losses are reduced, (*vii*) Immune from weed growth, (*viii*) Permits speedy construction by mechanical equipment, (*ix*) Economical in areas where natural shingle and sand are locally available in abundance within short leads, (*x*) Provides more sanitary conditions where passing through or near abadies, (*xi*) Lean proportion concrete 1:4:8 is also adequate provided properly graded aggregate is used and concrete is properly compacted and cured, (*xii*) Partial replacement of cement by surkhi is possible which results in economy besides improving durability of the concrete, (*xiii*) Low maintenance cost, and (*xiv*) Most resistant to erosion even at high velocities. As such suitable where gradient is steeper and velocity is high to economise in cross sectional area.

Disadvantages: (*i*) High initial cost where coarse and fine aggregates are not locally available within short leads, (*ii*) High consumption of cement, (*iii*) Skilled construction labour is needed, (*iv*) Susceptible to adverse subgrade conditions, (*v*) Elaborate concrete mixing, transportation equipment and quality control are needed, (*vi*) Prone to temperature cracking due to high coefficient of expansion and low extensibility, (*vii*) Limited resistance to rapid drawdown conditions and external hydrostatic pressure due to groundwater condition, (*viii*) Prone to climatic changes, (*ix*) Repairs costly and difficult, (*x*) Under high velocities finer material is eroded leaving behind coarse surface with high rugosity coefficient, (*xi*) Economical use of mechanical placers is possible only in case lining section is standardized, and (*xii*) Likely to be damaged by alkaline water.

(*ii*) *Precast Concrete Lining*

Precast lining has the highest durability, hydraulic efficiency and imperviousness. The precast slabs and tiles are more compact and dense than *in situ* lining. Thickness of lining is also less and it is economical to construct both from the standpoint of material and labour cost besides ensures early revenue return. The slab and tiles are cast manually or machine made out of well graded cement concrete 1:3:6. Usual sizes of slab are 50 × 50 cm and 50 × 25 cm, 40 × 40 cm, 30 × 30 cm and 25 × 25 cm. Each of these are manufactured in thickness 60, 50 and 40 mm. The tiles have their sides at right angles to the faces. The slabs are laid on well compacted and properly dressed subgrade. Precast lining is suitable (*i*) Where seepage losses are of no consideration, (*ii*) Speedy lining in a short closure is required, (*iii*) Where water table is high as water pressure is released through the joints between the slabs, and (*iv*) Coarse and fine aggregates are locally available in abundance, may not be very close to the canal.

Advantages: (*i*) Higher strength for equivalent thickness and smooth surface as precast slabs in perfectly uniform shape can be machine made from concrete properly vibrated and cured under controlled conditions, (*ii*) Form work is not required for laying lining, (*iii*) Low carriage charges of raw materials because slabs are cast near the site of availability of aggregates and water, (*iv*) Skilled labour is not required, (*v*) Drainage behind lining is not required as high water table pressure is released through joints of slabs, (*vi*) Joints filled with bituminous mastic instead of cement mortar improve flexibility, (*vii*) Larger surface coverage in short canal closure. Water can be released immediately after completing a heading of lining, (*viii*) Minimum curing of lining is required, precast slabs being already cured, (*ix*) Use of flyash and small quantity of sodium sulphate in the mix makes it highly resistant to sulphate attack besides economy in cost, (*x*) Superior to other types of lining being more durable, less pervious, more flexible, convenient in construction and economical in initial and maintenance costs, (*xi*) Repairs easy and economical. Failure of slab is rare. Joints given way are easily reset, (*xii*) More resistant to weathering, erosion and abrasion, (*xiii*) Permits use of various types of joints. Joints are uniform. Elaborate construction joints are not needed, (*xiv*) More flexible due to small size slabs and cracks are minute compared to in *in situ* concrete lining, (*xv*) Low cost, (*xvi*) More impermeable being denser in texture, (*xvii*) 50% saving in cement consumption, (*xviii*) Longer life because precast tiles are more durable and denser and least affected by salts in the soil or water, and (*xix*) Suitable for rock cuts in steep and roughly finished slopes.

Disadvantages: (*i*) Too light for hydel channel as thickness is half of *in situ* concrete lining, (*ii*) Susceptible to damage due to hydrostatic pressure when canals are closed, (*iii*) Not suitable for sharp curved lengths, casting being flat and rectangular, (*iv*) Low progress in lining operation because of hand placing of slabs and filling of joints, (*v*) Joints are the weakest points in lining, and (*vi*) Low resistance to rupture.

2. Tile Lining

Tile lining or brick lining, as it is generally called, is the most commonly used type of lining next to concrete lining. Tiles in themselves are not impervious but the sandwiched mortar layers between them impart imperviousness to tile lining. The degree of impermeability of the lining is approximately the same as that of the concrete lining. The lining is thus suitable in first cost, impermeability and expediency.

Tile lining generally consists of two layers of tiles with sandwich mortar in between. Double tile lining used on Bhakra and Rajasthan canals and conforming to IS: 10646-1991 is as shown in Fig. 6.7. It consists of two layers of tiles, 30 × 15 × 5 cm, with sandwiched cement mortar. Cement concrete coping of grade M 100 or double layer of tile with a mortar specified for the side slope is provided horizontally on the top of lining in order to check any leakage of rain water getting behind the lining. The subgrade has to be compacted and properly dressed.

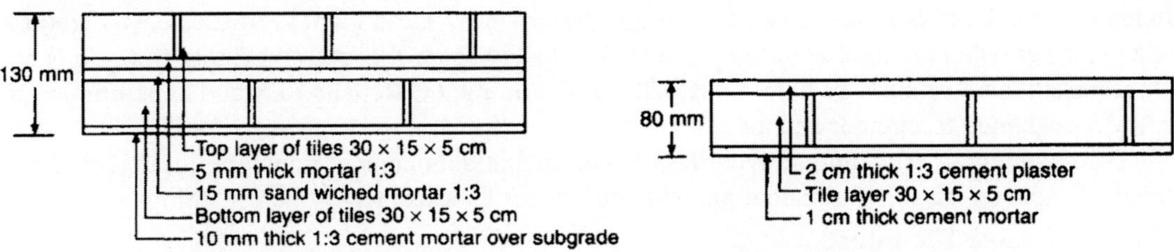

Fig. 6.7 Double tile lining. **Fig. 6.8** Single tile lining.

Tile lining may also consist of single layer of tiles termed as single tile lining (Fig. 6.8.). In large capacity canals, single tile lining in bed is primarily adopted for reasons of economy in the cost. In small capacity canals both the bed and sides are lined with single tile. Table 6.7 indicates the details of double and single tile lining.

Table 6.7 Double and single tile lining

Double tile lining	*Single tile lining*
(*i*) 50 mm (30 × 15 × 5 cm) tile in 1:3 cement mortar	
(*ii*) 5 mm cement mortar 1:3	50 mm tile in 1:3 cement mortar.
(*iii*) 15 mm sandwich mortar 1:3	10 mm cement plaster 1:3
(*iv*) 50 mm bottom layer tile in 1:3 mortar	12 mm cement plaster 1:5
(*v*) 10 mm 1:5 cement mortar on subgrade	Joint thickness 10 mm maximum

Fig. 6.9 indicates single tile lining as per IS: 3872-1966. Fig. 6.10 depicts single tile lining in bed and double tile lining on sides. Fig. 6.11 indicates brick lining adopted in Punjab.

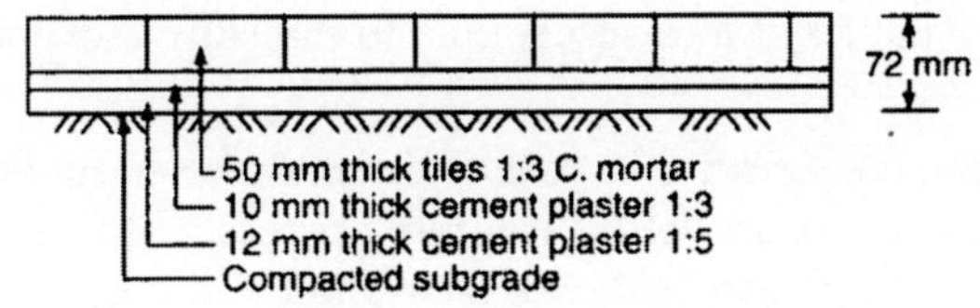

Fig. 6.9 Tile lining (IS : 3872)

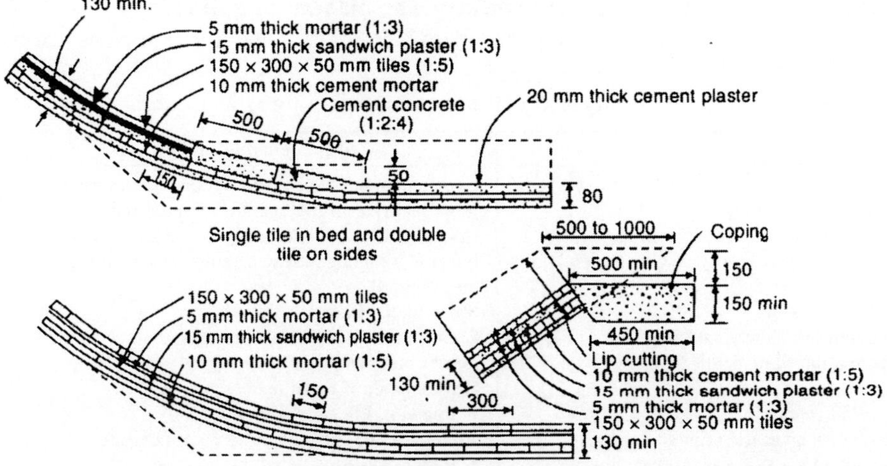

Fig. 6.10 Double tile in bed and sides.

Advantages: (*i*) Low initial and maintenance cost, (*ii*) Quicker construction, (*iii*) Economical where aggregates for concrete lining are not available, tiles can be burnt locally with minimum carriage costs, (*iv*) Elaborate construction equipment and quality control are not required, (*v*) Skilled labour as needed for concrete lining is not required, (*vi*) Lesser (40 per cent) consumption of cement than cement concrete lining. Saving is more in the case of single tile lining, (*vii*) No danger of undetected weak spots as thickness of lining is governed by the thickness of tiles, (*viii*) Natural safeguard against cracking due to the presence of closely spaced joints, (*ix*) Expansion joints are not required, there being large number of joints in tile lining, (*x*) Repairs easy and expeditious and (*xi*) More adapted to circular section.

Disadvantages: (*i*) Comparatively less impermeable than cement concrete lining, (*ii*) Comparatively low resistance to erosion, and (*iii*) Lesser life than cement concrete lining.

Single vs Double Tile Lining

Single tile lining has certain demerits compared to double layer tile lining such as (*i*) Structurally single layer tile lining is less stable on account of lesser self weight against differential head, (*ii*) Less durable and susceptible to failure, (*iii*) Remote chances of carrying out repairs in the bed if lining fails and canal is perennial, (*iv*) Smooth cement plaster on the top, as in Rajasthan canal, (Fig. 6.10) is liable to deteriorate and may not provide intended improvement in coefficient of rugosity in the long run, (*v*) Top cement plaster is prone to temperature cracks. In single tile lining the movement due to expansion under direct sun heat is much greater, and as such bigger cracks may develop in top plaster, (*vi*) Shear strength of single tile is less, (*vii*) Every chance of undetected faulty joint filling or flaking out of the plaster. In double tile lining, since joints are staggered in the two layers, faulty joint in bottom layer is covered by the top layer tile, of course sandwiched layer is there in between, (*viii*) Sandwiched layer which imparts imperviousness to tile lining cannot be provided in single tile lining, and (*ix*) More susceptible to moisture movement.

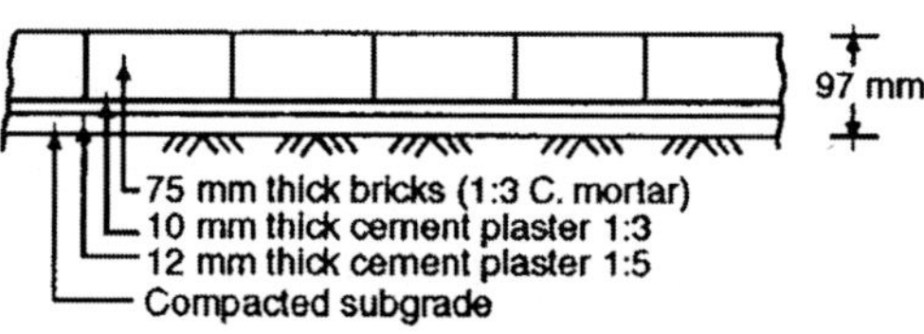

Fig. 6.11 Brick lining.

Shotcrete Lining

Shotcrete or gunited lining comprises thin coat of cement mortar applied to the rock surface under pneumatic pressure by dry mix as well as wet mix process. It consists of cement and sand admixture only. The lining may be constructed with or without reinforcement in the form of mesh or expanded metal. The pumped admixture coats over all the crevices in the rugged rock surface without leaving any gaps to give a lined surface. The maximum grain size of sand is limited to 5 mm. The thickness of lining usually varies from 3.5 to 5 cm. The life expectancy of the lining is over 20 to 30 years.

A comparison of shotcrete lining and cement concrete lining is as under:

Shotcrete lining	*Cement concrete lining*
Installation of necessary equipment for lining is simpler and less time consuming.	Even pneumatic placement of concrete lining is time consuming. Elaborate arrangement is required for concrete placement.
Consolidation of concrete is not required.	Consolidation of concrete by vibrators is essentially required.
Form work is not required, only staging for erection of platform is required.	Form work and centring arrangement to support form work are required.
Lining of thinner section.	More thick lining.
Difficult to control thickness of lining.	Lining usually of uniform thickness.
Higher comsumption of cement.	Lesser comsumption of cement.
Costly.	Economical; of half the cost.
Lesser life, 20-30 years.	Permanent, life over 60 years.
Best adapted for rock cut irregular surfaces.	Adaptable for soil as well as rock surfaces.
Fine dressing of subgrade not essential.	Subgrade has to be properly dressed.

Advantages: (*i*) Adapted to lining irregular and steep side slopes and where trimming to exact profile is difficult and costly, (*ii*) Specially indicated for resurfacing old canals, (*iii*) Most suitable for smaller canals, and (*iv*) Lighter equipment and small working labour is required.

Disadvantages: (*i*) High cost, roughly twice the cost of concrete lining of the same thickness, (*ii*) High consumption of cement, (*iii*) Unless suitably reinforced its resistance to rupture is not greater than that of concrete, (*iv*) Difficulty in controlling thickness of lining which may leave areas where thickness is less than the minimum required, and (*v*) Susceptible to damage by external water pressure easily due to small thickness.

Cement Mortar Lining

It is not a durable type of lining and is used in conjunction with some other protection material such as a layer sandwiched between two layers of tiles where it imparts imperviousness to the lining. Its thickness varies from 10-25 mm. Construction method is identical to that of concrete lining. Because of high consumption of cement it is costly and is not commonly used. However, 25 mm thick cement mortar lining is sufficient to reduce seepage losses by nearly 75%.

Asphalt Lining

It is a type of lining which besides being costly, absorbs heat and encourages weed growth causing rapid deterioration and as such has not found much use in India. It is of two types:

1. *Buried asphalt membrane.* Hot applied asphaltic, prefabricated asphalt materials, plastic film and layer of bentonite and provided with 30 cm cover of earth for protection. The lining generally consists of canvas or kraft paper impregnated with 5 mm layer asphalt and is laid directly over the subgrade and covered with earth.

2. *Asphaltic concrete lining.* It consists of a mixture of asphaltic cement and suitable aggregates mixed hot. The mix is placed by hand and equipment. Asphaltic membranes tend to flow down in summer.

Advantages: (*i*) Greater ability to withstand changes in subgrade, (*ii*) Can be used for repairing cement concrete lining by placing a resurfacing layer of asphaltic concrete, and (*iii*) Suitable where aggregate of good quality for cement concrete is not available but is suitable for asphaltic lining.

Disadvantages: (*i*) Velocity is limited to 1.5 m/s, (*ii*) Danger of weed growth resulting in puncturing of lining, (*iii*) Insufficient resistance to external hydrostatic pressure, and (*iv*) Danger of sliding during hot season.

Hydraulic Lime Concrete Lining

It consists of lime concrete. The use of this type of lining is indicated where kankar is available in abundance along the canal. Hydraulic lime made locally from kankar and mixed in suitably designed proportion with grit and ballast make a good lining. Kankar can also be used as coarse aggregate. On crushing, kankar grit can be used instead of sand. It is not as impermeable as cement concrete lining. Hydraulic lime in combination with cement results in a decrease in compressive strength of concrete in proportion to the quantity of hydraulic lime used. The thickness of lining varies from 10 to 15 cm for discharge up to 200 cumecs. The concrete mix is 1 : 1.5 : 3 of kankarlime: kankar grit or sand: kankar aggregate, stone aggregate or brick ballast.

Stone Pitched Lining

Also called boulder lining. The lining is useful and economical where stones of suitable size are locally available in abundance, subsoil water table is higher than the bed of canal, drainage to lining due to back pressure of high subsoil water level is avoided as pressure is released through its interstices, and prevention from erosion is also to be ensured. The stones should be of reasonably uniform size, rounded or subangular river cobbles or blasted rock pieces with sufficient base area to be stable. The stones shall be sound, hard and durable, free from cracks, seams and other defects to sustain weathering and water action. Permissible velocity in the canal is 1.5 m/s. The thickness of lining and stones dimensions are as under:

Canal discharge (cumecs)	*Thickness of lining (mm)*	*Average dimension along the longest axis (mm)*
Up to 50	150	150
50 – 100	225	225
100 and above	300	300

In laying lining subgrade of bed and sides is divided into compartments of not more than 15 m along the centre line of the canal by stone masonry or concrete ribs. The spacing of ribs across the centre line is such as to divide the canal bed and slope symmetrically about the centre line and in such a manner that ribs are provided at the junction of the slope and bed and at the upper extremity of the slope (Fig. 6.12). The ribs along the slope of the bank are kept continuous. A 15 mm thick filter layer is provided under the lining.

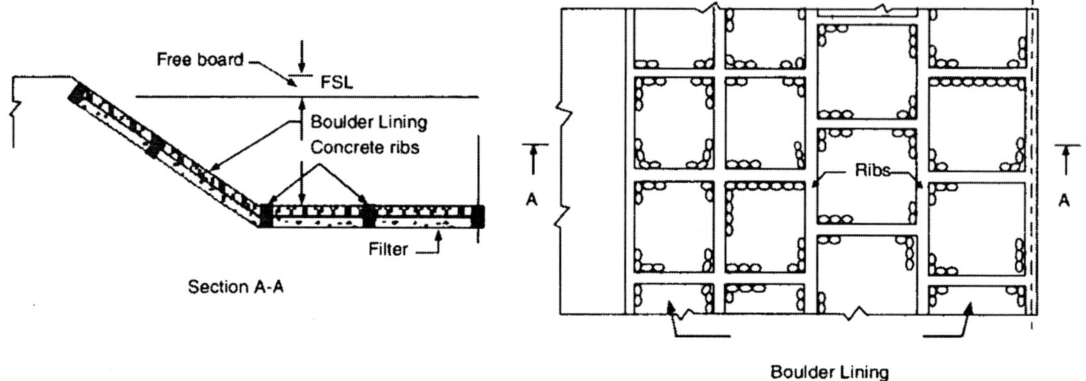

Fig. 6.12 Stone pitched lining.

Soil Cement Lining

Soil cement lining, as the name implies, is constructed by compacting soil cement mix at optimum moisture content. Economical use of cement and locally available sandy soil is made to make lining. Sandy soil or soil 100 per cent passing 10 mm I.S. sieve containing not more than 20 per cent material passing a 75 micron I.S. sieve but free from injurious salts such as carbonates and sulphates is suitable. This type of lining is specially suitable for small and medium sized channels and water courses where subgrade is sandy. Usually thickness of lining is 10-15 cm with velocity limited to 1 m/sec. More impervious surface can be obtained by providing a top layer of tile lining or black polyethelene film spread below the layer of soil cement lining.

Advantages: (*i*) Economical where subgrade is sandy or sandy soil is available for blending with cement, (*ii*) Suitable for minors and watercourses, and (*iii*) Reduces seepage at very low cost.

Disadvantages: (*i*) Short life, (*ii*) Maintenance cost is high, (*iii*) Prone to weed growth, (*iv*) Unsuitable for lining on expansive soils, and (*v*) It is not weather resistant.

Clay Lining

Clay lining is suitable for areas where expansive soils or high ground water table conditions are met with, because of flexibility and greater weight of the lining. Compaction is done in horizontal layers of 15 cm for 60 to 90 cm thick lining on sides normal to slope and about 60 cm for the bed. The usual thickness of lining is 30 cm protected with a 30 cm layer of silt. Clay being less porous makes impermeable and serviceable lining at low cost. The clay is weathered and pugged thoroughly after saturating it with water by tramping under human or animal feet.

Bentonite Lining

It consists of either as a membrane of 5 cm thickness covered with protective blanket of 30 cm or a mixture of sandy soil and bentonite well compacted to 5-7 cm thickness. On alternate wetting and drying cracks are developed in the lining. Bentonite earth with a mixture of lime, cement, pozzolana and coal ash is used with advantage. It helps reduce seepage losses at economical cost. Bentonite lining has the drawback that it deteriorates due to weathering, weed growth and cracking.

Bitumen Lining

Bitumen lining may be sprayed *in situ*, or prefabricated. Sprayed lining consists of a membrane of 6-12 mm thickness sprayed into place at a high temperature to form a watertight barrier

5 mm thick prefabricated sheets are also used. Protective covering is very important in bitumen lining. Bitumen concrete 5-7 cm thick gives satisfactory performance. Bitumen lining is better than asphaltic lining.

Combination Lining

Combination lining consists of a membrane in bed and brick, tile or concrete lining on the sides. It is adaptable where the canal is stable and is not susceptible to scouring.

Polyethylene Film Lining

The polyethylene film lining, usually black in colour, has nominal thickness not less than 1.50 microns. The film is laid over subgrade prepared below the designed bed level to the extent of cover thickness in strips perpendicular to water flow depending upon width of the film, width of bed and perimeter of the section. Longitudinal joints are avoided. The side slope usually vary from 1.5:1 to 2:1. The film laid over 12 to 25 mm sand layer is covered with earth cover of minimum thickness of 30 cm in bed and 60 cm on sides and compacted. In large canals, a masonry toe wall is provided (Fig. 6.13). The top cover is of pre-cast concrete tiles, in-situ-cement concrete, stone slabs or tiles.

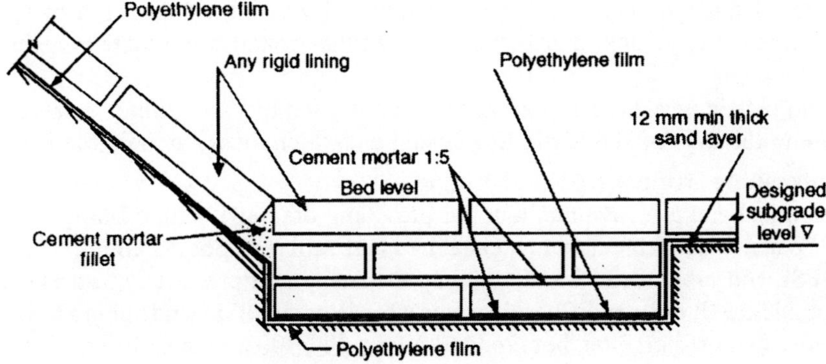

Fig. 6.13 Typical masonry toe wall.

6.10 LINING IN EXPANSIVE SOILS

Lining in expansive soils is avoided. Where unavoidable to traverse a reach of expansive clay, if in thin layer or in small pockets may be overexcavated and replaced with a suitable non-expansive soil and compacted. Swelling of clay can be controlled by loading the surface with 60 cm thick non-expansive compacted soil or gravel and filled to the grade of the underside of lining with good draining material. The excavated surface of expansive clay be given a coat of asphalt before loading.

In other than predominantly sandy reaches where dry bulk density of natural soil is not less than 1.8 g/cm^3, initial excavation is done to about 30 cm above the final section and cutting to final shape is done immediately before the lining.

Anti-salt Treatment: In reaches where salt contents are over 1.0 per cent or sodium sulphate is over 0.36 per cent, the subgrade is first covered with about 2 mm thick layer of bitumen obtained by evenly spraying bitumen at a rate of about 2.35 l/m^2. Crude oil at a rate of 0.5 l/m^2 is sprayed over it in advance of spraying bitumen to get a good bond between bitumen and soil.

6.11 DRAINAGE BEHIND LINING

Failure of canal lining is generally due to excess hydraulic pressure behind the lining resulting either from high watertable condition or pressure built up due to time lag in drainage of subgrade following drawdown of canal, hence the necessity for the provision of drainage behind lining. Drainage behind lining is not required when subgrade is free draining and there is no time lag in the dissipation of drawdown pore pressure in the backfill.

Drainage behind lining is provided depending mainly upon the position of watertable and type of subgrade. Pressure behind lining is built up in case the backfill is practically impervious ($K < 10^{-6}$ cm/s) or of poor draining soil (10^{-6} cm/s $< K < 10^{-4}$ cm/s) which cannot drain out at the same rate as the inflow. The inflow is due to (*i*) high water table, *i.e.*, above the canal bed level, (*ii*) saturation by canal water itself, (*iii*) saturation due to rainfall, and (*iv*) sudden drawdown conditions.

When the supplies are run for a long time, equilibrium is established between the full supply level in the canal and the free surface in the banks. Under sudden drawdown conditions created by closure or breach of the canal, the water level in the canal drops at a much faster rate than the lowering of the free surface in the banks and as a result lining is subjected to differential hydraulic head detrimental to its safety. Pervious soil in the canal banks overlying strata of low permeability is more likely to set up adverse pressure against the lining than pervious soil overlying highly permeable strata. High water table above canal bed level also causes uplift against lining. Water table is lowered by pumping to create working conditions for laying lining in bed. When pumping is stopped, water table again rises and exerts pressure on the lining unless proper drainage behind lining is provided. Drainage behind lining is not required where the backfill is of self draining material with permeability greater than 10^{-4} cm/sec and the ground water table is below bed level of the canal.

External hydrostatic pressure has a great effect on the stability of a lining. Drawdown rate of 30 cm/hr is safe in the case of single tile lined canal and 45 cm/hr for double tile lining.

Methods for providing Drainage Behind Lining

Most economical and effective methods for providing drainage behind lining are as under:

1. *Filter*: When water table is below canal bed and subgrade poor draining or it is between canal bed and FSL and subgrade is free draining, poor draining or practically impervious, provide 150 to 200 mm thick well designed filter. If water table is above FSL, and subgrade is free draining, provide 150-200 mm thick filter, but if poor draining 200-300 mm thick filter. In case subgrade comprises practically impervious, the subgrade is removed to a depth of 600 mm and replaced by sand, murum.

2. *Humps.* In high water table reaches, high crests of falls, drainage crossings and other masonry works or alternatively humps are provided to retain sufficient depths of water on the upstream to counteract the uplift pressures (Fig. 6.14). Pipes with necessary valves are provided to open or close them to facilitate draining of the reach upstream of its location for inspection. Provision of humps causes head loss in the canal.

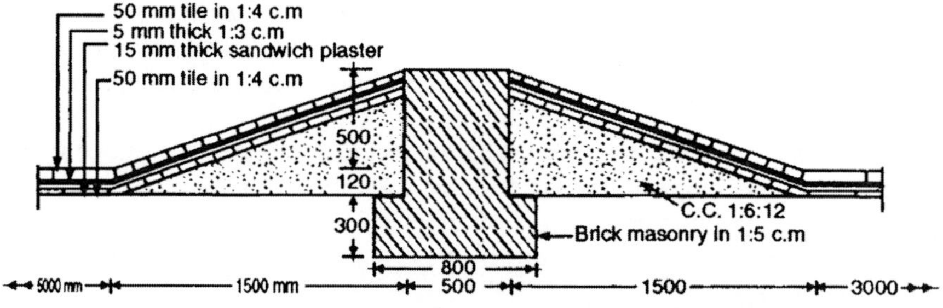

Fig. 6.14 Hump.

3. *Pressure Release Valves.* It is a device installed on slopes of lining or on the longitudinal transverse drains, if provided, with a valve opening into the canal to release excess hydrostatic pressure built up behind the lining. It is capable of opening at a differential pressure less than that which may damage the lining. It may be provided in pockets filled with graded filter underneath lining (Fig. 6.15). Pressure release valves are indicated in poor draining or practically impervious

subgrade and water table is below bed level, between canal bed and FSL, and provided on longitudinal and transverse drains (Figs. 6.15. and 6.16).

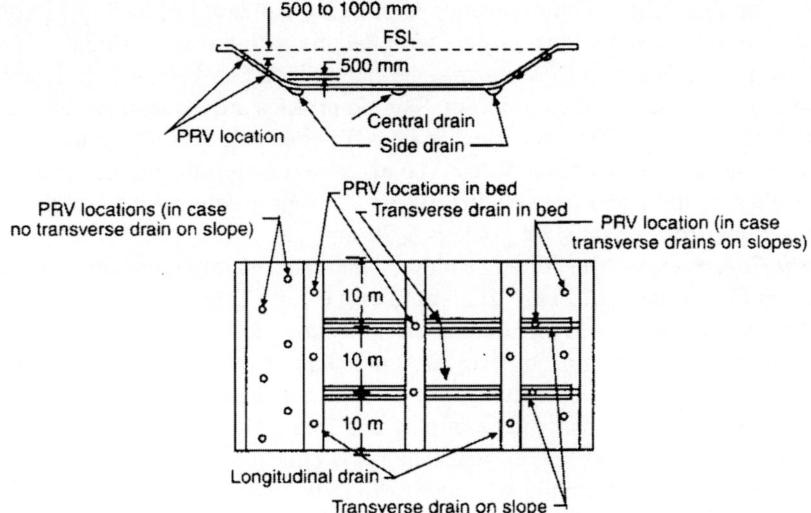

Fig. 6.15 Drains and PRV locations

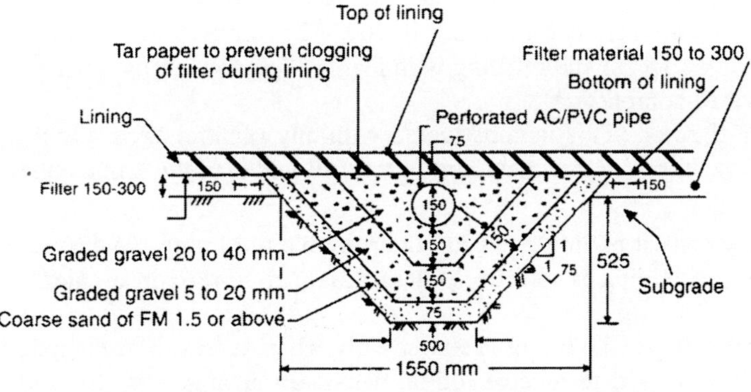

Fig. 6.16 Longitudinal/transverse drain (PRV not shown)

4. *Longitudinal and Transverse Drains:* These are provided; longitudinal in bed and sides with PRV and transverse drains with PRV on sides when water table is above canal FSL. Only longitudinal drains in bed and PRV on sides in other water table and subgrade conditions. The drain is trapezoidal with bottom width 500 mm, depth 525 mm with steep slopes (Fig. 6.15) provided with 150 mm asbestos or PVC perforated pipe embedded in graded filter. The longitudinal drains in bed are provided at least one for 10 m width, placed symmetrically with reference to centre line of the canal. The transverse drain is provided, where necessary, in the bed and on the side slopes up to free board level of the section as indicated for longitudinal drains. In general, transverse drains are provided at 10 m interval.

5. *Regulation of Supplies.* Regulation of supply in the lined canals is so devised that water level in the canal about to close is allowed to fall only at a rate slower than the rate at which the subsoil water level is expected to fall. The closure of canal as far as possible is arranged when water table is the lowest.

6.12 CONSTRUCTION OF LINED CANAL

Excavation. In reaches where the dry bulk density of the natural soil is not less than 1.8 g/cm^3, the channel section is excavated up to about 30 cm above the final grades in the first instance to allow for unequal settlement, weathering action and ravelling by equipment travel and subsurface preparation under the lining. The subsurface preparation to the finished grades below lining is to include only cutting and no filling. The excavation is done by standard excavation equipment such as rippers, shovels, draglines or bulldozers, etc., with the precaution that overbreaks, if any, are to the minimum. Over excavation up to 5 cm may be backfilled. The blasting is resorted to only when absolutely necessary and to the minimum. Sample profiles are made at an interval of about 20 m in the straight reaches and 10 m in curved reaches for checking the uniformity of side slopes. Concrete templates are laid on sample profiles. The excavated material suitable for being utilised in the filling reaches of the canal is used directly in the appropriate reaches of the canal. The tolerance in excavation from established grade is $\pm$ 20 mm.

Embankment Construction. The filling section of the bank mainly comprises two zones:

1. *Impervious Compacted Fill Zone.* It comprises the earth filling up to the top level of the lining from the natural surface level. The bank section is constructed wider by 50 cm to cater for a margin of probable improper compaction on the inner edge of the embankment. The additional compacted soil is removed (called lip cutting) prior to placing the lining and filter. The impervious material shall have sufficient percentage of clay and free from gravel and having coefficient of permeability not greater than 30 cm per year. The fill is placed at optimum moisture content in layers generally not more than 15 cm thick to obtain dry bulk density of not more than 95 per cent of the density at optimum moisture content. The number of coverage of compaction equipment is such as to give the specified compacted fill density. Rolling is done parallel to the centre line of the canal. In predominantly sandy reaches, compaction of the bed is done by over-saturating the bed by flooding it with water before lining is laid. The compaction of sides is done by overcutting the subgrade by 15 cm and refilling with lean mortar containing adequate quantity of lime or cement or by vibro-compactors.

2. *Random Fill Zone.* This is constructed with any random excavation material placed in layers not exceeding 30 cm in thickness and to the full width of the embankment.

Laying of Tile Lining

Tile lining may consist of single tile lining or double tile lining. To ensure proper workmanship and precision, the lining is laid in a phased manner, spread in a calendar of seven days, detailed as under:

First and Second Day. The lining is started only when at least 35 m length of the canal with subgrade is properly dressed to receive lining; necessary arrangement for sufficient number of tiles, proper soaking of tiles and mortar mixing shall be made. The tiles properly soaked in water for at least two hours shall be available for carrying out the work during each day. The water table, if high, is lowered by dewatering to at least 30 cm below the subgrade. The subgrade is then uniformly soaked with water to ensure that water penetrates to a depth of about 30 cm in sandy soil and about 15 cm in other soils but without making the subgrade slushy. Wetting of subgrade is continued in advance of laying tiles to avoid moisture absorption from the 10 mm thick mortar laid on the subgrade for laying tiles. Mortar is uniformly spread over the subgrade and tiles are laid in position quickly and properly so that vertical joints are completely filled with mortar. The tiles are firmly embedded in mortar. Hollows, if any, are rectified by relaying the defective portions with fresh mortar. Tiles are laid at right angles and parallel to the centre line of the canal in bed and sides respectively. Enough number of profiles both in the bed and side slope are laid each day so that when the work is started next day the marks are already in position. Construction points or panels at suitable width each in bed and the sides are perfectly rolled, levelled and laid with profiles every day to receive lining of the first layer on the next day.

Third Day — The layer of tiles laid are kept wet by sprinkling water over it to keep the mortar well wetted.

Fourth Day — The surface is kept wet and joints of tiles are carefully tested. Hollow joints, if any, are raked to a depth of 12 mm, loose mortar removed from sides and top of tiles and the joints properly refilled. Loose tile, if any, is removed and relaid.

Fifth Day — (*i*) For Single Tile Lining. Cement plaster 1:3, 20 mm thick is laid on the completed portion of the tiled bed. The surface of plaster is finished even and smoothened.

(*ii*) For Double Tile Lining. The second layer of tile may be laid in the bed or on the side slopes. The 15 mm thick sandwiched layer of 1:3 cement sand mortar is laid over the first layer of tiles. The plaster is well pressed and upper side made rough to ensure bond with the second layer of tiles.

Sixth Day — The cement sand plaster is kept well wetted. Single tile lining is complete and is checked for level with wooden templates and spirit levels. Subsequent to the laying of cement plaster and after 24 to 36 hours it is cured for at least 28 days by keeping lining submerged under 15 cm deep water and curing of side slopes is continued by water spraying.

Seventh Day — The top layer of tiles is laid in 1:3 cement mortar about 5 mm thick. This layer of tiles is laid firmly and properly to proper levels in bed and correct slopes on the sides with joints not more than 10 mm thick. The tiles used in the top layer are of best quality and laid with great precision to ensure smooth lined surface. Vertical joints are laid flush with cement mortar. The completed lining is checked for level with wooden templates and spirit level. The lining is cured for at least 28 days in the manner indicated for single tile lining in sixth day schedule.

Laying of Concrete Lining

The lining is started only when at least 35 m length of the canal with subgrade is properly dressed to receive lining and necessary arrangement for mixing of concrete and installation of form work has been made. Lining is laid with great precision in a phased manner described as under:

1. *Placing of Concrete.* The placing of cement concrete is started after the subgrade is moistened thoroughly, 1:3 cement slurry spread over it so that the soil does not absorb moisture from the freshly placed concrete. Hand placing of concrete is normally adopted. All concrete is placed directly in its final position within 20 minutes of mixing. The concrete is dumped and spread on the sides and bed of the canal in panels of not more than 3 m side with suitable joints in between. Screed guides are laid on the subgrade for placing concrete to proper thickness. For efficient placing of concrete on slopes, the use of weighted vibrated steel faced slip-form screed in the direction of movement are adopted. The joints are filled with mastic.

2. *Mixing.* The concrete is normally mixed in a mechanical mixer. The consistency of concrete shall be such that it barely stays on the slope. The slump of concrete varies from 6 cm to 7 cm, 5 cm in the case of machine mixing.

3. *Finishing.* The surface of concrete finished against forms shall be smooth and free from projections, honeycombing and other defects; variation within ± 10 mm.

4. *Curing.* Subsequent to the laying of concrete lining and after a period of 24 to 36 hours, the lining is cured for at least 28 days.

6.13 LINING OF WATER-COURSES

Water-courses lining is a new concept of water management to (*i*) Minimise water loss due to seepage, (*ii*) Effecting more equitable distribution of water and its better regulation and management, (*iii*) Making possible flatter slope, minimum being 1.5 cm/100 m and at the same time higher velocity and lesser cross section, (*iv*) Economise in land coming under it due to lesser section possible, and (*v*) Ensure stability, rigidity and maintenance. The various types of lining are as under:

1. *Brick Lining*: It consists of brick masonry in bed and sides, plastered inside (Fig. 6.17).

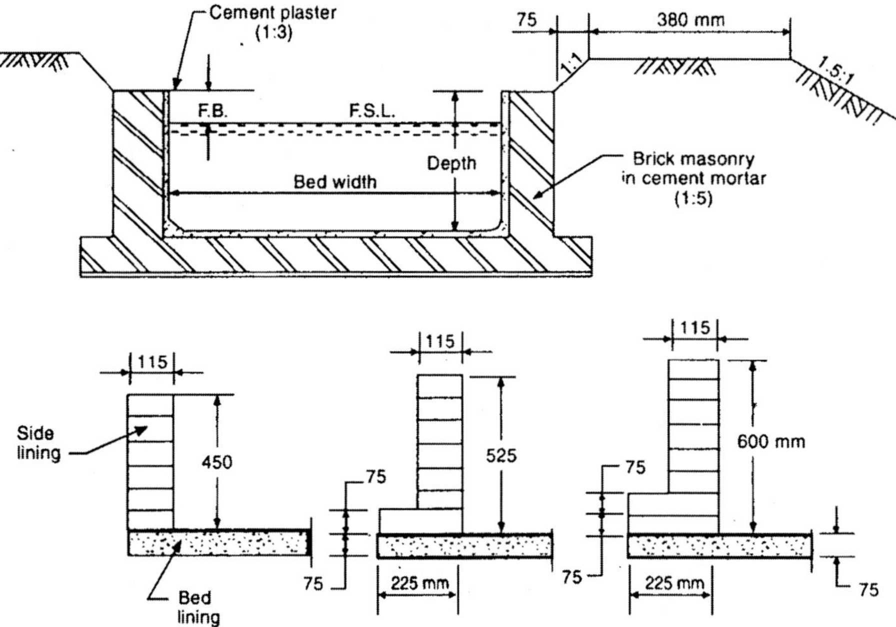

Fig. 6.17 Brick lining.

2. *Concrete Lining.* It consists of 5 cm thick 1:3:6 cement concrete laid in bed over a 100 microne LDPE film with 7.5 cm thick side walls. In case LDPE film is not used 7.5 cm thick concrete is laid (Fig. 6.18).

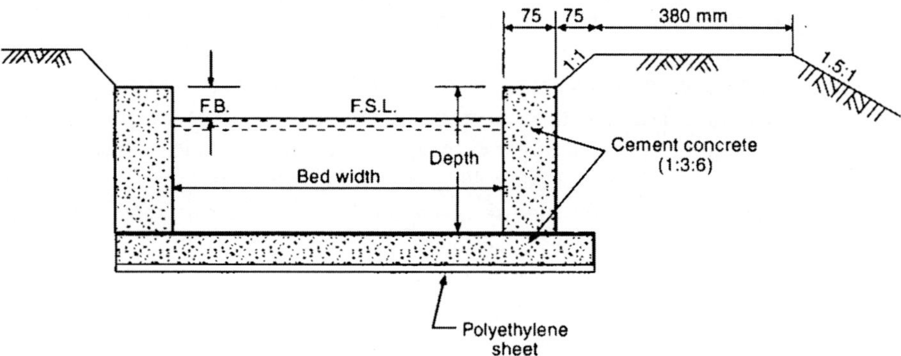

Fig. 6.18 Concrete lining.

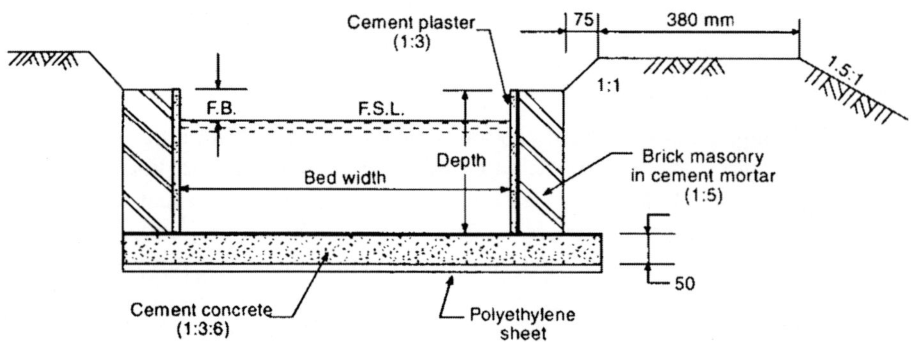

Fig. 6.19 Concrete and brick lining.

3. *Composite Concrete and Brick Lining.* It consists of 5 cm thick 1:3:6 cement concrete in bed, brick masonry walls in 1:5 cement sand mortar with 10 mm thick 1:3 cement sand plaster or in 1:4 cement sand mortar laid over 100 microne LDPE film (Fig. 6.19).

4. *Stone Masonry Lining.* It consists of 2.5 cm thick stone slab laid in 1:5 cement sand mortar. The masonry on sides is laid in 1:4 cement sand mortar.

6.14 OPERATION AND MAINTENANCE OF LINED CHANNELS

Various measures which should be taken for maintaining a lined channel in perfect condition are:

1. *Proper drainage of banks.* Proper drainage of banks is very essential so that pressure is not exerted on lining due to percolation of rain water from the bank. Proper cross slope (1 in 80) of the banks should be maintained.

2. *Distress to lining.* Lining, even of rigid type and conforming to best standards of design and construction, is susceptible to failure. The chief causes of distress to lining are settlement of subgrade, drawdown, development of cracks due to weathering, chemical action of subgrade with lining material due to improper filter action and inadequate drainage. The important amongst these is drawdown. The world's largest lined canal, Narmada main canal, has drawdown criteria which stipulates that the drawdown should not exceed 15 cm in one hour, 30 cm in two hours and 65 cm in 24 hours.

3. *Pressure relief arrangement.* Proper functioning of properly designed and built under-water drainage network is very essential. The humps and cross regulators should be well maintained and repaired for their efficient working.

4. *Repairs to cracks in lining.* The cracks in lining or at joints should be properly raked and then filled with cement mortar and bitumen to make the lining watertight. V-groove should be cut along the face of the crack before filling with sealing compound. The joint filler where found damaged or displaced should be filled with fresh filler material.

5. *Sand grouting behind lining.* The lining slabs generally fail due to formation of cavity behind the slabs. The cavity should be filled by filling sand behind the lining.

6. *Removal of weeds.* The weed growth on the lining is removed with a sharp edge and cracks/open joints are raked and filled with mortar and bitumen.

7. *Silting of channel.* The fluctuations in supplies in lined section should be regulated one. The channel should not be run at less than 50 to 70 per cent of the designed capacity particularly during high flow periods so that there is no silting in the bed which impairs its carrying capacity.

8. *Maintenance of inner slopes.* The inner slopes of canal banks in high cutting reaches should be turfed and grass grown not cut from roots.

SOLVED EXAMPLES

Example 6.1 *Design a lined distributary for 25 cumecs discharge, side slope 1.25:1, n = 0.018, slope 1/10,000.*

Solution. Let d is depth of circular lined section

$$V = \frac{1}{0.018}\left(\frac{d}{2}\right)^{2/3}(1/10{,}000)^{1/2}$$

$$V = 0.349\, d^{2/3}$$

Also $$V = Q/A = (25/1.925\, d^2) = 12.987/d^2 \; (i) \; d^2 \; (ii) \; 12.987/d^2$$

Equating (*i*) and (*ii*)

$$0.349\, d^{2/3} = 12.987/d^2$$

or $$d = 3.34 \text{ m.}$$

Example 6.2 *Design a lined canal for 300 cumecs discharge, slope 1/10,000, side slope 1:1, n = 0.018.*

Solution. Assume section 26 × 5.7 m

$$A = bd + 1.785\, d^2$$
$$A = 26 \times 5.7 + 1.785\, (5.7)^2 = 206.19 \text{ m}^2$$
$$P = b + 3.57\, d = 26 + 3.57 \times 5.7 = 46.35 \text{ m}$$
$$R = A/P = 206.19/46.35 = 4.45 \text{ m}$$
$$V = (1/0.018) \times 4.45^{2/3} \times (1/10{,}000)^{1/2} = 1.51 \text{ m/s}$$
$$\therefore \quad Q = 206.19 \times 1.51 = 311 \text{ cumecs, hence o.k.}$$

Example 6.3 *Determine financial forecast of lining an existing earthen canal with the data: total cost of lining = Rs. 1000 crore, Total discharge saved due to lining = 50 cumecs, and water allowance = 0.213 cumec per 1000 ha.*

Solution. Additional area to be irrigated with the water saved from seepage

= (50 × 1000)/0.213 = 2,34,742 ha

Net area irrigated at 60% intensity = 2,34,742 × 60/100 = 1,40,845 ha

Average additional benefit per hectare = Rs. 8000.

Net benefit at Rs. 8,000 per ha. = 1,40,845 × 8000

= Rs. 112.68 crore

Annual expenses

Interest on capital outlay at 10% $= \dfrac{1000}{2} \times \dfrac{10}{100}$

= Rs. 50 crore

Annual cost of lining assuming life of lining 60 years (= 1000/60) = Rs. 17 crore

Administration and working expenses at 0.5% (of Rs. 1,000 crore) = Rs. 5 crore

Total: = Rs. 72 crore

Benefit: Cost ratio = 112.68:72 = 1.56:1.

Example 6.4. *Work out the expenditure justifiable for lining a 50 km long canal of 100 cumecs capacity with 10 m bed width, 4 m depth and 1.25:1 side slopes. Assume one cumec of water valued Rs. 2 crore and 10% is the prevalent rate of interest. Ignore other benefits accruing from lining.*

Solution.

(*a*) Absorption losses in lined section (cumec/10^6 of wetted perimeter) $K = 0.349\,(Q)^{0.056}$

$K = 0.349\,(100)^{0.056} = 0.453$

Wetted perimeter of section with 1.25:1 side slope in 50 km $= (B + 3.85D) \times 50 \times 10^3$ m

$= (10.0 + 3.85 \times 4.0) \times 50 \times 10^3 = 1.27 \times 10^6$ m

Absorption losses $= (0.453 \times 1.27 \times 10^6)/10^6 = 0.575$ cumec

(*b*) Absorption losses in unlined section $K = 1.6\,(Q)^{0.0625}$

$K = 1.6\,(100)^{0.0625} = 2.134$

Wetted perimeter in 50 km (as per Lacey) $= [4.75\,(Q)^{0.5}] \times 50 \times 10^3$

$= [4.75\,(100)^{0.5}] \times 50 \times 10^3 = 2.375 \times 10^6$ m

Absorption losses $= (2.134 \times 2.375 \times 10^6)/10^6 = 5.068$ cumecs

Water saved due to lining = 5.068 – 0.575 = 4.493 cumecs

Value of water saved $= 4.493 \times 2 \times 10^7$ = Rs 9 crore

Capital investment justifiable on lining at 10% interest = 90 crore.

Example 6.5 *Design a lined canal section for a discharge of 50 cumecs at 1 in 10,000 bed slope. Assume suitable value of n and side slopes.*

Solution. Since canal capacity is below 85 cumecs, circular section is adopted. Assume 1.25:1 side slope and value of $n = 0.018$.

$$\tan\theta = \frac{1}{1.25} \text{ or } \cot\theta = 1.25$$

Therefore, $\theta = 0.675$ radians

Area, $A = \pi D^2 \dfrac{\theta}{\pi} + 2\dfrac{D^2}{2}\cot\theta$

$= D^2(\theta + \cot\theta)$

$= D^2(0.675 + 1.25) = 1.925\, D^2$

Wetted perimeter, $P = 2\pi D\dfrac{\theta}{\pi} + 2D\cot\theta$

$= 2D(\theta + \cot\theta) = 2D\,(0.675 + 1.25) = 3.85D$

Hydraulic mean depth, $R = A/P = \dfrac{1.925D^2}{3.85D} = 0.5D$

From Manning's formula, $V = \dfrac{1}{n} R^{2/3} S^{1/2}$

$$V = \frac{1}{0.018}(0.5D)^{2/3}\left(\frac{1}{10{,}000}\right)^{1/2} = 0.35\, D^{2/3} \qquad (i)$$

Also $Q = AV$

$50 = 1.925\, D^2 V$

or $V = 50/1.925\, D^2$ (*ii*)

From (*i*) and (*ii*)

$0.35D^{2/3} = 50/1.925D^2$

On solving, $D = 5.03$ m

Also $V = 50/1.925D^2 = 50/1.925\,(5.03)^2 = 1.03$ m/sec.

Example 6.6 *Design a lined canal for 250 cumecs discharge, side slope 1.25:1, bed slope 1 in 6666 and n = 0.018.*

Solution. Since discharge capacity is more than 85 cumecs, trapezoidal section of the lining is adopted. Assume permissible velocity 1.75 m/sec

$A = Q/V = 250/1.75 = 142.85\ \text{m}^2$

$V = \dfrac{1}{n} R^{2/3} S^{1/2}$

$1.75 = (1/0.018)\,(R)^{2/3}\,(1/6666)^{1/2}$

or $R = 4.15$ m

Also $P = A/R = 142.85/4.15 = 34.42$ m

Now $142.85 = BD + 1.925\, D^2$ (*i*)

Also $34.42 = B + 3.85D$ (*ii*)

Multiply (*ii*) by D and deduct (*i*) from it, we get

$1.925D^2 - 34.42D + 142.85 = 0$

or $D = 6.55$ m

From (*ii*) $34.42 = B + 3.85 \times 6.55$

or $B \simeq 9.25$ m

$f = 2.5\, V^2/R = [2.5\,(1.75)^2]/4.15 = 1.84$

Hence, Bed width $f = 9.25$ m, F.S. depth = 6.55 m.

Example 6.7 *A lined irrigation canal has bed width 10 m, depth 3m, side slope 1:1. If the maximum permissible velocity for the type of lining used is 1.5 m/s, find the maximum permissible* *bed slope at which the canal can be laid. Assume coefficient of Manning's roughness n = 0.013. Use Manning's equation.*

Solution. $A = 10 \times 3 + 1.785\,(3)^2 = 46.065\ \text{m}^2$

$A = 10 + 3.57 \times 3 = 20.71\ \text{m}$

Now, $P = 46.065/20.71 = 2.22\ \text{m}$

From Manning's formula, $1.50 = (1/0.013)\,(2.22)^{2/3}\,S^{1/2}$

or $S = 0.00013.$

QUESTIONS

1. (*a*) What is the necessity of canal lining? What are its advantages and disadvantages:
   (*b*) Describe various steps for the design of a lined canal. (*PU, Nov., 1980*)
2. (*a*) What is lining of a canal?
   (*b*) What is the necessity of lining a canal?
   (*c*) Enumerate the various types of lining.
   (*d*) What are the advantages of brick lining over concrete lining? (*PU, Nov., 1980*)
3. (*a*) What are the advantages of canal lining? What are the considerations for selection of a suitable type of lining?
   (*b*) Describe in brief different types of lining used for irrigation canals. (*PU, April, 1980*)
4. (*a*) How will you decide about the suitability of a canal lining material?
   (*b*) Design a lined canal for discharge = 150 cumecs, slope = 20 cm per km side slope = 1½:1, and limiting velocity = 1.5 m/s. Value of N in Manning's equation may be taken as 0.016. (*PU*)
5. (*a*) What are the factors that determine the FSL of a lined canal?
   (*b*) Explain the advantages of concrete lining over other types of lining. (*Kurukshetra Uni.,*)
6. (*a*) Design an irrigation lined canal to carry a discharge of 36 m^3/s. Mean diameter of average soil particle (m_r) is 0.464, side slope 1.25:1, bed width is zero. Assume missing data suitably.
   (*b*) What are the properties that determine the suitability of any material as canal lining? (*PU*)
7. (*a*) What are the considerations for selection of a suitable type of lining for a project?
   (*b*) Suggest the section for a concrete lined hydel channel to carry 36 m^3/s of water. Assume coefficient of rugoisty as 0.014. (*PU*)
8. What are the main advantage of a lined canal?
9. (*a*) Describe advantages and disadvantages of lining distributaries with polyethylene film.
   (*b*) Calculate velocity and discharge in a lined channel with the data; bed width = 25 m, depth = 5.5 m, side slope = 1.25:1, water surface slope = 1 in 10,000 and Manning's N = 0.018.
10. (*a*) Design a tile lined branch canal having a discharge of 500 cumecs with side slope of 1.5:1 and bed slope of 1 in 10,000. Make suitable assumptions where necessary.
    (*b*) Draw a neat sketch of designed section for a filling reach of this branch canal.
11. Describe different types of lining being done on channels and water courses. What are the factors that influence the choice of particular type of lining.

7

DRAINAGE ENGINEERING

7.1 INTRODUCTION

Drainage engineering is an essential component of irrigation engineering. Drainage is one of the important managements but is usually neglected. Drainage is as essential as irrigation and no water system is efficient without drainage. Additional water resulting from irrigation practices coupled with poor drainage result in raising of ground water from a considerable depth to within a fraction of a metre of the soil surface in a relatively short time causing waterlogging, salinity and alkalinity.

National Commission on Agriculture (1976) has defined an area to be waterlogged when the water table rises to an extent that the soil pores in the root zone of crop become saturated resulting in restriction of the normal circulation of air, decline in the level of oxygen and increase in the level of carbon dioxide. As a consequence yield of crops commonly grown on the land is reduced well below the normal for the land or if the land is not cultivated it cannot be put to its normal use because of high subsoil water table. In course of time, such lands turn saline or alkaline and ultimately become unfit for cultivation. The depth of water table at which such conditions arise depends on the height of capillary fringe and the type of crops grown, type of soil, quality of water, and the period for which the water table remains high. The yield of crops is adversely affected if the water table is within 90 cm (sugarcane), 60 cm (rice), 90 cm (gram and barley), 90 - 125 cm (wheat), 120 cm (fodder), 125 cm (maize, bajra and cotton), and 210 - 240 cm (Lucerene). Usually the land with subsoil water table within 1.5 m of the natural surface not prior to onset of monsoon is taken as waterlogged. In general, the areas with water table within 2 m below the ground surface are considered as prone to waterlogging, and with water table within 2 - 3m are viewed as critical areas. In fine grained soils, capillary rise may be appreciable causing saturation of root zone even with relatively low water table. In Pakistan, from revenue assessment angle, a land is said to be waterlogged when the yield is 25% or less of the normal yield.

Waterlogging has become more pronounced in India with the increase in the irrigated areas and shift in emphasis from protective irrigation to intensive irrigation as a means of attaining greater agricultural production. The problem exists in varying degrees in most parts of the country but it has taken a serious turn in the irrigated areas of Indo-Gangetic plains which have very flat countryside and inadequate outfalls. An area of about 2.46 million ha is affected by waterlogging and another 3.06 million ha by soil salinity and 0.24 M ha by alkalinity in irrigation command areas.

7.2 WATERLOGGING — ITS ILL EFFECTS

The waterlogged soils are rendered infertile because of ill effects of waterlogging to the soil, to the crops and to the environment.

Ill Effects on Soil

1. *Lack of aeration.* Due to presence of excess water, soil pores within the root zone of plants are saturated to cut off the normal circulation of air. The capacity of soil bacteria is affected resulting in retardation in the formation of organic acids to dissolve plant food material.

2. *Reduced soil temperature.* Waterlogged soil is slow to warm up. Low soil temperature restricts root development, depresses biotic activity in the soil resulting in lowered rate of production of available nitrogen hampering seed germination and seedling growth.

3. *Creation of salinization.* Capillary water coming to the surface brings up in solution harmful salts from the soil or those present in the groundwater. Thus, high salinization and deposits of sodium salts in the soil at or near the ground surface are created which may be toxic or lead to the formation of alkaline condition.

4. *Inhibiting activity of soil bacteria.* When soil structure is affected and tillage and cultivation of wet soil is resorted to, which tend to reduce normal biotic activity and root development.

5. *Denitrification.* Denitrification occurs because of the competition for nitrogen by the soil micro-organisms that thrive in saturated soil and reduction in numbers of nitrifying organisms due to lack of aeration.

6. *Retards cultivation.* Difficulty in carrying out normal cultivation operations in waterlogged soil.

Ill Effects on Crops

1. *Delayed cultivation operation.* Normal cultivation operations of tilling and ploughing are adversely effected due to presence of excess water in the soil.

2. *Growth of weeds.* Water-loving wild plants grow profusely and have competition with the crops, thereby affecting the growth of useful crops. Weed removal also entails extra investment on the part of the cultivator. In extreme waterlogged condition, only wild growth is there.

3. *Diseased crops.* Waterlogged conditions cause physiological disease to crops. Decay of roots, external symptoms on the foliage, fruits, etc. are common.

4. *Loss of cash crops.* Cash crops desired to be grown in the land cannot be cultivated. High subsoil restricts the use of land to crop like paddy.

5. *Low yield.* Maturity period of crops is reduced resulting in low yield.

6. *Oxygen depletion.* In saturated soil, plant roots are denied normal circulation of air; the level of oxygen declines and that of carbon dioxide increases resulting in wilting and ultimately death of plants.

Ill Effects on Environment

Waterlogging impairs sanitary conditions, promotes malaria and thereby results in unhealthy environment for human population, animals and plants in the area.

7.3 CAUSES OF WATERLOGGING

In the pre-canal stage, a state of equilibrium in the water table is maintained. The factors tending to raise the water table termed as 'inflow' are (*i*) Infiltration from rivers towards the bottom of the tough, (*ii*) Percolation of rainfall in saturated phase, (*iii*) Percolation from well or flooding irrigation, (*iv*) Subsoil flow from upper regions. The factors tending to lower the water table, termed as 'outflow' are; (*i*) Subsoil flow towards the low regions, and (*ii*) Soil evaporation from water table surface.

With the introduction of irrigation, a new source of constant inflow into the subsoil is introduced which results in continuous rise of water table. Thus, the sources of inflow and outflow are also changed. The sources of inflow are (*i*) Infiltration from rainfall in saturated phase, (*ii*) Subsoil flow from upper regions, (*iii*) Seepage from canals in saturated phase, and (*iv*) Infiltration from canal irrigation in saturated phase. The sources of outflow are (*i*) Infiltration into rivers, (*ii*) Subsoil flow to the lower regions, (*iii*) Increased soil evaporation, (*iv*) Transpiration by plants and (*v*) Seepage drains.

With the rise in water table, volume of outflow from the above sources increases until a stage is reached when outflow matches inflow and water table attains a state of equilibrium. The water table can be lowered in case inflow into the subsoil can be reduced or outflow increased.

Various causes of waterlogging are discussed below :

1. *Over irrigation.* Over irrigation and intensive irrigation result in waterlogging due to heavy percolation and rise of water table.

2. *Seepage from canals.* Excessive seepage from unlined canal system and water courses result in the rise of water table leading to waterlogging conditions.

3. *Impervious barrier.* There may be an impervious stratum below the top layer which obstructs the seepage from going down resulting in creation of false water table or perched water table leading to waterlogging.

4. *Inadequate drainage.* In irrigated areas, water losses from canal system and water courses continuously contribute to water table. If drainage is inadequate, waterlogging conditions are developed.

5. *Poor irrigation management.* Poor irrigation management by cultivators results in development of waterlogging.

6. *Obstruction of natural drainage.* Interception of natural drainage by the construction of canals, roads, railways, water courses, etc., results in development of waterlogging.

7. *Excessive rainfall.* Excessive rainfall, in the absence of adequate drainage, cause waterlogging. Rain, apart from irrigation, is the major source of waterlogging.

8. *Flat topography.* Flat terrain with depression lead to waterlogging as disposal of excess water is delayed resulting in increased percolation into the soil.

9. *Occasional spills by floods.* Occasional flooding of the countryside and flood water not quickly drained off give rise to water table resulting in waterlogging.

10. *Soils and subsoils.* High textured soil and black cotton soil are prone to waterlogging.

11. *Abandonment of percolation wells.* Advent of canal irrigation has resulted in abandonment of percolation well giving rise to water table.

12. *Land locked patches having no outlets.* In land locked patches waterlogging develops due to absence of outlets to drain excess rain water.

13. *Unsuitable cropping pattern.* Unsuitable cropping pattern also leads to waterlogging.

14. *Physiography of watershed.* Physiography of watershed, *i.e.*, shape, topography and drainage pattern has an important bearing on the drainage problem of the watershed. In area with high value of shape index, compactness coefficient and drainage density, the problem of waterlogging is not as serious as in the catchments where the values of these factors are low.

7.4 ANTI-WATERLOGGING MEASURES

The anti-waterlogging measures may be preventive as well as curative.

Preventive Measures : Preventive measures envisage reduction of inflow into the subsoil water table. Various measures in this context are:

1. *Arresting percolation from rainfall.* Quick disposal of storm water in excess of crop requirement is a must for canal irrigated lands.

2. *Arresting percolation from canals.* (*a*) Lining of canals to control excessive seepage from unlined canals is a very important preventive measure, (*b*) Lowering of designed full supply level in the canal, where possible, with a view to cut down seepage losses, (*c*) Provision of pustha on canal bank, and (*d*) Interception drains along the canals 15 to 30 m away from the canal bank, are effective to arrest seepage from the canals.

3. *Arresting percolation from irrigation in the fields.* (*a*) Restriction of irrigation. High water table areas may be restricted to receive kharif irrigation only and during rabi season, the cultivators may irrigate from open wells and tubewells. In high water table areas controlled intensity of irrigation or permiting only a fraction of the total culturable commanded area to receive irrigation water, (*b*) Lining of watercourses, (*c*) Economical use of water. Economical use of water by adopting khalkiari system of fields, levelling the fields and refraining from applying more water than the soil can hold, scientific cropping pattern, etc.

Curative Measures : Curative measures envisage extraction of excess water from the root zone of crops to restore the condition conducive for normal growth of crops. Various measures in this context are:

1. *Vertical drainage.* Pumping from subsoil through shallow tubewells, deep tubewells, seepage-*cum*-pumping stations, open well to lower water table and use the pumped water for irrigation as it is or by mixing with canal water. Its advantages over horizontal drainage is that it is quite flexible and effective depending on extent, disposition and characteristics of aquifer.

Surface drainage : The natural drainages *viz.* existing drains, streamlets can be made efficient after remodelling them. Artificial drainage is required to be constructed in command area having flat topography.

Regulating canal discharge : Seepage losses can be minimized if canal discharge is regulated during non-irrigational period. Overuse of canal water is curbed if canal supply is regulated together with pumping of ground water.

2. *Tree plantation:* Eucalyptus tree plantation is helpful in lowering water table, enhancing farmers income and creating healthy environment.

3. *Land grading:* Land grading is useful for betterment of waterlogged lands.

4. *Cropping pattern.* Vegetation and crops which have high rate of evapotranspiration may be patronized to serve as natural drainage system.

5. *Optimum water use.* Conjunctive use of surface and ground water be encouraged by digging wells, sinking shallow tubewells etc.

6. *Better irrigation methods.* Improvement in water application by giving only the optimum depth of irrigation water. Sprinkler irrigation and drip irrigation methods be introduced to reduce percolation losses.

7. *Sub-surface drainage.* Sub-surface drainage or deep buried field drains to intercept ground water constitute the most efficient and permanent type of anti-waterlogging measure.

8. *Puncturing underground barriers.* Geological formation such as buried ridges as may be interfering with the subsoil flow leading to waterlogging conditions may be punctured to lower water table.

9. *Efficient drainage system.* An efficient drainage system is essential for the quick disposal of the storm water and excess irrigation water.

10. *Removing obstruction in natural drainage.* The natural drainages should be improved by removing debris, etc. from the waterway as also providing adequate waterway under the road bridges, drainage crossings, etc.

Bio-drainage : Deep rooted trees like Cineraria draw a good quality of water from the ground. Eucalyptus tree has roots extending up to 3 m but transpire water at a high rate and thus works as a biological pump. It is as such suitable as anti-waterlogging measure to take care of seepage along canals. The bio-drainage is effective if there is adequate aeration in the root zone and the roots extend up to ground water reservoir and draw water from capillary zone.

11. *Reclamation of soil.* Reclamation of waterlogged soil by various reclamation methods is an important curative measure.

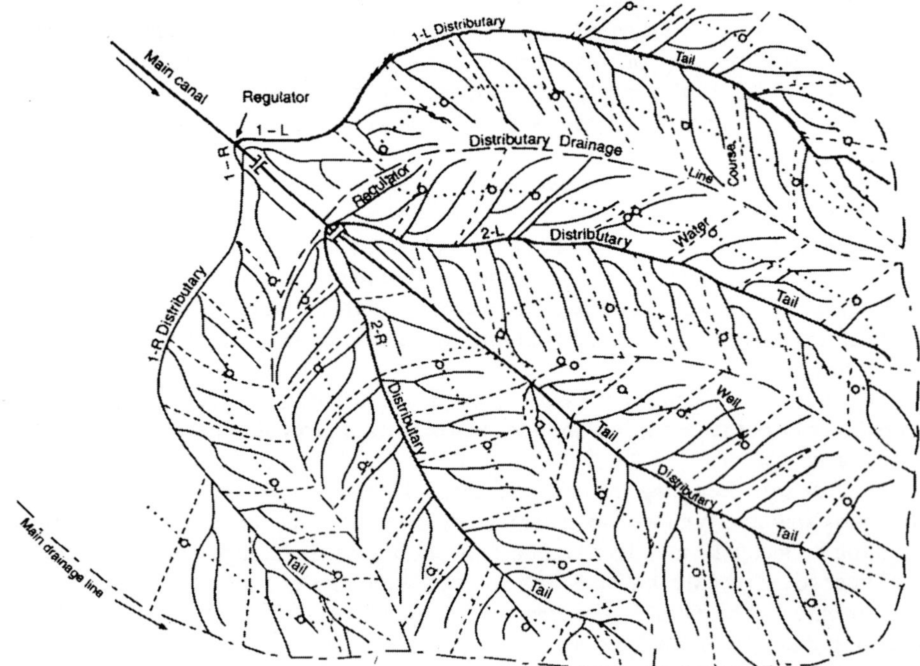

Fig. 7.1 Typical design of conjunctive use: lower half of each block by well irrigation.

A judicious combination of canal and tubewell irrigation system offers an ideal solution of the waterlogging problem in most areas (Fig. 7.1), besides lining of canals and provision of surface and sub surface drains. More problematic waterlogged areas are more intensively drained by providing underground drains and comprehensive programme of providing tubewells in the command area. Topographic, geological and soil surveys, nature of soils and sub-soils from point of view of water table and its fluctuations, quality of ground water, irrigation practices, farming activity, cropping pattern need to be considered in studying a waterlogged area.

7.5 SALINE/ALKALI SOILS

Sodic Soil : A sodic soil is defined as a soil having exchangeable sodium percentage of more than 15, a conductivity less than 4 mmhos/cm and a pH greater than 8.5. A saline-sodic soil is a sodic soil with an electrical conductivity of more than 4 mmhos/cm.

Saline Soil : The soil whose yield has been reduced due to the presence of toxic concentration of salts in root zone is called saline soil. Such a soil has an exchangeable sodium percentage of less than 15, a conductivity more than 4 mmhos/cm and a pH less than 8.5.

Thur: The deposition of sodium sulphate as white crystals on the surface of earth, usually in winter, is called Thur. Thur land, if not reclaimed early, deteriorates to alkalinity or rakkar.

Alkali Soil : Alkali soil is formed due to rise of salts from water table or the lower depths. The alkali soil is impermeable to water and air. The development of alkalinity is a gradual process and after the first stage of thur appearance has been reached, the longer the land is left before reclamation, the greater will be approach to the second stage deterioration termed as alkalinity.

The characteristics of saline, alkali soils are as under:

Type of soil	Electrical conductivity (micromhos/cm)	ESP	pH
Saline	> 4000	< 15	< 8.5
Saline-alkali	> 4000	> 15	8.5
Alkali	< 4000	> 15	8.5 - 10.0

Formation of Saline/Alkali soils

Various factors responsible for the formation of saline/alkali soils are (*i*) Capillary rise from saline ground water, (*ii*) Indiscriminate use of canal waters, (*iii*) Weathering rocks and the salts brought down from the up streams to the plains by river water and subsequent deposition along the alluvial materials, (iv) In-situ decomposition of soil material, and (*v*) Ingress of sea water. Climate, geology, topography and hydrology play either singly or in combination important part in determining the salt regime of a tract.

In command areas, this formation is due to (*i*) Excessive irrigation water which raises the ground water level sufficiently to permit concentration of salts from saline ground water at the surface by capillary action through intense evaporation, (*ii*) Excessive seepage from poorly maintained canals and lateral channels which run at higher level resulting in a high water table, (*iii*) Soil has high salt content, (*iv*) Poor drainage keeps salts in the surface soil and prevents the leaching of salts down below the root zone, and (*v*) Erratic use of irrigation water, *i.e.*, flooding at one time followed by intense drying and long breaks and when the total supply of water is limited and irregular which would also leave the salts in the root zone.

Remedial Measures : Salinisation in the command area is the effect of excessive irrigation. The remedial measures include improvement of surface water management in the form of canal lining, on farm management by large scale application of sprinkler and drip irrigation and improved sub-surface water management by increased water extraction for irrigation and through ground water drainage.

7.6 LAND RECLAMATION

Land reclamation is defined as the process of restoring to cultivation lands which have been rendered unculturable or have suffered reduction in the yield. Various considerations governing the methods of land reclamation are (*i*) Soil characteristics, (*ii*) Salt distribution and nature in soil, (*iii*) pH value and total soluble salt content in the soil, (*iv*) Irrigation water quality, and (*v*) Present state of salts *vis-a-vis* yield of crops.

Land Reclamation Process

The land reclamation process comprises:

1. *Reduction in salt content.* It is achieved by leaching and rice cultivation. Sufficient depth of water in rice cultivated land tends to maintain the salt at a safe depth below the surface from crop growth point of view. It is the first step of reclamation.

2. *Reduction in soil alkalinity.* Reduction in soil alkalinity is obtained by the roots of rice crops. Carbon dioxide generated by rice roots on the soil replaces the sodium which brings about improvement in soil.

3. *Restoration of nitrogen.* Nitrogen essential for plant food, is reduced during rice cultivation due to absence of air and excess water content of the soil. A leguminous crop is grown during rabi season to restore its nitrogen. Guara and San possess maximum percentage of nitrogen.

Land Reclamation Methods

Principal methods of land reclamation are as given below:

1. *Levelling.* Before any reclamation is attempted, the land is essentially perfectly levelled to ensure uniform application of water and prevention of accumulation of water in fields.

2. *Leaching.* Leaching is a prerequisite for reclamation of saline sodic soils. Leaching is the process of flooding the surface and deposits of salt with abundant irrigation water to a depth of 15-25 cm. The dissolved salts are drained off to deep seepage drain. The leaching operation may be repeated a couple of years later, if salt deposits reappear.

The leaching requirement is determined by the following equation:

$$LR = \frac{Dd_w}{Di_w} = \frac{Eci_w}{Ecd_w} \qquad (i)$$

Where, LR = leaching requirement expressed as a ratio or as per cent, Eci_w = electrical conductivity of irrigation water (mm hos/cm) Ecd_w = electrical conductivity of drainage water, mm hos/cm, Dd_w = equivalent depth of drainage water, Di_w = depth of irrigation water.

$$Di_w = D_c + D_{dw} \qquad (ii)$$

From (*i*) and (*ii*) eliminating D_{dw}

$$Di_w = \frac{D_c}{1 - LR}$$

For crops which can tolerate Ecd_w = 8000 micromhos/cm and Eci_w is 2000 micromhos/cm, the leaching requirement would be: $\frac{2000}{8000} \times 100 = 25\%$

Equation for water requirement *vis-a-vis* degree of salinity of soils is as under :

$$Y = N\,400\,X \pm 100 \qquad \text{Kovda's formula}$$

where, Y = quantity of water required for leaching (in mm depth of water) over the land, X = average salinity of soil (expressed in % of depth 1 to 2 m), N = coefficient depending on (*i*) permeability of soil, (*ii*) depth of occurrence, and (*iii*) degree of salinity of ground water of reclaimed field. Usually the value varies from 0.5 to 2.0; sandy silt 0.5, loam and deep located ground water 1.0, and clayey soil and shallow located saline ground water 2.0.

3. *Crop rotation.* Leaching is followed by a combination of crop rotation. The most efficient crop rotation for reclamation of alkali soils is manuring with sesabania-aculeata. Rice followed by gram come next. The rotation of 'Pani-*cum*-colonum' followed by 'berseem' is recommended for alkali soils having pH values not above 9.5. Sugarcane might also stand an alkalinity of 9.5 pH. In areas susceptible to waterlogging, light irrigation crops should be introduced. Restriction on area under sugarcane contributes substantially to prevention of waterlogging.

4. *Addition of Amendments.* Addition of amendments containing soluble calcium salts such as gypsum, acid or acid forming fertilizers is very useful for the improvement of alkali soils. For saline sodic soil use of gypsum results in not only higher yield of crops but also in improving the soil. Unlike fertilizers, gypsum application is not to be repeated every year.

5. *Surface drainage.* Drains wash out salts from the soil surface as also remove them from within the soil. Reduction in salt content by drains is very slow but they help in reclamation of saline soils.

6. *Subsurface drainage.* Provision of sub-surface drains is a very effective reclamation measure for areas with high salt content throughout the profile or saline underground water.

7. *Addition of waste products.* Addition of waste products such as molasses in groundnut hull, pressmud cakes, sawdust is useful to reduce soil salinity. Distillery waste being acidic is effective in reducing alkalinity.

8. *Dug-out ponds.* For permanent reclamation of waterlogged and saline-alkali soils, the excess runoff is collected in dug-out ponds of sufficient capacity in low lying areas. Either runoff from first rain of the season is not ponded or such waters with high sodium carbonate content could be used for irrigation with addition of gypsum.

9. *Breaking impervious pan.* The hard impervious subsoil pan should be broken by deep chiseling or deep ploughing with sub-soiler plough.

10. *Skimming drainage.* The brackish water overlies saline water in areas of rising water table. The better quality of upper water can be removed by skimming drainage, *i.e.,* installing partially penetrating wells just to pump out better quality water.

7.7 LAND DRAINAGE

Optimum agricultural production essentially requires artificial drainage of most of the soils having high water table, high salt content, low lying soils and soils at the lower ends of the irrigated fields. There are areas where irrigation may not be required during monsoon season though drainage is necessary. Surplus water is as harmful to crops as inadequate water. The excess water in the field due to over-irrigation or seepage from an adjacent channel has to be removed to help crop growth. Drainage is also required in delta areas when irrigation is extended there. The drainage requirement is determined by the excess moisture on aeration, soil temperature, structural stability of the soil, soil chemistry, biological activity and the overall problems of land and crop management. Efficient soil and water management can lessen or at times avoid the need for artificial drainage.

In canal irrigated areas, the implementation of drainage needs to be taken up simultaneously with the irrigation development so as to avoid the problems of waterlogging, and/or salinity development at a later stage. Thus the drainage as a means of disposal of excess water is necessitated due to various factor such as (*i*) Water accumulation on the land surface usually resulting from heavy precipitation and/or river bank spill combined with the deficiency of drainage capacity, (*ii*) Excess water on the land surface resulting from waterlogging or stagnation in depressions and low areas, (*iii*) Seepage water from canals, tanks and subsoil flow from higher ground, (*iv*) Rise of sub-surface water due to excessive irrigation and percolation from other sources, (*v*) Water stagnation due to obstruction in the natural flow of drainage such as construction of road, railway line, canal, etc. From plant growth standpoint drainage is essential because of the adverse effects of the excess moisture in the soil such as (*i*) *Damage to roots.* Saturated soil results in the stunted crops with yellow leaves. The plants die if the excess water remains for some time because of damage to roots caused by reduced supplies of oxygen and accumulation of carbon dioxide, (*ii*) *Poor aeration.* Poor aeration also results in accumulation of sufficiently high concentrations of reduced iron and manganese which becomes toxic to the crops, (*iii*) *Reduced soil temperature.* Saturated soil is slow to warm up under a given amount of heat input. Low soil temperature restricts root development, depresses biotic activity in the soil resulting in lowered rate of production of available nitrogen, (*iv*) *Denitrification.* Denitrification occurs because of the competition for nitrogen by the soil microorganisms that thrive in saturated soil and reduction in numbers of nitrifying organisms due to lack of aeration, (*v*) *Reduced uptake of plant nutrients.* High moisture level in the soil results in reduced uptake of plant nutrients because of limited root growth which restricts the volume of soil from which the plant may draw nutrients, (*vi*) *Difficulty in tillage operation.* The tilling and harvest with machinery becomes difficult and costlier.

The functional classification of drainage is as under :

1. **Surface Drainage :** It is the removal of excess water from the surface of land by providing drainage channel in the area.

2. **Subsurface Drainage :** It is the removal or control of ground water and removal or control of salts by means of water.

7.8 BENEFITS OF DRAINAGE

The benefits of drainage are (1) Improvement of the soil structure and increase in productivity of the soil, (2) Helps tillage operations due to increased soil tilth, (3) Lengthening of growing season, (4) Facilitates early ploughing and sowing of the crops. Crop period is thus increased resulting in higher crop yield, (5) More soil moisture is made available for crop growth due to extension of crop root zone into the soil, thereby ensuring vigorous plant growth, (6) Maintains proper aeration of upper soil layers, (7) Maintains higher soil temperature. The soil is kept warmer, (8) Reclamation of waterlogged lands. Harmful salts are leached off, (9) Maintenance of water table at a reasonable depth so that water cannot rise above the natural ground by capillary action, (10) Improvement in sanitary conditions of the area, malaria and weed control, and (11) Larger varieties of crops can be grown.

7.9 ESSENTIAL REQUIREMENTS OF A DRAIN

The essential requirements to be satisfied by a drain are (*i*) Admit all the flood discharge from the catchment, (*ii*) Quick and unobstructed flow towards the drain from the catchment, (*iii*) Capacity to carry away the received water to the outfall, (*iv*) Ideal outfall conditions, (*v*) Stable section with non-silting tendency and capable of avoiding sloughing of side slopes, (*vi*) Seepage and/or low discharge does not spread thin over the entire section, (*vii*) Low maintenance cost, and (*viii*) Low initial cost.

The drainage system of an area is just the reverse of the irrigation system ; the drainage collects water through small drains and outfall into major drains and ultimately into a river or the sea.

In irrigated areas, the drainage required is of three types :

Type I. Area with water table between 0 to 1.5 m. Drain is required 2.5 to 3.0 m deep. Natural drainage may be deepened so that 0.5 to 0.6 m depth of drain is in pervious strata.

Type II. Area with water table between 1.5 to 3 m and is likely to rise. Drain is required along the canal to the limit of 1.25 to 1.75 m or up to the pervious strata whichever is less.

Type III. Water table is below 3 m. Artificial drains are not required. The natural drainage may be trained to the depth of 1.5 to 1.75 m or up to the pervious strata.

Drains cater for storm water and seepage water. It is advantageous if the seepage water collected in the drain is pumped for irrigation in lower down reaches.

7.10 CLASSIFICATION OF DRAINS

The drains may be open drains or closed drains with further classification, as under :

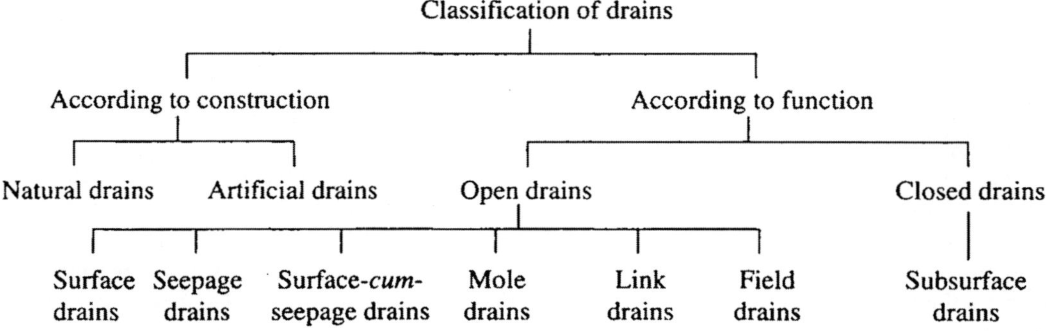

Classification According to Construction

Natural Drains : These are the lowest valley line between two ridges.

Artificial Drains : These are the constructed drains generally aligned along drainage line, sometimes taken across the valley to reduce length of the drain or to have proper outfall conditions.

Classification According to function served

1. **Open Drains :**

(*i*) *Surface Drains.* Surface drains are normally used for the removal of excess surface irrigation water or for the disposal of storm water. They remove water before it has entered the soil. Whether constructed for the purpose or not, deep surface drains with bed level below water table also cater for seepage water.

(*ii*) *Seepage Drains.* Seepage drains cater for the subsoil water. They are made deep enough to allow water table to drop in the drain and seepage water is carried away. They are of smaller section compared to surface drains. They help maintain aeration of root zone depths. Usually these are constructed along canal bank to drain directly into a natural outfall or into a carrier drain.

(*iii*) *Surface*-cum-*seepage drains.* They serve the dual purpose of seepage and storm water drain. During rainy season they carry storm water and seepage water in non-monsoon months. They have bed level below the water table. A cunnette is usually provided to cater for the small seepage water.

(*iv*) *Mole Drains.* Mole drains constitute valuable supplement to open drainage where they can be used. They are useful in equalising water levels between ditches for both drainage and for sub-irrigation. Mole drainage is a method of draining soil by means of mole drains. Mole drains are cylindrical drains formed in the subsoil by pulling a mole plough of 5 to 10 cm diameter by a tractor. The plough is pulled along the sloping ground with the mole shoe at a depth of 60 cm. They are spaced 3.5 to 5 m apart. A round channel is formed in the soil with cracks along the mole. The drainage water enters the mole through these cracks and is carried along the slope in sub-surface or open drains. They are suitable in clay and clay loam soils but impracticable in soils of coarse texture. Durability of the mole depends on the texture and structure of the soil. In general, mole drains are not satisfactory.

(*v*) *Link and Field Drains.* These are branch drains draining sub-catchment into the outfall drain. These are aligned along subsidiary valley lines.

Field drains are small drains draining individually or a group of fields into the link drains.

2. **Closed Drains :**

The sub-surface drains remove water which has entered the soil. They are usually laid 1 to 1.5 m below ground surface and at a suitable spacing and grade to lower watertable to greater depths.

7.11 SURFACE DRAINS — DESIGN CRITERIA

These are usually 1 to 1.5 m deep to cater for storm water and seepage water. They are suitable when (*i*) Large volume of either surface or subsoil water from land are to be catered, (*ii*) Slope is too slight to permit installation of the tile drains, (*iii*) Plant roots are likely to clog the tile drain, (*iv*) There is no satisfactory outfall for tile drain, (*v*) Low cost land is traversed. The design considerations are as under :

1. **Rainfall:** Intensity, frequency and duration of rainfall dictates the design discharge of a drain. Usually the maximum rainfall of 3 days duration is considered on economic considerations. For drain section rainfall corresponding to 5 year return, and 10 or 15 years for higher degree of protection is considered and for the masonry structures rainfall corresponding to 50-year return is adopted.

2. **Depth and duration of submergence:** The discharge to be catered also depends on the permissible depth and duration of submergence which in turn varies from crop to crop. As per IS:8835-1978 the period of disposal is limited to 7-10 days for paddy, 3 days for bajra, maize and cotton, 7 days for sugarcane and 1 day for vegetables.

3. **Economic considerations:** Economic considerations require excavated drain section of limited capacity.

4. **Environmental Aspects:** The negative effects on the environment, if any, must be considered and remedial measures included in the drainage scheme.

Other important considerations of design are soil drainage (water table), crop drainage and salinity drainage.

5. **Drain Capacity:** In determining drain section, peak rate of runoff, total volume of runoff as also distribution of runoff rate throughout the year is considered. IS:8835-1978 recommended run-off coefficient for different soils in plains is 0.7 for plateaus lightly covered, 0.55 for clayey soils, stiff and bare and clayey soils lightly covered, 0.4 for loam, lightly cultivated or covered, 0.30 for loam, largely cultivated and suburbs with gardens, lawns, roads, 0.20 for sandy soils, light growth and 0.05-0.20 for parks, lawns, meadows, gardens, cultivated area.

Boston Society formula $Q = C\sqrt{A}$ is also used. The existing drains may have design capacity $Q=C\sqrt{A}/5$ in forced reaches comprising closed drainage tracts in which drains cut across high land. In the case of new drains the capacity is determined by the formulae $Q=C\sqrt{A}/6$ for catchment area up to 260 sq km, $Q=C\sqrt{A}/4$ for C.A. 261-650 sq km and $Q=C\sqrt{A}/2$ for C.A. 651 sq km and above but bed width is excavated for $C\sqrt{A}/4$ and balance capacity from $C\sqrt{A}/4$ to $C\sqrt{A}/2$ is provided by deepening the drains. Here Q = discharge (cumecs), A = catchment area (sq km) and C = coefficient; value 3.5 for areas having annual rainfall 50 cm, 8.4 for rainfall 51-75, and 35 for rainfall 75-100 cm. For areas with rainfall above 100 cm, every case is considered individually on actual basis.

U.P. and Andhra Pradesh : Q = 0.11 cumec per sq km of catchment area.

Madhya Pradesh : Q = 0.22 to 0.44 cumec/sq km, 0.44 for C.A. up to 13 sq km and 0.22 for C.A. 79 sq km and above.

6. **Alignment:** Alignment of the drain is required to be such that it traverses through the lowest contours, *i.e.*, along the drainage line and length of the drain is minimum consistent with the requirement to drain off the lowest spots by either directly or through subsidiary drains. It should not cross irrigation canal or pass through village habitations as far as possible. Drains aligned down the slope are usually much more effective than those excavated normal to the direction of the slope of the ground. The reduced distances, RDs, are marked along the drain, the zero RD being at the outfall end and increasing upstream, *i.e.*, just the reverse of canals wherein RDs increase towards downstream.

7. **Water Surface Slope:** Water surface slope in the drain is governed by the general slope of the ground and outfall condition. Non-weeding velocity which is considered higher than non-silting velocity is provided. The slope is either kept constant or gradually decreasing towards the outfall in keeping with the increased discharge downstream. Slope is generally determined from Lacey's formula $S = 0.0003 f^{5/3}/Q^{1/6}$.

8. **Velocity:** Drains are designed by Lacey formula. Generally adopted velocities in the drains for firm loam and clay loam = 0.5-1.0 m/s, alluvial soil = 0.6-1.25 m/s, ordinary gravelly soil = 1.0-1.25 m/s, coarse gravel, shingle, etc. = 1.25-1.50 m/s. IS : 8835-1978 has suggested value of n as 0.025 in computing velocity.

9. **Side Slope:** Side slope of 1:1 is generally adopted for cutting. Usually considered side slopes are, loose rock and hard soil = 0.5:1, alluvial soil = 1:1, sandy soil and deep black cotton soil = 2:1, and very sandy soil = 3:1.

10. **Bed Width and Depth:** Bed width and depth corresponding to Lacey's formula, discussed under canals, is generally adopted. The discharge is divided into suitable reaches, considering that

the flow increases towards the downstream, without taking into account transmission losses applicable to canals. The full supply line is generally kept below the natural ground level, say by 0.6 m, so that there is no flooding outside and the drain caters for the drainage effectively but not higher than 0.3 m above average ground level at the starting point of the drain.

It is a good practice to dig a cunnette at the centre of the drain to cater for low flow and seepage. Cunnette section also helps in preventing weed growth because low flow is not made to spread thin over the entire section but is contained within the cunnete section. It also reduces maintenance which for the most period is required for the cunnette portion only.

In Punjab and Haryana States, drain section is dug for the discharge corresponding to $Q = C\sqrt{A}/4$ to $C\sqrt{A}/6$ and the maximum flood $Q = C\sqrt{A}$ is contained within the banks (Fig. 7.2). In Andhra

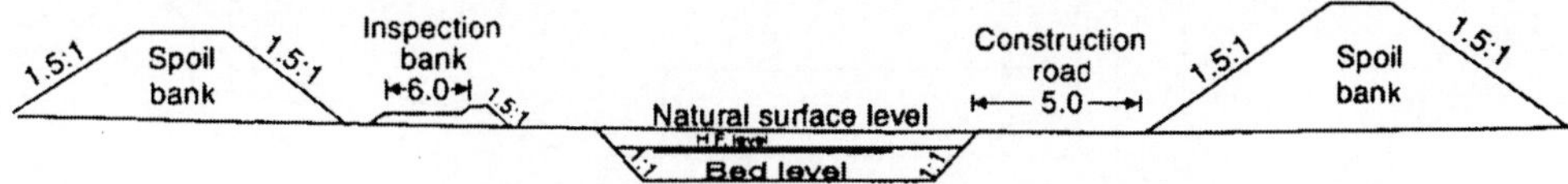

Fig. 7.2 Typical cross section of drains.

Pradesh, on Nagarjunasagar project, drains are excavated for ordinary flood discharge only at the lowest average level of the adjoining wet fields while the maximum flood discharge section will have its maximum flood level at 0.3 m above ordinary flood level with a very wide berm (Fig. 7.3).

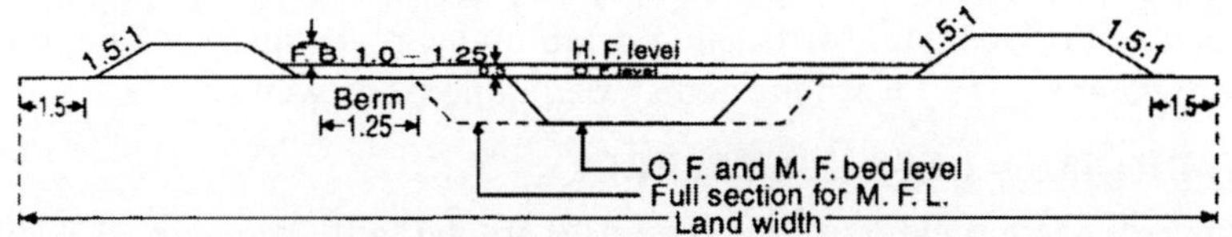

Fig. 7.3 Typical drain section (Andhra Pradesh)

Bank Width. The drains, in general, are not banked on both sides. As per IS:8835-1978 top of embankment is kept 1 m higher than design full supply level and 1 m minimum berm width. Bank is essentially required on one side and the side contributing flow may be left unbanked. Where heavy spilling may take place on both sides, continuous embankments on both sides are provided. Regulated inlets (Fig. 7.4 to 7.6) are provided, where necessary, to allow the outside water to enter into the drain. In diversion drains continuous banks on both sides of the drain are essentially provided.

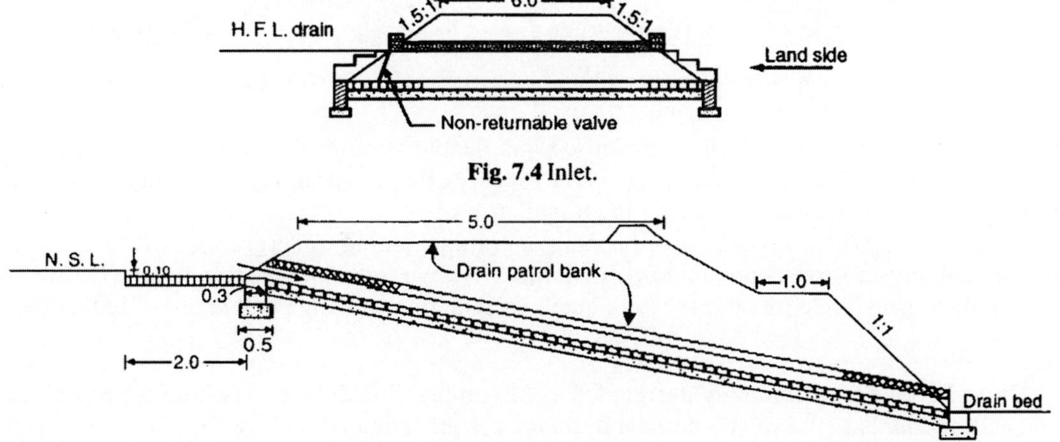

Fig. 7.4 Inlet.

Fig. 7.5 Chute inlet.

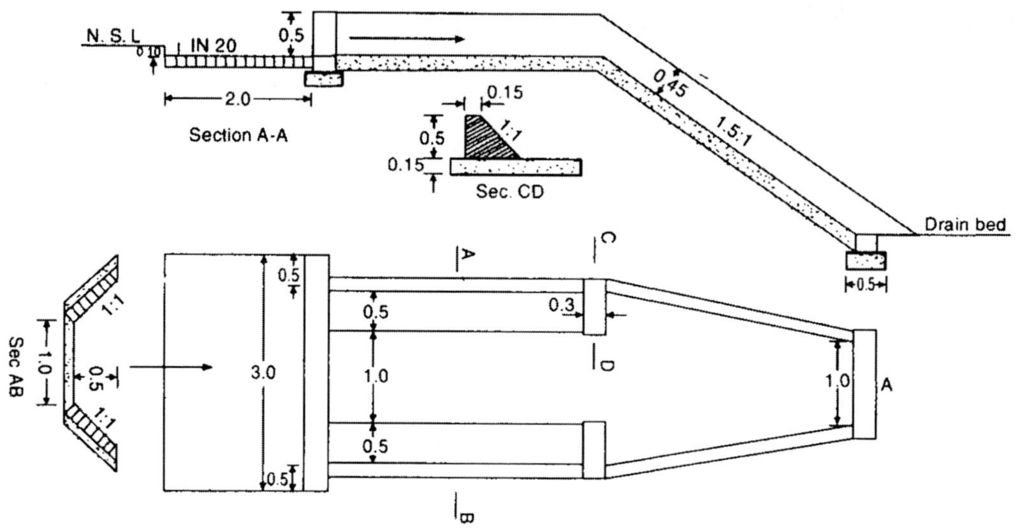

Fig. 7.6 Open trough shaped inlet on drains.

Minimum bank width is 2.0 m; general practice being 3 m width on non-patrol bank and 6 m on patrol bank side. In large capacity drains a boundary road 5 m wide is provided (Fig. 7.2).

Disadvantages : (*i*) Cause wastage of land, *i.e.*, the land brought under the drain cross section, (*ii*) Require bridges, etc. for passing drain under road, railway line, canal, etc., (*iii*) Require frequent cleaning, (*iv*) Harbour and spread obnoxious weeds.

7.12 CLOSED DRAINS – DESIGN CRITERIA

Closed drains are underground drains laid deep in the ground and then covered. Their use is indicated in high cost land because they do not occupy surface land and also cause no hindrance to the agricultural operations. They are located at a suitable depth and grade below the ground surface depending on the topography, existing water table and the extent of depression of water table required Fig (7.7).

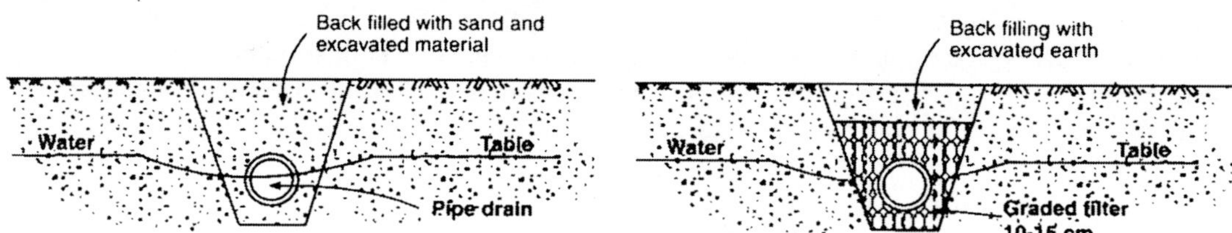

Fig. 7.7 Tile drain (without graded filter) in a previous soil.

Fig. 7.8 Tile drain (with filter) in a previous soil.

They are placed in permeable stratum. In low permeable stratum the drain is shrouded with filter material of high permeability (Fig. 7.8) to ensure effective drainage. They help lower the full saturation line adequately below the ground surface.

Advantages : (*i*) No hindrance to farming operations, being underground, (*ii*) Occupy no surface land, thus no area is put out of cultivation, (*iii*) Give root zone greater depth, (*iv*) Lower water table to greater depth, and (*v*) Permanent reclamation of saline and saline-alkali soils.

Tile drains

Closed drains are commonly designated as tile drains. Tile drains constitute the most efficient and permanent type of sub-surface drainage for the irrigated areas where the water table has permanently risen close to the ground surface. They are located at a suitable depth, usually 1 to 1.5 m, below the ground surface and at a suitable spacing and grade depending on the soil, climate and topography of the area. The centre of the tile drain is usually 0.3 to 0.6 m below the level up to which the water table is desired to be lowered below the root zone of the plants (Fig. 7.9). Free water enters through the open joints of the tiles. The water drained by tile drain

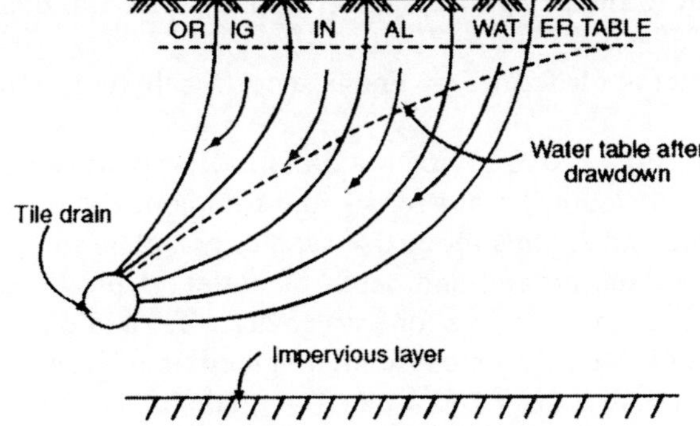

Fig. 7.9 Drawdown curve with a single tile drain.

is disposed of by gravity into deep surface drain or pumped out depending on outfall conditions. In the system, laterals remove water from the soil, sub-drains collect water from a group of laterals and empty into mains for disposal into outfall.

Tile spacing and depth. The drain depth is reckoned from the ground surface to the bottom of the tile. The deeper the drain the more is the drainable area per drain line and farther is the spacing of drains. The drains are closely spaced in clay soils and far apart in sandy soils. IS:10907-1984 recommended drains placed about 1.25 m deep are given in Table 7.1.

Table 7.1 Drain spacing in humid areas

Soil type	*Hydraulic conductivity*	*Spacing (m)*
Clay and clay loam	Very slow (< 1.3 mm/h)	9 to 21.5
Silt and silty clay loam	Slow to moderately Slow (1.3-2 mm/h)	18.5 to 30.5
Sandy loam	Moderate to rapid (20-250 mm/h)	30.5 to 91.5

Hooghoudt formula for spacing is as follows (Fig. 7.10).

where, S = spacing between tile drains, V = rate of discharge or rainfall per unit area of land surface, d = depth of bottom of drain above impervious layer, h = depth of water in tile drain, H = maximum height of water table above drain bottom, k = coefficient of permeability.

For practical purpose, $S^2 = 4KH(2d + H)/V$, considering drain is empty and $h = 0$.

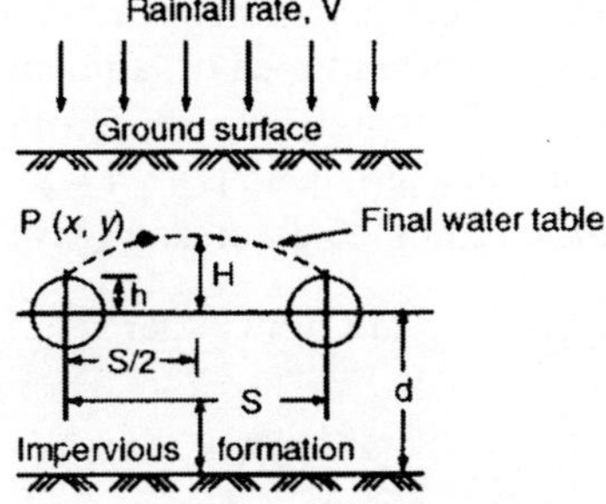

Fig. 7.10 Hooghoudt tile drain

$$S^2 = \frac{4K\left(H^2 - h^2 + d - 2hd\right)}{V}$$

Design of Tile Drains

The capacity of the drain flowing full is determined from the Manning's velocity formula taking roughness coefficient, $n = 0.018$. The diameter of the drain is computed by equating the capacity of the tile drain to the design runoff for the area. The tile diameter is given by the formula, $D = 0.1635\, D_c^{0.375} \times S^{-0.1875} \times A^{0.375}$
where, d = internal diameter of tile (cm), D_c = drainage coefficient (cm per day), A = drainage area (m^2), S = drain slope.

The minimum size of tile is 10 to 15 cm. Maximum velocity prescribed by IS:10907-1984 is 1.1 m/s for sand and sandy loam, 1.5 m/s for silt and silt loam, 1.8 m/s for silty clay loam, 2.1 m/s for clay and clay loam and 2.7 m/s for coarse sand or gravel. A suitable longitudinal slope, depending on the slope of the ground and depth of outlet, is provided. IS:10907-1984 has indicated chart for determining discharge for a given area. Tiles laid to little grade tend to fill up readily while those with steep grade cause high velocities of flow. The desirable range of working grade along the tile drain is 0.1 to 0.2%. The grade for 10 cm tile is 0.10%, for 13 cm tile is 0.07% and for 15 cm tile is 0.05%. The drainage coefficient with no surface water admitted directly into the drain is 5 to 10 mm/day, recurrence interval 5 years is recommended by IS:10907-1984.

The laying of tiles begins at the lower end of the line and progresses up grade. The tiles are laid true to the line of the trench and firmly bedded in the bottom of the trench and on grade. Joints between the tiles are kept open, shrouded with filter, to admit drainage water into the line. A gap of 3 mm in the case of silt, loam and clay soils and 6 to 10 mm for peat and muck is usually allowed. An approximate method for designing tile drain (refer Fig. 7.11) is as under:

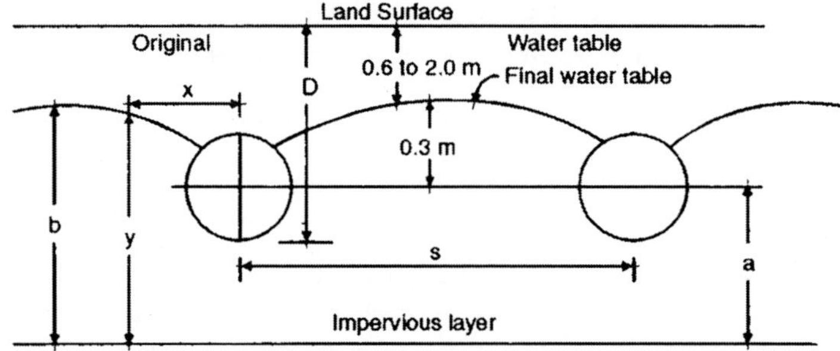

Fig. 7.11 Spacing of tile drains.

Where, S = spacing of tile drain (m), a = depth of impervious layer from the centre of the drain, b = maximum height of the drained water table above the impervious layer, x, y = x is any distance from the centre of the drain where height of water table above the impervious layer is y, and V = rate of rainfall or discharge per unit area of land surface.

It is assumed that the hydraulic gradient at a distance x from the drain is dy/dx, flow lines are parallel and cross sectional area of flow at a distance x is $y \times 1 = y$, and discharge q towards the drain is inversely proportional to the distance from the drain, and Q is the total discharge per unit length carried by the drain so that $\frac{1}{2}Q$ enters the drain from either side.

According to Darcy's law

$$Q = KIA$$

$$\text{or } q = Ky\frac{dy}{dx} \qquad (i)$$

Where, q is discharge per unit length of drain passing through y. K is coefficient of permeability of soil.

Now $$q = \frac{1}{2}Q \quad \text{when} \quad x = 0$$

and $$q = 0 \quad \text{when} \quad x = \frac{S}{2}$$

Therefore, $$q = \frac{1}{2}Q - \frac{1}{2}Q\frac{x}{S/2}$$

or $$= \frac{Q}{2S}(S - 2S) \qquad (ii)$$

Equating equations (*i*) and (*ii*)

$$Ky\frac{dy}{dx} = \frac{Q}{2S}(S - 2x)$$

or $$\frac{Q}{2KS}(S - 2x)\,dx = ydy$$

Integrating, we have

When $$x = 0, y = a, c = -\frac{a^2}{2}$$

$\therefore$ $$Q = \frac{SK\left(y^2 - a^2\right)}{\left(Sx - x^2\right)}$$

When $$x = \frac{S}{2}, y = b$$

Hence $$Q = \frac{4K\left(b^2 - a^2\right)}{S} \qquad (iii)$$

or $$S = \frac{4K\left(b^2 - a^2\right)}{Q} \qquad (iv)$$

Also $$S = 2\sqrt{\frac{k}{V}\left(b^2 - a^2\right)} \qquad \text{Dupuit formula}$$

Q depends on infiltration into the ground and is usually assumed as 1% of the average annual rainfall in 24 hours.

Equations (*iii*) and (*iv*) give discharge and spacing of the tile drain.

Advantages : Advantages of tile drains have been enumerated under closed drains.

Disadvantages : (*i*) High initial cost, (*ii*) Limited drainage capacity, (*iii*) Not open to inspection, being underground, (*iv*) Repairs costly and inconvenient, and (*v*) Requires steeper slope.

7.13 TILE DRAINAGE SYSTEM

Drainage system comprises a main drain, its branches and subsidiary drains. Tile system is so devised as to cater for all the wet areas that could eventually be drained into one main drain. Various tile drainage systems are briefed as follows:

1. *Natural system.* It consists of a system of drains, similar to trunk system, but covering a much larger area (Fig. 7.12). The main drains are located along the depressions or low spots to

conform to topography. Natural system is generally installed in areas of rolling or broken topography where drainage of isolated tract is required.

2. *Parallel system.* It consists of a system of drains with long parallel laterals emptying into a single main drain (Fig. 7.13). It is used in poorly drained soils having uniform texture and little slope.

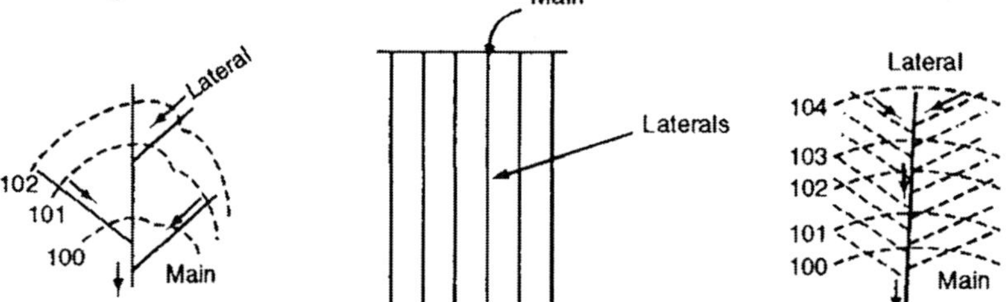

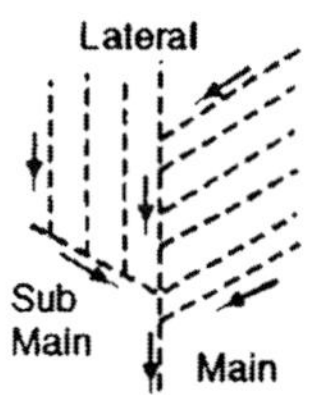

Fig. 7.12 Natural system. **Fig. 7.13** Parallel system. **Fig. 7.14** Herringbone system. **Fig. 7.15** Grid iron system

3. *Herringbone system.* This system of drains consist of a main or submain along the depression with parallel lines of field drains sloping towards the main drain and joining it at staggered intervals (Fig. 7.14). It is used for lands lying on both sides of a narrow depression and laterals must enter from both sides. It is less economical on account of double drainage occurring where the laterals and main join.

4. *Grid iron system.* In this system, the field drains are constructed in parallel lines along the direction of slope and join the main drain at its bottom (Fig. 7.15). It is used for flat land with a uniform slope.

5. *Double system.* It is a system of drains similar to that of herringbone system except that there are two main drains in this system on each side of the depression (Fig. 7.16). It is used when the bottom of the depression is wide. It is not generally used as conditions which it requires are seldom met with in practice.

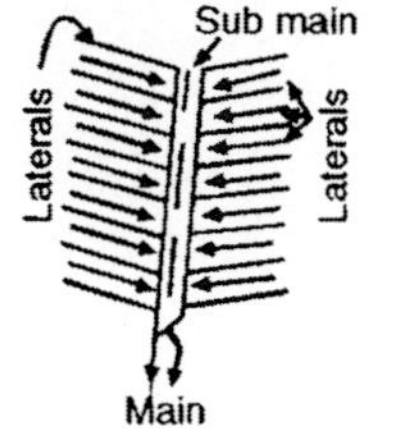

Fig. 7.16 Double main system.

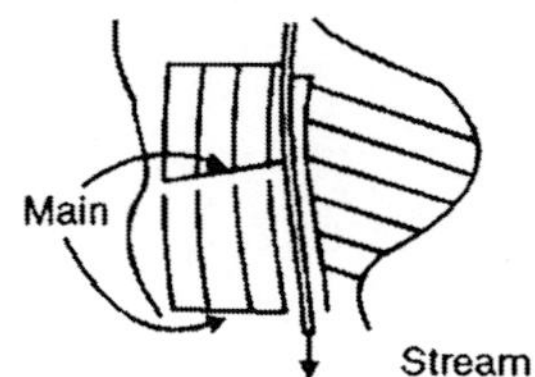

Fig. 7.17 Grouping system.

6. *Grouping system.* It is similar to the natural system except that a few laterals are provided in wet areas or ponds along the system (Fig. 7.17). It envisages collection of water from a number of small systems and discharging into the ditch through one outlet. It is used where the topography and wetness on the field vary and pattern of drainage must be changed to fit the different conditions.

7. *Random system.* In this system drains are laid more or less at random to drain the wet areas (Fig. 7.18). The main is located at natural drainage line and individual wet spots are connected through submains and laterals. Where wet spots are large the arrangement of the submain and laterals for each wet place may utilize one or more of the parallel systems to provide the required drainage. It is used in rolling areas that have scattered wet areas slightly isolated from each other.

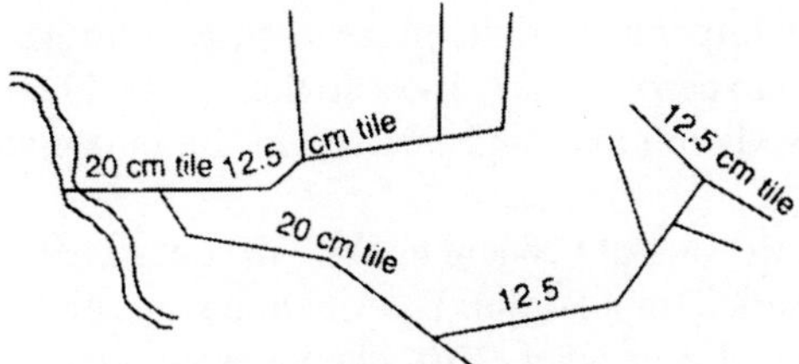

Fig. 7.18 Random system.

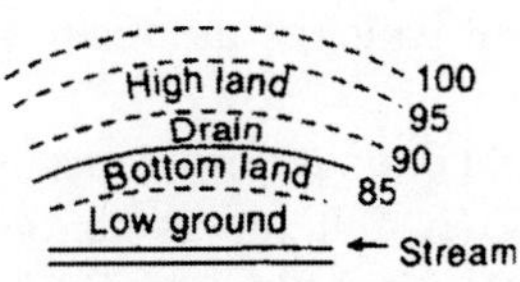

Fig. 7.19 Intercepting system.

10. *Intercepting system.* In this system, tiles are placed along the hillside to intercept the seepage water that follows the upper surface of an impervious subsoil to prevent it from reaching the bottom land (Fig. 7.19). It is used for draining seepage along hillsides.

11. *Composite system.* It is a combination of systems of tile drain arrangement such as the herringbone and grid iron systems (Fig. 7.20).

12. *Sink hole drainage system.* It is a system of drainage used to intercept seepage water, but has in addition wells dug at regular intervals to let the water come up from a lower stratum and enter the drain (Fig. 7.21).

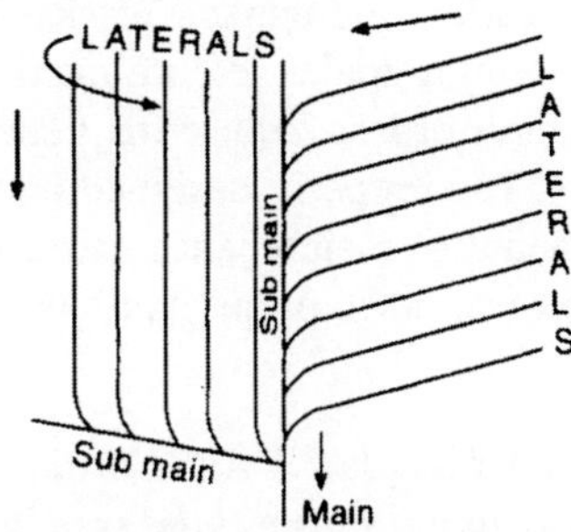

Fig. 7.20 Composite system.

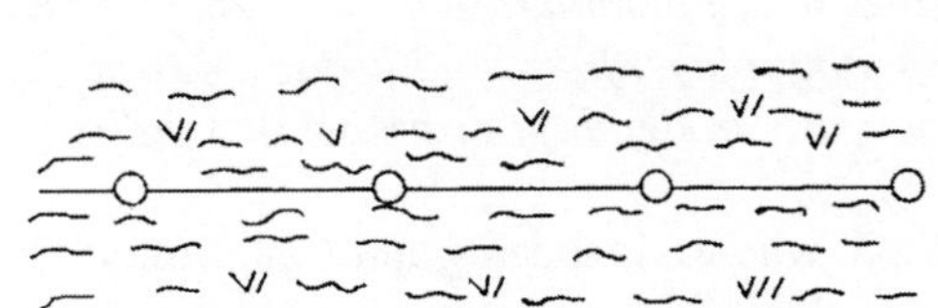

Fig. 7.21 Sink hole system.

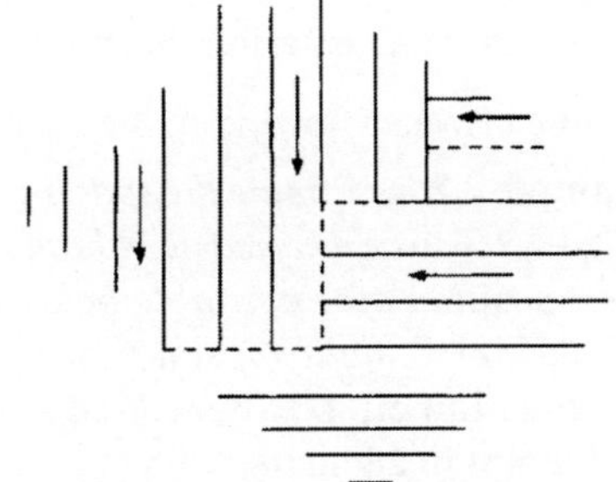

Fig. 7.22 Zigzag system.

13. *Zigzag system.* In this system field drains as well as mains are constructed zigzag (Fig. 7.22) to reduce high velocities.

7.14 OPERATION AND MAINTENANCE OF DRAINS

The drains designed on regime theory applicable to canals inherent maintenance problems, since these are not stable. Regular maintenance is required to keep them functional as designed. Various factors on which the frequency and degree of maintenance depends are amount of rainfall, climate and ground water conditions. In general, open drains require maintenance after a heavy storm. They suffer from operation and maintenance problems such as (*i*) They carry variable flow, unlike irrigation channels; maximum during heavy rains and normally very small discharge and as such are susceptible to erosion and silting up and do not maintain their section for long, (*ii*) They are infested with weed growth as these are run with low velocities for most of the time which choke their waterway, (*iii*) Repairs and maintenance difficult because of their location far away from roads and other means of communication, (*iv*) Rarely inspected since they are situated away from irrigation channels and generally not provided with inspection bank, (*v*) Roads, railroads and canals often cause obstruction to drainage as they cause an afflux in water level and create congestion, and (*vi*) Cross bunds are often put up across the drains to divert or pump out water for irrigation and not entirely removed afterwards, impairing their normal functioning. It is thus imperative that, in general, a drain has cunnette to cater for seepage water and the widened width to carry surface water so that the problem of weed growth which is the main maintenance problem is minimal and that too in the cunnette section only.

The general drainage problems in open drains are (*i*) Erosion of bed and banks, (*ii*) Settlement, (*iii*) Sloughing, (*iv*) Silting, vegetation and seepage. The buried (pipe) drains have maintenance problems like (*i*) Physical blockages, (*ii*) Organic or biological blockages, (*iii*) Chemical or mineral sealing and outlet restrictions. The work plan for maintenance include (*i*) Monitoring, (*ii*) Cleaning of section and banks, (*iii*) Root removal and cleaning and repair of outlet grills. The drains with minimum grades require maximum maintenance. Drains with gradients that result in velocities of 0.3 m/s or less require frequent cleaning.

The properly designed and installed subsurface drains generally require comparatively lesser amount of maintenance care for satisfactory operation. In early period of operation, these drains, however, need proper vigilance to increase their effectiveness as a drain.

SOLVED EXAMPLES

Example 7.1 *A tile drainage system draining 10 ha flows at a design capacity for three days following a storm. If the system is designed using a drainage coefficient of 2.0 cm, how much water will be removed during this period ?*

Solution. Water flow depth from drainage area in 24 hours = 2.0 cm

Volume of water entering the drain (per day) $= \frac{2.0}{100} \times 10 \times 10^4 \text{ m}^3 = 2000 \text{ m}^3$

Volume of water flowing in the drain within 3 days $= 3 \times 2000 = 6000 \text{ m}^3$

Example 7.2 *Determine the spacing of tile drain to cater for annual rainfall = 100 cm, drainage coefficient = 1% to be drained in 24 hours, depth of impervious layer from land surface = 10 m, depth of highest position of water table below land surface = 1.5 m, depth of drain below land surface = 2.0 m and permeability of soil = 6×10^{-6} m/s.*

Solution. Runoff, Q to be catered at 1% in area ($S \times 1$, where S is spacing and 1 m is longitudinal length of drain) in 24 hours.

$$Q = \frac{100}{100} \times \frac{0.01\,(S \times 1.0)\,\text{m}^3}{24 \times 60 \times 60}$$

$$Q = \frac{S}{8.64 \times 10^6} \qquad (i)$$

Spacing of drain, (S) $$= \frac{\left[4K\left(b^2 - a^2\right)\right]}{Q} \qquad (ii)$$

Putting values in equation, (*ii*) $S^2 = 4 \times 6 \times 10^{-6}\left[(10-1.5)^2 - (10-2)^2\right] \times 8.64 \times 10^6$

or $S = 41$ m

Example 7.3 *Determine the size of tile drain if the tile grade is 0.25%, rugosity coefficient is 0.018 and drainage area is 10 ha., drainage coefficient 1.0 cm.*

Solution. Runoff per second $= \frac{1.0 \times 10 \times 10^4}{100 \times 24 \times 60 \times 60} \text{ m}^3 = 0.0116$ cumec

Using Manning's formula, $0.0116 = \frac{1}{0.018} \times \Pi \frac{D^2}{4}\left(\frac{D}{4}\right)^{2/3}\left(\frac{0.25}{100}\right)^{1/2}$

$D = 20$ cm.

Example 7.4 *A surface drainage channel has to be designed for an area of 500 ha. If the drainage coefficient is 6 cm/day, determine the capacity of the channel required and a suitable section of the channel for the given capacity. Take side slope as 1:1. Use Lacey's formula with Lacey's silt factor, f = 1.0. What will be the Lacey's regime slope ?*

Solution.

Discharge to be drained $= \dfrac{500 \times 10^4 \times 6 \times 10^{-2}}{24 \times 60 \times 60} = 3.47$ cumecs

Lacey' $V = \left(Qf^2/140\right)^{1/6} = \left[(3.47 \times 1.0)/140\right]^{1/6} = 0.54\,\text{m/sec}$

$A = Q/V = 3.47/0.54 = 6.43\ \text{m}^2$

$P = 4.75\, Q^{1/2} = 4.75\,(3.47)^{1/2} = 8.85$ m

Now $A = BD + D^2$

and $P = B + 2\sqrt{2}\,D$ (with 1:1 side slope)

Putting values of A and P

$$6.43 = BD + D^2 \qquad (i)$$

$$8.85 = B + 2\sqrt{2}\,D \qquad (ii)$$

Multiply (*ii*) by D and deduct (*i*) from it:

$$1.828\,D^2 - 8.85\,D + 6.43 = 0$$

or $D \simeq 0.9\,\text{m}$

Putting D as 0.9 m in (*ii*), $B = 6.3\,\text{m}$

Lacey's slope, $S = 0.0003\,f^{5/3}/Q^{1/6} = (0.0003 \times 1)/(3.47)^{1/6} = 1/4100$

Example 7.6 *Design a drainage channel for a catchment area of 100 ha, slope 0.45% coefficient of runoff 0.70, intensity of rainfall 3 cm/hr, coefficient of rugosity 0.025.*

Solution.

Discharge = 2.78 CIA

$= 2.78 \times 0.70 \times 3 \times 1.0 = 5.84$ cumecs

$N = 0.0225\, f^{1/4}$

$0.025 = 0.0225\, f^{1/4}$

or $f = 1.52$

$R = 0.47\,(5.84/1.52)^{1/3} = 0.74$ m

$P = 4.75\,(5.84)^{1/2} = 11.48$ m

$$D = \frac{P - \sqrt{P^2 - 6.944\,RP}}{3.472}$$

$$D = \frac{11.48 - \sqrt{(11.48)^2 - 6.944 \times 0.74 \times 11.48}}{3.472}$$

$D = 0.85$ m

$B = P - 2.236 \times 0.85$

$B = 9.5$ m

$$S = \left(f^{5/3}/3340\right)^{1/6} = \left[(1.52)^{5/3}/3340\right]^{1/6}$$

= 0.45% as given

$\therefore$ Bed width = 9.5, depth = 0.85

$$V = \left(\frac{Qf^2}{140}\right)^{1/6} = \left[\frac{5.84(1.52)^2}{140}\right]^{1/6} = 0.68 \text{ m/s}$$

$$\text{Area} = (9.5 + 0.5 \times 0.85)\ 0.85 = 8.44 \text{ m}^3$$

$$\therefore \quad \text{Discharge} = 8.44 \times 0.68 = 5.74 \text{ cumec, and is acceptable}$$

QUESTIONS

1. Discuss waterlogging, its causes and control.
2. What is waterlogging ? What are its ill-effects ?
3. Discuss the various anti-waterlogging measures.
4. (*a*) What is waterlogging ? What are its causes ?
   (*b*) Write a note on various anti-waterlogging measures. [*PU, April, 1980*]
5. What are the measures to be taken in consideration while planning an irrigation project to safeguard against danger of waterlogging in an alluvial plain ?
6. (*a*) What are the effects of waterlogging ?
   (*b*) What measures are adopted to reclaim the waterlogged areas in Haryana ?
7. What are the causes of waterlogging ? What reclamation methods should be adopted ?
8. State briefly the possible causes of salinity and alkalinity of irrigated lands. Name some important standard tests and the units used in measuring salinity and alkalinity. Suggest briefly the remedial measures.
9. What do you understand by land drainage ?
10. Discuss the importance of drainage scheme in the agricultural development of a region.
    [*AMIE, Summer 1974*]
11. Explain why drainage of irrigated land is necessary ? Describe various methods employed for such drainage. [*AMIE, Summer 1975*]
12. Why is proper drainage of irrigated land necessary ? If an area of irrigated land is already waterlogged, what remedies would you suggest for reclaiming the land. Briefly discuss the methods of sub-surface drainage of irrigated land.
13. Discuss different types of drainage schemes provided in irrigation tracts as a precaution against waterlogging. Give salient features of design of an open drain system.
    [*AMIE, Summer 1975*]
14. Discuss whether it is possible to design the channel dimensions of a surface drain by Lacey's regime equation. Mention relative advantages and disadvantages of surface and sub-surface drains.
15. Compare relative efficiency of sub-surface drains and surface drains. Describe the situations in which sub-surface drains would be preferred over surface drains. [*P.U.*]
16. In a drainage system closed drains are placed with their centres 10.5 m above the impervious stratum and maximum height of the drained water spacing of the drains is 38 m and the drains carry 15% of the average annual rainfall in 24 hours. Find the average annual rainfall, $K = 5 \times 10^{-6}$ m/s. [*P.U.*]

PART - II

RIVER ENGINEERING

- **River Mechanics**
- **River Training**

8

RIVER MECHANICS

8.1 INTRODUCTION

Water is an integral part of man's life and environment. Man's habits and activities in all walks of existence and forms of social organisations are influenced more by the close association with water than with the land through which he earns his bread. Man has realised the importance of water since the dawn of civilisation. The ancient civilisation developed only along river banks where water was readily exploitable. Rivers, lakes and oceans greatly facilitated worldwide spread of population and commerce.

India is blessed with abundant water resources. The Himalayas in north have perpetual snow, rainfall and melting of snow give rise to mighty perennial rivers like the Indus, the Ganga and the Brahamputra. The important rivers of Central and Southern India which are entirely rainfed are the Mahanadi, the Narmada, the Tapti, the Godawari, the Krishna and the Cauvery.

8.2 CLASSIFICATION OF RIVERS

Rivers are classified on the basis of topography of river basin, stage, shape in plain, flood hydrographs and on location, illustrated below:

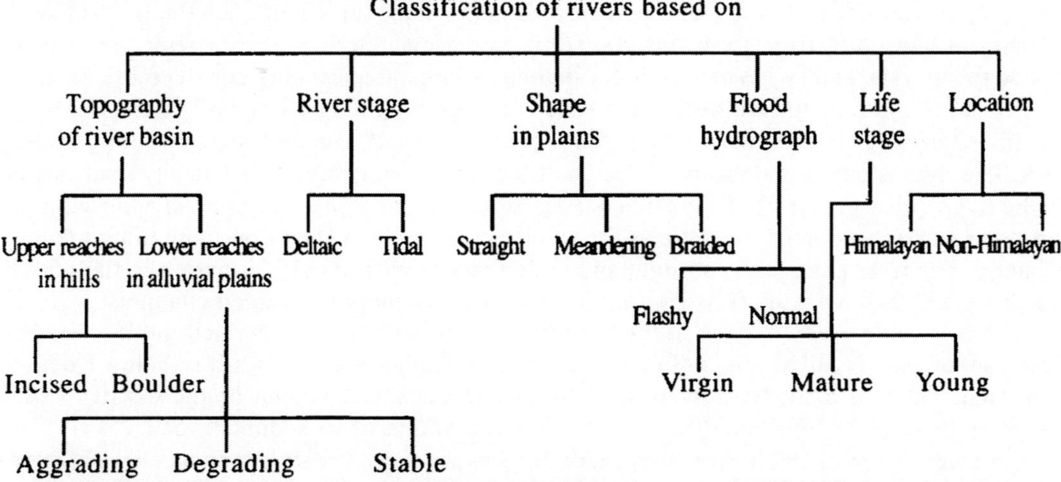

1. Classification Based on Topography of River Basin

1. *Rivers in Hills.*

Rivers in the upper reaches in hilly region are further classified into (*i*) Incised river, and (*ii*) Boulder river.

(*a*) *Incised river.* Incised river channel is formed by the process of degradation. The river passes down steep slopes and carves out for itself a deep channel through the bed of the valley floor, as opposed to one flowing on a flood plain. The chief characteristics of such a river are: (*i*) Well defined narrow cross section cut through hill gorges, (*ii*) Bed and banks highly resistant to erosion; rocky bed, (*iii*) Bed is lined with different sizes of boulders. Some material also rolls along the bed, (*iv*) Clear water; mostly snow water, (*v*) Swift flow, (*vi*) Steep bed slope with abrupt falls, (*vii*) Sediment transported is often dissimilar in character to that of the river bed, as most of it comes from the catchment due to denudation and soil erosion, and (*viii*) Pattern of meanders is not regular.

(*b*) *Boulder river.* It is the reach of the river when it emerges from the hills with a well defined cross section and non-submersible banks on either side. The river is characterised by its steepness and bed is a mixture of boulders, gravel, shingle and sand. The chief characteristics of the river in this reach are (*i*) It flows through sub-mountainous tract, (*ii*) It remains confined within the valley, as the sides are hard, (*iii*) Bed slopes are steep, usually 1 in 500 to 1 in 1000, (*iv*) Velocities are high but lesser than in hill reach, (*v*) Water is generally clear except in rainy season, (*vi*) Larger discharges are accommodated by the river digging in deeper in the valley, (*vii*) Boulders deposit and finer materials are carried to plains, (*viii*) It flows through wide, shallow bed and develops straight course, and (*ix*) Bed material is pervious giving rise to a strong subsoil flow.

2. *Rivers in Flood Plains*

Alluvial river. In this reach the river flows through deposits created by itself and is constantly building its flood plain by overflowing. The sediment carried, except for the wash load, is similar to that in its bed and banks. The chief characteristics of the river are (*i*) Meanders freely from one bank to the other; zig zag course. The meanders also travel down and during the course of adjustment form cutoffs as well, (*ii*) Flows through the alluvium created by itself, *i.e.*, the material carried is similar to that of the bed, (*iii*) The river runs on relatively flat ground and finer materials settle down, (*iv*) Material is deposited by the river during the alternate processes of spreading and receding. In the deposits laid by the river itself, the river carves its own deep winter channel — a flood-way or a Khadir width, (*v*) When once the river channel deviates at any point from its linear course, it sets a meandering pattern with the original course for its axis, and (*vi*) When in flood, inundates large area causing considerable damage.

Rivers in flood plains are further classified as aggrading, degrading and stable.

(*a*) *Aggrading river.* A river in the process of aggradation* is called accreting or aggrading river. The river is building up its bed. The chief reasons for aggradation are heavy load carried by the rivers, obstruction like a dam or a barrage across it raising the water level and flattening the slope and extension of the delta at the river mouth, sudden intrusion of sediment from a tributary. The river has usually straight and wide reaches with shoals formation in the middle which shift with floods, the flow getting divided into a number of braided channels.

(*b*) *Degrading river.* A river in the process of degradation†, *i.e.*, the bed getting scoured from year to year, is called a degrading river. It is found either above a cutoff or below a dam or a barrage. There is a sudden lowering of the water surface upstream of the cutoff♦ which increases its slope of flow, and there is a sudden diminution of its sediment load.

(*c*) *Stable river.* A stable river may be defined as the river stream which as a whole maintains its slopes, depths and channel dimensions without any noticeable raising or lowering of its bed. Such a river may denote a condition that may be temporary from a geological point of view but for practical engineering purposes may be accepted as one of considerable stability.

**Aggradation* — An increase or rise in specific levels of the bed (the bed level at specific discharge) of a channel at any site.

†*Degradation* — The lowering of specific bed level of a channel. It is the converse of aggradation.

♦*Cutoff* — A process by which an alluvial river flowing along curves or bends abandons a particular bend and establishes its main flow along a comparatively straighter and shorter channel.

2. Classification Based on River Stage

(i) Deltaic river. Delta river is rather a stage of a river than a type in itself. It is the reach of a river when it approaches the sea with very gentle slope and velocity, drops down the sediment and divides into channels on either side of the deposits resulting in the formation of deltas. The delta stage does not exist in rivers of Northern India.

(ii) Tidal river. It is the tail reach of a river where it joins ocean. When the ocean is in flood tide its water enter into the river and flows back during the ebb tide. The river thus flows in both directions depending on the tide or ebb stages of the ocean.

3. Classification Based on Shape in Plain

(i) Straight river. It is a river which has attained stable flow condition and carries forward its silt charge. It is straight in plan with maximum flow in the centre.

(ii) Meandering river. Alluvial rivers have meandering tendency; tortuous course with bends and curvatures.

(iii) Braided river. It is a river flowing in a very wide and shallow section characterised by small interlaced channels separated by bars or shoals. The river channel does not meander as a whole.

4. Classification Based on Flood Hydrograph

Flashy river. A river in which the rise and fall of floods are sudden is termed as flashy river. The flows collect rapidly from the steep slopes of the catchment and flood peaks occur soon after the rain. The flows subside as rapidly as they collect. Flood hydrograph is very steep indicative of sudden flood. A sizeable small flow is maintained for some time till the end of the rainy season. Characteristics of bed and banks are similar to those of rivers in alluvial plains.

Normal river. Rivers essentially have variable flows according to the season but the rise and fall may not be as that of a flashy type river. The flood hydrograph too is not steep indicating that flood has occurred gradually.

5. Classification Based on Life Stage

(i) Virgin river. A river which completely dries up in the arid zones, before joining another river or sea is termed as virgin river. Virgin rivers are found in Kutch and Rajasthan States.

(ii) Mature river. A river which flows through certain fixed limits is a mature river. The limits are called Khadir.

(iii) Young river. A river is young, in geological terms, when it is actively eroding its channel.

6. Classification Based on Location

(i) Himalayan rivers. Himalayan rivers take off from the Himalayas, the home of snow, and flow through alluvial plains. These rivers are fed by melting of snow during dry period and by rainfall in monsoon months. They are more or less perennial and carry heavy sediment loads because of soft, friable Himalayan rock. The north-eastern part is of seismic character and seismic disturbance loosen the friable rock leading to landslides and consequently increased sediment load. They include rivers of the Chenab basin and all rivers of north and north-eastern regions including, the Ganga and the Brahamputra, which are the principal rivers having their origin in the Himalayas. Ravi, Beas, Sutlej, Ghaggar, Gandak, Gomti and Kosi rivers also fall in this category.

(ii) Non-Himalayan rivers. Non-Himalayan rivers are not snowfed. They receive their supply chiefly from rainfall during monsoon and by regeneration of water during dry periods. They dry up practically in summer and as such are called non-perennial rivers. They are more stable than Himalayan rivers as they flow through non-alluvial soils. Non-Himalayan rivers transverse Central and South India. They originate from the Aravali, Vindhya, Satpura and Sahyadri ranges and run in all directions. They drain into the east by Mahanadi, Brahmani, Damodar, the Subarnarekha, Vaitarni, Godavari, Krishna, Pennar and Cauvery and to the west Sabarmati, Mahi, Tapti, Narmada and a large number of west flowing rivers draining the Western

Ghats. Narmada flowing towards the west and Krishna, Godavari and Mahanadi to the east account for the bulk of drainage of peninsular India.

8.3 BEHAVIOUR OF RIVERS

The behaviour of rivers is chiefly affected by silt-laden flow through the action of sediment deposition on river beds and raising them. The river in its course from its origin to the mouth at sea presents varying behaviours in the shape of bends, meanders, cutoffs, delta formation, interlacing channels, etc., discussed below :

Straight Reaches. In a straight river, section is of trough shape with the high velocity flow in the mid section. As a consequence water surface in the centre of the river is lower than that at the sides (Fig. 8.1) causing a transverse gradient from the sides towards the centre.

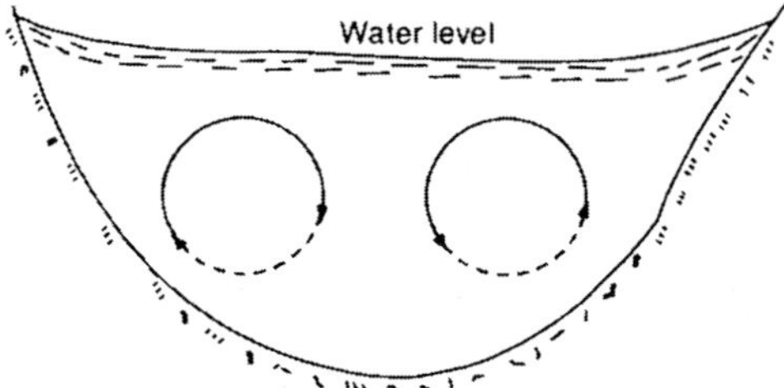

Fig. 8.1 Transverse currents in a regular cross section and straight channel.

Bends. Alluvial rivers tend to develop bends which are characterised by scouring and erosion on the concave side and shoaling on the convex side. The characteristic features of the flow when it negotiates a bend are (*i*) The flow is subjected to centrifugal forces and, as a consequence, there is a transverse slope of water surface from the inner bank towards the outer bank, (*ii*) Water, in order to keep its own level, tends to flow from convex side to concave side. However, this movement is prevented along the water surface by centrifugal forces, (*iii*) Top water dives in at the concave (outer) bank and flows along the bottom carrying therewith sand and silt to the convex (inner) bank, where it is deposited, (*iv*) There is a transverse current from the concave side, where the cross section is deep, to the convex side (Fig. 8.2), (*v*) This rotary action brings about erosion of concave bank and deposition of sediment towards convex side where shoaling takes place, (*vi*) When once a bend is established, the flow tends to make its curvature larger.

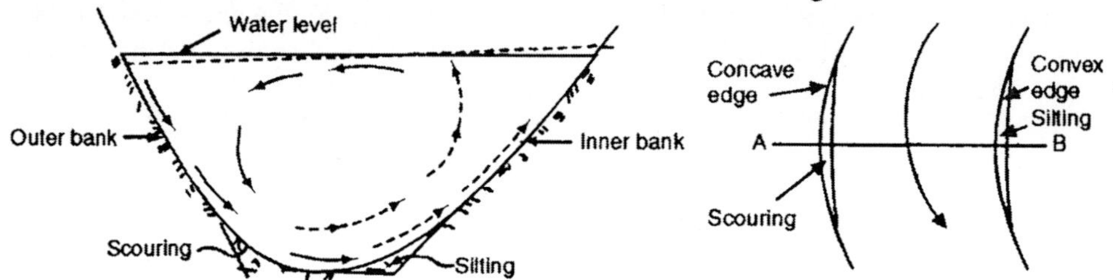

Fig. 8.2 Section AB—Helical movement of water on a curved channel.

8.4 RIVER REGIME THEORY

River development is nature's work spread over several hundred years. Principal river regime theories are discussed as under :

1. **Earth's Rotation Theory.** Earth moving round itself has a rotary motion. The flow in rivers running north or south does not acquire the same rotary motion as does the land and attack their west bank due to earth's rotation bringing them against the current. This assumption is, however, not tenable and the theory is not in use.

2. **Ellis's Tributary River Regime Theory.** This theory states that a river bends to the infalls of the tributaries owing to its bank being there blank and weak, the tributary forming a channel on that side. In order to control the river, some points should be fixed and neither any tributary enters it nor does any canal take off except at fixed points. The usual conditions stipulated are (*i*) No stream coming in on either side, a straight well defined channel, (*ii*) Stream coming in one side —

a well-defined channel on the other side only, and (*iii*) Streams coming in on both sides — an ill-defined channel and a shoal formation in midstream.

3. **Molley's River Regime Theory.** Molley, a Punjab irrigation engineer, presented river regime theory, as under :

1. Plan of the River (Fig. 8.3 (*a*)). (*i*) River bends alternately from one bank to another. The characteristics of each bend are that there is deep water at the crest of bend and shallow crossings or bars are formed in the lengths where the river changes over from one side to the other due to the deposition of silt in the bed, (*ii*) The bars are formed by the action of the side channels. The side channels have offtakes and outfalls (Fig. 8.3 (*a*)). The withdrawal of discharge into side channel at the off takes brings about reduction in velocity in the main stream resulting in deposition of silt forming bars, and (*iii*) Between the main river and the side channel there are islands which are formed when flood recedes and river separates into the main and side channels.

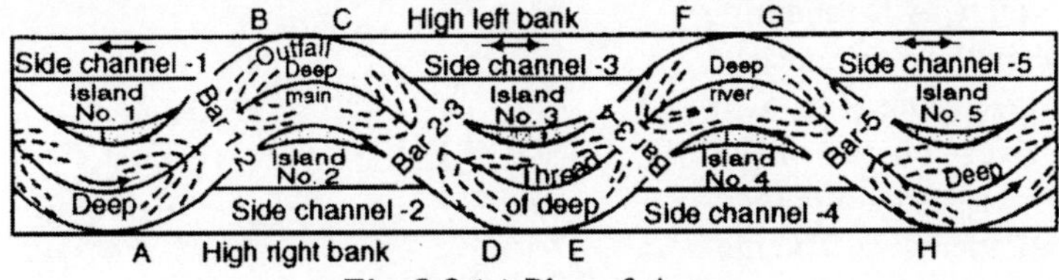

Fig. 8.3 (*a*) Plan of river.

2. Longitudinal Section of the River (Fig. 8.3 (*b*)) (*i*) It shows the conditions of surface levels. Mean gradients are in dotted lines while the actual ones are in firm lines, (*ii*) Low water longitudinal section—Due to the obstruction caused by the bars actual gradient compared to the means is raised above the bars to the extent indicated by *a* and is depressed by *b* to gain sufficient velocity in order to pass the bars, (*iii*) Flood longitudinal section—When the flood is rising, the bars are silting up and

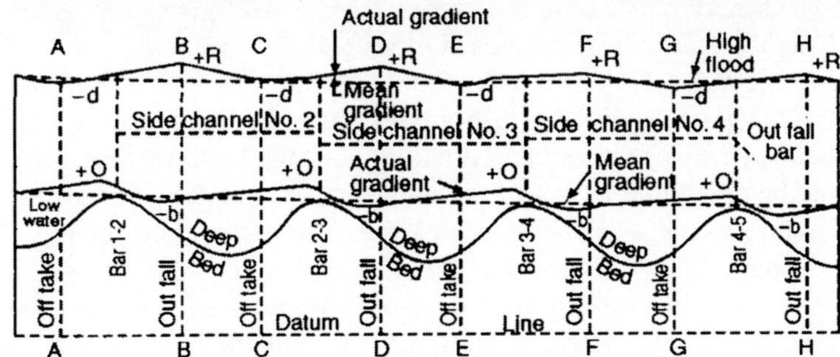

Fig. 8.3 (*b*) Longitudinal section of river.

discharge of side channel at the offtakes tends to increase. This tendency is resisted in the following two ways, as the floods continue rising, (*a*) Rise of 'R' and drop of 'd' at the off takes resulting in reduced gradient of side channel, and (*b*) Back flow at the outfall. This process continues till the flood covering the full section is reached and conditions depicted are attained.

4. **Oldham's River Regime Theory.** The four principles of Oldham's river regime theory are (*i*) Every stream tends to be in an equilibrium condition in which the velocity shall scour the bed, thereby reducing gradient, velocity and silt transporting power while a lesser velocity shall entail the silt deposition, reduction in section and increased velocity required to transport silt, (*ii*) Every stream is alternately collected into a single well-defined channel, called a reach, or spread forming a fan (Fig. 8.4). These fans can be seen when the river dries up in winter, (*iii*) The gradient of a stream is not uniform. The bed is flatter in reaches and steeper in

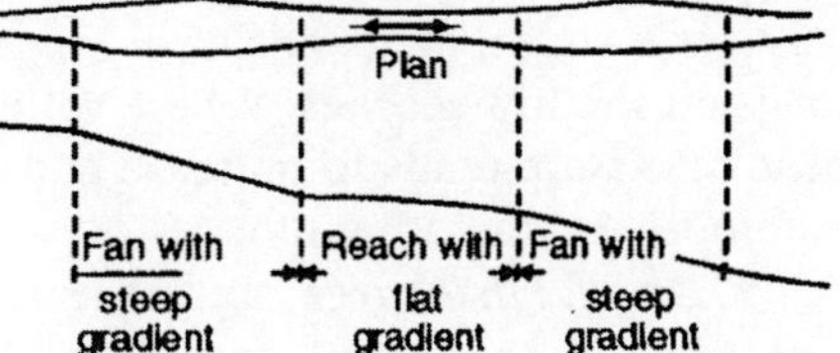

Fig. 8.4 Oldham's river regime theory.

fans relative to the mean gradient. Fans are shallow and have reduced velocity, and (*iv*) Both reach and fan work gradually upstream. Reach encroaches the upper fan by erosion and is encroached upon by the lower one by silting.

8.5 MEANDERING

Alluvial rivers are characterised by their meandering tendency. The term meandering is usually applied to any channel having a tortuous course, with bends of either regular or irregular shape and small or large curvature. However, the term applies only to letter S-channel pattern that is free to shift; as part of a slow migratory movement of the river's meander pattern as a whole down the valley. A meandering river on alluvial plain flows in consecutive curves of reverse order (Fig. 8.5). Meandering in a river consists of two consecutive loops, one flowing clockwise, the other anticlockwise. The mechanics of meandering is a transfer process in which the materials eroded from a concave bank are redeposited on the convex bank immediately downstream on the same side of the river. The curvature of flow established in a river due either to its own characteristics or to the impressed external forces propagates downstream by building up shoals on the convex side by means of secondary currents.

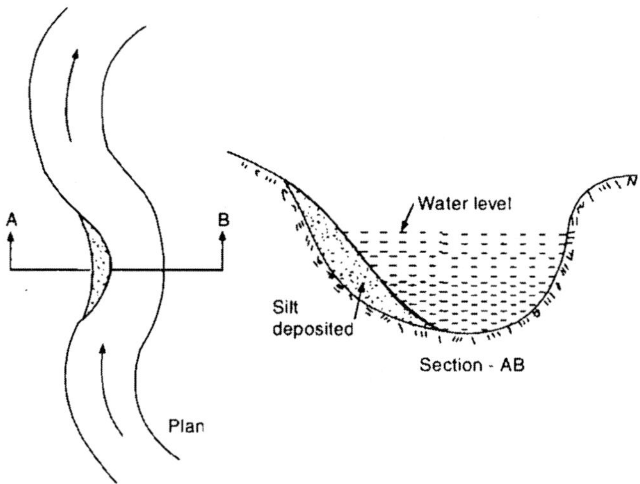

Fig. 8.5. Meander of a river.

Causes of Meandering : The initiation of meandering takes place due to one of the causes such as (*i*) Local disturbance such as variable bank material, obstruction, etc., (*ii*) Excessive energy of the streams, (*iii*) Excessive bank erosion, and (*iv*) Transverse oscillation. The cause of meandering may be attributed to any of the these factors but the primary cause of meandering is excess of total charge associated with excessive turbulence during floods. The silt charge in excess of that required for stability tends a river to build a steeper slope by depositing the charge on the bed. The increase in slope tends to increase the width of the river channel if banks are not resistant. Only a slight deviation from uniform axial flow is then necessary to cause more flow towards one bank than towards the other. Additional flow is immediately attracted towards the former bank, leading to shoaling along the latter, actuating the curvature of flow and producing, finally, meanders in its wake.

Factors Governing Meandering Pattern. Meandering of a river once established has a certain pattern and dimensions. The principal factors governing the meandering pattern are as follows :

1. *Valley slope:* Meander occurs downstream of reaches where the silt charge picked up in the upper reach deposit due to slope being too flat to carry the charge. Where valley slope is steep, as in the upper part of an alluvial valley, the meander length is longer in proportion to the river's width, the bends have longer radii of curvature and cross overs consist of relatively long and shallow reaches. Where valley slope flattens, as in the down part of an alluvial valley, bend curvature tends to increase and the meanders increase in size, assuming no decrease in erodibility of the alluvium.

2. *Bankful discharge:* Increase in bankful discharges as below the confluence of an important tributary, there are larger channel dimensions and a larger meander pattern.

3. *Bed load:* With considerable increase in bed load, there are large meanders, and a wider shallower channel. Considerable decrease in bed load results in narrower and deeper channel with less sinuous course on a lighter hydraulic gradient.

4. *Bed and side resistance:* In gravel and boulder bed no definite meander pattern is observed as the river has a tendency to form interlaced channels.

Classification of Meanders: Meanders are classified as either regular or irregular in shape and as either simple or compound, as described below:

1. *Regular meanders.* A train of bends of the same curvature and frequency (Fig. 8.6).

2. *Irregular meanders.* Deformed in shape and may vary in amplitude and frequency. They may occur because of variation in discharge along the channel by additions from tributaries, withdrawals by branches, lakes, etc., and non-homogeneity of the strata in which the channel is carved, *i.e.*, rock outcrop, stiff lime deposits and clay formations, weir, barrage, spurs, ghats, etc. Figure 8.7 shows deformed meander formed due to revetment and rock outcrops on the banks.

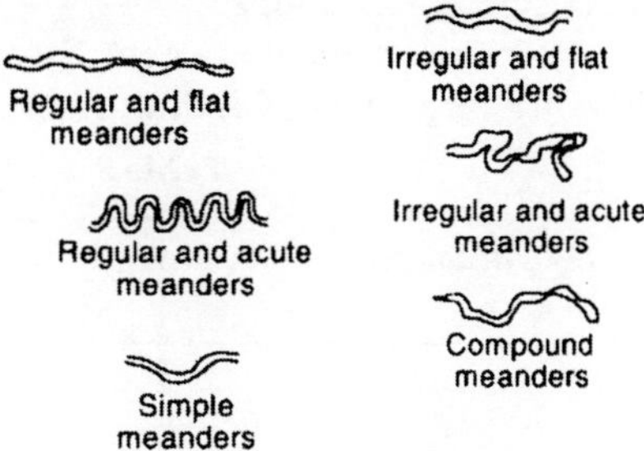

Fig. 8.6 Types of meanders.

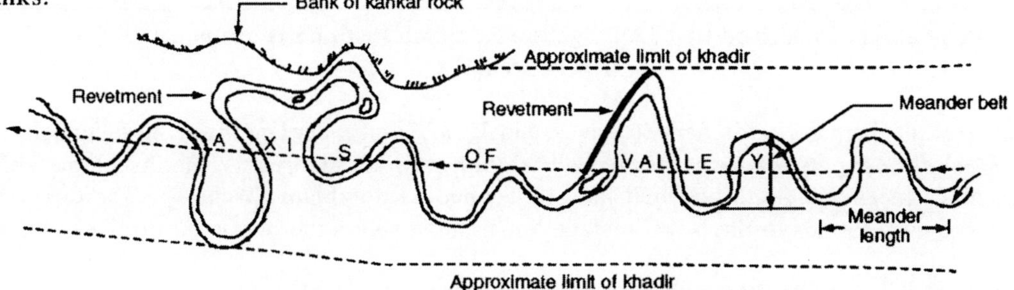

Fig. 8.7 Deformed meander due to obstruction.

3. *Simple meanders.* Bends of one dominant amplitude and frequency (Fig. 8.6).

4. *Compound meanders.* Bends of different amplitudes and frequencies (Fig. 8.6). They may develop if a river has more than one dominant discharge, each of which could independently influence meander characteristics.

Meander Parameters

The pattern and dimensions of a meander are as under:

Meander. Meander or full meander in a river consists of two consecutive loops, one flowing clockwise, the other anticlockwise (Fig. 8.8).

Meander belt (M_b). It is the distance between lines drawn tangential to the extreme limits of successive fully developed meanders (Fig. 8.8).

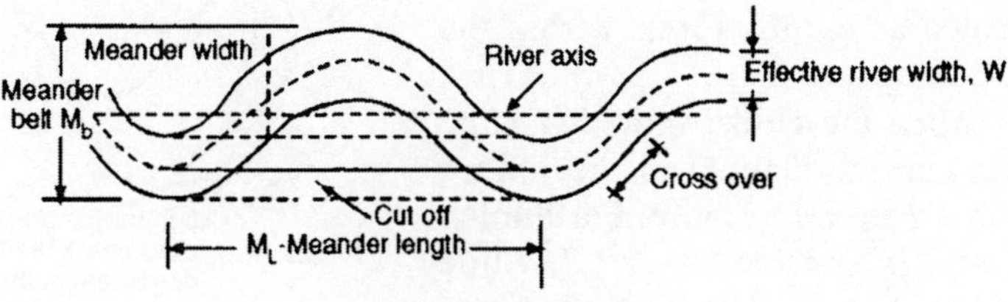

Fig. 8.8 Meander belt.

Meander length (M_l). It is the air distance along a meandering type river, measured from the tangent point of one curve to the tangent point of the next similar curve (Fig. 8.8).

Meander ratio (M_b/M_l). It is the ratio of meander width to meander length.

Tortuosity. It is the ratio of the length of river channel to the axial length of the river, *i.e.*, l/M_r

Crossover. It is the short straight reach of the river connecting two consecutive clockwise and anticlockwise loops.

The relationship between different parameters of a meander, established by Inglis on analysing large data of Indian rivers is given in Table 8.1.

Table 8.1 Relationship of meander parameters

Parameter	*Rivers*	
	In flood plains	*Incised rivers*
M_l	6.06 W	11.45 W
M_b	17.40 W	27.3 W
M_b	2.86 M_l	2.2 M_l
M_l	53.61 $Q^{0.5}$	46.01 $Q^{0.5}$
M_b	153.42 $Q^{0.5}$	102.16 $Q^{0.5}$
W	8.84 $Q^{0.5}$	4.53 $Q^{0.5}$

The relationship derived by U.P. Irrigation Research Institute is as under:

$$M_l = 14.36\, x^{0.125} Q^{0.7}$$

$$M_b = 7.62\, x^{0.125} Q^{0.7} x$$

where, x = silt charge, Q = discharge (cusecs), and M_l, M_b = meander length and meander belt (ft).

Meander Length-Dominant Discharge Relation. In natural streams, the discharge which determines meander length and belt may be termed as dominant discharge. The dominant discharge corresponds to the bankful stage but in most rivers the dominant discharge varies between $\frac{1}{2}$ and $\frac{2}{3}$ of the maximum discharge.

$M_l = 65.8\, (Q_{dom})^{0.5}$, assuming Q_{dom} as 2/3 maximum discharge.

8.6 CUTOFF

It is defined as the process by which an alluvial river flowing along curves or bends, abandons a particular bend and establishes its main flow along comparatively straighter and shorter channel. Cutoff ratio is the ratio of the length of arc to chord of a meander at which the cutoff generally takes place. Cutoff is essentially a characteristic of alluvial rivers and signifies development of river meandering to the extreme conditions in the shape of hair-pin bends. Under favourable condition, these bends become large loops with narrow necks. The narrowing of the neck reaches a limit when a break through occurs and a chute channel, known as cutoff, forms across the neck (Fig. 8.9).

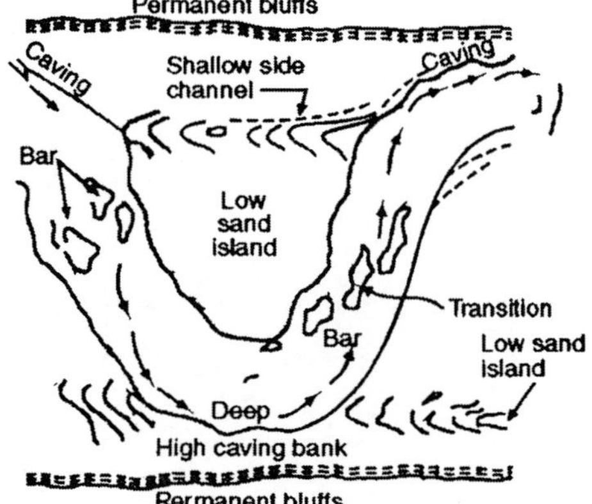

Fig. 8.9 Cutoff.

Cutoff occurs when the cutoff ratio l/M_l, *i.e.*, length of the main curved channel to the chord (meander length) is 1.7 to 3.0 or more. For Punjab rivers the cutoff ratio is considered as 1.7, while that for Indus in Sind (Pakistan) it is a little more than 2. Another important factor causing cutoff is too sharp a curvature for the discharge, *i.e.*, a small value of $r/\sqrt{Q}$ between 13 to 24, where r is the radius of curvature (m) and Q is the discharge (cumecs).

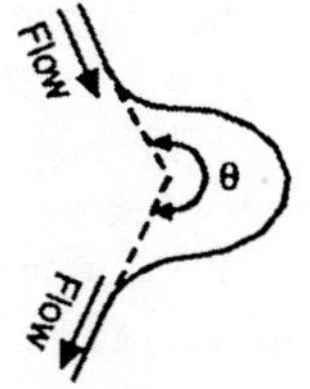

Fig. 8.10 Angle of swing

According to Chatley, cutoff occurs when the angle of swing (Fig. 8.10) is greater than 180° or, when arc/chord ratio exceeds 1.571. The maximum angle of swing, = 180° + 2 [Vers^{-1} (chord/2 radius)].

Effects of Cutoff: Various effects of cutoff are (*i*) With the establishment of a cutoff bank starts caving, new channels are formed, some old channels get silted up, until at last, probably during low stages, a temporary stability is established for a couple of months. The next flood again sets in an agitated period of adjustment until equilibrium is fully established, (*ii*) The action of cutoff is similar to that of channel improvement for stage reduction of flood, (*iii*) Cutoff also causes local changes in the direction of current, depending on local conditions, which may be occasionally radical, leading to bank caving both above and below the cutoff. As soon as the channel adjusts itself to new conditions imposed by cutoff, caving decreases to its normal proportions, (*iv*) Due to scour in the channel above the cutoff, flood waters are charged with bed load to their maximum carrying capacity, the bar* formation at the lower end of the cutoff channel is not washed away rapidly and a rise in flood stage in this region will continue for some time, and (*v*) There is a marked tendency to compensate for shortening the length of river, as a result of successful cutoff, by lengthening the channel elsewhere.

8.7 EFFECT OF DAM ON RIVER REGIME

With the introduction of a dam across a river, a reservoir is created, the effects of which are as follows :

1. Aggradation Upstream of Dam

As a result of dam the transportation capacity of the inflow into the reservoir is reduced by backwater so that sand and coarser sediments travelling on or near the stream bed tend to settle out continuously where the current is retarded as it flows into the reservoir and inclined downward in the direction of flow, the inclination increasing with the coarseness of the sediments. The deposition of sediment in a reservoir take place as bottom-set, fore-set and top-set beds (Fig. 8.11). The top-set beds are usually composed of coarse sediments and are generally sloped upstream at a low gradient from the edge of the fore-set bed, with slightly increasing steepness upstream and extend as far as the backwater extends upstream of the reservoir. The finer particles of silt and clay carried by the river in suspension settle down very slowly in the entire bottom of the reservoir. The distance to which the sediment is carried before settling at the bottom varies according to the rate and duration of inflow, size and settling rate of particles, etc.

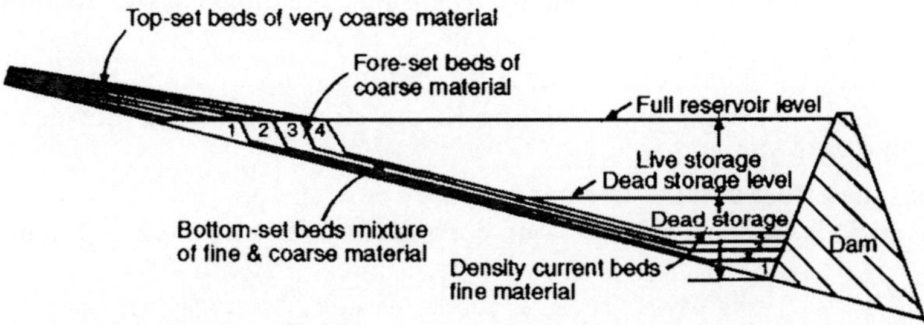

Fig. 8.11 Longitudinal section through reservoir showing various types of sediment deposits.

*Bar is an alluvial deposit or bank of sand, gravel or other material at the mouth of a stream or at any point in the stream itself which causes an obstruction to flow or to navigation.

The exposed delta deposits at the upstream end of a reservoir promote growth of vegetation in the upper reaches. However, no such growth is evidenced in the reservoirs in South India. The vegetation growth acts as a filter preventing inflow of sediment into the reservoir which eventually silts up the upstream channel and raises the bed levels thereby reducing the rate of silting of the reservoir to a certain extent, and also promotes considerable water loss from the reservoir due to transpiration.

2. Degradation Below a Reservoir

A dam generally causes degradation immediately downstream. The released, relatively silt-free, water which passes down over the dam has again the capacity for picking up the sediment load in the downstream reaches progressively lowering the bed levels and steepening the slope of the channel downstream. The effects of degradation, in general, are (*i*) Lowering of flood levels, (*ii*) Damage to foundations of structures, (*iii*) Materially affecting the functioning of dissipation devices such as stilling basin apron with blocks, bucket, etc. due to reduced tail water levels, (*iv*) Increase in available head at power installations due to lower water level in tail channel. It can be made use of by properly designing the power installation, and (*v*) Lowering of water table. The construction of dam reduces the sediment charge in the river downstream, reduces the dominant discharge and the fluctuations in discharge and charge, consequently the downstream channel tends to straighten up thus minimizing the meandering tendency, and tail water levels downstream are appreciably affected.

3. Reservoir Induced Seismicity

Reservoirs are considered to trigger earth tremors. There is no complete agreement on the exact cause of this seismic activity. However, an important measure is early installation of seismographs. Seismic observations are the most efficient means of advance alarm, as the main earthquake or quakes are preceded by a series of shocks of increasing intensity.

QUESTIONS

1. Name the important rivers of North, Central and Southern India. Classify the rivers on the basis of topography and flood hydrography.
2. Discuss the behaviour of rivers with special reference to straight reaches and bends.
3. Explain different stages of a river.
4. Discuss various river regime theories.
5. Define meandering of rivers. What are the causes of meandering ? Discuss the factors governing meandering. How does nature counter-balance the increased river length due to meandering ?
6. Why are the meanders of an alluvial river normally found to shift downstream ? Can they be stabilized ?
7. Define various patterns of meandering. Discuss meandering parameters with the help of a sketch.
8. What is cutoff ? What is its use as river training work ?
9. Explain fully the changes occurring in the river regime on the development of a natural or artificial cutoff.
10. Discuss the effect of dam on river regime.
11. Do reservoirs induce simicity — no correlation is found as yet? Discuss

9

RIVER TRAINING

9.1 INTRODUCTION

River training is an age-old practice resulting in incessant development and application of human ingenuity to correct vagaries of the rivers. It requires deep and precise study of river mechanism and behaviour discussed heretofore. River training has assumed considerable significance in India due to huge annual recurring damage caused by the floods; 80 per cent of which accounts for loss of crops. River training, in its broad aspects, covers all engineering works constructed on a river to guide and confine the flow to the river channel, and to control and regulate the river bed configuration for effective and safe movement of floods and river sediments. In essence, river training envisages training and stabilising a river within a suitable waterway and along a certain alignment for a variety of purposes. River training works involve large outlays and it is essential to select the type of the training work and materials of construction so as to make optimum utilisation of funds, and effective and economical utilisation of the available construction materials.

9.2 OBJECTIVES OF RIVER TRAINING WORKS

The various objectives of river training are (*i*) To guide the axis of flow at ordinary and low stages and safe passage of floods without overtopping the banks, (*ii*) To protect the banks from erosion and generally improve their alignment by stabilising the river channel, (*iii*) To train the river flow along a safe course, thereby avoiding damage by flooding or erosion of valuable lands, habitations, crops, factories, etc., (*iv*) To prevent outflanking of a bridge, barrage or weir by directing the flow in a defined stretch of the river, (*v*) To prevent river from changing its course, (*vi*) To confine a river channel which has become too wide by swinging from side to side and to reclaim the land from river bed, (*vii*) To check certain devastations like that of flash torrents, (*viii*) To trap bed load in areas of superfluous width, (*ix*) To transport efficiently bed load and suspended sediment load, (*x*) To provide sufficient depth of flow for safe navigation, (*xi*) To establish channel boundaries where braiding has created too wide a section divided into small channels separated by islands, and (*xii*) To correct disorderly banks or flow conditions.

9.3 CLASSIFICATION OF RIVER TRAINING WORKS

1. Classification Based on Purpose

1. *High Water Training:* River training aimed at flood protection is called high water training or *training for discharge*. It envisages provision of adequate waterway for safe passage of maximum flood by proper location, alignment and height of embankments for a given flood discharge without tending to change river bed conditions.

2. *Low Water Training:* Also termed as *training for depth*. It envisages to provide adequate water depths during low water periods in the river channel for navigation by concentrating flow in

a desired channel and closing other channels by the method of bandalling, *i.e.*, contracting the width of the river channel with the help of groynes, etc.

3. *Mean Water Training:* Also termed as *training for sediment.* It is by far the most important type of river training. It envisages rectification of the river bed configuration and efficient movement of suspended and bed load for maintaining the channel in good condition. The maximum aggrading capacity of a stream occurs in the vicinity of mean water or dominant flood discharge, and as such tends to change the river bed in accordance with that stage of bed flow. Mean water training includes river training for efficient sand exclusion from canals by correcting adverse river curvature to locate the canal offtake.

2. Classification Based on Structure Alignment

1. *Longitudinal structure:* These aim at guiding the axis of flow at ordinary and low stages, protecting the banks from erosion, generally to improve their alignment, trapping bed load in the areas of superfluous width, and establishing channel boundaries where braiding has created too wide a section divided into small channels separated by islands. Longitudinal training works are preferred to projecting training works in rivers carrying small bed loads, and with narrow channels having steep slopes and swift currents. In rivers with unstable bed conditions, longitudinal works are susceptible to damage by undermining.

2. *Projecting training works.* Projecting training works aim to protect the bank from which they project into the river by deflecting the current away from the bank. They are more suitable in rivers with unstable bed conditions. Cross connecting dikes to the bank at intervals add strength to withstand flood action and promote sedimentation in the closed-off channel spaces.

9.4 PRINCIPLES OF RIVER TRAINING

Two approximate theories, the theory of tractive force for bed load, and regime theory for suspended load are available for application to problems of river training.

1. Tractive Force Theory

Tractive force approach is essentially lane's approach based on the concept of critical tractive force for the incipient conditions of bed movement. The alluvial rivers carry sediments in suspension as well as bed load. The capacity of river channel is designed to cater for the discharge and sediment load by the following formula:

$$Qs = \frac{0.17}{m^{3/4}} \gamma^2 BR^2 S^2$$

Also $$Q = \frac{1}{n} BR^{5/3} S^{1/2} \text{ (Manning's formula)}$$

Where, Qs = silt bed charge (cumecs), m = dia. of sand (mm), γ = specific weight, B = channel width, R = hydraulic mean depth, and S = slope.

2. Regime Flow Theory

In incoherent alluvium, Lacey's regime theory is adopted for determining initial design dimensions, although there is a marked divergence when applied to large rivers due to omission of an important parameter, *viz.*, quantity of silt charge. The size and quantity of silt in suspension are important where silt charge in the flow predominates.

9.5 METHODS OF RIVER TRAINING

Planning and design of river training works is done by empirical methods and on the intuitive judgement of engineers. The method of river training applicable depends mainly on the type of river and sediment load, as under. Under certain cases, where river is very unstable, training is almost impossible and the situation has to be met as it arises.

1. *Stable rivers:* Stable rivers are characterised by stability of alignment, slopes and regime as they mould their beds to carry into sea almost all the silt brought down by them. They are amenable to river training to attain ultimate stability with the aid of training works such as spurs and guide banks.

2. *Aggrading rivers:* Aggrading rivers have inherent instability and are not equally amenable to river training. Stability cannot be imposed, *e.g.*, bank protection works may either be destroyed by severe erosion or get buried under deposition unless complementary measures like soil conservation in the upper reaches supplemented with check dams and dams on the tributaries are adopted for the most effective training and control of the rivers.

3. *Degrading rivers:* Degrading rivers too have inherent instability and are not easily amenable to river training. Stability cannot be imposed, *e.g.*, training works may be destroyed by undermining due to bed scour. Gradient control by constructing dams and weirs is a prerequisite to attempting any other modifications in the river. Local training works may be possible if their repercussions on river regime upstream and downstream are not drastic.

Model studies* is a very useful tool in the design of river training works. Generally mobile bed models are constructed. When a river has a predominant bed load movement, its simulation, both qualitatively and quantitatively, is possible in the model. However, similitude of the suspended load in models is not possible. The conformity between hydraulic models and prototype is subject to limitations.

9.6 PLANNING FOR RIVER TRAINING MEASURES

The training measures are planned with due regard to the limitations imposed by the type of river. The various steps involved are as follows:

1. *Alignment.* Alignment of training work is generally based on the layout and number of training structures involved. A better way to arrive at radius of curve is to study the photographs of alignment of a stable stretch of river on a similar grade and soil conditions in the vicinity of the erosion site and from this stable section measure the radius of curves. An attempt is then made to develop an alignment with curves of similar radius to those measured. Once the proposed alignment is decided, the methods of control to stabilise the channel to its alignment are considered.

2. *Cross Section.* The cross section is that built by the river itself, *i.e.*, when the channel is uniform and does not show signs of accretion or retrogression. The proposed section is either by river contraction or by river diversion and is compared with the normal cross section with regard to gauge and discharge relation, velocity of flow, and total sediment runoff. The section is modified by successive analysis to bring it in line with the normal cross section as far as possible.

9.7 TYPES OF RIVER TRAINING WORKS

River training works are of the types discussed as under:

1. Guide Banks

Also called *Bell's bund.* It is defined as a protective or training embankment constructed at the site of a barrage, weir, bridge, etc. to guide the river flow through the confined waterway without causing damage to the structure and its approaches. They are provided in the direction of flow, both upstream and downstream of the barrage on one or both flanks depending on requirement. The guide banks are usually provided in pairs, symmetrical in plan and may either be parallel or converge slightly towards the structure, extending a little downstream but largely on the upstream of the structure and curved inlands on both ends to provide a bell-mouth entry and smooth exit. The layout of training works for a barrage, weir or bridge are almost identical and consist of a pair of guide banks and armoured groynes. The guide banks are two heavily armoured embankments in the river in the form of a bell-mouth (Fig. 9.1). Simple approach embankments span the portion of the river between the normal river banks and the guide banks.

* For details refer "Hydrology and water resources engineering by the author".

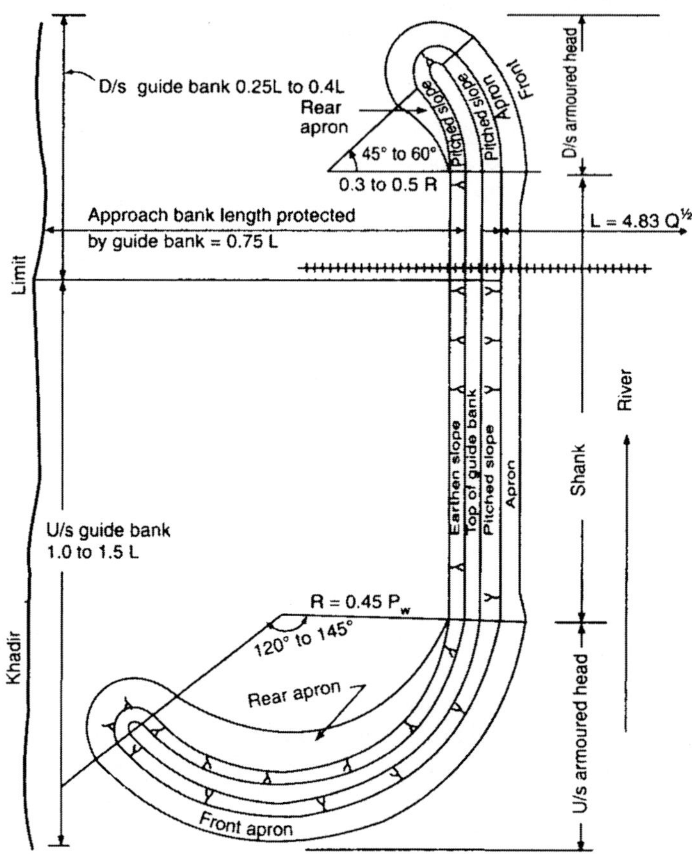

Fig. 9.1 Layout of guide bank.

Functions of a Guide Bank: The functions served by a guide bank are (*i*) Economical spanning of a wide river, (*ii*) Safe and expeditious passage of floods, (*iii*) Protect adjacent lands, upstream of a weir or barrage from spills due to afflux caused by the barrage construction, (*iv*) Control and confine the flow axially through a restricted waterway so that the river has no possibility to swing about and outflank the structure, (*v*) Prevent occurrence of cross-flow immediately upstream of barrage, and (*vi*) Protect the approach embankments (on either side of a bridge extending from the river bank to the guide banks) from direct attack by the river.

Effects of Guide Banks: The effects of guide banks are (*i*) Increase in the rate at which flood wave passes down the river, (*ii*) Increase in maximum discharge at all points downstream, (*iii*) Rise in the water surface elevation of the river during flood, (*iv*) Increase in the velocity and scouring action through the embanked section, and (*v*) Reduction in water surface slope of the river above the embanked portion.

Classification of Guide Banks

1. *According to Form in Plan*: The pair of guide banks may be parallel, divergent or convergent in plan upstream of the structure.

(*a*) *Parallel guide banks.* Where the river is likely to meander on both sides, parallel guide banks with suitable curved heads are essentially required to give uniform flow from the head of the guide bank to the work. Symmetrical and parallel guide banks (Fig. 9.2) are usually adopted unless local conditions warrant adoption of convergent or divergent banks; their suitability being indicated by model studies.

(*b*) *Convergent guide banks:* Convergent guide banks (Fig. 9.2) are rarely used. They have the disadvantage of excessive attack and heavy scour at the head and shoaling all along the shank rendering the end bays inactive.

(*c*) *Divergent guide banks.* Divergent guide banks exercise an attracting influence on the flow and are indicated where approaching flow is oblique to the structure or work is located at one edge of khadir, in order to keep the flow active in the spans adjacent to them. However, they provide relatively less protection to the approach embankment, under worst possible embayment*, as compared to the equal length of parallel guide banks (Fig. 9.3). The divergent guide banks thus require longer length in comparison to parallel guide banks for the same degree of protection to the approach embankment. They induce oblique flow onto the barrage and give rise to tendency of shoal formation in the centre due to larger waterway between curved heads.

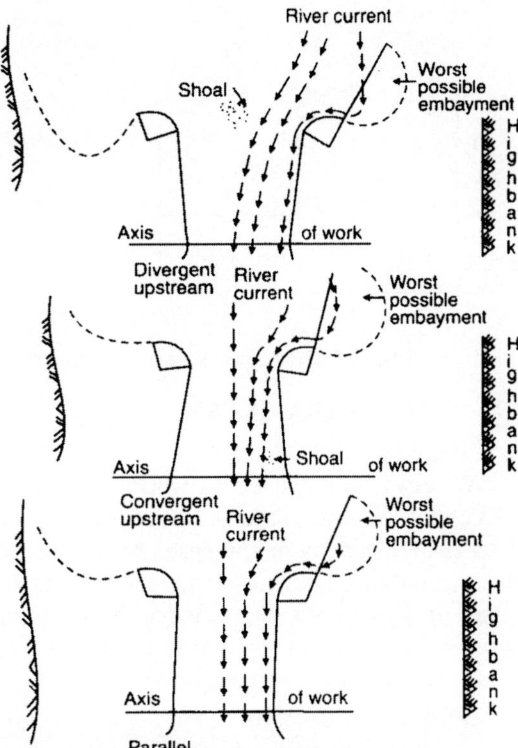

Fig. 9.2 Different forms of guide banks.

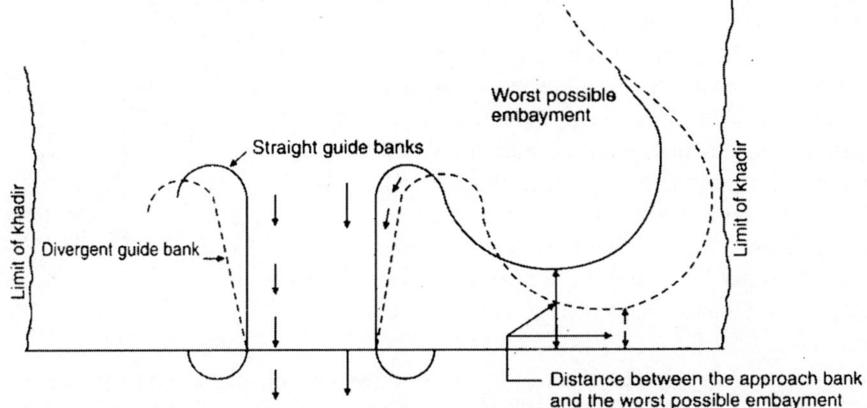

Fig. 9.3 Extent of protection provided by the straight and divergent guide banks.

2. *According to Geometrical Shape.* On the basis of geometrical shape, the guide banks are classified as straight and elliptical with circular or multi-radii curved head (Fig. 9.4). Any other shape, if warranted by site conditions, is determined with the aid of hydraulic model studies.

Elliptical guide banks are preferred to the straight ones on account of the considerations (*i*) More suitable for wide flood plain rivers compared to straight guide banks, (*ii*) Flow hugs the guide banks all along their length due to gradual change in curvature, while separation of flow occurs in straight guide banks after the curved head which leads to obliquity of flow, and (*iii*) Provide better control on development and extension of meander loop towards the approach embankment.

* Embayment. A localised widening of a river or channel brought out artificially or naturally.

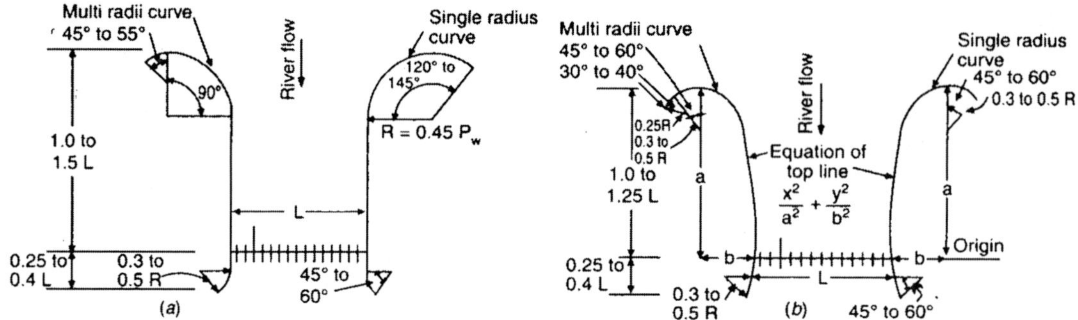

Fig. 9.4 Geometrical shapes of guide bank (*a*) straight guide bank, (*b*) elliptical guide bank.

9.8. DESIGN PARAMETERS

1. Waterway

Waterway or length of the work (L) is adopted as given by Lacey's stable width for rivers, *viz.* 4.83 $Q^{1/2}$. It is termed as active waterway. An additional span is generally provided to make up for the loss of waterway in end spans which are partially obstructed by the training works or khadir edge. In railway practice, active waterway is assumed equal to the length of span from centre to centre of adjacent piers, less twice the width of pier-foundation wells. Overall waterway, L, between guide banks is usually taken as 1.1 to 1.25 Lacey's waterway. Model studies are carried out to ascertain the waterway required.

2. Length of Guide Bank

The two important considerations for determining the length of guide banks for wide alluvial rivers are (*i*) Maximum obliquity of the current. In order to avoid heavy river action on the guide bank, the obliquity of flow to the river axis is limited to not more than 30° (Fig. 9.5), and (*ii*) Permissible limit to which the main river channel can be allowed to flow near the approach embankment in the event of the river developing excessive embayment behind the training works. The radius of the worst possible loop is ascertained from the data of the active loops formed by the river.

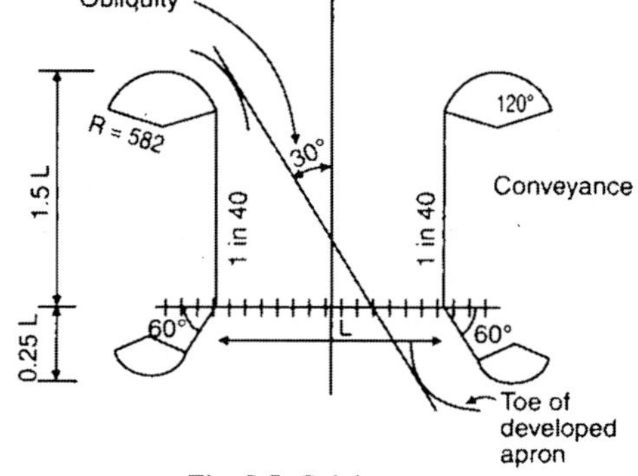

Fig. 9.5 Gale's pattern.

(*a*) The upstream length of guide bank is normally kept 1.0 to 1.5 L as per IS: 8408-1976. Spring recommended length 1.1 L and Gales 1.25 L for Q up to 20,000 cumecs and 1.50 L for Q above 20,000 cumecs.

(*b*) *Downstream length.* The river fans out on downstream of the work to attain its normal width. The function of downstream length of guide bank is to ensure that in fanning out the river does not threaten the approach embankments on either side of the work or afflux bunds or canal embankments. A short guide bank with sharp curved head is considered suitable for the purpose. The downstream length of guide banks is usually kept 0.25 L to 0.4 L as per I.S. 8408-1976. Spring recommended 0.1 L to 0.2 L, and Gales 0.25 L.

3. Curved Head and Tail

(*a*) *Circular head.* The upstream curved head is designed to guide the flow smoothly and axially to the structure to keep the end spans effective and active.

Upstream head. The upstream curved portion of the guide bank, termed as impregnable head, is provided with a suitable curvature usually of a radius as small as possible consistent with proper functioning of the guide bank. Unless indicated by model studies, the minimum and maximum radius is 150 m and 600 m respectively. The radii of curvature suggested by Spring is 180 to 250 m for rivers with velocities between 2.4 to 3.1 m/sec. Gales recommended 250 m for discharges of 7,000 to 21,000 cumecs, $R = 580$ m for discharges of 42,000 to 71,000 cumecs. For intermediate discharges R is determined by interpolation.

Degree of curve: The degree of curve in terms of maximum river discharge is 7° for discharge up to 14,000 cumecs, 5° for discharge 14,000–42,000 cumecs and 3° for discharge above 42,000 cumecs. Gales suggested a radius of 250 m (7°) for rivers with maximum discharge up to 21,000 cumecs and 580 m (3°) for discharge above 21,000 cumecs.

The curved head has sweep angle (central angle) varying from 120° to 145° according to the river curvature. For curved tail it normally ranges from 45° to 60° or 0.3 to 0.5 times the radius of upstream curved head.

(*b*) *Elliptical head and tail.* The shape is finalised on the basis of model studies. The elliptical curve is provided up to the quadrant of ellipse, followed by multi-radii or single radius circular curve. In the case of multi-radii curved head, the larger radius adjacent to apex of ellipse is generally kept 0.3 to 0.5 times the radius of the curved head for circular guide banks, with sweep angle varying from 45° to 60° and the smaller radius equivalent to 0.25 times the radius of curved head for circular guide banks with sweep angle of 30° to 40° (Fig. 9.4).

Maximum Scour Depth. The maximum depth of scour depends on the velocity of river flow which in turn is influenced by the intensity of flood, angle of attack, duration of flood and grade of silt charge carried by the river. Spring suggested three types of scour conditions for larger rivers flowing in alluvial plains, as under:

1. *Normal scour.* This is obtained in rivers with unobstructed flow in their straight reaches, generally 9 to 12 m below low river water level.

2. *Abnormal scour.* The scour of the order of 2 to 2.5 times the normal scour, *i.e.*, 18 to 30 m below low water level caused by the forward river flow in the vicinity of a rigid structure such as revetment, pitching, etc. is termed as abnormal scour.

3. *Extraordinary scour.* It is about 3 times the normal scour, 25 to 35 m below low water level. It results from a defective design of the curved head of the guide bank whereby the river flow cannot follow its face all along and separates from it to form strong swirls.

Since only a portion of the guide bank comes under attack and oversafe design would be wasteful, usually the apron is designed for a 50 to 100-year flood. The recommended values of probable scour depths along different parts of a guide bank are given in Table 9.1.

Table 9.1 Scour depths (IS: 6966–1973)

Location	*Range*	*Mean*
Nose of guide bank	2.0 to 2.5 R	2.25 R
Nose of divide wall	2.0 to 2.5 R	2.25 R
Transition from nose to straight portion	1.25 to 1.75 R	1.5 R
Straight reaches of guide banks (shank)	1.0 to 1.5 R	1.25 R

Empirical formula for scour depth

Various empirical formulae for estimating scour depth (R) are as given below:

Khosla	*Kennedy*	*Ishar Dass*	*Lacey*	*ISI*
$1.35\,(q^2/f)^{1/3}$	$1.11\,q^{0.61}$	$(q/c_1)^{2/3}$ $C_1 = V/R^{1/2}$ $C_1 = 1.13$ (F.P.S.)	$0.473\,(Q/f)^{1/3}$	$0.473\,(Q/f)^{1/3}$ when looseness factor is > 1 $1.35\,(q^2/f)^{1/3}$ when looseness factor is < 1

Cross Section

Guide banks are made of river bed material, *viz.* silt, sand or sand and gravel borrowed from the side facing the river. The salient dimensions of cross section of the guide bank usually provided are as follows:

1. Top width: Top width at shank usually adopted is 6 to 9 m at the formation level, the minimum being 4 m. Larger width is provided to permit laying of track to facilitate carriage of material and vehicles. At the nose, the width is suitably increased to enable vehicles to take turn and for stacking reserve stone.

2. Side slopes: Side slopes depend on the nature of the river bed material of which banks are made and their height. Side slope of 2 : 1 on river side and 2 : 1 to 3 : 1 on rear side is usually adopted.

3. Free board: A free board of 1 to 1.5 m above either the highest flood level (HFL) for a minimum 500-year frequency flood, or above the affluxed water level in the rear portion of the guide bank calculated after adding velocity head to HFL corresponding to 100-year flood/standard project flood at the upstream nose of the guide bank, whichever is higher, is adopted.

Water level at the back of guide bank near the abutment is appreciably higher than that on the river side on account of (*i*) Water is headed up at the rear by absorption of velocity head, and (*ii*) Water level falls continuously according to the surface slope. The free board is provided with respect to the maximum water level behind the guide bank near the abutment and not above the maximum water level in a river.

4. Apron: The guide bank is divided into two portions (*a*) the curved head and tail, and (*b*) shank or the straight portion. The curved head is more prone, compared to shank portion, to damage due to (*a*) concentration of discharge caused by collection of the overbank flow, and (*b*) direct attack of current if it hits obliquely. The shank portion is subjected to attack by parallel flow but sometimes due to obliquity of flow as well. The river attack results in formation of deep scour at the toe of the bank and erosion of the river side slope. Protection at the toe and extending a little in the river is provided in the form of a flexible stone cover, called apron, and river side slope as also rear slope of the curved portion is protected by stone pitching.

Apron or launching apron is the extension of slope pitching on the river side. It is laid horizontal at low water level from the standpoint of construction feasibility and is designed to launch itself due to its own weight, along the slope of the scour hole to provide a continuous unerodible cover up to the point of deepest scour to safeguard against further scooping out of the underlying bed material. The launching is reduced to the minimum with corresponding minimum loss of stones in case the apron is laid at the lowest possible level. The average slope assumed by the apron on launching is considered 2 : 1 for loose boulders or stones and 1.5 : 1 for concrete blocks or stones in well knit wirecrates. With regard to grade and size of material in river bed, the slope of launched apron recommended by IS: 6966–1973 is 1.5:1 for rivers in boulder reach, and 2:1 for rivers in alluvium sandy reaches, and 3:1 for rivers with very fine sand or silty bed.

(*a*) *Width of apron:* It is desirable to keep the scour at a safe distance from the side pitching so as to ensure a minimum sub-surface slope of 2:1. With this end in view, width of apron as prescribed by various engineers is as under:

Bell–1.25 times the scour depth below bed level at which laid.

Spring–1.5 times the scour depth for shank and 2 times for head.

Gales–1.5 times plus 5.0 to 8.0 m wide berm at the foot of slope pitching.

(*b*) *Thickness of apron.* The thickness of launching apron recommended by various investigators is as under:

Spring: Tapering wedge-shaped section with a thickness of *T*, *i.e.*, slope pitching thickness at the inner end and increased up to 2.76 *T* at the outer edge. Average thickness of launched apron is 1.25 *T* (Fig. 9.6.).

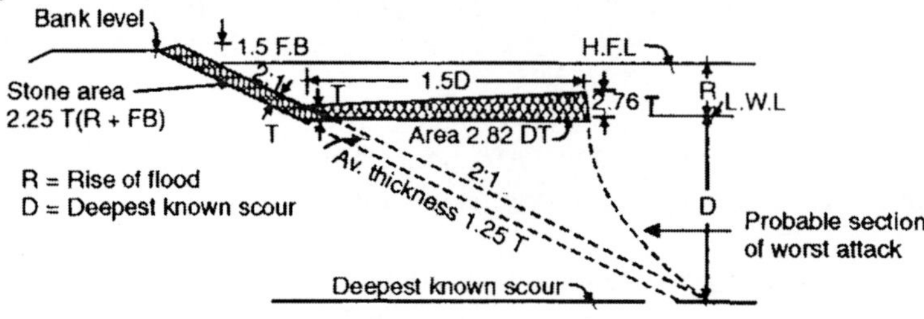

Fig. 9.6 Spring pattern.

Inglis: Equal to thickness of slope pitching.

Sethi: 1.33 to 1.50 times the slope pitching thickness at shank and head respectively (Fig. 9.7).

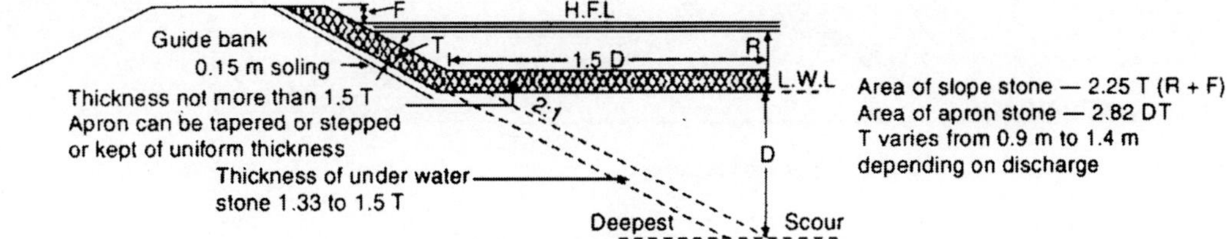

Fig. 9.7 Sethi design.

Rao: 1.5 T at the inner end and increased to 2.25 T at the outer edge (Fig. 9.8).

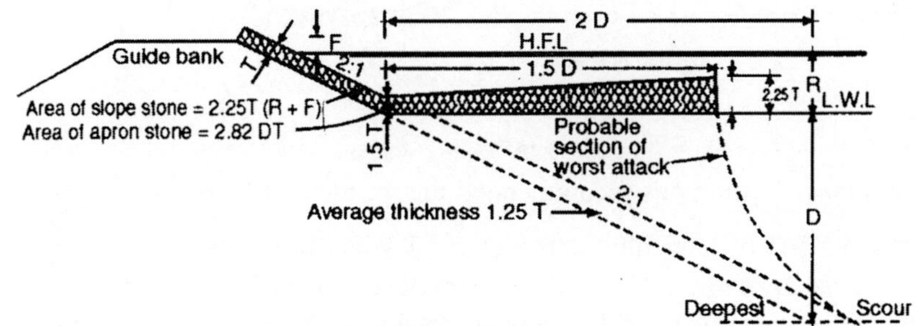

Fig. 9.8 Rao design.

Blench: Two times the size of stone required for slope pitching.

Gales: Uniform apron thickness 1.5 T (Fig. 9.9). His recommendations, Table 9.2, are more in use than Spring's.

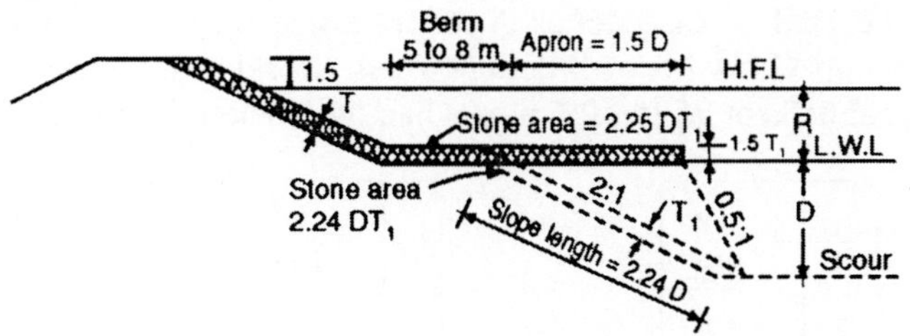

Fig. 9.9 Gales pattern.

Table 9.2 Gale's thickness of apron

Location	*Discharge (cumecs)*	*Thickness (m)*
Head	7,000 – 21,000	1.65
Body and tail		1.35
Head	21,000 to 42,000	1.90
Body and tail		1.65
Head	over 42,000	2.10
Body and tail		1.90

The thickness generally suggested is 1.4 to 2.0 times the size of individual stone used in stone pitching. The apron is designed thicker at the outer edge than at its inner end adjoining the toe of slope pitching.

Indian Standard recommendations are (*i*) Apron of uniform thickness of 1.5 T for the shank portion, (*ii*) Thickness of apron is increased gradually from 1.5 T to 2.25 T (Fig. 9.10) at the curved head where the apron has to cover a wider area, and (*iii*) Apron is extended from the shank to the head in a length L equal to one-fourth the radius of the curved head. A little apron is required on the rear side of the curved nose. The extreme line of the apron at the nose of the curved head is turned and ended in a length of about *L*.

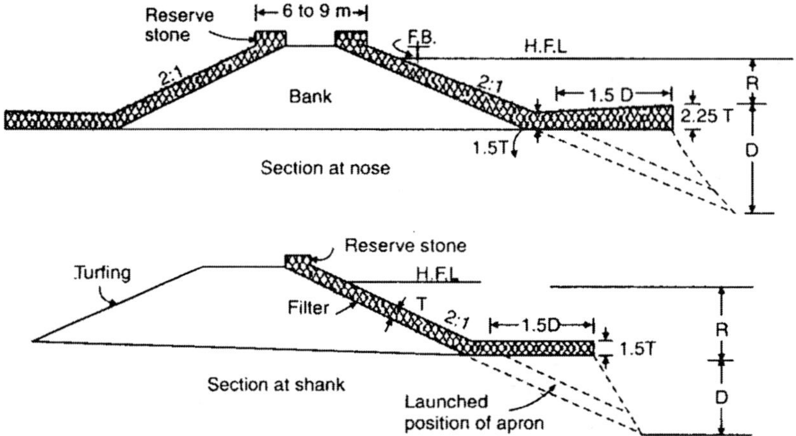

Fig. 9.10 Section of apron.

Quantity of stone. Spring has recommended the quantity of stone as 2.82 *DT* (*i.e.* $1.25\ T \times \sqrt{5}\ D$) at assumed slope of launching apron as 2 : 1 when full scour develops and the average thickness of launched apron as 1.25 *T*. I.S.I. recommendation is 2.25 *TD* with an apron width of 1.5 *D* and uniform thickness of 1.5 *T*. In Punjab barrages, the value of *T* is generally 0.75 m so that quantity of stone per metre run comes to 2 *D* approximately.

Size of stone. The size of stone for launching apron is determined from curves (Fig. 9.11). The curve on the right corresponds to stones surrounded by other stones of the same size whereas the curve on left corresponds to the case of isolated stone. Cement concrete blocks or smaller size stones in wirecrates may be used, in case stones of required size are not economically available. The crates are made of 4 mm G.I. wire with double knots, closely knit and securely tied. The thickness of launching apron is kept 25 to 50% more than the thickness of pitching of slope.

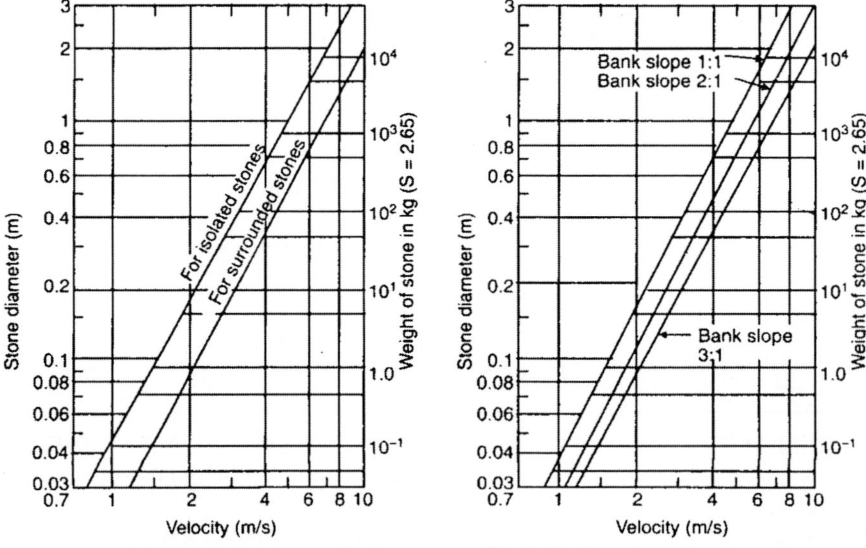

Fig. 9.11 Size of apron stone *vs.* velocity. **Fig. 9.12** Size of pitching stone *vs.* velocity.

Side pitching. The river side slope is protected against river action by suitable pitching consisting of dumped or hand placed stones and concrete blocks. The predominant flow characteristic which affects the stability of pitching is the velocity along the guide bank. Other factors are obliquity of flow and eddy action, waves, etc. The primary causes of failure of pitching are under estimation of the size of individual stone, inadequate thickness of the pitching, and inadequate filter underneath the pitching.

Size of stone. Various criteria are in vogue for selecting the size of stone for dumped pitching depending either on velocity of flow or wave height. USBR has recommended use of same size of stone pitching as required for apron for average velocity of flow. I.S.I. has specified curve (Fig. 9.12) for determining the size of stone required on sloping surface to withstand erosive action of flow. For hand pitching, the criteria is quite different because of interlocking of hand packed stones which offer great resistance to the movement of stones. In India, the pitching is usually placed on slopes manually. I.S.I. recommended size of stones are (*i*) Burnt clay brick on edge for velocity up to 2 m/s, (*ii*) Stone weighing 40 to 70 kg (0.3 to 0.4 m in diameter) for velocity 2.1 to 3.5 m/s and (*iii*) For velocity over 3.5 m/s cement concrete blocks of depth equal to thickness of pitching as discussed later, and size equal to or bigger than the size of stone required for dumped pitching. At higher velocities pattern grouting in blocks of required size may be done where necessary.

Thickness of pitching

Spring: Thickness of pitching recommended is as follow:

River bed material	*Thickness of pitching (m)*		
	River bed slope		
	0.00005	0.00019	0.00038
Very coarse	0.41	0.56	0.71
Medium	0.71	0.86	1.02
Very fine	1.02	1.17	1.32

Gales: 1.10 m both for shank and curved head over 15 to 25 cm thick ballast.

Inglis: $0.06\ Q^{1/3}$

Sethi: $T = K\,(Q/f)^{1/3}$, where, K is constant varying from 0.065 to 0.035; the higher value being for lower discharges. The thickness (T) includes soling.

CWC: Thickness varying from 30 to 60 cm.

USBR: Maximum thickness 90 cm, minimum 45 cm, depending on the size of stone required. It is usually kept as 1.5 to 2.0 times the size of stones subject to above restriction. Thickness of hand placed pitching is equal to one-half of the thickness for dumped stone pitching but not less than 30 cm.

I.S.I: As per I.S.I. recommendation, thickness of hand packed pitching is kept equal to the size of stone but not less than 25 cm. For velocities at which pitching stone of size greater than 40 cm is needed, cement concrete blocks of size 40 to 50 cm thickness can be used. The thickness of pitching is also computed by the formula $T = 0.06\ Q^{1/3}$ subject to upper limit of 130 cm.

The unarmoured slopes are covered with 0.3 to 0.6 m of earth and turfing of grass and sarkanda is provided on both sides to counteract wave action, river being on the front side and still water on the rear side.

Filter. Fine material of the embankment is subjected to sucking action by the flowing water and failure may take place. A well graded filter below the pitching is very essential. USBR has recommended thickness of filter equal to half the thickness of dumped pitching. CWC recommendation is for filter of thickness 15 to 30 cm, the larger thickness is used at locations where violent wave action is anticipated and where boulder is used instead of quarried stones or bricks. *IS*: 8408–1976 has recommended a graded filter 20 to 30 cm in thickness underneath pitching of guide banks.

9.9 EMBANKMENTS

Embankments, Levees or dykes may be defined as earthen banks extending generally parallel to the river channel and designed to protect the area behind them from overflow by flood waters. These are protected by spurs or pitching. Fig. 9.13 shows the various types of embankments provided at a barrage. The embankments thus mitigate the flood damage, confine and direct the flood flows and as such also serve the function of the river training work. Thus, the choice, the location, the alignment, the type, the shape and size of the embankment depends upon the flood, the protected area, the economics and the after-effects of such protective measures. Because of their comparatively lower initial costs and quicker benefits embankments constitute a major component of the flood control. The embankments may be classified, as under, depending on the purpose to be served:

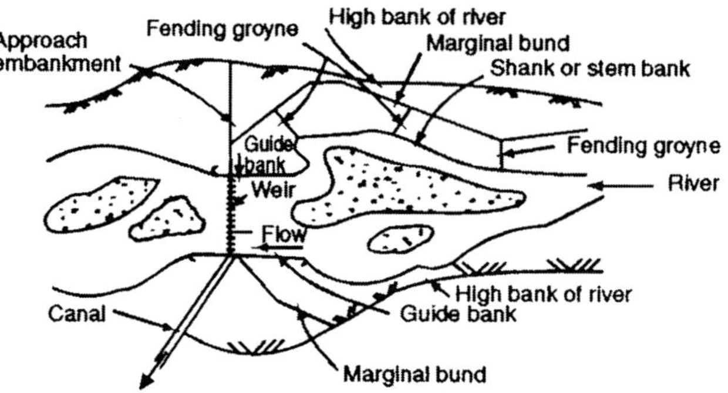

Fig. 9.13 Embankments provided at a barrage.

1. *Marginal Embankment:* Marginal embankments are built along both sides of a river upstream of a barrage or weir at a short distance from the margin (Fig. 9.13). Also termed as *Afflux embankments,* these are earthen embankments extending from both the abutments or the approach embankments, as the case may be, and connected on upstream to the high ground above affluxed highest flood level or to flood embankments, if existing. On the downstream, the embankments are taken to such a length as are necessary to protect the canal/approach embankments from high floods. The alignment of afflux embankments also influence the layout of guide banks and approach embankments. The alignment considered safe for a particular flood condition may become unsafe later due to change in the river course and as such proper training works need to be provided. The embankments are designed to contain water up to the highest affluxed H.F. level without spilling over them. The section of the embankments depends on the nature of the soil of river bed of which they are made and their height. Fig. 9.14 depicts typical cross sections of marginal embankments.

The marginal embankments apparently shut off flood effectively but are at best a temporary measure where river carries a heavy silt charge because they hasten raising of the river bed with consequent rise in flood levels and hence potential danger of breach of the embankments.

2. *Approach Embankment:* It is the simplest embankment that is constructed to connect the restricted waterway provided at a barrage or weir with the unbridged gap between the high river edges on either side. Alternatively, it is the embankment that is provided to approach the barrage or weir from the high river edges on both sides. The approach embankments are thus aligned in line with the axis of the barrage up to a point beyond the range of worst anticipated loop.

3. *Retired Embankment*: Retired embankments are provided to enclose the area which has no other means of irrigation except the river spills. Retired embankments are thus constructed at a distance from the river edge behind the existing embankment as a second line of defence. The embankments may be provided with reclamation bunds and projecting spurs to the main river channel where it is likely to attack the culturable lands and thus help grow post-inundation crops or forest plantation. The retired embankments by confining the flood waters between them raise HFL, maximum discharge in downstream, rate of flood wave travel downstream and velocity encouraging coming action betwen them. The upstream water surface slope tends to reduce.

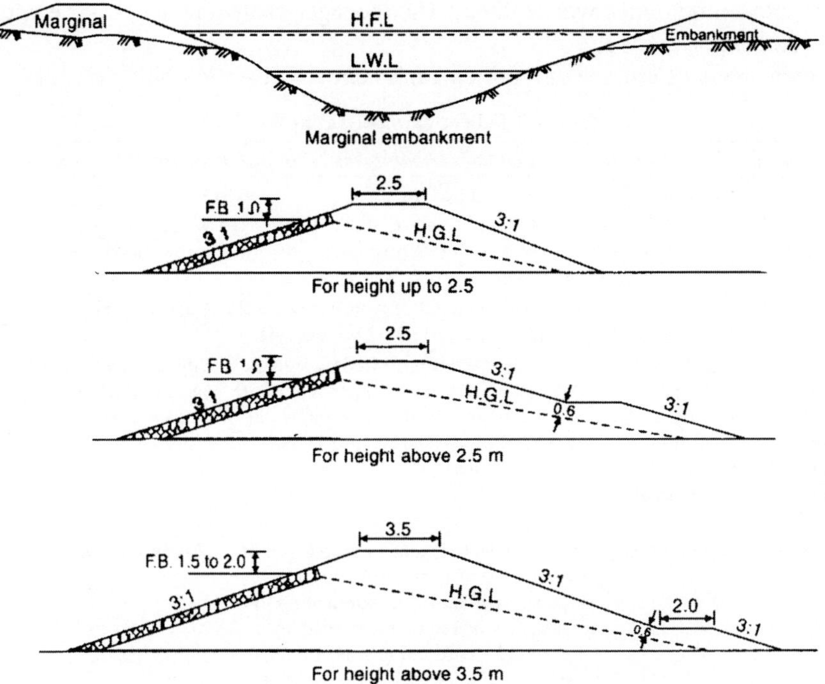

Fig. 9.14 Typical cross sections of marginal embankment.

Advantages: (*i*) Minimum interference with the natural rise in level of the countryside by silt deposition during floods, (*ii*) Maintenance of a large pool capacity, (*iii*) Moderation of the peak high floods by filling the reservoir pocket enclosed by the embankments, (*iv*) Temporary storage of irrigation water for use in canals later, (*v*) Low initial cost and (*vi*) Construction and maintenance is simple.

Disadvantages: (*i*) Reduction in normal spilled area of the river and consequent higher flood levels, (*ii*) Deprive the area beyond the protected lands of the manurial silts deposited by flood waters, (*iii*) The area behind the protected lands is denied raising by the natural operation of silting by the river, (*iv*) Interfere with the natural leaching of the protected lands and cause them to deteriorate by the rise of subsoil salts by shutting off inundation water, and (*v*) Sudden flooding of entire area if breached.

4. *Flood Embankments*: Flood embankments are constructed along both sides on high ground, sufficiently away from the river bank, aligning them more or less straight and leaving a certain fixed distance from the main river channel to minimise risk of river attack. The single river channel provided with embankments on sides is designed to provide for the same transporting capacity for water discharge as well as sediment charge by adopting same slope as existing. The flow depth is designed so that there are no spills.

Advantages: (*i*) Reclamation to maximum extent of flood-prone area, (*ii*) No need for guide banks for proposed railway and road bridges after the river is canalised. Thus, the cost of bridges will be reduced, and (*iii*) Helps in making river navigable.

Disadvantages: (*i*) High flood levels are raised appreciably due to restricted waterway, (*ii*) High flood discharge increases as valley storage is lost, (*iii*) Bridges, if any, need to be made safe for the afflux and foundations protected for deeper scour, (*iv*) Protection of the river side slope of the embankment may be needed at salient points to safeguard it from erosion, and (*v*) Embankment, if too near the river edge, is susceptible to breaching.

Design of Embankments

The essential conditions governing the design of cross section of embankments are (*i*) An adequate height to prevent overtopping, and (*ii*) A cross section sufficiently massive for security against dangerous seepage through the section. Seepage gradient is kept inside the body of the embankment with a minimum cover of 0.9 m. The seepage gradient in accordance with the nature of soil varies from 4 : 1 to 6 : 1.

Salient parameters of cross section of embankment are given in Table 9.3.

Table 9.3 Salient parameters of cross-section

Parameter	*Marginal embankment*	*Approach embankment*	*Retired embankment*	*Flood embankment*
Top width	6 to 9 m	6 to 9 m	6 to 9 m	6 to 9 m
Free board. 1 to 1.5 m above HFL for 1 in 500-year flood or above the affluxed water level in the rear portion of the embankment calculated after adding velocity head to HFL corresponding to 1 in 100-year flood/standard project flood at the upstream nose of the bank, whichever is higher.				
Side slopes. Depending on the nature of river bed material of which they are made and the height of the embankment, Slope of 2 : 1 is generally adequate on river side and 2 : 1 to 3 : 1 on rear side.				
Slope pitching.	Embankment is not subjected to strong river current. Slope pitching is not necessary, turfing or vegetal cover is provided. Where strong currents are likely to attack, paving of slope with some resistant material is provided.	River side slope is protected by suitable pitching calculated for 40 per cent of the design discharge. Filter is normally not required except in case approach embankment is heavy.	Embankment is not subjected to river current. Turfing is provided. Spurs and other protective measures are provided at salient points.	
Launched apron.	Apron is not necessary.	The scour depth is taken as 0.5 to 1.0 *D*. Apron of 0.5 *D* width at 2:1 launched slope is provided from abutment up to a distance covered by the curved head of the guide bank and beyond 1.0 *D*. The size and thickness of the stones as for guide banks.	Not necessary.	

Embankment Failure

Factors responsible for embankment failure are (*i*) Erosion of river side by wave wash during very strong current, (*ii*) Sloughing of the bank slope when saturated with water by floods of long duration, (*iii*) Sloughing of inland side slope caused by under seepage, (*iv*) Piping in sub-layers due to movement of ground water towards the river, which carries away material with it. Gradual reduction of cross section of the embankment takes place due to gradual removal of the embankment material through the joints or interstices of the protection revetment, (*v*) Toe of the bank undermined by eddies, currents, etc. followed by a collapse of overhanging material deprived of support, (*vi*) Overtopping due to a flood of a magnitude greater than the design flood, (*vii*) Insufficient cross section, leaks and cracks due to shrinkage of soil and rat-holes while breaches could result from overtopping of an embankment due to higher than design flood levels being obtained in a river as also due to river erosion. One of the important causes usually is lack of proper maintenance by way not only of annual repairs, but also inadequancy or slackness in watch and timely action during the flood season.

9.11 EMBANKMENTS AS RIVER TRAINING WORK

The embankments, discussed earlier, are fragile works indicating an inherent weakness of this method of river training and control. Their merits and demerits as river training work are as under:

Advantages: (*i*) Only means of preventing inundation on tidal plains, (*ii*) Low initial cost, subsequent raising may be possible, (*iii*) Easy construction, making use of locally available material and unskilled labour, (*iv*) Maintenance simple and economical, (*v*) Can be executed in parts, provided that the ends are properly protected.

Disadvantages: (*i*) Cause raising of high flood levels, (*ii*) Fragile works susceptible to boring by crabs and worms resulting in failure by piping, (*iii*) Need close supervision during floods and protected as soon as they are in danger by measures which are generally temporary, (*iv*) In the event of a breach, there is a sudden and considerable inflow of water which may cause damage and deposition of sediments rendering the area unproductive, (*v*) Susceptible to direct attack of river flow which can erode or undermine them, (*vi*) Low lying areas may become infested with malaria, if proper drainage facilities are not provided, (*vii*) In rivers carrying a considerable sediment charge, embankments exclude silt deposition on the flood plain, extension of the delta towards the sea is thus hastened, resulting in the increase of flood stages and the rate of aggradation of river bed.

9.11 BANK PROTECTION

Bank protection is an important part of river training works. The purposes of bank protection can be (*i*) Training of the river, (*ii*) Protection of adjacent agricultural land, (*iii*) Protection of urban lands and valuable properties threatened by river erosion, (*iv*) Protection of hydraulic structures such as weirs, barrages and bridges against the direction and nature of current, (*v*) Protection of flood embankment, and (*vi*) Affording facilities for water transportation.

River bank may be considered to consist of two sections, upper and lower. The upper bank, *i.e.*, the portion between high water level and low water level is subject to severe action of river in high stages due to (*i*) strong current impinging normal to the bank, and (*ii*) strong current acting along the bank. Lower bank is the portion of the bank below low water level. It is generally more susceptible to erosion. It acts as a foundation supporting the upper bank. Recession of the bank is caused by erosion of the lower bank, particularly at the toe. The recession is fast when there is a sandy substratum below; sand is washed away by a strong current and the overhanging bank collapses. When the river rises, the banks become partially or fully saturated with water and sloughing is caused.

Types of Bank Protection: Bank protection may be direct or indirect, discussed as follows:

Direct Bank Protection: It includes the work done on the bank itself, such as providing vegetal cover, pavement, revetment, grading of slope, etc. Three conditions exist, (*i*) *Mild current.* Turfing or vegetal cover, *e.g.*, growth of shrubs and willows is sufficient, (*ii*) *Strong current.* Paving of slopes with erosion resistant materials is necessary, and (*iii*) *Very strong current.* Protection has to be provided by stone revetment or various types of mattresses such as willows, lumber, asphalt or articulated concrete.

Indirect Bank Protection: It includes undermentioned works constructed not directly on the banks, but in front of them for reducing the erosive forces of the current, either by deflecting the current away from the banks or by inducing silt deposition against them.

Revetment: Revetment is placed along an existing or proposed river bank to prevent its recession by caving and erosion. The protection against erosion by revetment induces erosion of bed so that there is a possibility of deep current always hugging a revetted bank. It is used in places where siltation inducing structures cannot be provided. Generally, revetment is provided at places like highways, railway lines, etc. It provides riprap along the bank slope and toe protection.

Launching apron: It consists of a flexible stone apron to cater for maximum anticipated scour at the toe of the revetment. It launches as the scour develops. Full details of apron are discussed under "Guide banks".

Mattress: A mattress is provided extending 15 to 30 m beyond the toe of the side slope.

Curtain of sheet piles. An impregnable curtain of sheet piles is provided. It is very expensive and is rarely used.

Submerged sills: Submerged sills in front of the concave bank across the channel are provided to help stop deepening of the bed and thus protect the toe of the bank.

9.12 SPURS

Also called *groynes, spur dikes,* or *transverse dikes.* They constitute probably the most widely used training work. Spurs are the structures constructed on the river bank normal to the dominant flow direction or at an angle pointing upstream or downstream. A spur consists of a shank and a nose or head. Shank is a bund of adequate section which connects the nose of the spur with the highest point above H.F.L. on the river bank. The upstream face and nose of the spur are duly armoured with stone pitching and apron of proper dimensions. The apron is built with stones in wirecrates. The nose may have different angles and slopes to cater for local requirements. They can be used singly or in series as also in combination with other training measures. In case reach to be protected is long, single spur may not be enough and a number of groups may be required. The spacing, orientation and length of groynes is usually decided by model experiments.

The essential requirements of a spur are (*i*) Fairly stable and strong to withstand river action, (*ii*) Proper alignment and angle to meet the objective, (*iii*) Top level sufficiently above HFL, (*iv*) Strong armoured upstream head, adequately protected nose, head and shank against anticipated scour, and (*v*) Essentially tied to high ground.

Functions

Spurs serve one or more of the functions (*i*) Training the river along the desired course to reduce concentration of flow at the point of attack, guiding or deflecting the axis of flow, (*ii*) Creating a slack flow zone with the object of silting up the area in the vicinity of river bank. Silting up a minor stream in a meandering or a braiding river to open up a well defined channel so as to improve hydraulic efficiency of the river, (*iii*) Protecting an eroding bank by keeping the flow away from it, (*iv*) Contracting a wide river channel for establishing normal channel width usually for improving depth for navigation, (*v*) Promoting scour and sediment deposition, where desired, and (*vi*) Trapping bed load to build up new river banks.

Constraints

(*i*) Spurs can protect a reach of river bank to the extent of 2 to 2.5 times their own length protruding inside the river. They thus have limited use in that they cannot protect long reaches of the river economically, (*ii*) Spurs upset the equilibrium of river, (*iii*) Many times river starts attacking other points, particularly the opposite bank, and calls for providing spurs on the opposite bank, and (*iv*) They require heavy repairs after every flood.

Types of Spurs

(*i*) **Classification based on method and materials of construction**

1. **Permeable Spurs:** These spurs permit passage of restricted flow. Because of partial obstruction to flow they promote deposition of sediment as the flow is retarded. They are built of materials like ballies, bamboos, timber, brush, steel or wire discussed hereinafter. They slow down the velocities over a portion of the channel area and thereby induce rapid deposition of sediment. Heavy silt laden river water when so retarded, drops much of its suspended load, bed load to come to rest, thereby building up new bank lines. This side contraction causes a larger proportion of the discharge to flow in the main channel, thus increasing its water and sediment transport capacity. They are best suited to deep and narrow river transporting heavy sediment loads. They are not suitable for small rivers flowing on steep gradients as also the deep rivers carrying light sediment loads.

(*a*) *Tree Spur:* Tree spur consists of a 25 mm diameter wire rope firmly anchored at one end to the river bank and tied at the other end to a heavy buoy or a concrete block (Fig. 9.15). The length of stretched wire rope constitutes the length of tree spur. The entire leafy trees with abundant branches such as Babul are employed to increase sedimentation by retarding velocity of flow and intercepting sand and silt. A hole, about 0.3 m up the stem of each tree is drilled through which an iron ring is drawn and attached, either directly or by subsidiary pendant wire ropes, to the main rope. When trees become heavy with detritus they sink. Therefore, they are packed as closely as possible. Initially the spur is laid pointing upstream at an angle between 60° to 70°, so that when it launches and becomes sand-bound it assumes a position still facing slightly upstream.

(*b*) *Dip Trees Spur.* Dip trees spur consists of entire bamboo trees with branches whose bottom is tied to a heavy concrete block or small crate full of stones by means of a hook (Fig. 9.16). It is then launched in water. It keeps the tree at the launched position and sinks itself in whenever scour occurs. Rows of dip trees form the cross or longitudinal permeable spurs.

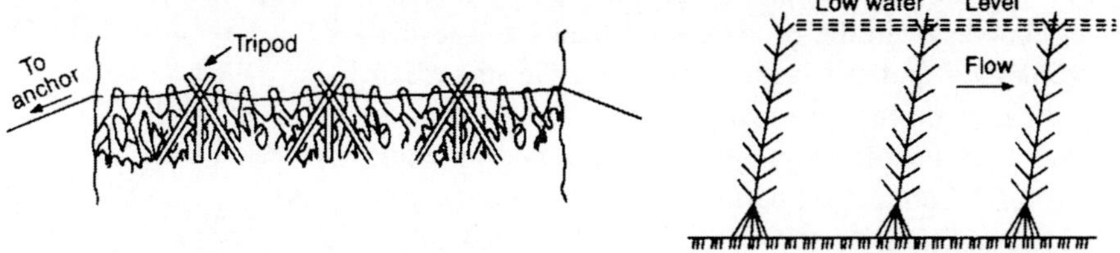

Fig. 9.15 Typical tree groyne. **Fig. 9.16** Dip trees.

Tree spurs serve the object of (*i*) Diverting or deflecting the current directly threatening erosion of a bund, and (*ii*) To deaden the flow in one bye-river near the bund in order to open out another bye-river away from the bund, and (*iii*) to silt up a channel of the river at its source by checking the flow in it.

(*c*) *Pile Spur:* The piles may be wooden balli, sheet pile or R.C.C. pile. They constitute the main verticals driven down 2.5 to 3 m about 6 to 9 m inside the river bed and in at least 2 to 3 similar rows. For ease in driving, the piles are tapered at the base and are fixed with heavy iron shoes. In between the main verticals, there can be two intermediates, embedded at least 1.25 m below the bed. Each row is closely intertwined either with brushwood branches, going in and out around each vertical, or by horizontal railings. The upstream row is braced back to the down-stream row by transverses and diagonals. Every other main vertical of the rear row is strutted, the struts are embedded a minimum of 2.5 m below bed. Between two rows, the space is filled with brushwood branches in alternate layers of 1.75 m thick brushwood weighted by 0.6 m thick stones or sand bags. Being a permeable spur, the whole space is not filled with stones. The filling is closely packed and tamped. The bed is protected both at upstream and downstream of the spur and around the nose with a fascined mattress or a stone apron 1.0 m thick over a width of 3 m and 6 m along the shank and nose respectively.

(*d*) *Bamboo Cage Spur:* Fig. 9.17 indicates typical cross section of bamboo cages flat bottomed permeable spurs constructed at Ninan-Nurpur reach of river Hooghly. They were made of full and half split bamboo members, jointed with nails. Each cage consisted of 0.75×0.75 m cubical boxes, top and bottom of which were covered with 1.75×0.75 m sized lids and contained about 50 bricks. Each cage when dumped in groups covered about 1.70 m^3 and remained about 50% permeable ensuring thereby fast construction. The interlocking between the cages increases their resistance to possible drifting and the spur is rendered gradually stable by deposition of silt.

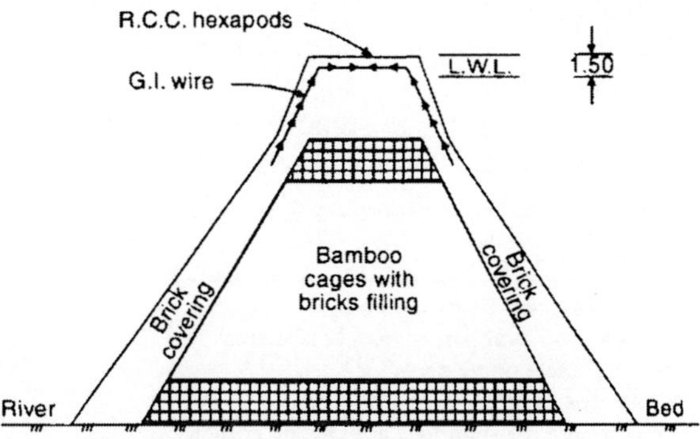

Fig. 9.17 Typical cross-section of bamboo cage spur.

(*e*) *Porcupines.* Bamboo porcupines have been used in India to induce siltation along the river banks. These are made of 3 to 6 m long bamboos tied together at the centre in the form of a space angle and are weighted by tieing some stones to the centre (Fig. 9.18). These add roughness to the channel in the overbank area and thereby deflect the eroding current away from the bank. The vegetation within the jacks take roots in the course of time to enhance their siltation action.

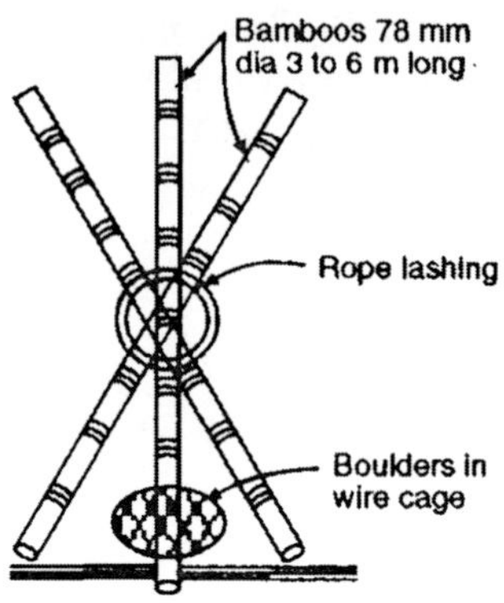

Fig. 9.18 Porcupine.

Advantages: The permeable spurs have the advantages (*i*) Construction is cheap, easy and rapid, (*ii*) Suitable where stone is difficult to obtain, (*iii*) More effective than impermeable spurs in the regulation of river courses or in the protection of banks and levees, especially in erodible-bed and silt laden rivers, (*iv*) Do not create big scour holes as the flow through the spur does not change abruptly like that passing an impermeable spur. Silt deposition is evenly and quickly effected, and (*v*) Submerged permeable spurs do not create turbulence and eddy conditions as strong as with submerged solid spurs, (*vi*) Offer flexibility in construction and maintenance and any alteration, if required at a later stage, is possible, and (*vii*) Effective and economical for deep and narrow rivers.

Disadvantages: (*i*) They are not suitable for upper reaches of rivers being not strong enough to resist shocks and pressures from debris, floating ice and logs, (*ii*) Submerged spurs are a danger to navigation during floods, (*iii*) Bamboos, timber, etc. employed for building permeable spur are liable to theft.

2. **Impermeable Spur:** Impermeable spurs do not permit passage of flow through them and as such constitute solid obstruction. They are constructed in masonry or rock-fill or earth core armoured with resistant material like stone, fascine mattress or sausages filled with stones. If slots are provided in the impermeable spurs at regular spacings, these are known as slotted spurs. They attract, repel or deflect the flow away from the banks along a desired course.

Choice of Type

Out of the various types of spurs discussed, the choice of type suitable for given site conditions or design thereof depends on various factors such as (*i*) River channel width at high, mean or low water, (*ii*) Depth of river channel, height of flood rise and nature of flood hydrograph, (*iii*) Fall and velocity of flow, (*iv*) Character of river bed material carried such as shingle, boulders, silt or sand, (*v*) Nature of available construction materials, (*vi*) Construction time and funds available, (*vii*) Nature of protection desired whether permanent, semi-permanent or temporary.

Impermeable spurs are suitable for confining a river to well defined channel during high and low water flows as also where stones are available in abundance. However, these are not suitable for boulder rivers with slow rate of silt deposition as also flashy rivers which rise and fall quickly and silting does not take place. They are also less effective in regulation of river course or in the protection of banks.

Permeable spurs are suitable for deep and narrow rivers or those carrying considerable suspended sediments. However, these are not suitable for small rivers having steep gradients.

Permeable *vs*. Impermeable Spurs

Permeable spur	*Impermeable spur*
Permits restricted flow through it.	Does not permit flow to pass through it.
Temporary or semi-permanent construction.	Permanent structure subject to vagaries of river.
Damping effect on velocity.	Deflecting or repelling action on velocity.
Suitable for sediment carrying rivers.	Indicated for sediment free rivers.
Flow through the spur does not change abruptly resulting in no serious eddies and scour holes.	Flow changes abruptly resulting in serious eddies and scour holes.
More effective in regulation of river courses or in bank protection and levees especially in silt-laden rivers.	Comparatively less effective in regulation of river course or bank protection.
Effective in deep and narrow rivers with banks of sand layers covered by clay.	Effective in deep and narrow rivers where depths are considerable.

Permeable spur	Impermeable spur
Submerged permeable spurs are preferable as they do not cause strong turbulent and eddy conditions.	Not preferred as they cause severe turbulent and eddy conditions.
Not strong enough.	Strong structure.
Generally make larger angle with the bank upstream.	Generally made to face 60° upstream.
Cheap, specially adaptable where stone is difficult to obtain.	Costly, adaptable only where stone is locally available or nearby.

(2) Classification based on height of spur below high water

1. *Submerged Spurs:* Submerged spurs are adaptable for narrow and deep rivers where solid spur above high water is substantial and also solid spurs cannot be extended lest they cause harmful conditions on the opposite bank or further downstream. Submerged permeable spurs are preferable to submerged solid spurs as the former do not create turbulent and eddy conditions as strong as with the latter. Submerged spurs are also preferred in situations where lower portion of bank is erodible. Submerged spurs inhibit undercutting of the bank and improve its stability against undermining. The height of submerged spur may be kept one-third the depth of flow at full discharge. At lower discharges, the submerged spurs act as spur of full height but there should not be adverse flow conditions.

A single submerged spur may not be as effective as a series of submerged spurs. A series of spurs at a spacing 5 times the length of spur are found to be effective in affording protection to the bank in straight reaches (Fig. 9.19).

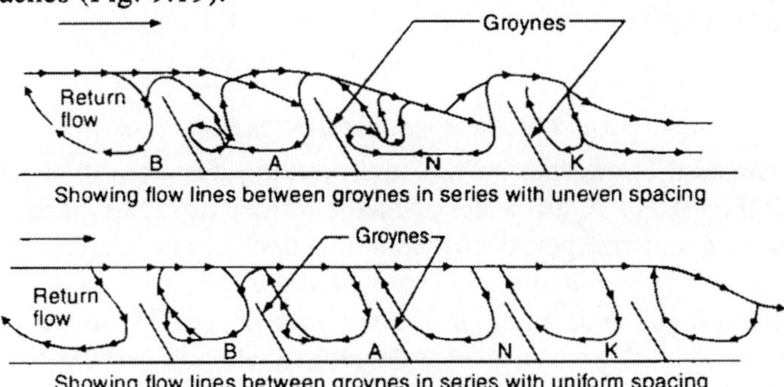

Fig. 9.19 Action of series of repelling spurs

2. *Non-submerged Spurs:* Non-submerged solid spurs are not suitable besides being costly for narrow and deep rivers because they create undesirable flow conditions. The cost of spur to a height above the high water level is high due to heavy nose protection required to prevent excessive scours. Solid spurs cannot be extended much as they create adverse flow conditions on the opposite bank or further downstream.

(3) Classification based on functions served

The spurs may be aligned perpendicular to the river bank, inclined upstream or downstream according to the function served by it.

(*i*) *Straight Spur.* It is constructed normal to the direction of flow and is only mildly repelling (Fig. 9.20). After a kick at the nose, the current generally tends to return to the bank. For giving orientation to the current to flow parallel to the bank or away from the bank, a *T*-head is constructed at the nose. The shank joins the *T* with two-third of the length protruding upstream. The shank upstream is exposed to parallel flow and some further erosion of the bank takes place in the initial stages before equilibrium conditions are attained. The nose may be of the same section as the shank or made into a bulkhead to withstand the river flow action.

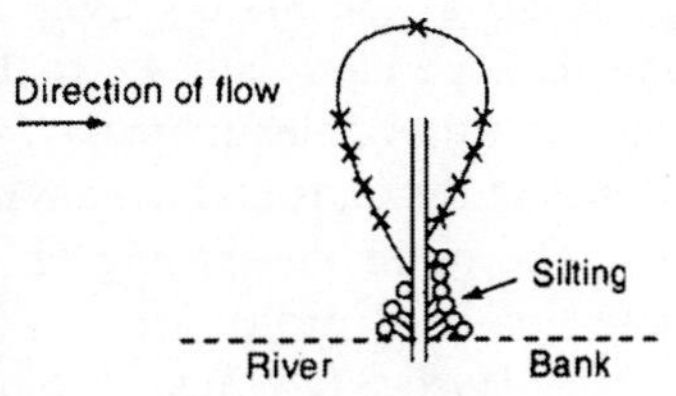

Fig. 9.20 Straight spur.

(*ii*) *Attracting Spur*: Also called Fending spur or Bankhead spur. It is the spur pointing downsteam (angle between 30° to 60°) to attract the river flow towards it (Fig. 9.21). The nose of the spur is constructed at an obtuse angle to the flow, the current hugs the side of the spur where scour holes are formed close to the bank to maintain deep current there. The main attack of the river flow is on the upstream face of the spur and requires more protection than that on the downstream. Such a spur is constructed not at the location where the river bank is desired to be protected but on the opposite bank to attract current towards the latter. Choice of site for the location of the spur is important for the success or failure of the spur depends on it. It is constructed at 0.4 of the meander length upstream of the spot to be protected.

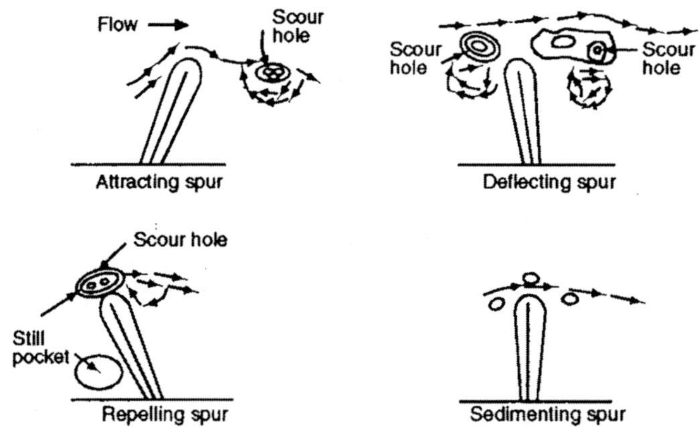

Fig. 9.21 Attracting and deflecting spurs.

(*iii*) *Repelling Spur.* It is the spur constructed pointing upstream to repell the deep current away from the bank (Fig. 9.21). A still water pocket is formed upstream of it, and suspended load brought down by the river channel gets deposited in this pocket. The current is kicked off the bank for considerable length. It is provided at the spot where the river bank is desired to be protected. It is so constructed that it subtends an angle varying from 60° to 75° to the direction of flow. A curvature of flow of nearly still pond pocket is developed wherein suspended load carried by the river flow gets deposited thus incidentally shields a portion of the shank. The spur especially the nose is exposed to greater river action and requires strong protection. It is more effective than attracting spur. When the nose of the spur is joined over a suitable curve to the shank, the spur takes the shape of an inverted hockey stick.

Deflecting Spur. It is a spur of short length which changes only direction of flow without repelling. It is also called holding spur as it tends to hold the river channel at that point, scouring action on the upstream side and silting along the bank on its downstream side.

(4) Special types

Special types of spurs (*i*) Denehy's spur, and (*ii*) Hockey spur.

1. *Denehy's Spur.* Also called T-Spur (Fig. 9.22). It is defined as a spur provided with cross spur at the head giving the entire spur a *T*-shape, which protects the main spur on the same principle as the main groyne protects the bank. The front perpendicular arm, about 100 m in length, is parallel to the current. A longer portion of the head is placed upstream of the spur shank so that it resembles more to a reversed *L* than a *T*. The head is sloped on both sides, with flatter slopes at the ends and is heavily armoured.

The optimum spacing of T-spurs is 800 m with river face length of 100 m or so for most effective bank protection.

2. *Hockey Type Spur.* It is defined as a spur shaped at its lower end like a hockey stick (Fig. 9.22). Hockey spur has attracting effect on the flow and is not useful for bank protection by repelling the current away from it. In this spur, the scour is nearly equal along the upstream and downstream

faces. A spur reverse in orientation to hockey spur is termed as inverted hockey spur The up-stream face and up to half length on downstream side requires apron protection.

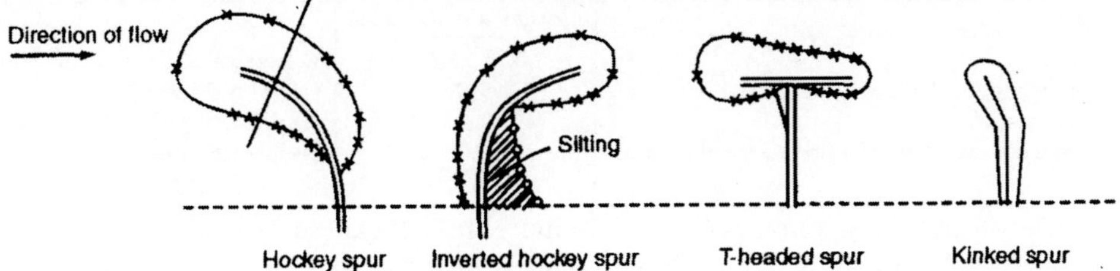

Fig. 9.22 Special types of spurs.

Design Parameters

Design discharge: 50 years flood.

Layout: The spurs are more effective when constructed in series as they create a pool of nearly still water between them which resists the current and gradually accumulates silt forming a permanent bank line in course of time. The repelling spurs are constructed with an inclination upstream which varies from 10° to 30° to the line normal to the bank. In T-shaped spur, greater length of the cross spur projects upstream and a smaller portion downstream of the main spur.

Length of groyne: It is not less than that required to keep the scour hole formed at the nose away from the banks; more than 2.5 ds assuming angle of repose of sand 2.5:1 and anticipated maximum depth of scour below bed 'ds'. Short length may lead to bank erosion on u/s and d/s of the groyne due to eddies formed at the nose and too long a groyne may restrict the river and get damaged. Normally effective length is not to exceed 1/5th of width of flow in the case of single channel. In wide, shallow and braided rivers, the protrusion of groyne in deep channel is not to exceed 1/5th of the channel width on which the groyne is proposed excluding the length over the bank.

Number of Spurs: The number of spurs to be provided in river training works depends upon the particular location to be protected, the river curvature, discharge intensities, sediment characteristics and control conditions. In cases where only one particular reach is attempted to be saved and the discharge is less than 2,800 cumecs a single spur may be enough. Where the reach to be protected is long and discharge intensity is more than 2,800 cumecs, one or more spurs may be provided, even one may suffice in certain cases.

Spacing: Generally a spacing of 2 to 2.5 times the effective length of spur at convex banks and equal to length at concave banks is adopted. T-shaped spurs are generally spaced 800 m apart with T-heads on a regular curved straight line. For wide river the ratio of spacing to length of spur is larger than that for concave banks. Spacing of permeable spurs is larger than that of impermeable or solid spurs.

Top level: It depends on whether the spur is submerged, partially submerged or non-submerged. In non-submerged, top level is 1.0 to 1.5 m (free board) higher than design flood level.

Top width: 3 to 6 m at formation level with slopes of sides and nose between 2 : 1 to 3 : 1.

Stone pitching size of stone,
$$W = \frac{0.02323 S_s}{K(S_s - 1)^3} V^6$$

where, W is weight of stone (kg), k is $\sqrt{\dfrac{1-\sin^2\theta}{\sin^2\theta}}$, S_s is specific gravity of stones, θ is angle of repose of protection material, θ is angle of side slope, and V is velocity (m/s). Thickness of pitching is equal to two layers of stones determined from velocity indicated above.

Launching apron: Size of stones, concrete blocks, crates, etc. for launching apron is determined as indicated above. Scour depth adopted is as under:

Location	*Maximum scour depth*	
Nose	2.0 to 2.5 D	D is depth of scour below HFL by Laceys' formula, it $D = 0.473\,(Q/f)^{1/3}$. Where q is known, preferable formule is $D = 1.33\,(q^2/f)^{1/3}$.
Transition from nose to shank and first 30 to 60 m in u/s	1.5 D	
Next 30 to 60 m in u/s	1.0 D	
Transition from nose to shank and first 15 to 30 m in d/s	1.0 D	

Thickness of launching apron is 25 to 50% more than thickness of pitching on slopes. The slope of launched apron may be 2:1 for loose boulders or stones and 1.5:1 for concrete blocks or stones in crates. 15 cm thick sand layer is provided on filter to prevent mechanical rupture of the synthetic filter (geotextile) used. The shape and size of apron is shown in Fig. 9.23.

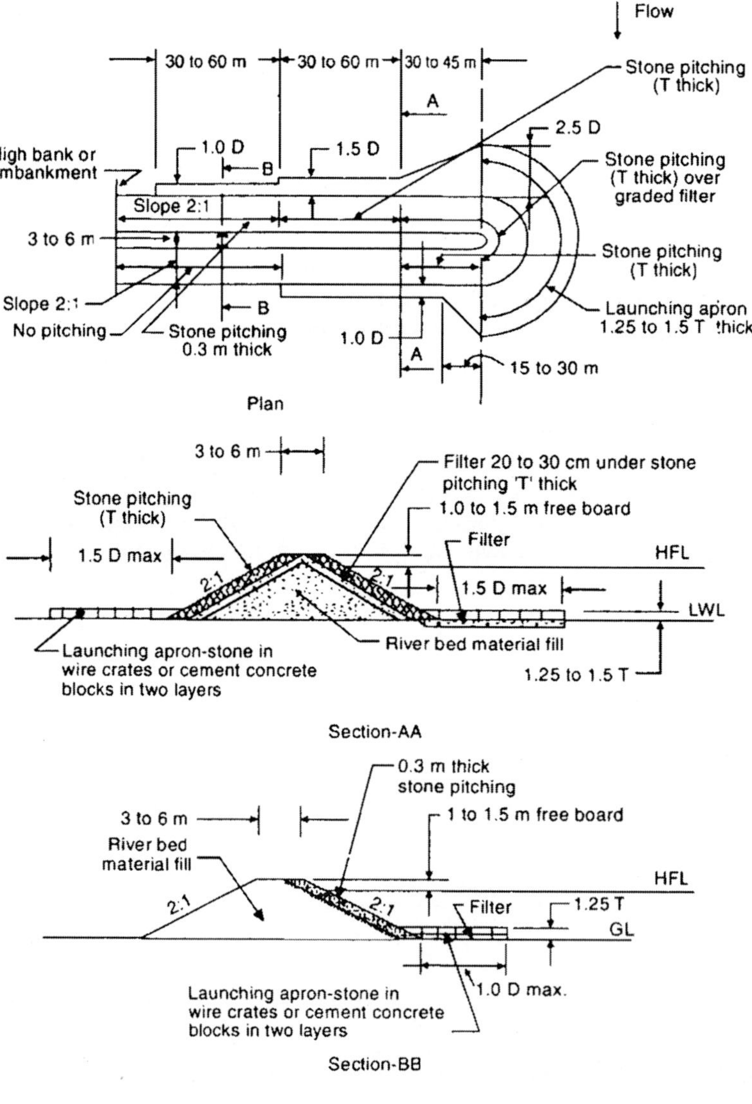

Fig. 9.23 Typical design of impermeable groynes.

9.13 ARTIFICIAL CUT-OFFS

A cut-off can be artificially induced with advantage as a river training measure. This is generally done when loop length exceeds 1.5 to 2.5 times the chord length of the meandering river channel and the artificial cut-off is made along the chord lines. The alignment of cut-offs is such that the cut at both ends is tangential to the main direction of the river flow approaching and leaving the cut. To promote such development the alignment of the pilot cut is preferably along a slight curve, the curvature being flatter than the dominant curvature of the river itself. The entrance to a cut-off is to have a bellmouth up to a double section in a length of about 180 m. Bellmouthing at the outlet is unnecessary as the cut develops at the lower end works progressively upstream. Where cuts are unlikely to develop on their own they are excavated to the main river section. A pilot cut is made as deep as possible, a deeper cut is helpful for rapid development. The width of pilot cut is relatively unimportant as in erodible beds, the flood gradually enlarges the pilot section into the full size cross section. A cut-off alters the river regime violently, may be for a short duration, as subsequently the river has a tendency to rebuild the slopes which characterised its original profile whenever it has a chance to scour its bottom. This process of reversion to the original slope may be retarded by tough bottom material. The artificial cut-off aims at establishing ultimately hydraulic gradients which the river can maintain without requiring correcting work at frequent intervals.

As a river training work, cut-offs are not enough by themselves. They correct the instability and inefficiency at sharp bends and loops, where much head is lost by excessive river length. But they do little to correct conditions in the reaches between these bends. As such extensive work between cut-offs is necessary to improve the alignment, width and depth of the channel by supplementary training works with a view to create a uniform river width, and establishment of a central river channel deep enough to maintain itself by normal scour action.

The objectives of artificial cut-offs as a river training work are (*i*) Improvement in flood carrying capacity of a river reach, (*ii*) Reduction of flood heights and flood periods, (*iii*) Shortening of navigation course, (*iv*) Correction of the instability and inefficiency at sharp bends and loops where excessive head is lost by longer river length.

Advantages: (*i*) No interruption to navigation, (*ii*) Valley storage during flood stages remains unimpaired, (*iii*) Eventual restoration of the normal hydraulic slope of the river, and (*iv*) Pilot cut enlarged to full sized river channel is economical because further on huge materials are eroded by the river itself.

9.14 PITCHED ISLAND

It is defined as artificially created islands in the river bed protected by revetment of stone pitching on all sides to direct and maintain the approach to some engineering work in the desired direction (Fig. 9.24). The underlying principle of pitched island is that river flow hugs *pacca* structures in its course. This principle is utilised in training the river centrally by constructing an artificial *pacca* island at the desired place. Where an island exists, the same is reshaped and pitched with advantage. Pitched island causes redistribution of the velocity and tractive force. The tractive force near the pitched island begins to increase rapidly after the construction of the island with the result that deep scour begins to form round the island and gradually draws the main river channel into itself and ultimately holds it permanently abandoning the old course. It essentially consists of a sand core protected on sides with stone pitching and protected at the toe by a falling stone apron. The general shapes adopted are traingular, elliptical or egg-shaped. Pitched island can be used either singly or in series.

The objectives of pitched island are (*i*) Correcting oblique approach upstream of hydraulic structures by training the river, (*ii*) Improving river channel for navigation, (*iii*) Stabilisation of river course by altering river flow pattern, (*iv*) Training the river reach away from bridges and weirs or relieving attack on marginal embankments, guide banks, and (*v*) Rectification of adverse curvature to effect sand exclusion. The general aspects of designs of pitched island are similar to those of guide bank. However, the length and shape of the island are better determined by model studies.

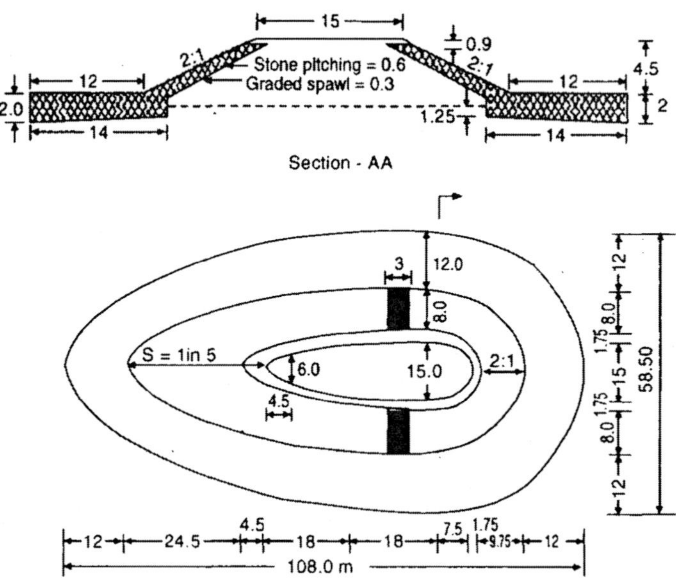

Fig. 9.24 Pitched island (Sulemanki weir Pakistan)

The salient parameters are (*i*) As deep scour occurs round the pitched island thereby creating there heavy concentration of flow, the stone pitching and apron are designed for maximum lacey's scour of 2 *D*, (*ii*) Since scour at the toe of a steeper slope is deeper than at the toe of flatter slope, sometimes different side slopes are given to the pitched island when it is used for distribution of concentration of flow, especially around the bends, and (*iii*) Usually steeper slope of 2 : 1 is adopted on the side on which a concentration of flow is desired, and flatter slope on the side where a throw-off becomes necessary.

9.15 MISCELLANEOUS METHODS

1. River Training Without Embankments

The method envisages building up river banks by driving stake fences on the river banks. During floods, the fences slacken velocity of flow and induce heavy silting on falling floods, banks form over these fences. Over the new banks, a second system of fences is selected and the process is repeated until a deep river channel is formed with natural high banks on both sides. This bank formation is continued gradually from upstream downwards. It is used only on hill streams carrying heavy sediment which is essential for its success. This method was evolved in Burma. This method of river training is not designed for facilitating flood flow nor to improve navigation. It is more correctly the systematic reclamation of low lands by guiding the spill of flood flow with stakes and fences to induce uniform silt deposition.

2. Bandalling

Bandalling evisages river training work designed to confine the low water flow in a single channel for maintaining required navigation depths. A bandal consists of a framework of bamboos driven into river bed, set 0.6 m apart by means of horizontal ties supported by struts 1.25 m apart. Bamboo matting is tied to this bamboo framework by means of coir ropes at water levels. The bamboos used on the framework are generally 3 to 6 m long and matting is 1 m wide, strengthened at the edges by strips of split bamboos (Fig. 9.26). Bandals are placed at an angle of 30° to 40° inclined downstream to check the flow and cause sand to be deposited parallel to and behind the bandals. Thus, river channel confined between the bandals is formed with sand banks on either side to which the whole river discharge is directed. After two or three weeks, deepening of the river channel is generally achieved. Work commences after the flood season as soon as the water levels begin to lower. The deepening in depth thus achieved subsists until the next flood period without requiring any attention during the same dry season.

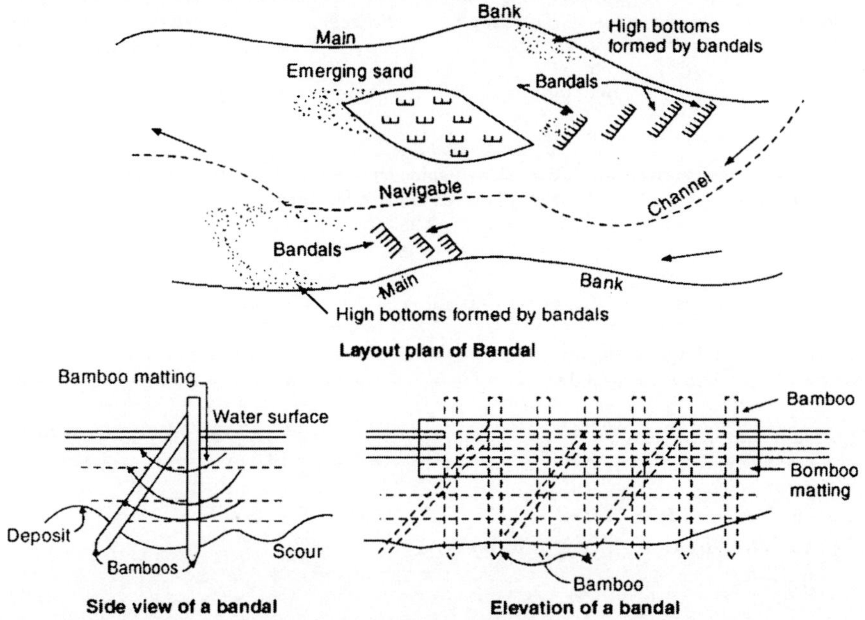

Fig. 9.26 Bandalling.

9.15 RIVER DIVERSION

River diversion as a river training measure is necessary in the case of barrages or weirs or where an actively eroding channel is to be diverted to an abandoned river channel when the eroding channel is eroding valuable land and threatening inhabited areas. The methods consist of diverting the creeks in falling winter discharges by means of stones dumped in crates from boats, stone crates advanced from one side or both sides, dropping stones from trestle bridges, pushing muck with big bulldozers, and sinking loaded boats and building crate bunds over the base thus formed and ultimately putting in earthen embankment. These methods entail exhorbitant cost and many times failures occur with complete loss of material and postponement of the next attempt to the following year.

Diversion of the changed main river channel into the old central channel with a balli-sarkanda bund is the recent innovation in the armoury of river training. As the balli-sarkanda bund is advanced and river discharge rises gradually more and more discharge is drawn into the central channel. The increased discharge dredges the channel mouth and eases the diversion. The back of the bund gradually silts up, because the velocity of percolating water decreases and drops silt. The last gap is closed in the falling season subsequently to complete the river diversion.

Advantages: (*i*) Economical, (*ii*) Locally available material is used, (*iii*) No skilled technicians or machinery are required, (*iv*) Work can be carried out in rising discharge and in all stages of the river, and (*v*) New channel is dug by the river itself.

SOLVED EXAMPLES

Example 9.1 (*a*) Calculate the width of waterway and scour depth for a discharge of 5000 cumes, Particle size 1.00 mm. Lacey's formula may be used.

(*b*) Design and sketch Bell's bunds including the launching apron to train the river.

Solution. Assume HFL 110 m, river bed level = 106 m

Lacey's f	*Lacy P*	*Overall water way*	*q*	*Lacey's scour depth, R*
$1.75\sqrt{dm} = 1.76 \times \sqrt{1.0}$ = 1.76	$4.83\ (5000)^{1/2}$ = 341 m	1.2 × 341 = 400 m	5000/400 = 12.5 cumecs	$0.473\ (5000/1.76)^{1/3}$ = 6.68 m

Velocity	Velocity head (ha)	Water level on rear side of guide bank.		
12.5/6.68 = 1.87 m/s	$(1.87)^2/2 \times 9.81 = 0.18$	HFL + ha	= 110 + 0.18	= 110.18 m.

Guide bank Top R.L.	*Height*	*U/s length*	*d/s length*	*U/s head radius*	*U/s head sweep angle*	*D/s head radius*	*D/s head sweep angle*
110.18 + 1.5 (F.B) ≃ 112 m	112–106 = 6 m	1.25L = 1.25× 400 = 500m	0.25 L = 0.25 × 400 = 100m	0.5 L 0.5 × 400 = 200m	130°	0.25L 0.25 × 400 = 100 m	60°

Top width	*Stone pitching*	*Apron*
6 m with 2:1 side slope on river side and 3 : 1 on rear.	River side of guide bank and rear slope of curved tail is stone pitched. Thickness $T = 0.06\ (Q)^{1/3} = 1.0$ m with 25 cm thick graded filter	Design thickness 1.0 × 1.25 = 1.25 m

	Nose	*Shank*	*Transition*
Scour depth, *D*	2.5 R – 4.0 = 2.5 × 6.68 - 4.0 = 13m	1.5 R - 4.0 = 1.5 × 6.68-4.0 = 6 m	
Apron width, 1.5 *D*	1.5 × 13 = 19.5 m	1.5 *D* = 1.5 × 6 = 9 m	
Apron thickness 2.25 *T* to 1.5 *D*	2.25 × 1.25 ≃ 3.0 m	1.5 *T* = 1.5 × 1.25 ≃ 2.0 m	Apron thickness increases from 2.0 m at shank to 3.0 m near nose.
Launched apron length	$\sqrt{5}\ D = \sqrt{5} \times 13 = 29$ m	$\sqrt{5} \times 6.0 = 13.5\ m^3$.	
at 2:1 slope stone/m length of apron	3.0 × 19.5 = 58.5 m^3	2.0 × 9 = 18.0 m^3	
Thickness of lauched apron	= 58.5/29 ≃ 2.0 m	18.0/13.5 ≃ 1.5 m	

Fig. 9.27 indicates the design of guide bank.

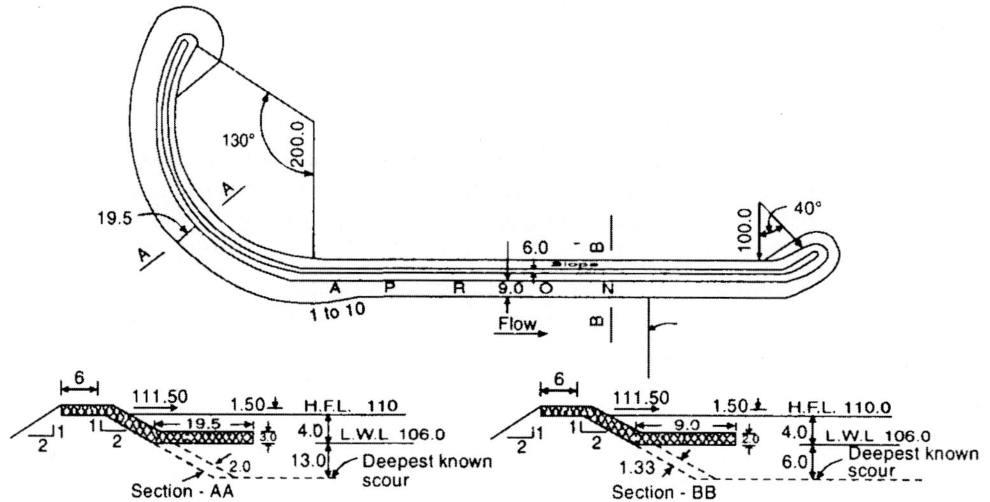

Fig. 9.27 Design of guide bank.

Example 9.2 Design guide bank for a road bridge across a river having maximum flood discharge of 65,000 cumecs, river bed slope of 0.08 m/km, mean diameter of medium sand in bed is 0.45, H.F.L. is 105.0 m and L.W.L. is 100.0 m.

Solution.

Active waterway	=	$4.83\ Q^{1/2} = 4.83\ (65{,}000)^{1/2} = 1231$ m
Overall waterway with 20% extra for piers, etc.	=	1231 × 1.2 = 1477 m
Overall waterway (*L*) provided with 13 spans of 100 m each and pier width of 3 m	=	(100 × 13 + 12 × 3) = 1336 m say 1400 m.

1. Guide bank
   (*i*) Upstream length = 1.1 L = 1.1 × 1400 = 1540 say 1550 m
   (*ii*) Upstream curved head = Provide a composite curvature of 3° curve, *i.e.*, with a radius of 580 m up to 90° sweep, then, a 5 ° curve, *i.e.*, a radius of 350 m between 90° to 120° sweep.
   (*iii*) Downstream length = 0.25 L = 0.25 × 1400 = 350 m
   (*iv*) D/s curved head radius (0.3 to 0.5 R) = Provide a radius of 240 m over a sweep of 90°.
   (*v*) Thickness of stone pitching on river side slope.

As per Spring for a slope 0.1 m/km and medium sand 75 cm. As per Gales, 130 cm inclusive of soling. By Inglis formula, $T = 0.06\ (65000)^{1/3}$, it comes to 240 cm. Adopt, T = 130 cm.

(*vi*) Lacey's f = $1.76\sqrt{0.45} = 1.15$

(*vii*) Depth of scour, (as per Lacey) $D = 0.473\ (Q/f)^{1/3} = 0.473\ (65000/1.15)^{1/3} = 18$ m

Particular	*Apron Design*	
	At head	*At shank*
D_{max} at H.F.L.	$2.5\ D \approx 45$ m	$1.5\ D \approx 27$ m
D_{max} below bed level	45 – 5 = 40 m	27–5 = 22 m
Apron thickness	1.5 T to 2.25 T = 2 to 3 m	1.5 T = 2 m
Volume of stone (assuming launching at 2 : 1 slope) Allowance for fanning out has to be considered.	Surface area to be covered by 3° curve of 90° sweep (refer section A. A)	$\sqrt{5} \times D_{max} \times$ apron thickness $= \sqrt{5} \times 22 \times 2 = 98$ m³

$$= \frac{\pi r}{2}\sqrt{5}\ D, \text{ where } r = \frac{599+679}{2} = 639 \text{ m}$$

$$= \frac{3.14 \times 639}{2} \times 2.24 \times 40 = 89889 \text{ m}^2$$

Volume = 89889 × 2.5 av. thickness = 224723 m³

(*viii*) Apron width = 1.5 D = 1.5 × 40 = 60 m; 1.5 × 22 = 33 m

(*ix*) Average length of laid apron $= \frac{\pi}{2}\left(\frac{r_2+r_3}{2}\right) = \frac{3.14}{2}\left(\frac{599+689}{2}\right) = 988 \text{ m}$

(*x*) Volume/m run $= \frac{224723}{988} = 227 \text{ m}^3$

(*xi*) Average thickness $= \frac{227}{60} = 3.78$ m; $\frac{98}{33} = 2.97$ m

(*xii*) Radius of d/s head = 240 m which permits movement of broad gauge rolling stock over it.

(*xii*) Width and thickness of apron. at d/s head and shank.

Fig. 9.28. Shows the section of guide bank and its cross sections.

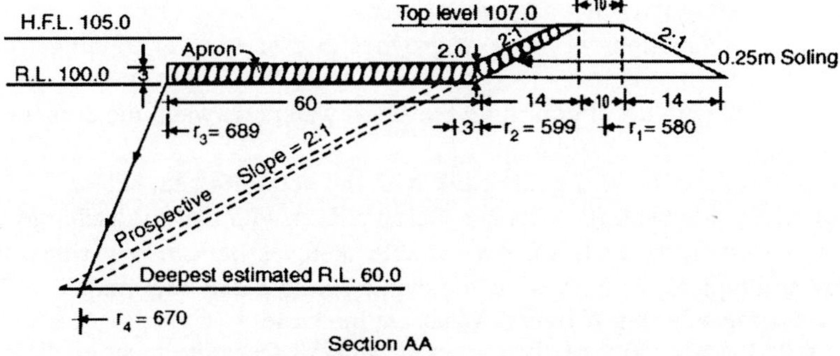

Fig. 9.28 Design of guide bank.

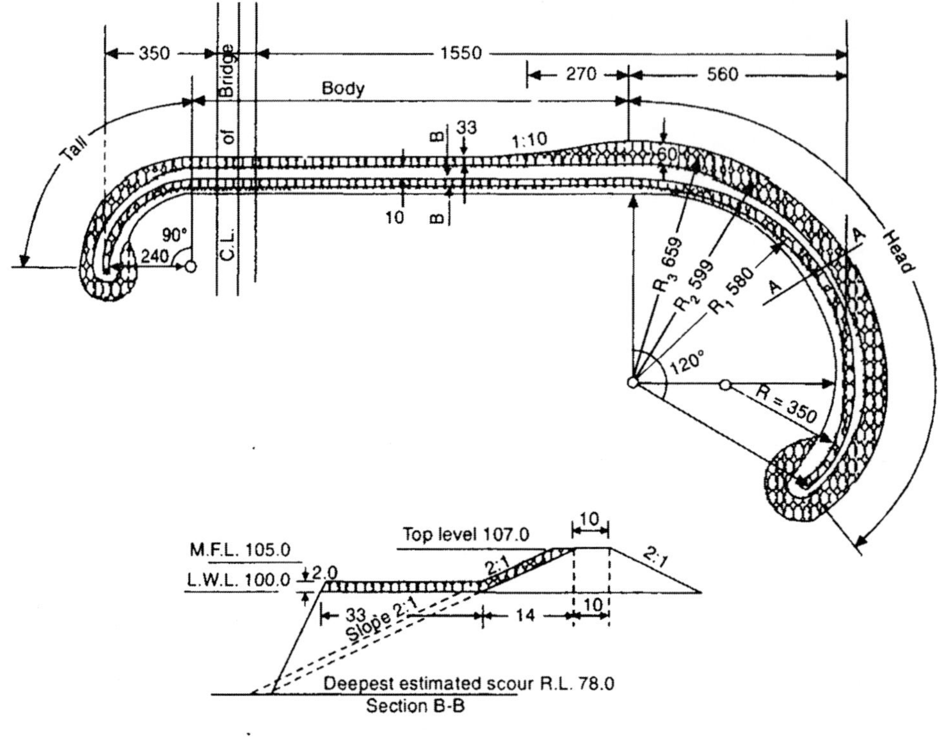

Fig. 9.28 Design of guide bank.

QUESTIONS

1. Define river training. Describe various types of river training and protection works.
2. What are the principles of river training? Discuss the statement "Planning for river training depends on the type of the river."
3. What do you understand by high water, low water and mean water training works?
4. (*a*) What is meant by river training works? Describe briefly their objectives.
   (*b*) Distinguish between:
      (*i*) Permeable spurs and impermeable spurs.
      (*ii*) Guide bund and marginal bund. [*Kurukshetra, University*]
5. (*a*) Distinguish between:
      (*i*) Permeable spurs and impermeable spurs.
      (*ii*) Aggrading river and degrading river.
   (*b*) Discuss the functions of spurs or groynes in case of river training works. Classify them with neat sketches. [*P.U.*]
6. (*a*) What are the effects of confining the flood waters between the embankments? State the limitations of embankments.
   (*b*) Show various parts of a guide bank with the help of a neat sketch.
   Design a guide bank for a bridge site on a river with design discharge as 8500 m³/s. Assume silt factor as 1.5. Draw suitable sketches indicating various dimensions in plan and typical sectional views. Assume suitable data required.
7. (*a*) What is meandering of rivers? What are its causes? [*P.U.*]
   (*b*) What is the necessity of river training works? Describe in brief different types of river training works. [*P.U.*]
8. (*a*) What is the necessity of river training works?
   (*b*) Explain the role of cut-offs in the meandering of rivers.
   (*c*) Sketch and explain the different types of spurs commonly used. [*P.U.*]
9. What is meant by guide banks? What are their functions and effects?

PART - III

CANAL STRUCTURES

- **Headworks**
- **Control Structures**
- **Cross Drainage Works**
- **Bridges and Culverts**
- **Canal Outlets**

10

HEADWORKS

10.1 INTRODUCTION

Canals draw supplies from rivers. Rivers do not have constant flow; the maximum being during monsoon months and minimum in winter months but the canals require continuous and dependable supplies for irrigating the command areas. For this purpose, perennial canals are essentially provided with permanent head known as headworks. Headworks comprise a weir or a barrage on the river and head regulators on the offtaking canals, the former for local pondage and raising of water level for diversion of supplies into the offtaking canals as well as passing regulated supplies in the river downstream while the latter is meant to draw the water supply into the canals. A weir is defined as a barrier constructed across a river to raise the water level in it course, with or without low shutters.

The functions of a barrage or a weir are (*i*) Divert water from river into canal, (*ii*) Effect pondage and raise water level in the river in the order to feed the offtaking channel during low supplies, (*iii*) Effect control over fluctuations in water level of the river, *i.e.*, impart permanency to the arrangement unaffected, within working limits, by vagaries of the river, (*iv*) Command the irrigable area by gravity flow, (*v*) Regulate the supplies into the offtaking canal and the river downstream, (*vi*) Pass extra supply into the canal when required and prevent flooding the canal during high floods in the river, and (*vii*) Control silt entry into the offtaking canal.

10.2 WEIR

Also termed as *anicut*; a weir is an obstruction constructed across a river to effect pondage and raise water level locally and to divert part or all the supplies into the canal. The weir is of height limited to 6 m and the volume and duration of pondage is comparatively small. Ponding of water is done against a raised crest and partly against shutters 1 to 2 m high (Fig. 10.4). Other salient features of a weir are (*i*) The crest level, usually high set, is determined by the permissible afflux under maximum floods, the discharge intensity per metre run and pond level, and (*ii*) The crest level is either constructed with top at pond level such that ponding is done against it, or at a lower level supplemented by shutters or counter-balanced gates. The shutters are dropped to pass floods. The raised crest has the disadvantage of causing excessive afflux during maximum floods and lack of control over the river during low floods. The essential difference between a weir and a dam apart from the structural is only in height and the quantum and period for which the supply is stored.

Advantages: (*i*) Low initial cost, and (*ii*) Simple in construction.

Disadvantages: (*i*) Lacks in speed and effective control possible with a gated weir (barrage), (*ii*) Operation of shutters involves considerable time and labour, (*iii*) Excessive afflux in floods, (*iv*) Silting upstream due to high set crest, and (*v*) Control over river flow can not be had fully as in the case of barrages.

10.3 CLASSIFICATION OF WEIRS

1. Classification According to Design of Floor

(*a*) *Gravity weir.* In a gravity weir, weight of the weir body and floor is more than the uplift pressure due to head of seepage water under the weir floor, *i.e.*, the weir depends on its weight for stability against uplift pressure due to seepage.

(*b*) *Non-gravity weir.* In a non-gravity weir, the weight of the floor is less but the weight of reinforced concrete floor of relatively lesser thickness together with the weight of the piers provides stability against the uplift pressure. The uplift is also resisted by the bending action of the reinforcement in the floor.

2. Classification According to Material of Construction

(*a*) *Rockfill weir:* It is the simplest type of weir constructed at Okhla on Yamuna river (Fig. 10.1). The weir body and a few intervening core walls are made of dry packed wall. The upstream and downstream rockfills are laid in the form of glacis. It depends on its weight and effective length for stability and hence is big in section. The stability of the weir is not amenable to exact theoretical treatment. It is suitable only in mountainous region where large quantities of stones are available within economical leads.

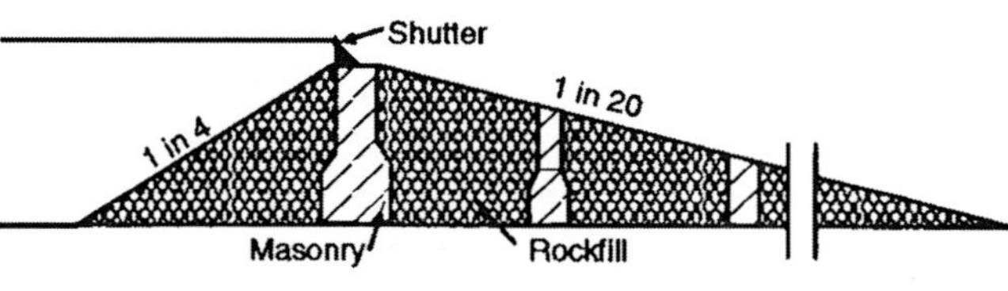

Fig. 10.1 Rockfill weir.

(*b*) *Concrete weir:* (Refer Sloping glacis weir).

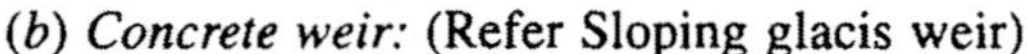

(*c*) *Masonry weir:* It is a weir constructed of stone masonry and usually fitted with shutters to effect pondage of water. The vertical drop wall and downstream floor are all made of masonry (Fig. 10.2). It is usually constructed on fine sand foundation. It is suitable in all site conditions except where drop is more than 5 m.

Fig. 10.2 Masonry weir.

3. Classification Based on Control of Surface Flow

(*a*) *Vertical drop weir.* It consists of a vertical drop wall or crest wall with or without shutters. Impervious floor is provided with cutoffs at upstream and downstream ends (Fig. 10.3). The cutoffs are protected followed by launching aprons both at the upstream and downstream. At the downstream end of impervious floor a graded inverted filter is provided to relieve uplift pressure. The raising of water level is against the raised crest. If provided with shutters, the raising is partly against the crest and partly against the shutters. It is suitable in submountainous region and for low falls on hard clay and consolidated gravel foundation.

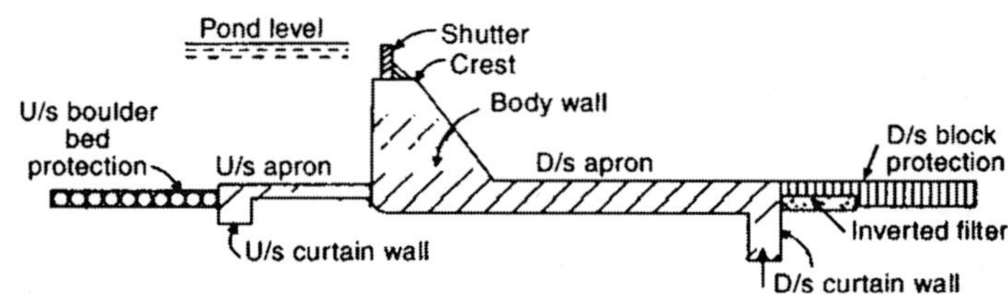

Fig. 10.3 Vertical drop weir.

(*b*) *Sloping glacis weir:* It is a weir of recent origin, based on the latest concept of subsurface flow. It is fitted with falling shutters. Cutoffs are provided at the upstream and downstream ends of the impervious floor besides under the crest. Hydraulic jump is formed on the downstream sloping glacis to dissipate the energy of overflowing water (Fig. 10.4). It is suitable for soft sandy foundation commonly met with at weir sites and is the type generally adopted.

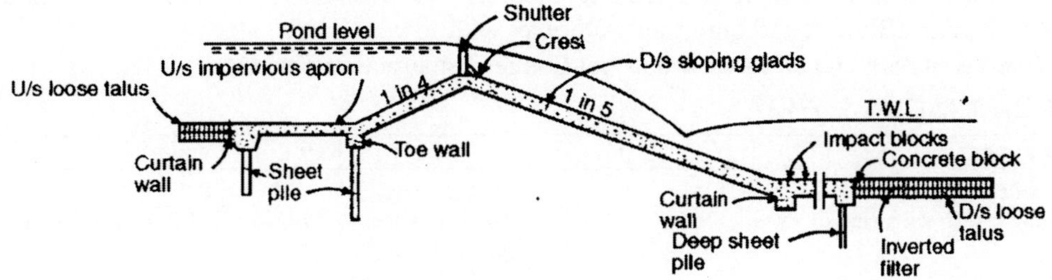

Fig. 10.4 Sloping glacis weir.

(*c*) *Barrage*: It is the modern concept of a weir. It is a low crest weir fitted with a series of gates over its entire length for creating the required pondage and for regulating the flood discharge. (Fig. 10.5)

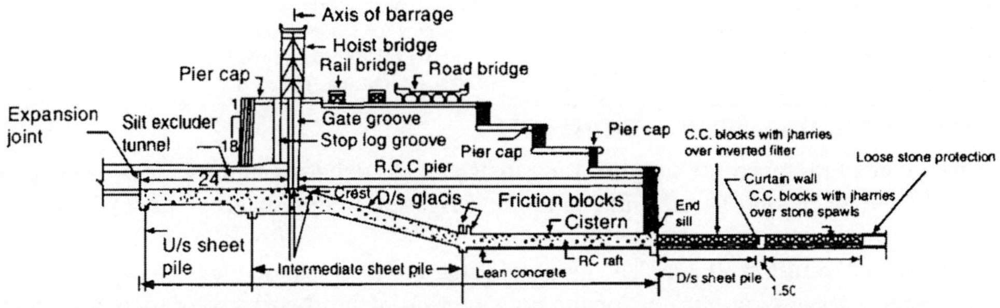

Fig. 10.5 Typical section of barrage undersluice with silt excluder bay.

4. Classification According to Function Served

(*a*) *Storage weir:* It is a weir permanently constructed for storage of water. Also termed as low dam. It is provided on Sutlej River at Nangal, called Nangal Dam, 29 m high and 305 m long barrage with a rail-*cum*-road bridge across it.

(*b*) *Waste weir:* Waste weir serves as spillway to escape floods which cannot be stored in the pond. It is constructed to ensure safety of the weir proper. Waste weir comes into operation when water level in the pond rises above the designed pond level.

(*c*) *Pickup weir:* A weir constructed at the offtake of a canal across the river to raise the water level for feeding the canal. The term is generally applied to a weir across the river on which there is a storage reservoir or a dam upstream. It is constructed as an adjunct to a reservoir.

(*d*) *Diversion weir:* Diversion weir is constructed as a constituent part of the headworks to raise water level in the river and divert supplies into the offtaking channel.

10.4 BARRAGE

A barrage is defined as a diversion structure across a river fitted with a series of gates over its entire length for creating the required pondage and for regulating the flood discharge to regulate the water surface level above it and pattern of flow upstream. Distinguished from a weir it is a gate controlled weir with low crest wherein ponding is done by means of gates. During high floods the gates are raised clear off the high flood level so that there is minimum afflux. A road bridge is usually provided across the river at a barrage at economical cost. The barrage is thus the obvious choice and is usually constructed in preference to a weir.

Advantages: (*i*) Better control on the river in floods, both inflow and outflow are regulated by means of operation of gates, (*ii*) Efficient control of the river channel leading to the undersluices, (*iii*) Better control over silt entry into the canal, (*iv*) Improved facilities of inspection and repairs, (*v*) Control over flow conditions, shoals formations and cross currents upstream of the barrage, and (*vi*) Lesser afflux due to optimum waterway with low set crest.

Disadvantages: (*i*) High cost, and (*ii*) Longer construction period.

10.5 BARRAGE *VS.* WEIR

Barrage	*Weir*
Low set crest.	High set crest.
Ponding is done by means of gates.	Ponding is done against the raised crest or partly against crest and partly by shutters.
Gated over entire length.	Shutters in part length.
Gates are of greater height.	Shutters are of smaller height, 2 m.
Gates are raised clear off the high floods to pass floods.	Shutters are dropped to pass floods.
Perfect control on river flow.	No control of river in low floods.
Gates convenient to operate.	Operation of shutters is slow, involve labour and time.
High floods can be passed with minimum afflux.	Excessive afflux in high floods.
Less silting upstream due to low set crest.	Raised crest causes silting upstream.
Longer construction period.	Shorter construction period.
Silt removal is done through undersluices.	No means for silt disposal.
Road and/or rail bridge can be constructed at low cost.	Not possible to provide road-rail bridge.
Costly structure.	Relatively cheaper structure.

10.6 SITE SELECTION FOR A BARRAGE

Rivers constitute source of supplies for canals and as such tapping them at a suitable stage is required for the diversion of water into canal for irrigation in command areas, hydel power generation, navigation, etc., as under:

Barrage in Mountain Stage: The mountain stage of river is suitable for a diversion structure for hydel power generation but not for the construction of headworks for canals due to (*i*) Low supply in the river due to lesser catchment, (*ii*) Being mountainous area, land suitable for irrigation is not available, rather the command area lies in plains far away. Moreover, due to high rainfall in the area, water for irrigation is not generally required, (*iii*) Away from command area and, therefore, more idle length of canal, (*iv*) Suitable location for headworks is not available, (*v*) River is more flashy associated with sudden rise and receding of flood necessitating more protection works downstream, (*vi*) Ground has steep gradient necessitating a number of falls in the canals, (*vii*) More cross drainage works are required to be constructed and as such high cost on construction of canals, (*viii*) Water silt-free is devoid of manurial value, and (*ix*) Costly structure as it has to resist high velocities and rolling of boulders which may damage floor and pier noses.

Certain positive points in favour of location of headworks in mountain stage are (*i*) Availability of sound rocky foundation for a weir at shallow depth, (*ii*) High head available for hydel power generation. Power generation also possible in the canal at the location of falls in it, and (*iii*) Economy in river training works.

Barrage in Boulder Stage and Trough Stage: Brief description of barrage in the two stages in comparative form is as under:

Barrage in boulder stage	*Barrage in trough (alluvial) stage*
River passes from rock to boulder stage.	River passes from boulder to alluvial stage.
Away from command, smaller area is available nearby for irrigation.	Nearer to the command area with smaller canal lengths.
Lack of demand for irrigation in head reach and hence long idle length of canals.	Serves large areas on both sides of the canal and there is little idle length.
Involves costly cross drainage works.	Small number of cross drainage works.
Excessive subsoil flow.	Comparatively low subsoil flow.
Silt factor is high which means economy in the requirement of cut-off.	Silt factor is low with higher cost on cutoff and protection works.
High seepage losses from canals in boulder tract.	Comparatively low seepage losses from offtaking canals.
Falls on canal may be utilized for hydel power generation.	Surplus head is usually not available.

Barrage in boulder stage	*Barrage in trough (alluvial) stage*
River more flashy.	River less flashy.
Low rate of silt charge.	Water silt laden and causes silting tendency in canal.
River section being not very wider, barrage length is shorter.	River section is very wide and barrage is of longer length.
Moderate river training works are required, river banks being high.	Extensive river training works are required, besides costly maintenance.
Availability of construction materials such as boulder and sand.	Stone is transported from long distances at heavy cost.
Comparatively better foundation strata.	Poor foundation resulting in costly structure.
Site not conveniently connected.	Site usually accessible.
Bed and banks composed of boulder and gravel.	Bed and banks made of alluvial silt & sand.
Cross-section is well defined.	Cross-section is wide between low banks wide apart.
Higher bed slope & Velocity.	Flat bed slope & low velocity.
High afflux poses no problem of inundation of land.	High afflux means extra construction cost & inundation upstream.

Barrage in Delta Stage: The river next passes from alluvial to this stage which is the last leg of its jounery to approach ocean. It is not suitable for the location of barrage due to extra wider section and shifting tendency of the river, besides lack of suitable compact command area for irrigation. The bed slope and velocity are too mild to carry forward the silt load which is deposited resulting in formation of delta.

The choice is thus limited to sub-mountainous or boulder stage and trough stage.

The essential requirements which a site shall qualify for locating a barrage, after having selected the reach of the river are (*i*) Narrow, straight and well defined river channel between high non-submersible banks, (*ii*) Easy accessibility by rail/road, (*iii*) Near to command area, (*iv*) Availability of construction materials and (*v*) Easy and economical construction.

10.7 INVESTIGATIONS FOR SITE SELECTION

Various investigations generally required to be conducted for the planning and layout of a barrage or weir are as follows:

Preliminary Investigation

1. *Map study*: Study of topographical and contour maps is made to ascertain the topography of the area, location of suitable sites and catchment areas upstream of these locations. River survey maps are prepared to assess the stability of the river at the site.

2. *Geology and foundation study:* Geological study is undertaken of the proposed sites to see any adverse geological features such as faults, fractured zones, shear zones, solution cavities, etc.

3. *Runoff and flood data:* It includes study of rainfall data in the catchment area, river gauge and discharge data in order to assess the available average weekly or 10 daily and monthly runoff and peak flows.

4. *Water requirement assessment:* An assessment is made of the water requirements for diversion into the canal system during various periods and the feeding levels is made.

5. *Availability of construction materials:* An assessment is made of the requirement *vis-a-vis* availability of the various construction materials at or in the near vicinity of the proposed sites and within economical leads.

6. *Communication facility:* Easy and all-weather means of communication to the proposed sites is investigated for easy accessibility and economical transportation of the materials and equipment to the sites under consideration.

7. *Environment and Ecology:* The effect of barrage on ecosystem especially on fish, wild life and human inhabitants adjacent to the structure is studied. Site selected should cause minimum environmental disturbances.

Detailed Investigations

1. *Topographic survey:* Topographical survey comprises (*i*) An index plan showing the entire catchment area upstream of the proposed barrage site with position of gauge and discharge sites,

raingauge sites, important irrigation works, road and railway crossing, if any, (*ii*) Contour plan of the area around the proposed barrage site extending up to 5 km on upstream and downstream sides with contour interval 0.5 m up to an elevation of at least 2.5 m above HFL, (*iii*) Cross section of the river at 2 km intervals up to pondage effect on upstream, (*iv*) Longitudinal section of the river to indicate observed water levels along the deep current. In the case of meandering river the survey is to cover at least two fully developed meanders on the upstream of the barrage axis and one meander length on the downstream or as may be required for detailed model studies. For model studies, cross sections are observed at closer intervals, (say 100 m) up to 2 km upstream and 1 km downstream of the proposed barrage or weir site, (*v*) The cross-levels in the river bed are spaced 10 to 30 m depending upon the topography of the river. The cross sections are extended on both banks up to 2.5 m above the HFL as far as possible, otherwise to an extent such that proper layout of guide and afflux bunds may be decided. In hilly streams cross sections at closer intervals are taken.

2. *Collection of hydrological data:* The hydrological data are collected to (*i*) Compute the design flood, (*ii*) Assess the available weekly or 10 daily and monthly runoff on a more realistic basis. For these studies it is necessary to obtain rainfall and runoff data. For the estimation of design flood the following data are collected:

(*a*) *Rainfall*–daily rainfall recorded at all the rainfall gauging stations in and around the catchment area for as many years as possible as also storm data, (*b*) Flood hydrographs for unit hydrograph, (*c*) Catchment characteristics for developing synthetic hydrograph, if adequate data are not available, (*d*) Peak flow data, (*e*) High flood marks, (*f*) Daily gauge and discharge data of the river at or near the proposed barrage site, higher flood gauges and observed discharges for as many years as possible is collected. A discharge site is established at a suitable point in the vicinity of the barrage site and the observed gauge discharge data are utilised to evolve a gauge-discharge curve for computing the discharges for the period for which only river gauge data are available, (*g*) The runoff data thus obtained is utilized for estimating dependable yield, (*h*) Data of sediment load carried by the river, its quality and quantity especially during flood season, for planning sediment excluding or ejecting works and to frame a suitable mode of regulation. In case the sediment charge brought by the water is excessive, due margin is given for sedimentation while fixing the pond capacity, particularly in case where pondage is proposed to be provided to meet diurnal power fluctuations.

3. *Surface and sub-surface investigations*: (*a*) Bore holes are driven at 100 m apart along axis of the proposed barrage, along divide wall and under abutment carried to hard rock level or to a depth of 20 m below the deepest river bed level depending on strata and structure. Trial pits are excavated to determine the depth of overburden comprising large size boulders. Where necessary geophysical method may be employed to locate the rock surface. Observations of water table in the area adjacent to the location of the barrage is also carried out for three-dimensional electrical analogy studies.

(*b*) *Log chute:* Statistics of logs, such as their numbers, sizes and periods in which they are handled, cost of handling and other relevant data are collected.

4. *Construction materials:* Survey of construction materials, their availability with lead for determining the type of construction and for preparing comparative estimates. Availability of hard stone may make masonry preferable to concrete.

5. *Diversion requirements*: Diversion requirements are worked out in accordance with the need of the project.

6. *Communication system:* Investigation includes dislocation of existing facilities and their relocation and additional facilities required during construction and operation.

7. *Other miscellaneous studies:* These include pond survey for the area submerged up to normal pond level or within the afflux bunds, as acquired, and all immovable properties coming

within it are recorded and valued and data for the necessity for the provision of fish pass for migratory fish are collected and advice from fisheries department in this regard is obtained.

8. *Environmental and ecological:* As discussed under preliminary investigations.

9. *Flood plain:* Aerial map of the flood plain indicating dominant river course.

10.8 LOCATION OF A BARRAGE

The location of a barrage is that which satisfies the requirements for weir proper, undersluices and canal head regulator, discussed hereinafter:

1. *River stage:* Mountain and delta stages are obviously not suitable for the location of headworks and the choice is, therefore, restricted to sub-mountainous or boulder stage and trough stage.

2. *River reach:* The river reach shall be straight and well defined so that velocities are uniform and cross sectional area is constant to ensure that the approach to the barrage is straight and distribution of flow is uniform. A slight curvature at the site is preferred from the standpoint of off-taking canal which when located on the downstream end of the outer curvature draws less sediment.

3. *Barrage alignment:* The barrage or weir is aligned at right angle to the river course in order to ensure (*a*) minimum length, (*b*) normal and uniform flow through all bays of the weir, minimum shoal formation and oblique flow. A skew alignment is avoided. The alignment is finalized after model studies.

4. *Protection embankment:* High, well defined, inerodible and non-submersible banks are suitable to ensure that the requirements of guide banks and flood protection embankment is the minimum.

5. *Foundation:* Good foundation and at minimum depth. Sometimes a favourable site has to be discarded due to large quantity of rock excavation involved in the construction of the work.

6. *Construction materials:* Availability of construction materials in required quantity and quality is one of the merits of a good site.

7. *Site accessibility:* Accessibility of site by rail and road to facilitate economical transportation of cement, steel, timber, fabricated gates and gearing equipment.

8. *Undersluices:* Location of undersluices shall be essentially in deep channel to ensure adequate supply to the canal head regulator under all stages of flow. When canals take off from both banks, a site with deep channels on both banks and low water in the centre is the most suitable.

9. *Head regulator:* Canal head regulator is constructed in conjunction with a barrage or weir. The alignment of headworks aims at providing a suitable curvature of flow to the canal to ensure least entry of silt into it. The head regulator is usually aligned at an angle of 90° to 110° to the barrage axis in order to minimize sediment entry into the canal which may be confirmed from model studies.

10. *Command area:* From irrigation point of view it is preferably so located that full command is obtained by a barrage of reasonable height. The overall cost of construction of headworks and the canal from the weir to the point where irrigation commences shall be as small as possible, consistent with the efficiency of the project, *i.e.*, nearest to the command, optimum command with minimum heading up of water level and low height of the headworks structure.

11. *Canal:* Good approach to the canal to enable water to flow smoothly into the canal without causing turbulence in the pocket. The alignment of canal shall be suitable and capable of giving good command. Idle length of the canal is the least with minimum number of cross drainage works.

12. *Silt ejector:* The site selected provides a suitable location for silt ejector and availability of levels for effective functioning of escape channel.

10.9 LAYOUT OF A BARRAGE—COMPONENT PARTS AND THEIR FUNCTIONS

Fig. 10.6 show the general layout and parts of a barrage; their functions are discussed as follows:

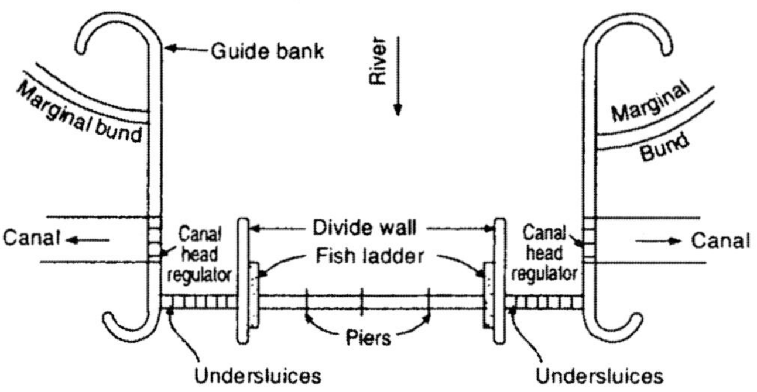

Fig. 10.6 Layout of a barrage.

Head Regulator: As the name implies it is the structure constructed at the offtake of the canal adjacent to the river undersluices. The functions served are (*i*) To admit water into the offtaking canal, (*ii*) To regulate the supplies into the canal, (*iii*) To indicate the discharge passed into canal from the designed discharge formula and observed head of water on crest, (*iv*) To control silt entry into the canal. Canal is completely closed during maximum floods lest heavy silt laden flood water silts up the canal.

River training works: It includes guide banks, marginal bunds, spurs, etc. It serves (*i*) To provide smooth, non-tortuous approach to the weir, (*ii*) To prevent the river from outflanking the weir, (*iii*) To prevent additional area being submerged due to afflux head (marginal bund), and (*iv*) To prevent erosion of river banks or protection works.

Divide well: It is a long wall or a groyne constructed at right angle to the weir axis. It extends up to the upstream end of canal head regulator. In case one canal takes off from either bank of the river, one divide wall is provided in front of each head regulator of the offtake. Its functions are (*i*) To isolate canal head regulator from main river flow for effecting silt exclusion since it creates a still pool. Silt deposits in the pocket and clear water enters the canal, (*ii*) To separate undersluices from the main weir since floor level of undersluices is at lower level than that of the main weir. It prevents turbulent action, (*iii*) To keep the cross current, if any, away from the canal, (*iv*) To concentrate scouring action of the undersluices for flushing out silt deposit in the pocket by ensuring a straight approach through the pocket, (*v*) To serve as a trap for coarser bed material, and (*vi*) To serve as one side wall of fish ladder.

Undersluices: Undersluices are the openings at low level in that part of the weir or barrage which is adjacent to the canal offtake. In case two canals take offtake on either side of the weir, two undersluices are provided, one on either side. The openings are controlled by gates. They form downstream end of the still pond, bounded on the two sides by divide wall and canal head regulator. The functions served are (*i*) To scour periodically silt deposited in the pocket, (*ii*) To maintain a clear and well defined river channel approaching the canal regulator. As the crest level is low, a deep channel develops towards them, (*iii*) To pass winter freshets and low floods and thereby reduce weir length, (*iv*) To lower highest flood level, (*v*) To control silt entry into the canal, (*vi*) To protect downstream floor from hydraulic jump action by filling the river bed downstream with the supplies passed through undersluices in case of sudden floods, and (*vii*) To impound receding floods.

Fish ladder: It is a narrow trough opening along the divide wall towards the weir side, provided with baffles so as to cut down the velocity of flowing water from upstream to downstream. Its functions are to facilitate natural migration of fish life across the weir; fish move from upstream to downstream in search of relatively warm water in the beginning of winter and return upstream for clear water before the onset of monsoons.

Weir or barrage: It is the obstruction constructed across the river. Its functions are to effect pondage, raise water level and divert water into canal.

Impervious floor: Impervious floor consists of upstream apron, upstream glacis, crest, downstream glacis and downstream apron. It serves to raise the water level and to take care of the disturbances caused by the hydraulic jump and uplift pressure.

Pervious floor: Upstream talus and downstream talus. Its functions are (*i*) To prevent scouring under the impervious floor, (*ii*) To serve as inverted filter, (*iii*) To check scour downstream, and (*iv*) To withstand high velocities downstream of hydraulic jump.

Sheet piles: In modern weirs, sheet piles upstream and downstream replace the curtain walls or cut off walls of masonry or well lines because these can be driven to any depth. It serves: (*i*) To protect the works against undermining of the foundation soil, (*ii*) To reduce uplift pressure under the weir, and (*iii*) To reduce the exit gradient.

Shutters or gates: Weirs are provided either with shutters or counter-balanced gates to maintain pond level. Better control is possible in a gated weir (barrage). Its functions include (*i*) To maintain pond level, and (*ii*) To raise the water level during low supplies.

10.10 POND LEVEL

The water level required to be maintained in the undersluices pocket in front of canal head regulator to feed the canal with full supply discharge is termed as pond level. Pond level is fixed 1 to 1.5 m higher than full supply level in the canal at head so that enough working head is always available for the modular working of the canal head regulator even if the canal bed is silted up in due course of time. The working head includes the head required to pass the design discharge into the canal and the head losses in the head regulator. It is also fixed in consideration of winter (low discharge period) storage. In ice formation regions, an ice cover of about 0.5 m may be added to the working head. Pond level in turn determines (*i*) crest level of the head regulator to feed the canal, (*ii*) crest level for the barrage and undersluices of the barrage, (*iii*) height of undersluices gates of the barrage, and (*iv*) height of shutters above raised crest level in the case of a weir.

10.11 UNDERSLUICES

Deep pockets of undersluices portions in front of the head regulator of the offtaking canal and long divide wall are required to separate the remaining river bays portion from the undersluices portion. It helps in keeping approach channel of the head regulator comparatively clear of the silt and to minimise the effect of main river current on the flow conditions in the regulator.

Undersluices are defined as openings in the weir adjacent to the canal head regulator. Two undersluices are provided in the weir in case there are two offtaking canals. A divide wall separates the main portion of the barrage from the undersluices portion of the barrage. The undersluices, with low set crest level than that of the normal portion of the barrage, are fitted with a series of gates, usually stoney gates, which can be easily operated from regulating platform at top for controlling the supplies through them.

Design Considerations

1. *Location:* The undersluices are located in the deepest channel to ensure adequate supply to the canal head regulator at all times. When canals take off from both the banks a site with deep channels on both banks and low water in the centre is most suitable.

2. *Discharge capacity:* It is fixed in consideration of the maximum of (*i*) twice the canal discharge so as to be effective in scouring action, or (*ii*) 10 to 20 per cent of the maximum flood

discharge at high floods, or (*iii*) maximum winter flood or fair weather freshets and low monsoon floods for obviating overtopping and/or operation of crest gates in the case of rivers.

3. *Crest level:* It is generally kept at general lowest bed level of the deep river channel consistent with the economy in cost of foundation and dewatering operation. Where a high pond level has to be maintained, such a low set crest is not necessary. Crest level is usually fixed up to 2 m below the crest level of the head regulator in case silt excluding device is provided, or 1 to 1.25 m if no silt excluding device is there, and 1 to 1.5 m below the crest level of the main weir.

4. *Undersluice's pocket:* The length of undersluice's pocket between the divide wall and the head regulator is kept 1.5 times the length of upstream divide wall. The undersluices are made wide enough so that the approach velocities are sufficiently lower than critical velocities to ensure maximum settling of suspended silt load. Where silt excluders are provided, the width of the pocket is determined by the velocity required in the pocket to induce siltation. Where the barrage is very wide, river sluices adjoining the pockets can be provided to cater for low floods and freshets thereby economising on the cost. However, the width of undersluices portion and length of divide wall are fixed on the basis of model studies.

5. *Floor:* The length and thickness of impervious floor is designed as for main weir from the considerations of formation of hydraulic jump and against uplift on permeable foundation but is made extra stronger keeping with the high discharge intensity. Downstream floor is depressed up to anticipated retrogression bed level downstream.

6. *Apron:* Adequate apron is provided in view of higher discharge intensity.

7. *Span:* Wider spans of undersluices are preferred so as to be effective in scouring action. Bridge decking is provided to facilitate raising or lowering of gates by other mechanical means.

10.12 DIVIDE WALL

Divide wall, as the name implies, is a long wall which separates the undersluice bays from barrage or weir bays (Fig. 10.5) and is aligned at right angle to the axis of the barrage. It extends both on upstream and downstream sides of the work, the portion on the upstream side being of longer length than that on the downstream side. On the upstream of a barrage, it isolates the pocket in front of canal head regulator from the turbulent main flow in the alluvial river in flood for effecting sand exclusion by creating a still pond from which discharge is drawn into the canal.

Divide wall in the upstream of the barrage or weir does not extend beyond upstream of the head regulator. For effective silt exclusion the nose of the divide wall is extended up to the upstream abutment of the canal head regulator and as near the barrage as possible. A divide wall of two-third full width of canal head regulator generally gives satisfactory flow conditions when only one canal takes off from the barrage. In the case of two off taking canals, on the same bank, the divide wall is taken up to the uppermost head regulator.

The divide wall is continued on the downstream side up to the end of apron. It is to (*i*) Ensure adequacy of tail water depth in the undersluice bays for the formation of hydraulic jump to facilitate proper dissipation of energy, (*ii*) Avoid cross flow in the close vicinity of the structure which may cause scours, and (*iii*) Provide side wall to the fish ladder. Model studies provide best guide for determining length of divide wall.

Design Considerations: Various design considerations are (*i*) It is founded on sound foundation. At least 30 m length from the nose is provided with well foundations (Fig. 10.7*a*) going sufficiently below the anticipated deepest scour. The remaining length protected by *pacca* floor is provided with open foundation (Fig. 10.7*b*). Fig. 10.8*a* and 10.8 *b* depict typical layout of a divide wall

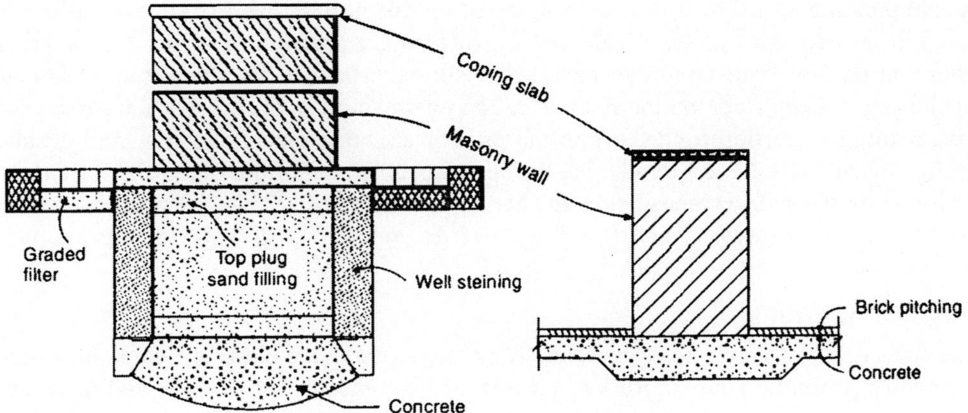

Fig. 10.7a Divide wall with well foundation. **Fig. 10.7b** Divide wall with open foundation.

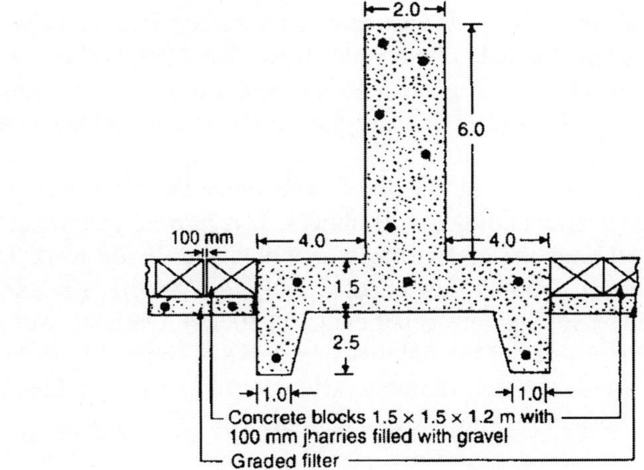

Fig. 10.8a Divide wall (Section A-B).

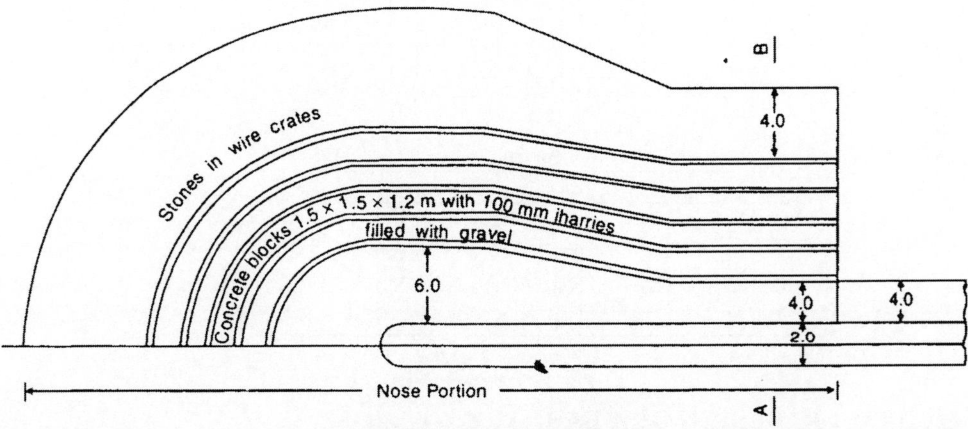

Fig. 10.8b Layout of divide wall.

Top width of the divide wall is usually 1.5 to 3 m and the bottom width is determined depending on height, water depth and differential pressure. Normally there is water on both sides of the divide wall but it is likely to be subjected to differential pressure (*a*) when flood discharge is passing through the undersluices and relatively little depth is there on the side of the main weir, a differential pressure of 1.0 m is usually considered on this account, and (*b*) due to difference of levels of silt on two sides of the divide wall. A differential pressure of about 2 m is generally considered in design. With a view to have better silt exclusion, a flat slope to nose is kept on pocket side and a steep slope on the river side. The upstream nose of the divide wall is provided with extra strong protection in the form of apron. For effective sand exclusion from canals, the nose of the divide wall should extend up tc the upstream abutment of the first offtaking canal. The foundation of divide walls extends below the bed up to a depth of 2.25 R below H.F.L at the nose portion and may be decreased progressively towards impervious floor of the barrage up to a minimum depth of main cut-offs provided.

10.13 FISH LADDER

The fish are always there in river water. They are on the move either in search of food or to their spawning grounds. They migrate to upstream in summer for cold water and in winter towards downstream for warm water. In pre-monsoon months of May and June they move upstream again in search of clear water. In the weirs and barrages provision in the form of fish ladder is made for the uninterrupted movement of the fish across the work. The fall or difference of head in the pool level upstream and tail water level downstream is broken up into small water steps by means of baffle walls which are constructed oblique to the flow (Fig. 10.9) and in which holes are provided so that fish can take rest after passing one hole and before moving to the next. The walls act as energy dissipators in order to have smooth water flow at a velocity not exceeding 2.5 m/sec along and across which the fish can travel.

Suitable location for the fish ladder is near the divide wall where there is some water throughout the year in the river bed downstream of the undersluices. It is normally constructed between the barrage and the divide wall near the undersluices or within the divide wall. The essential requirements of a fish ladder are (*i*) Smooth flow at sufficiently low velocity suitable for the fish migration. The slope of fish ladder is kept not steeper than 1 in 10, (*ii*) Availability of ample water throughout the year, (*iii*) Top and sides of the fishway are above ordinary high water level and (*iv*) Easy and safe passage of fish through the compartments of the fish ladder without any risk for the ascending fish to collide with the sides and upper end of each bay.

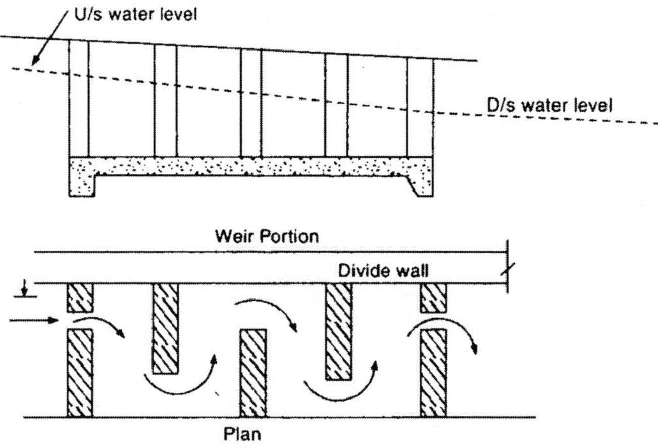

Fig. 10.9 Fish ladder.

10.14 HYDRAULIC DESIGN OF A BARRAGE

1. Waterway

Waterway or the length of the barrage is fixed to cater for the maximum flood discharge corresponding to 50 year frequency. In deep and confined rivers with stable banks, the overall waterway between abutments is approximately equal to the actual width of the river at the design flood. For meandering alluvial rivers for minimising shoal formations, Lacey's waterway, $P = 4.83\ Q^{1/2}$ with looseness factor of 1.2–1.0 and 1.0–0.6 is adopted for silt factor less than 1 and 1 to 1.5 respectively as per IS: 6966 (Part I) 1989. Permissible intensity of discharge and afflux and pond level are also considered.

Permissible Intensity of Discharge: The average intensity of discharge per metre of waterway between abutments is limited to a safe permissible value. High discharge intensity may be allowed in boulder reaches where coarse bed material can withstand higher velocities and large afflux poses no serious problem of inundation of land. In alluvial reach, the slopes are flat and bed material generally cohesionless sand and hence lower discharge intensity consistent with economy is adopted. Higher discharge intensity reduces the linear waterway but (*i*) Increases the velocity and resulting scouring necessitating sluices to be adequately safeguarded, (*ii*) Increases cost of protection works, (*iii*) Deepens the downstream floor of the barrage resulting in increased cost due to dewatering and increased thickness of the floor. Further, higher the pond level, narrower is the waterway and *vice versa*.

The advantages of restricted waterway are (*i*) Construction of weir itself at low cost; the weir length being short, (*ii*) Better river control, and (*iii*) Removes shoal formation immediately upstream of weir or barrage.

The disadvantages are (*i*) Excessive velocities which cause deep scour along the guide banks, (*ii*) Lower levels of upstream and downstream impervious aprons which ential higher pumping expenses to lay them, (*iii*) Higher cost of river training works, undesirably high afflux which either submerges a vast area or requires high afflux bund and consequent higher cost of river training works.

Wider waterway, however, has the advantages of (*i*) Low velocities which mean moderate scour along the guide bank, and (*ii*) Low afflux, and (*iii*) Less cost of river training works with the disadvantages (*i*) Slack velocities which promote formation of shoals with consequent non-uniform flow distribution, and (*ii*) increased cost of construction of barrage.

The waterway may be provided at one level over a barrage having a uniform crest level or partly at a lower level, as is the general practice, through the undersluices and partly at a higher level over the weir crest.

2. Retrogression: Progressive retrogression or degradation of the downstream river and levels as a result of construction of a weir or barrage causes lowering of the d/s river stages and has to be considered in the design of d/s cistern and exit gredient. A value of 1.25 to 2.25 m is adopted as retrogression for alluvial rivers at lower river stages. At design flood, the reduction of gauges due to retrogression may be considered to vary from 0.3 to 0.5m depending upon whether the river is shallow or confined during floods.

3. Afflux: The rise in the maximum flood level of the river upstream of the weir due to its construction is termed as afflux. The value of afflux corresponding to the design flood is important for the design of various components of barrage. High afflux results in a narrow weir but at the same time has the disadvantages of (*i*) Greater cost of river training works, (*ii*) Greater intensity of discharge per metre width of the weir, (*iii*) Greater depth of probable scour and hence additional length and thickness of launching aprons upstream and downstream, (*iv*) Greater depth of cutoffs both upstream and downstream of weir, (*v*) Greater afflux means greater difference of water level between upstream and downstream of the weir and greater dynamic action as also its effect on the location and depth of standing wave, (*vi*) Higher cost of gates, and (*vii*) Greater risk of failure due to outflanking.

Provision of higher afflux has the advantages of (*i*) Narrow weir at low cost, (*ii*) Makes the weir modular at all stages by formation of standing wave even at maximum flood level, and (*iii*) Helps operation of extractors and excluders. On the other hand provision of smaller afflux has the demerits of (*i*) Inordinately larger size and length of weir, (*ii*) Greatly reduced control on weir, and (*iii*) Use of extractors and excluders is ruled out.

In the design of weir afflux is limited to 1 m in upper and middle reaches of a river and in lower reaches with flat gradient afflux is limited to 0.3 m.

4. Free Board: Generally, a free board of 1.5 to 2 m above affluxed water level on the upstream side due to flood of 500 years frequency or the standard project flood and above the highest flood level (unretrogressed) on the downstream side is provided. However, if pond level is higher than water level for 500 years flood, the free board is provided over the pond level.

5. Floor and Bed Protection of Barrage Bays

Floor

The upstream floor level of a weir or a barrage bays (other than undersluices bays) is fixed at the general river bed level at or below the level of the crest. Usually the floor level is kept at 0.5 to 1.0 m higher than the upstream floor level of the undersluice bays.

Crest Level

The crest level is fixed in consideration of design flood, permissible afflux, discharge per metre run through the weir, and pond level. The crest level also varies with the type of the weir or barrage, *viz*., weir without shutters, weir with shutters and barrage.

For weirs without shutters, the crest level is kept at the pond level by the construction of a permanently raised crest.

For weirs with shutters, the crest level is kept not lower than 2 m below the pond level; 2 m being the usual maximum height of the falling shutters. Thus, the top level of the shutter corresponds to the pond level.

For barrages, a gate controlled crest with the provision of a roadway across the river is provided.

Discharge intensity is determined by the weir formula:

$$q = 1.7\, H_e^{3/2}$$

where, q = discharge per metre run (cumecs), $H_e = H + h_a$ (m), H = depth of water over the crest (m), h_a = head due to velocity of approach (m).

T.E.L. U/s = T.E.L. d/s + loss of head = T.E.L. d/s + afflux (*i.e.*, 1.0) + 0.3 m retrogression

= T.E.L. d/s + 1.3 m.

T.E.L. d/s = maximum flood level at the weir site before the construction of weir plus head due to velocity of approach minus 0.3 m to allow for possible drop in the maximum flood level due to retrogression of the downstream river bed level.

Discharge formula (IS: 6966–Part I–1989) is $Q = CLH^{3/2}$.

Where, Q is discharge (cumecs), C is coefficient of discharge (in free flow condition), value determined from Fig. 10.10 subject to determination by model studies, L is clear waterway of weir or barrage (m), and H is total head causing the flow (m).

$$\text{Drowning ratio} = \frac{\text{d/s water level} - \text{crest level}}{\text{u/s water level} - \text{crest level}}.$$

As per CBIP practice L is replaced by Lc which is equal to (L-k.n.H). Where, K is a coefficient (0.01 to 0.03) depending on shape of pier nose and n is number of end contractions.

The crest level is, however, kept higher than the crest level of the undersluice bays, the discharge relation for which is $q = 1.84\, H_e^{3/2}$. The crest is usually kept 2.0 m wide, flat at the top with a glacis of 2 : 1 to 3 : 1 slope to meet the upstream floor and a slope of 3 : 1 for the downstream glacis. Hydraulic jump type stilling basin with horizontal apron is provided for the dissipation of the surplus energy downstream.

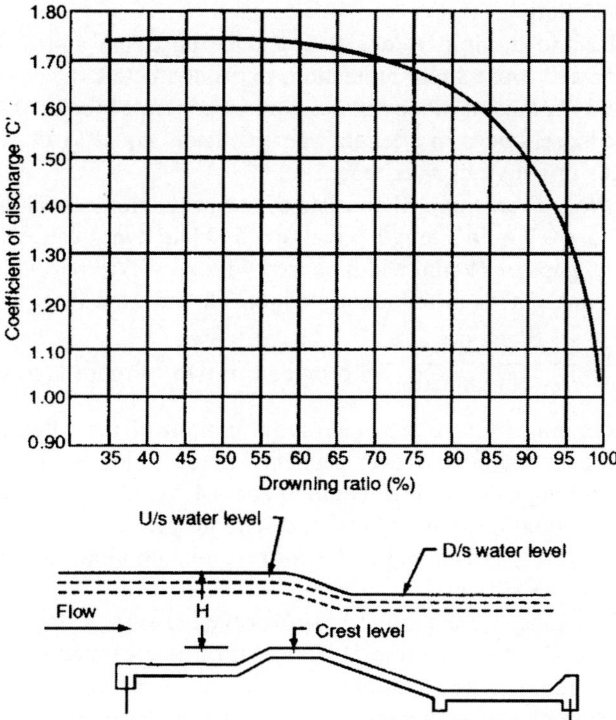

Fig. 10.10 Drowning ratio *vs.* coefficient of discharge curve (IS: 6966-1989)

Cutoffs

The cutoffs are usually of concrete, masonry or sheet piles depending on the type of soil and depth of cutoff. The depths of upstream and downstream cutoffs are provided to cater for scours upto 1 R and 1.5 R respectively, where R is the depth of scour below highest flood level subject to the condition that the depth of cutoff and floor length is sufficient to limit the exit gradient within safe values. The subject-matter has been dealt with in detail under design of barrages on permeable foundation.

The scour is likely to occur in erodible soils such as clay, silt, sand and shingle. In non-cohesive soils, the depth of scour is calculated by Lacey's formula, as under:

$$R = 0.473\,(Q/f)^{1/3}, \text{ when looseness factor is} > 1$$

or

$$R = 1.35\,(q^2/f)^{1/3} \text{ when looseness factor is} < 1$$

where, R is depth of scour below the highest flood level (m), Q is high flood discharge in the river (cumecs), q is discharge intensity (cumecs/m) and f is Lacey's silt factor = 1.76 $\sqrt{mr}$, (f is checked from Lacey's relationship, $f = 290\,(R^{1/2}\,S^{1/3})^{2/3}$. The extent of scour depth in a river with erodible bed material at various points along a weir or a barrage is given in Table 10.1:

Table 10.1 Scour depths (IS: 6966 (Part I)-1989)

Location	*Scour depth (m)*	
	Range	*Mean*
Upstream of impervious floor	1.25 R to 1.75 R	1.5 R*
Downstream of impervious floor	1.75 R to 2.25 R	2.0 R*
Noses of guide banks and divide wall	2.0 R to 2.5 R	2.25 R
Transition from nose to straight	1.25 R to 1.75 R	1.5 R
Straight reaches of guide bank	1.0 R to 1.5 R	1.25 R

* These are considered for designing flexible protection only. For cutoff, it is *IR* and 1.25 R for upstream and downstream respectively.

7. Flexible Bed Protection

Bed protection beyond the impervious floor on the upstream and downstream sides comprises block protection and loose stone protection, discussed below:

1. *Upstream Block Protection.* It comprises cement concrete blocks of size 1.5 × 1.5 × 0.9 m laid over 0.6 m thick loose stones in a length approximately equal to *D*, scour depth below the floor level for barrages in alluvium river reach.

2. *Downstream Block Protection.* It consists of bed protection consisting of cement concrete blocks of size 1.5 × 1.5 × 0.9 m in a length equal to 1.5 *D* laid over inverted filter for the grade of river bed material (with gaps of 75 mm width packed with gravel) in part length and block protection only with loose stone spawls in remaining length. The inverted filter, conforming to criteria, $\frac{D_{15} \text{ of filter}}{D_{15} \text{ of foundation}} \geq 4 \geq \frac{D_{15} \text{ of filter}}{D_{85} \text{ of foundation}}$ is provided in two or more layers and protected at the end by a toe wall extending up to 0.5 m below the bottom of the filter. At the end of block protection on the upstream and downstream, launched apron of loose boulder or stones (stone of size + 300 mm and weighing + 40 kg) in a length of about 1.5 *R* and 2 *R* respectively are provided. No allowances for concentration is made in the discharge per unit length in computing *R*. The slope of the launched apron is considered as 2:1 for rivers in alluvium sandy reaches and 3 : 1 for rivers with very fine sand or silty bed.

The thickness of the loose stone required to cover the launched slope is 1.25*T*, where *T* is the thickness of pitching obtained from Table 10.2. The thickness of stone at the inner edge corresponds to the quantity required for a thickness of *T* and with the material required for extra thickness of 0.25 *T* the thickness is increased towards the outer edge. The total quantity of loose stone worked out, as per criteria given above, is laid in a length of about 1.5 *D* to 2.5 *D*, the higher figure for flatter launched apron.

Table 10.2 Thickness of pitching (*T*) for loose stone protection (cm)

Type of river	*River slope in m/km*				
bed material	*0.05*	*0.15*	*0.20*	*0.30*	*0.40*
Very coarse	40	50	55	65	70
Coarse	55	65	70	80	85
Medium	70	80	85	95	100
Fine	85	95	100	110	115
Very fine	100	110	115	125	130

8. Canal Head Regulator

The canal head regulator is so aligned that silt entry into the canal is minimum. It is located at an angle of 90° to 110° with respect to the axis of the diversion work. Fig. 10.11 shows typical plan of canal head regulator. The crest level of the regulator is pond level minus head of water over crest (H_e) required to pass full supply discharge. It is kept at a level higher than the level of bed of the still water pocket and the crest level of the undersluices with a view to allow only water entraining finer sediment particles to enter into the offtake and arrest the entry of coarser sediment. In practice, the crest level is kept at more than 1/3 of full supply depth. The effect of retrogression is considered in the design of a barrage. A retrogression of 0.5 m increasing linearly up to 2.0 m at low discharges is usually considered in the design of barrages. When a silt excluder is provided, the crest level of head regulator is determined in conjunction with the design requirements of silt excluder

$$Q = C L_e H_e^{3/2}$$

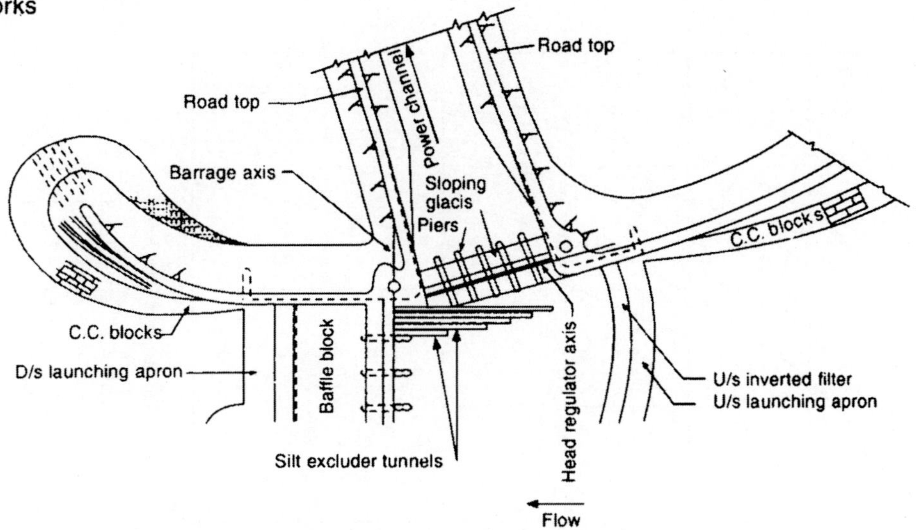

Fig. 10.11 Typical plan of head regulator with sediment excluder.

where, Q is discharge (cumecs), C is coefficient of discharge, value depends on head over crest, shape and width of crest, u/s and d/s slopes Z_1 and Z_2 respectively, height over u/s floor and roughness of its surface. Values of C for varied He/P and He/w for ungated flow and for $Z_1 = 0, Z_2 = 2$ and 3 are given in Fig. 10.12. The discharge reduction factor for varied submergence ratios H_d/He is obtained from Fig. 10.13, where H_d is depth of tail water level above the crest, L_e is effective waterway. When the outflow is controlled by partly open gates, the condition similar to sluice flow developes and required discharge is:

$$Q = (2/3)\ \sqrt{2g}\ C\ Le\ (H_1^{3/2} - H_2^{3/2})$$

where, H_1 and H_2 are total heads to the bottom and top of the orifice. The width of crest is kept (2/3) He minimum. The edge of crest is rounded off with a radius equal to He. The upstream face is generally kept vertical and d/s sloped at 2:1 or flatter.

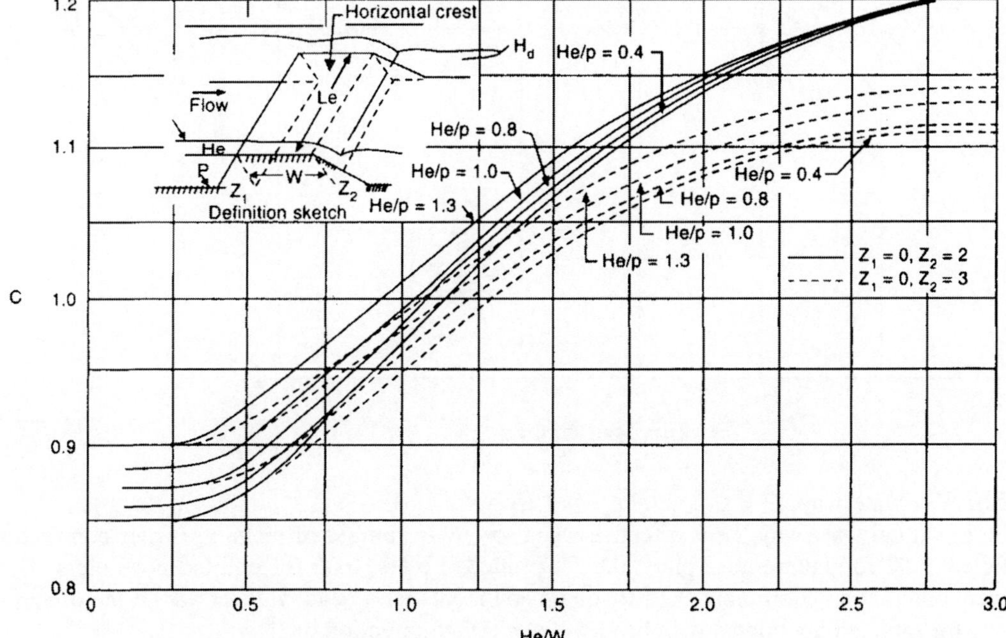

Fig. 10.12 Coefficient of discharge for varied H_e, P and W.

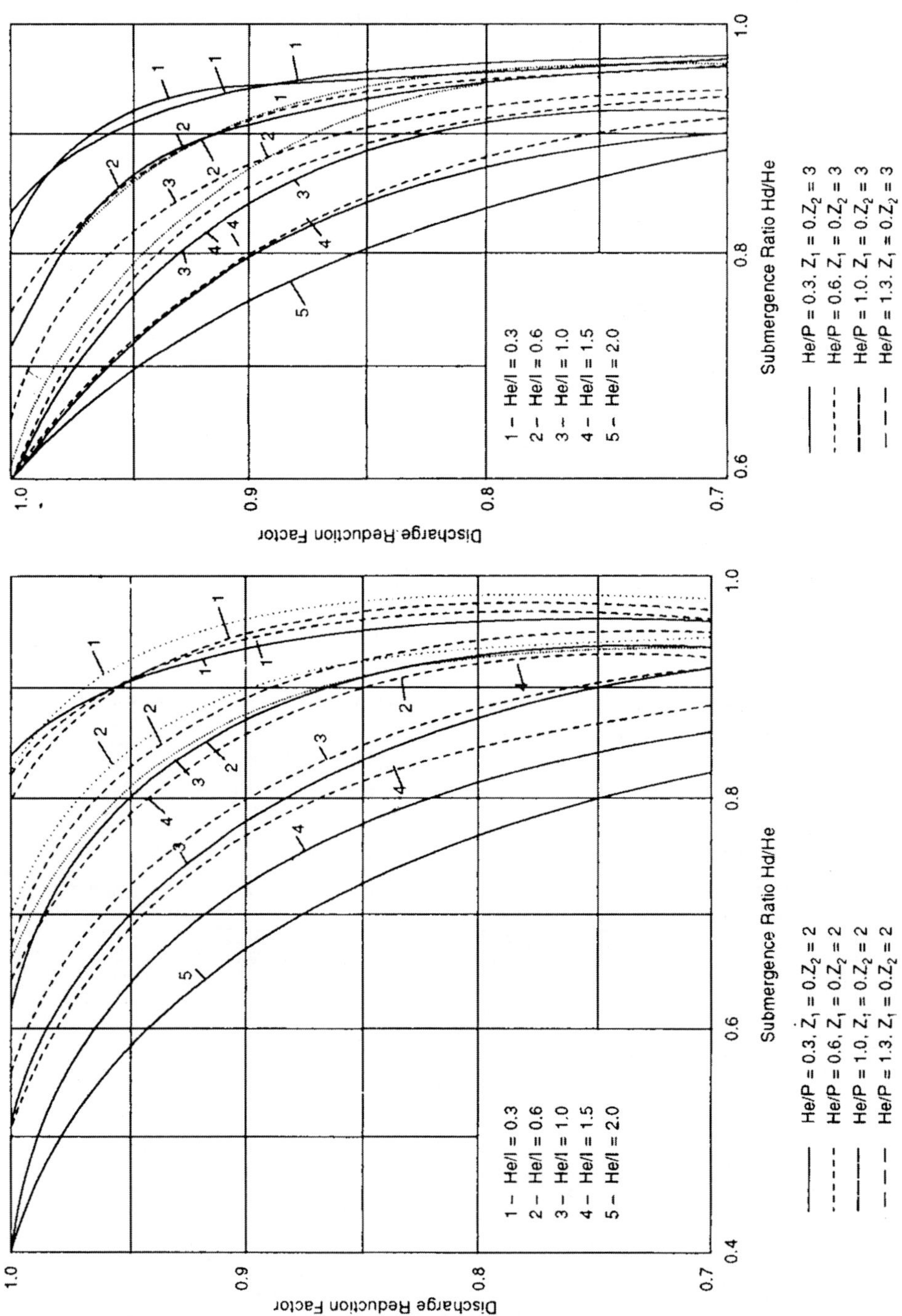

Fig. 10.13 Variation of discharge reduction factor for submerged flow (IS: 6531:1994)

Effective waterway; $L_t = L_e + 2\,(N\,k_p + k_a)\,H_e + W$

where, *Lt* is total waterway, *Le* is effective waterway, *N* is number of piers, k_p is pier contraction coefficient-0.02 for square nose piers, 0.01 for rounded piers, 0.00 for pointed nose piers, K_a is abutment contraction coefficient-0.2 for square abutments with head walls at 90° to the direction of flow, for rounded abutments with head walls at 90° to direction of flow:

$$0.5\,H_e \geq r \geq 0.15\,H_e,\; K_a = 0.1.$$

Approaches: At upstream inlet a smooth entry is ensured by providing circular, elliptical or hyperbolic transitions. The splay may be 1:1 to 3:1. At downstream side, straight, parabolic or hyperbolic transitions are provided. The splay may be 3:1 to 5:1. Wing walls are kept vertical up to end of impervious floor. But to effect economy, vertical up to toe of glacis and beyond which flares gradually to 1/2:1 at the end of impervious floor. In the remaining length, flare from 1/2:1 to actual slope of canal section.

Minimum floor length: It is sum of horizontal floor in d/s from surface flow considerations, length required to accommodate sloping glacis and crest and about 3m extra u/s of crest.

Depth of d/s cutoff: Minimum d/2 + 0.5, where d_2 is water depth (m) corresponding to full supply discharge.

U/s and d/s loose aprons: Inverted filter 1.5 to 2*D* long, *D* being depth of scour below bed and 1.5 *D* on downstream side.

9. Guide Banks

Guide banks help secure a suitable river approach to flow axially through the barrage, to prevent outflanking of structure, minimise cross flow through weir which may endanger the structure and protection works. It is a good practice to determine alignment, length and shape of guide banks by model studies. Principal considerations in fixing the parameters of guide banks are (*i*) The length and curvature of head of guide banks in wide alluvial rivers is kept such that worst meander loop is away from entire canal embankment and the approach embankment, (*ii*) In case the alluvial bank is close to barrage, the guide banks are tied to it by providing suitable curvature, if necessary, and (*iii*) The guide banks are tied to any outcrops of bed rock on the banks. The subject matter has been discussed in detail in River Training works.

10. Energy Dissipation

For barrages or weirs on alluvial foundations hydraulic jump type stilling basins are most effective means of dissipation of surplus energy (Fig. 10.14). The formation of standing wave depends on the discharge per unit width, afflux, shape of glacis, etc. Unlike level and smooth floor wherein the position of standing wave is unstable and cannot be closely predicted, on a sloping glacis its position is definite and can be closely predicted. The length of the jump is approximately five times the height. For the structure to be safe, the standing wave is confined on the sloping glacis of 3:1 slope and not permitted to form on the horizontal floor beyond the toe of the glacis. The elevation and length of downstream impervious floor thus depends on the position and height of the standing wave under various sets of conditions. Length of the horizontal floor is accordingly kept 5 (D_2-D_1) where D_1 is depth of water just upstream of the point of formation of the wave. The design procedure for locating the elevation and position of the standing wave is as follows:

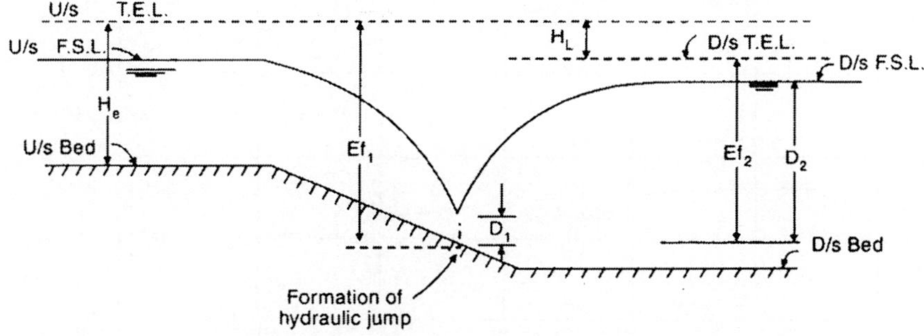

Fig. 10.14 Energy dissipation by standing wave.

1. By Use of Curves

(*i*) From known values of the intensity of discharge with concentration of flow, usually

20%, (q), head loss (H_L, *i.e.*, difference between u/s and d/s T.E.L. values) using Blench curves (Fig. 9.13) the value of E_{f_2} (energy of flow just d/s of the point of formation of the standing wave) are read out for various discharge values.

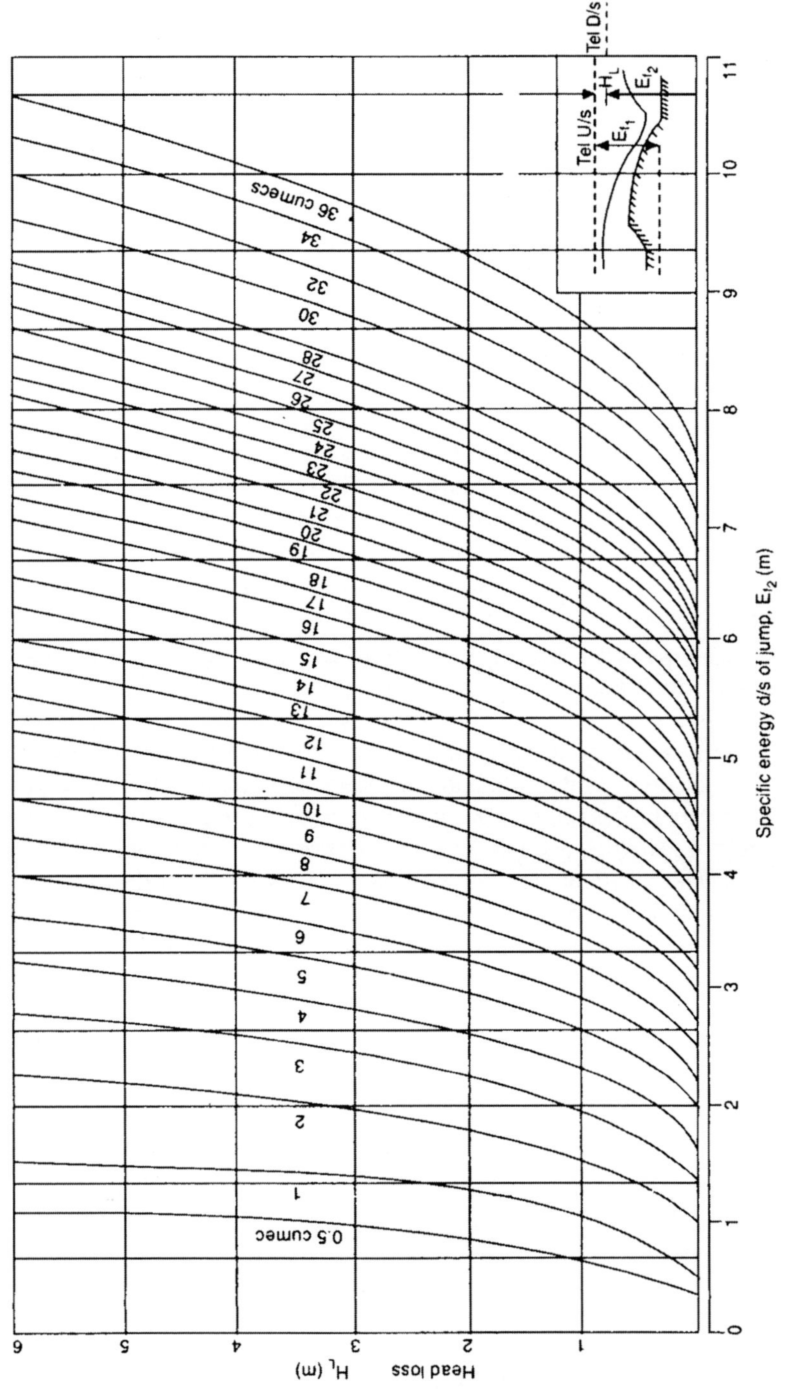

Fig. 10.15 Blench curves

(*ii*) E_{f_1} (*i.e.*, energy of flow just u/s of the point of formation of standing wave) = $E_{f_2} + H_L$. Hence, for various discharge values E_{f1} are calculated.

(*iii*) For the known values of E_{f1} and q, using Montague curves (Fig. 10.15), the values of D_1 and D_2, *i.e.*, hypercritical and subcritical depths of water before and after the hydraulic jump are read out.

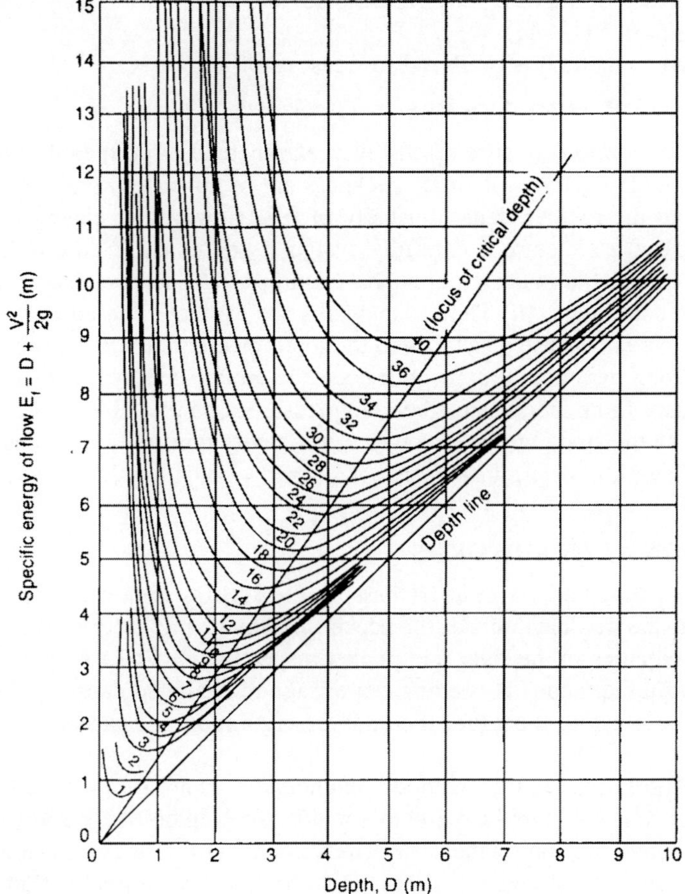

Fig. 10.16 Montague curves.

(*iv*) The required cistern level is equal to the retrogressed water level for that discharge minus D_2. The cistern length is 5 $(D_2–D_1)$.

The energy dissipation devices are generally in the form of (*i*) a continuous end sill or Rehbock's dentated end sill, (*ii*) arrows, (*iii*) raised blocks, (*iv*) a roughened or toothed floor, and (*v*) a combination of two or more of these forms. U.S.B.R. standaradised stilling basin designs discussed in the chapter 'Spillway' with one set of chute blocks at the toe of downstream glacis are usually provided.

2. By Analytical Method

(*i*) With known T.E.L u/s and d/s, an arbitrary cistern level for the particular discharge is initially assumed.

(*ii*) E_{f_1} = U/s T E.L.– assumed cistern level.

(*iii*) From the known values of E_{f_1} and q with concentration, D_1 is calculated from the relationship, $E_{f_1} = D_1 + q^2/2g\ D_1^2$.

(*iv*) Pre-jump Froude's member F is calculated from the relationship $F^2 = q^2/g\ D_1^3$.

(*v*) From the calculated values of D_1, and F_1, D_2 is calculated from the relationship,

$$D_2 = \frac{D_1}{2}(-1+\sqrt{8F^2+1}).$$

(*vi*) The required cistern level is equal to the retrogressed water level for that discharge minus D_2.

(*vii*) The various steps above are repeated till the assumed and calculated cistern levels are the same. The exact values of D_1 and D_2 are thus, known.

(*viii*) Cistern length is 5 (D_2-D_1).

Generally analytical method is preferred for reasons of accuracy.

10.15 BARRAGE ON CLAYEY STRATA

Where barrage foundation consists of clayey strata, instead of permeable foundation as discussed heretofore, the clay patches of small thickness are removed and replaced by sandy material and compacted to the relative density usually of 0.7. However, if clayey material is extensively present, the barrage structure is carefully designed and constructed on clayey foundation. The various factors considered in design of such a barrage are (*i*) The waterway to be provided is governed by the site conditions, (*ii*) The scour depths are estimated based on the soil properties overlying the clayey strata because resistance to the scour offered initially by the cohesion of clay is overcome over a long period and total scour could be anticipated, (*iii*) Adequacy of the structure is ensured against sliding, permissible bearing pressures and probable settlements due to load conditions imposed on the strata by the structure, (*iv*) Gravity type of floor is desirable instead of raft type, (*v*) Floor length is determined from surface flow considerations since exit gradient in clayey soil has no relevance.

10.16 REGULATION AT HEADWORKS

The regulation at headworks aim at (*i*) Passing the surplus high discharge carrying heavy sediment load through the weir bays outside the pocket, thereby giving V_R/V_p a value higher than 1, V_R and V_p denote the velocities in the river and pocket respectively, (*ii*) Passing comparatively silt-free water into the offtakes, and (*iii*) Feeding the offtake to designed capacity. From the consideration of sediment exclusion, the usual methods of regulation at headworks are discussed as under:

1. Still Pond Regulation: In this method, the undersluices are kept closed when the canal is running. In this way only the supply required for the canal enters the pocket and the balance escapes through the other section of the weir. The velocity in the pocket is thus very much reduced there being excessive waterway. In the river, slope towards the pocket flattens, decantation takes place and sediment in suspension settles down in the approach channel and relatively clear water enters the canal. The sediment is free to accumulate in the pocket. When it reaches to within about 0.6 m of the crest level of the canal regulator, the canal is closed for about 24 hours, undersluices are opened to scour out the deposited silt in the pocket. The canal is then opened and the undersluices are closed. The process is repeated when flushing off of the accumulated silt is again warranted, say one day each month.

Still pond regulation is the most efficient method for the headworks which are not provided with silt excluders. It is superior to other two methods described hereinafter and is generally resorted to as far as possible. The method, however, has the drawbacks of (*i*) Forced closure of the canal and consequent loss of irrigation, and (*ii*) Loss of discharge passed down the undersluices.

Another technique in the method is that the undersluices are kept closed and the river is allowed to develop a curved approach channel by escaping excess supply over the other portion (than undersluices) of the barrage or by providing another set of undersluices called *river sluices* in the middle of the barrage, adjoining the undersluices. The second pocket creates favourable curvature by forcing more discharge through the river sluices, outside the pocket divide wall. There is cross flow of the silt in the river channel section on a curve resulting in low silt charge on outside of the curve than on the inside. The relatively silt free water is further to undergo the still pond regulation before entering the canal.

2. Semi-open Flow Regulation: In this method of regulation, the undersluice gates are partially opened to escape the surplus water entering the pocket in front of head regulator while the canal is drawing the top water from the pond. It is considered that coarse sediment concentrate near the pocket floor and can be escaped through the undersluices. The separation of the water entering the pocket into two streams, one flowing towards the canal head and the other escaping down the undersluices maintains such a velocity in the pocket which causes the silt to remain in suspension and the turbulence so created enables even some of the coarser sediment to rise up against the crest of regulator and may find entry into the canal. It implies that the crest of canal regulator should be sufficiently high. Silt content of the canal water is observed daily and if it exceeds a certain limit, the accumulated silt in the pocket is flushed down by raising the undersluice gates. The method has the merit that the canal has not to be closed and the operation of feeding the canal and scouring of silt through undersluices is continuous.

An improvement in the method applicable to the canals taking off from both the banks of the barrage is *double wedge system* wherein gate opening of undersluices is minimum at centre, increasing gradually towards the divide walls. The wedge method may be either wedge from the right (maximum opening at the right end bay decreasing towards the left) or left or centre.

3. Open flow Regulation: In the method, the undersluices are kept open to escape supply in excess of the canal discharge. The canal draws top water from the pocket which implies that the crest of regulator is high. It is considered that the coarse sediment concentrates near the pocket floor and escapes through the undersluices. However, this method is conducive to comparatively more silt entry into the canal. It has the merit that the canal is not to be closed for scouring the sediment but is practicable only in case there is ample inflow *vis-a-vis* canal discharge.

10.17 EFFECT OF WEIR ON RIVER REGIME

Barrage or weir is essentially an obstruction in the normal flow of the river and effects changes in the river regime both upstream and downstream. The sequence of the effect of barrage on river regime may be (*i*) The river water carries its sediment load but as a result of construction of barrage, the supply drawn by the canal and the water past the weir is also relatively silt free resulting in silt deposition upstream, (*ii*) The heading up of water upstream of a barrage results in reduced velocity and consequent deposition of bulk of the sediment load in the pond resulting in formation of irregular shoals upstream of the barrage, (*iii*) The water surface slope of the river upstream of the barrage flattens, *i.e.*, accretion starts upstream and degradation or retrogression downstream for the first 4 or 5 years depending on the initial heading up over the natural water levels, (*iv*) The progressive silting increases tortuosity and formation of shoals in the upstream tends to rise as the bed levels at the barrage are fixed. The resistance to flow of river increases to overcome which increased head is required and as a consequence there is a tendency on the part of the river to regain its original slope, and (*v*) A stage is then reached when the pond absorbs no more silt burden and the excess of silt is passed down the barrage with the reduced discharge. It results in progressive silting up downstream, an increase in tortuosity and hence a recovery of the bed level downstream.

The accretion on the upstream and retrogression on the downstream is of great importance in the design of a barrage or weir. The accretion upstream if not duly provided for or controlled may necessitate subsequent raising of the afflux bund. Likewise the excessive lowering of water levels in the downstream, excessive than anticipated, and not controlled may make it difficult to maintain the formation of hydraulic jump at the toe of glacis resulting in major damage to the structure. The effects of degradation are (*i*) Lowering of flood levels, (*ii*) Damage to foundations of structures, (*iii*) Materially affecting the functioning of dissipation devices due to reduced tailwater level, and (*iv*) Lowering of watertable.

10.18 FAILURE OF A BARRAGE OR WEIR

Failure of a barrage or a weir may be caused singly or in combination due to the subsurface flow, surface flow, besides other factors briefed as follows:

1. Subsurface flow

(1) *Piping*: Piping or floatation failure is caused when the exit gradient exceeds the critical value for the soil and as a consequence the subsurface soil starts boiling and is washed away by the percolating water. The process of erosion progresses backwards towards the upstream side and results in the formation of a pipe under the floor causing its failure. Piping is caused due to (*i*) inadequate depth of downstream cutoff, (*ii*) Absence and/or non-functioning of inverted filter on the downstream of the floor, (*iii*) Short circuiting of seepage flow carrying the foundation soil through the cracks, and (*iv*) Cavities in the floor. The remedy lies in (*i*) Providing a deeper cutoff on the downstream and provision of properly designed inverted filter, or (*ii*) Increasing the floor length on the upstream to increase length of percolation and hence reduction in exit gradient but there should be no short circuiting of seepage path.

(2) *Eruption of floor due to excessive uplift pressure*: This failure occurs when floor thickness prόvided in trough portion is inadequate to counteract uplift pressure. The remedy lies (*i*) in increasing the thickness of floor with due consideration for adequate cistern level for proper energy dissipation, (*ii*) Provision of reinforced cement concrete floor raft instead of gravity section, (*iii*) Grouting of cavities formed, if any, and (*iv*) Increasing upstream floor length so as to reduce pressure especially over the trough region by increasing the seepage path.

2. Surface Flow

(1) *Formation of deep scours*: Failure occurs if cutoffs both on the upstream and downstream provided are of inadequate depth *vis-a-vis* deep scours caused by high intensities of discharge through the structure. This may be due to inadequate waterway, faulty gate regulation or improper energy dissipation. The remedy lies (*i*) In lowering the cistern level, (*ii*) Providing adequate energy dissipation arrangement, and (*iii*) Provision of the upstream and downstream cutoffs of adequates depths with requisite aprons.

(2) *Suction caused by standing wave*: Hydraulic jump may form outside the floor and the suction associated with it may cause failure of the floor. The remedy lies in taking cognisance of the retrogression on the downstream of the bed and provision of adequate floor thickness to counteract suction or negative pressure due to standing wave.

(3) *Faulty regulation:* Failure may occur due to faulty regulation of gates or high concentration of flow through the structure. It can be prevented by proper regulation and proper maintenance of gates mechanism.

(4) *Faulty construction*: It may be stratification of concrete layers in the impervious floor and absence of proper bond between the layers of concrete, inadequate cover over the reinforcement bars, improper concrete mix, improper foundation treatment and other construction and structural design defects.

Overtopping: Failure due to overtopping of the components of the diversion structure occurs mainly due to inadequate design flood and maximum flood for free board and afflux. Remedial, rather preventive measures, consist in the correct assessment of design flood for waterway and maximum flood for free board, also taking into account siltation aspect.

(5) *Outflanking*: Failures due to outflanking are related to (*i*) Wrong estimation of flood values and subsequent overtopping of afflux bunds, etc., (*ii*) Wrong location of the diversion structure, and (*iii*) Improper gate operation.

10.19 FOUNDATION DESIGN OF WEIRS AND BARRAGES

The design of a weir or a barrage consists of hydraulic design, and structural design. The hydraulic design implies hydraulic analysis of the conditions under which a weir or a barrage is to work which in turn means analysis of discharges, velocities, water levels, water slopes and subsoil formation. Thus the hydraulic design consists of hydraulic calculations for fixing various elevations such as high flood level, pond level, crest level, etc., waterway of the weir or barrage and that of undersluices, river training works appurtenant to the work. It also includes hydraulic calculations governing the design of the work in relation to the subsurface flow of water through the permeable subsoil as also for the flow of water over the surface. The design of work for subsurface flow and surface flow consists of the determination of hydraulic forces developed by water seeping through the foundation, and acting on the structure and configuration of the structure which is economical and has functional efficiency.

1. *Subsurface Flow:* A weir or a barrage founded on impervious foundation has the only force exerted by the uplift pressures. The impervious subsoil offers so much resistance to the flow of water that the head of water (due to the water stored against the work) is largely dissipated in overcoming friction before downstream end of the structure is reached. As such little significance is attached to the seepage force exerted by percolating water as it is not likely to damage the foundation rock. Since rocky foundations are not available in the rivers in plains where a weir or a barrage is usually constructed, the structure is usually constructed on pervious alluvial soil. The pervious river bed offers little resistance to the seeping water which may travel up to the downstream end of the structure without substantial loss of head. The percolating water may bring soil particles with it leading to failure of the structure. Thus, the governing factors in the design of the structure for subsurface flow are the exit gradient and uplift pressures which depend on the total head and the foundation profile of the work.

2. *Surface Flow:* In surface flow there is the additional factor of the dynamic impact and the designs in this respect depend on afflux, changes in regime of the river caused by the construction of the work, discharge intensity, flood, scour and position and depth of the trough of the standing wave.

In essence hydraulic design of the structure of a weir or a barrage on pervious foundation consists of general arrangement of the work and dimensioning of the weir profile, checking the profile for safety against uplifting of the floor and undermining. The substructure generally consists of continuous concrete foundation and floor, cutoffs, apron, block protection in the upstream and downstream of the work. The design of superstructure is usually based on common theories of structural engineering and is not discussed as such.

The principles of design of weirs on permeable foundation apply to that founded on loam, clay, gravel, etc. The general consideration in the design of weirs in pervious soils like sand is that the water forced through the soil due to head upstream of the weir has safe exit velocity as it emerges at the downstream end, *i.e.*, there should be safe length of the line of creep. Law of flow of water through permeable soil, as under, enunciated by Darcy signifies that the velocity of flow varies directly as the head and inversely as the length of path of flow, that is

$$V = KH/L$$

where, V = velocity, H = head, L = length of path of flow and K = a constant, called transmission constant.

Further investigations with regard to validity of the law in relation to weir design, led to the development of two important theories; Creep theory and potential theory. These theories assume the soil to be homogeneous and the depth of pervious stratum infinite.

10.20 CREEP THEORY

1. Bligh's Creep Theory

Water creeps along the base profile of the weir at its joint with the subsoil due to head of water across the work. The percolating water loses its head enroute and finally comes out at the downstream end of the work. According to Bligh, creep is more important and destructive in undermining a weir than percolation through the soil below it. Other salient features of the theory are (*i*) Hydraulic gradient line or slope is constant through the impervious floor length of weir. It is an imaginary line which joins upstream and downstream water levels, (*ii*) Water percolation starts from the upstream joint between the base profile of the floor and the subsoil and emerges out at the downstream end of the downstream cutoff. The total length traversed by the percolating water is termed as *creep length,* (*iii*) Percolation flow, instead of following the shortcut route indicated by long arrows in Fig. 10.17., hugs the line of contact of base profile of weir and permeable soil as indicated by dotted lines, (*iv*) Creep length is the sum of the horizontal as well as vertical lengths of creep, the length of creep along each cutoff being twice its depth as the water travels along both sides of the depth of cutoff, (*v*) Percolating water loses head enroute, proportional to the length of path traversed irrespective of the fact whether the path is horizontal, inclined or vertical. There is no distinction between horizontal, vertical or inclined path, (*vi*) No amount of sheet piling or cutoff can stop the percolation unless it extends up to the impermeable formation, (*vii*) Vertical cutoff at the downstream end of the floor is not a structural necessity, and (*viii*) Percolating water emerges at the downstream end of the floor with a velocity known as exit velocity. In order that the exit velocity is safe, the creep length should be sufficient. To ensure safety of the weir against piping and uplift pressure, Bligh suggested the formulae, discussed below:

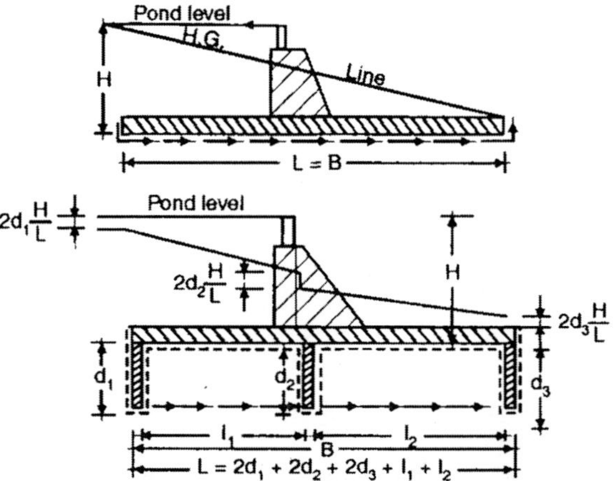

Fig. 10.17 Bligh's creep theory.

1. *Against Piping:* The weir is safe against piping if the percolating water has negligible upward pressure as it emerges at the downstream end of the floor, which implies that the length of creep should be sufficient to provide a safe hydraulic gradient according to the type of soil. The structure is safe if the average pressure gradient, defined by the ratio of the total head to the total creep length, does not exceed a critical ratio $1/c$, where c is a constant depending on the size of the bed sand. He gave the formula for creep length as follows:

$$L = CH$$

where, L = creep length $(2d_1 + l_1 + 2d_2 + l_2 + 2d_3)$, C = creep coefficient, value as under:

Type of soil	Gravel	Boulder or shingle & gravel & sand mixed	Coarse grained sand (central and south Indian rivers)	Fine micaceous sand (north Indian rivers)	Light sand and silt
Value of C	5	5–9	12	15	18

H = Total head over the weir, *i.e.* seepage head reckoned from the highest headed up water level upstream to the corresponding lowest water level downstream, *i.e.*, downstream bed level under nil supply condition.

For no piping, $H/L = 1/C$ or hydraulic gradient $\leq 1/C$.

2. *Against Uplift Pressure:* Computation of uplift pressures underneath the floor at various points is of primary importance. At any point of the weir, the uplift pressure due to subsurface

flow is represented by the ordinate between the hydraulic gradient line and the weir floor level. Corresponding to the ordinate H_1 at any point between HG line and bottom of floor the uplift pressure exerted by the water at that point is wH_1 (Fig. 10.18). Assuming floor thickness (t) and specific gravity of the floor material (ρ), the thickness of floor is given by

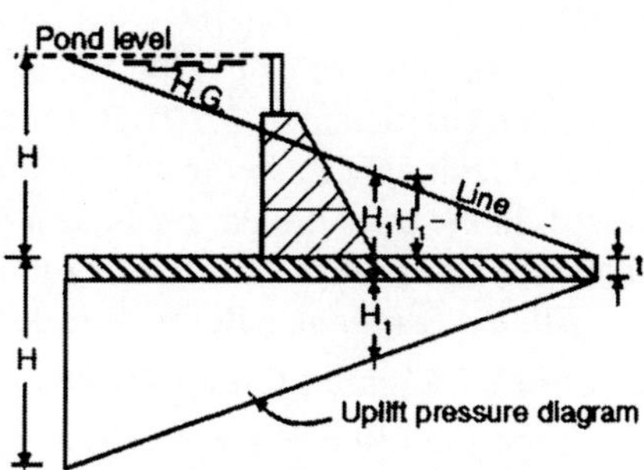

Fig. 10.18 Uplift pressure diagram

$$t = (H_1 - t)\,/\,\rho - 1$$

or $t = 4/3\,(H_1 - t)/\,\rho - 1$ assuming a factor of safety 1.33 where $H_1 - t$ = ordinate length between HG line and top of floor, $\rho - 1$ = submerged specific gravity of floor material.

It is evident that (*i*) If there is no water upstream of the weir crest or barrier there is no seepage head and consequently no uplift pressure, (*ii*) The uplift pressure upstream of the barrier, is more than counterbalanced by the weight of standing water on the floor and as such minimum floor thickness to resist wear, impact of flowing water, etc. shall do, (*iii*) The floor downstream of the crest is thicker to resist uplift pressure (assuming nil supply downstream), (*iv*) It is economical to provide as much of the total creep length upstream of the crest as possible. The minimum floor length is governed by the consideration of surface flow to resist the action of fast flowing water past the crest. Large floor length upstream of crest also reduces the uplift pressures on the downstream portion of the floor, and (*v*) Upstream sheet pile reduces the uplift pressure all over the floor whereas the downstream one increases them (Fig. 10.19).

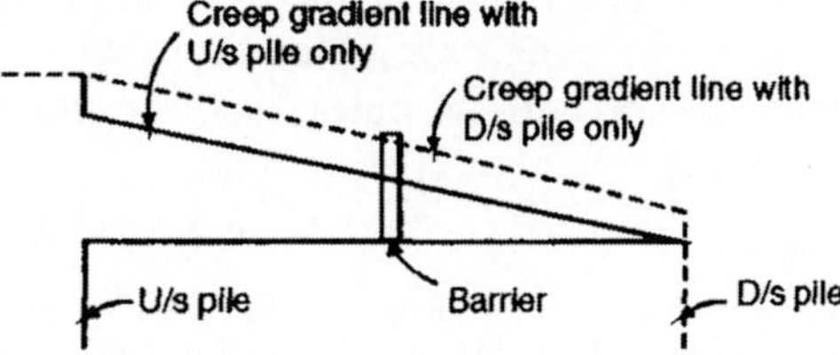

Fig. 10.19 Effect of d/s pile on creep gradient.

Limitations of Bligh's Theory: Because of its simplicity, Bligh's creep theory found general acceptance. Some works constructed on the basis of this theory failed while others stood, depending on the extent to which the importance of upstream and downstream vertical cutoffs was ignored. Limitations of this theory, revealed as a result of later investigations, are (*i*) Danger in application of this theory lies in the statement that the length of path of flow has the same effectiveness, length for length, in reducing uplift pressures whether it is along the horizontal or the vertical because loss of head does not take place in the same proportion as creep length, (*ii*) Uplift pressure distribution is not linear but follows a sine curve, (*iii*) It made no distinction between horizontal and vertical creep, (*iv*) It considered vertical cutoff at the downstream end not a structural necessity which is essential to prevent undermining, (*v*) Absolute necessity of providing sheet pile at downstream end is not emphasised. Undermining of floor starts from the downstream end and sheet pile at this end is absolutely essential, (*vi*) No distinction is made between the inner and outer faces of sheet piles whereas according to later investigations outer faces of end sheet piles are more effective than the inner ones. Similarly the role of intermediate sheet pile was not considered which if smaller in length than the outer ones are ineffective except for local redistribution of pressures, (*vii*) The theory holds good so long as the horizontal distance between cutoffs is greater than twice depth, (*viii*) Necessity of keeping exit gradient below critical value is not given due importance, and (*ix*) According to Bligh, the safe value of C cannot be determined from direct experiments but only from observations on actual structures.

Anomalies

(*i*) Assignment of coefficient of 2 to the depth of cutoff is far from being too large. Lane suggested coefficient of 3, (*ii*) Bligh's coefficients are applicable for a definite criterion of design and are significant in so far only as they define the relative efficiencies of vertical and horizontal creeps. In the piles driven too close a distance from one another, line of creep follows the straight shortcut path, instead of following the base profile and vertical depths of cutoffs.

Bligh's Formulae: Refer weir design section 10.24.

2. Lane's Weighted Creep Theory

Lane modified Bligh's creep theory and enunciated Lane's weighted creep theory. According to this theory (*i*) Weighted creep length is the sum of vertical (steeper than 45°) creep distances, (*ii*) Weighted creep head ratio is the weighted creep distance divided by the effective head, (*iii*) Safe weighted creep head ratio may be reduced by 10% if reverse filter, drains, weep holes and pipe drains are used as a measure against underseepage, and (*iv*) Upward pressures are estimated by assuming that the drop in pressure from head water to tail water along the contact line of the dam and foundation is proportional to the weighted creep distance. Thus, Lane's weighted creep theory, in essence, may be called Bligh's creep theory corrected for vertical cutoff and sloping faces. Lane suggested the following formula for computing effective creep length.

$$L_w = V + 0.33\,H, \text{ such that } L_w \text{ is greater than } CH$$

Where, L_w = weighed creep length, H = sum of all horizontal contacts and of all the sloping contacts less than 45° (to the horizontal), and V = sum of all vertical contacts greater than 45°.

Further, $L_w = C_l H$, to ensure safety against piping

Where, L_w = length of floor (m) and C_l = creep coefficient, value depending on the type of soil as given in Table 10.3.

Table 10.3 Lane's creep coefficient values

Type of soil	C_l	*Lane's safe hydraulic gradient ($1/C_l$) not exceeding*	*Type of soil*	C_l	*Lane's safe hydraulic gradient ($1/C_l$) not exceeding*
Very fine sand or silt	8.5	1/8.5	Coarse gravel including cobbles	3.0	1/3.0
Fine sand	7.0	1/7.0	Boulders with some cobbles and gravel	2.5	1/2.5
Medium sand	6.0	1/6.0	Soft clay	3.0	1/3.0
Coarse sand	5.0	1/5.0	Medium clay	2.0	1/2.0
Fine gravel	4.0	1/4.0	Hard clay	1.8	1/1.8
Medium gravel	3.5	1/3.5	Very hard clay or hardpan	1.6	1/1.6

Lane's theory though marked an improvement on the Bligh's theory is empirical and lacks the background for a rational basis for the design of weirs. The factor of safety of works designed on the basis of Lane's theory is uncertain. Cutoffs have to be provided to take care of scour at the downstream and also necessary filter layer to take care of unrelieved gradient at the tail.

Limitations of Creep Theory. The methods under this theory are not satisfactory for determining uplift pressure and for check against heaving. However, where there is possibility of loss of contact between the structure and foundations, and thin weak layers or seams of high permeability are present, only creep theory which is based on performance studies can safeguard against possible failure due to seepage erosion.

10.21 DESIGN BY FLOW NET

In the hydraulic design of structures involving subsoil flow and certain cases of surface flow a diagram known as the flow net is constructed. It furnishes the pattern of stream lines and the equipotential lines which form two systems of lines mutually cutting at right angles and dividing the area enclosed by the boundaries of flow into curvilinear squares. The flow net is very useful in design for computing uplift pressures, exit gradients, pore pressures and seepage quantities in subsoil flow and in surface flow for predicting cavitation effects, etc.

In constructing flow net, the hydraulic boundary conditions of the problem and their effect on the shape of the flow lines are considered. The bottom profile of the weir represents the top-most and the shortest stream line and the top surface of the foundation rock represents the lowest flow line and also the longest. The other stream lines lie between these two and their shapes must represent a gradual transition from one to the other. The upstream and downstream horizontal ground surfaces represent equipotential lines and hence the stream lines at inlet and exit should have vertically downward and upward directions respectively. Refering to Fig. 10.20 a chosen number of stream tubes, 4 in this case, are sketched. A system of equipotential lines are then drawn to meet the former lines at right angles and approximately forming squares. The shape of the net is rectified by trials by adjusting the spacing of stream lines and equipotential lines.

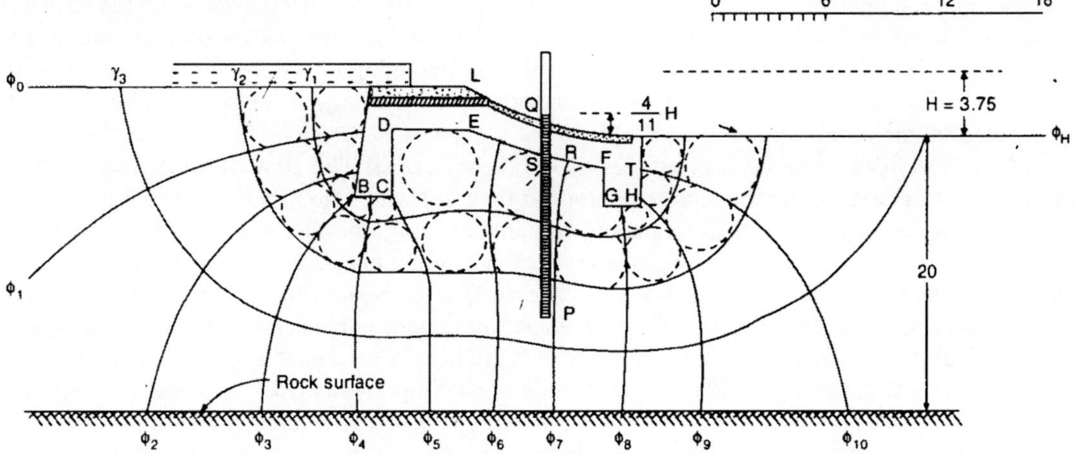

Fig. 10.20 Flow net for Godavari weir.

In using flow net for the design of a structure, the equipotential lines mean that the hydraulic potential for all points on that line is the same. With regard to Fig. 10.20, if the piezometer tappings are taken for a number of points along the equipotential line ϕ_7 all of them will rise to a common level. The exact height to which monometric column will rise indicated by Q is determined as follows:

The total potential or head is 3.75 m. In the flow net there are 12 lines or say 11 equipotential steps. Along the equipotential line ϕ_7 the residual head will be that due to the 4 potential drops to follow. The residual potential along $\phi_7 = (4/11)\ H = (4/11)\ 3.75 = 1.36$ m of water. Thus, the piezometers installed at different points along ϕ_7 will rise to the elevation of Q which is 15.3 m above the downstream sand face.

Uplift Pressure: In determining uplift pressures, the point where the various equipotential lies meet the bottom profile of the weir are noted. The equipotential line ϕ_7 meets the profile at R.

Since at this point the hydraulic potential is equal to 4/11 H and is represented by the elevation of Q, the uplift pressure at R is given by the manometric column SQ, where R and S are points on the same horizontal line. There will be equilibrium at R if the weight due to the thickness of the floor at R is either equal to or in excess of the uplift head at the point. Thus minimum floor thickness required is $3.3/2.24 \simeq 1.5$ m, 2.24 being the relative density of masonry and 3.3 is uplift head (in m) of water.

Exit Gradient: To ensure safety of a structure against piping, the pressure gradient at the exit or the floatation gradient should not be greater than unity theoretically but in practice a minimum factor of safety of 5 to 7 depending on the nature of the subsoil material is allowed. In Fig. 10.20, the equipotential line ϕ_{10} meets the weir profile at point T. The hydraulic potential at T = residual head at the point = $H/11 = 3.75/11 = 0.34$ m of water. The elevation of T below the downstream sand face is 2 m. Exit gradient = $0.34/2 < 1/5$ and hence safe.

10.22 POTENTIAL THEORY

The modern approach to the theory of weir design is that seepage takes place according to the theory of seepage flow throughout the underlying strata, and stability of granular particles depends on the limiting value of hydraulic gradient at the upper surface of the granular material which gradient to satisfy the consideration of equilibrium is to be smaller than the critical gradient. The important potential theory is due to Khosla. Dr. Khosla and his associates, on the basis of field experiments and hydraulic model tests, showed that the loss of head is nonlinear along the creep length. Therefore, creep theory is not reliable for predicting uplift pressures. He explained the nonlinear loss of head along the creep length satisfactorily through analytical results obtained from potential theory for simple weir profiles.

Main Principles: The main principles of the theory are (*i*) The flow of water through the subsoil does not creep along the bottom contour of the impervious floor but is in streamlines and hence susceptible of mathematical treatment, (*ii*) The creeping of water along the bottom of the impervious floor can be prevented by providing vertical cutoffs one each at u/s and d/s ends of the impervious floor, and a few intermediate vertical cutoffs in between, (*iii*) The undermining of the subsoil starts from the d/s end of the d/s impervious floor and if not checked it travels u/s towards the weir wall. It is, therefore, essential that the exit gradient is safe for the subsoil and a reasonably deep vertical cutoff is provided at the d/s end of the impervious floor to prevent undermining. The depth of the cutoff is to be safe against maximum anticipated scour and safe exit gradient calculated for the subsoil on which the weir is founded, (*iv*) Subsoil water has a certain residual force at each point along its flow through the subsoil and acts in the direction of flow, tangential to the path of flow and is proportional to the pressure gradient at that point. At the d/s end of the d/s impervious floor this force is obviously upwards and tends to lift up the soil particles if it is more than the submerged weight of the subsoil. Once the surface particles are disturbed, the resistance against upward pressure of water is further reduced, tending to progressive disruption of the subsoil. To safeguard against undermining, the exit gradient is not allowed to exceed a certain safe limit as per Table 10.4. It can be read from Fig. 10.21. Liberal factor of safety has to be allowed to critical value of exit gradient to obtain safe exit gradient for the safe design of structure.

Table 10.4 Safe exit gradient

Kind of subsoil	*Factor of safety*	*Safe exit gradient*
Shingle	4 to 5	0.25 to 0.20
Coarse sand	5 to 6	0.20 to 0.17
Fine sand	6 to 7	0.17 to 0.14

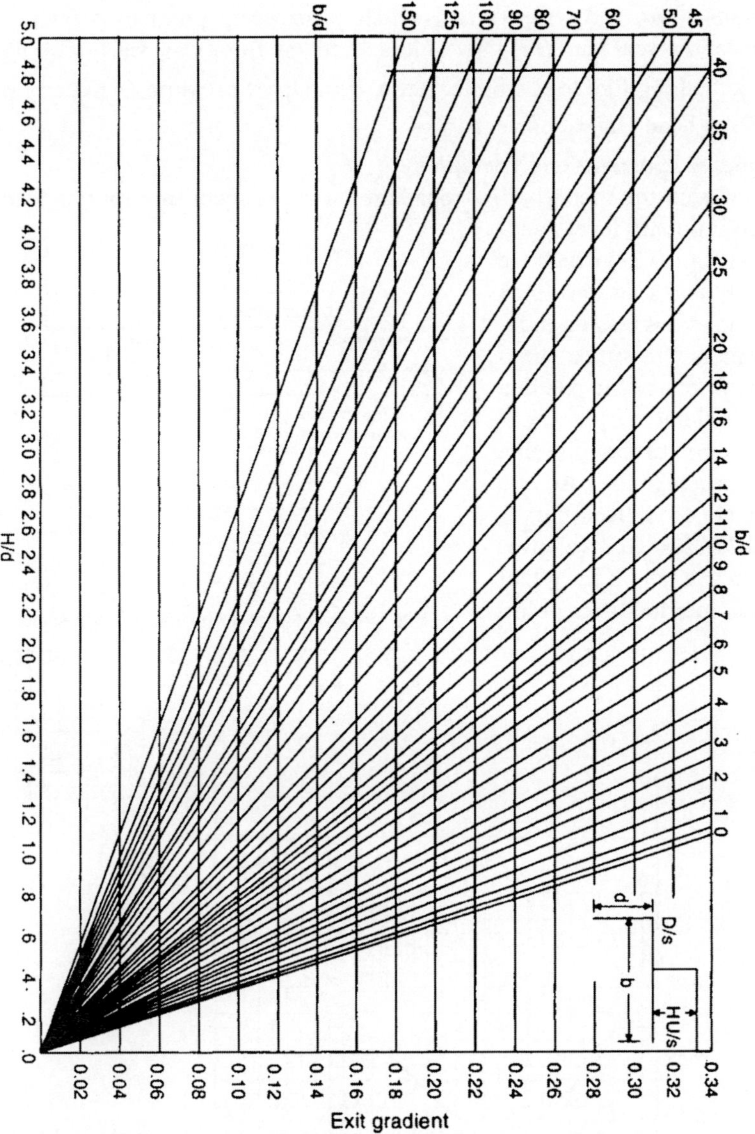

Fig. 10.21 Exit gradient.

Conclusions of Khosla's Theory. In repudiation of Bligh's creep theory, Khosla arrived at certain provisional but important conclusions (*i*) Outer faces of the end sheet piles were much more effective than the inner ones and the horizontal length of floor, (*ii*) Intermediate piles, if smaller in length than the outer ones, were ineffective except for local distribution of pressures, (*iii*) Undermining of floors started from the tail end. If the hydraulic gradient at exit was more than the critical gradient for the particular subsoil, the soil particles would move with the flow of water, thus causing progressive degradation of the subsoil, resulting in cavities and ultimate failure, (*iv*) It was absolutely essential to have a reasonable deep vertical cutoff at the downstream end to prevent undermining. As a result of further investigations, the facts established were (*a*) The flow of water through the subsoil is in streamlines and therefore, susceptible of mathematical treatment, (*b*) The ratio (ϕ) of uplift pressure (P), at any point along the base of particular weir founded on permeable soil to the total head (H) is constant and independent of head (H), class of subsoil so long it is homogeneous, upstream and downstream water levels, and temperaimpervious floor, and temperature, which varies from point to point in subsoil and in different seasons of the year, and (*iii*) Law of loss under the floor was nearly a straight line and for the sheet piles something like logarithmic. Ratio ϕ = Uplift pressure (P) at any point along the base of weir / Total head (H) = constant.

Khosla's Method of Independent Variables

Khosla introduced the theory of independent variables according to which the complicated weir profile is split up into its elementary components (Fig. 10.22) which are amenable to analytical solution. Each elementary component is treated independent of the others. Pressure at junction points of the floor and the pile line of that particular elementary component, the bottom point of that pile line and the bottom corners in the case of a depressed floor are read from curves (Fig. 10.23 to 10.24). Fig. 10.24(*a*) gives pressures at points *C, D* and *E* for a floor with pile at some intermediate point.

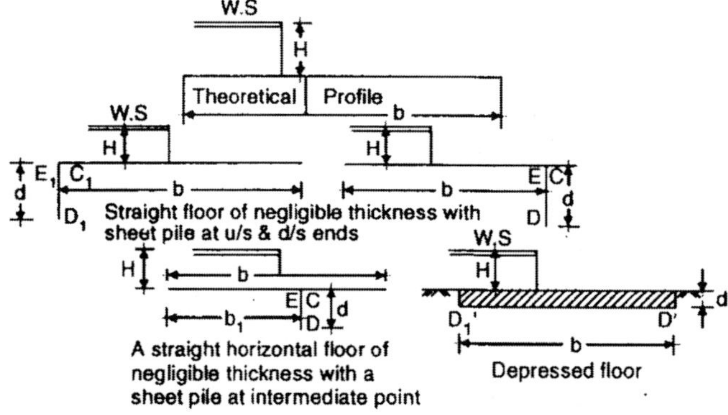

Fig. 10.22. Elementary profiles of Independent variables.

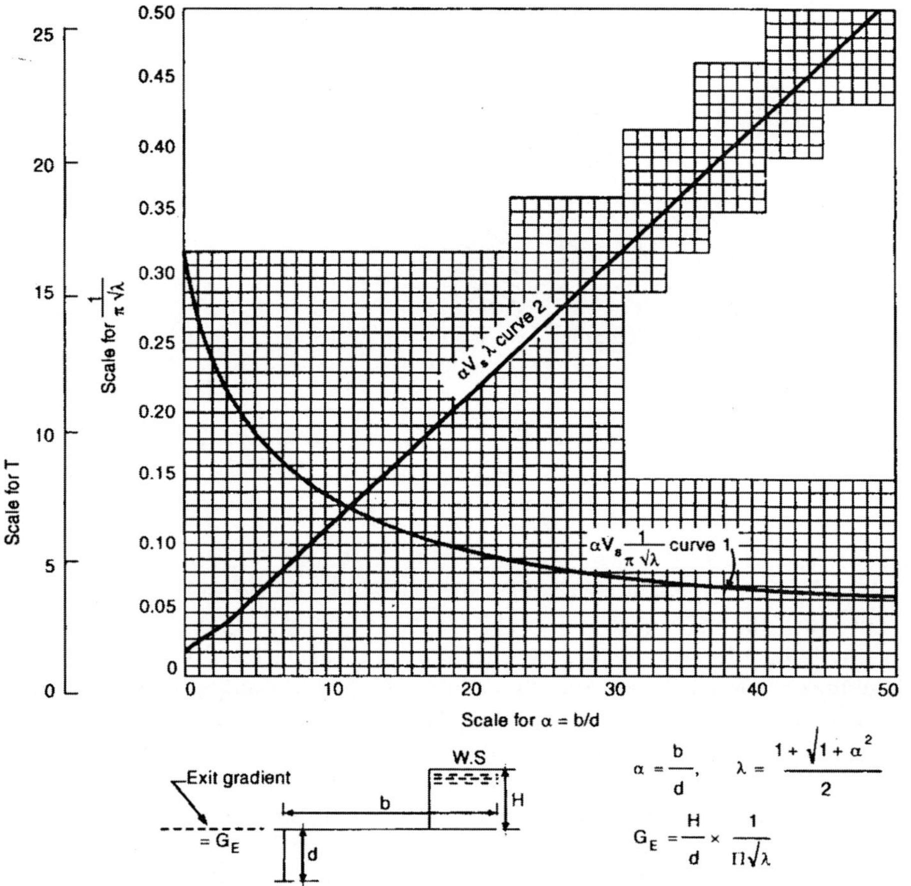

Fig. 10.23 Khosla's safe exit gradient curve.

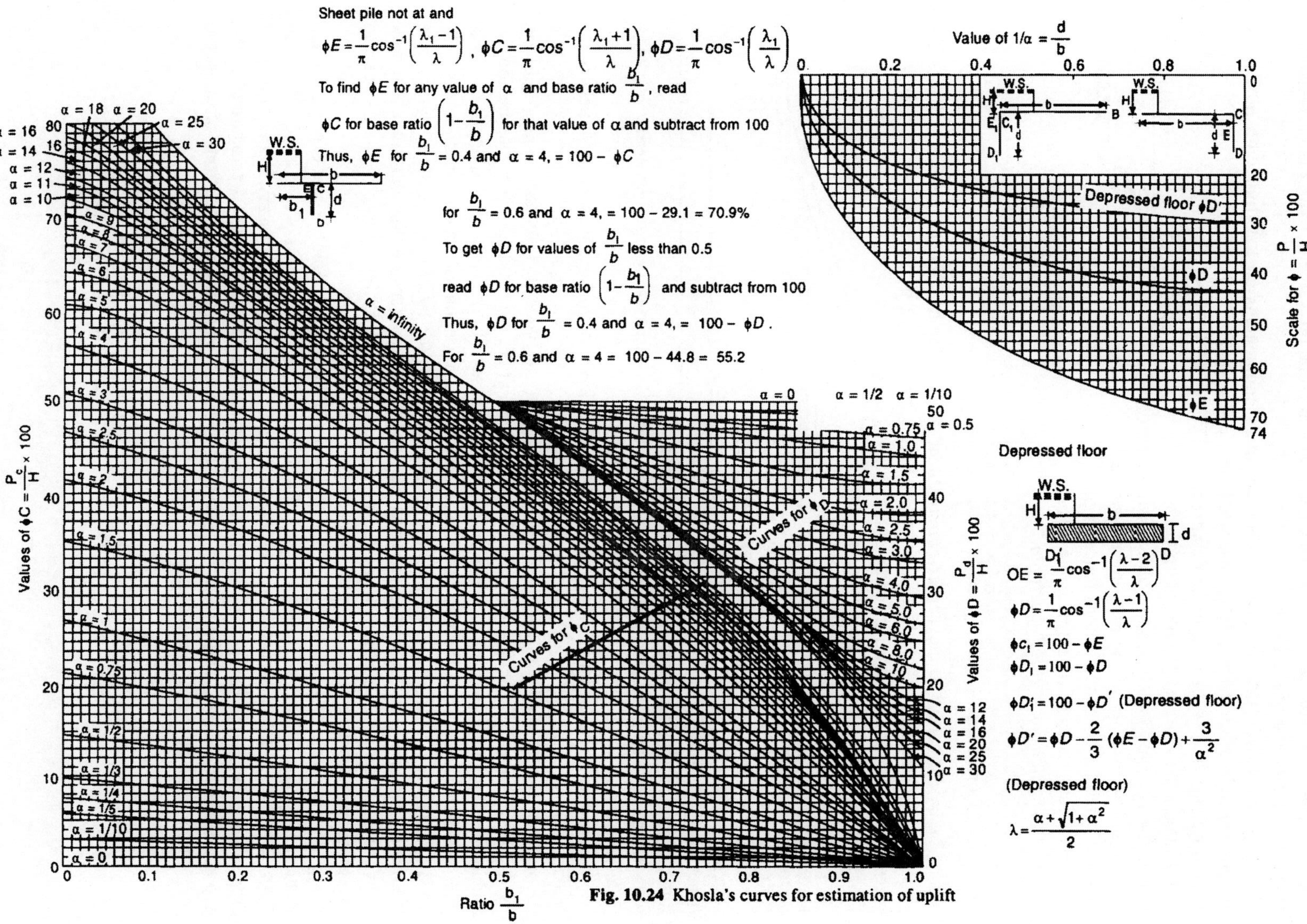

Fig. 10.24 Khosla's curves for estimation of uplift

Figs. 10.23 and 10.24(*b*) and give pressures at *C, D* and *E* and the value of exit gradient when pile is at downstream end of the floor.

In the usual weir profile (Fig. 10.25) elementary components are (*i*) floor with pile at upstream end, (*ii*) floor with pile at intermediate point, and (*iii*) floor with pile at the downstream end. In these the floor length of negligible thickness is the common factor and the piles form the variables. The readings at the junction points are then corrected for (*i*) mutual interference of piles, (*ii*) floor thickness and (*iii*) slope of the floor.

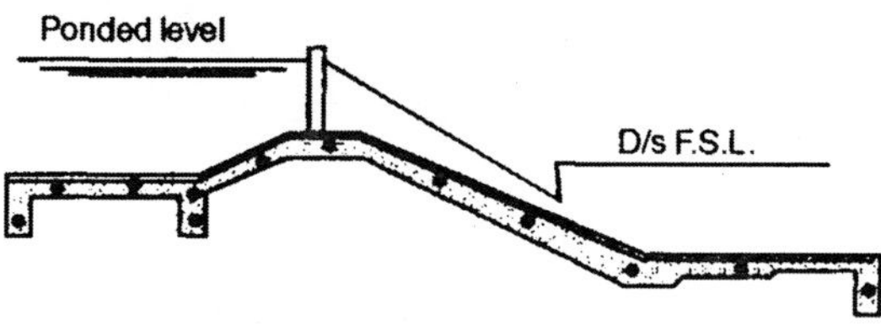

Fig. 10.25 Usual weir profile.

1. *Correction for mutual interference of piles*

$$\text{Correction } C = 19\sqrt{\frac{D}{b'}}\left(\frac{d+D}{b}\right)$$

where, C = correction at the corner, where sheet pile of depth d meets the floor, due to the presence of sheet pile of depth D. It is applied as a percentage of sheet pile of depth D and percentage of head. It is additive for points in the rear or backwater and subtractive for points forward in the direction of flow, b_1 = distance between the two piles, D = depth of pile whose influence has to be determined on the neighbouring pile of depth (d), d = depth of pile on which the effect of pile (D) is required to be determined. Both d and D are measured below the level at which interference is required to be determined (Fig. 10.26), and b = total floor length.

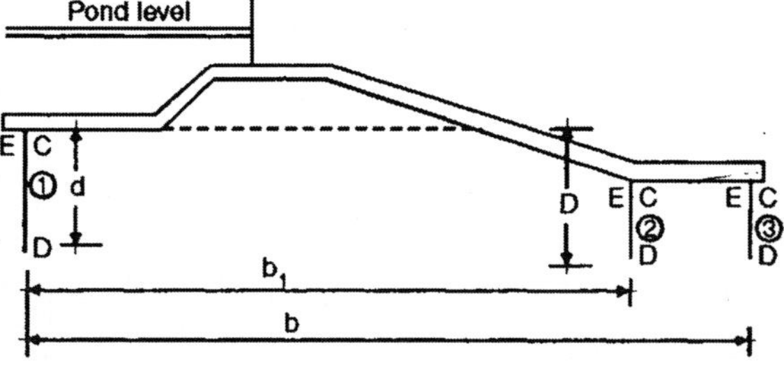

Fig. 10.26 Correction for mutual interference of piles.

Limitations: The equation gives results with in ± 2½% of those obtained by experiments. Mutual influence of piles is local and mainly extends to a distance equal to the depth of the pile beyond which it gradually falls off till the residual effect at twice that distance is negligible in most cases. Where the spacing of piles is close and depth great, the residual effect may be considerable. The above equation gives this influence for all cases. It takes note of the depth of either pile, the distance between the two and the floor length. The equation, however, does not hold good for the effect of an outer pile on an intermediate pile if the latter is equal to or smaller than the former and is at a distance less than twice the length of the outer pile. Subject to these limitations the equation is applicable for finding the influence of outer piles on intermediate ones and vice *versa*, irrespective of their depth or spacing.

2. *Correction for Floor Thickness:* The pressures at the junction points E and C pertain to the level at the top of floor, whereas, the junction of pile is at the bottom of the floor (E_1, C_1 in Fig. 10.27). To obtain the corresponding pressure at E_1 and C_1, it is necessary to determine the drop in pressure along each face of the sheet pile, considering linear variation of pressure from the hypothetical point E to D and also from D to C (Fig. 10.27). For different locations of piles, the corrections to be applied are as follows.

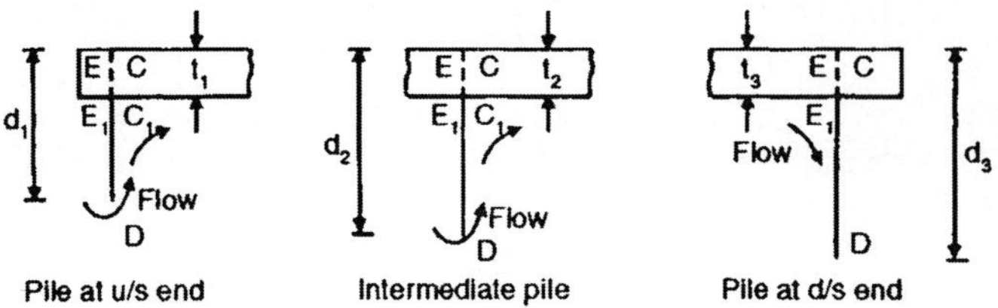

Fig. 10.27 Correction for floor thickness.

Correction for u/s pile
Corrected pressure at C_1

$$\phi_{C_1} = \phi C + \frac{\phi D - \phi C}{d_1} t_1$$

Correction for intermediate pile
Corrected pressure at E_1

$$\phi_{E_1} = \phi E - \frac{\phi E - \phi D}{d_2} t_2$$

$$Q_{C_1} = \phi C \times \frac{\phi D - \phi C}{d_2} t_2$$

Correction for end pile
Corrected pressure at E_1

$$\phi E_1 = \phi E - \frac{\phi E - \phi D}{d_3} \times t_3$$

Here, d_1, d_2, d_3 = depth of pile, and t_1, t_2, t_3 = floor thickness.

3. *Correction for Slope:* Due to sloping floor, a suitable percentage correction is to be applied. The correction is plus for the down and minus for the up slopes following the direction of flow. Values of correction for various slopes are given in Table 10.5. This percentage correction (c) is further multiplied by the proportion of the horizontal length of the slope to the distance between the two pile lines between which the sloping floor is located, *i.e.*, the correction applied is bc/b_1. The slope correction is applicable to the key point of pile line fixed at the beginning or the end of the slope.

Table 10.5 Slope correction

Slope (V : H)	*% Correction*	*Slope*	*% Correction (% of pressure)*
1 : 1	11.2	1 : 5	2.8
1 : 2	6.5	1 : 6	2.5
1 : 3	4.5	1 : 7	2.3
1 : 4	3.3	1 : 8	2.0

Exit Gradient. Exit gradient at the downstream end of an impervious floor length b and vertical cutoff d is given by $G_E = \frac{H}{d} \frac{1}{\pi\sqrt{\lambda}}$.

where, $\lambda = \left(1 + \frac{\sqrt{1+\alpha^2}}{2}\right)$, H = total head and $\alpha = b/d$.

For any value of α or b/d, the corresponding value of $1/\pi\sqrt{\lambda}$ is read off from the curve (Fig. 10.23). Knowing H and d, the value of G_E is calculated from the above equation. Thus, if $d = 0$, G_E is infinite. The value of exit gradient so calculated should be within safe limits.

Design procedure. The substructure of a typical weir on pervious foundation consists of the components, *viz.* (*i*) upstream pervious apron, (*ii*) upstream impervious apron, (*iii*) upstream pile or cutoff, (*iv*) downstream pile or cutoff, (*v*) intermediate pile or cutoff, (*vi*) downstream impervious apron, and (*vii*) downstream pervious apron. The design procedure of substructure of weirs on permeable foundations is as under:

The tentative dimensions are fixed from (*i*) Levels and lengths of upstream and downstream floors, based on surface flow considerations, (*ii*) Depths of upstream and downstream cutoffs are fixed on scour depth consideration, (*iii*) Deep cutoff under the main body wall of weir or two cutoffs at both ends of glacis are provided on site conditions and general practices, (*iv*) Total creep length is computed and checked to verify that the profile satisfies the criteria of creep theory. This step may be omitted if it is considered that (*a*) the soil is homogeneous, that is, there are no weak seams at shallow depths below the weir floor, and (*b*) perfect contact between the substructure and the foundation soil can be achieved, (*v*) nominal lengths of inverted filters or pervious aprons on upstream and downstream ends are provided, and (*vi*) relief holes and other arrangements for relieving uplift pressure are provided, where necessary. The weir profile is checked for safety against uplift of floor and undermining.

10.23 CUTOFFS

Cutoff is an impervious barrier constructed deep into pervious soil (*i*) To minimize the likelihood of undermining of foundation soil by increasing the creep length and thereby reducing velocity of percolation at exit, and (*ii*) To ensure safety of the structure behind it from scouring and undermining being deeper than the anticipated scour hole. The concentration factor, (*i.e.*, the factor by which the discharge per unit length of a barrage or weir assuming uniform distribution is required to be multiplied to get the design discharge per unit length for designing its various elements), is taken into account in fixing depth of cutoff. Cutoffs are, therefore, provided at both ends of the impervious floor, besides intermediate one at the lowest point of the weir glacis or under the crest. The functions served at each location are described below:

Functions of Pile Lines

1. *Upstream and Downstream Pile Lines.* Pile lines are required at the u/s and d/s ends of the weir floor in order to protect the structure against undermining of the foundation soil. The pressure under the d/s floor increases as the depth of d/s pile line increases. The u/s pile line is effective to destroy pressure head and reduce uplift pressures under the weir while the pile line at the d/s end is most effective to reduce exit gradient. The u/s pile line has little influence on the uplift pressures under the d/s floor but it effects a reduction of pressure under the u/s floor which may not be of any consequence in case of weirs where the load of water on top of upstream floor is much in excess of the uplift pressure.

Pile lines are also required at both ends of the weir floor to prevent failure by slipping of the subsoil into the scour holes by simple earth pressure. At u/s end, the seepage flow tends to restrain the soil from slipping due to earth pressure in the scour hole, while at the d/s end the force assists and accelerates slipping as these two forces shall act together. The restraint and acceleration at the u/s and d/s ends increases in direct proportion to the applied head. Failure at u/s end generally occurs at low heads when the restraint exercised by seepage flow against slipping is reduced while at the d/s end failure tends tó occur at high heads when the acceleration imposed by the seepage is increased.

2. *Intermediate Pile Lines.* Intermediate pile lines afford no protection against undermining of the floor at u/s or d/s ends or pressure distribution to reduce uplift pressures under the downstream floor but these serve as second line of defence to save the rest of floor and superstructure if the impervious floor at either the u/s or the d/s ends gives way by failure of end piles.

Depth of Sheet Pile

Scour holes occur not only during floods, but also at lower stages if there be parallel flow along the weir either upstream and downstream, or if there are swirls. The depth of u/s pile line is governed by the depth of the anticipated maximum scour only while that of the d/s pile line by the depth of scour in addition to exit gradient. The pile line at the u/s end is kept higher than that at the d/s end by the amount of afflux or the difference between u/s and d/s levels in maximum flood. While the top level of aprons is fixed according to Kennedy's scour depth, the bottom of pile lines should go deeper than the scour depth calculated from Lacey's formula.

10.24 WEIR DESIGN

Essentials of Weir Design

The two essentials of weir design, which in fact are interconnected, are (*i*) Residual or uplift pressure under the weir, and (*ii*) Exit gradient. The u/s pressures are kept low *consistent with* safety at the exit in order to have minimum floor thickness. To guard against undermining, the exit gradient is not allowed to exceed a certain safe limit, generally 1/5 to 1/7 of the critical exit gradient.

Design Procedure. Various steps involved in the weir design are outlined as follows.

Vertical Drop Weir

1. H.F. level—Local highest flood mark.
2. Maximum flood discharge—Design flood or of 50-100 years frequency. Free board to cater for a minimum of 500 years frequency flood or the standard project flood.
3. Minimum winter water level—From observed past gauge and discharge data of the river.
4. Waterway—Lacey's regime waterway of 4.83 $Q^{0.5}$ plus 10 to 20% extra.
5. Afflux—Usually adopted 1 m, range 1 to 1.5 m.
6. Total energy lines (T.E.L.)
   (*a*) T.E.L. d/s = H.F.L. at weir site before construction + head due to velocity of approach.
   (*b*) T.E.L. u/s = T.E.L. d/s + loss of head, *i.e.*, T.E.L. d/s + afflux + retrogression d/s (say 0.3 m).
   (*c*) H.F.L. u/s = T.E.L. u/s — head due to velocity of approach.
7. Discharge formula, q = 1.7 $H^{3/2}$
8. Crest level = T.E.L. u/s — H
9. Crest width, B = $H/\sqrt{\rho-1}$ usually 1.5 to 2.0 m or 0.6 of crest shutter height.
   Base width = $(H + \text{weir height})/\sqrt{\rho-1}$
   Side slope (batter) 0.4 : 1 on u/s and 1 : 1 on d/s.
10. Pond level = F.S.L. in canal + modular head for head regulator (say 1 to 1.25 m)
11. U/s floor level = As low as possible, usually at Kennedy's scour depth below H.F.L.
12. D/s floor level = d/s H.F.L.– D or d/s T.E.L. – E_{f2} whichever is lower.
13. Cutoffs
   (*a*) Bottom R.L. of u/s pile line = U/s H.F.L. — 1.5 R, where R is Lacey's regime scour depth
   (*b*) Bottom R.L. of d/s pile line = D/s H.F.L. — 0.3 m (for retrogression) — 2.0 R.
14. Length of impervious floor L = CH_p, where H_p is percolation head
   (*a*) D/s impervious floor length, l_2
   For weirs with crest shutters, $l_2 = 2.21C\sqrt{H/13}$
   For weirs without shutters, $l_2 = 2.21C\sqrt{H/10}$
   (*a*) D/s impervious floor length including d/s talus (length l_3)
   Weir with crest shutters, $l_2 + l_3 = 18C\sqrt{H_p/13 \times q/75}$
   Weir without crest shutters $= 18C\sqrt{H_p/10 \times q/75}$
   (*c*) U/s impervious floor length $l_1 = L-(l_2 + B + 2d_1 + 2d_2)$ where, B is base width of weir and d_1 and d_2 are depths of u/s and d/s cutoffs.
15. Thickness of impervious floor d/s, $= 1.33(H_p/\rho-1)$

where, 1.33 = factor of safety, H_p = residual percolation head at the point, ρ = specific gravity of floor material, say 2.4. Minimum thickness of u/s and d/s apron usually adopted is 0.6 m and 1.25m respectively.

Undersluices portion: $L_2 = 3.87\ C\sqrt{H_p/13}$ and $L_2 + L_3 = 27\ C\sqrt{H_p/13 \times q/75}$ with crest shutters, and that without crest shutters, $\sqrt{H_p/13}$ is replaced by $\sqrt{H_p/10}$.

The floor thickness is checked by Khosla's theory as well.

Pervious floor length, D/s $l_3 = (l_2 + l_3) - l_2$ as determined above

U/s, $l_4 = \frac{l_3}{2}$

Fig. 10.28 shows the section of vertical drop weir.

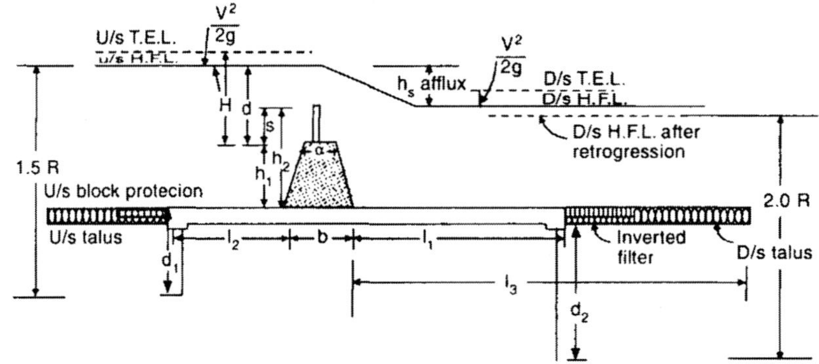

Fig. 10.28 Section of vertical drop weir.

Barrage with Sloping Glacis

1. H.F. level = Local highest flood mark.
2. Maximum flood discharge, (Q) = Design flood of 50-100 years frequency. Free board to cater for a minimum of 500 years frequency flood or the standard project flood.
3. Minimum winter water level = From observed past gauge data of the river.
4. Waterway
   (*a*) Barrage bays = 4.83 $(0.8\ Q)^{1/2}$
   (*b*) Undersluices bays = (*i*) 4/3 of offtaking canal bed width or (*ii*) 4.83 $(0.2Q)^{1/2}$ *i.e.*, for 10 to 20% of maximum flood discharge or (*iii*) twice canal discharge.

   Overall waterway = Undersluices waterway + divide wall width + fish ladder width + barrage waterway.
5. Discharge per metre width, $q = \frac{(0.8Q)}{4.83(0.8Q)^{1/2}}$ + 20% for concentration of discharge.
6. Lacey's regime scour depth, (R) = When looseness factor is < 1, $R = 1.35\ (q^2/f)^{1/3}$ and when looseness factor is > 1, $R = 0.473\ (Q/f)^{1/3}$.
7. Regime velocity, (V) = q/R
8. Velocity head, (ha)
   (*a*) Barrage bays = $V_b^2/2g$, where V_b is velocity through weir bays.
   (*b*) Undersluices bays = $V_u^2/2g$, where V_u is velocity through undersluice bays.
9. Afflux = Usually adopted 1 m
10. Discharge formula
    (*a*) Barrage bays $Q = 1.84\ (L - 0.1\ nH)\ H^{3/2}$
    (*b*) Undersluice bays $Q = 1.7\ (L - 0.1\ nH)\ H^{3/2}$
11. Crest level = T.E.L. u/s – H, Lowest water level in winter river section is also considered in the case of undersluice bays or 1 to 1.5 m below the crest level of barrage bays.
    Crest width = 2.0 m minimum
    Side slope (batter) = 2 : 1 to 3 : 1 for u/s glacis and 3 : 1 for d/s glacis.
12. Pond level = F.S.L. in canal + modular head for head regulator.
13. U/s floor level = As low as possible, usually at Kennedy's scour depth below H.F.L.

14. D/s floor level = H.F.L. – D, or d/s T.E.L. – E_{f2} whichever is lower.
15. Cutoffs
    (*a*) Bottom R.L. of u/s pile line = U/s H.F.L. — 1.5 R, where R is Lacey's regime scour depth.
    (*b*) Bottom R.L. of d/s pile line = D/s H.F.L — 0.3 m (for retrogression) — 2.0 R.
16. Length of impervious floor, L = u/s floor length + horizontal length of u/s glacis + crest width + horizontal length of d/s glacis + length of d/s floor.

    As per Khosla's theory, $G_E = \frac{H_p}{d} \frac{1}{\pi\sqrt{\lambda}}$, where, d is depth of d/s cutoff. G_E is exit gradient fixed from consideration of bed silt.
17. Thickness of impervious floor = Calculated from Khosla's theory with reference to standing wave and tested for maximum uplift pressures with maximum pond level u/s and no seepage d/s under condition of maximum retrogression of bed levels. Minimum values 1m up to crest, 1.0 to 2.0 m up to glacis toe, 2.0 to 1.5 m up to end of d/s floor.
18. Pervious floor length = Equal to Kennedy's scour depth (D) on u/s and 1.5 D on d/s, consisting of concrete blocks 1.5 × 1.5 × 0.9 m. Beyond block protection, launching apron in a length equal to 1.5 R and 2.0 R on u/s and d/s sides respectively. Thickness is the same as that of stone apron at this end, generally 1.25 T. In weirs constructed in Punjab it consists of 0.6 m deep concrete blocks over 0.6 m of graded stone on u/s and 1 to 1.25 m deep concrete blocks over 0.6 m of graded filter bed made up of shingle over the bed sand. The spaces between blocks are filled up with shingle.

SOLVED EXAMPLES

Example 10.1 A weir has 51 number of vertical gates span of each gate = 10 m, full reservoir level (u/s) = 110 m, crest level = 106 m, coefficient of end contraction for piers = 0.02, coefficient of end contraction for abutment = 0.1 and coefficient of discharge, C_d, (in Francis formula) = 1.70.

Compute the maximum flood discharge which can safely pass over the weir without exceeding the full reservoir level. Neglect velocity of approach.

Solution.

$$Q = 1.7\, L_e H^{3/2}$$

where,

$$L_e = L - (0.1 \times \text{No. of end contractions for abutments} + 0.02 \times \text{No. of end contractions for piers}) \times H$$

$$= 51 \times 10 - (0.1 \times 2 + 0.02 \times 2 \times 50)\, 4.0$$

$$= 501.2 \text{ m}$$

$$Q = 1.7\, L_e H^{3/2} = 1.7 \times 501.2 \times 4^{3/2} = 6816 \text{ cumecs.}$$

Example 10.2(*a*) Determine the pressures at points C_2 and E_2 as shown in Fig. 10.29 by Bligh's creep theory. Pond is full at upstream and no water at downstream. [*AMIE*]

(*b*) Determine floor thickness at E_2 and C_2, assuming specific gravity of concrete as 2.4.

Solution.

Total creep length, $b = 0.60 + 2 \times 6.0 + 15.8 + 2 \times 6.0 + 40.0 + 2 \times 10.3 + 0.6 = 101.6$ m

Hydraulic head, $d = 160.0 - 152.0 = 8$ m

Hydraulic gradient, $d/b = 8.0/101.6 = 0.08$

(*i*) Uplift pressure at E_2, 15.8 m from upstream

Creep length up to $E_2 = 0.6 + 6.0 \times 2 + 15.8 = 28.4$ m

Unbalanced head, $h_1 = 8\,(1 - 28.4/101.6) = 5.76$ m

Uplift pressure at $E_2 = wh_1 = 1 \times 5.76 = 5.76$ tonne/m^2

(*ii*) Uplift pressure at point C_2

Creep length $= 0.6 + 6.0 \times 2 + 15.8 + 6.0 \times 2 = 40.4$ m

Unbalanced head $h_2 = 8[(1 - 40.4/101.6)] = 4.80$ m

Uplift pressure at $C_2 = w \cdot h_2 = 1 \times 4.8 = 4.8$ tonne/m^2

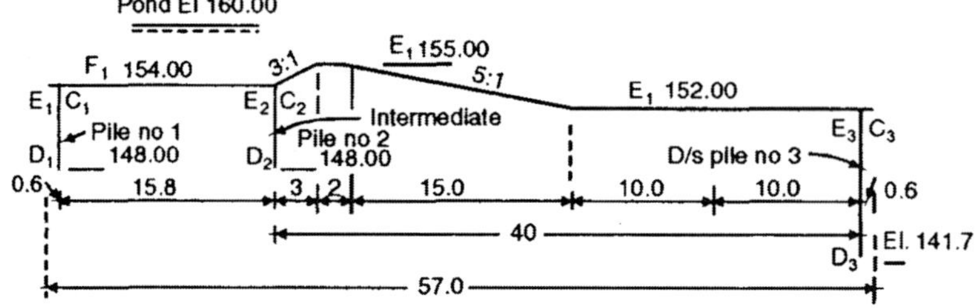

Fig. 10.29 Profile of barrage floor.

3) Floor thickness at $E_2 = 1.33\,(h/\rho - 1) = 1.33\,(5.76/2.4 - 1) = 5.5$ m

at $C_2 = 1.33\,(4.8/2.4 - 1) = 4.6$ m.

Example 10.3 Design a vertical drop weir by Bligh's theory for design flood discharge of 3150 cumecs, HFL before construction 193.0 m, average river bed level 186 m, Lacey's $f = 1.0$, FSL of canal 191.0 m coefficient of creep = 12.

Solution.

Pond level $= 191.0 + 1.0$ (canal working head) $= 192.0$ m

Waterway $= 4.83\,Q^{1/2} = 4.83\,(3150)^{1/2} = 271$ m

Discharge intensity, $q = 3150/271 = 11.62$ m^3/s

Regime scour depth, $R = 1.35\,(q^2/f)^{1/3} = 1.35\,[(11.62/1)^{1/3}] \simeq 7$ m

Regime velocity, $q/R = 11.62/7 = 1.66$ m/s

Velocity head, $ha = V^2/2g = (11.62)^2/2 \times 9.81 = 0.14$ m

D/s TEL $=$ HFL $+ ha = 193.0 + 0.14 = 193.14$ m

U/s TEL $=$ D/s TEL $+$ afflux $= 193.14 + 1.0 = 194.14$ m

U/s HFL $= 194.14 - 0.14 = 194.0$ m

D/s HFL with retrogression $= 193.0 - 0.5 = 192.50$ m

Discharge intensity $q = 1.7\,H_e^{3/2}$

or $H_e = (11.62/1.7)^{2/3} = 3.60$ m

$\therefore$ Crest level $= 194.14 - 3.60 = 190.54$ m

Height of shutter, $S = 192.0 - 190.54 = 1.46$ m

RL of bottom of U/s cutoff $=$ U/s HFL $- 1.5R = 194.0 - 1.5 \times 7.0 = 183.50$ m

Depth of u/s cutoff below bed, $d_1 = 186.0 - 183.50 = 2.5$m

RL of bottom of d/s cutoff $= 192.50 - 2 \times 7.0 = 178.50$ m

Depth of d/s cutoff below bed, $d_2 = 186.0 - 178.50 = 7.50$ m

Height of crest above bed level, $H = 190.54 - 186.0 = 4.54$ m

1. Weir Wall

Depth of water over crest, $d = 194.0 - 190.54 = 3.46$ m

Top width, $a = d/\sqrt{\rho} = 3.46/\sqrt{2.24} = 2.31$ m

also $a = 3d/2\rho = (3 \times 3.46/2 \times 2.24) = 2.32$ m

also $a = S + 1.0 = 1.46 + 1.0 = 2.46$ m, Adopt 3 m.

Bottom width: Condition 1 : Water headed up to crest level with no tail water,

$Hs = H + S = 4.54 + 1.46 = 6.0$ m

Overturning moment, $Mo = (w\,Hs^3)/6 = 1 \times (6.0)^3/6 = 36$ t.m

Moment of resistance, $Mr = \frac{w}{12}[\{(\rho + 1.5) H + 2.5S\} b^2 + a(\rho H - H - S) b$

$- \frac{1}{2} a^2 (H + 3S)]$

Putting values and equating to *Mo*

$$35.0 = \frac{1}{12}[\{(2.24 + 1.5)\ 4.54 + 2.5 \times 1.46\}b^2$$

$$+3(2.24 \times 4.54 - 4.54 - 1.46)b - \frac{1}{2}(3)^2 (4.54 + 3 \times 1.46)\}]$$

or $\quad 20.63\, b^2 + 12.51b - 472.14 = 0$

or $\quad b = 4.49$ say 5 m.

Condition 2: Flowing condition when weir is submerged:

Overturning moment, $\quad Mo = wh\, H^2/2$

When tail water is just at the crest, *d* and *h* are equal.

Head of water over crest, $d = \left[\frac{q^2}{(2/3c)^2 \times 2g}\right]^{1/3}$, neglecting velocity of approach, $c = 0.58$-coefficient of discharge

$$\therefore \quad d = \left[\frac{(11.62)^2}{[(2/3 \times 0.58)]^2 \times 2 \times 9.81}\right] = 3.58 \text{ m}$$

$$Mo = wh\, H^2/2 = [1 \times 3.58\ (4.54)^2]/2 = 36.89 \text{ t.m} \qquad (i)$$

$$Mr = \frac{wH(\rho - 1)}{12}(b^2 + ab) = \frac{1 \times 4.54\,(2.24 - 1)}{12}(b^2 + 3b) \qquad (ii)$$

Equating (*i*) and (*ii*)

$$0.47\, b^2 + 1.41b - 36.89 = 0$$

or $b = 8$m

3. Impervious and Pervious Aprons

Total Creep length, $\quad L = CH_s = 12 \times 6.0 = 72$ m

D/s impervious apron, $\quad l_1 = 2.21\, c\sqrt{Hs/13} = 2.21 \times 12\ \sqrt{6.0/13} = 18$ m

U/s impervious apron, $\quad l_2 = L - l_1 - (b + 2d_1 + 2d_2)$

$l_2 = 72 - 18 - (8 + 2 \times 2.5 + 2 \times 7.5) = 26$ m

Total length of d/s apron, $\quad l_3 = 18c\sqrt{Hs/13 \times q/75} = 18 \times 12\sqrt{6.0/13 \times 11.62/75}$

$l_3 = 58$ m

Length of filter + d/s launching apron $= l_3 - l_1 = 58 - 18 = 40.0$ m

Minimum length of inverted filter $= 1.5\, d_2 = 1.5 \times 7.5 \simeq 11$ m

Minimum horizontal length of launching apron $= 2.5\, d_2 = 2.5 \times 7.5 \simeq 19$ m

Total minimum length of both $= 11 + 19 = 30$ m

But $\quad l_3 - l_1 = 40$ m

Provide 1 m thick filter of stones over 0.6 m graded filter in 20 m length and launching apron in 20 m length.

Length of launching apron on 3:1 slope $= \sqrt{10}\, d_2 = \sqrt{10} \times 7.5 = 24$ m. Thickness in horizontal position, taking 1 m thickness = 24/20.0 = 1.2 m. Provide 1.5 m thick launching apron.

U/s concrete block protection and launching apron length = d_1 = 2.5 m say 5 m. Provide 1.5 × 1.5 × 0.9 m thick concrete blocks over 0.6 m thick gravel filter. With 1 m thickness, volume in 3 : 1 launched position = $\sqrt{10}\, d_1 \times 1 = 3.16 \times 2.5 \times 1 = 8\ m^3$.

Thickness in horizontal position = 8/5 = 1.6 m.

Thickness of impervious floor: Provide nominal 1m thickness to u/s of weir wall and 1.5 m below:

Floor Thickness

Residual pressure at point A (Fig. 10.30).

$$H_r = H_s - (Hs/L)x = 6.0 - (6.0/72)\,(2 \times 2.5 + 26 + 8) = 2.75\text{m}$$

Floor thickness $= 1.33\,(Hr/\rho - 1) = 1.33\,[(2.75/(2.24 - 1.0)] = 3.0$ m

Hr at 6 m from A = 6.0 – (6.0/72) (39 + 6) = 2.25 m

Floor thickness = 1.33 [2.25/(2.24 – 1.0)] ≃ 2.5 m

Hr at 12 m from A = 6.0 – (6.0/72) (45 + 6) = 1.75 m

Floor thickness = 1.33 [1.75/(2.25 – 1.0)] = 2.0 m.

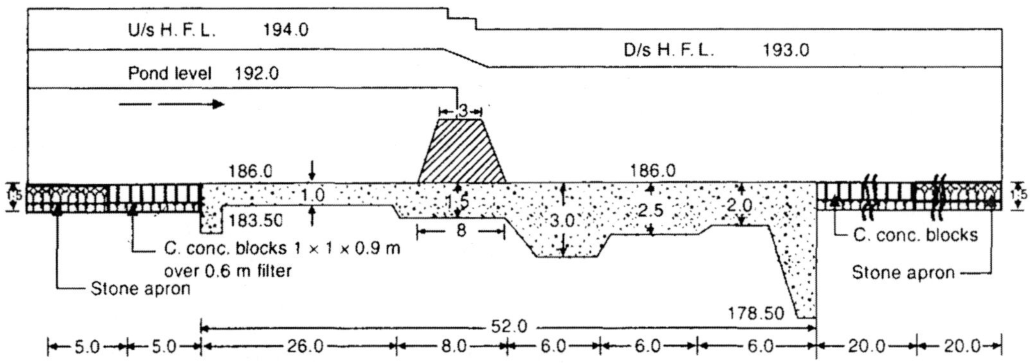

Fig. 10.30 Vertical drop weir

Example 10.4 Investigate stability of a weir section shown in Fig. 10.31 by Lane's weighted creep theory. It is founded on medium gravel. (*a*) Determine uplift pressure at point A, B and C.

Solution.

Weighted creep length = (Twice vertical depth of cutoff) + (1/3 horizontal floor length).
= (5 + 5 + 3 + 3 + 2 + 2) + 1/3 (3 + 6 + 9 + 21) = 33 m

Pressure head = Head water - tail water = 10.0 - 2.0 = 8 m

Weighted creep ratio = 33/8.0 = 4.125.

Lane's recommended ratio for medium gravel is 3.5 and as such the weir section is safe.

(*b*) *Uplift pressure*: The uplift pressure is determined by assuming that drop in pressure from headwater to tail water along the contact line of the weir and foundation is proportional to the weighted creep distance.

Uplift at $A = 8.0 - \left[\dfrac{8(5+5+3/3)}{33}\right] + 2.0$ *i.e.* tail water depth above foundation level.

$= 7.33$ m

Uplift at $B = 8.0 - \left[\dfrac{8(5+5+3+3+(1/3)\times 3)}{33}\right] + 2.0 = 5.88$ m

Uplift at $C = 8.0 - \left[\dfrac{8(5+5+3+3+1/3(3+6+9)}{33}\right] + 2.0 = 4.67$ m

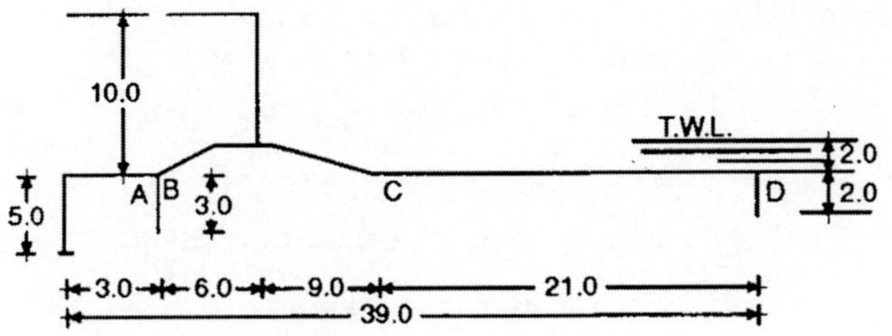

Fig. 10.31 Profile of weir floor.

Example 10.5 Check for exit gradient the floor profile given in Fig. 10.32 and also determine floor thickness at salient points.

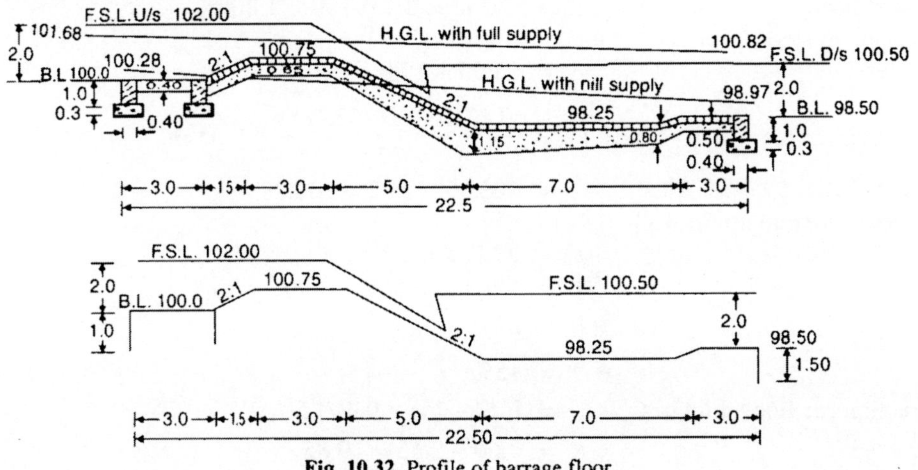

Fig. 10.32 Profile of barrage floor.

Solution. 1. Check against exit gradient

H = 100.75 – 98.50 = 2.25 m, b = 22.5 m and d = effective depth of downstream curtain wall excluding blocks and inverted filter (0.5 m thick) = 1.30 – 0.50 = 0.80 m

$$\alpha = b/d = 22.5/0.8 = 28.12$$

$$H/d = 2.25/0.8 = 2.81$$

Exit gradient, $= G_E = H/d\,\pi\sqrt{\lambda}$

Here,
$$\lambda = \left(\frac{1+\sqrt{1+\alpha^2}}{2}\right)$$

$$\lambda = \left(\frac{1+\sqrt{1+(28.12)^2}}{2}\right) = 14.57$$

$$\sqrt{\lambda} = \sqrt{14.57} = 3.82$$

Therefore,
$$G_E = \frac{H}{d\pi\sqrt{\lambda}} = \frac{2.25}{0.8\times 3.14\times 3.82} = 0.23$$

which is less than 0.3 depth of curtain wall as such it is OK. It can also be read from Fig. 10.21.

2. *Floor thickness*. The floor is to be checked for two critical conditions, *viz.* (*i*) canal closed, water headed up to crest upstream and downstream floor pumped dry, (*ii*) canal running full.

Uplift pressures and HFL.

(*i*) Canal closed and water headed up to crest level, downstream floor pumped dry. Water head = 100.75 – 98.50 = 2.25 m (*a*) Upstream curtain wall b = 22.5 m, d = 1.3 m

$1/\alpha = d/b = 1.3/22.5 = 0.06$

ϕE = 22% read from Fig. 10.24.

$\phi C_1 = 100 - 22 = 78\%$

ϕD = 15% read from Fig. 10.24.

$\phi D_1 = 100 - 15 = 85\%$

Correction for floor thickness $= \left(\dfrac{\phi D_1 - \phi C_1}{d}\right) t. = \left(\dfrac{85-78}{1.3}\right) 0.4 = 2.15$

Corrected ϕC_1 = 78 + 2.15 = 80.15

Pressure head = (80.15 × 2.25)/100 = 1.80 m

Downstream floor level = 98.50

H.G. level = 98.50 + 1.80 = 100.30 m

Correction for toe wall thickness $= \dfrac{0.40(100.30 - 98.95^*)}{22.5} = 0.02$ m

Corrected H.G.L. = 100.30 – 0.02 = 100.28 m

(*b*) Downstream curtain wall

b = 22.5, d = 1.3

$1/\alpha = d/b = 1.3/22.5 = 0.06$

ϕE = 22%

ϕD = 15%

Correction for floor thickness = [(22 – 15) × 0.4]/1.3 = 2.15

Corrected ϕE = 22.0 – 2.15 = 19.85

Pressure head = (19.85 × 2.25)/100 = 0.45 m

H.G.L. = 98.50 + 0.45 = 98.95* m

Correction for thickness of toe wall = 0.02 (as calculated for u/s toe wall)

Corrected H.G.L = 98.95 + 0.02 = 98.97 m

Slope of H.G.L. $= \dfrac{100.28 - 98.97}{(22.50 - 0.80)} = 0.06$ m/m

2. Canal Running Full, water head = 102.0 – 100.50 = 1.50 m

(*i*) Upstream toe wall

ϕC_1 = 80.15%

Pressure head = (1.5 × 80.15)/100 = 1.20 m

F.S.L. downstream = 100.50

H.G.L. = 100.50 + 1.20 = 101.70 m

H.G.L. of d/s face of toe wall $= 101.70 - \dfrac{0.4(101.70 - 100.80^{**})}{22.5}$

= 101.70 – 0.02 = 101.68 m

(*ii*) Downstream toe wall

Corrected ϕE = 19.85%

Pressure head = (19.85 × 1.5)/100 = 0.30 m

H.G.L. = 100.50 + 0.30 = 100.80** m

H.G.L. at u/s end of toe wall = 100.80 + 0.02 = 100.82

Slope of H.G.L. $= \dfrac{101.68 - 100.82}{(22.5 - 0.8)} = 0.04$ m/m

Calculations for floor thickness

Distance	*H.G.L. at the point*		*Floor level or water level*		*Pressure head*		*Max.*	*Floor thickness*
from toe wall d/s	*With nil supply*	*With full supply*	*With nil supply*	*With full supply*	*With nil supply*	*With full supply*	*pressure head*	*Col. vii/1.24 + 0.10*
(*i*)	(*ii*)	(*iii*)	(*iv*)	(*v*)	(*vi*)	(*vii*)	(*viii*)	(*ix*)
0.4	98.97	100.82	98.50	100.50	0.47	0.32	0.47	0.47/1.24 + 0.10 = 0.48, say, 0.80 m
3.0	99.13 (98.97 + 2.6 × 0.06)	100.92 (100.82 + 2.6 × 0.04)	98.25	100.50	0.88	0.42	0.88	0.88/1.24+0.10 = 0.81, say, 0.80 m
10.0	99.55 (98.97 + 9.6 × 0.06)	101.20 (100.82 + 9.6 × 0.04)	98.25	100.50	1.30	0.70	1.30	1.30/1.24+0.10 = 1.15m
15.0	99.85 98.97 + 14.6 × 0.06)	101.40 (100.82 + 14.6 × 0.04)	100.75	100.75	—	0.65	0.65	0.65/1.24+1.10 = 0.62, say, 0.65 m
22.1	100.28	101.68	100.0	102.0	0.28	—	0.28	0.28/1.24+0.10 = 0.33, say, 0.4 m

QUESTIONS

1. (*a*) Explain the different stages of a river, which stage is suitable for headworks.
   (*b*) Discuss the effect of construction of weir on regime of rivers.
   (*c*) Explain the criterion for fixing pond level and waterway. [PU.]
2. (*a*) Suggest suitable location of headworks on a river considering its different stages. What are the relative merits and demerits of its different locations?
   (*b*) State the functions of the following in a headworks with illustrative sketches:
   (*i*) silt excluder, (*ii*) divide wall, (*iii*) sheet piles. [PU.]
3. (*a*) Draw a neat sketch showing various components of a canal headworks. Briefly explain the functions of each component. [PU.]
   (*b*) What is afflux? Explain how will you calculate afflux in a weir.
4. (*a*) Sketch the general layout of a diversion headworks and explain the functions of various components.
   (*b*) Explain the criteria for fixing the pond level, and the crest level of a diversion headworks. [PU.]
5. (*a*) What is the difference between a weir and a barrage? Why is a barrage preferred to weir in modern times?
   (*b*) Differentiate between the following:
   (*i*) Scouring sluices and head sluices.
   (*ii*) Surplus weir and storage weir.
   (*iii*) Gravity and non-gravity weirs [PU.]
6. (*a*) Sketch the general layout of headworks. List all the components of headworks and explain their functions.
   (*b*) Calculate the design discharge for undersluices for given data:
   High flood discharge = 2500 cumecs
   Winter flood discharge = 400 cumecs
   Canal discharge = 300 cumecs. [PU.]
7. Explain Bligh and Lane's creep theories. How does Lane's theory differ from Bligh's creep theory?
8. What is exit gradient? How does it effect the design of a weir?

9. Explain the criteria adopted in designing various components of weir built on permeable foundation using Khosla's theory.
10. Discuss briefly how the surface and sub-surface flow across a barrage may eventually lead to failure of the barrage, if not properly controlled. What precautionary measures will you recommend to safeguard such failures? Illustrate your answer with neat sketches, wherever necessary.
11. Explain surface flow considerations involved in the design of thickness of the sloping glacis and downstream floor of a weir for different flow conditions.
12. What are the main causes of failure of weir on permeable foundations and what remedies would you suggest to prevent them?
13. Discuss Khosla's theory for design of weirs on permeable foundations. Enumerate the various corrections that are needed in its application.
14. State the fundamental difference between Khosla's theory and Bligh's creep theory for seepage flow below a weir.
15. Explain the design of a weir on permeable foundations for surface flow conditions.
16. What is meant by scour? What precautions are necessary against it in weir design?
17. Explain the design procedure of a vertical drop weir.
18. Explain Khosla's method of independent variable. How do you apply corrections for floor thickness, sloping floor and interference of piles?
19. (*a*) Describe with the help of suitable sketches Bligh's creep theory for the safe design of apron in an irrigation work. What would be the nature of gradient diagram according to this theory and why?
20. Discuss the limitations and anomalies of Bligh's method. [AMIE]
21. Briefly explain Khosla's theory and how it is used in design of weir on permeable foundations:
22. Following corrected ϕ values were computed from Khosla's curves in a barrage placed on permeable foundation:

U/s sheet pile —	$\phi E_1 = 100\%$,	ϕD_1 90%,	$\phi C_1 = 85\%$
Intermediate pile	$\phi E_2 = 80\%$,	ϕD_2 70%,	$\phi C_2 = 65\%$
D/s sheet pile	$\phi E_3 = 55\%$,	ϕD_3 45%,	$\phi C_3 = 0\%$

Distance between the u/s and intermediate piles is 20 m and that between the intermediate and d/s pile is 40 m. Assuming that the floor is horizontal throughout, draw the hydraulic gradient line for the subsoil flow. If the net head is 10 m, determine the thickness of d/s floor at a distance of 20 m and 30 m away from the intermediate sheet pile. [PU]

23. (*a*) Discuss development, utility and limitations of Khosla's theory.
    (*b*) Write a short note on Bligh's creep theory. [PU]
24. (*a*) Briefly explain the design basis of a weir.
    (*b*) Given Bligh's coefficient $C = 6$,
    (*i*) Draw hydraulic gradient line for the given section (Fig. 10.33).

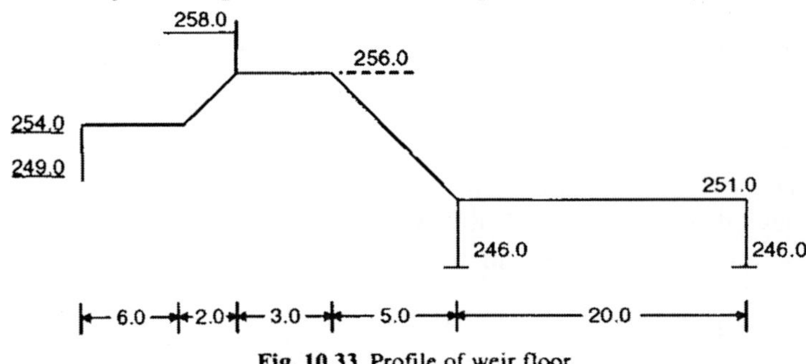

Fig. 10.33 Profile of weir floor.

11

CONTROL STRUCTURES

11.1 INTRODUCTION

The head regulator at the head of a canal cannot, in general, secure an accurate and automatic adjustment of the discharge let through it into the canal. Provision of control point in a canal system is a necessity to localise malfunctioning of the head regulator besides to cater for the diverse requirements such as (*i*) To divide the channel length in certain reaches so as to effect proper regulation of the supplies thereat as also to help attain designed conditions in the channel, (*ii*) To measure the discharge at various reaches of a channel, (*iii*) To effect proportionate distribution of supplies, (*iv*) To feed an offtake during low supplies in the parent channel, (*v*) To secure balancing depth for reasons of economy as also for the destruction of surplus energy, and (*vi*) To exclude silt charge which is higher than the carrying capacity of the offtakes.

Furthermore, canals drawing their water supplies from rivers draw excessive silt charge which tends to make the channel shallow and wide rendering it incapable to draw authorized discharge thereby jeopardizing the irrigation in the command. Heavy silt charge in hydel channels results in deterioration of the smooth lining surface and power generation equipment. The success of an irrigation project depends on the control of silt entering the channel taking off from a reservoir or a river. The silt exclusion is aimed at as the river can disposed of its silt load by passing down into the sea. Coarse stuff in the silt charge is transported as bed load and finer material as suspended load. Roughly, coarser particles are mainly confined to one-third depth above bed, while sand and clay particles are more or less uniformly distributed throughout the vertical. As such if bottom layers of flow are directed without disturbing the natural distribution of sediment load a large quantity of coarse sediment can be prevented from entering an offtaking canal. The various devices described below aim at exclusion of coarser particles.

11.2 SILT CONTROL MEASURES

Various measures to control silt entry are classified as preventive and curative. Preventive methods envisage exclusion of coarser sediment at the head of the canal before it enters it. Disposal of the silt which has found its way into the canal falls under curative measures.

Preventive Measures

Silt Control at Head Works

1. River Approach: Proper river approach conditions play a very important role on the control of sediment entry into the canal. Straight river approach channel is preferable. Where it is not available, the barrage is located outside the main river channel in some creek which is dry in winter and then the main river is diverted through the barrage, and the offtake is located towards the downstream end of the concave bend and as near to the point of contraflexture as possible.

2. Orientation of Headworks: Silt exclusion is influenced by the orientation of headworks. Sand charge decreases from midstream towards the bank in a straight channel. Offtake at right angle to the river does not contribute towards silt exclusion. Headworks located in concave curve towards downstream end is the natural and satisfactory method of controlling sand entry into the canal. A head regulator with angle varying from 90° to 110° with the axis of barrage functions satisfactorily from stand point of minimum silt entry.

3. Regulator Crest Level: Crest level of the canal head regulator set higher than the level of the bed of still water pocket and crest level of undersluices allows only top water entraining finer sediment particles to enter the canal and arrest the entry of coarser sediment.

4. Divide Wall: Judicious length of divide wall, usually one-half to two-third of the length of head regulator or up to the last head regulator in the case of two offtaking canals, is effective to control silt entry into the offtake. Long divide wall does not act as a trap for coarse bed material rather lessens the effect of regulation on curvature of flow.

5. Shape of Guide Banks: Control of silt entry into the offtake is secured by a suitable approach to the canal pocket through proper shape and alignment of the guide banks. Bottleneck and converging guide banks do not provide effective silt control as large islands are formed upstream of the pocket.

6. Canal Closure: Heavily sediment laden water in the river when in flood is better rejected down the barrage and denied entry into the offtake but this is rarely done in the case of irrigation canals and not feasible in hydel channels.

7. Double-wedge Regulation: Better silt exclusion in still pond system is secured by opening the barrage gates adjacent to the divide wall more than those farther away. Opening of gates is minimum in the centre, increasing gradually towards the divide walls, in case canals take off from both banks of the barrage.

8. Submerged Vanes: A submerged curved vane of one-third the depth of water, pointing downstream and projecting from a bank, produce a local concave curvature and helps deflect bed material away from the bank into the mid-stream. It entails very little obstruction to the natural waterway of river. Its demerits are (*i*) Vanes are not effective in case a deep channel is not formed or maintained along or near a bank at high river discharges and throughout the flood season, (*ii*) A single vane on one bank is liable to draw too large a share of top water to feed the canal and may possibly pull the whole river towards it necessitating counter measures to adverse the conditions thus created, (*iii*) Unless a vane is constructed sufficiently upstream, bed sediment thrown up in suspension may find its way into the canal, (*iv*) A vane designed for lower discharge may not be effective during higher or intermediate discharges, and (*v*) Accretion in the river may render the vane partially or completely ineffective.

9. Silt Excluder: A large quantity of sediment in suspension besides the saltation and bed load is denied entry into the canal through silt excluder constructed in the river pocket feeding the canal. Sand or sand gravel and boulders in the river pocket are thus excluded down the river.

Curative Measures

1. Silt exclusion in the canal. The silt that has found entry into the canal from the headworks is ejected from the canal through a silt extractor or silt ejector constructed in the head reach of the canal a little downstream from the offtake.

2. Silt control measures at the offtakes. The sediment exclusion in the distributary head offtaking from a canal aims at proportionate distribution of sediment according to the silt carrying capacity of the canal which should be achieved because the canal can not dispose of any excess silt, unlike river at headworks site.

(*i*) **Gibb's Groyne Wall.** A curved wall in extension of the downstream wing wall of the offtake is constructed to distribute the discharge and hence the silt of the parent channel in proportion to the discharge of the offtake and the parent channel downstream of the offtake. A hole 1 m in diameter or square is provided near the bottom of the curved wall to further reduce silt entry into the offtake.

(*ii*) **King's vanes.** Curved vanes impose an artificial curvature on the bed flow diverting it away from the offtake and thus help preventing silt from entering the offtake. King's curved vanes (Fig. 11.1) are made of plain or reinforced concrete.

The longest vane extends 0.5 to 1.5m beyond a line drawn at an inclination of 2:1 to the parent channel axis from the downstream end of the offtake. Their height is generally 1/3rd to 1/4th of the depth of water in the parent channel. Ordinarily width of parent channel covered by the upstream ends of vanes is half the width of offtaking channel. The recommended values of *Y* (refer Fig. 11.1) are given in the Table 11.1. Their thickness is kept the minimum as thinner vanes are more efficient. Their spacing is generally 1.5 times the vane height. The minimum radius of curved vanes is 7.5 m and 12 m maximum in small and average capacity channels. Larger the radius greater is the silt exclusion efficiency. The downstream end of first vane is removed from the drawdown effect of the offtake to the extent of 1.0 m in very small channels and 1.75 m for channels 15 m wide. The curved vanes are preceded by a straight length parallel to the centre line of the parent channel and finished to a V-shape. The straight portion has a slope of 3:1 and a cut water shape at upstream end. The downstream ends of curved portion of the vanes are tangential to lines which are at an angle not less acute than 27° with centre line of parent channel. The bed and sides of the parent channel are lined for a minimum length of 10 times the depth of flow or 15 m. The pitched bed level is generally kept 0.15 m higher than the normal silted bed level. Minimum number of silt vanes used is 2.

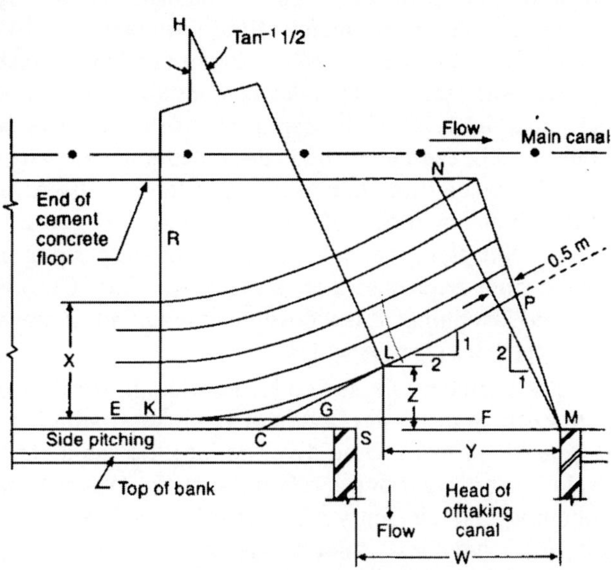

Fig. 11.1 King's vane.

Table 11.1 Minimum value of *X*, *Y* and *R*

Width (m) of offtake = *w*	0.60	1.2	1.8	2.4	3.0	3.6	4.6	6.0	7.6	9.0	10.6	12.0
X (m)	1.20	1.5	2.1	2.4	3.0	3.6	4.6	5.4	6.0	7.0	7.8	8.5
Y (m)	0.6	1.2	1.5	1.8	2.4	2.7	3.0	4.0	5.2	6.0	6.6	7.6
Z (m)	1.2	1.2	1.5	1.8	2.4	2.7	3.0	3.6	4.2	5.2	5.8	6.6
R (m)	9	9	10	12.0	18.0	21.0	24.0	30.0	35.0	44.0	50.0	57.0

Limitations : King's vanes are not suitable (*i*) If the discharge of the offtake is more than 1/3rd of the parent channel, (*ii*) When a small offtaking channel is situated between two large branches and has its bed at high level, and (*iii*) In canals where there is no fixed discharge and there is fairly large surging due to fluctuations in supplies.

(*iii*) **Cantilever skimming platform:** This device is provided in a head regulator. It consists of a slab cantilevered into the parent channel a little below the designed crest level of the head regulator to tangentially cut apart the top and bottom layers of water and to allow only the surface water to enter the offtake and as such reduces silt entry. Heavy sediment held by water in suspension is also made to drop into the lower layers by pitching the bed and side slopes of the parent canal in a small reach just upstream of the offtake such that the skimming platform is constructed at its downstream end. The platform is made wide enough to pass 25% extra discharge than the authorized discharge of the offtaking canal when there is minimum supply in the parent channel at which offtake is expected to run. The downstream edge of the platform is made at an angle of 60° to the centre line of the parent channel. The upstream edge may be parallel to the other edge but at 1.5 to 3.0 m upstream of the upstream edge of the offtake. It is suitable only where the parent channel is deep, say 2 m or more and the offtake is comparatively small.

(*iv*) **Silt escape:** This device consists of a weir with scouring sluices at a level below the designed bed level of the parent channel to remove bottom silt laden water into some natural drainage. Silt escape has been discussed in detail separately.

(*v*) **Two or three tier gates:** Silt entry is also reduced by temporarily raising the crest level of the offtake by means of two tier or three tier gates.

(*vi*) **Silt pits:** Internal borrow pits or silt pits dug in the bed of the canal serve to accumulate the silt from the canal water as the canal is run. During closure the silt is removed.

(*vii*) **Dredging:** It is a costly method of silt removal from perennial canals by dredgers.

11.3 SILT EXCLUDER

Silt excluder is a device located just in front of the canal head regulator and its function is to exclude the lower layers of river water which carry comparatively coarser particles and prevent them from being diverted into the canal. The basic requirements for satisfactory silt exclusion are (*i*) It draws relatively silt-free water from the top layers and excludes heavy laden bottom layers, (*ii*) Entry of water is smooth so as to avoid turbulence in water and agitating of silt, (*iii*) It provides a smooth surface by paving or by plastering the bed and the sides to reduce friction and give chance to the silt in the upper layers to settle down in the layers adjacent to the bed from where it is led away downstream, and (*iv*) Velocity is reduced in the pocket so as to attract silt load in lower layers. These requirements are met with by passing the slow moving bed water containing heavy charge of coarser grades of material through tunnels which have top level of the roof covering at the crest level of the canal head regulator. The water approaching the head regulator is cut apart without causing eddies; the top water enters into the canal and that beneath carrying silt-laden water is led through the tunnels to pocket gates in the barrage.

The salient parameters of the silt excluder are (*i*) About 20 per cent of the canal discharge is passed through the tunnels and the pocket gates to secure efficient silt exclusion, (*ii*) Each tunnel must take the same discharge as others in spite of their unequal lengths which is achieved by suitably adjusting the cross section at intermediate points such that the total loss of head in each is the same, (*iii*) The location of upstream mouths of the tunnels is staggered so that the length of the tunnels decreases in a series of steps, the longest tunnel being nearest to the head regulator, and (*iv*) The top slab of the excluder, besides being bell mouthed is projected beyond the noses of the tunnel divide wall by 0.3 to 1 m.

Salient Features

Approach: It is preferably kept straight to the mouth of the tunnels.

Tunnels: The tunnels are provided in the form of RCC barrels, normally 4 to 6 in number, depending on the available discharge for escapages, approach conditions and length of canal regulator. The excluder tunnels are located in front of the canal head regulator and their alignment kept parallel to the regulator . Any change in alignment, if necessary, is on smooth curves. The tunnel nearest to the head regulator is of the same length as that of the regulator. The consecutive tunnels are spaced at distances such that the mouth of the tunnels nearer to the head regulator comes within the suction zone of the succeeding tunnel so that no dead zone is left between the two to permit sediment deposition. Usually the mouth of two tunnels is kept 12 m apart. The tunnels at the inlet are suitably bellmouthed to minimise entry losses and improve suction. The bell mouthing is done within the thickness of divide wall on any suitable elliptical curve. The tunnels are normally of size 1.75 × 2.0 m. Their size depends on the number of tunnels, discharge available for escapage, self-clearing velocity of 3 m/s and 4-4.5 m/s in alluvial and boulder stages respectively. The roof of the tunnels is kept flush with crest of the canal regulator(s) and bed kept at the u/s floor level of the barrage or weir. All the tunnels outfall into stilling basin through one or two undersluice bays of the barrage next to the canal regulator. It is usually one in sandy reaches and two in shingle or boulder stage of the river. The tunnels are suitably throttled laterally or vertically or both, depending upon the conditions. The aim is to produce accelerating velocities in the tunnels; maximum at exit so that sediments once extracted are not deposited anywhere in the tunnels. The tunnels are preferably straight. Bend, if inevitable, may be of radius 10 to 15 times the tunnel width. All transitions to piers in bellmouthing at top or sides are preferably elliptical, the major axis being in the direction of flow and 2 to 3 times the minor axis.

The excluder tunnels are operated by undersluice gates. These are regulated either for the tunnels to run full bore or to remain completely closed.

The extracted sediments outfall into the river downstream of the barrage through undersluice bays; outfall channel as such is not required. In boulder or shingle bed river, a curved wall in the stilling basin is necessitated to avoid formation of big deposits of sediments there.

The escape discharge is 15 to 20 per cent of canal discharge. A minimum working head of 0.5–0.75 m and 1.0–1.25 m for excluders in sandy rivers and shingle or boulder river bed respectively is required.

Losses in Tunnels

(*i*) Friction loss

$$hf = V^2 L N^2/R^{4/3}$$

Where, hf is head loss (m), V is velocity (m/s), L is length of tunnel (m), N is rugosity coefficient, R is hydraulic mean depth (m).

(*ii*) Loss at bend

$$hb = f(v^2/2g)\,(\theta\ /180)$$

Where, hb is loss due to bend (m), f is $0.124 + 3.104\ (S/2r)^{1/2}$, g is acceleration due to gravity (9.81 m/s^2), θ is angle of deviation (degree), S is tunnel width (m), r is radius of bend along centre line of tunnel (m).

(*iii*) Transitional loss due to change of velocity in contraction

$$h_c = 0.1\ (V_2^2/2g - V_1^2/2g)$$

Where, h_e is transitional loss due to change of velocity in contraction, V_1 and V_2 are velocities before and after transition (m/s).

(*iv*) Expansion Loss

$$he = k(V_2^2/2g - V_1^2/2g)$$

where, k is constant; 0.2 for gradual transition, V_1 and V_2 are velocities in the expanded and smaller sections respectively (m/s):

Fig. 11.2 shows Narora silt excluder.

Advantages: (*i*) Head across the barrage for its operation is usually available, (*ii*) Economy is secured by the use of barrage gates, and (*iii*) Excessive sediment entry into the offtaking canal causing silting up and gradually reduction in carrying capacity is avoided, and (*iv*) Separate outfall channel is not required as is the case with silt ejector.

Disadvantages: (*i*) The structure being subject to river action has to be made structurally strong and hence costly, (*ii*) Difficulty in securing good approach condition, and (*iii*) Undersluice bays covered by the excluder cannot be used simultaneously for the passage of flood discharge.

Efficiency of Silt Excluder

Efficiency (E) of excluder is given by the relationship

$$E = \frac{(qs_1) - (qs_2)}{(qs_1)} \times 100$$

where, qs_1 is sediment proportion in canal without excluder, and qs_2 is sediment proportion in the canal with excluder.

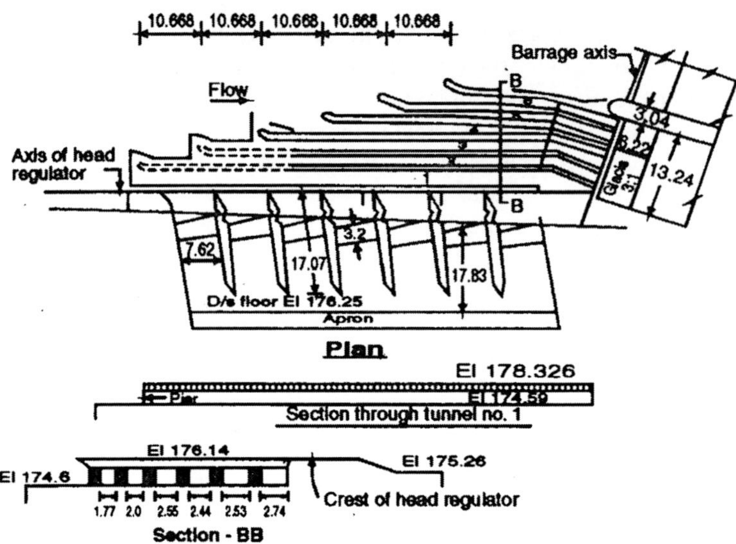

Fig. 11.2 Narora silt excluder.

11.4 SILT EJECTOR

Also called *sediment* extractor, it is a work constructed on the canal itself a little below the head regulator to remove the coarse sand that has found its way into the canal in spite of the silt excluder at the diversion weir, anicut or barrage. The escape discharge, usually 15-20 per cent of the designed canal discharge downstream of the ejector is fixed to ensure removal of desired sediment size and load as also to cater for flushing in individual tunnels. In principle silt extractor is similar to tunnel type silt excluder. The extraction of sediment is effected by causing sediment concentration in the bottom layers and separating them in such a way that there is least disturbance in the sediment distribution of the approaching flow.

Location: The location of the silt ejector is governed by the considerations (*i*) Not very close to the head regulator as the residual turbulence may cause the sediment load to remain in suspension and prevent its ejection to the desired extent, (*ii*) Not far removed from the head reach lest the sediment may settle down earlier and reduce the channel capacity upstream, (*iii*) Topographical considerations and availability of a suitable outfall channel, (*iv*) Suitable straight approach channel free from curves or kinks as otherwise it is likely to change the sediment concentration across the channel and disturb the uniform distribution of flow in front of the ejector. The length of straight reach upstream is fixed from the consideration of velocity, depth, grade of material to be removed and degree of turbulence, (*v*) Availability of sufficient working head, 1 m, in the canal u/s of ejector and outfall channel at the exit of ejector tunnel to extract the desired sediment.

Salient Features

1 **Approach channel.** A lined or pitched approach channel of length 3 to 4 times the depth is provided to increase the concentration of sediment in the bottom layers. The cross section of approach channel is increased by depressing the floor suitably, say about 0.3 m and/or increasing the width with a splay of 5:1 to reduce the velocity of flow. The reduced velocity is maintained for a sufficient length to achieve the desired silt concentration in the bottom layers. Fig. 11.3 shows a typical silt ejector.

Diaphragm. The diaphragm properly tied to supports to prevent its dislodging is so designed that least disturbance is caused in the sediment concentration attained in the bottom layers of flow upstream of the ejector tunnels. The diaphragm level, generally at the downstream bed level of the canal, is fixed with due consideration to (*i*) Desired sediment size to be trapped and extracted, (*ii*) Bed level and size of tunnels, (*iii*) Thickness of diaphragm, and (*iv*) Designed bed level of the canal downstream of the ejector.

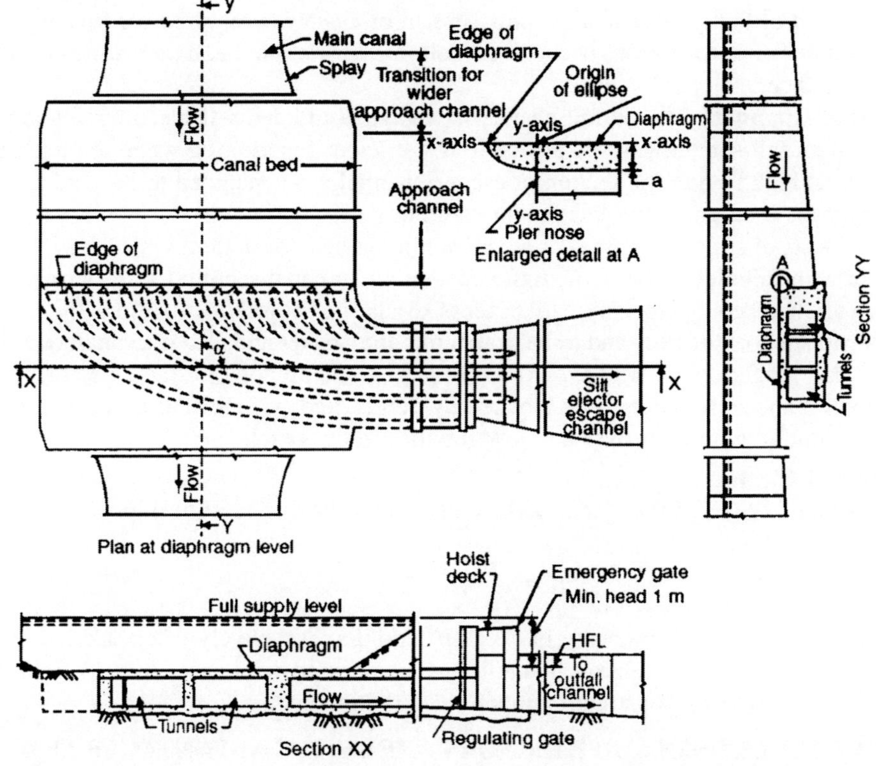

Fig. 11.3 Silt ejector.

The diaphragm is extended beyond the pier noses and the underneath of diaphragm is given a streamlined bellmouth shape conforming to the following equation as per IS : 6004.

$$\frac{x^2}{4a^2}+\frac{y^2}{a^2}=1, \; a \text{ is the thickness of diaphragm.}$$

Tunnels: Canal width at the mouth of the extractor is divided into three or four main tunnels by vanes gently converging towards the outfall on one side of the canal. The main tunnels are subdivided into four to six smaller compartments or sub-tunnels by vanes of radii varying from 3 to 4 times the width of sub-tunnels to avoid cross flow in the transition section and terminate in the straight portion towards the emergence of the main tunnels. The upstream noses of the vanes have cut water shape while the downstream end is fish tailed. The tunnel dimensions are fixed (*i*) To generate the velocities that would carry the size of sediment to be removed, (*ii*) The section of sub-tunnel at entry is so chosen that the velocity of flow at intake is slightly higher than the velocity of bottom filaments of water upstream of ejector, (*iii*) Sub-tunnels section up to their exit, where these end into the main tunnel, is reduced gradually to ensure 10-15 per cent increase in the velocity of emerging flow, (*iv*) At the exit of sub-tunnels the section of main tunnel is so designed that the velocities of combined discharge are not less than the velocities emerging out from sub-tunnels. Velocity of 2.5 to 6 m/s at exit of main tunnels is maintained but always less than the critical velocity, (*v*) Height and width of tunnels is fixed to facilitate inspection and equal losses in each tunnel respectively. Usually depth of main tunnels is kept about 1.75 to 2.25 m. Generally no slope is required for tunnel beds, (*vi*) At emergence, each tunnel is provided with a gate for regulation of discharge and gives maximum efficiency for full bore. Minimum operation head is 0.75 m; higher heads, however, give better efficiency, and (*vii*) For efficient working 15-20 per cent of the canal discharge downstream of ejector is required for flushing.

Losses in Tunnels: Losses occur in tunnels due to friction, bend, contraction and expansion as in silt excluder.

Outfall Channel: The outfall channel is designed to have self-cleaning velocity. Adequate head at the outfall shall always be available for efficient functioning when the natural stream is under normal high flood. The section of escape channel is so designed to have adequate capacity to transport the total quantity of sediment entering it.

Drawbacks of silt ejector are (*i*) Loss of minimum head of 0.75 m which may be of the order of 1.25 m with shingles on bed, (*ii*) Extra cost on wider canal section for the surplus water to be carried from head to ejector site, (*iii*) Excess discharge required to operate the ejector causes disproportionate load of bed sand to be drawn into the pocket and into the canal, (*iv*) Wider head-regulator and longer divide wall for extra discharge to be escaped through ejector, (*v*) Comparatively small tunnels are liable to be blocked by debris, and (*vi*) Difficulty in securing a working head at the outfall with high river and low supplies in the canal.

Efficiency of Ejector

The efficiency (E) of the ejector is given by the following relationship:

$$E = \frac{I_u - I_d}{I_u} \times 100$$

Where, I_u and I_d = silt intensity in canal u/s and d/s respectively of ejector.

Advantages: (*i*) Good approach conditions are secured with ease, and (*ii*) Extent of extraction can be improved by constructing more ejectors in the canal.

11.5 SILT EXCLUDERS AND EJECTORS—RELATIVE ADVANTAGES AND DISADVANTAGES

Generally most of the coarse silt is in the bottom layers of water and the essence of exclusion or removal of silt is to separate the top and bottom layers so that there is no disturbance and the distribution of silt remains normal. Further, the silt is carried by the eddies formed at the bed due to friction. If this friction is reduced by provision of a smooth cross section, concentration of the silt in the bottom layer can be increased. These conditions can be better realised in the case of canals than that in a river. Thus, silt ejector is likely to be more effective than excluder. For silt ejector it would, however, be necessary to carry the surplus escape discharge from the head to the silt ejector and also to provide an escape and a regulator. It would also not be desirable to allow the coarse silt to enter the canals.

Where sand is fine and uniformly distributed across a vertical, excluders and ejectors are unsatisfactory, as even a little turbulence at the tunnel mouth may throw sand in suspension and reduce chances of its exclusion.

11.6 SILT EXCLUDER *VS.* SILT EJECTOR

Silt excluder	*Silt ejector*
Costly structure.	Comparatively low cost structure.
Structure is heavy as it is subject to river action.	Structure is light being located on canal.
Good approach conditions are difficult to secure in a river.	Good approach conditions can be easily obtained in a canal.
Constructed in conjunction with canal head regulator.	Entirely additional structure on canal.
Work is concentrated at one place.	Separate outfall work may be necessary.
Extraction is secured only once, i.e. at the head regulator.	Multiple extraction is possible by having a number of ejectors in a canal system.

Silt excluder	*Silt ejector*
Working head for operation of excluder is always available.	Working head may not be available under low supplies.
Tunnels are not liable to be blocked being large in size.	Tunnels are liable to be choked by debris being smaller in size.
Canal head regulator is not required to be widened beyond normal capacity.	Canal head regulator is required to carry extra discharge meant for escapage.

11.7 CONTROL POINTS AND THEIR LOCATION

Larger number of control points on a channel are required for its efficient operation which in turn require certain amount of head (fall) for their working. It is, however, not advisable to introduce control points by flattening water surface slope in a channel as this would induce silting trouble. A control point requires certain working head for its working. A minimum drop of 10 per cent of the depth of the channel or 0.25 m whichever is higher is essential for modular working of a control point.

Location of control points is governed by various considerations such as (*i*) In major distributaries two control points are essentially required, first within 3 km downstream of the head regulator to localize the effect of any faulty design or functioning of the head regulator and/or improper selection of silt factor '*f*' and other design parameters of the canal, and the second about 3 km upstream of the tail, where the discharge is of the order of 0.50 cumec, so as to segregate the tail reach from the whole channel, because the silting and berming up is more pronounced in the tail reach which may upset the regime of the whole channel, (*ii*) One control point is required at every change in the channel section where the discharge is reduced appreciably due to drawoff by outlets. For this purpose a bridge is also considered as a control point, and (*iii*) At a point where the offtake has proportionately higher discharge than the parent channel downstream or has insufficient working head to work for proportionate distribution of supplies. In case drop is not available, control point in the form of an headless flume is provided to effect equitable distribution of supplies.

11.8 TYPES OF CONTROL STRUCTURES

Various types of control structures with specific functions served by each are as under:

1. *Regulating fall:* Regulating falls or headless regulating bridges are necessary below all offtaking channels, except when the working head for the offtake channel is large enough to feed it with its full supply discharge, when only 1/3rd supply is running in the parent channel, or the working head is more than half the designed depth of the parent channel.

2. *Fall (non-regulation):* Fall or drop is a work designed to secure the lowering of the water surface in a channel and for safe destruction of the surplus energy so generated.

3. *Rapid:* Rapid is a fall with long sloping glacis for the destruction of surplus energy.

4. *Meter flume*: Meter flume is a fall with additional provision for measuring discharge from the direct measurement of the depth of water flowing over its crest.

5. *Head Regulator:* Head regulator is the work constructed at the offtake of a channel to admit water into it from the parent channel. It is designed to pass not only the authorized discharge but also fair share of silt from the parent channel. It should be capable to draw authorised discharge with approximately 2/3rd supply in the parent channel. Minimum modular working head for a head regulator is 1/5th the depth of water over the crest which should always be available.

6. *Proportional distributor*: Proportional distributor is a cross regulator constructed on a supply channel in conjunction with the offtake head in such a manner that when supply fluctuates each offtake always draws a constant proportion of the supply. Cross regulator is not generally

provided in conjunction with the head regulator when the available working head in the offtake is more than half the full supply depth in the parent channel. Its primary function is thus to effect proportionate distribution of supply without any necessity for regulation. Regulation arrangement, operative from a regulating bridge in case the throat width is more than 0.6 m and the flume is not combined with a road bridge, is provided.

7. *Distributary head regulator:* As the name implies, it is the head regulator of a distributary or minor to divert supplies from a main canal, branch or distributary into the offtaking distributary or minor and serves the same functions as a head regulator.

8. *Escape:* Escape is a work through which surplus or excess supply may be removed from a canal to a drainage.

11.9 FALLS—LOCATION AND CLASSIFICATION

Fall is a work designed to secure the lowering of the water surface in a channel and for the safe destruction of the surplus energy so generated. It is also termed as a drop. Water surface level in canals is governed by the area to be commanded. Furthermore, water surface slope permissible in the canal system is fixed from the consideration of regime slope corresponding to the given discharge as per Lacey's theory, consistent with the nature of the soil comprising bed and sides of the canal. Natural surface slope may vary in nature to any limit depending upon the configuration of the country. In cases, where the country slope exceeds the canal slope, falls have to be provided (Fig. 11.4).

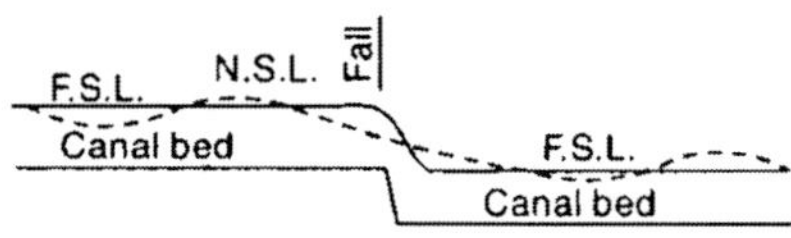

Fig. 11.4 Location of fall.

Location of Fall: Most economical section of a canal is represented by balancing depth. A suitable location of fall is thus a site where the depth of cutting falls short of balancing depth and banks go in embankment with the attendant disadvantages of high cost of construction and maintenance. Various considerations governing the location of falls are:

1. *In Main Canals and Branches:* Falls may be regulating or non-regulating. Regulating falls are located from the consideration of feeding an offtake immediately upstream of the work while non-regulating falls are fixed primarily from the consideration of economy in digging of the canal with regard to balancing depth and cost of fall itself, there being no direct irrigation.

2. *Distributaries and Minors:* Location of falls depends upon the command of the area to be irrigated. Fall located with outlets taking off immediately upstream of it with crest at the same level as that of fall embody the advantage of better command. In essence the location of a fall is indicated where full supply level rises appreciably above the natural ground level and the canal is completely in embankment.

A fall is combined with a bridge by either shifting its site to the bridge site or *vice versa* for reasons of economy. Fall on a parent channel with the head regulator of offtake is desirable from the standpoint of better working conditions for the offtake, proportionate distribution of supplies and economy in the construction of combined structure.

Classification of Falls: Falls are divided into the following four classes:

Class I: To maintain the depth-discharge relationship. In this class of fall the depth-discharge relation applicable for normal channel section is maintained at the fall. There is neither drawdown nor heading up of water as the water approaches the fall. Trapezoidal notch falls and low-crested rectangular notch falls belong to this class of fall. This class of fall is best suited to escapes which generally run with low surplus supplies. Rectangular notch fall can be designed as a meter.

Class II: To maintain a fixed supply level in the channel above the work. Falls which maintain nearly fixed water surface level upstream of the work are desirable (*i*) when a hydel power station is combined with a fall, (*ii*) when a subsidiary canal takes off at some distance upstream of the fall. High crested falls and siphon spillways are the principal types of this class of falls. High crested weir falls are constructed to the full bed width of the channel with low discharge intensity per metre width. These admit their use as meters.

Class III: To admit variations of the water level above the work. This type of fall admits of variation in the water level upstream of the work. It serves as a regulator. The variation in water level is accomplished in three ways, *i.e.*, by (*a*) sluice gates, (*b*) horizontal stop-logs (or karries), and (*c*) vertical needles or karries. All other types of falls which are designed from special consideration without any reference to the approach conditions such as rapids, well falls or cylinder falls belong to this class of falls.

11.10 TYPES OF FALLS

Fundamentally there are two types of falls, briefed as under:

1. **Vertical Drop Fall:** In these falls the nappe impinges in a pool downstream and the energy of flow is dissipated by the process of turbulent diffusion. This type is suitable for discharge up to 15 cumecs and drop up to 1.5 m and is not flumed.

(*i*) **Ogee fall:** All falls on Ganges canal are ogee falls; the first design of fall developed by Sir Proby Cantley. Despite provision of bed sloping gradually along successive convex and concave curves, (Fig. 11.5), this type of fall suffered from scouring in the canal on the upstream and downstream sides on account of heavy drawdown on the upstream side associated with lower depth, higher velocity and consequent bed erosion, and due to smooth downstream transit, kinetic energy is not dissipated but preserved causing scouring of bed below the fall to a depth sufficient to form standing wave.

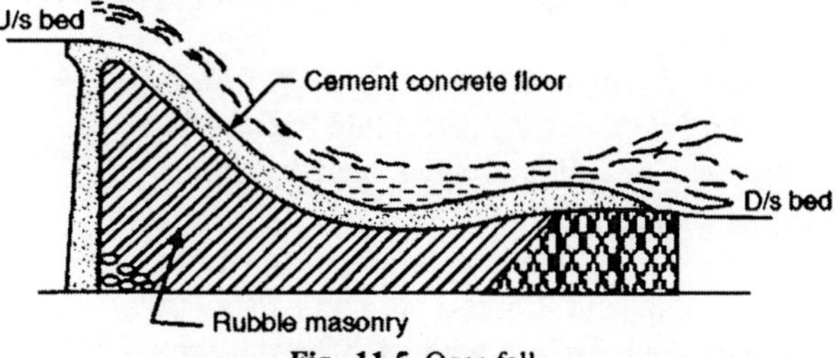

Fig. 11.5 Ogee fall.

(*ii*) **Rapid:** Next design evolved was of rapid essentially consisting of long galcis sloping at 1:10 to 1:20 (Fig. 11.6) with stone pitched sides as constructed on Western yamuna Canal to admit of timber traffic. As the destruction of kinetic energy was achieved through the formation of hydrualic jump on long glacis these worked well but were very expensive.

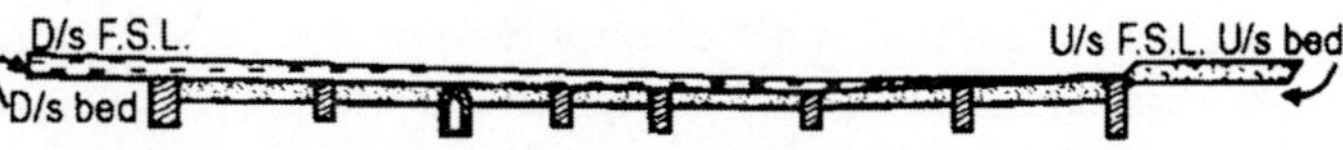

Fig. 11.6 Rapid fall.

(*iii*) **Stepped fall:** It was the modified form of rapid in that the long glacis was replaced by a floor in steps (Fig. 11.7) and essentially inherited the disadvantage of high cost. Later on, the efficiency of a vertical impact on the floor for energy dissipation came to be recognised and vertical falls with cistern were evolved.

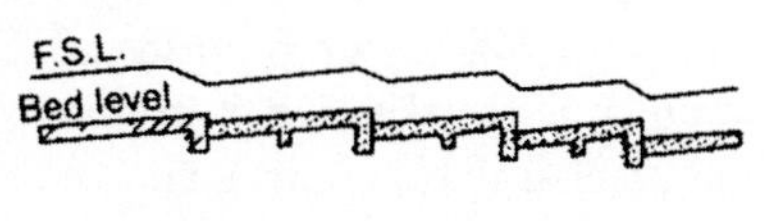

Fig. 11.7 Stepped fall.

(*iv*) **Vertical drop fall with cistern :** Development of vertical drop fall with cistern followed rapids. The cistern was introduced to destroy the surplus energy of water leaving the crest (Fig. 11.8). Grid consisting of baulks of timber were spaced few centimetres apart to expend partially the energy of the falling Jet. It suffered from getting clogged with rubbish, decay of timber necessitating their frequent replacement.

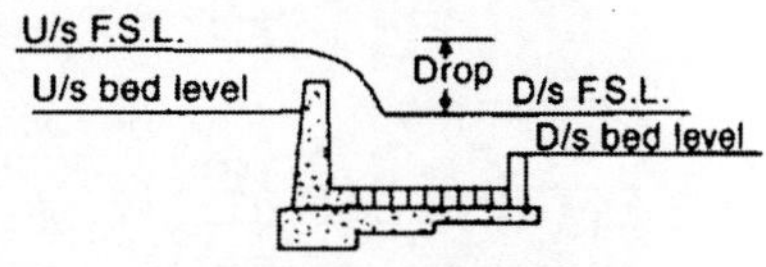

Fig. 11.8 Vertical drop fall.

(*v*) **Notch falls:** Notch fall consists of a high breast wall across the channel with one or more trapezoidal notches or openings (Fig. 11.9) to pass the discharge downstream. A smooth entrance and a flat lip projecting downstream was provided to splay the dropping jet. The crest was kept flush with upstream bed level of the channel to ensure no silting up upstream in the channel during low supplies. The notches were designed to pass the discharge corresponding to 1/4th and 3/4 of the full supply depth in the channel. Notch falls found extensive use in India and abroad but suffered from the drawback that they did not admit the use of discharge table connecting the gauge with the discharge.

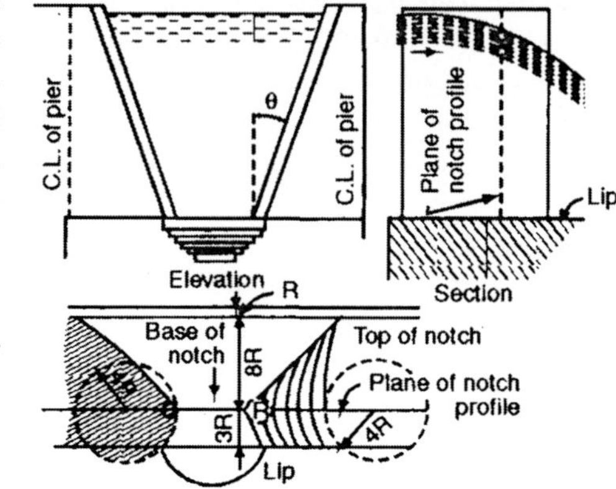

Fig. 11.9 Notch fall.

2. **Glacis Type Fall:** These are the falls in which a hydraulic jump is used for dissipation of energy on a suitably placed apron. These are of two types, briefed as under:

(*a*) *Straight Glacis with Baffle Platform and Baffle wall*: Also called Baffle fall. This type of fall was developed by Inglis and is called Inglis fall (Fig. 11.10). It is suitable for all discharges and drop more than 1.5 m and flumed. It consists of a straight sloping glacis with a baffle wall provided at a calculated height (1/10 of water depth) and distance from the toe of glacis to ensure formation of hydraulic jump on the baffle platform. At the end of the baffle wall, a cistern with a deflector at its downstream end followed by an apron and another cistern at the end are provided.

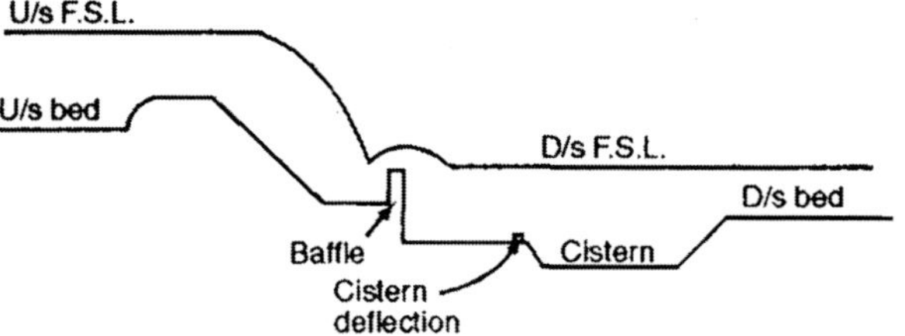

Fig. 11.10 Inglis fall.

(*b*) *Straight Glacis without Baffle Wall*: Montague fall. In it the straight glacis is replaced by a parabolic glacis, known as 'Montague profile' (Fig. 11.11). The formation of hydraulic jump on the glacis is assured. It is suitable for drop up to 1.5 m and discharge 60 cumecs and may be flumed if economical.

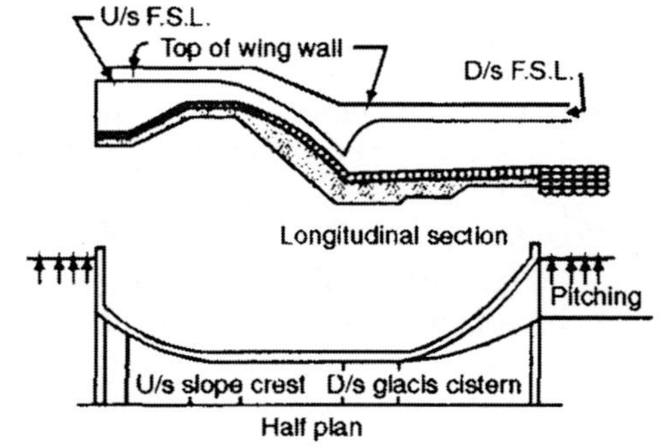

Fig. 11.11 Montague type fall.

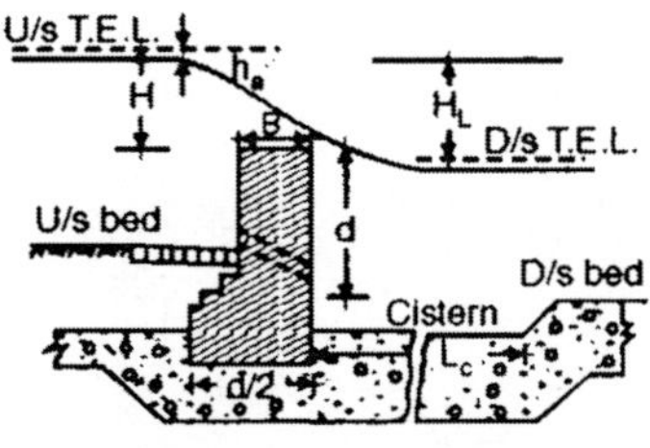

Fig. 11.12 Rectangular crest, sarda fall

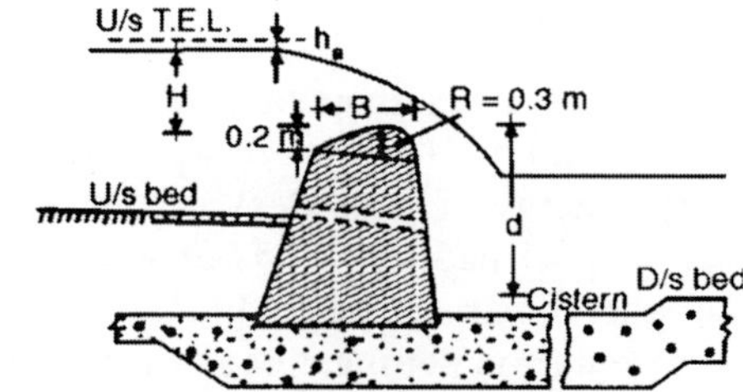

Fig. 11.13 Trapezoidal crest, Sarda fall.

(*a*) *Sarda fall:* Sarda type falls were constructed on Sarda cannal system in Uttar Pradesh which envisaged a raised crest in full channel width with vertical impact and no depressed cistern.

Model experiments conducted to improve this type of fall indicated the requirement of depressed cistern of length $5\sqrt{H \times H_l}$ and depth 0.25 $(H \times H_l)^{2/3}$.

Sarda falls are of two types, rectangular crest fall for discharge up to 14 cumecs (Fig. 11.12) and trapezoidal crest fall for discharge over 14 cumecs (Fig. 11.13). Design parameters for these two types of fall are as under:

Trapezoidal crest Sarda fall	*Rectangular crest Sarda fall*
1. Discharge formula $0.45\sqrt{2g}\,LH^{3/2}\,(H/B)^{1/6}$	$0.415\sqrt{2g}\,LH^{3/2}\,(H/B)^{1/6}$
$= 1.99\,LH^{3/2}\,(H/B)^{1/6}$	$= 1.84\,LH^{3/2}\,(H/B)^{1/6}$
2. Crest top width, $B = 0.55\sqrt{H+d}$ with u/s batter 1:3 and d/s 1:8.	$0.55\sqrt{d}$, base width $= \dfrac{H+d}{\rho}$

3. *Upstream protection*: (*i*) Wing walls are made segmental with a radius of 5*H* to 6*H* and subtending an angle of 60° at the centre and then continued tangentially into the berm. Wing walls are splayed straight at 45°, (*ii*) Bed pitching is laid at u/s bed level in 1:10 slope towards the crest in the length equal to full supply depth upstream.

4. *Downstream protection*: (*i*) Wing walls are kept vertical in a length 5 to 8 times $\sqrt{H \times Hl}$ from the crest, thereafter flared from vertical to 1.5:1 or 1:1 slope at the end of impervious floor; with average splay of 1 in 2.5 to 1 in 4. The wing walls are circular in plan which is tangential at the starting point of warping. These are taken up to the end of impervious floor, (*ii*) Side pitching consists of brick on edge supported on curtain walls one and a half brick thick and depth half of d/s depth or 0.5 m minimum. The pitching is curtailed at 45° from the end of bed pitching in plan or extended straight from the end of bed pitching, (*iii*) Bed protection consists of 20 cm thick brick over 10 cm ballast filling in a length indicated in Table 11.2. The dry brick bed pitching is laid level up to the end of wing wall and thereafter in a reverse slope of 1:10. The depth and number of curtain walls shall also be as given in Table 11.2, thickness being 0.3 m.

Table 11.2 Bed protection

H, Head over crest (m)	*Total length of d/s pitching (m)*		*Remarks*	*Curtain walls*	
				No.	*Depth (m)*
Up to 0.3	3.0	or	Sloping 1:10.	1	0.30
0.3 to 0.45	$3.0 + 2\,H_l$	3	Horizontal	1	0.30
0.45 to 0.60	$4.5 + 2\,H_l$	times	up to end of	1	0.45
0.60 to 0.75	$6.0 + 2\,H_l$	d/s	masonry	1	0.60
0.75 to 0.90	$9.0 + 2\,H_l$	depth	wings and	1	0.75
0.90 to 1.05	$13.5 + 2H_l$	(min.)	then sloping	2	0.90
1.05 to 1.20	$18.0 + 2H_l$		down at	2	1.05
1.20 to 1.50	$22.5 + 2H_l$		1:10	3	1.35

5. *Floor thickness*: Upstream floor is of thickness 0.3 to 0.4 m minimum. Downstream floor thickness is determined by Bligh's or Khosla theory; the minimum being 0.3 to 0.4 m. Vertical cutoff of 1 to 1.5 m depth is provided at d/s end of the floor.

(*b*) *Weir type fall:* Broad crested type meter flume falls are designed to bear a constant coefficient of discharge under variable heads so as to be a measure of discharge passing under various heads. With the formation of hydraulic jump at downstream of weir crest, the free fall formula holds good even up to drowning ratio of 85-90%. These are constructed for full designed bed width of channel. For further details refer Fig. 11.19.

(*c*) *Contracted fall:* Also called Glacis fall (Fig. 11.11), the fall with the flumed width, *i.e.*, contracted from the normal bed width of the channel is constructed as non-meter falls for economy in the cost of construction. Surplus energy after the formation of hydraulic jump on the glacis is destroyed further by means of friction blocks. CDO type fall developed in Punjab is the commonly used type (Fig. 11.19).

Design Procedure: (*i*) Throat width, $Bt = 0.6\ B$; rounded to nearest tenth of a metre, (*ii*) Depth of water over crest, H is determined from the formula $Q = C.Bt.\ H^{3/2}$; value of C is as under. It covers effect of velocity of approach.

Type of fall	*Valuce of C*			
	Discharge (cumecs)			
	0.015–0.055	*0.055–1.5*	*1.5–3.5*	*over 3.5*
Broad crested	1.66	1.675	1.69	1.705
Sharp crested	1.78	1.80	1.82	1.84
	The value of C is reduced by 2.5%, 3.5% and 5% for 45°, 60° and 90° offtake.			

(*iii*) Throat length, $Lt = 2\ H$, (*iv*) Radius of u/s approach wall = $2\ H^{3/2}$, starting from crest, the curve subtends an angle of 60° and continues tangentially by 0.6 m beyond surface width on either side of distributary of < 3 cumecs capacity and 1 m for higher discharges, (*iv*) Horizontal length of side curve and bed curve for joining crest with u/s bed, $La = \sqrt{3}\ H^{3/2}$, Radius of curve, $R = \dfrac{La^2 + x^2}{2x}$, where x is height of crest above u/s bed, (*vi*) Glacis has a slope of 2.5:1. Radius of curve joining crest with downstream glacis is 0.6 m, (*vii*) Length of cistern, $Lt = 4\ (D_2 - D_1)$, where D_2 is depth of water in the cistern and D_1 is hypercritical depth at hydraulic jump, (*viii*) Depth of cistern below d/s bed level = 7.5 cm minimum , (*ix*) Downstream expansion, $Le = \dfrac{3(B - Bt)}{2}$ or length of glacis + length of cistern, whichever is more. Radius of expansion, $R = \dfrac{Le^2 + (B - Bt)^2 / 2}{B - Bt}$, Bt = overall width including piers, if any, (*x*) Upstream bed protection: As per Khosla theory. Minimum thickness is 15 cm thick brick on edge over 15 cm thick concrete with toe wall of 1/3 F.S. depth, subject to minimum of 0.3 m depth over 0.15 m concrete, (*xi*) Downstream bed protection: Side pitching in a length 3 times the length of cistern (*i.e.* d/s F.S. depth + fall) consisting of dry brick pitching 20 cm thick supported on toe wall of F.S. depth/2 subject to minimum of 30 cm thick masonry over 15 cm thick concrete. The bed protection consists of brick bats of thickness, as under:

Brick bats	*Discharge (cumces)*				
	Up to 1	1–2	2–3.5	3.5–5.5	5.5–10
Thickness (cm)	15	25	30	45	60

Floor to be of minimum thickness of 20 cm masonry over 0.3 m thick concrete suitably increased at the point of hydraulic jump, (*xii*) Fall combined with bridge. Parallel sides of flume serve as abutments. Bed of distributary up to u/s end of abutment is provided with impervious floor, and (*xiii*) Free board = 0.5 m. For more details refer example 11.4.

11.11 UTILITY OF THE VARIOUS TYPES OF FALLS

The utility of the various types of falls is as follows:

Vertical falls: Unflumed vertical falls are suitable for discharges up to 15 cumecs and fall up to 1.5 m. These do not function satisfactorily if flumed.

Glacis falls: Straight glacis type falls are suitable for all conditions, if unflumed, but in that case the cost is high. Friction blocks are needed for destruction of energy downstream of the fall in the case of flumed falls.

Inglis fall: It functions satisfactorily either flumed or unflumed so long as it is not drowned.

Notch fall: Trapezoidal notch does not give satisfactory results as a meter but can be calibrated to work as meter provided the channel upstream of the work neither silts nor scours. Rectangular notch fall can serve as meter.

Broad crested weir: Highly successful as meter and admit of its use as such with a drowning ratio of 85–90% of the depth of water over the crest.

Sharp crested falls: Sharp crested rectangular falls can function as meter so long as free fall conditions are available. The crest of the fall should always be higher than the downstream bed level of channel.

The type of fall preferable is one which is most effective in the dissipation of kinetic energy for the given discharge and fall. However, vertical drop fall is not in general use. Glacis fall (CDO type) is mostly used.

11.12 CONTRACTED AND UNCONTRACTED FALLS–COMPARISON

Full-channel width fall	*Contracted fall*
Discharge intensity is low and hence serves as reliable meter.	Higher the contraction, higher is the discharge intensity and hence unsuitable as meter.
High set crest.	Low set crest.
High set crest causes silting up in the upstream.	With low set crest, it causes neither silting nor afflux.
Suitable for accurate gauging depth of water on crest as loss of head in entry is negligible.	Loss of head in entry and exit from the work is more but is of no consequence.
Energy dissipation devices downstream of the fall are generally not required.	Energy dissipation devices downstream of the fall are necessitated for complete destruction of surplus energy.
Economy in the cost of pitching as the action in the channel section (earthen) is very much reduced.	Elaborate bed and sides protection both on the downstream and upstream is necessitated.
Comparatively short wing walls required are conducive to hydraulic efficiency.	Longer wing wall appurtenant to fluming are required.
No complications of contraction and expansion.	Expansion and contraction are there due to fluming.
Uneconomical when combined with a bridge due to wider waterway.	Economical especially when combined with a bridge due to restricted waterway.

11.13 METER FLUME

The work designed for measuring discharge from the direct measurement of depth of water flowing over its crest is called meter flume. Discharge in a clear overfall flume depends upon one factor only, *i.e.*, the depth of water over the crest and as such the probable errors of discharge observation depends on one measurement only against three measurements in the case of a regulator, *viz.* (*i*) upstream water depth, (*ii*) downstream water depth, and (*iii*) area of the opening. The drawback of a clear overfall type is the material loss of head it involves but it affords better conditions for calculating the discharge and is essential for an efficient control of water distribution than a submerged one, which may require the measurement of both the upstream and downstream water levels under certain conditions.

Advantages: (*i*) Constant coefficient of discharge, (*ii*) Negligible loss of head between upstream and downstream sides of the flume, (*iii*) No afflux is caused as a result of provision of smooth transition, and (*iv*) Does not obstruct the passage of bed load and suspended silt charge and as such does not lead to silting trouble.

Design parameters of meter flume (uncontracted) are given in example 11.5.

11.14 REGULATORS

The regulators aim at admitting only the suspended sediments into the offtake in accordance with the carrying capacity (discharge) of the offtakes by dividing the sediment load carried by the flow in the parent channel into coarse particles in contact with the bed, and the finer suspended load comprising of particles in suspension. Thus the design of regulator is to ensure proportional distribution of supplies and sediments. Crest level of the regulator plays important role in sediment distribution as also in distribution of supplies. The crest level of offtakes is designed to run with full supply discharge with a water level 1/5th depth of water in the upstream parent channel below designed full supply level corresponding to approximately 2/3rd supply in the parent branch canal. Crest level of the regulator on the parent channel is kept the same as that of head regulator of the offtaking channel. The crest level is kept at least 0.15 m higher than the downstream bed level of the channel, parent or offtaking whichever is higher, to cater for any silting up downstream and thereby drowning the crest.

Canal regulating works may be (*i*) head of the offtaking channel, (*ii*) cross regulator combined with head regulator of the offtake, and (*iii*) cross regulator combined with the head regulators of two offtaking channels. The orientation of the regulator thus depends on the alignment of the offtaking channel with respect to the parent channel, as under. Alignment of the offtaking channel is considered to be that which makes zero angle with the parent channel initially and separates out in a transition.

1. Parent channel running straight and the offtaking channel takes off at 45°, 60° or 90°. It is the case when (*a*) offtake is of small discharge compared to that of the parent channel, and (*b*) it can be fed even during low supplies in the parent channel.

2. Parent channel bifurcates into two offtaking channels equally aligned. Here the discharge capacity of each is not of consequence, the head regulator has to be provided on both.

3. Parent channel running straight with two offtakes at 45°, 60° or 90°. In this case regulator on the parent channel is necessarily provided in conjunction with that of each offtake. It is termed as proportional distributor. A headless flume is provided on parent channel in case there is no fall between the full supply levels at upstream and downstream.

The supplies drawn by a canal from the river through the headworks is controlled by cross regulators and distributed amongst the offtaking channels through the head regulator.

Cross Regulator

The regulator constructed across a channel is called cross regulator. It is constructed just on the downstream of the offtaking channel to control distribution of supplies and silt between the parent and offtaking channels and also to maintain water level on its upstream as well as downstream, besides serving other functions described hereinfter. It is usually combined with a road bridge or a fall on the parent channel and is then also termed as a regulating fall. A cross regulator in conjunction with head regulator of the offtake is not required when the available working head in the offtake is more than half the F.S. depth in the parent channel that is the offtake is capable of drawing its discharge even during one-third supply in the parent channel (Refer example 11.6). The regulation at the cross regulator is made with the help of manually operated karries in grooves provided in the upstream portion of piers and mechanically operated gates when depth of flow is more than 2 m.

Functions: A cross regulator serves the functions (*i*) Effect equitable distribution of supplies amongst the distributing channels thereat, (*ii*) Raise water level when supply in the parent channel is low to feed offtakes, (*iii*) Provide means for cutting off supplies for repairs, etc. downstream by diverting the supplies to other canals or part of the same canal, (*iv*) Divide the canal in certain reaches for effective regulation, (*v*) Provide means of communication across the channels through the bridges provided over them, (*vi*) Provide means for absorbing fluctuations in various reaches of the canal system in order to avert breaches in tail reaches, (*vii*) Ensure safety of canal lining in the reaches where the subsoil water levels are high, (*viii*) Escape water from canals in conjunction with escapes, and (*ix*) Control water surface slope, in conjunction with falls, for bringing the canals to regime slope and section.

Design Parameters: (*i*) Discharge, $Q = C\,Bt\,H^{3/2}$

where, Q is full supply discharge (cumecs), C is coefficient of discharge, usual value 1.7, Bt is clear waterway (m) and H is head of water over crest *i.e.* F.S.L. u/s + head due to velocity of approach (ha)–crest level, (*ii*) *Crest profile:* The crest profile generally conforms to that of a fall. IS: 7114-1973 specified crest profiles for canals with discharge up to 10 cumecs and above are as under:

Discharge (cumecs)	*Crest width (m)*	*Radius of curve joining crest with*		*Glacis slope*		*Curve joining crest with d/s glacis*
		U/s glacis	*D/s glacis*	*U/s*	*D/s*	
Up to 10	2/3 H	H	1.5 H	curve of $R = (3H^3 - x^2)/2x$	2.5:1	0.6 m
Over 10	2/3 H	H	1.5 H	2:1	2:1	0.6 m
	H = depth of water over crest, x = height of crest above u/s bed level.					

Waterway. For a headless regulator, overall linear waterway is equal to bed width. On lined canal, it is kept equal to width of canal at full supply level. For regulator combined with a fall, the drowning ratio is greater thán 0.8 for submerged falls. For free falls, crest is kept higher than downstream bed level. The fluming ratio Bt/B is obtained from Fig. 11.14.

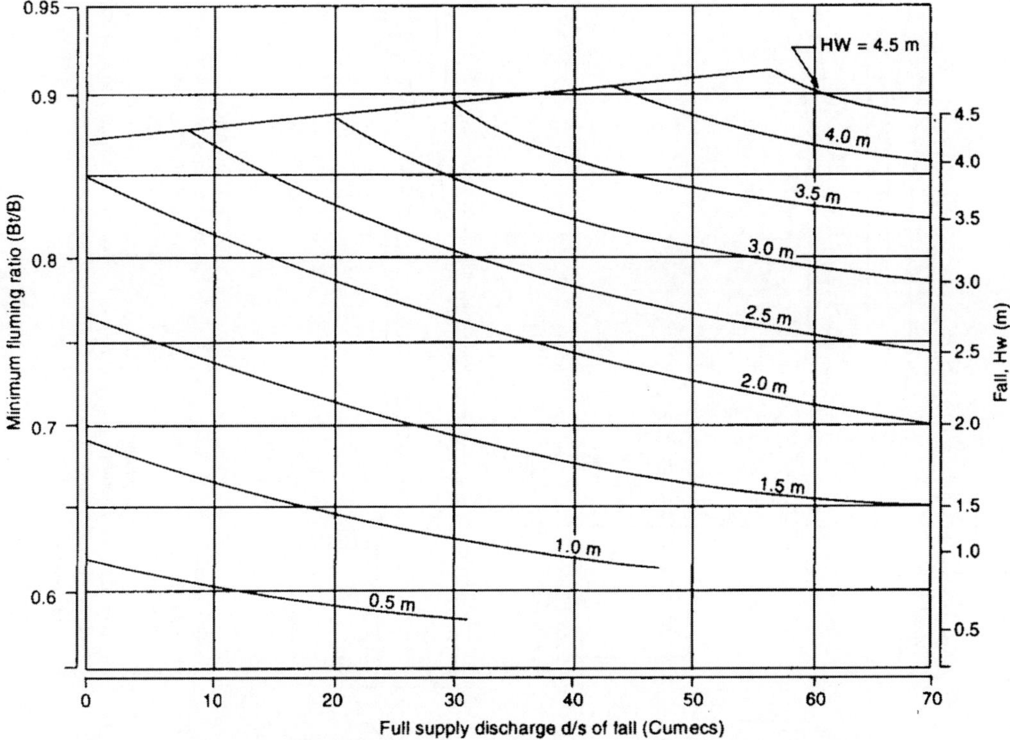

Fig. 11.14 Fluming ratios for falls combined with regulators (IS: 7114-1973)

Head Regulator of a Distributary

The silt selective head regulator is a device by means of which only silt conforming to the silt conductive power of the offtaking canal is allowed to be taken into it. It consists of approach chamber, regulator and the flume as shown in Fig. 11.15.

1. Approach Chamber

(*i*) *Depth of flow:* The depth of flow in the approach chamber is calculated from the relation:

$$r_1 = (Ha/du)^{1/3}$$

Where, r_1 is silt conductive power in offtaking canal, ha is depth of flow in the approach chamber (m), and du is depth in parent channel (m).

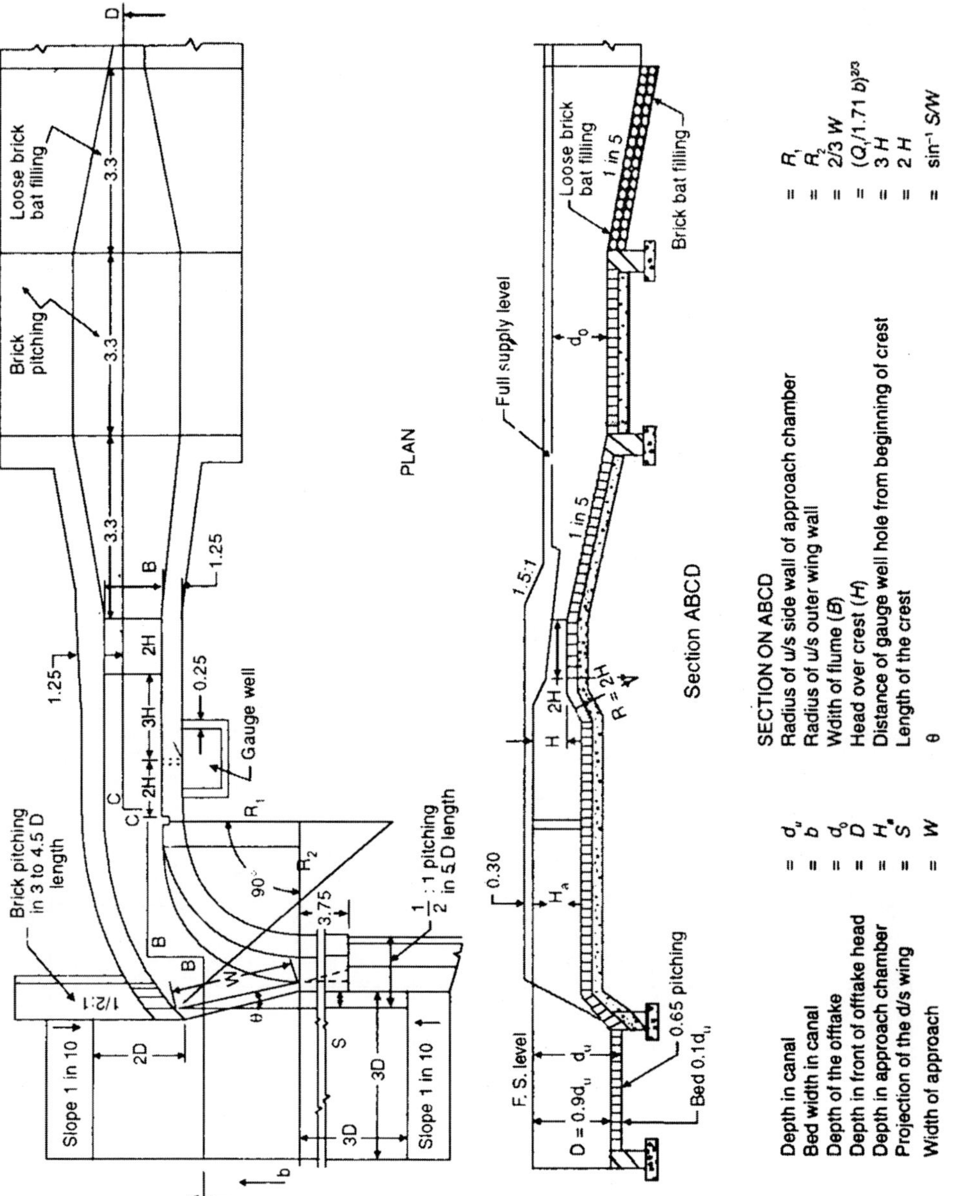

Fig. 11.15 Head regulator of a distributary (IS: 7495-1974).

(*ii*) Silt conductive power of offtaking canal,

$$R = (r_1)^{0.63}\ (\lambda)^{1/6}$$

where, R is CVR in offtaking canal/CVR in parent canal, and λ is depth in parent channel/depth in offtaking canal.

(*iii*) Width of approach chamber,

$$W = k\,(S \times du/Ha)$$

Where, W=width of approach chamber, $k = C$ coversine $\phi\frac{1}{2}\times$ coversine $\phi\frac{1}{6}$ a constant generally varying from 1.5 to 2.0 [C = constant = 2.0, $\phi = \sin^{-1}(S/W)$ that is, inclination of the approach to the central line of the parent channel, and ϕ is inclination of the offtake to the normal of the parent channel]; and S is setback of the upstream wing wall given by the formula:

$$S = \frac{Q_1}{Q_2}\,(b + d_u/2),$$

where, Q_1 is discharge in offtaking canal, Q_2 is discharge of the parent canal, and b is bed width of the parent canal.

(*iv*) *Radius of wings:* The radius of u/s side wall of the approach chamber is 3 times the depth of flow in the chamber.

2. Regulation

The width of crest of the regulator is 2/3 that of approach chamber. The straight approach channel length is 5 H, H being head over crest. Length of crest is $2H$. The bed of the approach chamber is joined to the top of crest by a curve of radius $2H$, and the crest is joined to the bed of the offtaking canal by a glacis of slope 5:1.

3. Fluming

The width of flume is widened to the bed width of the offtake through straight 5:1 splays starting from downstream edge of the horizontal portion of the crest.

11.15 CANAL ESCAPES

Canal escape is defined as a channel meant for the removal of surplus or excess water from the canal into nearby drainage. Despite control at head, a canal portion may be in distress due to breach, etc. The supply in the canal can be reduced from the head regulator. But first it shall take time to convey the message to the head for closure and secondly the supply already released from the head up to the breach point shall have to be endured but immediate relief is essential to control widening of the breach and averting further damage to crops. The necessity for the provision of escapes in a canal is due to other reasons as well, such as (*i*) Excessive rainfall in upper reaches of the canal and somehow admission of storm water into the canal may result in excessive rise in water level and discharge beyond the carrying capacity of the canal, (*ii*) Closure of canal outlets in downstream reach by the cultivators due to low demand or heavy rainfall meeting their water requirement, (*iii*) Faulty regulation at the canal head, and (*iv*) Non-withdrawal of supply by a lift offtaking canal due to breakdown in water lifting installation or sudden closure of an offtake due to breach in it. Escapes are thus safety valves for the canal system and are better constructed at intervals along the canals to cater for surplus water and control the water levels in the canals with a view to save the latter from breaching or outflanking.

The functions served by canal escape are (*i*) Safety valve to protect the canal against possible damage by flooding, (*ii*) Emptying of the canal reach, below the escape, for silt, or weed removal, repairs and maintenance, (*iii*) Removal of supply from the canal upstream of a breached site in order to avert damage to standing crops and property and habitations nearby as also to facilitate its expeditious closure, and (*iv*) Periodical flushing off the silt prone head reach of a canal through the escape.

Location: Most economical location for an escape is near suitable drains, depressions or river at low level from the canal or in combination with a canal aqueduct, the other desirable locations are (*i*) In a single bank canal, the unbanked length brings in storm water as it acts as long continuous surface inlet. Escapes may be provided at suitable locations in the vicinity, (*ii*) In head reach of a canal having high silting trouble, escape may be located within 5 km of the canal head to cater for periodical flushing of silt from the head reach, (*iii*) An inlet bringing in considerable discharge and sediment necessitates provision of an escape in the canal bank directly opposite to the inlet or more conveniently below for the disposal of drainage water and sediment, (*iv*)

In lift channel, upstream of the pumping station to escape the supply in the event of breakdown of pumping installation, and (*v*) In a power channel, escape is necessary upstream of penstocks. Escapes may be provided at 15 to 20 km in main canals and 10 to 15 km in the case of other canals.

Types of Escapes

1. Weir Escape

Also termed as surface escape. It is a weir or flush escape which is capable of disposing surplus water from the canal. In this type, the crest level of the escape is kept at full supply level of the canal to cater for the disposal of surplus water above that level in the canal. It may or may not be fitted with the crest shutters (Fig. 11.16). It is desirable that the bed of escape canal is below bed level of the canal and has non-silting, non-scouring velocity. The waterway through the escape is fixed considering (*i*) the discharge it has to cater which may be 50% of the maximum discharge of the canal or the difference between the maximum discharge of the canal at the site and maximum flow at the next lower escape, and (*ii*) depth of flow not exceeding 50% of free board in the canal. It is a useful type of escape at the tail of a canal or lift channel to cater for escaping excess supply when there is fluctuation in the pumping.

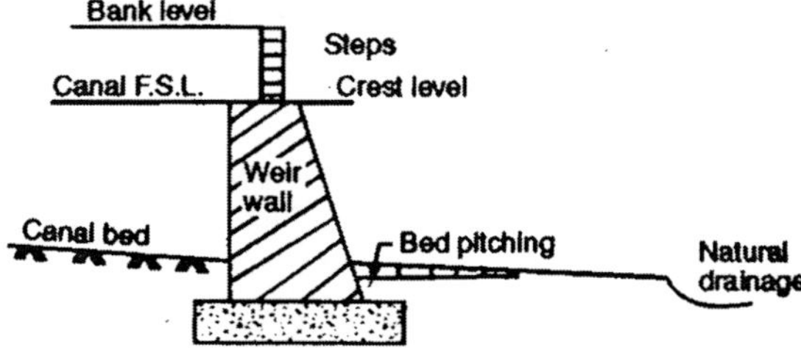

Fig. 11.16 Weir escape.

2. Sluice Escape

In this type, crest level of the escape is at or below the bed level of the canal (Fig. 11.17) to enable quick emptying of the canal or to act as scouring sluice for the removal of silt that has entered the canal upstream of the escape. The flow is controlled by gates of undershot type. The waterway through the sluice is fixed with a view to ensure quick emptying of the canal, considering the discharge, which is limited to a percentage of the design discharge of the canal at the site and crest level as low as possible depending on the permissible bed level of the escape channel which in turn is related to the outfall drainage levels. The escape channel has its bed level below the bed level of the canal and provided with adequate cross section and slope to carry the flow and sediment down below. Energy dissipation arrangement is provided to cater for all conditions of flow and operation. It is a useful type when quick emptying of the canal is required as also when an inlet brings in considerable sediment to be disposed of through scouring operation of the escape.

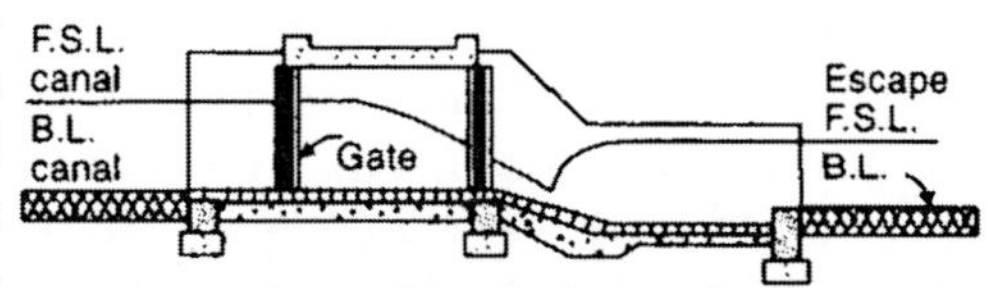

Fig. 11.17 Sluice escape.

SOLVED EXAMPLES

Example 11.1 *Data observed in a canal fall are FSL 110 m, bed level of canal 107 m, bed width 16 m, discharge 35 cumecs, crest length of fall 10 m (crest section is rectangular) and coefficient of discharge 1.7 $m^{1/2}$ s. Design fall crest.* *(AMIE)*

Solution.

Canal depth, (110–107) = 3.0 m

Sectional area = $(16 + 6) \times 3 = 66$ m²

Velocity of approach, $V_a = Q/A = 35/66 = 0.53$ m/s

$$h_a = V_a^2/2g = (0.53)^2/2 \times 9.81 = 0.014 \text{ m}$$

$$Q = C\,L\,E_f^{3/2}$$

$$35 = 1.7 \times 10 \times E_f^{3/2}$$

or $E_f = 1.62$ m

$H = E_f - h_a = 1.62 - 0.014 \simeq 1.61$ m

Therefore, crest level $= FSL - H = 110.0 - 1.61 = 108.39$ m.

Example 11.2 *Design a sarda type fall on a channel of 15 cumecs, bed width 18 m, depth 1.5 m, u/s FSL 101 m, bed level 99.5 m, NSL 90.0, fall 1.0 m, side slope 1/2:1.*

Solution.

It is a case of trapezoidal fall.

Discharge = 15 cumecs

Length of crest, Bt = Bed width = 18 m

Top width of crest, $B = 0.55\sqrt{D+H_l} = 0.55\sqrt{1.5+1.0} \simeq 1$ m

Discharge formula, $Q = 1.99\, L\, H^{3/2}\, (H/B)^{1/6}$

$15 = 1.99 \times 18 \times H^{3/2}\, (H/1.0)^{1/6}$

or $H = 0.59$ m

Velocity head, $ha = V^2/2g = [15/(18 + 1.5 \times 0.5)\, 1.5]^2/2 \times 9.81 = 0.014$m

TEL u/s $= 101.0 + 0.014 = 101.01$ m

Crest level $= 101.00 - 0.59 = 100.41$ m

Crest height above u/s bed level $= 100.41 - 99.50 = 0.91$ m

U/s approach radius $= 5$ to $6H = 6 \times 0.59 = 3.54$ m

Radius of u/s wing wall is 3.5 m subtending an angle of 60° at the centre.

D/s approach straight length $= 6\sqrt{H \times H_l} = 6\sqrt{0.59 \times 1.0} \simeq 5$ m

and then warped in 1:1 slope.

Hydraulic head, d = Crest level – d/s bed level

$= 100.41 - 98.50 = 1.91$ m

Exact top width of crest $= 0.55\sqrt{H+d} = 0.55\sqrt{0.59+1.91} \simeq 1$ m

Length of u/s floor with Bligh's $c = 10$ $= 1.91 \times 10 = 19$ m

D/s toe wall depth $= d/2 + 0.60 = 1.5/2 + 0.6 = 1.35$ m

U/s toe wall depth $= d/3 + 0.60 = 1.5/3 + 0.6 \simeq 1.0$ m

Horizontal floor length $= 19.0 - 2(1.35 + 1.0) \simeq 15$ m

U/s floor length = Total length – d/s floor length required

= 15 – 2 (water depth + 1.2) – 1.0

$= 15.0 - 2\,(1.5 + 1.2) - 1.0 \simeq 8$ m.

Cistern length $= 5\sqrt{H_l \times H}$

$= 5\sqrt{(1.0 \times 0.59)} \simeq 4$ m

Cistern depth $= [(H \times H_l)^{2/3})]/4$

$= [(0.59 \times 1.0)^{2/3}]/4 \simeq 0.2$ m

Cistern level $= 98.50 - 0.20 = 98.30$ m

Total d/s floor pitching length, $= 4.5 + 2\,H_l$ (From table 11.2)

level and then sloping in 1:10 $= 4.5 + 2\,(1.0) = 6.5$ m

U/s bed pitching. It is level at u/s bed level in 1:10 slope towards the crest. 20 cm drain holes are provided in the entire length at 3 m *c.c.* for drying the bed during closure.

Downstream side pitching. The side pitching in a length 3 × d/s depth = 3 × 1.5 = 4.5 m is warped from side slope 1:1 to 1.5:1. The pitching is curtailed at 45° angle from end of pitching in plan. Fig. 11.18 shows the Sarda fall designed as following.

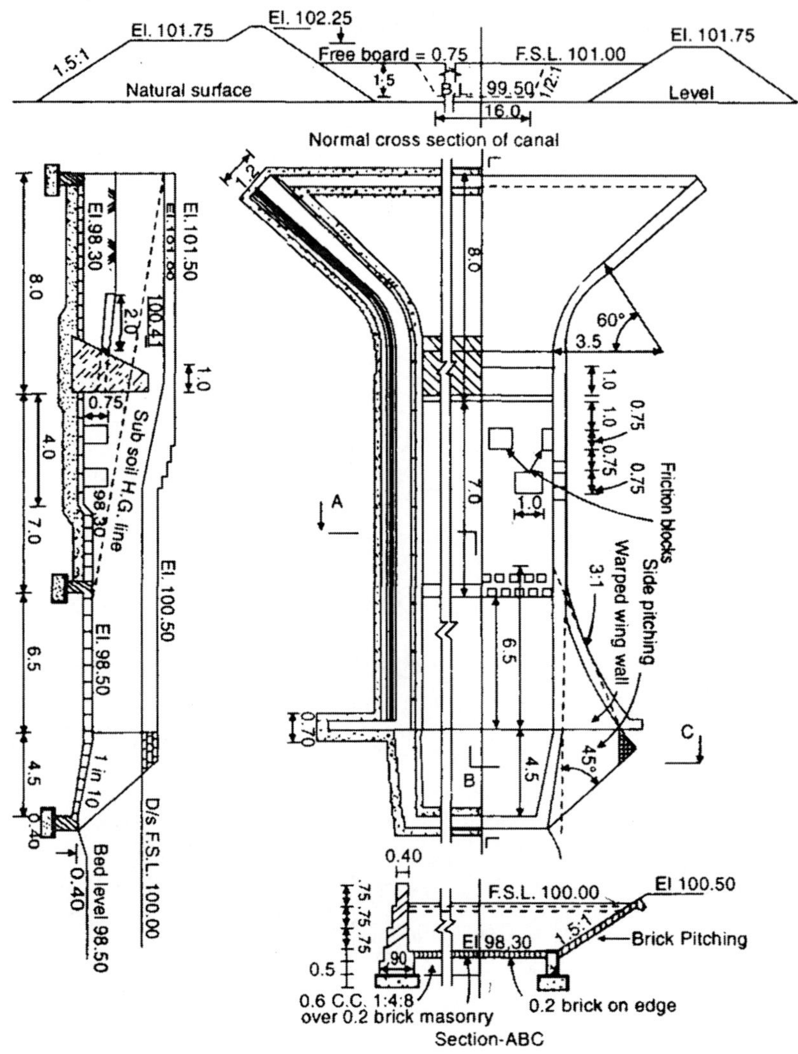

Fig. 11.18 Sarda type fall.

Example 10.3. *Design a Sarda type fall for an irrigation channel of 13 cumecs capacity, bed width 13 m, depth 1.5 m, side slope 1/2:1, F.S. level u/s 101.00 m, fall 1 m, soil is loamy-Bligh's coefficient 6.*

Solution. The discharge is below 14 cumecs and as such the fall is of rectangular type.

Crest length, L = 13 m, the same as channel bed width.

Assume top width of crest, B = 0.75 m

Discharge formula, $Q = 1.84\, L\, H^{3/2}\, (H/B)^{1/6}$

$13 = 1.84 \times 13 \times H^{3/2}\, (H/0.75)^{1/6}$

or H = 0.68 m

Velocity of approach, $V_a = Q/A = 13/(13 + 1.5 \times 0.5)1.5 = 0.63$ m/s

Velocity head, $V_a^2/2g = (0.63)^2/2 \times 9.81 = 0.02$ m

U/s T.E.L. = U/s F.S.L + velocity head

U/s T.E.L	$= 101.0 + 0.02 = 101.02$ m
Crest level	= U/s F.S.L.–H
	$= 101.0 - 0.68 = 100.32$ m
Height of crest above d/s F.S.L.	= Crest level - d/s F.S. level
	$= 100.32 - 100.0 = 0.32$ m
Height of crest above d/s bed level, d	= Crest level – d/s bed level
	$= 100.32 - 98.50 = 1.82$ m
Top width of crest	$= 0.55\ d^{0.50}$
	$= 0.55\ (1.82)^{0.5} = 0.75$ m, the same as assumed
Width of crest at base	$= (H + d)/\rho$
	$= (0.68 + 1.82)/2.24 = 1.25$ m
Length of cistern	$= 5\sqrt{H \times H_l} = 5\sqrt{0.68 \times 1.0} \simeq 4.25$ m
Depth of cistern	$= 0.25\ (H \times H_l)^{2/3} = 0.25\ (0.68 \times 1.0)^{2/3} = 0.19$ m
Impervious floor length	= C.d., as per Bligh's theory
	$= 6 \times 1.82 \simeq 11$ m
Minimum length of d/s floor	$= 2$ (F.S. depth + 1.2) + H_l
	$= 2\ (1.5 + 1.2) + 1.0 = 6.5$ m
Floor length u/s of crest	$= 11.0 - 6.5 = 4.5$ m
Maximum unbalanced head	$= 100.32 - 98.50 = 1.82$ m.

Floor thickness

Distance (D) from d/s toe (m)	Unbalanced uplift pressure $(U) = 0.19 + \frac{1.82 \times D}{11.0}$ (m)	Floor thickness $= \frac{U}{1.24}$ (m)
1.0	0.36	0.30, provide 0.20 m cement masonry over 0.3 m cement concrete.
2.25	0.57	0.46 say 0.50 m, provide 0.20 m thick cement masonry over 0.3 m concrete.
4.5	0.94	0.77 say 0.75 m, provide 0.20 m masonry over 0.55 m concrete.
6.5	1.28	1.04 say 1.0 m, provide 0.20 m masonry over 0.80 m concrete.

Upstream bed protection: 0.20 m thick pitching is laid at u/s bed level in 1:10 slope towards the crest in 1.5 m length *i.e.* equal to u/s depth. 3 No. 150 mm dia G.I. pipes are laid in the crest at u/s bed level to drain out the u/s bed during the canal closure.

Depth of u/s curtain wall = u/s depth/3 = 1.5/3 = 0.5 m, provide 0.40 m wide masonry curtain wall, 0.5 m deep over 0.3 m concrete base.

D/s side protection: Provide 0.20 m dry brick pitching over 0.1 m thick ballast in a length of 8m (*i.e.*, 6.0 + 2 HI = 6.0 + 2 × 1.0) resting on 0.4 m wide curtain wall 1.35 m deep (*i.e.*, d/s depth/2) + 0.60 = 1.5/2 + 0.6 = 1.35 m.

Fig. 11.19 shows Sarda fall designed as above.

Example 10.4 *Design a CDO (Punjab) type fall on a distributary of 3.05 cumecs discharge, bed width 6 m, F.S. depth 1 m, F.S.L. u/s 101.0 m, d/s 100.0, bed level u/s 100.0, d/s 99.0, natural surface level 99.5 m.*

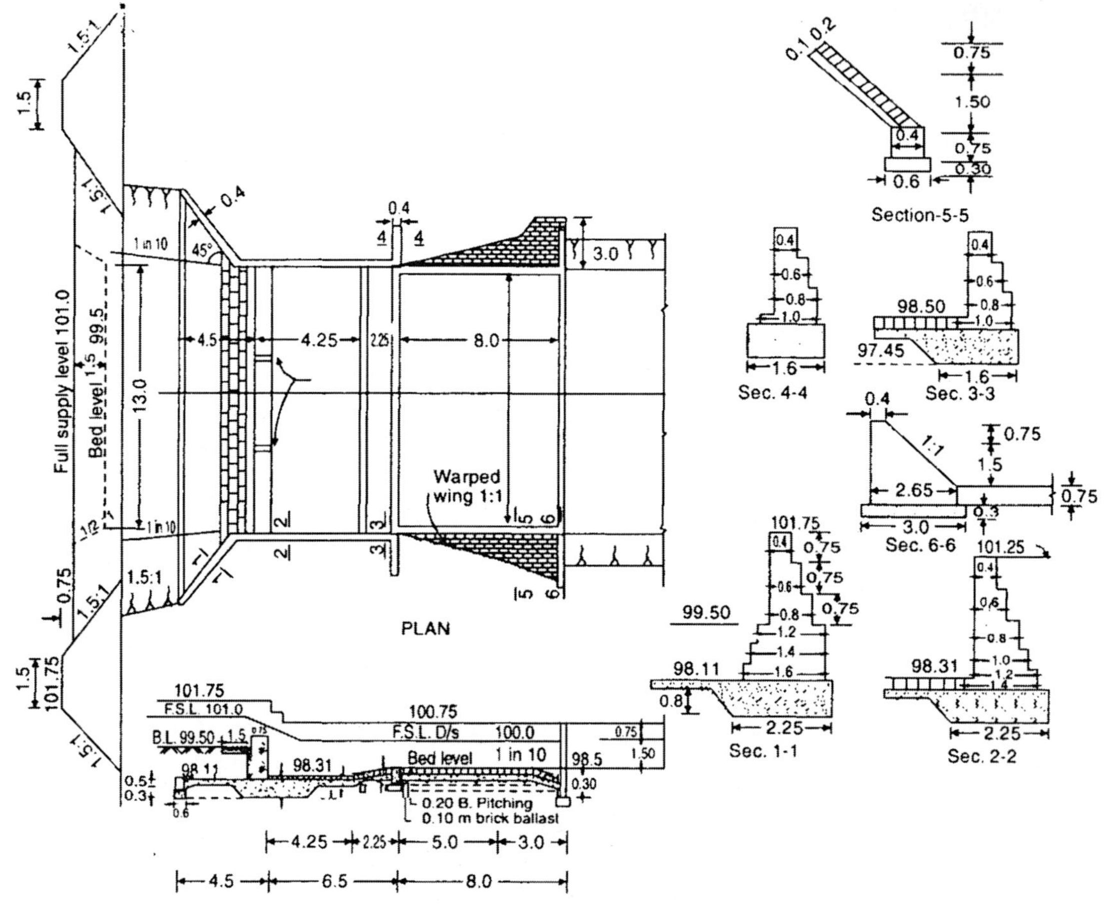

Fig. 11.19 Sarda fall.

Solution. Throat width, $B_t = 0.6 \times 6.0 = 3.6$ m

Discharge formula, $Q = C B_t H^{3/2}$

$$H^{3/2} = \frac{Q}{C B_t} = \frac{3.05}{1.69 \times 3.6} = 0.50 \text{ or } H = 0.63 \text{ m}$$

Crest level $= FSL - H = 101.0 - 0.63 = 100.37$ m

Length of u/s approach, $L_a = \sqrt{3}\, H^{3/2} = 1.732 \times 0.50 = 0.87$ m

Length of throat, $L_t = 2H = 2 \times 0.63 = 1.26$ m

Radius for u/s side wall $= 2 H^{3/2} = 2 \times 0.5 = 1.0$ m

Radius for curve joining crest with u/s floor $= (L_a^2 + x^2)/2x$, x being height of crest above u/s bed level $= 100.37 - 100.00 = 0.37$ m.

$\therefore$ $= [(0.87)^2 + (0.37)^2]/2 \times 0.37 = 1.21$ m

Radius for joining crest with d/s floor = 0.6 m

Cistern level = D/s bed level – 0.10 = 99.0 – 0.10 = 98.90; 0.1 m minimum.

Length of cistern $= 4\,(D_2 - D_1) = 4(0.65 - 0.18) = 1.88$ m (from Fig. 11.20) where D_2 is depth of water in the cistern and D_1 is hyper-critical depth at the point of hydraulic jump

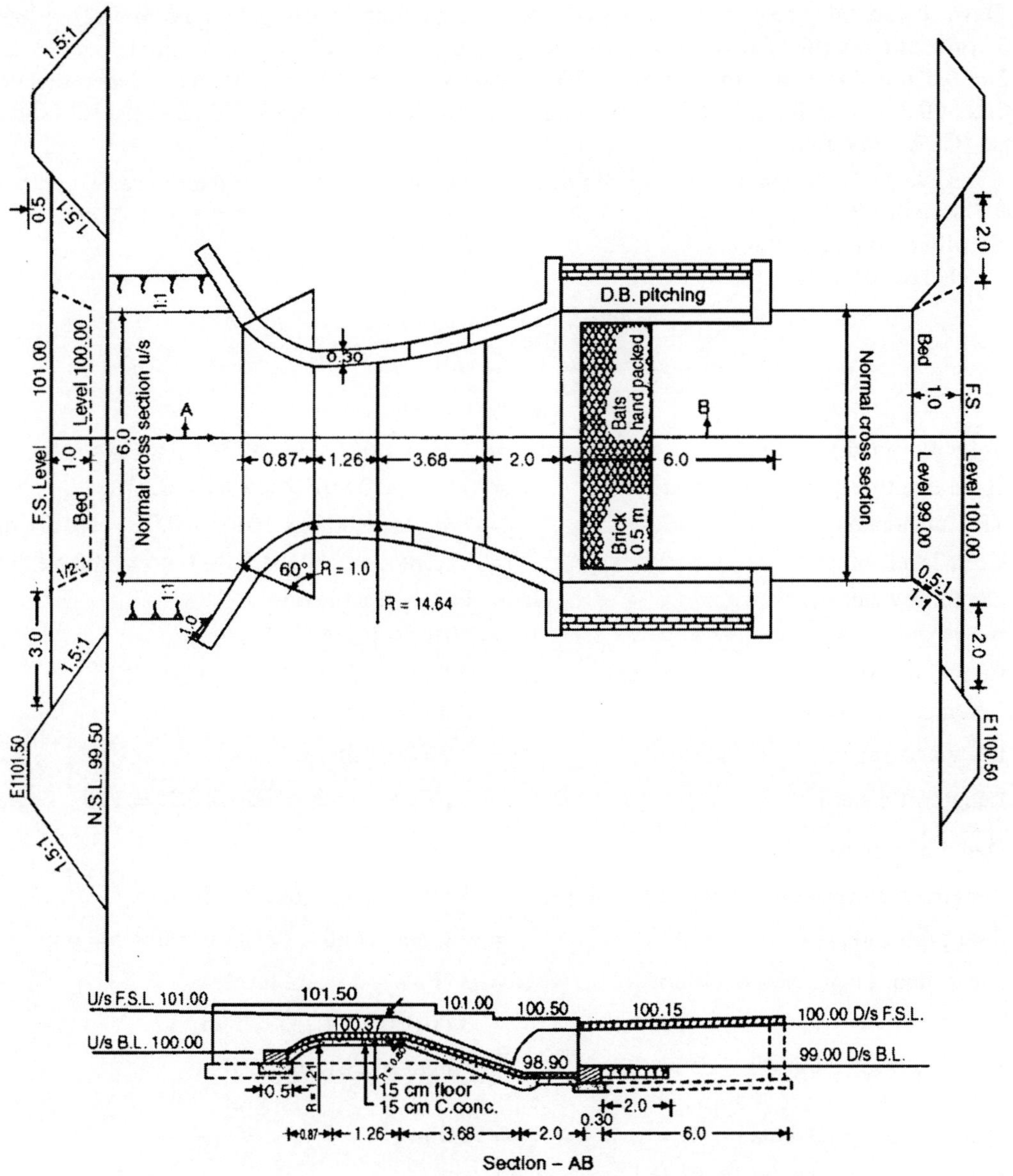

Fig. 11.20 Punjab (CDO) type fall.

or $D_n + H_w$ *i.e.* d/s FS depth + fall = 1.0 + 1.0 = 2.0 m, Adopt 2 m.

Length of d/s glacis = 2.5 (100.37 – 98.90) = 3.68 m

Length of d/s expansion, L_e = 3 $(B - B_t)/2$ or length of glacis + length of cistern, whichever is more

= 3 [(6.0 – 3.6)/2] = 3.6 m, or

or = 3.68 + 2.0 = 5.68 m,

Adopt 5.68 m.

Radius of d/s expansion, $R = [L_e^2 + (B - B_t)^2/2]/B - Bt$

$= [(5.68)^2 + (6.0-3.6)^2/2]/(6.0-3.6) = 14.64$ m.

Bed Protection. The upstream end of curved floor in the approach rests on a toe wall 0.3 m thick and depth equal to 1.0/3 = 0.33 m subject to minimum of 0.3 m of masonry floor over 0.15m thick cement concrete.

Downstream side protection consists of dry brick pitching in a length equal to 3 $(D_n + HW)$ *i.e.* 3 times the length of cistern = 6.0 m supported on toe wall of depth equal to $D_n/2$, *i.e.* 1.0/2 = 0.5 m subject to a minimum of 0.3 m masonry over 0.15 m concrete. Bed protection consists of 0.3 m thick brick bats in a length equal to length of cistern, *i.e.* 2.0 m. Fig. 11.20 shows Punjab (CDO) type fall.

Example 11.5 *Design a meter flume for a distributary of 3 cumecs capacity, bed width 6 m, depth 1.0 m, fall 0.25 m.*

Solution. Width of throat, $B_t = B = 6.0$ m

Discharge formula, $Q = C\,B_t\,E_f^{3/2}$

$$\therefore \quad E_f^{3/2} = \frac{Q}{C B_t}, \quad C = 1.7$$

or $E_f = 0.44$ m

Velocity in canal section, $V = Q/A = 3.0/(6.0 + 1.0/2)\,1.0 = 0.46$ m/s

Head due to velocity of approach $= V^2/2g = (0.46)^2/2 \times 9.81 = 0.01$ m

TEL upstream = FSL u/s + 0.01 = 101.0 + 0.01 = 101.01 m

Crest level = TEL u/s – E_f = 101.01 – 0.44 = 100.57 m

Depth of water over crest, H = FSL – crest level

H = 101.0 – 100.57 = 0.43 m

Length of throat, $L_t = 2.5\,H = 2.5 \times 0.43 = 1.08$ m

U/s approach radius $= 2.5\,H = 1.08$ m

U/s glacis radius $= 2.5\,H = 1.08$ m

Length of cistern $= 3\sqrt{E_f \times H_l} = 3 \times \sqrt{0.44 \times 0.25} = 1$ m

Depth of cistern = 1.0/6 = 0.15 m

Length of downstream glacis at 5:1 slope = 5 (100.57 – 99.60) = 4.85 m

Side protection d/s = 6 times length of cistern = 6 × 1 = 6 m

Pacca floor length below cistern = 3 (FSL – d/s bed level)

= 3 (101.0 – 99.75) = 3.75 m

Brick bat protection d/s = length of cistern = 1.0 m

Thickness of brick bats = 0.3 m

Fig. 11.21 shows meter flume fall as designed.

Example 11.6 *Design head regulator for the following data:*

	Parent channel	*Offtake*
F.S. discharge	*19.5 cumecs*	*0.45 cumecs*
F.S. level	*222.50 m*	*221.75 m*
Bed level	*221.0 m*	*221.20 m*
Depth	*1.5 m*	*0.55 m*
Bed width	*17.5 m*	*2.0 m*
W.S. slope	*0.15‰*	*0.3‰*
Lacey's f	*0.88*	*0.80*
Angle of offtake	—	*60°*

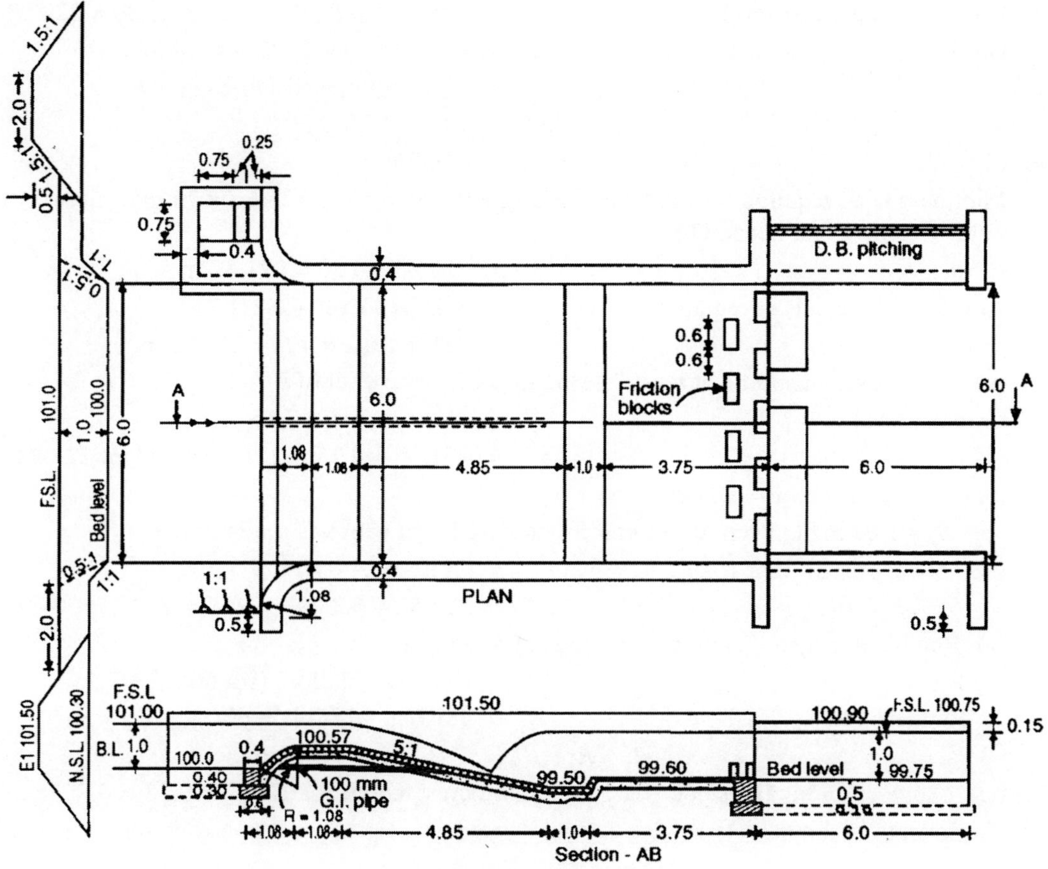

Fig. 11.21 Meter flume fall.

Solution. Parent channel

$$Q = Kd^{5/3}$$

$$19.5 = K\,(1.5)^{5/3} \text{ or } K = 9.92$$

When supply is 1/3rd of design discharge

$$Q = Kd^{5/3}$$

$$19.5/3 = 9.92\,(d)^{5/3} \text{ or } d = 0.78 \text{ m}$$

Water level with 1/3 supply = bed level + 0.78 = 221.0 + 0.78 = 221.78 m.

But F.S.L. of offtake is 221.75 m and as such the offtake can draw due share even with 1/3rd supply in parent channel. Therefore, no regulating work on parent channel is needed.

Check: Working head for offtake = 222.50 – 221.75 = 0.75 m

and F.S. depth/2 = 1.5/2 = 0.75 m, hence no regulatory work is necessary on branch canal.

Offtake

(*i*) Throat width, 0.6 *B* $B_t = 2.0 \times 0.6 = 1.2$ m.

Provide double set of karrie groove with 0.9 m wide R.C.C. regulating bridge. Provide 0.3 m wide gate groove for installation of gate, if necessary.

(*ii*) *Crest level.* It is necessary to feed offtake with design discharge with 2/3 supply in the parent channel *i.e.* with u/s supply level in parent channel at *D*/5 *i.e.* 1.5/5 = 0.3 m below design water level in parent channel.

U/s governing water level = U/s F.S.L.– 0.30 = 222.50 – 0.30 = 222.20 m

Discharge, $Q = C\, B_t\, H^{3/2}$, C = 3.5% less from 1.675 (refer contracted fall page 309)

$$0.45 = 1.62 \times 1.20 \times H^{3/2}$$

or $$H = 0.38 \text{ m}$$

Minimum head required for modular working = $0.2\,H = 0.2 \times 0.38 = 0.076$ m against 0.75m available working head, hence O.K.

Crest level = 222.20 – 0.38 = 221.82 m

Also lowest crest level required = U/s bed level + $0.2D$

= 221.0 + 0.2 × 1.5 = 221.30 m

Also crest level should be at least 0.15 m above d/s bed level of offtake.

i.e. crest level = d/s bed level + 0.15

= 221.20 + 0.15 = 221.35 m, adopt 221.82 m.

(*iii*) U/s approach of regulator

B_t = 1.20 m but keep u/s approach width as 1.5 m which is the minimum.

(*iv*) U/s approach floor level = Crest level – $H/4$

= 221.82 – 0.38/4 = 221.73 m

(*v*) Radii and tangent length of u/s wings of approach.

Radius $R_1 = (0.25\,D + 2.25)/2\,[\sin 60/2]^2 = 7.25$ m

Radius $R_2 = 1.5\,D + 1.75/2\,[\sin 60/2]^2 = 5.15$ m

$R_3 = R_2 + B_t = 5.15 + 1.5 = 6.65$ m

(*vi*) *Shape of crest*: U/s slope is 2:1, minimum crest length = $2\,H = 2 \times 0.38 = 0.75$ m.

However, to accommodate karries as well as gate grooves adopt 3 m as horizontal length of crest.

Length of 1st glacis = 2 (221.73 – 221.00) = 1.46 m

and length of 2nd glacis = 2 (221.82 – 221.73) = 0.18 m

(*vii*) Cistern level = 221.20 – 0.10 = 221.10 m; 0.1 m minimum.

(*viii*) Length of d/s glacis with 2.5:1 slope = 2.5 (221.82 – 221.10) = 1.80 m

(*ix*) Length of cistern = $(D_n + H_w)$ = (0.55 + 0.75) = 1.30 adopt 3 m minimum

(*x*) Depth of u/s curtain wall = $D/3$ = 1.5/3 = 0.5 m founded on 0.3 m concrete base.

(*xi*) Depth of d/s curtain wall = $D/2$ = 0.55/2 = 0.28 adopt 0.5 m founded on 0.3 m concrete

(*xii*) Length of 0.75 m thick u/s brick bats = D = 1.5 m

(*xiii*) Length of 0.15 m thick d/s brick bats = $2\,D$ = (2 × 0.55) = 1.0 m say

(*xiv*) Length of dry brick pitching = $3\,(D_n + H_w)$ = 3 × (0.55 + 0.75) = 4 m.

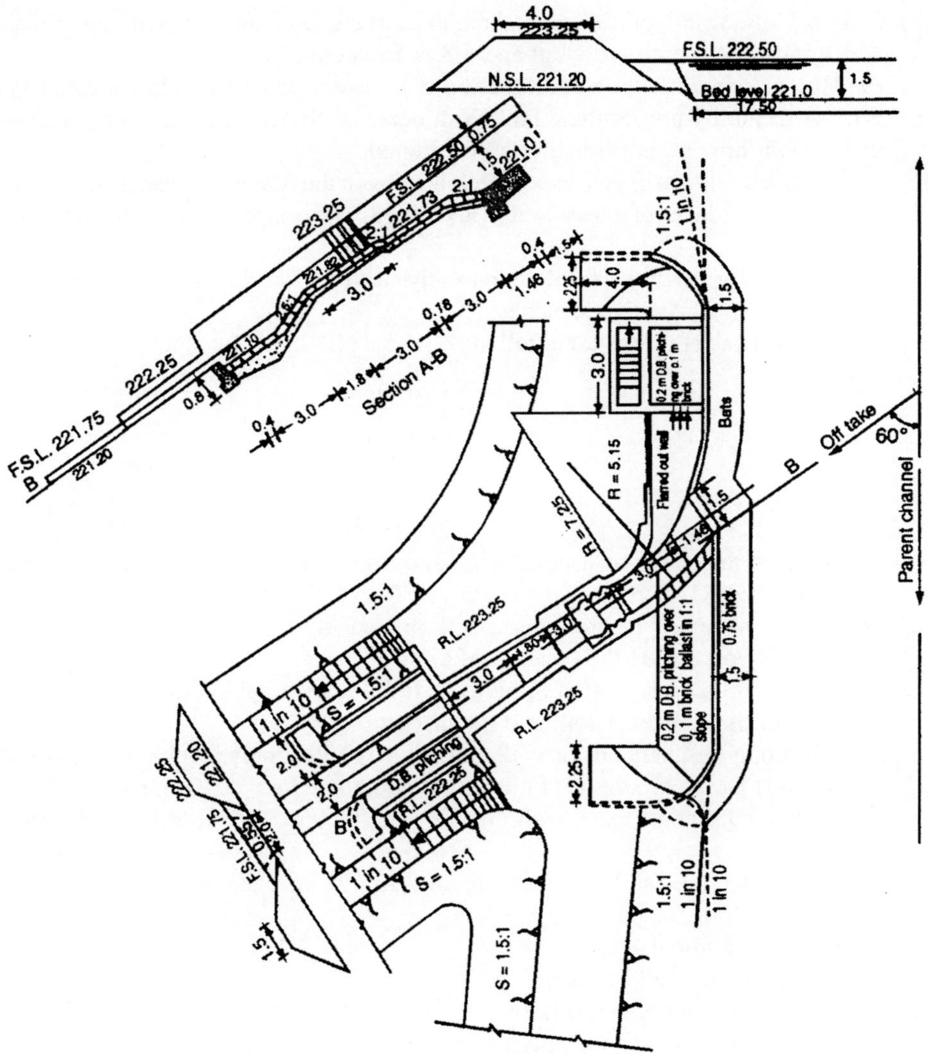

Fig. 11.22 Head regulator

QUESTIONS

1. Enumerate and explain briefly the different methods for control of entry of silt into canals.

   Give a neat sketch of any one silt exclusion work to show its principle of functioning.
2. Briefly describe a system of regulation that is practised to minimize sediment load entering a canal.
3. Compare silt excluder with silt ejector.
4. Explain the role of silt excluders and silt ejectors in prevention and control of silting problems.
5. With neat illustrative sketch state the functions of silt ejector.
6. Discuss briefly the principles and design of silt ejector.

   Explain the main difference aimed at in sediment exclusion at the canal head and distributary head.
7. (*a*) What is meant by canal falls? Why are falls constructed in modern canal systems?

   (*b*) What are roughening devices? Discuss their use in construction of falls.

(c) Design 1 m Sarda type fall on a channel carrying 20 cumecs discharge whose bed width and water depth are 14 m and 1.5 m respectively.

8. (a) Explain the design criteria and function of a cistern provided below a canal fall.
   (b) What is meant by proportional fall? With neat sketch, illustrate any proportional fall and explain how proportionality is maintained.
   (c) Discuss briefly how will you decide the following in the design of a canal fall (i) Fluming ratio, (ii) Crest level of breast wall, (iii) Length and shape of downstream expansive transition.
9. (a) Explain briefly various devices commonly used for dissipation of energy below a canal fall with illustrative sketches.
   (b) Following data refer to a canal fall site:

Item	*Upstream*	*Downstream*
Full supply discharge	50 cumecs	50 cumecs
Bed width	30 m	30 m
Full supply level	270.00 m	268.50 m
Bed level	268.00 m	266.50 m

   Recommend the fall to be used. Design and draw the following elements of the fall: (i) Length and RL of crest, (ii) Length and RL of cistern. [PU, No. 1979]
10. (a) What is the necessity of providing a fall on a canal?
   (b) How would you decide the location of a fall?
   (c) The following data relate to a Sarda type fall. Make suitable assumptions, if necessary, and design the crest wall and the cistern:
   (i) Full supply discharge us/ds = 10 cumecs, (ii) Full supply level = u/s/d/s = 101.50/100.50, (iii) Drop = 1 m, (iv) Full supply depth u/s/d/s = 1.5/1.5, (v) Bed level u/s/d/s = 100.00/99.00, (vi) Bed width u/s/d/s = 8.0/8.0 m, (vii) Soil is good loam. [PU, Dec. 1978]
11. (a) Design a Sarda type fall for the data; fall height = 1.7m, canal discharge = 25 cumecs, bed width = 20 m, canal depth = 1.6 m.
   (b) Draw its longitudinal section.
12. (a) Why we provide a fall on canal?
   (b) Enumerate different types of falls.
   (c) Data refer to fall site, full supply discharge us/ds = 50 cumecs, bed width u/s/d/s = 28.0 m, full supply level u/s/d/s = 150.00/148.50, bed level u/s/d/s 148.00/146.50 m. What type of fall would you recommend for this canal?
   Design (a) Length and RL of cistern, and (b) Length and RL of cistern. [PU, Nov. 1980]
13. (a) What are the functions of a head regulator and a cross regulator?
   (b) Describe the design criteria for head regulator and cross regulator.
   (c) Explain canal escapes. [PU, Nov. 1980]
14. With neat illustrative sketches, state the functions of the following: (i) Escape, (ii) Distributary head regulator, (iii) Silt ejector.

12

CROSS DRAINAGE WORKS

12.1 INTRODUCTION

Cross drainage work or drainage crossing is a structure constructed to negotiate an irrigation canal over, below or at the same level of a drainage or another canal. Irrigation channels, as a principle, are aligned along ridges or watersheds so that normally no drains are intercepted by them; all the drainage being away from the watershed towards the two drainages on either side of the ridge. However, in the head reach the canal taking off from a river has to run for a certain distance before it is able to mount the watershed and it is in this reach that the canal intercepts the drainages from the watershed towards the river (Fig. 12.1) requiring the provision a large number of cross-drainage works to carry the canal across them. Likewise, when a canal alignment is made to leave the watershed for a certain distance to cut across a loop drainage crossings are necessarily provided to cater for the drainage of the pocket intercepted between the watershed and the canal.

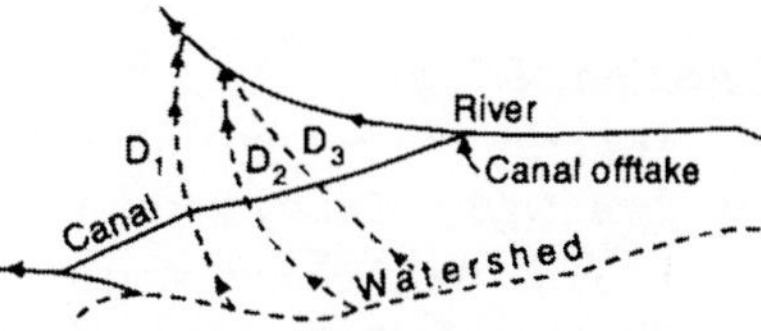

Fig. 12.1 Drainage in head reach of a canal.

12.2 MINIMUM NUMBER OF CROSS DRAINAGE WORKS

Since cross drainage works are very costly structures to construct and maintain, these are avoided, as far as possible, by (*i*) Judiciously changing the canal alignment to a point downstream of the confluence of two streams, thereby avoiding necessity for the construction of cross drainage works separately for each stream, (*ii*) Diverting one stream into another nearby drainage upstream of its crossing with the canal alignment. It shall be economical to divert small discharge carrying stream into the bigger one, and (*iii*) Shifting the bridge site to the nearest drainage crossing or *vice versa*, more particularly in the case of village road bridges. First, the combined structure shall be economical to construct, and secondly the upstream and downstream training works of the stream can be made to serve as approaches for the village road. In any case the drainage should cross the canal alignment at right angle for reasons of economy in the construction of cross drainage works.

12.3 SITE OF DRAINAGE CROSSING

The site of crossing is considered suitable if it meets the requirements (*i*) Requires minimum regrading of approach and tail reaches of the drainage, (*ii*) Sufficient headway is available for the superstructure of the aqueduct over the maximum flood level of the drainage, (*iii*) Suitable soil for foundation is available at reasonable depth, and (*iv*) Suitable existing topographical, geological and hydraulic conditions for the drainage work at reasonable cost.

12.4 CLASSIFICATION AND SELECTION OF TYPE

Cross drainage works are classified, as illustrated below, depending on the type of structure to negotiate the canal over, below or at the same level of the drainage.

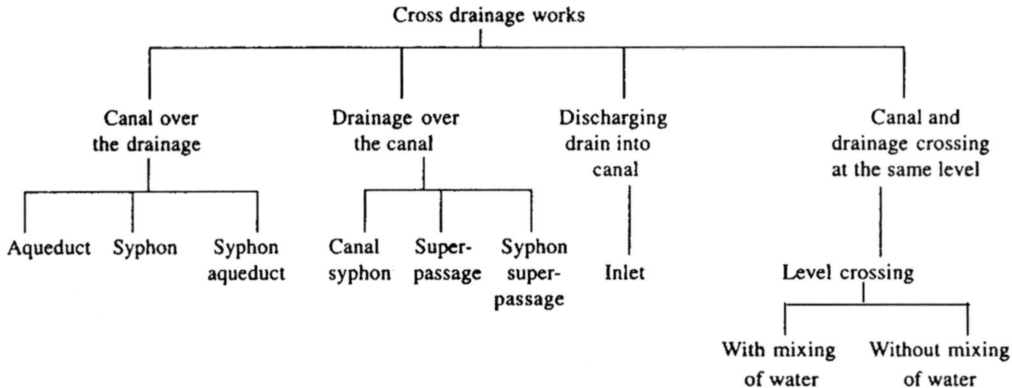

Selection of Type

The selection of type of drainage crossing suitable for the given conditions of drainage *vis-a-vis* canal is governed by various considerations, briefed as under:

1. *FSL of the Canal vis-a-vis HFL of the Drainage*: HFL of the drainage *vis-a-vis* FSL of canal dictates the type of cross drainage work suitable for the location. Aqueduct when canal bed level is well above HFL of the drainage. Superpassage when FSL of canal is well below the bed of the drainage. Syphon-aqueduct when bed level of the canal is at or below HFL of the drainage and bed of drainage is depressed.

2. *Topography of Terrian:* A right angled crossing is preferable. Possibility of diverting the drainage into nearby drainage or locating the crossing below the confluence of two drainages is studied with a view to explore the feasibility of avoiding drainage crossing on each drainage for economy.

3. *Regime of Drainage:* A stable and confined drainage section is preferable. Drainage carrying high sediment charge may rule out the possibility ot providing syphon-aqueduct because of susceptibility to choking.

4. *Foundation Strata:* Firm sub-strata close to the bed of the drainage is a prerequisite for the suitability of a site for the economical design and construction of the structure.

5. *Ratio of Discharges*: If ratio of canal discharge to design flood of drainage is substantially low, canal syphon is preferred.

6. *Head Loss*: Head loss permitted in the canal dictates the type of cross drainage work. Higher head loss means loss of command and restriction of head loss necessitates wider section and hence costly structure. In Sardar Sarovar canal system, the head losses considered for various structures are (*i*) road bridge (0.01 m), railway bridge (0.02 m), cross regulation (0.075 m), and superpassage (0.01 m). For other structures like aqueducts, canal syphons and drainage syphon, the head loss shall be more and based on actual calculations.

7. *Subsoil Watertable:* Subsoil watertable *vis-a-vis* dewatering requirement for laying foundation is studied to estimate the cost of dewatering. Dewatering is costly as well as time consuming.

8. *Economical Aspects*: Economy of various types of drainage crossings *vis-a-vis* alternative alignment of the canal is studied. Aqueducts are costly.

9. *Canal Alignment:* The relative bed levels of the canal and drainage may be altered with advantage by suitably changing the canal alignment in order that the point of crossing is shifted upstream or downstream of the drainage.

10. Availability of soil for making embankments for canal and drainage.

In essence, preference of one type of drainage crossing over the other is as follows:

Type	*Preference*	*Where generally suitable*
Syphon	Generally undesirable and avoided in case the drainage is of large capacity. Preferred when the drainage is of small capacity compared to canal.	HF level of the small drainage is much higher than the canal bed level and HF level and bed level of the drainage have to be depressed to pass underneath the canal. It is the case when the canal is running in partial cutting.
Aqueduct	Preferable when the drainage is wide and of higher discharging capacity like a river.	Canal in filling with bed level sufficiently above HFL in drainage. Flood discharge is large and of long duration.
Syphon-aqueduct	Generally undesirable.	When bed level of canal is at or below HFL of drainage channel, the depression of drainage bed is often more economical.
Superpassage	Avoided where practicable in comparison with syphon. Inferior compared to aqueduct also.	Stream passes over the canal in deep cutting with FS level well below bed level of drainage. Flood discharge is large and of long duration.
Level crossing	Preferred being economical in first cost as also if syphon or superpassage are costly and unsuitable. Broadly it is inferior to inlet.	When the stream and canal cross each other with almost the same bed levels and the maximum flood discharge is large but of short duration and intermittent.
Intet	Preferred to dispose of very small discharging drainage.	When small drainage has bed level at or above FSL in the canal in deep cutting. The maximum flood discharge is small.

In general, cross drainage works in the order of preference are (*i*) Inlet, (*ii*) Aqueduct, (*iii*) Superpassage, (*iv*) Syphon aqueduct, (*v*) Syphon, and (*iv*) Level crossing,

12.5 TYPES OF DRAINAGE CROSSINGS

1. Aqueduct

An aqueduct is defined as a drainage crossing in which the canal is carried over the drainage channel and the bottom of the canal trough, or the covering over the drainage openings, is above the high flood level in the drainage channel. An aqueduct is thus provided when the canal bed level is higher than H.F.L. of the drainage. Depression of canal or drainage bed level is not involved. Provision of hump in the canal section is sometimes made to obtain necessary headway for the drainage and thereby facilitate construction of an aqueduct. The canal being smaller, compared to the drainage, is usually flumed. The canal width is constricted but not less than 75 per cent of the bed width to create a velocity of 2.5 to 3 m per second in order to reduce the cost of crossing. The flumed trough is joined to the normal canal section with a transition of 2:1 and 3:1 on upstream and downstream sides but not flatter than 3:1 and 5:1 respectively. The flow should follow the boundaries of the transitions. Judicious waterway is provided for the drainage. Fluming from 60 to 90 per cent of Lacey's waterway may be adopted in alluvial regions. Economy in cost is effected by allowing velocity higher than 3 m/s through the works. Wing walls for the drainage are provided with 2:1 and 3:1 splays on upstream and downstream sides; the splay being not flatter than 3:1 and 4:1 respectively.

Types of Aqueduct

The aqueducts are broadly classified into the following two types, depending on the arrangement of canal passing over the drainage:

Barrel Type Aqueduct

In this type the canal continues over the drainage channel in its normal earthen section including the banks and earthen slopes (Fig. 12.2). It is suitable for a small drainage channel as no canal transitions are required. The saving made due to the absence of canal transitions more than compensate the increased cost due to longer length of drainage barrels. The barrel may be of R.C.C or masonry.

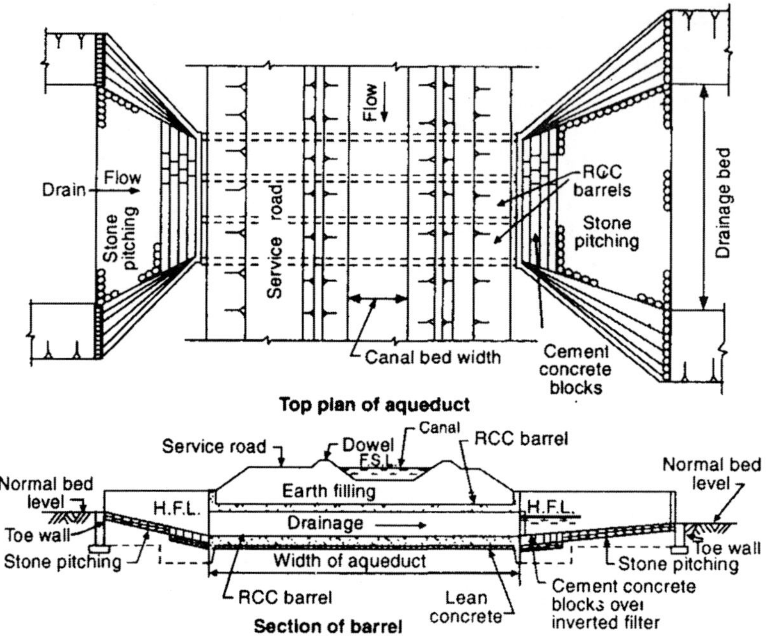

Fig. 12.2 Aqueduct (IS 7784-1995)

Another variant of this type, is wherein the outer slopes of the canal banks are replaced by retaining walls thereby reducing the length of drainage barrels. It may be suitable for intermediate capacity drainages.

The drainage channel is diverted towards the structure bv suitable training works like training walls, guide banks, spurs, etc. The canal banks adjacent to the structure are protected by turfing, pitching and launching apron, wherever necessary.

Arch Type Aqueduct

It consists of a trough for the canal supported on arches, which may be of masonry, or reinforced cement concrete. Alternatively, it may be described as an arched bridge with a platform at canal bed level and provided with two solid masonry retaining walls or earthen banks to contain the water flowing through the canal. A masonry breast wall on both upstream and downstream sides is provided to retain the bank slope. The arches may be semi-circular, funicular but generally segmental with a rise to span ratio of 1:4. The foundation for the drainage may consist of open foundation, well foundation and raft foundation and floor is generally of inverted arch masonry to distribute the load evently over relatively shallow foundation, as also for taking up the upward water pressure on the floor. Arch type aqueduct is suitable when there are unyielding foundations and adequate headway is available.

Trough Type Aqueduct

It is the form of aqueduct in which the earthen banks of the canal are discontinued over the aqueduct and the canal duly flumed is carried in a masonry or concrete trough, box, barrel, pipe (Fig. 12.3). The sides of the trough are connected on either side of the work to the earthen banks of the canal by means of transition walls. It is economical for a large river as the length of drainage culverts across the canal is small and saving made on this account offsets the increased cost of canal transitions.

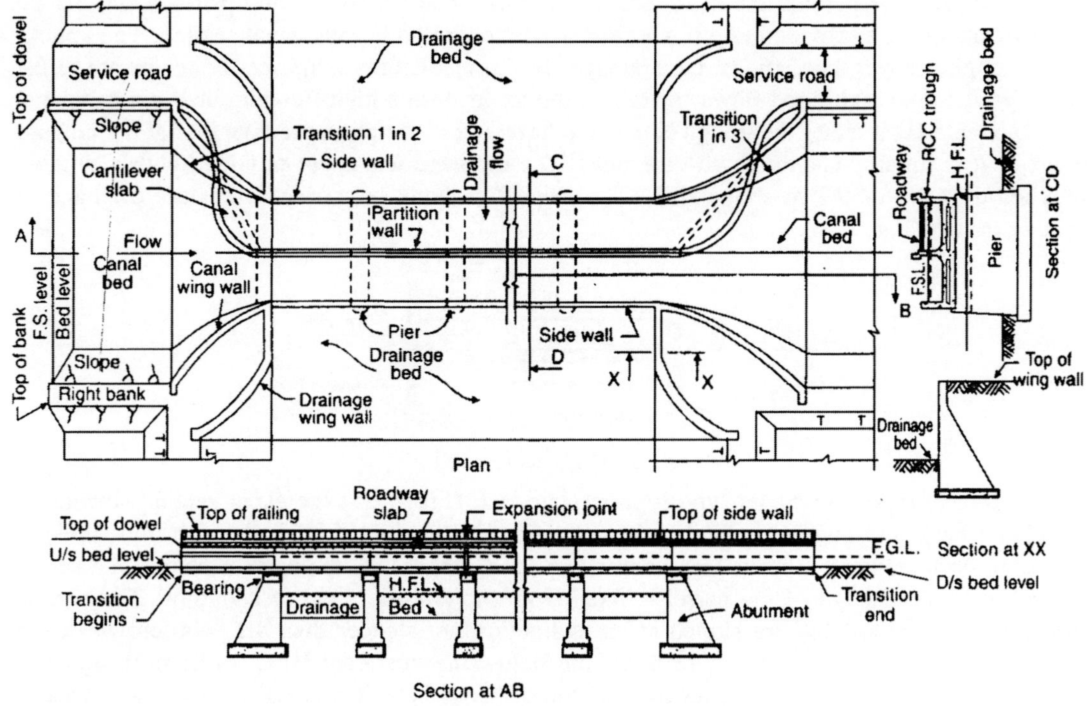

Fig. 12.3 Typical plan and section of trough aqueduct

Buttress Aqueduct

This type of aqueduct is best suited and generally economical when the foundations are not of rocky nature. In this type, the canal section is trapezoidal with piers and abutments on sides raised to support the inclined walls of the troughs.

Advantages of Aqueduct

An aqueduct is preferred when the drainage is of large capacity and carries heavy silt bed load. The advantages of this type of drainage crossing are (*i*) Canal is open to inspection, and (*ii*) Damage to canal by floods in the drainage are rare.

Disadvantages: (*i*) Elaborate training works are required for the drainage, (*ii*) Foundations of the work are susceptible to scouring during heavy floods in the drainage, (*iii*) Waterway of the drainage is susceptible to choking with trees and debris, etc, and (*iv*) Larger headway is required between canal bed level and HF level in the drainage.

2. Syphon-Aqueduct

A syphon aqueduct is defined as a drainage crossing in which the canal is carried over the drainage channel preferably at right angle and the drainage discharge is carried under pressure through the structure. The bed level of the drainage passes underneath a canal such that the headway is insufficient, the high flood level of the drainage is depressed as it passes under pressure under the canal trough; the bed level may or may not be depressed below its normal level but is generally lowered (Fig. 12.4). The syphon aqueduct barrels may be of prestressed concrete, RCC or masonry or a combination of these.

Types of Syphon Aqueduct

Two types of syphon aqueduct are trough type and barrel type.

1. **Trough type:** In trough type, the canal is carried over the drainage channel through a trough supported on piers (Fig. 12.4). The canal is flumed to not less than 75% of the bed width

keeping in view the permissible head loss in the canal. Transitions 3:1 on upstream and 5:1 on downstream side are provided to join the flumed section to the normal canal section. The bottom of the trough is lower than HFL of the drainage. The trough is thus subjected to the upward thrust acting on it due to high flood flows corresponding to the design high flood in the drain and canal dry. To counteract it either the trough is made to have a dead weight more than that of the upward thrust or it is suitably anchored with the piers. The drainage wings are provided with a suitable free board of 0.3 or 0.5 m above HFL. The inspection bank is carried across the drainage in another trough (Fig. 12.4).

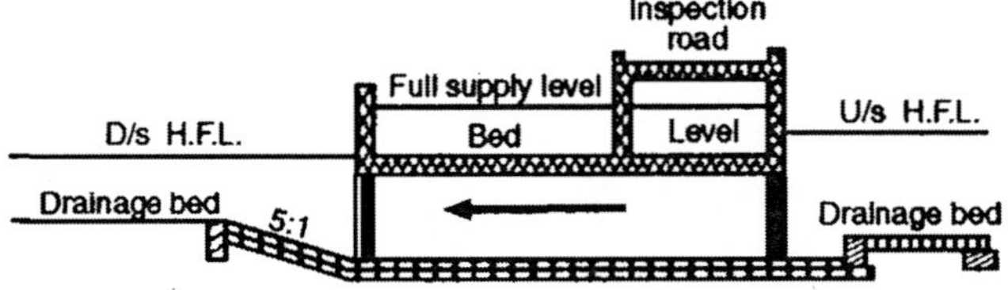

Fig. 12.4 Syphon aqueduct.

2. **Barrel type:** In barrel type syphon aqueduct, (Fig. 12.5) the canal section (including inspection bank and non-inspection bank with suitable covering to HGL through the banks) is flumed or unflumed and is carried over barrels designed to pass the drainage discharge at a velocity of 3 m/s. The length of the barrel is fixed from the consideration of economy. The barrels supporting the canal bank are sloped at an inclination not steeper than 4:1. The crown of the barrels at the entrance is kept 0.5 m minimum below the upstream HFL. Bellmouth entry is provided to reduce afflux. At outlet end, the top of barrel is kept depressed below the normal

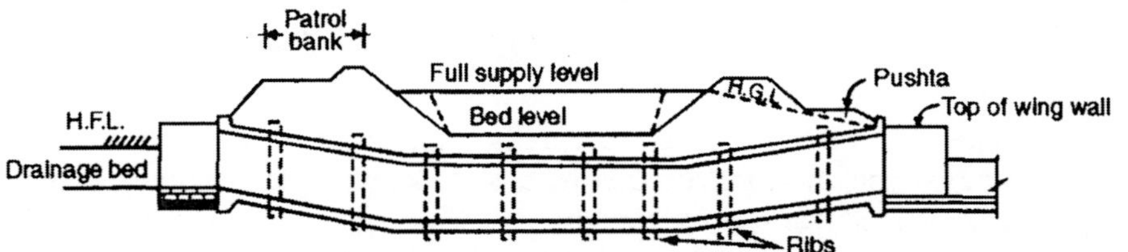

Fig. 12.5 Barrel type syphon aqueduct

downstream flood level in the drainage channel equal to the difference in velocity head at the exit end of the barrel and that in the drain on the downstream side. The drainage wings are provided in a smooth curve to give a splay of 3:1 on upstream and 5:1 on downstream side. A hyperbolic curve (Fig. 12.6) is provided to ensure smooth transition using formula

$$B_x = \frac{B_C \times B_F \times L}{L \times B_C - X(B_C - B_F)}. \quad \text{Mitra Hyperbolic Transition}$$

Fig. 12.6 Transition

Advantages: (*i*) Canal runs with its designed levels undisturbed, (*ii*) Canal is open to inspection, (*iii*) Preferred when the drainage is small compared to the canal capacity, (*iv*) Damage to the canal due to floods is rare.

Disadvantages: (*i*) Lowering of bed levels by a drop is not economical in the case of larger drainages, (*ii*) Higher cost of dewatering to lay foundations of the work, (*iii*) Foundations of the work are susceptible to undermining due to scouring during heavy floods.

2. Drainge Syphon

A drainage syphon is a cross drainage work in which the drainage is carried, with depressed H.F.L and bed, below the canal. The drainage is small with H.F.L higher than the canal bed. The important considerations in design of drainage syphon at Sardar Sarovar project are (*i*) Minimum size of drainage barrel is 2.5 × 2.5 m to facilitate maintenance, and easy flow through the barrel even under partial siltation condition, (*ii*) Permissible velocity for design flood is 4.5 m/s and 6 m/s for check flood (1.3 times design flood), and further checked that for 25 year frequency flood (dominent flood) the minimum velocity is 2.5 m/s to ensure non-silting, (*iii*) At exit end, the barrels are kept horizontal instead of in the form of inclined limb to avoid siltation within the barrels, (*iv*) The tail channel is regraded up to invert level of the barrel for a distance of 100 m on d/s followed by recovery slope of 1 in 15 in required length to ensure minimum clearance of 2/3 height of the barrel above the regraded bed. Fig. 12.7 shows typical drainage syphon on Sardar Sarovar Project.

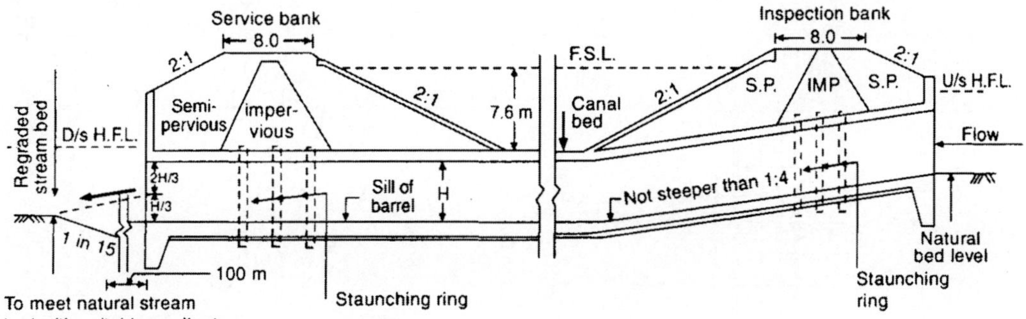

Fig. 12.7 Typical section of drainage syphon.

3. Canal Syphon

Canal syphon is a structure with a closed conduit, designed to run full and usually under pressure to carry the canal water under the drainage. When a canal passes underneath a drainage, such that headway between bed level of the drainage and full supply level of the canal is insufficient, the bed level of the canal is depressed below the normal designed level and the canal water is passed through inverted syphon under the drainage, rising again on the other side. A bridge is provided across the drainage for the patrol bank of the canal to permit running of the inspection vehicles. The conduit may be of (*i*) RCC single or multi-cell barrels, circular, rectangular, horse shoe shaped; circular or horse shoe are preferred or (*ii*) Masonry barrel with flat RCC or masonry arch roofing. Precast RCC pipes may be economical for syphoning small discharges. The barrel is horizontal under the bed of canal but sloping under the banks; the usual slope is 3:1 on the upstream side and 4:1 on the downstream side. Where the subsoil water is above or near the floor level of the syphon, in the case of masonry syphons an inverted arched floor is provided to be safe against the uplift pressure. A self-cleaning velocity of 6 m/s and 3 m/s is provided through RCC barrel and masonry barrel respectively. The barrel is founded at least 1/3 of overall height of barrel below the anticipated/design scour level corresponding to design flood. The top of the syphon is kept at least 30 to 75 cm below the drainage bed level. The floor level of the barrel at the entrance is kept at the bed level of canal but the top of the barrel is kept at least 15 cm or 1.5 times the change in velocity head below the designed FS level in the canal with a view to ensure that the conduit always runs under pressure and air does not enter the barrel. More precisely, it is fixed from the consideration of total energy line of the canal at the entrance to the barrel, as under as follows :

Invert level of barrel at entrance	=	TEL u/s–water seal (15 cm) – velocity head due to approach-height of barrel.
Invert level of barrel at exit	=	Invert level at the entrance-loss of head through the work.

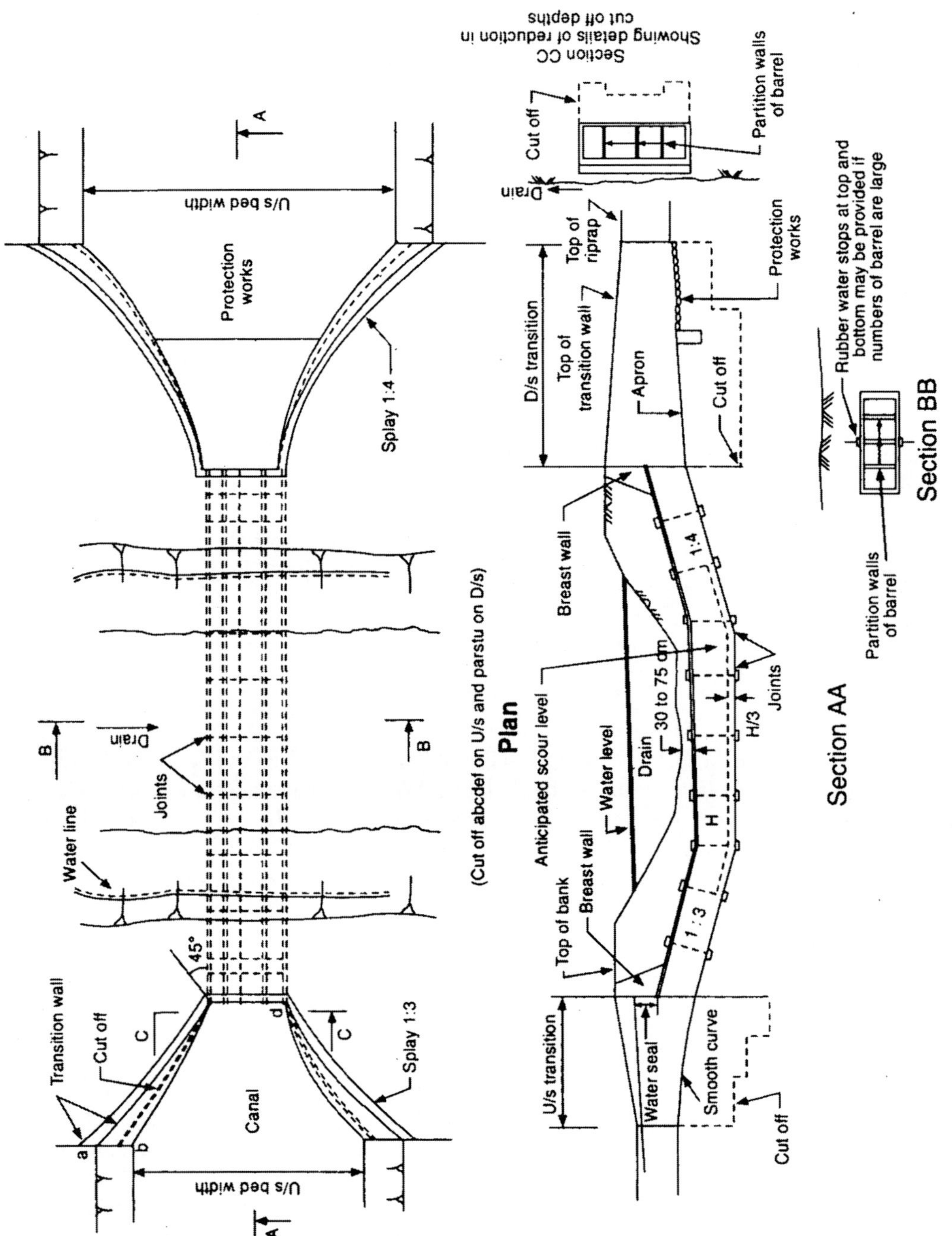

Fig. 12.8 Typical Canal Syphon

The canal section is flumed to not less than 70 per cent of canal bed width. For computing fluming ratio, the width of mid depth is taken as 100. In drainage channel with undefined course a fluming ratio of 70 to 90 per cent of Lacey's waterway is adopted. The flumed barrel section is joined to the canal bed in slope of 3:1 on upstream and 4:1 on downstream.

Loss of head through the structure, $H = h_1 + h_2$
where, H is total head loss, h_1 are losses at inlet and outlet and h_2 are losses in the barrel or pipe, due to skin friction.

1. Head loss at inlet or outlet of syphon, $h_1 = (1 + f_1)\ V^2/2g$
where, h_1 is entrance or exit loss (m), f_1 is coefficient for loss of head at entry; 0.08 for a bell-mouth entrance, 0.505 for cylindrical entrance with sharp edges (unsharped mouth of the same sectional area of the barrel), V is velocity in syphon (m/s) and g is acceleration due to gravity.

2. Head loss in the barrel. Loss of head due to friction in the barrel is estimated by Manning's formula, $S = (VN/R^{2/3})^2$. Hence, loss of head through barrel = length of barrel × slope.

The estimated losses are increased by 10 per cent for design purposes.

Advantages: (*i*) Preferable when the drainage is large compared to the canal capacity, (*ii*) Damage to the canal due to floods in the drainage is rare, and (*iii*) Canal section can be conveniently flumed to effect economy in the cost of the drainage crossing.

Disadvantages: (*i*) Canal being under the drainage is not available for inspection, (*ii*) Normal designed levels of the canal are disturbed entailing loss of head which shall have to be catered for, and (*iii*) Higher cost of dewatering to lay foundations.

4. Superpassage

Superpassage is defined as a drainage crossing in which the drainage channel is carried in a trough over the canal normally with full supply level of the canal below the bottom of the trough. It is reverse of an aqueduct. It is generally adopted when the high flood level of the drainage is higher than the full supply level of the canal. A modified version of the superpassage is 'superpassage syphon', the brief description of the two is as under:

1. **Superpassage:** Superpassage is the structure where waterway of the drainage is carried above the water in the canal. Canal passes under the drainage trough with normal FSL and bed level of the canal unchanged and requisite clearance is available between the bottom of trough and FSL of the canal. The drainage trough is carried over the piers constructed in the bed of the canal (Fig. 12.9).

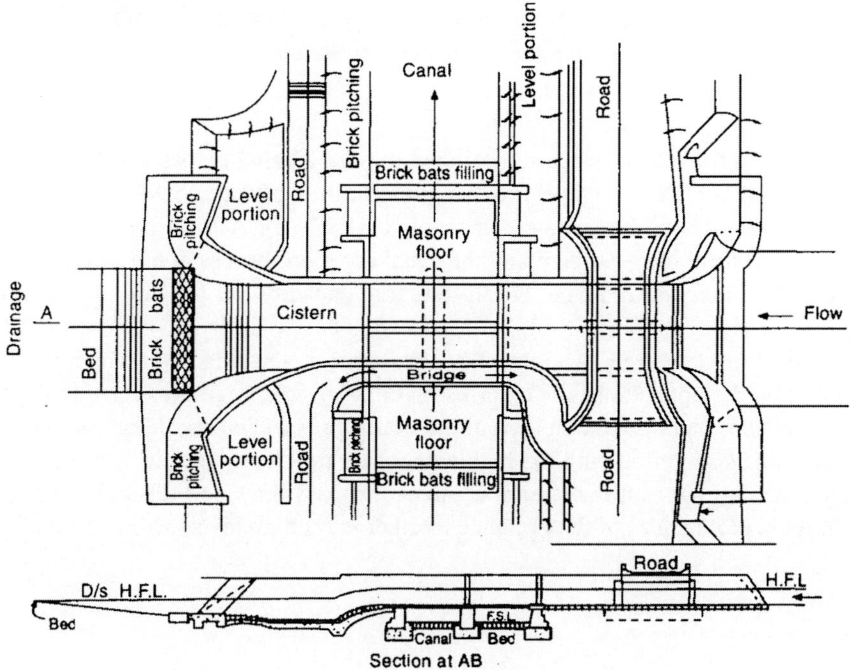

Fig. 12.9 Superpassage

Superpassage structure may be either of brick or stone masonry or RCC. RCC superstructure is preferred. In the case of masonry superpassage, masonry arches are provided with their crown at bed level of the drainage channel while that in RCC trough the slab is laid at the bed level of the torrent or slightly higher, if need be. The clearance for rectangular or arch openings is 50% of those indicated for aqueducts (Table 12.2). The trough is divided into compartments by low height divide walls for better distributed stream flow through the work.

2. **Superpassage Syphon:** Superpassage syphon combines the characteristics of a superpassage and a syphon at the same time in that the canal passes under the drainage with bed level depressed below its normal bed level and the trough bottom is below the FSL of the canal such that the canal water flows under pressure. Syphon superpassage is provided where the canal FSL is at or above the drainage bed. The canal is flumed for reasons of economy and a splay of 2:1 and 3:1 on upstream and downstream sides respectively are provided but the drainage is not flumed. Where the waterway of the drainage is restricted, suitable splay of 3:1 to 5:1 is provided on upstream and downstream sides. A velocity of 2 to 3 m/s is generally allowed.

Loss of Head

(*i*) *In upstream approach*: (*a*) due to convergence, $h_1 = 0.1\ (V_2^2 - V_1^2)/2g$, where V_2 and V_1 are velocities (m/s) at the starting and other end respectively of the upstream flared out walls, (*b*) due to friction loss, $h_2 = SL = V^2N^2L/R^{4/3}$ from Manning's formula, where S is the water surface slope, L is the length of approach, and V is average velocity $(V_2 + V_1)/2$ through the approach.

(*ii*) Through the trough, $h_3 = S.L. = V_2^2\ N^2\ L/R^{4/3}$, where S is water surface slope and L is length of trough and V_2 is the velocity through the trough.

(*iii*) *In downstream transition*: (*a*) due to divergence, $h_4 = 0.2\ (V_3^2 - V_4^2)/2g$, where V_3 and V_4 are velocities at the upper and lower ends respectively of d/s transition, (*b*) Due to friction, $h_5 = SL = V_5^2\ N^2 L\ /\ R^{4/3}$, where V_5 is average velocity, *i.e.*, $(V_3 + V_4)/2$ in the approach and L is length of the approach.

Loss of head through the structure = $h_1 + h_2 + h_3 + h_4 + h_5$.

Generally the drainage has steep slope and the fall available is more than the loss of head through the work. The extra fall is disposed of by providing a raised crest or hump followed by a glacis and a deep cistern for the dissipation of energy.

Disadvantages: (*i*) Canal is not open to inspection, and (*ii*) Drainage troughs are susceptible to silting up and it is difficult to remove the silt deposited therein.

5. Level Crossing

Type with Intermixing of waters

Level crossing is defined as a structure designed to pass a canal across another channel or a natural drainage at or about the same level involving intermixing of the two waters with regulating structures on the canal and on the drainage channel (Fig. 12.10). A level crossing is generally provided when the bed level of the stream and the canal are more or less at the same level at their junction and both the canal and drainage are of the large capacity but the drainage has high but intermittent floods, when syphoning of either of the two is costly or due to non-permissibility of loss of head through syphon barrels. An inlet (B) is provided on the side from which the stream is coming and an exit on the opposite bank (D). This involves an inlet regulator across the canal (A) and another at the farther bank across the exit of the drainage (C). The regulators on the canal are kept open during non-flood period with the regulator across the stream closed. On the other hand, during the floods the regulator on the stream is opened to pass the flood discharge downstream. Generally it is possible to dispose of the drainage in other ways than level crossing which is rarely provided.

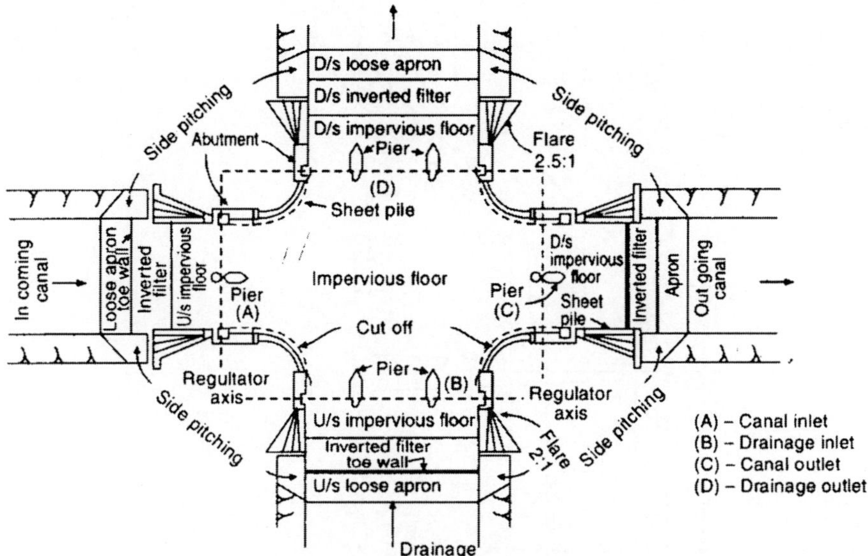

Fig. 12.10 Level crossing

Disadvantages: (*i*) Likely to disturb the regime of the canal by silting due to intermixing of two waters calling for frequent silt clearance, (*ii*) Canal capacity has to be increased to take care of drainage discharge, (*iii*) Costly due to the provision of controls at the inlet and outlet sides and perhaps a regulator in the parent channel if the discharge of stream is considerable, (*iv*) Higher maintenance cost, (*v*) Risk to safety of the canal on account of failure of human element in the operation of the structure.

Advantages: (*i*) It functions as a balancing reservoir to make best use of the waters of the stream and canal, (*ii*) Perennial discharge of the stream can be utilized to augment the canal supply in lean supply period, and (*iii*) Suitable for drainage having low floods of short duration and at considerable intervals.

Type Without Intermixing of Water

This type of level crossing is adapted only to lined canal because of its deeper depth. The drainage water is carried across the canal through a round barrel (Fig. 12.11) with attendant advantage of not allowing turbid drainage water to mix with canal water.

6. Inlet

Inlet is defined as a structure consisting of an opening in canal bank, suitably protected to admit upland drainage water into the canal. The canal section *i.e.* bed and sides are suitably pitched for a certain distance upstream and downstream of the inlet. Inlet type drainage work thus enables utilization of drainage water for irrigation. The drainage water admitted, if not to be utilized in the canal, or is large compared to the canal capacity, is escaped through an outlet which is another opening in the opposite bank of the canal with bed and sides suitably pitched or through existing escape downstream nearest to the inlet. The discharge escaped may not be equal to that admitted into the canal through the inlet, a part of it may be carried by the canal for irrigation. Inlet for the disposal of drainage water is economical than a superpassage. Inlets, depending on relative levels of the bed level of the drainage and full supply of the canal, are classified as flush inlet and weir inlet, described below:

Flush Inlet: When the bed level of the drainage is at or higher than the designed full supply level of the canal, the inlet provided is called flush inlet or freefall inlet.

Weir Inlet: When the bed level of the drainage is slightly below designed FSL of the canal, the inlet provided for the admittance of the upland water into the canal is termed as weir inlet. In

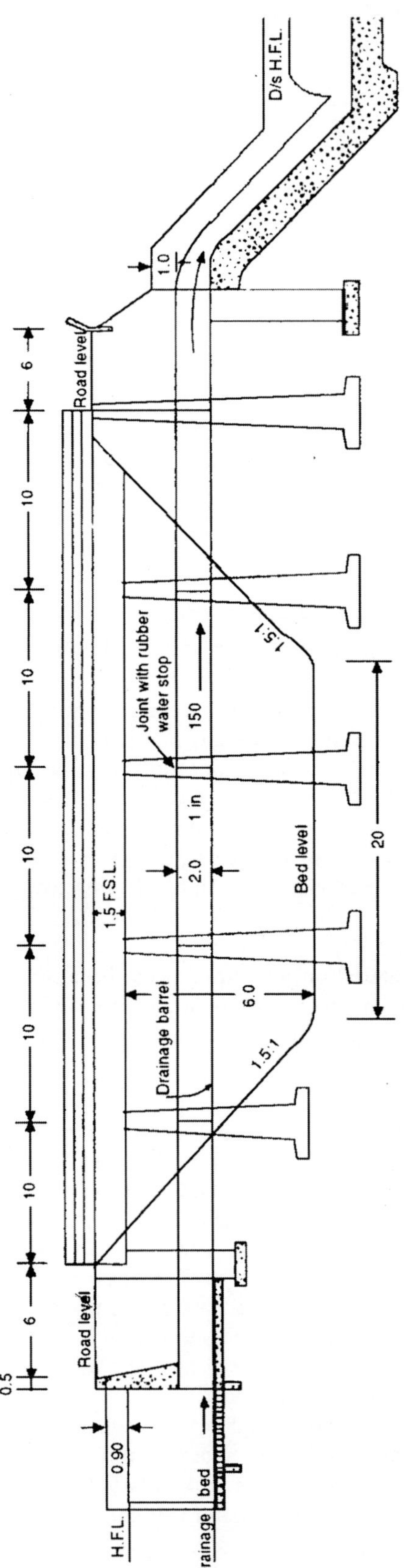

Fig. 12.11 Level crossing without mixing of waters

weir inlet a low weir wall is provided across the opening in the canal bank with crest level at the designed FSL of the canal. The drainage water thus spills over the crest into the canal.

Advantages: (*i*) Simplest type of cross drainage work, (*ii*) Minimum cost of construction, (*iii*) Minimum maintenance cost, (*iv*) Avoids incurring heavy expenses for providing a costly superpassage, (*v*) Enables utilization of the drainage water for irrigation.

12.6 SITE DATA

Collection of data, as under, is a prerequisite for the formulation of design of the cross drainage works:

1. *Index plan*: Index plan to a scale not smaller than 1 cm = 500 m indicating prominently the location of crossing, existing lines of communication, general topography of the area and nearby habitations.

2. *Catchment plan*: Part contour plan indicating contours at 0.5 to 1.5 m intervals, main drainage, tributaries of the drainage showing important topographical features extending 1.5 km upstream and 1.5 km downstream of the crossing, limits of the catchment area, course of drainage, general types of soil and land use-cultivated, uncultivated, forest areas, etc., storage-artificial or natural in the catchment, embankment breaches, if any, in the past and location of various raingauge stations.

3. *Site plan*: A site plan to a suitable scale showing details of site selected and extending upstream and downstream of the centre line of proposed crossing, direction of flow and angle of crossing, alignment of canal, etc.

4. *Geological plan*: Geological section at the centre line of the crossing extending 5 m below the lowest foundation level, indicating the nature of soil in bed, banks and approaches as determined through trial pits or bore-hole sections.

5. *Rainfall*: Highest annual rainfall, short duration intensity and frequency data of rainfall in the catchment.

6. *Seismic disturbances*: Likelihood of the site to seismic disturbances.

7. *Discharge and sediment data*: The observed gauge and discharge data of the drainage and flood gauges. Silt or sand samples at the bed level as also at the depth up to which the deepest scour is anticipated to ascertain value of Lacey's *f* and scour depth.

8. *Canal and drainage data*: The canal and drainage data required are as follows:

Sr. No.	*Canal data*	*Drainage data*
1.	Authorized discharge;	Highest observed flood discharge, catchment area and maximum rainfall and maximum rainfall intensity.
2.	Full supply level;	(*i*) Highest flood level, ordinary flood level and low water level.
3.	Bed level;	Deepest bed level.
4.	Normal cross section showing F.S depth, bed width, bank level, bank width, natural surface level and maximum subsoil water level;	A few cross sections up to about 300 m u/s and d/s of C.L. of the crossing at 10 to 50 m intervals, bed and banks of drainage channel, etc.
5.	Water surface slope, designed (and existing, if canal is in existence);	Actual water surface slope, as observed under the highest flood conditions on upstream and downstream of the work as also the average slope.
6.	Lacey's *f*, nature of bed material and value of rugosity coefficients N.	Silt factor and type of bed load.
7.	Approved longitudinal section of the canal showing nearest proposed bridge site and drainage crossing on the canal to ascertain the possibility of combining the bridge with drainage crossing or combining two-drainage crossings.	Longitudinal section of the stream from about 1.5 km u/s and 1.5 km d/s of the canal crossing indicating location of cross drainage work, HFL, ordinary flood level and low water level at suitably spaced intervals along the line of deep channel. Waterway provided at the drainage where it crosses other roads, etc. nearest to the work site.

12.7 HYDRAULIC DESIGN PARAMETERS

Design of a cross drainage work is formulated on the basis of the given data of the drainage and canal, outlined above. The design of the structure consists of hydraulic design and structural design. The scope of the book limits itself to hydraulic design. The hydraulic design envisages determination of maximum design flood, approaches and outfall channel, waterway, afflux, head loss through the structure, uplift pressure on roof of trough, uplift pressure on floor and floor thickness, and bank connections. Model studies is a useful tool for the finalization of hydraulic design of major cross drainage works. These studies take into account the impact of ancillary components of the structure, namely approaches, connections, afflux bunds, floor and protection arrangements.

1. *Design Discharge*: Design discharge, as under, is the estimated flood discharge used for the design of waterway, foundation of a particular work or its component works and connotes fixation of the flood magnitude after thorough consideration of the flood characteristics, frequencies and flood potentialities of the drainage basin:

Table 12.1 Design flood values (IS : 7784 (Part I) : 1993

Category of Structure	*Canal discharge (cumecs)*	*Drainage discharge (cumecs)*	*Frequency of Design flood (1 in yrs)*
A	0 – 0.5	0 – 150	25
B	0.5 – 15	above 150	50
C	15 – 30	0 – 100	50
D	above 30	0 – 150	100
		above 150	Actual but not less than 100

At Sardar Sarovar Project (Gujarat), cross drainage works with catchment up to 500 sq km have been designed for 100 year flood and checked for 1.3 times 100 year flood derived by rational formula. The structures draining larger catchments are designed for SPF and checked for PMF derived by either regional approach analysis or sub-basin-wise unit hydrograph approach.

2. *Waterway*: From the known maximum flood discharge, observed water surface slope, value of *f*, bed width and depth are calculated. Due consideration to the size of silt particles or bed load is given in determining value of *f*. Waterway for drainage crossings is fixed from hydraulic and economic considerations, particularly (*i*) design discharge, (*ii*) topography of site, (*iii*) existing stream section, (*iv*) slope of the stream in the vicinity, (*iv*) permissible afflux, and (*v*) construction and maintenance problems.

In plains, drainage channel is generally in alluvium and waterway, between two abutments, provided with rigid floor is 60 to 80% of Lacey's perimeter, $C\sqrt{Q}$, usual value of C is 4.83. In works with rigid floors, waterway may be further flumed within the permissible velocity through ventage. The maximum permissible velocity in stone masonry is 3-4 m/s, brick masonry 2-2.5 m/s, hard rock 4 m/s, murum 1.5-2 m/s and soil, silt 0.75–1.0 m/s. In sub-mountainous and mountainous area waterway is as within the width of existing stream, as under:

Central Water Power Commission Practice: As per Central Water Power Commission (India) practice, waterway for canal aqueducts is provided to pass a 50-years flood but their foundations and free board are provided for a flood of not less than 100-years return period. In cross drainage works, which carry highways or railways, waterway provided shall also satisfy the respective standard code of practice of highways or railways.

3. *Approaches*: A splay of 2:1 for upstream transition and 3:1 for the downstream expansion is provided.

4. *Clearance*: Clearance is defined as the vertical height between the designed flood level, including afflux of the stream or full supply level of the canal and the lowest point of superstruc-

ture. The clearance in the case of aqueduct and superpassage depends upon the relative levels of the canal and drainage. The minimum clearance taking into account allowable afflux, is given in Table 12.2.

Table 12.2 Minimum vertical clearance

Rectangular openings		*Arch openings*	
Design flood (cumecs)	*Clearance (m)*	*Design flood (cumecs)*	*Clearance (m)*
Below 3	0.45	Below 3	Rise or 0.8 m whichever is more
3 – 30	0.60	3 – 6	2/3 rise or 1.0 m whichever is more
30 – 300	0.90	6 – 21	2/3 rise or 1.25 m whichever is more
300 – 3000	1.25	21 and above	2/3 rise or 1.5 m whichever is more
3000 and above	1.50		

5. *Free Board*: Free board is the difference in levels between the maximum flood line including afflux and the top of the embankment or guide banks. In aqueducts and superpassages it is reckoned from HFL including afflux in stream or from FSL in canal to the formation level of guide bank or canal embankment. Minimum free board of 0.9 m is provided. It is increased suitably wherever heavy wave action is anticipated.

6. *Afflux*: Afflux is defined as the upstream rise of water level above the normal surface water in a canal or drainage channel caused by an obstruction resulting from contraction of normal waterway. The afflux is restricted to such a value that (*i*) the resulting velocity does not cause excessive bed scour in the drainage, (*ii*) there is no flooding of the country side. The area submerged depends on the slope of the country side. Afflux is calculated by the following methods:

(*i*) *Rational formula*: Depending on the flow conditions through the opening of the drainage crossing, afflux is calculated by weir formula or orifice formula. Weir formula is applicable when a standing wave is formed, otherwise orifice formula holds good. Approximately, when the downstream depth D_d above the crest is more than 80% of the upstream depth D_u, the weir formula does not hold good.

(*i*) *Weir formula* $$Q = 1.70\, C_w L H^{3/2}$$

where, Q = discharge through the openings (cumecs), C_w = coefficient of discharge accounting for losses in friction, its values are 0.94 for narrow opening with or without floor, 0.96 for wide opening with floor, and 0.98 for wide opening without floor, L = linear waterway (m), H = total energy head u/s of obstruction (m) *i.e.* $D_u + V^2/2g$, D_u = depth of flow upstream (m), $V^2/2g$ = velocity head, where V is the average velocity in the approach section worked out from the known width (W) of unobstructed section.

(*ii*) *Unwin's formula*: Unwin's formula is also used for the determination of afflux:

$$h = \left(1 + f_1 + f_2 \frac{L}{R}\right)\frac{V^2}{2g}$$

where, h = head loss (m), f_1 = coefficient of head loss at the entry; 0.505 for unshaped mouth, and 0.08 for bell mouth, R = hydraulic mean depth of barrel (m), f_2 = coefficient of value $a\,(1 + b/R)$, where a and b have the following values for different materials:

Nature of barrel surface	*a*	*b*
Smooth iron pipe	0.00497	0.025
Smooth cement plaster	0.00316	0.030
Ashlar or brickwork	0.00401	0.070
Stone masonry or stone pitching	0.00507	0.25

(*iii*) *Orifice Formula*: $$Q = C_o (2g)^{1/2} L D_d \left[\frac{(h + (1+e)V^2}{2g}\right]^{1/2}$$

where Q=discharge through openings (cumecs), Co = coefficient of discharge, L = liner waterway (m), D_d = depth d/s of obstruction (m), h = afflux (m), e = a factor accounting for recovery of some velocity as potential head on emergence from the cross drainage work openings, and V = average velocity in approach section (m/s). Values of 'C_o' and 'e' are given in (Figs. 12.12 and 12.13).

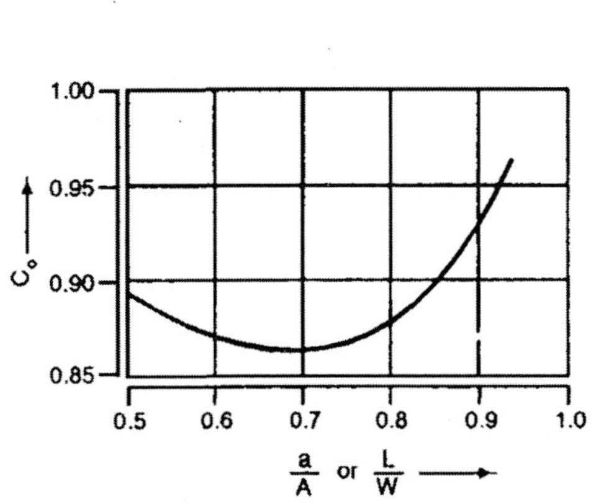

Fig. 12.12 Coefficient 'C_o' in orifice formula

Fig. 12.13 Coefficient 'e' in orifice formula

(*iv*) *Empirical formula*: When area of obstruction is not very large compared to original unobstructed area, the formula is

$$h = [(V^2/17.85) + 0.0152]\,[(A^2/a^2) - 1]$$

where, V = velocity in unobstructed drainage channel (m/s), A = unobstructed sectional area of drainage channel (m^2); and a = sectional area of structure (m^2).

7. *Loss of Head* (Energy Loss): Loss of head (H) through the structures in inlet, and outlet transitions and through the conduit is as under:

$$H = h_1 + h_2 + h_3 + h_4$$

where, h_1 = losses at inlet and outlet (for syphon), h_2 = losses at elbows or bends (for barrel), h_3 = losses due to transitions (other than syphon) and h_4 = losses due to skin friction (for barrel and trough).

Loss of head at inlet and outlet of syphons;

(*a*) At entrance, $h_1 = (1 + f_1)\,V^2/2g$

where, h_1 is loss at entrance or exit (m), f_1 = coefficient, value 0.08 for bell mouth entrance, 0.505 for cylindrical entrance with sharp edges (unshaped mouth of the same sectional area of barrel), V = velocity in syphon (m/s).

(*b*) Loss of head due to elbows or bends h_2 in barrels is determined as per IS : 2951 (part 2) : 1965.

(*c*) Loss of head due to skin friction in barrels or troughs. Losses in normal transitions of cross drainage works other than syphons are determined, as under:

Type of transition	*Losses at*	
	inlet	*outlet*
Streamline warped	$0.1\,[(v_1^2 - v_2^2)/2g]$	$0.2\,[(v_1^2 - v_2^2)/2g]$
Straight warped	$0.2\,[(v_1^2 - v_2^2)/2g]$	$0.3\,[(v_1^2 - v_2^2)/2g]$
Straight warped with bottom corner fillets	$0.3\,[(v_1^2 - v_2^2)/2g]$	$0.4\,[(v_1^2 - v_2^2)/2g]$
v_1 is velocity before transition and v_2 velocity after transition (m/s).		

Conduit: Loss is $S \times L$, where S is given by Manning's formula, value of n for concrete and brick masonry = 0.018, 0.025 for rubble masonry.

Scour Depth: Mean scour depth below check HFL is estimated by the formula:

$$D_{sm} = 1.34\,(q^2/ksf)^{1/3}$$

where, D_{sm} = mean scour depth (m) below the maximum flood level, ksf = silt factor for a representative bed material sample taken at the anticipated depth of scour below the bed of the stream given by 1.76 $(dm)^{1/2}$, q = discharge/m width (cumecs). Concentration of flow is not considered for overall waterway < 60 m.

Maximum scour depth for foundation

Piers and abutments in straight reach with individual foundations without floor protection	Floor protection works, raft foundations or shallow foundations.	
In vicinity of piers 2.0 *dsm*	In straight reach	1.27 dsm
Near abutments 1.27 *dsm* approach retained	At moderate bend	1.50 dsm
	At severe bend	1.75 dsm
2.0 dsm scour all round	At right angled bend	2.0 dsm

The foundation should rest on homogeneous, undisturbed and uniform subgrade of fairly dense type. On subgrade of different type, suitable joints are provided to avoid cracks due to differential behaviour of foundation soil. Comparatively lower bearing pressures are allowed on foundations on subgrades prone to appreciable settlement.

In works provided with impervious floor scour condition is not applicable and the foundation is usually taken to about 1.5 m below the floor level with a suitable cutoff for the floor. Passive resistance of the soil formation and the effect of scour determines the depth and type of foundation of works not provided with impervious floor. Foundation is usually taken to a depth 1.33 times the scour depth below the anticipated scour level or 1.05 times the scour depth estimated for probable maximum flood (PMF).

The foundation for the cross drainage works may be either open foundation, or well foundation or raft foundation. Open foundations are taken down to rock, and are economical if (*i*) the foundation is of shallow depth, *i.e.* up to 6 m depth, and (*ii*) subsoil watertable is not that high as to entail heavy expenditure on lowering the watertable in order to lay the open foundation. In open foundation, deep curtain walls are provided on either side of the impervious floor at bed level to protect against scouring action. Well foundations are generally dictated by the consideration of high subsoil watertable, depth of scour and low bearing capacity of the subsoil. Even well foundations may be costly at certain locations leaving the choice to mass concrete raft floor supporting the piers with deep cutoff at the upstream and downstream ends of the floor. The exit gradient of seepage water should be limited within the permissible limit.

Fluming: Fluming or constriction of the canal waterway is necessary to reduce the length of the trough or the width of the aqueduct in order to effect economy in the construction of cross drainage work. Fluming, on the other hand, entails extra transition wings for joining the flumed section to the normal section. The greater the extent of fluming, the greater is the length of transition wing walls required on either side of the work. Various considerations which influence the extent of fluming are (*a*) actual loss of head through the structure is not more than the fall available, (*b*) the cost of additional length of canal wings is balanced by the saving resulting from reduction in the aqueduct width, and (*c*) the velocity through the trough is not to exceed critical velocity $V_c = q/D_c$, where critical depth $D_c = (Q^2)^{1/3}/f$ because supercritical velocity, if generated, shall entail formation of hydraulic jump adding to the cost of the transition back to the normal section downstream. Generally, velocity in the trough is limited to 3 m/s. Fluming ratio is less than 70% of the bed width. For a canal, width at mid depth is taken as 100 per cent. In drainage channel in alluvium, a fluming ratio from 60 to 80 per cent of Lacey's waterway is adopted. The section of barrel selected is preferably in round number of metres.

Transitions: The approach and departure transitions, on upstream and downstream respectively, from normal canal section to the flumed section are designed to provide smooth entry and exit so as to have a streamline flow and minimum loss of head. The approach transition is usually not steeper than 30°, corresponding to a constriction of 2:1, and departure transition not steeper than 22½°, *i.e.* an expansion of 3:1 with curved and flared wing walls. With the change in section from trapezoidal in earthen channel to rectangular in the trough or flumed section, the depth of flow may not be the same in the normal and flumed section. Increased depth in the flume means reduced width but it also entails increase in uplift pressures requiring higher thickness of the roof and floor. Increased depth of flow thus does not result in appreciable economy. The canal transitions can be designed with water depth constant by Mitra's hyperbolic transition:

$$B_x = \frac{B_c B_f L}{L B_c - (B_c - B_f)x}$$

where, B_c = bed width of normal canal section, B_f = bed width of flumed canal section, B_x = bed width at any distance x from the flumed section, and L = transition length.

(*b*) *Chaturvedi's semi-cubical parabolic transition*: R.S. Chaturvedi, on the basis of his experiments, has proposed the following equation for the design of canal transition:

$$x = \frac{L_f B_c^{3/2}}{B_c^{3/2} - B_f^{3/2}} \left[1 - \left(\frac{B_f}{B_x}\right)^{3/2}\right]$$

Choosing various convenient values of B_x, the corresponding distance x can be computed from the above equation.

Transition with Varying Water Depth: Hind has suggested the following method for the design of transition when the depths of flow in the flumed sections are the same or varying.

Referring to Fig. 12.14 we have

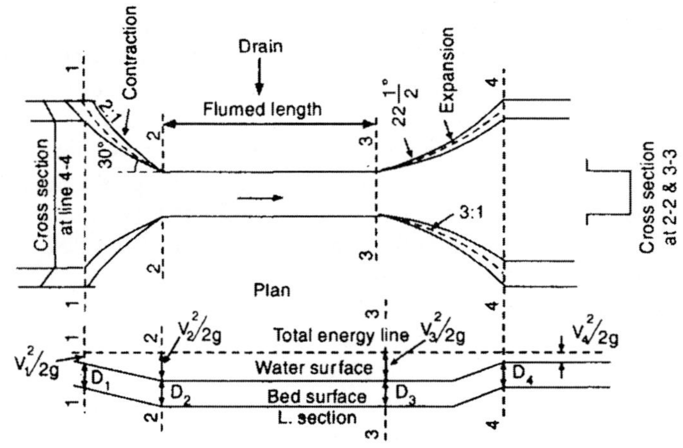

Fig. 12.14 Transition with varying water depth

Step I. Let D and V with appropriate subscripts refer to depth and velocities at the different sections. Also assume the elevation of bed at section 4 as 300.00. FSL at section 4–4 = 300.0 + D_4.

TEL at section 4–4 = 300.0 + D_4 + $V_4^2/2g$.

Step II. Between sections 4–4 and 3–3 there is negligible energy loss (0.1 to 0.3) $(V_3^2 - V_4^2)/2g$ due to expansion of stream lines and friction.

Step III. TEL at section 3–3 = El. of TEL at section 4–4 + energy loss between section 4–4 and 3–3.

With the known trough dimensions at section 3–3 and known V_3, energy loss can be computed.

Section 1 – 1 = normal canal cross section from which contraction starts

2 – 2 = end of contraction transition

2 – 2 to 3 – 3 = flumed section

3 – 3 = start of expansion transition

4 – 4 = end of expansion transition and start of normal cross section downstream.

FSL at section 3 – 3 = TEL at section 3 – 3 - $V_3^2/2g$.

Bed level at section 3 – 3 = FSL at section 3 – 3 – D_3.

Step IV. Between section 2-2 and 3-3 the channel flows in a trough of constant cross section and uniform flow. The only loss is, therefore, friction loss which can be computed by Manning's equation:

$$V = \frac{1}{n} R^{2/3} S^{1/2}$$

or
$$V = \frac{1}{n} R^{2/3} (H_l / L)^{1/2}$$

$$\therefore \quad H_l = (V^2 . n^2 . L)/R^{4/3}$$

which is the friction loss in the trough.

TEL at section 2-2 = TEL at section 3-3 + friction loss

FSL at section 2-2 = TEL at section 2-2 – $(V_2^2)/2g$

Bed level at section 2-2 = F.S.L. at section 2-2 – D_2

Since velocity ($V_2 = V_3$) and depth ($D_2 = D_3$) is constant in the trough, TEL, FS line and bed line are parallel to each other from section 2-2 to section 3.3.

Step V. Between section 2-2 and section 1-1 there is energy loss of 0.1 to 0.2 $(V_2^2 - V_1^2)/2g$ due to contraction of stream lines and negligible loss.

TEL at section 1-1 = TEL at section 2-2 + 0.2 $(V_2^2 - V_1^2)/2g$

Water surface elevation at section 1-1 = TEL at section 1-1 – $V_1^2/2g$

Bed level at section 1-1 = Water surface elevation - D_1.

In the contraction transition, section 1 to section 2, there is drop in water surface due to drop in energy line and increased velocity head at section 2-2. This drop in water surface is negotiated by a smooth curve tangential at both ends, *i.e.* by two parabolic curves, the first convex and the second concave upwards (Fig. 12.15).

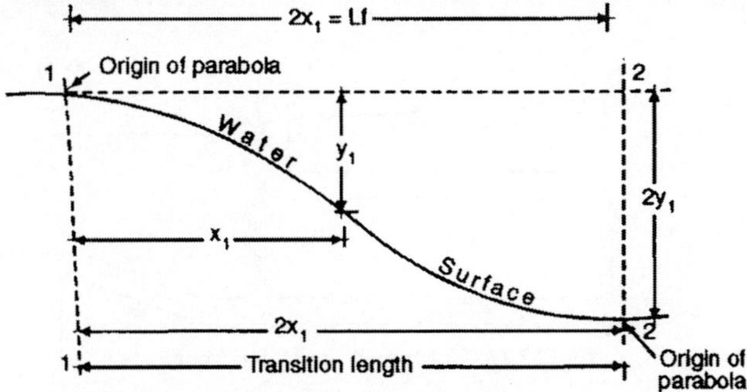

Fig. 12.15 Water surface curve for contraction transition

Let, length of contraction transition = $2X_1$

Drop in water surface = $2Y_1$

At the middle point of the transition, the distance from section 1-1 is X_1 and drop in water surface Y_1. Taking water surface at section 1-1 as the origin, the equation of the parabola is $y = cx^2 = (y_1/x_1^2)\, x^2$.

By substituting known values of y and x, value of c is determined, and hence the first parabolic curve is plotted downstream and downwards from the origin at section 1.1. Similarly the second parabolic is plotted by taking origin on water surface at section 2-2 and values of y are measured upstream and upwards from the origin at the water surface at section 2-2. Similarly the water surface in the expansion transition between section 3-3 and 4-4 is plotted. There is a rise in the water surface from section 3-3 and 4-4 (Fig. 12.15). This rise in water surface is negotiated with a smooth curve tangentially at both ends similar to that explained for contraction transition.

With the water surface profile over the whole length so plotted, the velocity head h_a is found by measuring the vertical distance between TEL and water surface line at any point. The velocity head is converted into equivalent velocity by the relationship $V = \sqrt{2g h_a}$. Thus velocity at each point is known. The cross sectional area at any point corresponding to the given discharge is determined by the relationship $A = Q/V$. By subtituting the known values of area (A), depth (D) and side slope (S) at any point in the equation $A = BD + SD^2$, the bed width (B) is determined in a trapezoidal channel section.

In the flared out wall transition, the side slopes are gradually brought to vertical from an initial side slope and accordingly the side slopes at any point are interpolated in proportion to the length of transition undergone. Thus, all the dimensions of the transition are determined.

12.8 UPLIFT PRESSURE

Uplift pressure acts on roof of work, or underside of trough and on floor of the work, described below:

Uplift Pressure on the Roof of Trough

1. *Aqueducts*: The bottom of the trough in an aqueduct is clear of the flow in the drainage and as such is not subjected to uplift pressure. The weights acting on the trough are dead weight of the trough itself and vertical load of water column in the trough. The side walls of the trough are subjected to the horizontal force exerted by the flow in the canal and wind load. The side walls are designed either monolithic with the bottom of the trough or the bottom slab rests on the side walls.

2. *Syphon-aqueduct*: Uplift pressure is exerted by the flow in the drainage on the underside of the trough as the tail water level below a syphon is higher than the underside of the trough. With the afflux worked out, HGL is drawn as illustrated in Fig. 12.16. The maximum uplift pressure occurs at the upstream end of the trough and the minimum at the downstream end of the trough. The uplift pressure at any other point on the trough is given by the ordinate between the

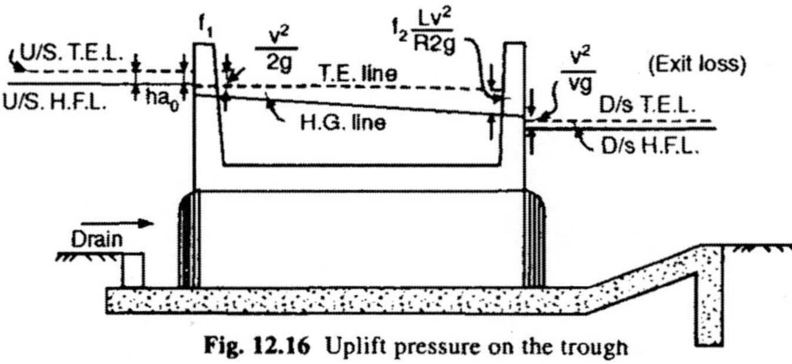

Fig. 12.16 Uplift pressure on the trough

HGL and the underside of the trough. The bottom slab of the canal trough is subjected to downward force due to dead weight of the trough and canal water in the trough, and uplift force is exerted by flow in the drainage, both acting in opposite directions so as to counteract each other. However, the worst conditions for which the slab is designed are:

Condition I: Canal dry and drainage passing the highest flood, *i.e.* maximum uplift pressure but little downward pressure due to dead weight of the slab only. The thickness of slab is sufficient to counteract uplift pressure to the extent that the bed level of the downstream channel is 1m or below the underside of the trough. But when downstream water level is high, the uplift pressure is so high that it may not be counteracted by increasing the slab thickness. The increase in slab thickness implies lowering the level of the underside of the trough and resulting in increased uplift pressure on the trough as well as flooring which is not desirable. It is, therefore, necessary to counteract the uplift pressure by providing the slab thickness as required for canal water and weight of the slab, and resisting the remaining uplift pressure by bending action by providing reinforcement at top; and the bottom reinforcement for the load of the canal trough. The trough is better anchored down to the piers.

Condition II: Canal running full (up to brim of trough considering silting up or heading up in the trough), and drainage dry. Maximum downward pressure is due to (*a*) water in trough, and (*b*) dead weight of trough but no uplift pressure. The trough is designed for the downward acting loads of canal water and trough itself with the slab reinforcement at the bottom. A box culvert is a better alternative.

Uplift Pressure on the Floor

The uplift pressure on the floor of the syphon-aqueduct is exerted by high subsoil watertable and seepage of canal water to the drainage.

1. *Due to Subsoil Watertable*: Uplift pressure due to subsoil acts in syphon aqueduct with the floor depressed below the drainage bed. Maximum uplift under worst condition occurs when there is no flow in the drainage and the subsoil watertable has risen up to the drainage bed and is equal to the difference between the bed of the drainage and the bottom of the depressed floor.

2. *Due to Seepage from the Canal*: Uplift pressure due to seepage from the canal to drainage is maximum under worst condition when the canal is running full but there is no flow in the drainage.

The flow pattern in this case is three dimensional which cannot be approximated to two dimensional flow there being no typical place across which the flow is practically two dimensional. The uplift pressure so acting at any point under the floor is complex and difficult. An approximation can be made by Bligh's creep theory in small works and by model studies in the case of large works.

Seepage Uplift Pressure Determination by Bligh's Creep Theory

The seepage flow occurs from either side of the canal trough and reappears in the bed of the drainage at either side of the impervious floor. The total creep length, *i.e.*, the total length traversed by the seepage flow from *a* and *b* and thence *c* (Fig. 12.17) is calculated to compute the residual pressure at any point which in accordance with Bligh's theory is in linear proportion to the total creep length. For example, residual uplift pressure at point *b* is equal to:

$$H_b = H - \left[\frac{H}{L} \times l\right]$$

where, H_b = residual uplift pressure at *b*, H = total seepage head, *i.e.*, level difference between full supply level in the canal and downstream bed level of the drainage, L = total creep length, *i.e.*. $ab + bc$, and l = creep length up to point *b*, *i.e.*, *ab*.

The floor of the syphon-aqueduct is designed for the sum of uplift pressure due to seepage head, plus uplift pressure due to static head. The total uplift pressure is resisted partly by the weight of floor thickness and partly by the bending action in the reinforcement.

Method for Reducing Uplift Pressure

The three methods available for reducing uplift pressure on the floor are:

1. *Extension of impervious floor*: Extension of the impervious floor in the canal in continuation and on either side of the canal trough in order to increase the creep length (Fig. 12.17).

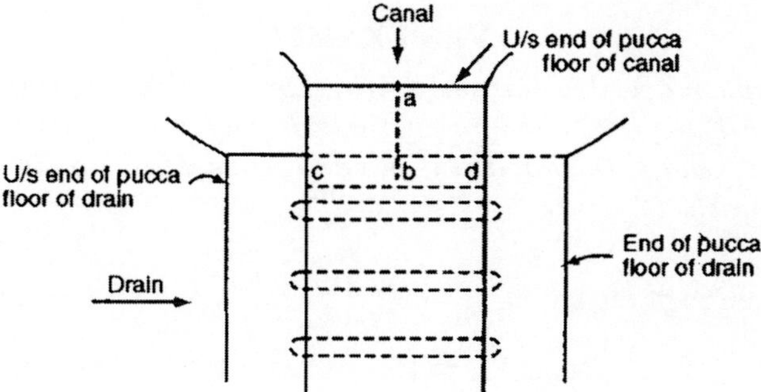

Fig. 12.17 Uplift pressure under the floor

2. *Provision of water cushion on the floor*: Arrangement is made to load the impervious floor, with a certain depth of impounded water. The uplift pressure is counteracted to the extent of the weight of the column of impounded water over the floor.

3. *Provision of drainage holes*: Provision of drainage holes in the floor is made to release the uplift pressure. Inverted filter is also necessarily provided between the subsoil and the impervious floor with a view to prevent the soil particles from getting lifted up with the seepage water. The relief holes are essentially provided with flap valve to allow seepage water to release into the drainage but prevent the drainage water from getting down.

12.9 IMPERVIOUS FLOOR DESIGN

The impervious floor is designed on the basis of the theory of exit gradient. The safe value of exit gradient for alluvial permeable soil is taken as 0.3. The floor thickness is determined for the maximum uplift pressure, caused at any point under the maximum pressure head available. In case of superpassage maximum pressure head is determined under the highest flood conditions, while for a syphon or an aqueduct, the maximum pressure head occurs when the canal runs full supply and the drainage is dry. Hydraulic gradient is assumed as 1 in 4 for clay soils and 1 in 4 to 5 for shingle, 1 in 5 to 6 in coarse sand and 1 in 6 to 7 for fine sand. Pressure head is determined at various points along the profile of the floor. Floor thickness is determined at any point by dividing the effective pressure head so obtained by $\rho - 1$, where ρ is the specific gravity of the material used for the construction of the floor and 75 mm is added to the floor thickness obtained to allow for wearing, etc.

Cutoffs: The upstream and downstream cutoff walls at the end of impervious floor are taken to a depth of 1.33 times the scour depth estimated below HFL or 1.5 times the scour depth for probable maximum flood.

12.10 BANK CONNECTION

Bank connections are provided to increase the length of the seepage path and control the exit gradient. The bank connections consist of canal or land wings and drainage or water wings. The layout of the wing walls depends upon the extent of constriction provided to the canal and drainage section at the work and the general arrangement of the work.

1. *Canal Wings*: Normal section of the canal is trapezoidal but is rectangular at the trough. Canal wing walls on upstream and downstream of an aqueduct are provided to facilitate this change in section of the canal from trapezoidal to rectangular and extend up to the end of upstream transition to retain and protect the earthen embankment of the canal.

2. *Drainage Wings*: The drainage wings are provided on upstream and downstream of the drainage at the aqueduct to (*a*) retain and protect the earthen slope, (*b*) form guide walls to the water entering and leaving the work, (*c*) join the work to guide banks, and (*d*) provide a vertical cutoff for seepage water from the canal percolating into the bed of the drainage. They are designed to provide smooth entry and exit, to be safe against the deepest anticipated scour in the drainage when the wall extend sufficiently into the top of guide banks.

SOLVED EXAMPLES

Example 12.1 *A syphon aqueduct has single barrel of 3 m diameter, length 90 m, discharge capacity 25 cumecs, f in Darcy-Weisbach formula 0.013, coefficient of (i) bend loss (2 bends) 0.10, (iii) in expansion at outlet 0.20 and (iii) in contraction at inlet 0.10. Determine afflux. Neglect velocity of approach.*

Solution. Velocity through the barrel, $V = Q/A = 25/\left(\frac{22}{7}(3)^2/4\right) = 3.54$ m/s.

1. Loss at inlet, $= (0.1)\ V^2/2g = 0.1\ [(3.54)^2/2 \times 9.81] = 0.06$ m

2. Loss at bends, 2 No. $= 2 \times 0.1\ (V^2/2g) = 2 \times 0.06 = 0.12$ m

3. Friction loss in pipe $= 4\,flV^2/2gd = \dfrac{4\times0.013\times90\times(3.54)^2}{2\times9.81\times3} = 1.0$ m

4. Loss at exit $= (0.2)\ V^2/2g = 0.12$ m

Total loss through the work or afflux $= 0.06 + 0.12 + 1.0 + 0.12 = 1.30$ m.

Example 12.2 *Design a suitable cross drainage work for a canal having discharge 35 cumecs, bed width 25 m, F.S. depth 1.75 m, F.S. level 201.75 m, bed level 200.0 m, Lacey's f = 1.0. The drainage has catchment area 50 sq km. average annual rainfall 100 cm, H.F. level 200.50 m, bed level 198.0 m, natural surface level 199.0 m, Lacey's f = 1.5.*

Solution. The highest flood level in the drainage is higher than the bed level but below full supply level of the canal and as such it is a case of syphon-aqueduct. As the size of drainage is large, trough type syphon-aqueduct shall be economical and hence adopted. The canal section shall be flumed for reasons of economy and taken in a concrete trough over the drainage.

1. Drainage

(*i*) Peak discharge

Peak discharge, $Q = CA^{0.5}$, as per Boston society formula and value of $C = 35$ for area of 100 cm rainfall.

$\therefore \quad Q = 35\sqrt{50} = 247$ cumecs

(*ii*) Approach and outfall channels

Lacey's wetted perimeter,

$$P = 4.83\ Q^{0.5} = 4.83\ (247)^{0.5} = 75 \text{ m}$$

$\therefore$ Velocity, $V = (Qf^2/140)^{1/6} = (247\ (1.5)^2/140)^{1/6} = 1.26$ m/s

Area, $A = \dfrac{Q}{V} = \dfrac{247}{1.26} = 196$ m^2

Hydraulic mean depth, $R = 2.5\ V^2/f = 2.5\ (1.26)^2/1.5 = 2.65$ m

Also, $R = A/(B + 2.828D)$, with 1:1 slope

$$2.65 = 196/(B + 2.828\ D)$$

or $(B + 2.828\ D) = 74$ (*i*)

Also $A = 196 = BD + D^2$ (*ii*)

On solving (*i*) (*ii*)

Bed width = 66 m and F.S depth = 2.85 m

(*iii*) Drainage barrels

Velocity through masonry barrels = 2.5 m/sec (assumed)

$\therefore$ Area of barrels required $= 247/2.5 = 99$ m^2

Provide 7 barrels of 6×2.5 m

Actual velocity through the barrel $= 247/7 \times 6 \times 2.5 = 2.35$ m/sec.

Overall waterway, with 6 piers of 1.25 m each = 49.50 m.

2. Canal waterway

Bed width of 25 m is proposed to be flumed to 15 m with 2:1 and 3:1 Splay in contraction and expansion respectively

Length of contraction transition $L_f = 2(25 - 15)/2 = 10$ m

Length of expansion transition $= 3(25 - 15)/2 = 15$ m

In the transitions, the side slopes of the canal section shall be warped from the original slope of 1.5 : 1 to vertical.

3. Design of levels at different sections.

(*a*) At section 4-4

Area of section with 1.5 : 1 side slopes = $(B + 1.5\ D)\ D$

$= (25 + 1.5 \times 1.75)\ 1.75 = 48.34\ \text{m}^2$

Velocity, $V = Q/A = 35/48.34 = 0.72$ m/s

Velocity head, $ha = V_a^2/2g = (0.72)^2/2 \times 9.81 = 0.026$ m

R.L. of T.E.L $= 201.75 + 0.026 = 201.776$ m

(*b*) At section 3-3

Area of trough, $A = 15 \times 1.75 = 26.25\ \text{m}^2$

Velocity, $V = Q/A = 35/26.25 = 1.33$ m/s

Velocity head, $ha = V^2/2g = (1.33)^2/2 \times 9.81 = 0.09$ m

Loss of head in expansion from section 3-3 to 4-4

$= 0.3\ (V_3^2 - V_4^2)/2g$

$= [0.3\ (1.33)^2 - (0.72)^2]/2 \times 9.81 = 0.019$ m

R.L. of T.E.L. at section 3-3 = T.E.L. at 4-4 + head loss

$= 201.776 + 0.019 = 201.795$ m

R.L. of water surface at section 3-3 = T.E.L. – velocity head

$= 201.795 - 0.09 = 201.705$ m

R.L. of bed (to maintain constant depth)

$= 201.705 - 1.75 = 199.955$ m

(*c*) At section 2-2

Hydraulic mean depth, $R = A/P = \dfrac{15 \times 1.75}{15 + 2 \times 1.75} = 1.42$ m

Velocity, $V = \dfrac{1}{n} R^{2/3} S^{1/2}$ (Manning's formula).

or Slope $S = \dfrac{V^2 . n^2}{R^{4/3}}$

$= \dfrac{(1.33)^2 (0.018)^2}{(1.42)^{4/3}}$, assumed $n = 0.018$

$= 0.00036$ m/m

Head loss due to friction $= 49.5 \times 0.00036 = 0.018$ m

R.L. of T.E.L. at section 2-2 = T.E.L. at 3-3 + friction loss

$= 201.795 + 0.018 = 201.813$ m

R.L. of water surface at section 2-2 $= 201.813 - 0.09 = 201.723$ m

R.L. of bed at section 2-2 $= 201.723 - 1.75 = 199.973$ m

(*d*) At section 1-1

Loss of head in contraction from section 1-1 to section 2-2

$= [0.2\ (V_2^2 - V_1^2)]/2g$, here $V_2 = V_3$ & V_1 & V_4

$= 0.2 \left[\dfrac{(1.33)^2 - (0.72)^2}{2 \times 9.81}\right] = 0.013\text{m}$

R.L. of T.E.L. at section 1-1 = R.L. of T.E.L. at section 2-2 + loss in contraction

$= 201.813 + 0.013 = 201.826$ m

R.L. of water surface at 1-1 = R.L. of T.E.L. – $V_1^2/2g$

$= 201.826 - (0.72)^2/2 \times 9.81 = 201.80$ m

R.L. of bed at 1-1 (to maintain constant depth)

$= 201.80 - 1.75 = 200.05$ m.

4. Design of canal transitions

(a) Contraction transition.

The transition shall be designed according to Mitra formula:

$$B_x = \frac{B_c\,B_f\,L}{L B_c \times x(B_c - B_f)}$$

or

$$B_x = \frac{25\times15\times10}{10\times25 - x(25-15)} = \frac{375}{25-x}$$

Value of B_x for various value of x varying from 0 to 10, reckoned from section 2-2 are:

x	0	2	4	6	8	10
B_x	15	16.3	17.86	19.24	22.06	25

(b) Expansion transition

$$B_x = \frac{25\times15\times15}{15\times25-(25-15)x} = \frac{5625}{375-10x}$$

Values of B_x corresponding to various value of x reckoned from section 3-3 are:

x	0	3	6	9	12	15
B_x	15	16.30	17.86	19.74	22.06	25

5. Design of Trough

The trough shall be of monolithic construction in reinforced cement concrete. It shall have flumed waterway of 15 m partitioned into two compartments, one carrying roadway 5 m wide and the other is 10 m wide separated by 0.4 m thick partition wall. The outerwall and bottom slab shall be tentatively 0.5 m thick. Thus, the outer width of trough or the length of syphon barrel is 16.4 m (*i.e.* 15 + 0.40 + 2 × 0.50). A free board of 0.75 m shall be provided in the trough as in the case of a branch canal. As such height of trough shall be 2.5 m (*i.e.* 1.75 + 0.75 m).

6. Loss of head through syphon barrels.

By Unwin's formula:

$$h = \left(1 + f_1 + f_2\frac{L}{R}\right)\frac{V^2}{2g}$$

where, h = loss of head (m), f_1 = coefficient of head loss at entry, *i.e.*, 0.505 for unshaped mouth, f_2 = $a\left(1+\frac{b}{R}\right)$, where a = 0.00316 and b = 0.030 for cement plastered surface

$$= 0.00316\left(1+\frac{0.030}{1.36}\right) = 0.00323$$

L = length of barrel, *i.e.*, 16.4 m.

R = (6 × 2.5)/(6 + 2 × 2.5) = 1.36 m

Substituting these values in the formula,

$$h = \left(1+0.505+0.00323\times\frac{16.4}{1.36}\right)\frac{(2.35)^2}{2\times9.81} = 0.43\text{ m}$$

D/S H.F.L. = El. 200.50 (given)

U/S H.F.L. = D/S H.F.L. + loss of head

= 200.50 + 0.43 = 200.93 m

7. Uplift on the roof of barrel

R.L. of bottom of trough = Bed level of canal - slab thickness

= 200.0 – 0.50 = 199.50 m

Loss of head at entry of barrel = $0.505\ V^2/2g = 0.505\ (2.35)^2/2 \times 9.81 = 0.14$m

Uplift on the roof = U/s H.F.L. – head loss – level of roof slab underside

= 200.93 – 0.14 – 199.50 = 1.29 m of water

= 1.29 t/m²

The trough slab 0.5 m thick exerts downward load of 1.2 t/m² (*i.e.*, 0.50 × 2.4) which counter-balances the upward pressure. However, suitable reinforcement shall be provided at the top of the slab to resist bending action. Since the trough as a whole is heavier than the total uplift pressure no anchorage to piers is necessary. The trough slab is to be designed for the full water load of the canal when the drainage is dry and as such suitable reinforcement shall be provided at the bottom of the slab.

8. Uplift on the floor of barrel

(*a*) Static Uplift.

R.L. of barrel floor = R.L. of bottom of trough slab-height of barrel

= 199.50 – 2.5 = 197.0 m

R.L. of bottom of floor = 197.0 – 0.75 = 196.25 m, assume floor thickness 0.75 m

Drainage bed level = 198.0 m

∴ Static uplift on floor = 198.0 – 196.25

= 1.75 m

(assuming subsoil water level up to bed level of drainage)

(*b*) Seepage head.

The seepage head shall be maximum when the canal is running full and drainage is dry.

∴ Total seepage head = F.S.L. in canal — bed level of drainage.

= 201.75 – 198.0 = 3.75 m

The residual seepage head at the centre of the first barrel is to be calculated. The seepage line, according to Bligh's theory shall follow the path abc.

Neglecting the floor depression or floor thickness, the various lengths are as under:

(*i*) Length of u/s transition = 10 m

(*ii*) Half of barrel = 3m

(*iii*) End of drainage floor from centre of barrel, assuming total length of 30 m = 15 m

∴ Total creep length = 10 + 3 + 15 = 28 m

Creep length up to centre of first barrel = 13 m

∴ Residual seepage head at head $B = 3.75\left(1-\frac{13}{28}\right) = 2.01$ m

Total uplift = static head + seepage head

= 1.75 + 2.01 m = 3.76 m = 3.76 t/m²

Uplift resisted by 0.75 m thick floor due to its
self weight $= 0.75 \times 2.4 = 1.8$ t/m²

∴ Balance to be resisted by the
bending action $= 3.76 - 1.8 = 1.96$ t/m².

This shall be taken care of by suitable reinforcement provided at the top of the floor.

9. Cutoff and Protection Works.

Cutoff:

Scour depth $D_{sm} = 1.34\ (q^2/f)^{1/3}$

$$D_{sm} = 1.34\left[\frac{(247/6\times7)^2}{1.5}\right]^{1/3} \simeq 3.75 \text{ m}$$

Upstream cutoff

R.L. of bottom of u/s cutoff = u/s H.F.L.–1.5×3.75 = 200.93–6.0 = 194.93m

R.L. of bottom of d/s cutoff = 200.50 – 2 × 3.75 = 193.0 m

Length of u/s protection consisting of 40 cm brick pitching = 2 (u/s bed level – bottom R.L. of u/s cutoff)

= 2 (198.0 – 194.93) ≃ 6.5 m

Length of d/s protection consisting of 40 cm brick pitching resting on 0.4 m wide toe wall 1.5 m deep. = 2.5 (198.0 – 193.0) = 12.50 m

Fig. 12.18 shows the section of syphon aqueduct.

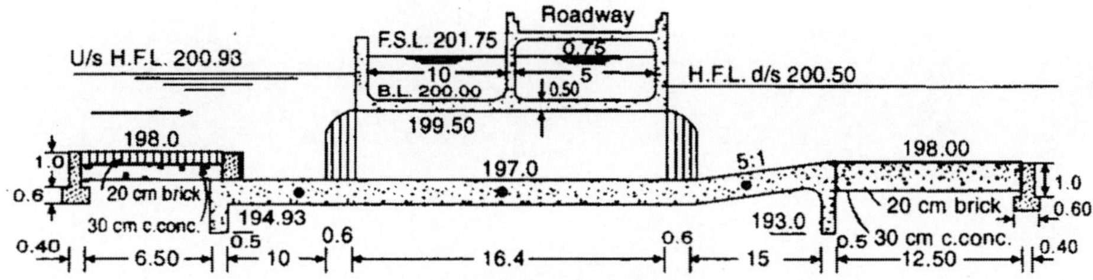

Fig. 12.18 Syphon -Aqueduct

QUESTIONS

1. (*a*) What are different types of cross drainage works.
   (*b*) Classify aqueducts and explain under what conditions each one is suitable.
   (*c*) Draw a neat sketch to illustrate syphon-aqueduct. [PUM, Nov. 1980]
2. (*a*) Explain the necessity of a cross-drainage structure.
   (*b*) How do you classify cross-drainage works?
   (*c*) Suggest suitable cross-drainage works, under the following situation and draw neat sketch to illustrate the same:

	Canal data	*Drainage data*
FSL/HFL	125.00 m	124.00 m
Bed level	120.00 m	122.00 m
Discharge	100 cumecs	20 cumeces

[PU]

3. (*a*) What are the different types of cross drainage works in common use?
   (*b*) Draw neat sketch to illustrate syphon-aqueduct.
   (*c*) State how will you determine a syphon aquedue (*i*) Contraction of canal waterway, (*ii*) Waterway for the drain, and (*iii*) Thickness of floor for the drain in the transition reach. [PU]
4. (*a*) Discuss the factors governing the suitable type of a cross drainage work.
   (*b*) Describe various steps for the design of a syphon-aqueduct. [PU]

5. (*a*) Give neat sketches of suitable designs of aqueducts for each of the crossings: (*a*) A major canal over a small drainage,
   (*b*) A canal carrying low discharge over a large drainage,
   (*c*) A major canal over a large drainage.
6. (*a*) What is the necessity of providing cross drainage works?
   (*b*) Distinguish between, (*i*) Aqueduct and syphon-aqueduct, (*ii*) Syphon and superpassage, and (*iii*) Level crossing and inlet. [PU. Nov., 1977]
   (*c*) Discuss the design criteria for a syphon aqueduct.
7. (*a*) During the construction of a canal, which types of cross drainage works shall possibly come across. Briefly state the site conditions under which each one is used.
   (*b*) Describe various steps for the design of a syphon-aqueduct. [PU. Nov. 1980]
8. (*a*) State under what circumstances you will recommend use of cross drainage structures (*i*) syphon, (*ii*) inlet, (*iii*) aqueduct. Draw a neat sketch to illustrate any one of them.
9. (*a*) Give neat sketch of suitable designs of aqueducts for each of the crossings; (*i*) A major canal over a small drainage, (*ii*) A canal carrying low discharge over a large drainage, and (*iii*) A major canal over a large drainage. Explain in detail the method of design of fluming required in the crossing at (*iii*) above.
10. Discuss various types of cross drainage works used in canal systems. What considerations govern the selection of the different types of works mainly depending upon the levels of the canal and the drainage. Illustrate by drawing a neat sketch of each type of structure.

13

BRIDGES AND CULVERTS

13.1 INTRODUCTION

Bridges and culverts are required to be constructed whenever a canal or a drainage crosses an existing road or *vice versa*. The bridges are classified on the basis of type of existing road crossed and bridge vent height *vis-a-vis* high flood level of the drainage, as under :

Classification Based on Type of Road Crossed

(*i*) Arterial road bridge (AR bridge) — crossing an arterial (highway) road.
(*ii*) District road bridge (DR bridge) — on a district road.
(*iii*) Village road bridge (VR bridge) — on a village road.
(*iv*) Foot bridge (F bridge) — on a pedestrian path.

Classification Based on Vent Height of Bridge *vis-a-vis* HFL

(*i*) *High level bridge.* A bridge which can safely pass the highest anticipated flood without submergence thereby providing an all weather communication across the drainage.

(*ii*) *Submersible bridge.* It is a bridge which is designed to be overtopped in floods with the formation level so fixed that disruption to traffic is for not more than 3 days at a time or for not more than 6 times in a year.

(*iii*) *Causeway.* It is a road crossing with a dip to pass floods over it. The causeway may or may not have vents to pass low water flow.

13.2 SELECTION OF SITE

The essential data of the road and drainage required for the design of a bridge are as under:

1. *Road data.* It includes (*i*) Type of road, (*ii*) Metalled width, (*iii*) Road level, (*iv*) Normal cross section of the road, (*v*) Road width provided on other nearest bridges on the road, (*vi*) Index plan, scale 1/50,000 showing proposed location of the bridge and the angle of crossing of the road with the stream, and (*vii*) Site plan extending not less than 100 m upstream and downstream of the centre line of the crossing.

2. *Drainage data.* (Refer Cross Drainage Works).

3. *Site data.* The site selected for a bridge should have the characteristics (*i*) The river reach should be practically straight to minimize the possibility of an oblique approach of the river and to obtain a uniform distribution of flow throughout the waterway. On a meandering river, bridge should preferably be located in transition reaches (connecting two bends) rather than on the bends, (*ii*) Minimum width between khadir/banks to reduce extent of possible embayment at the back of the guide banks and to permit shorter guide banks, (*iii*) Stable river reach. Desirable location is at the nodal point where the river has been permanently flowing, irrespective of the changing meanderings in the past, (*iv*) Main deep channel of the river flowing along a bank is a preferable site, (*v*) Permanence of the bank material. The bank where

bridge is to be located should be firm and inerodible, (*vi*) Alluvial river bed consists of strata of erodible sand to permit deepening of bed to gain adequate waterway after the bridge construction, within design limits.

13.3 DESIGN PARAMETERS OF A BRIDGE

Designed Maximum Discharge : Designed maximum discharge is computed from (*a*) Empirical formula applicable to the region or by a rational method, (*b*) Hydraulic characteristics of the stream such as cross sectional area, slope of the stream for calculating velocity of flow, and (*c*) Maximum observed discharge at the site or nearest to the proposed bridge site. For further details regarding computation of the designed maximum discharge from catchment area refer "Cross Drainage Works".

Waterway: As indicated in Cross Drainage Works.

Span Length : Number of spans is kept odd, *i.e.*, either one, three, five or seven whichever is economical. Where found economical, use may be made of two spans. For economical span, cost of decking for one span excluding wearing coat, railing and wheel guard shall approximately equal the cost of one pier inclusive of foundations. Span length in the case of skew bridges is determined along the axis of the road. The angle of crossing shall not be sharper than 60°, in exceptional case 45°. Village road bridges and foot bridges have right angle crossing.

Free Board : Vertical clearance or free board as per Indian Road Congress specifications is specified in Table 13.2.

Table 13.2 Vertical clearance

Discharge (cumecs)	*Vertical clearance (m)*
Below 0.3	0.15
0.3 to 3.0	0.5
3 to 30	0.6
30 to 300	0.9
300 to 3000	1.25
over 3000	1.5

In practice, a minimum free board of 0.6 m above HFL in a drainage is provided. In the case of irrigation channels, the free board above designed full supply level provided in main canal and branches is 0.75 m, and 0.5 m in distributaries and minors.

Maximum Scour Depth : Refer details discussed under "Cross Drainage Works". In the case of irrigation canals scour depth is taken as $D/3$ below the designed bed level or natural surface level whichever is lower, where D is the designed full supply depth in the canal. The top level of the foundation concrete of the abutment or pier is kept at $D/3 + 0.5$ m below the designed bed level.

Depth of Foundation : *Deep foundation.* Maximum depth of foundation below HFL is 1.33 D, where D is the anticipated maximum depth of scour below the designed HFL including that on account of possible concentration of flow.

Shallow foundation. In inerodible strata, the foundation is taken to a comparatively shallow depth below the bed surface provided the foundation bearing stratum is practically incompressible and is prevented from lateral movement and is also protected against scour.

Decking Slab : The decking is designed for Indian Road Congress class A loading or class AA loading whichever is severer. The decking may be of reinforced cement concrete, precast cement concrete or arch made of masonry or reinforced concrete. Precast slab is indicated for long span bridge. Arch decking made of masonry is not commonly used.

13.4 CULVERT

A culvert is a small bridge with a span up to 6 m between the faces of the abutments or extreme ventage boundaries. A bridge and a culvert is distinguished from each other on the basis of span length. The critical span at which a bridge becomes a culvert varies from 1.5 to 8 m. The minimum vent dimensions of a culvert is 0.75 m to facilitate inspection and carrying out of repairs as also to be safe from choking due to debris, etc. The length of the vents or barrels of a culvert is kept equal to the full formation width of the road at the top, so as to provide for two way traffic. The parapet of suitable design is kept 0.3 m minimum above the road surface Fig. 13.1. Water passages (barrels or vents) may be round, arched but usually rectangular. Small culverts carrying high fills usually have a square or rectangular barrel. For channels of small capacity RCC hume pipe culverts are provided.

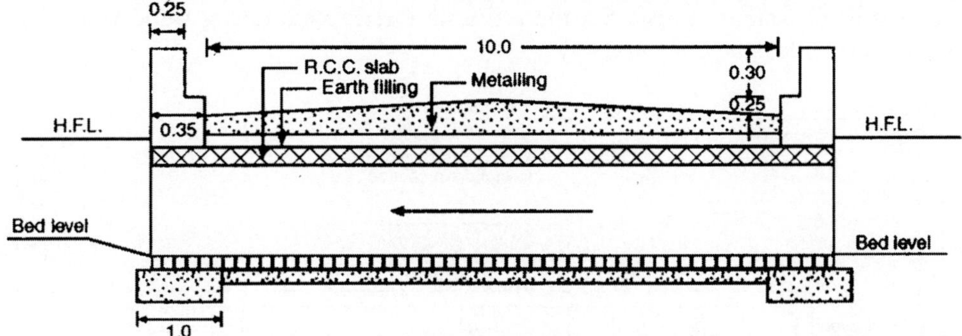

Fig. 13.1 Culvert.

A culvert serves a drainage area of small size to cater for surface runoff of rain, groundwater and melted snow (where applicable). The waterway depends on (*a*) Catchment area, (*b*) Amount of rainfall which is likely to fall in the catchment within a certain period, (*c*) Character and slope of ground, (*d*) Water flow velocity through the culvert, (*e*) Normal stream flow, (*f*) Afflux upstream against the embankment before overtopping it or damaging the property behind it by flooding and probable damage from overtopping. The upstream water level at the time of maximum design discharge and the downstream water level, which is generally taken as the surface level of the unobstructed flow of the stream, gives the operating head H_r. The greater the heading up of water at the upstream end, greater is the head causing the flow and lesser is the opening or span of the culvert. Under the maximum design flood the culvert opening shall run full and if v is velocity through the barrel, then operating head, H_l, is equal to the sum of the losses through the culvert, *viz.* entrance loss, friction loss in the barrel and velocity head in the culvert. Therefore,

From Manning's formula, $$V = \frac{1}{n} . R^{2/3} S^{1/2}$$

or $$S = \frac{n^2 V^2}{R^{4/3}}$$

∴ $$H_l = K \frac{V^2}{2g} + \frac{n^2 V^2}{R^{4/3}} L + \frac{V^2}{2g}$$

or $$H_l = \frac{V^2}{2g} \left(K + \frac{2n^2 Lg}{R^{4/3}} + 1 \right]$$

where, V = velocity through the barrel (m/s), K = coefficient –0.505 for a square edged entrance and 0.05 for a bellmouth entrance, and L = length of barrel (m).

Discharge through culvert = AV

It is a good practice to lay the culvert on the bed level of the stream or in a smoothly dug ditch and to use the original stream channel for the culvert passage and to have the culvert cross the road at right angle. The requirement of strength of roof slab is the same as for high level bridges.

13.5 CAUSEWAY

A causeway is a dip in the road which allows flood water to pass over it. A causeway having vents for low water flow is called *vented causeway.* A causeway is adapted to a shallow stream of small discharging capacity which is dry most of the time but occasionally carries such large volumes of water that adequate bridge would be expensive. The road may cross the stream at its natural grade but no provision is made for passing water under it. A causeway may be satisfactory for light or occasional traffic but the traffic is interrupted during flood flow. Fig. 13.2 indicates the usual design of a causeway.

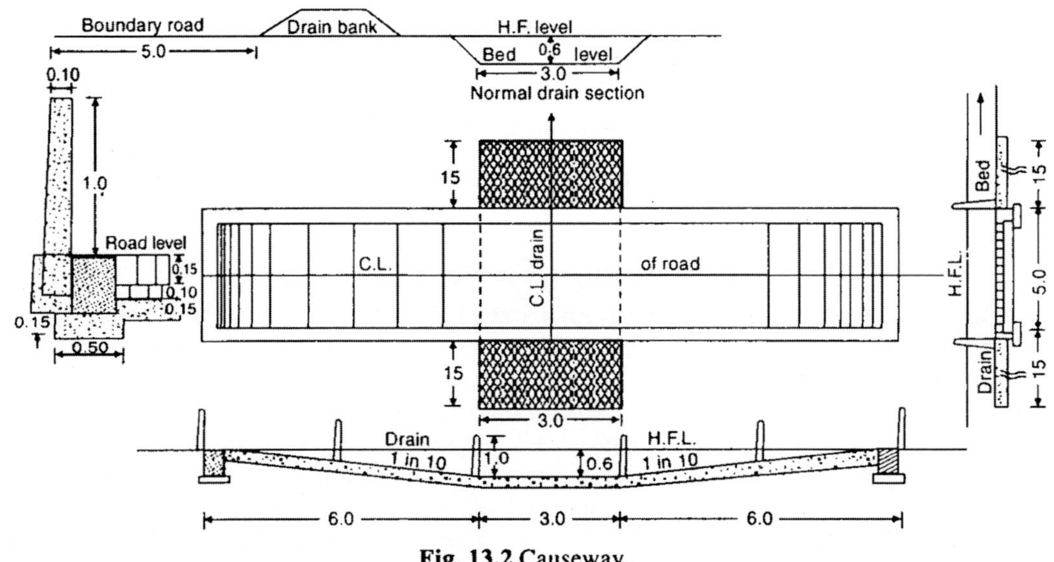

Fig. 13.2 Causeway.

QUESTIONS

1. Differentiate between a bridge and a culvert.
2. (*a*) Define economical span lengths for bridges.
   (*b*) Discuss the statement, reducing the waterway of a bridge will reduce cost of superstructure but will increase the cost of substructure.
3. Enumerate the essential data required for the design of a bridge over a drainage or a canal.
4. How will you determine the maximum design discharge for a bridge over a stream.
5. Discuss the waterway usually provided for an irrigation canal and a drain under various types of bridges.
6. (*a*) Discuss causeway and vented causeway.
   (*b*) Give a line sketch of a causeway.
7. Distinguish between normal scour depth and maximum scour depth with regard to a bridge design. How does it help in determining the depth of foundation of a bridge.

14

CANAL OUTLETS

14.1 INTRODUCTION

An outlet may be defined as a device constructed in a canal to deliver designed discharge from a government channel to the field channel or water course (cultivator's channel). It is a work which connects a government channel with the water course. Outlets are generally not provided on the main canal and branches, but are installed in such cases in the ditch distributaries constructed along the canals. In exceptional cases, when provided on main canal and branches these are called direct outlets. Since equitable distribution of the canal supplies is dependent on outlet, it must be such that it not only passes a known and constant quantity of water, but must also essentially be a measure of the discharge.

14.2 OUTLET CHAK

The area of land which an outlet irrigates is called outlet area or outlet chak. The extent of area to be irrigated by an outlet is usually limited by drainages on either side. The location of an outlet is fixed at the highest point in the chak adjacent to the canal.

Factors Determining Size of Chak : The factors which determine the size of an outlet chak are (*i*) Discharge required for the chak shall be such as can be conveniently and efficiently handled by the cultivators, (*ii*) Holdings of the people in the area, (*iii*) Nature and slope of the area to be irrigated, (*iv*) Permissible intensity of irrigation and water allowance because of upper and lower limits of 0.09 cumec and 0.03 cumec for which an outlet can be provided, (*v*) Killa line or village boundary shall be the chak boundary where possible, (*vi*) Length of water course to irrigate a chak shall not exceed 3 km, (*vii*) National highway or road crossings are avoided as far as possible, (*viii*) Natural drainage is the boundary of the chak. The water course shall not be required to cross natural drainage.

The modular sizes of chak are indicated in Table 14.1:

Table 14.1 Modular chak sizes

Delivery rate (lt/s/ha)	*Modular chak size range (ha)*			*Delivery rate (lt/s/ha)*	*Modular chak size range (ha)*		
	Heavy soils	*Medium soils*	*Light soils*		*Heavy soils*	*Medium soils*	*Light soils*
	A. 0–1% Land slopes			*B. 1–3% Land slopes*			
1	2	3	4	1	2	3	4
0.5	36-72	42-84	45-90	1.0	15-30	17-34	18-36
0.6	30-60	35-70	37-75	1.2	13-26	14-28	15-30
0.7	25-50	30-60	32-64	1.4	11-22	12-24	13-26
0.8	22-44	26-52	28-56	1.6	10-20	11-22	11-22
0.9	20-40	23-46	25-50	1.8	9-18	9-18	10-20
1.0	18-36	21-42	22-45	*C. 3-5% Land Slopes*			
1.2	15-30	17-35	19-38	0.5	20-40	22-45	25-50
1.4	13-26	15-30	16-32	0.6	17-34	19-38	21-42
1.6	11-22	13-26	14-28	0.7	14-28	16-32	18-36

(*Contd.*)

(*Contd. table 14.1*)

B. 1-3% Land slopes				C. 3-5% Land slopes			
1	2	3	4	1	2	3	4
0.5	31-62	34-68	36-72	0.9	11-22	12-24	14-28
0.6	26-52	28-56	30-60	1.0	10-20	11-22	13-26
0.7	22-44	24-48	26-52	1.2	8-16	9-18	10-20
0.8	19-38	21-42	22-44	1.4	7-14	8-16	9-18
0.9	17-34	19-38	20-40	1.6	6-12	7-14	8-16
				1.8	5-10	6-12	7-14

At Sardar Sarovar Project, the basic delivery capacity is fixed as 0.60 and 0.70 lt/s/ha of net CCA. A value of 0.45 lt/s/ha is used for regions with heavy clay soils. The regional irrigation intensities by seasons are 60-80% Kharif and 60-75% Rabi of net CCA served.

Maximum Length of Water Course : The maximum length of water course, in general, shall not be more than 3 km. In Orissa state, field channel length is taken as 80 m per hectare on the average. The entire command is divided into small units known as service areas (200–500 ha), limit of which is generally the revenue village and hence termed village service area VSA, which in turn is divided into smaller units called 'Chaks' of 25 to 60 ha. A chak is to have 5 to 8 sub-chaks, each having area of 6 to 10 ha and 8 to 10 farm holdings. The VSA is to receive water through one outlet from a distributary within each service area, lined minors are to carry water to different chaks for proportionate flow delivery to a chak. Whereas in a chak lined sub-minors are to carry water to sub-chaks and within a sub-chak unlined farm channels are to carry water to individual farms. The full chak flow will be rotated from farm to farm and from sub-chak to sub-chak during a rotation period of 7 days. Each form is allotted fixed time for drawing water proportionate to the areas. In Orissa state size of sub-chak is 5 to 8 hectares and 5 to 8 number of beneficiaries.

Water course longer than 3 km has the disadvantages : (*i*) Longer the water course, higher the discharge, rendering maintenance and control of water course difficult for the cultivators, (*ii*) Losses of water in the water course shall be more owing to inability of the cultivators to handle higher discharge, (*iii*) More absorption losses if for the same discharge length of water course is over 3 km, and (*iv*) Inefficient use of water.

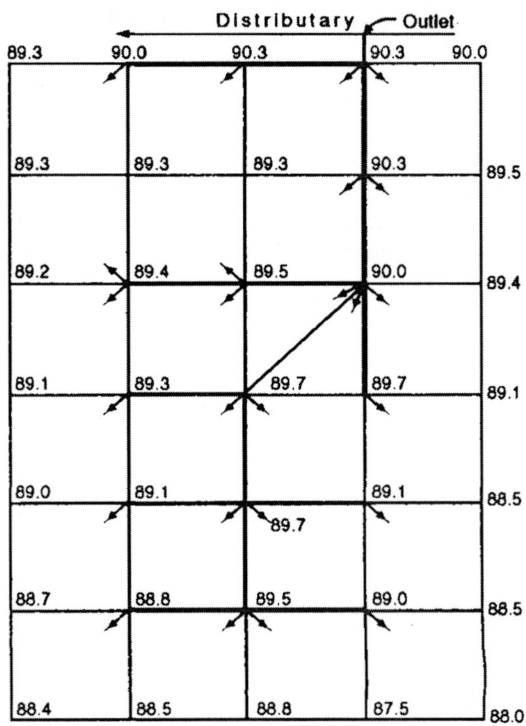

Fig. 14.1 Typical alignment of water course system.

14.3 MINIMUM WATER LEVEL REQUIRED AT AN OUTLET HEAD

A chak is essentially a miniature irrigation scheme in all respects. In the compact block constituting the area to be irrigated by an outlet, select a suitable representative critical level to be irrigated in the chak. Measure the length of water course, which is essentially aligned along the sides of rectangles. Care is exercised in selecting critical level because a lower ground level at longer distance from the outlet head shall need higher water level at outlet head to irrigate it than a comparatively higher ground level nearer the head as illustrated in Table 14.2. Allowing a minimum water surface slope of 1/5000 in the water course and a field command of 7.5 cm, the minimum water level required in the distributary at each outlet head is calculated in the tabulated form (command statement) as described in Table 14.2.

Table 14.2 Command statement

S.No.	*R.D. (m)and side of outlet (R or L)*	*Critical level in the chak (m)*	*Distance from outlet head via. water course alignment (m)*	*Loss of head due to slope in water course (1/5000)*	*Field command (m)*	*Minimum water level required in water course at outlet head (m) (col. 3 + 5 + 6)*	*Minimum working head required at outlet (m)*	*Minimum water level required in distributary (m) (col. 7 + 8)*
1	2	3	4	5	6	7	8	9
1	300-L	250.10	3000	0.60	0.075	250.775	0.15	250.925 (say 250.93)
		250.50	500	0.10	0.075	250.675	0.15	250.825 Adopt the highest of two, *i.e.*, El. 250.93 m

Field Command

A depth of 7.5 cm is sufficient for one watering. It follows that 0.028 cumec (1 cusec) flowing for a period of 3 hours is sufficient to give one watering to an area of 0.4 ha (1 acre). The extent to which the actual area irrigated in a given time falls short of this standard is an indication of the amount of water lost by absorption in field channel and water course and by giving excessive depth of water in the field.

14.4 OPTIMUM SIZE OF AN OUTLET

The optimum size of an outlet constitutes a discharge which is conveniently and economically (involving no avoidable loss of water in the water course and in the field) handled by the cultivators and time taken to irrigate a field is reasonable. The factors governing the size of an outlet are size of the outlet chak and permissible discharge. With these considerations, an outlet of 0.06 cumec authorised discharge is the best. The discharge shall not, in general, exceed 0.09 cumec and be less than 0.03 cumec.

The disadvantages of an outlet of small discharge compared to that of higher discharge are : (*i*) Higher water losses in water course, (*ii*) Higher spread loss which occurs in the field. An outlet with 0.015 cumec discharge may irrigate only 1/3rd the area irrigated by an outlet of 0.03 cumec, (*iii*) The longer the time taken to irrigate a field, the greater is the amount of water absorbed in that part of field already irrigated while applying minimum irrigation to the remainder, (*iv*) Low velocity in water course causes silting up requiring frequent clearance, (*v*) Greater number of outlets cause trouble in distribution of water and regime of channel, (*vi*) Cost is comparatively more due to greater number of outlets, water course culverts, etc.

Because of these obvious reasons it appears desirable to have outlets of maximum discharge which in turn have the demerits: (*i*) Practical difficulty of irrigators to handle maximum discharge in the water course and in the field while opening and closing nakkas, (*ii*) Turn (Wari) of an irrigator is of very short duration. Warabandi difficulties arise particularly for shareholders of small holdings, (*iii*) More areas catered by an outlet may involve lands belonging to different parties or villages that cannot pull on with one another, and (*iv*) Length of water course may exceed the optimum limit of 3 km.

14.5 DEFINITIONS

Available working head. The minimum difference between full supply level in canal and delivery water level in water course.

Setting. The ratio of the depth below full supply level of the crest of an outlet to the supply depth of the channel at that point.

Minimum modular working head. The difference of water level or pressure between supply and delivery sides, which is the minimum necessary to enable a modular or semi-modular outlet to work as designed.

Modular limits. The extreme values of any factors at which a module or semi-module ceases to be capable of acting as such.

Flexibility. The ratio which the rate of change of discharge of the outlet bears to the rate of change of discharge of the supply or parent channel.

Depression ratio. The ratio between the depression and the height of the opening of an orifice outlet.

Modular range. The range of conditions between the said limits, within which a module or semi-module works as designed.

Minimum modular head ratio. The ratio between the minimum modular head and the depth of upstream water level on the crest of an outlet.

Sensitivity. The ratio that the rate of change of discharge of an outlet bears to the rate of change in level of the distributary water surface, referred to the normal depth of the channel.

Sensitiveness. The variation of discharge of a semi-module for 30 mm variation of supply level.

Submerged orifice. An orifice which in use is drowned by having the tail water higher than all parts of the opening.

Water course. The term is applied to an irrigator's channel, taking its supply from a government channel, from which fields are irrigated directly.

Efficiency. The efficiency of an outlet is the ratio of the head recovered to the head put in, expressed as a percentage. The efficiency is more as the working head required for the functioning of an outlets is less.

14.6 ESSENTIAL REQUIREMENTS FOR AN OUTLET

An outlet must satisfy the essential conditions (*i*) It must be strong enough and should not have moving parts liable to derangement by silt or weeds or requiring periodic attention, (*ii*) Interference by the cultivators must be difficult and, if made, should be easily detectable, (*iii*) It should draw its fair share of silt carried by the parent channel, (*iv*) It should be possible for an outlet to work efficiently with a small working head. The larger the working head, higher the water level required in the parent channel and higher is the cost of the distributing system, (*v*) It should be simple in construction, (*vi*) The cost should not be high, (*vii*) Subsequent alterations should not be costly, (*viii*) Within the working limits, the discharge should be constant irrespective of the water levels in the watercourse, and (*ix*) It should be proportional in discharge distribution that is draws more or less corresponding to the varrying supply in the channel.

14.7 CLASSIFICATION OF OUTLETS

The outlets are classified into three types, as under :

1. **Modular Outlet :** Also called *modules* or *rigid modules*. It is the outlet whose discharge is independent of the water levels in the distributary and the water course within reasonable working limits. The outlet discharge is thus constant within the working limits. The loss of head through the outlet is more. This type of outlet is either with moving parts or without moving parts. In the latter case, these are called rigid modules. Outlets with moving parts have disadvantages (*i*) Expensive being not simple to design and construct, and (*ii*) Liable to derangements due to increase in friction, rusting of the moving parts and any obstruction in the working of the parts caused by silt and weeds in the canal water.

Gibb's Module. It is the most common type of modular outlet. It essentially consists of a bell-mouth inlet pipe curved upwards from which water is led into an eddy chamber, semi-circular in plan, around which water flows giving rise to a free vortex flow (Fig. 14.2). A number of baffles are

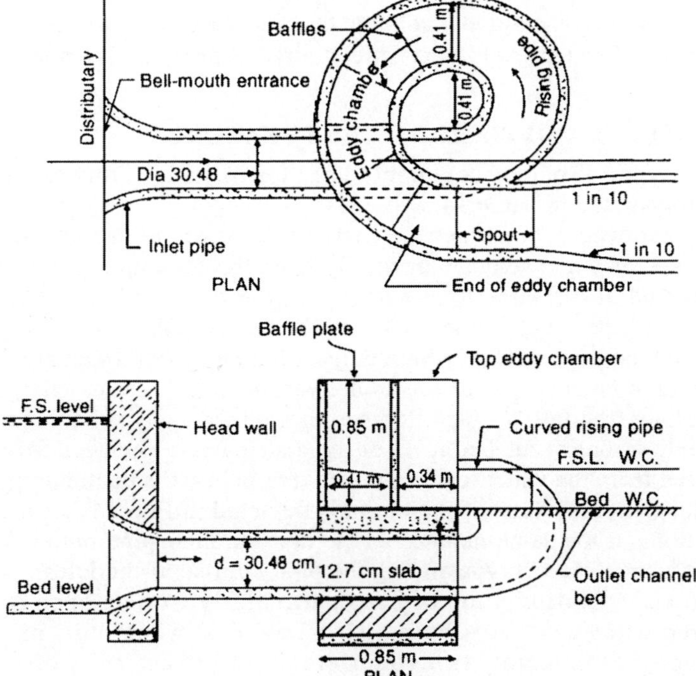

Fig. 14.2. Gibb's module outlet.

provided in the eddy chamber with thin lower ends sloping against the direction of flow at the required height above the bottom to prevent increased discharge passing through the module. The number of baffles coming into action depends on the variation in head. With increase in the head causing flow, the water banks up at the outer circumference of the eddy chamber and impinges against the baffles imparting an upward, rotational direction of flow to the water, which spins round in the compartment between two successive baffles and finally drops on the incoming stream of water thereby dissipating excess energy and keeping the discharge constant. The disadvantages of the outlet are that it is costly and easily tempered with.

Discharge formula

$$Q = r_0 \sqrt{2g}\,(y_1 + h_0)^{3/2} \left\{ \frac{m^2 - 1}{m^3} \log_e m + \frac{1}{m} \log_e m - \frac{m^2 - 1}{2m^2} \right\}$$

where, Q = discharge (cumec), r_0 = radius of outer semi-circle of eddy chamber, r_1 = radius of inner semi-circle of eddy chamber, $m = r_0/r_1$, *i.e.*, ratio of outer to inner radius, y_1 = water depth at inner circumference, and h_0 = head at outer circumference.

The formula holds good only for Gibb's standard design in which $m = 2$, $h_0/D = 1/7$, where D is the difference of level measured from the minimum water level in the distributary to the floor of eddy chamber.

2. **Semi-modular Outlet :** It is the outlet the discharge of which is independent of water level in the watercourse but dependent on the water levels in the distributary so long as the minimum working head required for its working is available. The outlet discharge thus increases with rise in the water level in the distributary. Properly designed semi-modules can distribute more or less equitably variations in the supply in distributary.

Adjustable proportional module, open flume outlet and free fall pipe outlet belong to this class. In free fall outlet the exit end of pipe is placed higher than the water level in the water course (Fig. 14.6) such that the working head is the difference between water level in the distributary and centre of the pipe at exit end.

3. **Non-modular Outlet :** It is the outlet whose discharge is a function of the difference in levels between the water surfaces in the distributing channel and the water course. Variations in either effect the discharge. The outlet discharge thus increases with rise in water level in the distributary or lowering of water level in the water course. Pipe outlet is a non-modular type of outlet.

14.8 TYPES OF OUTLETS AND THEIR SALIENT FEATURES

Three important types of outlets commonly used for irrigation from canals and their salient features are summarised in Table 14.3.

Pipe-cum-Semi-module. It combines the characteristics of pipe outlet and open flume or APM outlet. It is a type of outlet selected primarily from the economic point of view through wide and high bank and at the same time embodying the merits of open flume or APM. It essentially consists of a pipe set at bed level, to lead water from the distributary into a cistern 0.6 to 0.9 m square, at the other end of which is installed an orifice fixed at a higher level to ensure modularity and of correct dimensions to deliver the authorized discharge of the outlet. It may be a pipe working free-fall, an open flume with its crest at any suitable level, or an APM set anywhere above, at or below the bed level of the distributary. The head for calculating the discharge is measured from the water level in the cistern below the upstream pipe, allowing a loss of head of 3 cm. General layout of the outlet is depicted in Fig. 14.3 (*b*). If the orifice is submerged, it functions in the same manner as the non-modular pipe outlet. With its orifice working free-fall, it is the simplest type among the pipe-cum-semi-modules.

Advantages : (*i*) Low consumption of head particularly with an open flume, (*ii*) Ease in adjustability even when canal is running, (*iii*) Low cost when built in wide and high banks, (*iv*) High degree of immunity from interference due to certainty of early detection, (*v*) Large range of modularity, and (*vi*) Constant coefficient of discharge.

Table 14.3 Salient features of three types of outlets.

Features	*APM*	*Open Flume*	*Pipe outlet*
Structural	Essentially consists of an orifice provided with a gradually expanding flume on the downstream side. The critical velocity is exceeded in the orifice and the discharge is thus independent of the water levels in the water course. Structurally, it is a long throated flume fitted with a roof block, capable of vertical adjustment, introduced into the upstream end of the parallel throat [Figs. 14.3 (a) & (b)]. Cast iron base, cast iron roof block and check plates on either side form the nucleus around which masonry is built. The roof block is fixed to the check plates by bolts which can be removed and depth adjusted.	Structurally open flume is a simple smooth weir with a throat constricted sufficiently long to ensure that the controlling section remains within the parallel throat for all discharges up to the maximum. A gradually expanding flume is provided at the outfall to obtain the maximum recovery of head (Fig. 14.4 & 14.5). The advantage of rigidity is embodied by the installation of a roof block. The entire work is built in brick masonry, but the controlling section is generally provided with cast iron or steel bed and check plates.	It is the simplest type of outlet and essentially consists of a pipe placed horizontally and at right angle to a distributary. The pipe is generally laid at designed bed level of the distributary provided with face and tail masonry walls (Fig. 14.6). A pipe outlet with exit end of the pipe submerged in the water in water course works as a non-modular outlet.
Flexibility	Flexibility of the outlet depends on the relation H to D. Proportionality ($F = 1$) is secured when $H = 0.3\ D$ *i.e.*, the bottom of the roof block is submerged below the full supply level by 0.3 depth of water in the canal. With a rise in the FS level, the flexibility is reduced and the outlet becomes sub-proportional. Similarly, with a fall in the FS level the flexibility is increased and the outlet becomes hypercritical.	Flexibility ($F = 1$) is secured by setting the outlet at $0.9D$. If the crest is kept higher than this, the outlet becomes more flexible, *i.e.*, hyper-proportional, and if lower, it tends towards rigidity. With a fall in FSL in the distributary the flexibility increases and with rise in the water level it decreases.	If the water level in the water course drowns the pipe at exit, the discharge passed is a function of the water level in the distributary and the water course and the variations in either effect the discharge, characteristic of a non-modular type of outlet.
Modularity	Formation of hydraulic jump renders the outlet independent of water level in water course, making it semi-modular outlet. So long as standing wave is steady and remains clear of the exit of the orifice, the discharge coefficient does not alter.	Formation of hydraulic jump renders the outlet independent of water level in water course, making it semi-modular outlet.	In case of free-fall condition, the discharge passed is independent of water level in the water course and under this condition it is semi-modular outlet. The pipe may be laid horizontal or tilted in accordance with the above condition.

Features	*APM*	*Open flume*	*Pipe outlet*
Adjustability	Adjustability is secured by raising or lowering the roof block. This is fixed by a couple of bolts which in turn are secured from tempering by masonry key. The key can be broken out, the roof block adjusted and the key rebuilt within a couple of hours most economically and without injury to the rest of structure.	Adjustability can only be secured by dismantling one side wall to increase or decrease the width of flume or by raising or lowering the crest to decrease or increase *H*. In sum, adjustment means partial dismantlement and reconstruction.	Adjustability is secured most economically by fixing a socket of reduced size to the pipe to reduce the discharge. Enhancement in supply can be secured by installing another pipe for the increase in discharge alongside previous one.
Silt conduction	An APM set at 0.6 depth does not draw its due share of silt. Silt conduction increases as the setting is increased, as under: Setting — Silt conduction 0.6 — 99.5% 0.8 — 109.7% 1.0 — 121.9% For proportionate silt distribution, *h/D* should be between 0.375 to 0.48.	Open flume is either deep and narrow or is shallow and wide in which case it fails to draw its fair share of silt. Generally this type of outlet does not cause silting above the work. If the channel has silting tendency, the crest is set near the bed level of the distributary from silt conduction point of view.	Usually the pipe is placed at the bed of the distributary to draw fair share of silt charge. In general pipe outlet draws off more silt than any other type of outlet at the same level. Complete control on silt conduction is exercised in the case of pipe-cum-semi-module outlet.
Efficiency	Minimum working head required for adequate working of the outlet, viz., 0.75 *h* (approximate) is a measure of its efficiency. This type of outlet requires higher working head for modular working which necessitates the use of open flume and pipe outlets where head available is less.	It has greater efficiency as it can work with just a fraction of the head on the outlet, the minimum being 6 cm.	Pipe outlet requires the minimum working head and has higher efficiency. It is capable of passing discharge, given suitable diameter at a minimum working head of 3 cm.

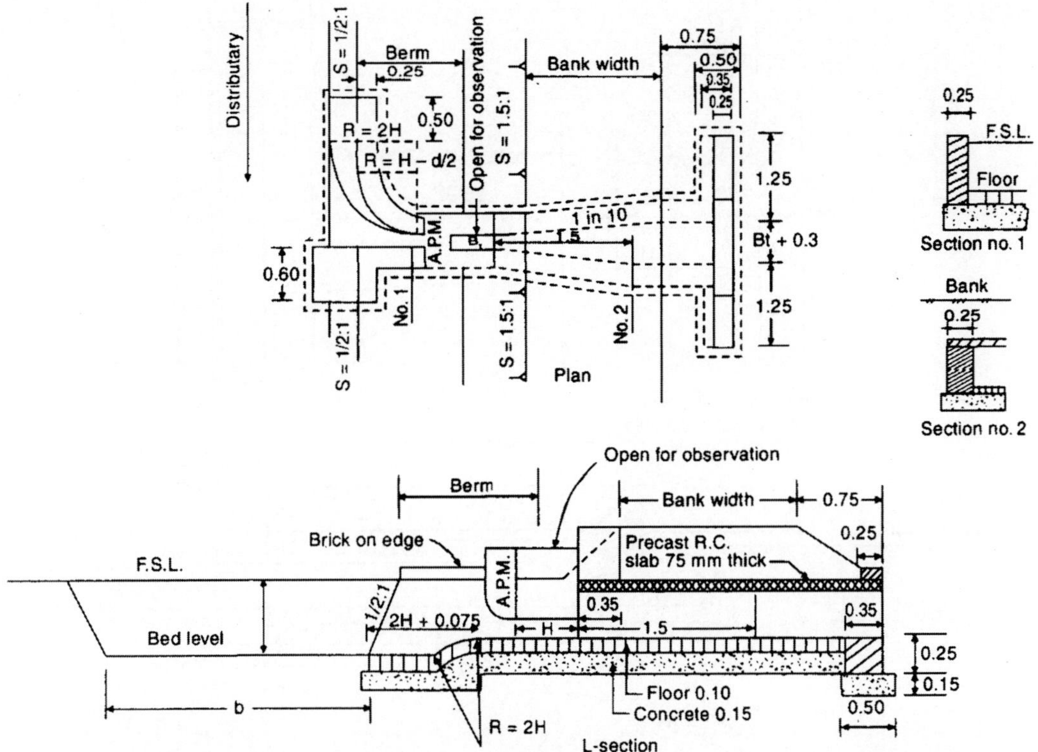

Fig. 14.3 (*a*) APM outlet.

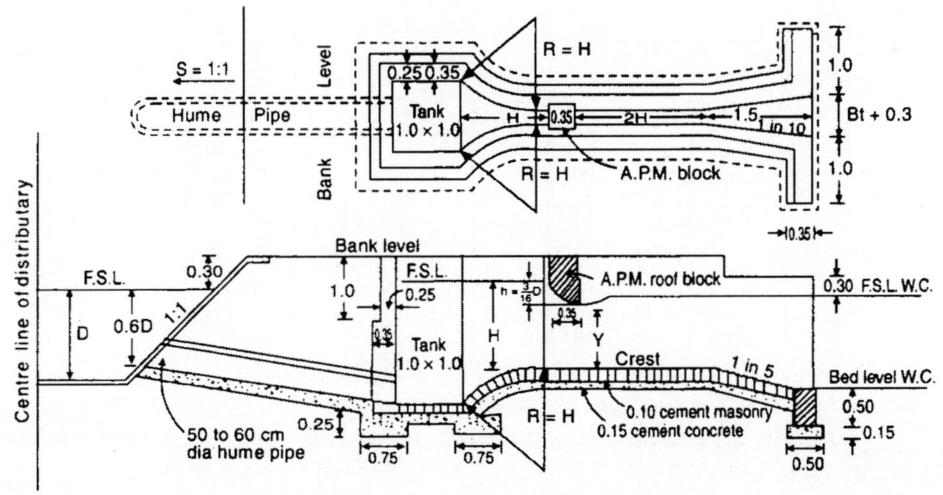

Fig. 14.3 (*b*) APM outlet on lined distributary

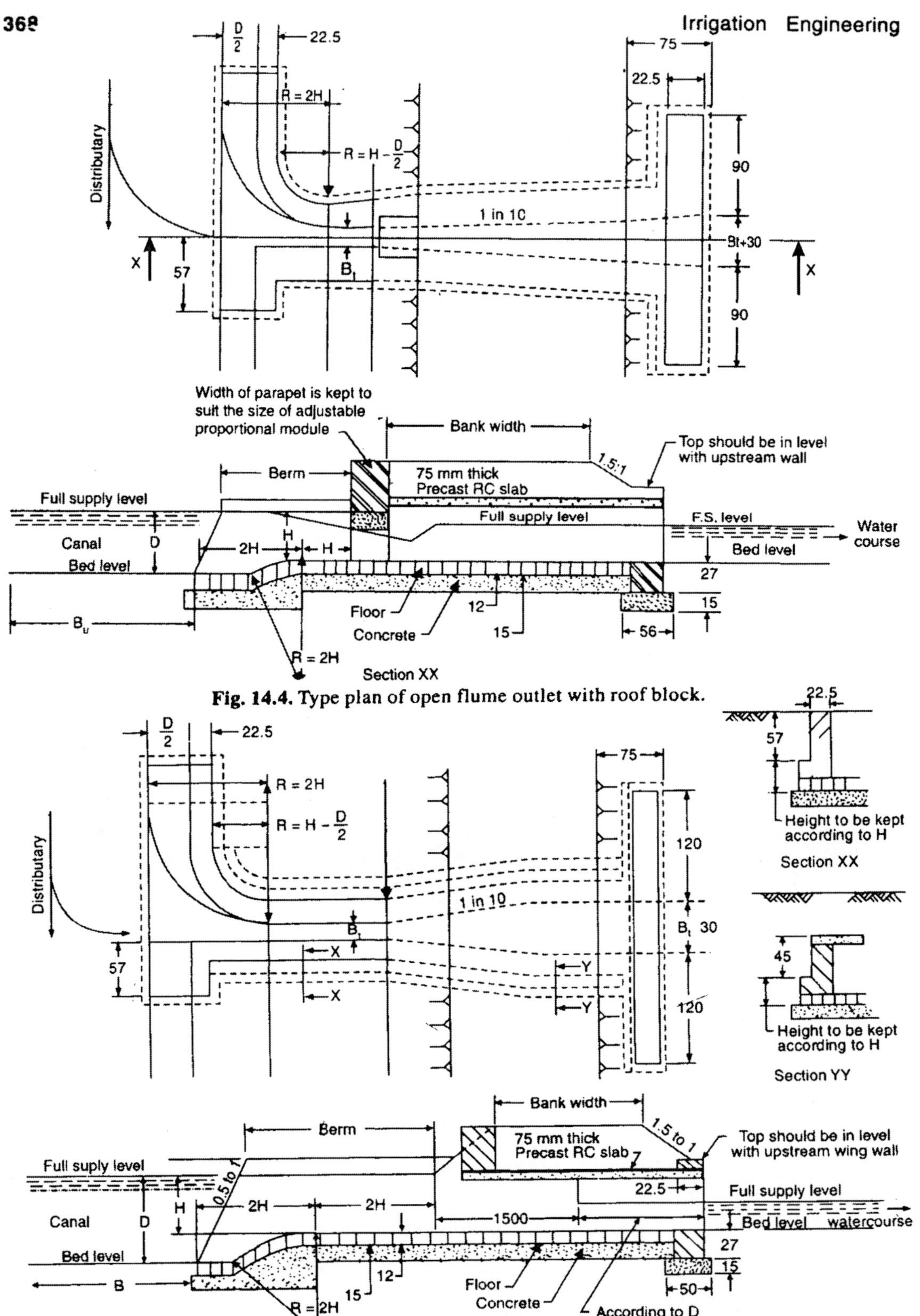

Fig. 14.4. Type plan of open flume outlet with roof block.

Fig. 14.5. Type plan of open flume outlet for distributary above 0.6 m depth and *H* is less than full supply depth (Dimensions in cm).

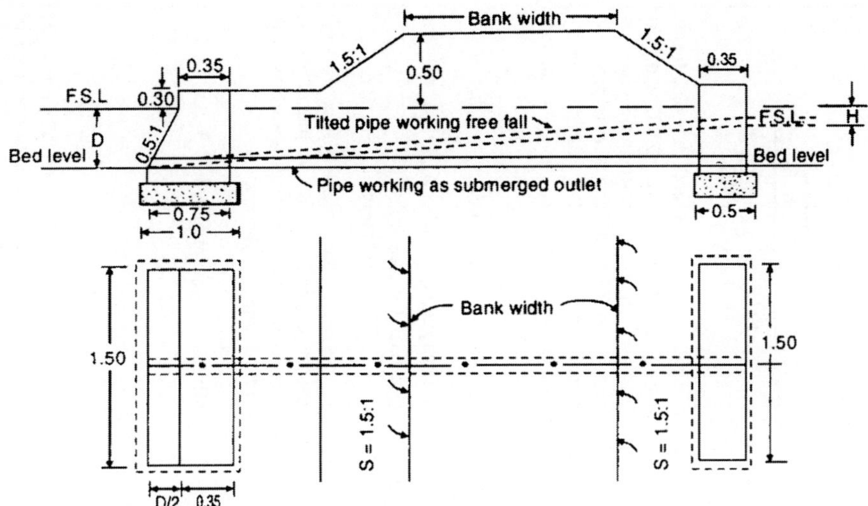

Fig. 14.6 Pipe outlet.

14.9 DESIGN PARAMETERS OF OUTLETS

The design parameters of three common types of outlets are discussed in Table 14.4.

14.10 MERITS AND DEMERITS OF THREE TYPES OF OUTLETS

APM Outlet. *Advantages* : (*i*) Best type of outlet if required working head is available, (*ii*) Most economical in adjustment either by raising or lowering the roof block or crest, and (*iii*) It is one of the best type of outlets to adopt when set near the bed level of the canal.

Disadvantages : (*i*) Costlier type of outlet, (*ii*) Requires more working head, (*iii*) Not suitable at or near the tail of a canal or immediately upstream of control points where it cannot work as safety valve to absorb temporary excess with a view to save the tail reach or control point from flooding, (*iv*) Channels fitted with APM (at 6/10 setting) silt up badly, and (*v*) Waterway is either deep and narrow which could get blocked easily or is shallow and wide in which case it does not draw its fair share of silt.

Open Flume Outlet. *Advantages* : (*i*) Cheaper than APM outlet, (*ii*) Does not cause silting above the work, except when low supplies are run for considerable periods. The silt, as may have accumulated, gets swept away during high supplies, (*iii*) Water surface at control point is relatively steady during small fluctuations in supply, (*iv*) Useful in distributing the supply proportionately amongst the outlets at tail, (*v*) Best suited for tail reaches and within 300 m upstream of control points such as falls to prevent flooding of tail and control points, as the outlet can overdraw, (*vi*) Low working head is required which means avoiding undue raising of FSL level in the channel resulting in economical construction of canal system, and (*vii*) As proportional distributor, offers a control point with minimum loss of head.

Disadvantages : (*i*) Adjustments are costly as they mean dismantlement and reconstruction of one side wall, if not fitted with roof block, (*ii*) Being either deep and narrow it is easily blocked or is shallow and wide, it is hyper-proportional and also fails to draw its fair share of silt, and (*iii*) Capable of overdrawing if water level in the canal is raised.

Pipe Outlet. *Advantages* : (*i*) Simple and economical to construct, (*ii*) Lowest working head, 30 mm, required to pass discharge, (*iii*) Suitable for small discharges, (*iv*) Draws fair share of silt, (*v*) Best as temporary outlet or for use on new canals in the first instance, (*vi*) Economical adjustment in reduction of diameter by fixing suitable reducing socket. Increase in supply can be secured by installing an auxiliary pipe for additional discharge, and (*vii*) High reuseable value of pipe on dismantling of the outlet.

Disadvantages : (*i*) Coefficient of discharge varies from outlet to outlet and at different times at the same outlet, (*ii*) Non-modular type of outlet, (*iii*) Discharge can be increased by deepening the water course and thereby lowering the water surface level in it and increasing working head.

Table 14.4. Design parameters of the three types of outlets.

Description	*APM*	*Open flume outlet*	*Pipe outlet*
Discharge formula	$Q = CB_t y h^{1/2}$ or $Q = 4.03\ B_t y h^{0.5}$	$Q = CB_t H^{3/2}$	$Q = CA\sqrt{2gh}$
Coefficient of discharge (C)	$C = 4.03$	$C = 1.60$ for B_t 6 — 9 cm $= 1.64$ for B_t 9 — 12 cm $= 1.66$ for B_t over 12 cm	$C = \frac{1}{2 \times 10^5}\sqrt{\frac{d}{f\left(L + \frac{1.5\,d}{400\,f}\right)}}$ where d is diameter of pipe (cm), L is pipe length (m) and f is coefficient value =0.005 and 0.01 for clean and slightly encrusted iron pipes respectively. $f = 0.0075$ for earthen ware pipes, h* is difference of water levels in the distributary channel and watercourse (m), g is acceleration due to gravity (m/s[2]).
Gullet or throat width	Base plate and roof block are manufactured in standard sizes 6, 7.5, 10, 12, 15, 20, 24 and 30 cm.	Minimum $B_t = 6$ cm	
Setting H	1.0 to 0.6 depth of distributary. h/D is 0.375 to 0.48 for proportionality. h/D is 0.8 or less for modularity. $H = h + y$.	1.0 to 0.6 depth of distributary. $H = 0.9$ for proportionality.	Free fall, $Q = 2.75\,A\sqrt{h}$ Submerged, $Q = 3.54\ A\sqrt{h}$ $C = 0.62$ for free fall $= 0.80$ for submerged
Minimum modular working head	$0.82\ h - B_t/2$ or $0.75\ h$ approximately. Depression head or depth of submergence of roof block below designed FSL in the distributary, *i.e.*, head measured from FSL in distributary to crest level of the outlet (H) minus depth of orifice y. For exact proportionality $h = y$, thus fixing $H = 0.6$ depth. Clearance between bottom level of roof block & crest level of outlet, y is kept not more than $0.5\ H$ and less than $0.3\ H$.	$0.2\ H$, minimum 6 cm. H = depth of water over crest.	Submerged — Setting generally at bed level of distributary. Free fall — centre line of pipe at d/s end is set at a distance equal to working head below FSL in the distributary. Minimum head required is 3 cm. * $h = 0.5\ V^2/2g + 4flv^2/2gd + V^2/2g$ h = difference in water level in distributary and watercourse. g = acceleration due to gravity A = cross sectional area of pipe (m[2]) V = velocity in pipe (m/s)

14.11 COMPARISON OF THREE TYPES OF OUTLETS

APM	*Open flume*	*Pipe outlet*
Costly type of outlet.	Cheaper than APM outlet.	Simple and economical.
Requires sufficient minimum working head for modularity.	Working head required is minimum amongst semi-modules.	Lowest working head required to pass discharge.
Best adapted for head reaches to feed tail properly.	Best suited for tail reaches & upstream of control points.	Best as temporary outlet or for use on new canals in the first instance.
Not suitable for small discharges.	Not suitable for small discharges.	Indicated where discharge is very small.
Most economical in adjustment either by raising or lowering the roof block or crest.	Adjustments are costly as these mean dismantlement and reconstruction of one side wall, if not fitted with roof block.	Most economical adjustment in reduction or increase in discharge.
Semi-modular type of outlet; draws fixed discharge within working limits.	Semi-modular type of outlet capable of overdrawal if FS level in canal rises.	Non-modular type of outlet. Only under free-fall condition, works as semi-modular outlet.
Discharge drawn is independent of water level in water course.	Discharge drawn is independent of water level in the water course.	Dishcharge passed is a function of difference of water levels in distributary and water course.
Does not draw due share of silt unless set below bed level.	Does not draw fair share of silt unless set at bed level of distributary.	Draws fair share of silt.
Discharge can be increased by cultivators by inserting at the downstream side of roof block wooden plank, thereby forming an air tight roof in continuation of roof block.	Discharge can be increased by 10-20% by inserting a horizontal lamina, upper portion works as a weir and the lower one as orifice, as also by heading up of water level in the distributary.	Discharge can increased by cultivators drowning free-fall pipe by heading water level in water course and by increasing working head by deepening the water course or heading up water in the distributary.

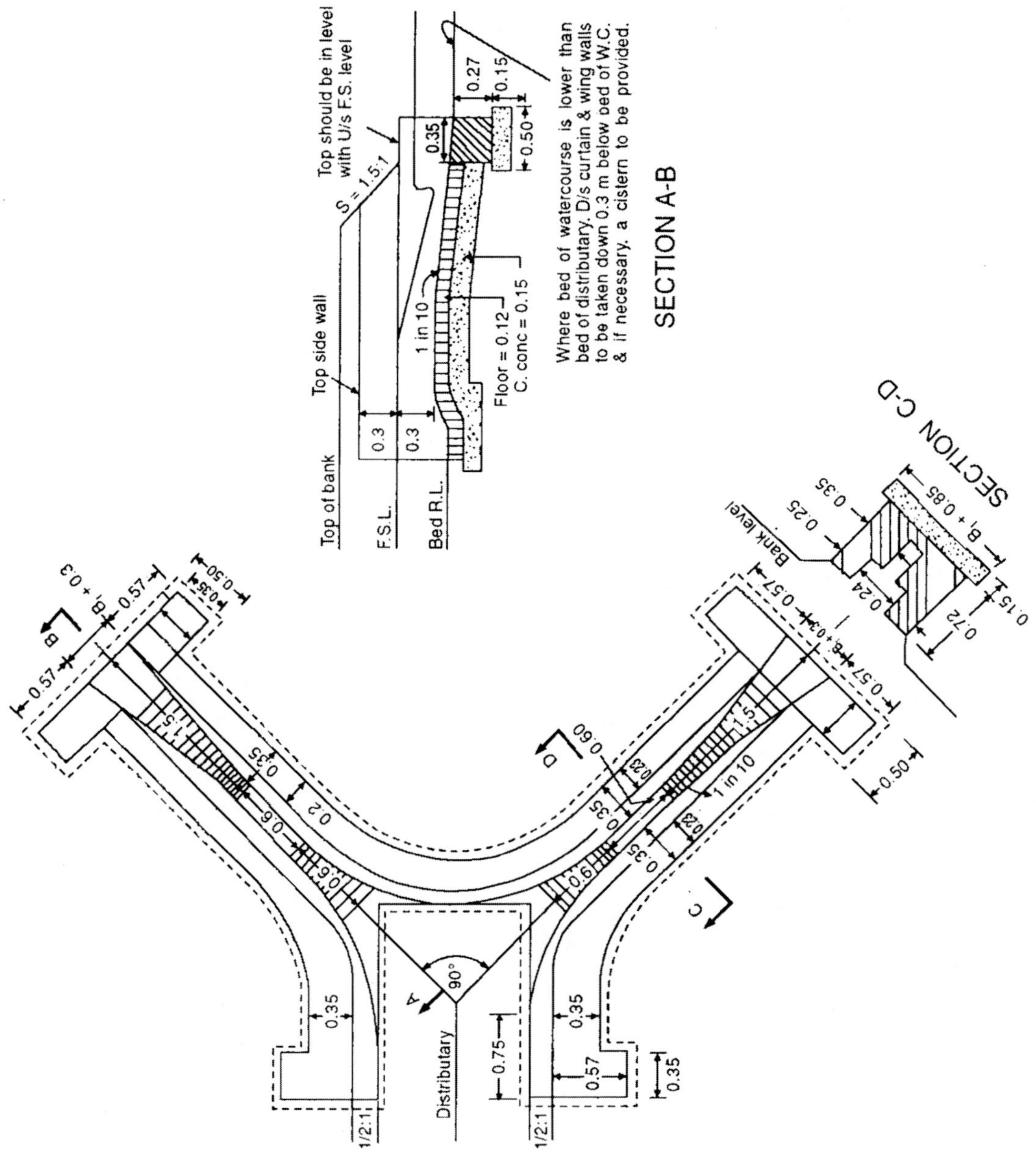

Fig. 14.7 Two outlets tail cluster.

14.12 TAIL CLUSTER

The work constructed at the tail end of a channel for the distribution of supplies there is called tail cluster. As the name implies generally two or more outlets are clustered at the tail with aggregate discharge over 0.07 cumec and the side offtakes having equal angles with respect to the centre line of the distributary. The offtaking angles are either 45°, 60°, 90° or straight in line with the channel alignment. Tail cluster is thus a combination of two, three or four open flume outlets (Fig. 14.7). So that the fluctuations in the distributary may effect them equally, the crest level of all the outlets is kept at the same level and proportionality is secured by adjusting throat width (B_t). Depth of water on the crest is kept 0.3 m and value of C is 1.66 in the open flume discharge formula. The discharge formula thus simplifies to $Q = 0.27\ B_t$. The angle of offtake and lower approach velocity effect in the side offtakes being not appreciable are ignored

The tail clusters are useful in distributing the supply proportionately and in easily absorbing any excess that reaches the tail. The gullet width is widened at full supply level to facilitate passing down any excess supply reaching the tail without endangering safety of the structure and the canal reach upstream. The low working head, 6 cm, required for the modular functioning of the tail outlets avoids undue raising of the full supply level in the distributary resulting in economy in the construction of the canal system.

14.13 SELECTION OF CLASS OF OUTLET

Following considerations govern the selection of class of the outlet :

Modules. Use of modules is limited to direct outlets subject to variations in supply as on a branch canal. On a distributary system the number of modules should not exceed a small precentage of the total, as by their very nature modules cannot absorb any fluctuations in the parent channel and thus the channels would be either flooded or would run dry in the tail reach.

Semi-modules. Semi-modules, if properly designed, can be made to distribute more or less equitable distribution of the supply in the distributing channel. In case water levels in the head reach rise due to silting, the semi-modules overdraw and tail runs short. On account of the immunity from variations in discharge with varying water levels in the watercourse, the semi-modules are the best class of outlets to be used in practice.

Non-modular outlets. Non-modular outlets are avoided as far as possible. Their use is justified only when the working head available or the discharge of the outlet is so small that a semi-modular outlet cannot be a designed.

14.14 SELECTION OF TYPE OF OUTLET

In the selection of the type of outlets to be used in a distributary system, it is essential to examine expected fluctuations in water level in the canal, available working head for each outlet, location of the outlet head-middle or tail reach of the canal, and channel condition whether in regime or new canal. In the selection of the particular type for any situation, the factors considered are (*i*) Outlet should draw authorized discharge within the working head available, (*ii*) Outlet should draw its due share of silt, (*iii*) It should be immune from tempering with, (*iv*) It should admit easy and economical adjustment, and (*v*) Its cost is low.

Pipe Outlet : Non-modular (pipe) outlets are avoided as far as possible. Their use is justified only when the working head available or discharge of the outlet is so small that a semi-modular outlet cannot be designed. Pipe outlets are provided in the newly constructed channels and subsequently replaced by semi-modular outlets as soon as the distributary attains its regime. When additional supply is to be given to the irrigators during Kharif, Imdadi or temporary outlets of this type are given for specific period only.

Open Flume : On account of immunity from variation in discharge with raising of water level in a water course, the semi-modules are the best class of outlets to be used in practice. Following type of outlet may be adopted for different conditions such as open flume: (*i*) Open flume as far as possible restricted to reach above control points. Whenever possible, outlets are clustered above, within 300 m of a control point. In such cases, open flume outlets having their crest level the same as that of control point are provided. (*ii*) At tail cluster, open flume outlets are useful in distributing the supply reaching there proportionately and in easily absorbing any excess that reaches the tail thus avoiding chances of flooding tail reach, (*iii*) Where available working head is too small to design an APM, as OF requires just a fraction (20%) of the head of water over the crest of the outlet, and (*iv*) Where bank section is heavy, a pipe-cum-O.F is preferred on account of low cost, ease of adjustment and low consumption of head.

APM : Under all other conditions, APM outlets are preferred, provided they can be designed with their crest at or near bed level and sufficient working head is available. Reachwise the following types are suggested :

Head Reach : APM outlets set below 8/10 depth or even up to 9/10 depth in head reach of a distributary so that excess supply at head is conveyed to the tail reaches, where it is most wanted.

Middle Reach : Proportional outlet, viz., open flume set at 9/10 depth and APM at 6/10 depth.

Tail Reach : At tail open flume outlets are most suitable for proportionate distribution of the supply reaching there. Also due to low working head required, it avoids undue raising of full supply level resulting in economy in the construction cost of distributary system.

14.15 TURNOUT

Turnout or Nakka is a structure capable of directing entire stream of water from the watercourse to the chaks or farm holdings. A steel Nakka plate generally used gets rusted with time and also stolen. Circular concrete turnout (Fig. 14.8) consists of a square panel with a circular lid

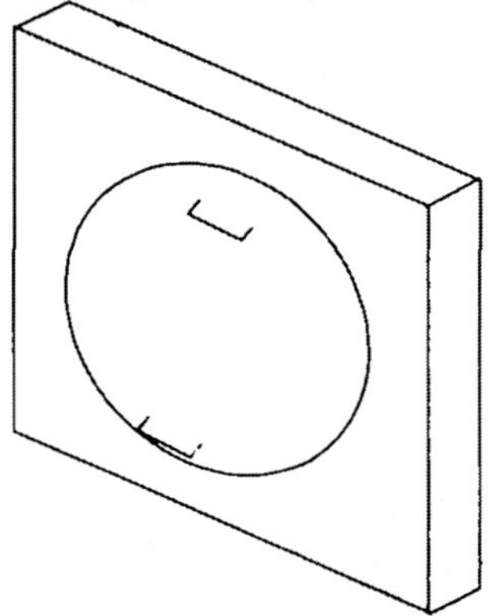

Fig. 14.8. Isometric view of circular Nakka.

cast in reinforced concrete 1:2:3. The reinforcement of the panel and lid consists of 4 No. 10 guage wires. Two handles of 5 mm rod are provided for handling the lid. The sizes of Nakka for various discharges are as given in table 14.5 :

Table. 14.5 Sizes of Nakks

Size of Nakka	*Bed slope*			
Diameter (cm)	*0.15‰*	*0.2‰*	*0.25‰*	*0.30‰*
		Discharge (cumecs)		
30	0.035	0.040	0.045	0.048
40	0.047	0.054	0.060	0.065
45	0.052	0.061	0.068	0.074
52.5	0.064	0.074	0.082	0.091

14.16 DERIVATION OF RELATION TO SHARE PROPORTIONALITY

1. Flexibility

General formulae for discharge of an outlet and discharge through an irrigation channel are as under :

$$q = ch^m \quad (1)$$

$$Q = kD^n \quad (2)$$

where, q = discharge of outlet, h = head acting on the outlet, Q = discharge of canal, D = full supply depth of canal, m = outlet index, and n = channel index.

Differentiating (1), we get

$$dq = mch^{m-1}dh \quad (3)$$

Dividing (3) by (1)

$$\frac{dq}{q} = \frac{mch^{m-1}}{ch^m}dh = m\frac{dh}{h}$$

Differentiating (2) with respect to *d*, we get

$$dQ = nkD^{n-1}dD \qquad (4)$$

Dividing (4) by (2) $\dfrac{dQ}{Q} = \dfrac{nkD^{n-1}\,dD}{kD^n} = \dfrac{ndD}{D}$

∴ Flexibility, $F = \dfrac{dq/q}{dQ/Q} = \dfrac{mdh/h}{ndD/D} = \dfrac{m}{n}\dfrac{dh}{h}\dfrac{D}{dD}$

Since discharge passed by a semi-module and hence head acting on the outlet varies with the depth of water in the channel $dh = dD$.

Therefore, $F = \dfrac{m}{n}\dfrac{D}{h}$

2. Proportionality

For the outlet to be perfectly rigid or module, its flexibility, $F = 0$. For the outlet to be proportional *i.e.*, the rate of change of discharge of the outlet is equal to the rate of change of discharge of the channel, flexibility, $F = 1$.

Also, $F = \dfrac{m}{n}\dfrac{D}{h}$, ∴ $F = \dfrac{m}{n}\dfrac{D}{h} = 1$ or $\dfrac{h}{D} = \dfrac{m}{n}$ (5)

where, h/d = setting of the outlet, and m/n = outlet index/channel index.
$m = 1/2$ for pipe outlet, as its discharge is proportional to $h^{1/2}$, and $n = 3/2$ for weir type outlet (OF), discharge being proportional to $H^{3/2}$, and $n = 5/3$ for a trapezoidal channel based on Lacey's theory, discharge being proportional to $D^{5/3}$.

Putting values in (5)

$$h/D = \frac{1}{2}\Big/\frac{5}{3} = 3/10, \qquad \text{also} \qquad h/D = \frac{3}{2}\Big/\frac{5}{3} = 9/10$$

Thus for proportionality, pipe outlet should be set at 3/10 of the depth below full supply water surface in the channel and weir type outlets at 9/10 the depth.

3. Sensitivity

Also, $S = (dq/q)/d_G/D$ or $S = (dq/q)/(dD/D)$, (6)

assuming $dG = dD$ *i.e.*, change in outlet gauge reading = change in depth of flow in canal.

Flexibility, $F = dq/q/dQ/Q$

But $\dfrac{dQ}{Q} = \dfrac{n}{D}dD$ ∴ $F = \dfrac{dq/q}{n\,(dD/D)}$ (7)

From (6) and (7)

$$S = nF = \frac{5}{3}F \text{ for wide trapezoidal channel.}$$

14.17 VISUAL TEST OF MODULAR WORKING OF OUTLETS

APM : Visual test of the modular working of the outlet is the formation of a standing wave and so long as the wave is steady and remains clear of exit of the orifice the discharge coefficient is constant, subject of course, to the satisfaction of the conditions at the control section upstream of the jump that is (*i*) y is not more than $0.67\,h$ (that is why under designed conditions y is restricted to $0.5\,h$), (*ii*) filaments beyond the control section should be either convergent, adherant or rectilinear, and (*iii*) jet leaving the orifice is adherant, *i.e.*, in contact with the bed.

Open Flume : Presence of a steady standing wave below the outlet constitutes a visual test of the outlet to show that it is taking its authorized discharge under full supply condition. So long as the wave is steady, the discharge through the outlet is independent of the water levels in the water course.

Pipe Outlet : Free overfall pipe outlet is a semi-module but discharge is increased 15 to 20 per cent by heading up the water to the top level of the pipe to drown it. Visual examination can show whether the outlet is working as a free-fall as designed or made to drown by the cultivators to increase the discharge by heading up the water level in the water course up to top level of the pipe.

SOLVED EXAMPLES

Example 14.1 (*a*) *Calculate the minimum water level required in channel in front of the outlet head with the data : critical level to be irrigated = 105.0 m, length of water course = 3 km, and working head of outlet = 0.25 m.*

(*b*) *Calculate authorized full supply discharge of an outlet with the data : area of chak = 300 ha, kharif : rabi ratio = 30 : 60, full supply factor at distributary head = kharif 700 ha/cumec, rabi 1700 ha/cumec, and absorption in distributary = 10%.*

Solution.

(*a*) Critical level in the chak to be irrigated = 105.0 m
Command in the field = 0.075 m
Water level required to irrigate it (105.0 + 0.075) = 105.075 m
Difference of water level at head and tail of water course
(3 × 0.20) @ slope 0.20 ‰, = 0.6 m
Water level required at head of water course (105.075 + 0.60) = 105.675
∴ Minimum water level required in the channel at outlet head
(105.675 + 0.25 working head given) = 105.925 say 105.93 m

(*b*) Area of chak (assumed C.C.A.) = 300 ha
Area under kharif crops @ 30% = 90 ha
Area under rabi crops @ 60% = 180 ha
Authorized discharge for the outlet at distributary head
during kharif at F.S.F. 700 = 90/700 = 0.13 cumec
AFS discharge of outlet at outlet head at 10% absorption = 0.1287 cumec
AFS discharge of outlet at distributary head
during rabi FSF 1700 = 180/1700 = 0.106 cumec
AFS discharge of outlet at outlet head at 10% absorption = 0.106 – 0.011
= 0.095 cumec

Example 14.2 *Design an outlet for the data : discharge = 0.05 cumec, FSL in distributary = 105.0 m, FS depth in distributary = 1.0 m, El. of highest area to be commanded = 104.0 m, and distance of critical point to be irrigated from outlet head = 2000 m.*

Solution. The available working head for the outlet is worked out from command statement, as under :

Command statement

El. of critical area (m)	*Distance from outlet head (m)*	*Loss of head @ 0.20‰ in watercourse*	*Field command (m)*	*Minimum water level required at outlet head (m)*	*FSL in distributary (m)*	*Working head available*
104.0	2000	0.4	0.075	104.475	105.0	0.525 m

Outlet Design. Since ample working head is available, APM outlet can be provided.
Let setting be 90%, such that $H = 1.0 \times 0.9 = 0.9$ m

Now $Q = 4.03\, B_t y \sqrt{h}$

Let $y = 0.4$ m, $h = 0.5$ m and $B_t = 0.06$ m

Therefore, $Q = 4.03 \times 0.06 \times 0.4\,(0.5)^{1/2}$
= 0.07 cumec, more than required discharge.

Limiting value of h = Available working head/0.75 = 0.525/0.75 = 0.7 m

Now discharge with $y = 0.2$, $h = 0.7$, $Q = 4.03 \times 0.06 \times 0.2\,(0.7)^{1/2} = 0.04$ cumec which is less than required discharge.

Now, increase B_t to next standard size 0.08 m

Therefore, $Q = 4.03 \times 0.08 \times 0.2\,(0.7)^{1/2} = 0.053$ cumec

which is nearly the required discharge of 0.05 cumec, hence O.K.

Minimum modular working head, *mmwh* $= 0.75 \times 0.7 = 0.525$ m

which is the same as available working head, hence O.K.

Example 14.3 *Design an APM outlet for 0.05 cumecs discharge in a distributary of 0.5 m depth and 0.3 m is working head.*

Solution.

With 90% setting $H = 0.9 \times 0.5 = 0.45$ m

Discharge formula, $Q = 4.03\, B_t y \sqrt{h}$

Assume $y = 0.2$ m, $h = 0.25$ m and $B_t = 0.06$ m

Now, $Q = 4.03\, B_t y \sqrt{h}$

$= 4.03 \times 0.06 \times 0.2\,(0.25)^{0.5} = 0.024$ cumec

which is less than required discharge of 0.05 cumec.

Now, assume $B_t = 0.12$, $y = 0.22$ and $h = 0.23$

Now, $Q = 4.03\, B_t y \sqrt{h}$

$= 4.03 \times 0.12 \times 0.22\,(0.23)^{1/2} = 0.05$ cumec

which is the required discharge.

Minimum modular working head $= 0.75\, h = 0.75 \times 0.23 = 0.17$ m

which is less than available working head of 0.3 m, hence O.K.

Example 14.4 *Design an open flume outlet for the data as given in example 14.3.*

Solution. With 90% setting,

$$H = 0.9 \times 0.5 = 0.45 \text{ m}$$

Discharge formula $Q = C\, B_t\, H^{3/2}$

Putting known values $0.05 = 1.6\, B_t \times (0.45)^{3/2}$

or $$B_t = \frac{0.05}{16 \times 0.30} = 0.10 \text{m}$$

Minimum modular working head $= 0.2\, H = 0.2 \times 0.45 = 0.09$ m which is less than available working head.

Note. Mark the difference that an open flume requires 0.09 m working head to work whereas an APM for the same data requires 0.17 m of working head to function.

Example 14.5 *Design a pipe outlet for the data in example 14.4.*

Solution.

$$Q = C\, A\, \sqrt{2gh}$$

or $$0.05 = 0.62 \left(\frac{22}{7}\frac{d^2}{4}\right)\sqrt{2 \times 9.81 \times 0.30}$$

or Pipe diameter, $d = 20$ cm

The centre of the pipe should be at a level 0.30 m below F.S.L. in distributary.

QUESTIONS

1. Define chak. How is the size of a chak fixed ?
2. Define term outlet. What are the essential requirements of a good outlet ?
3. What is the optimum size of an outlet ? What are the objections to large and small size outlets ?
4. Describe three classes of outlets, their relative merits and demerits. Outline the three types of outlets to suit different reaches of a canal.

5. Why is the installation of an adjustable proportional module preferred by irrigation department but unpopular amongst irrigators ?
6. Give discharge formula for three types of outlets.
7. Define sensitivity and proportionality of an outlet. Establish the relation between sensitivity and flexibility of an outlet.
8. Draw neat sketches illustrating each of the following types of canal outlets.
   (*i*) Non-modular, (*ii*) semi-modular, and (*iii*) rigid module
9. (*a*) State briefly how you will fix up the location and capacity of canal outlet.
   (*b*) A crump semi-modular outlet is to be built in a distributary canal having its FSL 100.0 m and bed level = 98.5 m. Determine the RL of crest and the corresponding width of throat (contracted section) of the outlet required to pass a discharge of 100 *l*/sec through the outlet. Flexibility of the outlet is unity. Coefficient of discharge (C_d) of the outlet is 1.7 m/sec. Neglect velocity head in the distributary canal. [*AMIE*]

Solution. Unity flexibility implies that the outlet is proportional, *i.e.*, the rate of change of discharge of outlet is equal to the rate of change of discharge of the distributary,

Therefore, $F = \frac{m}{n}\frac{D}{h}$

where, $F = 1$ (given)

m = outlet index, value, 3/2 for crump open flume outlet

n = canal index, value 5/3

Hence $1 = \frac{3/2}{5/3}\frac{D}{H}$

or $H = \frac{3}{2} \times \frac{3}{5} D$

or setting, $H = \frac{9}{10} D$

Distributary depth $= 100.0 - 98.50 = 1.5$ m

Therefore, $H = 0.9 \times 1.5 = 1.35$ m

RL of crest of outlet $= \text{FSL} - H$

$= 100.0 - 1.35 = 98.65$ m

Outlet discharge = 100 *l*/sec = 0.1 cumec

Now, by discharge formula $Q = CB_t H^{3/2}$

$0.1 = 1.7\, B_t (1.35)^{3/2}$

or $B_t = 37$ mm, which is not suitable, B_t being less than 60 mm

Adopt 60% setting *i.e.*, $H = 1.5 \times 0.60 = 0.9$ m

Now, $0.1 = 1.7\, B_t (0.9)^{3/2}$

or $B_t = 6.9$ cm, hence O.K.

Minimum modular working head $= 0.2\, H = 0.2 \times 0.9 = 0.18$ m

If a pipe outlet is to be fixed $Q = 2.75\, A\sqrt{h}$, assuming free-fall as ample working head is available.

Assume 25 cm dia pipe $0.1 = 2.75 \frac{\pi}{4} (0.25)^2 \sqrt{h}$

or $h = 0.55$ m

Centre line of pipe is fixed at El. 100.0 – 0.55 = El 99.45 m in the distributary.

10. (*a*) What are tail clusters ? Draw sketch to illustrate the same.
    (*b*) Design and draw an APM outlet from the following data :
    (*i*) RL of bed of distributary = 150.00 m.
    (*ii*) RL of bed of field channel = 150.00 m.

(*iii*) Full supply depth of distributary = 0.95 m.
(*iv*) Bed width of distributary = 6.0 m.
(*v*) Full supply discharge of distributary = 2.7 m^3/sec.
(*vi*) Full supply depth of field channel = 0.30 m.
(*vii*) Width of outlet opening = 0.16 m, and
(*viii*) Discharge of field channel = 0.05 m^3/sec. [*PU*]

11. (*a*) State requirements for an ideal outlet.
(*b*) Distinguish between a modular, a non-modular and a semi-modular outlet. Give one example of each type.
(*c*) Explain tail clusters with the help of sketches. [*PU*]

12. Define :
(*a*) Sensitivity (*b*) Drowning ratio (*c*) Proportional outlet.
(*d*) Modular limits, and (*e*) APM and OF.

14. Design an an open flume at tail for a minor of discharge 0.030 cumec and F.S depth 0.3 m.

Solution :

$$Q = CB_t H^{3/2}$$
$$0.030 = 1.64 \times B_t (.30)^{3/2} \quad \because \text{ at tail } H = 0.30 \text{ m}$$
$$\text{or } B_t = 0.11 \text{ m}$$
$$\text{Check flexibility, } F = (m/n)(D/H)$$
$$F = [(3/2)/(5/3)](0.30/0.30)$$
$$F = 0.90 < 1$$

It means that the outlet is sub proportional. The variation in discharge of the outlet will be less compared to the variation in discharge of fluctuation in F.S.L. of the minor.

Sensitivity, $S = n \times F = (5/3)(0.90) = 1.5 < 5/3$. It implies that the variation in discharge will be 50% more compared with the fluctuations in the water depth.

15. Determine size of tail cluster having two outlets of 0.032 cumec and 0.030 cumec at the tail.

Solution :

Assume $Bt = 0.12$ m and $C = 1.64$, H being 0.30 m

$$\therefore \quad 0.032 = 1.64\,(Bt)\,(0.30)^{3/2}$$
$$\text{or} \quad Bt = 0.12 \text{ cm}$$
$$\text{Now} \quad Bt_1/Bt_2 = Q_1/Q_2$$
$$\text{or} \quad Bt_1 = (0.030/0.032)\,(0.12) = 11 \text{cm}$$

Hence tail cluster will have two outlets of gullet width size of 12 cm and 11 cm.

PART - IV

HYDROLOGY

- **Precipitation and its Measurement**

15

PRECIPITATION AND ITS MEASUREMENT

15.1 INTRODUCTION

Hydrology is the science which deals with the water of the earth in all its states, *i.e.* occurrence, distribution and circulation, through the unending hydrologic cycle of precipitation, consequent runoff, stream flow, infiltration and storage, eventual evaporation and reprecipitation. It is concerned with the atmosphere, the medium through which water is brought to and removed from the earth's surface. The study of hydrology concerns itself with three different forms of water *i.e.* (*i*) above land as atmospheric water or precipitation, (*ii*) on land surface as surface water or runoff, and (*iii*) below the land surface as subsurface water or groundwater.

Engineering hydrology includes the science which deals with the estimation of runoff and its transmission from one place to another, besides the properties of water in nature. For an engineer, surface and groundwaters are of direct importance but owing to lack of data of stream flow, it is usually necessary to study precipitation, evaporation and percolation to indirectly determine the availability of surface and groundwaters. For planning and management of water resources development projects, an engineer has to determine (*i*) Water resources potential of a basin, (*ii*) Dependable yield for irrigation and hydel power generation, municipal and industrial water supply, (*iii*) Probability of floods, their pattern and magnitude, (*iv*) Sediment yield and transportation, (*v*) Hydrometeorological data, (*vi*) New sources of water for irrigation and other purposes, (*vii*) Streamflow forecasting by the use of precipitation data, (*viii*) Floods, droughts and control measures thereof, (*ix*) Water control, navigation, erosion control and pollution control of rivers and streams and (*x*) Operation and maintenance of river valley projects,

15.2 HYDROLOGIC CYCLE

Earth's water sources are mainly rivers, oceans, lakes and underground sources which get water from rain. The rain in itself depends on evaporation from surfaces of rivers, oceans and lakes. The moisture in turn is lifted and is eventually condensed and falls back as precipitation to the earth's surface. The precipitation and evaporation continues for ever. Thus the earth's total available supply of moisture and water, except for some deep groundwater supplies, is constantly in circulation. It is a gigantic system operating in and on the ground surface, the oceans and the atmosphere surrounding the earth. It has no beginning or ending. This unending process of circulation and redistribution of water by atmosphere and earth is called *hydrologic cycle.* The hydrologic cycle is defined as a phenomenon relating to the circulation of water from the sea, through the atmosphere to the land, and thence often with many delays, back to the sea or ocean through various stages and processes, for example, precipitation, interception, runoff, infiltration, percolation, groundwater storage, evaporation and transpiration. Also there are many short circuits of the water that are returned to the atmosphere without reaching the sea. The hydrologic cycle is illustrated in Fig. 15.1.

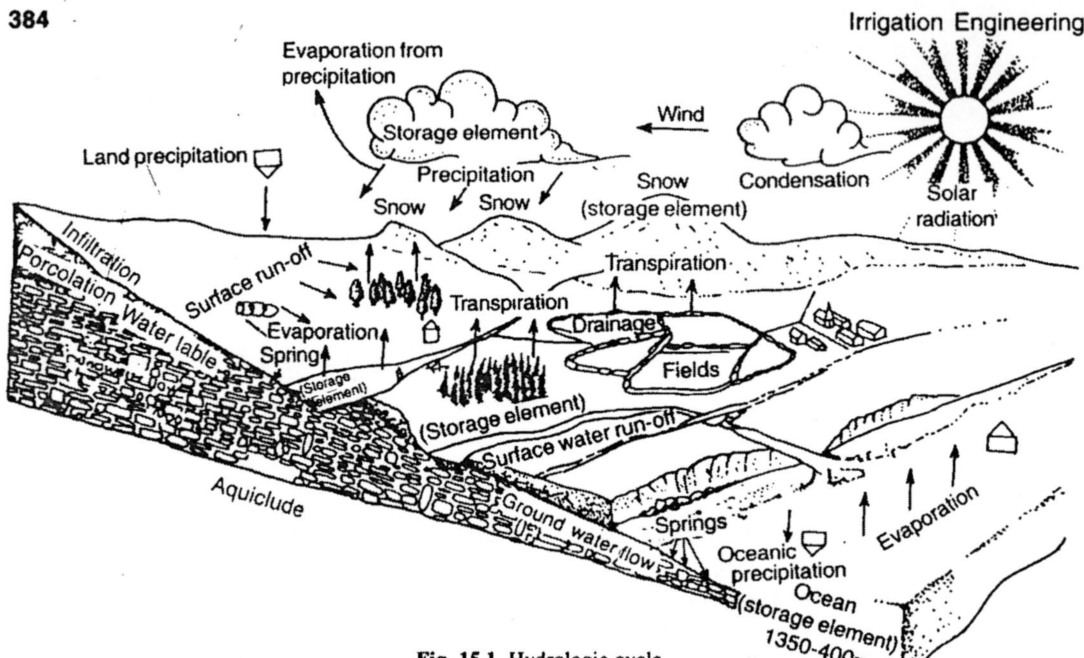

Fig. 15.1 Hydrologic cycle.

This cycle continuously keeps a balance between the waters of the earth and moisture in the atmosphere. It is represented by the following simplified relation:

$$P = R + E + S$$

where, P = preciptiation. It is the total supply of water derived from the atmosphere in the form of rain, snow, dew, moist, frost, hail, sleet, etc, R = runoff. It is that portion of precipitation which is not absorbed by the deep strata but appears in the streams after meeting persistent demands of evapotranspiration including interception and other losses. It includes other delayed runoff that has been temporarily detained as snow cover or stored in natural lakes or swamps, E = evaporation. It is the quantity of water that is evaporated, S = storage. It is the change in soil storage on or below the surface of the earth within the region.

15.3 PRECIPITATION

It is the general term for all forms of moisture emanating from clouds and falling to the surface of the earth. It is thus the total supply of water derived from the atmosphere in the form of rain, snow, mist, frost, hail, sleet, etc. It is usually expressed as depth of liquid water on a horizontal surface in a day, month or year and designated as daily, monthly or annual precipitation. Precipitation is thus the sum of the amounts of liquid precipitation *i.e.* rainfall and liquid equivalent of frozen precipitation, *i.e.* snow, hail, sleet, freezing rain.

Occurrence of Precipitation

For precipitation to occur, the three essential requirements are (*i*) Some mechanism is required to cool the air sufficiently to cause condensation and droplets growth, (*ii*) Condensation nuclei are also necessary, but they are usually present in the atmosphere in adequate quantities, and (*iii*) Large scale cooling is needed for significant amounts of precipitation. It is achieved by lifting of the air. In order for precipitation to take place, it is first necessary for a portion of the atmosphere to be cooled until the air is saturated with water vapour. Condensation thereupon occurs, resulting in the formation of minute droplets which are visible collectively in the form of clouds or fog. These droplets will not in themselves form into rain drops and may remain indefinitely as droplets in a condition of colloidal suspension. Rain drops, or other forms of precipitation, result when the minute droplets of moisture collect on particles in the air termed hygroscopic nuclei. The commonest hygroscopic nuclei are salt particles from oceans and products of combustion in the atmosphere over cities. The exact form of precipitation, whether as rain drops, snowflakes or the ice particles depends upon the temperature of the air in which the precipitation is formed and the temperature of lower air layers through which it falls to the earth. It is necessary for moist air to be cooled to the saturation point (dew point) before condensation can occur.

Types of Precipitation

1. *Convective Precipitation*: It is the precipitation resulting from the upward movement of air that is warmer than its surrounding. The effect may be caused by the temperature difference between hot, bare fields and cool woods, between land and lakes than the surrounding country-side. Convective precipitation is generally of a showery nature with rapid changes of intensities.

2. *Orographic Precipitation*: It is the precipitation caused by dynamic cooling of air as an air current rises over a mountain barrier. When moisture bearing winds usually blowing from ocean to land surface are forced up by the influence of coastal mountain ranges, the cooling and condensation processes take place and precipitation occurs along the windward slope of mountains. The precipitation in Himalayan region is due to orographic ascent of air mass laden with moisture content because of long travel over ocean. Orographic precipitation is ordinarily of low intensity.

3. *Cyclonic Precipitation*: It is the precipitation associated with the passage of depressions or cyclones. It results from the lifting of air converging into a low pressure area. Cyclonic precipitation may be heavy and prolonged. Most of the winter rains in Punjab and Haryana are due to cyclones. Cyclonic precipitation may be divided into Frontal precipitation, and Non-frontal precipitation.

Frontal precipitation: Cyclonic storms and their resulting precipitation are associated with the planes of contact (frontal surfaces) between the air masses of different temperatures and moisture contents. The boundary of discontinuity between the two air masses is termed as front. When a moving warm moist air overrides a cold air mass or is forced up by a cold air wedge, it cools down at high altitudes resulting in precipitation. A front may be of two types (*i*) warm front, or (*ii*) cold front depending whether there is active or passive ascent of warm air mass over the cold air mass. Warm front precipitation, generally light to moderate, is formed in warm air moving gradually upward over a wedge of cold air. Cold front precipitation, generally of showery nature, is formed in warm air which is forced upward by an advancing wedge of cold air.

Non-frontal precipitation: It results when a moving cold air mass meets a rather stationary moist warm air mass. The warm air mass, being lighter, is lifted up over cold air which cools down at higher altitudes resulting in precipitation.

4. *Precipitation Due to Turbulent Ascent*: Air mass is forced up due to increased friction of earth surface after its travel over the ocean and friction on condensation results in precipitation. Winter rainfall in Tamil Nadu is mainly due to this process.

15.4 RAINFALL

Rainfall is the total liquid product of precipitation or condensation from the atmosphere as received and measured in a raingauge. A rainy day is defined as the day when 2.5 mm or more rainfall is recorded. Rainfall excess or net rainfall or effective rainfall is the part of the rainfall that appears as runoff in the streams. Rainfall intensity is the amount of rain occurring in a unit interval of time. It is generally expressed in mm per hour.

Average rainfall over India is computed as 120 cm. The rainfall is erratic and ill distributed. Practically the whole of Assam and its neighbourhood, the Western Ghats and the adjoining coastal strip and parts of Himalaya receive heavy rainfall of more than 200 cm. Some places in the Khasi, Jayantia hills of Meghalya receive the heaviest rainfall in the world. Cherrapunji records a rainfall of 1142 cm and in contrast Rajasthan, Kutch and high Ladakh plateau receive rainfall as low as 10 to 50 cm in a year. Rainfall also varies during different parts of the year due to the difference in seasonal temperature between land and sea. Even during monsoons in Northern India, some regions remain deficient. The rainfall in India is seasonal in character and varying in geographical distribution over the country. In major parts of the country almost 85 to 90 per cent of the total rainfall is contributed by the South-West monsoon and is received only in four rainy months of June to September more precisely in about 100 hours in a year. The balance precipitation occurs during the North-East monsoon of November and December only in the coastal belts of Tamil Nadu, Andhra Pradesh and parts of Orissa. Some parts of the country receive hot weather rainfall between March and May due to thunderstorms activity, called North-Westers. The regions with rainfall less than 40 cm form 12 per cent of total geographical area and 35 per cent or little over one-third of India has a rainfall below 75 cm, only 8 per cent receive more than 250 cm of rainfall.

Rainfall Measurement: Rainfall is measured on the basis of the vertical depth of water accumulated on a level surface during an interval of time, if all rainfall remained where it fell. It is measured in an instrument called raingauge. Raingauge measures the quantity of rain that falls at a given place and time usually 8.30 A.M. Rainfall is measured in mm. An amount less than 0.1 mm is recorded as a trace. The sum total of all the measurements during the previous 24 hours is entered as the rainfall of the day.

Selection of Site: The raingauge is installed at a site selected with the considerations that (*i*) It is true representative of the area for which the raingauge station is to give rainfall, (*ii*) It is in an open place, (*iii*) It is in a level ground, not upon a slope or terrace, (*iv*) Away from trees, buildings or other obstructions by a minimum distance of four times but never less than twice the height of obstruction, (*v*) Site is shielded from strong wind. The ground should not fall away steeply on the side of the prevailing wind, and (*vi*) Easily accessible at all times. In hills minimum level space required is 6 × 6 m shielded from high winds.

Network of Gauges : The rainfall varies from place to place and as such a number of raingauges are set up at various locations. A network of raingauges is the simplest and most common means of measurement of rainfall. The number of stations are required to be adequate to represent reliable picture of the intensity and duration of rainfall in the area. The network of raingauges design is considered as saturated when in a network any addition of gauge to it will not appreciably alter the rainfall assessed. It is in regard to important project wherein maximum accuracy is desired. The optimum network design is that in which by interpolation between values of different stations, it is possible to determine with sufficient accuracy the characteristics of the basic hydrometeorological elements at any point of interest. In planning network it has to be ensured that it may not contribute towards faulty assessment of water resources and flood producing potential. As per IS : 4987-1994 for optimum network in plains one raingauge station up to 500 km^2 is sufficient while in hilly areas, with general average elevation one km above sea level, one raingauge in 250 to 400 km^2 is sufficient. In predominently hilly areas with heavy rainfall, one raingauge in not more than 150 km^2; the denser stations being preferable. In areas where optimum network may not be feasible, network may be designed wherein at least one raingauge may represent 750 km^2 in plain areas 100-250 km^2 in hilly areas and 1,500–10,000 km^2 in arid areas.

15.5 OPTIMUM NUMBER OF RAINGAUGE STATIONS

Optimum number of raingauge stations required to compute average rainfall in the basin within 10% error is determined by the procedure as under:

Total rainfall from *n* number existing gauges (R)	Mean rainfall (m)	Sum of squares of all gauges (SS)	T	Square of standard deviation (S^2)	Coefficient of variation (Cv)	Optimum number (N) of gauges within 10% error	Additional raingauges required
$P_1 + P_2 .... + P_n$	$\frac{R}{n}$	$P_1^2 + P_2^2 + ... P_n^2$	$\frac{R^2}{n}$	$\frac{SS - T}{n - 1}$	$\frac{100\sqrt{S^2}}{m}$	$\left(\frac{C_V}{10}\right)^2$	$(N - n)$

15.6 TYPES OF RAINGAUGES

The two principal types of raingauges are non-recording and recording raingauges, as under:

1. Non-recording Raingauge: It is also termed as non-automatic raingauge. It essentially consists of a collector to intercept the sample of rainfall to be measured and a receiver consisting of a base and a bottle to collect and store the rainfall until measured. The collector is extended above ground level while the receiver is fixed partly below ground level. The bottle is of a capacity sufficient to hold the extremes of rainfall likely to occur in 24 hours at a location. The bottles are of standard capacity of 2 litre, 4 litre and 10 litre. The usual time to measure rainfall is 8.30 A.M. Standard raingauge (Fig. 15.2) as per IS: 5225–1969 is adopted in India. The design provides for four raingauges of 100, 200, 400 and 1000 mm capacities.

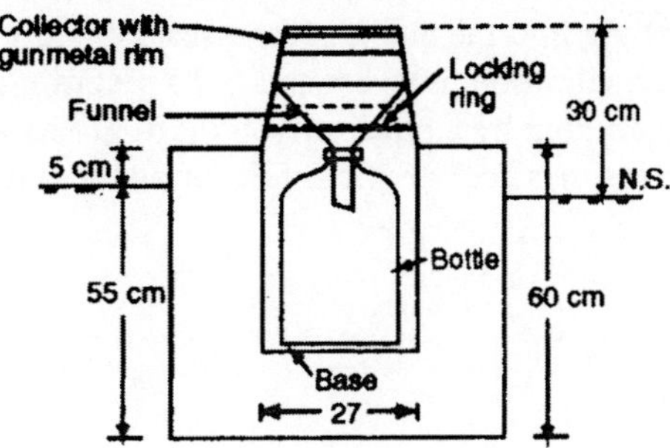

Fig. 15.2 Standard raingauge

Symon's raingauge extensively used earlier in India consists of a receiving funnel 127 mm in diameter, the base, the body and the receiver with a brass rim 1.25 cm high. The rain falling in the receiver is funnelled into a 175 cm capacity container and measured. The funnel shank is inserted into the neck of the bottle which is 75 to 100 mm in diameter. The rainfall is measured by a measuring glass which can be read to the nearest 0.1 mm. Recently India meteorological department has introduced fibreglass reinforced polyester raingauge which is an improvement over the Symon's gauge. Non-recording raingauge gives only the total depth of rainfall for 24 hours and not its intensity and duration during different time intervals of the day. In remote areas, storage gauge is used to measure total seasonal precipitation. It consists of a vertical 60 cm diameter steel pipe of sufficient length to place its 20 cm catch ring above maximum accumulated snow. To convert the snow which falls into the gauge to liquid state an antifreeze solution is placed in the receiver.

2. Recording Raingauges: Recording raingauge, an improvement over the non-recording raingauge, is automatic which gives record of rainfall. Since the raingauge represents cumulative rainfall, it is also termed as integrated raingauge or continuous raingauge. It is specially adapted to inaccessible places in hilly areas where it is not possible to visit the station daily. The mechanical arrangement in the recording raingauge records rainfall for extended periods in the form of a pen trace on a clock-driven chart. The graph of the cumulative rainfall is termed as mass curve of the rainfall (Fig. 16.11). The chart indicates (*i*) total amount of rainfall since the record is started at the station, (*ii*) duration of rainfall, *i.e.* times of starting and ending of the rainfall, and (*iii*) intensity of the rainfall for any given period.

The raingauge is installed on a concrete or masonry platform 60 × 60 × 35 cm, sunk to a depth of 27.5 cm, truly horizontal and at a height of exactly 75 cm above the ground surface.

The installation of self-recording gauge stations is fixed in consideration of (*i*) Economical feasibility, (*ii*) Necessity for obtaining short duration intensities, and (*iii*) Requirement of the project. It is desirable to equip at least 20% of the raingauge stations with recording gauges to know intensities of rainfall. Self-recording raingauge may also be installed in conjunction with a non-recording gauge at a distance of 3 m away from the latter to serve as a standard for checking and adjusting the recordings of self-recording gauge.

Advantages: (*i*) Only means to get rainfall data from inaccessible hilly areas, (*ii*) Automatic recording of rainfall, free from error due to human judgement, (*iii*) Records intensity of rainfall at any time besides total rainfall, (*iv*) Duration of rainfall is also indicated, and (*v*) Large capacity of the gauge.

Limitations: (*i*) High initial cost, (*ii*) Recording may be interrupted due to development of any fault in mechanical as well as electrical recording mechanism or mechanical defects.

The various types of recording gauges are as under:

1. Weighing Bucket Raingauge: Weighing bucket gauge is used for recording rainfall as well as snow. It consists of a receiver bucket supported by a spring or lever balance (Fig. 15.3*a*). The catch from funnel empties into a bucket mounted on a weighing scale. The weight of the rain falling into the bucket depresses the entire mechanism. This movement is transmitted to a recording pen through levers and links recording the increased weight of the bucket and its content on a chart held by a clock-driven drum giving a trace of accumulated amounts of rainfall with time. It gives mass curve of rainfall for as long as one week.

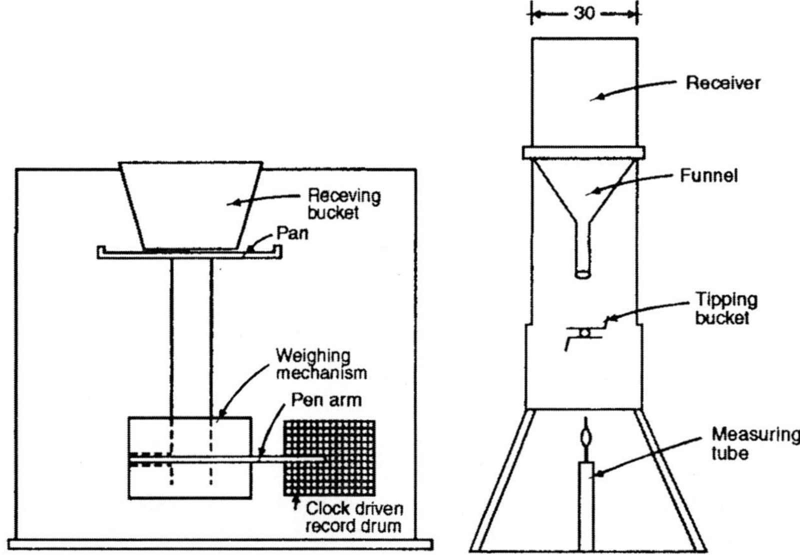

Fig. 15.3 Weighing bucket raingauge. **Fig. 15.4** Tipping bucket raingauge.

Advantages: (*i*) Simple mechanism, and (*ii*) Suitable for digitalising of output signal for the transmission of telemetry.

Disadvantage: The main disadvantage is that rainfall of less than the capacity of bucket compartments is not recorded.

2. Tipping Bucket Raingauge: It consists of a sharp edge receiver, 30 cm in diameter, and a funnel. The rain water collected in the receiver is funnelled into a twin compartment tipping bucket (Fig. 15.4). When 0.25 mm of rain collects to fill one compartment, it tips and empties the water into a storage reservoir and simultaneously second compartment of the bucket moves under the funnel to receive rain. Thus the bucket is tipped by each 0.25 mm of rain. As the bucket is tipped, it actuates an electrical circuit causing a pen to mark on a revolving drum. The movement of the tipping of bucket can be transmitted electronically over distances to a control room to record rainfall measurement directly and in this case graph paper or drum is not installed in the gauge. It is ideally suited for digitalising of the output signal.

3. Float Recording Gauge: It consists of a rectangular float chamber, carrying a funnel and a drum on either side on its top and a light hollow float at its bottom provided with a self-starting siphoning arrangement (Fig. 15.5). The funnel receives the rain and conducts it into the chamber. The float gets raised as the rainwater level in the chamber rises. This movement is transmitted through a suitable mechanism to the pen which moves on a revolving chart actuated by a clock-work. When the chamber gets filled, thereby causing the float to touch the top, siphonic action starts and the accumulated water is drained off. The clock mechanism revolves the drum once in 24 hours. Rewinding is needed once in a week when the chart fixed around the drum is also replaced with a new one.

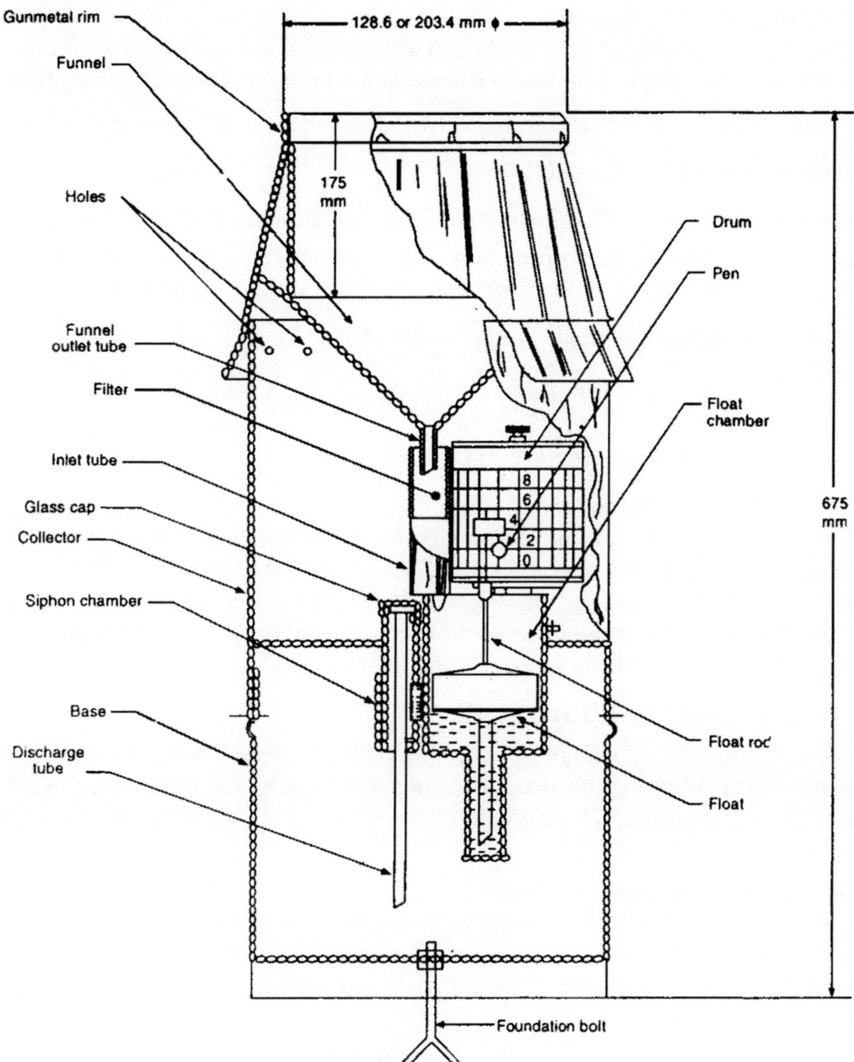

Fig. 15.5 Float type raingauge.

4. Automatic Radio Reporting Raingauge: Automatic radio reporting raingauge signals rainfall data over radio at regular intervals or on command. Rainfall is collected in a receiving funnel and measured by a tipping bucket raingauge. The tipping bucket gives electrical pulses equal in number to the millimetres of rainfall collected in the gauge. These electrical pulses are stored by a data register which are converted into coded messages by a recording unit and impressed on the transmitter during the broadcast. A programme clock switches the coding and programming units as well as radio transmitter at predetermined instants of time. The programming unit governs the sequence of operations during a broadcast which lasts three minutes. The rainfall data are recorded as a coded message beginning with a repeated 'call sign' of the station followed by the coded information on the total rainfall recorded. The signals also indicate whether it is raining at the time of broadcast and the intensity of rainfall.

15.7 ESTIMATION OF MISSING RAINFALL RECORD

Rainfall records of certain stations suffer a break because of non-observation of the rainfall by the observer or failure of the instrument installed. The missing rainfall record is approximated as follows:

1. From the record of Three Stations

(*i*) *Arithmetic Mean Method*: This method is applicable if the normal annual rainfall at the missing station *X* is within 10 per cent of the normal annual rainfall at the adjoining three stations *A, B* and *C*. Arithmetic means of the rainfall of the three stations is taken as the rainfall at the station with missing record *i.e.* $P_x = \frac{1}{3}(P_A + P_B + P_C)$.

(*ii*) *Normal Ratio Method*: The method is applicable if the normal annual rainfall at any of the nearby operative stations *A, B, C* differ from that at the station *x* with missing data, value P_x, by more than 10 per cent. In this method the rainfall values at three stations are weighted by the ratios of the normal long term (annual) rainfall value (*N*).

$$P_x = \frac{1}{3}\left[\left(\frac{N_X}{N_A}\right)P_A + \left(\frac{N_X}{N_B}\right)P_B + \left(\frac{N_X}{N_C}\right)P_C\right]$$

N_A, N_B or N_C differs from N_X by more than 10 per cent.

2. From the Record of Two Stations

Rainfall values of two or even one adjoining station, where three adjacent stations are not there, may be used to estimate the missing rainfall record of a station.

Ratio Method: In this method the missing record of rainfall is computed from the ratio of the rainfall for the particular year, value in respect of which is missing at one station, to that of the rainfall of complete years at both stations, as illustrated in example 15.11.

15.8 AVERAGE RAINFALL OVER A BASIN

Where there is only one raingauge station, the rainfall recorded there is considered applicable over the entire area. Where more than one raingauge stations are established in a basin, the computation of the average rainfall which will represent the whole basin is done by the following methods:

1. Arithmetic Unweighted Mean Method

It is the simplest method of determining the average depth of rainfall over a basin by taking simple arithmetic mean of the rainfall gauges recorded at various stations. The method gives good results in flat regions if the gauges are uniformly distributed and the rainfall at various stations do not vary widely from the mean.

$$\text{Average rainfall over a basin, } (P_{av}) = \frac{P_1 + P_2 ... + P_n}{n} = \frac{\Sigma P}{n}$$

where, $P_1, P_2 \ldots, P_n$ = rainfall recorded at the stations, and n = number of raingauge stations.

2. Thiessen Polygon Method

This method is used for non-uniform raingauge network by assigning weights to station data in proportion to the space between the raingauge stations. Thus the method simply assumes linear variation of precipitation between the raingauge stations and assigns equal segment of the area to the nearest station. It is also called weighted mean method. It is more accurate than arithmetic mean method and is commonly used for medium size catchment areas.

To draw Thiessen polygon (*i*) Plot location of raingauge stations on a map of the basin, (*ii*) Draw straight lines between the adjacent raingauge stations, (*iii*) Draw perpendicular bisectors on these connecting lines in such a way that the bisectors enclose areas referred to as Thiessen ploygon. Around the edge of the basin, where parts of polygons extend beyond the basin boundary, only the portion of polygon inside the drainage area is used. All points within one polygon will be closer to its raingauge than to any of the others. The gauge recorded at the station is then considered to represent the rainfall within the appropriate areas, (*iv*) Multiply the area of each

Thiessen polygon by the raingauge value of the station enclosed, (*v*) Find the total area (ΣA) of the basin, and (*vi*) Determine the average rainfall over the basin from the equations.

$$P_{av} = \frac{A_1P_1 + A_2P_2 + ... + A_nP_n}{A_1 + A_2 + ... + A_n} = \frac{\Sigma(A \times P)}{\Sigma A}$$

where, P_{av} = average depth of rainfall in the basin, P_1, P_2, ..., P_n = rainfall depths at various stations, and A_1, A_2, ..., A_n = areas surrounding the raingauge stations.

Limitations: The method has the limitation of (*i*) Inflexibility, as a new Thiessen diagram is required every time there is a change in the gauge network, (*ii*) Orographic influence is not considered, (*iii*) Assumption that raingauge represents the area which is close to it may not be valid when the rainfall is influenced by topography or results from intense convection.

A refinement of the method envisages that the area governed by each station is expressed as a percentage of the whole area. The stations outside the catchment could also be used for assigning weights to marginal stations inside the catchment.

$$R_{av} = P_AR_A + P_BR_B + P_CR_C$$

where, P_A, P_B, P_C = areas of stations *A*, *B* and *C* expressed as a percentage of the total basin area, and R_A, R_B, R_C = point rainfall at stations *A*, *B* and *C*.

As an illustration, assume that three raingauge stations *A*, *B* and *C* represent 20%, 40% and 40% of the total basin area. Then the average rainfall over the basin = 0.2 R_A + 0.4 R_B + 0.4 R_C.

3. Isohyetal Method

It is probably the most accurate method for the determination of average depth of rainfall over a large basin and provides a means of considering the orographic (mountain) effect. The method takes into account the actual special relationship of the raingauges and that of the nearby gauges and makes corrections accordingly. In this method, the stations and rainfall values are plotted on a map to a suitable scale. Then the contours of equal precipitation *i.e.* isohyetes* are drawn. In constructing the isohyetal maps, full use of the orographic effects of elevation on rainfall and storm movements can be made so that the final map may represent a more realistic pattern of rainfall. The procedure of isohyetel method is to (*i*) Plot rainfall at each station, (*ii*) Prepare isohyetal map of the area enclosed, (*iii*) Areas between successive isohyetes are measured by means of a planimeter, (*iv*) Each of these areas is multiplied by the average rainfall intensity between the successive isohyetes, and (*v*) Average rainfall is determined from the relation:

$$P_{av} = \frac{A_1\dfrac{(P_1+P_2)}{2} + A_2\dfrac{(P_3+P_4)}{2} + ... + A_{n-1}\dfrac{(P_{n-1}+P_n)}{2}}{A_1 + A_2 + ... A_n}$$

or

$$P_{av} = \frac{\Sigma\,[A(P_1+P_2)/2]}{\Sigma A}$$

Advantages: (*i*) Most accurate method when the stations are large in number and is used by an experienced analyst, and (*ii*) Enables interpretation of all available data.

15.9 MEAN ANNUAL RAINFALL

The term 'mean annual rainfall' is synonymous with 'rainfall' at a place. Mean annual rainfall is determined by taking simple average of the total rainfall of several consecutive years at the place. Since the annual rainfall varies at the same station, rainfall records for a considerable number of years are required. The greater the number of years considered, the more accurate is the mean value of rainfall so obtained. In India, a rainfall cycle of 35 years is considered. The

---

*An isohyete is a contour, joining points of equal rainfall in the basin. It has the properties that (*i*) Two different isohyetes do not cross each other, (*ii*) Each isohyete closes on itself, and (*iii*) Isohyete of higher value indicates higher rainfall.

value is recomputed every decade. According to Alexander Binnie, mean rainfall worked out from 35 years of rainfall record is within 1.8% of the true value. If the rainfall record is for a period shorter than 35 years, the error in the computed mean value will be as given in Table 15.1.

Table 15.1 Error in average rainfall value computed from record of period less than 35 years.

Period (years)	5	10	15	20	25	30	35
Error in mean rainfall (percentage)	14.9	8.2	4.8	3.3	2.8	2.3	1.8

U.S. Weather Bureau has considered somewhat longer periods of rainfall record. According to it, about 40 years observations are required to give results within 1.5% of the true mean.

IS 4987:1994 has recommended the following number of years of rainfall record:

Catchment layout	Island	Shore	Plain	Mountains
No. of years	30	40	40	50

Studies by Central Water and Power Commission, India show that 33 to 40 years of rainfall record is necessary to predict a 100-year flood and 80 to 90 years record is required to predict a 250-years flood.

15.10 RAINFALL DEFICIENCY

The mean annual rainfall in India shows great variation in as much as it is 12 cm in desert areas of Rajasthan and 1150 cm in Assam. In general, the coefficient of variation is less than 20% in west coast and Assam while in Rajasthan, Punjab, Haryana and Gujarat it is more than 40% and 30% in the rest of the country. Deficiency of 30 to 45 per cent below normal is called large deficiency, between 45 to 60 per cent serious deficiency and over 60 per cent disastrous deficiency. In a particular year, the rainfall may not be equal to the mean annual rainfall. The ratio of actual rainfall in a particular year at a place to the corresponding normal rainfall is termed as *index of wetness*. It indicates the abnormal or deficient rainfall. A deficient year is that in which the rainfall is less than the mean annual rainfall. Normal rainfall is the annual rainfall averaged from record of 35 years. A good year, also called wet year, is the year in which rainfall is more than the mean rainfall. According to Blendford, wet year is 1.24 to 2.54 × average annual rainfall and as per Binnie, wet year is 1.51 × average annual rainfall. For two consecutive years, maximum rainfall of both the years is equal to 2 years × average annual rainfall × 1.35 and for the three consecutive years as 3 years × average annual rainfall × 1.27. A dry year, also called bad year, is the year in which rainfall is less than the average annual rainfall. According to Blendford, dry year is 0.27 to 0.78 × average annual rainfall and as per Binnie, dry year is 0.60 × average annual rainfall. For two consecutive years, dry rainfall year = 2 years × average annual rainfall × 0.69. For three consecutive years, dry year = 3 years × average annual rainfall × 0.75.

An area with normal rainfall less than 40 cm is called arid, 40–75 cm semi-arid and above 75 cm as humid area.

15.11 ADJUSTMENT OF RAINFALL RECORD

The inconsistency of the record of a station arises due to (*i*) Change in location of raingauge, (*ii*) Change in exposure, and (*iii*) Changes in instrument or observational procedure. The consistency is tested by comparing its accumulated annual or seasonal rainfall with the concurrent values of mean rainfall of a group of adjoining stations termed as base stations. A group of minimum 10 base stations in the immediate surrounding with similar environment of a station the record of which is being tested (say X) is selected. The data of rainfall (annual or monthly) of the station X and the average long-term rainfall of the base stations are arranged in reverse order, starting with the latest record and ending with the entry of the oldest record. The accumulated rainfall of the station X and the accumulated values of the average of group of base stations are determined starting from the latest record and plotted for various consecutive time periods (Fig. 15.6). A change in the slope of the double mass curve (if it persists for more than 5 years) indicates change in the rainfall regime of the station X. The rainfall values of station X beyond the point of change of slope is corrected to make the record compatible with the new regime of the station by applying the following correction factor:

$$P_C = P_o(S_c/S_o)$$

where, P_C is corrected rainfall at station X at any time t, P_o is original recorded rainfall at station X at time period t, S_c is corrected slope of double mass curve, and S_o is original slope of the double mass curve.

Fig. 15.6 shows change in regime in the year 1980. Correction factor 80/110 = 0.7273 is to be applied to records prior to 1980.

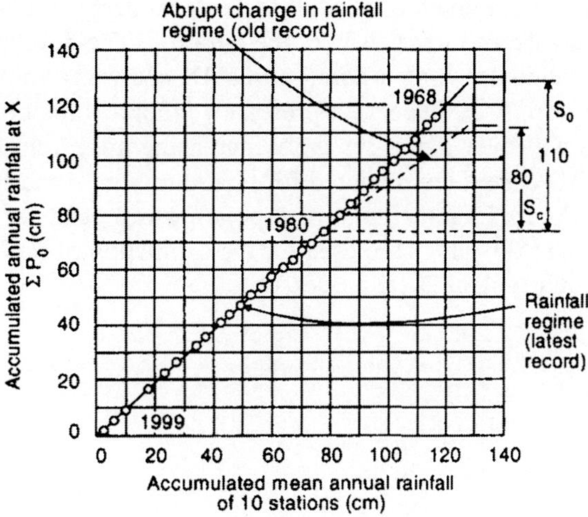

Fig. 15.6 Double mass curve

15.12 RAINFALL FREQUENCY ANALYSIS

Rainfall frequency analysis is defined as the rainfall which can be expected to be equalled or exceeded on the average once every N-years, the N-year rainfall. *Exeedence interval* is the average number of years between the occurrence of an event and a greater event. The *recurrence interval* is the average number of years within which a given rainfall event will be equalled or exceeded. Frequency of occurrence of maximum and minimum rainfall is determined to predict the future rainfall trends for the design of river protection works, drainage works, etc.

Various methods for determining recurrence interval are:

1. California method (1923): $T = N/m$
2. Hazen's method (1930): $T = 2N/2m-1 = N/m - 0.5$
3. Weibull method, $T = (N + 1)/m$
4. Gumbel's method, $T = \dfrac{N}{m+C-1}$

where, T = recurrence interval, N = number of years of record, m = serial number of a particular event arranged in descending order so that maximum rainfall is at top to find out maximum rainfall value and in ascending order to determine minimum rainfall value, and C = Gumbel's correction, values are as under:

m/N	1.0	0.9	0.8	0.7	0.6	0.5	0.4	0.3	0.2	0.1	0.08	0.04
C	1	0.95	0.88	0.845	0.78	0.73	0.66	0.59	0.52	0.40	0.38	0.28

California method, also expressed as $a = y/b$, where a = recurrence interval, y = years of rainfall record, and b is ranking of rainfall, has the merit of simplicity. It is a conservative method from the point of view of safety, but the method is not applicable if it is required to determine rainfall having recurrence interval of more than the total number of years of which the records are available. Hazen's method is widely used for giving most probable recurrence interval. Hazen's formula assigns a recurrence interval of $2N$ to the maximum event on record. Since the actual frequency of occurrence of the maximum event in N-years is not known, it is assumed to be $2N$. But no statistical method can lengthen the period of record. Gumbel's method has the best theoretical basis and is the latest best contribution to the theory of frequency analysis.

Depth-Area-Duration Curves

The estimation of severe floods for the hydrological design of dams require the determination of the maximum amount (depth) of rainfall for various durations over areas of various sizes.

Maximum depth of rainfall for a storm decreases as the area increases; for a given area the maximum depth increases with the duration. The development of relationship for the three parameters is called depth-area-duration (DAD) analysis to prepare DAD curves for the region under consideration. A DAD curve represents a graphical relation between a progressively decreasing depth of rainfall over a progressively increasing area from the centre of storm outward to its edges. The DAD curves are prepared for total duration of the storm or selected durations of 6, 12, 18, 24 and 36 hours incremental; corresponding isohyetel for these time intervals being available. Maximised DAD curves are used for the estimation of probable maximum precipitation (PMP). Fig. 15.7 shows typical DAD curves.

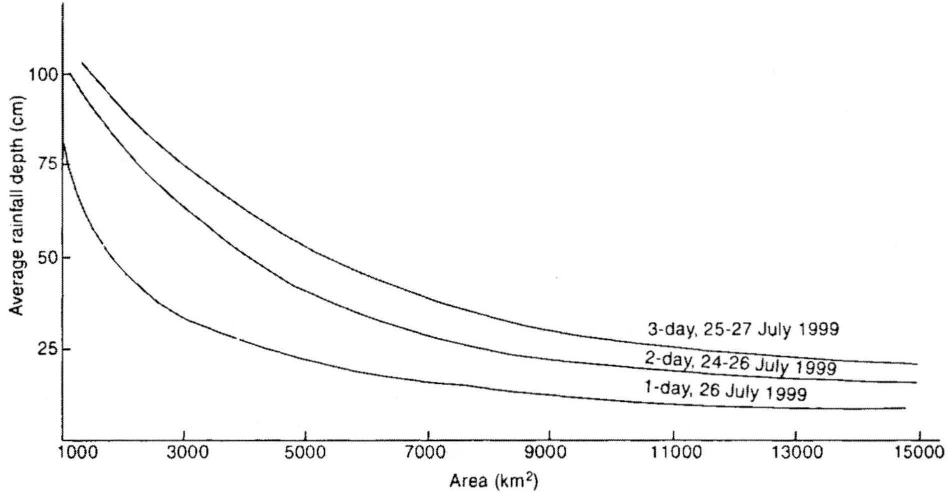

Fig. 15.7 Typical depth-duration-area curves.

15.13 RAINFALL INTENSITY-DURATION RELATIONSHIP

Rainfall intensity is defined as the rate at which rainfall occurs, expressed in depth units per unit time, usually mm/hr or cm/hr. It is the ratio of amount of rainfall occurred in unit duration of the rainfall. The rainfall intensity of 2.5 mm/hr is termed as light intensity, 2.5 to 7.5 mm/hr as moderate intensity and over 7.5 mm/hr as of heavy intensity. The greater the intensity of rainfall the higher is the runoff and vice versa.

The rainfall intensity is usually the average intensity over the rainfall duration. In a small period, suitably selected within a total storm duration, the rainfall intensity can be much higher than the average intensity during the total duration of the storm. Thus greater the intensity of rainfall, shorter is the duration for which rain continues that is the intensity is an inverse function of its duration. Fig. 15.7 shows a curve of maximum rainfall intensities for various durations. The rainfall intensity with time is graphically indicated in a hyetograph in the form of bar graph (Fig. 15.8). It is useful in determining maximum intensities of rainfall during a storm. The values for rainfall intensity are read from intensity-duration-frequency curves using rainfall duration equal to the time of concentration. The rainfall intensity is used in the design of soil and water conservation works and preparation of hydrographs of maximum probable and maximum possible flood flows in the design of spillway capacity.

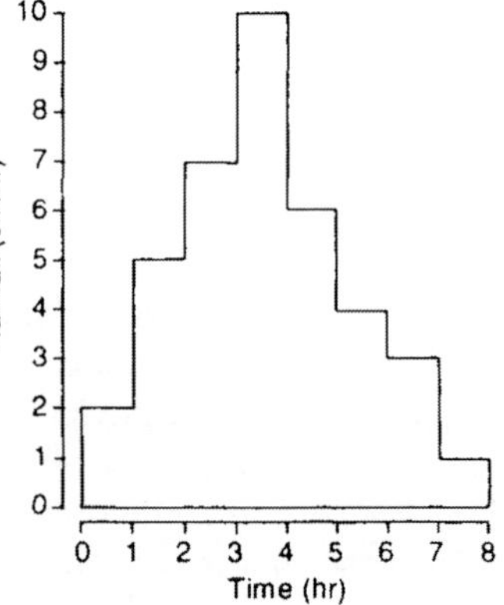

Fig. 15.8 Hyetograph

15.14 RUNOFF AND FACTORS EFFECTING IT

When rain falls, before runoff can occur, the precipitation must satisfy the demands of evaporation, interception, infiltration, surface storage, surface detention and channel detention. Runoff is thus the flow collected from the drainage basin and appearing at an outlet of the basin. It includes surface runoff received into the streams after rainfall as perennial or intermittent flow, delayed runoff that enters the streams after passing through portion of the earth, and other delayed runoff that has been temporarily detained as snow cover or stored in natural lakes or swamps. The various factors effecting the runoff are discussed as under:

1. Precipitation Characteristics

(*i*) *Type of precipitation*: Precipitation may be rainfall, snowfall or both. Rainfall immediately produces runoff. Snowfall is a delayed runoff temporarily detained as snow cover.

(*ii*) *Rainfall intensity*: Rainfall intensity, *i.e.* the amount of rain occurring in an unit interval of time, generally expressed in mm per hour, has pronounced effect on the runoff. Rainfall intensity influences both the rate and volume of runoff. Higher the rainfall intensity, greater is the runoff.

(*iii*) *Frequency of intense rainfall*: More frequent the intense rainfall, higher is the runoff.

(*iv*) *Duration of rainfall*: Storm of short duration may produce no runoff, whereas a storm of the same intensity but of long duration results in runoff. More the duration of rainfall, more prolonged is the runoff, as the soil gets saturated and consequently reduced infiltration loss and humid atmosphere slows evaporation loss.

(*v*) *Distribution of rainfall*: Distribution of rainfall over the basin has great effect on runoff. When rainfall is concentrated over the basin and the entire watershed contributes, the runoff is maximum. An intense storm on one portion of the catchment results in higher runoff than a moderate storm over the entire watershed.

(*vi*) *Direction of storm movement*: A storm moving in the direction of stream flow results in shorter period of runoff and higher peak than a storm moving upstream.

(*vii*) *Snow*: Snow retards runoff from rainfall and also tends to store a part of such rainfall. Snow accumulates on land surface till rise in temperature melts it to produce runoff.

2. Drainage Basin Characteristics

(*i*) *Size of catchment*: A small area tends to have higher discharge per square kilometre than a large area. Peak flow decreases as the area increases because intense storm is usually over a small area and period of surface runoff increases with area. The square, broad and short watersheds (Fig. 15.9) have almost symmetrical drainage pattern but shorter distance to the outlet.

(*ii*) *Shape of catchment*: Runoff is influenced by the shape of catchment area (Fig. 15.9). Long narrow watersheds are likely to have lower runoff rates than compact watersheds of the same size because runoff from the former does not concentrate as quickly as it does from the latter and long catchments are less likely to be covered by intense storm. Fan-shaped catchments (Fig. 15.9) are indicative of higher runoff because all the tributaries are approximately of the same size and peak flow from the tributaries tends to reach the main stream roughly at the same time.

Fern-leaf catchment gives lesser runoff because the tributaries are generally of different lengths and meet main stream at regular intervals thereby the discharges are likely to be distributed over a large period of time.

(*iii*) *Slope of catchment*: Catchment areas with flat slope and without surface outlets have lower runoff than areas with steep slope and well defined drainage pattern. Runoff from a long slope is generally slower but lasts longer after the rainfall ceases. Effect of slope is less pronounced during the period of prolonged and intense rainfall when runoff usually becomes constant and reverse is the case when rainfall is of short duration.

(*iv*) *Orientation of catchment*: Orientation, *i.e.* general direction of slope of a catchment, with respect to prevailing winds, storm movement and sun's position has considerable effect on the runoff.

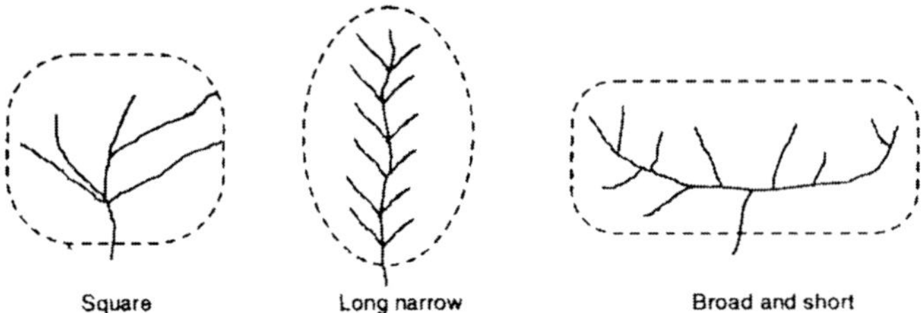

Fig. 15.9 Shapes of catchment.

3. Geological Characteristics

Type of soil and its permeability greatly influence runoff. Runoff is less if soil and subsoil are pervious as seepage is more. Runoff is more from rocky or impervious soil because absorption is least. Fault zones, fissures and cracks result in rain water finding its way through these openings and surface runoff is less. Seepage water may flow to the stream during low flow if it has outlet within the stream catchment.

4. Meteorological Characteristics

Temperature, humidity and wind effect runoff. High temperature, low humidity and strong wind result in higher evaporation loss and reduced runoff.

5. Geographical Characteristics

(*i*) *Altitude*: Higher catchments usually have steep slope and are less pervious and as such produce greater total runoff though dry season runoff is much lower due to less ground storage. Snow at higher altitudes tends to retard runoff from rainfall and store a portion of such rainfall.

(*ii*) *Drainage net*: Drainage net has an important effect on runoff. More dense and efficient the drainage network, the more flashy is the stream flow and vice versa. Catchments without surface outlets have lower runoff than areas with steep, well defined drainage patterns.

(*iii*) *Nature of catchment surface*: The vegetation and cultivation greatly influence runoff. Thick vegetation and well maintained soil reduce runoff to some extent. Effect of vegetation on low water flow is relatively small compared to that of geological and topographical characteristics of the catchment. Cultivation results in decreased surface runoff.

6. Storage Characteristics

Lakes, reservoirs, check dams, swamps, etc. act as flood moderators and reduce peak flow. They also tend to increase water losses due to evaporation.

7. Other Factors

These include urbanisation. Urbanisation on individual catchment basis leads to an increase in the area of impervious surfaces and alteration of natural drainage patterns resulting in increased volumes and rates of water runoff, increased peak flood over a shorter period of time and consequent high potential for flooding.

15.15 ESTIMATION OF RUNOFF

Various methods of estimation of runoff from a catchment are as under:

1. Rainfall-Runoff Coefficient

Runoff coefficient is the ratio of volume of stream flow above base flow to the volume of rainfall. It is expressed as $R = KP$.

Where, R is runoff (cm), P is rainfall (cm), K is runoff coefficient; usual values of K are as follow:

Runoff coefficient (IS: 8835–1978)

Type of soil	*Runoff coefficient*	*Type of soil*	*Runoff coefficient*
Plateau lightly covered	0.70	Loam, largely cultivated and suburbs with gardens, lawns, macademised roads	0.30
Clayey soils stiff and bare and clayey soils lightly covered	0.55	Sandy soils, light growth	0.20
Loam, lightly cultivated or covered	0.40	Parks, lawns, meadows, gardens, cultivated area	0.05-0.20

2. Runoff Percentage

(*i*) *Binnie's percentage*: Alexander Binnie's runoff percentages, based on data of two rivers in Madhya Pradesh, are as under:

Annual rainfall (cm)	*Runoff (%)*	*Rainfall (cm)*	*Runoff (%)*
50	15	90	35
60	21	100	38
70	25	110	40
80	29		

(*ii*) *Barlow's percentage*: T.G. Barlow's runoff percentages, based on studies of catchments mostly under 130 km^2 in Uttar Pradesh, are as under:

Class	*Description*	*Runoff (%)*
A	Flat, cultivated and black cotton soils	10
B	Flat, partly cultivated various soils	15
C	Average	20
D	Hills and plains with little cultivation	35
E	Very hilly and steep, with hardly any cultivation	45

These percentages are for average type of monsoon and are modified by the application of following coefficients according to the nature of season:

Nature of season	*Class of catchment*				
	A	*B*	*C*	*D*	*E*
Light rain, no heavy downpour	0.70	0.80	0.80	0.80	0.80
Average or varying rainfall, no continuous downpour	1.0	1.0	1.0	1.0	1.0
Continuous downpour	1.50	1.50	1.60	1.70	1.80

3. Tables and Curves

1. *Strange's Tables and Curves*: Strange gave tables and curves, based on data of Maharashtra, between rainfall and runoff. Strange's table and curves give runoff for daily rainfall for good, average and bad catchments and surface conditions, viz., dry, damp and wet prior to rain (Table 15.2).

Table 15.2 Daily runoff and yield according to Strange runoff percentage and yield

Daily rainfall (mm)	*Percentage*	*Original stage of ground surface*				
		Dry	*Damp*		*Wet*	
		Yield (mm)	*Percentage*	*Yield (mm)*	*Percentage*	*Yield (mm)*
5	—	—	4	0.2	7	0.35
10	1	0.10	5	0.5	10	1.0
20	2	0.40	9	1.8	15	3.0
25	3	0.75	11	2.75	18	4.5
30	4	1.20	13	3.9	20	6.0
40	7	2.8	18	7.2	28	11.20
50	10	5.0	22	11.0	34	17.0
60	14	8.46	28	16.8	41	24.6
75	20	15.0	37	27.75	55	41.25
100	30	30.0	50	50.0	70	70.0

For good or bad catchments, add or deduct up to 25 per cent of yield. Fig. 15.10 depicts the Strange's curve.

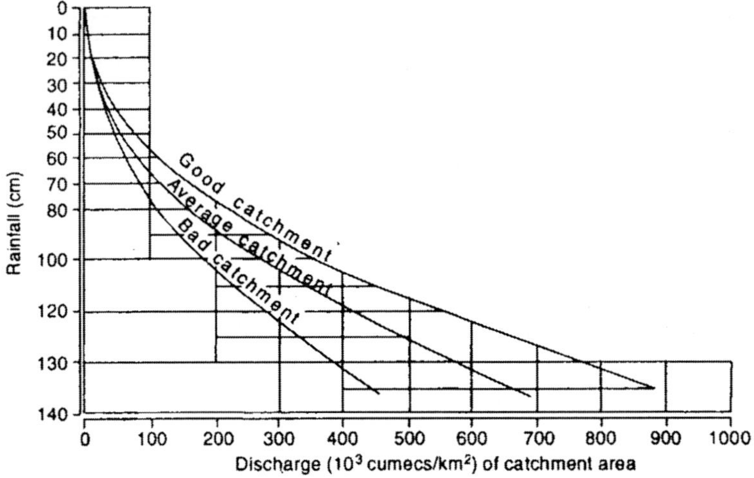

Fig. 15.10 Strange's curve

4. Empirical Formule

1. *Lacey's Formula*

$$R = \frac{P}{1 + \frac{304.8}{P}\left(\frac{F}{S}\right)}$$

where, R is runoff, P is rainfall, F is monsoon duration factor varying between 0.5 – 1.5 and S is catchment factor depending upon slope; value varies from 0.25 for flat cultivated areas to 3.45 for very hilly areas.

2. *Khosla's formula (1960)*

$$R = (P - T/3.74)$$

where, T is temperature (°C), P is rainfall (cm) and R is runoff (cm).

The formula gives fairly good results for the annual yield for preliminary studies.

3. *Inglis and Desouza's formula (1929)*

$$R = 0.85\,P - 30.5 \text{ for ghat areas of western India}$$

$$R = \frac{(P - 17.8)\,P}{254} \text{ for Deccan plateau.}$$

5. Infiltration Method

It is the most reliable method of estimation of runoff. Infiltration is defined as the entrance of water into the soil through the interstices or pores of the soil. Infiltration occurs both prior to and during the occurrence of surface runoff but decreases with time until a minimum rate of infiltration is reached. In rainfall, infiltration begins at high rate and decreases to a minimum rate as rain continues. At any instant, infiltration capacity of a soil is the maximum rate at which water enters the specified soil in given condition. The rate at which infiltration actually takes place during a rainfall, expressed in depth of water per unit of time, usually in mm per hour, is known as *infiltration rate.* It is equal to the infiltration capacity or the rainfall rate whichever is less. The limiting value of infiltration capacity is controlled by the soil permeability and can be determined experimentally by subjecting an experimental plot to rainfall rate in excess of infiltration capacity and by measuring surface runoff (Fig. 15.11). Infiltration capacity curve (Fig. 15.12) is the curve showing the infiltration rate at any period during a specified storm if the rainfall intensity were to equal or exceed the capacity at that instant.

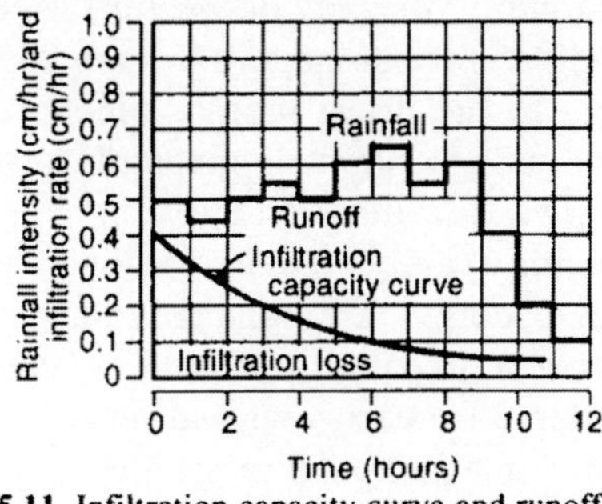

Fig. 15.11 Infiltration capacity curve and runoff

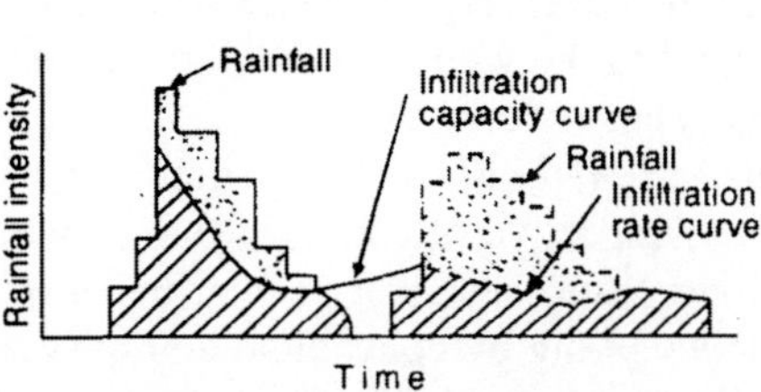

Fig. 15.12 Typical infiltration rate and infiltration capacity curve.

Infiltration index is the average rate of loss such that the volume of rainfall in excess of that rate will be equal to the direct runoff. Runoff from large area is determined by using infiltration indices. Three common types of infiltration indices are:

1. ϕ*-index*: The average rainfall intensity above which the volume of rainfall equals the volume of observed runoff and equals total basin recharge divided by duration of rainfall provided rainfall intensity is continuously in excess of basin recharge.

The index (Fig. 15.13) is termed as ϕ-index:

$$\phi = \frac{\text{Basin recharge}}{\text{Rainfall duration}}$$

Fig. 15.13 Typical ϕ-index and runoff.

2. *W-index*: *W*-index is the average rate of infiltration during the time the rainfall intensity exceeds the infiltration capacity, that is

$$W = (P - R)/t_r \text{ cm/hr}$$

where, t_r is duration of rainfall (hr), P is precipitation (cm), R is runoff (cm). Essentially, *W*-index is equivalent to ϕ-index minus the average rate of retention or depression storage.

3. W_{min}*-index*: With very wet conditions, when the retention rate is at minimum, the values of *W*-index and ϕ-index are almost identical and *W*-index is termed as W_{min}-index.

The infiltration index can be used to estimate runoff coefficient K from the following relation:

$$K = (P - W)/P$$

where, P = rainfall intensity (cm/hr).

The use of infiltration indices does not constitute a rational application of infiltration theory and results obtained are considered empirical.

15.16 FLOW DURATION CURVE

The flow duration curve is a very useful tool for studying the stream flow variability for the water power development. The stream flow data may be daily, ten daily or monthly values. A flow duration curve of average daily discharge gives more accurate results than those obtained from monthly average values. The flow duration curve is a plot of stream flow (Q) in ascending order of magnitude as ordinate against the percent of time the flow was equalled or exceeded as abscissa. The exceedence probability P_r is worked out. The availability of power from a hydro plant can also be represented by a power duration curve exactly on the lines as flow duration curve, the discharge ordinate in this case is replaced by horse power per metre of head or for a given head to give power from the plant.

The flow duration curve is developed from as long a stream flow data as is available or by considering only the year which has the lowest flow rate out of the entire flow record. The two methods used for drawing flow duration curve with entire stream flow record are as under:

(*i*) *Calendar year Method*: In this method the mean monthly (or daily or weekly) values of flow rates of each year are first separately arranged is ascending order of magnitude and tabulated in that order. Then the average flow values corresponding to the driest month, second driest month and so on up to the wettest month are determined by taking arithmetic mean of all values of the same rank. These average values are then plotted to give flow duration curve.

(*ii*) *Total period Method*: In this method, the mean monthly (daily or weekly) flow rates for the entire period irrespective of calendar year in which they occurred, are arranged in ascending order of magnitude. With these values the flow duration curve is prepared. Alternatively continuous ranges of flow varying by convenient amount are considered for each year and separately the number of days during which the daily stream flow is recorded in a particular range are noted. The number of days, for all the years, are considered and those having the daily flow in the same range are added and percent of time the daily flow occurred in a particular range is determined. The lower value of the range of flow rate is then plotted against the corresponding percent of time to give flow duration curve. The total period method gives more correct results than calendar year method which averages out extreme values.

15.17 FLOOD ESTIMATION

Flood is defined as the flow pattern in a stream constituting a distinct progressive rise culminating in a peak or summit together with the recession that follows crest. Flood is the maximum rate of runoff that is unusually high stage of the river. The highest peak discharge in a water year is termed as annual flood.

Flood estimation is essential for the design of irrigation, hydel power and flood control projects. On proper selection of the design flood depends the safety and cost of such structures. An excessive flood implies unnecessarily costly structures and too small a design flood, on the other hand, involves a high risk not only of the total failure of the structure and services rendered by it, but also safety of persons and property located downstream. In this context certain definitions of floods are as follows:

Design flood: It is the flood adopted for design purposes. It may be the probable maximum flood or the standard project flood or a flood corresponding to some desired frequency of occurrence or historical or any other flood depending on the standard of security to be provided consistent with its purpose.

Flood peak: The highest value of the stage or discharge attained during a flood.

Flash flood: It is a flood of short duration and abrupt rise with a relatively high peak rate of flow, usually resulting from a high intensity of rainfall.

Flood frequency: Flood frequency is defined as the number of time a flood above a given magnitude is likely to be equalled or exceeded in given number of years on the average.

Standard project flood: It is the flood discharge that may be expected from the most severe combination of meteorological and hydrological factors that are considered reasonably characteristic of the region and is computed by using the standard project storm.

Maximum intensity of flood: Also called momentary flood peak. It is maximum instantaneous rate of flow during a flood.

Maximum known flood: The highest flood which has occurred within the living memory.

Probable Maximum flood: (*PMF*) It is the flood discharge that may be expected from the most severe combination of critical meteorological and hydrological conditions that are reasonably possible in the region and is computed by using the probable maximum storm which is an estimate of the physical upper limit to maximum precipitation for the basin. It is used in the design of major hydraulic structures for dams for which virtually complete security from potential flood is sought.

Extraordinary flood: A flood with magnitude much higher/larger than the design flood.

Return period flood: A flood with a return period of 'T' year ('T' year flood) is defined as a flood that is expected, on the average, to be equalled or exceeded once, in 'T' years. Thus the probability of occurrence of a flood equalling or exceeding the 'T' year flood is $1/T$.

Methods of Flood Estimation:

Various methods of estimation of flood discharge are as under:

1. Factor of Safety to Maximum Historical Flood

It is the earliest method of estimating flood discharge, obtained by application of a suitable factor of safety, usually 1.5, to the observed maximum flood at site or at nearby site on the stream. An extension of the method is that from high flood marks the cross sectional area of the river is computed, water surface slope determined and flood discharge calculated by slope area method. Coefficient of rugosity is assumed or computed by substituting data from a few measured floods.

2. Rational Method (1989)

Rational method correlates runoff with rainfall, as follows:

$$Q = 2.78\ CIA$$

where, Q is flood flow (cumecs), C is runoff coefficient (Table 15.3), I is average intensity of rainfall (cm/hr) lasting for a critical period of time t_c (t_c is time of concentration) 0.00032LS, where, L is maximum length of travel of water (m), S is slope equal to H/L, where H is difference in elevation (m) between the remotest point on the basin and the outlet and A is drainage area (sq. km).

It is assumed that the rainfall is uniform and constant over the catchment area and the slope is also uniform. In this method the catchment area is divided into a number of subareas and with known time of concentration of different subareas runoff contribution from each area is computed. This method is applicable to small catchments.

Table 15.3 Values of coefficient, C (Richards)

Type of catchment	*C*	*Type of catchment*	*C*
Rocky and impermeable	0.8 – 1.0	Cultivated absorbent soil	0.3 – 0.4
Slightly impermeable, bare	0.6 – 0.8	Sandy soil	0.2 – 0.3
Cultivated or covered with vegetation	0.4 – 0.6	Heavy forest	0.1 – 0.3

3. Empirical Formulae

Dicken's formula (1865), $Q = CA^{3/4}$

where, Q = discharge (cumecs), A = catchment area (km^2), C = coefficient; values are northern India plains = 6, northern India hilly = 11-14, central India = 14 – 28, western India = 22 – 25, Maximum value = 35, Actual observed value in Northern India = 22. Dicken's formula is used in central and northern India.

Ryve's formula (1884), $Q = CA^{2/3}$

where, C = 6.8 within 80 km of east coast, 8.5 within 80-160 km of east coast, 10.2 limited area near hills, and 40 actual observed value.

The formula is used in Tamil Nadu and parts of Andhra Pradesh and Karnataka states.

Inglis's formula, $Q = \dfrac{124\,A}{\sqrt{A+10.4}} \simeq 124\,A^{1/2}$

The formula is used in Maharashtra state for small catchments.

Ali Nawaz Jang Bahadur's formula, $Q = C(0.3906A)^{[0.925\,-\,(1/14)\log 0.3906\,A]}$

where, C = coefficient, values, north India = 60, South India = 48 and Maximum value = 86.

The formula gives better results than Dicken's and Ryve's formulae for areas below 25,000 sq km.

Fuller's formula: $Q_{max} = CA^{0.8}\,(1 + 0.8 \log T)\,(1 + 2.67\,A^{-0.3})$

where, Q_{max} = maximum flood (cumecs) having recurrence interval T years, T = number of years in which maximum flood is considered, A = area of drainage basin (sq km), and C = constant, value varying from 0.026 to 2.77.

Creager's method: Creager's flood frequency method is used in case sufficient data are not available

$$Q = CA^{0.5}\left[2-1.374\left(e^{-0.04A^{-0.3}}\right)\left\{1-\frac{\log_e^{0.1T}}{3}\right\}+\frac{\log_e^{0.1T}}{3}\right]$$

where, Q = discharge (cumecs) for a period of T years, A = area (sq km), and C = constant of value 1405 for areas most favourable to large floods.

Percentage increase method: Central Water and Power Commission formulae for obtaining percentage increase (P) to the maximum known flood in a record of N-years, to evaluate 100 years flood are:

For North and Central Indian rivers $P = 50 - 0.4\,N$

For South Indian rivers $P = 50 - 0.45\,N$

where, P = percentage increase to be made to obtain design flood, and N = number of years for which known flood is the maximum (minimum 25 to 30 years).

The empirical formulae have regional application and do not give clue to the probable frequency at the computed flood figure.

4. Envelope Curve

Creager Envelope Equation: Justin, Creager and Hinds obtained following equation, for the envelope curve of maximum floods:

$$Q = (A/2.95)^{0.894\,(A/2.95)^{-0.048}}$$

Kanwar Sain envelope curve: Fig. 15.14 indicates envelope curves for South, North and Central India rivers.

The method is better than empirical formulae in that selection of coefficient based on judgement is eliminated. The order of frequency of floods derived is somewhat greater than the longest period of record.

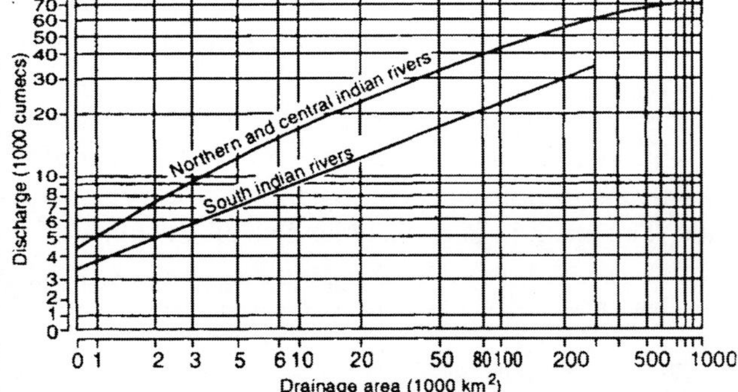

Fig. 15.14 Envelope curve

5. Probability or Flood Frequency Method

The probability method requires data for fairly long period; minimum of 30 years. In this method, record of peak flood discharges are interpreted in terms of future probabilities of occurrence to determine the recurrence interval of floods of given magnitude. The recurrence interval, also called return period, is the average interval of time within which the magnitude X of a flood event will be equalled or exceeded once. The recurrence interval can be determined, if a particular flood discharge has a serial number m, by the following methods:

1. *California method (1923)*: $T = N/m$
2. *Hazen's method (1930)*: $T = 2N/2m-1$
3. *Weibull method (1839)*: $T = (N+1)/m$
4. *Gumbel's method*: $T = \dfrac{N}{m+C-1}$

where, C = Gumbel's correction. It depends on m/N ratio, given in Table 15.4, T = recurrence interval (years), n = number of events *i.e.*, years of record, and m = order of rank of event, highest flood (m = 1) and $m = n$ for the lowest flood events.

Table 15.4 Gumbel's C

m/N	1.0	0.9	0.8	0.7	0.6	0.5	0.4	0.3	0.2	0.1	0.08	0.04
c	1.0	0.95	0.88	0.845	0.78	0.73	0.66	0.59	0.52	0.40	0.38	0.28

In order to find any design flood of desired frequency a plot of Q vs T (or F) is prepared on a probability graph. Unknown flood for any value of T is determined from the plot.

Let $Q_1, Q_2, ... Q_n$ comprise a series of extreme values of floods, the probability of occurrence of a value equal to or less than Q is given by the relation,

$$P = e^{-z} \quad ...(i)$$

where,

$$Z = e^{-y} \quad ...(ii)$$

such that

$$Y = a(Q - Q_f) \quad ...(iii)$$

where,

Q_f = value of flood discharge having highest frequency

$= Q_{av} - 0.45\,\sigma x$

Q_{av} = average value of discharge

$$\sigma x = \sqrt{\frac{(Q_1 - Q_{av})^2 + ... + (Q_n - Q_{av})^2}{N}}$$

$$a = \text{constant} = 1.28/\sigma x$$

Recurrence interval of Q in years, $T = 1/(1 - P)$

For given data, find σx, Q_{av} and calculate Q_f.

From Eqs. (*i*), (*ii*) and (*iii*) plot P and T for various values of y.

Then, for any desired value of T, Q is found out by extrapolation.

7. Flood Hydrograph

It is a better method for flood estimation, discussed hereinafter:

15.18 HYDROGRAPH

Discharge hydrograph is a graphical representation of the discharge versus time at a particular point of a stream. It shows the time distribution of total runoff at the point of measurement. Fig. 15.15 shows a typical single peaked simple hydrograph resulting from an isolated storm. A hydrograph consists of:

(*a*) *A rising limb or concentration curve*: It corresponds to the rising flood. It is concave upwards rising slowly in the early stages of flood but more rapidly towards the end of the rise.

(*b*) *A peak or crest segment*: It is the highest instantaneous value of the hydrograph and extends from the point of inflection on the rising limb to a similar inflection point on the falling limb. The peak occurs at the end of the rain at some time fixed by the characteristics of the catchment, if rainfall is uniform, while under non-uniform rate of rainfall with the highest rates early in the storm, the peak may occur before the end of rain. Hydrograph of some catchments from a single, relatively short period of rainfall have two or more peaks which may merge in rainfall of long duration. Multi-peaked hydrographs may result from rainfall non-uniform in time.

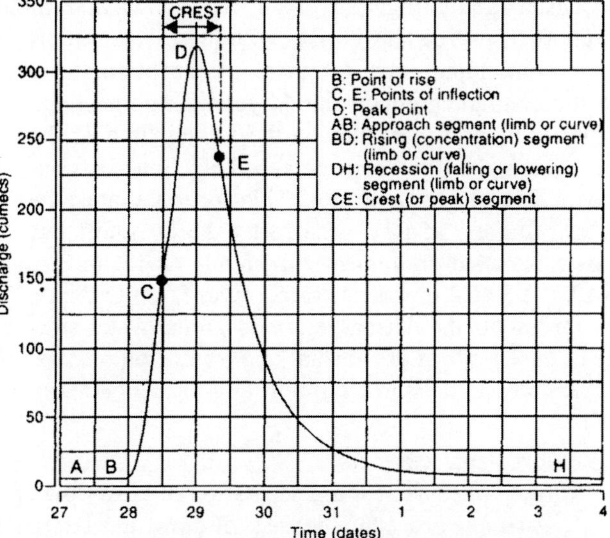

Fig. 15.15 Typical single peaked hydrograph.

(*c*) *Falling limb or recession curve*: It is the descending portion of a hydrograph. It is the falling limb beyond the point of contraflexure of a hydrograph after a flood event. The point of inflection on the falling limb is assumed to mark the time at which the surface inflow to the channel system ceases. The recession curve is convex and is reverse of that of the rising limb.

Components of Hydrograph

A hydrograph represents following three types of flow:

1. *Surface runoff*: It represents the rainfall which flows overland and channel flows after satisfying all the losses, *e.g.* evapotranspiration, infiltration and filling up the depressions, etc. It includes channel precipitation representing the precipitation which falls directly on the water surface of the stream and is the first water from a storm to reach a stream. Most of the surface runoff is discharged into the stream during the rainfall and is a major contributor to flood peak.

2. *Subsurface flow*: It is the runoff that moves through upper soil layers and returns to surface or appears in streams without entering the water table in the zone of saturation. It thus represents that portion of rainfall which has not passed down the water table, but is discharged from the area as subsurface flow into the stream. Unlike surface runoff, subsurface flow reaches the stream later than cessation of rainfall and spreads over a long period of time.

3. *Groundwater flow*: It represents that part of the runoff which consists of water that passes into the earth and enters the zone of saturation and is later discharged into a spring or seepage water. When water level in river rises, the flow spreads on the banks and seeps down to be retained as bank storage (Fig. 15.16). When the river falls, the bank storage flows back into the river thereby contributing flow under the period of no rainfall (Fig. 15.17). Thus through bank storage surface water becomes groundwater.

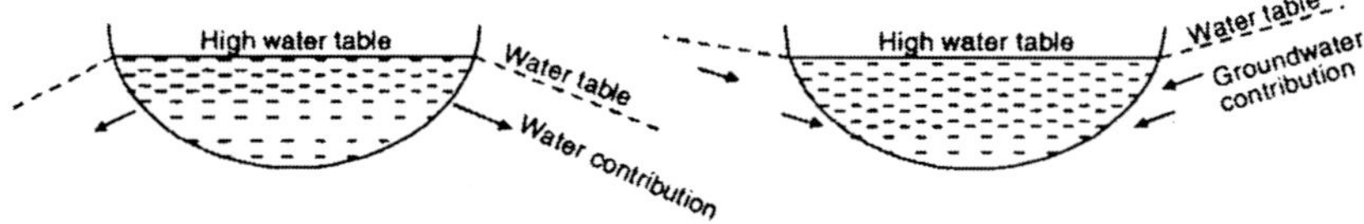

Fig. 15.16 River contributing to bank storage. **Fig. 15.17** Bank storage contributing to river flow.

Discharge hydrograph (Fig. 15.18) shows that initially there is contribution of groundwater to runoff. The stream level rises with increase in runoff due to rainfall in the area and there is contribution from the stream to the groundwater. Beyond the point B, the point of sharpest curvature on the falling limb of the hydrograph, the flow represents the base flow, *i.e.* continuous flow due to snow melt. This portion, that is, the recession after the flow created by direct runoff has ceased, is called *groundwater depletion curve.* In Fig. 15.18, (*a*) $EACB$ represents direct runoff, (*b*) Ordinate MN represents the discharge due to groundwater storage or base flow, at any instant and (*c*) LM represents the flow due to the surface runoff and influent stream both.

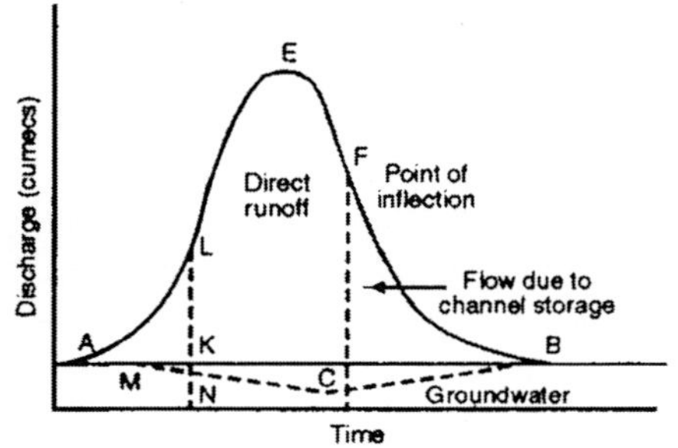

Fig. 15.18 Components of discharge hydrograph.

Base Flow Separation

Three procedures for the separation of base flow are as under:

1. A simple procedure consists of drawing a straight line from the point of rise to an arbitrary point on the lower portion of falling limb of the hydrograph line *bd* (Fig. 15.19). An arbitrary point is roughly located at a time $N = 0.84A^{0.2}$ days after the peak of hydrograph. A is the drainage area in sq km.

2. Another simple procedure is to extend the recession prior to storm to a point under a little beyond the peak of the hydrograph, line *bc* (Fig. 15.19). From this point, a straight line is drawn to the arbitrary point on the hydrograph determined as above.

3. The third procedure makes use of composite base flow recession curve. Curves aa' and bb' (Fig. 15.20) corresponding to recessions of lower end and approach segment of the hydrograph

are first drawn to match the composite curve. The connecting curve cc' is then drawn arbitrarily to join the two portions aa' and bb' smoothly in such a way that its crest point lies below the point of inflection.

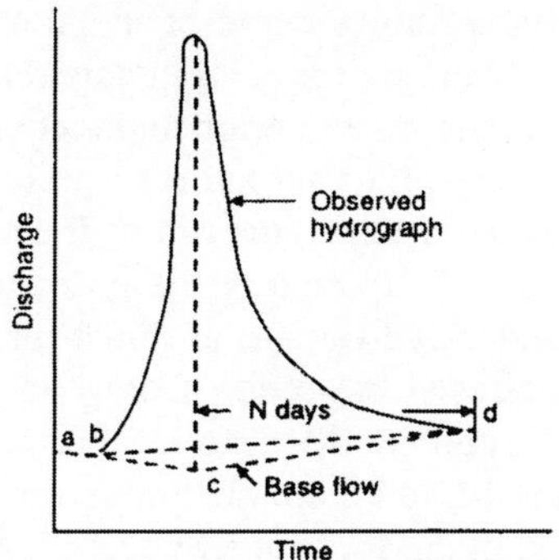

Fig. 15.19 Base flow separation by simple method.

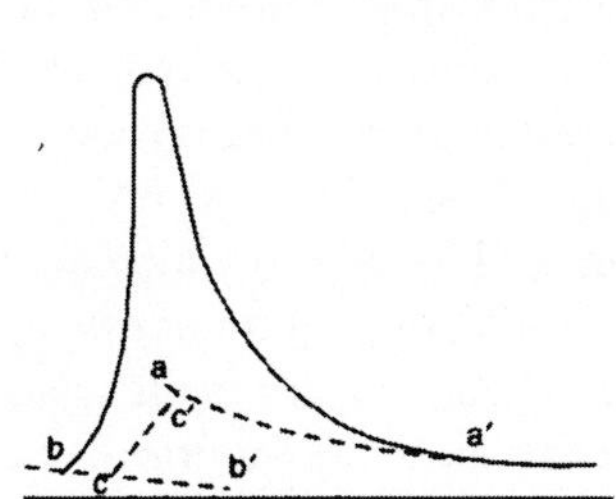

Fig. 15.20 Base flow separation by composite base flow recession curve.

15.19 UNIT HYDROGRAPH METHOD

Unit hydrograph method is widely used for flood estimation and flood prediction. Unit hydrograph of a drainage is a hydrograph of direct runoff resulting from 1 cm of effective rainfall of a unit duration generated uniformly over the basin area at a uniform rate. The unit rainfall duration refers to the smallest period of time or duration applicable to and commensurate with the catchment in which the rainfall can be assumed to be uniform. It should not exceed the period during which the design storm rainfall is uniform in intensity in various portions of the basin under study. The best unit duration is about one-fourth of the basin lag time, *i.e.*, the time from the middle of the block of excess rainfall to the peak of the observed hydrograph. It may vary from one day in the case of large catchments to a few hours for small catchments, but usually taken as 6, 12 or 24 hours. For example, a 6-hour hydrograph results from 6-hour unit rainfall duration, meaning thereby a discharge hydrograph produced by surface runoff from a storm of 6-hour duration and yielding a rainfall excess of 1 cm.

Principles: (*i*) Effective rainfall is uniformly distributed within its duration or specified time period, (*ii*) Effective rainfall is uniformly distributed throughout the entire area of the drainage basin, (*iii*) The base of all unit hydrographs (period of surface runoff) from all unit storms (of unit duration irrespective of intensity) is approximately the same, (*iv*) For all unit hydrographs, if the total period of surface runoff is divided into any number of equal time intervals, the percentage of total runoff that occurs during each of these periods is approximately the same, irrespective of the magnitude of total runoff or the coordinates of unit hydrograph are approximately proportional to the volume of surface runoff resulting from rainfall of unit duration, irrespective of the amount of such rainfall, and (*v*) The hydrograph of direct runoff from a basin due to a given period of effective rainfall reflects the combined physical characteristics of the basin (including infiltration, surface detention and storage).

The basic premise of unit hydrograph principle is that R cm effective rainfall over the basin in the same unit duration will produce a direct runoff hydrograph of the same time base as the unit hydrograph but having ordinates R-times those of the unit hydrograph. Thus a rainfall excess of 2 cm within the unit duration will produce a runoff hydrograph having ordinates twice as great as those of unit hydrograph.

1. Unit Hydrograph from Simple Hydrograph

The most direct method of deriving a unit hydrograph involves the analysis of a simple hydrograph resulting from an isolated unit storm that produces reasonably uniform effective rainfall rates within its duration. The procedure of unit hydrograph construction is (*i*) From the past discharge records, select an isolated, well defined and single peaked flood rises of short duration, (*ii*) Runoff volume for all such floods are computed roughly from the daily average flows and those having considerable volume are selected, (*iii*) The instantaneous hydrograph for each selected floods is plotted using the gauge records and suitable rating curve of the site, (*iv*) Each hydrograph is then analysed; various steps involved are (*a*) Analyse the rainfall data related to the hydrograph and prepare mass curves and the hyetograph for the storm period under consideration, (*b*) Separate the base flow from the stream discharge hydrograph to get hydrograph of the direct runoff. The observed hydrograph may have to be modified to eliminate the runoff from preceding or subsequent rainfall, (*c*) Find the ordinates of direct runoff subtracting the ordinates of base flow from the total ordinates, (*d*) Calculate the volume under hydrograph of direct runoff (in cm of depth over the basin) by planimeter or by counting squares (1 cm runoff from 1 sq km = 2.78 cumec hours), (*e*) Calculate the ordinates of unit hydrograph by dividing the ordinates of the direct runoff hydrograph by its volume as computed in step (*d*) to obtain the unit hydrograph, and (*f*) Compute the rainfall excess quantities and plot the data in the form of hyetograph. The period of excess rainfall as indicated by the hydrograph is the unit duration for the unit hydrograph derived in step (*e*). A rough idea of the period of excess rainfall can also be had from the inspection of mass rainfall curves.

As a general rule, it is desirable to derive unit hydrographs from effective rainfalls of periods equal to about one-fourth of the peaking time in order that the variations in rainfall are reflected by the hydrographs. If unit duration of the unit hydrograph derived above is appreciably different from the desired unit duration, the unit hydrograph may be adjusted to the desired duration by S-hydrograph method. A tolerance of 25% of the desired duration is ordinarily accepted without serious error.

2. Unit Hydrograph from Complex Hydrograph

Where simple hydrographs from isolated unit storm are not available for deriving unit hydrograph, a complex hydrograph (hydrographs) resulting from a storm of varying intensity is studied. In case of two peaked complex hydrograph resulting from sufficiently separated storms, the two storms are separated by the use of a direct runoff depletion curve and base flow separated. The unit hydrographs are then developed from each storm. However, in the case of extended period of rainfall wherein hydrograph cannot be separated into portions contributed by each rain burst, the complex hydrograph is analysed by the method of successive approximation or Collin's method*.

3. S-Hydrograph (S-Curve)

When a unit hydrograph of a desired unit duration cannot be developed by analysis of available data, it is derived by conversion of the available unit hydrograph. In general, if a unit hydrograph of a given unit duration is available, unit hydrograph of other durations can be derived by the application of the principle of superposition. If other durations are integral multiples of the given duration, the computations are relatively easy. For example, to derive a 6-hour unit hydrograph from a 2-hour unit hydrograph, three 2-hour unit hydrographs with their initial points separated by 2-hour are summed up to yield a hydrograph of 3 cm of runoff during a 6-hour period. If the ordinates of this summation graph are divided by 3, a 6-hour unit hydrograph is obtained. A general method based on the principle of superposition is the S-hydrograph method.

The S-hydrograph is a graph showing the summation of the ordinates of a series of unit hydrographs spaced at unit rainfall duration intervals. It represents the hydrographs of average rate of rainfall excess of the unit duration continued indefinitely. This method is quite suitable when uniform spells of rainfall have been recorded at greater time interval than the unit duration

* For details, see author's *Hydrology and Water* Resources Engineering.

applicable to the catchment. However, it is not so suitable for smaller catchment areas where rainfall is recorded at shorter intervals not exceeding the unit duration.

The S-hydrograph is essentially a hydrograph produced by a continuous effective rainfall at a constant rate for an indefinite period. It can be shown that the ordinates of the S-hydrograph ultimately approach the rate of effective rainfall at a time of equilibrium. If I cm per hour is the constant rate of continuous effective rainfall and A is the drainage area in sq km, the discharge at the time of equilibrium is equal to 2.78 AI cumecs. The S-hydrograph can be constructed graphically by summing up a series of identical unit hydrographs spaced at intervals equal to the duration of effective rainfall from which they were developed.

4. Synthetic Unit Hydrograph

Synthetic unit hydrograph is a unit graph developed on the basis of estimation of coefficients expressing various physical features of a catchment. Synthetic unit hydrograph is developed, based on known physical characteristics of the basin, where adequate rainfall, runoff data are lacking. Among several known methods of synthetic unit hydrographs, the one developed by Synder is most commonly used having the following relations:

Lag time, $t_p = C_t (L.\ L_c)^{0.3}$ (*i*)

Standard duration of net rain, $t_r = t_p/5.5$ (*ii*)

Peak flow, $Q_p = (2.78\ C_p.\ A)/t_p'$ (cumecs) (*iii*)

Time base, $T = 3 + 3\ (t_p/24)$ (days) (*iv*)

Peak flow/sq km, $q_p = Q_p/A$ (*v*)

For any duration (t_r') other than standard duration (t_r), the basin lag time

$$(t_p') = t_p + \frac{(t_r' - t_r)}{4}$$

$$= \frac{21}{24}t_p + \frac{t_r'}{4}$$

Time to the beginning of rising limb $= \dfrac{t_r'}{2} + t_p'$

where, t_p = lag time from mid-point of effective rainfall duration t_r to peak of a unit hydrograph (hours), t_r = standard duration of effective rainfall (hours), Q_p = peak discharge of unit hydrograph for standard duration t_r, C_p = coefficient; value ranging from 4 to 5, C_t = a coefficient value ranging from 0.2 to 2.2, L = basin length (km) measured along the water course from the basin upstream boundary to the gauging station, L_c = distance (km) along the main water course from the gauging station to a point opposite the centroid of basin, and T = base length of unit hydrograph for t_r in days.

For sketching the unit hydrograph, U.S. Army Corps of Engineers relations for widths of unit hydrograph at 50% and 75% of the peak discharge are:

$W_{50} = 5.6/q_p^{1.08}$ and $W_{75} = 3.21/q_p^{1.08}$, q is peak discharge (cumecs/km²).

15.20 STREAM GAUGING

Discharge, also termed as rate of flow, is defined as the volume of water flowing through a cross-section in a unit time such as cubic metres per second (cumecs). The objects of discharge measurement are accurate assessment of the available and dependable natural water resources of a river, their tributaries and streams for the development of irrigation and power potential, flood control, water supply, navigation, etc., and issue of flood warnings, regulation and equitable distribution of supplies in a canal system, determination of seasonal and annual variations in runoff and estimation of water losses and regeneration in different reaches of the river and planning for inland navigation.

The need for extensive collection of stream flow data on a scientific basis is well recognised. The World Meteorological Organisation has laid down norms, as given below, for the minimum network of stream gauging stations:

Terrain	*WMO norms (1 station for)*
Flat region	3000 – 10,000 km^2
Hilly region	1,000 – 5,000 km^2
Arid region	No norms

The stage or gauge is the elevation of water surface of the stream measured relative to a datum which can be mean sea-level on any arbitrary datum connected to the former.

15.21 SELECTION OF GAUGE AND DISCHARGE SITES

The water depth is read at gauge site and the rate of flow is determined at discharge site. The essential requirements of a gauge site are, (*i*) It offers a place where, once fixed, the gauge is permanent, *i.e.* immune from shifting due to silting or scouring, and (*ii*) Registers always the same elevation of water surface corresponding to the given discharge.

Types of Gauges

The gauge may be a direct or indirect. Direct gauges are staff gauges and indirect gauges are self-recorder type and crest stage type.

1. *Vertical staff gauge*: Vertical staff gauge is preferably an enamel plate gauge graduated to a millimetre, the figures being in English. The markings are clear and sufficiently accurate with bold line indication at every 20 mm interval. The graduations may be made either by etching or by painting or enamelling so as to read centimetres, decimetres and metres distinctively, the smallest graduation being a centimetre. The staff gauge is generally fixed at or near the edge of water in the river so as to be easily readable and accessible and at the same time not unduly exposed to the face of the current.

2. *Inclined gauge*: The inclined gauge shall fit closely and be solidly anchored to the slope of the natural bank of the river channel. It is calibrated on the site by precise levelling. Inclined gauge may be constructed to one continuous slope or may be a compound of two or more slopes. It is often convenient to construct a flight of steps alongside the inclined gauge to facilitate taking readings.

3. *Hook gauge*: A hook gauge is used in a vertical well, the top of which is so constructed as to permit accurate and easy reading relative to a cross bar or other form of reference datum.

4. *Temporary gauge*: The temporary gauge may be wooden or made of steel plate, but the latter is preferred as enamelled metal staffs give a much longer life and are available in various lengths of sections and graduated to desired precision. Temporary enamelled iron gauges 75 mm wide in sections of 2 m, secured and screwed to a wooden stack or sal balli are suitable where river has a shifting channel.

5. *Crest stage gauge*: It is a device to obtain the record of flood crests at sites where recording gauges are not installed. A variety of such gauges are used. It includes small floats which rise with the stage but are restrained at the maximum level and water soluble paints at bridge pier side shielded from rain. The portion of the paint coming in contact with water is washed out to indicate high water level.

Discharge Site:

Gauge reading and discharge observation are complementary operations and as such the site selected for the installation of the gauge shall preferably be suitable for discharge observations as well. If site conditions are such that the two sites have to be distinct, these shall be as close as possible. Besides, the two essential requirements indicated for gauge site, the discharge measurement site in plains shall meet the additional requirements such as (*i*) At or as close to the gauge site as possible in the case of velocity measurement by current meter or near the middle of a straight reach where velocity measurement is done by a float, (*ii*) Located in a straight reach of such length as is the least of 3 to 4 times the river width or 0.75 km during high floods, both upstream and downstream of the normal cross section where velocities are to be measured, (*iii*) The flow in the straight reach is streamlined, steady, free from cross currents, reverse slope and eddies and showing less departure from normal flow at all stages of the river, (*iv*) The water flow is preferably in a single channel. Two straight channels are, however, preferable to one defective channel, (*v*) Site is free from aggradation and degradation, (*vi*) Located where the greatest range of fluctuations in river stage occurs, (*vii*) Reasonably uniform cross section of the river channel in the selected reach, (*viii*) Site, as far as possible, is free from trees and obstruction as is likely to interfere with flow and clear vision during observation, (*ix*) Section line be as nearly as possible at a right angle to the direction of the flow, (*x*) Site is free from the tendency of formation of vortex or development of backward flow, (*xi*) Located preferably away from bridges or other hydraulic structures which are likely to affect the flow of water, (*xii*) Located neither on the upstream of a confluence nor in its close proximity, (*xiii*) Reasonably stable bed and banks for affording a regular gauge-discharge relationship, (*xiv*) The banks are high enough to contain HFL and have more or less regular side slope to permit fixation of sloping gauges to record both low supplies and high floods at the same location, (*xv*) Easily accessible at all times of the year.

Besides the above, the discharge measurement sites in the hills shall meet the requirements (*i*) Straight reach of minimum length of 150 m above section line, (*ii*) Free from projections in the bed and sides, (*iii*) Minimum 150 m removed from incoming torrent upstream and downstream from the section line, (*iv*) A bridge without piers or rope bridge, and (*v*) Nearest to a proposed dam site.

15.22 SEGMENTATION

The river cross section is divided into suitable compartments or vertical sections and total discharge is computed as the sum of the discharges in various sections on the basis of the depths and velocities observed at these verticals. Fixing the numbers and positions of verticals is termed as segmentation (Fig. 15.21), minimum 1/20 of water width and discharge between two verticals is not more than 10% of the total discharge. Table 15.5 indicates the intervals at which the depths of water are measured along the cross section on rivers of different widths as specified in IS: 1192-1981.

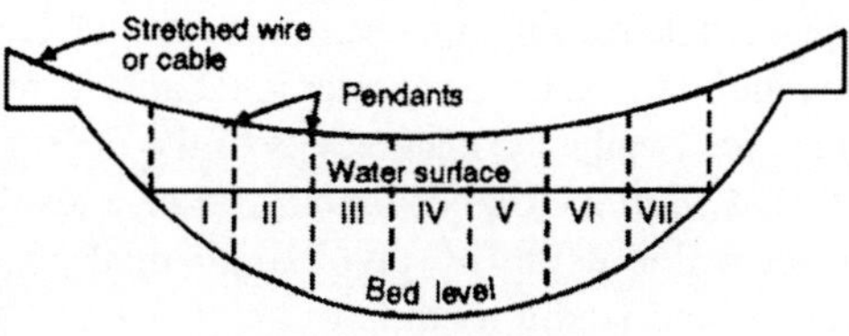

Fig. 15.21 Segmentation.

Table 15.5 Width of interval for measuring depths

Rivers channel width (m)	*Observation points (No)*	*Segment width (m)*
Less than 15 m or in channels where river bed changes abruptly	15	1.5
Between 15 to 90	15	6.0
Between 90 to 180	15	15.0
Over 180	25	Variable

The verticals for depth and velocity measurements are kept the same according to Table 15.5. In velocity measurement the maximum spacing between adjacent verticals is so maintained that the mean velocity does not differ by more than 20% with respect to the lower value of the two velocity measurements.

15.23 DEPTH MEASUREMENT

Direct Methods

1. *Wadding or suspension rod*: Wadding or suspension rods are used for the measurement of depths when the velocities in the river are of the order of 1 m/s and depths are wadable and small up to 1 m.

2. *Sounding rod*: When observations by wadding rods are found difficult, sounding rods are used for the measurement of depths up to 6 m under medium velocities. The sounding rod may be either (*i*) a wooden rod oval in shape 80 × 20 mm and 2 or 3 m long and provided with a metal base plate of 150 mm in diameter and 5 mm thick, or (*ii*) a long bamboo, usually 38 mm in diameter and provided with a 100 mm diameter base plate with a flange to hold the bamboo rod.

The sounding rods are graduated in metres, decimetres and centimetres with a suitable paint or etched and painted with white, black and red paint. The number in metres and decimetres and their graduations, are precisely marked in red enamel paint and checked with a steel tape. The lettering size is usually not less than 2.5 cm.

3. *Lead lines*: Also called *log lines,* these are used for the measurement of water depth greater than 6 m provided the current is slow. For the current meter suspension, lead line is used when the depth is more than 2.5 m for rod suspension. It consists of a cable of 2.5 mm diameter to the lower end of which is attached a hanger for placement of the current meter and sounding weight during observations. Weight of 7 kg to 14 kg is necessary to avoid deflection of cable from the true vertical. Tags attached to the lead line indicate metres. A core of insulated copper wire in the log line enables electrical circuit to be made for counting ticks of the revolution of the current meter when using battery and headphones.

4. *Reel line and cranes*: When the depths are great and velocities are high to permit use of log line, a cable line is lowered by means of a crane to measure water depth. The arrangement, either mounted on a boat or used from bridge parapet, consists of a hand-operated reel fixed to a pedestal and carrying 3 mm diameter hard steel flexible cable passing over pulley at the other end.

Indirect Method

Echo sounder: Echo sounders of indicator as well as recorder type are used for accurate and rapid measurement and recording of water depth where the channel is of sufficient depth and the velocity is high. The echo sounder of recorder type records a continuous graph of the stream bed during the traverse on moving paper chart drawn by a stylus. At 30 to 40 points along the cross section a vertical mark is made on depth sounder chart to serve as observations point. In using echo sounder of indicator type a single reading may be made for each depth measurement but regular calibrations of the sounder are required. The echo sounder is capable of recording as small a depth as 0.3 m but with less accuracy. In greater depths the accuracy is of the order of 10 mm per metre depth.

Theodolite: Use of theodolite or a level fitted with stadia wires and a levelling staff is made to determine depth by pivot point method, linear measurement method, angular measurements method and stadia method.

15.24 DISCHARGE MEASUREMENT

The methods for discharge measurement are direct and indirect. The important direct methods are velocity-area method using current meter, floats and chemical dilution method. The indirect methods include slope-area method and using flow-measuring structures such as gated structures, flumes, weirs etc.

1. Velocity Measurement by Current Meter

The velocity varies approximately as a parabola, from zero at the bottom of the channel to a maximum at or near the water surface at top. The maximum velocity is observed to occur at 0.15 or more precisely between 0.10 to 0.20 of the depth below water surface at the vertical. The ratio of maximum velocity to the mean velocity is approximately 1.2. The mean velocity on any vertical section occurs at a depth 0.6 of the total depth. A general idea regarding the various methods applicable under various flow conditions is given below:

1. *Velocity distribution method*: This method is used when observations are required to be accurate, or for the purpose of calibration for determining the coefficients to be used for application to the results obtained by other methods. The average of velocities at two-tenths and eight-tenths depth below the water surface equals the mean velocity in a vertical, the velocity at six-tenths depth below the water surface being very nearly the mean in vertical.

2. *One point method*: In this method velocity is observed at 0.6 depth below water surface to give the mean velocity within 5 per cent accuracy and as such is generally used for shallower depths. Use of this method is also indicated where the current meter, if lowered to 0.8 depth, would interfere with projections from bed, grass, weed growth, etc. Correction factor applicable is 0.92 for depths less than 1 m.

3. *Two points method*: This method gives more consistent and accurate results except veloc-

ity distribution method. Accuracy of observation is within 2%. Velocity is measured at 0.2 and 0.8 of depth of water and the average of two is taken as the mean velocity in the vertical.

4. *Three points method*: Use of this method is preferred where velocity distribution in a vertical is abnormal to give accurate results with two point method. The velocity is measured at 0.2, 0.6 and 0.8 of the depth of water at each vertical and mean of the three values is taken as the mean velcoity in the verticals. Alternatively 0.6 depth measurement is weighted and $V_{mean} = 0.25\ (V_{0.2} + 2V_{0.6} + V_{0.8})$.

5. *Six points method*: In this method velocities are observed at 0.1, 0.2, 0.4, 0.6, 0.8 and 0.9 of the depth of water at each vertical. The velocity observations at each position are plotted in graphical form and the mean velocity determined with the aid of a planimeter.

6. *Two-tenth method*: The use of this method is indicated when the current is too fast to permit lowering of current meter at 0.6 or 0.8 depth. The velocity at each vertical is measured at 0.2 of the depth and converted into mean velocity by the application of a correction coefficient.

7. *Subsurface velocity method*: This method is analogous to two-tenth method. The velocity is measured just below the water surface. It is used where either the river is in flood or too swift and deep and there are disturbances at the surface or where it is not possible to make observations at 0.6 depth or at 0.2 depth.

8. *Integration method*: This method is rarely used in practice. In this method the current meter is lowered to the bed and then raised to the surface at a uniform rate not more than 5% of the mean velocity of flow in the cross section but not greater than 0.04 m/s and the results are averaged. The number of revolutions and the time taken in lowering and raising it are noted. Two complete cycles are made in each vertical and if the results differ by more than 10%, the measurement is repeated. The mean velocity in the vertical is calculated by multiplying the average number of revolutions per second with the current meter calibration coefficient.

9. *Surface velocity method*: This method is used only when the river is in flood. The current meter is submerged either to a constant depth below the water surface or to a certain percentage of depth. The velocity observed with a float is the surface velocity. The mean velocity is usually taken as 0.89 of the observed surface velocity.

Current Meter

Current meter is the instrument used for velocity measurement. It is simple in construction, rugged and most reliable. Broadly current meters are of two types, depending upon the orientation of their revolving axle as under:

1. Vertical-axis current meter

A Gurley or Price current meter essentially consists of the head, tail, hanger, recording or indicating device and suspending device. The head is just a yoke carrying a wheel of 12.7 cm diameter with six conical cups attached to a horizontal frame. This wheel revolves in an anti-clockwise direction on a shaft which passes into the cummutator box having arrangements to electrically show the revolutions by either single or ponta strokes. The tail consists of a stem with two vanes, one adjustable and one rigidly fixed. One of the vanes has an adjustable weight. The tail balances the head and retains the meter parallel to the current of flow. Each rotation of the rotar operates a contact mechanism which energises a headphone as well as electric flash counter. The recording device consists of a telephone and battery equipment. The latest current meters employ electro-magnetic counters with digital or analogue displays. The accuracy of the instrument is about 1.5% at threshold value and improve to about 0.3% at speeds in excess of 1.0 m/s. In deep waters meter can be used by cable suspension and in shallow waters by wadding rod. The limitation of the instrument is that it cannot be used where there are appreciable vertical components of velocities.

2. Horizontal-axis meter

It consists of a propeller mounted at the end of an horizontal shaft which lends it the name of propeller current meter. The rotation about horizontal axis is accomplished by means of screws or propeller shaped blades. It is a rugged instrument unaffected by oblique flows of 15°. Its accuracy is about 1% at threshold value and about 0.25% at a velocity of 0.3 m/s and above.

Rating of Current Meter

The current meter is precisely rated in still water. It is mounted on a carriage which moves on rails along a straight channel or rotates about a central pivot in a circular basin. The rating is done by moving the carriage through still water tank. The speed of the carriage is computed from the time required to travel a known distance. From several such observations of the runs at various speeds a curve is plotted showing the relation between meter contacts per unit time and water speed. The relation between the revolutions per second of current meter cups and velocity of water is given by the equation:

$$V = aN + b$$

where, V = velocity of flow (m/s), a and b = constants; value 0.65 and 0.03 respectively and N = number of revolutions of meter per second.

2. Velocity Measurement by Floats

Velocity measurement by floats is resorted to when it is not possible to use current meter due to unsuitable velocities or depths or quick observation is to be made. It is a very simple method. The surface velocity of flow is ascertained from the time noted during which a float traverses a known distance called float run. The method is suitable for canals or other regular channels in long straight reaches which are uniform in cross section and grade, the flow is free from cross currents and eddies, bed is free from weeds and flow velocity does not exceed 3 m/s and wind velocity is small. Three cross sections are selected along a straight reach of the channel spaced far apart so as to give a travel time of 20 seconds or less in high velocities. Velocity measurements by floats are less accurate (+ 10%) than current meter observations. The surface velocity is reduced to mean velocity by the application of a suitable reduction coefficient, usually taken as 0.89, 0.85 as per IS: 3911-1994. A reduction factor F is applied as function of R, the ratio of immersed depth of float to water depth to convert the float velocity to the mean velocity, as under:

$R \leq$	0.10	0.25	0.50	0.75	0.95
F	0.86	0.88	0.90	0.94	0.98

Floats commonly used are as under:

Surface float: It consists of (*i*) circular disc, (*ii*) cylindrical stem and (*iii*) metallic floats (Fig. 15.22). The circular disc has overall dimensions approximately of 150 × 40 mm, streamlined at edges and having at its top centre a square shaped recess approximately 40 × 40 × 25 mm deep made in the centre to engage the lower end of a vertical rod or stem 200 mm long and 20 mm in diameter, the bottom 25 mm length being 40 × 40 mm square to fit squarely in the recess. The upper end of float supports a rectangular flag 100 × 50 mm. The cylindrical wooden float is identical to circular disc float except that the circular disc there is not streamlined but is vertical at the edges. The metallic float is of barrel type approximately 300 mm in diameter at the centre decreasing uniformly to about 200 mm at the two ends and approximately 300 mm in length. The ends are sealed to make the float watertight.

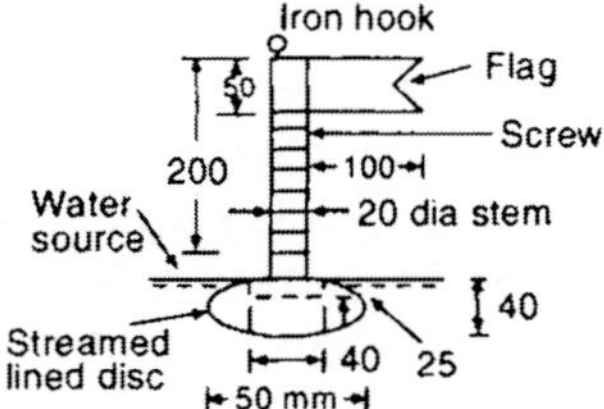

Fig. 15.22 Surface float.

Double float: It consists of a weighted subsurface float; a hollow metal sphere or cylinder suspended by means of a thin cord of known length from another float at the surface (Fig. 15.23*a*). The depth of subsurface float is regulated by the length of an adjustable cord. The velocity is obtained by timing the surface float over the float run. It gives the mean velocity across a vertical when the centre of the sphere is kept at about 0.2 depth above the bed level by adjusting the cord length. Twin float (Fig. 15.23*b*) is a variety of double float. It consists of two hollow metal spheres connected with a thin cord such that the upper sphere is afloat and the lower one so weighted as to keep the float vertical. Mean velocity is given by twin float if the lower sphere is just clear off the bed level.

Velocity rod: It consists of a square or round rod of 25 to 50 mm in diameter or width depending on its length. It is weighted to the bottom with lead to keep it afloat in a vertical position and submerged to the required depth (Fig. 15.23(*c*)). The depth of submergence varies from 0.91 to 0.97 times of the water depth.

The use of velocity rods is restricted to canal discharge observation. Because of the limitation of the use of long rods, the method is applicable to water depths up to 4.5 m.

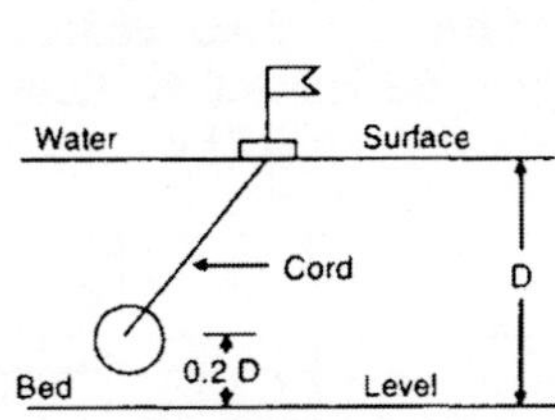

Fig. 15.23(*a*) Double float.

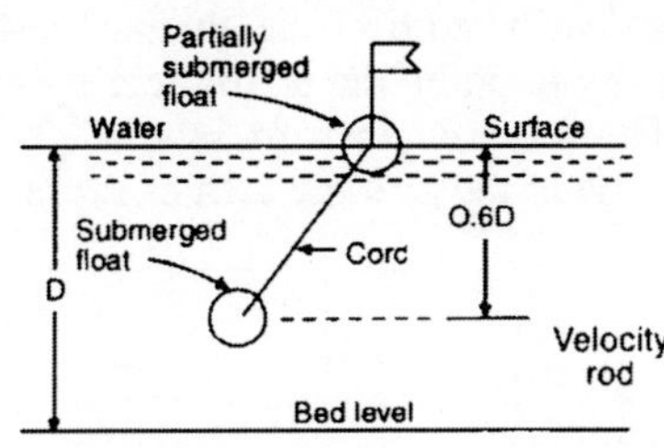

Fig. 15.23 (*b*) Twin float.

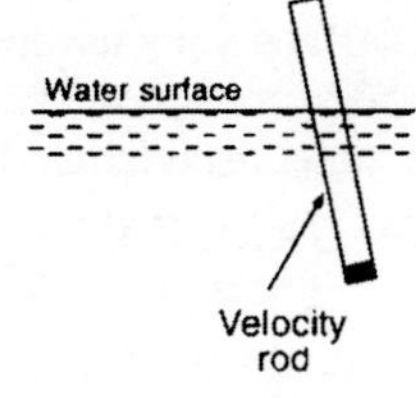

Fig. 15.23 (*c*) Velocity rod.

15.25 DISCHARGE COMPUTATION

The various methods of discharge computations are briefly described as under :

1. *Velocity area Method*: It is the most accurate method, making use of observed depth and velocity measured by current meter or float. The two methods of discharge computation which is essentially a product of velocity and area are as under:

(*i*) *Mid-section Method*: The cross section is considered to be made up of a number of compartments each bounded by two adjacent verticals. Mean velocity is observed at 0.6 depth at verticals. Referring to Fig. 15.24, $Q = \bar{V}_1 d_1 \left[\frac{b_1 + b_2}{2}\right] + \bar{V}_2 d_2 \left[\frac{b_2 + b_3}{2}\right] + \ldots\ldots\ldots + \ldots\ldots\ldots$ The flow at the edges is omitted.

(*ii*) *Mean section method*: Width of strip is the distance between two verticals. Mean depth is the average of the depths in two verticals.. Velocity is.the average value of mean velocities in two adjacent verticals (Fig. 15.24). Discharge in each strip $q = b\left[\left(\frac{d_1 + d_2}{2}\right)\left(\frac{\bar{V}_1 + \bar{V}_2}{2}\right)\right]$ and $\Sigma q = Q$. The first and last verticals are sited as near to the banks as possible, since discharge in end segments is omitted.

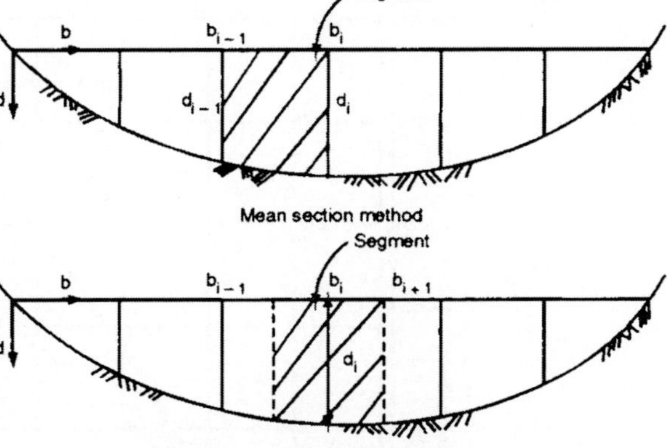

Fig. 15.24 Mid-section method.

2. Slope area Method: This method provides a rough estimate of the discharge and is used when actual measurements by velocity area method are not possible. In this method the average cross sectional area is multiplied with the velocity derived from observed water surface slope. The cross sectional area is measured at three locations, two at the ends and one in the centre of a straight river reach of length at least 10 times the river width or 300 m whichever is more.

Mean cross sectional area, $A = (A_1 + 2A_2 + \ldots . + (2A_{n-1}) + A_n)/2(n - 1)$.

Where, $A_1, A_2, \ldots, A_n$ are the cross sectional areas measured at various locations.

Mean wetted perimeter,

$P = (P_1 + 2P_2 + \ldots + 2P_{n-1} + P_n)/2\,(n - 1)$.

Water surface slope is computed from gauges installed at least in three cross sections, one on either river bank, making a total of six gauges. The water surface drop or slope in the reach shall not be less than 150 mm, the minimum being 30 mm. Mean velocity is determined from the observed water surface slope by Chezy formula $V = C\sqrt{RS}$ or Manning's formula. $V = \frac{1}{n} R^{2/3} S^{1/2}$.

Discharge computation

Each float is released far enough the cross section to attain a constant velocity before it reaches the first cross section. The time at which the float crosses each of the three cross sections are noted by stop watch. The floats are uniformly distributed over the channel width. Repeated observations are made. The surface velocity determined by dividing the travel time by the distance between cross sections is converted into mean velocity. The water depth is measured. The discharge is the sum of the flow in each panel ascertained by multiplying the average area of cross section with mean velocity.

15.26 GAUGE DISCHARGE CURVE

The estimation of discharge from gauge record for periods for which no discharge observations are available is best done by using a proper stage-discharge curve. The curve is obtained by plotting an adequate number of observations of discharges along the abscissa against gauges along the ordinate, suitably distributed throughout the whole range of stage (Fig. 15.25).

During floods, the rate of rise of gauge on a rising flood is generally higher than the rate of fall on a falling flood thereby passing a higher discharge at a particular gauge on the rising gauge than on the falling gauge. The loop in rating curve reverts to a single curve at medium or low flood stage (Fig. 15.26).

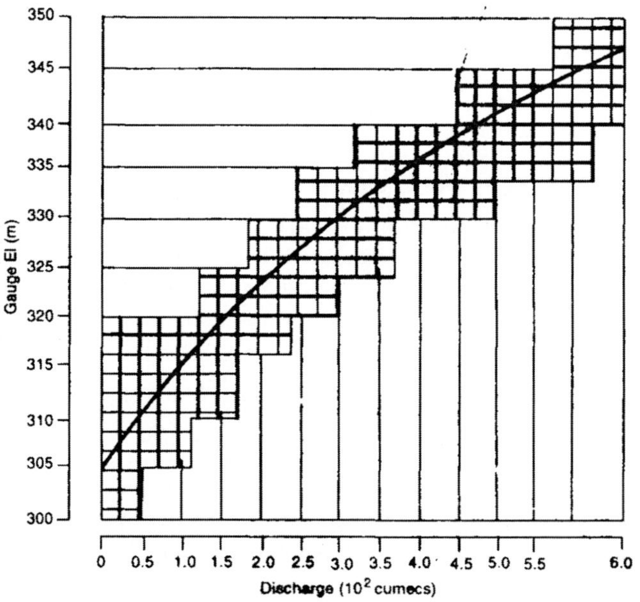

Fig. 15.25 Gauge-discharge curve

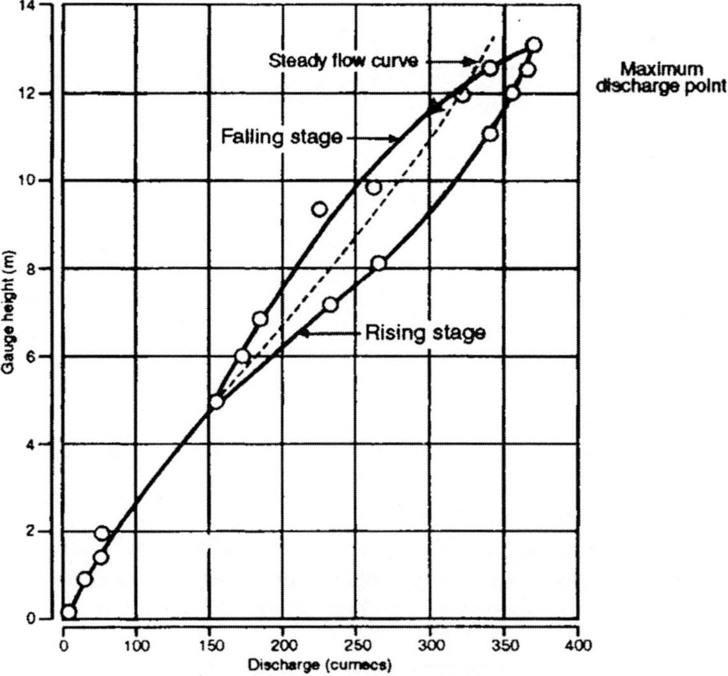

Fig. 15.26 Loop rating curve.

SOLVED EXAMPLES

Example 15.1 *Fig. 15.27 shows a typical river basin with 4 raingauge stations and 4 outside as per IS: 4987-1994. Determine adequacy of the gauges and allocate additional gauges so that percentage error is less than 10%.*

Solution:

(*i*) *Adequacy of raingauges*

Total rainfall, $R = 800 + 540 + 445 + 410 = 2195$ mm

Mean rainfall, $m = R/4 = 2195/4 = 548.75$ mm

Sum of squares, $SS = (800)^2 + (540)^2 + (445)^2 + (410)^2 = 1297725$

$T = R^2/4 = (2195)^2/4 = 1204506$

Square of standard deviation, $S^2 = (SS-T^2)/4-1 = (1297725-1204506)/3 = 31073$

Coefficient of variation, $C_v = [100\,(S^2)^{1/2}]/m = [100 \times (31073)^{1/2}]/548.75$

$= 32.12$

Optimum number of gauges, $N = (C_v/P)^2 = (32.12/10)^2 = 10.32 \simeq 11$

Additional gauges required $= 11 - 4 = 7.$

(*ii*) *Allocation of additional gauges*: (*i*) Draw isohyets at equal intervals (380, 510, 640, 770 and 900 mm) taking into consideration rainfall outside the catchment (Fig. 15.27), (*ii*) Measure the area bounded by two consecutive isohyets within the catchment with a planimeter. The proportional areas of the different isohyetal zones marked are $X_1 = 0.1229$, $X_2 = 0.3760$, $X_3 = 0.3392$, $X_4 = 0.1021$, and $X_5 = 0.0598$, (*iii*) 11 gauges are to be allocated to different zones in proportion of their areas. Thus the No. of raingauges in different isohyetal zones are $X_1 = 1.4$, $X_2 = 4.1$, $X_3 = 3.7$, $X_4 = 1.1$ and $X_5 = 0.7$, (*iv*) Additional No. in each zone are:

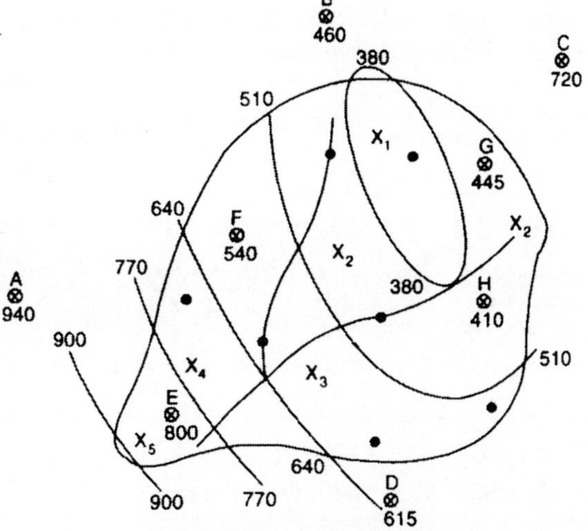

Fig. 15.27 Distribution of raingauges.

Zone	X_1	X_2	X_3	X_4	X_5	*Total*
Existing No.	0	2	1	0	1	4
Additional No. of gauges	1	2	3	1	0	7
Total No.	1	4	4	1	1	11

The additional raingauges are indicated by circles in Fig. 14.27.

Example 15.2 *The annual rainfall (cm) at a place for a period of 10 years from 1971 to 1980 are respectively: 30.3, 41.0, 33.5, 34.0, 33.3, 36.2, 33.6, 30.2, 35.5, 36.3. Determine the mean and median values of rainfall for the place.*

Solution: Mean value of rainfall

$$P = \Sigma P/n = (1/10)\,(30.3 + 41.0 + 33.5 + 34.0 + 33.3 + 36.2 + 33.6 + 30.2 + 35.5 + 36.3) = 34.39 \text{ cm}$$

Median rainfall: By arranging the rainfall data in ascending order 30.2, 30.3, 33.3, 33.5, 33.6, 34.0, 35.5, 36.2, 36.3, 41.0, No. of items = 10 *i.e.* Even median value of rainfall (*i.e.* average of central two items) = (34.0 + 33.6)/2 = 33.8 cm.

Note: Had the number of items been odd, the middle item gives the median value.

Example 15.3 *Compute the average precipitation by the arithmetic average method and Thiessen polygon method from the following data:*

Station No.	Precipitation (mm)	Area (sq km)	Station No.	Precipitation	Area
1	30.8	45	3	32.0	30
2	34.6	40	4	24.6	38

Solution:

By arithmetic average method, $P = (30.8 + 34.6 + 32.0 + 24.6)/4 = 30.50$ mm

By Theissen polygon method,

$$P = \frac{P_1A_1 + P_2A_2 + P_3A_3 + P_4A_4}{A_1 + A_2 + A_3 + A_4}$$

$$= \frac{30.8\times45+34.6\times40+32.0\times30+24.6\times38}{45+40+30+38}$$

$= 30.49$ mm.

Example 15.4 *Estimate the total volume of rainfall water received in m^3 over a basin of 4444 ha for the following observations:*

Station	A	B	C	D	E
Polygon area (ha)	518	777	906	1495	748
Observed rainfall (mm)	267	198	142	114	81

Solution: Average depth of rainfall over catchment.

$$= \frac{P_aA_a + P_bA_b + P_cA_c + P_dP_d + P_eA_e}{A_a + A_b + A_c + A_d + A_e}$$

$$= \frac{267\times518+198\times777+142\times906+114\times1495+81\times748}{518+777+906+1495+748} = 146.67 \text{ mm}$$

Total volume of rainfall water over the basin = $(4444 \times 10^4 \times 146.67/10^3) = 6.52 \times 10^6$ m^3.

Example 15.5 With the information given below, calculate the mean rainfall over the area:

Steps between isohyets (cm)	Area (km^2)	Steps between isohyets (cm)	Area (km^2)
127.0 – 139.7	181	165.1 – 177.8	311
139.7 – 152.4	388	177.8 – 190.5	104
152.4 – 165.1	337	190.5 – 203.2	26

Solution:

Average precipitation (P)	Area (A)	Product (P.A.)	Average precipitation (P)	Area (A)	Product (P.A.)
(127.0 + 139.7)/2 = 133.35	181	24,136	(165.1 + 177.8)/2 = 171.45	311	53,321
(139.7 + 152.4)/2 = 146.05	388	56,667	(177.8 + 190.5)/2 = 184.15	104	19,152
(152.4 + 165.1)/2 = 158.75	337	53,499	(190.5 + 203.2)/2 = 196.85	26	5,118
				Total A = 1347	P.A. = 211,893

Average precipitation, $P_{av} = (2,11,893)/1347 = 157.31$ cm

Example 15.6 *Determine mean rainfall for the catchment from the rainfall data as under:*

Station	Station weight	Rainfall (mm)	Station	Station weight	Rainfall (mm)
A	0.04	400	D	0.15	500
B	0.06	300	E	0.27	240
C	0.08	200	F	0.40	175

Solution:

Weighted mean rainfall $= (0.04 \times 400 + 0.06 \times 300 + 0.08 \times 200 + 0.15 \times 500 + 0.27 \times 240 + 0.40 \times 175) = 259.8$ mm.

Example 15.7 *The sketch (Fig. 15.28) below represents catchment area with the precipitations observed in a year. Calculate the mean precipitation by Thiessen method and check the result roughly by the arithmetic mean method.*

Solution: Draw right bisector of EA and ED to cut AD at U and V respectively (Fig. 15.29), $AG = 5$ km, $GU = 3.75$ km, Join LM, MN, NO and OL, $ML = 11.31$ km.

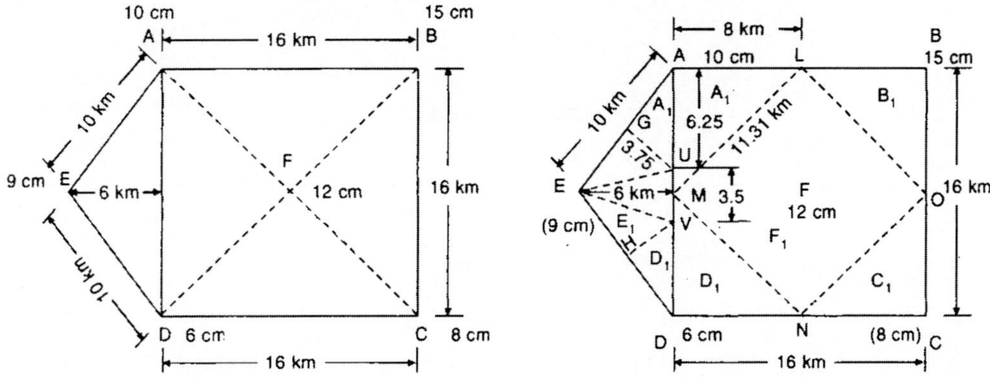

Fig. 15.28 **Fig. 15.29** Thiessen polygon.

Station	*Area*	*Rainfall (cm)*	*Figure*	*Area (km²)*			*P.A.*
A	A_1	10	*AGU*	(5 × 3.75)/2	= 9.375		
			AML	(8 × 8)/2	= 32.0	41.375	413.75
B	B_1	15	*LOB*	(8 × 8)/2	= 32.0		480.0
C	C_1	8	*ONC*	(8 × 8)/2	= 32.0		256.0
D	D_1	6	*MDN*	(8 × 8)/2	32.0		
			HDV	(5 × 3.75)/2	= 9.375	41.375	248.25
E	E_1	9	*EGU and EHV*	2(5 × 3.75)/2	= 18.75		
			UEV	(3.5 × 6)/2	= 10.5	29.25	263.25
F	F_1	12	*LMNO*	11.31 × 11.31	= 128		1536.0
				Total A	= 304		Σ PA = 3197.25

$$P_{av} = \frac{\Sigma P.A.}{\Sigma A} = \frac{3197.25}{304} = 10.52\,cm$$

By arithmetic mean method, $P_{av} = (10 + 15 + 8 + 6 + 9 + 12)/6 = 10$ cm.

Example 15.8 *The catchment area of a basin is in the shape of a regular hexagon of side 40 km. 7 raingauges located one at each end of the hexagon and one at the centre recorded precipitation of 10, 20, 30, 40, 50, 60 and 70 cm respectively. Using Thiessen polygon method, compute the average precipitation over the basin.* *(AM.I.E)*

Solution. By drawing right bisectors of the sides, the Thiessen polygon is completed (Fig. 15.30).

From $\Delta\, JJE'$

$$JE' = JE \sin 30° = 20 \times ½ = 10$$

$$EE' = JE \cos 30° = 20 \times \frac{\sqrt{3}}{2} = 17.32$$

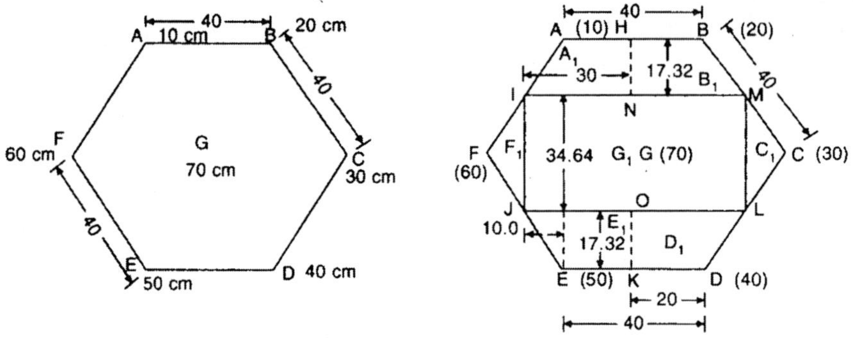

Fig. 15.30 Thiessen polygon.

Area	*Rainfall (cm)*	*Figure*	*Area (km²)*		*P.A.*
A_1	10	*AHNI*	$\frac{20+30}{2} \times 17.32$	= 433	4,330
B_1	20	*HBMN*	Same as of *A*	= 433	8,660
C_1	30	*MLC*	$\frac{34.64 \times 10}{2}$	= 173.2	5,196
D_1	40	*OKDL*	Same as of *A*	= 433	17,320
E_1	50	*JEKO*	Same as of *D*	= 433	21,650
F_1	60	*FJI*	Same as of *C*	= 173.2	10,392
G_1	70	*IJLM*	34.64 × 60	= 2078.4	1,45,488
				ΣA 4156.80	ΣPA 213036

$$P_{av} = \Sigma PA / \Sigma A = 213036/4156.80 = 51.25 \text{ cm.}$$

Example 15.9 *The following table gives the data of an isohyetal map. The total area of the basin is 300 sq km. Compute the average depth of rainfall over the basin.*

Isohyet (cm)	*40*	*35*	*30*	*25*	*20*	*15*	*10*
Basin area covered by isohyet (sq km)	—	*17*	*45*	*75*	*165*	*215*	*300*

Solution: (*P.U.*)

Mean isohyets (P)	*Area (A)*	*P × A*	*Mean isohyets (P)*	*Area (A)*	*P × A*
40 – 35 = 37.5	17	637.5	25 – 20 = 22.5	90	2025.0
35 – 30 = 32.5	28	910.0	20 – 15 = 17.5	50	875.0
30 – 25 = 27.5	30	825.0	15 – 10 = 12.5	85	1062.5

Average rainfall, $$P_{av} = \frac{\Sigma P \times A}{\Sigma A} = \frac{6335}{300} = 21.12 \text{ cm.}$$

Example 15.10 *The annual average rainfall of four stations A, B, C and D in an area is 105, 122, 95 and 102 cm respectively. In a particular year station C did not function and the rainfall figures of A, B and D were 90, 97 and 82 cm respectively. Compute the missing precipitation at station C.*

Solution:

	A	*B*	*C*	*D*
Annual average rainfall (cm)	105	122	95	102
Annual rainfall (cm)	90	97	—	82

Since annual rainfall at any of the stations (122 cm at station *B* in this case) differs from that at the station with missing data (95 cm at station *C*) by more than 10 per cent, normal ratio method is applicable.

$$\therefore \quad P_c = (N_x/3)\,[P_A/N_A + P_B/N_B + P_C/N_C]$$

$$= (95/3)\,(90/105 + 97/122 + 82/102] = 77.78 \text{ cm}$$

Example 15.11 *Estimate the missing rainfall record at station B for the year 1976 from the following data:*

Year	*Annual rainfall (cm)*		*Year*	*Annual rainfall (cm)*	
	Station A	*Station B*		*Station A*	*Station B*
1971	*138.9*	*150.5*	*1974*	*132.3*	*147.4*
1972	*145.9*	*154.7*	*1975*	*126.2*	*145.0*
1973	*146.9*	*159.0*	*1976*	*143.8*	—

Solution:

$$\frac{\text{Rainfall at station } A \text{ for 1976}}{\text{Rainfall at station } B \text{ for 1976}} = \frac{\text{Rainfall at station } A \text{ from 1971} - 75}{\text{Rainfall at station } B \text{ from 1971} - 75}$$

$\therefore$ Rainfall at station *B* for 1976

$$= \frac{\text{Rainfall at station } B \text{ from 1971} - 75}{\text{Rainfall at station } A \text{ from 1971} - 75} \times \text{rainfall at station } A \text{ for 1976}$$

$$= \left(\frac{150.5+154.7+159.0+147.4+145.0}{138.9+145.9+146.9+132.3+126.2}\right) \times 143.8 = 157.6 \text{ cm.}$$

Example 15.12 *The maximum values of 24 hours rainfall at a raingauge expressed in cm from 1930 to 1959 are given below:*

12.7	*13.2*	*12.8*	*11.6*	*16.9*	*17.2*	*14.0*	*14.2*	*17.8*	*18.8*
11.7	*13.3*	*13.6*	*13.9*	*14.4*	*14.7*	*8.4*	*12.5*	*11.2*	*20.7*
19.1	*19.7*	*18.9*	*17.4*	*15.8*	*14.9*	*18.3*	*17.7*	*18.6*	*19.2*

Estimate the maximum rainfall having a recurrence interval of (i) 10 years, and (ii) 50 years.

Solution: For the estimation of maximum rainfall, arrange the given rainfall data in descending order, as follows:

Rainfall (cm)	*Ranking (m)*	*Recurrence interval T = N/m – 0.5*	*Rainfall (cm)*	*Ranking (m)*	*R.I. (N/m – 0.5)*	*Rainfall (cm)*	*Ranking (m)*	*R.I. (N/m – 0.5)*
20.7	1	30/1 – 0.5 = 60	17.4	11	2.86	13.6	21	1.46
19.7	2	30/2 – 0.5 = 20	17.2	12	2.61	13.3	22	1.40
19.2	3	30/3 – 0.5 = 12	16.9	13	2.40	13.2	23	1.33
19.1	4	30/4 – 0.5 = 8.57	15.8	14	2.22	12.8	24	1.28
18.9	5	30/5 – 0.5 = 6.67	14.9	15	2.07	12.7	25	1.22
18.8	6	30/6 – 0.5 = 5.45	14.7	16	1.94	12.5	26	1.18
18.6	7	30/7 – 0.5 = 4.62	14.4	17	1.82	11.7	27	1.13
18.3	8	30/8 – 0.5 = 4.0	14.2	18	1.71	11.6	28	1.09
17.8	9	30/9 – 0.5 = 3.53	14.0	19	1.62	11.2	29	1.05
17.7	10	30/10 – 0.5 = 3.16	13.9	20	1.54	8.4	30	1.02

Rainfall with 10 years frequency read from Fig. 15.31 = 19.15 cm
Rainfall with 50 years frequency read from Fig. 15.31 = 20.45 cm

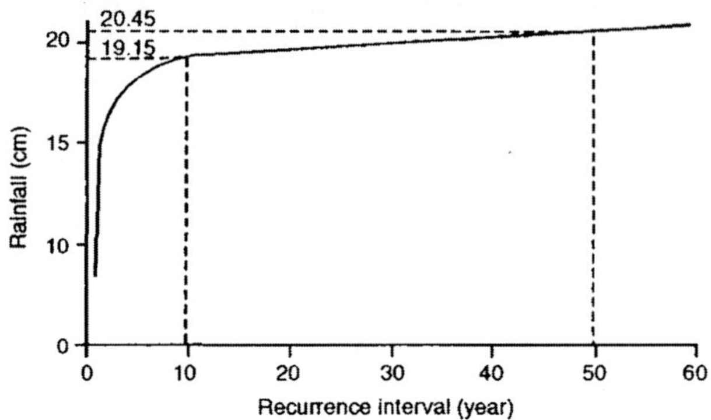

Fig. 15.31 Rainfall frequency

Example 15.13 *Maximum values of 24 hour precipitation (mm) at a raingauge station are 140, 113, 132, 115, 130, 118, 127, 127, 123, 121. Estimate maximum and minimum precipitation having a recurrence interval of 5 and 15 years. Explain limitations of the method used, if any.*

Solution:

Maximum rainfall			*Minimum rainfall*		
Rainfall (mm)	*m*	*T = 2n/2m-1*	*Rainfall (mm)*	*m*	*T = 2n/2m -1*
140	1	20	113	1	20
132	2	6.67	115	2	6.67
130	3	4.0	118	3	4.0
127	—	—	121	4	1.86
127	5	2.22	123	5	2.22
123	6	1.82	127	—	—
121	7	1.54	127	7	1.54
118	8	1.33	130	8	1.33
115	9	1.18	132	9	1.18
113	10	1.05	140	10	1.05

* Using higher value of *m*, when two rainfall events of the same value.

Fig. 15.32 indicate the maximum value of precipitation with 5 and 15 years frequency as 131 mm and 137 mm respectively. The minimum values with these recurrence interval are 117 mm and 113.75 mm (Fig. 15.32).

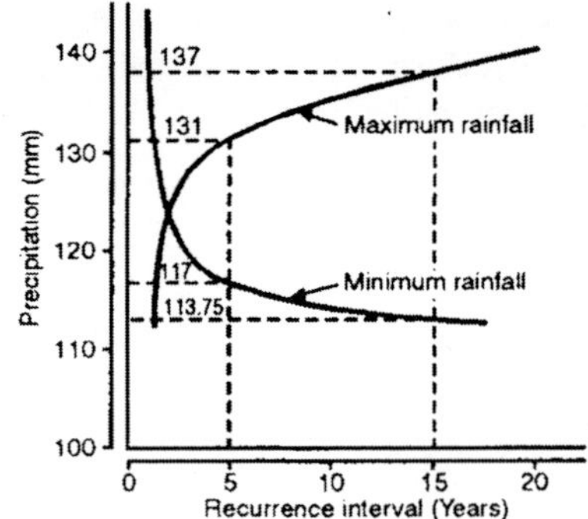

Fig. 15.32 Maximum & minimum precipitation.

Since the actual frequency of occurrence of maximum event in 21 years is not known and no statistical device can lengthen the period of record, the assumption of recurrence interval of $2n$ to the highest event on the record is the limitation of the Hazen's method used.

Example 15.14 *Estimate the return period of a rainfall whose probability of exceedence is 5%. Also determine the probability that this rainfall may occur in the next 5 years.*

Solution:

Probability of exceedence $P = 1/T = 5\%$

or $T = (1 \times 100)/5 = 20$ years

The probability that the rainfall will not occur in 5 years

$= (1 - 5/100)^5 \times 100 = 77.38\%$

The probability that the rainfall will equal or exceed in 5 years

$= (1 - 0.774) \times 100 = 22.62\%.$

Example 15.15 *Estimate the magnitude of rainfall the probability of whose exceedence would be 1 in 10,000 years from the observed values of annual rainfall (in cm) from 1961 to 1985. What shall be rainfall whose return period is 1 in 1,000 years?*

95	87	113	98	78	125	140	105	88	67	115	103	127
99	102	65	91	130	132	68	60	58	50	82	100	

Solution:

S.No. (m)	Rainfall (cm)	Exceedence probability F = (m/N + 1) × 100	S.No. (m)	Rainfall (cm)	Exceedence probability F = (m/N + 1) × 100	S.No. (m)	Rainfall (cm)	Exceedence probability F = (m/N + 1) × 100
1	140	(1/25 + 1) × 100 = 3.85 and so on				18	82	69.23
2	132	7.69	10	102	38.46	19	78	73.08
3	130	11.54	11	100	42.31	20	68	76.92
4	127	15.38	12	99	46.15	21	67	80.77
5	125	19.23	13	98	50.00	22	65	84.62
6	115	23.08	14	95	53.85	23	60	88.46
7	113	26.92	15	91	57.69	24	58	92.31
8	105	30.77	16	88	61.54	25	50	96.15
9	103	34.62	17	87	65.38			

The exceedence probability values, as above, are plotted on probability paper (Fig. 15.33). The rainfall values for return periods of 1 in 10,000 years (exceedence 0.01 per cent) read from the figure comes to 1850 mm and that for 1 in 1,000 years (exceedence 0.1 per cent) as 1700 mm.

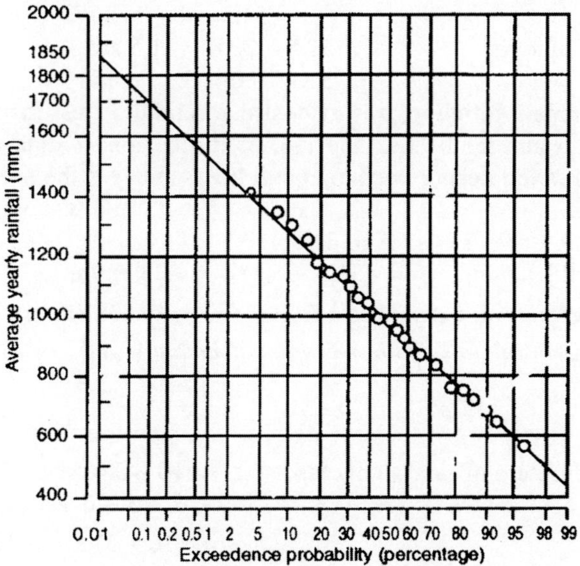

Fig. 15.33 Exceedence probability.

Example 15.16 *The rates of rainfall for successive 20 minutes period of a 140 minute duration storm are 2.5, 2.5, 10.0, 7.5, 1.25, 1.25 and 5.0 cm/hr. Taking the value of ϕ-index as 3.2 cm/hr, estimate the net runoff (surface runoff only) in cm. Also determine the value of W-index.*

Solution:

Time (minutes)	20	40	60	80	100	120	140
Rate of rainfall (cm/hr)	2.5	2.5	10.0	7.5	1.25	1.25	5.0
ϕ-index	3.2	3.2	3.2	3.2	3.2	3.2	3.2
Net runoff (cm/hr)	—	—	6.8	4.3	—	—	1.8

Plot the rate of rainfall against time and superimpose the curve of ϕ-index at 3.2 cm/hr (Fig. 15.34). The area above the curve of ϕ-index represents the net rainfall, while that below the curve is the infiltration.

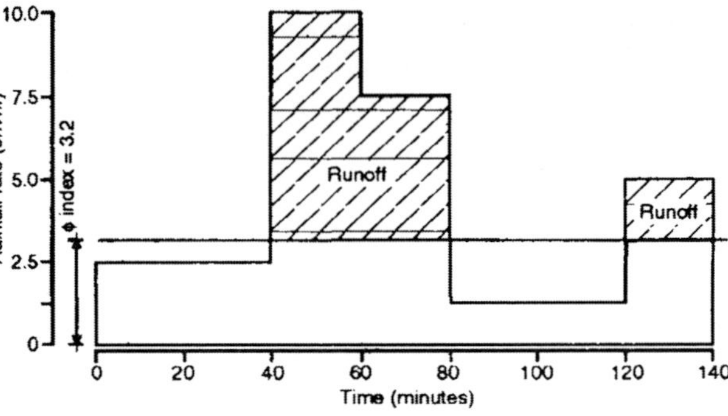

Fig. 15.34 ϕ-index and runoff

Net runoff, $R = (20/60)\ (6.8 + 4.3 + 1.8) = 4.30$ cm

Total runoff, $P = (20/60)\ (2.5 + 2.5 + 10.0 + 7.5 + 1.25 + 1.25 + 5.0) = 10.0$ cm

W-index, $W = (P - R)/t_r$ cm/hr $= (10.0 - 4.3)/140/60 = 2.44$ cm/hr.

Example 15.17 *In a 10-hour storm rainfall, depths occurred over a catchment are:*

Hour	*1*	*2*	*3*	*4*	*5*	*6*	*7*	*8*	*9*	*10*
Depths (cm/hr)	*1.0*	*1.5*	*5.0*	*6.0*	*10.5*	*8.5*	*9.0*	*7.0*	*1.5*	*1.5*

Surface runoff resulting from the storm is equivalent to 20 cm of depth over the catchment. Determine (i) Average infiltration, and (ii) Average rate of infiltration.

Solution: Total rainfall in 10 hours = 1.0 + 1.5 + 5.0 + 6.0 + 10.5 + 8.5 + 9.0 + 7.0 + 1.5 + 1.5 = 51.5 cm

Total infiltration in 10 hours = 51.5 – 20.0 = 31.5 cm

Average infiltration = 31.5/10.0 = 3.15 cm/hr

With 3.15 cm/hr average infiltration, it is evident that rainfall is less than average infiltration during first, second, ninth and tenth hour and rate of infiltration is equal to the rainfall. The rainfall is more than infiltration during third to eighth hours. Let f is the average rate of infiltration during these 6 hours.

$$1.0 + 1.5 + 6f + 1.5 + 1.5 = 31.5$$

or Average rate of infiltration, $f = (31.5 - 5.5)/6 = 4.33$ cm/hr.

Example 15.18 *Determine ϕ-index for a basin of 500 sq km. The direct runoff of a flood was 5 cm and equivalent uniform rainfall depth over the catchment was 15 cm. The time distribution of rainfall was as under:*

Time (hr)	*5 – 6*	*6 – 7*	*7 – 8*	*8 – 9*	*9 – 10*
Rainfall (cm)	*2*	*3*	*6*	*4*	*1*

Solution: Assume duration of rainfall excess = 10 – 5 = 5 hr

Total rainfall = 2 + 3 + 6 + 4 + 1 = 16 cm

ϕ-index = (16 – 5)/5 = 2.2 cm/hr.

This value of ϕ-index renders rainfall in 5-6th hr, and 9-10th hr ineffective being less than 2.2 cm/hr. The duration of rainfall excess as such is 3 hr (not 5 hr assumed).

Infiltration = [(3.0 + 6.0 + 4.0) – (5.0)] = 8 cm

$\therefore$ ϕ-index = 8/3 = 2.67 cm/hr.

The rainfall excess thus occurs during 6-7th hr to 8-9th hr equal to 0.33, 3.33 and 1.33, aggregating to 4.99 say 5 cm which is the same as given, hence ϕ index of 2.67 cm/hr is o.k.

Example 15.19 *Find out the ordinates of a storm hydrograph resulting from a 3 hour storm with rainfall of 2.0, 6.75 and 3.75 cm during subsequent 3 hours intervals. The ordinates of unit hydrograph are:*

Hr	*3*	*6*	*9*	*12*	*15*	*18*	*21*	*24*	*3*	*6*	*9*	*12*	*15*	*18*	*21*	*24*
OUH (cumecs)	*0*	*110*	*365*	*500*	*390*	*310*	*250*	*235*	*175*	*130*	*95*	*65*	*40*	*22*	*10*	*0*

Assume an initial loss of 5 mm, infiltration index of 2.5 mm/hr and base flow of 10 cumecs.

Solution:

1. Rainfall excess
   (*a*) During 1st three hours = 2.0 – (0.25 × 3) – 0.5 = 0.75 cm
   (*b*) During 2nd three hours = 6.75 – (0.25 × 3) = 6.0 cm
   (*c*) During last three hours = 3.75 – (0.25 × 3) = 3.0 cm
2. Rainfall excess as ratio of unit rainfall of 1 cm during subsequent 3-hour intervals is 0.75 cm, 6 cm and 3 cm.
3. Runoff due to 0.75 cm rainfall starts from 3 hour, because of 6 cm rainfall from 6 hour, and 3 cm rainfall from 9 hour. The computation of ordinates of storm hydrograph is tabulated in Table 15.5.

Table 15.5 Storm hydrograph

Time (hr)	*Ordinate of (3 hr) U.H. (cumecs)*	*Rainfall excess (cm)*	*Surface runoff from rainfall excess during successive unit periods (cumecs)*				*Base flow (cumecs)*	*Ordinate of storm hydrograph (cumecs)*
			0.75 (col. 2 × 0.75)	*6.0 (col. 2 × 6.0)*	*3.0 (col. 2 × 3.0)*	*Total*		
1	2	3	4	5	6	7	8	9
3	0	0.75	0				10	10.0
6	110	6.0	82.5	0		82.5	10	92.5
9	365	3.0	274.0	660	0	934.0	10	944.0
12	500		375.0	2190	330	2895.0	10	2905.0
15	390		292.5	3000	1095	4387.5	10	4397.5
18	310		232.5	2340	1500	4072.5	10	4082.5
21	250		187.5	1860	1170	3217.5	10	3227.5
24	235		176.0	1500	930	2606.0	10	2616.0
3	175		131.5	1410	750	2291.5	10	2301.5
6	130		97.5	1050	705	1852.5	10	1862.5
9	95		71.0	780	525	1376.0	10	1386.0
12	65		49.0	570	390	1009.0	10	1019.0
15	40		30.0	390	285	705.0	10	715.0
18	22		16.5	240	195	451.5	10	461.5
21	10		7.5	132	120	259.5	10	269.5
24	0		0	60	66	126.0	10	136.0
3				0	30	30	10	40.0
6					0	0	10	10.0

Example 15.20 *Given below are the observed flows (cumecs) from a storm of 6 hour duration on a stream with a drainage area of 316 sq km. Assume a constant base flow of 17 cumecs, derive and plot a 6-hour duration unit hydrograph.*

Time (hr)	*0*	*6*	*12*	*18*	*24*	*30*	*36*	*42*	*48*	*54*	*60*	*66*	*72*
Flow	*17*	*113.2*	*254.5*	*198*	*150*	*113.2*	*87.7*	*67.9*	*53.8*	*42.5*	*31.1*	*22.64*	*17*

Solution: Separating base flow of 17 cumecs, surface runoff is as under:

Time (hr)	0	6	12	18	24	30	36	42	48	54	60	66	72
Surface run-off (cumecs)	0	96.2	237.5	181	133	96.2	70.7	50.9	36.8	25.5	14.10	5.64	0

$$\text{Runoff depth} = \frac{6\times60\times60}{3}[0 + 0 + 2(96.2 + 181 + 96.2 + 50.9 + 25.5 + 5.64) + 4(237.5 + 131 + 70.7 + 36.8 + 14.1)]$$

$$= \frac{20.7308\times10^6}{316\times10^6}\times 100 = 6.56 \text{ cm.}$$

Ordinate of unit hydrograph = 96.2/6.56 = 14.66 and so on.

Time (hr)	0	6	12	18	24	30	36	42	48	54	60	66	72
OUH	0	14.66	36.20	27.59	20.27	14.66	10.78	7.76	5.61	3.89	2.15	0.86	0

Fig. 15.35 shows the unit hydrograph.

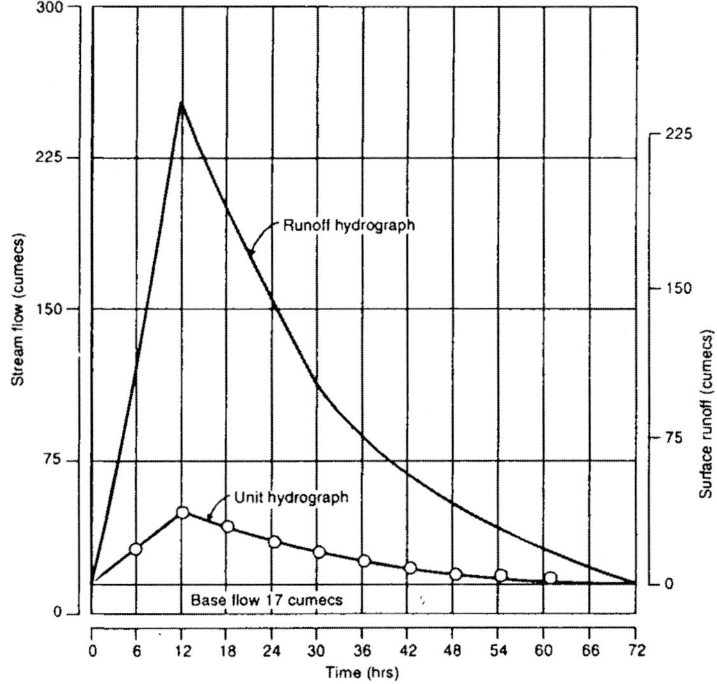

Fig. 15.35 Unit hydrograph.

Example 15.21 *Given below are the ordinates of a unit hydrograph for a storm of 4-hour duration. Find ordinates of flood hydrograph when the maximum flood observed was 4000 m^3/s and base flow was 250 m^3/s.*

Time (hours)	*0*	*4*	*8*	*12*	*16*	*20*	*24*
Surface runoff (cumecs)	*0*	*1500*	*1200*	*600*	*220*	*80*	*0*

Solution: Maximum surface runoff = 4000 – 250 = 3750 cumecs.

Rainfall excess = 3750/1500 = 2.5 cm

Time	0	4	8	12	16	20	24
O.U.H. of surface runoff	0	1500	1200	600	220	80	0
O.S.R. = 2.5 × O.U.H. (cumecs)	0	3750	3000	1500	550	200	0
Base flow	250	250	250	250	250	250	250
O.F.H. (cumecs)	250	4000	3250	1750	800	450	250

Example 15.22 *From a small catchment (230 km^2) flows for a 6-hour storm are known. Determine rainfall excess and draw unit hydrograph.*

Date	*Time*	*Flow (cumecs)*	*Base flow (cumecs)*
10 August	*0000*	*15*	*15*
	0600	*190*	*10*
	1200	*305*	*5*
	1800	*227*	*7*
11 August	*0000*	*148*	*8*
	0600	*94*	*9*
	1200	*61*	*11*
	1800	*35*	*13*
12 August	*0000*	*15*	*15*

Solution:

Date	Time (hrs)	Flood ordinate (cumecs)	Base flow ordinate (cumecs)	Direct runoff ordinate (cumecs)	Ordinate of U.H. (col. 5/9.36) (cumecs)
(1)	(2)	(3)	(4)	(5)	(6)
10 August	0	15	15	0	0
	6	190	10	180	19.23
	12	305	5	300	32.05
	18	227	7	220	23.50
11 August	0	148	8	140	14.96
	6	94	9	85	9.08
	12	61	11	50	5.34
	18	35	13	22	2.35
12 August	0	15	15	0	0

Runoff $= [(6 \times 60 \times 60)(180 + 300 + 220 + 140 + 85 + 50 + 22)$

$= 21.535 \times 10^6$ m^3

Rainfall excess $= (21.535 \times 10^6/230 \times 10^6) \times 100 = 9.36$ cm

Unit hydrograph is shown in Fig. 15.36.

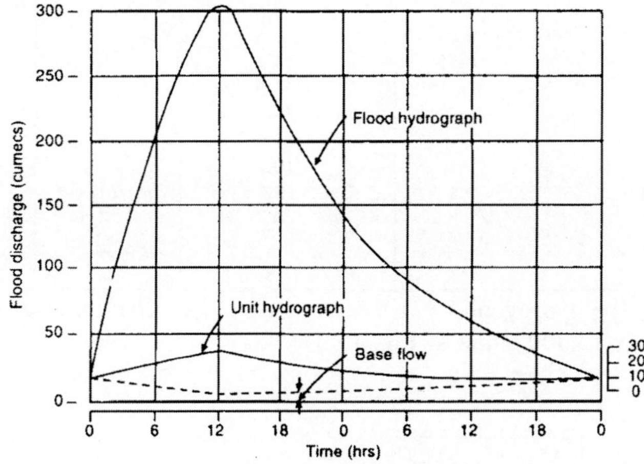

Fig. 15.36 Unit hydrograph

Example 15.23 *The stage values of a flood resulting from a uniform effective rainfall of 4 hours duration were measured at the catchment outlet, at two hourly intervals. The area of the catchment is 2200 km² and the rating curve is represented by the expression $Q = 100 (H - 0.18)^2$ where, Q = cumecs, and H = stage in metres*

Time (hrs)	*Stage (H)*	*Time (hrs)*	*Stage (H)*
0	0.50	12	2.80
2	2.95	14	3.20
4	4.45	16	1.85
6	4.60	18	1.20
8	4.25	20	1.0
10	3.30	22	0.50

Compute the ordinates of 4 hour unit hydrograph. Assume constant base flow of 10.2 cumecs.

Solution:

Time	0	2	4	6	8	10	12	14	16	18	20	22
Stage (H)	0.50	2.95	4.45	4.60	4.25	3.30	2.80	3.20	1.85	1.20	1.0	0.50
Flood discharge $Q = 100 (H - 0.18)^2$	10.2	767.3	1823.3	1953.6	1656.5	973.4	686.4	912.0	278.9	104.0	67.2	10.2

Base flow	10.2	10.2	10.2	10.2	10.2	10.2	10.2	10.2	10.2	10.2	10.2	10.2
(ODR) hydrograph	0	757.1	1813.1	1943.4	1646.3	963.2	676.2	901.8	268.7	93.8	57.0	0

ΣODR = 9120.6

$$\text{Runoff excess} = \frac{2\times60\times60}{2200\times10^6}\ (9120.6)\times100 = 2.98 \text{ cm}$$

O.U.H.	0	254.06	608.42	652.15	552.45	323.22	226.91	302.62	90.17	31.48	19.13	0

Example 15.24 *The following are the ordinates of a 12 hour unit hydrograph.*

Time (hr)	*0*	*12*	*24*	*36*	*48*	*60*	*72*	*84*
Flow (10^2 cumecs)	*0*	*16*	*29*	*26*	*14*	*7*	*1.5*	*0*

If the successive 12 hour rainfall excesses are 1.5 cm, 3.0 cm and 0.75 cm for the catchment, plot the resulting flood hydrograph on a graph sheet and mark the lag time and flow expected for the river.

Solution:

Time (hrs)	*O.U.H (10^2 cumecs)*	*Rainfall excess*	*Surface runoff from rainfall excess in successive periods (10^2 cumecs)*			*O.F.H. (10^2 cumecs)*
			0.5	*3.0*	*0.75*	
0	0	1.5	0			0
		3.0				
12	16	0.75	24	0		24
24	29		43.5	48	0	91.5
36	26		39	87	12	138
48	14		21	78	21.75	120.75
60	7		10.5	42	19.5	72
72	1.5		2.25	21	10.5	33.75
84	0		0	4.5	5.25	9.75
96				0	1.125	1.125
					0	0

The lag time from the hydrogaph is 21.6 hours (from C.G. of rainfall to C.G. of peak) and peak flow for the river is 13,800 cumecs (Fig. 15.37).

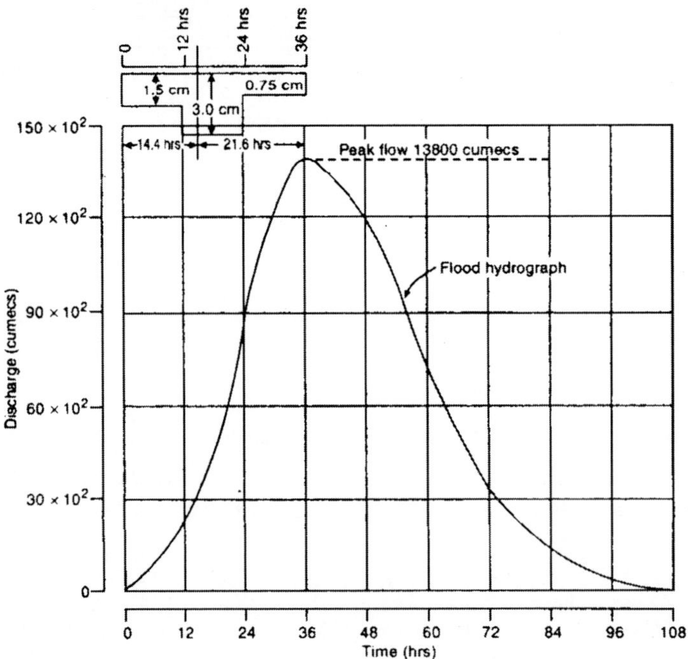

Fig. 15.37 Flood hydrograph.

Example 15.25. *Determine a 6 hr synthetic unit hydrograph for a catchment of 50 sq km, length of main stream 10 km, distance of outlet to centroid of basin is 4 km $C_t = 1.5$ and $C_p = 0.69$.*

Solution:

Basin lag, $t_p = C_t (L_c . L)^{0.3} = 1.5 (4 \times 10)^{0.3} = 4.54$ hr.

$t_r = t_p /5.5 = 4.54/5.5$

$= 0.83$ hr

$t_{PR} = t_p + (t_R - t_r')/4$

$= 4.54 + (6 - 0.83)/4$

$= 5.83$ hr

t_R' (given) $= 6$ hr

or $Q_p = 2.78 \, C_p A/t_p'$

$= (2.78 \times 0.69 \times 50)/5.83$

$= 16$ cumecs

Base period length,

$T = 3 + 3 \, t_{PR}/24$

$= 3 + (3 \times 5.83)/24$

$= 3.73$ days $= 90$ hr

Time from beginning of

rising limb to peak $= t_p' + \dfrac{t_R}{2}$

$= 5.83 + 6/2 = 8.83$ hr.

$q_{PR} = 16/50$

$= 0.32$ cumec/sq km

$W_{75} = 3.21/(q_{PR})^{1.08} = 3.21/(0.32)^{1.08} = 11$ hr

$W_{50} = 5.6/(q_{PR})^{1.08} = 5.6/(0.32)^{1.08} = 19$ hr

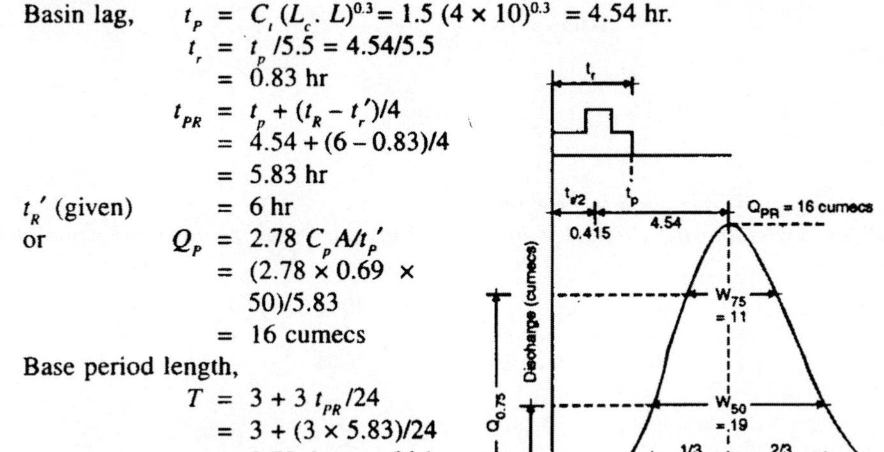

Fig. 15.38 Synthetic unit hydrograph

Fig. 15.38 indicates the unit hydrograph.

Example 15.26 *(a) A flood of a certain magnitude has a return period of 20 years. What is the probability of its exceedence?*

(b) What is the probability that a flood of this magnitude may occur in next 15 years and 10 years.

Solution: (*a*) There is 5 per cent chance of 20 years flood being equalled or exceeded in 20 years.

(*b*) Probability of this flood to occur in next 15 years

$= 1.0 - (0.95)^{15} = 0.54$

Thus there exists 54 per cent chance for this flood to occur in next 15 years.

Probability of this flood to occur in next 10 years $= 1.0 - (0.95)^{10} = 0.40$.

Thus there exists 40 per cent chance for this flood to occur in next 10 years.

Example 15.27 *Following velocities were recorded in a stream with a current meter. Find the discharge per unit width of stream near the point of measurement. Depth of flow at the point was 5 m.*

Depth above bed (m)	*0*	*1*	*2*	*3*	*4*
Velocity (m/s)	*0*	*0.5*	*0.7*	*0.8*	*0.8*

Solution: $V_{0.2} = 0.8$ m/s, $V_{0.8} = 0.5$ m/s and $V_{0.6} = 0.7$ m/s

(*a*) By two point method

Discharge per unit width $= 5 \times (0.8 + 0.5)/2 = 3.25$ cumecs

(*b*) By three point method

$$\frac{1}{2}(V_{0.2}+V_{0.8}) = (0.8 + 0.5)/2 = 0.65 \text{ m/s}$$

$$V_{0.60} = 0.7 \text{ m/s (given)}$$

Mean velocity $= \dfrac{0.65+0.70}{2} = 0.675$ m/s

Discharge per unit width = 0.675 × 5 = 3.375 cumces

(*c*) By one point method (at $V_{0.6}$)

Discharge per unit width = 0.7 × 5 = 3.5 cumecs.

Example 15.28 *Calculate the discharge of river, given the following measurements made with a flow meter.*

Distance from one bank (m)	*0*	*10*	*20*	*30*	*40*	*50*	*60*	*70*
Depth of bed (m)	*0*	*0.7*	*1.2*	*1.5*	*1.8*	*1.5*	*0.9*	*0*
Average velocity V, (m/s)	*0*	*0.15*	*0.24*	*0.30*	*0.36*	*0.33*	*0.24*	*0*

Solution:

Discharge (*i.e.*, $d \times v$)	0	0.11	0.29	0.45	0.65	0.50	0.22	0

Total discharge = 10 [(0 + 0)/2 + 0.11 + 0.29 + 0.45 + 0.65 + 0.50 + 0.22)]
= 22.2 cumecs.

Example 15.29 *Compute the stream flow for the measurement data below:*

Distance (m)	*0*	*0.6*	*1.2*	*1.8*	*2.4*	*3.0*	*3.6*	*4.2*	*4.8*	*5.4*	*6.0*	*6.6*
Depth, d (m)	*0*	*0.3*	*1.29*	*2.16*	*2.55*	*2.22*	*1.68*	*1.41*	*1.05*	*0.63*	*0.42*	*0*
Velocity, $V_{0.2}$ *(m/s)*	*0*	*0.42*	*0.57*	*0.78*	*0.87*	*0.81*	*0.75*	*0.69*	*0.63*	*0.54*	*0.45*	*0*
$V_{0.8}$	*0*	*0.21*	*0.36*	*0.54*	*0.60*	*0.30*	*0.51*	*0.45*	*0.39*	*0.33*	*0.30*	*0*

Solution:

$(V_{0.2} + V_{0.8})/2$	0	0.32	0.46	0.66	0.74	0.55	0.63	0.57	0.51	0.44	0.37	0
Discharge ($d \times V_{av}$)	0	0.10	0.59	0.43	1.89	1.22	1.06	0.80	0.54	0.28	0.16	0

(*a*) Total discharge by trapezoidal rule
= 0.6 [(0 + 0)/2 + 0.10 + 0.59 + 1.43 + 1.89 + 1.22 + 1.06 + 0.80 + 0.54 + 0.28 + 0.16] = 4.84 cumecs

(*b*) By simpson rule
= 0.6/3 [0 + 4 (0.10 + 1.43 + 1.22 + 0.80 + 0.28) + 2 (0.59) + 1.89 + 1.06 + 0.54) + 0.16] + 0.6 (0.16 + 0)/2 = 4.75 cumecs.

Example 15.30 *Determine discharge from the following current meter observations. Current meter equation,* $V = 0.3\,N + 0.04$.

Distance from one bank (m)	*Depth (m)*	*Current meter observations*		*Distance from one bank (m)*	*Depth (m)*	*Current meter observations*	
		No. of revolutions (R)	*Time (sec)*			*No. of revolutions (R)*	*Time (sec)*
2	0.5	50	160	10	1.8	140	70
4	0.7	70	150	12	1.5	110	80
6	1.2	80	100	14	0.9	90	140
8	1.5	120	80				

Solution:

Distance from one bank (m)	*Depth (m)*	*No. of revolutions (R)*	*Time (t)*	*N = R/t*	*Av. velocity (0.3 N + 0.04)*	*Discharge 2dv*
2	0.5	50	160	0.31	0.13	0.13
4	0.7	70	150	0.47	0.18	0.25
6	1.2	80	100	0.80	0.28	0.67
8	1.5	120	80	1.50	0.49	1.47
10	1.8	140	70	2.0	0.64	2.30
12	1.5	110	80	1.38	0.45	1.35
14	0.9	90	140	0.64	0.23	0.41
						Total discharge = 6.58 cumecs

Example 15.31 *A river cross section measurements by surface float method are given as under. Estimate the discharge passed through the section at the gauging site, assuming velocity factor of 0.89.*

S.No.	*Width of section (m)*	*Mean depth in each section (m)*	*Surface velocity, observed (m/sec)*	*S.No.*	*Width of section (m)*	*Mean depth in each section*	*Surface velocity, observed (m/sec)*
1	24	1.60	0.80	7	30	1.70	0.98
2	30	2.27	1.00	8	30	2.40	0.68
3	30	2.43	1.07	9	30	1.40	0.68
4	30	2.30	0.98	10	30	0.30	0.46
5	30	1.20	0.94	11	30	0.40	0.40
6	30	1.20	0.75	12	30	1.00	1.00

Solution in tabular form is as under :

S. No.	*Section width (b)*	*Mean depth (d)*	*Area (b.d.)*	*Surface velocity (V_s)*	*Mean velocity $V_s \times 0.89$*	*Discharge (area × V_m)*
1	24	1.60	24 × 1.6 = 38.4	0.80	0.71	27.26
2	30	2.27	30 × 2.27 = 68.1	1.00	0.89	60.61
3	30	2.43	30 × 2.43 = 7.29	1.07	0.95	69.26
4	30	2.30	30 × 2.30 = 69.0	0.98	0.87	60.03
5	30	1.20	30 × 1.20 = 36.0	0.94	0.84	30.24
6	30	1.20	30 × 1.20 = 36.0	0.75	0.67	24.12
7	30	1.70	30 × 1.70 = 51.0	0.98	0.87	44.37
8	30	2.40	30 × 2.40 = 72.0	0.68	0.61	43.92
9	30	1.40	30 × 1.40 = 42.0	0.68	0.61	25.62
10	30	0.30	30 × 0.30 = 9.0	0.46	0.41	3.69
11	30	0.40	30 × 0.40 = 12.0	0.40	0.36	4.32
12	30	1.00	30 × 1.0 = 30.0	1.00	0.89	26.70
						Total discharge = 420.14 cumecs

QUESTIONS

1. Define hydrology. What role does hydrology play in the water resources planning?
2. (*a*) Discuss engineering application of hydrology.
   (*b*) Define hydrologic cycle and explain it with the help of a sketch. Discuss the statement "Hydrologic cycle has no beginning or ending."
3. Describe the 'hydrologic cycle' and give the hydrologic equation. Explain the terms precipitation, stream flow and evaporation. How are they measured? List different methods of finding average depth of rainfall over an area.
4. Briefly discuss the role of various components of hydrologic cycle with the help of a suitable sketch.
   (*a*) What are the various factors that effect the rainfall characteristics?

(b) What is the intensity of rainfall and annual rainfall average?

5. (a) How would you obtain the average rainfall over a catchment area?
   (b) Discuss the various methods of obtaining peak flows.
   (c) How long range runoff from a catchment area is estimated?
6. How is the rainfall at a place measured? What are the factors that effect rainfall at a place?
   (a) What points will you keep in view while selecting a raingauge station?
   (b) Give details of recording and non-recording types of raingauges.
7. What is the practical usefulness of rainfall data analysis?
8. Define raingauge. Name various types of raingauges and discuss with the help of sketch non-recording type raingauge.
9. Discuss the merits and demerits of recording and non-recording raingauges. Explain weighing bucket raingauge.
10. Name the principal sources of errors in rainfall measurements.
11. How will you compute average rainfall over a basin? Which method is superior?
12. Explain how rainfall is measured at a place.
13. Describe different types of recording and non-recording raingauges. What is the difference between continuous and discontinuous type gauges? Draw neat sketches to illustrate any one of them. Explain how will you compute rainfall in catchment from gauge readings.
14. Discuss the various methods of measurement of rainfall.
15. Describe the various methods of calculating mean rainfall from point rainfall records in a given area.
16. (a) Give optimum number of raingauge stations required as per B.I.S. recommendation.
    (b) Determine optimum number of raingauges in a basin from the following data so as to limit the percentage error within 10%.
    Existing stations = 5 No.
    Average rainfall at the stations, = 90, 80, 54, 45, 41, cm. 8 No. **Ans.**
17. (a) Values of annual precipitation at a raingauge station, expressed in cm per year in chronological sequence from 1956 to 1967, are indicated below:

36.5	29.0	56.2	82.0	27.8	23.4
71.2	48.3	31.4	18.1	29.0	65.6

Using Hazen's or any other method, estimate the maximum and minimum value of precipitation which has a recurrence interval of 5 years.

(b) Maximum 66 cm, Minimum 27 cm **Ans.**

Compute the rainfall at station *X* from following data:

Station	*X*	*A*	*B*	*C*
Storm rainfall (cm)	X	12.5	14.5	16.9
Annual rainfall (cm)	115	125	145	130

12.65 cm **Ans.**

18. What is runoff? What are the factors that effect the runoff from a catchment area.
19. List the factors effecting runoff.
20. List the methods of working out runoff from a drainage area and describe any two methods in detail.
21. Discuss the various methods of determining runoff from a drainage area? State three important formulae.
22. Explain the effect over runoff of (a) thick vegetal cover, (b) snow cover, (c) grass, (d) permeable soil and (e) uneven land.
23. (a) What do you understand by catchment area and runoff?

(*b*) Explain the various factors that influence the runoff from a catchment area.
(*c*) How will you measure rainfall?

24. (*a*) What are the different factors effecting the runoff from a catchment area?
(*b*) How will you estimate the flood from a basin where the stream runoff data are scarce?
25. Define fan-shaped and fern-shaped catchments. Discuss their chief characteristics.
26. What is infiltration? Define (*a*) infiltration capacity, and (*b*) infiltration rate. How would you use infiltration capacity curve to calculate runoff from a small catchment?
What is the difference between *W*-index and ϕ-index.
Explain the different methods of measuring infiltration.
27. Explain with sketches, if necessary, the following methods for calculating net storm rain: (*a*) Infiltration capacity method, (*b*) *W*-index method.
A three-hour storm occurs over a 62 sq km area. From the following data, estimate the net storm rain for the whole area and its hourly distribution.

Sub-area (km^2)	ϕ-index (cm/hr)	Hourly rain		
		1st	2nd	3rd
14.0	2.0	1.5	5.0	0.8
20.0	3.5	1.5	5.0	0.8
28.0	1.0	1.5	5.5	1.5

28. Describe the terms as used in hydrologic practices (*i*) Infiltration index, (*ii*) Time of concentration, (*iii*) Runoff, (*iv*) Isohyetals.
29. Discuss the Rational method for the qualitative determination of urban storm water run-off. What are the limitations of the method?
30. What is flood? Describe the various methods of flood estimation.
31. What are the various factors which effect the peak flood.
32. List various formulae to obtain the peak flow and discuss their limitations.
33. Distinguish between maximum flood and spillway design flood.
34. What is the importance of calculating the maximum flood discharge, total runoff from a river for the design of a water resources project?
35. What are the various purposes for which calculation of maximum probable flood are necessary? What methods are used for the calculation of flood discharge? Describe briefly.
36. Explain briefly how peak flood discharge can be estimated from frequency analysis of stream flow data.
37. Find the values of C and n in the empirical formula, $Q = CA^n$ for flood discharge, Q in m^3/s from a catchment for area A km^2, from the following data:

A (km^2)	4	10	50	100	200
Q (cumec)	100	158	354	500	706

($C = 50$, $n = 0.5$) Ans.

38. Describe briefly how you will determine the maximum flood by using (*i*) rational formula, and (*ii*) Dicken's formula.
39. Determine the net runoff and total rainfall and *W*-index for the following data when ϕ index = 3 cm/hr.

Time of rainfall (minutes)	30	30	30	30	30
Rates of rainfall (cm/hr)	2.0	2.5	12.0	8.0	2.0

Net rainfall = 7.0 cm, total rainfall = 13.25 cm, Windex = 2.5 cm/hr. Ans.

40. Describe the methods of river gauging in different situations.
41. What are the requirements of a site for area velocity method of measurement of discharge in a river?

42. (*a*) Write a note on the velocity measurement by current meter.
    (*b*) How will you measure the discharge of a stream with the help of float?
43. Define current meter. Indicate two principal types of current meters generally used for determining velocity in deep and shallow depths.
44. Discuss discharge computation under area-velocity method.
45. How does slope-area method of discharge observation differs from area velocity method? Where and why the former is used for discharge computation?
46. What is echo sounder? Outline the principle of working of echo sounder. What are its limitations?
47. With neat illustrative sketch explain how the discharge can be measured by area-velocity method using current meter?
48. Describe an appropriate method for measuring discharge of a major river.
49. Explain the working principle of a current meter. Give a stepwise outline of determining stream discharge using a current meter.

PART - V

DAM ENGINEERING

- Reservoirs and Dams Planning
- Earth and Rockfill Dams
- Gravity Dams
- Arch and Buttress Dams
- Discharging Facilities

16

RESERVOIRS AND DAMS PLANNING

16.1 INTRODUCTION

A reservoir is created with the impounding of runoff from the catchment upstream by the construction of a dam across a river or stream. Storage is done during the period when the flow is in excess of the demand for release during the lean supply period so as to maintain continuous hydel power generation, besides meeting requirements of irrigation, water supply, etc. The reservoir thus provides (*i*) Protection from flood as the flood water escaping downstream is moderated so as to be compatible with the flood carrying capacity of the river channel downstream, (*ii*) Water stored is utilised for irrigation, power generation, water supply, navigation, fishing, recreation, etc., (*iii*) Improvement in climate, (*v*) lessening river pollution, (*v*) Check on spread of diseases due to improved water supply and sanitation, and (*vi*) Reduction in river section downstream of the dam thereby making available considerable land from river bed for cultivation.

The reservoirs, on the other hand, have certain disadvantages, viz., (*i*) Submergence of fertile valley lands, (*ii*) Displacement of large population from reservoir area and their resettlement cost at new areas, (*iii*) Reduced capacity to flush pollution and salt from esturary, (*iv*) Possible adverse effects on the ecology of the project area, (*v*) Entraping of fertile sand which has manurial value for cultivation in command area. Lack of silt adversely effects fishing, and (*vi*) flooding of forest and displacement of wild life.

16.2 SELECTION OF SITE FOR A RESERVOIR

Various considerations in the selection of site for a reservoir are (*i*) Good runoff from the catchment area with minimum percolation losses, (*ii*) Runoff free from excessive silt load, (*iii*) Soil in the catchment area does not contain harmful soluble salts and minerals, (*iv*) Water tight reservoir perimeter. There should be no serious leakage from the reservoir basin, (*v*) High and steep side hills to ensure deep reservoir with maximum storage but minimum surface area and hence minimum evaporation losses and lesser possibility of weed growth. Steep sided reservoirs are, therefore, ideal but are rarely available, (*vi*) Geology of the region *i.e.*, type of rock and soil besides climate suitable to ensure minimum erosion of the soil. Impervious rock formation is the best, (*vii*) Rock formation having dip in the upstream direction is preferable. Storage water is led out of the basin if the rock is dipping downstream, (*viii*) Ideal dam site *i.e.*, narrow gorge with sound foundations to locate dam and rapid widening of the valley above dam site to afford greater average volume per metre of height and length of the dam, (*ix*) Reservoir basin barren so that cost of land acquisition and jungle clearance is the minimum, (*x*) Minimum submergence of forest and cultivated lands and habitations so that the uprooted population is the least and the land acquisition and compensation charges are bare minimum, (*xi*) Remote from seismic zone as far as possible, (*xii*) Nearest to the commanded area to ensure minimum length of canal system and hence minimum water conveyance losses, (*xiii*) Minimum disruption to rail-road communication in the region, (*xiv*) Exclusion of tributaries bringing heavy silt-laden water, (*xv*) Availability of construction materials for the dam within minimum leads, (*xvi*) Smallest possible drainage area to yield the required supplies as rate of silting increases directly as the ratio of drainage area to reservoir capacity, and (*xvii*) Absence of objectionable vegetation and marshy land which may otherwise impart bad odour, taste and colour to the impounded water. Reservoir sites possessing all the characteristics, as above, are rare to find.

16.3 CLASSIFICATION OF RESERVOIRS

1. Classification on the Basis of the Purpose Served:

Storage Reservoir

Also termed as conservation or impounding reservoir. Storage reservoir stores surplus water during the period of excess flow for a considerable period of time so as to maintain continuous supply for irrigation, hydel power generation, municipal water supply, industrial purposes, navigation etc. during the period of lean supply in the river but when demand is keen.

Flood Control Reservoir

Flood control reservoir holds a part volume (in excess of the safe carrying capacity of the river channel downstream) temporarily during floods and releases later, when flood recedes, as rapidly as the channel capacity permits. Flood control reservoir thus reduces flood stage and consequently avoids damage downstream. Construction of reservoir solely for flood control measures is not advisable except in very special cases due to the high cost of construction. Flood control reservoirs may be either retarding reservoirs or detention reservoirs.

(*i*) *Retarding reservoirs*: Retarding reservoirs are those at which no gates are provided in the outlets to regulate the releases but the discharging capacity of the outlets and spillway is so fixed that it is not in excess of the flood carrying capacity of the river channel downstream. With the rise in reservoir level, the amount of water released is such as would not create flooding in the areas downstream. They are preferred on small rivers. The suitable location of a retarding reservoir is immediately upstream of the city to be protected from floods or above the confluence of two or more streams.

(*ii*) *Detention reservoirs*: Detention reservoirs have gated outlets and store water for a relatively brief period of time so as to provide greater flexibility in the operation of reservoirs. They are specially suitable when the area under control increases in size and protected area is widespread.

Advantages: (*i*) Profitable venture and prime necessity, (*ii*) Maximum benefits from the stored water by way of irrigation, power, flood control, navigation, public water supply, etc., and (*iii*) Lot of indirect benefits such as development of fisheries and tourism.

Limitations: (*i*) High initial cost, (*ii*) Cost becomes disproportionate to the results aimed at in case the capacity required for flood control cannot be utilized completely or partly, (*iv*) Usually involve submergence and consequent uprooting of population and consequent their rehabilitation and resettlement problems.

2. Classification on the Basis of Regulation:

Distribution Reservoir

It is a reservoir connected with water supply project of a city or a town. It is usually of limited storage capacity, used primarily to cater for fluctuations in demand which may occur over short periods of several hours to several days and also as local storage to take care of emergency in the event of break in main supply line or failure of the pumping plant, etc.

Single Purpose Reservoir

It is the reservoir to serve only one purpose which may be flood control or irrigation or power generation, municipal water supply, recreation etc.

Multipurpose Reservoir

Also termed as multiuse reservoir. In this reservoir, the storage and release cater for a combination of two or more purposes such as irrigation, hydel power generation, flood control, public water supply, navigation, recreation, fisheries, etc. A reservoir which is designed for one of the purposes put provides other incidental benefits because every reservoir serves or should serve more than one purpose on account of indirect benefits that may result from its operation, is also referred to as multipurpose reservoir. Keeping with the diverse purposes served by a multipurpose reservoir, it is usually of large capacity so as to (*i*) Reserve a certain minimum storage capacity at all times for flood control, (*ii*) Provide steady supply for kharif and rabi crops, (*iii*) Generate maximum firm power . Usually, these enormous benefits accruing from a multipurpose project repay the cost of building the dam within a couple of years.

Balancing Reservoir

It is a reservoir, usually of limited capacity, located downstream of (or subsidiary to) a main reservoir to (*i*) Store the water let down from the reservoir in excess of that required for irrigation or additional power generation, (*ii*) Provide flexibility of operation to the distribution system, and (*iii*) At certain locations primarily for permitting regulated supply to the power penstocks with a view to cater for the fluctuations in requirement of water supply to the turbines depending on power load.

Auxiliary Reservoir

Also termed as compensatory reservoir. This is a reservoir which supplements and absorbs the spill of a main reservoir:

System of Reservoirs

These consist of a group of single or multipurpose reservoirs which may be operated in an integrated manner for optimum utilisation of water resources of the river system.

3. Classification on the basis of Water Quality:

Fresh Water Reservoir : It is the reservoir the water of which is replenished at least once, preferably twice in a year by the continuous cycle of inflow and outflow due to low capacity-inflow ratio. It provides quality water nearest to the fresh water.

Stored Water Reservoir : It is the reservoir which stores almost entire inflow, the bulk of which is received during monsoon. Evidently the capacity-inflow ratio is high. The quality of water may not be comparable to the fresh water reservoir also possibly due to bad odour and taste imparted by contamination.

16.4 STORAGE ZONES OF A RESERVOIR

Various zones of storage in a reservoir, illustrated in Fig. 16.1 are discussed below:

Dead Storage

Dead storage in a reservoir, about 10% of the gross storage capacity, is provided to cater for sediment deposition by the impounded sediment-laden waters as very small portion of the sediment passes down through dam outlets and for the minimum drawdown in the case of power projects. It is equivalent to the volume of sediment expected to be deposited in the reservoir during the designed life of the reservoir, usually taken as 100 years. The level attained by the reservoir corresponding to the dead storage is termed as dead storage level. It is the level below which a reservoir is not susceptible to release water by the in-built outlet means. In power projects, releases are allowed up to minimum drawdown level only instead of dead storage level so as to maintain the minimum head required for power generation.

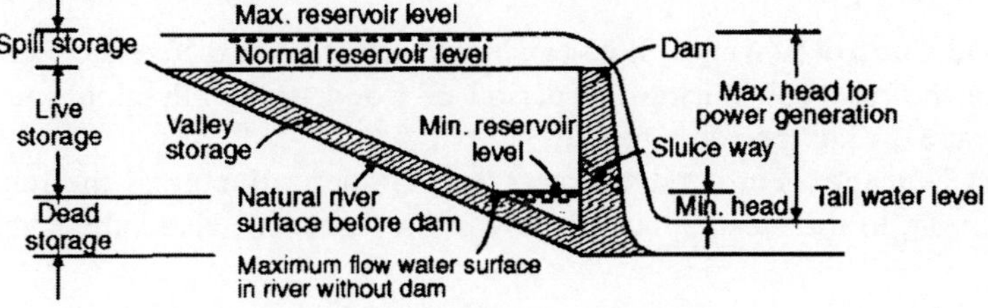

Fig. 16.1 Storage zones of reservoir.

Live Storage

It is the storage capacity of the reservoir above dead storage level which constitutes useable portion of the total storage. It is thus the difference of gross storage capacity and the sum of dead storage capacity and inactive capacity for hydro power generation. Inactive capacity is the storage capacity exclusive of dead storage below which evacuation is not contemplated because of minimum irrigation and power load requirement. However, in practice dead storage capacity and inactive capacity are considered the same. Live storage has to be sufficient so that the project is successful for (*i*) 75% of its life period in an irrigation project, (*ii*) 90% of its life period in an hydel power project and (*iii*) 100% of its life period in water supply project.

Flood Storage

Flood storage is the storage contained between the maximum reservoir level and the full reservoir level. It is the storage space provided in a reservoir for storing flood water temporarily to moderate the releases downstream. Flood storage varies with the spillway capacity for a given design flood. When the spillway capacity is equal to the inflow design flood, there is no flood storage and the full reservoir level and maximum reservoir level are the same. Effective flood storage is the difference between the amount of water stored in the reservoir during the period of flood peak reduction and the amount of natural valley storage during the same period.

Induced surcharge storage

The storage between FRL and MWL of a reservoir which may be induced by regulating the outlet gates after the reservoir is filled up to full reservoir level.

Bank Storage

It is the storage absorbed and stored in the bed and banks of a stream, lake, in the voids in the soil cover in a reservoir and becomes available in whole or in part as seepage water when the water level drops down in the reservoir. Bank storage increases the reservoir capacity over and above that given by the elevation storage curve.

Valley Storage

Rivers in dry season flowing within its banks have small water flow compared to that when in floods. Valley storage is the storage in the river or stream in floods after it has overflowed its banks. It is an important element in the design of large size flood control reservoirs where the valley storage at the time of flood is of significant proportions in regard to the reservoir storage volume.

Conservation Storage

Water impounded in a reservoir for conventional uses such as irrigation, power generation, industrial uses, municipal water supply, etc. is termed as conservation storage

Carry Over storage

Also called 'over year storage'. The storage left over in a reservoir at the end of the depletion period of a year which is available for use in later years. Alternatively, it is the storage collected during surplus years for making up deficiencies in dry or lean year.

Cyclic Storage: Water stored in a reservoir during periods of more than average supply, and the releasing of it for use during periods of insufficient supply; also called 'over-year storage'.

Effective Power Storage: Storage between minimum and maximum reservoir levels for power generation.

Effective Flood Control Storage: Storage which represents the difference between the amount of water stored in the reservoir during the period of flood peak reduction and the amount of natural valley storage during the same period.

Coefficient of Storage: For surface waters, a coefficient to express the relation of storage capacity in a reservoir, to the mean annual inflow in the reservoir; also called 'reservoir factor'.

In multipurpose reservoirs, five basic storage zones of reservoir space for use in operating the reservoir for various functions are indicated in (Fig. 16.1 *a*) and discussed as under:

1. *Spill Zone*: It is the storage space above flood control zone between FRL and MWL as is occupied mostly during high floods.

2. *Flood Control Zone*: It is the storage space earmarked as temporary storage for absorbing high flows for alleviating downstream flood damages.

3. *Conservation Zone*: It is the storage zone between FRL and dead storage level used for conservation of water to meet various future demands.

4. *Buffer Zone*: It is the storage space above dead storage level which is used to satisfy only very essential water needs in case of extreme situation.

5. *Dead Storage Zone*: It is the lowest zone meant to absorb sediment inflow. It is not susceptible to release by in-built outlet means.

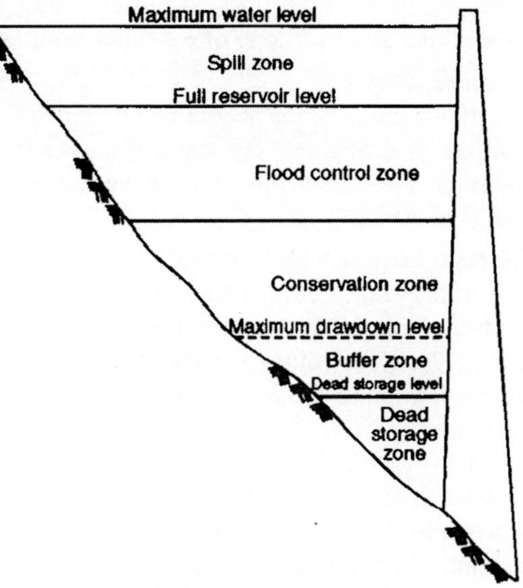

Fig. 16.1 (*a*) Storage zones in multipurpose reservoir.

Active Capacity: It is the storage available for project purposes; usually between the lowest allowable level of release (minimum drawdown level) and the highest controlled water surface (static full pool level); also called 'effective storage', 'useful storage', 'usable storage', 'effective capacity', or 'useful capacity'.

Area Capacity Curves: The graph of area of water spread and the storage volume of reservoir as function of elevation (above lowest river bed level).

16.5 DESIGN CAPACITY OF RESERVOIRS

The design capacity of storage reservoirs is to ensure that the water will be replenished often enough; at least once, preferably twice, in a year to classify it as fresh water reservoirs and accordingly the drainage area must be large enough. The design capacity of a reservoir is fixed in consideration of the various factors such as (*i*) Long-term precipitation records for the catchment, (*ii*) Long-term runoff data at or near the site, (*iii*) Sediment yield into the reservoir from catchment, (*iv*) Area and capacity curves, (*v*) Trap efficiency, (*vi*) Losses in the reservoir, (*vii*) Maximum requirement of water for different uses to be met from the reservoir and direct flows of the river irrespective of the availability of assured supplies, (*viii*) Consideration of dependability, (*ix*) Upstream use, and (*x*) Density current aspects and location of outlets. Usually the storage capacity is determined by mass curve method.

Mass Curve*

It is a curve of the accumulated total flow (or rainfall) against time. The characteristics of the curve are that (*i*) It rises continuously as it shows accumulated inflow, (*ii*) The rise is steep if the rate of inflow is high. Steepness of the curve indicates the rate of inflow at that time interval, (*iii*) It is horizontal if there is no inflow during a certain period. The relatively dry periods are indicated as hollows on the curve, (*iv*) Any point on

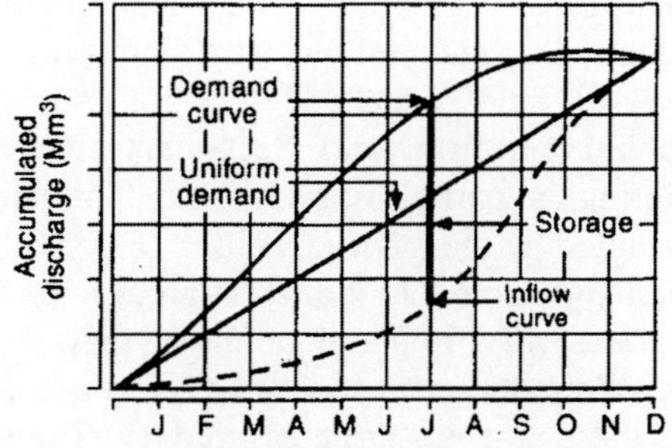

Fig. 16.2 Mass Curves.

*For details refer "Hydrology and Water Resources Engineering" by the anthor.

it represents the total flow, in million cubic metres, from the beginning of the period up to the given time, and (*v*) Slope of a tangent to it at this point indicates the discharge in cumecs.

Demand Curve

It is a curve of the accumulated demand with time. In other words it is a plot of accumulated outflow against time. The demand curve is a straight line, rising from the origin, if there is a constant rate of demand. In case there is a variable rate of demand, the demand curve is curved one (Fig. 16.2).

Storage Capacity Determination

The storage capacity of a reservoir to meet the demand of continuous supply is determined with the help of the observed discharge data of a stream or a river on which the dam is to be built. The flow values for the driest years in as long a period as is available, say 25 to 30 years; minimum 10 years is considered. The storage capacity is determined by the following methods:

1. *Analytical Method*: The inflow and demand values in each month are determined. The demand includes prior rights, if any, evaporation losses, rainfall etc. The deficit and surplus of water which is the departure of inflow volume from the demand volume is determined. The maximum value of cumulative excess of demand over the inflow represents the minimum storage necessary to meet the demand. The cumulative excess inflow volume starting from each demand withdrawal from storage is also determined which indicates the filling up of the reservoir and the volumes in excess of storage to be spilled over as also whether the reservoir fills up after a demand and when.

2. *Mass Curve Method*: In this method the storage capacity is determined from mass curves of inflow and demand (Fig. 16.11). An examination of the demand curve and inflow curve drawn to the same scale would indicate when storage is required. When inflow rate is low, *i.e.*, the inflow curve is flat but at that time if the slope of demand curve is more, it indicates that storage is required to meet the demand.

3. *Bar Graph Method*: In this method, the average inflows of the driest year are plotted as ordinates against time as abscissa to get a stepped graph known as bar graph (Fig. 16.12). The area in the diagram represents the total volume of water which enters the reservoir in that year. Likewise, the average demand is then plotted on the graph, straight line if demand is uniform and stepped in the case of non-uniform demand. The area of maximum deficit *i.e.*, between the demand and inflow represents to scale the minimum storage capacity of the reservoir. The storage required to meet the maximum uniform demand that can be met with without loss of water over spillway is given by the area of surplus over the demand.

16.6 SAFE YIELD

Reservoir yield is the amount of water that can be released from a reservoir in a certain interval of time. The time interval is a month or a year for large conservation reservoirs. Reservoir yield, depending upon inflow, vary from time to time. Firm or safe yield is the maximum amount of supply or yield guaranteed during a critical dry period. Firm power generation is determined corresponding to firm yield. Secondary yield is the quantity of water available in excess of safe yield during the periods of high flows. The safe yield from a reservoir of a given storage capacity is determined by mass curve method (Fig. 16.11) by (*i*) Drawing inflow mass curve, (*ii*) Drawing mass demand curve. On it draw straight lines from a common origin to represent the demands at various rates, (*iii*) From high points on mass inflow curve, draw tangents such that their maximum departure from the mass curve does not exceed the specified reservoir capacity, (*iv*) The slopes of each of these tangents are measured, (*v*) The slope indicates the yield as can be attained in each year from the reservoir of a given capacity, and (*vi*) The slope of the most flat demand line is the safe yield.

16.7 FLOOD ABSORPTION CAPACITY OF A RESERVOIR

The reservoirs are generally constructed to store a small percentage of maximum annual runoff from the catchment. In storage reservoirs, the flood waters are impounded for release during the periods of low flow and as such there is little guarantee that they will have sufficient available capacity of flood absorption at the incidence of a flood for effective flood control. Flood control reservoirs are operated so as to store flood water temporarily and release as rapidly as possible as downstream river channel permits without causing any damage. The multipurpose reservoirs are effective as they have large capacity.

The floods may be disposed of by (*i*) Absorption in the reservoir to fill it, (*ii*) Storage as flood storage, between maximum reservoir level and normal reservoir level, and (*iii*) Moderated and routed through spillway. A correct estimate of the flood storage capacity of a reservoir is essential so that depletion of the reservoir in premonsoon period is so devised as to enable that the reservoir may be full in the monsoon season.

16.8 DENSITY CURRENTS

Density current is defined as gravity flow of fluid under, over or through another fluid of approximately equal density or the density of which differs by a small amount from that of the primary current. Further, it is essential that two fluids are miscible and that the density difference be a function of differences in temperature, salt content and/or sediment content of the two fluids, but independent of pressure intensity and elastic properties of the fluid. A sediment laden inflow with density greater than the reservoir water descends to the lower water layers and moves there in the direction of the dam. Such underflows are one of the common types of density currents. Fig. 16.3 shows the formation of such currents.

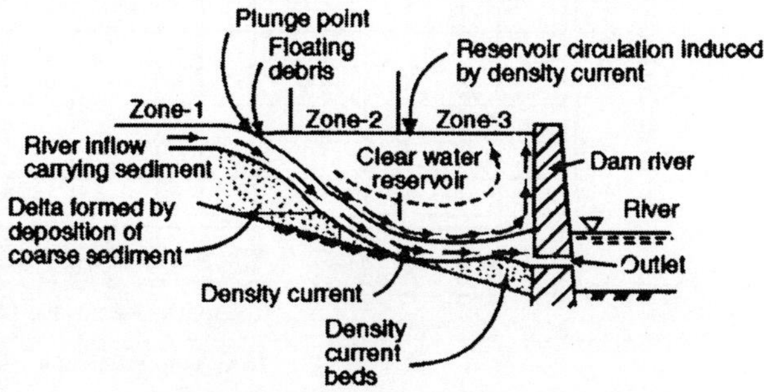

Fig. 16.3 Formation of density current.

Density currents of importance in reservoirs are those that transport sediments from upper and higher parts of the reservoir to its lower reaches especially to the dam. The inflowing muddy water may plung directly, on entry in the reservoir, with the clear still water of the reservoir and partial mixing occurs due to the losses involved in the change of momentum caused by the reduction in velocity of flow. The rate of silting of reservoirs can be reduced if the density currents are vented through the outlets, and sluice gates properly located and operated in the body of the dam so that a major portion of the sediment laden water entering the reservoir can be ejected downstream of the dam in the form of density currents before if has a chance to settle down.

16.9 TRAP EFFICIENCY

Reservoir sedimentation is measured in terms of trap efficiency. Trap efficiency is defined as the ratio of sediment retained in the reservoir to the sediment brought by the water into the reservoir. The trap efficiency primarily depends on the sediment load characteristics, and the detention time of the inflow, method of reservoir operation and age of reservoir. The detention time depends on storage capacity-inflow ratio (C/I); the larger the C/I, more is the detention time and higher is the percentage of incoming sediments trapped in the reservoir and higher is the trap efficiency. Furthermore, normal ponded reservoirs with sediments always submerged have lower compaction rate than those with periodic drawdowns. The trap efficiency decreases with age as in a majority of reservoirs the rate of silting generally falls off with the age.

Trap efficiency can be computed based on the actual inflow and outflow of the sediments assessed from the sediment observations made at the entry into and released downstream immediately below the outlets. The volume occupied by the sediments assessed is computed from the known specific weight of the sediment in actual conditions at the reservoir. While allocating space for dead storage in most cases, the trap efficiency is considered at least 95% and only on rare and special occasions it falls below 90%. Observations at Bhakra and Beas reservoirs have revealed that the trap efficiency is almost 99% and 97% respectively.

The two methods for determining trap efficiency of a reservoir are Brune's method and Churchil method. In Brune's method, trap efficiency is related to inflow of water into the reservoir and its storage capacity. Fig. 16.4 gives the generalized trap efficiency based on actual observations from existing reservoirs. From the curves the amount of sediment which a new reservoir will trap can be read against known value of the capacity-inflow ratio. It indicates that trap efficiency is low for reservoirs having small capacity-inflow ratio and it gradually increases as capacity-inflow ratio increases.

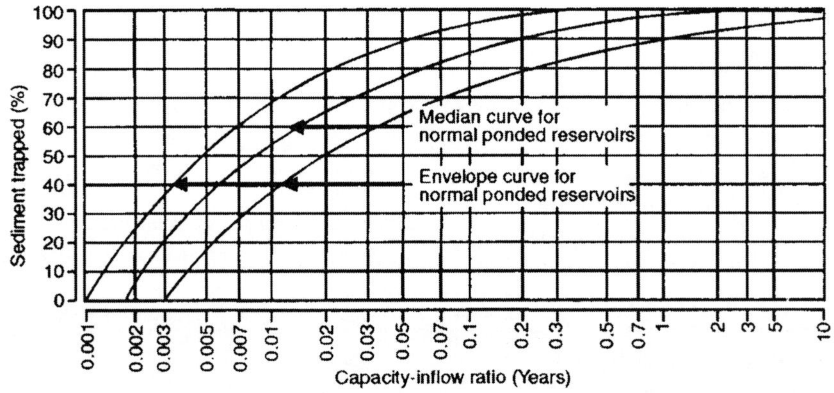

Fig. 16.4 Trap efficiency (After Brune)

Churchil curve (Fig. 16.4*a*) for trap efficiency period of detention and velocity through reservoir as controlling factors. It is a relationship between percentage of incoming sediment passing through reservoir and sedimentation index of a reservoir. The sedimentation index is defined as the ratio of the period of detention to the mean velocity through the reservoir.

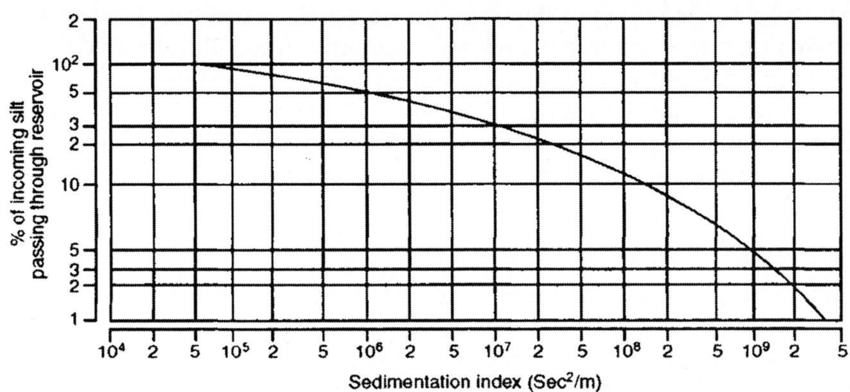

Fig. 16.4 (*a*) Churchil's curve-percent sediment passing vs sediment index.

16.10 RESERVOIR SEDIMENTATION

The river water carries silt. The flow impounded in the reservoir during monsoons is heavily silt laden. The fine sediment in the river water is carried in suspension, as suspended load, while the larger solids move on or near the bed as bed load. When the river water reaches a reservoir, its velocity and turbulence are reduced and as a consequence the bed load generally gets deposited along the thawleg or pre-reservoir course, but the suspended sediments which take a long time to settle down spread more evenly all over the reservoir bed, and frequently the thickest deposit of such sediment are below the deepest standing water nearer the dam. A small amount of sediment deposit is passed down along with the supply released through the outlets. The bed load and a part of suspended sediment is deposited in the reservoir bed if the reservoir discharges the surplus freely for long time which is the case when runoff of the river is large in comparison to the reservoir capacity, *i.e.*, small capacity-inflow ratio. However, if the capacity-inflow ratio is large, *i.e.* the reservoir is large in comparison to the annual runoff, all the water-borne sediment brought by runoff is deposited in the reservoir bed.

The probable rate of silting in a reservoir is estimated on the basis of (*a*) Sediment deposits surveyed over long period in existing reservoirs with similar catchment characteristics, (*b*) Observations of sediment load carried by the river and tributaries and that in the outflow through the gates, (*c*) Predictions based on total erosion, and (*d*) Remote sensing technique.

Sediment Load

Sediment load consists of suspended load, and bed load. The sequence of variation of sediment in a vertical is that (*i*) Large particles moving along in contact with the bed of the stream constitute bed load, (*ii*) Immediately above the bed, a thin layer comprises the materials which are relatively coarse with high concentration, carried either in suspension by saltation or both, and (*iii*) Above it and extending up to the water surface, comprises a thick layer of relatively fine grained materials and their concentration diminishing with height carried in suspension.

Suspended Load

The suspended load which comprises the greatest part of total sediment load is important in reservoir silting. The suspended load is very fine and is generally carried away by the river to the delta or by irrigation channels to the fields. Term wash load is often used to designate very minute particles approaching colloidal sizes, which always remain in suspension and hence are seldom found on the bed of channels. The suspended load is classified in terms of sediment diameter* into three grades; coarse sediments are particles above 0.20 mm diameter, medium sediments are particles between 0.20 to 0.075 mm, and fine sediments have particles below 0.075 mm diameter.

Periodic samples from the river water are taken at various discharges and the suspended sediment concentration is measured. A sediment rating curve**, which is a plot of sediment concentration against the discharge is prepared and is used for approximate estimation of suspended sediment load.

Bed Load

Bed load denotes the particles moving on or near the bed. Movement of particles takes place by rolling, sliding and hopping, *i.e.*, saltation depending on the velocity of flow. Generally the bed load is taken as a certain percentage of suspended load which is susceptible to sufficiently accurate measurement. This percentage usually varies from 10 to 20 per cent of the total suspended load depending upon the nature of bed material, its grade, the concentration of suspended material etc. A percentage of 15 is the more commonly acceptable figure.

Distribution Pattern of Sediments in Reservoirs

The pattern of sediment distribution in a reservoir is very important in planning and design of conservation structures, *e.g.*, (*i*) To ascertain how high shall sediment accumulate at the dam during a given period to fix the crest level of outlets and penstock gates, etc., (*ii*) For allocation of different storages, as silt is distributed throughout the reservoir, (*iii*) To estimate region where delta will form and consequent increase in back water levels after the reservoir comes into operation, (*iv*) To locate spots for recreational facilities such as swimming and boating. The sediment distribution pattern in a reservoir depends on many factors, as under, but reservoir operation has an important control over other factors.

---

* Sediment diameter is defined as the diameter of a sphere of the same specific weight and the same terminal settling velocity as the given particles in the same sedimentation fluid.

**For details refer "Hydrology and Water Resources Engineering" by the author.

(*i*) *Slope of the river*: In a river with steeper slope, the length of the reservoir for the dam height is shorter and the sediment deposition takes place in lower reaches of the reservoir. The sediment deposition takes place in higher elevations in reservoirs located at rivers with flat slope.

(*ii*) *Length of reservoir*: The large sediment deposition takes place in lower reaches in reservoirs of shorter length but at higher elevations in reservoirs of longer length.

(*iii*) *Constriction in reservoir*: Where there are constrictions in a reservoir, deposition is more in the upper reaches. The deposition is flatter in constricted reach and does not follow the natural slope of the river.

(*iv*) *Size of sediment*: The size of sediment greatly controls the pattern of sediment deposition. The larger the portion of large size particles, the greater is the sediment deposition at higher elevations. If fraction of fine particles is more, more deposition takes place near the dam.

(*v*) *Capacity-inflow ratio*: A small reservoir on a large river passes most of its inflow so quickly that the finer sediments do not get time to settle in the reservoir but are discharged downstream. A large reservoir may retain water for a number of years and permit almost complete deposition of suspended sediment.

(*vi*) *Reservoir operation*: The mode of reservoir operation greatly influences the pattern of sedimentation. Single purpose reservoirs are generally operated at a relatively constant level and as such the primary pattern of deposition is not much disturbed. The multipurpose reservoirs, on the other hand, are operated within a wide range of elevations to cater for various demands in different parts of a year which causes significant secondary variations in the primary pattern of deposition because of flushing down of some sediment deposited in live storage at higher elevation due to seasonal drawdown in the reservoir. Multiple deltas are formed. Large volume of sediment deposits near the minimum drawdown level.

(*vii*) *Vegetal growth*: Vegetal growth helps in trapping more sediment at higher elevations thereby accentuating delta formation.

(*viii*) *River inflow pattern*: In a river which is in flood in early part of the monsoon, more sediment is trapped in the reservoir and deposited in the lower elevation as compared to that when major flood occurs at the end of monsoon period.

(*ix*) *Sediment load*: The sediment load brought by a river water varies greatly in quantity and quality as it depends on the characteristics of the catchment with regard to soil cover and climatic conditions which in turn vary widely during the life of a reservoir. The sediment trapped being a function of retention period reduces with the age of reservoir. The sedimentation pattern is greatly affected by these variations.

(*x*) *Shape of reservoir*: In a reservoir regular in shape the suspended sediment deposits uniformly along the direction of flow, with decreasing depth away from the dam. There is a significant variation in the depth of bottom set beds if the reservoir is irregular in shape.

(*xi*) *Outlets*: Outlets of adequate capacity and located at lower elevations tend to eject out density currents, thereby reducing the extent of deposition of density currents near the dam.

(*xii*) *Sediment-reservoir volume ratio*: If sediment inflow is large compared with the reservoir capacity, the sedimentation is heavy and the reservoir has very short useful life.

However, the pattern of sediment deposition in a reservoir is a combined effect of all these factors.

Factors Effecting Silting of Reservoirs

Principal factors controlling rate of silting in a reservoir are (*i*) Quality, quantity and concentration of the sediment brought down by the river, (*ii*) Percentage of the silt intercepted by an upper reservoir or all the silt from the river that reaches the reservoir, (*iii*) Percentage of the silt reaching the reservoir trapped or what proportion passes through which in turn depends upon methods of reservoir operation and nature of spillway and other discharging facility in operation, (*iv*) Degree of consolidation *i.e.*, weight of silt per unit volume, (*v*) Length of reservoir, (*vi*) Runoff volume and peak discharge from catchment and subcatchments, (*vii*) Ratio of reservoir capacity to annual runoff, (*viii*) Exposure of deposited material, (*ix*) Depth and age of sediment deposited, (*x*) Size and shape of reservoir. In general, the unit rate of sediment production decreases with the increase in watershed area for land use tends towards greater uniformity in larger watersheds, (*xi*) Steepness of thawleg. Steep slopes give rise to higher velocity and hence higher silt-charge, (*xii*) Growth of vegetation at the head of reservoir. Erosion is reduced if there is adequate vegetation cover in the catchment area, (*xiii*) Nature of soil of the catchment area. Clayey and coarse grained sandy soils are less erodible than silty soil. Soils are more erodible when dry than moist. Igneous rocks are less easily erodible than the sedimentary or metamorphic rocks, (*xiv*) Type of rainfall and snowfall in the catchment. High intensity rainfall of short duration yields higher run off and more silt charge. Most of the sediment is brought during freshets in monsoon season, (*xv*) Effective annual rainfall, (*xvi*) Mean monthly and annual temperature in the watershed area, and (*xvii*) Earthquakes generate additional amount of silt. During earthquake huge amount of debris are displaced and carried by the river which are ultimately deposited in the reservoirs.

Effects of Sedimentation

The principal adverse effects of sedimentation of a reservoir are (*i*) Reduction of active storage capacity which may reduce the capability of a reservoir to deliver the benefits which could have been delivered by the reservoir but for sedimentation. The progressive reduction of the storage capacity affect the output, (*i*) Reduction in the availability of firm water and power supply, (*ii*) May block the outlet intakes hampering operation of gates especially those which are not frequently operated, (*iii*) Increased passage of sediments in irrigation canals, power houses etc which may become serious with the passage of time, (*iv*) Increase in flood levels upstream may cause additional submergence, formation of marshy lands, (*v*) River regime at entry to the reservoir is affected. Delta formation and braided river pattern may result and this may be unsightly, (*vi*) With large sediment deposits, humps are formed in reservoir bed. The deleterious effect of hump formation is early reduction of live storage capacity, (*vii*) Turbidity of the reservoir water is increased resulting in deterioration of water quality detriment to fish survival.

Annual Rates of Silting of Reservoirs

The earlier designs of reservoirs in India were based on the silting rate of 0.036 Mm^3 per 100 sq km of catchment areas as suggested by Dr. Khosla, Application of Dr. Khosla's recommendations to Indian reservoirs had limitations, firstly because only 5 points out of 38 points on which he developed his enveloping curve pertained to data of Indian reservoirs and secondly the catchment areas of these reservoirs were very big but reservoirs of very small capacities. Dr. Khosla equation is $Y = 0.323/A^{0.2}$, where Y is sedimentation rate (*ha m*/100 sq. km/year) and A is catchment area (sq. km). Recent surveys carried out on some of Indian reservoirs have revealed that the actual rate of silting is much higher. Dr. Joglekar's annual rate of silting enveloping curve is in the form

$$Y = 0.597/A^{0.24}$$

where Y = annual silting rate (Mm^3 per 100 sq. km), and A = catchment area at dam site (sq. km). The curve indicates that the silting rate falls down as the catchment area increases.

Rate of reservoir sedimentation falls off with the advance of years, though no correlation is established as yet. The progressive reduction in the volume of deposits is due to consolidation and shrinkage as a result of the superimposed load and periodical exposure of deposit to weather, fall in trap efficiency and mainly due to the progressive formation of delta at the mouth of main river as well as in the tributaries where most of the silt is trapped above reservoir level. As a consequence lesser amount of sediment is received into reservoir basins. Recent surveys conducted in India have shown reduced rate of silting over a few decades of commissioning.

Results of sedimentation surveys of Indian reservoirs have revealed that the rate of sedimentation for major catchments vary between about 4.75 to 14.25 ha m annually per 100 sq km. and the reservoirs are losing a capacity of half a per cent to one per cent annually. IS: 6518–1972 has recommended provision of sedimentation at the rate of 0.1 to 0.2 ha m/year/sq km of drainage area for the entire economic life of the reservoir.

Estimating Silt Load into a Reservoir

The estimate of sediment flow into a reservoir is based upon stream flow sampling and by silt survey method.

1. Silt Sampling Method: It is an indirect method of estimation of sedimentation of a reservoir consisting of measurement of inflows and outflows with sediment sampling. Suspended silt load reaching a reservoir through the river and its tributaries that flow directly into the reservoir is observed by means of silt sampling by taking samples at 0.6 depth where boats are plied. Bed load is also measured. During monsoons, where there is a great turbulence, surface samples may be collected at boils. On tributaries also samples may be collected from the boils. The suspended load is determined in sediment laboratories. The suspended silt load is calculated by the following formula representing total quantity of silt in cu m in 24 hours:

$$\text{Suspended load} = \frac{S \times 86400 \times D}{P}$$

where, S = silt (kg/m^3), D = discharge (cumecs), and P = density of silt (kg/m^3).

The unmeasured portions of the load are usually the bed load and the load that is passing in the unsampled zone *i.e.*, about 7 cm to 15 cm from the bed which cannot be sampled by the sampler. Where judgement shows this error is small, a figure of 5 to 10 per cent may be added to account for unsampled load.

The difference of sediment inflow into the reservoir including estimated bed load and the sediment outflow therefrom give the quantity of silt deposit during the period of analysis in gravimetric terms and its conversion into volumetric units requires estimation of the average unit mass of the deposited sediments. The systematic samples of the deposits are taken in an undisturbed state from all over the reservoir bed and a correlation between the average dry unit mass and the fractional composition of different grains such as clay, silt and sand is determined. The entire depth of the sediments is sampled and its volume weight measured accurately. Consolidation continues to occur throughout the entire sediment depth.

2. Capacity Survey Method: It is a direct measurement method carried out periodically to assess the volume of deposit as also its location in a reservoir, usually when a major reservoir losses 3 per cent of its capacity. The two methods are range line method and contour method. In range line method, the reservoir area is surveyed and range monuments, bench mark pillars etc. are fixed. The reservoir area is divided into suitable sections called silt ranges, along the main river and its tributaries that flow directly into the reservoir. The ranges may be generally parallel to each other fixed at 0.75 to 2 km interval depending on the length of the reservoir. The ranges are kept closer in upper reaches where movement of deposits is more frequent due to drawdown and filling operations of the reservoir. Range monuments and bench mark pillars are fixed on both the banks of the main river and its tributaries for observing cross-sections. Cross-sections of the river bed and its tributaries are observed along the range lines by sounding by means of echo-sounding equipment. The echo-sounding equipment, operated from a motor launch which runs along the predetermined line or cross-sections defined by monuments, determine water depths by utilising supersonic sound pressure waves and precision timing. It gives a permanent and continuous record on a calibrated chart of depths of water through which the sound waves are sent. To determine the value of silt deposit the observed cross-sections are superimposed on the corresponding previous cross-sections to indicate the extent of silt deposit at these cross-sections. The total quantity of silt deposited in the reservoir is computed from the value of silt deposited at each cross-section. The method does not indicate origin of the sediment yield subcatchmentwise.

In contour method, a contour map of the reservoir with suitable scale and contour interval is prepared from which the capacity of the reservoir is computed by planimeter. The difference in capacity between this survey and the earlier one indicates the loss of capacity due to sediment deposition during the intervening period. The loss of capacity above the reservoir level at which the survey was made is also estimated by aerial survey and remote sensing techniques. The contour method is suitable for all types of reservoir shapes whereas range method is limited to relatively straight reaches.

Control of Reservoir Sedimentation

The sediment yield from the watershed, the trap efficiency of the reservoir and the ultimate specific weight of deposited sediment fix the volume of sediment which is to be catered in the design of a reservoir to avoid premature loss of storage dedicated to a specific purpose for a given period of time. The allocation of dead storage space for sedimentation is not a measure of sediment control because sediments do not confine to the allocated dead storage space only but encroach upon live storage simultaneously. As stated earlier, every reservoir is to ultimately silt up, but the life of each reservoir can be considerably prolonged if proper remedial measures are adopted. The methods for the control of reservoir silting can be classified as preventive and curative, as follows:

Preventive Measures

1. Reservoir site selection: A site where sediment inflow is low is preferable than which receives high amount of sediment. The possibility of selecting suitable alternate reservoir site, particularly in the case of smaller reservoirs, should be fully investigated.

Reservoir in seismic area may be avoided, if possible, as earthquake brings in additional silt to the reservoir.

2. Reservoir off the main river channel: Possibility of location of reservoir off the main river channel and creating an auxiliary reservoir on the main channel and transferring silt clear water to the main reservoir through a diversion channel may be fully explored.

3. Design of reservoir: (*i*) *Storage capacity-watershed inflow ratio*: In designing a reservoir the ratio of storage capacity-watershed inflow ratio, design and location of outlet works are important for the rate of silting. In general, where the capacity of the reservoir is a high percentage of the annual yield such reservoirs suffer a high rate of silt accumulation. Thus if economics permit, the life of reservoir is enhanced by increasing the capacity-inflow ratio to such an extent as to have the annual loss of 0.25 per cent only. The ratio between the capacity of the reservoir and the size of the drainage area is one of the most important factors governing the annual rate of accumulation of sediment.

(*ii*) *Outlet work*: Adequate outlet works shall be provided in the design of a dam for the release of water and sediment, such as sluices at various elevations which discharge finer sediments before they can settle to the bottom. Undersluices have, however, limited effectiveness because water carves out a channel near the sluices leaving most of the sediment deposition undisturbed.

(*iii*) *Reservoir operation*: The rate of sedimentation decreases with the use of reservoir. The final solution depends on the economics involved.

(*iv*) *Dam raising in stages*: Instead of providing huge final storage capacity in the first instance raising of the dam subsequently may be considered so as to adjust the capacity-runoff ratio after some capacity has been lost by sedimentation. In the case of some reservoirs, where evaporation is a critical factor, it may be economical to build a substantial lower dam initially and to raise it at intervals until its ultimate height for the given original capacity so that long useful life may result.

(*v*) *Density currents*: Venting out of sediment laden inflow through outlets as density currents without readily mixing with the reservoir water and maintaining its identity for a considerable time can reduce sediment deposition.

4. Control of sediment inflow: Sediment inflow can be controlled by the construction of (*i*) check dams on the tributaries bringing highly silt charged waters, (*ii*) contour trenches on the steep slopes of the hills in the catchment area, and (*iii*) vegetation screens provided by promoting growth of vegetation in the catchment and at the mouth of the reservoir through which the flood waters must pass before entering the reservoir.

5. Rejecting peak floods: In case project economics permit, the reservoir may be filled only after passing the peak flood as provided in Aswan dam in Egypt.

6. Watershed area: Catchment area should not be excessive. It shall be just sufficient to fill the reservoir in an average year with a marginal carry over storage. A reservoir to have a safe life of 100 years, according to Brown, should hold at least 3.6 ha m of water per sq km of watershed area.

7. Control of sediment deposition: The deposition of sediment in a reservoir can be controlled to a certain extent by operating gates or other outlets in the dam in such a manner as to permit selective withdrawal of water having sediment content higher than the average. If the water could be escaped at those levels where the concentration of suspended sediment is the highest, considerable amount of sediment is passed down without settling into the reservoir.

8. Watershed erosion control: This includes all those measures which are effective in preventing or delaying the movement of silt from the point of origin in the watershed to the reservoir; watershed management and afforestation in the catchment area are very effective to reduce the silt flow into the reservoir besides other erosion control and soil conservation measures such as regrassing and grass land cultivation processes, gully control, wild life protection, reservoir shore line protection, highway and railway erosion control, etc.

Curative Measure

Removal of silt deposits: Sediment deposits can be removed by excavation, dredging and sluicing with hydraulic and mechanical agitation. Dredging is expensive and is not economically feasible for most water storage projects. Sluicing with hydraulic or mechanical agitators for stirring up the silt deposits along with the flood flushing may be possible in small reservoirs.

16.11 LIFE OF A RESERVOIR

The term life of reservoir denotes the period during which whole or a specified fraction of its total or active capacity is lost.

Every reservoir is ultimately to silt up. In earlier practice, the life of a reservoir was considered as number of years when the storage allocated as dead storage is filled up. It was based on the assumption that all the sediment load first deposits into dead storage and then encroaches on live storage. This concept is not in line with the actual process of sedimentation of reservoirs as heavier material begins to deposit, on impinging the reservoir backwater, a portion of which in live storage. Thus capacities reserved for the irrigation, hydel power and flood control are encroached upon simultaneously with the deposition of finer material in dead storage zone.

Central Water Commission (India) practice is to fix minimum drawdown level a little above the zero elevation which is likely to develop after 100 years of operation. Generally, the life of a reservoir in India is adopted as 100 years and to provide for silt storage space (dead storage) equal to the estimated total volume of silt deposition during its entire life. In this context, there are centain relevant terms, as under.

Full Service Time: For a specific purpose, the period or notional period for which a reservoir provided or is was expected to provide, a part of the full planned benefits inspite of sedimentation. Full service time of an irrigation project shall not be less than 50 years after the start of operation as per IS : 12182-1987.

Economic Life: If at any point of time, the benefits likely to accrue in further operation of a reservoir compare unfavourably under the relevant economic criteria with the further costs involved in the operation and maintenance of the system but excluding any element to cover the past costs incurred, the reservoir is said to have reached the end of economic life.

Feasible Service Time : For a specific purpose, the period or notional period for which the reservoir provided or is/was expected to provide a part of planned benefit in respect of storage in the reservoir being impaired by sedimentation. It is estimated as the time after which the new zero elevation of the reservoir would equal the sill of the outlet relevant for the purpose. It is not less than 100 years after the start of operation.

16.12 RESERVOIR LOSSES

Considerable quantity of stored water in a reservoir is lost due to evaporation, seepage and leakage, briefed as under:

1. Evaporation Losses

Evaporation losses from a reservoir depend on water spread area of the reservoir, and the rate of evaporation which varies directly with temperature of air and water and wind velocity and inversely with humidity. The evaporation losses are expressed in cm of water depth. The various methods of determining evaporation loss are:

(*i*) Analytical Method: (*a*) *Water budget or storage equation method*

$$E = P + I - O \pm U \pm \Delta S$$

where, E = evaporation (mm depth) of water area for convenient time interval, P = precipitation on water surface, I = surface inflow, O = surface outflow, U = underground inflow or outflow into or from reservoir, ΔS = change in storage, which is negative for increase in storage and positive for decrease in storage.

(*b*) *Mass (vapour) transfer method*

$$E = \frac{-10K^2 \rho(q_2 - q_1)(u_2 - u_1)}{(\log_e z_2 / z_1)^2}$$

where, E = evaporation (mm), K = Von Karman's constant $\simeq$ 0.41, ρ = density of air (g/cc), q_2, q_1 = specific humidities at heights Z_2 and Z_1, which are taken as 2m and 1m respectively above the water surface, and U_2, U_1 = wind speeds (cm/s) at heights Z_2 and Z_1, respectively.

(*ii*) Empirical Formulae: (*i*) Rohwer's Formula

$$E = 0.771 \times \text{pan evaporation}$$

or $$E = 0.771\ \{0.075\ (1.465 - 0.000594\ B)\ (0.44 + 0.0733\ W)\ (e_s - e_d)\}\ \text{mm}$$

where, E = reservoir evaporation (mm), B = bar reading (millibars), W = wind speed (km/h), e_s = vapour pressure of saturated air at temperature of water surface (millibars), and e_d = vapour pressure of saturated air at temperature of dew point (millibars).

(*iii*) Energy Budget Method

$$R = H + LE + G$$

where, R = net radiation flux received at the surface (cal/cm^2) measured by a net radiometer, H = sensible heat flux (cal/cm^2), LE = latent heat flux, where L is latent heat of vaporisation (cal/g) and E is evaporation (mm), and G = heat flux in the soil or water (cal/cm^2).

H is difficult to determine. Bowen ratio $\beta = H/LE$ is evaluated from the equation

$$\beta = \frac{T_s - T}{e_s - e} \times 0.61 \times \frac{P}{1000}$$

where, T_s = water surface temperature (°C), β = air temperature (°C), e_s = saturation vapour pressure at T_s in hectopascal (hpa), e = vapour pressure of air at T in hpa, and P = station level pressure (hpa).

$\therefore$ Evaporation, $$E = \frac{R - G}{L}\left(\frac{1}{1+\beta}\right).$$

(iv) Pan Evaporimeter Method: Observation of evaporation loss at or near the reservoir site with pan evaporimeters is essential for the estimation of reservoir evaporation loss. Network of evaporation stations facilitates the determination of average and extreme values of parameter at any point in the region covered by the network by interpolation of the observed data.

Selection of Site

The pan evaporation measurement is usually associated with meteorological observations and as such the site which meets the requirements of the latter is suitable for the installation of pan evaporimeters as well. However, essential considerations in the selection of a site for pan evaporimeters are (*i*) Nearer to the reservoir site, (*ii*) Fairly level ground, preferably on an open plateau or terrace, (*iii*) Open with no obstruction casting any shadow on the pan except during sunrise and sunset. Trees, buildings etc. are not closer to the site than 10 times their height, (*iv*) Where radiation or sunshine observations are to be used, there shall be no substantial objects to the east or west subtending an elevation angle of more than 30° with respect to the level of the radiation instrument or sunshine recorder, (*v*) The highest groundwater level is not less than 2 m from the ground surface, (*vi*) Not susceptible to flooding or waterlogging even during heavy rains, (*vii*) Located upwind of the reservoir, along the most prevalent direction of the high winds and at a distance away from drifts of spray from water waves, (*viii*) Nearer to the source of water for cleaning and daily filling of the evaporimeter, and (*ix*) Easily accessible.

Types of Evaporimeters

Pan evaporation is an instrument for measuring the amount of water lost by evaporation per unit area in a given time interval from a shallow container. The pan evaporation values are used to interpret the evaporation losses from the water surface of reservoirs. The evaporation pans may be installed above or in the ground or may float on water surface of the reservoir. The commonly used four types of evaporation pans are discussed as follows:

B.I.S. Class A Pan (Modified)

Class A pan standarised by Bureau of Indian Standards and used in India (Fig. 16.5) consists of a large cylindrical pan 1220 mm in diameter and 255 mm deep and a fixed point gauge. It is made of 20 gauge copper sheet tinned inside and painted white outside. It is also made from fibre glass reinforced ploystер and covered with a wire mesh. A stilling well provides an undisturbed water surface around the point of fixed point gauge for taking measurement. The pan rests on a white painted wooden grill about 100 mm above the ground surface and is covered with wire netting of standard mesh. A thermometer is fixed to the side of the pan to measure surface temperature of water. The fixed point gauge serves as a reference point for the free water surface in the pan. The amount of water added to the pan from a graduated cylinder till the water level touches the reference point divided by the time interval gives the amount of water lost by evaporation from the pan in the given interval of time. Due allowance for the rain falling into the pan is made from the catch of nearby raingauge. The coefficient used to obtain evaporation from the free water surface of the reservoir from pan evaporation may vary between 1.10 and 0.90 for lake evaporation of the order of 4-5 mm/day and between 0.75 and 0.65 for reservoir evaporation over 10 mm/day and 0.80 for the transition months.

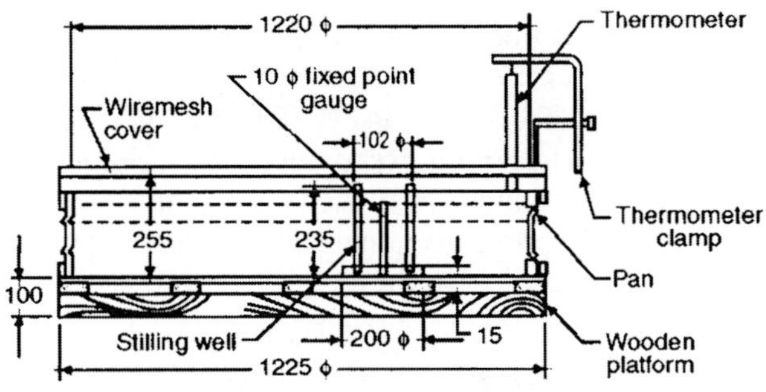

Fig. 16.5 Class A pan (Modified)

Advantages: (*i*) More data are available from this type of pan, (*ii*) Its coefficient has relatively little geographic variation, and (*iii*) Reasonably free from drifting dirt and debris.

Class A U.S. Weather Bureau Land Pan

It is circular 1207 mm in diameter and 254 mm deep (Fig. 16.6). It is made of galvanised iron or monel metal where corrosion is apprehended. The pan rests on a wooden open platform 150 mm above the ground to permit free circulation of air below the pan. The pan is filled within 51 mm of the rim. Evaporation measurements are made by hook gauge or refilling to a fixed point. The coefficient used is 0.7.

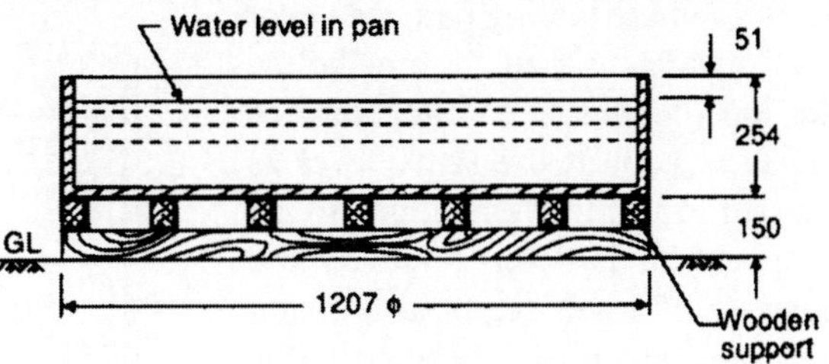

Fig. 16.6 U.S. weather Bureau class A pan.

Colorado Sunken Pan

It is 914 mm square and 457 mm deep (Fig. 16.7). The pan is sunk in the ground with rim 51 mm above the ground level. It is made of unpainted galvanised iron. The water level in the pan is maintained at or slightly below the ground level. A hook gauge fixed to an angle iron frame by means of a base plate and bolted is placed on the top of the pan to take measurements of water surface in the pan. The measurement can also be made by means of a fixed point gauge. The hook can be read up to 0.02 mm. The coefficient used is 0.89.

Fig. 16.7 Colorado sunken pan.

Advantages: (*i*) Radiation and aerodynamic characteristics are similar to those of a reservoir. However, it is desirable to use a large pan to minimise the relative effect of heat exchange taking place between the pan and the soil, and (*ii*) The evaporation from the pan agrees closely with that from a floating pan of the same dimensions.

Disadvantages: (*i*) Difficult and costly to instal, clean and repair, (*ii*) Collects more trash, (*iii*) Leaks occurring in the pan are not easy to detect, and (*iv*) Height of the ground vegetation surrounding it effects evaporation from the pan.

US Geological Survey Floating Pan

The floating pan has been developed to simulate conditions similar to those of the surrounding water in the reservoir. The pan is 900 mm square and 450 mm deep. It is made of galvanised iron. It is supported on drum floats in the centre of a raft 4.25 × 4.87 m and kept afloat in the reservoir. The water level in the pan is supposed to be at the same elevation as that of the surrounding body, sides of pan projecting 75 mm above. The floating pan is not generally used; where used it is essential to instal a land pan also to supplement data for the missing periods in the records of floating pan. The pan coefficient is 0.8.

Advantages: (*i*) More nearly simulate the conditions pertaining in the reservoir, (*ii*) Not affected by drifting soil and snow (if applicable).

Disadvantages: (*i*) Less accessible for taking measurements than land pans, (*ii*) Measurements may be vitiated by splashing of water which occurs frequently due to wave action, and (*iii*) Installation and observational expenses are more.

GGI-3000 Pan (Russian): It is a cylindrical tank 618 mm diameter 600 mm deep at tank wall and 685 mm at centre with a conical base made of galvanised iron sheet (Fig. 16.8). The tank is sunk in ground with the rim approximately 75 mm above. A similar sunken tank with a funnel is installed alongside for measurement of precipitation.

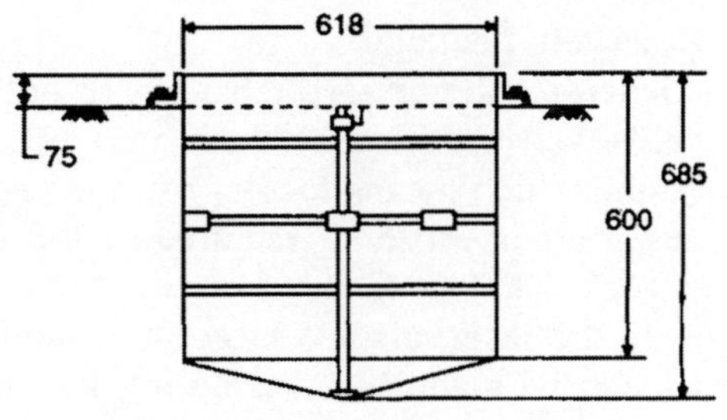

Fig. 16.8 GGI-3000 Pan (Russian).

Class A US Weather Bureau Floating Pan: The land pan (Fig. 16.9) used as floating pan has water level inside the pan 75 mm below the top of pan. The water surface inside is kept at the same level as that of outside. The amount of evaporation is determined by the quantity of water required to bring the water level up to a fixed index point in the centre of the pan.

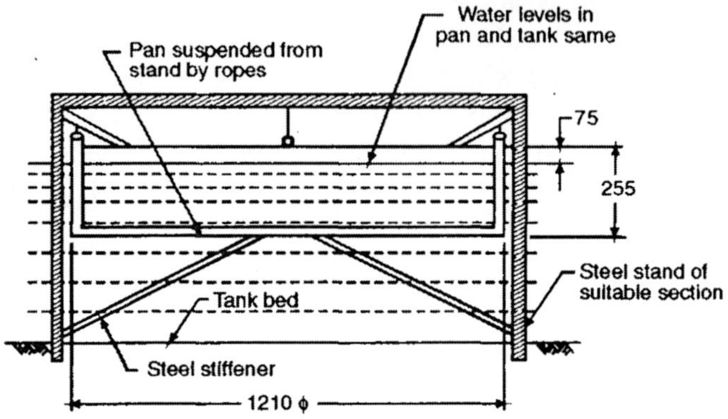

Fig. 16.9 US weather bureau class A floating pan.

Evaporation Losses

In case observed data are not available from the neighbouring reservoirs for application with due correction, the evaporation losses given in the Table 16.1 may be assumed. Evaporation in hot months of May and June is two to five times than in winter months of December and January. For the Indian subcontinent loss of water by evaporation generally varies from 150 to 200 cm depth of stored water annually; a maximum of 350 cm in parts of Rajasthan and Saurashtra and a minimum of 50 cm in parts of Jammu and Kashmir.

Table 16.1 Reservoir evaporation loss

Month	*Evaporation Loss (cm)*		
	South and Central India	*North India*	*Design assumption at pong dam (including seepage losses)*
January	10	7	6.5
February	10	9	7.4
March	18	13	11.9
April	23	16	17.7
May	25	27	20.5
June	18	24	15.4
July	15	18	9.7
August	15	14	8.4
September	15	14	7.7
October	13	13	6.1
November	10	9	5.2
December	10	8	5.5
Total	182	172	122

2. Seepage Losses

Seepage losses are difficult to estimate and are not important in most cases. Seepage losses are generally more when the reservoir is underlain by porous strata having ample outlets beneath the surrounding hills or under the dam. Absorption loss may be significant in the initial stages but gradually reduces as the soil pores become saturated.

3. Reservoir Leakage

Generally the reservoir banks are not so permeable as to cause leakage of any significance. However, serious leakage may occur where the reservoir rim is porous or cavernous limestone which means loss of storage capacity. Where the banks have continuous seams of porous strata or made of fractured rock formation, pressure grouting is done to seal the fractured rock. The water lost from the reservoir to the groundwater through seepage and passages in the reservoir bed is not amenable to measurement. However, an assessment of the losses can be made by considering the inflow, the outflow, and the precipitation over the reservoir and evaporation loss and then working out the unaccounted loss of water due to seepage and absorption. The accuracy of the determination evidently depends upon the precision with which other connected observations are made.

16.13 FLOOD ROUTING

Flood rounting is the process of determining reservoir stages, storage volumes, and outflow rates corresponding to a particular hydrograph of inflow. The maximum reservoir water level is obtained by routing the design flood through the reservoir and spillway. The routing* is determined by (*a*) initial reservoir stage, (*b*) design-flood hydrograph, (*c*) rate of outflow including the flow over the crest, through the sluices or outlets and through power units, and (*d*) incremental storage capacity (each stage). The two general methods of flood routing are integration method and step by step method.

Integration Method

In integration method, a given flood hydrograph is replaced by an equivalent flood of uniform intensity. The available storage curve is represented by an empiricial formula and it is necessary to know only the exponent of the spillway discharge curve.

Integration method provides a more direct solution to the continuity equation, while step by step method gives satisfactory results particularly for small head water reservoirs.

Step by Step Method

In this method, increase in storage and rate of outflow resulting from the volumes of inflow during successive short increments of time are computed. Increments of outflow are computed for periods of time sufficiently short to warrant the assumption that the mean of the inflow rates and the mean of outflow rates at the beginning and end of the intervals would closely approximate the average rates for the respective periods. The procedure of flood routing is illustrated in Example 16.11.

16.14 ENVIRONMENTAL EFFECTS OF RESERVOIRS

The creation of reservoirs result in far-reaching changes in the ecosystem. The major effect is on the land which is inundated and on the aquatic environment. Sedimentation, soil erosion, stratification, adverse effect on fish and proliferation of aquatic weeds are some of the major disruptions in the ecosystem which may involve economic loss. The effect on terrestrial environment are generally felt in case of forests, wild life, groundwater, climate and agriculture. Human environment is affected in respect of alteration of human settlement and occupational patterns etc., and water borne diseases. An increasing constraining factor in hydropower development has been the conflict between the expanding electric energy demands and the concern for environmental effects of reservoirs. Compared to high head dams, low head hydropower projects are less destructive to aquatic life, and cause less damage to the general environment and other aesthetic factors. However, an understanding of the interrelationship between water reservoir and natural ecosystems of the region is essential with a view to not only preserve the existing environments but also to further improve its quality. Such goals as the improvement of natural conditions and environments are adequately achieved by the construction of multipurpose reservoirs. The long-term use of reservoirs has clearly proved that in a variety of climatic zones, the water reservoirs form the bases of stable ecosystems matched with natural environment. The environmental problems arising from water reservoirs and remedial measures are as given below:

Fish: Free passage of migratory fish to and from their spawning grounds is disturbed due to the obstruction created by the construction of dam. Species and habitat of fish in the downstream river reach are also effected. Creation of reservoir provides conducive environments favouring reproduction of several fish species. The river discharges below the reservoir are higher during

*For details refer Hydrology and Water Resources Engineering by the author.

dry season than under natural pre-storage condition. This provides for favourable conditions for the development of fishery in the downstream river reach.

Flora and fauna: Deforestation of the reservoir submerged area and consequent displacement of wild life population is inevitable. Area submerged in a reservoir constitutes an insignificant percentage of catchment area upstream of the dam. Moreover, more mature ecosystem, such as primary forests, is least affected by the reservoir and its surrounding. However, to enhance flora and fauna, it is essential in particular to develop national parks, game reserves, forest reserves and forest management.

Diseases: Reservoirs are considered to contribute to the spread of water borne human diseases. Planning for reservoir should include consideration of the water quality standards for every use of water and the control of water quality.

2. Physical Impact

Reservoir induced seismicity: Reservoirs are considered to trigger earth tremors. While there is no complete agreement on the exact cause of this seismic activity, a general belief is that earthquakes are caused by the filling of reservoir at site where natural stresses in the underlying rock mass have developed to a state very close to rupture. An important measure is early installation of seismographs to provide information on the seismic potential of the region before construction and monitor the post-construction behaviour.

Aquatic nuisance plants: Proliferation of nuisance aquatic vegetation is associated with reservoirs. Concern, however, is limited to those aquatic plants which are larger than microscopic algae. Early planning and action can avoid some of the hazards posed by aquatic plants to public health, fisheries and navigation.

Silt: The river water deposits silt into the reservoir which in turn triggers a number of environmental problems. Silting may be reduced by placing outlets in the dam at such points that allow some of the silt to escape.

In conclustion, it is evident that future multipurpose irrigation and flood control projects should take into account, right at the planning stage, appropriate steps for the protection, preservation and development. With suitable environmental protection measures, irrigation and hydropower plants are considered harmless from environmental viewpoint as compared to other energy sources.

16.15 WATERSHED MANAGEMENT

Water storage reservoirs created at a very high cost are natural assets of great importance. One of the important problems in their operation and maintenance is the loss of storage capacity due to continuous deposition of water borne sediments. The soil erosion originates in the catchment, is transported through the streams and rivers and deposited in the reservoirs. The reservoirs provide space for the accumulation of such sediments but they do not confine themselves in the allocated space, called "dead storage" and simultaneously encroach upon the live storage at a rate higher than the design assumption which means not only their inability to provide sustained water supply and power generation, but most importantly reduction in their operational life. Soil erosion and consequent sedimentation of reservoirs is a complex process dependent on a number of factors such as geology, climate, landscape features, soil characteristics which are not readily influenced by human resources but land disturbance which includes land management practices like tillage, overgrazing, mining, urban development, logging etc. Construction activities and tourism are the major factors under human control. The river ecosystems are increasingly subjected to greater stress due to these human activities. Their adverse effect is more pronounced in the catchment area. It requires soil and land use survey, catchment stabilisation and watershed management which includes soil conservation, afforestation and engineering measures in moderating floods and reducing sediment discharge, maintenance and enhancement of the ecology and land productivity. Effective watershed management requires three level approach *i.e.*, state level, regional level and individual farm level.

The benefits of watershed management are (*i*) Reduction in runoff and damaging floods, (*ii*) Quality management of land resources of soil, water and vegetation for sustained availability and development, (*iii*) Maintenance and improvement of water quality, (*iv*) Soil fertility conservation and improvement in agricultural production, (*v*) Prolonging life of storage reservoir by reduction in sediment inflow, (*vi*) Wild life and forest protection, (*vii*) Improvement in environment, and (*viii*) Restoration and revegetation of seriously eroded areas.

The integrated watershed management was launched in the country during Fifth Plan (1974-78) and soil conservation measures in the catchment area of a few river valley projects were also taken up as a centrally sponsored scheme. The results obtained from earlier treatments have shown that soil conservation measures result in considerable reduction in silt load.

16.16 DAMS

A dam is a hydraulic structure constructed across a river or a stream to impound part of runoff from the catchment upstream of the dam. Streamflow is variable within very wide limits from season to season largely conforming to the pattern of rainfall and snow melting in snow fed rivers. The streamflow also varies from year to year. The rate at which water is required to meet the demands of irrigation, hydropower, water supply, flood control rarely conform to the pattern of water availability. Hence the need to conserve or impound water during wet periods for use during lean periods. This is achieved through construction of a dam.

The construction of dams for creating storage reservoirs has been practised for at least 6000 yeas. Evidence is found of the construction of dams in ancient times. In pre-independence era diversion weirs of small earthen dams not exceeding 15 to 20 m high mainly in the nature of tank bunds were constructed. Prior to 1947, 118 dams were constructed in different parts of the country, 87 dams for irrigation and 31 dams for irrigation-cum-generation of power. Most of these dams were earthen embankments or masonry gravity dams. Only 30 dams were 30 m and above in height, the maximum height was 66 m. The construction of large dams received a great fillip in the post-independence era. Large storage dams such as the Bhakra, the Hirakud, the Tungabhadra and Damodar Valley were constructed. Other major hydroelectric projects constructed are Machkund, Pykara, Kundah, Pong dam and Ramganga projects. The number of large dams constructed or under construction now stand at over 4300. These dams are of wide variety from masonry dam, concrete dam earth dam, rockfill dam, arch dam to hollow gravity dam but 90 per cent of which are earth dams. The pace of construction of dams is likely to accelerate further as they are vital for irrigation and hydel power generation. Major limitation in construction of dams is lack of suitable sites for storage. Good dam sites are few in the Indian subcontinent. Those which are available have to be harnessed optimally and their potential fully utilized. Surface storages are economically very attractive and lend themselves to multipurpose benefits. The national perspective plan now formulated envisages building up of as many storage dams as possible to conserve water for beneficial use as well as minimizing flood damages and transferring them to drought prone and backward areas in need of even drinking water as highlighted by the severe drought conditions in Gujarat and Rajasthan states in the first year (2000 AD) of the new millennium. Thus, a multipurpose and multi-objective development of water resources for irrigation, hydropower generation, flood control, navigation and environments preservation is aimed at.

16.17 INVESTIGATIONS AND SURVEYS

The object of investigation and studies for a dam site is to evolve a suitable location, design and construction of the dam structure which is safe, expeditious, economical environment friendly involving minimal dislocation of human, wild life and forests for sustainable development, and at the same time effects the desired amount of impounding of water.

Engineering Survey

Reconnaissance Survey: It is a general survey carried out visually to examine feasibility of the project, topographical features of the area, possible sites for locating dam, submergence of land, rehabilitation problem and environmental aspects.

Preliminary Survey: Preliminary survey aims at obtaining sufficient details to establish merits and demerits of various sites, the most economical and suitable dam site, tentative cost and accruing benefits therefrom. It also includes trial borings, topographical surveys, geological and hydrological studies, available construction materials etc.

Final Survey: Final survey is conducted in respect of the finally selected site to collect more precise data for detailed designs and precise estimates of the construction cost, besides, (*i*) The dam site topographical survey plan to cover an area sufficient to accommodate all possible arrangements of the project (dam), spillway, outlet works, diversion works, etc. For a large structure, a scale of 1/1000 with contours at intervals of 2 m are adequate. The plan will cover an area at least up to about 200 m upstream and 400 m downstream of the dam site and extend well beyond the abutments, (*ii*) The reservoir submergence plans may be prepared to scale of 1/15000 with contours of 2 to 3 m interval depending upon the size of the reservoir. Area capacity curves and the tables are prepared to an elevation high enough to cover the anticipated maximum reservoir level, (*iii*) Command area survey plans to a scale of 1/15000 with contour interval of 0.5 m, (*iv*) For a barrage structure, detailed survey maps to cover the area under the barrage and appurtenant works. The plan may be to a scale of 1/4000 with levels in a 30 m grid, (*v*) River surveys in a length of about 10 km upstream and 10 km downstream of the diversion structure. Cross-sections are taken 300 m apart extending sufficiently above HFL, (*vi*) Along the canals alignment, L-section shall be prepared with levels at 50 m intervals and cross-sections 100 m apart extending to 100 m on either side of the centre line of the canal, (*vii*) For the location of power station, survey plan shall cover an area sufficient to include alternative station layout and shall give contours at 5 m interval, (*viii*) L-section covering power channel, penstock and tail race channel shall give levels at 50 m interval. Cross-sections shall be taken at 100 m interval extending sufficiently on both sides, (*ix*) For tunnels, L-section along tunnel alignment and contour plan with 50 m interval including covering about 100 m on either side of the alignment and also up to contours corresponding to tunnel grade, if applicable, indicating location of adits, (*x*) For surge tank contour plan of surge tank area at 5 m contours interval, (*xi*) For underground power house, contour plan at 1/4000 showing contour interval at 15 m covering location of access tunnels, tail race tunnel and switchyard, (*xii*) Survey connected with cross drainage works shall include L-section and cross-section of the drains along with details of catchment area, high flood discharges, observed HFL along with canal data at the site of crossing, (*xiii*) Plane table surveying for the selection of suitable sites for residential colony, workship complex, etc, (*xiv*) Estimation of the quantum of excavation, fill placement, concreting work involved, etc., (*xv*) Computation of the quantum of land acquisition and compensation for the land and property to be submerged, (*xvi*) Determination of the type and height of the dam, (*xvii*) Implication of riparian law to be observed, (*xviii*) Areal survey as the reservoir is usually large and the areal survey will be good enough for computing the capacity at various levels and land properties to be submerged, (*xix*) Determiantion of capacity: (*a*) Area of water spread or reservoir area is computed in million cubic metre or hectare metre by measuring the area at definite contours by a planimeter. A curve showing the water spread at various elevations, called area-capacity curve (Fig. 16.10) is prepared for reference, (*b*) *Capacity*: Volume of storage (V) or capacity of the reservoir by taking contour area at equal intervals and totalling up by formula, as follows:

1. *Contour Area Interval Method*: The capacity between successive unit elevations is worked out arithmetically by taking the average submergence area and multiplying by the unit height.

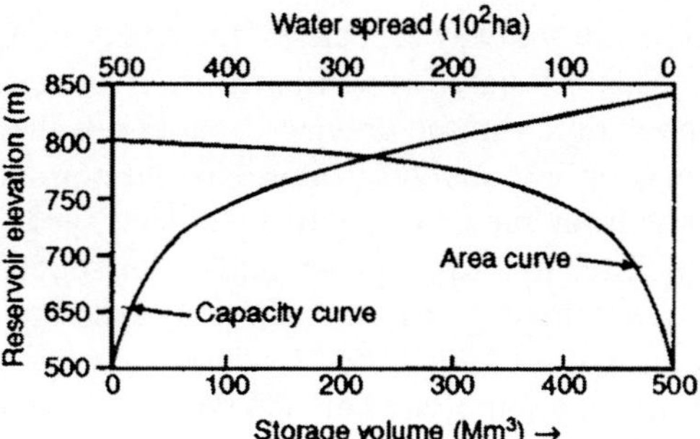

Fig. 16.10 Area-capacity curves.

2. *Cone formula*: 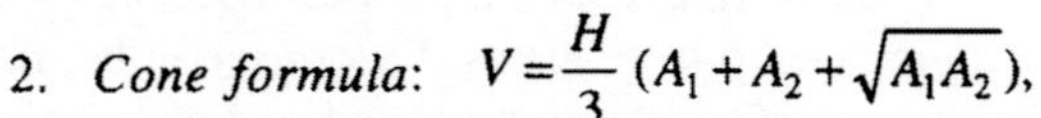 $V = \frac{H}{3}(A_1 + A_2 + \sqrt{A_1 A_2})$,

where A_1 is area at one contour, A_2 is area at another contour (sq km) and H is contour interval (m).

3. *Modified Prismoidal Formula*: $V_s = (2H/6)(A + 4B + C) - V_y$. Where, V_s is capacity between middle and top contours, H is contour interval between A and B contour, A is area of the bottom contour, B is area of middle contour, C is area of top contour and V_y is capacity between middle and bottom contours previously determined. The capacity below the lowest contour is computed by contour area interval method which provides V_y. After finding the volume below the lowest contour, the formula is used progressively for each succeeding higher contour.

4. *Average End Area Length Method*: $V = (L/2)(A_1 + A_2)$ where, V is capacity at elevation 'X' between two successive range lines, A_1, A_2 are areas of successive cross sections at elevation 'X', and L is distance between range lines.

Geological Investigations

Geological investigations costing hardly 0.25 to 1.0% of the total construction cost of the project, are essentially required to give the data such as (*i*) Geological and structural features, viz., folds, faults, fissures, etc., their nature and extent, (*ii*) Water tightness of the reservoir basin, (*iii*) Presence or otherwise of solution channels, sink holes, etc., (*iv*) Existing and potential slide areas, (*v*) Assessment of valuable mineral deposits in the reservoir, (*vi*) Groundwater condition, (*vii*) Seismic conditions of the region with reference to geological map of the vicinity, and (*viii*) Properties of foundation soil, hidden weak spots and shear zones, etc.

Foundation Investigations

Foundation investigation include to determine properties of foundation soil, hidden weak spots and shear zones, etc., and depth of overburden. Location and thickness of weathered, altered or otherwise softened zones and their characteristics and structural weaknesses and discontinuities shall be investigated and contours of bed rock presented. Usual methods of exploration comprise (*i*) Exploratory bore holes, excavation of open pits and trenches, (*ii*) Drifting or tunnelling into the side of the valley and (*iii*) Core drilling.

Meteorological Studies: These include (*i*) Watershed plan showing prominent orographic features, normal annual isohyets, location of raingauge stations in and around the catchment, gauge and discharge sites and interstate boundaries, (*ii*) Assessment of weighted rainfall in the catchment, (*iii*) Collection and evaluation of data and frequencies of heavy rainfall in the catchment and its neighbourhood, (*iv*) Determination of standard project storm or maximum probable storm, rainfall depths, and (*v*) Mean monthly temperature, maximum and minimum monthly temperatures, monthwise average of wind velocities (km/hr), relative humidity percentage, and percentage of cloud cover.

Hydrological Investigations: These include (*i*) Gauge and discharge observations extending for at least five years at project site, (*ii*) Hourly gauges and discharges during flood occurrence and survey map of flood marks of past floods, (*iii*) Determination of observed river hydrographs to fix the spillway capacity of the reservoir, (*iv*) Criteria to be followed in the estimation of design flood, (*v*) Rainfall-runoff correlation studies, critical rainfall condition in regard to the intensity and duration in the drainage area above the dam site, (*vi*) Estimation of annual yields and their break up into monthly, monsoon and non-monsoon runoff, (*vii*) Actual 10 daily observed runoff pattern for runoff of the river projects, and (*viii*) Estimated yields reliabilities of 75%, 90% and 100% for irrigation, power and water supply projects respectively.

Sediment Studies: Sediment observations are carried out for a minimum period of 3 years of the suspended load, bed load and natural soil conditions, including catchment characteristics from the point of erosion. Chemical and petrographic analysis of river water are also done.

Construction Material Investigations: A map showing the location of the sources of construction materials such as rock, coarse and fine aggregates, pozzolana as if proposed to be manufactured locally, soil for use in earth dam, dykes, etc., geological and related characteristics of aggregates for concrete and masonry dams is prepared. It is very essential to locate in advance quantities sufficiently in excess of the estimated requirements.

Communication: Map shall be prepared to show the facilities required during construction and completion phases of the project such as (*i*) Existing and proposed roads, water and rail routes with information on load and size limitations, (*ii*) Telephone and telegraph lines, (*iii*) Sources for obtaining power with transmission line routes, and (*iv*) Rerouting of the communication system where existing system is disrupted due to construction of the project.

Environmental Considerations: Before the project is undertaken surveys and investigations are carried out to assess its historical and cultural repercussions, and determine its impact on navigation, fish culture, wild life, recreational aspects and overall ecology of the region affected.

16.18 SELECTION OF DAM SITE

The various considerations which govern the final selection of a dam site, the general location having been chosen as a result of investigations are (*i*) Good catchment upstream of the dam site as shall contribute adequate supply of water to meet the intended water demand, (*ii*) Minimum length of dam for maximum storage for a given height. Narrow gorge opening out upstream to admit large storage in a deep reservoir less susceptible to weed growth and minimum evaporation losses is preferable, (*iii*) Minimum height of dam for the maximum intended storage, cost of dam is proportional to square of the dam height, (*iv*) Suitable foundations as for high concrete or masonry dams, solid rock foundation is essential, (*v*) Major portion of the dam is on high ground compared to river basin for economy in cost and better drainage, (*vi*) Availability of suitable location for spillway. A deep gorge for the dam and a flank for the spillway separated by a high hillock is the best site for earthen dam, (*vii*) Availability, locally or within economical leads of suitable construction materials such as concrete ingredients, pervious and impervious soils and stones (for rip rap) in the case of concrete and earthen dams respectively, (*viii*) Ensure adequate storage capacity of the reservoir, (*ix*) Rim of dam watertight. Reservoir basin should also be reasonably watertight. Cup shaped basin with fairly watertight bed and sides is preferred, (*x*) Stable banks and adjoining hill slopes, (*xi*) Minimum construction and maintenance cost, (*xii*) Availability of suitable site for economical construction of residential colony, office, construction equipment and workshop complex, (*xiii*) Requirements of coffer dam, diversion and dewatering operations. The site should be suitable for diversion of river during construction period, (*xiv*) Minimum value of land and property compensation. The land to be submerged should be of low value not densely wooded as far as possible, (*xv*) Location of dam and appurtenant works in one state, if possible, to avoid interstate disputes, (*xvi*) Good quality water for intended uses and essentially carrying minimum sediment load, high silt contributing tributaries are avoided, as far as possible, and (*xvii*) Healthy locality. Very cold climate and heavy rainfall impede construction activities.

16.19 SELECTION OF TYPE OF DAM

Various considerations in the selection of a suitable type of dam are as under:

Topography

It is the principal governing factor in the selection of type of dam suitable for a dam site. A narrow river flowing between high, rocky banks, *i.e.,* U-shaped valley is suitable for a concrete overflow dam. A narrow V-shaped valley of top width less than one-fourth of its height with a separate site for spillway is suitable for arch dam. A low wide gorge with a separate spillway site is suitable for an earthen dam. However, earthen dam can be constructed in almost every topography except in general narrow conyons with steep abutments. If the length of a dam is very long and its height is low, earthen dam is the obvious choice. Gravity dam is suitable if length is less and height is more.

Geology and Nature of Foundation

Geological character, thickness, permeability and inclination of the underlying strata, existing faults and fissures determine the foundation conditions of the dam. The selection of type of dam is governed by the nature of foundations, as under:

1. *Sound rock foundation*: On good rocky foundation with no faults or fissures any type of masonry or concrete dam can be constructed. On poor rock foundation having seams and fractures, foundation treatment such as removal of disintegrated rock and grouting to seal the seams and fractures is necessary.

2. *Gravel foundation*: Well compacted gravel foundation is suitable for low gravity dam, earthen dam and rockfill dam. However, gravel foundations have high water percolation for which effective cutoffs or seals have to be provided.

3. *Silty and fine sandy foundation*: Such foundations are subject to settlement, seepage and toe erosion. Properly designed low concrete gravity dam and earthen dam can be constructed but not a rockfill dam.

4. *Clay foundation*: Clay foundations are subject to long range consolidation under the dam weight. Earthen dam with proper foundation treatment can be constructed. Unconsolidated clay with high moisture content is not suitable for concrete gravity dam or rockfill dam.

5. *Non-uniform foundation*: Non-uniform foundations of rock and soft material are not satisfactory for building a dam on it. A dam can, however, be built with special design features.

Spillway Location

A suitable site for the required size and type of the spillway too governs the type of dam to be constructed. Spillway site and type is influenced by the magnitude of the flood to be surplused. As such where large capacity spillway is required to cater for high magnitude flood, combining the dam and spillway into one structure in the form of concrete overflow gravity dam is considered. Non-overflow type dam like earth dam or rockfill dam is suitable where spillway can be located outside the limit of the dam by excavating through one or both of the abutments and chute spillway is most commonly used. Tunnel spillway is used in case site for chute spillway is not available.

Safety Consideration

The type of dam suitable for the given foundation and site condition is also governed by the safety consideration. Loss of life and property arising from dam incidents is an indication of the severity in some cases. Safety of dam is of paramount importance.

Availability of Construction Materials

Main materials required for the construction of dams of various types are (*i*) sand, gravel, crushed stone as concrete aggregate, (*ii*) soils for embankments, and (*iii*) rock for embankments and riprap. Economical availability of these materials determines the type of dam to be constructed. Where good concrete aggregate is available, concrete dam is suitable. Where suitable soils are available, earthen dam is suitable. Rockfill dam is suitable where rocks for embankment and riprap are available. Availability of excavated material from spillway in adequate quantity sometimes suggests earthen dam.

Earthquake Zone

For dams to be constructed in earthquake region, earthquake forces have to be considered in the design of dam. Type of dam suited to resist the earthquake shocks without damage are earthfill dam and concrete gravity dam.

Purpose and Economics

Purpose for which a dam is to be built such as continuous and dependable supply from storage dam for irrigation and hydel power, control of floods by detention dam and to regulate the river flow, or diversion dam or a weir without storage features too suggest the type of dam suitable for construction.

River Diversion

Suitable location for river diversion and coffer dam is important. Incorporation of upstream and downstream coffer dams in the main dam section, if possible, results in saving in cost of dam construction as in the case of earthen dams.

Construction Schedule

Within limited and specified time available for dam construction, the use of excavated material from spillway for dam embankment ensures economy and expeditious construction.

Aesthetic Considerations

Aesthetic considerations too determine the type of structure especially when meant primarily for recreational use. As also because dams are important tourist attractions.

Life of Dam

Dams, like other structures, are subject to decay and deteriorate with passage of time. Concrete or masonry dams have long life compared to earth and rockfill dams.

A summary of the character of site conditions vis-a-vis type of dam suitable for the situation is as under:

Character of site	*Type of dam*
(*i*) Foundation character	
(*a*) Impervious solid rock such as schist and gneiss.	Any type
(*b*) Gravel and coarse sands	Earth-rockfill dam
(*c*) Grit, sand and silt clay	Earthen dam
(*d*) Very fine and uniform sands	Earthen dam
(*e*) Clay, plastic clay	Earthen dam with very flat slopes
(*f*) Silt and fine sand	Low gravity dam or earthen dam
(*g*) Heavy over-burden on rock layer	Earthen or buttress dam
(*ii*) Character of gorge	
(*a*) V-shaped narrow gorge	Arch dam
(*b*) U-shaped narrow gorge	Overflow concrete dam
(*c*) Low-wide gorge	Earth/rockfill dam
(*iii*) Character of construction materials	
(*a*) Ingredients of concrete, *i.e.*, sand, shingle and cement-locally available	Concrete dam
(*b*) Sand, shingle and cement involve long distance carriage	Buttress dam
(*c*) Pervious and impervious soils and stone (for riprap) locally available	Earth/rockfill dam
(*d*) Stone of required specific gravity and size available	Stone masonry dam
(*iv*) Spillway requirement	
(*a*) Spillway location separate from main dam	Earthen dam
(*b*) Adequate width for dam only	Gravity dam with overflow spillway
(*c*) Large spillway requirement	Overflow concrete dam
(*v*) Character of the region	
(*a*) Seismic zone	Earth dam and concrete gravity dam
(*b*) Excessive uplift pressure exerted	Arch dam

16.20 CLASSIFICATION OF DAMS

Dams are classified on the basis of hydraulic design, structural design, function served, construction materials used and capacity, as discussed below:

1. Classification Based on Hydraulic Design

Overflow dams: Also called spillway dam. It is designed to pass surplus flood water over the crest which is kept lower than the top of the main dam. It is generally made of concrete or masonry.

Non-overflow Dam: A non-overflow dam is designed to pass surplus flood waters without overtopping the dam. This type of dam can be constructed of a wide variety of materials such as earth, rockfill, concrete, or masonry. In practice, overflow dam and non-overflow dam are sometimes combined, main dam as non-overflow dam made of earth or rockfill and the overflow dam (spillway) made of concrete or masonry.

2. Classification Based on Structural Design

Gravity Dam: It is the dam which depends on self weight to resist the action of water stored on the upstream side. It is a rigid type of dam and is generally constructed of concrete, approximately triangular in section. In plan it may be straight or curved. Curved one has the merit of economy in cost and safety. It is best suited to sites with relatively sound rock foundation. It is designed as an overflow spillway crest dam. The gravity dam may be solid gravity dam or hollow gravity dam; the former is composed of solid masonry or concrete, the latter has hollow spaces in the dam body.

Arch Dam: It is a curved dam convex upstream that depends principally on arch action for its stability. The water load is mostly transferred by arch action to the canyon walls or abutments. It is best suited where the ratio of width to height is not great and the abutments are necessarily strong, the foundation may or may not be so strong. It is also a rigid type of dam made of concrete or masonry.

Buttress Dam: Also called hollow dam. It is constructed of concrete. This rigid type of dam consists of a water supporting upstream face or deck, usually reinforced concrete slab, supported by buttresses generally in the form of equally spaced counterforts that transmit water load and weight of deck to the foundation.

Low Dam: It is a dam which is designed so that the resultant of all forces passes through the lower middle third point and the maximum compressive stress at the toe is within permissible limit.

High Dam: It is a dam in which the maximum permissible compressive stress is exceeded if the resultant of all forces is to pass through the middle third point. The resultant is, however, maintained nearer to the centre of the base to avoid excessive stress by flattening downstream slope and a batter at the upstream face.

Classification Based on Function Served

Storage Dam: It is a common type of dam constructed to impound surplus flood waters for subsequent supply in the lean period. They serve diverse purposes like hydroelectric, irrigation, water supply, etc. It may be constructed of a wide variety of materials such as concrete, stone, earth and rockfill.

Diversion Dam: It is a dam of smaller height. It is constructed primarily for irrigation or hydroelectric development to raise water level of the river and not for storage or equalization of flow. When used for diversion from river to a water conducting system or canal it is called barrage or weir. Generally it is overflow type of dam or provided with separate spillway section in the case of rockfill dam.

Detention Dam: It is primarily constructed to reduce flood peaks by impounding the flood flow in excess of the capacity of spillway or outlet works of the dam and thereby minimize the effect of sudden floods as also to trap sediments. The storage may be temporary for release consistent with the carrying capacity of the downstream channel or for a longer period to recharge the underground water supply. Besides the overflow spillway, outlets with automatic control of rate of releases within safe limits are provided in detention dam. In general all reservoirs have some detention effect.

Check Dam: It is a low height dam provided for soil conservation and checking uncontrolled transportation of sediments to lower reaches.

Diversion-cum-storage Dam: It is a dam serving the dual purpose of storage and diversion.

Coffer Dam : It is a temporary dam constructed to segregate the dam construction area (working area) from the river flow.

Multipurpose Dam : It is a dam which serves more than one purpose like irrigation power, flood control etc.

Single Purpose Dam : It is a dam to serve a single purpose such as irrigation or hydro power etc.

4. Classification Based on Construction Materials Used

Rigid Dam: Rigid dam is constructed of rigid materials such as concrete or masonry. Gravity dam, arch dam, and buttress dam fall in this category.

Non-rigid Dam: It is constructed of selected earth or rockfill. Earthen dam, rockfill dam and earth and rockfill dam fall in this category.

5. Classification Based on Capacity

Large Dams: All dams above 15 m height, measured from the deepest foundation level to the crest, as also the dams between 10 and 15 m height provided they comply with at least one of the conditions that (*i*) Crest length of dam is not less than 500 m, (*ii*) Reservoir capacity formed by dam is not less than 10 million cu m, (*iii*) Maximum flood discharge is not less than 2000cu m/s, (*iv*) Dam has specially difficult foundation or of unusual design, and (*v*) Volume of earthwork exceeds 0.75 million m^3.

Small Dams: All dams below 15 m in height and storage capacity between 0.5-10 Mm^3.

16.21 TYPES OF DAMS

Various types of dams are Gravity dam, Arch dam, Buttress dam, and Earth and Rockfill dam. The comparative merits and demerits of various types of dams are as follows:

Gravity Dams

Advantages: (*i*) Suited to narrow gorges with steep side slopes where earthen dam, if constructed, might slip, (*ii*) Can be built to any height provided sound foundation is available, (*iii*) Low maintenance cost, (*iv*) Suited to areas with heavy rainfall, (*v*) Portion of dam can be designed as overflow section to pass surplus flood waters, where suitable location for spillway is not available, (*vi*) Outlets can be provided at various elevations to cater for discharging facilities, (*vii*) Relatively more strong and stable than earthen dam, (*viii*) Outlets at low elevation retard sedimentation and help prolong life of the reservoir by passing sediments past the dam, (*ix*) Rapid progress of construction with modern mechanical techniques of concrete mixing and placement, (*x*) Section can be made thinner to economize in cost, and (*xi*) Failure, if any, is not sudden.

Disadvantages: (*i*) Requires very sound rock foundations. Unsuitable for wide gorge or permeable foundations on which earth dam can be constructed, (*ii*) More costly than earthen dam of same height, specially where foundations are deep, (*iii*) Require elaborate construction plant and machinery besides skilled labour, (*iv*) Huge requirement of costly construction materials such as cement, reinforcement and shingle, (*v*) Unless provided for originally in the design, subsequent raising of dam is not possible, and (*vi*) Not desirable to locate in seismic zone, as far as possible.

Arch Dams

Advantages: (*i*) Base rock foundation may not be very sound as pressure transmitted to the foundation is less. Hence suitable for moderate foundations, (*ii*) Particularly suitable for narrow V-shaped gorges where the depth of water stored is greater than the width of gorge, (*iii*) Cheaper than solid gravity section for the given height as it has thinner section and as such requires less concrete, (*iv*) Base width is much less than gravity dam and as such uplift pressure on the base is not so important.

Disadvantages: (*i*) Requires abutments to be of very sound rock to resist arch thrust, (*ii*) The aggregate cost of dam and spillway may make it uneconomical compared to other types of dam even if the length to height ratio is within economical limit (5 : 1 or 6 : 1), (*iii*) Requires skilled labour, (*iv*) Requires elaborate form work, and (*v*) Unsuitable for wide gorge.

Buttress Dams

Advantages: (*i*) Less massive than gravity dam; requires 60% lesser concrete per metre length than solid gravity type, (*ii*) Lesser pressure on the foundation, the dam being of lesser weight and as such foundation unsuitable for solid dam may be suitable for this type. There is better distribution of the compression stress on the foundation, (*iii*) Power house is located economically in the space between buttresses, (*iv*) Unlike gravity or arch dam, full use of the structural strength of concrete is made, (*v*) Decreased uplift, problem of uplift is not so serious if the foundation is not continuous, (*vi*) More suited where carriage cost of cement and other construction materials is prohibitive, (*vii*) Better conditions for stability against sliding, (*viii*) Specially adapted where further increase in reservoir capacity is contemplated, further raising of dam being possible, (*ix*) Inspection and maintenance easier. Back of upstream face and foundations between buttresses are available for inspection. Subsequent drilling and grouting can be done, if required, (*x*) Higher speed of construction because of larger exposed area and thinner section, and (*xi*) Can be used on permeable foundations also with suitable modifications.

Disadvantages: (*i*) Reduced stability against lateral thrust and earthquakes. Increased danger of action of frost, (*ii*) Require richer mix of concrete, reinforcement and complicated shuttering work, (*iii*) More susceptible to damage or failure due to sabotage or military attack, (*iv*) Highly skilled labour is required for construction, (*v*) Sensitive to even moderate deterioration of concrete in thin upstream face concrete surface, (*vi*) More close supervision necessary to ensure careful construction and to avoid unusual exposure conditions, and (*vii*) Not suitable where labour cost is high.

Earth and Rockfill Dams

Advantages: (*i*) Adaptable to any type of foundation. Rockfill dam, however, imposes some restriction on the quality of foundations, (*ii*) Permits subsequent raising, (*iii*) Adaptable for sites in seismic zone, (*iv*) Economical in first cost, (*v*) Permits use of wide variety of available materials, (*vi*) Comparatively unskilled labour can be employed, and (*vii*) Modern excavation, hauling and compaction equipment has added to its economic and speedy construction.

Disadvantages: (*i*) Susceptible to damage by floods or burrowing animals, (*ii*) Cannot be used as overflow dam. Separate spillway structure add to its cost, (*iii*) High maintenance cost, (*iv*) Unsuitable for very heavy rainfall regions, (*v*) Requires subsequent close supervision. Instrumentation necessary to study the behaviour of dam during and after construction add to its cost, (*vi*) Rapid siltation of reservoir as outlets at low level to exclude silt cannot be provided in earthen dam, (*vii*) Failure is sudden and without warning, (*viii*) Unsuitable for narrow gorges with steep side slopes, and (*ix*) Cannot be constructed for greater heights.

16.22 APPORTIONMENT OF COST OF A MULTIPURPOSE RESERVOIR PROJECT

River valley projects are essentially multipurpose for irrigation and hydel power generation besides indirect benefits which accrue from them such as drinking water, flood protection, pisciculture, navigation, tourism etc. The cost of construction of the project is apportioned on the basis of the uses for which the reservoir is planned. The main objective of the allocation is to distribute the cost on equitable basis amongst the various purposes, usually to determine, (*i*) Apportionment of cost between the participating states, (*ii*) Apportionment of revenue between the participating states, (*iii*) The sale price of power or water, and (*iv*) To determine revenue return of irrigation and power departments, and main beneficiaries on the basis of apportionment of cost. The basic principle to ensure equitable allocation to the various purposes is that the cost allocated to a purpose is (*i*) Not more than the benefits to be achieved by that purpose, (*ii*) Not more than the cost of an alternative project built for that purpose alone, and (*iii*) Not less than the cost of the items meant for the specific (exclusive) use of that purpose.

Common Allocation Procedure

The different methods for cost allocation generally have a common four step procedure *i.e.*, (*i*) To determine specific costs which serve each of the purposes, (*ii*) To assign such specific costs to various purposes from the capital cost and to find out the remaining joint cost, (*iii*) To distribute joint cost equitably to various purposes, and (*iv*) To sum up, for each purpose, the specific cost and the share of joint cost chargeable to that purpose.

1. Alternative Cost Method:

In this method, the alternative cost of a new single purpose project is split up into two parts, namely, the specific cost equal to that in the multipurpose project, and the balance. The joint cost of the multipurpose project is then distributed in the ratio of the balance. The allocated joint cost is then added to the respective specific cost to arrive at the total cost allocated to each function.

2. Alternative Justifiable Cost Method:

This method is similar to the alternative cost method, the only difference being that the specific cost is subtracted from the justifiable cost. The joint cost is then distributed in the ratio of the balance as in the alternative cost method and the cost allocated to each purpose/function is worked out by adding the respective specific cost to the allocated joint cost.

3. Bearability Concept:

This concept suggests that the net revenue from the minor advantage (gross revenue minus the operation, maintenance and depreciation on both the specific cost and share of the joint cost of the minor advantage) shall be equal to the interest charges on the capital investment on the minor advantage, both specific cost and share of joint cost chargeable to minor advantage.

4. Benefits Methods:

In this method, the cost is allocated to each function by its specific cost plus a share of the joint cost in direct proportion to the estimated net benefits accruing to that function.

5. Ceiling Allocation (or Priority of Use) Method:

In a multipurpose project the various purposes to some extent compete with each other for the use of the water or storage space. The functions have different timing for the periods for optimum storage and release of water and thus all of them cannot be served in an optimum manner. If some of the functions are served proportionately more than others, the priority of use method will give special consideration to such priorities. The priority of use method assigns the specific cost entirely to the individual purpose and assigns the remaining joint costs to project purposes in a descending order of priority. Thus, the top priority purpose is charged with joint costs equal to the lesser of two amounts, namely (*i*) the benefits, minus specific costs assigned to that purpose, or (*ii*) the cost of the most economical alternative minus specific costs assigned to that purpose. The remaining joint costs are then allocated in the same manner to the other functions in order of priority. In this way, the higher priority functions are assigned the maximum allocation represented by alternative joint costs for these purposes, thereby relieving the lower priorities from sharing the joint costs.

6. Equal Apportionment Method:

In this method, the joint cost is distributed equally among the different purposes served by multipurpose project. It is based on an arbitrary assumption that an equal portion of the joint cost is reasonably chargeable to each of the purposes.

7. Separable Cost:

Also termed as remaining benefits method. The basic principle of distribution in this method is the same as in alternative justifiable cost method. It is a refined form of that method in so far as the concept of specific cost for each purpose is replaced by the concept of separable costs. Separable cost for each purpose is subtracted from the justifiable cost and the remainder which is really the 'remaining alternative costs' is termed in this method, as the 'remaining benefits' for the purpose. The joint cost, which in this case is the difference between entire project cost and the sum of separable cost, is then distributed in the ratio of remaining benefits for each purpose/function. When separable cost for all purposes/functions consists of specific cost only the separable cost-remaining benefits method is identical to the alternative justifiable cost method in computation and results. It is rational and sufficiently accurate method but it is difficult to work out the benefits or its alternative costs.

8. Use of Facilities Method:

This method is based on the premise that the joint costs are proportional among the various functions according to their amount of use of the joint facilities. The usual two approaches for assessing the use for each purpose are the 'capacity approach' and 'water released approach'. In the 'capacity approach', the joint cost is distributed in the ratio of storage assigned to different purposes, whether or not those storages are actually used. Two or more purposes may make use of the same reservoir capacity alternately or simultaneously. In the 'water released approach', the joint cost is distributed in the ratio of total quantity of water released for each purpose. It is a simple but not an accurate method.

9. Vendibility Method:

The vendibility theory assumes that the products of a multipurpose project are sold in an entirely free competitive market. The method assumes that the revenues from all benefits of a multipurpose project shall at least be equal to the total cost of the project. The revenues of each benefits should be sufficient to cover the specific costs and in addition should contribute towards the joint cost in such a manner as to render the entire project remunerative with revenues realized from each of the benefits at competitive prices. In this method, a particular purpose, the revenues from whose benefits are not sufficient to meet its costs are subsidized by other purposes the benefits from which are produced at more competitive price.

The method of dividing joint cost is (*a*) The specific cost is subtracted from the estimated revenues for that purpose, (*b*) The remainder is capitalized at the prevailing rate of interest, and (*c*) The remainder is the share of the joint cost or is the upper limit of the share of the joint cost for this particular purpose.

10. Incremental Method:

In this method the separable costs are allocated to their respective functions and the total joint cost to one basic function considered the basic or principal function.

SOLVED EXAMPLES

Examples 16.1 *The yield of water from a catchment area during each successive month is (1.4, 2.1, 2.8, 8.4, 11.9, 11.9, 7.7. 2.8, 2.52, 2.24, 1.96, 1.68) × 10^6 m^3. Determine the minimum capacity of a reservoir required to allow the above volume of water to be drawn off at a uniform rate assuming that there is no loss of water over the spillway.*

Solution. 1. *By Analytical Method*:

Total inflow: $= (1.4 + 2.1 + 2.8 + 8.4 + 11.9 + 11.9 + 7.7 + 2.8 + 2.52 + 2.24 + 1.96 + 1.68) \times 10^6\ M^3 = 57.4 \times 10^6\ m^3$

Monthly demand $= (57.4 \times 10^6)/12 = 4.78 \times 10^6\ m^3$

Storage required $= [(8.4 + 11.9 + 11.9 + 7.7) \times 10^6] - [(4.78 \times 4) \times 10^6]$

$= 20.78 \times 10^6\ m^3$

2. *By Mass Curve Method*: The step by step procedure is as follows:

(*i*) The cumulative inflow works out as (1.4, 3.5, 6.3, 14.7, 26.6, 38.5, 46.2, 49.0, 51.52, 53.76, 55.72 and 57.40) × 10^6 m^3.

The mass inflow curve (dotted line in Fig. 16.11) is drawn by plotting the cumulative inflow as ordinate against time (months) as abscissa.

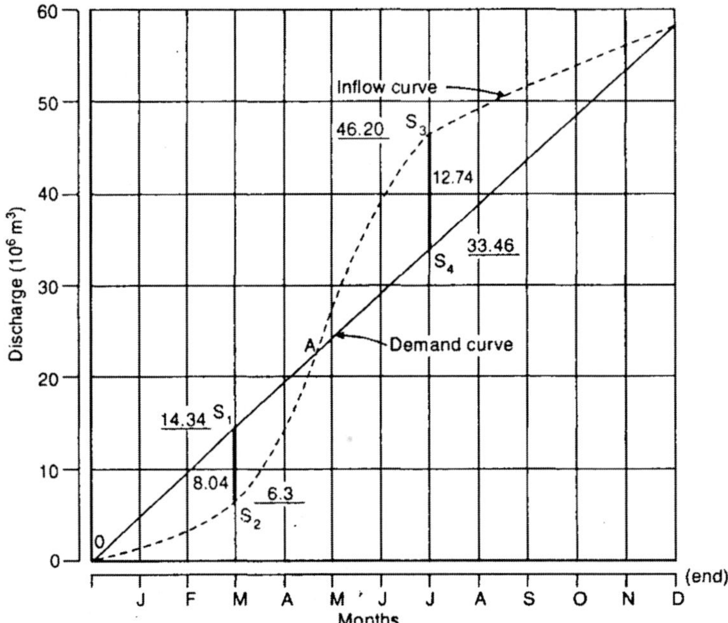

Fig. 16.11 Mass curve.

(*ii*) The monthly demand = $(57.40 \times 10^6)/12 = 4.78 \times 10^6$ m³. The demand curve is drawn as straight line for the uniform monthly demand.

(*iii*) It is evident from the curve that (*a*) S_2 is the lowest point on the depression or trough of the mass inflow curve indicating that the inflow is just equal to the outflow, (*b*) S_3 is the highest point on the hump of the curve indicating maximum reservoir level, (*c*) from 0 to S_2 the slope of the mass inflow curve is less than that of the demand curve indicating that inflow into the reservoir is less than outflow from it and the reservoir in emptying. S_2 represents the minimum reservoir level, (*d*) from S_2 to *A* the slope of mass inflow curve is more than that of the demand curve indicating that inflow is more than outflow and the reservoir water level is rising during the period S_2 to *A*, (*e*) at *A* the reservoir water level is the same as it is in beginning *i.e.*, corresponding to point 0, (*f*) $S_1 - S_2$, *i.e.*, the vertical through S_2 cutting demand curve at S_1 represents, to scale, the minimum initial storage required to cater for deficit during the period 0 to S_2, At S_2 reservoir has fallen to the minimum water level (*g*) from *A* to S_3 the slope of the mass inflow curve is more than that of the demand curve indicating that the water level in the reservoir is rising. S_3 represents full reservoir level, (*h*) $S_3 - S_4$, *i.e.*, the vertical through S_3 cutting the demand curve at S_4 represents the storage, between the initial water level, corresponding to point 0, and full reservoir level, and (*i*) from S_3 upwards, the slope of the mass inflow curve is less than demand curve indicating that the inflow is less than the outflow and the reservoir is again falling.

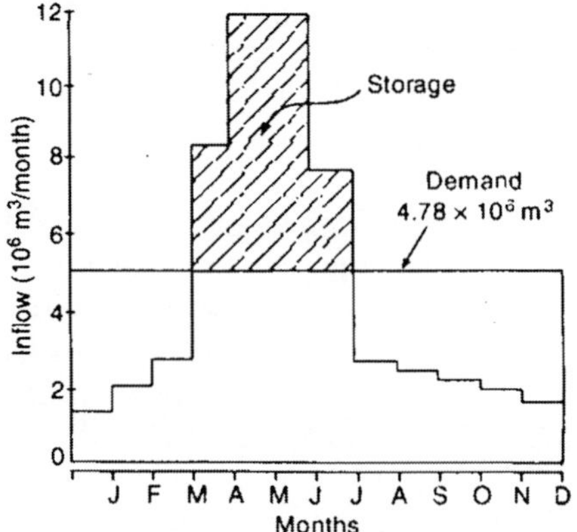

Fig. 16.12 Bar graph.

The minimum storage capacity required to meet the demand

$$= (S_1 - S_2) + (S_3 - S_4)$$
$$= (8.04 + 12.74) \times 10^6 \text{ m}^3$$
$$= 20.78 \times 10^6 \text{ m}^3.$$

Initial storage = 8.04 Mm³.

3. *By Bar Graph Method*: Bar graph (Fig. 16.12) area shows that total volume of 57.4×10^6 m³ has entered the reservoir in the year and required minimum storage is 20.83×10^6 m³ which is the area hatched be-

tween the bar graph above the average demand graph of 4.78×10^6 m³ drawn as a horizontal line. It also represents maximum cumulative excess of outflow over inflow during the year which is also the maximum cumulative deficit of water. The value of storage 20.83×10^6 m³ slightly differs from that obtained by mass curve method. It is because of the curvilinear variation of inflow volumes in mass curve as against linearly from one month to the other in this method.

Example 16.2 (*a*) *The runoff data for a river along with the probable demands (M cu m) is given below. Can the flow in the river cater to the demands? If so, how?*

(*b*) *What is the maximum uniform demand that can be met and what is the storage capacity required to meet the demand.*

Month	*J*	*F*	*M*	*A*	*M*	*J*	*J*	*A*	*S*	*O*	*N*	*D*
Inflow	*135*	*23*	*27*	*21*	*15*	*40*	*120*	*185*	*112*	*87*	*63*	*42*
Demand	*60*	*55*	*80*	*102*	*100*	*121*	*38*	*30*	*25*	*59*	*85*	*75*

Solution: (*i*) *By Analytical Method*: Evaporation losses and prior water rights of the downstream, if any, are not given and as such ignored. Solution in tabular form is as under:

Month	*Inflow* (*M m³*)		*Demand* (*M m³*)	*Deficit* (*M m³*)	*Cumulative deficit* (*M m³*)	*Cumulative excess flow* (*M m³*)
January	135		60	+ 75		75
February	23		55	− 32	32	—
March	27	Total	80	− 53	85	—
April	21		102	− 81	166	—
May	15	=	100	− 85	251	—
June	40		121	− 81	332	—
July	120	870	38	+ 82		82
August	185		30	+ 155		237
September	112		25	+ 87		324
October	87		59	+ 28		352
November	63		85	− 22		
December	42		75	− 33	55	427

The maximum cumulative deficit of 332 M m³ as in the month of June is the storage required to meet the indicated demand. The column of cumulative excess flow shows that the reservoir shall be filled again in the month of October.

2. *By Bar Graph Method*: Fig. 16.13 shows the bar graph for the inflow and demand. The minimum storage is indicated by the portion of the bar graph between demand and inflow (shown by dots) which works out to 332 M cum.

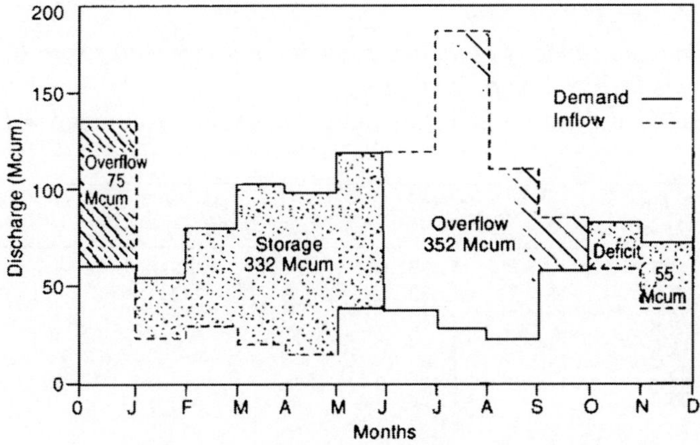

Fig. 16.13 Bar graph.

(*b*) The maximum uniform monthly demand that can be met with from cumulative inflow of 870 M. cum is 72.5 M cum (*i.e.*, 870/12). The storage required to meet this demand is 276.5 Mm^3

i.e., [(135 + 120 + 185 + 112 + 87) – (72.5 × 5)]

Example 16.3 *Evaporation loss (cm) has been measured by a pan evaporimeter (pan coefficient = 0.7) placed near a reservoir area 5.2 sq km. After a year the surface area of the reservoir reduces to 2.2 sq. km. Determine the annual quantity of water lost by evaporation when the pan evaporation loss (cm) over the year are known;*

Month	*Jan*	*Feb.*	*Mar.*	*Apr.*	*May*	*June*	*July*	*Aug.*	*Sept.*	*Oct.*	*Nov.*	*Dec.*
Evaporation loss	*11.5*	*11.0*	*13.2*	*12.8*	*13.2*	*16.2*	*14.1*	*11.2*	*10.2*	*12.0*	*12.4*	*11.6*

Solution:

Average surface area of reservoir (by cone formula) $= \frac{1}{3}[(A_1 + A_2 + \sqrt{A_1 + A_2}\,]$

$= \frac{1}{3}[5.2+2.2+\sqrt{5.2\times2.2})] = 3.59\ \text{km}^2$

Annual evaporation loss $= (11.5 + 11.0 + 13.2 + 12.8 + 13.2 + 16.2 + 14.1$

$= + 11.2 + 10.2 + 12.0 + 12.4 + 11.6) = 149.4$ cm

Annual water loss $= 3.59 \times 10^6 \times 149.4/100 \times 0.7 = 3.75 \times 10^6\ \text{m}^3$.

Example 16.4 *An evaporation pan 1.2 m in diameter was used to find out evaporation from a reservoir. The pan was initially filled up with water to a depth of 8 cm. During the period of observation a rainfall of 4 cm was recorded. To keep the water level the same in the pan an amount of water equivalent to 2 cm depth had to be removed. At the end of the period of observation, the depth of water in the pan was found to be 8.5 cm. If the pan coefficient is 0.75, estimate the evaporation from the surface of reservoir.*

Solution:

Net addition to water in pan (4.0 – 2.0) = 2 cm

Water depth in pan, had there been no losses = 8 + 2 = 10 cm

Water depth observed in pan = 8.5 cm

∴ Evaporation loss from pan = 10.0 – 8.5 = 1.5 cm

∴ Evaporation loss from reservoir = 1.5 × 0.75 = 1.125 cm.

Example 16.5 *The flows from a certain stream in each successive month (30.4 days) are:*

Months	*1*	*2*	*3*	*4*	*5*	*6*	*7*	*8*	*9*	*10*	*11*	*12*	
Flow	*3.0*	*3.6*	*6.0*	*19.6*	*25.2*	*25.2*	*21.6*	*9.9*	*7.8*	*7.2*	*6.6*	*6.3*	*M m^3*

Determine the minimum capacity of the reservoir if the above water is to be drained off at a uniform rate and none is lost by flow over spillway.

Solution: For 100% utilisation of inflow, uniform monthly demand = (Σ inflow)/12 = 142/12 = 11.83 M m^3.

Month	*Inflow (M m^3)*	*Demand (M m^3)*	*Deficit (M m^3)*	*Cumulative deficit (M m^3)*
1	3.0	11.83	8.83	8.83
2	3.6	11.83	8.23	17.06
3	6.0	11.83	5.83	22.89
4	19.6	11.83		22.89
5	25.2	11.83		22.89
6	25.2	11.83		22.89
7	21.6	11.83		22.89
8	9.9	11.83	1.93	24.82
9	7.8	11.83	4.03	28.85

Month	Inflow (M m³)	Demand (M m³)	Deficit (M m³)	Cumulative deficit (M cumec)
10	7.2	11.83	4.63	33.48
11	6.6	11.83	5.23	38.71
12	6.3	11.83	5.53	44.24

The storage capacity required for 100 per cent utilisation of inflow is 44.24 million m³.

Example 16.6 *Monthly flow rate in M m³ of a river are given below. What would be the maximum storage capacity when 85 cumecs of water is drawn at a uniform rate for power generation? If the reservoir capacity is fixed at 710 M cu m, what uniform rate of withdrawal is possible? Estimate also the percentage change of the generation capacity. Assume a month is equal to 30.4 days, if needed.*

Jan.	*Feb.*	*Mar.*	*Apr.*	*May*	*June*	*July*	*Aug.*	*Sept.*	*Oct.*	*Nov.*	*Dec.*
290	*237*	*164*	*132*	*79*	*66*	*171*	*580*	*790*	*497*	*303*	*290*

Solution:

Month	Inflow (M m³)	Demand (M m³)	Deficit (M m³)	Cumulative deficit (M m³)	Excess (M m³)	Cumulative excess (M m³)
Jan.	290	223*			67	
Feb.	237	223			14	81
Mar.	164	223	59	59		
April	132	223	91	150		
May	79	223	144	294		
June	66	223	157	451		
July	171	223	52	503		
Aug.	580	223			357	357
Sept.	790	223			567	924
Oct.	497	223			274	1198
Nov.	303	223			80	1278
Dec.	290	223			67	1345

* Demand = 85 × 30.4 × 24 × 60 × 60 = 223 Mm³.

Reservoir capacity required = 503 M m³. It is also obtained from Fig. 16.14 by drawing a line parallel to demand line with slope of 85 cumecs. Uniform rate of demand met with if storage capacity is 710 M m³, as read from Fig. 16.14, is 101 cumecs.

$$\text{Percentage increase in generation capacity} = \frac{101-85}{85} \times 100 = 19\%.$$

Cumulative excess shows that the reservoir shall be full in September.

Example 16.7 *The following table gives the mean monthly inflow and demand from a reservoir in cumecs:*

Month	*J*	*F*	*M*	*A*	*M*	*J*	*J*	*A*	*S*	*O*	*N*	*D*
Inflow	*35*	*23*	*27*	*21*	*15*	*40*	*120*	*185*	*112*	*87*	*63*	*42*
Demand	*60*	*55*	*80*	*102*	*100*	*121*	*38*	*30*	*25*	*59*	*55*	*35*

The mean monthly evaporation and rainfall (cm) are as under:

Evaporation	*5*	*6*	*11*	*19*	*26*	*23*	*13*	*10*	*11*	*10*	*7*	*5*
Rainfall	*2*	*1*	*1*	*—*	*—*	*—*	*50*	*40*	*30*	*10*	*—*	*2*

Determine storage required to meet the demand. Assume runoff coefficient of reservoir submerged area as 0.33. Average reservoir area is 30 km².

Solution:

Inflow volume (I) = I × No. of days in a month × 8 64 × 10⁴ m³

(in January) = 35 × 31 × 8.64 × 10⁴ = 93.7 M m³ (*i.e.*, col. 3 of solution in Table (page 470) and so on

Volume of demand (D) = 60 × 31 × 8.64 × 10⁴ = 160.7 M m³ (col. 7) and so on

Volume of evaporation loss $= (E/100) \times$ reservoir area (m^3)

$= (E/100) \times 30 \times 10^6 = 0.3$ E M m^3

Rainfall volume $= (R/100) \times (1 - 0.33)\ 30 \times 10^6 = 0.2$ R Mm^3

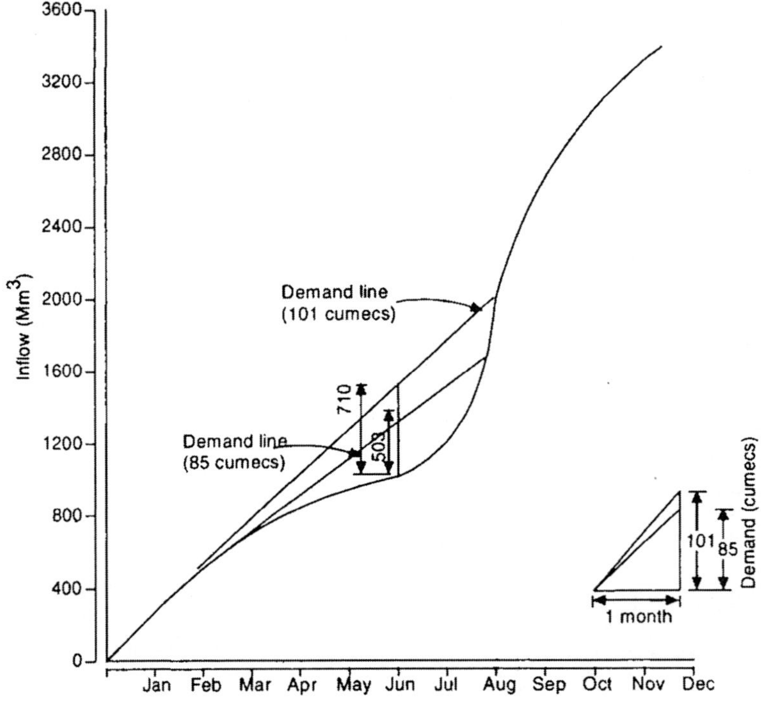

Fig. 16.14 Storage capacity

Table below shows the computation of storage capacity.

The minimum capacity of the reservoir required is 960.2 M m^3 which is the cumulative deficit. The reservoir shall be full again in November and flow in excess of 960.2 M m^3 spills over till end December.

Month	*Mean monthly inflow (cumecs)*	*Monthly inflow volume (M m³)*	*Rainfall volume = 0.2 R (M m³)*	*Total inflow (col. 3+ col. 4) (M m³)*	*Withdrawal* Evaporation = 0.3 E (M m³)	*Withdrawal* Demand (M m³)	*Withdrawal* Total (Col. 6 + Col. 7) (M m³)	*Deficit (col. 8 - col. 5) (M m³)*	*Cumulative deficit*	*Surplus (col. 5 - col. 8)*	*Cumulative surplus*
1	2	3	4	5	6	7	8	9	10	11	12
J	35	93.7	0.4	94.1	1.5	160.7	162.2	68.1	68.1		
F	23	55.6	0.2	55.8	1.8	133.1	134.9	79.1	147.2		
M	27	72.3	0.2	72.5	3.3	214.3	217.6	145.1	292.3		
A	21	54.4	—	54.4	5.7	264.4	270.1	215.7	508.0		
M	15	40.2	—	40.2	7.8	267.8	275.6	235.4	743.4		
J	40	103.7	—	103.7	6.9	313.6	320.5	216.8	960.2		
J	120	321.4	10	331.4	3.9	101.8	105.7			225.7	225.7
A	185	495.5	8	503.5	3.0	80.4	83.4			420.1	645.8
S	112	290.3	6	296.3	3.3	64.8	68.1			228.2	874.0
O	87	233.0	2	235.0	3.0	158.0	161.0			74.0	948.0
N	63	163.3	—	163.3	2.1	142.6	144.7			18.6	966.6
D	42	112.5	0.4	112.9	1.5	93.7	95.2			17.3	983.9

Example 16.8 *Work out the life of a reservoir before its capacity is reduced to 20% of the initial capacity from the following sedimentation data. Catchment area is 1,000 sq km,. reservoir capacity is 10,000 ha m, average annual flow amounts to 15 cm of runoff, average annual sediment inflow is 0.36 million tonnes.*

Solution: Average annual inflow, $I = \dfrac{15 \times 1000 \times 10^6}{100 \times 10^4} = 15000$ ha m

Reservoir capacity, $C = 10{,}000$ ha m

Capacity inflow ratio, $\dfrac{C}{I} = \dfrac{10000}{15000} = 0.67$

Average annual sediment inflow (assuming density of sediment deposits as 1.2 gm/cc) $= \dfrac{0.36 \times 10^6 \times 10^3}{1.2 \times 10^3 \times 10^4} = 30$ ha m

The capacity inflow ratio decreases as the sedimentation increases and trap efficiency also decreases.

Capacity (ha m)		*Capacity inflow ratio, C/I*	*Trap efficiency (percent)*		*Annual sedimentation (ha m) (30 × col 5/100)*	*Loss of capacity (ha m) (10,000 × 20/100)*	*No. of years in which capacity is lost (col. 7/col. 6)*
%	*Volume*		*for C/I (read from Brune's curve)*	*Av. for increment*			
1	2	3	4	5	6	7	8
100	10,000	0.67	98	97			
80	8,000	0.53	96	95	29.1	2000	69
60	6,000	0.40	94	92	28.5	2000	70
40	4,000	0.27	90	87	27.6	2000	72
20	2,000	0.13	84		26.1	2000	77
							Total 288 yrs.

Example 16.9 *Calculate the life of a reservoir from the following data:*
Dead storage capacity $= 1.0 \times 10^9 m^3$
Average annual suspended load in dead storage $= 0.02 \times 10^9 m^3$
Assume coarse silt 1%, medium silt 15% and fine silt 84%.

Solution. (*i*) Suspended sediment load deposition

Average annual suspended load in dead storage $= 0.02 \times 10^9$ m^3
Coarse silt at 1% $= 0.0002 \times 10^9$ m^3
Medium silt @ 15% $= 0.003 \times 10^9$ m^3
Fine silt @ 84% $= 0.0168 \times 10^9$ m^3
Total $= 0.02 \times 10^9$ m^3

Assume that all coarse silt, 80% of medium silt and 20% of fine silt is deposited in the reservoir. Further assume that fine silt is deposited in a semi-fluid state with 45% silt and 55% water.

Silt deposited in the reservoir:
Coarse silt $= 0.0002 \times 10^9$ m^3
Medium silt $= 0.8 \times 0.003 \times 10^9$ $= 0.0024 \times 10^9$ m^3
Fine silt $= 0.2 \times 0.0168 \times 10^9 \times 100/45$ $= 0.0075 \times 10^9$ m^3
Total $= 0.0101 \times 10^9$ m^3

(*ii*) Bed load deposition, assume 10% of total load = 0.002×10^9 m³

Total sediment load deposited $= (0.0101 + 0.002) \times 10^9 \text{ m}^3$

$= 0.0121 \times 10^9 \text{ m}^3$

Life of reservoir $= \dfrac{1.0\times 10^9}{0.0121\times 10^9} = 83$ years

Example 16.10 *Determine trap efficiency of a reservoir of 100 × 10⁶ m³ and 85 × 10⁶ m³ capacity at FRL 600 m and mean operating pool elevation 580 m, average inflow 2000 × 10⁶ m³/yr, reservoir length at MRL 25 km.*

Solution: 1. By Brune's Method

Capacity-inflow ratio, C_1/I $= 100 \times 10^6/2000 \times 10^6 = 0.05$

Trap efficiency for $C/I = 0.05$ = 76 per cent, read from Fig. 16.4 median curve

2. By Churchlli's Method

Average inflow $= 2000 \times 10^6$ m³/yr = 63.42 m³/sec

Retention period C/I $= 85 \times 10^6/63.42 = 1.34 \times 10^6$ sec

A (average cross sectional area = C/L $= 85 \times 10^6/25 \times 10^3 = 3400$ m²

Velocity = I/A = 63.42 m³/sec/3400 m² = 0.0187 m/sec

Sedimentation index = Retention period/velocity

$= 1.34 \times 10^6/0.0187$

$= 7.166 \times 10^7$ sec²/m

Percentage of incoming sediment passed down as read from Fig. 16.4*a* for the above sedimentation index is 16 per cent. As such trap efficiency = 100 – 16 = 84 per cent.

Example 16.11 *IS: 5477 (Part-IV)-1971 has specified an example on step by step method of reservoir flood routing for fixing maximum water level (modified Puls method).*

Solution: *Data required:* (*a*) Reservoir level vs spillway capacity (outflow) and storage capacity (Fig. 16.15), (*b*) full reservoir level, (*c*) Inflow hydrograph (Fig. 16.16), (*d*) Initial outflow, and (*e*) Initial storage (or initial reservoir level).

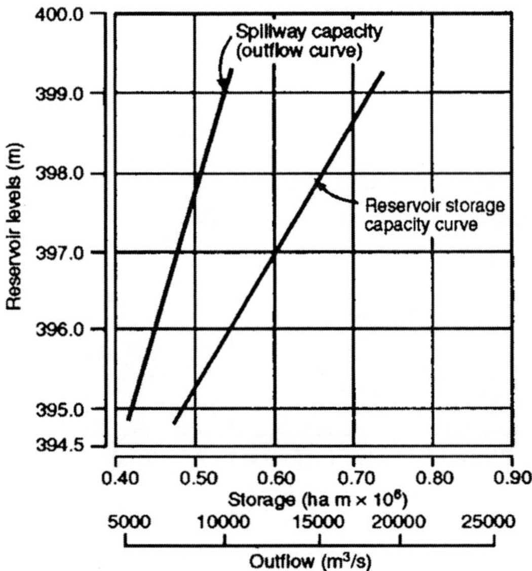

Fig. 16.15 Reservoir levels vs spillway and storage capacities

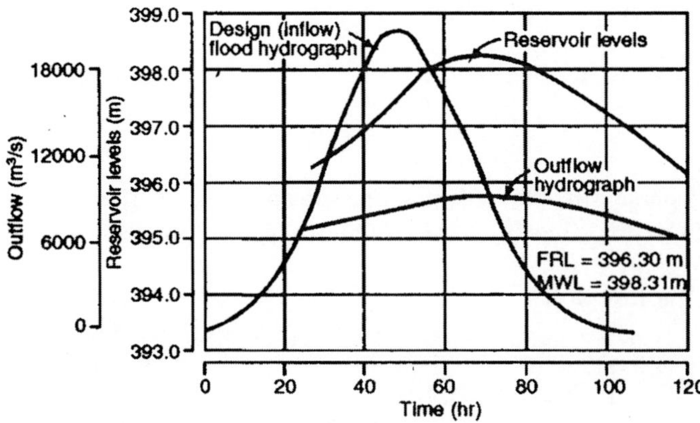

Fig. 16.16 Inflow and outflow hydrograph and time vs reservoir levels for the design flood hydrograph.

Basic equation for routing flood through reservoir is given below:

$$t\left(\frac{I_1+I_2}{2}\right)-t\left(\frac{O_1+O_2}{2}\right)=S_2-S_1=\Delta S$$

where, t = time interval, I_1 = initial inflow (cumecs), I_2 = final inflow (cumecs) after t hour, O_1 = initial outflow (cumecs), O_2 = final outflow (cumecs) after t hours, S_2 = final gross storage (cumec h) after t h, S_1 = initial gross storage (cumec h), and ΔS = incremental storage in time t h.

Storage capacity is usually given in hectare metres which is converted into cumec h or cumec day, if the time interval t is expressed in hours or a part of a day.

Procedure: Taking known values on one side

$$\left(\frac{I_1+I_2}{2}\right)+\left(\frac{S_1}{t}-\frac{O_1}{2}\right)=\left(\frac{S_2}{t}+\frac{O_2}{2}\right)$$

1 cumec day = 24 cumec h = 8.64 ha m

S ha, m = 2.771 S cumec h

S/t ha, m = 2.771 S/t cumec h

Taking the time interval t as 6 hours

$$S/t = 0.463\ S$$

Table 16.2 is prepared from Fig. 16.15.

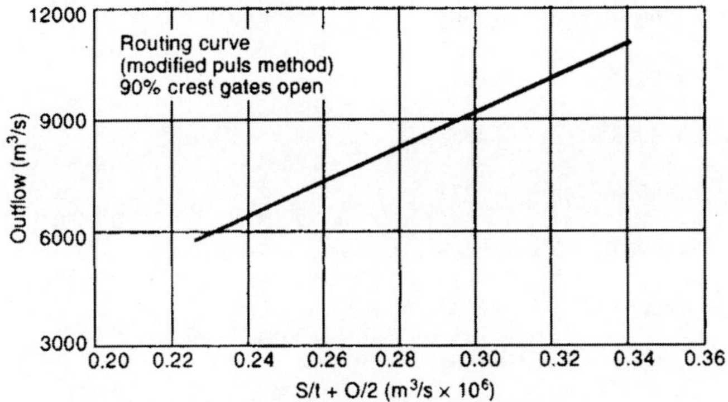

Fig. 16.17 $S/t + O/2$ vs outflow curve.

By means of Table 16.2, Fig. 16.17 is prepared.

Table 16.2 Reservoir level vs $S/t + O/2$ (Routing period 6 hours)

Reservoir level (m)	*Storage (S) (ham × 10⁶)*	*S/t = 0.463S (cumec × 10⁶)*	*Outflow (O) (cumec × 10⁶)*	*O/2 (cumec × 10⁶)*	*S/t + O/2 = (col. 3 + col. 5) (cumec × 10⁶)*
(1)	(2)	(3)	(4)	(5)	(6)
394.8	0.4800	0.2223	0.0055	0.0028	0.2251
396.3	0.5565	0.2577	0.0073	0.0037	0.2614
397.8	0.6413	0.2970	0.0089	0.0045	0.3015
399.3	0.7353	0.3405	0.0111	0.0056	0.3461

Referring to Table 16.3 col. 1 and 2 are obtained from the given inflow hydrograph (Fig. 16.16). The values given in col. 3 are the average of successive inflows. Column 4 is obtained by subtracting the initial $O/2$ value from the initial S/t value (which is either known or read out from Fig. 16.15 knowing the initial reservoir level). Column 6 is obtained by referring to Fig. 16.17 for values of col. 5. Corresponding to the values of col. 6, the values in col. 7 are obtained by referring to Fig. 16.15. With the values given in cols. 6 and 7, the outflow hydrograph and the curve of reservoir level with time are drawn in Fig. 16.16.

Table 16.3 Routing of design flood through reservoir (routing period-6 hours)

Time *h*	*Inflow* $m^3/s \times 10^6$	*Mean inflow* $m^3/s \times 10^6$	*(S/t-O/2)*	*(S/t + O/2)*	*Outflow* $m^3/s \times 10^6$	*Reservoir level m*
(1)	(2)	(3)	(4)	(5)	(6)	(7)
26	0.0071			0.2614	0.0071	396.30 FRL
		0.0097	0.2523			
32	0.0123			0.2620	0.0073	396.42
		0.0147	0.2547			
38	0.0171			0.2694	0.0076	396.76
		0.0182	0.2615			
44	0.0193			0.2797	0.0081	397.14
		0.0194	0.2714			
50	0.0196			0.2908	0.0086	397.59
		0.0189	0.2822			
56	0.0182			0.3011	0.0091	397.98
		0.0165	0.2919			
62	0.0149			0.3084	0.0095	398.21
		0.0130	0.2990			
68	0.0110			0.3120	0.0096	398.31 MWL
		0.0092	0.3024			
74	0.0074			0.3116	0.0096	398.28
		0.0057	0.3021			
80	0.0041			0.3078	0.0094	398.19
		0.0032	0.2984			
86	0.0023			0.3016	0.0091	397.98
		0.0017	0.2895			
92	0.0011			0.2913	0.0088	397.67
		0.0009	0.2852			
98	0.0006			0.2861	0.0084	397.43
		0.0006	0.2777			
104	0.0005			0.2783	0.0080	397.09
		0.0005	0.2703			
110	0.0004			0.2708	0.0077	396.79
		0.0004	0.2630			
116	0.0004			0.2654	0.0074	396.58
		0.0004	0.2564			
122	0.0004			0.2568	0.0071	396.22

Example 16.12 *Apportion the cost among the irrigation and power components of a multipurpose river valley project. Total cost of project is Rs. 1000 crores, Cost of dam portion is Rs. 700 crores, Cost of irrigation system is Rs. 300 crores, Cost of power specialities like penstock, trashrack structures, etc. is Rs. 100 crores, Cost of irrigation specialities is Rs. 50 crores, Height of dam is 150 m, Live storage is 2000 M cu m, Storage utilized for power is 1200 M cu m, and Storage utilized for irrigation is 800 M cu m.*

Single purpose dam:

(i) for power generation, (1200 M cu m. live storage) = *Dam height 135 m, cost Rs. 700 crore*

(ii) for irrigation (800 M cu m. live storage) = *Dam height 105 m, cost Rs. 500 crore*

Solution:

(*i*) By Equitable Justifiable Method

Cost of common civil works (700-100-50) = Rs. 550 crore

According to this method, the amount is to be apportioned among power and irrigation facilities in the ratio of cost of dam estimated, assuming it was constructed for providing irrigation and power benefits separately.

Hence, share cost debitable to power $= \text{Rs. } \dfrac{550 \times 700}{700 + 500} = 320$ crore

Share of cost debitable to irrigation $= \text{Rs. } \dfrac{550 \times 500}{700 + 500} = 230$ crore

Total cost to be borne by irrigation = Rs. 230 + 50 + 300 = 580 crore

Total cost to be borne by power = Rs. 320 + 100 = 420 crore

Total cost of transmission lines and the power house are not included in the above analysis as it is to be borne by power.

(*ii*) By Equal Apportionment Method

	Irrigation	*Power*	*Total*
Specific cost	50	100	= Rs. 150 crore
Share of joint cost $\dfrac{(1000 - 100 - 50)}{2}$	425	425	= Rs. 850 crore
Total allocation	475	525	= Rs. 1000 crore

Example 16.13 *Apportion the cost of a multipurpose project, total cost is Rs. 1000 crore.*

Estimated use — storage capacity

(a) For flood control = *5000 M cu m.*

(b) For power = *2500 M cu m*

(c) For irrigation = *2500 M cu m*

Separable cost, flood control = *Rs. 100 crores*

Irrigation = *Rs. 500 crores*

Power = *Rs. 300 crores*

Total = *Rs. 900 crores*

Solution.

Joint cost (total — separable cost)

= 1000 – 900 = Rs. 100 crores

For flood control, share of joint cost

(in ratio of use) = (100 × 5000)/1000 = Rs. 50 crores

For irrigation, share of joint cost

(in ratio of use) = (100 × 2500)/10000 = Rs. 25 crores

For power, share of joint cost,

(in ratio of use) = (100 × 2500)/10000 = Rs. 25 crores

Therefore, the total cost chargeable to:

Flood control	= 100 + 50	= Rs. 150 crores
Irrigation	= 500 + 25	= Rs. 525 crores
Power	= 300 + 25	= Rs. 325 crores
	Total	= Rs. 1000 crores

QUESTIONS

1. Define reservoir. Discuss various types of reservoirs. Distinguish between single purpose and multipurpose reservoirs.
2. What are single purpose and multipurpose projects? Under what conditions do you go for multipurpose projects?
3. What do you understand by storage zones of a reservoir? Discuss live storage and dead storage. Does dead storage constitute a loss in a reservoir feeding a hydel power plant?
4. Define mass inflow curves and demand curves. Explain the procedure for finding storage capacity of a reservoir in order to meet a particular rate of demand.
5. Where are storage schemes necessary? Describe different methods used for arriving at the storage capacity of a reservoir.

6. What is a storage elevation curve for a reservoir? How is it obtained from the contour plan of a reservoir site?
7. Discuss the factors which are considered for the selection of site for a proposed dam. It is assumed that the type of dam has already been selected for the project.
8. Explain causes of silting of reservoirs and method for its control. Discuss various methods of reservoir sediment control.
9. Describe a method for routing flood water through a deep reservoir using the fundamental relation between inflow, outflow (discharge) and storage. Take $Q = CLH^{3/2}$ for the spillway.
10. What is flood routing? Write a short note on the methods of flood routing.
11. What do you understand by flood routing through a reservoir? Briefly explain the procedure of routing of a given flood hydrograph through a reservoir knowing elevation versus storage and elevation versus outflow curves of the reservoir. Explain graphical method of flood routing.
12. Discuss the geological and topographical features which effect the selection of type of dam.
13. The following table gives the mean monthly runoff of a river. Draw a mass curve and find the capacity of the reservoir for 100% utilization of water at constant regulated flow.

Month	*Runoff* (m^3/s)	*Month*	*Runoff* (m^3/s)
April	80	October	30
May	40	November	20
June	9	December	40
July	10	January	50
August	20	February	60
September	40	March	120

14. Define the term "dam". What are single purpose and multipurpose projects? Under what condition is each type provided?
15. (*a*) What preliminary investigations should be conducted for the selection of a dam site?
    (*b*) Under what conditions would you adopt a gravity dam or earthen dam?
16. (*a*) What are the preliminary investigations involved in the planning of an irrigation project?
    (*b*) Explain the concept of a multipurpose project. Name some of the important projects in Northern India.
17. Define gravity dam and earthen dam. Differentiate between gravity and earthen dams.
18. What is meant by a multipurpose river project? Enumerate its different aspects.
19. (*a*) Explain the concept of a multi-purpose project.
    (*b*) What do you understand by planning of an irrigation project? Describe the various stages of its development.
20. (*a*) Discuss the various factors governing the selection of a dam site.
    (*b*) Enumerate the various considerations in the selection of type of dam with specific emphasis on topography.
21. Which type of dam you will recommend under the following conditions:
    (*a*) Foundation consists of gravel and coarse sand.
    (*b*) V-shaped narrow gorge.
    (*c*) Large wide gorge.
    (*d*) Ingredients like sand and shingle are locally available.
    (*e*) Spillway location separate from main dam.
    (*f*) Seismic zone.

22. Discuss the relative advantages and disadvantages of (*i*) gravity dam, (*ii*) arch dam, (*iii*) buttress dam, and (*iv*) earth and rockfill dams.
23. Differentiate between the following:
    (*i*) Overflow dam and non-overflow dam.
    (*ii*) Low dam and high dam.
    (*iii*) Rigid and non-rigid dam.
    (*iv*) Small dam and large dam.
24. (*a*) What are the factors which effect the selection of type of a dam?
    (*b*) Under what conditions are each of the following preferred:
    (*i*) Arch dam, (*ii*) Buttress dam, (*iii*) Earth dam, (*iv*) Masonry gravity dam.
25. Enumerate the different types of dams. What are the factors which effect the selection of a particular type of dam.

17

EARTH AND ROCKFILL DAMS

17.1 INTRODUCTION

Earthen dams have been in use since ancient times. In the earlier days these dams were limited to smaller heights due to lack of scientific knowledge. With the advancement in knowledge of soil mechanics and development of instrumentation to monitor the post-construction behaviour of a dam, earthen dams have been constructed to greater heights which were considered impossible earlier. Nurck dam in USSR, over 300 metres high, is the highest earth-cum-rockfill dam in the world.

In India, during the period 1870-1947, about 188 dams of low and medium heights were constructed, including about 70 earthen dams of which 19 were of a height above 30 m. In the post-independence era, since 1947, 31 earth and rockfill dams of varying heights have been constructed. There are as many as 4300 dams now, about 90 per cent are earth/rockfill dams. The longest is about 21 km long Hirakud dam and the highest is 260.5 m high Tehri dam under construction in Uttar Pradesh. To keep pace with the ever-increasing demands of water for irrigation, drinking water supply and hydel power generation etc., more and more earthen dams, in preference to other types of dams, are expected to be constructed in the country in times to come, due to the reasons (*i*) The number of dam sites favourable for masonry or concrete structures are constantly decreasing, (*ii*) The construction materials like cement and steel required in huge quantities for building concrete or masonry dams are getting costlier day by day, (*iii*) Earthen dams with recent advancement in the technology of design and construction and the resultant stability, adaptability and economy are becoming more commonly adopted type of dam, (*iv*) Earth dams are made of locally available materials, like clay, gravel, sand and silt. The two basic requirements to be satisfied by an earthen dam are (*i*) Imperviousness, and (*ii*) Stability under all conditions of operation. Evidently these two conditions can not be satisfied by sandy soils and clay soils alone, necessitating their suitable blending for use in embankment construction of an earthen dam.

17.2 TYPES OF EARTHEN DAMS

The type of embankment to suit various site conditions depends on (*i*) Quality and quantity of available materials, (*ii*) Foundation conditions and cutoff requirements, (*iii*) Types of available earthmoving equipment, (*iv*) Diversion considerations, (*v*) Construction schedule, (*iv*) Climatic conditions, and (*vii*) Safety considerations. Various types of earthen dams are as under:

1. Homogeneous Dam

Homogeneous dam, as the name implies, is constructed almost entirely of a single soil material even though it may have an internal drainage system and upstream slope protection. It may be composed of relatively pervious sands and sand-gravel mixtures but is usually made of impervious or semi-pervious soils. It is used when (*a*) Only one type of soil is locally available in required quantity, (*b*) The dam is of very low height so that it is economical to construct it homogeneous rather than zoned, or (*c*) It is uneconomical to construct a zoned embankment type dam due to paucity of free-draining material. Fig. 17.1 indicates homogeneous earthfill dam (Kuttiyadi dam) in Kerala state.

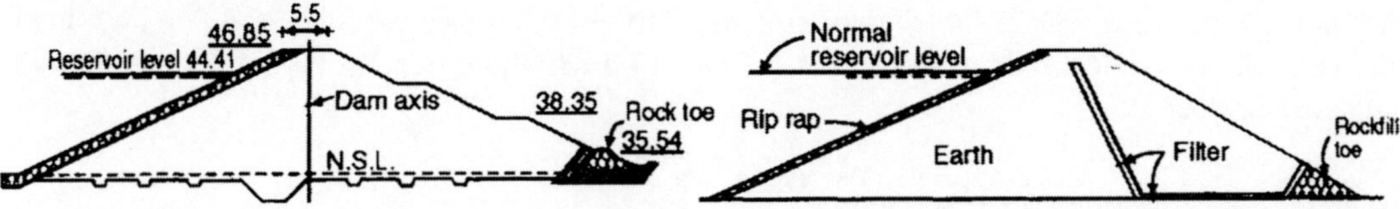

Fig. 17.1 Homogeneous earthfill, Kuttiyadi dam (Kerala)

Fig. 17.2 Modified homogeneous dam.

A purely homogeneous dam is essentially of low height and of relatively bigger section with flatter slopes so as to be safe against piping. The finer the soil, the higher the dam and flatter are the slopes. The essential requirements of a homogeneous dam are (*a*) The material comprising the dam is sufficiently impervious so as to be effective in providing barrier to the passage of water and develop required shear strength on compaction, maintain most of it on saturation with reservoir filling, (*b*) The side slopes are sufficiently flat (2 : 1 to 4 : 1) for stability. Upstream slope is safe against sudden drawdown of the reservoir and the downstream slope to resist sloughing when saturated to a high level, and (*c*) Seepage that emerges at the downstream due to long term storage in the reservoir, regardless of the flatness of the slope and impermeability of the soil, is taken care of by providing rock toe. It is on this account that instead of purely homogeneous section a modified homogeneous section wherein an internal drainage system is provided to keep the phreatic line well within the embankment is preferred (Fig. 17.2).

2. Rolled Fill Dam

Rolled fill dam embankment is constructed by controlled placement of earth material in thin layers and compacted by means of sheep foot rollers and other compaction machinery. The earth is spread in layers and water is sprinkled over the layer. Only additional moisture needed on the fill to replace moisture loss by evaporation, in order to maintain the material at desired moisture content, is added by light sprinkling of water. The compaction is done by sheep foot rollers hauled by crawler tractors properly weighted so as to ensure the requisite pressure. The compaction is also carried out by heavy pneumatic rubber tyred rollers. A number of coverages are given to ensure compaction of the material. The placement and compaction is carried out at the optimum moisture content. But method of rolling and optimum moisture content varies with the types of soils placed; better compact a soil at somewhat below optimum water content than above.

3. Zoned Dam

It is the most common type of rolled dam in which use is made of more than one types of the soil material. It essentially consists of a central impervious core zone followed by transition zones of comparatively pervious material and finally covered by free draining pervious outer zones on both sides. The impervious core is protected on both sides by filter, draining and transition zones. Two shells of sufficiently thick material, rockfill, ensure a high degree of permeability. Thus the permeability increases from the centre towards each slope. The entire dam, except the impervious core, forms part of drainage system. Filters allow the water to pass through them and retain the fine materials. Transition zones do the same thing with respect to the filters. The shells allow passage of filtered water through them. However, drainage system in the form of a horizontal drain or rock toe is provided. Fig. 17.3 shows a typical zoned earth dam.

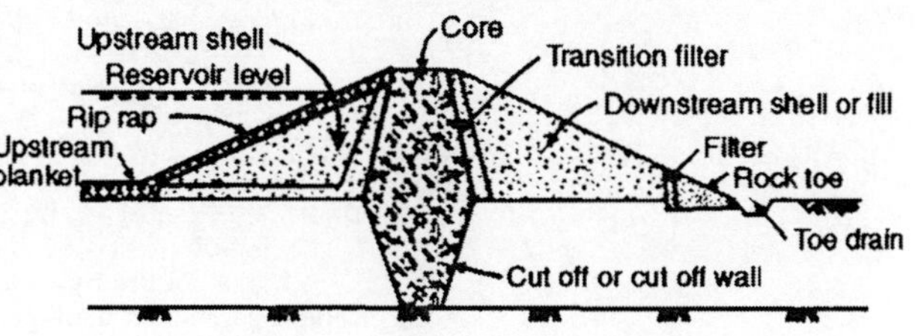

Fig. 17.3 Typical zoned earth dam

A dam is termed as of zoned embankment type if the horizontal width of the impervious zone at any elevation equals or exceeds the height of the embankment above that elevation in the d· and is not less than 3 m. Fig. 17.4 (*a*) depicts zoned earthen Pong dam on Beas river. Fig. 17.4 (*b*) is the general layout of the dam. Figs. 17.4 (*c*) and 17.4 (*d*) show general layout and section of Ramganga project.

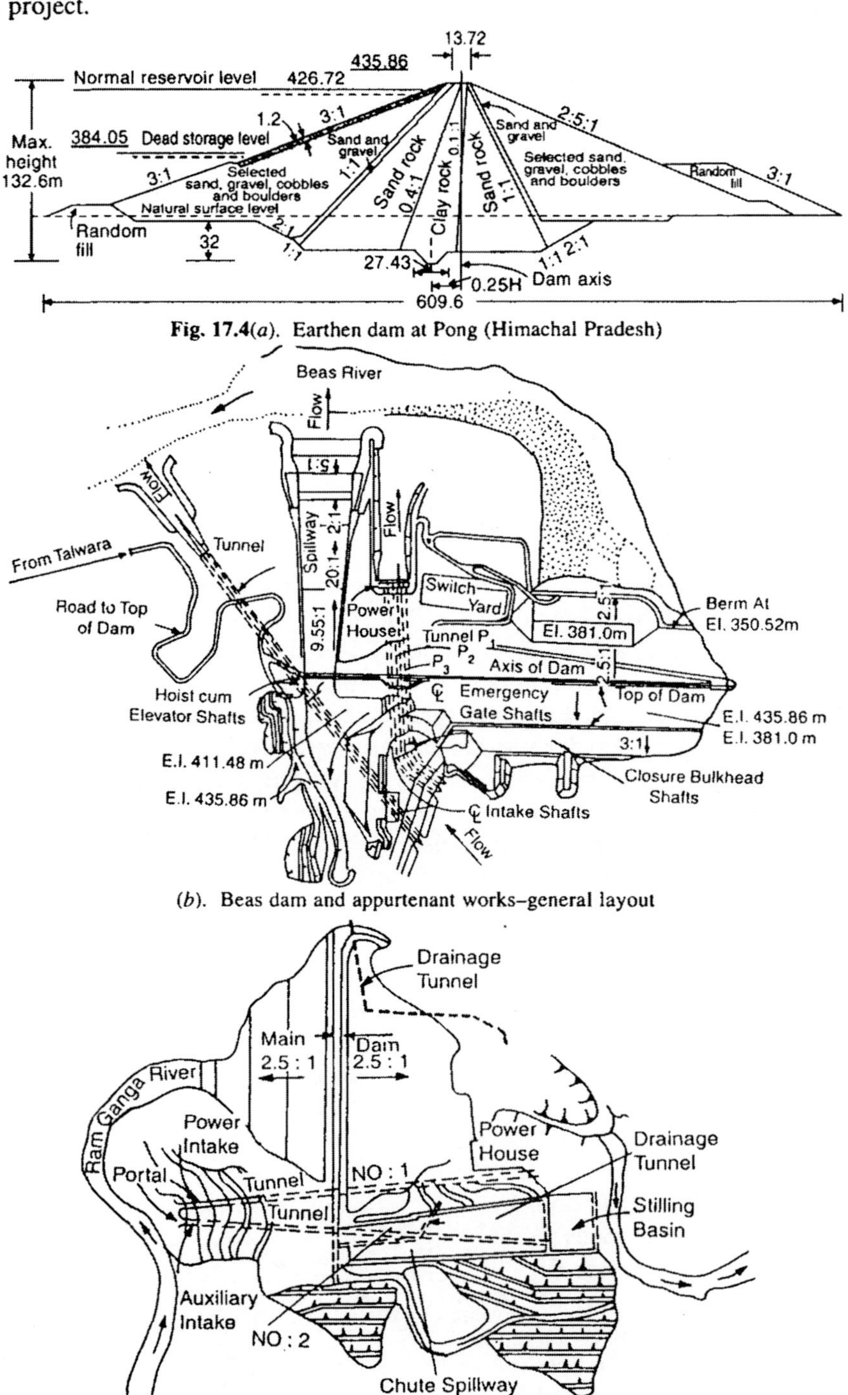

Fig. 17.4(*a*). Earthen dam at Pong (Himachal Pradesh)

(*b*). Beas dam and appurtenant works–general layout

(*c*). Ramganga earthen dam–general layout

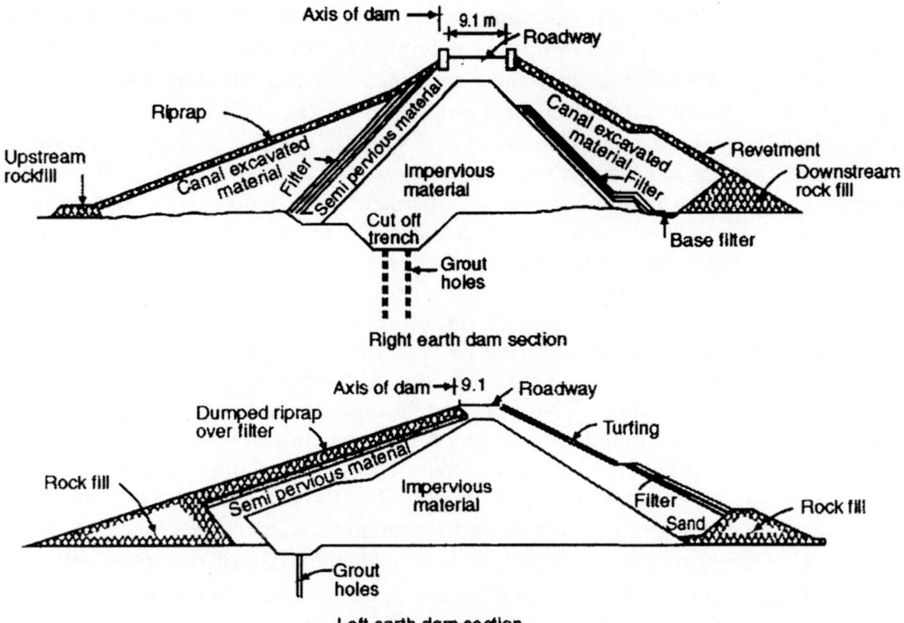

Fig. 17.4(*d*). Ramganga dam section

Significance of Various Zones in Zoned Dams

Earthen zoned dams are constructed of a variety of the available materials in different quantities and locations within the fill, discussed as under:

1. **Impervious zone:** Within the earthfill the distribution of materials is to be such that permeability and coarseness increases towards the outer slope. Keeping this consideration in view selected clay soils are used in impervious zone. In Pong dam impervious zone (Fig. 17.4(*a*) comprises (*a*) earth material containing more than 50 per cent claystone, and (*b*) earth material containing more than 50 per cent sandstone. The material is of homogeneous nature with a coefficient of permeability not greater than 10^{-6} cm/s.

The impervious zone of earthfill provides the water barrier. The material must be sufficiently impervious to preclude excessive water loss through the dam and must not consolidate or soften excessively on saturation by water from the reservoir. To satisfy the design criteria of imperviousness, low consolidation and resistance to softening, the maximum possible density is desirable. Core zone forms 40-70 per cent of the area of cross-section of a dam. The use of wider core zone cannot be regarded as the most efficient impervious membrane in the body of the dam. From the point of view of stress and deformation conditions in the core zone, more favourable conditions can be obtained from the core by giving it some inclination towards the upper side

2. **Filter zone:** It is impracticable to stop all seepage through the dam. As such an inverted filter is provided abutting both the upstream and downstream slopes of the core to collect seepage emerging out of the core and thereby keeping the downstream casing relatively dry. Filter is constructed of selected free draining sand and gravel with not more than 10 per cent of plus 75 mm size material as at Pong dam. It permits passage of water but prevents the displacement of fine soil particles which constitute the core. Provision of downstream filter is important for lowering the phreatic line and to prevent the movement of fine materials.

3. **Permeable zone:** Pervious zones provide an outer shell of high strength to impart stability and protect the impervious core, secure favourable hydraulic conditions of drainage, and act as filters between materials of greatly differing grain sizes. Within the pervious zone, individual loads are directed so that the material is graded in coarseness towards the outer slopes. The inner pervious zone provides stability against sudden drawdown and support and protect the impervious zone, while the outer pervious zone acts as a drain to control the line of seepage. Free drain-

ing material is essentially required in upstream pervious zone. The downstream zone may be of non-free-draining material in case free-draining material is not available. On Pong dam, pervious zones comprise selected free draining sand and gravel, cobbles and boulders with maximum size such that no stone protrudes outside the compaction lift of 60 cm.

4. **Toe weights:** To make the foundations of low strength suitable to support a dam, fills for weight only are often provided at the upstream and downstream toes of the dam. These fills act to restrain the tendency of the foundation materials to displace under load. Excavation from foundations or appurtenant works as is unsuitable or in excess of the requirements for structural zones of the dam is used for stabilizing fills at the toes of the dam.

5. **Riprap:** Riprap is used in earth dams for protecting earthen slopes from erosion by water currents and waves. The size of riprap depends on the reservoir fetch, the wind velocity and its direction. Riprap is dumped or hand placed, the latter to obtain shearing strength by interlocking of the angular fragments. The material consists of selected boulders.

6. **Connection with foundation:** The most widely used method of connecting the dam with foundation is a cutoff trench made in the upper layer of the foundation deposits and backfilling it with impervious clay soils. The depth of cutoff is of the order of 5-20 m working to about 15-45 percent of the total height of dam and width at bottom of cutoff from 0.1 H to 0.3 H. Such a design, especially if the cutoff is extended up to impervious layer in the foundation, is safe and reliable.

Advantages: (*i*) Optimum utilization of the available materials of different permeabilities, (*ii*) Zoning of the embankment permits use of steeper slopes because stronger materials are placed where they are most effective to resist shear stresses and drainage system controls and reduces pore water pressure, and (*iii*) Economy in cost.

4. Diaphragm Dam

It consists of a central impervious thin diaphragm of earth, cement concrete or bituminous concrete surrounded by earth or rockfill. The diaphragm extends through the cutoff to the impervious foundation to check seepage through the dam. It may be placed either centrally or at the upstream face as a blanket. The inclined diaphragm provides slightly better stability against earthquake. Earthen blanket on the upstream slope is not preferred, except for small dams, because it has to be buried to protect against erosion by wave action and as such it is not available for inspection or repairs. When the diaphragm is of earth, by definition, it has thickness of the impervious core at any elevation less than 3 m or less than the height of the embankment above the corresponding elevation. However, if the diaphragm thickness equals or exceeds these limits, it is considered to be of zoned embankment type. Earthen core diaphragm is preferred. The core has base width of 0.3 to 0.5 times the water head. Concrete diaphragm is susceptible to rupture due to settlement of the dam or its foundation.

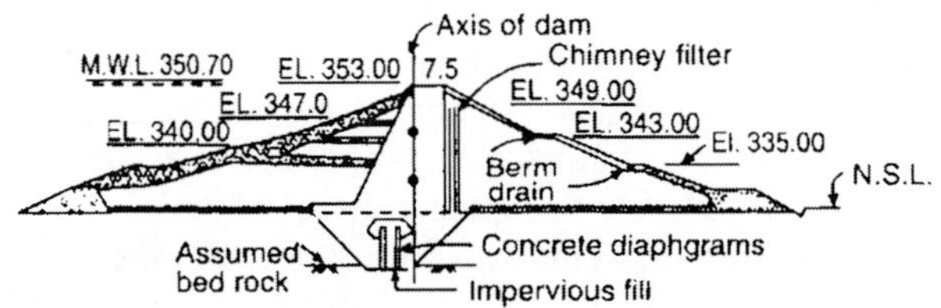

Fig. 17.5. Earth dam with concrete diaphragm

5. Hydraulic Fill Dam

In hydraulic fill dam, as the name implies, the soil material is excavated, transported and placed by using water (Fig. 17.6). Along the outer edges of the embankment, pipes–called flumes–

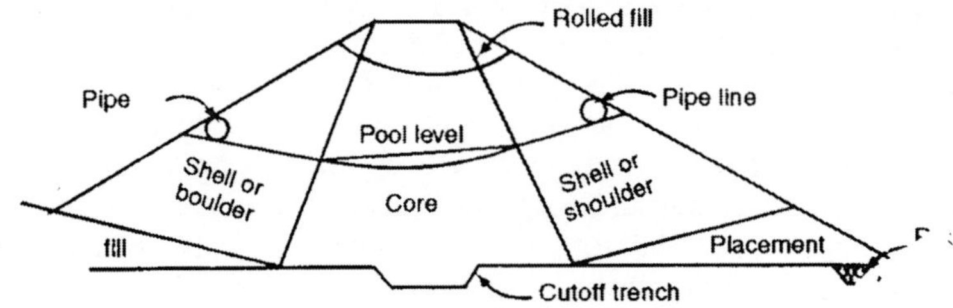

Fig. 17.6 Hydraulic fill dam

are laid at a suitable falling gradient. The material at the borrow pits is mixed with water to form slush and is pumped into these pipe flumes. The slush is discharged through the outlets in the flumes at suitable intervals along their length. The coarser materials of the slush settle soon after the discharge at the outer edge while the finer materials settle at the centre forming an impervious central core. Compaction is not done. However, high pore pressures develop in the core material. The drainage from the core is slow and as such settlement of embankment occurs over long periods. The hydraulic fill is suitable for sandy and gravelly materials. Hydraulic fill dams are not constructed now-a-days.

17.3 DESIGN CRITERIA FOR EARTH DAMS

The essential requirements of design of an earth dam are safe and stable structure at a minimum construction and maintenance cost. The essential design criteria are (*i*) Safe against overtopping during design flood by providing adequate spillway and outlet capacity, (*ii*) Spillway be of sufficient capacity to pass peak flood, (*iii*) Safe against overtopping by wave action by providing adequate free board, (*iv*) Side slopes stable during construction and under all conditions of reservoir operation. Upstream slope is safe against rapid drawdown conditions while the downstream slope is safe against sloughing, (*v*) Side slopes upstream and downstream are flat enough so that the shear stress induced in the foundation is within the shear strength of the material comprising the foundation with a suitable factor of safety, (*vi*) Upstream slope is protected against erosion by wave action while the crest and downstream slope is protected against erosion due to wind and rain. Horizontal berms at suitable intervals in upstream and downstream faces may be provided for this purpose, (*vii*) Downstream slope is safe during steady seepage under full reservoir condition, (*viii*) Portion of the dam downstream of the impervious core is properly drained by the provision of suitable drainage system, (*ix*) Seepage flow through the dam and foundation is so controlled that there is no danger of fine particles getting washed out from the downstream by the efflux of seepage. Moreover, quantity of seepage loss is restricted to the minimum, (*x*) There is no possibility of free flow of water from upstream through either the dam or the foundation, (*xi*) Adequate impervious core to act as water barrier. Top of impervious core is maintained higher than the maximum reservoir level, (*xii*) Seepage line, *i.e.*, phreatic line is well within the downstream face so that no sloughing of the slope takes place, (*xiii*) Seepage flow through the embankment, foundation and abutments is controlled so that no internal erosion takes place. The amount of water lost through seepage must be controlled so that it does not interfere with planned project functions, (*xiv*) Transition filters are incorporated to satisfy the two essential conditions of filters, viz., to prevent choking of the filter by soil and have minimum of head loss in the filter, (*xv*) Dam as a whole is earthquake resistant. The seismic conditions of the region are investigated with reference to geological map of vicinity. India has been divided into five seismic zones. Horizontal seismic coefficient for static design in different zones, is taken as under:

Zone	Horizontal seismic coefficient h
I	0.01
II	0.02
III	0.04
IV	0.05
V	0.08

Vertical seismic coefficient, where applicable, is taken as half of horizontal seismic coefficient as indicated above.

17.4 SECTION OF DAM

Top Width

The crest width of an earth dam has no appreciable bearing on the embankment or foundation stability. Besides width of roadway to be provided on the top of dam, the other considerations are nature of the embankment materials and the minimum allowable percolation distance through the embankment at normal reservoir level, height of the dam embankment, importance of the dam, practicability of the construction and required width to provide embankment mass for resistance to earthquake shock.

Empirical relations

$W = H/5 + 3.0$ for low dams

$W = 0.55H^{1/3} + 0.2\,H$ for dams lower than 30 m

$W = 1.65(H + 1.5)^{1/3}$ USBR practice for dams higher than 30 m

$W = (5/3)\,H^{1/2}$ common practice

where, W = top width (m), and H = maximum height of dam (m).

In general, the crest width varies from 6 to 12 m; 13 m is sufficient for high dams. The minimum width for low dams is 6 m.

Free Board

It is vertical distance between the horizontal crest level of the dam embankment without camber and the reservoir level. Parapet wall, usually 1 m high, provided on the top of dam does not form part of free board. It is estimated considering size of reservoir, method of operation of the reservoir, particularly wave height and wave length, height of wind set-up above still water level is the free-board reference elevation, slope of the dam and roughness of the pitching. Free board is provided so that the waves generated by wind on reservoir water are not able to overtop the dam. Since a reservoir has maximum and minimum reservoir levels, correspondingly the free board is designated as minimum and normal free board, as under:

1. *Minimum Free board*: It is the difference in elevation between crest level of the dam and the maximum reservoir level resulting from the design inflow flood and spillway and outlet facilities functioning as designed. Minimum free board is provided so that the dam embankment is not overtopped. Free board in consideration of wave action requires determination of wave height generated by winds in the reservoir depending on wind velocity, duration of wind, fetch, water depth and reservoir width (Figs. 17.7 and 17.8). Maximum reservoir level and half to two-third of wind velocity are considered for the computation of free board. Minimum free board is 1.5 m. Fig. 17.9 shows the relation between wave period (T_s) and related factors. Design wave height (H_0) is taken as 1.27 times the significant wave height (H_s). Table 17.1 indicates the wind velocity relationship of land to water.

Table 17.1. Wind velocity relationship-land to water

Effective fetch in km (Fe)	*1*	*2*	*4*	*6*	*8*	*10 and above*
Coefficient Q	1.1	1.16	1.24	1.27	1.30	1.31

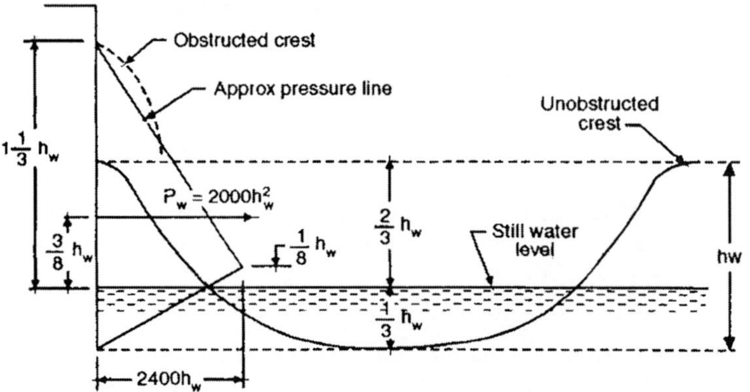

Fig. 17.7 Wind pressure and wave diagram

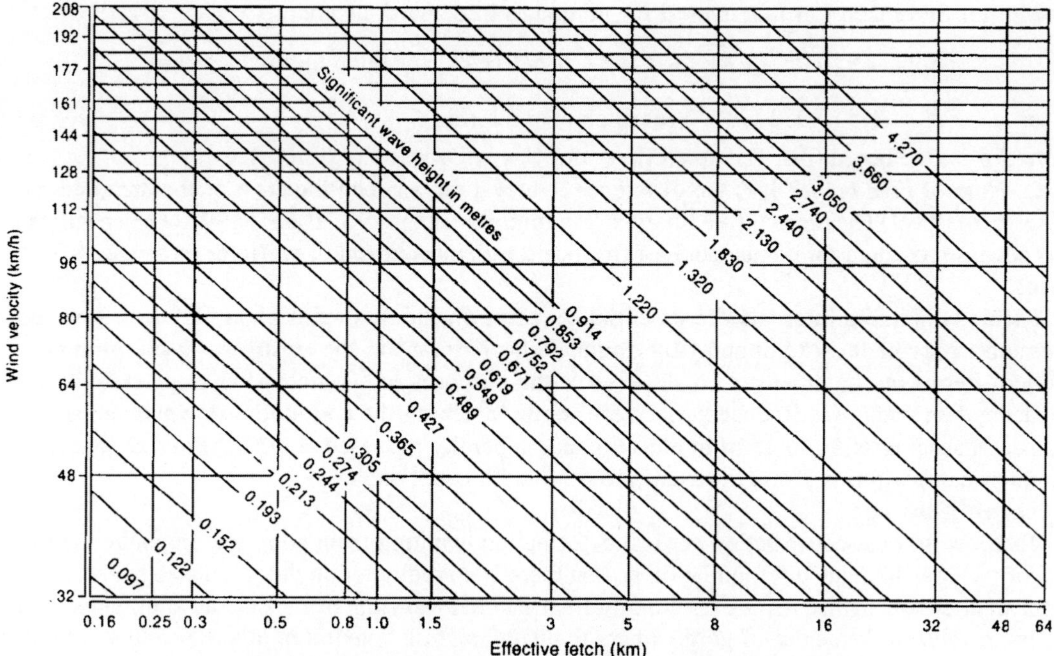

Fig. 17.8 Correlation of significant wave heights (H_s) with related factors

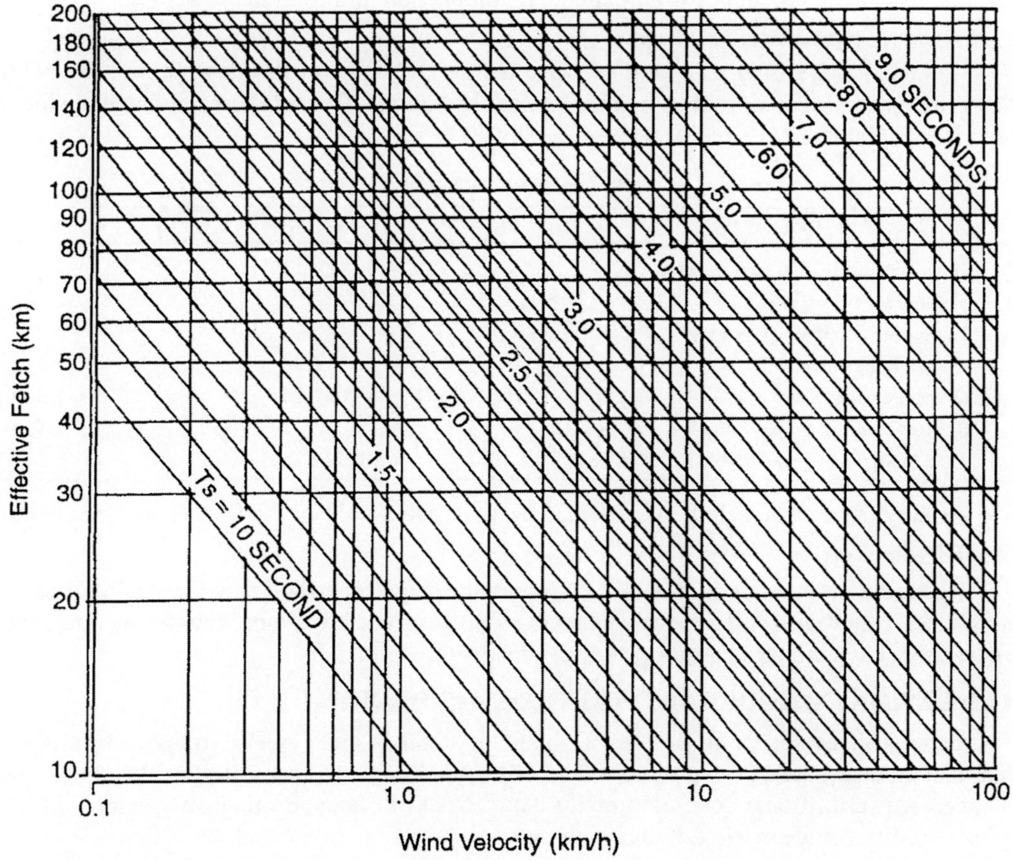

Fig. 17.9 Relation between wave period (T_s) and related factors

Molitor Stevenson formula is used for minimum free board, as under:

$$FB = 1.5\,[0.032\,\sqrt{FV} + 0.763 - 0.271\,(F)^{1/4}] \text{ for fetch less than 32 km}$$

or $$FB = 1.5\,(0.032\,\sqrt{FV}\,) \text{ for fetch over 32 km}$$

where, *FB* is free board (m), *F* is fetch (km) and *V* is wind velocity (km/hr).

2. *Normal Free board*: It is the difference in elevation between the crest of the dam and the full reservoir level. The normal free board is 2 m minimum as per IS: 10635: 1993. The specification contains procedure for computation of free board for embankment dams (refer Example 18.3).

Berms

Berms at suitable intervals are essential. A berm on the upstream about 1.5 m below the minimum reservoir level to support the foundation of riprap on the upstream is the minimum required. Berms on the downstream slope serve the purposes of (*i*) Increasing slope stability by increasing dam width, (*ii*) Reduce surface erosion by breaking the continuous downstream slope. Berms at a spacing of 10 to 15 m in elevation are generally provided, and (*iii*) Provide roadway. Minimum berm width is 3 m. However, 5 to 6 m width is desirable.

Settlement Allowance

To allow for post-construction vertical deformation resulting from compression and/or settlement of the embankment and foundation so that there is no reduction in designed free board, extra height of the dam in the form of a longitudinal camber varying from zero at abutments to a maximum value at the centre of gorge where the dam height is maximum and the settlement will be the most is provided. A provision of 1 to 2% of the embankment height above designed top level is provided to account for embankment compaction and foundation settlement.

Upstream and Downstream Slopes

Upstream slope is to take care of wave action and sudden drawdown condition, while the downstream slope is required to be safe against sloughing. Terzaghi's recommended side slopes are given in Table 17.2.

Table 17.2 Recommended side slopes

Type of material	*U/s slope*	*D/s slope*
Homogeneous well graded material	2.5 : 1	2 :1
Homogeneous coarse silt	3 : 1	2.5 : 1
Homogeneous silty clay or clay		
(*i*) height less than 15 m	2.5 : 1	2 : 1
(*ii*) height over 15 m	3 : 1	2.5 : 1
Sand or sand and gravel with clay core	3 : 1	2.5 : 1
Sand or sand and gravel with RCC core wall	2.5 : 1	2 : 1

Usually the upstream slope is provided as 3 : 1, and 2.5 : 1 or 3 : 1 on the downstream slope.

Slope Protection

Upstream slope is required to be protected against destructive wave action and sudden drawdown condition, while the downstream slope is protected against rain wash, wind, and tail water as discussed later in detail (see Section 17.11).

17.5 DESIGN CONSIDERATIONS IN SEISMIC REGION

Earthen dams appear to possess an inherent stability and reserve strength that make them resistant against shock of earthquake. The best suited dams to resist earthquake shocks without damage are earthfill and concrete gravity dams. Rockfill dams are relatively stable and possess great capacity for absorption of shock energy.

The dams situated in earthquake zones are likely to be subjected to additional stresses and deformation on account of earth acceleration. The additional factors to be considered are (*i*) Stability of embankment slopes under the extra forces set up by lateral and vertical accelerations, (*ii*) Settlement of loose or poorly compacted fill or foundations, (*iii*) Cracking of impervious fill, and (*iv*) Liquefaction of deposits of loose sand in the foundation of the dam. The design consideration to increase safety in regions of seismic activity is generally based on stability analysis of slopes with earthquake conditions in accordance with IS: 7894-1975. Usually it envisages a more conservative approach to the main elements of the dam, discussed as under:

1. *Extra Free Board*: It is a good practice to provide extra free board in the design to take care of (*i*) Any settlement of the crest of the dam, (*ii*) Raising of water level in the reservoir due to landslides in the reservoir basin, (*iii*) Movement of reservoir bottom, and (*iv*) Waves of greater height developed in the reservoir in an earthquake.

2. *Thick Core*: A thicker core is provided to resist piping action. When an earthquake may tend to twist the dam alignment; the additional width of the core will minimize the influence of possible localized effects. Dams with central concrete core wall are more vulnerable to damage by cracking.

3. *Foundation Treatment*: Dam founded on soil is more severly shaken in an earthquake than when founded on rock because both acceleration and amplitude of the ground movement are larger in the soil than in a rock. As such all loose or soft soil from the foundation should be excavated.

4. *Cohesionless Graded Filter Upstream*: Earthquake may develop cracks in the central core made of very tough cohesive soil. A cohesionless graded filter with appreciable content of fine sand placed upstream of the core has the advantage that sand from it will tend to flow into the cracks of the core and partially seal them.

5. *Cohesionless Graded Filter Downstream*: Excessive leakage in the core due to earthquake may develop excessive leakage and progressive piping through embankment. A suitably designed cohesionless graded filter placed just downstream of the core, extending up to the top of the dam, provides protection to the dam embankment against this leakage and piping.

6. *Very Pervious Downstream Zone*: Core of the dam if cracked by earthquake may become very pervious and if the filter is also ruptured large quantity of water may get into the downstream. Provision should, therefore, be made to cater for discharging the maximum anticipated leakage rapidly. For the purpose a massive downstream zone of large quarried rock or screened gravels are provided.

7. *Flatter Slopes Near Top of Dam*: The top of dam is susceptible to greater damage during an earthquake as it vibrates with considerably greater amplitude than the base. To render the dam safe by making longer the path of seepage through cracks which may develop, either the crest width is increased or the slope is made flatter at the top.

Analysis of Dam Section

The various considerations in the analysis of earth dam section under earthquake conditions are (*i*) When an earthquake occurs with a full reservoir behind the dam, vibrations of the dam induce dynamic water pressures which must be taken into account in design, (*ii*) The earthquake shaking may induce water waves in the reservoir. If there is fault displacement through the reservoir, it may generate relatively large water waves, the effect of which should be included in the design, (*iii*) Stability of soil and rock slopes surrounding the reservoir should be considered, (*iv*) For downstream slope, the most severe case for stability is if an earthquake occurs when the reservoir is full. The dam should be checked for stability against earthquake forces combined with reservoir full conditions, and (*v*) For upstream slope the two most severe cases for stability are sudden drawdown condition, and steady state (reservoir full) with earthquake. The dam is checked for stability for these two conditions. The simultaneous occurrence of sudden drawdown and earthquake is considered a remote possibility.

Sudden Drawdown Condition:Resisting forces for the impervious material are calculated using submerged weight of the material within the downstream range and moist weight above the maximum water surface level. The driving forces are calculated for the saturated weight within the drawdown range. Resisting and driving forces for the shell materials above dead storage level are computed for moist weight. All materials below the drawdown level are submerged and, therefore, all the resisting and driving forces are calculated on the basis of submerged unit weight of materials.

Steady State (reservoir full) with Earthquake: An additional horizontal inertia force due to horizontal earthquake acceleration is applied to the fill mass above the plane of slip acting in a direction towards the reservoir. The unit weight of the soil mass for computing the inertia force is the saturated weight of the soil mass. The submerged weight of the soil mass is used for determining the resisting and driving forces at the plane of slip. The reservoir thrust due to static head of water is reduced by hydrodynamic effect of the horizontal earthquake shock. The reservoir thrust thus obtained acts on the face of the impervious zone above the slip circle.

Factor of Safety: Factor of safety is considered as the ratio of forces resisting the movement to those tending to produce the movement. A minimum factor of safety of unity is accepted as adequate for ensuring stability of the upstream and downstream slopes against seismic condition and 1.5 for non-seismic conditions. The recommend (IS: 7894-1975) factors of safety for different conditions of stability of earth dam are as under:

Condition of analysis	*With earthquake*	*Without earthquake*
Steady seepage condition	1.0	1.5
Drawdown condition	1.0	1.3
Construction condition	1.0	1.3
	D/S slope (reservoir empty condition)	U/S slope (initial reservoir filling)

Earthen dams vibrate when subjected to ground motion during an earthquake, requiring thereby a dynamic analysis of the structure. The present design procedure, as under, is based on the assumption that the portion of the dam above the rupture surface is rigid, as stipulated in IS : 1893-1984:

Seismic force on soil mass: The seismic coefficient depends on the dam height and the lowest point of the rupture surface, as under:

(*i*) Fundamental period of the structure, $T = 2.9\ H_t \sqrt{\rho / G}$ where, T = fundamental period of the earth dam (in s), H_t = dam height above toe of the slopes, ρ = mass density of the shell material, and G = modulus of rigidity of the shell material.

(*ii*) Determine S_a/g for this period and 10% damping from average acceleration spectrum curves (Fig. 18.6).

(*iii*) Compute design seismic coefficient, α_h from the relation $\alpha_h = B\, I\, F_0\, S_a/g$. where, β = a coefficient depending upon soil-foundation system; value 1.0, I = a factor dependent upon importance of the structure; value 3.0, F_0 = seismic zone factor for average acceleration spectra; value as given in Table 18.1.

2. *Equivalent uniform seismic coefficient*: For checking slope failure with the lowest point of the rupture surface y below top of dam, the value of equivalent uniform seismic coefficient is taken as:

$$\alpha_y = (2.5 - 1.5\,y/H)\alpha_h., \; H \text{ is total dam height}$$

3. *Stability of upstream slope*: It is tested with FRL with horizontal forces due to earthquake acting in upstream direction and vertical forces due to earthquake, taken as one half of the horizontal, acting upwards. The factor of safety, assumed as 1, is tested only for failure surface which passes through the lower half of the dam.

4. *Stability of downstream slope*: In this case, the only exception from the testing of upstream slope is that the horizontal force due to earthquake is considered as acting in the downstream direction.

For prolonged construction period or prolonged empty reservoir condition, the earthquake forces are included in the stability analysis and calculated on 50% of the value obtained for upstream and downstream slopes mentioned above.

With a view to study behaviour of high dams under earthquake shaking, a full dynamic analysis is carried out using material parameters based on cyclic loading tests. Further, full dynamic analysis taking into consideration pore pressure built up during earthquake shaking, redistribution of pore pressures with time and potential for loss of strength at damaging strain levels is carried out in the case of coarse grained materials susceptible of loosing strength substantially.

17.6 PHREATIC LINE IN EARTH DAM

Phreatic or seepage line is defined as the line within a dam embankment below which there are positive hydrostatic pressures in the embankment. The hydrostatic pressure on the line itself is atmospheric or zero. Above the line, there is a zone of saturation due to capillary action which is neglected. Phreatic line separates the saturated region of flow from dry or moist zone of soil mass. Above the phreatic line soil is dry (moist) and below the line it is submerged. The determination of its locus in an earth dam is necessary to assess the soil properties while testing the stability of the earth dam embankment. The characteristics of phreatic line are (*i*) The location of phreatic line depends only on the cross-section of the dam embankment. Its position is not influenced by the permeability of the material comprising the dam so long as the material is homogeneous, (*ii*) The phreatic line for homogeneous material is a parabola except at ingress (starting point) and egress (emerging point). At the ingress it is normal to the upstream face since the upstream face is an equipotential line. At egress, it meets the discharge face tangentially if no filter arrangement is provided. The basic parabola needs correction at ingress if a horizontal filter is provided, and at lower end also if rock toe is provided, (*iii*) The focus of base parabola lies at the break out point of the bottom flow line, where the flow emerges out from relatively impervious medium to a highly pervious medium, (*iv*) The presence of a pervious stratum below the dam does not influence the position of phreatic line, and (*v*) In a zoned dam with impervious core, the effect of outer pervious shells is neglected and the focus of the base parabola is located at the toe of the core.

Determination of Phreatic Line

The phreatic line in a dam embankment can be determined by analytical method, graphical method, and experimental method. Among the analytical and graphical approximations, graphical method by Casagrande is widely adopted because of its simplicity. In this method, the phreatic surface begins normal to the upstream face at the water surface and in a short distance thereafter follows a parabola whose focus is at the beginning of the filter and which passes through a point in the water surface at a distance 0.3 time the horizontal face length in front of the dam. The procedure for locating the phreatic line is as under:

1. Dam with a Filter

In Fig. 17.10 *AB* is the upstream face and *BL* (equal to *L*) is its horizontal projection. Locate a point *C*, on water surface, at a distance equal to 0.3 *L* from *B*, *i.e.*, $BC = 0.3L$. *F*, the starting point of the filter is the focus of basic parabola. The filter length usually kept is 25% of the distance of toe of dam to the centre of dam crest. With point *C* as centre and *CF* as radius draw an arc cutting the line *LB* extended in *D*. Draw a vertical tangent to the curve *FD* at *D* such that *DE* is the directrix, *CD* being equal to *CF*. Point *G* is located midway between *F* and *E*. Thus a base parabola *CC'G* is drawn with focus at *F*. The base parabola is now corrected by eye, a short transition curve (*BB'*) being drawn to connect point *B* with base parabola. The transition curve should start normal to the upstream face at the point of intersection. *BB'C'G* now represents the phreatic line. Equation of base parabola—with *F* as focus and any point *P* (*x*, *y*) on the parabola (such that the distance of the point *P* from focus is the same as that from the directrix), is as under:

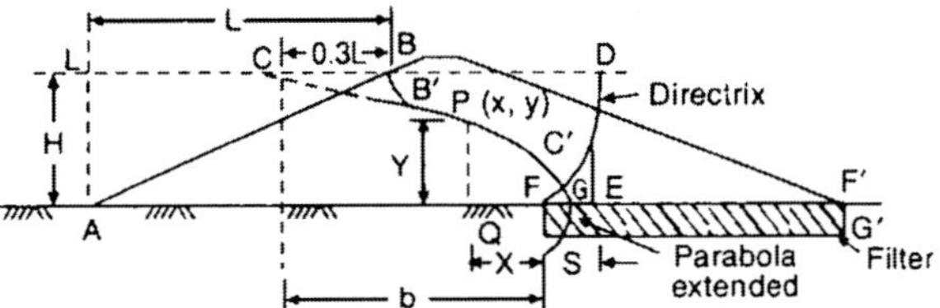

Fig. 17.10 Determining phreatic line in a dam with horizontal drainage filter (Casagrande's method)

$$PF = QE \text{ (from the property of parabola)}$$

$$\sqrt{x^2 + y^2} = QF + FE$$

or $$\sqrt{x^2 + y^2} = x + S \; (S = \text{focal distance } FE) \qquad ...(1)$$

Squaring both sides

or $$x^2 + y^2 = x^2 + 2Sx + S^2$$

or $$x = \frac{y^2 - S^2}{2S}$$

or $$y^2 = 2Sx + S^2$$

which is the equation of the base parabola. Considering point *C* with coordinates (*b*, *H*) on the parabola such that $CF = b$, substituting in equation (1) above.

$$\sqrt{b^2 + H^2} = b + S$$

or $$S = \sqrt{b^2 + H^2} - b$$

At *C'*, $x = 0$, $y = S$, thus vertical ordinate *FC'* at *F* is *S*.

For determining discharge through the dam consider unit width of dam and let *q* is the seepage discharge per unit width of the dam.

From Darcy's law

$$q = K.I.A.$$

$$q = K \frac{dy}{dx} (y \,.\, 1) \qquad A = \text{saturated depth } y \times \text{width}$$

or $$q = K\frac{dy}{dx}y \quad ...(2)$$

But equation of parabola

$$= \sqrt{x^2+y^2} = x + S$$

or $$x^2 + y^2 = (x + S)^2$$

or $$y^2 = (x + S)^2 - x^2$$

or $$y = \sqrt{(x+S)^2 - x^2}$$

Hence $$y = \sqrt{S^2 + 2Sx}$$

Also, $$\frac{dy}{dx} = \frac{S}{(2Sx+S^2)^{1/2}}$$

Substituting the value of y in equation (2) above

$$q = K\frac{S}{(2Sx+S^2)^{1/2}}\sqrt{(S^2+2Sx)}$$

or $$q = K\,S$$

The discharge can now be calculated from known coefficient of permeability K and focal distance S.

For determining q in terms of distance b and height H,

$$S = \sqrt{x^2+y^2} - x$$

at $$C,\ x = b \text{ and } y = H$$

Therefore $$S = \sqrt{b^2+H^2} - b$$

Hence $$q = KS$$

$$q = K(\sqrt{b^2+H^2} - b)$$

2. Dam without Filter

In this case, the focus of parabola is at the toe of the dam. The phreatic line is determined in the same manner as that for a dam with a filter. The base parabola BJG cuts the downstream slope at J and extends beyond the downstream slope of the dam (Fig. 17.11) up to point G, vertex of the parabola. The phreatic line will, however, emerge out at K meeting the downstream slope tangentially there. The portion KF is termed as discharge face and always remains saturated. The correction JK = (Δa) by which the parabola is to be shifted downward is found by the value of $\frac{\Delta a}{a+\Delta a}$ as given by Casagrande (Table 17.3 for various slopes α of discharge face with horizontal). Also $(a + \Delta a) = S/(1 - \cos\alpha)$ and $\Delta a = [(a+\Delta a)(180-\alpha)]/4$.

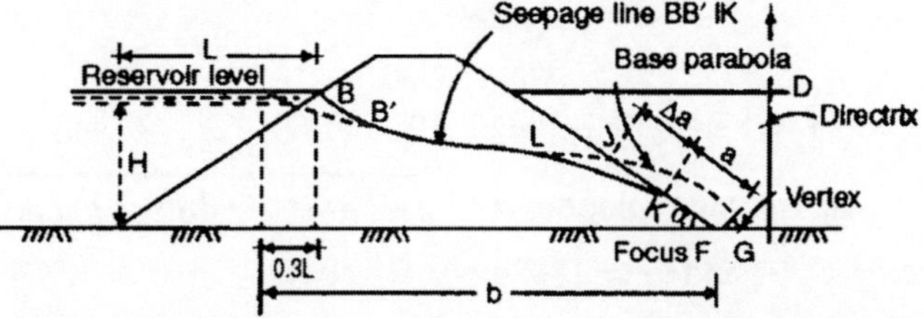

Fig. 17.11 Determining phreatic line in a dam without drainage filter

Intermediate values of α are read from Fig. 17.12. Value of α varies, $< 90°$ where no drainage is provided, 90° or more in rockfill toe, and 180° where horizontal filter is provided.

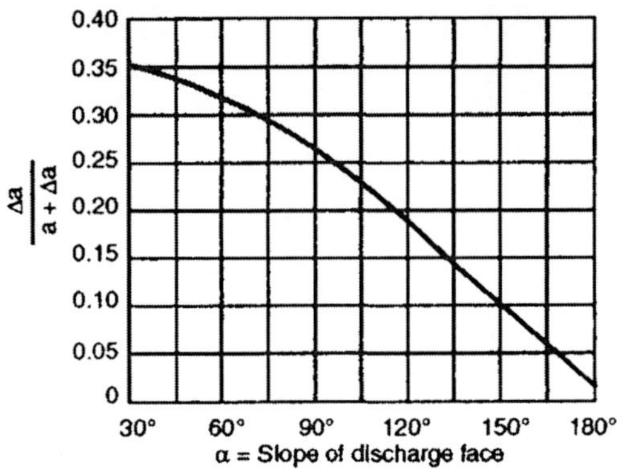

Fig. 17.12 α vs (Δa / a + Δa)

Table 17.3 Casagrande values of Δ a/(a + Δ a)

α (*degrees*)	$\frac{\Delta a}{a+\Delta a}$
30	0.36
60	0.32
90	0.26
120	0.18
135	0.14
150	0.10
180	0

Value of a position of point *K*, *i.e.*, where the phreatic line intersects the downstream slope, is determined by Schaffernak and Van Iterson relation, as under:

(*a*) When α < 30° (Fig. 17.13)

$$a = \frac{d}{\cos\alpha} - \sqrt{\frac{d^2}{\cos^2\alpha} - \frac{H^2}{\sin^2\alpha}}$$

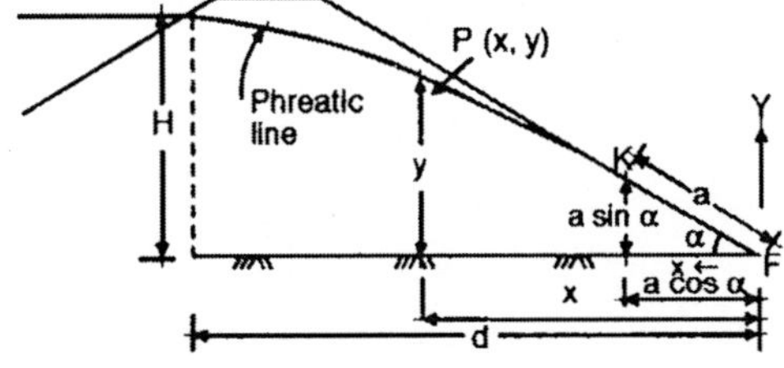

Fig. 17.13 Schaffernak's method.

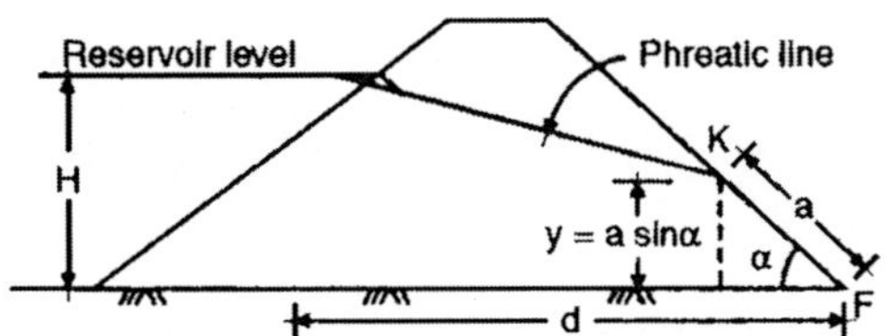

Fig. 17.14 Slope α = 30° to 60°

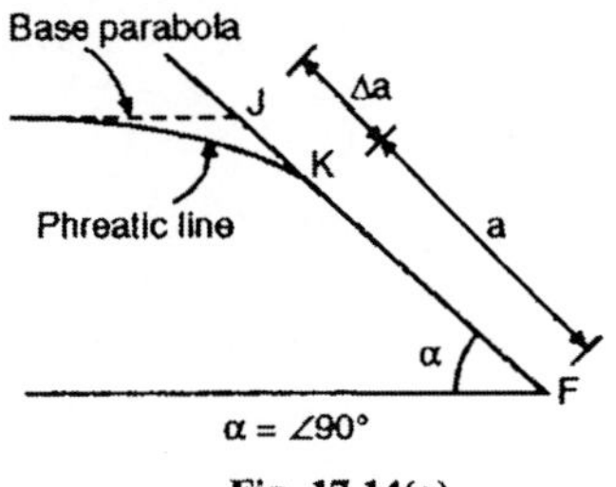

Fig. 17.14(*a*)

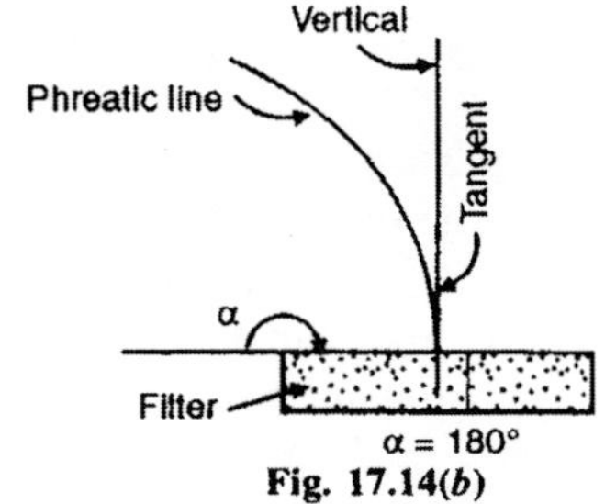

Fig. 17.14(*b*)

(*b*) When α is between 30° and 60° (Fig. 17.14).

Casagrande relation, $a = \sqrt{H^2 + d^2} - \sqrt{d^2 - H^2 \cot^2\alpha}$. Figs. 17.14(*a*) & (*b*) indicate when α = < 90° and α = 180° respectively.

3. Dam with Rock Toe

Focus of base parabola is located at the heel (the inner toe—Fig. 17.15) and basic parabola is drawn in the usual manner. The value of α is measured clockwise from the horizontal base.

Correction for $\frac{\Delta a}{a+\Delta a}$ is read from Fig. 17.15 and applied.

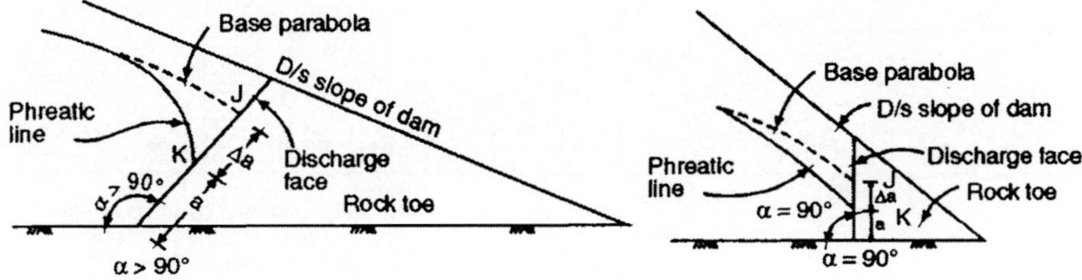

Fig. 17.15 Phreatic line in dam with rock toe

4. Zoned Dam

In zoned dam Casagrande's method is used for the location of phreatic line in the impervious central core at steady state condition (Fig. 17.16). The drop in head in the upstream pervious shell is neglected and the focus of base parabola is located at the downstream toe of the core. The point of entry of seepage line is the maximum reservoir level and the point of exit of phreatic line is considered at height $H/3$ (H is the head of reservoir water above ground level) on downstream boundary between central core and downstream pervious zone. The line joining these two points is taken as the position of phreatic line at steady state condition. Where inclined filter at the downstream slope of the core is provided in the dam embankment, the point of exit may be considered at height $H/2$ on downstream boundary between central core and downstream pervious zone. For designing upstream slope at sudden drawdown condition, the position of phreatic line in upstream pervious zone as well as in the central core is assumed to be without any change from that at steady state condition.

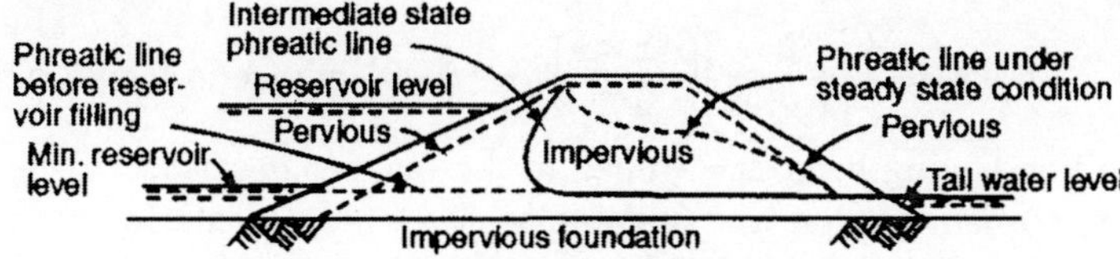

Fig. 17.16 Phreatic line in zoned dam

IS: 7894-1975 recommended approximate method for drawing the phreatic line for the discharge face of the core is that instead of drawing a basic parabola, the phreatic line is obtained by joining the entrance point and breakout point on the discharge face. The breakout point is located on discharge face at a height $0.5h$ from base of the dam, h being the head on upstream above the impervious base. At the entrance a short transition curve is introduced.

17.7 FLOW NET

Flow through the dam embankment composed of homogeneous material can be represented by a flow net. A flow net consists of two sets of curvilinear lines at right angles to each other. The Flow lines are the paths along which the seepage flows through the soil comprising the embankment. The direction of seepage is always perpendicular to the equipotential lines. Equipotential lines represent contours of equal head (potential).

In dam embankment, construction of flow net is complicated from the necessity to adjust by trial and error the shape of the free-surface line in the dam body, in addition to take account of a surface seepage of unknown length along the rear face. The flow net for a homogeneous dam is shown in Fig. 17.17. Here AE represents the phreatic line and EC the surface of seepage. However, the fact that the length EC has the same property as the line AE, that is the difference in elevation of consecutive equipotential lines meeting it should be equal, makes the flow net determinate.

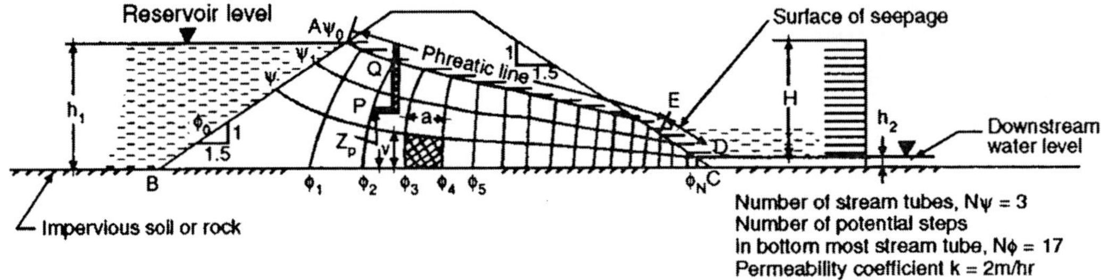

Fig. 17.17 Flow net for a homogeneous embankment.

Referring to Fig. 17.17, P is a point on the equipotential line meeting the phreatic line AE in Q. As such the piezometric tappings along the line ϕ_2 should all indicate the elevation Q. Although the hydraulic potential of all points along ϕ_2 is constant, the actual hydraulic pressure varies from point to point with the elevation. The pore pressure at the point P is given by the manometric column PQ. Here, $H = h_1 - h_2 = 12$ m, where, h_1 is depth of upstream water above base of embankment and h_2 is depth of downstream seepage water above the same datum. Elevation of P above tail water, $Z_p = 6$ m. Since there are 17 equipotential steps in the net and P is a point on the potential line ϕ_2, it follows that residual potential at P = 15/17 of the total head. Accordingly pore pressure at P

$$PQ = PQ\,[(h_1 - h_2) \times 15/17] - Z_p$$
$$= (12 \times 15/17) - 6 = 4.59 \text{ m of water}$$

Quantity of flow,

$$Q = K\frac{H}{N\phi}N\psi, \quad k = 2\text{m/hr},\ N\phi = 17 \text{ and } N\psi = 3,\ H = 12\text{m}$$

$$\therefore \quad Q = \left(\frac{2}{60\times 60}\right)\left(\frac{12\times 3}{17}\right) = 0.001 \text{ cumec/m length of embankment.}$$

The phreatic line having been determined in the manner described earlier, flow net can be drawn by plotting (*i*) dam section to a suitable scale — same scale for horizontal and vertical directions, (*ii*) drawing flow net by free hand sketching and making suitable adjustments and corrections by observing the properties of flow net *i.e.*, (*a*) The flow lines and equipotential lines intersect at right angles to each other, (*b*) The distance between the flow lines is made equal to the distance between the potential lines, so as to form a series of squares, (*c*) The rate of flow between each pair of flow lines, called *flow channels*, is the same and the energy drop between any two successive potential lines is the same, (*d*) Smaller the dimensions of the field, greater will be the hydraulic gradient and velocity of flow through them, (*e*) In a homogeneous soil, every transition in the shape of curves is smooth, being either elliptical or parabolic in shape.

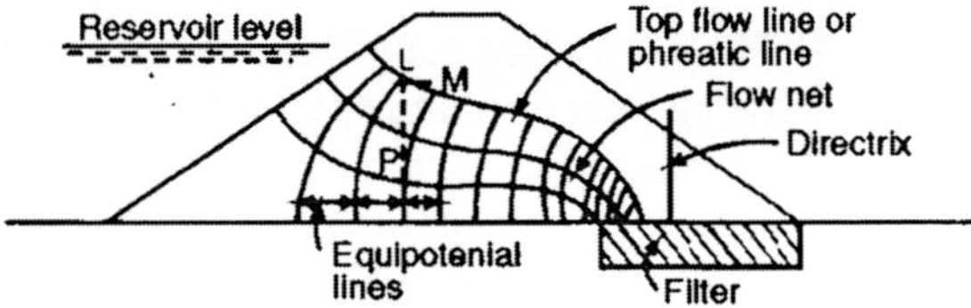

Fig. 17.18 Flow net by graphical method.

A limited number of flow lines and equipotential lines are drawn for convenience. The seepage discharge and pore pressures can be determined from flow net.

1. *Seepage Discharge*: The seepage rate (q) can be determined from the flow net using Darcy's law. The relation for discharge passing through a flow net are as follows:

(*i*) *Isotropic soils,* $q = K_H\ N_f/N_d = (K_H/N_d)$ (number of flow channels)

($K_H = K_V$, *i.e.*, horizontal permeability = vertical permeability)

(*ii*) *Non-isotropic soils,* $q = \sqrt{K_H K_V}\ \frac{HN_f}{N_d}$

where, q = rate of seepage, K = coefficient of permeability, H = total pressure head causing flow, N_f = number of flow channels in flow net, and N_d = number of equal potential drops in flow net.

In non-isotropic soils, wherein horizontal and vertical permeabilities differ in the dam section which is generally the case, the flow net is drawn in the manner as for isotropic soils. The dam section is drawn to the same vertical scale but to transformed horizontal scale. The transformed horizontal scale is actual horizontal dimensions reduced by multiplying with $\sqrt{K_V / K_H}$, where K_V is permeability in vertical direction and K_H is permeability in the horizontal direction. For example, if horizontal permeability is four times the vertical permeability, the horizontal dimensions of the dam shall be reduced by half for plotting.

2. *Pore Pressures*: Under steady state seepage, pore pressures at any point in the dam embankment is equal to the hydrostatic head due to reservoir minus the head loss in seepage through the embankment up to that point. The pore pressure at the point in the body of the dam can be found from flow net.

Pore pressure at point P (Fig. 17.18) is equal to difference in elevation between M and P, where M is the point of intersection of phreatic line with the equipotential line through P. Pore pressure at P is equal to LP, where LM is a horizontal line through M. Fig. 17.19 shows flow net (designed) for Pong dam.

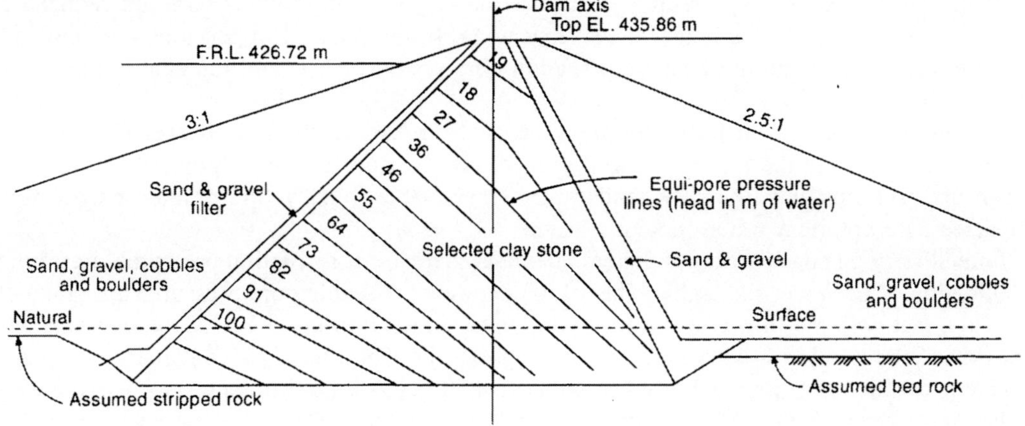

Fig. 17.19 Flow net

17.8 CONSTRUCTION PORE PRESSURES

The soil mass is composed of voids filled with water and air. When the moist soil mass is loaded without escape of air or water, part of the load causes soil grains to deform but without significant changes in their solid volume. This load is carried on the soil skeleton as effective stress and the remaining part of the load is carried by stress in the air and water contained in the voids and is termed as pore water pressure.

High pore pressures develop during construction (before filling of reservoir) of earth dam as a result of dam material compression due to the overlying fill and are not dissipated or are only partly dissipated. The layers of the soil are compacted with optimum moisture content. The pressure of the layers goes on changing with the increase in height of dam embankment and the water which fills the voids of the soil gets subjected to this extra pressure. Since there is no drainage to this water, a stress known as pore water pressure is developed in the dam embankment. This condition is not critical if the rate of raising of dam is less than about 15 m per year because the residual pore pressures in the dam and foundation are expected to be negligible except in highly clayey foundation with high water table. Construction pore pressure may control the design of dam as they may exceed the pore pressures as are likely to be developed because of seepage from reservoir. Construction pressures equal to 50 per cent of the total weight of the overlying embankment are common in the central portion of the impervious core where the overburden pressures are the highest and the effect of drainage to the adjacent pervious zone is least. The magnitude of the development of pore pressure during construction depends on placement water content, conditions of drainage, compressibility and permeability of the fill material, rate of fill placement and break in construction, overburden pressure etc.

1. *Water Content*: Average water content at which the embankment is constructed is the primary factor for the development of pore water pressures. The magnitude and distribution of these pore pressures depend primarily on construction water content of the embankment and natural water content in the foundation. Substantial pore pressures are found to have developed in the embankment in which the average water content exceeded 0.6 per cent below the standard proctor optimum. In the embankments compacted a few percent drier as also where the natural water content of the soil from borrow areas is always smaller than the optimum water content during construction no significant pressures develop. The closer the compacted fill is to saturation higher are the pore pressures under the given conditions of loading and compressibility.

2. *Drainage*: Magnitude and distribution of construction pressures depend on the rate at which dissipation may occur due to internal drainage. In the interior of thick impervious core, very little dissipation of pore pressures take place as a result of drainage during construction. However, construction pore pressures at the outside edges of impervious core are reduced by drainage of pore water to the adjacent pervious zone. In thin central impervious core embankment, appreciable dissipation of the construction pore pressures occur in the central part of the core.

3. *Soil type*: Development of pore pressures is also related with the soil type. The construction pore pressures are the highest in well graded, clayey sands and sand-gravel-clay mixtures, least in uniform silts and fine silty sands and inappreciable in very fine grained clayey soils, compacted near optimum water content.

Immediately on construction of the embankment there are construction pore pressures due to consolidation of fill under the embankment load. However, sizeable pore pressures are gradually dissipated. But with the filling of the reservoir, water enters into the pores of the dam embankment thereby developing a new pattern of pore pressures. The initial air contained in embankments with coefficient of permeability less than 0.1×10^{-6} cm/s is not influenced by seeping water but the embankment with coefficient of permeability more than 10^{-4} cm/s have all the initial air washed out of voids within a short span of time.

17.9 STABILITY ANALYSIS

An earthen dam has to be safe during construction and operation of the reservoir. It is recognized that in the life of a dam critical periods, for the stability of the dam are as under :

1. During Construction Condition, with or without partial pool

Pore pressures which develop in the embankment or foundation during construction as a result of embankment material compression due to overlying fill are not dissipated or are only partly dissipated and the factor of safety for upstream and downstream slope may be lowest during construction, particularly in case rate of dam raising is over 15 m per year. The construction pore pressures may exceed the pore pressures likely to be developed by seepage from the reservoir and consequently may control the design of dam. The pore pressures are computed by Hilf's method while examining the stability by effective stress method.

Hilf's formula,
$$U = \frac{P_a \cdot \Delta}{V_a + 0.02 V_w - \Delta}$$

where, U = induced pore pressure (kg/cm^2), P_a = absolute atmospheric pressure (kg/cm^2), Δ = embankment compression, in per cent of original total embankment volume, V_a = volume of free air in the voids of the soil directly after compaction, in per cent of original total embankment volume, and V_w = volume of pore water, in per cent of original total embankment volume.

The theoretical pore pressure at which all the air is forced into solution, that is when the dam becomes completely saturated which is the condition when the compression of the specimen Δ becomes equal to the initial volume of air in the voids V_a, is determined by the equation:

$$U = \frac{P_a V_a}{0.02 V_w}.$$

2. Reservoir Partial Pool (for upstream slope)

In initial partial filling of the reservoir it is assumed that a steady seepage condition has developed at the intermediate stages. The stability of the upstream slope is investigated for 1/3rd to 2/3rd height of full reservoir level and minimum value of factor of safety is determined. Reduction, if any, in the effective normal stresses where the pore pressures developed during construction are not dissipated before a partial pool condition can develop are considered in the analysis.

The zones above phreatic line drawn for upstream water level are considered as moist for determining resisting and driving forces and zones below it are taken with their submerged weights for working out both resisting and driving forces.

Analysis for partial pool condition may be carried out only in the case of high dams where the range of drawdown is small as compared to the height of dam.

3. Full Reservoir Condition

With full reservoir condition for long enough, seepage water percolates through the embankment, the pressure in the pore water in the downstream portion of the dam reaches its highest value. The downstream slope, under this condition, may have its lowest factor of safety against sliding. Only the downstream portion of the dam is required to be analysed for stability at normal reservoir level with fully developed phreatic line. Under steady state seepage conditions, the pore pressure head at any point is equal to the hydrostatic head due to the water level in the reservoir less the head loss in seepage through the dam fill to that point.

4. Sudden Drawdown Condition (for upstream slope)

After the reservoir is under operation for some time and seepage water has penetrated the embankment, drawdowns unprecedented either in the rate or in the extent of water surface lowering may cause sliding of the upstream slope. Sudden drawdown condition corresponds to the subsequent lowering of reservoir level rate faster than pore water can dissipate. Sudden drawdown to an intermediate level than complete drawdown induces unbalanced seepage forces and excess pore water pressures is a very critical condition for homogeneous dams and high dams where the range of drawdown is small as compared to dam height. Sudden drawdown causes reduction in resisting forces in comparison to actuating forces *i.e.*, (*i*) removal of supporting upstream pressure acting on the upstream slope above the drawdown level, (*ii*) the saturation line remaining higher as drainage is not as rapid as drawdown and pore pressures in the slope may not get dissipated immediately after the drawdown. This condition becomes critical if the material of the upstream portion of the dam are not free draining.

1. *Upstream Slope*: The analysis for upstream slope is done for drawdown condition from full reservoir level in the case of earth dam with gated spillway and a level corresponding to mid-level between full reservoir level and maximum water level to lowest water level in case of earth dams with ungated spillways. For the core (impervious) material, full pore pressures are allowed for the zone lying in the drawdown range. For a zone of random material with the properties intermediate between core and the shell material provided in between upstream shell and dam core, pore pressures for sudden drawdown condition are allowed for in the same way as for the core. Drawdown pore pressures can be determined by drawing flow nets, as also from the following formula:

Drawdown pore pressure

$$U = \gamma_w [h_c + h_r (1 - n) - h]$$

where, U = drawdown pore pressure at any point within the core, γ_w = density of water, h_c = height of core material at the point, h_r = height of shell material at the point, n = specific porosity of shell material, *i.e.*, volume of water draining out from the shell per unit volume, and h = drop in head under steady seepage condition at the point.

The weight of material is considered moist for computing driving and resisting forces for all zones above phreatic line and for zones in drawdown range core material and non-free draining material is considered as saturated and free draining material as moist for driving forces and pore pressure is adopted as given above corresponding to coefficient of permeability or by the formula given above for resisting forces. For all zones below drawdown level including foundation zone below drawdown level, the weight of material is considered buoyant for computing driving and resisting forces.

2. *Downstream Slope*: Analysis for downstream slope is carried out under sudden drawdown condition from maximum tail water level to minimum tail water level.

5. *Steady Seepage Condition*: Analysis for downstream slope is also carried out under the critical condition of steady seepage developed when water level is maintained at a constant level for sufficiently long time (say one month) and seepage lines are established in the earth dam section. In the analysis, existence of tail water and drawdown effects, if any, are taken into account. The stability of downstream slope is examined by effective stress method.

6. *Steady Seepage with Sustained Rainfull*: Stability of downstream slope is also analysed, where there is possibility of sustained rainfall, on arbitrary assumption that a partial saturation of shell material takes place due to sustained rainfall. In the analysis the shell and other material lying above the phreatic line is considered as moist for calculating driving forces and buoyant for resisting forces. Allowance for pore pressure in the analysis is made in terms of buoyancy of the material or by drawing flow nets. The saturation for downstream shell material is assumed (*a*) 50 per cent if coefficient of permeability is 10^{-4} cm/s or less, (*b*) zero per cent if coefficient of permeability is 10^{-2} cm/s or more, and (*c*) percentage varies linearly from fifty to zero for the coefficients of permeability lying between 10^{-4} cm/s and 10^{-2} cm/s.

7. *Earthquake Condition*: Earthquake forces are not normally included in stability analysis for the construction stage or reservoir empty condition. However, earthquake forces are included where the reservoir is to remain in empty condition for prolonged periods.

The stability of upstream and downstream slopes are analysed as discussed in section 17.5.

17.10 METHODS OF STABILITY ANALYSIS

The methods of analysing the stability of the slope, depending on the profile of the failure surface, are discussed as follows:

1. Circular Arc Method

Also known as Swedish or slip circle method. In this method, the surface of the rupture is assumed as cylindrical or in the cross-section by an arc of a circle. The method is generally applicable for analysing slopes of homogeneous earth dams and dams resting on thick deposits of fine grained materials. The assumptions made in this method are (*i*) No shearing stresses act across the plane of the cross-section and the analysis is treated as two dimensional, (*ii*) Section of dam analysed is of unit thickness, (*iii*) The sliding mass is divided into a convenient number of slices and each slice is assumed to act independently of its adjoining slices and the forces acting on the sides of a slice have no influence on the shear resistance which may develop on the bottom of the slice, and (*iv*) Shear strength of the various zones along the potential failure surface is mobilized simultaneously.

In the method a possible circular failure surface through the embankment and foundation (if it is not firm and through which failure is anticipated) is assumed. The trial sliding mass is divided into a number of vertical slices, usually 10 to 15, of preferably equal width, depending on the width and profile of the sliding mass, number of various zones included in the sliding mass and the accuracy required. For zoned embankment and stratified foundations with different properties, where an arc of the potential failure surface passes through more than one type of material, the vertical ordinates of the slices for each zone or part of foundation are obtained by locating the slice at each such dividing point. Trial surface computations are made of the shear force needed for equilibrium and the strength forces available. Fig. 17.21 (*a*) depicts assumed failure surface. Fig. 17.21 (*b*) shows the forces acting on a typical slice in the analysis of downstream slope for steady seepage condition.

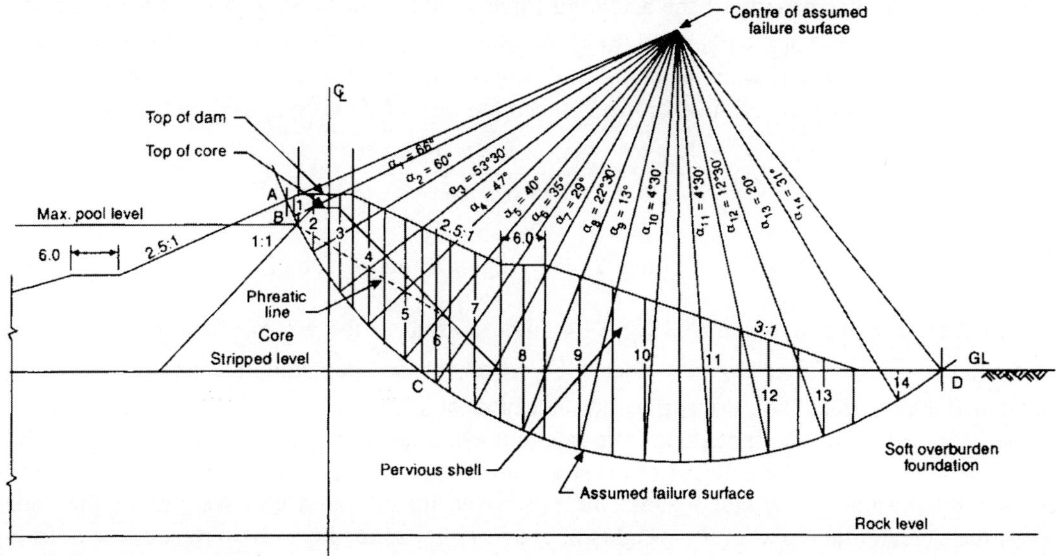

Fig. 17.21(*a*) Circular arc method.

Procedure for the analysis of the driving forces and resisting forces by arithmetical method is as follows:

Total weight W of the slice of width b is equal to the areas of various zones falling in the slice multiplied by the respective appropriate unit weights (soil plus water). The weight W acts vertically downward through the centre of gravity of the slice. The two components of W, *i.e.*, (*i*) the force normal to the arc of the slice, $N = W \cos \alpha$, and (*ii*) the force tangential to the arc of the slice, $T = W \sin \alpha$ are determined after resolving weight W in the radial and tangential direction, α being the angle made by the radius of failure surface with the vertical at the centre of the slice. Influence of pore water pressure is not accounted for separately in the analysis because it is already included in the test results by total stress method of analysis. However, allowance for pore pressure is made separately in the analysis by the effective stress method. Pore water pressure U acting on the arc of the slice results in an uplift force which reduces the normal component of the weight of the slice. The net or effective downward force acting on the curved bottom boundary of the slice is the total weight minus the upward force due to pore water pressure. The effect of pore pressure on resisting forces is accounted for by assuming buoyant weight of the material lying below the phreatic line. Component of shearing resistance due to internal friction is therefore $(N - U) \tan \phi$, where ϕ is the angle of shearing resistance of the material at the base of the slice, and $(N - U)$ is the effective

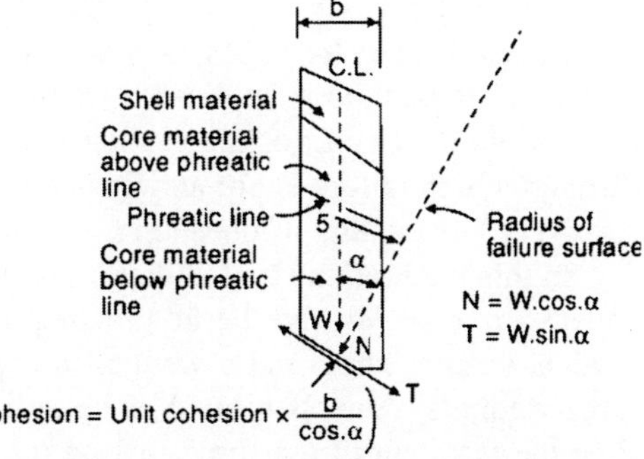

Fig. 17.21.(*b*) Analysis of forces acting on slice 5

normal load N_e. Another force acting at the bottom of the slice and which opposes the movement of sliding mass is the shearing resistance offered by the material due to its cohesion C, and is equal to the unit cohesion, c, multiplied by the length of the bottom of the slice and is approximately equal to $c \times b/\cos\alpha$. In practice, the length of the arc is measured accurately as the expression $b/\cos\alpha$ shall not give the length of the arc of the slice when b is not infinitely small. The total resisting or stabilizing force S developed at the bottom of the slice is equal to $(c \times b/\cos\alpha)$ + $(N - U)\tan\phi$. The driving or the actuating force T due to the weight of the slice is equal to $W \sin\alpha$. Similar forces are worked out for all the slices considered for a potential failure surface. Results of these computations are tabulated and sums of the resisting and driving forces obtained. Factor of safety, (FS) against sliding for the assumed failure surface is computed by the equation:

$$FS = \frac{\Sigma S}{\Sigma T} = \frac{\Sigma(C+(N-U)\tan\phi)}{\Sigma W \sin\alpha}$$

where, S = resisting or stabilizing force, T = driving or actuating force, $C = c \times \dfrac{b}{\cos\alpha}$, where c is unit cohesion and b is width of slice, N = force normal to the arc of slice, U = pore water pressure, ϕ = angle of shearing resistance, W = weight of the slice, and α = angle made by the radius of failure surface with the vertical at the centre of slice.

Seismic forces are also included in the calculations because they reduce the margin of safety or may even bring about the collapse of the structure. The total weight of the sliding mass considered for determining horizontal seismic force is based on saturated unit weights of the zones below, the phreatic lines and moist weights above it. If the zone above the phreatic line is freely draining, drained weights are considered for that zone.

2. Sliding Wedge Method

This method is generally applicable where it is considered that the failure surface may be best approximated by series of planes rather than smooth continuous curve as for circular arc method. The method is applicable where (*i*) One or more than one horizontal layers of weak soil exist in the upper part of the foundation, and (*ii*) Where the foundation consists of hard stratum through which failure is not anticipated and dam resting on it has a core of fine-grained soil with relatively large shells of dense granular material.

In this method, the trial sliding mass is divided into two or three segments, the top segment is called *active wedge* and the bottom segment the *passive wedge*. The middle wedge, in case of a three-wedge system, is termed as central block. First the resultant of the forces acting on the active wedge and the passive wedge are determined. These resultants acting on the central block along with other forces on the block gives a closed polygon for stability. Fig. 17.22 shows arbitrary failure surface A, B, C, D assumed for analysis. Here, C_A is developed cohesive force of active wedge, C_{CB} is developed cohesive force of central block, C_P is developed cohesive force of passive wedge, E_A is resultant force of the active wedge, E_P is resultant force of the passive wedge, F_A is resultant of normal and frictional force of active wedge, F_{CB} is resultant of normal and frictional force of central block, F_P is resultant of normal and frictional force of passive wedge, W_A is total weight of active wedge, W_{CB} is total weight of the central block, and W_P is total weight of the passive wedge. Vertical boundaries are assumed be-

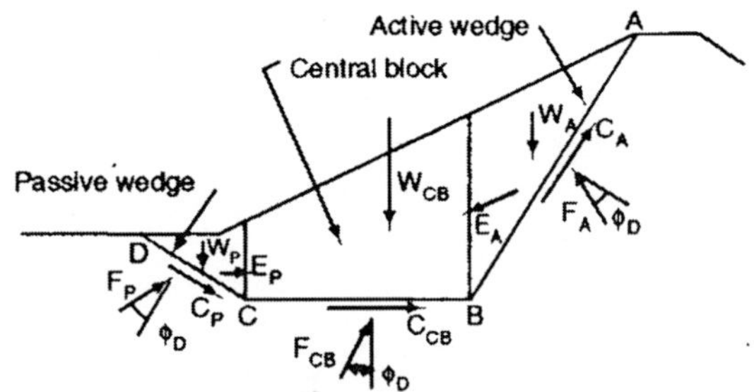

Fig. 17.22 Wedges in embankment section.

tween the central block and active and passive wedge. Total weight W of the segment is equal to the area multiplied by the appropriate gross unit weight of the material. Shear strength developed at the bottom of the wedge is assumed to be equal to:

$$SD = CD + (ND - U) \tan \phi D$$

where, C = total cohesive force, CD = developed cohesive force = C/FS, FS = trial factor of safety, ϕ = angle of internal friction, ϕ_D = developed angle of internal friction required for equilibrium, $\tan \phi D = \tan \phi / FS$, SD = developed shear strength, ND = normal force acting on the bottom when the wedge is at a state of equilibrium and not at failure, and U = pore water pressure.

The forces on each segment are considered separately. The developed values of cohesion, CD, and developed angle of internal friction, ϕ_D, along the failure surface are controlled by the assumed trial factor of safety. The magnitude of resultant earth forces E_A and E_P also depend on the value of trial factor of safety. However, the direction of forces acting on the vertical boundaries must be assumed. If the vertical boundary between the active wedge and the central block or passive wedge is located at or below the centre of embankment slope, the direction of the force E_A is assumed to be parallel to slope. If the vertical boundary is located near the top of the embankment slope horizontal direction for E_A should be assumed. For a vertical boundary between the centre and top of the slope, an intermediate angle may be selected. The direction of the force E_P in the vertical boundary between the passive wedge and the central block should be assumed parallel to the outer embankment slope if the boundary is located near the centre of the slope. If it is located at or near the toe of the slope the direction of E_P is assumed as horizontal. The resultant forces acting at the vertical boundaries of the active and passive wedges are determined by constructing force polygon (Figs. 17.23 and 17.24). These forces are finally incorporated in the force polygon for the central block (Fig. 17.25). A condition of equilibrium will generally not be obtained on first trial and several trial analysis with different factors necessary to close the polygon (Fig. 17.25) is denoted by ΔEH. The force ΔEH is assumed to act horizontally and its magnitude and sign vary with the trial value of factor of safety. A plot is made of ΔEH versus the trial values of factor of safety as shown in Fig. 17.26 to determine the value of factor of safety for which ΔEH is zero. This factor of safety is required to balance the forces for the sliding surface being analysed. Various trial locations of the active and passive wedges are required to determine the minimum value of factor of safety.

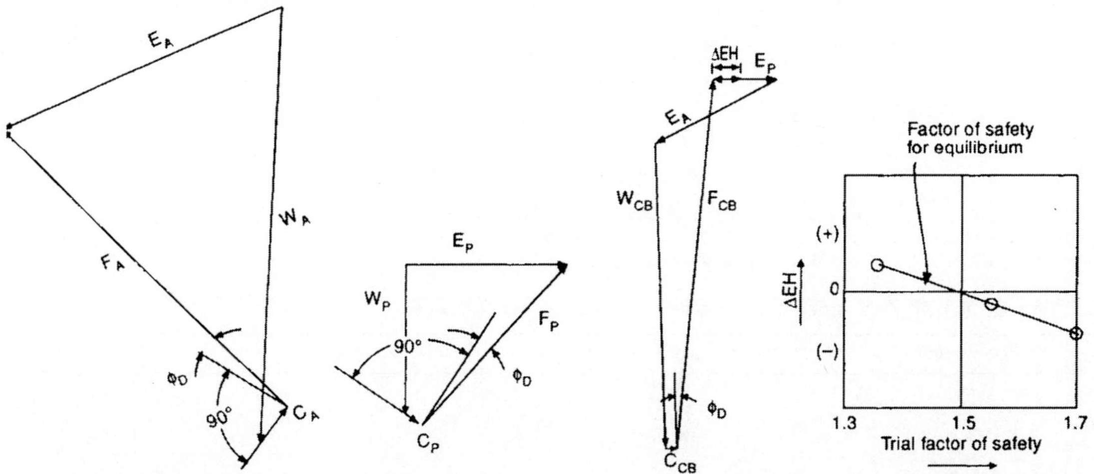

Fig. 17.23 Force palygon for active wedge.

Fig. 17.24 Force polygon for passive wedge.

Fig. 17.25 Force polygon for central block.

Fig. 17.26 Trial factor of safety vs ΔEH.

Rectangular Plot Method

A simplified rectangular plot method has been suggested by Alam Singh, as under:

$A_1, A_2, \ldots$ = boundary ordinates of the slices

b = width of each slice

mb = width of the last slice

Total weight of the sliding section

$$\Sigma W = b\left[A_1 + A_2 + \ldots \left(\frac{1+m}{2}\right)A_n\right] \quad \text{(in terms of area)}$$

The weight is taken corresponding to the end ordinates of each slice. The weight W of each slice is resolved into two components, say a normal component N and tangential component T (Fig. 17.27). Then, the N-rectangle is plotted by measuring various N-values (*i.e.*, $N_1, N_2, \ldots$) as abscissa and constant width b as ordinate. Similarly, T-rectangle is constructed (Fig. 17.27(*a*). Then ΣN = area of N-rectangles and ΣT = area of T-rectangles. Thus, if there are n-slices, the total number of ordinates will be $(n-1)$, the ordinate at the end of the last strip being zero. The product of areas of N and T rectangles and specific weight of the soil gives ΣN and ΣT respectively.

Area = $AN \cdot x^2$, if the scale of drawing is 1 cm = x metres and γ = unit weight of soil:

Volume, $V = AN \cdot x^2 \times 1 \text{ m}^3$

$$\Sigma N = AN \cdot x^2 \gamma \text{ kg}$$

$$\Sigma T = AT \cdot x^2 \gamma \text{ kg}$$

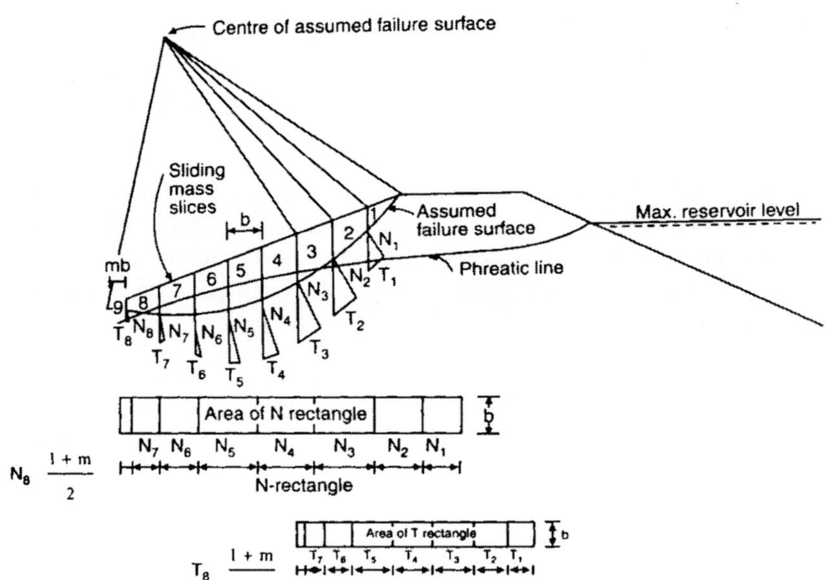

Fig. 17.27(a). Rectangular plot method.

The values of ΣN and ΣT can be worked out as follows:

Slice number	*Wt. of each slice (W)*	$N = W \cos \alpha$	$T = W \sin \alpha$
1	W_1	N_1	T_1
2	W_2	N_2	T_2
3	W_3	N_3	T_3
.	.	.	.
.	.	.	.
n	W_n	N_n	$-T_n$
		Sum ΣN	ΣT

Location of Centre of Slip Circle: For computing stability of side slope of a dam, it is necessary to find the least factor of safety which may occur on any possible surfaces of slippage or slip circles.

Location of critical slip circle as has least factor of safety is determined by trial and error which entails lot of time. With a view to reduce the number of trials Fellenious construction is used for locating the line on which the centre of the critical circle is most likely to lie in a homogeneous section. Various circles with centre on this line are tried to find one with minimum factor of safety. The position of the line on which the centre of the critical circle lies, according to Fellenious construction, depends only on the height and slopes of the embankment. In Fig. 17.27(*b*) the centre of critical slip circle lies on *PQ*. Coordinates of point *Q* are *H* downwards from toe and 4.5 *H* horizontally away. The other point *P* is located with the help of directional angles α and β (Table 17.4). The line from *PQ* is the line on which the centre of critical circle lies. Trial centres are selected on the line *PQ* and factor of safety corresponding to each centre is calculated from the equation, as under:

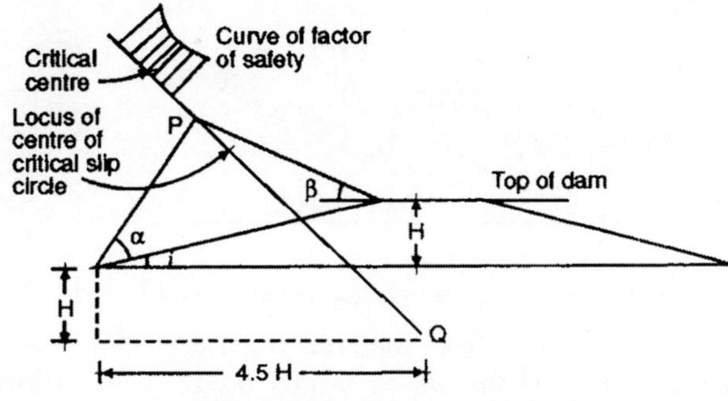

Fig. 17.27(*b*). Fellenious method.

$$F = \frac{M_r}{M_d} = \frac{CL + \tan\phi \Sigma N}{\Sigma T}$$

Various factors of safety obtained are plotted as ordinates on corresponding centre to obtain a smooth curve. The critical centre is the centre corresponding to the lowest factor of safety.

Table 17.4 Location of directional angles

Slope *H : V*	*slope* *angle (i)°*	*Directional angle (degrees)*	
		α	β
0.58 : 1	60°	29	40
1 : 1	45°	27.5	37
1.5 : 1	33° 41′	26	35
2 : 1	26° 34′	25	35
3 : 1	18° 26′	25	35
4 : 1	14° 2′	25	36
5 : 1	11° 19′	25	37

17.11 SLOPE PROTECTION

1. Upstream Slope Protection

Upstream slope protection is provided for the protection to the embankment material against erosion due to wave action, velocity of flow, rain wash, wind action, etc. by placing a protective layer, called riprap.

(*i*) *Riprap*: The cost of slope protection forms an appreciable percentage of the cost of the embankment and as such various considerations such as availability of suitable materials within reasonable distance, height of embankment and steepness of slope, fetch and wave height, drawdown, aesthetic and economics govern the choice of type of riprap. The riprap is provided from the top of dam embankment extending 1.5 m below the minimum reservoir level in a berm (Fig. 17.28) or is properly keyed. Berm is essentially required (*i*) To prevent undermining of the lower edge of the riprap where waves hit against the bottom of the riprap and filter under low reservoir condition, (*ii*) To provide a foundation support to the riprap to prevent it from slipping down the slope, and (*iii*) To provide a horizontal working surface for the construction equipment. Fig. 17.29 shows the end of riprap where there is no berm.

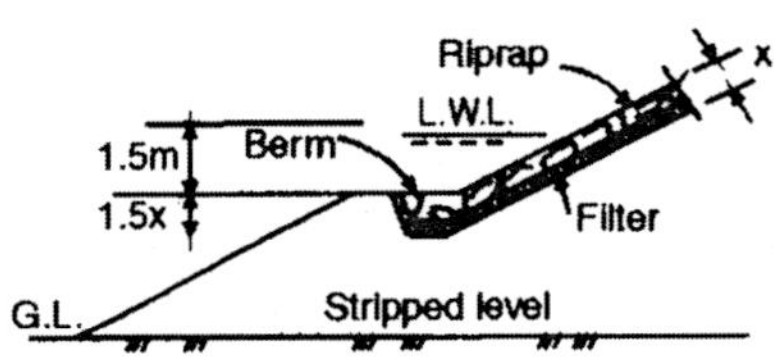

Fig. 17.28 Pitching with berm below low water level.

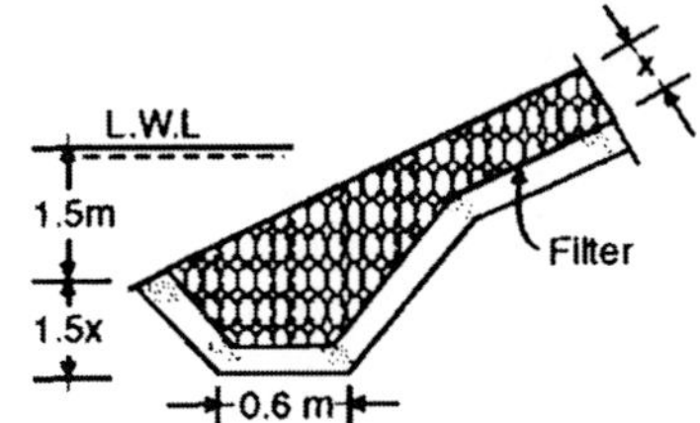

Fig. 17.29 Pitching with no berm below dead storage level.

Two layers of filter material consisting of coarse and fine is essentially provided under the riprap to prevent the waves from eroding and washing out the underlying compacted material through the voids in the riprap. Provision of adequate filter under hand placed riprap is all the more necessary as the latter is generally poorly graded due to predominance of one size material. The thickness of filter depends on intensity of wave action, riprap gradation, gradation of embankment and cost of filter. Minimum thickness of filter is 25 cm; generally it is in two layers of 15 cm thickness each. IS: 8237-1985 specifies that the gradation requirement for the filter material with respect to riprap material should conform to the criteria that D85 size of the coarse filter material is not less than 1/10 of D15 size of riprap material. With respect to embankment material, the gradation of the filter is that D15 size of filter material should not exceed 5 times the D85 size of the retained embankment material.

Types of Riprap

1. ***Hand Placed Riprap***: It consists of stones, boulders, cement concrete blocks or soil cement blocks of suitable size placed by hand in more or less definite pattern with a minimum amount of voids and with reasonably even and uniform top surface or alternatively panelwise concrete slabs or precast concrete interlocking type blocks. It has the drawback that the wave run-up height on the slope for the same wave height and embankment slope is about 1.3 to 1.5 times more than that for dumped riprap. The riprap consists of one man stone laid on edge. It is preferably laid in one course such that the layer thickness is the same as the stone size. However, at least 80 per cent of the area of riprap should have stones weighing more than 45 kg, spread uniformly in area.

Where stones of required size are not available to cover the entire thickness of riprap, the same may be laid in two layers. Alternatively, as adopted at Pong dam, boulders pitched on end are grouted over a portion of the area so that the stones are not dislodged by wave action (Fig. 17.30) or alternatively riprap is laid in two layers and grouted by pouring fluid cement mortar worked into the riprap.

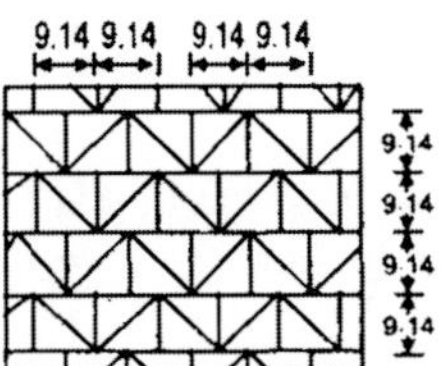

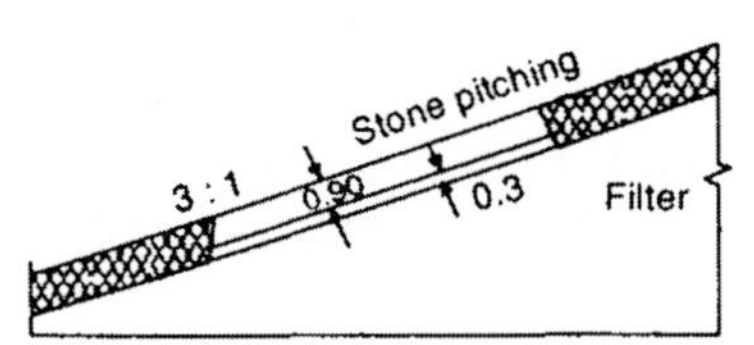

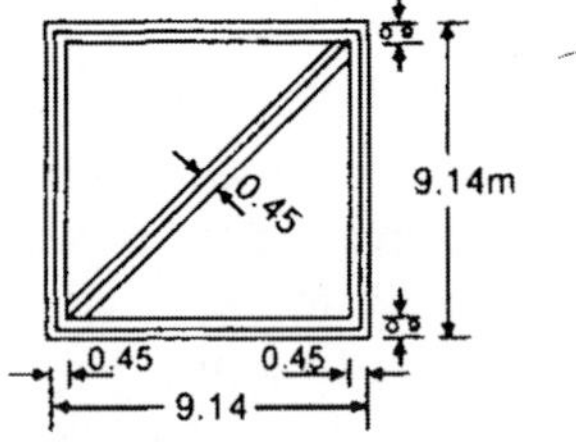

Fig. 17.30 Slope protection at Pong dam.

The minimum thickness of riprap is 30 cm. IS: 8237-1985 has specified minimum thickness of riprap as per Table 17.5.

Table 17.5 Minimum thickness of riprap

Wave height	*Hand placed riprap*	*Dumped riprap*	
(m)	*Minimum thickness (cm)*	*Min. average rock size D50 (cm)*	*Minimum riprap thickness (cm)*
Less than 1.0	30	30	60
1.0 – 2.0	45	40	75
Larger than 2.0	60	70	100

The advantages of hand placed riprap are (*i*) Thickness of hand placed riprap is one-half of the thickness required for dumped riprap, and (*ii*) Better appearance of the finished surface.

The *disadvantages* of this type of riprap are (*i*) Not so flexible as dumped riprap since it cannot adjust itself as well to local settlements and is not suitable where considerable settlement is expected, and (*ii*) Stone must be of excellent quality.

2. *Dumped Riprap*: Dumped riprap consists of boulders or blasted rock free from quarry fines and dumped in place by mechanical means over a properly graded filter, two layers, coarse and fine each of 20 cm thickness. It may be dumped over the upstream face from the embankment level as the embankment is being raised up or after the embankment has been completed. The full thickness of riprap is dumped in one layer and levelled off so as to uniformly distribute the large rock pieces and the smaller rocks filling the intervening spaces, there being no large hollow spaces left and reasonably smooth surface and uniform layer of specified thickness is obtained. The essential requirements of the riprap are (*i*) Individual stone is of sufficient weight to resist displacement by wave action, (*ii*) Filter below the riprap is not washed out through the voids in the riprap layer, (*iii*) Filter under the riprap is able to prevent any erosion of underlying embankment material.

Dumped riprap is considered superior to hand placed riprap as it provides excellent wave protection than hand placed riprap of the same thickness. Moreover, there is no appreciable effect of embankment slope on the wave ride but it is considerable for hand placed riprap, varying from 1.1 h_w to 2.0 h_w (h_w denotes wave height) for embankment slopes varying from 6:1 to 2:1 respectively.

The minimum thickness of dumped riprap is 45 cm. IS: 8237-1985 has specified the minimum thickness and average rock size as per Table 17.5.

USBR recommendation for thickness of riprap and gradation on 3:1 slope is as per Table 17.6. For 2 : 1 slope, the nominal thickness (except 90 cm thickness) should be increased by 15 cm and the corresponding gradation used. 1 m thickness of dumped riprap is recommended for major dams. Filter of crushed rock or shingle graded from 5 to 90 mm with thickness equal to one-half the thickness of riprap but not less than 30 cm is recommended.

Table 17.6 Thickness and gradation of riprap on 3 : 1 slope.

Reservoir fetch (km)	*Nominal thickness (cm)*	*Gradation, % age of stones of various weights (kg)*			
		Maximum size	*At least 25% greater than*	*45 to 75 per cent between*	*Not more than 25 per cent less than*
1.5 and less	45	45	15	0.5 – 1.5	0.5
4	60	65	30	1.5 – 30	1.5
8	75	115	45	2.5 – 45	2.5
16	90	230	90	5 – 90	5.0

The *advantages* of dumped riprap are (*i*) It provides excellent wave protection, (*ii*) Is very durable, (*iii*) Usually less costly, (*iv*) Low maintenance cost, (*v*) Mechanized laying requires a minimum of expensive hand work, (*vi*) Failure has been found in 5% cases due to improper size of stones against 30% in the case of hand placed riprap, and (*vii*) Preferable for high dams.

The *disadvantages* of dumped riprap are (*i*) Individual stone must be of sufficient weight to resist displacement of wave action, and (*ii*) Final surface is not of pleasing appearance as that of hand placed riprap.

Concrete Pavement: Concrete pavement is used for the protection of upstream slope of small dams. The pavement extends from the crest of dam to 1.5 m below the minimum reservoir level terminating on a berm and against a concrete curb extending at least 50 cm below the undersurface of paving. The embankment surface directly under the slab is well compacted to provide a uniform subgrade for the slab with a view to minimize subsequent cracks due to differential settlement caused by reservoir water pressure. Monolithic slabs are 20 to 30 cm thick with continuous steel in each direction equal to 0.5 per cent of concrete area. The thickness recommended by USBR for dams up to 15 m height is 20 cm. The concrete paving is constructed in blocks but paving monolithic to the greatest extent possible, with every measure adopted to prevent access of water and consequent development of hydrostatic pressures beneath the concrete, is preferred.

The disadvantages of the method are (*i*) Due to smooth surface waves run up high along the slab, (*ii*) Costly, and (*iii*) Suitable only when the settlement within the embankment after construction is insignificant.

Precast Concrete Blocks: Concrete blocks in the form of precast tetrahedrons, either dumped or placed in symmetrical pattern, are used where suitable riprap is not economically available. Concrete tetrahedrons each side 1.5 m, corresponding to stone riprap of 0.9 m diameter (Fig. 17.31) is provided. It is costly in first cost besides maintenance cost is also high.

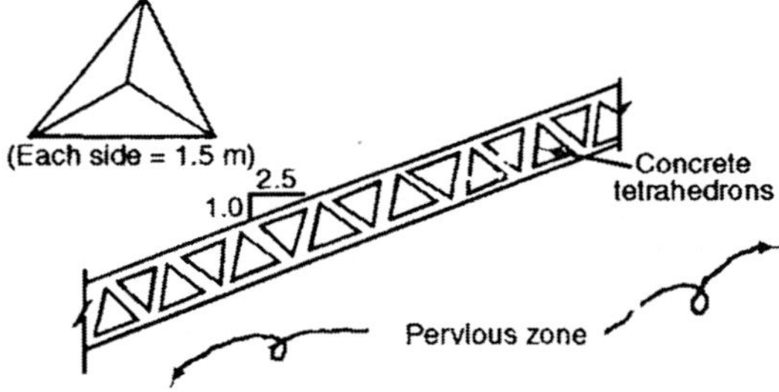

Fig. 17.31. Concrete tetrahedrons

Soil Cement Slope Protection: The soil cement dam facing requires soil cement to be placed and compacted in stair-step horizontal layers 2-3m wide. Usual soil cement slope protection is 1 m thick normal to slope in 1.5 cm thick horizontal compacted layers (Fig. 17.32). It is not suitable for high dams with long reservoir fetch. The layers interfere with drainage under drawdown conditions and excessive built up water pressures have to be relieved with provision of drains under the layer with outlets.

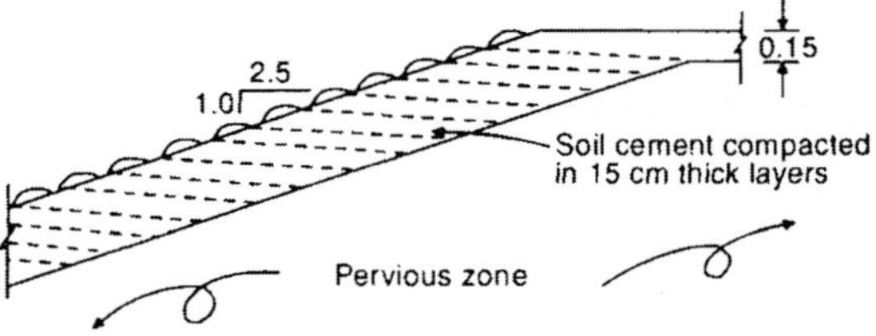

Fig. 17.32 Soil cement slope protection

2. Downstream Slope Protection

It is provided for protection to the dam embankment material against erosion due to rain wash, wind and tail water. Usual type of protection provided is as under:

1. *Turfing*: Turfing is provided on the entire downstream slope in the regions where average annual rainfall is less than 200 cm.

2. *Riprap*: Hand placed riprap 30 cm thick is provided on the downstream slope of the dam embankment in the regions where the average annual rainfall is more than 200 cm. Where tail water occurs every year, a rock toe up to the height of normal tail water is provided to support hand placed riprap (30 cm thick) up to 1 m above the highest tail water level.

17.12 SEEPAGE IN EARTH DAMS

In earth dam, the entire dam, except for the core, forms part of the drainage system. Filters, made of select materials, allow the water to pass through and in turn the transition zones do the same thing with respect to the filters. Finally the shells are draining elements and the filtered water is channelled through them downstream as far as the river or towards the reservoir. Thus in fill dams, drainage flows are generally expelled through their own filters or through the permeable downstream shell. Those from the abutments are channelled to the river bed in order to avoid surface erosion.

Seepage through the earth dams is unavoidable. The reservoir water seeps through the body of the dam or through its contact with foundation and abutments. The adverse effects of seepage in earth dam are (*i*) Loss of reservoir storage capacity, (*ii*) Sloughing of downstream toe due to saturation and progressive development of local shear failures, (*iii*) Weakening of soil strength as finer soil particles are moved through coarse particles resulting in internal erosion or piping failure, (*iv*) Foundation blow out due to excessive uplift, and (*v*) Reduction in effective strength of soil as seepage causes pore pressures.

For the location of zones of seepage through reservoir foundation or dam embankment, American Groundwater Consultants of new Maxico have developed a geophysical method known as ZETA-SP. The method is based on the principle that water moving through porous earthen materials creates an electrical field, the magnitude of which is proportional, among other things, to the velocity of the water through the earthen materials of the dam embankment or foundation of a reservoir.

In this method, two electrodes are placed side by side in water at one edge of the reservoir so that they are at zero potentials. To begin measurements, while one electrode is kept in that position, the other is moved to the remote shore and its potential relative to the stationary electrode is measured at that point and at regular intervals as it is drawn across the bottom of the reservoir towards the stationary electrode. The process is repeated from a number of different points around the shore so that a potential grid of the reservoir bottom is derived. Zones of maximum potential are indicative of seepage (Fig. 17.33).

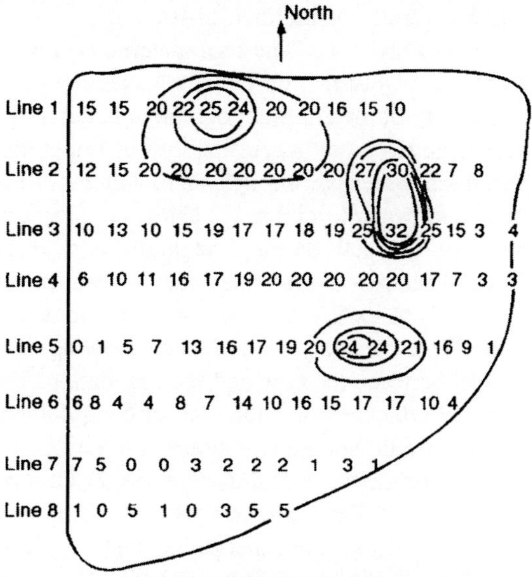

Fig. 17.33 Potential seepage zones.

Seepage Control Measures

The purpose of seepage control measures is to (*i*) Contribute to the stability of the dams and their foundations, (*ii*) Effect reduction of the loss of water, (*iii*) Avert failure of the dam structure by piping, and (*iv*) Excessive pore pressure in the dam. A suitably designed internal drainage system is essentially required to meet these requirements (Fig. 17.34).

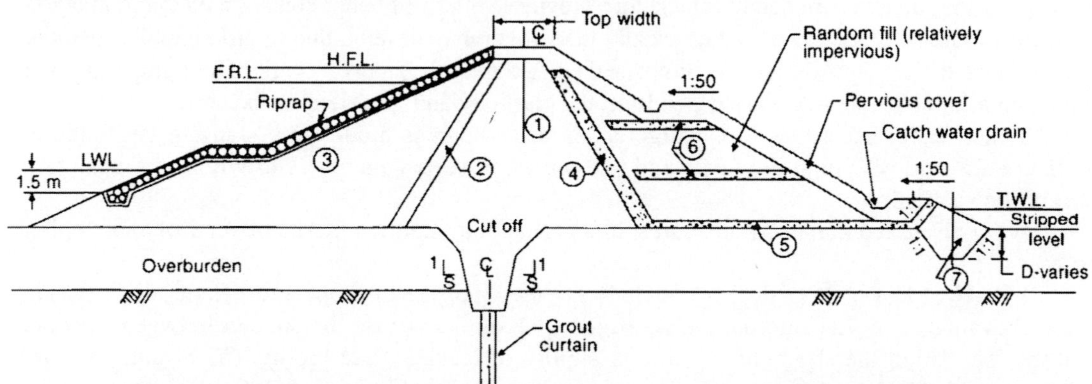

1. Central impervious core, 2. Transition zone, 3. Upstream pervious shell, 4. Inclined filter, 5. Horizontal filter, 6. Horizontal filters at intermediate level, 7. Rock toe and toe drain.
(The horizontal filters at intermediate levels are also provided in upstream shell, if the material is of impervious nature).

Fig. 17.34 Section of earth dam showing seepage control measures.

1. Control of Seepage through Embankment

(*i*) ***Impervious zones***: It consists of a watertight core (coefficient of permeability over 10^{-6} cm/s) protected on both sides by filter, draining and transition zones. Fig. 17.3 shows the most common section of an earth dam. The core depicted is vertical and symmetrical. However, in practice it is asymmetrical and slanted but this arrangement does not modify the manner in which the drainage elements next to it function. Impervious earth core checks the seepage through the embankment and also enables the portion of dam downstream of the core to act as a drainage zone.

***(ii) Core*:** The core provides improbable barrier within the dam body. It may be located either centrally or inclined upstream.

(*a*) **Thin core:** The core material normally has less shear strength than the rest of the embankment and may further develop appreciable pore pressures. The location of core depends mainly on the availability of materials, topography of site, foundation condition, diversion considerations etc. Thickness is governed by (*i*) Availability of impervious and filter materials, (*ii*) Resistance to piping, (*iii*) Permissible seepage through dam. Minimum thickness is 3 m. Top level of core is 1 m above, the maximum water level. From stability consideration thinner core is better. A thin core with a filter zone and pervious shells develops negligible pore pressures. A core is considered thin if its width is less than 15 to 20 per cent of the water head. Theoretical minimum volume for a thin core dam is obtained when the core is so located that the upstream and downstream slopes have approximately the same factor of safety. Thin core may be moderately slanting or central. A central core provides higher pressures at the contact between the core and the foundation. Thereby reducing the possbility of leakage and piping. Core with thickness less than 10 per cent of the head is not used except where a large leakage through the core would not lead to failure of the dam.

The *advantages* of thin core are (*i*) Thin core with pervious shells has less unit cost of placement, (*ii*) Smaller embankment volume, (*iii*) Expeditious construction as lesser time is required for constructing thin impervious zone, (*iv*) Lesser requirement of compaction equipment, and (*v*) Develops negligible pore pressures.

The *disadvantages* of thin core are that contact area between the core of the embankment and rock is less and there is likelihood of leakage through thin core. However, this can be taken care of to some extent by increasing the width of the core at junction with the rock to give wider contact area.

(*b*) **Thick core:** A core with a width of 30 to 50 per cent of the water head is considered as thick core. It is usually central and has steeper upstream slope and flatter downstream slope. It is suitable for any type of the soil and dam height. Thick core provides wider and better contact with the foundation and offers more resistance to piping especially that which may develop due to differential settlement cracks. Since the piping resistance of compacted embankment is largely a result of soil properties, the minimum allowable core thickness depends on the plasticity and gradation of the core.

An ideal core, according to Cedergeren, is that which is moderately slanting, sufficiently wide to eliminate very high gradients and yet confining seepage stresses to within the upstream part of the section.

Having discussed thin and thick cores, it is pertinent to mention merits of vertical and sloping cores.

(*c*) **Vertical Core:** A vertical core dam has a steeper upstream slope and a flatter downstream slope. Advantages of vertical core are (*i*) Higher pressures exist on the contact between the core and the foundation thereby reducing the possibility of leakage, and piping, (*ii*) Slightly greater thickness of the core is obtained for a given quantity of impervious soil, (*iii*) Central core permits carrying out of additional grouting from the crest of the dam after the dam is completed and without lowering the reservoir.

(*d*) **Sloping Core:** A sloping core dam has steeper upstream and downstream slopes. Advantages of sloping core are (*i*) Placement of main downstream portion of the embankment first and core placed later, (*ii*) Foundation grouting can be carried out while the embankment is being placed, (*iii*) Filter layers on either side of the core can be made thinner and placed more conveniently than vertical core dam, (*iv*) Reduces the pore pressures in the downstream part of the dam and thereby increases its safety, and (*v*) Allows use of relatively large volume of random material on the downstream.

(iii) Transition Zone or Casing **:** Transition zone or transition filter of free draining materials not subject to cracking on direct exposure to atmosphere are usually provided on both sides of the core. The transition zones minimise the possibility of failure by internal piping. The minimum thickness of this zone is 3 m. The filter is usually taken up to the top of impervious core and at the base it may, if necessary, be connected to horizontal filter. Fig. 17.4 indicates the transition zone provided at Pong Dam.

***(iv) Rock Toe Filter*:** Zoned dam embankment has pervious shell on the downstream which can serve as a drain but collection and disposal of drained water at the toe is required. In homogeneous dams it is all the more essential. Rock toe or toe filter (Fig. 17.35) has the function to facilitate drainage of seepage water and also to improve the stability of the downstream face. The rock toe does not extend far enough into the dam embankment and the seepage line may come close to the downstream. A minimum cover of 1 m between the seepage line and downstream slope is essential. The inner slope of rock toe is provided at 1:1 slope. The height of rock toe, depending on (*i*) The availability of material, (*ii*) Minimum allowable cover over the phreatic line, and (*iii*) Tail water level, is usually kept 15 per cent of head of water; maximum and minimum limits being 4 m and 1 m respectively. Rock toe is suitable for dams of low to moderate heights.

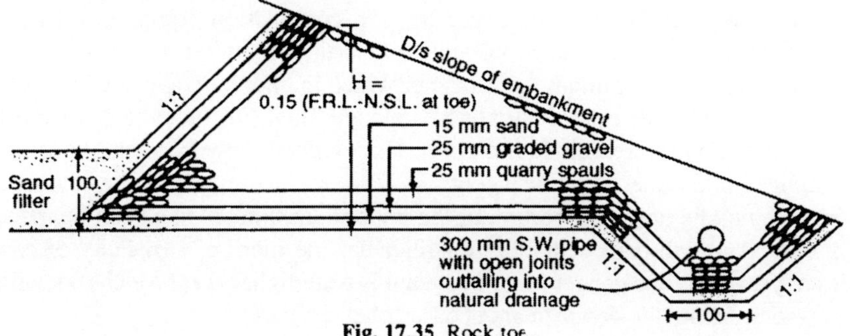

Fig. 17.35 Rock toe.

***(iv) Horizontal Drainage Filter*:** When the material available for the shells, because of its quality and granulometry, cannot assure a high degree of permeability, the transition zones are dispensed with. Instead a horizontal filter is provided under the downstream shell (Fig. 17.36) to collect seepage from the inclined or vertical filter or from the body of the dam including foundation to minimise possibility of piping along the dam seat and carry it to rock toe and toe drain. This horizontal filter is usually provided in two layers; the upper, a finer one, acts as a filter to retain the fine grains of the shell and the lower, coarser one, allows the water passing through the upper one to reach as far as the downstream toe outside of the dam. On occasions, because of quality of the material to be used for the shells, horizontal filters at various levels are incorporated in the shells to reduce pore pressures during construction, sudden drawdown or due to prolonged rainfall. These are placed at a vertical interval of 5 to 10 m and connected to upstream or downstream pervious shells to drain out water. It is usually provided at ground level on downstream of core in a zoned dam or in the downstream part of a homogeneous dam. It is best adapted for relatively homogeneous pervious foundations (Fig. 17.36).

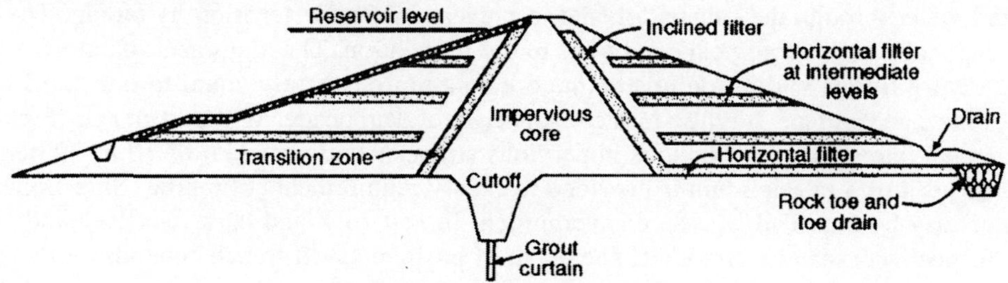

Fig. 17.36 Horizontal drainage filters.

The filter extends inside the embankment downstream from the impervious zone where it is more effective in controlling the seepage and reducing the pore pressures to the extent of 25 to 100 per cent of the distance from toe to the centre line of the dam or 3 times the height of the dam.

The disadvantages of the horizontal drainage filter are (*i*) It gives more leakage because of shorter seepage path, (*ii*) It tends to stratify the dam embankment and consequently more pervious in the horizontal direction than in the vertical, and (*iii*) It does not intercept stratification in the embankment. There may be relatively pervious layer or layers through which seepage water emerges on the downstream face.

(*vi*) ***Inclined Drainage Filter*:** Inclined or vertical drainage filter is provided where larger amount of foundation settlement or where a sharp reduction in pore pressure in the embankment is to be effected. Inclined filter is usually provided along the downstream face of the impervious core (Fig. 17.36) to collect seepage coming out of core and thereby keeping the downstream shell relatively dry. It extends up to FRL but up to the top of the core in seismic regions. At the base it is connected to horizontal filter or inner cross drains which is ultimately connected to the rock toe and toe drain. The filter is composed of free draining material of required permeability. The minimum thickness of the inclined filter layer depends on (*i*) Drainage requirements, (*ii*) Allowance of intermixing with adjoining zones depending on compaction equipment, (*iii*) Minimum width required for the facility of compaction, and (*iv*) Earthquake effects.

***Chimney Drain*:** Vertical or chimney drains are provided in high homogeneous dams to (*i*) Intercept all layers, (*ii*) Remedy the defects of horizontal drainage filter, (*iii*) Free the downstream portion of the dam from seepage with reservoir full, and (*iv*) Help reduce drawdown and construction pore pressures. The chimney drain may be vertical (Fig. 17.37) or inclined upstream or downstream. With an upward inclined drain, the upstream portion of the impervious embankment can be made to act as a relatively thin core. An earth dam with chimney filter has the merit of simplicity of construction. Here the core is not provided. The greater part of the dam is watertight except for chimney filter and its horizontal prolongation below the downstream shell or blanket filter.

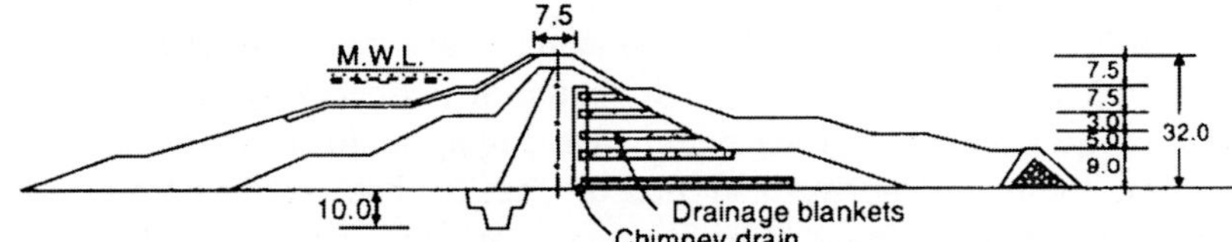

Fig. 17.37 Earth dam with chimney drain.

(*vii*) **Toe Drain:** It is provided at the downstream toe of an earth dam to collect seepage from horizontal filter or inner cross drains and through the foundation and to carry it away for disposal. It is usually provided as a part of rock toe. The drain is usually 1.0 m wide at bottom, 1.5 m deep with 1:1 side slopes. 300 mm diameter open jointed pipe at bottom may be provided (Fig. 17.35).

2. Foundation Seepage Control

(1) **Impervious cutoff:** Cutoff is a core in the foundation. The cutoff into the rock is useful for creating a seal between the dam core and foundation when the rock is relatively soft with decrease in permeability with depth below the surface. The functions of cutoff are to reduce loss of stored water through the foundation and to prevent subsurface erosion by piping. The type of cutoff depends on the geological conditions of the foundation. The alignment of cutoff is so fixed that its centre line is within the upstream core and approximately equal to one-third the base width of the dam from toe. It is also located ¼th height of dam upstream from dam axis (Fig. 17.4(*a*)) and is keyed into rock or continuous impervious strata. A bottom width of 10 to 30 per cent of hydraulic head or 4 m minimum is provided to satisfy requirements of piping. Side slopes of 1:1 or flatter may be provided in case of overburden. In soft rock and hard rock ½:1 and ¼:1 side slopes respectively may be provided. The depth of positive cutoff trench depends on the geological features influencing the configuration of the impervious substratum and profile of the unweathered mass of bed rock. The location and number of cutoff trenches depend on the nature

of foundation materials and their geological structure but usually a single trench is excavated either at the upstream, central or at downstream.

1. *Upstream Cutoff*: It minimizes seepage pressure in the greater part of the embankment and thereby increases stability of downstream half of the dam. However, it is vulnerable to rupture if there is partial failure in upstream face of the embankment.

2. *Central Cutoff*: It is less effective in reducing seepage pressure throughout the embankment but is less vulnerable to rupture and less likely to weaken the foundation.

3. *Downstream Cutoff*: It has no advantage and is dangerous as it increases pore pressures and reduces stability of downstream half of the embankment.

4. *Partial cutoff*: Where positive cutoff is not possible, partial cutoff with upstream blanket and pressure relief measures on the downstream is provided. It is best suited for horizontally stratified foundations with relatively more pervious layer near top. The depth of the cutoff in deep pervious alluvium depends on permeability of substrata.

5. *Steel sheet pile cutoff*: Steel piling is used singly or in combination with a cutoff trench in silty, sandy and fine gravel foundation, In boulders the sheet pile may not penetrate, besides interlocks may get damaged. Sheet pile cutoffs are found to be of low effectiveness. They are expensive and allow considerable leakage and as such are rather ineffective.

6. *Conrete cutoff diaphragm*: R.C.C. diaphragm walls provide an economical alternative to arrest the seepage through the foundation of earth dam. A single or a double-diaphragm is used. In two lines of diaphragms, the alluvium contained between them is grouted by tubes with sleeves. Limitations of the concrete diaphragm can be overcome by installing precast elements within slurry trenches filled with plastic concrete. At earth dam on Mahanadi river, two rows of 60 cm thick R.C.C. walls, 3 m apart (Fig. 17.5) have been provided in preference to open cutoff. The diaphragms are keyed about 1 m into the base rock and the junction between the two is grouted for water tightness.

7. *Grout curtain*: Grouted cutoff are produced by injecting cement, clay, chemicals or a combination of these materials into the voids of the sediments of pervious soils. Grouted cutoffs are generally effective when seepage occurs through zones or layers of coarse materials such as gravel, boulders and talus. Depth of grouting does not normally exceed half the head of water.

The curtains are usually constructed in a vertical plane. They may be provided at an inclination (not more than 30°) where the nature of rock indicates that grout holes at inclination will intersect more of the potential leakage cracks. The holes can be drilled up to about 60° with the vertical and grouted without appreciable cost and labour.

8. *Slurry trench*: In this method a trench cutoff is excavated through pervious deposits down to suitable impervious materials by a backhoe or dragline. The trench is kept open (without bracing) and filled with a bentonite-slurry, (specific gravity 1.2 to 1.8). The bentonite slurry, retained in the trench above existing groundwater level prevents the caving of the trench walls. After a sufficient length of the trench has been excavated and its bottom suitably prepared, back filling begins.

(2) *Upstream Blanket*: Upstream impervious blanket combined with relief wells in the downstream section is provided where positive cutoff is costly. An impervious horizontal blanket of compacted material is provided upstream of the dam extending into the reservoir in continuation of the impervious core of the dam embankment. It is used to increase the path of percolation in the foundation of dam comprising of unconsolidated and permeable materials. It thus provides a less expensive method of lengthening the path of percolation in preference to the expensive alternative of flattening the upstream slope of the dam with consequental increase in the cost of riprap. It helps to reduce, rather than cutoff, seepage altogether.

The length of the upstream blanket is determined by the desired reduction in the percolation gradient. Usually the material of the blanket is so tight in relation to the pervious stratum that it is not necessary to consider flow through the blanket in determining the desired length of the upstream blanket which may be computed from the relation:

$$L = \frac{khd - pqb}{pq}$$

where, L = length of upstream blanket (m), k = mean horizontal permeability coefficient of pervious stratum, h = gross head on the upstream blanket (m), d = depth of pervious stratum (m), p = fraction of flow under dam without the blanket to which it is desired to reduce seepage by means of blanket, q = flow under dam without a blanket per metre of dam, = $k(h / b)d$, and b = length of impervious portion of base of dam (m).

Thickness of the blanket varies with the permeability of the material comprising the blanket and the head of water to which it is subjected to. It is determined from the relation:

$$t = (K_b / k_s)\, b\, (B/d)$$

where, t = thickness of blanket at the point under consideration (m), k_b = permeability of blanket, K_s = average permeability of stratum, b = distance from point under consideration to upstream toe of blanket, B = length of blanket from upstream toe to impervious section, and d = depth of pervious stratum (m).

Generally adopted dimensions of the blanket are; length 10 times (15 times in the case of fine sand or silty foundation) the depth of water and thickness 1.5 to 3.0 m.

The limitations of the method are that (*i*) The blanket is susceptible to erosion by high velocity flow, and (*ii*) Placement of blanket under water or dewatering of the area for laying blanket is costly.

(3) **Drainage Tunnels:** Drainage tunnels are small size tunnels provided either in the body of dam or outside. They are excavated inside the rock of the foundation and abutments, or of a combination thereof to intercept the seepage water as may have seeped into the rock comprising the right and left abutments under high reservoir conditions. The tunnels are made of concrete and of dimensions enough for a man to enter and work; usually 1.25 to 1.5 m wide and 2 to 2.5 m high and provided with drainage arrangement. They serve the purpose of access for grouting, as and when needed, and collecting and discharging drainage water. Fig. 17.4 (*b*) shows drainage tunnels provided at Ramganga project.

(4) **Trenches:** If the dam foundation is relatively impervious compared to the material comprising the dam embankment, an artificial drainage system is essentially required. The drainage is usually provided by digging trenches perpendicular to the dam axis which are filled up with broken stones—larger stones at bottom and smaller ones at top 15 to 25 cm. Terra-cota sewer pipes with open joints surrounded with gravel are laid.

(5) **Pressure Relief Wells:** Relief wells provide an effective pressure relief system for low dams founded on pervious strata of sand and gravel overlain by an impervious stratum and where the cutoff is partial or upstream blanket is provided to control the pressure developed below the toe of dam. In case the thickness of top impervious stratum is equal to the reservoir head, an appreciable factor of safety against piping exists. However, if the thickness of top impervious stratum is less than the reservoir head but is too thick for treatment by drainage trenches, or if the pervious stratum is sandwiched between two impervious stratum somehow gets connected to water in the reservoir pressure relief wells are essentially required. The detrimental effect of outcropping of the underseepage at the toe of the dam is most dangerous when it develops in the form of uncontrolled 'boils'. Pressure relief wells provide a very effective control over excessive uplift pressures due to deep seepage. They ensure that excess hydrostatic pressures do not develop in the downstream portion of the dam which could lead to piping and also that uncontrolled seepage flowing downstream of the dam is reduced.

A relief well is a small drainage well, 50 to 90 cm in diameter, sunk near the downstream toe of earth dam, a pipe, about 10 to 20 cm in diameter having narrow slots, placed in the centre and surrounded by graded filter which permits the ingress of seepage water into the well and allows it to rise to the outfall (relief) level where the pressure gets reduced (Fig. 17.38). Synthetic filters, like geofabrics, have the advantage of reducing the diameter of relief well to 20 cm or so.

The slotted pipes are provided only in pervious layers and blind pipes in impervious layer portions. The filter is laid up to about 120 cm above the top of the slotted length. The annular space against blind pipe is back-filled with impervious soil or cement sand mortar to prevent an upward flow of water through the graded filter media. The relief well is surrounded and covered by a manhole. The relief well has a relief level or discharge level 1 or 2 m below the ground level from where the under seepage water is carried away for disposal.

For efficient working of relief well system, the wells shall penetrate to the full depth of pervious strata. However, if the pervious strata are very deep, relief wells partly penetrating the pervious strata are provided.

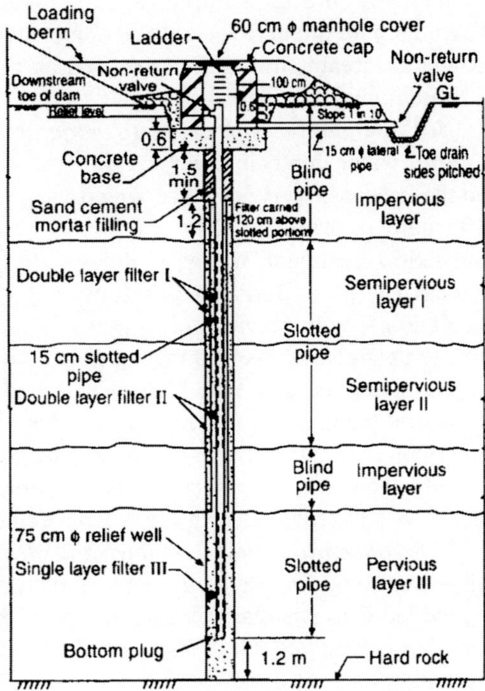

Fig. 17.38 Typical installation of a relief well (IS: 5050-1992).

The limitations of the method are (*i*) Excessive volume of seepage may require a large number of wells, (*ii*) Uncertainty of actual values of coefficient of permeability of pervious stratum, (*iii*) Uncertainty of seepage boundaries, (*iv*) Wells need protection from back flooding, (*v*) Regular maintenance and inspection is desired, and (*vi*) Well screen may be obstructed by chemical deposits or get corroded.

17.13 DAM SECTION TO SUIT AVAILABLE MATERIALS AND FOUNDATIONS

Earthen dams can be made out of a variety of available materials to suit various foundation conditions. A brief resume of the earth dam designs suitable under a combination of available fill materials and foundation is as under:

1. ***Clay Silt, Gravel or Coarse Sand are Available***

(*i*) *Foundation is impervious to greater depth*: The foundation is impervious to a depth greater than the reservoir water depth. It represents ideal foundation conditions as also suitable availability position of the fill materials. A zoned dam with impervious (clay silt) central core and outer shells of available gravel or coarse sand is suitable design section of the dam (Fig. 17.39 *a*). The core may be vertical or inclined (Fig. 17.39 *b*).

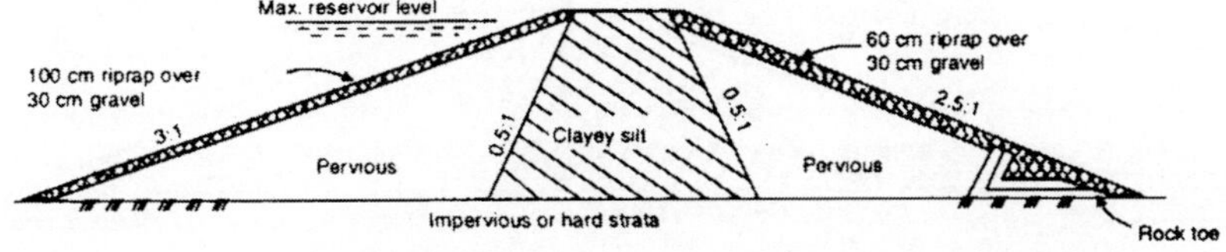

Fig. 17.39(*a*) Zoned dam with impervious (clayey silt) zone.

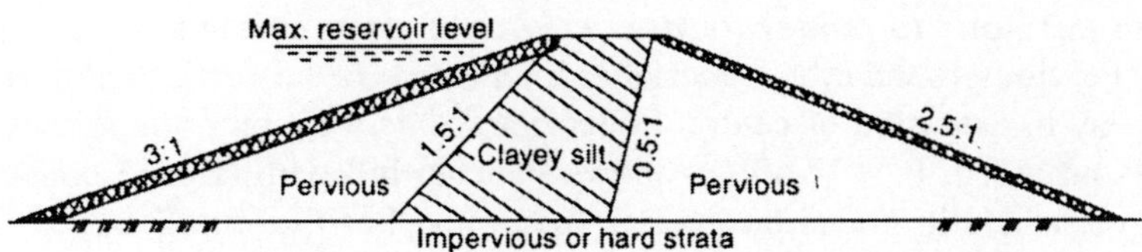

Fig. 17.39(*b*) Zoned dam with inclined impervious (clayey silt zone).

Inclined core is favoured. Positive cutoff trench extends into the impervious rock and is backfilled with impervious soil and compacted. There is no problem of seepage or seepage forces. Foundation treatment required depends on the nature of impervious layer as regards its stability and settlement.

(*ii*) *Foundation impervious to moderate depth*: The depth of impervious foundation is less than the water depth in the reservoir and the layer below is pervious, zoned earth dam is provided (Fig. 17.40). Foundation treatment, if any, is independent of the underlying pervious foundation. The impervious layer may serve as upstream blanket so that seepage losses are not excessive. However, the natural blanket is not assumed to be perfect or so extensive as to prevent seepage.

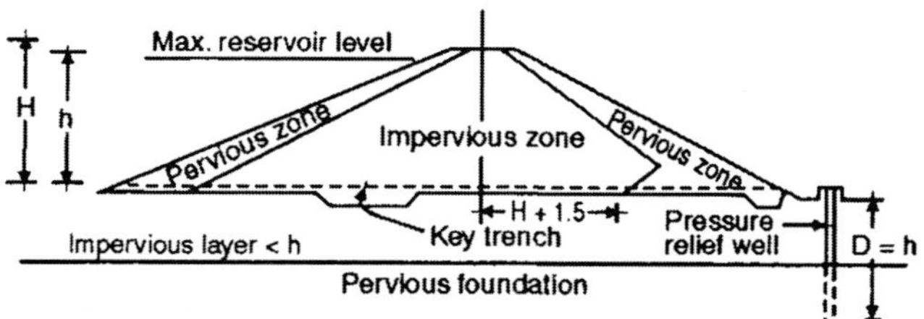

Fig. 17.40 Earthen dam on foundation pervious to moderate depth.

(*iii*) *Foundation impervious to shallow depth*: Zoned dam with impervious (clayey silt) central core is provided with a drainage trench.

(*iv*) *Foundation comprises fine sand alluvium*: Fig. 17.41 shows the section of Maithon dam (Bihar) based on fine sand alluvium of maximum depth of 21 m. A single row of steel sheet piles is provided 6 m upstream of dam axis and extending 6 m into the bed rock and top impervious blanket is considered cheaper and convenient than positive cutoff trench backfilled with impervious soil.

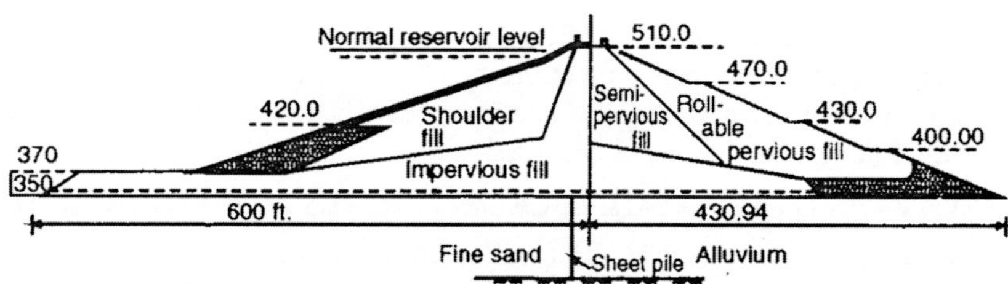

Fig. 17.41 Maithon dam.

(*v*) *Foundation pervious to deeper depth*: Zoned embankment with large impervious core is provided, provision of piling or curtain cutoff is not practicable owing to deeper depth of pervious foundation material. In order to limit seepage losses, upstream blanket is provided extending $10H$ to $15H$ from the upstream of dam so as to be contiguous with impervious core. The thickness of horizontal filter is required to be adequate. The blanket is also essential to avoid warping of the downstream slope. A toe drain is provided (Fig. 17.42).

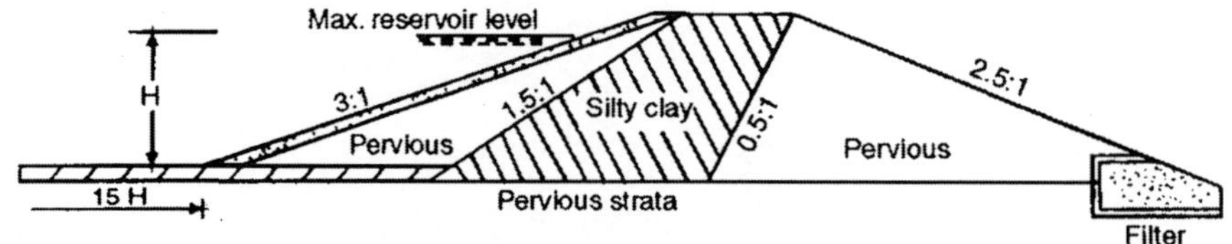

Fig. 17.42 Zoned dam based on deep pervious strata.

(*vi*) *Foundation pervious to moderate depth*: Zoned dam embankment is provided (Fig. 17.43 *a* & *b*). Pervious foundation material is excavated in the form of key trench and backfilled with silty clay in extension of central impervious zone. Minimum impervious zone is provided. Rock toe is provided. Fig. 17.43 (*c*) shows Panchet hill dam (Bihar) constructed on 15 m thick alluvium of sand, silt, gravel on gneissic bed rock. Positive cutoff trench backfilled with impervious material is infeasible and a single row of steel sheet piles cutoff is provided.

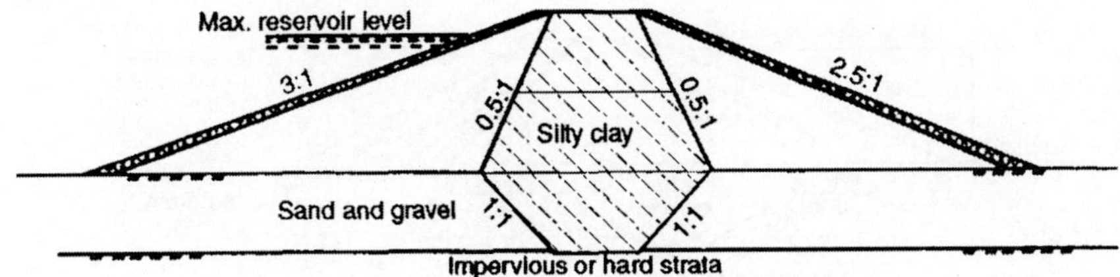

(*a*) Zoned dam based on pervious strata to moderate depth.

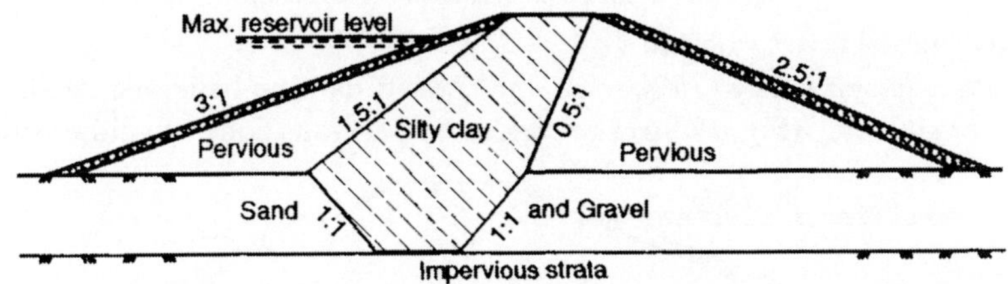

(*b*) Zoned dam (u/s inclined core) based on pervious strata to moderate depth.

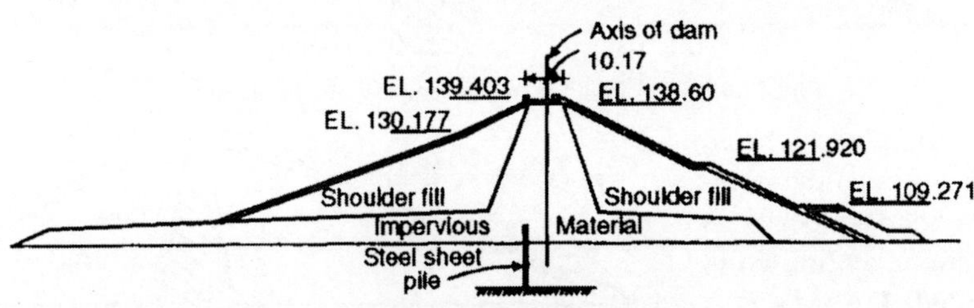

(*c*) Earthfill dam—Panchet hill dam.

Fig. 17.43 Dam sections on pervious strata to moderate depth

(*vii*) *Foundation pervious to shallow depth*: Zoned dam embankment is provided with a cutoff trench extending to the impervious stratum. Cutoff trench is very effective for avoiding excessive seepage losses and possibility of piping. Filter is not required if the downstream portion of the embankment is of sand and gravel similar in gradation to the foundation. Fig. 17.44 shows Badua dam (Bihar) based on coarse sand to a maximum depth of 15 m. Positive cutoff trench is excavated and backfilled with impervious material.

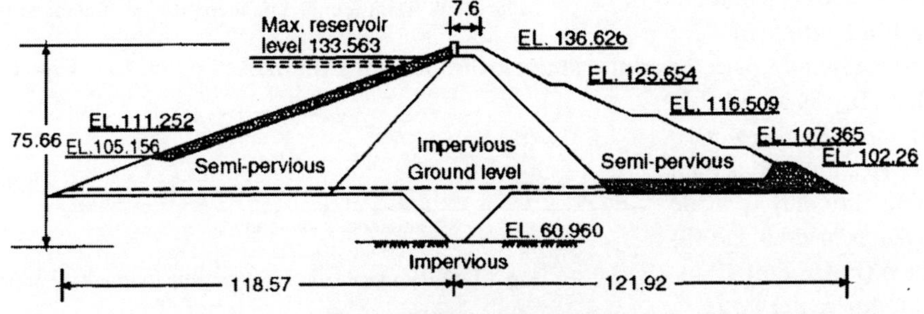

Fig. 17.44 Badua dam (Bihar).

(*viii*) *Stratified foundation*: Zoned dam embankment is provided. Sheet piling or cement grout curtain wall is replaced by partial cutoff trench (Fig. 17.45) extending into the foundation so that depth to top of uppermost pervious layer which is not cutoff is not less than reservoir head. The trench replaces horizontal drainage blanket.

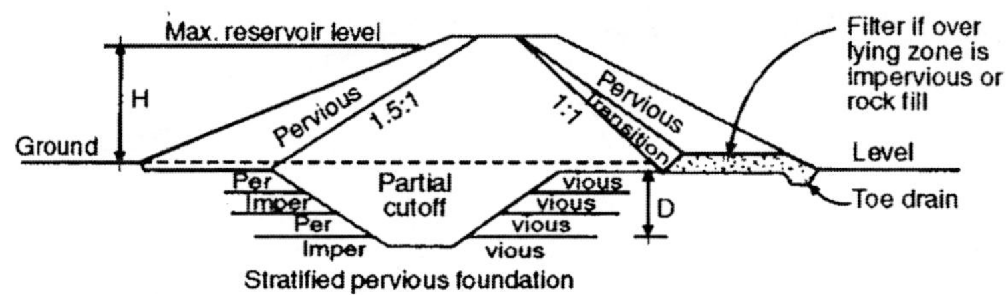

Fig. 17.45 Zoned dam on stratified foundation.

2. **Pervious Material is Available**

(*i*) *Foundation impervious to deeper depth*: Zoned dam embankment with a thin central impervious core of silty clay (Fig. 17.46) surrounded by pervious zones on either side is provided.

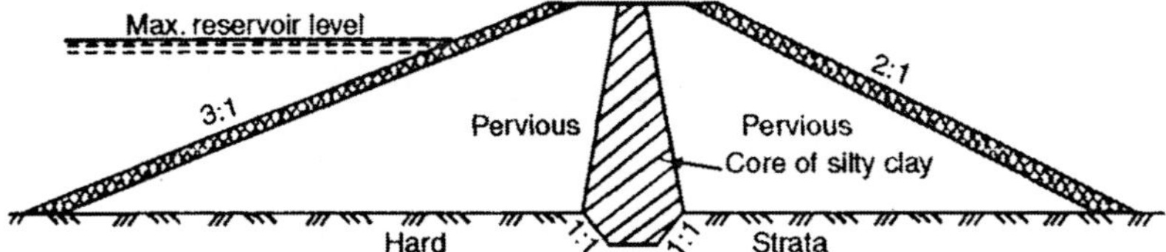

Fig. 17.46 Dam founded in deep impervious foundation.

(*ii*) *Foundation pervious to moderate depth*: Zoned dam embankment is provided. Impervious central core of silty clay limited in section is provided (Fig. 17.47).

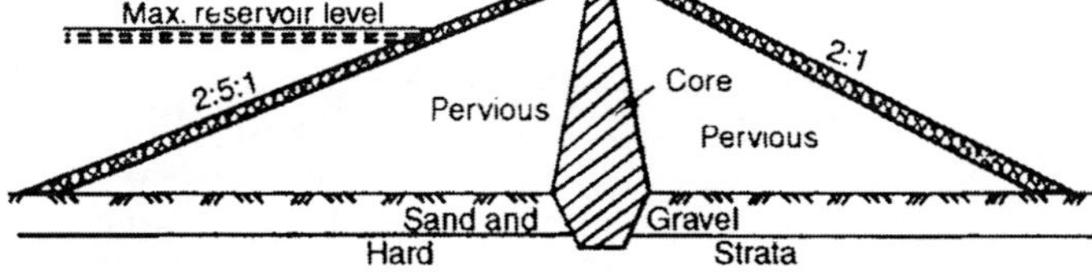

Fig. 17.47 Dam founded in foundation pervious to moderate depth.

(*iii*) *Foundation pervious to shallow depth*: Zoned dam embankment is provided. Key trench, through the pervious foundation of shallow depth, is provided and central core zone extends through it to the impervious strata.

(*iv*) *Foundation pervious to deeper depth*: Zoned dam embankment with a thin central impervious core extended upstream in the shape of impervious blanket is provided (Fig. 17.48).

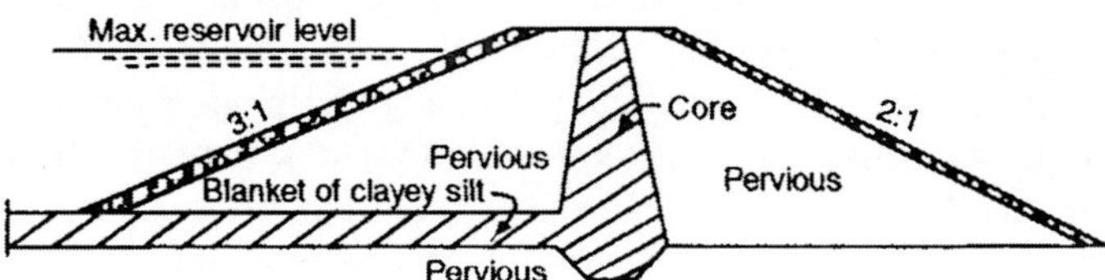

Fig. 17.48 Dam founded in deep pervious foundations.

3. **Silty clay is Available**

(*i*) *Foundation impervious to deeper depth*: Homogeneous dam of the available silty clay is made with the provision of drainage filter and rock toe (Fig. 17.49).

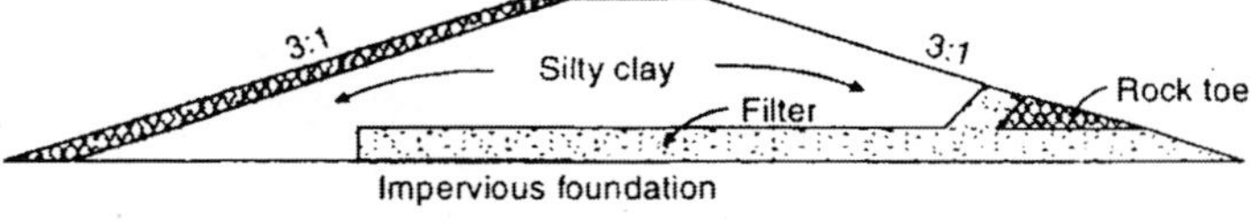

Fig. 17.49 Homogeneous dam on impervious foundation.

(*ii*) *Foundation pervious to moderate depth*: Homogeneous dam of silty clay is made with the provision of sheet pile cutoff and long horizontal filter (Fig. 17.50).

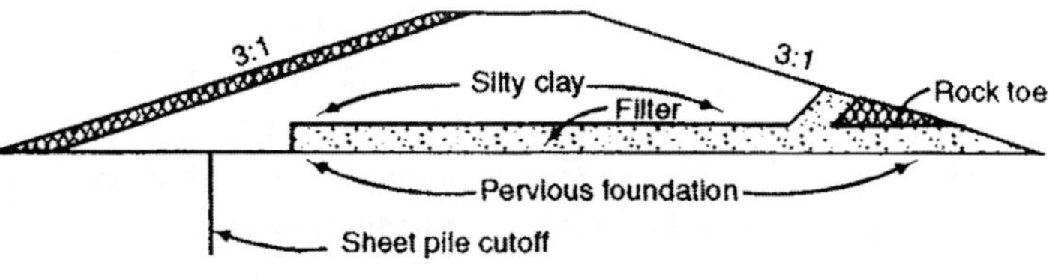

Fig. 17.50 Homogeneous dam based on pervious strata to moderate depth.

(*iii*) *Foundation pervious to shallow depth*: Homogeneous dam embankment with filter extending upstream by a distance of H + 1.5 m from the centre line of dam is provided.

17.14 EARTH DAM CONSTRUCTION

1. Excavation: The area of excavation is cleared of loose and semi-detached rocks as may become loosened or unstable later and jeopardise working conditions in the excavation area. The next step is removal of overburden which includes earth, gravel, boulders and soft disintegrated rock. Overburden is stripped to the rock foundation. Rock surfaces including all pockets or depressions are cleared of soil or rock fragments. The rock excavation may be classified into two categories as under:

Excavation Under Impervious Zone: Area under impervious zone includes core trench; the excavation for which is carried out to sound rock level. Feather edging of the rock band in transverse direction or longitudinal direction is avoided. Excavation of core trench is followed by the excavation of key trench which is generally about 10 m wide, 10 m deep with 1 : 1 side slopes depending on the dam height and foundation conditions. The excavation is done by the standard excavation equipment such as rippers, power shovels, draglines, bulldozers, scrappers. Blasting for the excavation, where necessary, is done under controlled conditions to minimize overbreak and shattering of the foundation. Blasting is generally not done in the last 2 to 3 m excavation in core trench and for the excavation of key trench.

Excavation Under Pervious Zones: The stripping is limited to expose the actual bed rock. Unsuitable or unstable materials, if any, are removed. In river bed portion the excavation is limited to expose clear sand, gravel and cobble material of grading similar to that of the shell zone material. The excavated material suitable for use in impervious zone, filter zone and pervious zones is stock piled separately.

2. Fill Placement: The stripped foundation is moistened; wet surface or standing water over the foundation is not desirable. The moistened surface is compacted by a tamping roller before placing first layer of fill material so as to ensure proper bond. Scarification of the foundation by harrows may sometimes be necessary to ensure proper bond. Local pockets are filled and compacted to bring to the general level. Hand tamping is done in pockets that cannot be compacted by the rollers. Proper bond between successive layers of fill placement is ensured.

Zoned dam essentially involves fill placement in impervious zone, filter zones and pervious zones. The fill is compacted by rollers; the usual type is sheep foot roller. From the standpoint of imperviousness, maximum practicable compaction is desirable. Over-compaction is not desirable especially in the case of clays for which only moderate compactive effort is needed to ensure impermeability. The compaction is done at optimum moisture content. The most efficient method of compaction is by tamping rollers. When the material is either perfectly dry or when it is nearly saturated with water it is compacted by vibratory rollers.

At Beas dam, 132.6 m high (from the deeped foundation level) zoned dam with a central impervious zone of mixed clayshale and sandrock material surrounded on both sides by filter zones and pervious gravel shell zones, the fill placement and compaction specifications adopted are briefly as follows:

The moisture content of fill materials immediately prior to their compaction in the dam embankment is determined in accordance with the requirements regarding moisture conditioning. Prior to and during compaction operations, the fill material is conditioned so that the optimum practicable moisture content required for compaction purposes is uniformly distributed throughout the layer.

Impervious zone: The area under fill placement is so divided that one area is normally under placement, second available for compaction and the third under observation for field tests, etc. The moisture content of the impervious zone is within 2 per cent on the drier side to 1 per cent on the wetter side of the optimum. The compaction of the layers in blocks is carried out by a train of six set of sheep-foot rollers moving in lanes parallel to the axis of dam, with sufficient overlap, to ensure the rolling of the layer by 12 coverages. To guard against ineffective rolling at the turning points, *i.e.*, at the ends of the blocks additional coverages are given parallel to the axis of the dam. The roller exerted a unit pressure of 35 kg/cm^2. The rollers work in tandem two in front and one at the rear covering the space between the first two drums, forming a rolled length of 4.9 m and moving parallel to the axis of dam. The impervious material is compacted such that not more than 5% results of density tests of impervious core have densities less than 98 per cent of maximum USBR Standard Proctor density, the density values of 97.5 per cent and above being considered as 98 per cent. The compacted densities are neither to fall below 95 per cent nor exceed 104 per cent of USBR Standard Proctor density.

Filter zone: The filter zone material is placed in two 0.3 m lifts, each of which is watered sufficiently by water wagons prior to the total lift of 0.6 m being rolled. The material is compacted by not less than four coverages of 5 tonne vibratory rollers to achieve a minimum fill dry density of 2.0 gm/cc.

Pervious zone: Pervious fill is placed dry in continuous approximately horizontal layers not more than 0.6 m in thickness. Rock pieces larger in maximum dimensions than the applicable layer thickness is removed to the outer 15 m of the zone measured horizontally and is so embedded in the fill as not to leave excessive surrounding voids. The material is compacted by not less than four coverages of 5 tonne vibratory rollers to achieve a minimum fill dry density of 2.16 gm/cc and a shear angle of 35°.

17.15 FAILURE OF EARTHEN DAMS

Dams, like other engineering structures, are susceptible to failure. But failure in the case of a dam causes loss of life and property for hundreds of kilometres downstream. Dams have failed in all parts of the world. During the last two centuries over 600 dam failures have been reported world wide, *i.e.* more than 3 per year on an average. A study of the causes of dam failures has yielded the following break up:

Causes	*Failure (%)*	*Causes*	*Failure (%)*
Foundation problems	40	Acts of war	3
Inadequate spillway	23	Embankment slips	2
Poor contruction	12	Defective materials	2
Uneven settlement	10	Incorrect operation	2
High pore pressure	5	Earthquakes	1
		Total	100

International Committee on Failures and Accidents to large Dams have classified the risks as follows:

Type of dam	*Gravity*	*Arch*	*Buttress*	*Earth or rockfill*	*Percent*
Origin of risk					
1. Geology at site	4	12	10	10	36
2. Hydrology of river	1	2	3	15	21
3. Construction	1	1	3	5	10
4. Operation, maintenance	1	1	1	4	7
5. Reservoir stability	1	2	4	3	10
6. Deterioration	2	2	9	3	16
Total	10	20	30	40	100

The causes of failures of earth dams are discussed as below:

1. Hydraulic Failure: It is the surface erosion of the dam by water. Causes of hydraulic failure are:

(1) *Overtopping*: It is one of the major causes of dam failure. It is caused by (*i*) Exceptionally large flood impinging the reservoir, (*ii*) Spillway is of insufficient capacity, (*iii*) Faulty opera-

tion of spillway gates or non-operation of spillway gates and hoists, outlet gates and operating equipment, (*iv*) Insufficient freeboard is also a cause of overtopping.

(2) *Wave Erosion*: Winds develop waves near the top water surface which tend to notch out the soil from the upstream slope in case adequate slope protection in the form of riprap is not provided. Progressive cutting and sloughing may destroy the free board and lead to overtopping and failure. Rollers are developed in the waves which try to scoop out the earth (Fig. 17.51). Waves also cause upstream slips (Fig. 17.52).

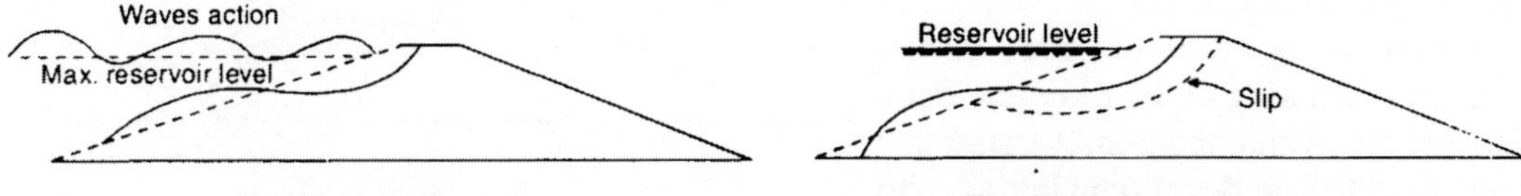

Fig. 17.51 Roller action. **Fig. 17.52** Upstream slip.

(3) *Gullying*: Gullies formed on downstream slope of the embankment due to heavy rain and consequent surface runoff may lead to failure of the dam. Turfing of the slope and provision of berms at certain intervals and proper drainage system is essential to prevent formation of gullies.

(4) *Toe Erosion*: The downstream toe of the dam may be eroded due to tail water or cross-currents that may come from spillway buckets or from exit areas of outlets leading to failure of the dam. The layout of the outlets and spillway should be such that cross flow does not travel towards the dam embankment and suitable riprap on the downstream slope up to a height slightly above the tail water level is provided.

(5) *Cracking Due to Frost Action*: In low temperature regions frost in the upper portion of the dam may cause heaving and cracking leading to seepage and consequent failure of the dam. Extra free board of 1.5 m may be provided to take care of it.

2. Seepage Failures: (1) *Piping*: Seepage through the dam is inevitable. However, uncontrolled excessive seepage through the body and foundation of the dam embankment leads to piping or sloughing and consequent failure of the dam (Fig. 17.53 *a*). Water seepage through the dam has harmful effects as follows:

(*i*) Seepage water generates erosive forces tending to dislodge particles from soil structure causing rearrangement or migration of the fines to voids between larger grains.

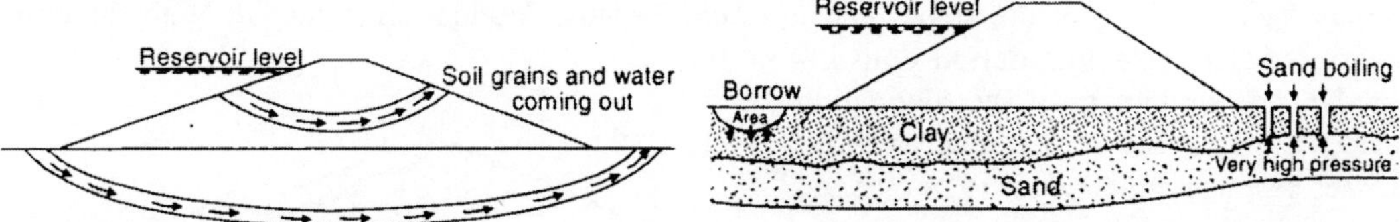

Fig. 17.53(*a*) Seepage or piping through dam embankment and foundation. **Fig. 17.53**(*b*) Piping due to excess exit gradient and due to removal of u/s blanket near the dam.

(*ii*) The flow with its associated differential pore pressure tend to lift portions of soil mass causing boiling (Fig. 17.53 *b*). The internal pore water pressures which cause deep slides are the result of seepage from the reservoir through or under the dam.

(*iii*) Formation of an open conduit through the soil due to internal erosion of the soil mass progressing backwards from the exit. Piping may develop through or under the dam embankment leading to its failure. Settlement is also caused by piping (Fig. 17.53 *c*).

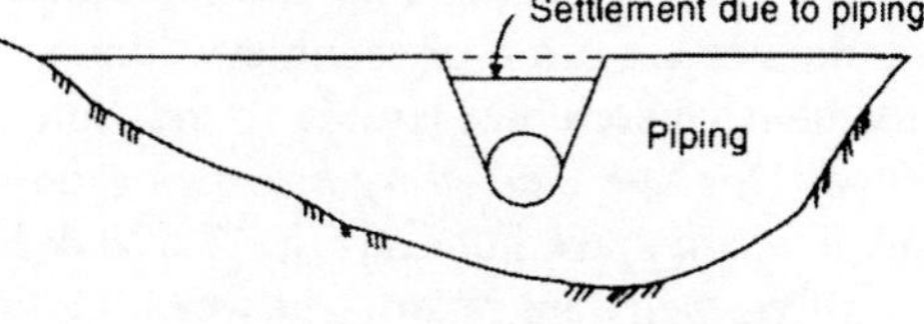

Fig. 17.53c. Piping failure.

(*iv*) Weakening of the soil mass and consequent failure by shear due to reduction in strength. The internal pressures in the soil water may reduce that part of the soil strength that is developed by internal friction causing weakening of soil mass and failure by shear.

(*v*) Uncontrolled seepage causes saturation of the downstream portion of the embankment that brings in failure.

(2) *Sloughing*: Progressive sloughing or ravelling is closely related to piping. Sloughing takes place when the downstream portion of the dam is saturated due to presence of highly pervious layers in the body of embankment or choking of the filter toe drain. Under full reservoir condition, the saturated downstream toe erodes producing a small slide (Fig. 17.54). This slide leaves a relatively steep face which becomes saturated with seepage water from the reservoir and slumps again forming a slightly higher and unstable face. The sloughing process continues till the remaining portion of the dam embankment is too thin to withstand the water pressure and breaks, causing complete failure.

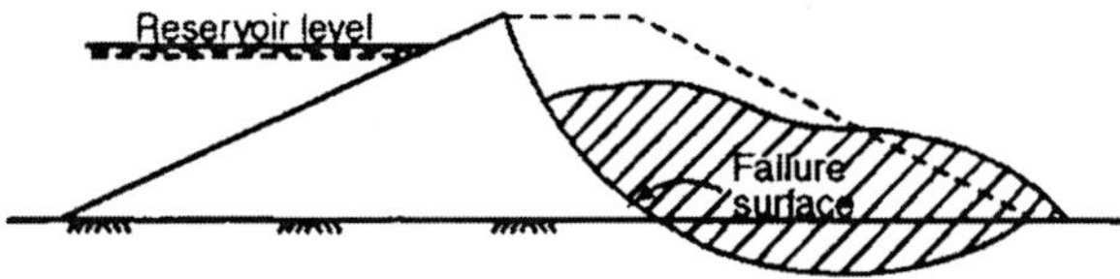

Fig. 17.54 Deep d/s slides during steady seepage.

3. Structural Failures: Structural failure occurs when stresses in the soil produced by the weight of embankment and the load of water exceed the soil strength, besides other causes discussed as follows:

(1) *Foundation slides*: Foundation problem is the leading cause for dam failures. Foundation slides causing failure of the dam may be (*i*) When the foundation of the dam is of fine silt or soft soil, it can slide wholly, (*ii*) When a soft and weak clayey seam or fissured rocks exist under the foundation, the dam can slide over it causing failure. The top of embankment cracks and subsides (Fig. 17.55), (*iii*) Excessive water pressure on confined sand and silt seams in the foundation may reduce the shear strength of the soil rendering it incapable of resisting the induced shear stress causing failure of dam foundation without warning, (*iv*) Loose sand foundation may fail by liquefaction (flow slides), (*v*) Expansion of soil on saturation may lead to lifting of the slopes causing failure of the foundations, and (*vi*) With the construction of dam, the foundation consolidates. Uneven settlement of the foundation produces cracks in the body of dam and thus increases seepage which leads to piping and failure of the dam follows.

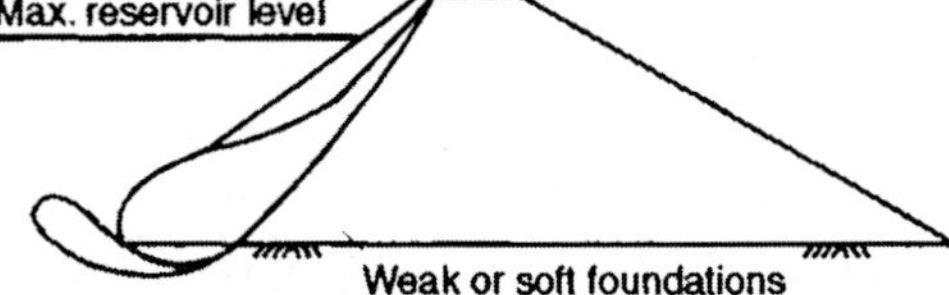

Fig. 16.55. Failure due to subsiding of u/s face.

(2) *Failure by spreading*: Failure by spreading may be caused when fill is located above stratified deposits containing layers of soft clay (Fig. 17.56).

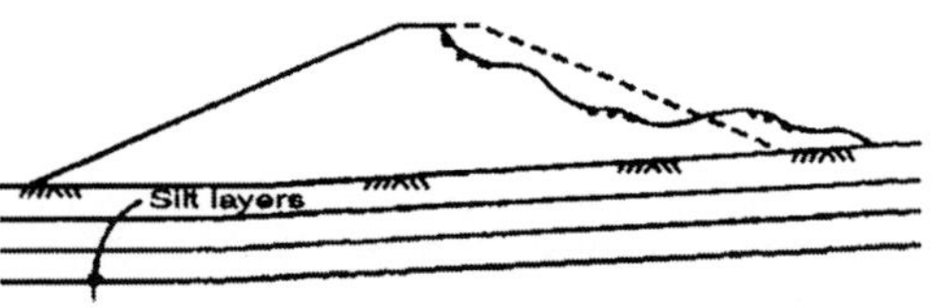

Fig. 17.56 Failure by spreading.

(3) *Damage caused by soluble materials*: Leaching of natural deposits of water soluble materials is a source of trouble in dams. Leaching of soluble materials may cause excessive settlement where considerable foundation leakage exists.

(4) *Damage caused by burrowing animals*: Burrowing animals like muskrats burrow embankments to make homes/(Fig. 17.57*a* & *b*). Many a muskrats by digging holes may honeycomb a small earthen dam redering it weak. Ground squirrels normally dig only in dry soil and stop at the point where seepage is encountered. If the water level in the reservoir is very low for a number of years, the squirrels may enter the upstream slope below the high water level and dig completely

through a small earthen dam. Failure by burrowing animals is not likely to occur in modern earthen dams due to dense compacted core material and adequate crest width and free board.

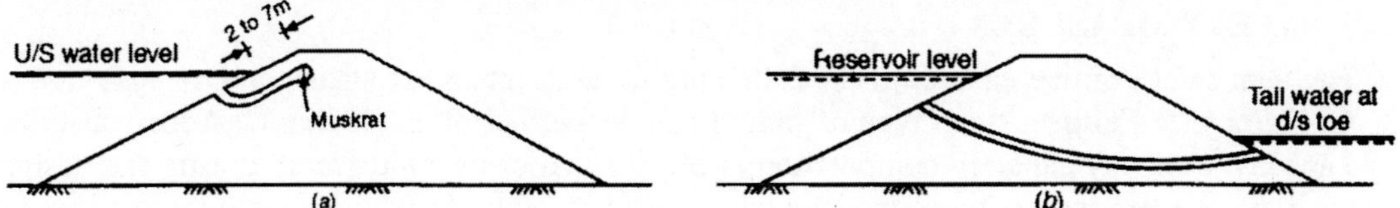

Fig. 17.57 Holes made by Muskrats.

(5) *Slope protection failure*: Riprap on the slope generally rests over a layer of gravel or filter layer. Water waves on the surface of the reservoir may beat repeatedly against the slope just above the reservoir level. The wave energy is dissipated in turbulent action on and within the rocks of riprap layer. The waves may pass through the voids of riprap tending to wash away the filter layer, settling the riprap layer and exposing the embankment directly to wave action (Fig. 17.58) or the stone comprising the riprap, if not heavy, may be washed out of the layer by the hydraulic forces generated by the waves.

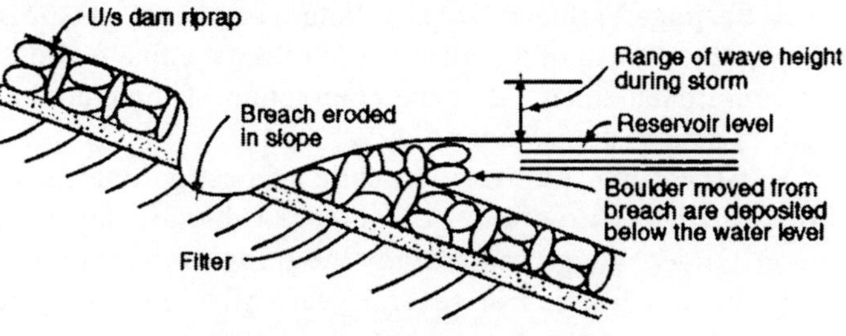

Fig. 17.58 Typical failure of riprap.

(6) *Slopes failure due to construction pore pressures*: Excess pore pressures develop during and immediately after construction when a dam is of relatively impervious compressible soil. When permeability is low, less than 10^{-6} cm/sec, there is no substantial drop in the pore pressure in the central zone of the dam by the end of construction (within usual range of 2 to 4 years).

(7) *Sliding:* The most important cause of failure is sliding. It may occur slowly or suddenly with or without forewarning. Such a failure causes a portion of the earth fill to slide downstream and outwards with respect to the remaining part generally along a well-defined slide surface. At the time of failure, the average shearing stress exceeds the average shearing resistance along the sliding surface.

(*a*) *Upstream slope slides during sudden drawdown*: Sudden drawdown following a long period of high reservoir level is the most critical condition for the slide of upstream slope (Fig. 17.59). When upstream slope is steep, soil used in dam construction is poor, and compaction is not adequate and drainage is poor, sudden drawdown of water causes sliding of upstream face of the dam.

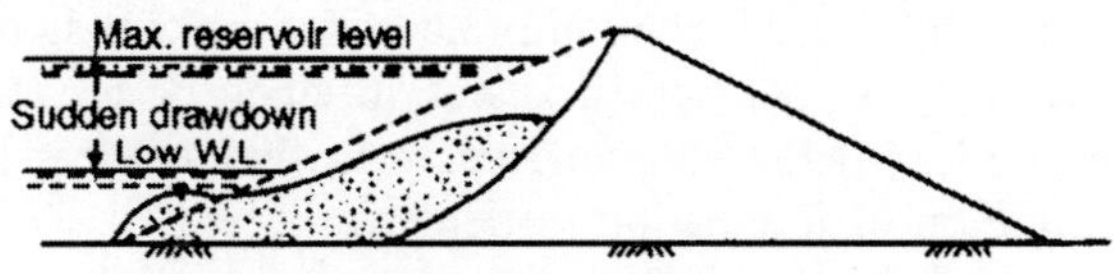

Fig. 17.59 Failure due to sudden drawdown.

(*b*) *Downstream slope slide*: For the downstream slope of the dam, the critical condition is when the reservoir is full and percolation is at its maximum rate. The direction of seepage tends to decrease stability, *i.e.*, the pore water pressure acting on the soil mass below the saturation line reduces the effective stress that resists shearing.

4. Failure by Earthquakes: The potential hazard to a dam from an earthquake depends on how large the earthquake is and how near to dam site it is. Main hazards to a dam from an earthquake are surface faulting under the dam, strong ground shaking, water waves in the reservoir produced by earthquake ground motions or land slides and rock falls and pervasive ground deformations associated with nearby faulting which may manifest in cracking at dam top and central core, settlement of dam crest, shear failure at the base of dam, liquefaction of loose and saturated soil mass in the lower portion of the dam, and overtopping due to high water waves generated in the reservoir.

17.16 SAFETY MEASURES

The dam safety can be ensured if the following failure aspects are taken care of:

1. Hydraulic Failure: Such type of failure can be averted by providing (*i*) Adequate spillway capacity. Spillway capacity computation is often a matter of controversy during the design stage and the resultant provision of inadequate spillway capacity. Provision of auxiliary or emergency spillway may be considered for dam safety, (*ii*) Adequate free board so that the dam safety is not endangered by overtopping during high floods, and (*iii*) Proper maintenance of gates so that they are always operative and do not get clogged.

2. Seepage Failure: Seepage failure can be taken care of by (*i*) Providing filter at the toe to minimize movement of the material, (*ii*) Seepage line is well within the body of the dam, and (*iii*) Provision of settlement, after the completion of the dam, be made from a normal 1 per cent of height of dam to a maximum of 6 per cent.

3. Structural Failure: In spite of best geological and foundation investigations done prior to dam construction, geological problems may arise such as induced seepage, earth tremors, slides, gougy seams and sloughing in the vicinity of dam and reservoir area surface during the construction or several years after the reservoir filling. Periodic geotechnical inspection is essential for early detection and resolution of potential problems, besides provision of adequate riprap and its maintenance and drawdown within permissible limits.

4. Earthquake Failure: Necessary provision shall be made in the design of a dam to account for additional forces due to earthquake.

For more details refer dam safety (Appendix-I)

17.17 ROCKFILL DAM

A dam with the diaphragm comprising mostly of rock materials is termed as rockfill dam. Term rockfill dam also refers to the dams constructed entirely of rock. Alternatively, a rockfill dam is defined as an embankment which uses rock of various sizes to provide stability and an impervious membrane to impart watertightness. The mass stability is developed by the friction and interaction of one particle on another rather than by any cementing agent binding the particles together. The water pressure is resisted only by the weight of the rock.

Rockfill dams are less flexible than earthen dams but more flexible than gravity dams. The foundation requirements are more rigid than earthen dams but not so strict and rigid as for gravity dams. The foundation requirements for rockfill dams are (*i*) Essentially the foundation consists of hard durable rock which is not softened or eroded appreciably by water percolating from the reservoir, (*ii*) Foundation free from faults, shear zones or other structural weakness, and (*iii*) Minimum of foundation settlement.

Rockfill dams are suitable for areas where there is scarcity of earthfill materials but durable, hard and sound rock such as unweathered igneous and metamorphic rocks are available for the construction of the dam. Sedimentary rocks and shales which slake in the presence of air are not suitable. A rockfill dam is not suitable when the normal operation of the reservoir does not permit periodic inspection of the impervious membrane constructed on the upstream slope. It is also essential that the dam is not overtopped in time of flood and as such spillway, usually of side channel type, of adequate capacity is provided.

17.18 CLASSIFICATION ROCKFILL OF DAMS

Rockfill dams may be classified on the basis of location of impervious membrane, as (*i*) Rockfill dam with upstream impervious membrane, and (*ii*) Earth core type rockfull dam as illustrated in Fig. 17.60. The components of the dams are also depicted therein.

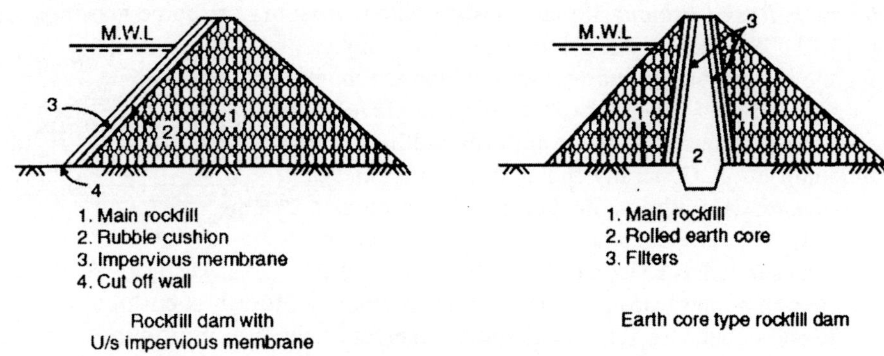

Fig. 17.60 Types of rockfill dams

17.19 DESIGN PARAMETFRS

Top width: The top width is normally fixed according to working space required at top, the minimum being 6 m.

Side slope: The upstream slope usually has hand placed stones and as such slope of ½ : 1 to 1 : 1 is adopted. For dams over 50 m height, the slope adopted is 1.3 : 1. The downstream slope made of dumped rock usually corresponds to the natural slope of the rock varying from 1.3 : 1 to 1.5 : 1. For high dams, berms are provided.

Base width: The overall base width usually ranges between 2.5 to 3 times the dam height.

Settlement: The settlement of compacted rockfill after placing the rubble cushion and the impervious facing is generally 1% of the height. It takes place during construction and up to the reservoir is filled. The horizontal movement of rockfill also takes place on filling of reservoir and like vertical settlement it continues over a long time.

17.20 DAM SECTION

A rockfill dam section comprises upstream impervious membrane, upstream rock cushion, downstream rockfill and upstream cutoff. The basic requirement is to ensure (*i*) Safety against overtopping, (*ii*) Stability and (*iii*) Safety against internal erosion.

1. *Upstream Impervious Membrane*: This watertight membrane is constructed on the upstream slope where it is available for inspection and repairs during the drawdown condition of the reservoir. The other advantages of upstream membrane are it has (*i*) Greater margin of safety against shear failure because of low pressures in rockfill, larger rockfill mass to resist water pressure and water pressure having larger downward component, (*ii*) Dam can be raised by dumping rock on the downstream side and then extending the membrane upward on the sloping surface, (*iii*) Permits grouting of the rock below cutoff wall at the upstream toe while embankment is being constructed, and (*iv*) Dam has comparatively lesser volume of fill.

The membrane usually consists of asphaltic concrete or cement concrete slabs with suitable joints duly filled with bitumen filler. Reinforced cement concrete membrane is the most common type because of its durability and ease of construction. The joints between the slabs are 25 to 50 mm wide in horizontal direction and 50 to 75 mm wide in vertical direction. Horizontal joints are avoided as far as possible. The joints are filled with a plastic or compressible material. Thickness of concrete slab is about 1% of water head or 30 cm minimum reinforced with steel equal to 0.5 per cent of concrete area, in both directions. Impervious membrane if of earth core and placed centrally is designed as in case of earth dams. Asphaltic concrete membrane is economical and more flexible than concrete slab.

2. *Upstream Rock Cushion*: It may consist of stone masonry or rubble masonry or pervious zone (Fig. 17.61). The masonry provides a smooth, compact bedding for the impervious membrane. In low dams, rubble masonry is replaced by a zone of graded sand and gravel or quarry fines. The pervious zone has a uniform width of 5 m to facilitate compaction. Generally rubble cushion is thicker at bottom than at top. A minimum thickness of 3 m normal to the slope is desirable. The upstream face against which the membrane rests usually has a slope of 2 : 1. The functions of rock cushion are to support upstream impervious membrane, to act as cushioning in stabilising upstream face and equalising settlement and to transfer water load uniformly to the loose rock mass filled at its back.

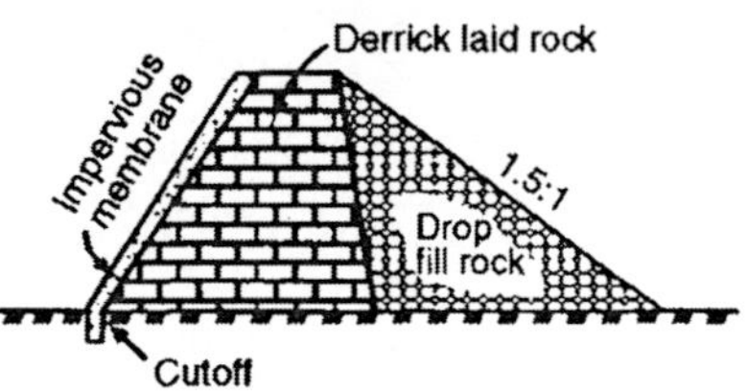

Fig. 17.61 Rockfill dam.

3. *Downstream Rockfill Zone*: It provides the structural support to the impervious membrane or core of the dam by its weight and internal stability against water pressure from the reservoir. The downstream rockfill zone (Fig. 17.62) serves to minimize total settlement and possibility of damage to the impervious membrane. The rockfill is dumped on the embankment and spread in layers with a maximum thickness of 1 m. The rockfill is to have minimum number of large voids and is compact. The downstream slope usually adopted is natural angle of the rockfill, *i.e.*, 1.4 : 1. In seismic regions, the slopes of rockfill are flattened beyond the angle of repose of the material to include the ratio of probable earthquake acceleration to the gravity acceleration. The requirements of crest width and free board are the same as indicated for earth dams.

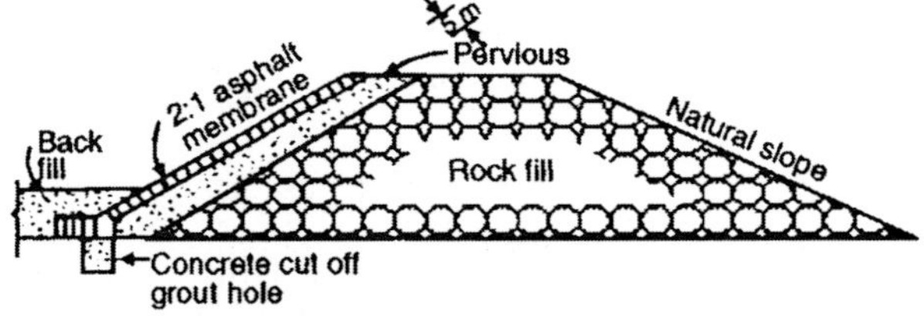

Fig. 17.62 Rockfill dam (d/s rockfill).

4. *Earth core*: Earth core, either vertical or inclined upstream, forms an impervious barrier in an earth core type rockfill dam (Fig. 17.60).

5. *Cutoff*: A watertight cutoff is provided at the upstream toe of the dam along with the contact of the impervious membrane with the foundation and abutments to prevent seepage under the dam and provide support for the weight and thrust of the impervious membrane. The contact length on foundation should be adequate to provide a minimum seepage path, 1/20 to 1/10th of head of water but not less than 3 m depending on rock quality. It should provide at least 1 m rockfill under membrane to permit the membrane to deflect normal to its face. Grouting is done regardless of the apparent quality of the foundation rock. The dam section is checked against sliding caused by water pressure on the upstream. The resisting force comprises vertical component of water and reduced weight of the rockfill due to buoyant effect. The coefficient of sliding varies between 0.6 and 1.0. Fig. 17.63 shows rockfill dam, Hidkal dam in Karnataka.

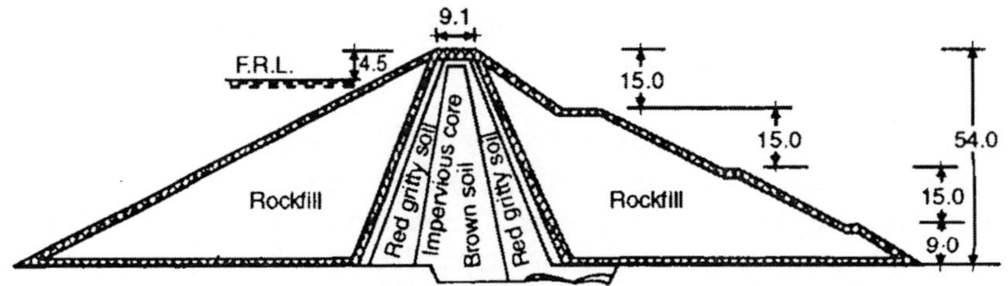

Fig. 17.63 Rockfill dam—Hidkal dam (Karnataka).

17.21 ADVANTAGES AND DISADVANTAGES OF ROCKFILL DAMS

Advantages of rockfill dam are (*i*) Cheaper where suitable hardrock is available, (*ii*) Suitable where suitable materials for earth dam are not available, (*iii*) Economical in remote locations where cost of cement for concrete dam is high, (*iv*) Suitable where foundation is not suitable for concrete dam, (*v*) Can be constructed with relatively unskilled labour, (*vi*) Can be raised subsequently, if so required, without much difficulty.

Disadvantages: (*i*) Time taken in construction is usually more than that required for a concrete dam. This is a crucial factor, (*ii*) More construction equipment is required, (*iii*) Foundation requirements are more rigid than earthen dam, and (*vi*) High maintenance cost.

SOLVED EXAMPLES

Example 17.1 *Determine the seepage line through the dam section (Fig. 17.64). Coefficient of permeability of soil material comprising the dam is 5 × 10⁻⁴ cm/sec.*

Solution.

$$LB = 20 \times 3 = 60 \text{ m}$$
$$AB = 0.3 \times 60 = 18 \text{ m}$$

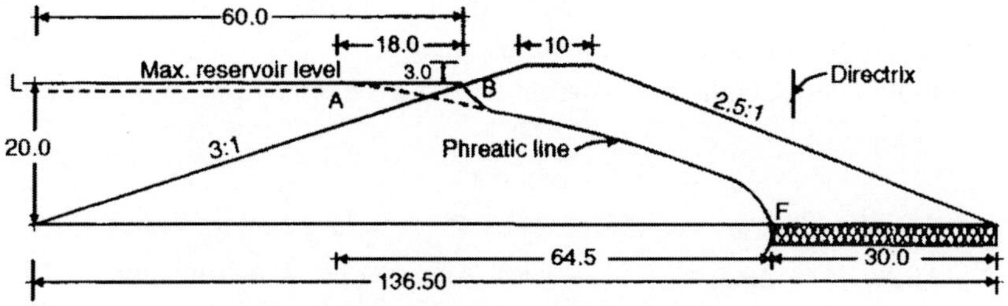

Fig. 17.64 Seepage line through earthen dam.

Coordinates of any point (x, y) on a parabola of equation:

$$\sqrt{x^2 + y^2} = x + s$$

Coordinates of point F are 64.5 *i.e.*, (136.50 – 30 – 60 + 18) and 20, *i.e.*

$$\sqrt{(64.5)^2 + (20)^2} = 64.5 + S$$

or $$S = 3.03$$

Vertex C of the parabola is 1.52 m *i.e.*, $S/2$ beyond F and directrix D is at 3.03 from F

Now, equation of parabola $$y = \sqrt{2Sx + S^2}$$
$$= \sqrt{2 \times 3.03 \times x + 3.03 \times 3.03}$$
$$= \sqrt{6.06x + 9.18}$$

Coordinates of parabola for various values of x are as under:

x	0	10	20	30	40	50	60	64.5
y	3.03	8.35	11.42	13.82	15.86	17.67	19.31	20.0

Fig. 17.64 shows the plotting of phreatic line.

Seepage discharge $$q = KS$$
$$= (5 \times 10^{-4} \times 10^{-2}) \times 3.03$$
$$= 15.15 \times 10^{-6} \text{ cumec per m length of dam.}$$

Example 17.2 *Draw seepage line through the dam if there is no drainage filter at the end. The central section consists of relatively impervious soil ($K = 0.2 \times 10^{-4}$ cm/s) and the outer shells are of pervious soil ($K = 100 \times 10^{-2}$ cm/s). Assume value of c i.e., $\Delta a/a + \Delta a = 0.34$.*

Solution. Due to very large difference in the permeability of the pervious upstream and downstream shells and the central impervious section, these shells have no influence on the position of the phreatic line and hence the seepage flow through the central section is only to be considered.

$$AB = 60 \times 0.3 = 18 \text{ m}$$

From equation of basic parabola

$$x = (y^2 - S^2)/2S$$

$$S = \sqrt{b^2 + H^2} - b$$

From Fig. 17.65 $\quad b = 54$ m and $y = 20$ m

$$s = \sqrt{b^2 + H^2} - b = \sqrt{(54)^2 + (20)^2} - 54 = 3.58 \text{ m}$$

Now, $\quad x = (y^2 - S^2)/2S$

or $\quad x = [y^2 - (3.58)^2]/2 \times 3.58$

or $\quad y^2 = 7.16\,x + 12.82$

or $\quad y = \sqrt{7.16x + 12.82}$

The coordinates of parabola for various values of x are as under:

x	0	5	10	15	20	30	40	50	54
y	3.58	6.97	9.19	10.96	12.49	15.09	17.30	19.26	20.0

(*iii*) Correction at the points of ingress and outgress.

The base parabola intersects the downstream face at a distance $(a + \Delta a)$ along the face from the point A.

$$a + \Delta a = S/(1 - \cos\alpha)$$

$$a + \Delta a = 3.58/(1 - \cos 45°), = 12.22 \text{ m}, \ \alpha \text{ being } 45°$$

$$\Delta a/(a + \Delta a) = 0.34 \text{ (given). Also can be read from Fig. 17.12.}$$

or $\quad \Delta a = 0.34\,(a + \Delta a) = 0.34 \times 12.22 = 4.15$ m.

∴ Location of point $k = 12.22 - 4.15 = 8.07$ m measured along d/s face from point G.

The seepage line at the downstream end is completed by making a short transition curve from point K to the base parabola. The upstream end of the seepage line is given a short transition curve as indicated in Fig. 17.65.

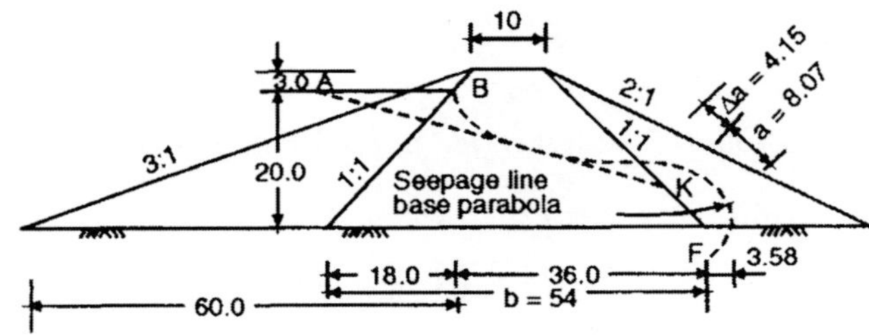

Fig. 17.65 Seepage through zoned dam.

Seepage discharge:

The seepage discharge per unit width of dam through impervious central section is given by the relation $q = k.s = 0.2 \times 10^{-4} \times 10^{-2} \times 3.58 = 0.72 \times 10^{-6}$ cumec.

The seepage flow through downstream pervious shell is also the same.

Example 17.3 *Determine seepage through foundation of the dam section shown in Fig. 17.66.*

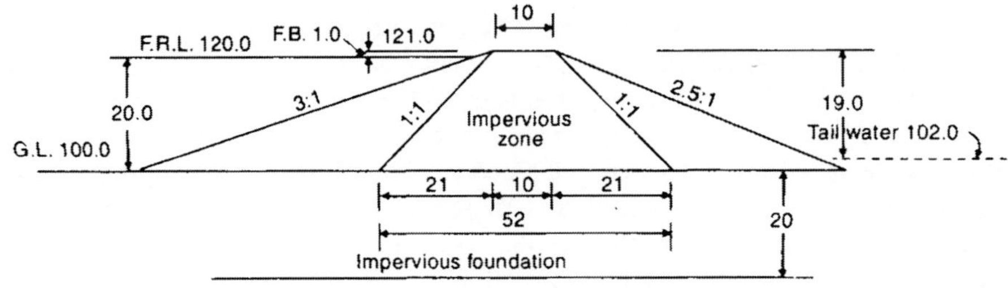

Fig. 17.66 Seepage through dam foundation.

Solution. Darcy's formula, $Q = KiA$

where, Q = discharge, k = coefficient of permeability of foundation $= 0.2 \times 10^{-6}$ m/s

i = hydraulic gradient $= h / l$

$i = (120.0 - 102.0)/52 = 0.346$

A = gross area of foundation through which flow takes place, $= 1 \times 20$ sqm for a dam foundation width of 1 m

Now, $Q = 0.2 \times 10^{-6} \times 0.346 \times 20$

$= 1.38 \times 10^{-6}$ cumec

Example 17.5 *A flow net is plotted for a homogeneous earthen dam of height 25 m and length 2000 m with free board 2 m. The results obtained indicate number of potential drops as 10 and number of flow channels as 4. The dam has a horizontal filter of 30 m at the downstream end and the coefficient of permeability of the dam material is 5×10^{-4} cm/sec. Calculate the discharge through the dam.*

Solution.

$$q = KH\frac{N_f}{N_d} = \frac{5\times10^{-6}\times23\times4}{10}$$

$= 46 \times 10^{-6}$ cumec/m length of dam

$Q = 46 \times 10^{-6} \times 2000 = 0.092$ cumec.

Example 17.6 *Determine the length and thickness of the upstream impervious blanket for an earthen dam of the dimensions; top width 5 m, base width 100 m, free board 3 m, minimum water head 16 m, depth up to impervious foundation 25 m, permeability coefficient of pervious foundation is 3 cm/minute, and desirable reduction in seepage 80%. Neglect seepage through the blanket.*

Solution. Seepage through dam foundation by Darcy's law.

$$q = KIA = K\frac{H}{B}d \qquad ...(i)$$

Seepage through dam with blanket, reduced to 20%,

$$Q = K \times \frac{H}{(B+L)}d$$

or

$$Q = (KHd)/0.2\,(B + L) \qquad ...(ii)$$

Equating (*i*) and (*ii*), $\frac{1}{B} = \frac{1}{0.2(B+L)}$ or $L = 4B$

Length of u/s blanket or $L = 4B = 4 \times 100 = 400$ m

Ratio of path of percolation to head $= (400 + 100)/16 = 31$ which is > 10, hence O.K.

Thickness of blanket $= 10\%$ of head $= 0.1 \times 16 = 1.6$ m

Provide 2 m thick blanket.

Example 17.7 *In a slip circle analysis of downstream slope of a dam during steady seepage, the section of the dam was drawn to a scale 1 cm = 5 m and the results obtained were; area of N-rectangle = 15.25 sq cm, area of T-rectangle = 6.5 sq cm, area of U-rectangle = 5.2 sq cm, and length of arc = 12.5 cm. The properties of the materials of the dam are; effective angle of friction = 26°, unit cohesion = 0.2 kg/km², unit weight of soil, $\gamma = 2$ g/cm³. Determine factor of safety for the slope.*

Solution. Let scale be 1 cm = x cm, x being = 500 cm

Consider 1 cm length of the dam

$$\Sigma N = A_N (x^2)\ \gamma$$
$$= 15.25 \times (500)^2 \times 2 \times 10^{-3} = 7625 \text{ kg}$$
$$\Sigma T = A_T (x^2)\ \gamma$$
$$= 6.5\ (500)^2 \times 2 \times 10^{-3} = 3250 \text{ kg}$$
$$\Sigma U = A_U (x^2)\ \gamma_w$$
$$= 5.2\ (500)^2 \times 1 \times 10^{-3} = 1300 \text{ kg}$$

Length of arc, $\hat{L} = 12.5 \times 500 = 6250$ cm

Factor of safety $= (C\hat{L} + \Sigma(N - U)\tan\phi')/\Sigma T$

$$= \frac{(0.2 \times 6250) + (7625 - 1300)\tan 26^\circ}{3250} = 1.33.$$

QUESTIONS

1. Explain how the following parameters effect design of an earthen dam: (*i*) optimum moisture content, (*ii*) C and ϕ value of soil, (*iii*) permeability of soil, (*iv*) sudden drawdown of reservoir.
2. Illustrate with neat sketch the following parts of an earthen dam and state their functions briefly:
   (*i*) Rock toe, (*ii*) horizontal drainage blanket, (*iii*) cutoff, (*iv*) riprap.
   (*b*) Give a few typical cross-sections of earthen dams to suit locally available material and foundations.
3. Show with the help of simple outline sketches suitable broad designs of earth dam for different available materials and the governing geology of the site.
4. Briefly discuss the checks that are required to be made to investigate the stability of an earthen dam.
5. Discuss control of seepage in earthen dams.
6. Explain the various steps that can be taken in the design and in the construction of an earthen dam to reduce seepage from it.
7. Discuss hydraulic fill dam.
8. Discuss cutoff trench.
9. Give a suitable design for a 20 m high dam for a site where both clay silt and sand gravel are available in abundance and where foundation is pervious to a depth of 10 m. Assume suitable data. Give reasons favouring the suggested design.
10. Differentiate between horizontal and vertical piping in earth dams. Suggest permanent remedial measures to check vertical piping.
11. (*a*) Illustrate with reasons a typical cross-section of an earthen dam to the adopted where fine grained weak soil is available for embankment and unconsolidated impervious silty clay as foundation. Explain the basic components of the same.
    (*b*) Explain: (*i*) Design requirements for filter.
    (*ii*) Phreatic line in an earthen dam.
12. (*a*) What are the different methods for reducing seepage in earth dams?
    (*b*) Check whether the free board provided in Fig. 17.67 is adequate. The longest straight expanse of water upstream of dam is 36 km and the maximum wind velocity is 81 km/hr.

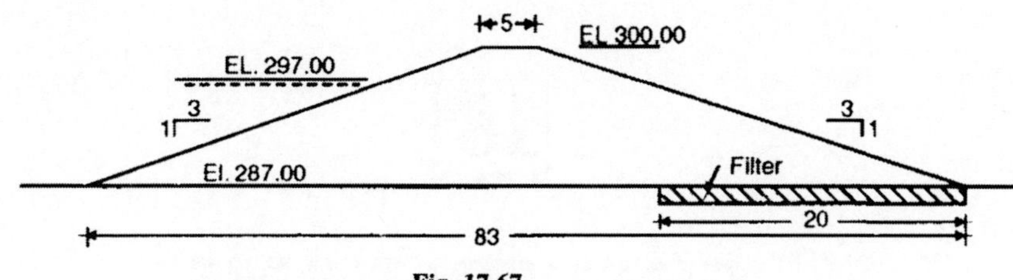

Fig. 17.67

13. (*a*) Deduce that the seepage velocity vector at the point of emergence of the top-seepage flow line in an earth dam of homogeneous section is directed along the line of the downstream slope.
(*b*) Explain how this point of emergence in (*a*) above is determined. What modifications could result in pulling down the top seepage line and keeping it well within the body of the dam?
(*c*) An earth dam impounding 15 m head of water and provided with an impervious clay core and u/s blanket is constructed on pervious foundations as per Fig. 17.68. Determine the seepage loss through the foundation per m length of the dam.

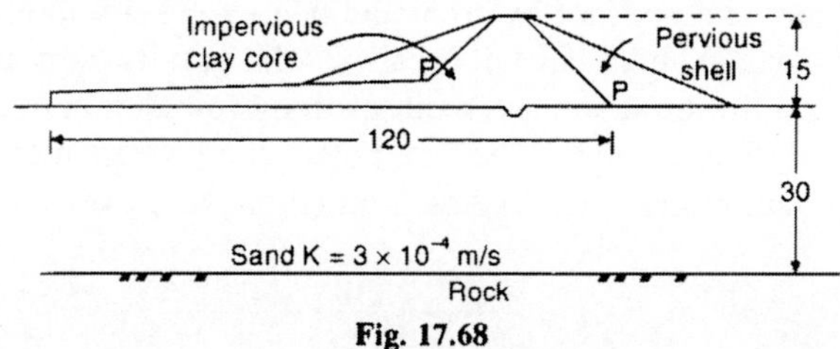

Fig. 17.68

Solution: Seepage loss/m length of dam $= K \dfrac{H}{L} d$

$$= 3 \times 10^{-4} \times \frac{15}{120} \times 30$$

$$= 11.25 \times 10^{-4} \text{ cumec.}$$

14. A zoned earth dam has to be built on a river bed with highly permeable gravel and sand to a depth of 20 m Height of dam above river bed is 50 m. Draw a suitable section of the dam you would propose and give brief specifications for different materials.
15. Enumerate and discuss briefly the various factors that determine the free board to be given for an earthen dam.
16. (*a*) What is a flow net? What are its salient properties?
(*b*) What points are to be noted while sketching a flow net?
(*c*) A homogeneous earthen dam is 21.5 m high and has a free board of 1.5 m. A flow net was constructed and the following results were noted:
Number of potential drops = 12
Number of flow channels = 3
The dam has a horizontal filter of 15 m length at its drawdown end. Calculate the discharge per metre length of the dam if the coefficient of permeability of the dam material is 2.7×10^{-3} cm/s.

Solution:

$$q = 2.7 \times 10^{-3} \times 10^{-2} \times (21.5 - 1.5) \times \frac{3}{12}$$

$$= 1.35 \times 10^{-6} \text{ cumec.}$$

18

GRAVITY DAMS

18.1 INTRODUCTION

Gravity dam is defined as a structure so proportioned that it derives its stability from its own weight to resist the external forces. A gravity dam is most durable and requires little maintenance. It may be constructed of masonry or concrete. Concrete gravity dams are greatly preferred and their construction is common. A gravity dam is generally straight in plan and termed as straight gravity dam. However, it may also be slightly curved in plan. Nevertheless, the external forces are resisted by the weight of the dam and not by arch action. In a gravity dam no bending stress is introduced at any point of the dam and as such it is called solid gravity dam. Gravity dams transfer the load to foundations by cantilever action and require strong foundation rock to resist the weight of the dam. They can be built to any height. They are most suited across narrow gorges with steep side slopes where earthen dam might slip. Bhakra Dam (Fig. 18.1) over river Sutlej is a straight concrete gravity dam.

Fig. 18.1 Bhakra dam.

Various dam parameters are discussed as under:

1. *Dam axis*: The line of upstream face of the dam or the line of crown of the dam, if upstream face is sloping, is taken as reference line for the purpose of layout. It is called *base line* of the dam or *dam axis*.

2. *Structural height*: It is the height between the top of dam and the lowest foundation level, exclusive of narrow fault zones, if any.

3. *Hydraulic height.* It is the height between the highest controlled water surface and the lowest point in the original river bed level at the axis of the dam.

4. *Dam length.* It is the length of the dam measured along the axis of the dam at the level of top of dam, *i.e.* from one abutment to the other.

18.2 SELECTION OF DAM SITE

Various considerations in the selection of a good dam site have been discussed in section 15. Those specific to gravity dams are (*i*) A narrow gorge, opening upstream, at the dam site, (*ii*) Strong rock foundation to safely withstand static and dynamic forces including earthquakes, (*iii*) Strong and water tight abutments, (*iv*) Stable side slopes of abutments, (*v*) Suitable location for power house and spillway, (*vi*) Availability of construction materials nearby, (*vii*) Accessibility by rail and road, (*viii*) Availability of electric supply for construction, and (*ix*) Dam height is reasonable for the required storage capacity.

18.3 FORCES ACTING ON A DAM

Various forces acting on a dam as considered in design are as under:

Dead Load: It comprises the major resisting force. It includes weight of the concrete or masonry or both plus that of appurtenances such as piers, gates and bridges. Usually unit length of the dam is considered. The cross-section of the dam is divided into several triangles and rectangles. Weight of each triangle and rectangle and their point of application at respective centre of gravity are computed. The resultant of all these downward forces is thus found out by taking moments of the component forces which constitute the total weight of the dam acting at its centre of gravity. Unit weight of concrete and masonry is usually taken as 2400 kg/m^3 and 2300 kg/m^3 respectively.

Reservoir and Tail Water Loads: Water pressure corresponding to full reservoir level constitutes the major external force acting on the dam. The water pressure acts horizontally and its intensity varies triangularly with zero intensity at the water surface and equal to $WH^2/2$ at the base, where W is the unit weight of water (1000 kg/m^3) and H is the water depth upstream. The resultant force $WH^2/2$ acts at $H/3$ from the base of the dam (Fig. 18.2). When the upstream face is partly vertical and partly inclined (Fig. 18.3), the resultant water pressures comprise (*i*) horizontal component $P = WH^2/2$ acting at $H/3$ from the base of the dam, and (*ii*) vertical component P_1 which is the weight of the column of water *ABCDA* acting at *CG* of the area. In low overflow dams, the dynamic effect of the velocity of approach is also considered.

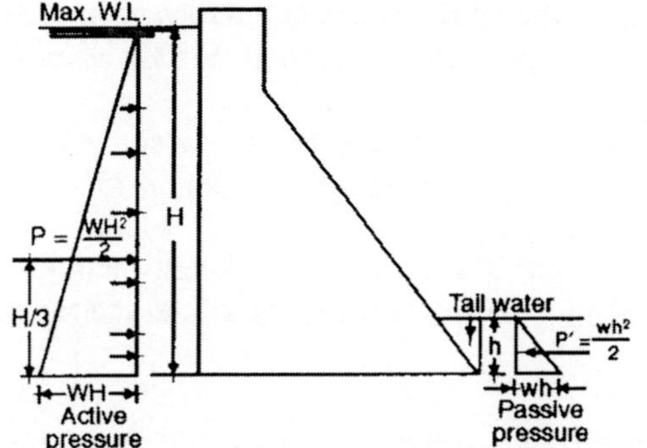

Fig. 18.2 Water pressure acting on dam with u/s face vertical and tail water.

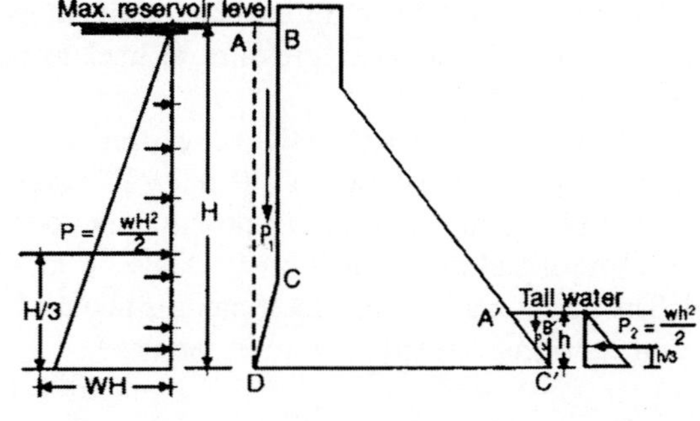

Fig. 18.3 Water pressure acting on dam with u/s face partly inclined and tail water.

Tail Water Pressure. Tail water pressure, adjusted for any retrogression, is taken at full value for non-overflow sections and at a reduced value for overflow sections depending on the type of energy dissipation arrangement provided and the anticipated water surface profile downstream. The full value of corresponding tail water is, however, considered in the case of uplift. The tail water of height h on the downstream side exerts both horizontal and vertical pressures; horizontal pressure, $P_2 = wh^2/2$ acting at $h/3$ from the base and vertical pressure, P_3 = weight of column of water *A'B'C'* (refer Fig. 18.3).

3. Uplift Pressure: It is the second major external force acting upward due to seeping water. The uplift forces occur as internal pressures in pores, cracks and seams within the body of the dam, at the contact surface between the dam and its foundation and within the foundation. The pressure intensities at the heel and toe of the dam are taken equal to the hydrostatic pressures at the heel and toe joined by a straight line in between (Fig. 18.4). Drainage galleries are sometimes provided in the dam body to relieve uplift pressure. Accordingly, the uplift pressure distribution in the body of the dam is assumed to have an intensity which at the line of the formed drain/drainage holes exceeds the tail water head by one-third the differential between reservoir level and tail water level. The pressure gradient is then extended linearly to heads corresponding to reservoir level and tail water level (Fig. 18.5). The pressure is assumed to act over 100 percent of the area. The uplift criteria in case of dams founded on compact and unfissured rock is as specified above.

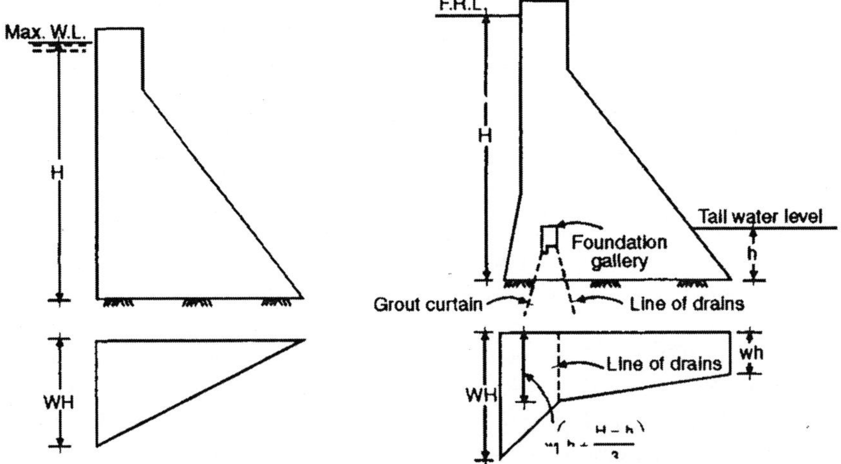

Fig. 18.4 Uplift pressure diagram without tail water. **Fig. 18.5** Uplift pressure diagram with drainage gallery and tail water.

Under worst loading conditions *i.e.* drains inoperative, the uplift is considered to vary linearly from maximum flood water pressure at heel to tail water pressure (or zero if no tail water is present) at the toe of dam.

Earthquake Forces: The effect of earthquake is equivalent to imparting horizontal and vertical acceleration to the foundation of the dam. The vertical acceleration is generally not considered in the design of a dam as it may not pose any serious damaging force.

1. Horizontal acceleration. (*i*) Dams up to 100m high: The seismic coefficient method is used. The horizontal seismic coefficient is taken as 1.5 times seismic coefficient, αh, at the top of the dam reducing linearly to zero at the base.

$$\alpha h = B\, I\, \alpha_0$$

Where, B = a coefficient depending upon soil foundation system; value 1.0, I = a factor depending upon the importance of the structure; value 3.0, α_o = basic horizontal coefficient (Table 18.1).

Table 18.1 Value of basic seismic coefficient and seismic zone factors in different zones.

Sl. No.	*Zone No.*	*Method*	
		Seismic coefficient, α_o	*Response spectrum, Fo*
(*i*)	V	0.08	0.40
(*ii*)	IV	0.05	0.25
(*iii*)	III	0.04	0.20
(*iv*)	II	0.02	0.10
(*v*)	I	0.01	0.05

(*ii*) Dams above 100m height. The response spectrum method is used. In the method, the response acceleration coefficient is obtained from the natural period and damping of the dam structure and the design value of horizontal seismic coefficient is computed from the relation:

$$\alpha h = B\ I\ F_o\ Sa/g$$

where, F_o = seismic zone factor for average acceleration spectra as given in Table 18.1 and Sa/g = average acceleration coefficient as read from Fig. 18.6 for appropriate natural period and damping of the structure.

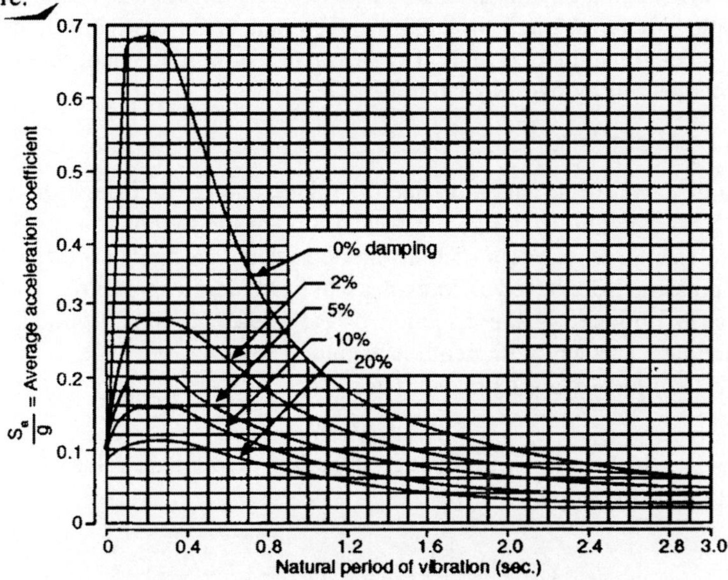

Fig. 18.6 Average acceleration spectra (IS: 1893-1984).

(*a*) The fundamental period of vibration, $T = 5.55\,\frac{H^2}{B}\sqrt{\frac{W_m}{gE_s}}$, where, H = height of dam (m), B = base width of dam (m), W_m = unit weight of dam material (kg/m^3), g = acceleration due to gravity (m/s^2), and E_s = modulus of elasticity of material (kg/m^2). With the period T, and for a damping of 5%, the design horizontal seismic coefficient αh is obtained.

(*b*) The base shear, V_B and base moment M_B are computed from the formulae:

$$V_B = 0.6\ W\alpha h \qquad M_B = 0.9\ Wh_1\alpha h$$

where, W = total weight of masonry or concrete in dam (kg), h_1 = height of C.G. of dam above the base (m).

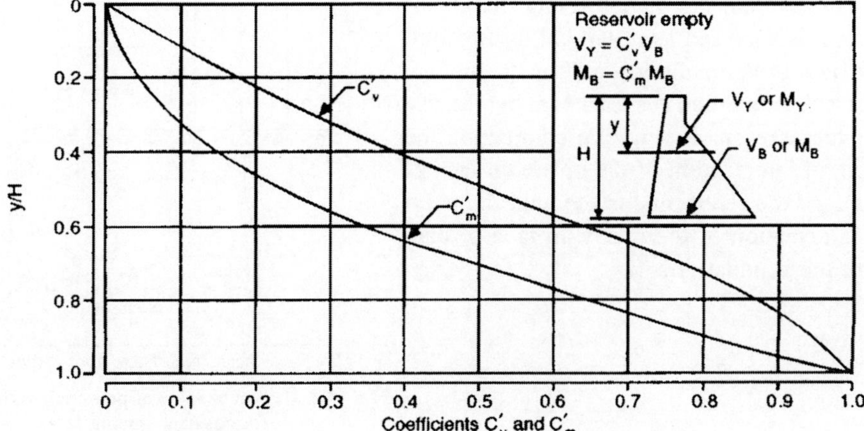

Fig. 18.7 Values of C'_v and C'_m along the height of dam.

(*c*) For any horizontal section at a depth y below top of dam, shear force, V_y and bending moment M_y is obtained from:

$$V_y = C'_v V_B \qquad M_y = C'_m M_B$$

where, C'_v and C'_m are obtained from Fig. 18.7.

Horizontal acceleration causes intertia force and hydrodynamic pressure :

Inertia force. It is computed by the principle of mass times acceleration acting through the centre of gravity of the section, regardless of its shape and in a direction opposite to the direction of ground acceleration. It is assumed to act from upstream to downstream or downstream to upstream to get worst combination for design. It causes an overturning moment about the horizontal section adding to that caused by hydrodynamic force.

$$\text{Inertia force} = \frac{W}{g}(\alpha g) = W\alpha$$

where, W = weight of dam, g = acceleration due to gravity, and α = acceleration coefficient.

Hydrodynamic pressure. Due to horizontal acceleration of the foundation and dam there is an instantaneous hydrodynamic pressure (or suction) exerted against the dam in addition to hydrostatic forces. The extra pressure so exerted is known as hydrodynamic pressure. The direction of hydrodynamic force is opposite to the direction of earthquake acceleration. The variation of hydrodynamic earthquake pressure with depth is as under:

1. *Dam with Vertical or Sloping Upstream Face.*

$$P = C_s \alpha_h wh$$

where, C_s = coefficient which varies with shape and depth. Approximate value of C for dams with vertical or constant upstream slope:

$$C_s = \frac{C_m}{2}\left[\frac{y}{h}\left(2-\frac{y}{h}\right)+\sqrt{\frac{y}{h}\left(2-\frac{y}{h}\right)}\right]$$

P = Hydrodynamic pressue (kg/m^2) at depth y from the reservoir water surface, α_h = design horizontal seismic coefficient (Refer earthquake forces, Page 527) , w is unit weight of water (kg/m^3), h is depth of reservoir (m), C_m = maximum value of C_s (read from Fig. 18.8).

2. *Dam with Combined Vertical and Inclined Face*

(*a*) If upstream face is inclined but height of vertical portion of upstream face is equal to or greater than one-half the total height of dam, it is analysed as if vertical throughout.

(*b*) If height of vertical portion of upstream face is less than one-half the total dam height, use the pressure on sloping line connecting the point of intersection of the upstream face of dam and the reservoir surface with the point of intersection of the upstream face of dam with the foundation.

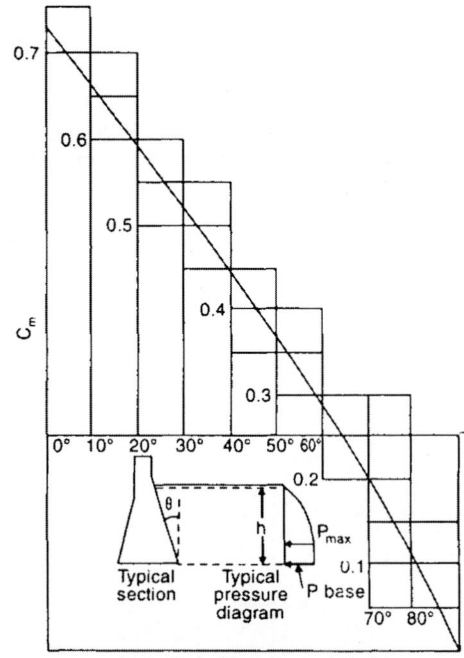

Fig. 18.8 Maximum value of pressure coefficient, C_m, for constant sloping faces.

The approximate values of total horizontal shear and moment due to hydrodynamic pressure for any height y of the dam from the water surface is given by the equation $V_h = 0.726\ Py$, where V_h is hydrodynamic shear (kg/m) at any depth. The moment of this force about the centre of gravity of a section at a depth y from the water surface is given by the equation:
$M_h = 0.299\ Py^2$, where M_h = moment (kg m/m).

Effect on Uplift Forces. It is determined as a function of the hydrostatic pressure of reservoir and tail water against the faces of dam. The water pressure is changed but the change is not considered effective in producing a corresponding increase or decrease in the uplift force. The duration of earthquake is too short to permit development of pore pressure in the concrete or rock foundations.

Effect on Silt Load. The increase in silt pressure due to earthquake is determined by considering hydrodynamic forces on the water up to the base of the dam and ignoring the weight of silt.

Effect on Vertical Component of Reservoir and Tail Water Load. The hydrodynamic pressure as it acts normal to the face of dam has a vertical component of this force if the dam face against which it is acting is sloping; the magnitude at any horizontal section is:

$$W_h = (V_2 - V_1) \tan \theta$$

where, W_h is increase (or decrease) in vertical component of load (kg) due to hydrodynamic force, V_2 is total shear (kg) due to horizontal component of hydrodynamic force at the elevation of the section under consideration, V_1 is total shear (kg) due to horizontal component of hydrodynamic force at the elevation at which the slope of the dam face commences and θ is angle between face of dam and vertical.

Moment. Moment due to vertical component of reservoir and tail water load is obtained by determining the lever arm from the centroid of the pressure diagram.

2. Vertical Acceleration. The effect of vertical earthquake acceleration is to change unit weight of water and concrete or masonry of the dam. The vertical seismic coefficient is assumed as a proportion of the horizontal seismic coefficient at any point in accordance with seismic and geological conditions.

Vertical acceleration may act either downward or upwards.

Acting downward. When acceleration is acting downward, the foundation tends to move downward from the body of the dam, thereby reducing the effective weight and stability of the dam. It represents worst condition for design purposes. The inertia force, $F = \alpha W$ acting upwards decreasing it momentarily. The downward weight is reduced weight of dam material = $W_c\,(1 - \alpha)$ and reduced weight of water = $W\,(1 - \alpha)$.

Acting upwards. When the acceleration is acting vertically upwards, it tends to lift the foundation upwards, *i.e.*, closer towards the dam body thereby increasing effective weight of dam and developed stress is increased. Increased weight of dam material = $W_c\,(1 + \alpha)$ and increased weight of water = $W\,(1 + \alpha)$.

The vertical earthquake acceleration is taken as 0.75 times the αh at top and reducing linearly to zero at the base in the both seismic coefficient method and response spectrum method of design.

Earth and Silt Pressures: Earth pressures have a minor effect on the stability of the dam and are ignored. Silt gets deposited against the upstream face of the dam. Dam is thus subjected to silt pressure in addition to the water pressure. Silt is treated as saturated cohesionless soil having full uplift and whose value of internal friction is not materially changed on account of submergence. Silt pressure and water pressure exist together in a submerged fill and silt pressure on the dam is reduced in the proportion that the weight of the fill is reduced by submergence. The horizontal silt pressure

$P_s = W_s \dfrac{h^2}{2}\left(\dfrac{1 - \sin \beta}{1 + \sin \beta}\right)$ is assumed to be equivalent to that of a fluid weighing 1360 kg/m^3. Vertical silt and water pressure is determined as if silt and water together have a density of 1925 kg/m^3. β is the angle of internal friction taken as 30° for sand, gravel, clay and silt, W_s is unit weight of silt or earth submerged in water, and h is the height of silt submerged in water above the base of dam.

Wind Pressure: Wind loads on dams are ignored.

Ice Pressure: Ice pressure is not encountered in the design of dams in India. Ice if formed on the water surface of the reservoir expands and contracts with change in temperature. The dam face is subjected to force due to expansion of ice. Ice pressure is taken as 25000 kg/m^2 applied to the face of dam over the anticipated area of contact of ice with the face of the dam as per IS:6512-1984.

Wave Pressure: Upper portions of the dam are subjected to impact of waves. Wind pressure depends on the height of wave. The wave pressure diagram is approximated by triangle 1-2-3 in Fig.17.7. The total wave force P_w is given by the area of triangle or $P_w = 2000\ hw^2$. The centre of application is at a height of 0.375 hw above the still water level. *T* Saville method recommended by IS : 6512–1984 for the computation of wave height is discussed in section 17.4 (Page 484).

(*i*) Maximum unit pressure, $P_w = 2400\ h_w$ kg/m^2 acting at 0.125 h_w above the still water level.

(*ii*) Total wave force, $P_w = 2400H_w \times 1/2 \times 5/3\ H_w, = 2000\ H^2{}_w$ acting at 0.375 h_w above the still water level assuming pressure distribution is triangular, and height equal to $5/3H_w$.

(*iii*) When a maximum wind velocity is known, the same is used for FRL condtion and 2/3 times for MWL condition.

Thermal Loads: Stresses in the dams are affected due to temperature variation in the dam body. The cyclic variation of air temperature and solar radiation on the downstream side and reservoir temperature on the upstream side also affect the stresses in the dam. Even deflection of the dam is maximum in morning and goes on reducing to a minimum value in the evening. Magnitude of deflection is generally less when spillway is working than when it is not working. While considering the thermal load, temperature gradients are assumed depending on location, orientation, surrounding topography etc.

Reaction of Foundation: Some water under pressure percolates between the dam and foundation, through horizontally bedded joints under the foundation and/or through pour joints in the dam. Part weight of the dam is supported by water and the direct foundation reaction is correspondingly reduced which effect is termed as uplift. The resultant of all horizontal and vertical forces including uplift must be balanced by an equal and opposite reaction at the foundation consisting of the total vertical reaction and the total horizontal shear and friction. Further, for the dam to be in static equilibrium, the force is so located that summation of moments is zero.

The greatest load, out of the various loads mentioned above, is the weight of the gravity dam as it is the only active force that resists failure of the dam. The safety of the dam is roughly proportional to the density of concrete or masonry it is made of.

18.4 COMBINATION OF FORCES FOR DESIGN

A dam is designed for the most adverse combination of load conditions as have reasonable probability of simultaneous occurrence. Such load combinations are : (*i*) Load Combination A – Construction condition, dam completed but reservoir empty and no tail water, (*ii*) Load Combination B – Normal operating condition with reservoir full (elevation) and normal dry weather tail water, normal uplift, ice and silt (if applicable), (*iii*) Load Combination C – Flood discharge condition. Reservoir at maximum flood level, all gates open, tail water at flood elevation, normal uplift and silt (if applicable), (*iv*) Load Combination D-Combination A, with earthquake, (*v*) Load Combination E – Combination B, with earthquake but no ice, (*vi*) Load Combination F – Combination C, but with extreme uplift (drains inoperative), (*vii*) Load Combination G – Combination E, but with extreme uplift (drains inoperative).

18.5 MODES OF FAILURE — SAFETY CRITERIA

A dam has to be stable against overturning, sliding, compression or crushing, and tension. Criteria for structural stability of a gravity dam in this context is as under:

Safety against Overturning: The dam has to be safe against overturning at any plane within the dam, at the base, or at any plane below the base. The overturning of a dam may take place if the resultant of all the forces acting on the dam passes outside the base. Practically before a dam overturns bodily, other failures such as crushing of toe material, cracking of upstream material and increase in uplift and sliding may occur. A gravity dam is thus considered safe against overturning if the criteria of no tension on the upstream face, the resistance against sliding as well as the quality of concrete/masonry of the dam and foundation is satisfied assuming the dam and foundation as a continuous body. A dam safe for the criteria of maximum compressive stress and sliding, usually has a factor of safety of 1.5 to 2.5 against overturning. The factor of safety is defined as the ratio of stabilizing moments to the overturning moments about the toe of the dam. Earthquake forces are not considered as contributing to overturning tendency.

Safety against Sliding: The horizontal loads including horizontal components of the loads acting on a dam are resisted by frictional or shearing forces along horizontal or nearly horizontal planes in the body of a dam, on the foundation or on horizontal or nearly horizontal seams in the foundation. It follows that the total magnitude of the forces tending to induce sliding shall be less than the minimum total available resistance along the critical path of sliding, *i.e.*, along that plane or combination of planes which mobilizes the least resistance to sliding. The sliding resistance is a function of the cohesion inherent in the materials and at their contact and the angle of internal friction of the material at the surface of sliding. The stability of a dam against sliding is evaluated by comparing the minimum total available resistance along the critical path of sliding (that is along that plane or combination of planes which mobilises the least resistance to sliding) to the total magnitude of the forces tending to induce sliding.

The factor of safety against sliding (F) is calculated, as under, on the basis of partial factor of safety in respect of friction ($F\phi$) and partial factor of safety in respect of cohesion (F_c) as per Table. 18.2.

Table 18·2 Partial factors of safety against sliding

Sl. No.	*Loading condition*	$F\phi$	F_c		
			For dams and the contact plane with foundation	*For foundation*	
				Thoroughly investigated	*Others*
(*i*)	A, B, C	1.5	3.6	4.0	4.5
(*ii*)	D, E	1.2	2.4	2.7	3.0
(*iii*)	F, G	1.0	1.2	1.35	1.5

$$F = \frac{1}{P}\left[\frac{(w-u)\tan\phi}{F\phi} + \frac{CA}{F_c}\right]$$

where, F = factor of safety against sliding, minimum 1.0, w = total mass of dam, u = total uplift force, tan ϕ = coefficient of internal friction of the material, C = cohesion of the material at the plane considered, A = area under consideration for cohesion, $F\phi$ = partial factor of safety in respect of friction, F_c = partial factor of safety in respect of cohesion, and P = total horizontal force.

Safety against Crushing: Safety against crushing is ensured if the compressive stresses produced are within the allowable stresses.

$$\text{Normal stress at toe} = \frac{W}{B}\left(1+\frac{6e}{B}\right)$$

$$\text{Normal stress at heel} = \frac{W}{B}\left(1-\frac{6e}{B}\right)$$

When pressures at toe and heel are both compressive, (Fig. 18.11) the maximum compressive stress occurs at the toe (when eccentricity) $e = B/6 = \frac{2W}{B}$ *i.e.* $\frac{W}{B}\left(1 + \frac{6}{B} \times \frac{B}{6}\right)$, and the corresponding compressive stress at heel will be zero. This maximum compressive stress shall be less than the allowable compressive stress. The excessive stresses at toe and heel can be brought within permissible limits by providing fillets sloping at 1 : 1 at the upstream and 2 : 1 at the downstream. The height of fillets is given by the relation:

$$h_{heel} = 6H - 0.658\,H^2 - 1.07$$

$$h_{toe} = 6.5H - 1.1H^2 - 0.9$$

where h = fillet height (m), and H = dam height (in 100 m)

The strength of concrete is to satisfy early load and construction requirements and at the age of one year it should be 4 times the maximum computed stress in the dam or 140 kg/cm^2, whichever is more. The allowable working compressive stress in any part of the dam is usually taken as 70 kg/cm^2 for concrete.

Safety against Tension: A gravity dam is usually designed so that no tension is developed in the dam body and for that the resultant must lie within the middle third. IS: 6512-1984 recommends that no tensile stress occurs at the upstream face of the dam for load combination B. Normal tensile stresses, however, may be permitted in other load conditions but not exceeding values given in Table 18.3. In the case of high dams, tensile stress up to a maximum of 5 kg/cm^2 is sometimes permitted for reasons of economy and keeping in view that such worst loading condition may occur rarely and for a very short duration.

Table 18.3 Permissible tensile stress

Load combination	*Permissible tensile stress*	
	Concrete	*Masonry*
C	0.01 *fc*	0.005 *fc*
E	0.02 *fc*	0.01 *fc*
F	0.02 *fc*	0.01 *fc*
G	0.04 *fc*	0.02 *fc*

fc = cube compressive strength of concrete/mortar for masonry

18.6 PRINCIPAL AND SHEAR STRESSES

The principal and shear stresses are determined by considering an elementary triangular section at either the heel or toe of the dam section such that stress intensities may be assumed to be uniform on its face. With reservoir full, the stresses at the toe are greater and as such stress analysis is done for the downstream face. Consider a triangular section *ABC*. (Fig. 18.9), *AB* and *AC* are the principal planes, mutually at right angles, plane *AB* will also have only a normal stress on it, and will be the other principal plane. Let the small lengths *AB*, *BC* and *AC* be d_r, d_y and d_s respectively. Let p = intensity of water pressure acting on face *AC*, σ = principal stress on plane *AB*, τ = shear stress along face *BC*, P_n = normal stress on face *BC*.

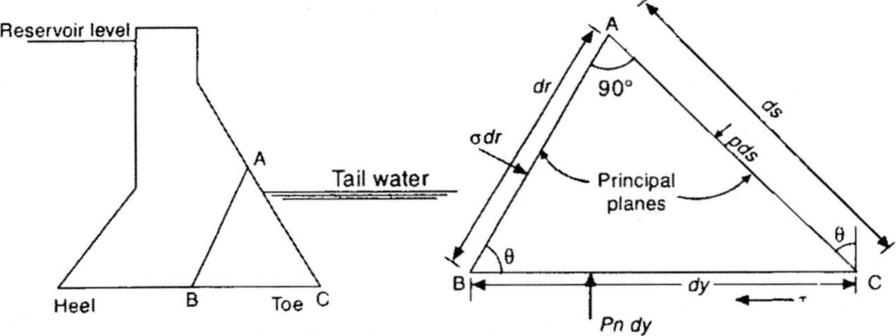

Fig. 18.9 Principal and shear stresses.

Consider unit length of the dam, the normal forces on the planes AB, BC and CA are respectively, σd_r, $p_n\, d_y$ and $p.ds$. Resolving all the forces in the vertical direction, we get :

$$p_n\, d_y = p.ds.\sin\theta + \sigma\, dr\cos\theta$$

But $$d_r = d_y\cos\theta \text{ and } d_s = d_y\sin\theta$$

$\therefore$ $$p_n\, d_y = p.d_y.\sin^2\theta + \sigma\, d_y\cos^2\theta$$

or $$p_n = p\sin^2\theta + \sigma\cos^2\theta$$

Hence $$\sigma = (p_n - p\sin^2\theta)/\cos^2\theta$$

or $$\sigma = p_n\sec^2\theta - p\tan^2\theta \qquad \ldots(i)$$

The principal stress relationship given by equation (*i*) is applicable to both upstream and downstream faces of the dam. However, the worst condition for the downstream face will be when there is no tail water, and hence p will be zero. In such case, the principal stress σ_1 is given by

$$\sigma_1 = p_n\sec^2\theta$$

Considering hydrodynamic pressure intensity P_e of tail water due to an earthquake, the principal stress at downstream face is given by

$$\sigma = P_n\sec^2\theta - (P - P_e)\tan^2\theta$$

Similarly considering hydrodynamic force to u/s, principal stress at d/s face is given by

$$\sigma = P_n\sec^2\theta - (P + P_e)\tan^2\theta$$

However, for u/s face σ_1 is always less than P. Hence σ is minor principal stress and P is major principal stress for the upstream side.

Shear Stress: Resolving all the forces in horizontal direction, we get:

$\tau d_y = \sigma\, d_r\sin\theta - p\, ds\cos\theta$

$\tau = \sigma\sin\theta\, d_r/d_y - p\, d_s/d_y\cos\theta$

Substituting $\sin\theta$ for d_s/d_y and $\cos\theta$ for d_r/d_y

$\tau = (\sigma - p)\sin\theta\cos\theta$

Substituting σ from Eqn. (*i*), we get

$\tau = (P_n\sec^2\theta - p\sec^2\theta)\cos\theta\sin\theta$

or $\tau = (P_n - p)\tan\theta.$ (*ii*)

The equation (*ii*) gives shear stress at the *d/s* face only. For u/s side, the magnitude of τ remains the same but the direction is reversed. Hence for upstream, $\tau = -(P_n - p)\tan\theta$ ...(*iii*)

The shear stress at downstream face, if tailwater is neglected, is given by

$$\tau = P_n\tan\theta$$

Considering the hydrodynamic pressure due to earthquake, the shear stress of *d/s* is given by

$$\tau = [P_n - (P - p'_e)\tan\theta]$$

Similarly, shear stress for u/s face $= \tau = -[P_n - (P + p_e)]\tan\theta$.

18.7 STABILITY ANALYSIS

Various methods for stability analysis of a gravity dam are discussed as under:

1. Gravity Method or Two Dimensional Method: It is an approximate but simple method. Preliminary analysis of the gravity dams are made by this method and final precise design is carried out by one of the other methods.

In this method, the dam is considered to be made up of a number of parallel vertical cantilevers of unit width, each of which acts independent of the adjoining cantilevers. The dam transfers the load to the foundation by cantilever action. The loads are resisted entirely by the weight of individual cantilevers of unit length.

The various assumptions made in this method are (*i*) The dam is made up of a number of cantilevers, each 1 m thick, acting independently of the others. However, this assumption is very nearly true only in case the dam is constructed in a wide U-shaped valley and transverse joints are not keyed or grouted, (*ii*) No load is transferred to abutments by beam action, (*iii*) Foundation and abutments behave as a single unit, the joint being perfect, (*iv*) No movements of foundation are caused due to transference of loads, (*v*) Materials in the foundation and body of the dam are isotropic and homogeneous, (*vi*) Stresses developed in the foundation and body of the dam are within elastic limits, and (*vii*) The general distribution of stresses is not affected by small opening made in the dam body and they only produce local effects.

(*i*) *Graphical Method:* The procedure adopted is (*i*) Entire dam section is divided into a number of horizontal sections at some suitable intervals (Fig. 18.15) such as where the slope changes, (*ii*) For each section the sum of all horizontal forces (ΣH) and all vertical forces (ΣV) acting above that section are determined and their line of action is drawn graphically, (*iii*) The resultant force R is determined and its line of action is located graphically, (*iv*) The above process, *i.e.*, steps (*i*) to (*iii*), is done for each section, and (*v*) A line joining all the points where the individual resultants cut the individual sections is drawn. This line represents the resultant force and shall lie within the middle third so that no tension develops. The procedure is adopted for reservoir full and reservoir empty conditions. The lines of resultant obtained in both the cases shall lie within the middle third so that no tension develops in the body of dam.

(*i*) *Analytical Method:* The procedure adopted is (*i*) Consider unit length of the dam, (*ii*) Calculate all the vertical forces acting on the dam, viz., self weight of the dam, weight of water acting on the inclined face, uplift pressure and inertia forces due to vertical acceleration. Compute their algebraic sum (ΣV), (*iii*) Calculate all the horizontal forces acting on the dam and their algebraic sum (ΣH) and the horizontal pressure due to hydrodynamic pressure, (*iv*) Calculate the lever arm of all these forces from the toe, (*v*) Calculate the sum of overturning moments (ΣM_o) and the sum of righting momehts (ΣM_R) at the toe, (*vi*) Calculate the algebraic sum of all those moments ($\Sigma M = \Sigma M_R - M_o$), (*vii*) Calculate the location of resultant force by determining its distance from the toe, $\bar{x} = \dfrac{\Sigma M}{\Sigma V}$, (*viii*) Calculate the eccentricity e of the resultant R from the centre of the base width b, *i.e.* $(b/2 - \bar{x})$. It must be $< b/6$ for no tension to develop, (*ix*) Calculate the normal stress at the toe $P_n = \dfrac{\Sigma V}{b}\left(1 + \dfrac{6e}{b}\right)$, (*x*) Calculate the normal stress at the heel $P_n' = \dfrac{\Sigma V}{b}\left(1 - \dfrac{6e}{b}\right)$. Normal compressive stresses are sometimes calculated ignoring uplift, (*xi*) Calculate the principal stress (σ) and shear stresses (τ) at the toe and heel.

$\sigma_{toe} = P_n \sec^2\theta - (P - P_e)\tan^2\theta$ considering hydrodynamic pressure (ρ_e) exterted by tail water during earthquake.

$\sigma_{heel} = P_n \sec^2\theta - (P + P_e)\tan^2\theta$

$\tau_{toe} = (P_n - P)\tan\theta$

$\tau_{heel} = -(P_n - P)\tan\theta$

$\tau_{toe} = P_n \tan\theta$, if tail water is neglected

$\tau_{toe} = [P_n - (P - P_e)]\tan\theta$, considering hydrodynamic pressure due to earthquake

$\tau_{heel} = -[P - (P + P_e)]\tan\theta$, considering hydrodynamic pressure due to earthquake

(*xii*) Calculate factor of safety against sliding $= \mu \dfrac{\Sigma V}{\Sigma H}$ it must be greater than unity, (*xiii*) Calculate factor of safety against overturning $= \dfrac{\Sigma M_B}{\Sigma M_b}$, (*xiv*) Calculate shear friction factor $= \dfrac{(\mu \Sigma V + bq)}{\Sigma H}$, where, ΣV is the net vertical force, inclusive of uplift. It must be greater than 3 but less than 5. The procedure is followed for reservoir full and reservoir empty conditions.

2. Trial Load Twist Method: In this method, the dam is divided into a number of vertical cantilever elements and horizontal beam elements. It is assumed that the horizontal elements of the dam behave as a beam and transfer the water pressure acting upon it to the abutments by beam action. Since the gorge across which the dam is built has both sides rising steeply, height of the dam and hence successive cantilever elements go on reducing in height from the centre towards the sides and their deflections also go on reducing. Each cantilever element is thus dragged on its one face to greater deflections by deeper adjacent cantilevers on the valley side and is held on by the shorter cantilever on the hill side, thereby developing a twist effect on each cantilever. The water thrust is shared by vertical cantilevers and horizontal beams such that the deflections at the common points on the dam faces are equal. More precisely (*i*) The vertical cantilever elements resist the external load by gravity action through bending and shear along horizontal planes, (*ii*) The horizontal beam elements resist bending, and (*iii*) The twisting effect resists twisting moments caused by shearing of horizontal elements and irregular rotation of vertical elements. The portion of the pressures taken by each element is determined by so dividing the load distribution between them that the deflection of a common point, considered on any element, is the same.

3. Slab Analogy Method: The method is based on the mathematical analogy between distribution of stresses in a dam section and deflection of a slab, similar in shape as a cantiliver section including a large block of foundation, by forces and moments around its edges which are bent into a form which corresponds to the stresses at dam surface. The dam is considered as a slab and analysis is similar to that of the calculation of stresses in bridge slabs. It is a cumbersome method.

4. Lattice Analogy Method: In this method, the dam is considered to be made of diagonally braced square frames. The solution involves adjustment of the lattice by a systematic relaxation of restraints at the joints. With the completion of the adjustment of the lattice to remove restraint, the strains are computed from the relative displacements between successive joints and there from stresses are determined.

5. Experimental Methods: 1. *Direct Method or Three Dimensional Models.* In this method, geometrically similar three dimensional models are made of elastic material to proportionate size and loading of the prototype. The models are located similar to the prototype and are acted upon by various pressures proportionate to the prototype. The high water pressure acting on the prototype is simulated by a liquid of high specific gravity such as mercury or the effect is obtained by the jetting action of water. The deformations, structural action, and stress conditions in the prototype are predicted by measurements of displacements of the model. The method is useful for the analysis of high dams.

2. *Indirect Method or Photoelastic Models.* In this method, model is made of elastic, transparent, isotropic materials like bakelite, celluloid and glass which are free from initial or residual stresses. Stresses in the photoelastic model are determined by means of visible optical effects produced by passing polarized light through the model while it is under load. The instrument used for making visible the effects of stresses in the model is known as photoelastic polariscope. The studies employing photoelastic models are limited to conditions of plain stress or strain and for determining regions of stress concentration.

18.8 ELEMENTARY PROFILE OF A GRAVITY DAM

The elementary profile of a gravity dam subjected only to the water pressure on the upstream face is a right angle triangle having zero width at the top, *i.e.*, water level where water pressure is zero and a maximum base width (B) at the bottom where the water pressure is maximum. Thus the elementary profile of the dam is of the same shape as that of the hydrostatic pressure distribution diagram (Fig. 18.10 (*a*)).

When the reservoir is empty, the only force acting is self weight (W) of the dam acting at a distance $B/3$ from the heel. It is the maximum possible inner-most position of the resultant so that no tension develops and provides the maximum possible stabilizing force against overturning without causing tension at toe under empty dam condition. If any triangular profile other than the right angled one is provided, its weight will act closer to the upstream face to provide a higher stabilizing force but will cause tension to develop at the toe.

Vertical stress distribution.

Forces acting on the dam are:

(*i*) Weight of dam, $W = \frac{1}{2} BH \rho w$

where, ρ = unit weight of dam material (2400 kg/m^3 of concrete), and

w = unit weight of water (1000 kg/m^3)

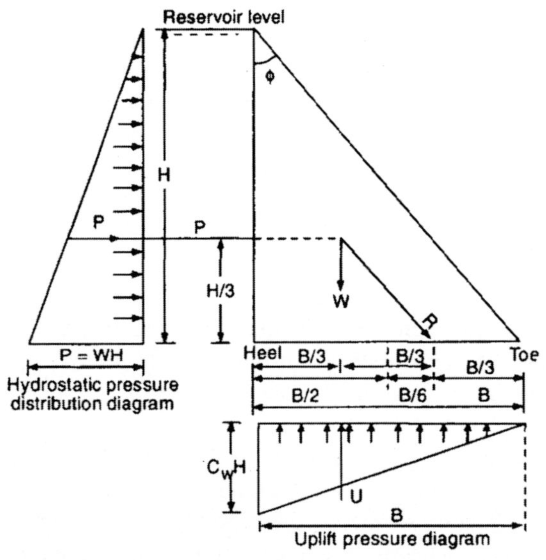

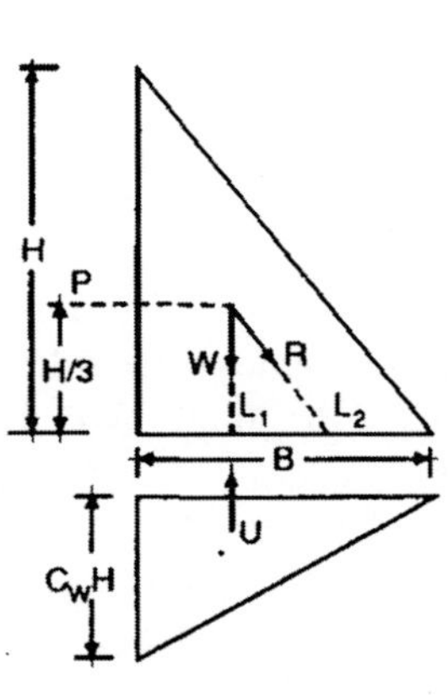

Fig. 18.10. (*a*). Hydrostatic pressure distribution diagram. (*b*). Elementary profile.

(*ii*) Water pressure, $P = \frac{1}{2} wH^2$, acting at $H/3$ from the base

(*iii*) Uplift pressure, $U = \frac{1}{2} cwBH$

where, c = coefficient of uplift pressure intensity, its value is *1*.

Normal stress at toe $P_{max} = \frac{W}{B}\left(1 + \frac{6e}{B}\right)$, without considering uplift pressure.

$$P_{max} = \frac{(W - u)}{B}\left(1 + \frac{6e}{B}\right)$$, with uplift pressure duly considered

Normal stress at heel $P_{min} = \frac{W}{B}\left(1 - \frac{6e}{B}\right)$, without uplift pressure

$$P_{min} = \frac{(W - u)}{B}\left(1 - \frac{6e}{B}\right)$$, with uplift pressure duly considered.

Eccentricity, $e \le \frac{B}{6}$.

When reservoir is empty

The maximum compressive stress at the heel is equal to $2W/B$ which is derived by putting $e = B/6$ in the relation

$$P_{max} = \frac{W}{B}\left(1 + \frac{6e}{B}\right),$$

and corresponding normal stress at the toe is zero. The resultant should act at inner third ($B/3$ from the heel). Fig. 18.11 shows the normal stress distribution at the base.

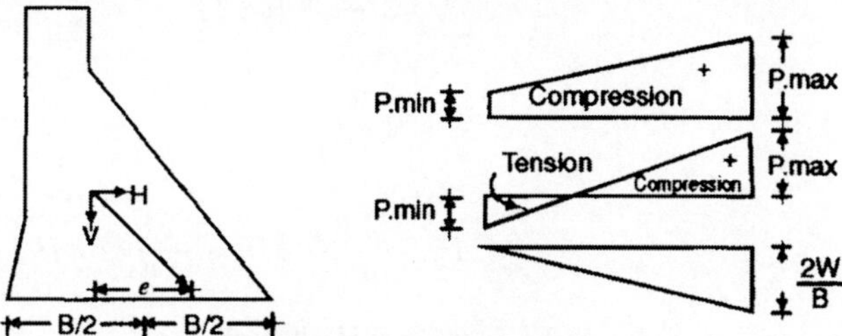

Fig. 18.11 Stress distribution

When the reservoir is full: For no tension to develop (*a*) The resultant should pass at outer third point. Taking moments about L_2 and equating it to zero (because moment of R about L_2 is zero), we get

$$P\frac{H}{3} + U\frac{B}{3} - W\frac{B}{3} = R \times 0 \text{ (refer Fig. 18.10b)}$$

or $$\frac{wH^2}{2}\frac{H}{3} + \frac{1}{2}cwBH\frac{B}{3} - \frac{1}{2}BH\rho w\frac{B}{3} = 0$$

Multiplying both sides by $6/wH$, we get

$$H^2 + cB^2 - B^2\rho = 0$$

$$B^2(\rho - c) = H^2$$

$$B = \frac{H}{\sqrt{\rho - c}}$$

Also $B = H/\sqrt{\rho}$, when uplift pressure intensity (c) is not considered.

(*b*) For stability against sliding, the horizontal forces causing sliding ($\Sigma H = P$) are balanced by the frictional forces [$\mu\Sigma V$ or $\mu(W - \mu)$] opposing it. Hence

$$P = \mu(W - U)$$

or $$\frac{1}{2}wH^2 = \mu\left(\frac{1}{2}BH\rho w - \frac{1}{2}cBwH\right)$$

or $$B = \frac{H}{\mu(\rho - c)}$$

or $$B \ge \frac{H}{\mu(\rho - c)}$$

Also, $$B = \frac{H}{\mu\rho}$$ when uplift is not considered.

Base width B should be maximum of the two values given by the above two conditions.

Normal stress, $$P_n = \frac{(W - u)}{B}\left(1 + \frac{6e}{B}\right), \text{ at the toe}$$

Under limiting condition $e = \frac{B}{6}$, therefore

$$P_n = \frac{W - u}{B}(1 + 1)$$

$$= 2\frac{W - u}{B}$$

or $$P_n = \frac{2}{B}\left(\frac{1}{2}BH\rho w - \frac{1}{2}cBwH\right)$$

or $$P_n = wH(\rho - c)$$

$$P_n = wH(1 - 1), \text{ at heel}$$
$$= 0$$

Principal stress near the toe (σ) is the maximum normal stress in the dam, that is,

$$\sigma_1 = P_n \sec^2\theta - P' \tan^2\theta$$

or $\sigma_1 = P_n \sec^2\theta$, when there is no tail water ($P' = 0$)

$$\sigma_1 = wH(\rho - c)\left[\left(\frac{B}{H}\right)^2 + 1\right]$$

But, $$B = \frac{H}{\sqrt{\rho - c}} \text{ or } B^2 = \frac{H^2}{\rho - c} \text{ or } \frac{B^2}{H^2} = \frac{1}{\rho - c}$$

or $$\sigma_1 = wH(\rho - c)\left[\frac{1}{\rho - c} + 1\right]$$

or $$\sigma_1 = wH(\rho - c + 1)$$

Shear stress (τ) at a horizontal plane near the toe,

$$\tau = P_n \tan\theta$$

$$\tau = wH(\rho - c)\tan\theta$$

$$\tau = wH(\rho - c)\left(\frac{B}{H}\right)$$

But, $$B = \frac{H}{\sqrt{\rho - c}} \text{ or } \frac{B}{H} = \frac{1}{\sqrt{\rho - c}}$$

$\therefore$ $$\tau = wH(\rho - c)\frac{1}{\sqrt{\rho - c}}$$

$$\tau = wH\sqrt{\rho - c}$$

The principal stress and shear stress at the heel are zero since normal stress at the heel is zero.

Low and High Gravity Dams

Low Dam. A low dam is of limiting height such that the resultant of all forces passes through the middle third and the maximum compressive stress at the toe does not exceed the permissible limit, *i.e.*,

$$\sigma_1 = wH\left(\rho - c + 1\right) = f$$

or limiting height,

$$H = \frac{f}{w(\rho - c + 1)}$$

or

$$H = \frac{f}{w(\rho + 1)}, \text{ ignoring uplift to be on the safe side.}$$

The limiting height, for the usual stress of dam material, viz. $w = 1000$ kg/m^3 $\rho = 2400$ kg/m^3 and $f = 30$ kg/cm^2 (or 300 tonne/m^2), gives

$$H = \frac{300}{1.0\,(2.4 + 1)} \approx 88 \text{ m}$$

High Dam: A dam the height of which exceeds the limiting height of low dam is termed as high dam. In a high dam the maximum permissible compressive stress is exceeded if the resultant of all forces were to pass through the middle third; to avoid excessive stresses the resultant is maintained still near the centre of the base for which purpose the downstream slope is flattened and the upstream slope is also provided with a batter. In this case, the resultant of water pressure and self weight of the dam passes out of the middle third at certain sections in the dam body indicative of development of tension. The dam profile has, therefore, to be adjusted by adding extra slope or batter on the upstream and downstream sides to limit the compressive stress within the permissible limit (Fig. 18.13).

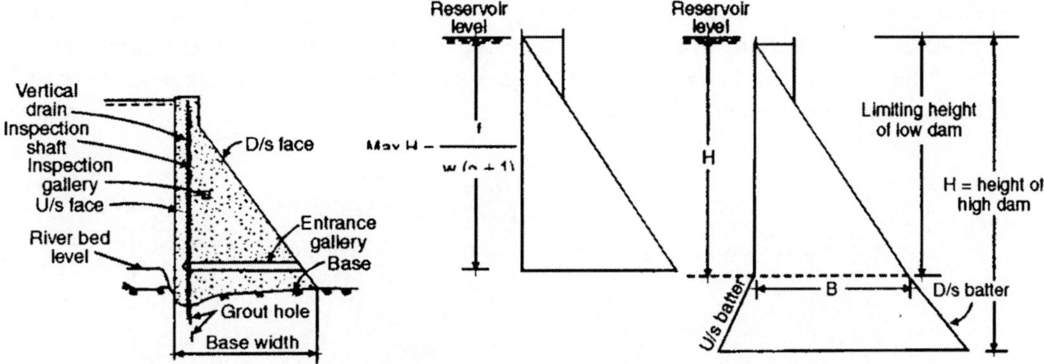

Fig. 18.12. Gravity dam of triangular section.

Fig. 18.13. Low dam and high dam.

Practical Profile of a Dam

The elementary profile of a dam composed of a triangle with apex at the maximum water surface is only a theoretical one and has to be modified from practical considerations; (*a*) Dam has to be raised above the maximum reservoir level to provide suitable free board to guard against overtopping by wave action, (*b*) Top of dam has to be of a certain width, instead of zero width of the elementary triangular profile, to provide suitable roadway.

The modifications due to (*a*) and (*b*) above cause the resultant to shift towards the heel crossing the inner middle third (under empty dam condition) indicative of development of tension at the toe. Batter has, therefore, to be provided on the upstream face to avoid development of tension. Figs. 18.12 and 18.13 show the profile of low dam and high dam respectively modified from practical considerations. Various considerations in fixing the parameters of the dam section are as under:

1. *Top Width.* The top width of the dam is generally dictated by the requirement of roadway to be provided. The most economical top width is 14% of the dam height. It is also taken equal to 0.55 $H^{1/2}$, where H is the maximum water depth. Usually the width provided varies from 6 to 10 m.

2. *Free Board.* The upstream parapet of the roadway above the top of dam is usually a solid wall but does not form part of free board. The free board provided is maximum of (*i*) $1.33h_w$, (refer example 18.3), (*ii*) 3 to 4%, usually 5% of the dam height, whichever is more.

3. *Base Width.* The base width of the dam shall be safe against overturning, sliding, and no tension in dam body. The base width (B) is determined from the equation, $B = H/\sqrt{(\rho - c)}$. The upstream face is kept vertical up to a height H_1 to be determined by trial. Approximately H_1 is equal to $2a\sqrt{(\rho - c)}$, where a is top width of the dam. Below height H_1 batter is provided to upstream and downstream faces to increase the base width with a view to ensure that no tension develops in the body of the dam. For this, the resultant force shall be located within the inner middle third and outer middle third for reservoir in empty and full conditions respectively.

18.9 DESIGN OF GRAVITY DAM

The design of a dam has to be economical and at the same time satisfying various requirements of stability. The first step in the design of gravity dam is to decide whether it is low dam or high dam. If the height (H) is within $f/w(\rho + 1)$, it is a low dam and is designed as such. Usual dimensions of a low dam are given in Fig. 18.14. For the height of dam exceeding H, it is high dam and is designed as follows:

1. Step by Step Method: The economical design of a gravity dam by this method envisages division of the dam section in various zones (Fig. 18.15). The successive zones are designed to meet the requirements of stability.

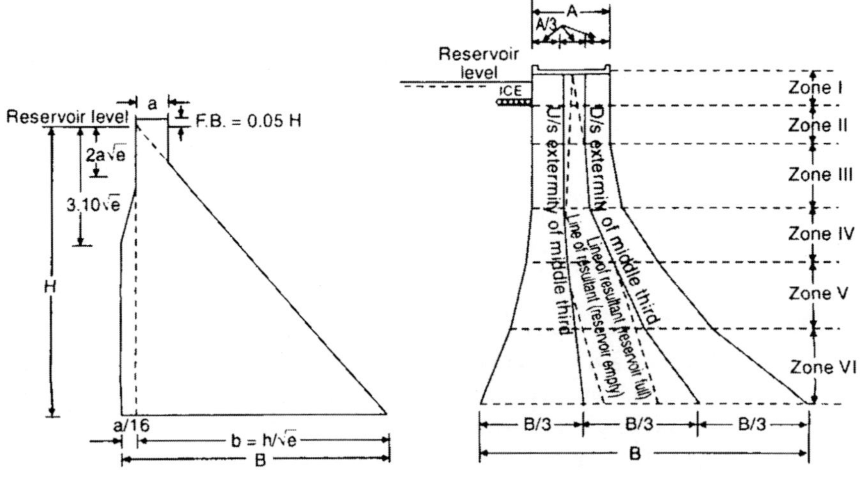

Fig. 18.14. Practical profile of gravity dam.

Fig. 18.15. Zoning of dam.

Zone I. It is the top portion of the dam above maximum reservoir level or above the bottom of ice sheet, if there is ice. If ice pressure exists, the section is checked for sliding. Normally the height of the section is equal to free board provided at the dam. The width is determined by practical consideration or economy for the section as a whole. The upstream and downstream faces are vertical.

Zone II. The limiting height for the section is the horizontal plane at which (*i*) the resultant cuts the inner middle third when the reservoir is empty and (*ii*) the resultant cuts the outer third under full reservoir condition. The width is generally more than required and as such both upstream and downstream faces are kept vertical.

Zone III. The limiting height for this section is the plane at which the resultant under full reservoir condition cuts the outer third. The resultant under empty reservoir condition remains within the inner middle third and as such the upstream face continues to be vertical.

Zone IV. The upstream face also begins to slope in this section to limit the resultant under empty reservoir condition within the inner middle third. The limiting height for the section is determined by the position of the resultant under full reservoir condition so as to limit the stress at downstream toe within the maximum permissible stress.

Zone V. Usually the maximum permissible stress is reached at the downstream face first and as such the downstream face of the dam is flattened so as to limit the stress at downstream toe within the maximum permissible stress. In this section, the resultant under full reservoir condition intersects well within the middle third while the resultant under empty reservoir condition intersects at the upstream extremity of the middle third.

Zone VI. In this section, there are maximum pressures at both upstream and downstream faces under reservoir empty and reservoir full conditions respectively. The limiting height for the section is the plane at which the inclined pressure at toe just reaches its maximum value. The resultant under reservoir empty and reservoir full condition remains well within the middle third.

Zone VII. With increase in the height of dam, batters of both upstream and downstream faces increase, but at a certain plane the maximum compressive stress at the toe exceeds the working limit. It is desirable to eliminate this section by revising the entire design. Thus the height of the dam is limiteo to zone VI by effecting changes in the upper zones or by using superior material in zone VI. Fig. 18.16 shows the zoning of masonry in Nagarjunasagar dam.

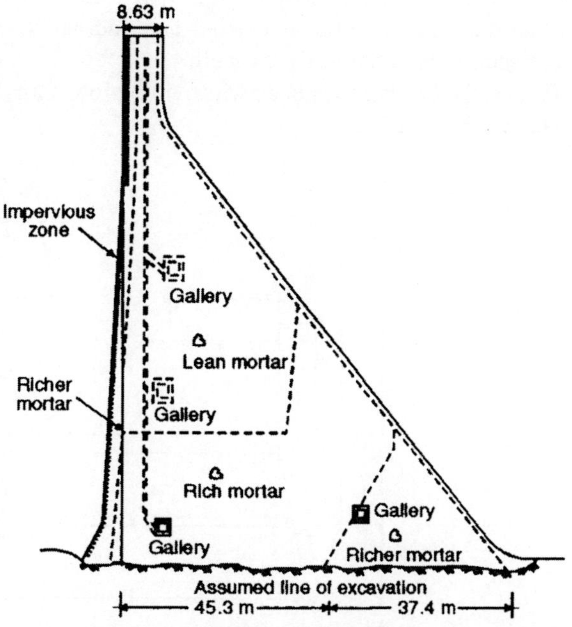

Fig. 18.16 Zoning of masonry in Nagarjunasagar dam.

2. Single Step Method: The step by step method is eminently suitable for low dams as it effects substantial saving in the material. Beyond zone IV, *i.e.*, high dam, it is observed that the upstream and downstream slopes work out to be of unusual shape, *i.e.*, upstream batter is of steep slope while the downstream slope is convex outwards necessitating revision of the design. In redesigning a single step method, as given below, is adopted :

(*i*) The dam in top certain height is kept vertical. This height H_1 is determined by trial. Approximately H_1 is equal to $2a\sqrt{(\rho - c)}$, where a is the top width of the dam, (*ii*) Beyond this

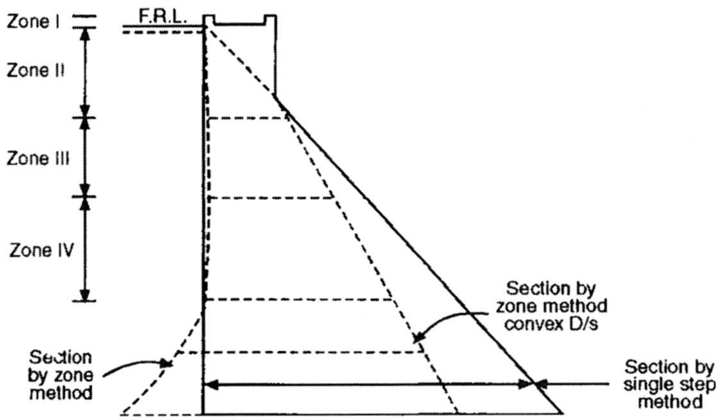

Fig. 18.17 Comparison of dam profiles by single step and zone methods.

height, both upstream and downstream faces are given suitable batter so that maximum inclined pressure at the two faces reach their maximum values simultaneously under empty reservoir and full reservoir conditions, and (*iii*) The section is checked for stability requirements. In this method, the dam section is understressed except at the base.

3. Strips Method: In this method the dam below the limiting height of low dam is divided into a number of suitable strips (Fig. 18.18). Each strip is designed to determine its dimensions such that the maximum normal stress or principal stress does not exceed the allowable value (f) and the section is economical as well.

First strip. Let B be the base width of the low dam. Base width B_1 of the first strip of high dam is given by the relation:

$$B_1 = \sqrt{\frac{wH_1^3}{f}\left(1 + \frac{w^2 H_1^4}{4W_1^2}\right)}$$

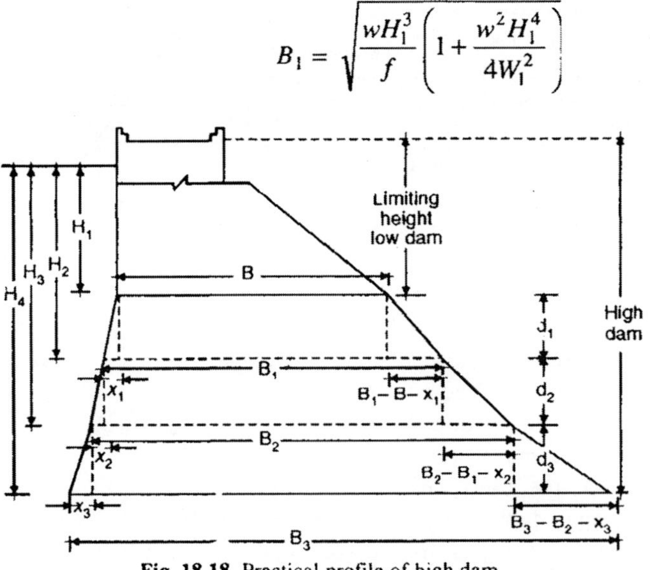

Fig. 18.18. Practical profile of high dam.

Projection on upstream face, $x_1 = \frac{pwd_1}{24}\{(3B^2 - B_1^2 + 6x_1(B + B_1) + 2BB_1)\}$

$$-\frac{wx_1}{12}(H + H_1)(2B_1 - 3x_1) - W\left[\frac{(B_1 - B)}{3} - x_1\right] = 0$$

where, H_1 = height of dam portion from maximum water level to the base of first step, d_1 = depth of first step, W_1 = total vertical load of dam and water above the base of first step and W = total weight of dam portion and water on the top of the first step.

Second strip,
$$B_2 = \sqrt{\frac{wH_2^3}{f}\left(1+\frac{w^2H_2^4}{4W_2^2}\right)}$$

$$x_2 = \frac{\rho w d_2}{24}\{(3B_1^2 - B_2^2 + 6x_2(B_1 + B_2) + 2B_1B_2)\}$$

$$-\frac{wx_2}{12}(H_1 + H_2)(2B_2 - 3x_2) - W_1\left(\frac{B_2 - B_1}{3} - x_2\right) = 0$$

Similarly, the other strips up to the base of the dam are designed. Suffix 2 is replaced by 3, 4, for third, fourth, ...strips respectively.

18.10 FOUNDATION TREATMENT

The gravity dams are massive structures transmitting tremendous loads on the foundation. The foundation rock should be strong enough to safely withstand pressures exerted on it under various conditions of loading. Ideal foundations for dams are rare to find, hence the necessity for building dams on less favourable sites after suitable treatment of the foundation usually by grouting in order to obtain a watertight foundation and help in eliminating uplift. The purposes of foundation treatment are thus to (*i*) Prepare stable support for the dam, (*ii*) Safeguard against sliding, (*iii*) Safeguard against uplift, and (*iv*) Safeguard against seepage through the junction of dam and its foundation. The foundation treatment envisages preparing the rock surface, grouting the foundation, and drainage.

1. Preparing Rock Surface. The weathered rock mass and overburden is removed to reach sound bed rock. Blasting is done carefully so as not to shatter the rock mass. The excavation is done in steps to provide a toe hold so that sufficient bed rock downstream from the dam is available to resist the sliding forces. Where geology exhibits extensive discontinuities like major faults or seams or shattered rock zone, or clay filled joints or bedding planes, these are removed and backfilled with concrete grouting. The final rock surface is properly cleaned and roughened by the use of high velocity air-water jets, wet sand blasting, stiff brooms, or other effective means.

2. Foundation Grouting. Foundation grouting is the process of injecting grout consisting of cementitious material into the foundation of dam to act as a binder and fill the voids for improving the stability and impermeability of the pervious foundation. Grouting has made it possible to undertake construction of dams on relatively unsound foundations. Various grouting techniques for foundation grouting are consolidation grouting, curtain grouting, and contact grouting.

(*a*) *Consolidation grouting:* It is also termed as area grouting or blanket grouting. The objectives of consolidation grouting are to strengthen the rock, to stop water passage through the disintegrated rock, to increase bearing strength of the strata and to seal off major crevices. Consolidation grouting is done in an area, usually entire dam foundation, by drilling shallow holes, 3 to 15 m deep, on a grid pattern at a spacing of 5 to 30 m. The drilled holes are grouted by a mixture of cement and water termed as grout at a low pressure of not higher than 3.5 kg/cm^2. This comparatively shallow and low pressure grouting is done for the general consolidation of the foundation before concreting of the dam section and is followed by high pressure grouting after some concreting of the dam has taken place.

(*b*) *Curtain grouting:* It is a high pressure grouting done to form a deep and approximately vertical grout curtain of moderate thickness at the head of the dam to reduce the passage of water to the required degree, thereby reducing the development of uplift pressures on the dam due to hydrostatic head. Generally a single line of closely spaced grout holes parallel to the axis of dam is sufficient but in inferior rock foundation the grouting may consist of a number of grout lines. The spacing of the holes may vary from 1.25 to 1.5 m. The depth of grouting holes may vary from 25% of hydrostatic head above the rock surface in the case of strong rock foundation to 70% in the case of inferior rock foundation. U.S.B.R. formula for depth of grout hole $D = 0.33H + C$ where, D = depth of hole (m), H = dam height (m), and C = constant, value varying from 7.5 to 25.

The grout holes may also be inclined. Inclined holes are considered to be more effective to intercept principal joints in the rock. The orientation of fractures in the foundation rock determine the inclination of the holes. The grouting pressure is the maximum that the rock can withstand without lifting the foundation strata. Grouting is done from the foundation gallery or from upstream face of the dam or from tunnels driven into the foundation rock below the dam.

(*c*) *Contact grouting:* It is done in the abutment areas of concrete dams to form a fully bonded contact between the concrete structure and the adjacent rock. It is also used where seepage paths along the contact are produced due to shrinkage of concrete on setting or deflection of the loaded structure.

18.11 GALLERIES IN GRAVITY DAMS

Galleries are openings or passageways left in the dam body. They may be provided parallel or normal to dam axis at various elevations. The galleries are interconnected by steeply sloping passages or by vertical shafts fitted with lifts. The shape and size of the gallery depends on the size of the dam and the function served.

The functions for which the galleries are provided are:

1. Drainage. To cater for the drainage of dam section by intercepting seepage from the water face and carry it away from the downstream face.

2. Inspection. To provide access to the interior of the mass comprising the dam with a view to inspect the structure and study the structural behaviour of the dam in post-construction period.

3. Drilling. To provide access for carrying out drilling and grouting of foundations, etc.

4. Operation of gates and control equipment. To provide access to mechanical equipment for the operation of gates and control equipment.

5. Post-construction grouting. To provide space for header and return pipes for post-construction grouting of longitudinal joints of the dam. Also to provide access for grouting the construction joints which cannot be done from the face of the dam.

Classification of Galleries

Galleries can be classified on the basis of the function served as also according to their alignment. On the basis of function served, the galleries are classified as drainage gallery, inspection gallery, grouting gallery, access gallery, foundation gallery and gate gallery. On the basis of alignment, galleries are classified as parallel to the axis and normal to the axis of the dam, as under.

1. Gallery Parallel to Dam Axis: (*a*) *Foundation gallery.* A foundation gallery generally extends along the length of the dam near the rock surface conforming in elevation to the transverse profile of the canyon; in plan it is nearer and parallel to the axis of the dam. The gallery is of a size suitable for carrying out drilling operations from inside it. Its usual size is 2.0×2.5 m. It caters for (*i*) The drainage of water percolating from upstream face or seepage through the foundations in the dam, (*ii*) Provides space for drilling and grouting of holes from the floor of the gallery for the main grout curtain, and (*iii*) For drilling foundation drain holes after foundation grouting has been completed.

(*b*) *Drainage gallery.* It is a supplementary drainage gallery located parallel to the crest at about 2/3rd of the base width from upstream face and extending usually only through the deepest portion of the dam. It serves for intercepting seepage from the water face and conducting it away from the downstream face and drilling and draining the downstream portion of the foundation.

(*c*) *Inspection gallery.* It is a gallery to provide access to the interior of the mass comprising the dam with a view to inspect the structure and study the structural behaviour of the dam in post-construction period. It also caters for the drainage of seepage water. It is rare that a gallery is provided purely for inspection. Foundation, drainage, gate and grouting galleries also serve as inspection galleries. Bhakra dam has galleries aggregating in length to 5 km.

(*d*) *Gate gallery.* It is a gallery made in a dam to provide access to and to house the mechanical equipment required for the operation of gates in outlet conduits, power penstocks or spillway crest. Location and size of the gallery depends on the need and size of the mechanical equipment.

(*e*) *Grouting gallery.* It is a gallery to locate the supply, return and vent headers of the grout pipe system, as also the piping system for artificial cooling of the blocks terminates in this gallery.

2. *Gallery Normal to Dam Axis:* Adit or access gallery is a horizontal gallery connecting the gallery system in the dam with the downstream face or features outside the dam such as power house or gate house.

3. *Vertical and Inclined Shafts:* These are vertical openings provided in the dam at suitable places for the purpose of location of elevator, hoisting equipment, stairway for connecting galleries, measuring devices, headers of artificial cooling system. The provision of a gallery in the body of a dam changes the normal stress pattern and stress concentrations occur around corners etc. Such stresses are ascertained to find out tension and compression zones to provide suitable reinforcement to counteract them. The corners are essentially rounded smoothly to minimize these stresses.

Figure 18.13 shows various galleries in a dam as also instruments installed.

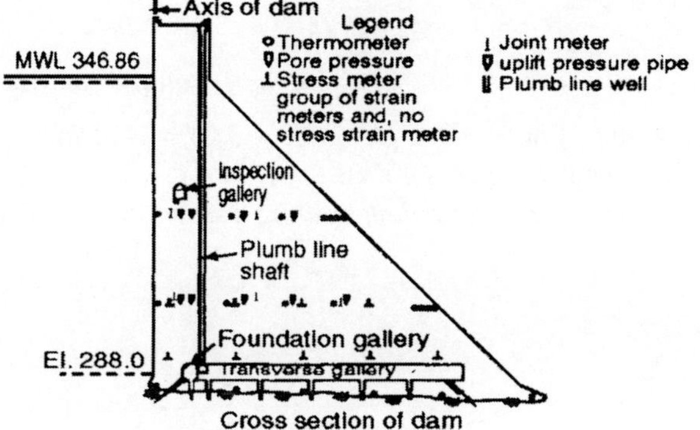

Fig. 18.19 Typical galleries in a concrete/masonry dam.

SOLVED EXAMPLES

Example 18.1 *What should be the maximum height of elementary profile of a dam, if the safe limit of stress on the masonry should not exceed 350 tonnes/m^2? Assume weight of masonry 2.4 tonnes/cum. (a) Determine base width also. (b) Determine H and B, if uplift intensity factor is 0.67 and factor of safety is 2.*

Solution. The limiting height of elementary profile of a masonry dam,

$$H = \frac{f}{w(\rho + 1)}$$

Putting given values

$$H = \frac{350}{1(2.4 + 1)} = 103 \text{ m}$$

Base width

$$B = \frac{103}{\sqrt{e}} = \frac{103}{\sqrt{2.4}} = 66 \text{ m}$$

(*b*)

$$H = \frac{350}{2 \times (2.4 - 0.67 + 1)} \simeq 64 \text{ m}$$

$$B = \frac{64}{\sqrt{2.4 - 0.67}} \simeq 49 \text{ m}$$

Example 18.2 *(a) Draw uplift pressure diagram for a dam holding 50 m water depth at upstream vertical face, top and bottom width 10 m and 30 m respectively. Uplift may be considered to be acting on 60% of the area of section. Tail water depth is 5 m.*

(b) Also if there is a drainage gallery at 6 m from face.

Solution. (*a*) Uplift at heel = 50 × 0.6 = 30 m

Uplift at toe = 5 × 0.6 = 3.0 m

Pressure diagram is shown in Fig. 18.20.

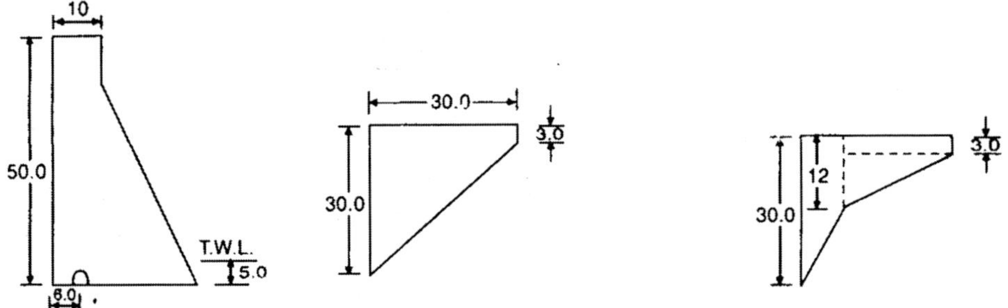

Fig. 18.20. Uplift pressure diagram. **Fig. 18.20(*a*)** Uplift pressure diagram with drainage gallery.

(*b*) Uplift at drainage gallery line = 3.0 + 1/3 (30.0 – 3.0) = 12.0 m.

Pressure diagram is indicated in Fig. 18.20(*a*).

Example 18.3 *Determine free board for a dam holding a reservoir with FRL 146.3 m, maximum wind velocity 140 km/h.*

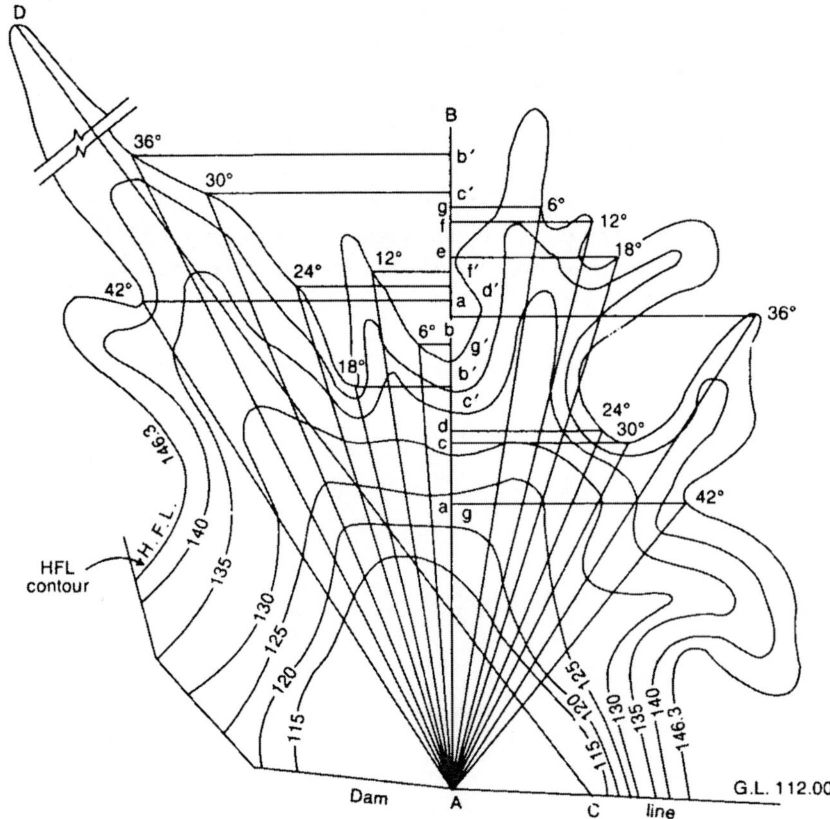

Fig. 18.21 Computation of effective fetch (F) and average depth (D).

Solution. 1. Referring to Fig. 18.21. Maximum reservoir fetch (CD) is 11 km.

2. Select a line *AB*, with *A* on dam axis and *B* on *FRL* contour in Fig. 18.21 to cover the maximum reservoir water spread area within 42° angle on either sides of line *AB*. Assume maximum wind velocity acting along *BA*. Draw 7 rays at 6° interval on either side of *AB* and compute effective fetch, as under:

Computation of fetch of water along fetch line C – D				*Calculation of effective fetch*			
Sl.No.	FRL (m)	GL (m)	Depth (m)	α°	cos α	x_1 (km)	x_1 cos α
1	146.3	146.3	0	42	0.743	Aa = 3.0	2.23
2	146.3	140.0	6.3	36	0.809	Ab = 4.85	3.92
3	146.3	135.0	11.3	30	0.866	Ac = 3.65	3.16
4	146.3	130.0	16.3	24	0.914	Ad = 4.80	4.39
5	146.3	125.0	21.3	18	0.951	Ae = 5.50	5.23
6	146.3	120.0	26.3	12	0.978	Af = 5.90	5.77
7	146.3	115.0	31.3	6	0.995	Ag = 6.0	5.97
8	At dam	112.0	34.3	0	0.0	AB = 4.50	4.50
	Total		147.1	6	0.995	Ag′ = 4.70	4.68
	Average depth = 147.1/8 = 18.3 m			12	0.978	Af′ = 5.30	5.18
				18	0.951	Ae′ = 4.20	3.99
				24	0.914	Ad′ = 5.25	4.80
				30	0.866	Ac′ = 6.20	5.37
	Effective fetch = $\frac{\Sigma x_1 \cos\alpha}{\Sigma \cos\alpha} = \frac{68.28}{13.512} = 5$ km.			36	0.809	Ab′ = 6.55	5.30
				42	0.743	Aa′ = 5.10	3.79
					Σ cos α = 13.512		Σx_1 cos α = 68.28

3. Wind velocity over water = 140 × 1.27 = 175 km/hr (Refer Table 16.1).
4. Significant wave height (H_s) = 2.8 m (From Fig. 16.8).
5. Design wave height (*H*) = 1.27 × H_s = 1.27 × 2.8 = 3.6 m (From Table below).

Percentage of waves exceeding specific wave height H	0.4	2	4	8	13
Ratio = sp. height/significant ht. = (*H/Hs*)	1.67	1.40	1.27	1.12	1.00

6. Wind setup = $V^2F/6200$ D = $[(175)^2 \times 11]/62000 \times 18.3$ = 0.3 m.
7. Computed free board = 1.33 (step 5) + (step 6) = 4.8 + 0.3 = 5.1 m
8. Similar calculations are done with reference to MWL for 2/3 wind velocity.
9. Minimum free board, above MWL = 1.0 m.
10. Top of dam is the height of FRL + 3.0, MWL + step 8 and MWL + step 9.

Out of free board computed 1 m may be provided as solid parapet.

Example 18.4 *A concrete gravity dam has maximum water level 305.0 m, bed level 225.0 m, top R.L. of dam 309.0 m, d/s face slope starts at R.L. 300.0 m, d/s slope 2 : 3, tail water is nil, upstream face of dam is vertical, centre line of drainage gallery is 8 m d/s of u/s face, uplift pressure is 100% at heel, 50% at line of gallery and zero at toe, weight of concrete is 2.4 t/m³. Considering only weight, water pressure and uplift, determine (i) Maximum vertical stresses at the toe and heel of dam, (ii) Major principal stresses at toe of dam, and (iii) Intensity of shear stress on a horizontal plane near the toe.*

Solution.

Height of dam = 309.0 – 225.0 = 84 m

Depth of water = 305.0 – 225.0 = 80 m

1. Top width of dam:

   14% of the height = 0.14 × 84 ≈ 12 m or 0.55 $H^{1/2}$ = 0.55 $(80)^{1/2}$ ≈ 5 m

Adopt top width of dam as 12 m

2. Bottom width of dam = 12.0 + (300.0 – 225.0) × 2/3 = 62 m

(*a*) Consider 1 m length of dam and *w* = 1 t/m^3. (Referring to Fig. 18.22)

Designation	*Force*	*Lever arm (m)*	*Moment about* toe (*t.m*)
Weight of dam			
$W_1 = 12 \times 84 \times 1 \times 2.4$	= 2419 t	50 + 12/2 = 56	135464
$W_2 = 50 \times 75/2 \times 1 \times 2.4$	= 4500 t	50 × 2/3 = 33.33	149985
$W = W_1 + W_2$	= 6919 t		ΣM_1 = (+) 285449
Uplift $U_1 = 40 \times 8/2$	= 160 t	54 + 8 × 2/3 = 59.33	9493
$U_2 = 40 \times 8$	= 320 t	54 + 8/2 = 58.0	18560
$U_3 = 54 \times 40/2$	= 1080 t	54 × 2/3 = 36.0	38880
ΣU	= 1560 t		ΣM_2 = (–) 66933 t.m

Water Pressure

$P = \frac{Wh^2}{2} = \frac{1}{2} \times 1 \times 80 \times 80 = 3200$ $\quad \frac{80}{3} = 26.67$ $\quad \Sigma M_3 = (-)\ 85344$ t.m

$V = W - u = 6919 - 1560 = 5359$ t $\quad M = M_1 - M_2 - M_3$

$= 285449 - 66933 - 85344$

$= (+)\ 133172$ t.m

Position of resultant from toe, $\quad \bar{x} = \frac{M}{V} = \frac{133172}{5359} = 24.85$ m

Eccentricity, $\quad e = \frac{62}{2} - 24.85 = 6.15$ m

Normal compressive stress at toe, $\quad P_n = \frac{V}{B}\left(1 + \frac{6e}{B}\right)$

$$= \frac{5359}{62}\left(1 + \frac{(6 \times 6.15)}{62}\right) = 138\ \text{t/m}^2$$

Normal compressive stress at heel, $\quad P'_n = \frac{V}{B}\left(1 - \frac{6e}{B}\right)$

$$= \frac{5359}{62}\left(1 - \frac{(6 \times 6.15)}{62}\right) = 35\ \text{t/m}^2$$

(*b*) Principal stress at toe, $\quad \sigma = P_n \sec^2 \alpha - P' \tan^2 \alpha$

= Here, *P'* is zero, there being no tail water,

$\tan \alpha = 2/3$

$$\sec^2 \alpha = 1 + \tan^2 \alpha = 1 + \left(\frac{2}{3}\right)^2 = \frac{13}{9}$$

Therefore, $\quad \sigma = P_n \sec^2 \alpha = 138 \times 13/9 = 199\ \text{t/m}^2.$

(*c*) Intensity of shear stress on a horizontal plane near toe, $\quad \tau_0 = (P_n - P') \tan \alpha$

$= (P_n - 0)\ 2/3$, *P'* is zero there being no tail water

$= 138 \times 2/3 = 92\ \text{t/m}^2$

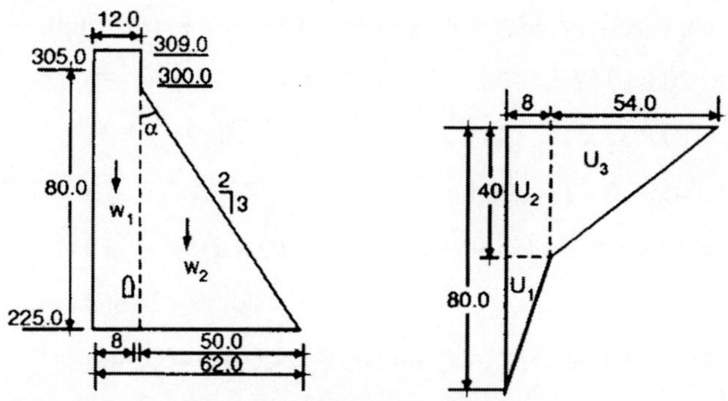

Fig. 18.22 Uplift pressure diagram.

Example 18.4 *A concrete gravity dam has maximum reservoir level 150.0 m, base level of dam 100.0 m, tail water elevation 110.0 m, base width of dam 40 m, location of drainage gallery 10 m from u/s face which may be assumed as vertical. Compute hydrostatic thrust and uplift force per metre length of dam at its base level. Assume 50% reduction in net seepage head at the location of the drainage gallery.* (AMIE)

Solution. Consider 1 m length of dam (Fig. 18.23).

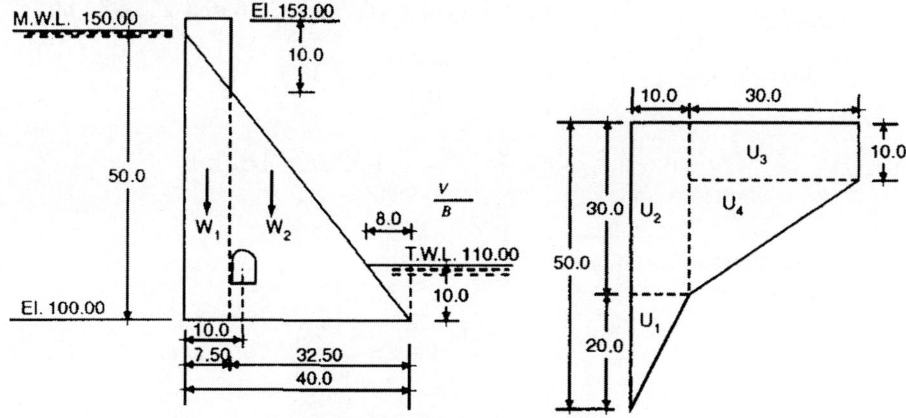

Fig. 18.23 Uplift pressure diagram.

Free board = 5% of dam height = 0.05 × 50 = 2.5 say 3 m

Dam height = 50 + 3 = 53 m

Top width = $0.14H$ or $0.55H^{1/2}$, where H is height of the dam

$= 0.14 \times 53 = 7.5$ or $0.55\sqrt{53}\ 1/2 = 4$ m, adopt 7.5 m

Designation	*Force (t)*	*Lever arm (m)*	*Moment about toe (t.m)*
Weight of masonry			
$W_1 = 7.5 \times 53 \times 1 \times 2.4$	= 954	$\left(32.5 + \frac{7.5}{2}\right) = 36.25$	34583
$W_2 = 32.5 \times \frac{43}{2} \times 1 \times 2.4$	= 1677	$\left(32.5 \times \frac{2}{3}\right) = 21.67$	36341
Weight of tail water			
$W_3 = 8 \times \frac{10}{2} \times 1 \times 1$	= 40	$\left(8 \times \frac{1}{3}\right) = 2.67$	107

$W = 954 + 1677 + 40 = 2671\ \text{t} \quad M_1 = 34583 + 36341 + 107 = +\ 71031\ \text{t.m}$

Uplift pressure at drainage gallery with 50% reduction = 10.0 + 1/2 (50.0 – 10.0) = 30 t

Uplift pressure			
$U_1 = 20 \times 10/2 = 100$ t	= (30 + 10 × 2/3) =	36.67	3667 t.m.
$U_2 = 10 \times 30 \times 1 = 300$ t	(30 + 10/2) =	35	10500 t.m.
$U_3 = 10 \times 30 \times 1 = 300$ t	(30/2) =	15	4500 t.m.
$U_4 = 20 \times 30/2 \times 1 = 300$ t	(30 × 2/3) =	20	6000 t.m.
U = 1000 t			$M_2 = -24{,}667$ t.m.

$V = W - U = 2671 - 1000 = 1671$ t, $M = M - M_2 = 71{,}031 - 24{,}667 = 46{,}364$ t.m.

Water pressure, u/s face $P = \dfrac{wh^2}{2} = \dfrac{(1 \times 50 \times 50)}{2} = 1250$ $\quad \dfrac{50}{3}$ $\quad$ (–) 20,833

d/s face, $P' = \dfrac{wh'^2}{2} = 1 \times 10 \times \dfrac{10}{2} = 50$ $\quad \dfrac{10}{3}$ $\quad$ + 167

Pressure acting downstream = 1250 – 50 = 1200, M_3 = (–) 20833 + 167
= (–) 20,666

Net moment = 46,364 – 20,666 = 25,698 t.m.

Position of resultant from toe, $\bar{x} = \dfrac{M}{V} = \dfrac{25698}{1671} = 15.38$ m

Eccentricity, $e = \dfrac{B}{2} - \bar{x} = \dfrac{40}{2} - 15.38 = 4.62$ m

Normal compressive stress at toe, $P_n = \dfrac{V}{B}\left(1 + \dfrac{6e}{B}\right)$

$$= \frac{1671}{40}\left(1 + 6 \times \frac{4.62}{40}\right) = 70.73 \text{ t/m}^2$$

Normal compressive stress at heel $= \dfrac{V}{B}\left(1 - \dfrac{6e}{B}\right)$

$$= \frac{1671}{40}\left(1 - \frac{6 \times 4.62}{40}\right) = 12.82 \text{ t/m}^2$$

Maximum principal stress at toe, $\sigma = P_n \sec^2 \alpha - P' \tan^2 \alpha$, here $P' = 10$

$\tan \alpha$ = 40/50 and $\sec^2 \alpha$ = 1.64

Therefore, $\sigma = 70.73 \times 1.64 - 10(40/50)^2 = 110$ t/m²

Intensity of shear stress on a horizontal plane near toe,

$$\tau_0 = (P_n - P') \tan \alpha$$

$$\tau_0 = (70.73 - 10)40/50 = 49 \text{ t/m}^2$$

Example 18.5 *A concrete dam can be assumed to be trapezoidal in section having a top width of 2 m and bottom width of 10 m. Its height is 12 m and the upstream face has a batter of 0.9. Give an analysis of the stability of the dam for the base section for overturning and sliding in the full reservoir condition assuming no free board allowance but allowing for uplift pressures. Assume* *uplift intensity factor as 100%. Also determine the compressive stress at the toe and the heel and major principal and shear stress developed at the toe. Assume weight of concrete to be 2.4 t/m³, unit shear strength of concrete to be 140 t/m², and the coefficient of friction between concrete and foundation soil to be 0.7.*

Solution. Consider 1 m length of dam and 1 t/m³ weight of water (Fig. 18.24).

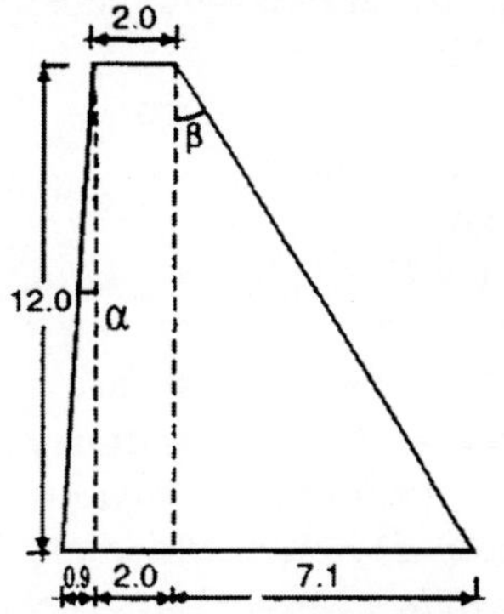

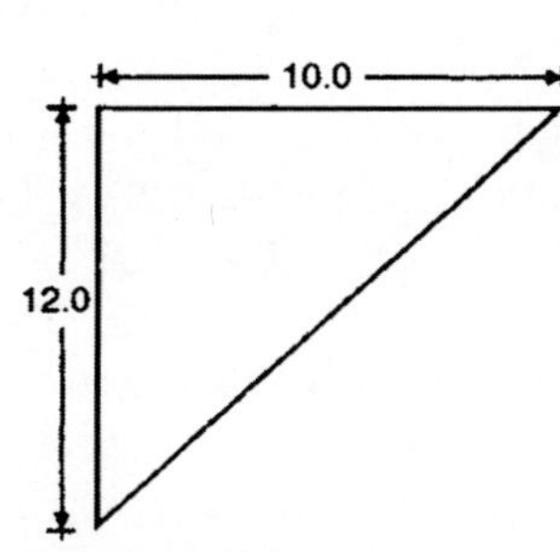

Fig. 18.24 Uplift pressure diagram.

Designation	*Force (t)*	*Lever arm (m)*	*Moment about toe (t.m)*
Weight of concrete	$\frac{(0.9\times 12)}{2}\times 1\times 2.4=13$	(9.1 + 0.9/3) = 9.4	122
	$2\times 12\times 1\times 2.4=58$	(7.1 + 2/2) = 8.1	470
	$\frac{(7.1\times 12)}{2}\times 1\times 2.4=102$	(7.1 × 2/3) = 4.73	482
Water weight	$(0.9\times 12)/2\times 1\times 1.0=5$	(9.1 + 0.9 × 2/3) = 9.7	49
	W = 178 t		M_1 = 1123 t.m
Uplift	(12 × 10 × 1 × 1)/2 = 60 t	10 × 2/3 = 6.67	M_2 = – 400
Horizontal water pressure	(1 × 12 × 12)/2 = 72 t	12/3 = 4	M_3 = – 288
			Total = (–) 688

Vertical weight $V = W - U$ = 178 – 60 = 118 t M = 1123 – 688 = + 435 t.m

Factor of safety against overturning = $+M/-M$ = 1123/688 = 1.63 > 1.5 hence, safe

Factor of safety against sliding $= \mu\Sigma V/\Sigma H$ = (0.7 × 118)/72 = 1.15 > 1, hence safe

Shear friction factor $= (\mu V + Bq)/\Sigma H$

= (0.7 × 118 + 10 × 140)/72 = 20.59 > 5, hence safe.

Stress analysis

Position of resultant $\bar{x} = \Sigma M/\Sigma V = 435/118 = 3.69$ m

Eccentricity, $e = B/2 - \bar{x} = 10.0/2 - 3.69 = 1.31$ m,

which is $< \frac{b}{6} < \frac{10}{6} < 1.67$, hence safe

Compressive stress at toe $P_n = V/B\,(1 + 6e/B)$

P_n = 118/10 (1 + (6 × 1.31/10) = 21.07 t/m²

Compressive stress at heel $P'_n = \Sigma V/B\,(1 - 6e/B)$ = 118/10 (1 – 6 × 1.31/10) = 2.53 t/m²

$$\tan\alpha = (0.9)/12.0 = 0.075$$
$$\tan^2\alpha = (0.075)^2 = 0.006$$
$$\sec^2\alpha = 1 + \tan^2 = 1 + 0.006 = 1.006$$
$$\tan\beta = 7.1/12 = 0.59$$
$$\sec^2\beta = 1 + \tan^2\beta = 1 + (0.59)^2 = 1.35$$

Principal stress at toe, $\sigma = P_n \sec^2\beta - P' \tan^2\beta$, but P' is zero there being no tail water.

$= 21.07 \times 1.35 = 28.44$ t/m^2

Shear stress at toe, $\tau_0 = P_n \tan\beta - P' \tan\beta$, but P' is zero

$= 21.07 \times 0.59 = 12.43$ t/m^2

Principal stress at heel, $\sigma h = P_n' \sec^2\alpha - P\tan^2\alpha$

$= 2.53 \times 1.006 - 1 \times 12 \times 0.006 = 2.47$ t/m^2

Shear stress at heel, $\tau_0 h = -(P'_n - P)\tan\alpha$

$= -(2.53 - 12.0) \times 0.075 = +0.71$ t/m^2

Example 18.6 *A triangular section gravity dam 40 m high has water stored up to top and base width is 30 m. Analyse the dam section to determine (i) Factor of safety against sliding, (ii) Factor of safety against overturning, (iii) There is no tension anywhere along the base of dam and (iv) Maximum compressive stress in the dam body is within safe crushing strength of the material (10 kg/cm^2). Assume coefficient of friction between the base and foundation as 0.7 and uplift pressure intensity coefficient as 0.45.*

Solution. Assume unit weight of material of dam as 2.4 t/m^3

Weight of dam, $W = 1/2\, BHwG$

$= 1/2 \times 30 \times 40 \times 2.4 = 1440$ t

Hydrostatic pressure, $p = 1/2\, w H^2$

$= 1/2 \times 1.0\,(40)^2 = 800$ t

Uplift pressure, $U = 1/2\, C_u H Bw$

$= 1/2 \times 0.45 \times 40 \times 30 \times 1.0 = 270$ t

Net vertical force, $V = W - U = 1440 - 270 = 1170$ t

Overturning moments $(\Sigma M_o) = P \cdot H/3 + U \times 2/3 \times B$

$= 800 \times 40/3 + 270 \times 2/3 \times 30 = 16067$ t.m

Stabilising moments $(\Sigma M_s) = W \times 2/3 \times B$

$= 1440 \times 2/3 \times 30 = 28800$ t.m

(*i*) Factor of safety against sliding $= (\Sigma V \times 0.7)/\Sigma H$

$= (1170 \times 0.7)/800 = 1.02 > 1$, hence o.k.

(*ii*) Factor of safety against overturning $= (\Sigma M_s)/\Sigma M_o$

$= 28800/16067 = 1.79 > 1.5$, hence o.k.

Sum of moments of all the forces about toe, $\Sigma M_t = \Sigma V \times 2/3 \times B - P \times h/3$

$= 1170 \times 2/3 \times 30 - 800 \times 40/3$

$= 12{,}733$ t.m

Distance of resultant from toe, $\bar{x} = \Sigma M_t/\Sigma V$

$= 12{,}733/1170 = 10.88$ m

Eccentricity, $e = B/2 - \bar{x}$

$= 30/2 - 10.88 = 4.12$ m

(*iii*) For no tension to develop any where along the base

$= e$ should be $< B/6$

$= 4.12 < 30/6$, hence o.k.

(*iv*) Maximum compressive stress at toe,

$$P_n = \frac{V}{B}(1 + 6e/B)$$

$$= \frac{1170}{30}\left(1+\frac{6\times 4.12}{30}\right)$$

$= 71.1 \text{ t/m}^2$

$= 7.1 \text{ kg/cm}^2 < 10 \text{ kg/cm}^2$, hence o.k.

EXAMPLE 18.7 *Determine minimum safe base width of an elementary profile of a gravity dam holding 50 m depth of water, assume specific gravity of dam material as 2.4, uplift factor as 0.4. Coefficient of friction between concrete and foundation soil 0.7. Also find out principal and shear stresses at toe of dam.*

Solution. Minimum base width

For no tension, $B = \dfrac{H}{\sqrt{\rho - c}} = \dfrac{50}{\sqrt{2.4-0.4}} \approx 35.4$ m

Against sliding, $B = \dfrac{H}{\mu(\rho - c)} = \dfrac{50}{0.7\,(2.4-0.4)} \approx 36$ m ; adopt 36 m

Weight of dam, $W = 1/2\,(B\,H\,\rho\,w)$

$= 1/2 \times (36 \times 50 \times 2.4 \times 1) = 2160$ t

Uplift pressure $U = C\,w\,B\,H/2$

$U = 1/2 \times 0.4 \times 1 \times 36 \times 50 = 360$ t

Net vertical force, $V = W - U = 2160 - 360 = 1800$ t

Hydrostatic pressure, $P = w\,H^2/2 = 1\,(50)^2/2 = 1250$ t

Sum of moments about toe, $M = V \times 2/3\,B - P \times H/3$

$= 1800 \times 2/3 \times 36 - 1250 \times 50/3$

$= 22{,}367$ t.m

$\bar{x} = M/V = 22{,}367/1800 = 12.43$ m

Eccentricity, $e = B/2 - \bar{x} = 36/2 - 12.43 = 5.57$ m

which is $< B/6 < 36/6 < 6.0$, hence safe.

Normal stress at toe, $P_n = V/B\,(1 + 6e/B)$

$= 1800/36\,(1 + (6 \times 5.57/36) = 96 \text{ t/m}^2$

$\sec^2\theta = 1 + \tan^2\theta = 1 + (36/50)^2 = 1.518$

Principal stress at toe, $\rho_1 = P_n \sec^2\phi = 96 \times 1.518 = 146 \text{ t/m}^2$

Shear stress at toe, $\tau = P_n \tan\phi = 96 \times 0.72 = 69 \text{ t/m}^2$

EXAMPLE 18.8 *Determine pressure, position and wave height for a dam 100 m high, direct fetch 24 km, wind velocity 100 km/hr.*

Solution.

Wave height, $Hw = 0.0322\sqrt{FV} + 0.763 - 0.271\sqrt[4]{F}$

$= 0.0322\sqrt{24\times 100} + 0.763 - 0.271\sqrt[4]{24}$

$= 1.74$ m

Free board required $= 1.5\,hw = 1.5 \times 1.74 \approx 3$ m

Wave pressure against dam/m width $= 2000\,hw^2 = 2000\,(1.74)^2$

$= 6056$ kg

Centre line of pressure is acting at 0.375 *hw i.e.*, 0.65 m above reservoir level.

Maximum unit pressure acting at 0.125 hw

$= 2400\,hw = 2400 \times 1.74 = 4176$ kg/sq m.

EXAMPLE 18.9 *Design practical profile for a gravity dam holding a water depth of 88 m, wave height 2 m, specific gravity of concrete 2.4 and permissible compressive strength of 300 tonne/m^2, top width of dam is 5 m.*

Solution.

Free board $= 1.5\ hw = 1.5 \times 2 = 3$ m

or 5% of $H = 0.05 \times 88 \approx 4.4$ m, adopt 4 m.

Limiting height of dam $= f/w(\rho + 1) = 300/1(2.4 + 1) \approx 88$ m

It is thus a low gravity dam

Base width, $B = \dfrac{H}{\sqrt{\rho - c}} = \dfrac{88}{\sqrt{2.4 - 0}} \approx 57$ m

Upstream batter, = top width, $a/16 = 5/16 = 0.31$ m

Total base width $= 57 + 0.31 = 57.31$ m

Vertical height of dam $= 2\,a\sqrt{\rho} = 2 \times 5\sqrt{2.4} = 15.5$ m

Height upto end of batter $= 3.1\,a\sqrt{\rho} = 3.1 \times 5\sqrt{2.4} = 24$ m

The profile of the dam is shown in Fig. 18.25.

EXAMPLE 18.10 *Design a gravity dam of 95 m height, data given as in example 18.9.*

Solution. It is a case of high gravity dam, the height being over 88 m. The portion between 88 m and 95 m may be divided into 2 strips of 4 m and 3 m.

Ist strip

Approximate base width, $B_1 = \left(57 + \dfrac{(57 \times 4)}{88}\right) \approx 60$ m

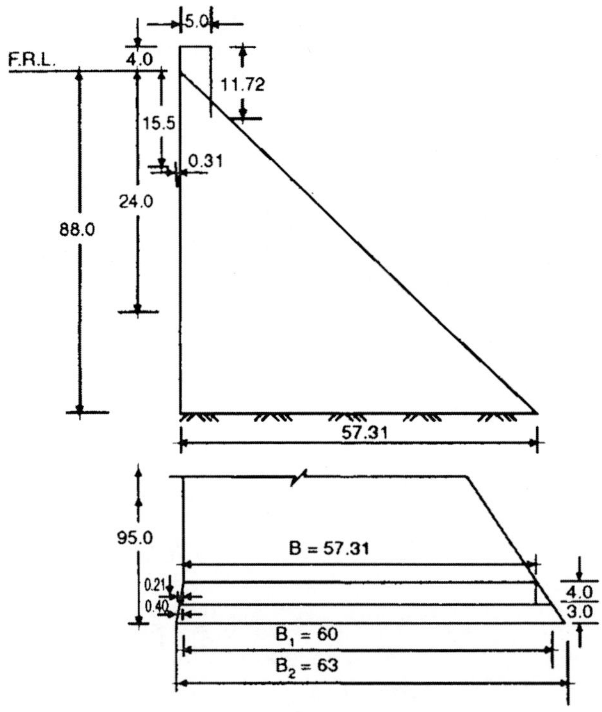

Fig. 18.25 Practical profile of low and high dams.

Weight of dam above the strip $= 2.4 \times 1\left[5\left(\dfrac{4.0 + 11.72}{2}\right) + \left(\dfrac{57 \times 88}{2} + \dfrac{64 + 72.5}{2} \times 0.31\right)\right]$

$= 6164$ tonne

Approximate weight of strip $= 2.4 \times 1\ (57.31 + 60)/2 \times 4 = 563$ t

Weight of water acting on u/s face $= 1 \times 1\ (24 + 15.5)/2 \times 0.31 = 6$ t

Total weight on 1st strip $= 6164 + 6 = 6170$ t

Total weight of dam above base of strip, $W_1 = 6170 + 563 = 6733$ t

Exact width of strip, $$B_1 = \sqrt{\frac{w H_1^3}{f}\left(1 + \frac{w^2 H_1^4}{4 W_1^2}\right)}$$

$$= \sqrt{\frac{1\,(92)^3}{300}\left[1 + \frac{1\,(92)^4}{4\,(6733)^2}\right]}$$

$= 60$ m

Projection on u/s side $$x_1 = \frac{\rho\, w\, d_1}{24}\left[3B^2 - B_1^2 + 6x_1\,(B + B_1) + 2BB_1\right]$$

$$-\frac{w\, x_1}{12}(H + H_1)(2B_1 - 3x_1) - W\left(\frac{B_1 - B}{3} - x_1\right) = 0$$

$$= \frac{2.4 \times 1 \times 4}{24}\left[3(57.31)^2 - (60)^2 + 6x_1\,(57.31 + 60)\right.$$

$$\left. + 2 \times 57.31 \times 60\right] - \frac{1\, x_1}{12}(88 + 92)\,(2 \times 60 - 3\,x_1) - 6170\left(\frac{60 - 57.31}{3} - x_1\right) = 0$$

or $x_1 = 0.21$ m

II strip

Weight of 1st strip $= 2.4 \times 1\left(\dfrac{57.31 + 60}{2}\right)4 = 563\,\text{t}$

Weight of water acting on u/s face of 1st strip

$= 88 \times 0.21 + 1/2 \times 4 \times 0.21 = 19$ t

Approximate base width of 2nd strip

$= 60 + (57 \times 3.0)/88 + (0.21 \times 3.0)/4.0 = 62$ m

Approximate weight of strip $= 2.4 \times 1\ (60 + 62)/2 = 3.0 = 439$ t

Approximate weight of water acting on u/s face, if $x_2 = x_1$

$= (92 + 95)/2 \times 0.21 = 20$ t

Total weight of dam and water on bottom of 2nd, strip,

$W_2 = 6733 + 19 + 439 + 20 = 7 \geq 11$

Exact base width $$B_2 = \sqrt{\frac{w H^3{}_2}{f}\left(1 + \frac{w^2 H^4{}_2}{4 W^2{}_2}\right)}$$

$$= \sqrt{\frac{1(95)^3}{300}\left[1 + \frac{1(95)^4}{4(7211)^2}\right]}$$

$= 63$ m

$$\text{Projection on u/s side } x_2 = \frac{\rho w d_2}{24}\left[3B_1^2 - B_2^2 + 6x_2\,(B_1 + B_2) + 2\,B_1 B_2\right]$$

$$-\frac{w\,x_2}{12}(H_1 + H_2)(2\,B_2 - 3x_2) - W_1$$

$$\left(\frac{B_2 - B_1}{3} - x_2\right) = 0$$

$$= \frac{2.4 \times 1 \times 3}{24}\left[3(60)^2 - (63)^2 + 6x_2\,(60 + 63)\right.$$

$$\left. + 2 \times 60 \times 63\right] - \frac{1 \times x_2}{12}(92 + 95)$$

$$(2 \times 63 - 3x_2) - 6733 \times \left(\frac{63 - 60}{3} - x_2\right) = 0$$

or $x_2 = 0.40$ m

Projection on other side = (63 – 60 – 0.40) = 2.6 m.

The profile of the dam is shown in Fig. 18.25.

Example 18.11 *Determine the stability of concrete dam shown in Fig. 18.25. Neglect uplift and tail water.*

Solution.

Description	Forces (tonne)			Lever arm	Moment of force
	Horizontal	Vertical		(m)	(t.m)
	pH	W	Pv		
1. Dam Portion	El 202 – 197				
Dam weight 2.4 × 5 × 5		60		1.5 + 5/2 = 4	240 t.m
Water pressure $1(3)^2/2$	4.5			3.0/3	4.5

Reaction to R at a point of incidence is equal and opposite to R. Let the resultant R cut at x metres from vertical face of dam or $(x + 1.5)$ from point A. The moment of PH about A is zero.

Reservoir Empty

Total moment of vertical load

W = 240 t.m

Total vertical load = 60 t

Taking moments about A

60 $(x + 1.5)$ = 240

or x = 2.5 m which is > 5.0/3, hence safe.

Reservoir Full

Total moment of vertical load

W and PH = 244.5 t.m.

Total vertical load = 60 t

Taking moments about A

60 $(x + 1.5)$ = 244.5

or x = 2.58 m which is within middle-third *i.e.*, (5.0 × 2/3) of this portion, hence safe.

2. Dam Portion El 197 – 185

			W	Pv	Lever arm		Moment
Dam weight =	2.4 (0.5/2 × 12)	=	7.2		1.0 + 0.5 × 2/3 = 1.33	=	10
	2.4 × 5 × 12	=	144		1.0 + 0.5 + 5/2 = 4	=	576
	2.4 × 6.5/2 × 12	=	94		1.0 + 0.5 + 5 + 6.5/3 = 8.67	=	815
Water weight =	0.5/2 × 12	=		3	1.0 + 0.5 × 1/3 = 1.17	=	4
	0.5 × 3	=		1.5	1.0 + 0.5/2 = 1.25		2
Water pressure	$1\,(15)^2/2 = 112.5$				15/3		563

Reservoir Empty

Total moments of vertical load

W = 240 + 10 + 576 + 815 = 1641 t.m

Reservoir Full

Total moments of vertical load W and PV and pH

= 1641 + 4 + 2 + 563 = 2210 t.m

Total vertical load W	Total vertical load W and PV
$= 60 + 7 + 144 + 94 = 305$ t	$= 305 + 3 + 1.5 = 309.5$ t
Taking moments about A_{185}	Taking moments
$= 305\ (x + 1.5 - 0.5) = 1641$	$309.5\ (x + 1.5 - 0.5) = 2210$
or = 4.38 m which is > 12/3, hence safe.	or $x = 6.14$ m which is $< 12 \times 2/3$, hence safe.

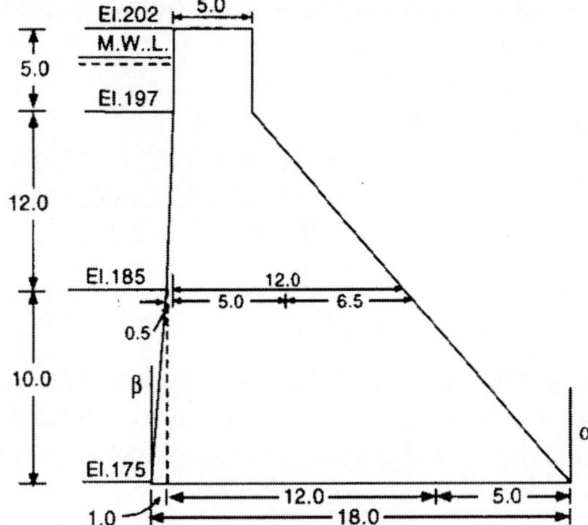

Fig. 18.26 Dam section.

3. Dam portion El 185 – 175

Dam weight =	2.4 (1.0/2 × 10)	12		1.0 × 2/3 = 0.67	= 8
	2.4 (12 × 10)	288		1.0 + 12/2 = 7	=2016
	2.4 × 5/2 × 10	60		1.0 + 12 + 5/3 = 14.67	= 880
Water weight =	1 (1.0/2 × 10)		5	1.0 × 1/3 = 0.33	= 2
	1 × 1.0 × 15		15	1.0/2 = 0.50	= 7.5
Water pressure	$1(25)^2/2 = 312.5$			25/3 = 8.33	=2603

Reservoir Empty	*Reservoir Full*
Total moments of vertical load	Total moments of *W, PV & PH* =
$W = 1641 + 8 + 2016 + 880 = 4545$	$= 4545 + 2 + 7.5 + 2603 = 7157.5$ t.m
Total vertical load	Total vertical load W and PV
$W = 305 + 12 + 288 + 60 = 665$	$= 665 + 5 + 15 = 685$ t
Taking moments about	Taking moments
$A_{175} = 665 \times x = 4545$	$685 \times x = 7157.5$
or $x = 6.83$ which is > 18/3, hence safe.	or $x = 10.45$ which is $< 18 \times 2/3$, hence safe.

4. Check for inclined P_{max} and P_{min}. This is checked for each portion of the dam, as above. But it is being done for the entire dam section.

Reservoir Empty	*Reservoir Full*
Total vertical load = 665 t	Total vertical load = 685 t
Eccentricity	Eccentricity = x – B/2
e = B/2 – x = 18/2 – 6.83 = 2.17 m	= 10.45 – 18/2 = 1.45 m
Vertical P_{max} at heel = $W/B\ (1 + 6e/B)$	Vertical P_{max} at toe = $W/B\ (1 + 6e/B)$
= 665/18 (1 + 6 × 2.17/18)	= 685/18 (1 + 6 × 1.45/18)
= 64 t	= 56 t
Inclined P_{max} at heel = $64 \times \sec^2 \beta$	Inclined P_{max} at toe = $56 \times \sec^2 \alpha$
$= 64\ [(1+1.0/10)^2]$	$= 56\ (1 + (5/10)^2)$
= 65 t.	= 70 t.

5. Check for shear stress

Reservoir Empty	*Reservoir Full*
P_s heel = Vert. $P_{max} \times \tan \beta$	P_s toe = $56 \times 5/10 = 28$ t/m^2
$= 64 \times 1.0/10 = 6$ t/m^2 < 60 safe.	< 60 safe

It shows that there is no danger of toe being sheared off.

6. Check for sliding. The check may be done at various levels of dam portions above. However, the entire dam is being checked against sliding under full reservoir condition, as under:
Total vertical load at base = 685 t
Total horizontal pressure at base = 312.5 t
For safety against sliding, $P_H < \mu (W + Pv)$
$= 312.5 < 0.67 \times 685$, assuming μ as 0.67
$= 312.5 < 459$, hence safe.
Factor of safety $= 685/312.5 = 2.19$, o.k.

QUESTIONS

1. (*a*) What preliminary investigations should be conducted for the selection of a dam site?
   (*b*) Under what conditions would you adopt a gravity dam and an earthen dam?
2. (*a*) Prove that the maximum shear stress intensity in an elementary profile is twice that the average.
   (*b*) Determine the safe height of a dam having elementary profile for the maximum compressive stress of 140 t/m^2 with an uplift intensity factor of 0.67.
3. How will you check the stability of the gravity dam section for reservoir full condition.
4. Define gravity dams. Enumerate the points considered while selecting site for the construction of a gravity dam.
5. Give the list of the forces affecting the stability of a gravity dam.
6. State the design criteria of a concrete gravity dam and explain briefly how you will determine the various factors of safety against possible failures.
7. Discuss the main principles of design of a masonry gravity dam. Explain how the design profile is evolved.
8. Explain how uplift considerations affect the design of a gravity dam. What measures can be adopted to reduce the undesirable effects due to uplift in such cases?
9. Write a brief note on the necessity and methods of foundation treatment of dams.
10. Discuss the evolution of the final profile of a gravity dam from its elementary triangular profile and explain the main principles of its design.
11. How are pore pressures and uplift pressures controlled in rigid dams?
12. What is meant by elementary profile of a gravity dam and how is it determined? What should be the maximum height of elementary profile of a dam, if the safe limit of stress on the masonry were not to exceed 150 t/m^2? Assume weight of masonry 2.4 t/m^3.
13. (*a*) Show that the base width of an elementary profile of a gravity dam is h/e where h is height of elementary profile and e is the specific gravity of masonry.
    (*b*) How and why elementary profile of gravity dam is modified to give practical profile?
14. A concrete gravity dam 15 m in height has a top width of 6 m and a free board of 3 m. Upstream face of dam is vertical, while downstream face has a slope of 0.8 (*H*): 1 *V* right from the top of the dam. Determine (*i*) factor of safety against overturning, and (*ii*) location and magnitude of maximum pressure on foundation. Neglect all other forces except those due to hydrostatic pressure, full uplift pressure and self weight. Take unit weight of concrete as 2.4 t/m^3.

15. State the design criteria of a concrete gravity dam and explain briefly how you will determine the various factors of safety against possible failures.
16. Discuss the main principles of design of a masonry gravity dam. Explain how the design profile is evolved.
18. Differentiate between a low gravity dam and a high gravity dam. What do you understand by the limiting height of a low dam?
18. Explain the step by step method for analysing the stability of gravity dam.
19. Name the galleries usually provided in gravity dams. Discuss the statement: it is rare that a gallery is provided purely for inspection purpose.
20. Discuss safety criteria of a gravity dam.
21. What do you understand by stability analysis of a gravity dam? Discuss two methods generally used.
22. Discuss strips method of design of practical profile of a high gravity dam.

19

ARCH AND BUTTRESS DAMS

19.1 INTRODUCTION

An arch dam is just a curved beam, the ends of which are restrained and the way in which the loads are resisted is termed as *arch action.* Arch dam may thus be defined as a curved masonry or concrete dam, convex upstream, which depends principally on arch action for its stability; itself resisting part of pressure of the reservoir by its own weight. The arch mostly transfers horizontally the water load to the canyon walls or abutments depending on the amount of curvature and as a result considerable economy is effected in the quantity of material used compared to gravity dams. Since the arch can be made to abut against the walls of the canyon under the hydrostatic loads acting on its extrados face, arches are efficiently employed as structural members in dams where the valley is narrow, the usual recommended ratio of crest length to dam height being 5 or less. For the arch action to be more effective, the rock constituting ends of arch should not yield. It follows that the abutments should be of sound rock. Uplift on the base of dam is not of importance in arch dams. Fig. 19.1 shows section of Idikki arch dam.

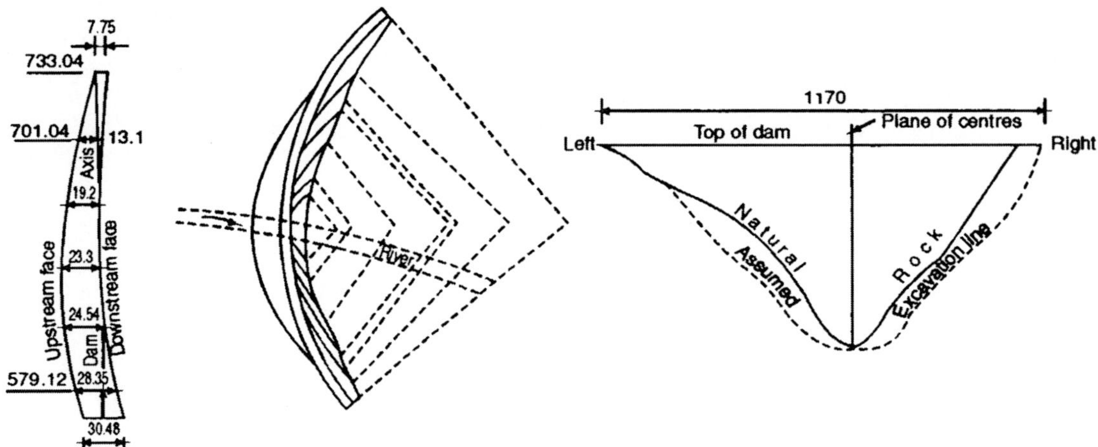

Fig. 19.1 Idikki arch dam.

19.2 TYPES OF ARCH DAMS

Arch dams are classified on the basis of thickness, symmetry with respect to crown section or characteristics of extrados and intrados curves but mainly as constant radius, variable radius or constant angle types.

1. Constant Radius Arch Dam: A constant radius arch dam is an arch dam, the arch centre of which for the upstream face, the downstream face and the centre line are all coincident with the axis centre at all elevations (Fig. 19.2). In simple types the upstream face is vertical so that its radius at all elevations is constant but thickness is proportional to the height. As a result the central angles of the arch rings of the intrados vary at various elevations, *i.e.*, goes on reducing from top to base, the maximum being at the top of dam and a certain minimum at the bottom of the dam (Fig. 19.2). This type of dam is triangular in cross-section with increased thickness at the crest. It is particularly adapted to U-shaped canyon where cantilever action carries a relatively large proportion of water load at lower elevations. Form work required for construction is simple. However, it is less economical compared to constant arch dam.

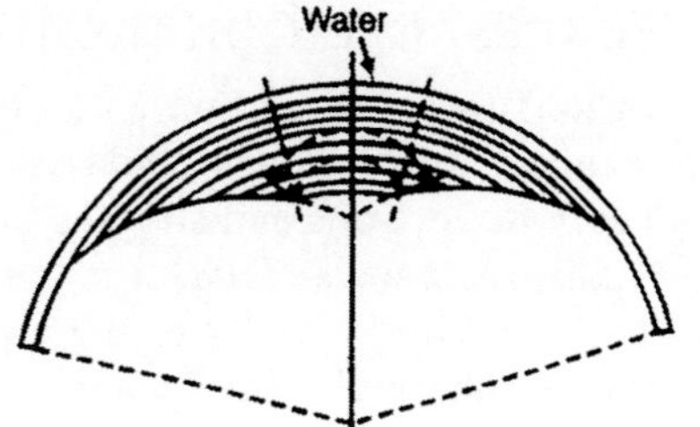

Fig. 19.2 Constant radius arch dam.

2. Variable Radius Arch Dam. Also termed as variable centre arch dam. In this type of dam, both the radius and central angle are not constant (Fig. 19.3). It is a compromise between the constant radius and the constant angle arch dam. The characteristic of this type of dam is variable radius, *i.e.*, radii of extrados surface or arch rings corresponding to upstream face and intrados surface or arch rings corresponding to downstream face, vary at various elevations; maximum at the top and minimum at the base. The radii gradually decrease as the depth below the crest increases. The central angle is as large and as nearly constant as possible to obtain maximum arch efficiency at all elevations. Central angle is different for different arches and usually varies from 180° to 150°. The upstream face is generally vertical or even overhanging at the upstream side near abut ments and that at the downstream side near the crown. The centres of the horizontal arch rings at various elevation do not lie on a single vertical line, rather the centres of different radii of extrados or intrados fall on a smooth curve which has lent it the name of variable centre arch dam. This type of dam is suitable for narrow V or U-shaped valleys as arch action can be depended at lower levels.

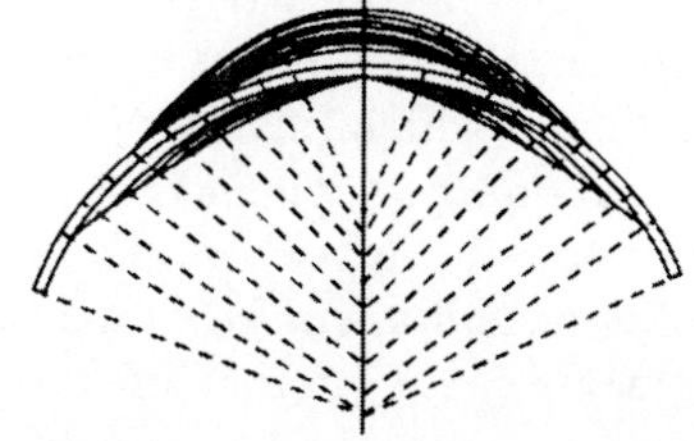

Fig. 19.3 Variable centre arch dam.

Advantages. (*i*) Greater arch efficiency, (*ii*) Saving in concrete; over 50% saving in concrete compared to constant radius arch dam, (*iii*) Useful to meet foundation requirements, and (*iv*) Preferred than constant radius arch dam.

3. Constant Angle Arch Dam. It is a type of arch dam in which all the arch rings have a constant angle (Fig. 19.4). It is thus an improvement on variable radius type arch dam in that the central angle of the horizontal arch rings at all elevations is constant. The central angle of the arch for most economical arch section is usually 133°. In a U-shaped valley, central angle of about 133° can be adopted but a V-shaped valley gives very low angle at the bottom. The constant angle arch dam is not easy to adapt to existing contour lines. However, it is most economical, 30 to 40 per cent saving in concrete compared to constant radius arch dam.

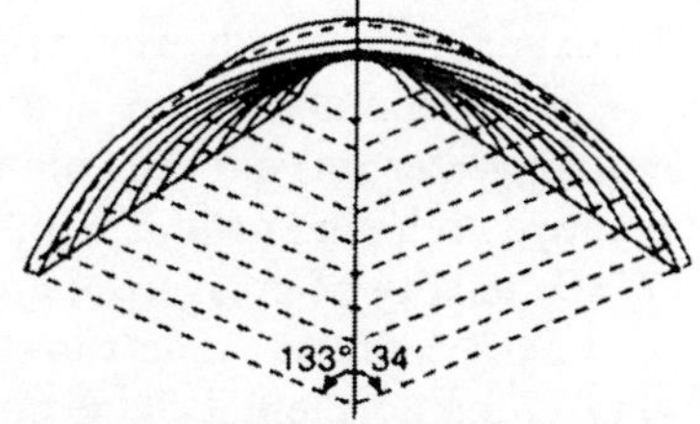

Fig. 19.4 Constant angle arch dam.

Modern dams are often neither constant angle, nor constant radius dams, but thin shells adapted to the shape of the valley.

19.3 FORCES ACTING ON ARCH DAM

Various forces acting on an arch dam are the same as for gravity dam but their relative importance is different in arch dams, as under. The temperature changes are very important as they cause deflections and stresses in arch dams.

1. *Reservoir water load.* It is the principal live load acting on the dam. Most of the water load is carried horizontally to the abutments by arch action depending on the amount of the curvature of the arch. The balance water load is transferred to the foundation by cantilever action.

2. *Uplift pressure.* Unlike gravity dams, uplift pressure in arch dams is not so important because arch dams are very thin in section. However, if cracks occur, uplift in cracks results in increase in downstream deflection, changes in load distribution accompanied by increases in maximum compressive stresses in the arch and cantilever elements.

3. *Ice pressure.* Ice pressure causes heavy concentrated loading on top arch, more precisely along the arch element at the level of the ice. The internal stresses caused by ice pressure are important in arch dams.

4. *Temperature changes.* Internal stresses caused by temperature changes are important, their effect being to (*i*) move the dam upstream during summer and (*ii*) to move the dam downstream during winter, thereby augmenting reservoir loads. In stress analysis, the latter condition, *i.e.*, moving of the dam downstream during winter is very important as it acts in the direction in which the reservoir water load acts.

5. *Yielding of abutment.* In order that the arch section is more effective, the ends of the arch shall not yield. It follows that the rock of the canyon constituting the abutments of the arch shall not yield under any force and movement developed at that section. Slight yielding of abutments cause internal stresses in the arch.

19.4 DESIGN OF ARCH DAM

The various methods commonly used for the design of arch dam are as under:

1. Thin Cylinder Theory. In thin cylinder theory of arch design it is assumed that (*i*) The arch is cylindrical, the restraint being provided in its radial movement, (*ii*) Stresses in the arch are approximately the same as in a thin cylinder of equal outside radius, (*iii*) Pressure on the ring is equal to the full hydrostatic pressure at the elevation, (*iv*) Weight of concrete and water is directly carried to the foundation, (*v*) Horizontal water thrust is taken up entirely by the arch action, and (*vi*) Stress distribution across arch thickness is uniform. Fig. 19.5 shows the section of arch as part of a cylinder of mean radius *r*, thickness *t* and central angle θ.

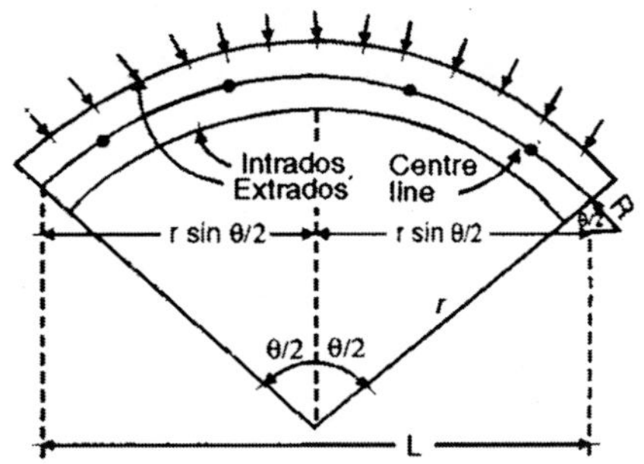

Fig. 19.5 Thin cylinder theory.

Horizontal radial hydrostatic pressure, $P = wh$ (t/m^2)

where, P = horizontal pressure acting in radial direction, w = unit weight of water (1 t/m^3), and h = depth of water (m) over the plane in consideration. The total hydrostatic force (PH) acting on the entire arch ring of unit height is given by :

PH = intensity of pressure × projected area = $wh\ 2r \sin \theta/2$,

$= 2w.h.r. \sin \theta/2$; it acts in the downstream direction along the axis of symmetry of the arch ring in the horizontal plane and is balanced by the components of reactions at the abutments in the upstream direction.

If R is the reaction at each abutment, its component acting upward is $2R \sin \theta/2$.

For equilibrium, $2R \sin \theta/2 = 2.w.h.r.\sin \theta/2$

or $$R = whr$$

This reaction at the abutments is equal to the maximum permissible compressive stress σ introduced in the arch. Therefore,

$$\sigma = R/t = \frac{whr}{t}$$

or
$$t = \frac{whr}{\sigma}$$

Also
$$t = \frac{w.h.r_i}{\sigma - wh}, \text{ where } r_i \text{ is intrados radius}$$

$$t = \frac{w.h.r_m}{\sigma - 0.5\,wh}, \text{ where } r_m \text{ is mean radius}$$

Central angle. The central angle of the top arch influences the curvature of the entire dam. The central angle is selected from the consideration of economy. For a minimum volume of concrete in the dam, the surface of each individual arch element is made small consistent with low stresses:

$$V = \theta rt, \ = \frac{whr^2\theta}{\sigma}, \ t \text{ being} = \frac{whr}{\sigma}$$

$$l = 2r \sin \theta/2$$

$$r = l/2 \sin \theta/2$$

Therefore,
$$V = \frac{\theta l. w. h. l}{2 \sin \theta/2\ \sigma 2 \ \sin\ \theta/2}$$

$$V = \frac{\theta l^2\ wh}{4\sigma \sin^2\ \theta/2}$$

$$V = \frac{whl^2\ \theta}{4\sigma \sin^2\ \theta/2}$$

For minimum volume,
$$\frac{dV}{d\theta} = 0$$

or
$$\frac{dV}{d\theta} = \frac{\sin^2(\theta/2) - \theta \sin(\theta/2)\cos(\theta/2)}{\sin^4(\theta/2)} = 0$$

or
$$\tan \theta/2 = \theta$$

or
$$\theta = 133°\,34'$$

Thus the most economical angle is 133° 34′. For practical reasons of economy in cost of shuttering, the angle is usually less than most economical theoretical value and may vary between 110° to 125°. A large central angle has the merit that reaction at the support is correctly oriented with respect to the sides of the gorge so that the mass of rock supporting the reaction is sufficient for transmission of the force.

Limitations. The cylinder theory has the limitations that (*i*) The arch sections are not thin cylinders, (*ii*) The arch ring is not free to move at the abutments and the fixing reduces movements and shears, (*iii*) Shearing and bending stresses, temperature stress and ice thrust besides stress due to rib-shortening and that due to yielding of the abutments are not taken into consideration, and (*iv*) The formula for thin cylinders yields unreliable results and is used only as a rough approximation.

2. Thick Cylinder Theory. Contrary to the thin cylinder theory wherein the stress distribution at a cross-section at any level is considered constant due to small thickness of the arch as compared to the radius, in thick cylinder theory of design of arch dam the actual variation of stress is considered. All the radial cross-sections are assumed to be subjected to similar stress-whereas the formula gives higher stress at the intrados at all sections, yet in bottom arches this is true only near the abutments, while at the crown the stresses are reversed. For top arches the conditions are quite the opposite.

3. Elastic Arch Theory. In this theory it is assumed that the entire external load acting on the dam is transferred horizontally to the abutments through arch action only and that the entire dam consists of a number of horizontal arch rings placed one over the other, their ends fixed and each ring is independent of the other. The cantilever action is neglected. The thin cylinder theory has certain defects as it takes into consideration only uniformly radial pressure, but the elastic arch theory takes into account variable loads along the length of the arch as also ice thrust, etc. In elastic theory of design of arches, additional stresses due to rib shortening, temperature changes, shrinkage and yielding of abutments are also considered. The ends of the arch dam are fixed and when water pressure acts radially from upstream, side rib shortening effect takes place which causes additional stresses. Similarly variation in temperature, shrinkage and yielding of abutment cause additional stresses.

4. Trial Load Analysis. The assumption in thin cylinder theory and elastic theory that any horizontal arch ring is independent of the arch ring below and above it is not true. The trial load method developed by USBR gives reasonably satisfactory approximation. The trial load analysis hinges on the assumptions that (*i*) Foundation deformations are independent of the shape of the foundation surface, (*ii*) Rock formations in foundation and concrete in dam are homogeneous and isotropic. Modulus of elasticity of concrete and rock foundation is the same for compression and tension, (*iii*) Water pressure on reservoir walls dose not cause differential movements at dam site, (*iv*) Direct stresses vary linearly from upstream to downstream faces, (*v*) Plane sections before bending remain plane after bending, (*vi*) Hooke's law is applicable, and (*vii*) Temperature strains are proportional to temperatures.

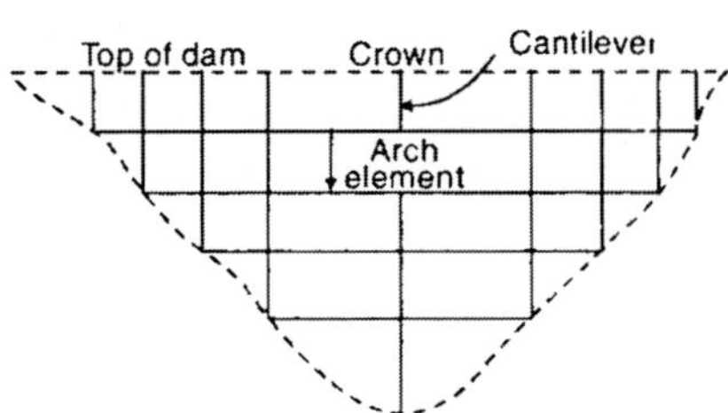

Fig. 19.6 Trial load analysis

The dam is assumed to be made up of two systems of elements, viz. horizontal arches transmitting thrust to abutments and vertical cantilevers fixed at the foundation, both the horizontal and vertical elements are considered a unit distance apart and representative units are selected for analysis (Fig. 19.6). Each system occupies the whole dam structure and the loading is assumed to be divided between them in such a manner that equal movements result in all directions at points of intersection of these elements. The horizontal component of water load is resisted jointly by arch action and cantilever action. The arches are considered to carry part of water pressure and total load due to other forces, while the cantilever takes the remaining water pressure and total load due to other forces. The distribution of load between the arch system and cantilever system is determined by trial and error. The stresses in the dam are computed on the basis of final load distribution. The dam dimensions fixed by the forementioned methods are analysed by this method and modified to arrive at final design. The trial load method is widely used in USA for arch dam designs; the computations are programmed for high speed electronic computers.

5. Finite Element Method. The method has the underlying principle to represent a continuous 2-dimensional system as an assemblage of finite 2-dimensional flat elements interconnected at a finite number of nodal points. The elements may be rectangular as in cylindrical shells or triangular as in curved shells. In the analysis, the stiffness of individual elements are first determined and then superimposed to obtain total stiffness.

Preliminary Design of Arch Dam

Salient dimensions of double curvature arch dam as per USBR practice are as under:

(*i*) Thickness at crest (T_C) $= 0.01\,(H + 1.2\,L_1)$ m

(*ii*) Thickness at base (T_B) $= 3\sqrt{0.0012\,H\,L_1\,L_2\left(\dfrac{H}{122}\right)^{\frac{H}{122}}}$ m

(*iii*) Thickness at 0.45 H, $T_{0.45H}$ $= 0.95\ T_B$

(*iv*) Crown cantilever projections

(*a*) At crest

Maximum u/s projection, (USp) $= 0.0$

Maximum d/s projection (DSp) $= T_c$

(*b*) At base

USp $= 0.67\ T_B$

DSp $= 0.33\ T_B$

(*c*) at 0.45 H

USp $= 0.95\ T_B$

DSp $= 0.0$

Concrete volume (v) $= V_1 + V_2$ m³

where,

$$V_1 = 0.000177\ H^2.L_2\left[\frac{(H + 0.8L_1)^2}{L_1 - L_2}\right]$$

$$V_2 = 0.0108\ H.L_1\,(H + 1.1\ L_1)$$

The upper end limits of above equations are :

$$30 \le H \le 370$$
$$30 \le L_1 \le 1825$$
$$4.5 \le L_2 \le 370$$

Fig. 19.7 illustrates the dimensions of a double curvature arch dam as per USBR practice. L_1 is straight line distance at crest level between abutments excavated to sound rock and L_2 is the straight line distance at 0.15 H above base, between abutments excavated to sound rock.

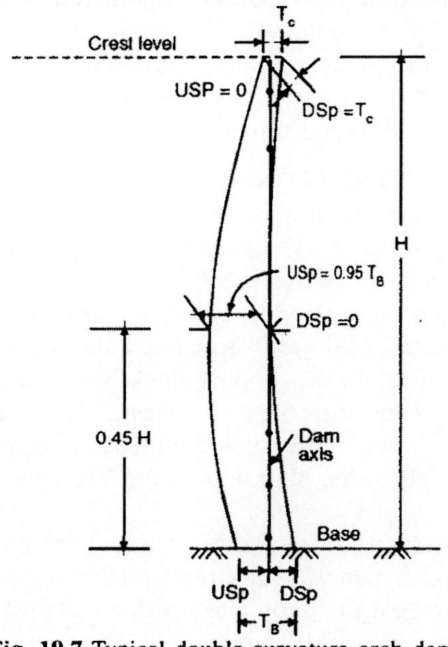

Fig. 19.7 Typical double curvature arch dam.

19.5 BUTTRESS DAMS

In a gravity dam the quantity of concrete is determined by the dead weight required for the stability of the dam, rather than by the strength of the concrete. Thus in gravity dams there are lower stresses in the dam body in central part of the dam and the high strength concrete there is not stressed as it could be, and hence lesser efficient use of concrete strength, large uplift force on dam base and hence increased dam section for stability. A buttress dam envisages to reduce the total volume of concrete by concentrating the material where the stresses are higher and other undesirable features of gravity dam but has higher cement content per cubic metre of concrete than a gravity dam. On steep hill slopes on both sides the buttresses are subjected to severe torsion effects which present formidable design problems requiring heavy anchorages and reinforcement for torsion moment and shear. Spillway arrangement would be difficult and costly.

A buttress dam is defined as a dam consisting of a relatively thin water supporting facing or deck supported by buttresses generally in the form of equally spaced triangular walls or counterforts that transmit the water load and deck weight to the foundations. The upstream face being inclined, usually at an angle of about 45°, water pressure exerted by it add to the stability of the dam. Buttress dams are adaptable to overflow and non-overflow conditions. In overflow dam, another deck at the downstream face is provided to guide the falling flow. The essential components of a buttress dam are:

(*i*) *Deck.* The upstream sloping membrane or deck which supports the water load and transmit it to a series of buttresses or counterforts. The water tight deck is continuous at the foundation, where cutoff is provided to reduce or prevent seepage.

(*ii*) *Buttresses.* The buttresses or counterforts are provided at certain intervals at right angles to the axis of the dam to support the deck and for transference of the load to the foundation. Corbels help transmit load from the upstream deck to buttresses.

(*iii*) *Lateral braces.* A system of cross bracing between buttresses is provided to effect lateral stiffness and resistance to buckling of counterforts.

(*iv*) *Foundation.* Mat foundation provides distribution of the pressure onto the soil.

Advantages. (*i*) Thin section with one-third to one-half economy in concrete compared to gravity dam of similar height, (*ii*) No problem of uplift or foundation drainage, (*iii*) Can be constructed on weak foundation, and (*iv*) Vertical component of water resting against upstream sloping deck add to the stability of the dam.

Disadvantages. (*i*) The thinner section is more vulnerable due to deterioration of concrete, (*ii*) Susceptible to damage or destruction by sabotage during war, (*iii*) Saving in concrete is largely offset by more reinforcement required, and (*iv*) Higher cost of form work.

19.6 FORCES ACTING ON BUTTRESS DAM

Forces acting on buttress dam are similar to those acting on gravity dam and arch dams except that (*i*) Downward component of water pressure is greater, (*ii*) Uplift is less due to lateral escape of the uplift pressure under the buttresses. Uplift from water head on rock foundation is usually neglected. However, the foundation is drilled for drainage where the rock is liable to uplift pressure on horizontal seams. Full uplift from tail water is considered, (*iii*) Ice pressure is not important as ice tends to slide up the sloping deck, (*iv*) Wind pressure is important if a diagonal wind of high velocity can reach the downstream side. It is, however, of little importance in the case of well braced or double walled buttresses, and (*v*) Because of small mass relative to rigidity of individual units, slab and round head buttress dams are efficient in resisting earthquake loadings.

The buttress dams are analysed for stability in a manner similar to gravity dams. The design element is taken as a full panel, rather than a slice of unit thickness. The essential design procedure envisages fixing the economical spacing and upstream slope of buttress dam, design of deck and buttresses and overall stability check.

19.8 TYPES OF BUTTRESS DAMS

1. Rigid Buttress Dams. Rigid buttress dams have rigidity of construction. The upstream sloping deck is made monolithic with the buttresses. Rigid buttress dams are suitable as high dams. The spacing of buttresses and layout of arches can be made to get best advantage of the foundation. Two types of rigid buttress dams are (*i*) multiple arch dam, and (*ii*) multiple dome buttress dam.

(*i*) *Multiple Arch Buttress Dam.* It comprises a series of inclined arches supported by buttresses. The water load is transferred from the arches to the foundation through the buttresses. Each arch does not constitute a separate structure due to the thrust at arch springing which are balanced. The two adjacent arches may abut on a common buttress (Fig. 19.8 *a*), or abut on independent walls — the buttress being two walled interconnected by horizontal connections (Fig. 19.8 *b*). The arch may be of uniform thickness or of variable thickness. The main considerations are slope of upstream face, central angle of each arch (between 150° to 180°) and buttress spacing.

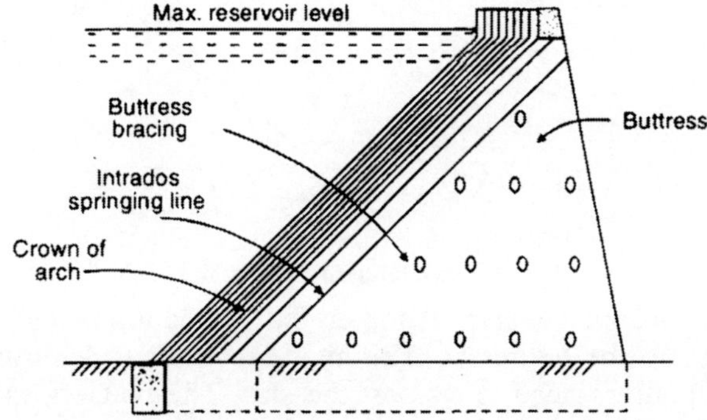

Fig. 19.8 (*a*) Multiple arch dam with adjacent arches abutting on a common buttress.

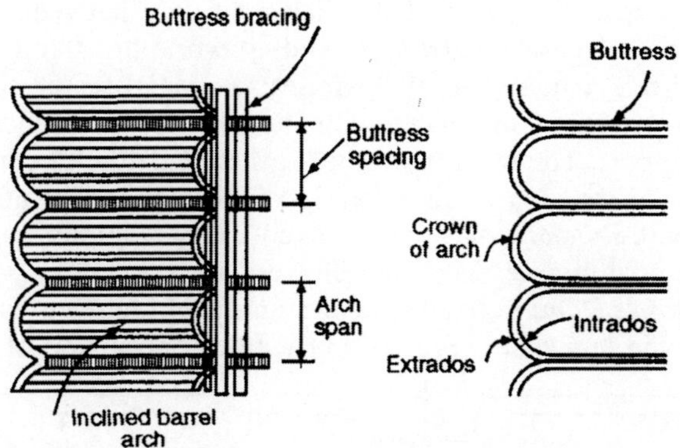

Fig. 19.8 (*b*) Multiple arch dam with buttress interconnected horizontally.

Advantages. (*i*) Economical in cost, (*ii*) Relatively stronger than flat slabs. Thinner deck permits a wide economical spacing of buttresses, (*iii*) Most suited for high dams, (*iv*) Suited for all types of foundations and better adapted to temperature changes.

Disadvantages. (*i*) Buttresses are thicker and more substantial than flat slab buttresses, (*ii*) Thin arches respond rather rapidly to climatic changes, (*iii*) Variation in pressures due to alternate filling and emptying of the reservoir causes frequent reversal of movements tending to reduce the life of the dam, and (*iv*) Unequal settlement of foundation is objectionable. Stability of any unit depends upon the adjacent unit so that any settlement in one may effect other units.

(*ii*) *Multiple Dome Buttress Dam.* It is buttress dam in which massive buttresses spaced far apart support a set or series of domes on which the water rests (Fig. 19.9). The arches act as water retaining membranes stretching across the length of the dam. The arches are under compression.

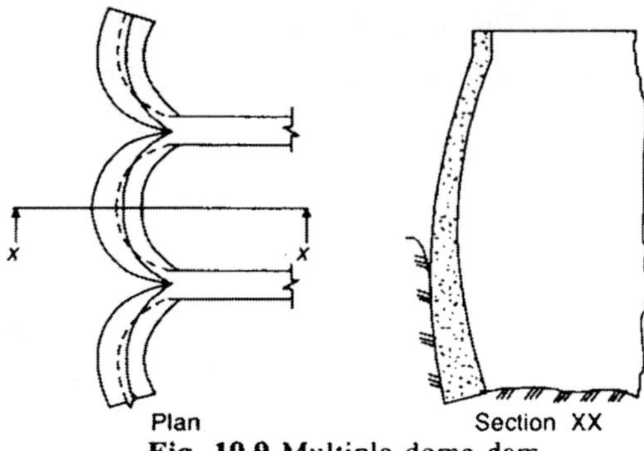

Fig. 19.9 Multiple dome dam.

2. Deck Slab Buttress Dams. Deck slab buttress dam (Fig. 19.10) is made of a flat reinforced concrete slab spanning across a series of parallel buttresses which rest directly on the foundation or upon a concrete slab resting on the foundation material. The water pressure is transmitted by the slab to the buttresses. The main layout considerations are spacing of the buttresses and shape of buttress head to support the slab. The buttress spacing is governed by the consideration of economy, *i.e.,* which gives minimum thickness of slab and buttress and concrete at all sections is fully stressed. The buttress spacing, depending on the height of dam, varies from 5 to 15 m. The upstream slope of the slab may vary between 35° to 45° with the horizontal. Too steep a slope increases tendency to sliding while too flat a slope causes a very high inclined stress at the heel. Articulation of flexibility, a special feature of this type of dam, is provided by not rigidly attaching the slab to the buttress. It is provided by having keys between the slab and the support. The recesses in the keys are filled with compressible material. It is to allow for expansion of the slab in directions parallel to base line of the dam and along the slope. The flat slab buttress dam may be of fixed or continuous deck slab buttress or cantilever deck slab buttress or simple deck slab buttress dam.

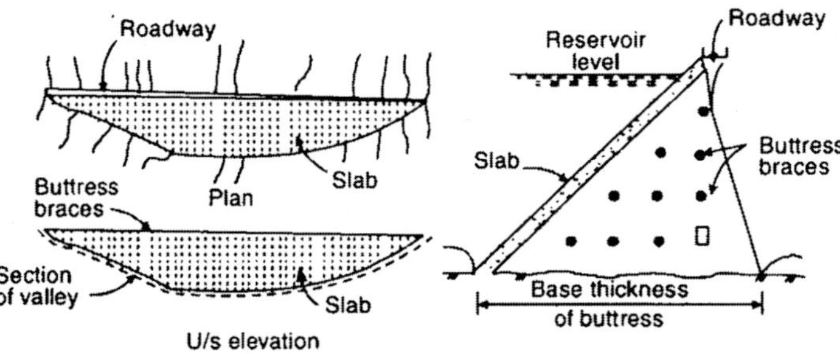

Fig. 19.10 Flat slab deck dam.

(*i*) *Fixed Deck Slab Buttress Dam:* In fixed or continuous deck slab buttress dam, the deck slab is cast monolithically with the buttresses (Fig. 19.11) with the provision of expansion joints at

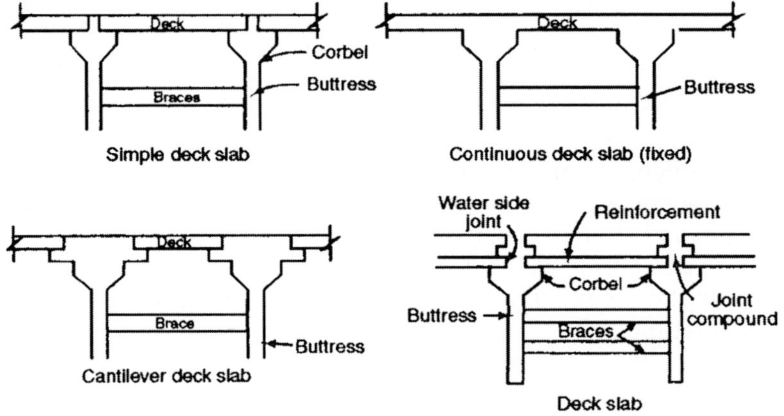

Fig. 19.11 Various types of deck slab buttress dams.

convenient places. The stresses are there both at the upstream and downstream faces of the deck slab. Significant stresses are also there due to temperature variation, foundation settlement, etc. In view of reversing bending moments, reinforcement is provided both on the upstream and downstream faces of the slab.

(*ii*) *Cantilver Deck Buttress Dam:* In this type of dam, as the name implies, the slab is laid monolithic with the buttress but cantilvered at both ends, with proper joints at both ends or at the centre (Fig. 19.11). The water load is transferred to the supporting buttress by cantilever action of the deck and reinforcement is provided at upstream face in view of cantilever action. It is subjected to high stresses due to temperature, shrinkage and foundation settlement. It is suitable for low dams.

(*iii*) *Simple Deck Slab Buttress Dam:* Also termed as free deck type or Amburson dam (Fig. 19.12). The reinforcement is placed at the downstream face of the deck slab and as such is not susceptible to rusting. The influence of temperature, shrinkage and foundation settlement is less. This type of dam is suitable for a wide valley with poor foundation. It requires thicker slab.

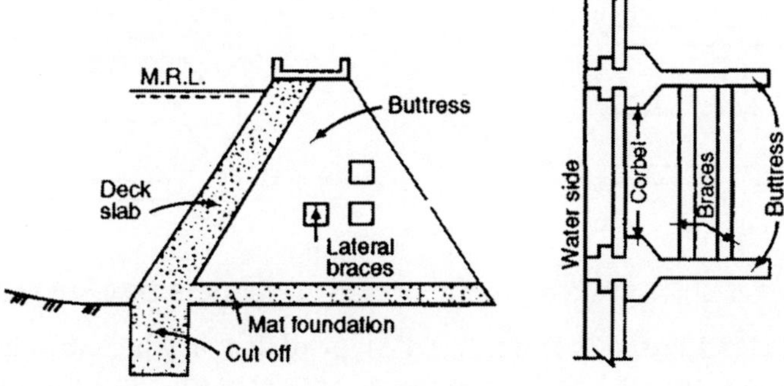

Fig. 19.12 Deck slab buttress dam.

3. Bulk Head Buttress Dam. It is a type of buttress dam in which the face slab is replaced by flaring the upstream edge of the buttresses to span the distance between buttress walls. It consists of a series of buttresses with massive head forming the water supporting member. The massive heads and buttresses, though independent structural units, behave as one monolithic unit for stability of the dam section. The flaring portion may be of different geometrical shapes.

Advantages: (*i*) Simple in construction, (*ii*) Water pressure is compressive and no reinforcement is required in the deck, (*iii*) Free from appreciable uplift, (*iv*) Stability against overturning is achieved with less concrete than in a gravity dam, (*v*) Minor relative settlement is tolerable as each unit is structurally independent of others, and (*vi*) Buttress can be spaced to suit the availability of good rock material in the foundations.

Disadvantages: (*i*) More form work is required, and (*ii*) Excavation is more if foundation overburden is large.

Types of Bulk Head Buttress Dam

1. Massive Head Buttress Dam. It consists of a series of buttresses with upstream massive heads flat as in deck slab abutting each other. The head and buttress together resemble letter T (Fig. 19.13).

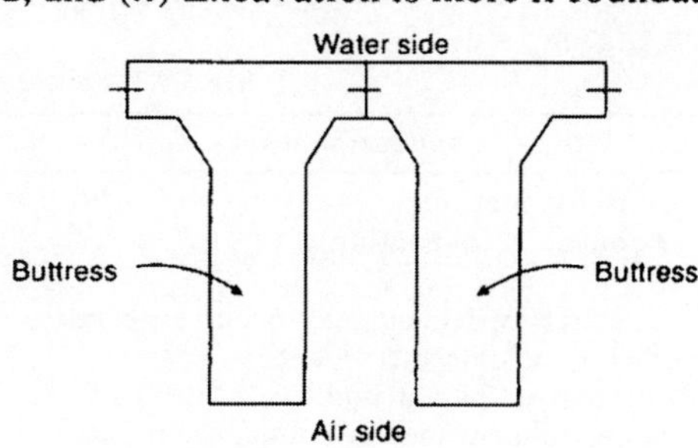

Fig. 19.13 Massive head buttress dam.

2. Round Head Buttress Dam. In this type of dam, the massive heads forming the water supporting member are circular (Fig. 19.14); the shape of round head is such that water pressure acts radially and is compressive. The purpose of round head is to eliminate reinforcement but buttresses are made massive and unreinforced. At the two adjacent massive heads water tightness is achieved by providing two layers of copper seals with an asphalt seal in between.

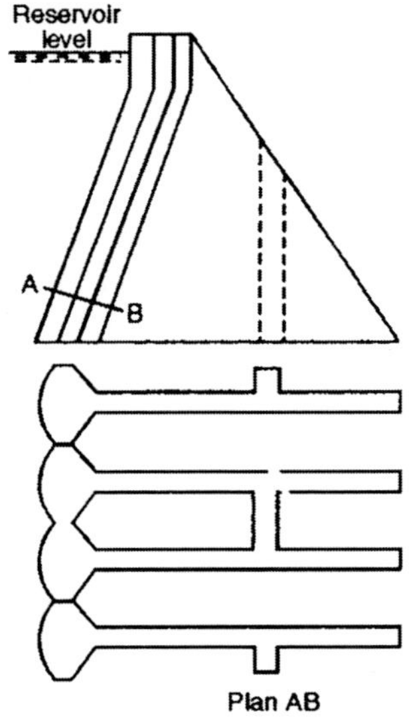

Fig. 19.14 Round head buttress dam.

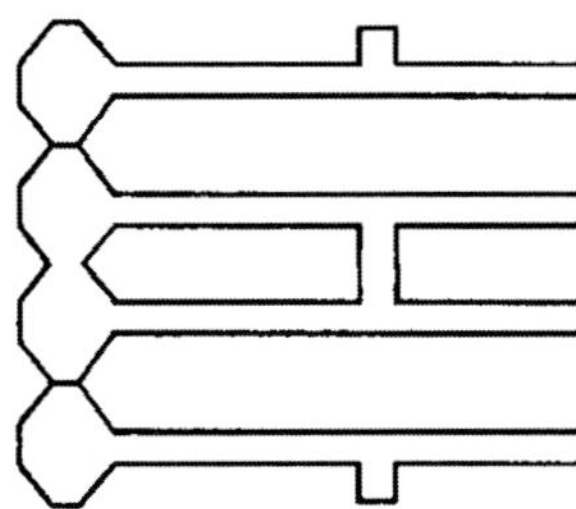

Fig. 19.15 Diamond head buttress dam.

3. Diamond Head Buttress Dam. It is the type of dam in which the massive heads are sharp and like a polygon (Fig. 19.15). The pressures act perpendicularly to the surface of the diamond head and are compressive.

4. Miscellaneous Types. (*i*) *Columnar Buttress Dam.* It consists of a continuous flat slab supported on a number of inclined columns rather than buttresses. The columns are stiffened by a two-directional system of struts.

(*ii*) *Truss Buttress Dam.* It is similar to columnar buttress dam with the exception that the deck slab is supported on reinforced concrete trusses.

19.8 SELECTION OF TYPE OF BUTTRESS DAM

The selection of various types of buttress dam depends on:

1. *Foundation conditions.* For good foundation any type of buttress dam is suitable. The type of dam suitable for various foundation rocks is roughly as stated in Table 19.1.

2. *Requirement of spillway, irrigation or power outlets.* Spillway can be conveniently provided at the top and outlets at the base of the buttress dam. The control equipment can be housed in the chamber constructed inside the buttresses.

3. Other considerations are economy governed by the availability of construction materials, skilled labour, etc.

Table 19.1 Foundation rock and dam type

Foundation rock	*Suitable type of dam*
Soft foundation	Any of buttress types dam with spread footing.
Permeable foundation	Slab and buttress dam with bottom slab under buttresses. Percolation and uplift are considered.
Unequal settlement may occur especially during seismic disturbances	Slab and buttress dam with buttresses strutted.
Jointed or faulted rock	Slab and buttress dam.
Good foundation, no unequal settlement	Multiple arch dam.
Variety of foundations	Massive head buttress dams.

QUESTIONS

1. Define arch dam. Discuss various types of arch dams and their merits and demerits.
2. For a fixed combination of span, loading and permissible cylinder theory stress, prove that the economic arch dam is one which subtends a constant angle of 133° 34′ at centre.
3. What are the situations to which the various types of dams are most suited, viz. arch dam, earth dam, and rockfill dam? What are the basic principles of their design?
4. What are arch and rockfill dams? Illustrate with sketches and mention site conditions favourable for construction of such dams.
5. What are arch and buttress dams? Illustrate with sketches and mention site conditions favourable for construction of such dams.
6. Discuss the various forces which act on an arch dam.
7. What is buttress dam? Discuss the functions served by its component parts.
8. (*a*) Explain the theory of cylinder action in the case of an arch dam.

   (*b*) Design a 100 m high constant angle arch dam by thin cylinder theory for a valley 20 m wide at the bottom and 220 m wide at 100 m height. The permissible compressive stress is 500 t/m^2.

Solution :

Central angle = 133° 34′ (assumed)

Chord length, l of top arch = 220 m

Radius of intrados, r_i = 119.68 m

($0.544\, l = 0.544 \times 220$)

$$\text{Thickness of arch, } t = \frac{Pr_i}{\sigma - P} = \text{cal.5} + \text{cal.6}$$

COMPUTATIONS

Depth of arch (m)	l (m)	r_i (m)	$P = wh$ (t/sqm)	Pr_i (t/m)	$\sigma - p$ (t/sqm)	t (m) cal.5÷cal.6	$r_e = r_i + t$ (m)	r_m (m) cal.8−t/2
1	2	3	4	5	6	7	8	9
0	220	119.68	0	0	500	0	119.68	119.68
10	200	108.8	10	1088	490	2.22	111.02	109.91
20	180	97.92	20	1958.40	480	4.08	102.0	99.96
30	160	87.04	30	2611.20	470	5.56	92.60	89.82
40	140	76.16	40	3046.40	460	6.62	82.78	79.47
50	120	65.28	50	3264.00	450	7.25	72.53	68.91
60	100	54.4	60	3264.00	440	7.42	61.82	58.11
70	80	43.52	70	3046.40	430	7.08	50.60	47.06
80	60	32.64	80	2611.20	420	6.22	38.86	35.75
90	40	21.76	90	1958.40	410	4.78	26.54	24.15
100	20	10.88	100	1088.00	400	2.72	13.60	12.24

9. (*a*) What are the advantages of buttress dams over other types? [P.U.]

   (*b*) Show with the aid of a neat sketch the general dimensions for buttress corbel and flat slab deck.
10. Discuss cylinder theory for arch and obtain an expression for the thickness or arch ring in terms of external radius, height of dam and allowable stress.
11. What are arch dams? Illustrate with examples and mention site conditions favourable for construction of such dams.

20

DISCHARGING FACILITIES

20.1 SPILLWAY

The various discharging facilities provided in a storage dam are spillway, intake and outlet works. Spillway is the escape provided in conjunction with a dam to dispose of surplus flood water from a reservoir. The surplus water is drawn from the reservoir and conveyed back to the river downstream or some natural drainage. Spillway is provided at storage dam and diversion dam to serve the designed purpose. For a storage dam, spillway is provided to release surplus flood water which cannot be contained in the allotted storage space in the reservoir. For a diversion dam, spillway is provided to bypass the flood water in excess of that turned into the diversion system. The spillway thus acts as a safety valve in the reservoir against failure of dam due to overtopping. It implies that the spillway is of adequate capacity to cater for spillway design flood. It is particularly applicable to earth and rockfall dams where overtopping is disastrous, whereas concrete dams may be able to withstand moderate overtopping. Many failures of dams have been caused by improperly designed spillways or spillways of insufficient capacity. The cost of providing spillway of adequate capacity is only moderately higher than that of one which is obviously too small but the consequence of an inadequate spillway may be catastrophic. Besides safety of the dam, the spillway governs (*i*) the length and height of dam, (*ii*) maximum flood levels attained both upstream and downstream of the dam, (*iii*) submergence in the upstream, and (*iv*) areas susceptible of flooding downstream.

The inlet and outlet works are devices to release water from the reservoir for irrigation and /or power generation etc.

20.2 ESSENTIALS OF A SPILLWAY

The essentials of a spillway are (*i*) Adequate capacity to serve as moderation of floods, (*ii*) Provide for safe and regulated release of the surplus water in excess of the reservoir capacity, (*iii*) While conveying the excess flows downstream the tail water maintained is such that the purpose and protection of dam and its appurtenant works are fully ensured, (*iv*) Hydraulically and structurally the spillway's bounding surfaces are adequate and are erosion resistant to withstand high scouring velocities created by the drop of flows from reservoir surface to tail water, and (*v*) Device for the dissipation of energy at the bottom of the dam is provided.

The spillway size and type depends on the best combination of storage and spillway capacity to accommodate the selected inflow design flood which in turn depends on (*i*) The character of the flood hydrography, (*ii*) Effects of various dam and spillway combinations on probable increase or decrease or damages above or below the dam, (*iii*) Relative costs of increasing spillway capacity, and (*iv*) Use of combined outlet facilities to serve as control of releases and control or passage of floods.

Spillway capacity. Apart from serving the primary purpose of safely passing a predetermined inflow design without any damage to its structure and other components of the dam thereby reducing the hazard to negligible or acceptable level, the spillway capacity is fixed so as to fulfil one or more of the additional purposes such as (*i*) To provide a small surcharge to compensate the effect of reservoir in decreasing the naturally available valley storage and in decreasing the travel time of the flood waves from the natural conditions so as not to change the natural flood regime on the downstream in a detrimental manner, (*ii*) To change the flood regime downstream to reduce the frequency of flood damage through provision of surcharge storage, (*iii*) To moderate the floods passed downstream by using a permanent flood control capacity or a joint use capacity in the reservoir, (*iv*) To control the backwater upstream of the reservoir so that the frequency of flooding of an upstream structure or property is reduced to acceptable limits, and (*v*) To economise the dam design by adopting an appropriate combination of surcharge storage versus spillway capacity.

Inflow Design Flood. The inflow design flood depends upon various factors such as safety of the dam against overtopping, performance of energy dissipation arrangements, upstream submergence of land and downstream flood hazard. The inflow design flood for dam safety is 100 year flood for small dams, standard project flood for intermediate dams and probable maximum flood for large dams. The floods of large or smaller magnitude may be used if the hazard involved in the eventuality of a failure is particularly high or low. The relevant parameters, in addition to size, for considering hazard are (*i*) Distance and location of the human habitations on the downstream after considering the likely future developments, (*ii*) Maximum hydraulic capacity of the downstream area at a level at which catastrophic hazard is not apprehended, (*iii*) For important projects dam break studies are done to decide whether PMP need to be used. In case damage to present or future human settlements is apprehended, PMP may be used.

20.3 SPILLWAY COMPONENTS

Spillway may be constructed as an integral part of the dam or as independent structure but essentially comprises components discussed as under:

Approach Channel : It is an entrance structure to draw water from the reservoir and convey it to the control structure. The approach channel may be a straight or curved in plan. Its banks may be parallel, convergent, divergent or combination of these and may be vertical or sloping. In order to ensure minimum head loss through the channel (which has the effect of reducing spillway discharge) and to obtain uniformity of flow over the control structure (which too has the effect of reduction in spillway discharge), the entrance velocities are limited and channel curvatures and transitions, if any, are gradual.

Control Structure : It consists of an overflow crest provided with a bridge and gates to regulate and control the outflows from the reservoir. The control limits the outflows below fixed reservoir levels and also regulated release when the reservoir rises above that level. The length of the crest is finalized from economic considerations and hydraulic model studies as for a given level of the crest, the height of dam increases if the crest is made of short length. (Fig. 20.1)

Discharge Carrier : It is the waterway provided to convey the flows released from the control structure to the river below the dam, except in the case of arch dam wherein the discharge falls free from the crest or where the flow is released directly along the abutment hill side to cascade down the abutment face. The structure of the discharge carrier may be the downstream face of a concrete dam, an open excavated channel, a closed cut and cover conduit placed through or under a dam or a tunnel excavated through an abutment. The dimensions of the discharge carrier are primarily governed by hydraulic requirements besides geologic and topographic characteristics of the site which influence the profile, shape of cross-section, width, length etc. The profile may be variably flat or steep, cross section may be variably rectangular, trapezoidal, circular or of other shape and the waterway may be wide or narrow, long or short.

Terminal Structure : At the downstream end of the discharge carrier where flow from reservoir to downstream river level manifests itself in the form of high velocities enough to cause scour, erosion and subsequent damage to the adjacent structures, a terminal structure usually in the form of stilling basin is provided for dissipation of residual energy.

20.4 Types and Design of Spillways

The site, shape, size and type of spillway and the arrangement of its components depend upon various factors, as stated below:

1. *Topographical.* Site conditions such as steepness of terrain traversed by the spillway control and discharge carrier, type and amount of the excavated material, possibility of utilisation of the excavated material for dam embankment, chances of scour of the bounding surfaces and the need for lining, permeability and safe bearing capacity of the foundation, stability of the excavated slopes, and geological site conditions.

2. *Hydrological.* Inflow and reservoir storage conditions such as inflow discharge, its frequency and shape of hydrography, reservoir capacity at various levels, and length and height of crest.

3. *Purpose of Storage.* Purpose of storage determines whether the spillway is required to be gated and the type of gates.

4. *Type of Dam.* Type of dam governs the spillway design flood. Spillways of certain types are eminently suitable for earth and rockfill dams, while the others are adaptable for concrete dams.

5. *Outlet Facilities.* Possibility of combined outlet facilitites to serve more than one function, such as control of release and control or passage of floods.

6. *Downstream Flow conditions.* Certain types of spillway greatly alter the shape of flood hydrograph downstream of the spillway. Siphon spillway, for example, gives rise to a wave travelling downstream in the river which is detrimental to navigation and fishing.

7. *Safety Considerations.* Spillway of inadequate capacity or improperly designed spillway may cause failure of the dam.

8. *Economic.* Several studies of alternative types of spillway are undertaken to finalize the most economical and hyraulically efficient spillway suited to the discharging requirement and site conditions. Furthermore, various combinations of spillway crest width and height are carried out to obtain a most economical solution. Fig. 20.1 indicates the result of economic studies carried out in respect of spillway at Ramganga Project. The most economical arrangement for all combinations was found to have crest length of 64 m, crest elevation at 351.28 m and normal reservoir level as 365.3 m.

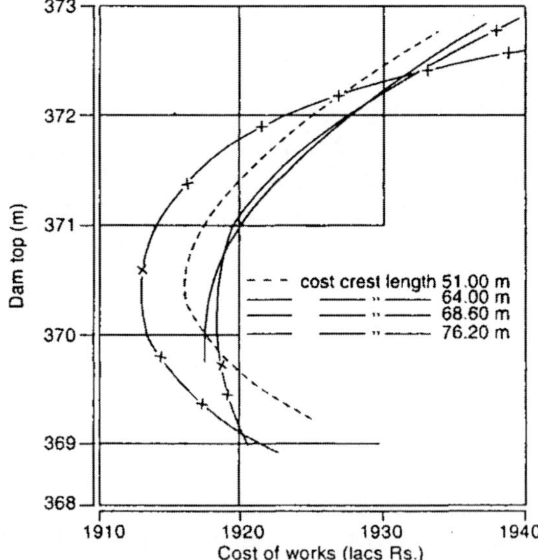

Fig. 20.1 Economics of spillway arrangements.

Types of Spillway

Various types of spillway, depending on the type of the structure constructed for surplussing the flows, are as follows:

1. Overflow Spillway. Also termed as Ogee Spillway or rollway. It comprises a control weir which is ogee or S-shaped. It is a section of a dam designed to permit water to pass over its crest. It is widely used on gravity, arch, buttress and hollow dams. Fig. 20.2 shows overflow spillway section of Nagarjunasagar dam. The ogee spillways are classified as (*i*) high spillway, and (*ii*)

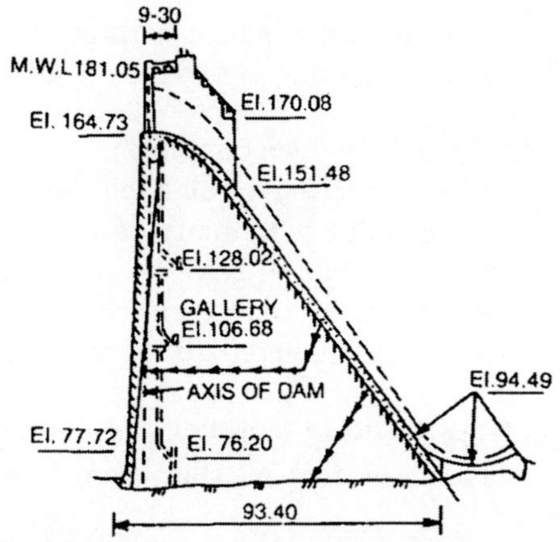

Fig. 20.2 Spillway section of Nagarjunasagar dam.

low spillway depending on whether the ratio of the height of the spillway measured from the river bed to the design head is greater than and/or less than or equal to 1.33 respectively. High spillway is an ogee spillway which is located within the dam body at the deepest portion of the valley and the design head is generally a fraction of the design head of the spillway. The velocity of approach is neglected in this case. A low spillway is built as a control structure for a chute spillway; the design head of which is either equal to or a multiple of the height of the spillway. Hydrodynamically if the downstream and upstream floors of a spillway are low enough so as not to effect the pressure distribution over the crest region, the spillway is called a high spillway, otherwise low spillway.

The ogee shape conforms closely with the profile of the aerated lower nappe of flow from a sharp crested weir. The upper profile of curve at the crest may be made either broader or sharper than the nappe. Broader profile helps in the stability in the crest due to hydrostatic pressure but reducing thereby the coefficient of discharge. The sharper crest increases the effective head and thereby increases the coefficient of discharge.

Downstream profile equation is

$$x^{1.85} = 2.0\, H^{0.85}\, y$$

where, H = design head over crest (m), and x, y = coordinates (refer (Fig. 20.2.)

Upstream profile equation. The profile is tangent to the vertical face with zero slope at the crest to ensure that there is no discontinuity along the surface of flow.

$$y = \frac{0.724(x+0.27H)^{1.85}}{H^{0.85}} + 0.126H - 0.4315H^{0.375}(x+0.27H)^{0.625}$$

x, y are the coordinates (refer Fig. 20.3). It extends up to the point 0.126 H downstream from the crest point.

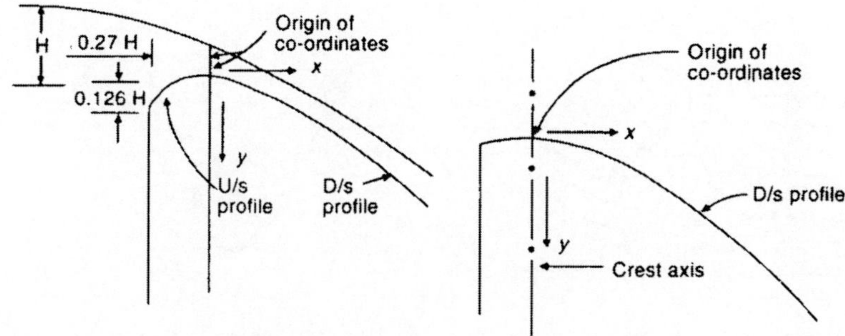

Fig. 20.3 U/s and d/s profiles of ogee spillway.

Discharge through gated spillway at partial gate openings $Q = 2/3\sqrt{2g}\,CL\,(h_1^{3/2} - h_2^{3/2})$, where, h_2 = head on upper edge, h_2 = head on lower edge, C = coefficient of discharge determined by model studies.

Discharge formula, $Q = CLH_e^{3/2}$ in fully open condition.

where, Q = discharge (cumecs), C = coefficient of discharge; usual value is 2.2, H_e = total head of flow on crest including the head due to velocity of approach = reservoir level – crest level + $V_a^2/2g$, L = effective length of crest = $L' - 2(NK_p + K_a)H_e$, L' = net length of overflow crest (m), N = number of piers, K_p and K_a = contraction coefficients of piers and abutment. Values of K_p are 0.01 and 0.02 and K_a are 0.10 and 0.20 for round and square shaped piers and abutments respectively.

2. Free Overfall or Straight Drop Spillway. In this type of spillway, the flow drops freely from the crest. It is suitable for thin arch dam or deck overflow dam or to a crest which has a nearly vertical face. The crest is occasionally extended in the form of an overhanging lip to direct small discharges away from the face of the overfall section (Fig. 20.4). To counteract vacuum which gets created in the underside portion of the falling jet, the underside of the nappe is sufficiently ventilated. To prevent a pulsating fluctuating jet, a secondary dam of low height is sometimes provided on the downstream side to maintain standing water for dissipation of energy of the falling water.

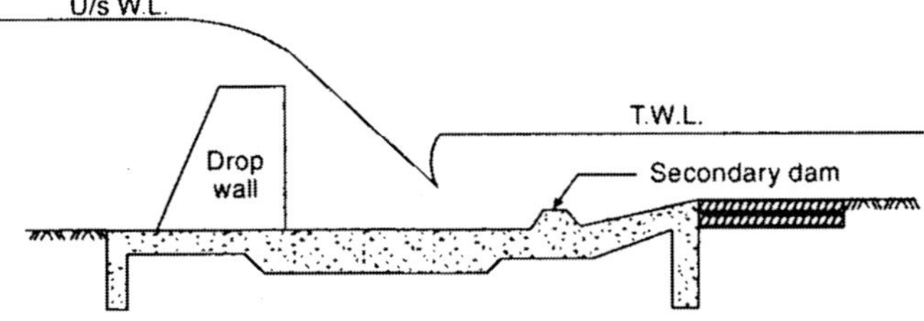

Fig. 20.4 Straight drop spillway.

3. Chute Spillway. Also called trough spillway. It is defined as a steep open channel, normally lined, conveying water usually at supercritical velocity from the reservoir to the waterway downstream of the dam. It is the usual type of spillway provided in earthen dam, rockfill or for dams over which it is not desirable or possible to pass floods. The chute spillway is used where ample room is available for locating it, usually either along a dam abutment (Fig. 20.5a) or through a saddle in the rim of the reservoir (Fig. 20.5b). When located at one end of the dam, it is called *Flank weir* and when in the river bank separated from the main dam, it is termed as *saddle weir.* The spillway is isolated from the dam having a flat crest of low height and normal to the centre line of the dam. The

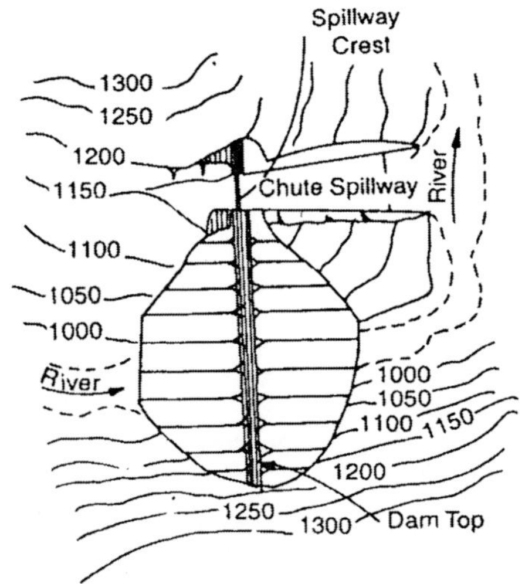

Fig. 20.5 (*a*) Chute spillway.

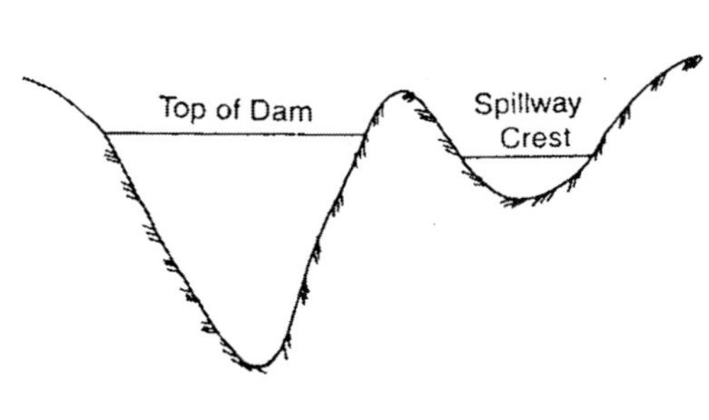

Fig. 20.5 (*b*) Saddle spillway.

length of crest is finalized from economic considerations and hydraulic model studies as for a given level of the crest, the height of the dam increases as the crest is shortened in length. Control at the crest is provided by gates, while stilling basin is constructed at the downstream end for dissipation of surplus energy as also to avoid undermining.

Advantages: (*i*) Simple in design and construction, (*ii*) More suitable for earth and rockfill dams, (*iii*) Adaptability to almost any type of foundation conditions, (*iv*) Overall economy is often achieved by making use of the spillway excavation in dam embankment, (*v*) Location is isolated from the main dam or the highest section of the dam, and (*vi*) Can be designed for any discharge intensity and is capable to pass even more than design flood.

Disadvantages: (*i*) Normally avoided on embankment, (*ii*) Essentiality of space for keeping the spillway basin away from the dam paving, and (*iii*) Straight alignment is required as efficiency is adversely effected if there are too many bends in the chute necessitated because of dam topography.

4. Side Channel Spillway. Also called *lateral flow spillway.* It is a spillway in which the initial and final flows are approximately at right angle to each other. The distinguishing feature is that the flow after passing over the spillway weir or ogee crest is carried away remote from downstream toe of dam by a channel running essentially parallel to the spillway crest (Fig. 20.6.). The discharge characteristics of the spillway are similar to that of an ordinary overflow weir except that at a high discharge it may partly submerge the crest. The discharge characteristics are dependent on the selected profile of the weir crest. The crest is usually a concrete gravity section. The channel receiving water from the spillway is designed so that its section at each point is adequate to carry the discharge accumulated up to that point without submerging the crest to any appreciable extent. The velocity distribution is irregular and average velocity is not the same as mean velocity obtained by dividing the discharge with area. However, mean velocity is generally used in computation of cross-section.

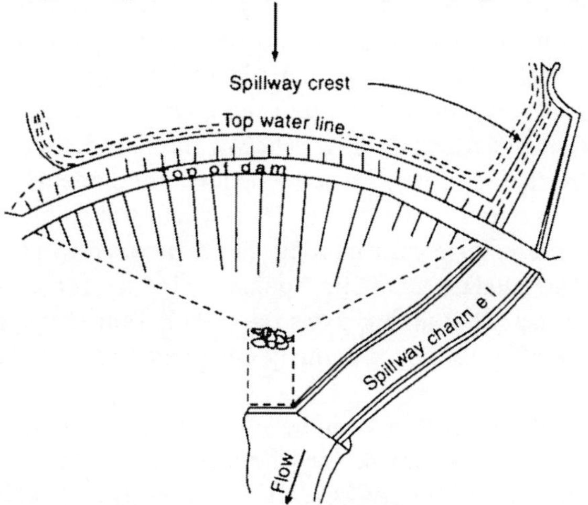

Fig. 20.6 Side channel spillway.

Side channel spillway differs from chute spillway in that water after flowing over the crest turns through 90° to flow parallel to the spillway crest. It is suitable for earth or rockfill dams in narrow canyons (*i*) Where space required for a chute spillway of adequate crest length is not available, (*ii*) Where direct overflow is not allowed, (*iii*) Where spillway discharge is to be connected to a narrow discharge channel, and (*iv*) Sides are steep and rise to a length greater than the dam.

Advantages: The side channel spillway, though neither hydraulically efficient nor inexpensive, has certain advantages which makes it adaptable to certain spillway layout, such that (*i*) Where long overflow crest is required to limit the surcharge head and discharge intensity, (*ii*) Where abutments are steep and percipituous and discharge channel is narrow, and (*iii*) Where the control is desired by the narrow side channel.

Limitations: (*i*) Not efficient hydraulically, and (*ii*) Not suitable for high discharging capacity.

5. Tunnel spillway. Also called *closed conduit spillway.* It is a special adaptation of shaft spillway. It is a spillway in which a tunnel is used as an outlet channel to carry the surplus waters around or under the dam. Closed conduit spillway requires either a very large size conduit or a number of conduits of suitable diameter to cater for high flood. Spillway tunnels are generally straight but may be curved along adequate radius. The tunnel is designed to flow three-fourths full. Tunnel spillway is suitable for a dam site in narrow canyons where there is danger to open channel from snow or rock slides. It may be appropriate at dam sites in wide valleys where abutments rise gradually and are at a considerable distance from the river channel. It has the limitations of discharging capacity and being underground is not open to inspection.

6. Shaft Spillway. Shaft spillway essentially consists of an overflow control weir, usually circular, vertical flaring tunnel turned through right angle at the lower end and connected to a horizontal tunnel (Fig. 20.7.). The surplus water, over and above the storage capacity of the reservoir, approaches in radial direction the horizontally positioned circular crest at the top of the vertical flaring tunnel to spill over the crest negotiating the inner surface of the shaft to find its safe exit downstream through tunnel or closed conduit through or around the dam. In concrete gravity dams, the tunnel is through the body of the dam while in earthen dams it is below the foundation. The shaft spillway is made of three components: (*a*) overflow control weir, (*b*) vertical transition, (*c*) closed discharge conduit. For low dams with a small shaft height, no special inlet design is necessary but on big dams the inlet is funnel shaped and is referred to as Morning glory. This type of spillway is variously designated as drop-inlet spillway or morning glory or glory hole spillway.

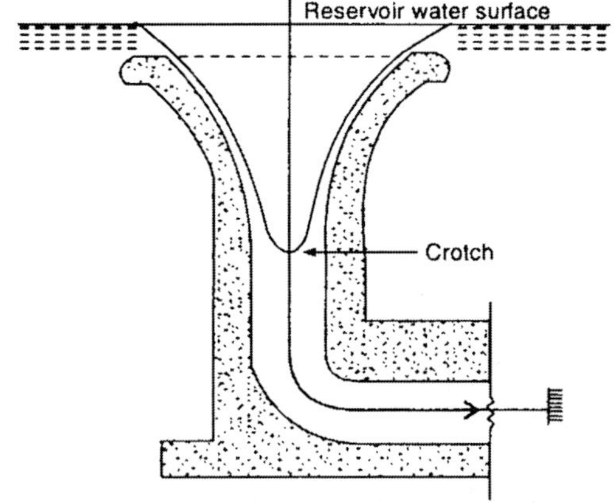

Fig. 20.7 Shaft spillway.

Shaft Spillways can be adopted advantageously at dam sites in narrow canyons where the abutments rise steeply or where a diversion tunnel or conduit is available for use as the downstream leg of the spillway. It is an ideal type of spillway where maximum spillway overflow is to be limited.

The discharge characteristics of the shaft spillway vary with the range of heads. As the head increases, the control shifts from the crest flow to the orifice flow in the inlet transition. and then to the full pipe flow in the downstream portion of the discharge carrier.

Advantages: (*i*) Suitable for dam sites in narrow canyons, (*ii*) Also suitable for low design discharge and operating heads, (*iii*) With a few modifications, it can very well combine in it the functions of diversion tunnel, the scouring sluice and reservoir outlet, (*iv*) There is little increase in capacity beyond the designed heads, should a flood larger than the design flood occurs. It would not be a disadvantage if the shaft spillway is used as a service spillway in conjunction with an auxiliary or emergency spillway.

Disadvantages: (*i*) Discharge capacity is limited, the known maximum design discharge is 1250 cumecs only, (*ii*) Large diameter of the intake (funnel) as depth of overflow is small, (*iii*) Radial flow at inlet is not established (in the absence of a guide vane) which results in unsymmetrical flow down the shaft causing considerable turbulence, (*iv*) Spiral flow in the funnel causes turbulence, (*v*) Heavy turbulence occurs at the elbow, *i.e.*, bottom of vertical shaft where the bend starts. Precautions are essential to eliminate chances of piping along its contact surface and unequal settlement, (*v*) May not be successful if flood higher than estimated occurs, (*vi*) Cavitation may occur if a gradual transition from vertical shaft and horizontal conduit is not provided, (*vii*) Inlet is prone to clogging especially at junction point, if floating debris from the reservoir are not prevented from entering the shaft.

Types of Shaft Spillways

1. *Standard Crested Shaft Spillway.* It is a circular spillway having its shape at the top conforming to the standard crest. The water begins its free fall immediately upon leaving the crest (Fig. 20.8a). In shaft spillway, the crest profile conforms to the shape of the lower nappe of a jet flowing over a sharp-crested circular weir (Fig. 20.8b). The discharge characteristics for flow over the crest are expressed by the relation $Q = CLH^{3/2}$.

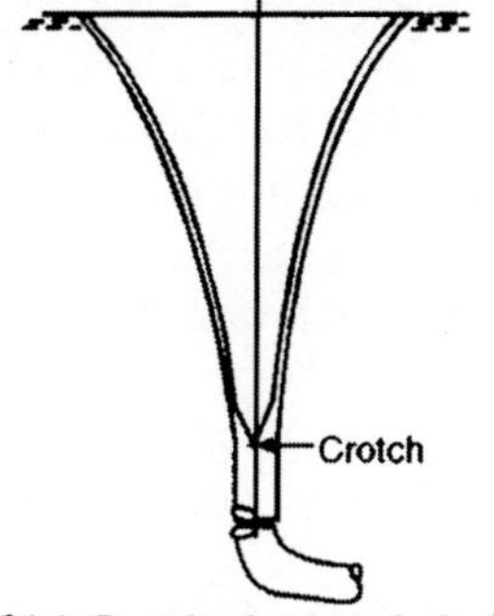

Fig. 20.8(*a*) Standard crested shaft spillway.

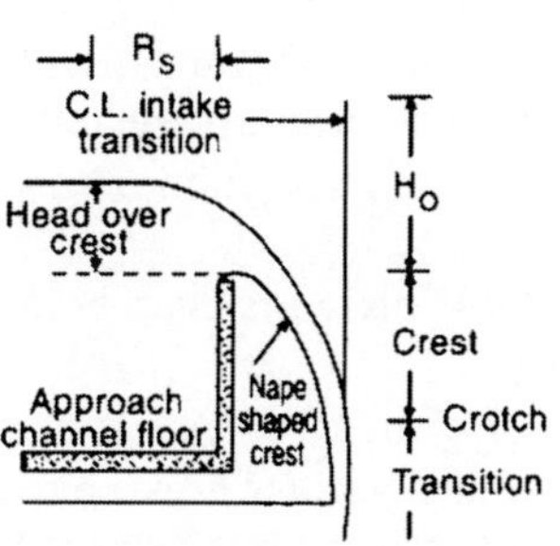

Fig. 20.8 (*b*) Profile of flow through standard crested shaft spillway.

where, Q = discharge (cumecs), C = coefficient of discharge, value 2.0 to 2.5. It changes with different definitions of L and H. For L taken at the outside periphery of the overflow crest and H measured to the apex of the overflow section, the equation takes the form, $Q = C_0 (2\pi R_s) H_0^{3/2}$. Value of C_0 for a circular crest differs from that of a straight crest because of the effects of submergence and backwater pressure caused by the joining of the converging flow, L = length of crest related to some specific point of measurement such as length of the circle at the apex, along the periphery at the upstream face of the crest or along some other chosen reference line, and H = head of water measured either to the apex of the under nappe of the overflow to the spring point of the circular sharp crested weir or to some other established point on the overflow.

2. *Flat Crested Shaft Spillway.* It is circular spillway composed of three parts, sloping weir section, free falling section and the vertical shaft section. The water is caused to approach the crest of the free falling section on flat slope before beginning its free fall (Fig. 20.9). The occurrence of sub-atmospheric pressure should be minimum. It requires a smaller diameter for the tunnel which is about 40% of standard crested spillway. The crest profile is of the shape conforming to the path of free falling jet. The discharge characteristics for flow over the crest are expressed by the relation $Q = CLH^{3/2}$.

where, C = coefficient of discharge, value 1.84 as for broad crested weir.

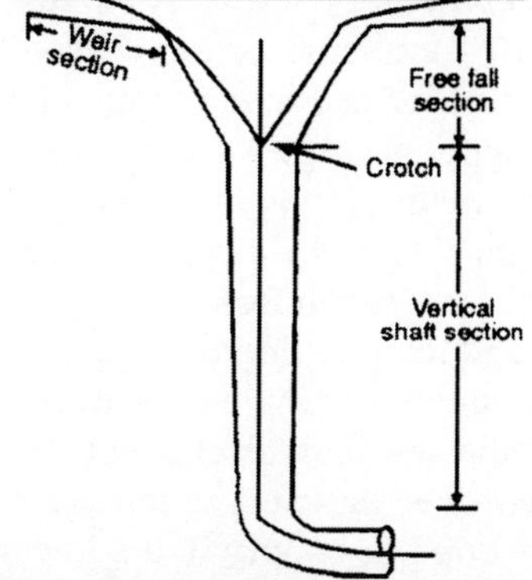

Fig. 20.9 Flat crested shaft spillway.

It is characteristic of flat crested shaft spillway that for heads greater than ¼th the crest diameter, the crest becomes submerged and causes a vortex in the spillway which because of tangential velocity reduces the discharge capacity of the spillway. The formation of the vortex is reduced to some extent by the provision of radial piers at the crest of the spillway. The spillway is suitable in rock excavations located on the knoll of abutments when the crest is almost flushed to the floor of the approach channel. It is preferred for reasons of economy due to smaller funnel diameter where the shaft is to be excavated in rock.

Standard Crest Spillway vs Flat Crest Spillway

Standard crest spillway	*Flat crest spillway*
(*i*) There is no weir section, water begins its free fall immediately on leaving the crest.	There is weir section which precedes free fall section. Water is caused to flow over the flat slope of the crest before the beginning of free fall.
(*ii*) The circular crest profile is shaped to the lower nappe of a sharp crested weir and there is a vertical shaft section below it.	The free falling section is of the shape conforming to the path of free falling jet and vertical shaft section which is completely full of water.
(*iii*) Crotch section is deep below in the vertical shaft.	Crotch section is much above in the vertical shaft.
(*iv*) Coefficient of discharge is higher.	Coefficient of discharge is low.
(*v*) Smaller diameter crest, about 75% of the flat crest for passing the same discharge.	Larger outer diameter of crest.
(*vi*) Larger diameter of the funnel.	Smaller diameter of the funnel *i.e.*, about 4o percent of that required for standard crest.
(*vii*) Vertical shaft section is somewhat smaller in diameter as head available for vertical velocity is little greater.	Vertical shaft section is negligibly larger as the head available for the vertical velocity is little lesser.
(*viii*) Well suited when the crest is projected like a tower above the floor of the approach channel.	Well suited where excavated in rock, located on the knoll of abutment, crest is almost flush with approach channel floor.
(*ix*) More excavation is involved, the funnel diameter being large.	Economy in rock excavation, the funnel diameter being smaller.
(*x*) Occurrence of sub-atmospheric pressure may be tolerated along the overflow section.	Occurrence of sub-atmospheric pressure should be minimum.

7. Siphon Spillway. It is defined as a spillway which works on the principle of siphon and is automatic in action. A siphon spillway carries the surplus waters from a reservoir past the dam by siphonic action. It consists essentially of a closed conduit of two legs of unequal lengths, formed in the shape of an inverted U-tube so positioned that the inside of the bend of the upper passage ways *i.e.*, the crest of the siphon is at the normal reservoir level. The shorter leg, called inlet, is in reservoir while the long leg, termed as discharging leg, is on the downstream side of the dam. Working on the principle of siphon, once the pipe is filled with water it shall continue to flow so long as the reservoir level is higher than the exit end of the discharge leg and the inlet leg is not exposed. Initial discharges from the spillway are like the discharge from the weir. Siphonic action takes place after the air in the bend over the crest is exhausted and the water is discharged through siphonic action. With the rise in reservoir level, water flows over the crest to initiate siphonic action and continuous flow is maintained by suction effect due to gravity in the lower leg of the conduit.

The essentials of a good siphon spillway are (*i*) Efficiency, (*ii*) Simple and economical in design and construction, (*iii*) Economical in maintenance, (*iv*) Automatic working, (*v*) Quick and knock free priming, (*vi*) Low depth of priming, (*vii*) Steady flow after priming, (*viii*) Self cleaning velocity in order to avoid deposition of silt and debris, and (*ix*) No entrainment of air.

Advantages: (*i*) Discharges full capacity flows at relatively low heads, (*ii*) Operating head is much more than the head of water over the crest, the discharging capacity per unit length of crest is disproportionately higher than other types of spillways, (*iii*) Non-overflow section of the dam can be kept lower for the same height of overflow section as the rise of flood over the crest is smaller, and (*iv*) Positive and automatic operation without mechanical device and moving parts.

Disadvantages: (*i*) High initial cost, (*ii*) Inability to pass appreciably higher discharges than the designed capacity even under raised reservoir levels, (*iii*) Inlet leg prone to clogging by accumulation of debris, (*iv*) Incapable to pass ice and floating debris, (*v*) Water may freeze in inlet leg retarding siphonic action, (*vi*) Minute crack in the cover interferes with its functioning by sucking in air, (*vii*) Requires more sound foundation to resist vibration disturbances, (*viii*) Siphonic action once initiated continues till the reservoir level falls down to the level of inlet leg causing loss of storage, unless air vent is there to break siphoning action, (*ix*) Likelihood of occurrence of sudden surges and stoppage of outflow in the downstream river stage as a result of erratic make and break action of siphon, (*x*) Possibility of clogging of the siphon passageway and breaking of siphon vents with logs and debris, (*xi*) Possibility of freezing inside the inlet and air vents of the siphon in cold climates, and (*xii*) It cannot be used for vaccum heads higher than 8m and there is danger of cavitation damage.

Types of Siphon Spillway

1. *Saddle Siphon.* Also called *hood siphon.* It is defined as a siphon having its duct formed of an U-shaped hood. The inlet lip of the hood is submerged below the full reservoir level and that of the outlet lip terminates in a water seal formed by various devices. A saddle siphon is provided in concrete dams only and is built monolithic with the dam with crest at full reservoir level (Fig. 20.10). The operative head is reckoned from the reservoir level to the tail water

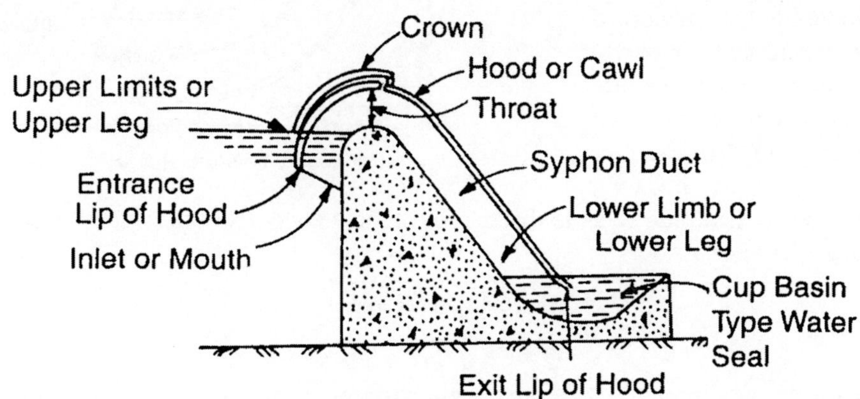

Fig. 20.10 Saddle siphon.

level of the level or the exit end of the discharging leg, whichever is higher. The inlet leg is always kept submerged below FRL so that it is not blocked by ice or floating debris. Since siphonic action once started continues till the reservoir level falls below the level of the inlet leg resulting in loss of storage an air vent consisting of a bent pipe connected to the siphon with its open end in the plane of the full reservoir surface, is provided to break siphoning action as the reservoir level falls just below the crest level when air vent is exposed and air finds entry into it. Likewise, when the reservoir level rises above FRL air vent is sealed due to submergence and priming takes place to initiate siphoning action.

Discharge formula. Discharge through the saddle siphon is given by the relation:

$$Q = CA\sqrt{2gH} \qquad (i)$$

where, Q = discharge (cumecs), C = coefficient of discharge, average value 0.65, A = area of cross-section at crown, and H = operating head, *i.e.*, (*i*) reservoir level — centre of free fall outlet, (*ii*) reservoir level — *d/s* tail water level in case the outlet is submerged. Due allowance for losses: (*i*) at entry = 0.5 $V_A^2/2g$, (where, V_A is velocity at inlet A), (*ii*) at upper and lower bends B and C = 0.1 ($V^2/2g$) each, and (*iii*) in lower leg (calculated as in the case of pipe) is made while calculating operating head.

Discharge in terms of radii of crest

$$Q = R_1 L \sqrt{2gh_0}\ \log_e R_2/R_1 \qquad (ii)$$

or

$$Q = 12.45\ R_1 L \log_e R_2/R_1 \text{ (at maximum safe head of 7.92m)}$$

or

$$Q = 28.7\ R_1 L \log R_2/R_1$$

where, R_1 = radius of crest (m), R_2 = radius of crown (m), and L = length of throat (m)

2. *Volute Siphon.* A volute siphon essentially consists of a dome* with a funnel having volutes** either on its inner curved surface or on the outside of vertical pipe near its top, placed underneath, leaving an annular space around, with a vertical pipe taken down the funnel to pass the discharge through the dam. The dome is supported by a cylindrical drum resting on pillars leaving an annular space alround the funnel. Over the main dome, a deprimer dome is attached (Fig. 20.11). The lip of the funnel is kept at normal reservoir level and number of volutes are fixed in the funnel to induce a spiral motion to the water flowing along them. Vertical pipe from the funnel is bent at right angle through the dome so as to discharge the flow away from the dam.

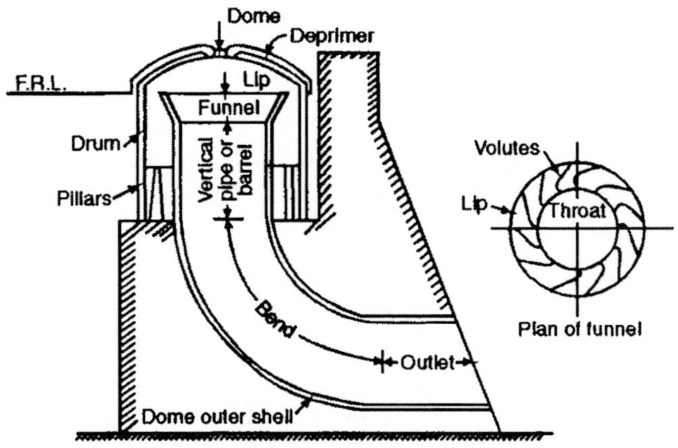

Fig. 20.11 Volute siphon.

Discharge formula

$$Q = CA\sqrt{2g(H - H_L)}$$

or

$$Q = CA\sqrt{2gH}$$

where, Q = discharge (cumecs), A = cross sectional area of pipe (m²), C = coefficient of discharge, H = maximum operating head (m), and H_L = head loss through the siphon (m).

With the rise in reservoir level, water seals the air entry as it spills over the circumference of the lip of the funnel and flows along the volutes with a spiral motion forming a forced vortex near the top of the vertical pipe. The vortex so formed induces suction which sucks all the remaining air creating priming very quickly. The siphonic action thus sets in. With the fall in reservoir level, air enters through the deprimer domes and breaks the siphonic action.

8. Emergency Spillway. Also called *auxiliary spillway.* It is an excavated or natural channel, usually removed some distance from a dam, provided to permit the release of extraordinary

**Dome.* A hemispherical covering with springing at a level slightly higher than the lip level and resting over the drum such that at no point the distance measured in direction normal to curvature be less than the annular space between the drum and the lip (Fig. 20.11).

***Volutes.* These are the curved vanes on projections placed on the inner sloping surface of the funnel shaped top of the barrel (called internal volute) or on the outer side of the barrel near its top (called external volute).

flood flows or flood discharge beyond the capacity of the service (main) spillway (Fig. 20.12). It is operated rarely. An inexpensive structure is usually aimed at. Control gates are seldom

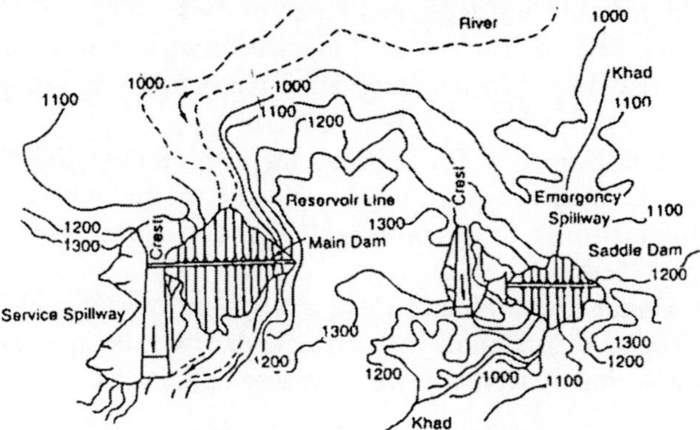

Fig. 20.12 Emergency spillway.

provided rather a low earthen embankment is used to allow the water surface of the reservoir to rise above the crest of the emergency spillway. The embankment is kept lower than the main dam, *i.e.*, at or near the normal reservoir level but sufficiently high to stand normal floods. It is intended to breach during abnormal floods only, the dam itself remaining safe.

An emergency spillway is, therefore, an auxiliary spillway which is called upon to operate infrequently. Emergency spillway is relevant to earthen dams. Fig. 20.13 is the auxiliary spillway provided at Ramganga project.

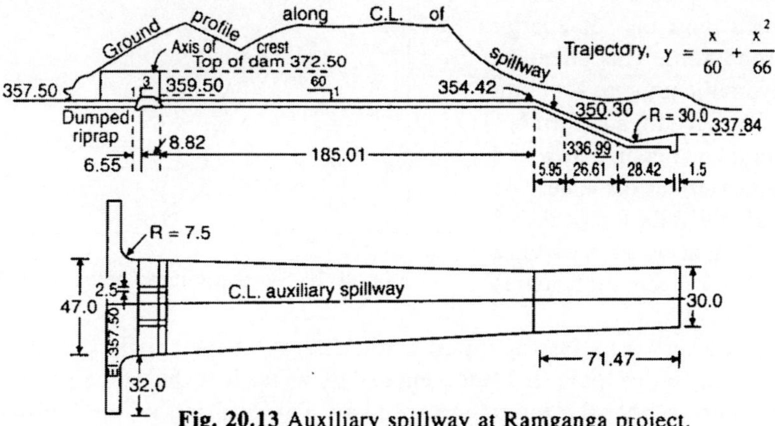

Fig. 20.13 Auxiliary spillway at Ramganga project.

20.5 ENERGY DISSIPATION BELOW SPILLWAYS

Since spillways are high structures, the water flowing over them has considerable velocity and kinetic energy due to conversion of potential energy into kinetic energy when it reaches the river bed and this has to be destroyed before the water is allowed to flow over the river bed, because otherwise there will be considerable erosion of bed at the toe of the spillway endangering the stability of the structure. The devices provided for destroying the energy of water flowing over the spillways are termed as 'dissipators'. The energy dissipation arrangement depends on the consideration of the nature of the device to be employed and the character of the river bed downstream.

Selection of Energy Dissipators

Various considerations governing the selection of energy dissipator are discussed below:

1. *Type of dam and spillway.* In earthen dam a chute spillway can be provided with hydraulic jump basin with or without baffles or end sills. In an arch dam, free fall spillway with

stilling pool of high depths is normally provided. In the case of a dam where water flows through a tunnel outlet, the energy is best dissipated by jet diffusion. Likewise, for high dams upturned buckets with high tail water depth can be employed.

2. *Tail water elevation.* The position of tail water elevation vis-a-vis jump height dictates the energy dissipating arrangement to be provided. Five conditions are generally met with (Fig. 20.15).

3. *Nature of foundation.* Nature of foundation met with dictates the dissipating arrangement. In sound rock upturned bucket or ski-jump bucket is preferred with much lesser length of stilling basin and in alluvial soil a long apron with hydraulic jump type stilling basin with baffles and end-sill is provided.

4. *Velocity of flow.* Under high velocity flow, cavitation may occur if the baffles in the stilling basin are provided. Jet diffusion or hydraulic jump type stilling basin or roller buckets are preferred under suitable foundation and water course conditions.

Types of Energy Dissipators

Various types of energy dissipators can be classified as under:

1. *Energy Dissipation by Hydraulic Jump.* Whenever a jet of water moving at a velocity higher than the critical velocity and at a depth lower than the critical depth meets with a body of water flowing at velocity lower than the critical velocity and at a depth greater than the critical depth, changes in velocity and depth take place resulting in great turbulence and adrupt rise in the surface at the line of junction accompanied by a loss of energy under appropriate conditions. This phenomenon is known as hydraulic jump or standing wave (Fig. 20.14). The surging and turmoil, accompanied by aeration, occur below the standing wave for a measurable distance. The standing wave invariably falls forward and covers the jet with a film of water ofvarying thickness which alters the characteristics of the standing wave itself.

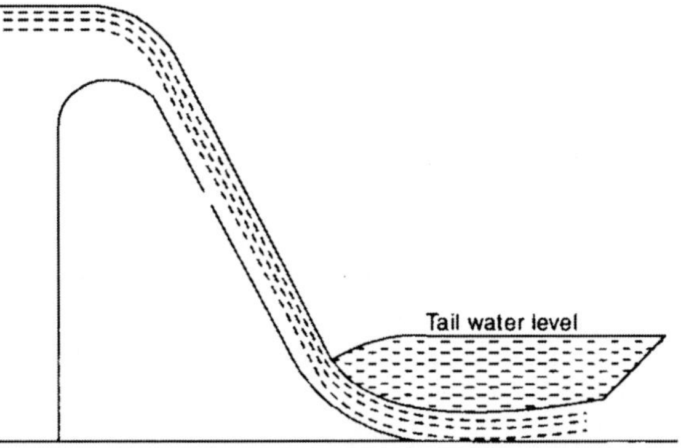

Fig. 20.14 Hydraulic jump below a spillway.

The hydraulic jump serves to destroy a part of the energy of the flowing water. Arrangement has to be provided to dissipate this huge energy of water lest the velocity of the water causes large scale scouring near the toe of the spillway. The position of the hydraulic jump depends on the depth and velocity of the water leaving the bucket at the toe of the spillway and the normal depth and velocity in the river for the particular discharge and this location can be calculated if these factors are known. The post-jump depth D_2 for a given discharge is computed from the following relation:

$$D_2 = \frac{D_1}{2} + \sqrt{\frac{2q^2}{gD_1} + \frac{D_1^2}{2}}$$

where, D_2 = post-jump depth (m), D_1 = pre-jump depth (m), and q = discharge intensity per unit width.

The efficiency of the hydraulic jump in dissipating energy vis-a-vis protection in the stilling basin depends on the height of tail water (D_1) in relation to post-jump depth D_2 for the given discharge intensity. The tail water level depends on the hydraulic dimensions and slope of the river channel below. The tail water depth for different discharges is determined from actual gauge-discharge observations and velocity measurement for wide range of discharges. A tail water rating curve (Fig. 20.15) gives the tail water depths corresponding to the discharges. On the same

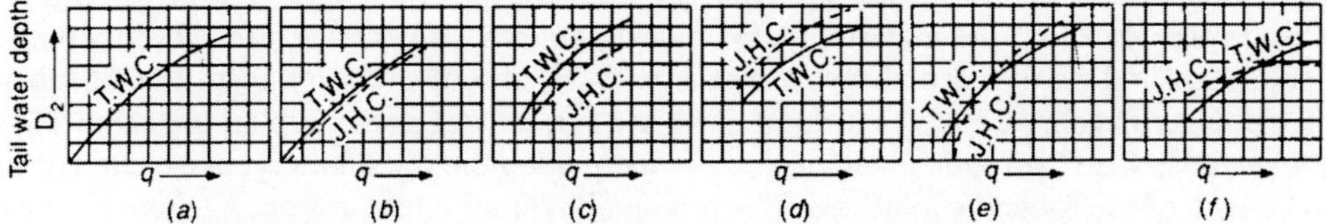

Fig. 20.15 Relationship between jump height curve and tail water curve.

graph, a plotting similarly done between discharge (q) and depth (D_2) gives jump height curve or D_2 curve. There are five conditions governing the relationship between the jump height curve (JHC) and tail water curve (TWC) and the corresponding energy dissipation arrangement required, as under:

1. *JHC and TWC coincide of all discharges (Fig. 20.15b).* It is the ideal condition wherein post-jump depth is available in the channel for the formation of a perfect hydraulic jump at the toe of spillway at all discharges. Protection in the form of a simple concrete apron of length $5 (D_2 - D_1)$ is generally considered sufficient (Fig. 20.16) as further confirmed from model tests.

2. *JHC is lower than TWC at all discharges (Fig. 20.15c).* In this condition the tail water depth is much in excess of sequent depth required for the formation of hydraulic jump, *i.e.*, the jump is completely drowned by TWC and very little energy is dissipated. As a consequence there is a tendency of high velocity flow to dive under the tail water and travel a long distance along the bottom forming only a very imperfect jump. High velocity flow continues for a considerable distance in the river channel unless dissipation arrangement is provided to lower the depth of tail water at the point of hydraulic jump. The dissipation arrangements generally adopted are:

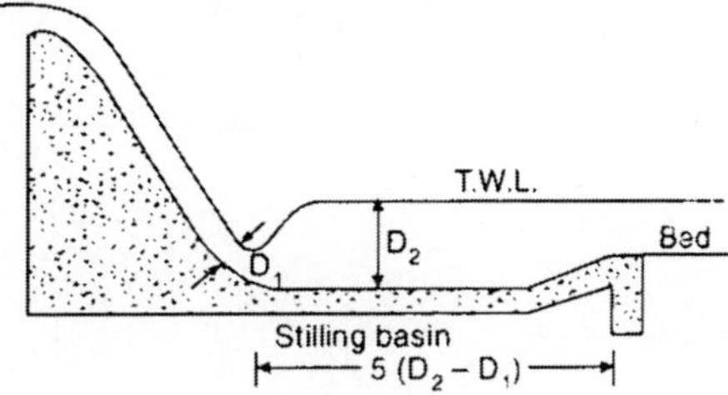

Fig. 20.16 Stilling basin.

(*a*) *Sloping apron.* A sloping apron is provided above the river bed level (Fig. 20.17) so that hydraulic jump is formed on the sloping glacis where depth D_2 lesser than tail water depth at toe is available. It is a commonly adopted method.

(*b*) *Roller bucket.* The roller bucket dissipator suitable under the condition is solid roller bucket type and slotted roller bucket type. The solid or slotted roller types of bucket are suitable when the tail water depth is in excess of the sequent depth of hydraulic jump, both measured with respect to channel bed level and economical when tail water depths are equivalent to about 1.1 times or more of the sequent depth of hydraulic jump as also where good rock levels are available.

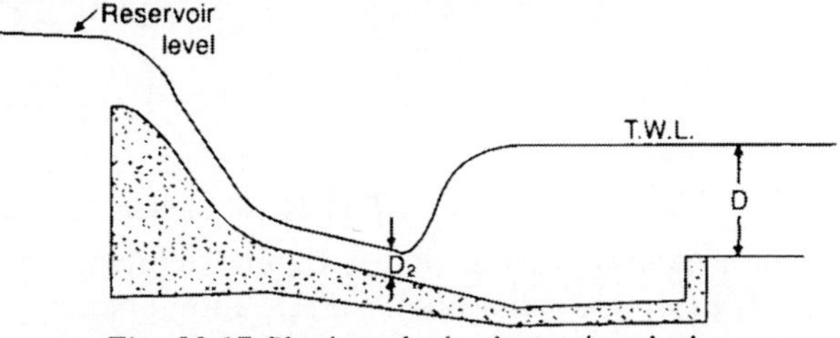

Fig. 20.17 Sloping glacis above river bed.

(*i*) *Solid Roller Bucket.* In an upturned solid bucket the dissipation of energy occurs as a result of interaction of two complementary rollers; one in bucket proper, called the surface or dissipating roller which is anticlockwise and the other downstream of the bucket, called the ground or stabilizing roller, which is clockwise (Fig. 20.18). The ground roller tends to move the

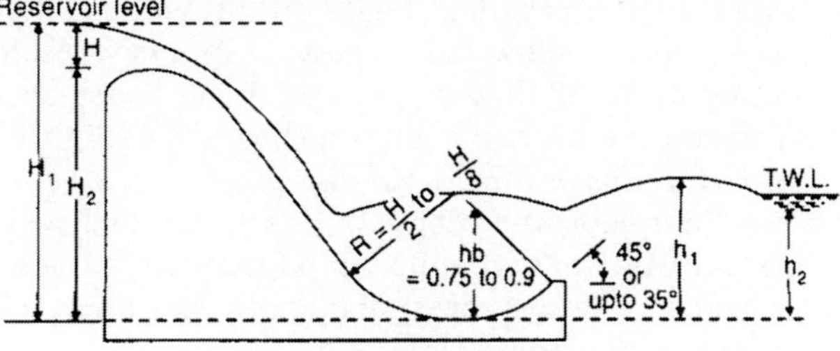

Fig. 20.18 Roller bucket.

scoured material towards the toe of the spillway thereby preventing formation of serious scour. Sometimes under the action of the surface roller the scoured material may enter the ground roller. The ground roller picks-up the material from downstream bed and carries it to the lip where it is partly deposited and partly carried away downstream by the residual jet from the lip, thereby keeping the material continuously in motion. Both the surface and ground roller are required to be well formed, otherwise hydraulic phenomenon of sweep out or heavy submergence occurs depending upon which of the rollers is inhibited. The principal hydraulic features are bucket invert elevation, radius of bucket and slope of the bucket lip or bucket lip angle.

(*ii*) *Slotted Roller Bucket.* It is an upturned bucket with teeth in it (Fig. 20.19 & 20.20). A part of the flow passes through the slots, spreads laterally and is lifted away from the channel bottom by a short apron at the downstream end of the bucket. The flow is thus dispersed and distributed over a greater area providing less violent flow concentrations than that occur with the solid roller bucket. The velocity distribution just downstream of the bucket is more akin to

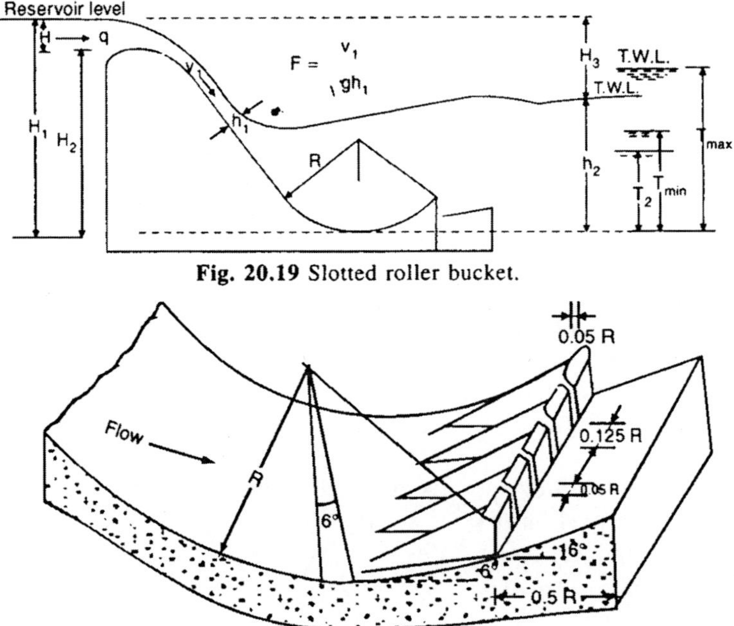

Fig. 20.19 Slotted roller bucket.

Fig. 20.20. Slotted roller bucket details.

that in a natural stream, that is higher velocities at the surface and lower velocities at the bottom. In slotted roller bucket dissipation of energy occurs by lateral spreading of jet passing through bucket teeth in addition to the interaction of two complementary rollers as in solid roller bucket. It is an improvement over the solid roller bucket particularly for smaller range of tail water depths. In slotted roller bucket the sweep out occurs at slightly higher tail water elevation with solid roller bucket and if the tail water level is extremely high, the residual submerged jet close to the bucket surface may dive from the apron lip to scour the channel bed downstream, a condition unlikely to occur with a solid roller bucket.

(*iii*) *Raised End Sill.* A raised sill provided at the end of a level apron below the bucket of spillway results in increasing depth of flow on the apron and hence encourages formation of hydraulic jump upstream of it (Fig. 20.21).

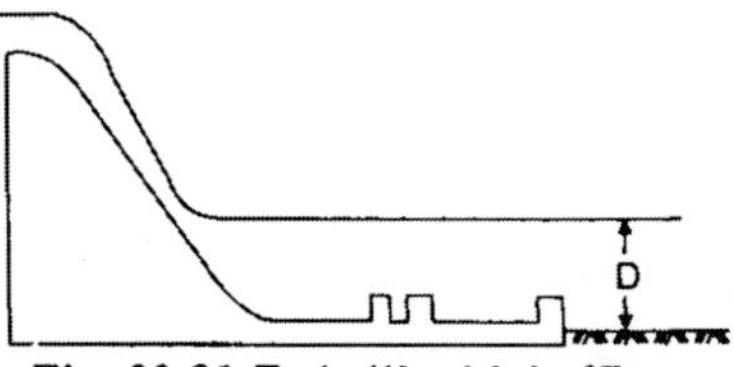

Fig. 20.21 End sill with baffles.

3. *JHC is above TWC at all discharge (Fig. 20.15d).* In this condition tail water depth is lesser than that required for the formation of hydraulic jump for all discharges. The jump tends to sweep across the apron at a high velocity and attacks the bed downstream. Dissipators provided are:

1. *Sloping glacis with depressed cistern.* Lowering of floor level downstrem of the dam so as to make the tail water depth in stilling basin equal to the jump height for all discharges. It is done in three ways: (*i*) A sloping glacis with a depressed cistern at a level lower than the river channel is provided in order to increase the water depth to ensure formation of hydraulic jump on the glacis of the spillway, (*ii*) A depressed floor but rising towards the downstream end, and (*iii*) A depressed floor but sloping away from the toe of the dam.

2. *Low secondary weir and baffles.* A low secondary weir, a baffle or a row of friction blocks of energy are provided to dissipate residual energy. The location, spacing size and shape of the baffles is better determined by model tests.

3. Stilling basin with baffles or sills at river bed level.

4. *Skijump bucket.* A sharply upturned bucket or skijump bucket provides dissipation of energy by deflecting the jet of water above the bucket into the air to fall some distance downstream into the river (Fig. 20.22). Obviously the skijump bucket type dissipator requires sound rocky bed to withstand part of the energy dissipation taking place by impact, whereas some of the energy is dissipated in air by diffusion and aeration.

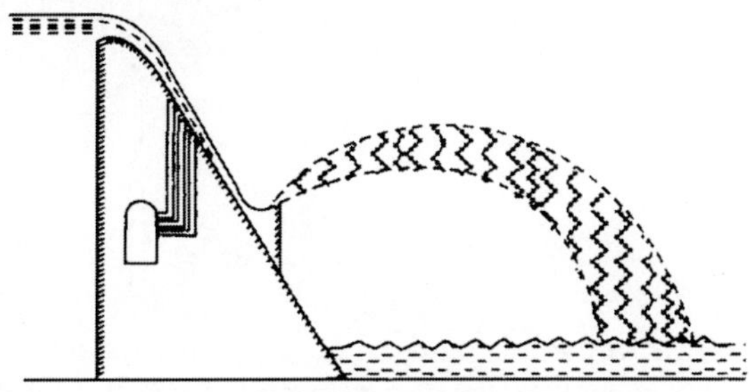

Fig. 20.22 Skijump.

4. *JHC is lower than TWC at lower discharges but lower at higher discharges (Fig. 20.15e).* In this condition the hydraulic jump is drowned at low discharges and at high discharges tail water depth is inadequate than required for the formation of hydraulic jump. As such it is necessary to provide sufficient depth of tail water for the formation of hydraulic jump at high flows. Solution lies in the construction of secondary dam (Fig. 20.23) or sloping glacis partly above and partly below the river bed with horizontal cistern of adequate length and end sill. At low discharges, the hydraulic jump is formed on the sloping glacis above the rive bed level and lower down, below the river bed level where adequate water depth is available, at higher discharges (Fig. 20.24). A bucket type dissipator is also suitable provided rock below is sound to act as roller for low discharges and ski-jump at higher discharges.

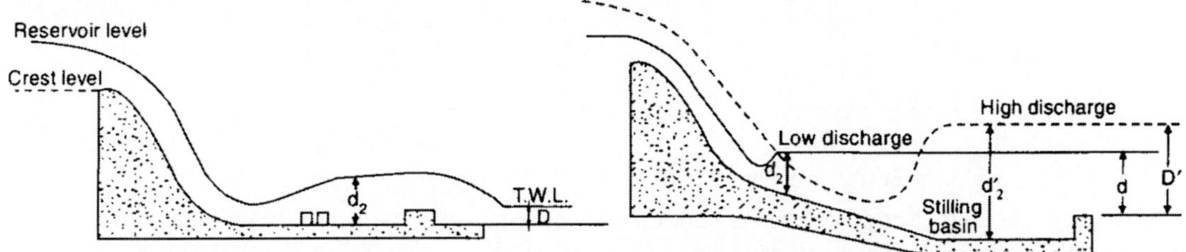

Fig. 20.23 Secondary dam.

Fig. 20.24 Sloping apron partly above and partly below d/s bed level

5. *JHC is above TWC at lower discharges but lower at higher discharges (Fig. 20.15f).* In this condition there is insufficient tail water depth at low discharges while at higher discharges the hydraulic jump is drowned, *i.e.*, the reverse of condition 4. The solution lies in creating enough water depth to make the jump form on the apron at low discharges while it is formed on the glacis above the river bed level at higher discharges. The dissipating arrangement comprises (*i*) Stilling basin with a low secondary dam, and (*ii*) Stilling basin with baffle piers or some form of dentated sill.

20.6 HYDRAULIC JUMP TYPE STILLING BASINS

Hydraulic jump type stilling basins are useful as energy dissipators downsream of spillways. These are of two types. (*i*) Stilling basin with horizontal apron, and (*ii*) Stilling basin with sloping apron. Their suitability and design parameters are discussed as follows :

Stilling Basin with Horizontal Apron

1. *Suitability as energy dissipator:* (*i*) When TWC approximately follows JHC or is slightly above or below it, the requisite depth is obtained on a proper apron near or at the ground level so that it is quite economical, and (*ii*) Recommended for spillways on weak bed rock conditions and weirs and barrages on sand or loose gravel.

2. *Classification:* (*i*) Basin I-Froude number of incoming flow is less than 4.5, generally encountered on weirs and barrages, and (*ii*) Basin II-Froude number of incoming water is greater than 4.5, a general feature for dams.

3. *Basin floor elevation:* Knowing H_L and q; D_c, D_1 and D_2 are determined from the following formula:

$$H_L = (D_2 - D_1)^3 / 4D_1D_2$$

$$D_c = \left(\frac{q^2}{9}\right)^{1/3}$$

$$D_2 = -\frac{D_1}{2} + \sqrt{\frac{2q^2}{D_1 g} + \frac{D_1^2}{4}}$$

With known D_1 and D_2, elevation of basin floor is calculated by either deducting the specific energy at pre-jump section from the total energy line at that section or that at post-jump section from the downstream total energy line. H_L is equal to the difference in *u/s* and *d/s* TE lines, where chute is not longer than hydraulic head or stilling basin is directly *d/s* from crest, (*ii*) where chute length exceeds the hydraulic head, D_1 is first determined by taking H_L equal to the difference between *u/s* and *d/s* TE lines. Next calculate friction losses in the chute by Manning's formula and determine the more accurate value of H_L and consequently that of D_1. The process is repeated till the desired accuracy is achieved.

4. *Basin length.* Stilling Basin-I. Length of basin is determined from curve (Fig. 20.25). The basin is provided with an end sill preferably dentated one. The raising of the basin floor is not done above the level required from sequent depth consideration or 15% of D_2 and chute blocks and basin blocks are provided.

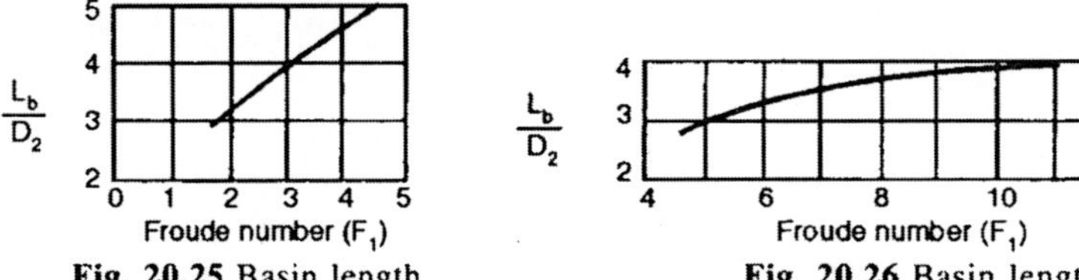

Fig. 20.25 Basin length. **Fig. 20.26** Basin length.

Stilling Basin-II. Length of basin is determined from curve (Fig. 20.26). Chute blocks and end sill are provided. Maximum raising of basin floor is not to exceed 15% of D_2.

5. *Basin appurtenances.* Stilling basin I — chute blocks, basin blocks, end sill are provided as given in Fig. 20.27.

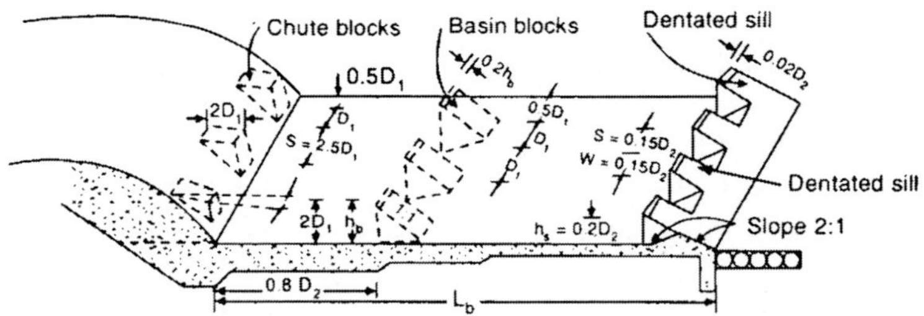

Fig. 20.27 Appurtenances for stilling basin type I.

(*i*) Chute blocks — Height = $2D_1$ on glacis slope, top length = $2D_1$ and spacing = $2.5D_1$, space left along each wall = $0.5D_1$.

(*ii*) Basin blocks — Height in terms of D_1 from (Fig. 20.28), width and spacing equal to the height. *U/s* face of basin blocks is vertical and in one plane. A half space adjacent to the walls. *U/s* face of basin blocks is set at a distance 0.8 D_2 from the *d/s* face of chute blocks.

(*iii*) End sill — Height of dentated end sill = 0.2 D_2. Maximum width and spacing of dent as per Fig. 20.28. A dent adjacent to each side wall is preferred. In narrow basins, reduce the width and spacing but in the same proportion. It is not necessary to stagger the end sill dents with reference to chute blocks.

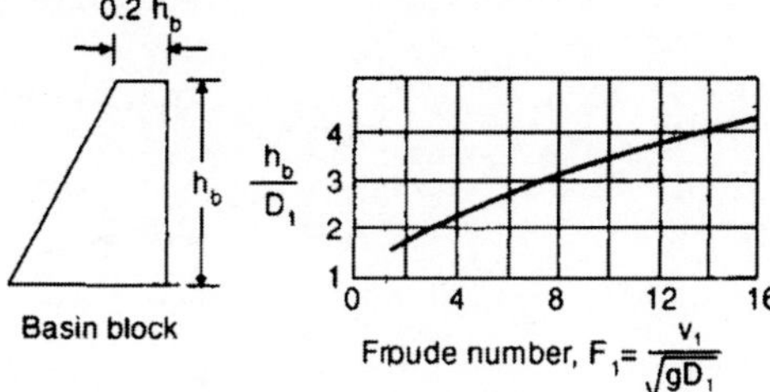

Fig. 20.28 Basin blocks.

Stilling Basin II — Chute blocks, basin blocks and end sill are given in Fig. 20.29.

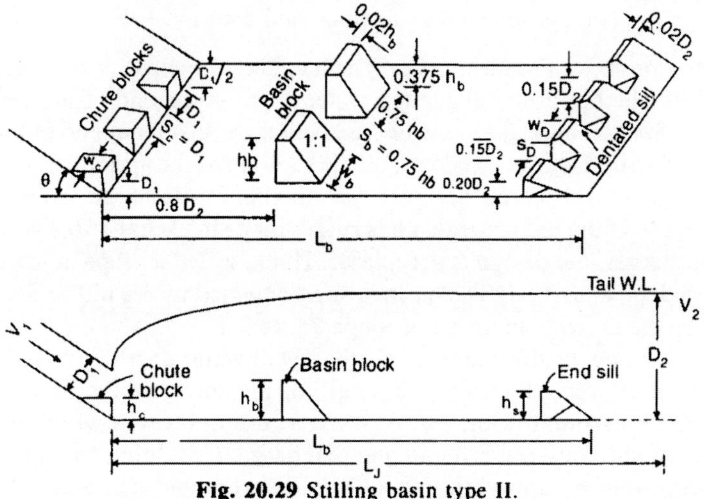

Fig. 20.29 Stilling basin type II.

(*i*) Chute blocks — Height, width and spacing of the chute blocks is kept equal to D_1. Width and spacing is varied to eliminate fractional blocks. A space equal to $0.5D_1$ is preferable along each wall.

(*ii*) Basin blocks — Height in terms of D_1 as per Fig. 20.28. Width and spacing is kept 0.75 height. The blocks are placed at distance of 0.8 D_2 *d/s* from chute blocks.

(*iii*) End sill — same as for stilling basin I.

Stilling Basin with Sloping Apron: When TWC is too deep as compared to the conjugate depth D_2, the jet left at natural ground level continues to go as a strong current near the bed forming a drowned jump which is harmful to the river bed. Efficient jump is formed at suitable level on the sloping apron.

Basin III. Where TWC is higher than D_2 curve at all discharges (Fig. 20.30).

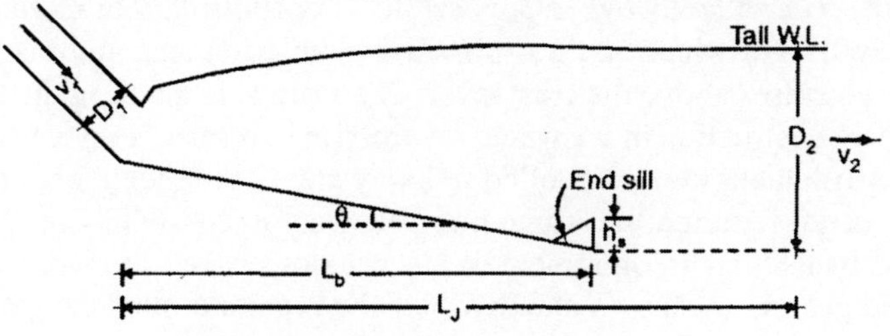

Fig. 20.30 Stilling basin type III.

Basin IV. Where tail water depth at maximum discharge exceeds D_2 considerably but is equal to or slightly greater than D_2 (Fig. 20.31) at lower discharges.

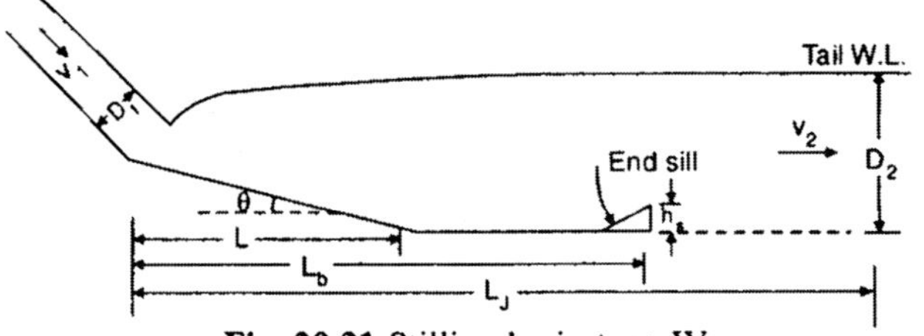

Fig. 20.31 Stilling basin type IV.

Design procedure

Basin III. (*i*) Assume a certain level at which the front of jump will form for the maximum tail-water depth and discharge, (*ii*) Determine D_1 from the known upstream total energy line by applying Bernoulli's theorem and calculate F_1. Then find out conjugate depth D_2 from equation $\frac{D_2}{D1} = \frac{1}{2}\left(\sqrt{1+8F_1^2} - 1\right)$, (*iii*) Assume a certain slope and determine the conjugate depth D'_2 and length of the jump for the above Froude number. The lenght of the apron is kept equal to 60% of the jump length, (*iv*) Test whether the available tail water depth at the end of the apron matches the conjugate depth D'_2. If not, change the slope or the level of the upstream end of the apron or both. Several trials may be required before the slope and the location of the apron are compatible with the hydraulic requirement, (*v*) The apron designed for maximum discharge is tested at lower discharges, say ¼, ½ and ¾. If the tail water depth is sufficient or in excess of the conjugate depth for the intermediate discharges, the design is acceptable. If not, a flatter slope at lower apron level is tried or Basin IV is adopted, and (*vi*) The basin is supplemented by a solid or dentated end sill of height equal to 0.1 to 0.2 D_2 with an upstream slope 2:1 to 3:1.

Basin IV. (*i*) Determine the discharge at which the tail water depth is most deficient, (*ii*) For the above discharge, determine the level and length of the apron on the basis of criteria given for stilling basin with horizontal floor, (*iii*) Assume a certain level at which the front of jump will form for the maximum tail-water depth and discharge, (*iv*) Determine D_1 from the known upstream total energy line by applying Bernoulli's theorem and calculate F_1. Then find out conjugate depth D_2 from equation $\frac{D_2}{D1} = \frac{1}{2}\left(\sqrt{1+8F_1^2} - 1\right)$, (*v*) Determine a suitable slope, by trail and error, so that the available tail-water depth matches the required conjugate depth D'_2. (*vi*) Determine the length of the jump for the above slope. It the sum of the lengths of inclined portion and horizontal portion is equal to about 60% of the jump length, the design is acceptable. If not, fresh trials may be done by changing the level of the upstream end of the formation, and (*vii*) The basin is supplemented by a solid or dentated end sill of height 0.05 to 0.2 D_2 and upstream slope of 2:1 to 3:1.

20.7 SPILLWAY GATES

Controlled spillways are provided with means to control the outflow from the reservoir which lend them the name "*Gated Spillways*". Opposed to it, uncontrolled or ungated spillways are the simplest form of spillways which are not provided with gates but automatically release water whenever the reservoir rises above the crest level. The storage is made against the raised crest and because of permanent obstruction in the waterway the rise in reservoir level is much more than that in gated spillway. Advantages of uncontrolled spillway are: (*i*) Attendant or operator for the regulation of the supply is not required; the siphoning being automatic with rise in reservoir level, (*ii*) Gate operating mechanism and requirement of electric power are not involved, and (*iii*) Elimination of the gate and gearing means no maintenance and repairs cost of the gates.

Gate is a welded or rivetted steel frame consisting of end vertical griders with horizontal girders spaced between them depending on the design water pressure and dimensions of the gate and provided with a skin plate on the upstream side (*i*) to hold water against it, (*ii*) to protect the steel frame from corrosion by reservoir water, damage by ice or heavy debris, (*iii*) to minimize downpull, and (*iv*) to facilitate maintenance. The skin plate may be on downstream side, in the case of fixed wheel gate, or in exceptional cases on both upstream and downstream sides.

The installation of gates in spillway have the following advantages: (*i*) Additional storage is made available and at the same time dam safety is ensured by raising gates full during high flood impinging reservoir at full reservoir level, (*ii*) Height of permanent crest is low with consequent economy in the height of dam for the same useful storage, and (*iii*) Economy in land acquisition.

Types of Spillway Gates

1. Non-automatic Gates.

Radial Gates. Also called tainter gates. These gates are extensively used as spillway gates and in certain cases as outlet gates in a dam. A radial gate, as the name implies, is a curved gate in the shape of a portion of a cylinder rotating about a horizontal axis. Normally the water is against the convex side of the gate but sometime water load is applied on the concave side of the gate also. The upstream skin plate bent to an arc with convex surface of the arc on upstream side attached to the suitably spaced stiffeners either horizontal or vertical or both supported by radial arms. If horizontal stiffeners are used, these are supported by suitably spaced vertical diaphragms which are connected together by horizontal girders transferring the load to the two end vertical diaphragms. The end beams are supported by radial arms, emanating from the trunnion hubs located at the axis of the skin plate cylinder. If vertical stiffeners are used, these are supported by suitably spaced horizontal girders which are supported by radial arms. The arms transmit the water load to trunnion anchorage girder. Suitable seals are provided along the curved ends of the gate and along the bottom. Each gate is pivoted on two trunnion assemblies anchored in the downstream portion of the spillway piers (Fig. 20.32) such that the gate is capable to rotate about the fixed horizonatal axis.

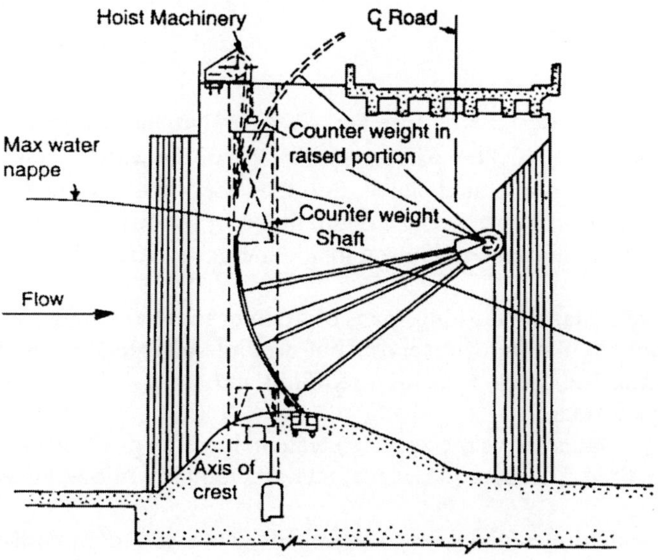

Fig. 20.32 Radial gate.

Advantages: (*i*) Widely used as spillway crest gates, (*ii*) Absence of any slots or grooves thereby providing better hydraulic flow efficiency, (*iii*) Trouble free operation, (*iv*) Less liable to vibration and thus better suited to work at part gate openings, (*v*) Absence of wheel and wheel assemblies, (*vi*) Better coefficient of discharge compared to vertical lift gates, (*vii*) Trunnion bearings remain above water level thereby facilitating better maintenance, (*viii*) Can be operated with comparatively smaller capacity hoists, (*ix*) Simple fabrication requires no precision finish except at the bearings, (*x*) Easily accessible for maintenance, (*xi*) Economical for large size flood gates, (*xii*) Risk of jamming is very much reduced and the crack open force is not great, and (*xiii*) Adapts itself easily to automatic control.

Drum Gates. Also termed as *sector gates.* These gates (Fig. 20.33a and b) are completely enclosed on all the three faces and at ends to form a water tight vessel. It resembles the quadrant of a circle formed by skin plate on the upstream attached to internal bearings. The gate rotates about the hinge generally provided at the upper edge so that the buoyant force aids in its lifting. A recess is made in the crest of the spillway to accommodate the gate in lowered position; the skin plate surface serving as a lid over the recess conforming to the shape of the ogee crest.

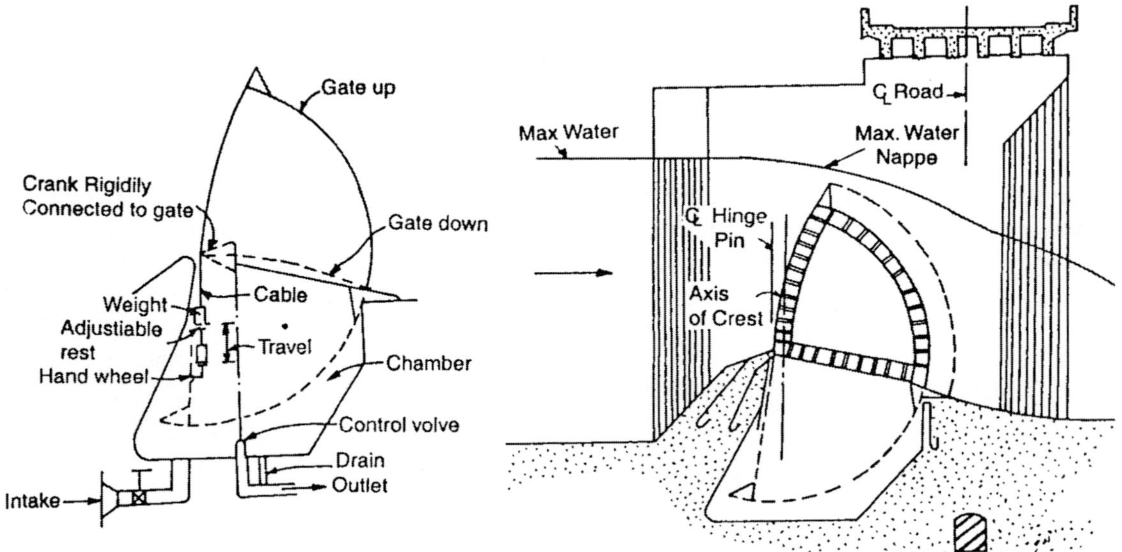

Fig. 20.33 (*a*) Drum gate. **Fig. 20.33** (*b*) Diagramatic layout of drum gate.

The gates are usually operated by water pressure. When the water is admitted through the inlet pipe, controlled by hand operated sluice, into the recess the hollow drum gate is forced upwards to the raised position. When surface regulation of reservoir is to be done by drum gate, the gate is lifted by control equipment housed inside the piers. These controls are fairly complicated.

Drum gates are well adapted for long spans of 40 m or more and under medium heights of 10 m or so for surface regulation of reservoir but are not suitable for low dams having low structures on account of the deep excavation required and the liability of the flooding of the recess in which the gate floats.

Disadvantages: (*i*) Gates are very costly, (*ii*) Gate in lowered position is liable to clogging, and (*iii*) Controls required for operation of the gate are complex in operation and are costly.

2. Vertical Lift Gates.

Earlier to development of radial gates, vertical lift gates were commonly used. A vertical lift gate is a reactangular steel frame consisting of skin plate on the upstream supported by suitably spaced beams and girders. The gate spans horizontaly between the grooves provided in piers and abutments and move vertically in its own plane. The grooves termed as guide grooves are lined with channel section of suitable size with a view to provide smooth bearing surface. Large friction forces are developed between the gate and the downstream guides due to hydrostatic force caused by the water against the gate and the lifting mechanism operated from the top must be able to overcome these high frictional forces. The frictional forces can be reduced by placing cylindrical rollers between the bearing surfaces of the gates and guide grooves. Vertical lift gates are simple to instal and do not require lengthy piers as for radial gates. The gates are designed as wheeled gates rather than slide gates to enable lowering by gravity and for lower hoisting capacity requirement. The common types of vertical lift gates are discussed below:

Slide Gate. The slide gate, as the name implies, operates vertically by sliding inside gate frame. The gate leaf is usually made of cast steel or of welded construction of structural steel. It consists of an upstream skin plate with all the sides closed except the downstream side. The leaf is adequately stiffened by horizontal and vertical ribs. The gate frame is made of cast steel or alternatively fabricated from structural steel.

Fixed Wheel Gate. It is the modern type of vertical lift gate. On the basis of the water head, the fixed wheel gates are classified as high head gate for water head of 30 m and above, medium head gate for 15 to 30 m, and low head gate for up to 15 m of water head.

The gate consists of a skin plate supported by horizontal beams which in turn are supported by vertical box type girders at both the sides. Wheels are mounted inside the vertical girders and move on the roller tracks fixed inside the gate grooves. The wheel transmit the water load acting against the gate to the track and also reduces the amount of frictional force and thereby permit use of hoists of small capacity. The bottom of the gate is shaped from the consideration of (*i*) Satisfactory performance, (*ii*) Freedom from harmful vibration under all conditions of operation, and (*iii*) Minimising downpull. The gate is normally designed to close under its own weight. For this purpose, the net downward closing force is not to be less than 25% of all frictional forces opposing the closure of the gate. Rubber seals are provided at the top and bottom to seal the opening between the upstream skin plate of the gate and the sides of the pier grooves. The type of rubber seal depends on the class of gate. IS:4622-1967 specifies the structural design requirements of fixed wheel gates.

Disadvantages: (*i*) Low coefficient of discharge. Gate grooves come in the waterway with consequent decrease in the coefficient of discharge necessitating increased net spillway width, (*ii*) Operational difficulties as the gate is liable to get jammed during operation due to extraneous matter getting wedged between the gate and slots, (*iii*) Regulation not possible as it is less suitable at partial openings and is more liable to vibration, and (*iv*) Maintenance difficult as for the replacement of seals, the gates have to be taken out of the groove which is time consuming and costly operation.

Roller Gate. Also called *stoney gate.* It is the oldest type of vertical lift gate. Roller gate is a steel cylinder spacing across the piers to the height of the opening. The gate slides on roller trains placed between the gate and roller track (Fig. 20.34 a and b). The rollers are neither attached to the gate nor to the guide groove but are placed independent of the gate and guide groove to roll vertically between the two when the gate is moved. The arrangement eliminates axle friction and only roller friction is developed and the entire water load is transmitted from a bearing strip on the gate to the roller track on the guide. The gate moves on the circumference of the roller and as such its speed is always twice the speed of roller trains. Roller gate is suitable for long span and moderate height.

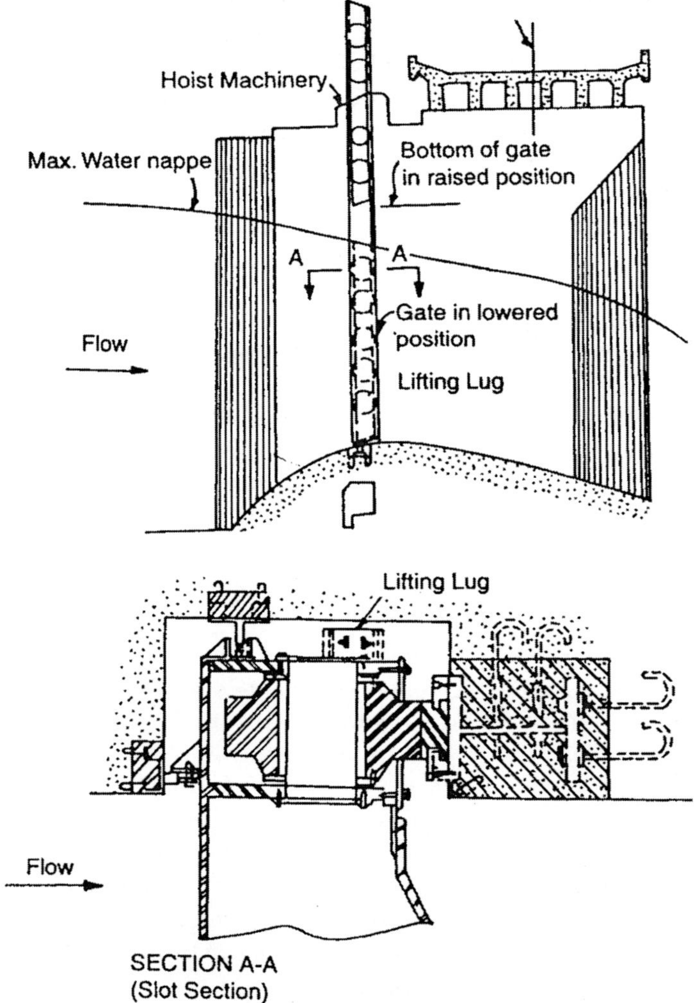

Fig. 20.34 *(a)* & *(b)* Roller gate.

Disadvantages: (*i*) Old type of gate; presently used at barrages only, and (*ii*) Not suitable for operation in rivers carrying large quantity of silt and debris.

20.8 INTAKE AND OUTLET WORKS

The stored water in a reservoir is essentially drawn to meet the intended purposes of irrigation, hydro power generation, public water supply, etc. through the discharging facility termed as intake structure. An intake structure may take on many forms depending on the functions it is to serve, the range in fluctuation of reservoir water levels, quality of water drawn, discharge to be released, frequency and amount of reservoir drawdown, trash removal requirement, wave action in the reservoir and other such considerations such as sedimentation control devices etc. The general requirements of an intake structure are:

(*i*) *Structural stability:* The intake structure is stable to resist water and wave thrust besides wind pressure when reservoir is empty as also against the shocks of earthquakes.

(*ii*) *Hydraulic efficiency:* There is smooth entry into the water conductor system to ensure gradual transformation of static head to conduit velocity so as to involve minimum hydraulic losses.

(*iii*) *Velocity limitation:* The velocity through trashrack gates and ports is within economic and safe limits.

(*iv*) *Operational efficiency:* The intake and the equipment are such as to prevent/minimize ice, floating trash and coarse sediment entering the water conductor system to ensure good operational efficiency.

The main components of an irrigation intake structure are (*i*) Trashrack and supporting structures, (*ii*) Anti-vortex devices, (*iii*) Bellmouth entrance with transition and rectangular to circular opening, and (*iv*) Gate slot closure devices with air vents.

20.9 Functions of Intakes

Intake structures serve to permit withdrawal of water impounded in the reservoir over a predetermined range of reservoir levels to the outlet. The outlet may release water at a retarded rate than the incoming flow from intake or release the impounded water at controlled rate consistent with the requirements of downstream such as feeding the power turbines, feeding the irrigation channels, discharging heavily silt laden inflow , depleting the reservoir to facilitate inspection, as flood control regulator to release water temporarily impounded in flood control storage space, to evacuate in anticipation of incoming flood in the reservoir or combination of multipurpose requirements. The other functions served by an intake are to support necessary auxiliary appurtenances such as trashrack, fish screens and bypass devices, etc., to provide temporary diversion openings and to house installations such as bulkhead or stoplog closure devices.

20.10 Location of Intakes

The various factors influencing the choice of location of an intake structure are (*i*) Type of edifice that is storage reservoir or run-of-river scheme, (*ii*) Location and type of dam/weir, (*iii*) Type of water conductor system that is canal or tunnel, and (*iv*) Topographical features of the river. Intakes are so located and their withdrawal level is so designed as to draw quality water that is clean and free from sediments so as not to injure the power generating equipment, discussed as under:

1. *Quality water.* Location of intake is required to be such as to draw the best quality water from the reservoir. Depth of water at intake is important. Quality of water varies at different levels in a reservoir and it is necessary to draw water from different elevations of the reservoir at different seasons of the year for which multiple-level intakes are frequently provided.

2. *Optimum water utilisation.* Intake is located in the deepest part of the impounding reservoir to enable full utilisation of the capacity of the reservoir and to protect intake from sediments in the reservoir. In reservoirs with wide variations in the water surface elevation, the intake is better located at the lowest stage so that one inlet is always submerged and operative to draw supply and minimum operating head is always available.

3. *Minimum silt entry.* In order to prevent silt from the reservoir being carried into the outlet system, location of intake at low points or pockets in the reservoir is avoided. Bottom sediments are kept out of the intake if the lowest entrance ports are sufficiently above the reservoir floor.

4. *Wave action.* The intake structure is placed sufficiently removed from the currents that might threaten the safety of the structure. Waves exert pressure on the superstructure of the intake. Waves may also stir up the bottom sediments if the intake is not so located deep in the water below the water surface. Locations where winds may drift debris into the intake are avoided.

5. *Ice action.* Ice may push down and clog ports. Ice troubles are reduced in case entrance velocities are kept between 7 to 10 cm/s. Low velocities do not transport ice and hold the entry of leaves and debris into the intake as also fish can escape from being drawn into by intake currents.

6. *Type of dam.* Usual location of intake in an earthen dam is in deep water near the upstream toe of the dam while in concrete or masonry dam, the intake may be in the dam structure itself.

7. *Sewage pollution.* The intake is located to avoid sewage pollution. Locations where stagnant water or algae can accumulate are unsuitable for intake structure. Navigation adds the hazard of pollution from passing ships. It is more relevant to river intake for public water supply.

8. *Shore location.* Usual location of the intake for public water supply is some distance away from the shore to avoid turbid water from surface wash along the banks.

20.11. Type of Intakes

Depending on the function served and the range in reservoir head under which it is to operate, the discharging capacity and frequency of reservoir drawdown, intakes for hydroelectric projects may be relatively simple submerged intakes or more elaborate structure raised as a tower above maximum reservoir level. Intakes may also be classified as low pressure and high pressure intakes though there is no clear line of division between the two classes of intakes; the essential requirements of the two types being the same, but may differ in details and type of equipment on account of the difference between the depths of water in which they are to be operated. Broadly the intakes are classified as (*i*) Run-of-the river type, and (*ii*) Reservoir type.

1. Run-of-River Intake. In run-of-river power plants, intake is appurtenent to power house and draws water from the river without any appreciable storage upstream of the diversion structure (Fig. 20.35). Characteristic of river flows, the intake is designed to withstand high peaks and short duration flood flows and high sediment loads. The bellmouth entrance is essentially provided with trashrack.

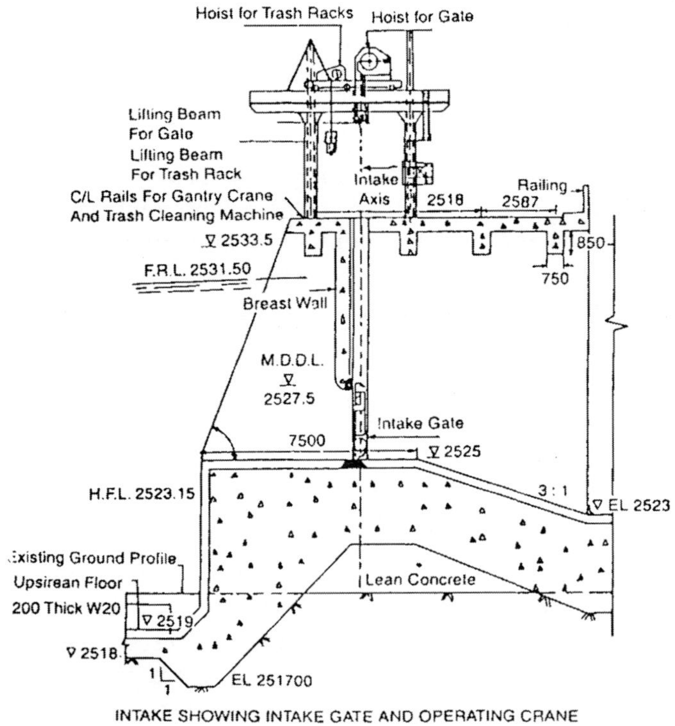

Fig. 20.35. Power plant intake showing gate and operating crane.

Canal Intake. It is also a variant of the run-of-river intake, that is provided adjacent to the diversion weir/barrage to admit water into the canal (Fig. 20.36). It is designed to function under low-heads and the topography and geology permits straight reach suitable for it. Sediment excluder is an essential component of the intake. The crest of the intake is generally raised to prevent entry of coarse fraction of bed load into the canal. Trashrack on the upstream and

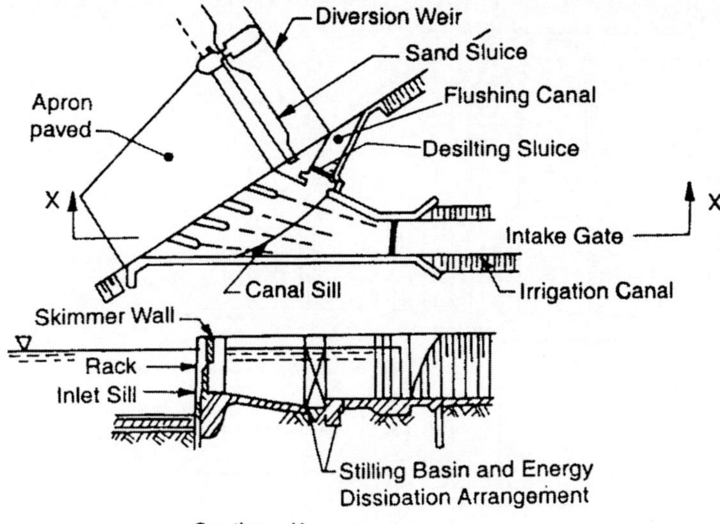

Fig. 20.36. Canal intake.

stilling basin and energy dissipation devices on the downstream of the intake are provided. In certain locations desilting tunnels may also have to be provided upstream of the intake.

2. Reservoir type Intakes. Intake towers classified as submerged, dry and wet intakes fall in this category.

(*i*) *Submerged Intake.* An intake structure which remains entirely under water during its operation is termed as submerged intake. It is provided where the structure serves only as an entrance to the outlet conduit and where ordinarily cleaning of trash is not required. It consists of simple concrete blocks or rock filled cribs usually square or octagonal to support the end of withdrawal conduit. The conduit entrance may be inclined, vertical or horizontal in accordance with the intake requirements. An inclined intake is provided in small installations where the gate is placed and operated on the upstream slope of a low dam with low entrance velocities. Submerged intake requires no attendance but partially submerged intake as the name implies, gets exposed in upper part during low reservoir level. The objection to submerged intake is that it is not easily accessible for inspection and repairs. Figs. 20.37 and 20.38 show the intake provided at Pong dam.

The submerged intakes remain entirely under water and possess the advantages (*i*) Simple structure, less costly to construct, (*ii*) No obstruction to navigation and little obstruction to river flows, (*iii*) Little danger from floating material and a minimum of trouble from ice, (*iv*) Trashrack cleaning is not generally required, and (*v*) Particularly suitable as public water supply intake from a river.

(*ii*) *Intake Tower.* An intake tower is used to draw water from reservoir in which there are huge fluctuations in water level or quality water is to be drawn at the desirable depth or both. It consists of an elaborate exposed or tower like structure rising above maximum reservoir level and closely located to the dam body or the bank of the stream so as to be approached by a connecting bridge of minimum span or at a distance as is reached by boat. The intake towers consist of circular concrete structure provided with openings or ports for water entry fitted with trashracks to prevent the entry of debris and ice large enough to injure the equipment and gates that control the flow through intake into the feeding conduit outlet. They are designed to

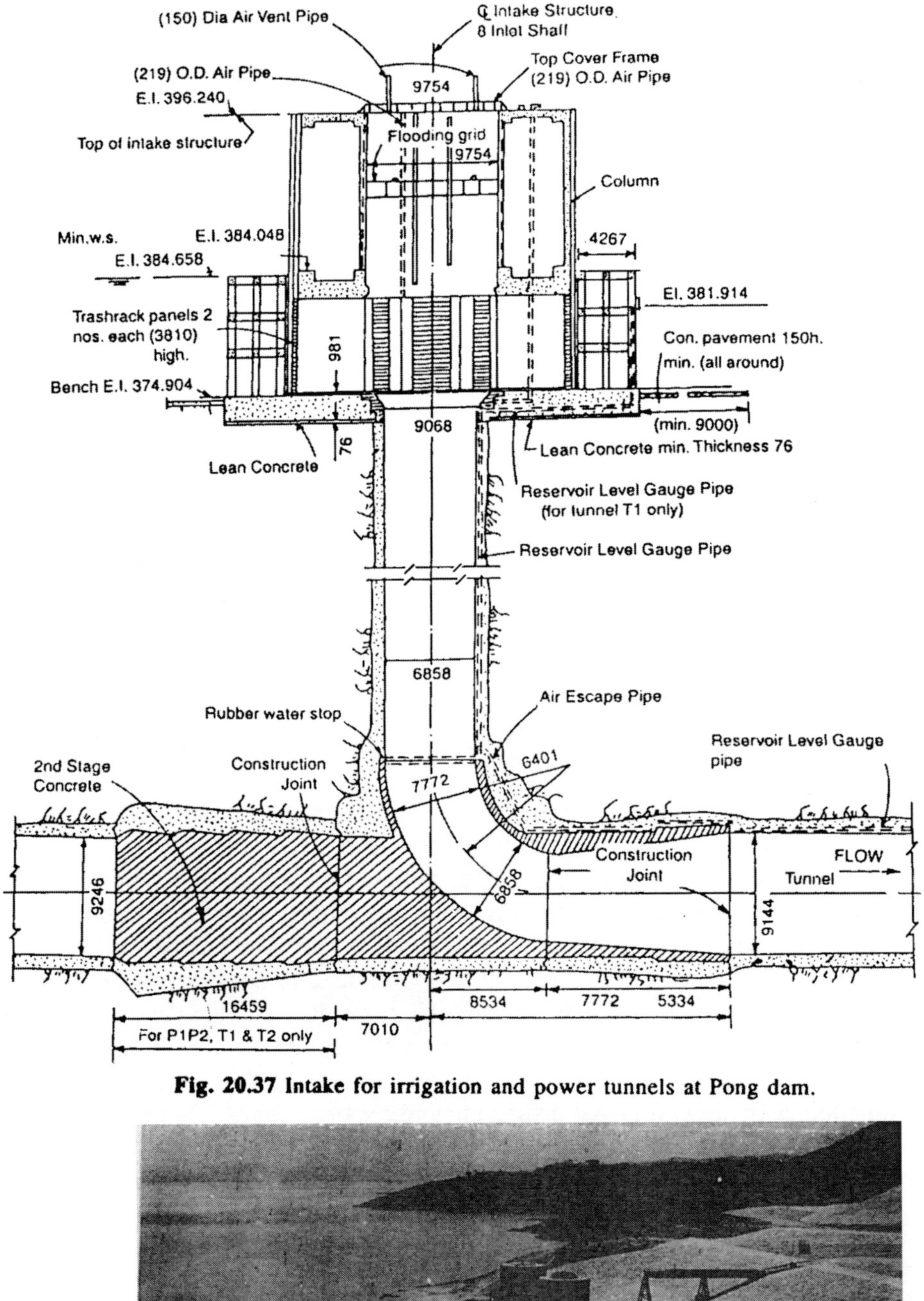

(150) Dia Air Vent Pipe
℄ Intake Structure 8 Inlet Shaft
9754
Top Cover Frame
(219) O.D. Air Pipe
(219) O.D. Air Pipe
E.I. 396.240
Top of intake structure
Flooding grid
9754
Column
Min.w.s.
E.I. 384.048
E.I. 384.658
4267
El. 381.914
Trashrack panels 2 nos. each (3810) high.
981
Con. pavement 150h. min. (all around)
Bench E.I. 374.904
9068
(min. 9000)
76
Lean Concrete min. Thickness 76
Lean Concrete
Reservoir Level Gauge Pipe (for tunnel T1 only)
Reservoir Level Gauge Pipe
6858
Air Escape Pipe
Rubber water stop
Reservoir Level Gauge pipe
2nd Stage Concrete
Construction Joint
7772
6401
Construction Joint
FLOW
Tunnel
9246
6858
9144
16459
8534
7772
5334
For P1P2, T1 & T2 only
7010

Fig. 20.37 Intake for irrigation and power tunnels at Pong dam.

Fig. 20.38 Intakes at Pong dam (Partially submerged).

safely withstand the severe combinations of hydrostatic pressures, waves, winds, debris and earthquake pressures. Usually 50 percent of the trashracks are considered clogged to visualise the worst designed conditions for the intake structure. However, there is no risk of damage due to clogging of debris or ice in the case of intakes the ports of which are to remain submerged at all operating reservoir levels. It has the merit that best quality water available at different depths at different seasons of the year can be drawn through ports at different elevations. Large size port openings protected with trashracks are made to minimize the entrance velocity in order to avoid clogging of the port as also to avoid difficulty with entrained air at low velocities. The two types of intake towers are as under :

(*i*) *Dry Intake Tower.* In dry intake tower (Fig. 20.37) the entry ports are directly connected with the withdrawal conduit and water is directly drawn into conduit through the gated entry ports and there is no water inside the tower when gates are in closed position. Dry intake tower has the merit that the interior of the tower being dry is made accessible for inspection and operation besides that the water can be withdrawn from any given selected reservoir level by opening the port at that level. However, dry intake tower is massive in structure, than wet intake to withstand additional buoyant forces to which it is subjected when the port gates are closed.

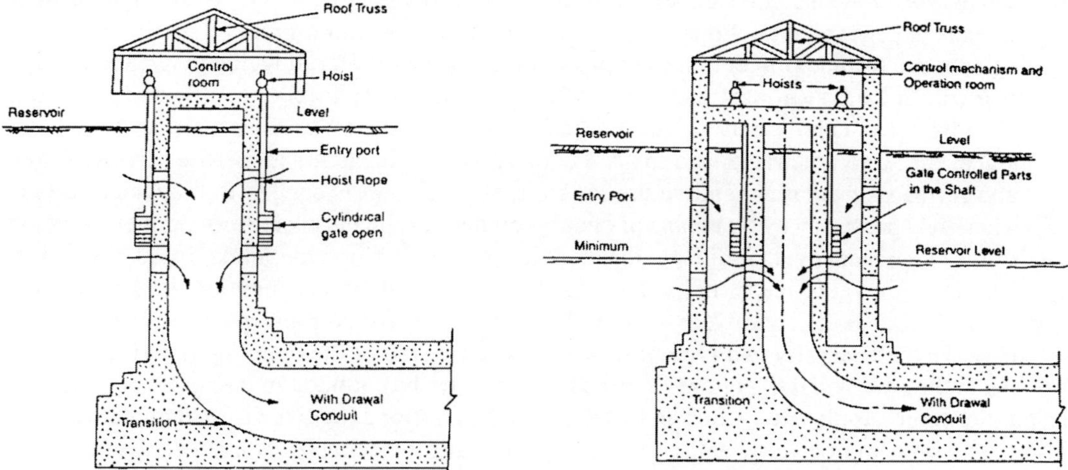

Fig. 20.39 Dry intake tower. **Fig. 20.40** Wet intake tower.

(*ii*) *Wet Intake Tower.* A wet intake tower has entry ports at various levels and the vertical shaft is filled with water upto reservoir level (Fig. 20.40). It differes from the dry intake tower in that the water enters from the ports into the tower and then into the withdrawal conduit through separate gated openings. As such it consists of a circular shell made of concrete filled with water upto reservoir level, housing another inside shaft directly connected to the withdrawal conduit. It is less costly to construct and is usually not subjected to flotation and certain other stress may not be the consideration.

3. Low Pressure Intake. It is the usual form of intake used for relatively smaller drawdowns as in the case of hydel power plants of small units catering for daily and weekly water surface variations. Low pressure intake usually performs the functions of a dam and is designed accordingly based on the theory of stability of a dam. The maximum velocity through the gross cross-sectional area of the trashrack is about 0.75 m/sec which may be of the order of 1.5 m/sec in the case of large units and mechanical racks of wider rack bars spacing. The racks are liable to complete and sudden plugging with ice or debris and are designed considering complete stoppage of flow through the racks. The gross losses through the racks are usually considered as 0.03 to 0.15 m.

4. High Pressure Intake. High pressure intake, in general, is used in a reservoir which serves as a storage reservoir and head water for hydel power generation involving considerable drawdown under normal operation conditions. The intake is constructed in a variety of forms depending on the type of dam whether of concrete or earthen dam. In concrete dam, high pressure intake varies little in detail from low pressure intake except in the type of gates installed. In earthen or rockfill dam, the intake is generally a tower built near the foot of the upstream slope of the dam, the structural details of which vary greatly from the low pressure intake. In some cases shafts of outlet tunnels are also used as intake. Considerable high velocities 3 to 4 m/s through gross area of trashracks are adopted.

20.12. Trashracks

Trashrack is defined as a screen or grating provided at the entrance of intake to prevent entry of debris, jungle, etc. Trashracks usually consists of trash sections 1.5 to 2 m wide and not too long for handling, made up of mild steel flats on edge 5 to 15 cm centre in both directions depending on the maximum size of debris to be excluded from entering the intake. Coarse trashracks are provided near the ports to prevent large drift, such as cakes of ice, roots, tree and timber from being drawn into the intake. In some part of the intake fine trashracks are provided to protect the machine and machine apparatus through which water flows. The spacing of trashracks is kept so that net opening between them is at least 5 mm less than the minimum opening between turbine runner blades to prevent clogging by material which may pass through the trashrack. In cold regions, trashrack is oftenly clogged with frazile ice. Electrical heating for small trashracks is provided to prevent ice formation on the racks. Electrical heating is not economical for large trashracks and instead compressed air is passed along the screens at low level and formed air bubbles remove the ice from the rack and is carried to the water surface from where it is passed away by means of chute provided for the purpose. Another method is the provision of vigorous reversal of flow through the port made when the intake is installed as an expedient for removal of ice from intake ports. The floating debris accumulated, as are denied entry into the intake, are removed with the help of power driven rack-rakes. Various factors which govern the necessity and type of trashrack and the size of the openings are (*i*) Nature of the trash in the reservoir, (*ii*) Nature of utilisation of water impounded in the reservoir, whether for public water supply, irrigation or hydel power generation, (*iii*) Need for the exclusion of small trash from the outflow, and (*iv*) Accessibility or otherwise for removing accumulated trash.

Design Considerations for Trashracks

Trashracks are closed spaced for excluding small trash when the outlet consists of small conduit with valve controls, and are generally more widely spaced in large sized conduit with large slide gate controls. The trashracks also vary in shape depending on the arrangements provided for their mounting on the intake structure. For a deep inlet intake, trashracks usually of the shape of a cage surmounting the entrance are arranged in an open box placed in front of a vertical entrance or positioned along the front side of a tower structure.

In the case of intakes which are to remain submerged at all operating levels of the reservoir, there is little risk of damage due to clogging of debris or ice, etc. and the trashrack may consist of widely spaced structural members without rack bars such as struts and beams so placed as to exclude only the large trash and other floating debris.

The required area of trashrack is governed by the limiting velocity through the rack and of course the nature of the trash to be excluded. In trashracks which are inaccessible for cleaning, usually the water velocity through them is limited to 0.75 m/s., while in racks accessible for cleaning a velocity of up to 1.5 m/s. is adopted. Where maximum loss value is considered, 50 percent of the rack area is considered as clogged which means twice the velocity through the trashrack. The water velocity through trashrack is usually limited to 1 m/sec. to reduce the pressure loss. The trashrack loss is usually neglected.

20.13. OUTLET WORKS

Outlet works comprise discharging facility for conveyance of water from a dam for power generation and irrigation. Outlet work for power generation is termed as penstock and that as a means for releasing water to meet the demand of irrigation water downstream is called irrigation outlet. Outlet works serve to regulate the releases at a retarded rate as per doynstream needs or a combination of multipurpose requirements such as releasing surplus inflows in conjunction with spillway if the outlet empties downstream into a river. The outlets are usually placed sufficiently below minimum reservoir level to provide necessary head for effecting flows. Occasionally the outlet may be placed at a higher level to deliver water to a canal. The irrigation outlet capacities are determined from reservoir operation studies, based on the consideration of a critical period of low runoff when reservoir storages are low and daily irrigation demands are at its peak. In a multipurpose reservoir, the irrigation outlets are operated rarely, the release being usually made through penstocks to serve for power generation as well as to meet irrigation demand.

The layout of the irrigation outlets is influenced by various considerations relating to the hydraulic requirements, site adaptability, the interrelationship of the outlet works to the diversion requirement during construction and other appurtenant works of the dam. An outlet work for a low dam may consist of an open channel or cut-and-cover structure placed at the dam embankment. For a low earthfill dam, the outlet may consist of single or multiple units of buried pipes placed through or under the dam embankment. For a high earthen dam, the outlet is usually carried through, under or around the dam as a cut-and-cover conduit or through the abutment as a tunnel. For a concrete dam, the outlet work is usually carried through the dam as a formed conduit or a sluice or a pipe embedded in the concrete mass. The irrigation outlet tunnel like penstock tunnels, usually comprise intake structure, tunnel with gated control, stilling basin and exit channel.

In earthen dam, the simple type is 900 mm minimum diameter pipe so as to permit entry of a man into it for clearance. Projecting collars are provided to increase the length of seepage path by atleast 25 percent in order to reduce the seepage along the outside of the conduit. Thus if N is the number of collars and P is the projection of each collar, $2NP$ is greater than $L/4$, where L is the length of the sluiceway from upstream to downstream end. In order to limit the hydraulic losses to the minimum, the interior surface of the sluiceway pipe of tunnel is made smooth and free from any projections or cavities and suitably painted. The entrance to an irrigation tunnel is designed to produce an acceleration similar to that in a jet issuing from a sharp edged orifice. The surface is formed to natural contraction curve and the tunnel is assumed to the size of orifice jet at its maximum contraction. The normal contraction with coefficeint of contraction C_c as 0.6 and 0.7 is used in high head and low head installations respectively in order to reduce the height of opening. The coefficients of discharge and loss coefficients for typical entrances for conduits are as given in Table 20.1.

Table 20.1. Coefficient of discharge and loss coefficient.

Type of entrance	*Coefficient*			*Loss coefficient*		
	Maximum	*Minimum*	*Average*	*Maximum*	*Minimum*	*Average*
Gate in thin unsuppressed contraction	0.70	0.60	0.63	1.80	1.00	1.50
Gate in thin wall-bottom contraction and sides suppressed	0.80	0.68	0.70	1.20	1.50	1.00
Gate in thin wall-corners rounded	0.95	0.71	0.82	1.0	0.10	0.50
Square-cornered entrances	0.85	0.77	0.82	0.70	0.40	0.50
Slightly rounded entrances	0.92	0.79	0.90	0.60	0.18	0.23
Fully rounded entrances $r/D = 0.15$	0.96	0.88	0.95	0.27	0.08	0.10
Circular bell-mouth entrances	0.98	0.95	0.98	0.10	0.04	0.05
Square bell-mouth entrances	0.97	0.91	0.93	0.20	0.07	0.16
Inward projecting entrances	0.80	0.72	0.75	0.93	0.50	0.80

Discharge formula:

$$Q = Cd.\ A\sqrt{2g\,H_{eff}}$$

where, Q = discharge (cumecs), A = area of the outlet sluice (m^2), H_{eff} = effective differential head acting on the sluice, *i.e.*, total head-head losses (m), and C_d = coefficient of discharge.

Head loss through conduit: The head loss through conduit comprises the loss caused by trashracks, conduit entrance, friction in conduit, transition and bends, gates, etc.

1. **Trashrack loss.** Trashrack loss depends on velocity, as under:

Velocity through trashrack (m/s)	*Head loss (m)*
0.15	0.006
0.30	0.03
0.45	0.09
0.62	0.15

2. **Transition loss.** Head loss in straight pipe, square or rectangular section conduit is determined by Darcy's formula and Scobey's formula:

(*i*) *Darcy's formula*

For pipe conduit, $$h_f = \frac{4flv^2}{2gd}$$

For square or rectangular conduit: $$h_f = \frac{flv^2}{8Rg}$$

where, h_f is friction loss through the conduit (m), l is length of the conduit (m), R is hydraulic mean radius (m), V is velocity through pipe (m/s), d is diameter of pipe (m), f is a loss coefficient, usual value is 0.014, $f' = 64/R_e$ for laminar flow and $0.316/R_e^{0.25}$ for turbulent flow, R_e = 3000–100000. $f = \frac{1}{R'^{1/2}} = 2.0 \log_{10} R f'^{1/2} - 0.8$ for R_e exceeding 100000. R_e is Reynold Number and g is acceleration due to gravity (m/s^2).

(*ii*) *Scobey's formula.* $$h_f = \frac{K_c V^{1.9}}{d^{1.1}}$$

where, K_c = loss coefficient depending on pipe material and interior condition of pipe.

3. **Bend loss.** The bend loss can be calculated from the formula:

$$h_b = \frac{KV^2}{2g}$$

where, h_b = is head loss resulting from bend and K is bend loss coefficient. In case of pipes it depends on shape of bend, deflection angle and ratio of radius of bend to pipe diameter.

4. **Contraction loss.** Head loss due to contraction,

$$h_c = \frac{K}{2g}\left(V_1^2 - V_2^2\right)$$

where, V_1 is velocity in contracted section, V_2 is velocity in normal section and K is a loss coefficient. It varies from 0.1 for gradual contraction to 0.5 for abrupt contraction.

5. **Exit loss.** When no recovery of velocity head will occur, *i.e.*, when the release from a pressure conduit discharge freely, or is submerged or supported on downstream from exit, head loss is $1.0\left(\frac{V^2}{2g}\right)$, V is the exit velocity.

20.14. Classification of Outlet works

Outlet works are classified according to their purpose, their physical and structural arrangement, or their hydraulic operation, as follows:

1. **Classification according to their purpose**

(*i*) *River outlet.* An outlet work which emplies directly into a river.

(*ii*) *Canal outlet.* An outlet which discharges into a canal.

(*iii*) *Pressure pipe outlet.* An outlet which delivers water into closed pipe system.

2. **Classification according to their physical and structural arrangements.**

(*i*) *Open channel.* It consists of a conventional open flume or rectangular channel with a gate similar to that of a spillway. Open channel outlet work is suitable for a low dam for release of water to a canal or to the river.

(*ii*) *Cut and cover conduit.* Cut-and-cover conduit outlet is suitable where the water-way is too small to justify a minimum size tunnel or where the foundation conditions are not suitable for a tunnel. It may pass through or under the dam. It is constructed before the placement of dam embankment and designed to safely withstand the load of fill-placement and internal hydrostatic pressure loading equal to full reservoir head when running as pressure conduit.

(*iii*) *Tunnel.* It is closed conduit waterway. It is suitable for high earthfill dams, where open channel outlet cannot be provided. It may be carried through. under or around the dam as a cut-and-cover conduit or as a tunnel through dam abutment. Diversion tunnel constructed to facilitate the construction of the dam is best utilized as outlet tunnel. Tunnel outlet, being not in direct contact with the dam embankment, provides a safe and durable means of release of water from the reservoir than a cut-and cover-conduit and open channel structure. Tunnels may be free-flow tunnel or pressure tunnel.

3. **Classification according to their hydraulic operation**

(*i*) *Gated outlet.* It is outlet provided with operating gates and regulating gates to control and regulate the outlet flow. The gates may be located at the upstream end of the conduit, at an intermediate point along its length or at the downstream end of the structure.

(*ii*) *Ungated outlet.* Ungated outlet flow in the conduit, as in the case of detention dams, is similar to that in a culvert spillway.

SOLVED EXAMPLES

Example. 20.1. *An ogee spillway with vertical upstream face has design discharge of 3000 cumecs, crest length of 150 m, normal reservoir level is 700.0, average river bed level is 650.0 m and coeffficient of discharge is 2.2. Determine the crest level and profile of the overflow spillway section.*

Solution. Assume an high overflow spillway, *i.e.*, height of dam is greater than 1.33 H so as to neglect the velocity of approach.

$$Q = CLH_e^{3/2}$$

$$H_e = \left(\frac{Q}{CL}\right)^{2/3}$$

$$H_e = \left(\frac{3000}{2.2 \times 150}\right)^{2/3} = 4.36 \text{ m}$$

Effective length of spillway, $L_e = L - 2\,(N.K_p + K_a)H_e$. Assuming 10 spans of 15 each, number of piers = 9, and piers of 90° cut water nose and rounded abutments so that $K_p = 0.01$ and $K_a = 0.1$

$\therefore$ $$L_e = 150 - 2(9 \times 0.01 + 0.1)\,4.36 = 148.34 \text{ m}$$

Now $$3{,}000 = 2.2 \times 148.34\,(H)^{3/2}$$

or $$H = 4.39 \text{ m}$$

Crest level $$= 700.0 - 4.39 = 695.61 \text{ m}$$

Height of dam, $$H_d = 50.0 - 4.39 = 45.61 \text{ m}$$

Effect of head due to velocity of approach,

$\frac{H_d}{H} = \frac{45.61}{4.39} > 1.33$, hence it is a high spillway and velocity of approach is neglected. For effect of tail water elevation, $\left(\frac{H_d + d}{H}\right) = \left(\frac{45.61 + 4.39}{4.39}\right) > 1.7$, hence coefficient of discharge is not affected by tail water conditions.

Downstream Profile: Tangent point co-ordinates, assuming slope of d/s glacis as 0.7:1

Hence, $$\frac{dy}{dx} = \frac{1.0}{0.7}$$

D/s profile relation, $$y = \frac{x^{1.85}}{2H^{0.85}} = \frac{x^{1.85}}{2(4.39)^{0.85}}$$

$$y = \frac{x^{1.85}}{7.03}$$

or $$\frac{dy}{dx} = \frac{1.85x^{0.85}}{7.03} = \frac{1.0}{0.7} \text{ or } x = 7.31$$

Co-ordinates of d/s profile

x	y	x	y
1	0.14	5	2.79
2	0.51	6	3.91
3	1.09	7	5.21
4	1.85	7.31	5.64

U/s Profile: U/s profile relation,

$$y = \frac{0.724\,(x + 0.27H)^{1.85}}{H^{0.85}} + 0.126H - 0.4315H^{0.375}\,(x + 0.27H)^{0.625}$$

$$= \frac{0.724\,(x+0.27\times 4.39)^{1.85}}{(4.39)^{0.85}} + 0.126\times 4.39 - 0.4315(4.39)^{0.375}\,(x+0.27\times 4.39)^{0.625}$$

$$= 0.21\,(x+1.19)^{1.85} + 0.55 - 0.75\,(x+1.19)^{0.625}$$

x extends up to $0.27H = 0.27\times 4.39 = 1.19$ m,

and $y = 0.126H = 0.126\times 4.39 = 0.55$ m

Co-ordinates of u/s profile

x	y	x	y
–0.25	0.02	–1.0	0.29
–0.50	0.07	–1.19	0.55
–0.75	0.15		

Example. 20.2 *Determine the discharge through a chute spillway of ogee crest; length of spillway is 250 m, height of spillway crest above u/s approach channel is 10 m, width of approach channel is 250 m, and depth of water over spillway crest is 5 m.*

Solution.

$$Q = CLH_e^{3/2}$$

$$Q = 2.5\times 250(5)^{3/2}$$

assuming coefficient of discharge as 2.5, neglecting velocity of approach

$$Q = 6988 \text{ cumecs}$$

Velocity of approach, $$V_a = \frac{Q}{BD} = \frac{6988}{250(10+5)} = 1.86 \text{ m/s}$$

Head due to velocity of approach, $$h_a = \frac{V_a^2}{2g} = \frac{(1.86)^2}{2\times 9.81} = 0.18 \text{ m}$$

$$H_e = H + h_a = 5.0 + 0.18 = 5.18 \text{ m}$$

Discharge through spillway, $$Q = CLH_e^{3/2}$$

$$= 2.5\times 250\,(5.18)^{3/2} = 7369 \text{ cumecs}$$

Example. 20.3 *A siphon spillway has height of the throat 1.5 m (distance from crest to summit) and width of throat 4.0 m. At the design flow, the tail water elevation is 7 m below the summit of the siphon, and the head water elevation is 2 m above the summit, (i) Taking coefficient of discharge as 0.6, and treating the siphon as a mouthpiece, determine the capacity of siphon, (ii) Determine the head that would be required on an ogee spillway 3.8 m long to discharge this flow, if coefficient of discharge is 2.25, and (iii) What length of this ogee weir would be required to discharge the same flow with a head of 2.2 m on the crest, assuming the coefficient of discharge to be constant?* *[AMIE]*

Solution. Capacity (discharge) of siphon spillway

Area of throat $= 1.5\times 4 = 6 \text{ m}^2$

(*i*) Head on siphon $= 7 + 2 = 9$ m

Therefore $$Q = C_d A\sqrt{2gh}$$

$$= 0.6\times 6\sqrt{2\times 9.81\times 9} = 47.84 \text{ cumecs}$$

(*ii*) Ogee spillway discharge,

$$Q = CLH^{3/2}$$

$$47.84 = 2.25\times 3.8\times H^{3/2}$$

$$H = (47.84\times 2.25\times 3.8)^{2/3} = 3.15 \text{ m}$$

(*iii*) Now $$Q = 2.25\,L\,(2.2)^{3/2}$$

$$47.8 = 2.25L\,(2.2)^{3/2}$$

or $$L = 6.5 \text{ m}$$

Example. 20.4. *Calculate discharge over a spillway, crest length 100 m, wherein water flow depth changes from 1 m to 4 m in hydraulic jump formation in stilling basin.*

Solution.

$$\frac{D_2}{D_1} = \frac{1}{2}\left(\sqrt{1+8\,F_1^2} - 1\right)$$

$$\frac{4}{1} = \frac{1}{2}\left(\sqrt{1+8\,F_1^2} - 1\right)$$

or $$F_1 = 3.16 \text{ m}$$

Further $$F_1 = \frac{V_1}{\sqrt{gD_1}}$$

or $$V_1 = F_1\sqrt{gD_1}$$

$$= 3.16\sqrt{9.81 \times 1.0} = 9.89 \text{ m/s}$$

Spillway discharge, $$Q = V_1 \times D_1 \times L$$

$$= 9.89 \times 1.0 \times 100 = 989 \text{ cumecs}$$

QUESTIONS

1. Define spillway. What is the necessity of a spillway? Mention functions of a spillway.
2. Enumerate various types of spillways and how they are selected for individual conditions. Describe in detail the most widely used type.
3. Describe ogee spillway and sketch an ogee profile. How it is designed?
4. Define chute spillway. Discuss the design principles involved in the chute spillway. Why is it preferred to ogee and other types of spillways?
5. Discuss the various component parts of a chute spillway.
6. What do you understand by siphon spillway? Enumerate two types of saddle siphon spillway and explain functions of its various parts.
7. What do you understand by hooded type siphon spillway? Describe its discharge equation.
8. What is meant by priming and depriming element of siphon? Discuss the devices used for early priming in a saddle siphon spillway.
9. Define energy dissipation. Discuss various methods used for energy dissipation.
10. Define spillway gates. Discuss various types of spillway gates. What are their relative merits and demerits?
11. Discuss with neat sketches the various spillway gates. [*AMIE*]
12. What are the various types of spillways? Explain in detail the design principles of an overflow spillway. [*PU*]
13. (*a*) Write short notes on (*i*) ogee spillway, (*ii*) side channel spillway.
    (*b*) What is the purpose of providing gates on spillway?
14. (*a*) Name various types of spillways and explain briefly any one of them.
    (*b*) A discharge of 1750 m^3/sec has to be passed over an ogee spillway with a coefficient of discharge of 2.25 at a head of 3.8 m. What should be the effective length of the spillway? Neglect velocity of approach. [*PU*]
15. (*a*) What are the factors which determine the type of gate to be installed over a spillway crest?
    (*b*) Briefly describe the design principles that are involved in the design of a radial gate. [*PU*]

16. (*a*) What are the spillways and why are they provided?
    (*b*) Name the various types of spillways.
17. (*a*) Describe with neat sketches the different types of spillways provided for dams and mention where each of these is suitable.
    (*b*) From the following data ascertain if the tail water depth is adequate for the formation of a standing wave on the downstream side of an overflow spillway. Also mention what type of stilling basin is indicated for this spillway:
    Design head over the spillway = 10.5 m
    Height of the crest of the spillway = 77 m
    Maximum tail water depth = 24 m
    Coefficient of discharge = 2.1 [*AMIE*]
18. Discuss outlet through an earthen dam.
20. Explain with neat sketch the functions of different parts of a reservoir intake.
20. Describe with the help of neat sketches a typical intake structure of outlet through an earthen dam over the permeable foundation. Dam height approximately 33 m.
21. The design and construction of the outlet arrangement to an earthen dam is the most important and difficult features of work; discuss the statement with the aid of neat sketch. Enumerate precautions to be taken for such construction.
22. Discuss types of outlets through an earthen dam.
23. What are the general features of a river intake? Describe them briefly with a neat sketch.
24. State various factors to be considered in locating suitable site of intakes on rivers and reservoirs.
25. Sate various factors to be considered for location of suitable intake sites on rivers for irrigation and water supply. Sketch a typical reservoir intake.
26. What are the main difference in the outlets through earthen and masonry dams.
27. Write a note on outlet works in storage dams clearly explaining different type, their need and typical features.
28. Sketch an outlet through an earthen dam. Indicate its different components.
29. What are the considerations in locating an intake on a river and in reservoir?
30. What is the difference between a reservoir intake and river intake.

PART - VI

WATER POWER ENGINEERING

- Water Power Engineering

21

WATER POWER ENGINEERING

21.1 INTRODUCTION

Water power has been one of the earliest sources of energy and is still attractive being cheap, clean and pollution free. In multipurpose projects use of water for irrigation, as well as hydro power generation is aimed at besides other benefits. In India the power development started with the installation of 1960 kw hydroelectric station in Darjeeling in Assam in 1897 and 4500 kw station at Sival Samudram in Karnataka in 1902. By 1947, at the time of independence, the hydro power generation capacity increased to 508 MW. The hydel power installed capacity as on March 1999 is 22,443 which is nearly one-fourth of the estimated potential of 84,000 MW at 60 percent load factor. According to fourth national power plan 58,000 MW hydro power addition is needed over the next 15 years. For details on the subject refer "Water Power Engineering" by the author.

Hydro-power is generated by the conversion of potential energy of high head water into electrical energy. The water is made to fall on the blades of turbines to run them by the conversion of hydraulic energy into mechanical energy which in turn is converted into electrical energy by directly coupling the shaft of turbine with the generator. The generator armature, thus rotated produces electricity.

21.2 CLASSIFICATION OF HYDRO POWER PLANTS

1. Classification Based on Hydraulic Characteristics

1. *Run-of-River Plants.* Run-of-river plant generates electricity with the stream flow as it comes and as such utilises the minimum flow. Evidently the stream should be perennial with minimum flow adequate to generate sufficient power. The full power output of such a plant can not be considered as firm. However, power plants on hydel canals which fall in the category of run-of-river plants because no pondage is envisaged in their cases give firm power, since water supply is fairly constant. Such power plants may be of medium head (above 10 m) and low head (2 to 10 m). On canal falls of low heights, bulb type units available for operation under heads in the range of 2 to 10 m are used to generate power (Fig. 21.1).

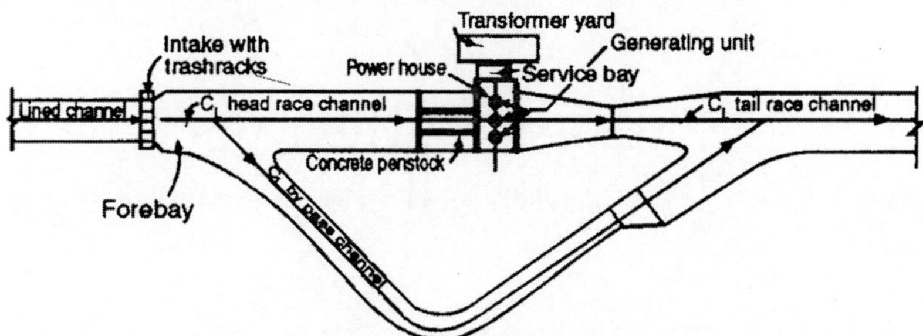

Fig. 21.1 Run-of-river canal fall power plant.

2. *Storage Plants.* In storage plants, substantial reservoir storage is provided by the construction

of a dam across a river to create high head. Storage is done during wet season when flow is in excess of demand to maintain continuous hydel power generation. Storage plants are designed for small as well as large heads with estimated yield for reliability at 90 percent. The power output is largely firm.

3. *Cascade Plant.* In cascade plant, available water is used more than once in a series of hydel power plants constructed to utilize the available drop in the head. Such hydel stations may either have separate storage reservoir or a common upstream reservoir or can simply be a run-of-river cascade with no significant storage.

4. *Pumped Storage Plant.* It is a peak head plant in which power is generated during the period of peak load from the water stored in an upper reservoir. The water is pumped back from a lower reservoir into the upper reservoir during the periods of low energy demand. Thus same water is reused. The machine used as turbine during generating electricity is used as a pump during off-peak load hours to pump water back into the upper reservoir. Similarly the machine used as a generator is used as a motor (Fig. 21.2). It embodies the advantages of improved load factor of the power system as also economy in the cost of the power plant. The various purposes of a pumped storage plant are (*i*) To improve power factor of the network, (*ii*) To control network frequency, (*iii*) To serve as emergency reserve, (*iv*) To meet weekly and seasonal peak load demands, and (*v*) To cover daily peak power demands and to improve capacity factor of base-load thermal plants. The potential of pumped storage power assessed is 96,524 MW

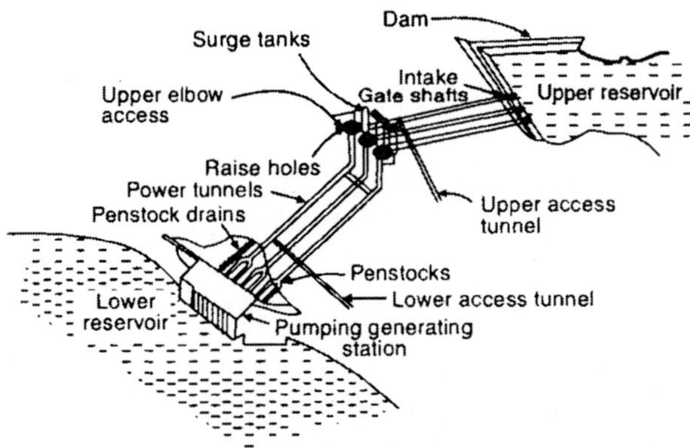

Fig. 21.2 Pumped storage plant.

5. *Tidal Power.* Water at the time of high tide when stored in a basin at a high level can be made to fall into sea during low tide through turbines to produce tidal power. The principle components of a tidal power plant are power house, dikes to form basin or basins and sluice ways from the basin to the sea and vice versa (Fig. 21.3). Tidal power generation is, however, to

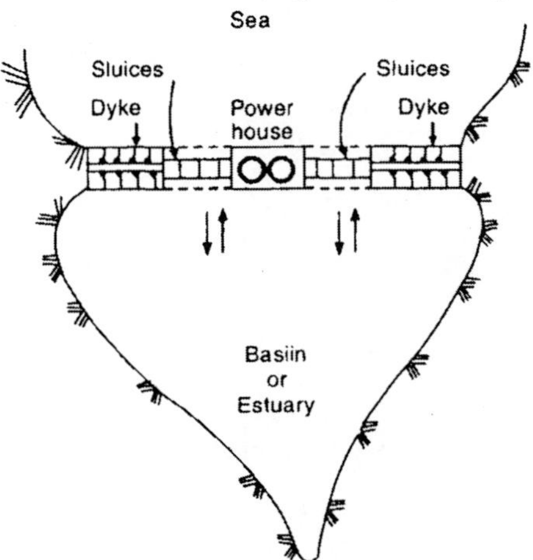

Fig. 21.3 Tidal power plant.

be justified by the minimum reliable peaking power it generates as an alternative to a standby thermal plant of similar capacity. India has tidal power potential of 10,000 MW along coast line of 1600 km. The tidal power is everlasting and is not influenced by the vagaries of nature such as failure of monsoons. It does not involve any extra submergence or uprooting of population. There is no loss of forest cover.

2. Classification Based on Unit capacity

In India, the unit rating based on generation capacity is Micro 1 to 100 kw, Mini 101 kw, up to 2000 kw; individual unit 101 to 1000 kw, Marco 1001 kw to 5 MW and Small up to 15MW, recently increased to 25 MW. As per latest estimates by Central Electricity Authority (India) total small hydel potential in the country is 6,782 MW including 564 MW from canal falls over 3m.

The advantages of such plants are (*i*) Low gestation period of upto 3 years, (*ii*) Small financial investment, (*iii*) Negligible submergence of land, (*iv*) Fairly constant cost of generation, (*v*) Large life, low operation-maintenance cost and special adaptability to remote control operation, (*vi*) Operation is made simple by microprocessor, (*vii*) Reliable and flexible, and (*viii*) Attractive proposition in remote mountainous areas to meet local energy demand and in plains through development of canal falls.

3. Classification Based on Available Head

1. *High head plant.* The power generation is made from high falls of over 300 m or by transfer of water from one river to another river or stream across the watershed to make use of high fall so achieved. Impulse turbines are generally used.

2. *Medium head plant.* It is a plant which operates under a head between 30 to 300 m. Francis, Propeller, Kaplan turbines are generally used.

3. *Low head plant.* It is a hydel power plant with operating head less than 30 m. It is run-of-river scheme. Francis, Propeller or Kaplan turbines are generally used.

4. Classification Based on Location

1. *Surface power plant.* Also termed as over ground plant *i.e.*, located over the ground surface.

2. *Underground power plant.* It is a power plant built underground *i.e.*, located in a cavity in the hill with no part of the structure exposed outside.

5. Classification Based on Functional Basis

1. *Base load plants.* These are the run-of-river and storage power plants which are designed to run at uniform load and generate nearly constant power for long periods. Base load plants are generally large in capacity and load factor is also high.

2. *Peak load plants.* As the name implies, these plants run at their peaking capacity during peak hours only. They thus work for a few hours of the day, when the load is more than the average generation. The stored water in the reservoir is passed down in the running of the plant and fresh storage is done again. This cycle is repeated. Evidently supply in the river at the plant site is meagre to run it round the clock.

6. Classification Based on Nature of Project

1. *Power project.* In mountainous regions, hydro electric stations are constructed only for power generation. Long transmission lines are required for connecting to the load centres in plains besides high cost on construction of rail/road communication for the transportation of men and material for the construction of power plant.

2. *Multipurpose project.* In multipurpose project, power is generated besides other functions such as irrigation, flood control, public water supply etc.

21.3 ADVANTAGES OF HYDRO POWER

Thermal and nuclear power plants are sometimes given precedence simply because of their shorter construction period. While development of thermal and nuclear power is no doubt imperative, hydel power generation deserves priority because of its obvious advantages such as (*i*) Cheapest source of energy, (*ii*) Indigenous, inexhaustible, perpetual and renewable energy resource. Hydro-power plants make use of water but do not consume water, (*iii*) Pollution free power generation. No foreign by-products like that of thermal effluents are produced, (*iv*) Operate at extremely low cost, (*v*) Afford incidental benefits of flood control, irrigation, public water supply, navigation, fishing, tourism, (*vi*) Afford a degree of independence from costly and at times unreliable supply of imported fuel, (*vii*) Technically more reliable than many thermal plants, (*viii*) Plant equipment is simple and less complicated than thermal, nuclear or even diesel power generating equipment, (*ix*) Better service operation flexibility; the operation of plant can be matched with load requirement *i.e.*, they can operate both as base load and peaking station, (*x*) Possible to build power plants of high capacity, and (*xi*) Efficiencies of about 95% to 98% are commonly achieved.

Limitations: (*i*) Susceptible to vagaries of nature such as drought, (*ii*) Longer construction period which means a disadvantageously freezing of the capital invested, (*iii*) Break down in hydro-power equipment may result not only in proportionate reduction in power generation but also, particularly in the case of run-of-the-river plants, in letting the precious water run waste, (*iv*) High initial cost, (*v*) Long term flow data is essential for proper assessment of irrigation and power benefits, (*vi*) Loss of large land due to submergence in the reservoir so created, (*vii*) Non-availability of suitable sites for the construction of dam, (*viii*) Displacement of large population from reservoir area and their rehabilitation, (*ix*) Environmental aspect; reservoir *vs* river ecology, and (*x*) High cost of transmission system for remote sites.

21.4 SELECTION OF SITE FOR HYDRO POWER PLANT

In order to ensure preparation of sound power projects, it is necessary to have thorough and systematic investigations. The selection of site for power plant is considered from hydrological, topographical and geological investigations. The subject matter has been dealt with in detail under "Dams".

21.5 ESSENTIAL DATA FOR WATER POWER STUDIES

The essential stream flow data required for water power studies of a storage projects is (*i*) Daily, weekly or monthly flow data over a long period of time for plant capacity and estimated output which is a function of the average flow as well as its distribution during a year as shown by flow-duration curves. The curve for power site forms the basis for the determination of plant capacity and available power at all times, (*i*) Maximum or flood flows for impoundment, spillway capacity and safety of the power plant. Unit hydrograph is a useful tool for it, (*ii*) Minimum or low water flows as a basis for primary power availability and size of auxiliary, if used. The flow analysis is best done by probability studies. Usually, a daily record of flows of say 10 years or more provides for most of the above, with perhaps the exceptions of the extremes of high and low flows which may be predicted with fair accuracy for a 10 year record of flow by taking into account precipitation and flow data of the nearby streams. The total period method gives more correct results than the calendar year method which averages out extreme values.

Power Duration Curve. The available power from a run-of-river plant wherein head is presumed to be constant at an average value is represented by a power-duration curve which exactly corresponds to a flow duration curve. However, for a precise power-duration curve the head corresponding to any discharge is required. The average monthly power from the plant is determined by the formula $P = 9.8\ QH\ h$ KW.

The power duration curve is then drawn by arranging the monthly power values in ascending order and following the same procedure as for flow duration curve.

Storage is a very important factor in operating a hydro power unit. The reservoir size required to satisfy the specified flow demand is accomplished by mass curve analysis (refer Reservoir and Dam Planning).

* For details refer Water Power Engineering by the author.

21.6 REQUIREMENTS OF WATER FOR HYDEL POWER

Hydro-power is generated with the fall of water through a head or fall. Thus the two basic elements of hydro-power development are (*i*) river or stream flow, and (*ii*) available head or fall.

Let a discharge of Q cumecs falls through a net height H m.

Rate of work done $= QwH$ kg. m/s.

Horse power generated (Theoretical) $= \dfrac{Q\omega H}{75} = 13.33\, QH$ taking w = 1000 kg/m^3 and metric H.P. = 75 kg. m/s.

Actual H.P. generated $= \dfrac{(1000\, QH\eta)}{75} = 13.33\, QH\eta$, h being the overall efficiency of the system *i.e.*, of generator, turbines etc.

Electrical energy $= 13.33\, QH\text{h}\ '\ 0.736$ kilowatts

(taking 1 metric H.P. = 0.736 kw)

$= 9.8\, QH\text{h}$ kw

or $= 8\, QH$, taking h as 80%.

The head available at a power plant is determined as the difference between reservoir level and tail race water level. For a given amount of available power, the greater the head as compared with the discharge, the more economical is the power generated. Power generated from high head is more economical than that from low head.

21.7 DEFINITIONS

Firm power. Also called *primary power.* It is the minimum continuous power which can be generated for 100 percent of time. It corresponds to the minimum river flow with due consideration of the pondage and load factor effects. In the case of storage reservoir it corresponds to the minimum outflow at minimum head, the condition usually met with just before the filling period.

Secondary power. It is the additional power, over and above the firm power, which can be generated during the periods of surplus flow.

Installed capacity. It is the total generating capacity of all the turbine-generator units installed in the power plant.

Dependable capacity. It is the load carrying capacity of the power plant with respect to the characteristic of the load to be supplied during a specified time interval and the period. It is determined by the power factor, capability and the load on the power plant.

Load factor. It is the ratio of the average load over a certain period to the maximum or peak load during the same period (Fig. 21.4). The load factor may be daily, monthly or annual depending on the period chosen such as one day, month or year. The load factor of an industrial area is higher than that for domestic area, the power demand by industry being more uniform. A high load factor is indicative of the better utilisation of the installed capacity and consequently the unit generating cost is less. A low load factor means that the generating capacity is not being fully utilised and as such the generation cost is high.

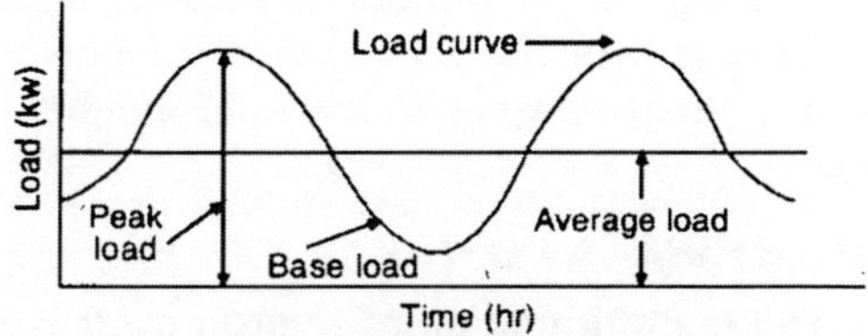

Fig. 21.4 Load curve

Capacity factor. Also called *plant factor.* It is the ratio of average load to the installed capacity of the plant. It is a measure of plant use whether the plant is generating power as per installed capacity.

Utilisation factor. It is the ratio of water actually utilised for power generation in the plant to that available in the river. With constant head, utilisation factor is also the ratio of power utilised to the power available. The utilisation factor equals capacity factor if there is always sufficient water to run the plant.

Plant factor. Also termed as capacity factor or plant use factor. It is the ratio of the average output to the installed capacity of the power plant.

$$\therefore \quad \text{Plant factor} = \frac{\text{Average output or load}}{\text{installed capacity}}$$

$$= \frac{\text{Average load}}{1.1 \times \text{peak load}}, \text{ taking reserve capacity of 10\% of peak load.}$$

The capacity factor is identical with load factor when the maximum load just equals the plant capacity. However, load factor is different if the plant is not used to its full capacity.

Power factor. It is the ratio of actual power (in kilowatts, kw) to the apparent power (in kilo volt amperes, KVA). The actual power generated depends on the power factor as the electrical machines are generally rated in KVA.

Average load. It is the total load produced divided by the number of hours in the day *i.e.*, average of the load for the period considered. It is expressed in kilowatt hours (kwh), *i.e.*, 1 kw of power delivered for 1 hour.

Peak load. It is the maximum instantaneous load or a maximum average load over a specified period of time. It determines the capacity of generating units.

Base load. It is the energy required to be supplied continuously for most of the time and accordingly to be generated.

Terms relating to water head

Gross head. It is the maximum head between the elevations of the reservoir water level and tail race water level. Gross head varies as the fall in reservoir level with the withdrawal of water and rise in tailrace level are often different.

Net head. It is the effective head available for power generation which is the gross head less all the losses in the water conductor system, including penstocks (usually the loss of head is taken as 2 to 5 percent of the gross head) *i.e.*, the effective head under which turbine operates.

Design head. It is the head at which the turbines installed in a power plant are designed to give the maximum efficiency at the rated speed.

Rated head. It is the head at which the turbine installed in a power plant produces the rated output at the rated speed and full gate opening.

Storage. It is the storage capacity of the storage reservoir to meet long term requirements of power plant.

Pondage. It is the small amount of storage to equalise daily, weekly or short term fluctuations in water requirement of power plants. Pondage factor is the ratio of the total inflow hours in a week to the total number of hours of power plant running during the same period.

21.8 THERMAL POWER

The main limitation of hydro power is its dependence on rainfall and reservoirs. Due to variability of rainfall, there may be periods when the reservoir feeding the power plant is not filled up specially during the dry years resulting in acute power shortage.

Thermal power, the largest energy source, is predominantly coal based. This natural resource is getting depleted day by day. The installed capacity of thermal power plants which was about 28,000 MW by the end of sixth plan is approximately 78,300 MW by the end of the century. The presently known coal reserves in the country are of the order of 1,10,000 million tonne. The consumption of coal is expected to be of the order of 1000 million tonne per year in the coming years. With all this the presently known coal deposits may last only for nearly a century. Besides thermal plants cause serious pollution and ecological hazards. It is thus imperative to go slow on thermal expansion and exploit hydro-power to the maximum possible extent.

The constraints of thermal power are: (*i*) Serious pollution and ecological hazards associated with it, (*ii*) Complicated thermal equipment, (*iii*) Poor industrial relations in generating stations, and (*iv*) Complicated operation and maintenance of thermal plants.

Hydro thermal mix. The mix in terms of storage hydro, run-of-river hydro, thermal, nuclear etc. is important as without adequate hydel back up the overall cost of meeting the power demand is more expensive besides for conservation of coal reserves of the country. It is estimated that every 1000 MW of hydro capacity releases normally 3 to 4 million tonne of coal per annum. The most desirable generation mix is considered as hydel 41%, thermal 56% and nuclear 3% whereas at present it is 22% hydro 2% nuclear and 76% thermal. To achieve the ideal mix, rapid increase in hydro power generating capacity is imperative.

21.9 HYDRO VS. THERMAL POWER

The relative usefulness of two sources of power is determined in consideration of availability of raw materials, distance from the plant, unit cost of energy, load factor, dependable capacity, unit rate, cost benefit and other parameters over a long period. A comparison of hydro and thermal power is as under.

Hydel power	*Thermal power*
Cheapest energy source.	Costly energy source.
Water is a renewable source. Production of power does not consume water.	Coal consumed is a non-replenishable source and is being fast consumed.
Pollution free generation of power.	Serious pollution and ash disposal problems.
Plant equipment is simple. Operation is also simple.	Plant equipment and operation are complicated.
High efficiency (90–95%)	Low efficiency (40–50%).
Nonconsumable small quantity of water is required for cooling system.	Large amount of water is consumed besides that for cooling purposes.
High extra cost on dam and appurtenant works rehabilitation of oustees and afforestation.	Construction of dam and appurtenant works is not involved.
Longer life, over 50 years.	Low life, 25 years.
Minimum stabilisation period of 2 to 3 months.	Longer stabilisation period of 6 to 12 months.
Smaller operating work force make them immune from labour problems.	Large operating work force and that engaged in mining and transportation of coal.
Incidental benefits of flood control, navigation, pisciculture, recreation etc.	No such incidental benefits.
Low capacity rating machines.	High capacity rating machines.
Longer gestation period.	Shorter gestation period.
Better service operational flexibility.	Lesser operational flexibility.
Better matching of operation of power plant with the load requirement.	Complicated matching of operation vis-a-vis load.
Generating sets can run and pick up load in almost no time to start and to stop.	Considerable time is required to run and pick up load.
Operation is dependent on water which in turn is susceptible to vagaries of nature.	Operation is dependent on coal susceptible to shortage, strikes and inferior quality supply.
Sites at isolated and inaccessible locations add to high cost on construction of dam and appurtenant works.	Located usually near the load centre, the essential requirement being availability of coal and water in adequate quantities.

High cost on long transmission lines as also high transmission losses.	Long transmission lines are not involved.
Huge area is submerged in the reservoir besides that for dam and appurtenant works in large storage power plant.	Land required for the power plant is less but large additional area is required for disposal of ash and coal production.

21.10 COMPONENTS OF HYDRO POWER PLANTS

1. Water conductor system

Forebay: It is the pondage at the beginning of a penstock leading water to turbine or the enlarged end of hydel channel in the case of run-or-river power plant. Its primary function is to store water temporarily when the load on the plant is reduced and to supply the initial water required when load is suddenly increased. The foreby thus absorbs the short interval variation of intake of water into the turbine corresponding to the fluctuations in power load. The foreby is provided with intake structure to direct water to penstock and a bypass channel or spillway to dispose of the excess water.

Intake structure: It is the structure meant to draw water from the reservoir or forebay to feed penstock. It essentially consists of trashracks, rakes to clear the trashracks and penstock regulating gates with their hoist mechanism.

Penstock: It is the conduit that conveys water from the intake structure to the power plant. It essentially consists of large diameter steel pipes to carry water under pressure. The penstocks may be buried underground, exposed pipes supported on pedestals or embedded at the lower end to feed several turbines (Fig. 21.5). The long penstocks are usually branched at the lower end to

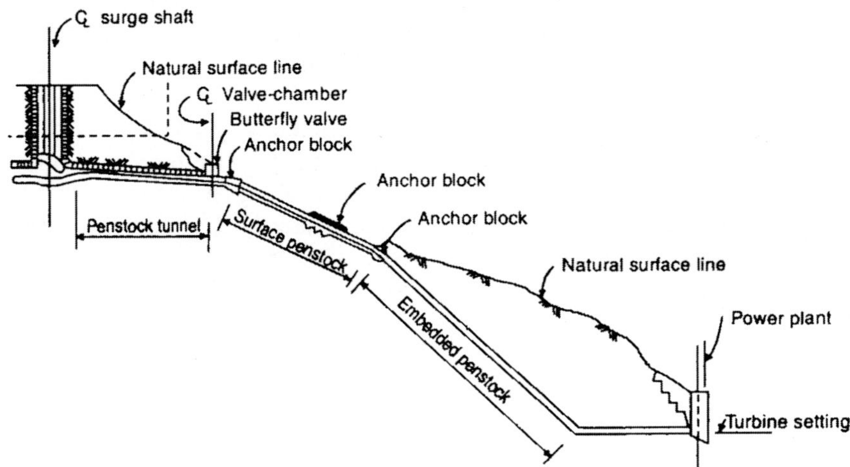

Fig. 21.5 Types of penstocks.

feed several turbines. Gated control is provided at the entrance in foreby to permit inspection and repairs of penstocks. Downstream from the gate, an air vent and stand pipe connecting the top of the penstock with open air is provided. The air vent permits entry of air into the penstock when the gates are closed and water drawn off through turbine as otherwise serious negative pressures may be exerted on the penstocks tending to collapse them.

Surge tank: Penstocks are subjected to water hammer pressure due to fluctuations in load. The short length penstocks at medium head plants are designed to take up this extra pressure but long penstocks are usually provided with a surge tank to absorb the water hammer pressures and to provide water to meet sudden load increase. The surge tank is thus a sizeable water receptacle interposed between power plant and high pressure penstock on the downstream side and low pressure tunnel and reservoir on the upstream side. Its main function is to intercept and dampen high internal pressure generated in the high pressure penstock due to water hammer and not allow it to transmit it into the low pressure tunnel. Besides, it serves as a storage basin *i.e.*, to absorb the excess discharge when load on the turbine decreases and the rejected water flows into the surge tank and thereby raising its level to absorb the flow and when the load increases it supplies the extra needed water to the turbine.

Power plant: The power plant is a multistoreyed structure wherein the hydro electric power is generated. It consists of substructure, intermediate structure and superstructure. The substructure in the usually used reaction turbines is the concrete block extending from foundation to the turbine level which is integrally cast while concreting water bringing scroll casing and water exit draft tube. The intermediate structure extends from the top of draft tube to the top of generator foundation. The superstructure is the portion of the structure above generator floor level to house generator, excitors, control room, service bay, overhead travelling crane etc. (Fig. 21.6). The power house may be a surface structure or underground structure. The underground power plant is housed in a large cavity excavated into a mountain where on surface space for the structure is not available.

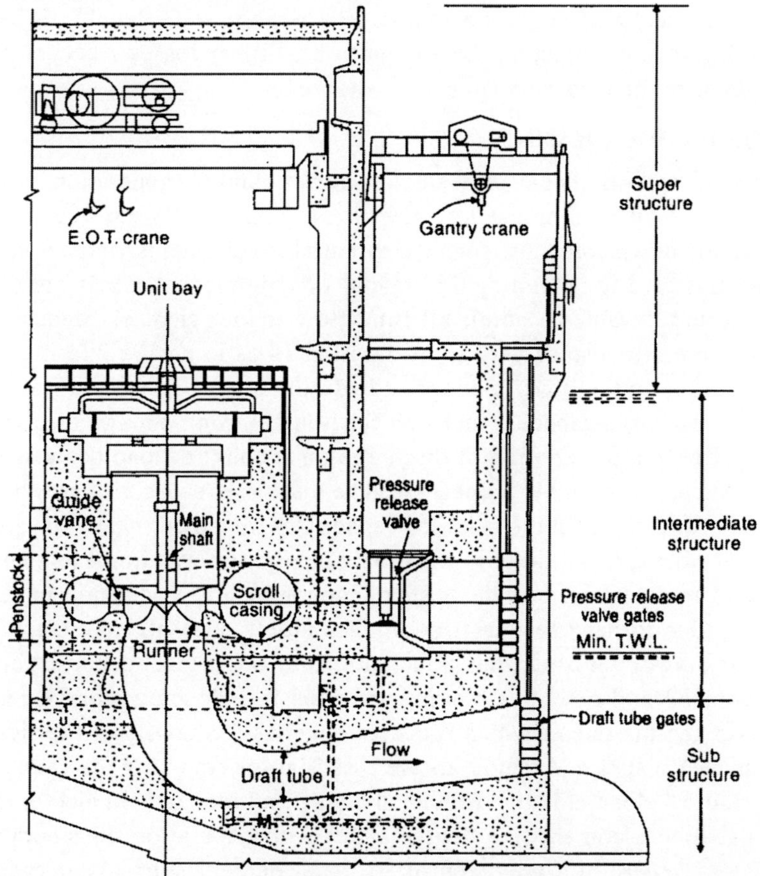

Fig. 21.6 Typical power plant structure.

Scroll casing: The water from penstock is delivered to the scroll casing which surrounds the turbine runner and guide mechanism. It is a spiral shaped steel or concrete conduit to distribute water evenly around the guide vanes and at the same time avoid formation of eddies. The concrete scroll casing is used usually in low head units.

Turbines: Turbines are the machines which convert hydraulic energy into mechanical energy which in turn is used in running an electric generator by directly coupling the shaft of the turbine with the generator. In general, a turbine consists of wheel, called runner, which is provided with specially designed curved blades or vanes so arranged that water flows over them. The high velocity water strikes the runner and makes it to rotate. The rotation is passed on to generator which is directly coupled to the shaft of the turbine. The generator armature so rotated produces hydro-power.

Draft tube: It is a conduit, straight, conical or with a circular section or elbow shaped tube with gradually increasing area; the shape changing from circular at the runner section to rectangular at the tail race. It serves the function to achieve recovery of velocity head at runner outlet which otherwise would have gone waste as an exit loss and also allow the turbine to be set at higher elevation to facilitate maintenance. The elbow type draft tube is commonly used for large units. The draft tubes are not required for impulse turbines.

Tail race: It is the exit channel carrying water from draft tube to river. It is designed so that silting or scouring of the channel does not take place. The silting raises tail water level and results in reduction in gross available head at the plant. The scouring of the tail race lowers tail water level and the process may aggravate to a stage where the water level falls below the top of draft tube leading to unsatisfactory flow conditions, higher losses, reduced turbine efficiency and may damage the turbine blades by cavitation.

21.11 DESIGN OF HYDEL CHANNEL

The hydel channel essentially carries water to a power plant for generation of power and as such forms important component of an hydro electric project. Usually the water carried by it is also used for irrigation downstream. As such the essential requirements which it must satisfy in addition to those indicated for planning and layout of irrigation channels are (*i*) Section is structurally strong and durable to admit all time flow in the channel because closure for repairs, if any, are rare and also of very short duration so as to ensure interruption in power generation for a brief period only, (*ii*) Section is hydraulically efficient with relatively flatter slope for the same discharge capacity, and (*iii*) Section is strong enough to take care of the phenomenon of surges due to variation of discharge on account of load demand or rejection. Hydel channel, as such, is preferably a lined one. The discharge capacity is based on the water power studies made for arriving at the installed capacity of the power plant. Usually discharge capacity is primarily fixed from the consideration of irrigation requirements. The cross section, bed slope etc. are determined on the basis of economic studies considering the optimum cost of energy lost due to head loss in friction. The side slopes of the channel section are designed to suit the drawdown conditions. Fluctuations in water level occur depending on the pattern of load demand and extent of storage provided in balancing storage reservoir. The effect of drawdown and the rate at which it occurs govern the stability of the side slopes and lining of the channel. Escapes and regulators are essential for emptying the channel in case of emergency or periodical closure, if any. Provision of silt excluder and silt ejector is essentially made to remove sediment content to a degree safe for the operation of generating units as given in Table 21.2. Generally a provision of 15 to 20 percent extra discharge for flushing purpose is made.

Table 21.2 Safe particle size for generating units.

Head (m)	*Largest particle size (mm)*
Below 100	0.5
100 to 300	0.2
above 300	0.1

SOLVED EXAMPLE

Example 21.1 *A medium hydro electric plant working at 30 m head at 85% efficiency has variable load, as under:*

Time (hr)	*Load (MW)*	*Time (hr)*	*Load (MW)*
0–3	*15*	*12–15*	*60*
3–6	*.20*	*15–18*	*70*
6–9	*90*	*18–21*	*63*
9–12	*50*	*21–24*	*12*

Determine: (i) Minimum average daily discharge required to supply the load, (ii) Storage required to produce peak load.

Solution:

(*i*) Total load at 3-hrs interval = 380 MW

Average load, $\frac{(380 \times 3)}{24} = 47.5$ MW

Discharge required to supply average load, $Q = \frac{47.5 \times 10^3 \times 75}{10^3 \times 30 \times 0.85 \times 0.736} = 190$ cumecs

$\left(\frac{QwH}{75} \times \eta \times 0.736 = 47.5 \times 10^3 \text{ kw}\right)$.

Discharge required to supply the load, $Q = \frac{(190 \times \text{load})}{47.5} \approx 4 \times \text{load}$.

(*ii*) Storage capacity required is calculated as under:

Time (hr)	*Load (MW)*	*Discharge required (cumecs) = 4 ´load*	*Deficit (from 190 cumecs)*	*Cumulative deficit*
0–3	15	60		
3–6	20	80		
6–9	90	360	170	
9–12	50	200	10	180
12–15	60	240	50	230
15–18	70	280	90	320
18–21	63	252	62	382
21–24	12	48		

The storage required = 382 cumecs for 3 hours
= 382 ´ 3 ´ 60 ´ 60
= 4.126 M m^3.

QUESTIONS

1. How will you work out the general requirements of water for a hydel power project? Explain with suitable examples. [*AMIE*]
2. Briefly discuss the economic aspects of hydro-power and thermal power.
3. How will you estimate the water requirement for a hydro-power project knowing distribution of electrical load? What improvements will be achieved if hydro-electric plants are used to generate power for peak loads only. [*AMIE*]
4. Discuss the relative merits and demerits of hydro-power as compared to other power sources.
5. List in detail the general and special surveys needed for hydro-electric power projects.
6. Write note on water requirements of a hydel-power project.

7. Briefly discuss the economic aspects of hydro-power vis-a-vis thermal power. [*AMIE*]
8. How is the water requirement for 'hydel-project' calculated? [*AMIE*]
9. What are various methods of classification of hydro power plants.
10. What do you understand by flow duration curve and power duration curve?
11. Discuss various types of surge tank. Why are they provided?
12. Briefly describe various types of turbines.
13. What are the various components of a power plant? What are their functions?

PART - VII

IRRIGATION WATER MANAGEMENT

- Irrigation Water Management

22

IRRIGATION WATER MANAGEMENT

22.1 INTRODUCTION

Because of the limited availability of utilized water and the growing need of water for irrigation and other purposes the need for water resources can hardly be overemphasized.

Irrigation management assumes special importance in view of the fact that irrigation alone needs about 75% of the total water requirement.

The object of irrigation management is to supply and apply the right amount of water at the right place and at the right time.

In the absence of data on employment & livelihood as a result of introduction of an irrigation system, the performance of irrigation system is gauged by the following:

(i) Area irrigated as proportion of the area planned to be irrigated.

(ii) Water logging

(iii) Tail end deprivation

(iv) yield

22.2 OPTIMUM USE OF WATER

Optimum water requirement of a crop is the quantity of water required during its growth period that results in maximum yield the optimum quantity of water include water supplied by precipitation as well as water delivered by irrigation.

The supply of water more than optimum requirement may not result in increase in yield of most crops but also decreases the quality of the product.

22.3 NEED OF OPTIMUM USE OF WATER

1. If the supply of water to the crop is more or less than its optimum requirement, it will adversely affect the crop yield supply
2. Supply of less water will cause the plant to spend extra energy to get moisture from the soil, which would otherwise has been used in its growth.
3. Supply of surplus water will expel the air from soil pores and will prevent free circulation of fresh air, which is essential for food preparation of the plant.

Therefore for the maximum crop yield, it should be supplied with optimum quantity of water.

22.4 FACTORS AFFECTING CROP WATER REQUIREMENTS

1. Depth of ground water table
2. Slope of the ground
3. Climate conditions of the region
4. Intensity of Irrigation
5. Texture and structure of the soil

6. Moisture storage capacity of the soil
7. Type and quantity of manure applied to the fields
8. System of Irrigation used.

22.5 INADEQUACIES OF IRRIGATION MANAGEMENT

From the print of view of performance, the management of the canal irrigation systems in India is far from satisfactory. The major inadequacies are as follows:

1. Insufficient planning & preparation at the stage of execution of the project which results in longer construction time and escalated project cost
2. Involvement of more than one ministry / department and poor coordination among them.
3. Non-responsive, authoritarian, and poor administration resulting in increased malpractices.
4. Lack of interaction between engineering and agricultural experts.
5. Log between creation of potential and its utilisation.
6. Improper assessment of personnel, equipment, and other facilities for proper operation and maintenance of reservoir and canal systems resulting in erratic supplies and inequitable distribution of available water.
7. Higher conveyance losses.
8. Absence of conjuctive use of ground water and surface water.
9. Insufficient drainage, excessive seepage and waterlogging.
10. Poor On-farm management.
11. Absence of former's participation in the management.
12. Lack of communication facilities in the command area.
13. Poor extension services - lack of pilot projects, demonstration farms, etc.
14. Problems related to land settlement & rehabilitation of displaced persons, and
15. Recovery of the project cost.

22.6 OBJECTIVES AND CRITERIA OF GOOD CANAL IRRIGATION MANAGEMENT

There we several conventional measures to improve the performance of canal irrigation systems. Some of these measures are

(i) Lining of canals and field channels
(ii) On farm Development
(iii) Farmer's organisation
(iv) Warbandi system of Water Distribution
(v) Charging farmers volumentrically for water
(vi) Educating farmers in water use management.

The most common criteria generally accepted for judging and performance of an irrigation system

(1) Productivity
(2) Equity
(3) Stability

22.7 PRODUCTIVITY

Productivity is defined as the ratio of output & input. The output can be water delivered area irrigated, yield, or income and the input can be water in the root zone, at the farm gate, at the outlet or at upstream points in the system including the point of diversion or storage.

Improved water supply influences the adoption of high yielding agricultural practices by farmers which justified the productivity criterion of performance.

22.8 EQUITY

Equity in canal irrigation systems implies equality, fairness and even-handed dealing in matters of allocation and appropriation of irrigation water there can be several ways to decide the equality of supplies to different farmers. In the method of prior appropriation, whoever first exploits in resource establishes a right to continue to do so. Thus head reach farmers and early corners to an irrigation project establish their right over irrigation water even if it means less water or no water to tail end farmers or late comers.

In the second method of proportionate equality the supply of water is in proportion to the size of the land holding as in the warbandi system of North West India.

Attempts to improve equity are usually limited to achieving the supply of water in proportion to the size of the land.

22.9 STABILITY

Stability of productivity as well as equity are important stability can either be short term stability or Long term sustainability.

The short term or interseasonal stability refers to the variations in productivity and equity between irrigation seasons, and is a function of climate, water supply, storage and control, system management, and other factors such as pests, disease, availability of labour & other inputs. It can be measured by comparing performance between seasons.

The long term sustainability refers to the prevention or minimising of adverse physical changes such as water logging, leaching of nutrients from soils, salinity, erosion, silting, the 'mining' of ground water, and infestations with weeds. Sustainobility can be monitored by measuring ground water levels, salinity, erosion or silting through inspecting works and by measuring long-term trends in productivity & equity.

Well-being is a broad objective achieved through productivity, equity, and stability there are several aspects of well being which must be borne in mind at all stages of project with both positive and negative effects. These aspects include health, nutrition, amenity & psychological factors such as freedom from domination, feeling of participation etc.

22.10 METHODS FOR IMPROVING CANAL IRRIGATION MANAGEMENT

1. Cropping patterns
2. Conjuctive use
3. Channel capacity
4. Canal lining
5. Regulators & Escapes
6. Canal outlets
7. Main System Management
8. Night Irrigation
9. Water Delivery system
10. Irrigation scheduling
11. Irrigation methods
12. Use of waste water.
13. Conservation of water on the field
14. Water logging
15. Soil Reclamation
16. Water charges and pricing of the Agricultural output
17. Command Area Development Programme
18. Farmer's Participation

1. CROPPING PATTERN

a. Cropping pattern is described in terms of the area under various crops at different periods of a year.
b. An Optimum cropping patterns for an area can ideally be determined by using System Analysis.

One such analysis is carried out for the Gomti-Kalyani doab under Sarda Sahayak command has recommended a cropping patterns which is expected to increase the net annual benefit to Rs

23.67 crore from the existing benefit of Rs 11.57 crore.

2. CONJUCTIVE USE

Conjuctive use means that water lifted from below the ground is used in conjuction with canal waters. It results in the coordinated, combined, and creative exploitation of ground water & surface water so as to minimise the dislocation caused by nature's inconsistent rainfall pattern.

Conjuctive use implies use of surface water during periods of above normal precipitation for irrigation & other activities to the extent possible & letting the balance reach the ground water storage which would be utilized for supplementing surface water supplies during years of sulnormal precipitation.

3. CHANNEL CAPACITY

The discharge capacity of the channel should be decided on the concept of evopo-transpiration rather than kor period.

4. CANAL LINING

a. Living of canals is a means to reducing the seepage losses from canals.
b. To such, the lining of field channels is the most beneficial.
c. Besides, it involved no dislocation in the operation of an existing system.
d. In order to prevent damage to lining, the slope of lined channel is reduced.
e. This reduce the sediment carrying capacity of an existing channel which is being lined.

5. CANAL REGULATORS AND ESCAPES

a. For ensuring proper distribution of irrigation water according to the adopted management policy, a suitable number of regulators & Escapes must be provided on channel network.

b. Canal Escapes are needed for the safety as well as for regulating canal supplies in areas which have received excess rainfall.

6. WATER LOGGING

Water logging results in lower yields, loss of lands for useful activities and health hazards. To eliminate or control waterlogging one or more if the following, remedial measures have usually been used:

(i) Reducing inflow through lining of canals
(ii) Removing ground water through pumping
(iii) Removing surface & ground water through drainage
(iv) Educating farmers in water management &
(v) Conjuctive use

22.11 WATER LOSSES:

Water enters the main canal at the headworks through the head regulator and flows through the distributors and the water courses & finally reaches the field. Throughout its journey, there is continuous water losses which have to be accounted for the design of channels.

These losses are especially in unlined canals. In some cases, the losses may be as high as one quarter to one half of the water diverted.

In North India, the water losses are usually found in practice are follows:

(a) Losses in the main canal & branches	=	15 to 20%
(b) Losses in the major & minor distributaries	=	6 to 8%
(c) Losses in the field channels	=	20 to 22%
(d) Losses in the application of water	=	25 to 27%
(e) Water actually utilized by crops as consumptive use	=	28 to 29%

From the above figures. It is obvious that 41 to 50% losses occur in the channels.

The losses in irrigation channels are mainly of the following types:

1. Absorption losses
2. Percolation (or seepage) losses
3. Evaporation losses.
4. Transpiration losses

22.12 ABSORPTION LOSSES

Absorption losses occur because of absorption of water by soil surface canal wetted perimeter. When the water table is at a considerable depth below the canal bed, the infiltrating the soil below the canal bed in unable to reach the ground water reservoir below the table.

Absorption losses are independent of the seepage head. These losses depend upon the water head h from the water level of the canal to the bottom of the saturated zone and the capillary head h_c for the soil at the boundary of the saturated zone and the capillary head h_c for the soil at the boundary the saturated zone. In general absorption losses depend mainly upon the depth of water in the canal X type of soil.

22.13 PERCOLATION (OR) SEEPAGE LOSSES

Percolation losses are usually much greater than absorption losses. They may be as high as 3 times or more of the absorption losses. Whether the losses will be by absorption or by percolation will depend primarily on the nature of soil strata and the level of the water table.

Both the absorption loss & percolation loss are initially large because the water is utilized for filling the pores of the soil. With the passage of time the losses decrease & an equilibrium is finally reached. Moreover the silt carried by the canal water gets deposited in the canal & reduces the soil permeability & hence the seepage.

Absorption & percolation losses from the canal mainly depend upon the following factors.

(i) **Permeability of soil**

The greater is the permeability of the soil in the bed & banks of the canal, the greater are the losses.

(ii) **Depth of water**

The greater is the depth of water, in the canal, the greater are the losses.

(iii) **Velocity of water**

The losses decrease with an increase in the velocity of flow in the channel.

(iv) **Amount of silt**

The losses decrease with an increase in the amount of silt carried by the canal.

(v) **Temperature of water**

The losses increase with an increase in temperature of water because permeability of the soil is increased.

(vi) **Age of the channel**

The losses are large in newly constructed channels and they reduce as silt deposited with the passage of time & a relatively impervious silt layer is formed.

(vii) **Level of the channel bed**

The lossers depend upon the level of the channel bed w.r.t natural surface or ground level. The losses are more when the canal is in heavy filling.

(viii) **Position of the water table**

The losses depend upon the position of water table w.r.t the canal bed the losses are more when the water table is high.

EVAPORATION LOSSES

Evaporation losses depend upon the water surface area of the canal, relative humidity, wind velocity, temperature & various other factors. In hot & dry summer months, the evaporation losses are high, but they seldom exceed 10% of the total losses Generally, evaporation losses is less than 1% of the total water entering the canal head.

There are no suitable practical measures to reduce the evaporation losses from the canal. For design issue, generally no separate consideration of evaporation losses in made.

TRANSPIRATION LOSSES

Transpiration losses occur through the vegetation & weeds in the canal. Losses are usually a small percentage of the total losses in an unlined channel. These losses can considerably decreased by keeping the canal banks free of vegetation.

ON FARM DEVELOPMENT

The main objective of on development is to develop a proper water distribution system to obtain maximum crop production per unit of water from a unit of land in unit time.

At farmer's level, the major aspects of wasteful use of irrigation waters are:

1. Wastage arising due to practicing field to field irrigation in the absence of field channels.
2. Wastage due to escape of water to drains during transit from the outlet to the field because of lack of supervision of flow of water in field channels.
3. Wastage due to surface run off of irrigation water because of lack of proper field levelling
4. Wastage due to over irrigation

COMPONENTS OF ON FARM DEVELOPMENT

A. **Community terms**

1. Formation of field channels
2. Formation if Internal Irrigation channels
3. Drop structure & distribution boxes in field channels
4. Crossing in field channels whenever required
5. Field drains
6. Drops in field drains
7. Crossing on field drains wherever required

B. **Terms of Individual farmers holding**

1. Providing bunds to form plots in the holding.
2. Land shiping to a planned grade for cultivation of light irrigated crops.
3. Land levelling to form level terraces for growing rice.
4. Efficient water use on the farmers' field.

22.14 IMPACT OF WATER USERS ASSOCIATION FORMATION

Public irrigation systems are predominant in our country. The experience is that irrigation efficiency is low & the expected increase in agricultural production has not materialized. It is therefore, necessary to design performance indices to monitor, evaluate and improve farmer management systems in major & miner irrigation projects.

The following indices could be considered

1. Conveyance Efficiency
2. Operations efficiency
3. Micro level disnet distribution efficiency
4. On-farm application efficiency
5. Crop MVC on the area irrigated, their water requirements, and efficiency in meeting these requirements.
6. Area water unit ration.
7. Equity
8. Financial & Organizational Viability

Other important aspects of performance are predictability, reliability, timeliness, matching crop water requirements & departures from the planned schedules of deliveries necessitated by climatological & Environmental factors. In this connection, the following indicators need to be considered Productivity : Physical & value, leading to better health standards, marketing & communication Better farming practises.

Soil & Water health (water logging, salinity)
Environmental effects, better health standards

FARM APPLICATION EFFICIENCY

The farmers have now started preparing fields more systematically & adopting the border method of irrigation all crops except sugarcane. The Application efficiency has improved considerably, through precise measurements have not been taken.

FIELD CHANNEL AND MINOR EFFICIENCY

The conveyance efficiency below the outlets has also improved, as shown earlier. Updating of the field channels would improve efficiency still further. Preparing the cut-off statement of outlets on the minor & letting out the discharge from the outlet fully to one farmer at a time will reduce the seepage losses in the minor. This concept is understood by the farmers in principle, but needs to further particed & improved upon.

MATCHING WATER SUPPLY WITH CROP REQUIREMENT

The farmers have realized the need for matching water supply with crop water requirement & have undertaken experiments in supplying water at high frequency, within the general schedules of the main canal.

The results are encouraging & the yield of groundnut have increased by providing two additional light waterings at the pegging & flowering times.

Similarly the sunflower crop, with proper furrows & light irrigation of 30 to 40 mn at the crop level, has shown increased productivity.

CONJOINT GROUND AND SURFACE WATER

The irrigation coverage indicated that the farmers are utilizing the ground water from the recharge of vain & canal water on priority, and the balance quantity is demanded from the available canal water. The guarantee of supplying water as well as payment on the volumetric basis has induced the farmer to do so.

ADVANTAGES TO WUAO/ FARMERS

1. Enhanced organizational strength for negotiating with the I.D. on water rights, timely supply, allocation of water, improvement in Main System Management MSM etc.
2. Better maintenance of the distributary leading to better water availability.
3. Possibility of organizing high value crop programmes due to flexibility of timing & frequencies
4. Possibility of getting irrigation frequency suitable for crops, soil & crop water requirements.
5. Better access to information about water availability & other irrigation policies & other programmes of the ID.
6. Possibility of constructing en-route storages, and service reservoirs for increasing frequency of supply.
7. Any Anementation of ground water from deep percolation, own offs & nallas in the command.

OBJECTIVES OF TRAINING

1. To help professionals to understand the principles, processes and specific field problems, & to appraise them on how the system presently works.
2. To provide field experience & knowledge, and develop skills in monitoring methodologies & procedures in inter-disciplinary responsibilities & actions.
3. To understand the process of team work
4. To help professionals appreciate positive attitudes in relation to involving the farmers in

decision making

5. To prepare them for their roles as project managers.

22.15 PARTICIPATORY IRRIGATION MANAGEMENT

Participatory Natural Resource Management is based on the concept of communities having very close relationship with resources provided by nature. Local communities have their own emotions and attachments with these resources. Such knowledge can be revealed through participatory interactions for learning & actions.

According PIM has objectives to explore relationship of local communities & natural resources as mentioned below:

1. To scale up involvement of villages by learning & listening about their perception, experiences & capabilities.
2. To generate primary data & information for immediate & future us
3. To learn about the impacts of previous & on-going policies & programmes & to formulate new ones.
4. To investigate trends & ascertain conditions of the issues at hand
5. To cross check reliability & validity of data collected from other sources.
6. To augment trained manpower in different categories of persons involved in the development process, whether from the Government, NGOs, donar agencies, researchers, extension agents & scientists.
7. To formulate & conduct research studies on use of PRA & suggest improvement in its methodology.

NEED FOR PIM

the participatory approach is an informal way of learning from local people about the problems, & opportunities in the given area PIM is both an attitude & methodology which produces authentic information of the village.

PIM is requires for :

1. Sustained change & the need for accurate & timely information.
2. Advocating that people themselves are 'solution agents' for their need & problems.
3. Reducing 'normal professional bias' & anti-poverty bias towards people.
4. Cutting down time-consuming long methods of survey which give results after considerate time. The method is cost effective, accurate & timely & information emanates from within the stokeholed.

PRINCIPLES OP PIM

1. Optimal Ignorance

The principle of optimal ignorance is to minimise cost & time. It is applied by investigation at or to know what is worth to know. One should know enough to serve the purpose. Associated with this is seeking appropriate inprecision or avoiding precision that is not necessary. The matter not desirable should be ignored to the possible extent to avoid confusion.

2. Seeking Diversity

PIM is to look for different analysis through diverse events. Various processes and deliberating relationships in local communities explore the facts in local situation.

QUESTIONS

1. Describe the following
   (i) Participatory Irrigation Management
   (ii) Objectives of Water Management
2. Define "Participatory approach"
3. Discuss the roles and responsibilities of farmers and governmental agencies in the farmers organisation.

APPENDIX I

DAM SAFETY

India's progress in dam building is favourably comparable to the technologically advanced countries in the world who too have developed their water resources potential on priority basis. India stands fourth in the world if only the large dams are considered. Dams are national assets of great importance. Safety of dams is a very important aspect to ensure continued optimum utility to the nation over the centuries to come as an endless input for food production, industrial, socio-economic and environmental improvement as also for ensuring public confidence in the river valley projects as monuments of prosperity to the posterity.

The dams impounding billions of cubic metres of water above a populous community have to be safe on two counts; firstly the appalling loss of life and property that would be threatened by failure justifies usual conservation in design and secondly the loss of investment on the project and its return in production, shattering the economy and normal life would be too hard a blow to be absorbed by any community developing or developed. The catastrophic incidents of recent dam failures in the country and abroad involving considerable loss of life and property have caused new awareness and concern for the safety of dams amongst dam builders and the general public. The case histories of majority of dam failures have revealed that these were not designed, constructed and maintained in accordance with the recognised standards. Old dams built with the techniques then available may not be safe with the current standards. Also dams like other structures are subjected to decay and deterioration with the passage of time. The dam, reservoir, spillway, tunnels and power house have distinct inherent constructional, operational and maintenance features. Monitoring post-construction behaviour of these permanent structures through the embedded instruments vis-a-vis design assumptions, settlement of dam embankment, upstream riprap, downstream slope and toe of the dam, proper working of all the mechanical and electrical installations connected with operation of gates, vigilance on inflow conditions and correct determination of inflow, outflow discharges, slides occurring in the reservoir area, prompt and dependable flood warning system, close pre-monsoon and post- monsoon inspections, besides the routine ones, dedicated surveillance of all the connected civil structures are the important parameters of dam safety.

To avoid catastrophics, new codes, rules and criteria for design, inspection and behaviour of the existing and new dams have been evolved in many countries of the world. Guidelines have been prescribed to achieve optimisation fully viewing the concept of safety. The procedures and practices of safety have to be strictly adopted from the conceptual stage to its completion. Even with a sound design and quality construction, the problems of dam safety can not be considered to have been resolved to entirety. It is necessary to keep the edifice under observation continuously even during the maintenance phase. It underlines that adequate instrumentation in the dam and appurtenant works is provided so that it will give timely information about the happenings on the dam and alert the maintenance engineers for possible action to be taken to keep the dam under safe conditions as also to regularly monitor the information generated from them and analyse the behaviour of the structure as constructed on a regular basis.

Dam Safety Organisation: Keeping in view the importance of dam safety in our country, a dam safety organisation was established in 1979 in the Central Water Commission, New Delhi to assist the state governments in various activities in dam safety. The subject of dam safety is multi-disciplinary, cross sectorial and all pervasive in character. The technicalities involved are pretty complex and do not lend themselves to uniform adaptation or application under all circumstances. The dam safety organisation has outlined Phase-I and Phase-II investigations for dam safety, as under:

Phase-I Investigation: The phase-I investigation is to develop an assessment of the general condition with respect to safety of the project based on available data and a visual inspection to determine any need for emergency measures and conclude if additional studies, investigation and analysis are necessary and warranted. A review is made of pertinent, existing and available engineering data relative to design, construction and operation of the dam and appurtenant structures including electrical and mechanical operating equipment and measurement from inspection and performance instruments and devices, and a detailed systematic visual inspection is performed of these features relating to the stability and operational adequacy of the project, flood warning system etc. An evaluation is made of the general condition of the dam, including, where possible, the assessment of the hydraulic and hydrologic capabilities and the structural stability, post construction changes and seismic stability, environmental impact of the reservoir, the effect of impounding of water in the reservoir on seismicity of the area, and crustal deformation, if any, in reservoir rim due to loading of water in the reservoir.

Phase-II Investigation: It is supplementary to phase-I investigation for additional in-depth studies, investigation and analysis to evaluate safety of the dam. Included, as required, are additional visual inspections, measurements, foundation exploration and testing, materials testing, hydraulic and hydrologic analysis of spillway design flood, maximum outflow, peak for project maximum flood, and structural stability analysis in the light of the data generated since the commissioning of the dam, landslide activities, sedimentation, potential hazard areas and preparation of report thereof.

Documentation of Engineering Data and Write Ups: Proper record and upkeep of engineering data and write ups on the lessons learnt during the construction and operation of the project are very important for evaluation of safety of the dam and structures. Design manuals, operation and maintenance manuals outlining the criteria of operation, and maintenance to (*i*) monitor the performance of the structure vis-a-vis design assumptions, (*ii*) observe data of pore pressure, stresses and strains, (*iii*) provide a continuing surveillance of the structure to warn of any development which may endanger its safety and indicate distress spots for immediate remedial measures, (*iv*) review of structural safety, and (*v*) provide information for future design criteria for high dams.

Parameters to be Measured/Observed: Various parameters which should be measured are as under:

In Earth Dam: Pore pressure, settlement (vertical movement, horizontal movement), tilt, drainage/seepage, soil pressure and seismic activity.

In Concrete Dam: Uplift pressure, total pressure, displacement and settlement, tilt, stresses and strains, temperature, elasticity modulus, joint movements and seismic activity.

Reservoirs: First filling: Initial filling of a reservoir is the first test of a dam to perform its intended functions. Reservoir filling criteria are established on a dam by dam basis. In general the objective is to provide a planned programme with adequate time for monitoring and evaluating performance of the dam and its foundation as the reservoir is being filled for the first time. The major factors to be considered in establishing initial filling criteria are type of dam, geology of foundation and reservoir, and landslide potential of reservoir rim, hazard potential, inflow characteristics, flood patterns, release provisions for the project requirements, flood release and emergency evacuation, instrumentation, safe capacity of downstream channel and characteristics of reservoir storage. The filling rates for concrete dams are much less restricted and not generally specified for the bottom half of the depth of the reservoir impounded by the dam. Broadly the stage-wise filling is filling of reservoir upto middle depth of dam level in first stage, and further to crest of spillway in second stage with controlled and specified rate of filling, in the third stage from above spillway crest level to full reservoir level in stages restricted to 0.3 m in 48 hours.

In earthern dams and also concrete dams having high earthen flanks, the first stage is the same as specified for concrete dams. The second stage upto spillway crest level is done in two stage; at a rate not exceeding 3 m per fortnight and filling temporarily stopped at 50 per cent elevation from middle depth of dam level to crest of spillway. Further filling may be decided in gradual sub-stages of 2 to 3 m depending on height of dam. The rate of filling in third stage shall be restricted to 0.3 m in 48 hours; the filling temporarily held at half of the height between full reservoir level and spillway crest level.

Evacuation Time: Where population growth in downstream areas is not controllable, the hazard potential at the initial stage and its subsequent increase cannot be assessed. The evacuation time may be assumed as under:

Depth of evacuation (from initial reservoir level) (percent)	*Degree of hazard or risk*		
	High	*Significant (Evacuation time in days)*	*Low*
25	20	30	50
50	40	50	70
75	80	90	100

The evacuation period would generally be within the overall requirement to drawdown the reservoir within a period of one to four months, allowing sufficient time for carrying out inspection and repairs, before the water level rises due to higher inflows of monsoon.

Instrumentation: Dams are monitored to assess their safety as well as performance during construction and in post construction period especially during earthquake, floods, maximum filling etc. At that time assessment of the strength of the dam is essentially required to compute its safety. Instruments are embeded/installed to ensure that performance of the dam and appurtenant structures is consistent with the design assumptions. The instruments generate the data base of the various parameters which are used as a guide to check input parameters as also to serve as an indicator for the health of the dam.

Various types of instruments are basically classified as (*i*) hydraulic and mechanical type, and (*ii*) electronic type. The electronic type instruments are further classified as (*i*) Vibrating wire type, (*ii*) Unborded strain gauge type (Carlson type), and (*iii*) Bonded strain gauge type. The hydraulic and mechanical type of instruments are very easy to handle, accurate, reliable, rugged and economical. The hydraulic type instruments, however, suffer from the effect of time lag on measurement of pore pressures. Automation is not possible. The electronic type instruments are accurate, sensitive, stable, easy to operate, require least maintenance and do not suffer from the effect of time lag. The mortality rates and cost are high as compared to hydraulic type instruments. USBR hydraulic types of instruments such as twin tube piezometers and mechanical type instruments viz telescopic cross arms, base plates, surface settlement plugs, inclinometers, embedded soil extensometers, earth pressure cell, band marks etc. are widely used in the country.

Instruments at Pong Dam: At Pong dam, the 132.6 m high earth core-cum-gravel shell dam constructed in 1974, two main types of instruments viz (*i*) USBR type piezometers are installation for measuring pore pressure developing in dam embankment and foundation, (*ii*) Settlement installation for observing displacement of the dam embankment and foundation have been provided. In addition, there is a slope indicator for measuring lateral movements in the dam alongwith two strong motion accelerographs, one at maximum dam section and the other near downstream toe of the dam for recording seismic activity in the area (Fig. 1). Since the area is seismically active, five seismic observatories have been set up around the periphery of the reservoir for monitoring earthquake and reservoir induced seismicity due to impoundment of water in the reservoir.

Instrumentation In Gravity Dams

The various instruments installed in concrete/gravity dams are (*i*) Strain measuring instruments (strain meters), (*ii*) Temperature measuring device (Temperature gauges), (*iii*) Pressure/stress measuring instrument (stress meter), (*iv*) Load on reinforcement bars (Re-bar load gauge), (*v*) Pore water pressure measuring device (Piezometer), Joint measuring device (Joint meter), (*vi*) Horizontal deformation measuring instruments (Electro level gauge and pendulums), (*vii*) Vertical movement measuring instrument (Inclinometer system), (*viii*) Rock movement (Boxhole extensometer), and (*ix*) Water leakage measuring instruments (Water leakage system).

At Bhakra dam, 225 m high straight gravity mass cement concrete dam, twin aims of instrumentation provided are (*i*) providing a means of verification of the pattern of stresses and strains in the dam body and thus afford an assurance of its continued structural stability, and (*ii*) supplying useful data in respect of stress, strain, deflection, temperature, settlement, uplift pressure, seepage etc. for evolving design on more economical and sounder basis for future dams. The instruments provided for making various types of measurements/observations are briefed as under:

Strain: The strain caused in massive concrete dam due to volume changes under the influence of temperature, chemical action, moisture changes and stress is measured by Carlson type strain meters embedded in the dam. Both stress meters and strain meters have been installed to make measurement of both stress and strain independently.

Stress: The stresses developed vis-a-vis design stresses in various portions of the dam under the loading condition are measured by Carlson stress meters.

Temperature: The temperatures are measured at a number of points within the dam as well as on the boundary by means of resistance thermometer to monitor the effect of temperature changes on the stress and volume change in the dam. The distribution of temperature within the reservoir and its annual variation is measured by resistance thermometers embedded flush with the upstream face of the dam.

Contraction joint openings: To evaluate openings of the longitudinal and transverse joints, joint meters have been installed.

Silt pressure: Specially designed stress meters embedded flush with the upstream face of the dam have been installed to supply information regarding actual silt load coming on the dam.

Deflection: Deflection measurements are done by various methods, namely internal measurements through observations of plumb lines and tilt meters, external measurements through observation by geodetic triangulation on targets on the downstream face of the dam and transverse markers at top of dam.

Settlement: Settlement bench marks have been provided in various galleries of the dam, on abutment rocks along periphery of the reservoir to monitor settlement.

Uplift pressure: Uplift pressure pipes have been provided at various locations for continued observance of uplift pressure beneath the dam.

Negative pressure: The working of river outlet at high heads usually result in creation of negative pressures, the magnitude of which is assessed through piezometers installed in outlets.

Seismicity: Seismological observations are carried out through seismological instruments installed in the body of the dam and outside it with the assistance of India Meteorological Department, New Delhi.

APPENDIX-II

SOME IMPORTANT DAMS IN INDIA

In India there are over 4300 dams either completed or under construction. A brief description of a few important dams is given below:

BHAKRA NANGAL PROJECT

Bhakra Nangal Project comprises, Bhakra dam, Nangal dam and Nangal hydel channel with its two power plants and a network of irrigation canal system.

Bhakra Dam: Bhakra dam across Sutlej river is located at the foot of Shivalik hills in Himachal Pradesh. It is one of the world's highest straight gravity concrete dams. The dam is founded on sound rock 57.9 m below the river bed. Spillway fitted with 4 No radial gates of 5.24 × 14.5 m size is provided at the top of the dam to surplus 5, 587 cumecs of flood discharge. The dam is flanked on both sides by two giant power plants and in the rear is beautiful Gobind Sagar reservoir. Work on the dam was undertaken in 1948 and completed in 1963. However, storage in the reservoir commenced in 1958. The dam involved concreting to the tune of 4.13 million m^3, sufficient for laying 2.4 m wide road around the globe at the equator. Bhakra dam is a symbol of India's progress as described by the then Prime Minister, Jawaharlal Nehru. The value of the agricultural produce per annum is over Rs. 500 crores and Rs. 635 crores as cost of out turn from major industries.

The Bhakra dam is a 225.55 m high concrete gravity dam with top width of 9.14 m and base width of 190.5 m; 402 m including apron and heel claystone plug. A network of galleries with a total length of about 5 km for the purposes of drainage, inspection, operation and maintenance etc. have been provided in the body of the dam. The reservoir created by the dam is 160 sq km in area, 97 km in length with gross storage capacity of 9621 million cu m at El.513.58 m; out of which live storage is 7191 million cu m at El. 445.62 m. It has a catchment area of 56,880 sq km. The reservoir submerged 371 villages and displaced a population of 36,000. There are 10 No. steel penstocks of 4.57 m diameter each for feeding water to power houses which have installed capacity of 1200 MW; left bank power plant has 5 units of 108 MW each (uprated) and right bank power plant has 5 units of 132 MW each (uprated). The maximum generating head is 156 m and 158.5 m on left and right bank power plants respectively with a normal designed head of 122 m. 16 No. irrigation outlets, in two tiers-8 each at El.1320 and El.1420 m are provided for increased irrigation water releases. The benefits from the dam include new irrigated area of 26 lac hectares, improvement in irrigation in 9 lac hectares, electrification in 128 towns and 13,000 villages besides energising numerous agricultural tubewells and industrial units.

Figure-1 shows upstream and downstream views of Bhakra dam.

Nangal Dam: It is a concrete dam located about 13 km downstream of Bhakra dam on the Sutlej river. The reservoir created upstream of the dam serves as a balancing reservoir for smoothening the diurnal and weekly fluctuations of the Bhakra power plants. The dam is 29 m high to head up river supplies by 16 m and divert them into Nangal hydel channel and Anandpur hydel channel through independent canal head regulators. A shingle excluder in front of the head

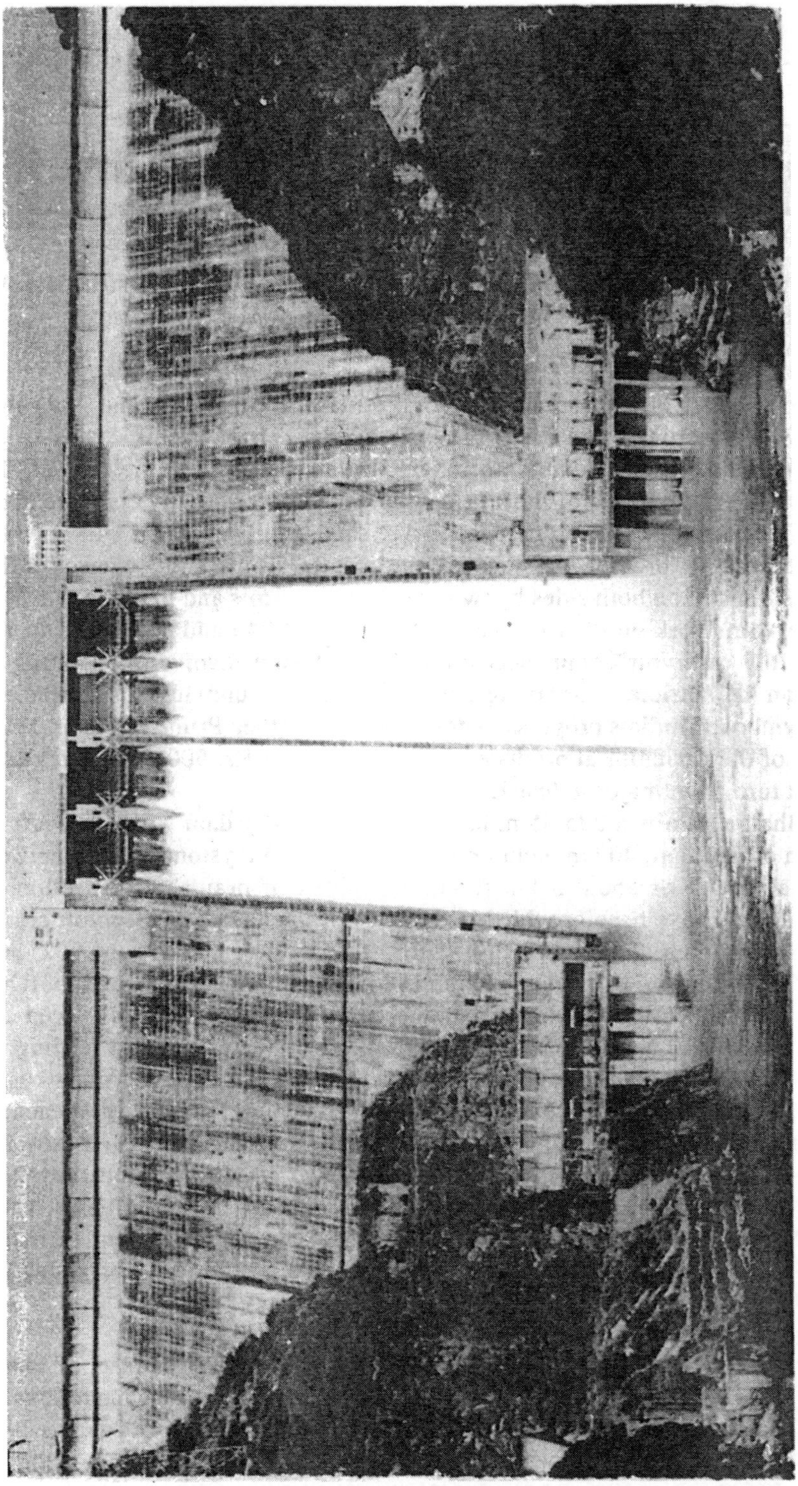

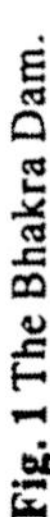

Fig. 1 The Bhakra Dam.

regulator as also covering two bays of the dam and consisting of six tunnels has also been provided to give complete control over the quality of water drawn from the reservoir. The dam is 305 m in length, 26 spans of 9 m each, and provided with gates. An inspection gallery in the foundation of the dam has been provided which runs across the river.

Nangal Hydel Channel: It is a concrete lined hydel channel of 354 cumecs capacity and 64 km long. Two power plants are located on this channel at Ganguwal and Kotla with installed capacity of 77 MW; 2 units of 24 MW and one unit of 29 MW at each power plant from the canal falls of 28 m. Nangal hydel channel ends into Bhakra main line canal which is a brick lined canal to carry water for irrigation to Punjab and Haryana States and drinking water to Chandigarh, Panchkula and Mohali. It is 174 km long with 1110 km length of main canal and branches, 3400 km length of distributaries and 4800 km length of network of canals. The Anandpur hydel channel too is of identical capacity and likewise feeds two similar power plants; designed to carry water to Haryana state canals.

BEAS PROJECT

It comprises two units: Unit-I—Beas Sutlej Link Project and Unit-II—Pong Dam.

Beas Sutlej Link Project: The Beas Sutlej Link is a power-cum-irrigation project intended to divert the Beas waters into the Bhakra reservoir through a system of tunnels and an open channel. There is a fall of 320 m at its tail end and about 122 m at Bhakra for generation of electric power. The river supplies so diverted have enabled extension of irrigation to arid areas in the south and south west of Punjab and Haryana besides improving irrigation facilities on the existing Bhakra canals command areas.

The capacity of the link canal is 213 cumecs with provision for additional 42 cumecs in the upstream portion of the water conductor system for exclusion of silt through silt ejectors in the open channel. Dehar Power Plant works in a combined hydro-cum-thermal grid of the region and any discharge surplus to the power requirement is diverted to the Sutlej through a by pass channel . The project comprises Pandoh dam, water conductor system of tunnels; an open channel and Dehar power plant.

Pandoh Dam. It is a diversion dam across Beas river in Himachal Pradesh having gross and live storage capacity of 4100 and 1856 ha m respectively. It consists of 76 m high earth-cum-rockfill dam, 12 m at top and 268 m at the base, a chute spillway fitted with 5 No. radial gates of 12 × 13 m size to surplus 9939 cumecs of flood discharge and intake structure for water conductor system transfers 563.5 cumecs of Beas water into Sutlej river at Dehar where a fall of 320 m is available to generate 990 MW of power; 6 units of 165 MW each. The water conductor system comprises 13 km long Pandoh-Baggi tunnel, 12 km long Sundernagar hydel channel followed by 12 km long Sundernagar Sutlej tunnel with a balancing reservoir of 2130 × 457 m size and 370 ha m capacity. The surge shaft provided in the penstocks is underground and of differential type with 22.9 m dia main shaft and 7.6 m dia riser shaft 125 m high, the biggest in Asia.

Pong Dam

It is one of the highest earth core-cum-gravel shell dams in the country. It is located on Beas river at Pong at the foot hills of Shivalik in Himachal Pradesh. Primarily it is an irrigation project to meet the irrigation water requirements of Rajasthan, Punjab and Haryana States besides generation of sizable power. It comprises (*i*) an earth-cum-rockfill dam 132.6 m high from the deepest foundation level, 1951 m long at top across the gorge and having a width of 13.7 m at the top and nearly 610 m at the base. The total earthfill placed in the dam is 35 million cum which is about 9 times the volume of concrete placed in the Bhakra dam, (*ii*) five concrete lined tunnels each of 9 m diameter and having an aggregate length of 5000 m. Initially these tunnels were used for the diversion of the river to enable work on the main dam being taken up. Two of these tunnels are utilized as outlet tunnels and the remaining three as penstock tunnels, (*iii*) a 42 km long reservoir to store 8570 million m^3 of water; out of which 7290 million m^3 is the live storage, (*iv*) a 102 m wide concrete chute spillway fitted with six radial gates of 14.48 × 12.34 m size with a capacity of 12,375 cumecs is provided on the left side of the main dam. The design flood discharge for the spillway is 33,555 cumecs, and (*v*) a power plant with an installed capacity of 360 MW; 6 units of 60 MW each, with maximum and minimum head of 95m and 48m respectively, six penstock branches bifurcated from three penstocks are of 5 m diameter and 108 cumecs discharge capacity to feed each turbine which is of vertical shaft Francis type.

The work on the dam commenced in 1965 and was completed in 1974. Water stored behind the dam provides perennial irrigation to an area of about 1.2 million ha in Rajasthan (on full development) and 0.4 million ha in Punjab and Haryana States. An additional annual production of about 2 million tonnes of foodgrains, besides substantial quantities of cash crops results. The annual increase in production is of the order of Rs 750 crores. 207 MW of power is produced at four power houses located on Mukerian hydel channel which utilises the water from Pong dam through Shahnehar barrage constructed downstream of the dam.

Fig. 2 shows the panoramic view of Pong dam and reservoir

Fig. 2. Pong dam and reservoir.

RAMGANGA DAM: 126 m high earth and boulderfill Ramganga dam is on Ramganga river in Uttar Pradesh. A power house located at the foot of the dam has an installed capacity of 198 MW. The dam forms a deep and wide reservoir which has water spread of 78 sq km to store 2443 million m^3 of water; out of which live storage is 2196 million m^3. It is designed to store all the water coming down the river in nine out of twenty years and surplus down the water in the remaining eleven years through a chute spillway. The spillway has a maximum discharge capacity of 8,467 cumecs plus 1,926 cumecs through auxiliary spillway.

The water released from the reservoir flows down the river for 26 km where a 347 m long barrage is constructed to divert the supplies into 11 km long link channel which in turn empties into Kho river. A barrage across the Kho river diverts the supplies into a 71 km long feeder channel which discharges into Ganga river opposite Garhmukteshwar. The augmented supplies of Ganga river are picked up at Narora barrage and utilised on lower Ganga canal. The irrigation is extended to an area of 0.659 million ha.

RANJIT SAGAR DAM: Thein dam, rechristened as Ranjit Sagar Dam, is an earth-core-rockfill dam 160 m high from deepest foundation level, 14 m wide at top and about 692 m at the base, 565 m long involving fill placement of 15 million cu m on Ravi river in Punjab. The four 12 m dia about 800 m long diversion tunnels are used to convey water for power generation and as irrigation outlets. All the tunnels together can pass a discharge of 9050 cumecs. The controlled gated ogee type spillway has a clear waterway of 133 m fitted with radial gates of 15.25 × 10.67 m size, having maximum discharging capacity of 18,000 cumecs. The reservoir behind the dam shall be 22 km long with a water spread of 83 sq km to store about 3280 million m^3 of water. It will submerge an area of 8760 ha. The power plant has installed capacity of 600 MW; 4 units of 150 MW each from hydraulic head of 121 m. The firm power is 207 MW. The increase in annual irrigation shall be to the tune of 0.35 million ha area in Punjab and Jammu and Kashmir States. The project will generate an income of about Rs 350 crores from additional agriculture produce and about Rs 240 crores by way of firm electric power production; besides fish production of about Rs 50 crores annually.

SARDAR SAROVAR PROJECT: The Sardar Sarovar dam, the third highest concrete gravity dam in the country and life-line of Gujarat State is a joint venture of the States of Gujarat, Madhya Pradesh, Maharashtra and Rajasthan. It is the largest water resources development project in the country. The dam is 163 m high, 1210 m long under construction on river Narmada near Navagam village. A reservoir with a live storage capacity 5500 million cubic metres, extending about 214 km in Narmada valley with water spread of about 37 sq km will be created behind the dam. It will submerge 37,533 ha of land, of which 13,386 ha are classified as forest. It will displace 40,727 families of 245 villages spread over the states of Gujarat, Maharashtra and Madhya Pradesh. Each oustee will get minimum 2 ha of land besides a plot for house 500 sq m free of cost. It has designed installed capacity of 1200 MW, 6 units of 200 MW each in underground power house, besides 250 MW at canal head power house, 5 units of 50 MW each. The spillway shall have surplusage capacity of 87,000 cumecs; the third largest in the world. The irrigation system consists of 532 km long Narmada main canal with a capacity of 1133 cumecs at its head and 66,000 km long distribution canal network for an annual irrigation of 1.91 million hectares land, besides drinking water to 40 million people.

Compensatory forests will be raised in non-forest area of 4650 ha, in lieu of 4523 ha forest land diverted for the project. In addition 12,729 ha of degraded forest lands will be afforested in the catchment area besides 110 ha and 125 ha in rainfed and irrigation land respectively near dam vicinity.

TEHRI DAM: The Rs 3800 crore Tehri hydro project is under construction below the confluence of Bhagirathi and Bhilaganga rivers near Tehri town in Uttar Pradesh. The earth and rockfill Tehri dam, 260.5 m high, is the highest in Asia. The catchment area of the dam is 7691 sq km; out of which 2735 sq km is snow bound. The reservoir shall have water spread of 42 sq km to store 3540 million cu m of water, out of which live storage is 2615 million cu m. 2583 ha of forest land is affected and in lieu of that 4595 ha of non-forest land has been taken up for afforestation. The reservoir will submerge Tehri town and 23 villages fully and another 72 villages partially. 10 villages will also be shifted for the construction of new Tehri town etc. Total population affected is 46000. A minimum of 0.8 ha of land with a ceiling of 5 ha is allotted to each displaced family. An underground power plant shall generate 1000 MW of power(4 × 250 MW) in first stage and another 1000 MW with a reversible pump turbine in a separate cavity in the second stage and 400 MW in third stage at 103 m high concrete dam at Koteshwar, 22 km downstream of Tehri dam site. The design inflow flood is 15,540 cumecs which is the maximum probable flood. The outflow routed through 95 m wide chute spillway is 11,712 cumecs. The irrigation facilities shall be extended to an additional area of 2.7 lakh hectares besides stabilising irrigation in 6.4 lakh hectares area.

APPENDIX III

SOME TYPICAL DESIGN PROBLEMS

Example 1. *Design a barrage with the following data: maximum flood discharge = 3150 cumecs, high flood level (before construction) = 193.0 m, average river bed level = 188.45 m, Lacey's f = 1.5, and full supply level of offtaking canal = 191.0 m.*

Sollution.

Pond level with 1 m working head of canal head regulator	= 191.0 + 1.0 = 192.0 m
Lacey's regime width	$= 4.83\ Q^{\frac{1}{2}} = 4.83\,(3150)^{\frac{1}{2}} = 271\text{m}$
Provide overall waterway, 3 bays of 10 m each in undersluices, 12 bays of 12 m each in river sluices, piers of 2 m width & fish ladder 5 m & divided wall of 3 m width	= 208 m *i.e.* (3 × 10 + 2 × 2.0 + 12 × 12.0 + 11 × 2.0 + 5.0 + 3.0)
Looseness factor	= 208/271 = 0.77
Clear waterway provided	= 3 × 10 + 12 × 12 = 174 m
Intensity of discharge u/s of barrage	= 3150/208 = 15.14 cumecs/m
Scour depth, R (for looseness factor < 1)	$= 1.35\left(q^2/f\right)^{\frac{1}{3}} = 1.35\left[(15.14)^2/1.5\right]^{\frac{1}{3}} = 7.21\text{ m}$
Velocity of approach,	Va = *q*/*R* = 15.14/7.21 = 2.10 m/sec
Head due to velocity of approach,	ha = $Va^2/2g = (2.10)^2/\ 2 \times 9.81 = 0.22$ m
Afflux (assumed)	= 1.0 m
D/s TEL	= 193.0 + 0.22 = 193.22 m
U/s TEL = D/s TEL + 1.0	= 193.22 + 1.0 = 194.22 m
U/s H.F.L.	= 194.22 – 0.22 = 194.0 m
Crest level — of undersluices	= 188.45 m *i.e.* deepest river bed level
— of river sluices	= 188.45 + 1.0 = 189.45 m
Head of water over crest of — undersluices	= 194.22 – 188.45 = 5.77 m
— river sluices	= 194.22 – 189.45 = 4.77 m
Discharge — through undersluices	$= 1.7\,(L - 0.1\ nH)H^{3/2}$
(20% of 3150 = 630 cumecs)	$= 1.7(3 \times 10.0 - 0.1 \times 3 \times 2 \times 5.77)(5.77)^{3/2} = 625$ cumecs
— through river sluices	$= 1.84\ (L - 0.1\ nH)H^{3/2}$ since H is 4.77 m which is > 1.5 times crest width of 2 m, (it is sharp crested weir)
	$= 1.84\ (12 \times 12.0 - 0.1 \times 2 \times 12 \times 4.77)(4.77)^{3/2} = 2541$ cumecs
Total discharge	= 625 + 2541 = 3166 cumecs against 3150 cumecs required, hence crest level and waterway assumed are correct.
Drowning ratio through river sluices	$= \dfrac{\text{d / s w.L} - \text{crest level}}{\text{u / s w.L} - \text{crest level}} = \dfrac{193.0 - 188.45}{194.0 - 188.45}$
	= 0.82, hence O.K.

DESIGN OF UNDERSLUICES

The discharge intensity and head loss is determined under maximum flood discharge, pond level flow with all gates open, without and with 20 percent concentration and 0.5 m retrogression.

Case 1. High flood Condition

(*a*) *Without concentration and retrogression*

$$q = 1.7H^{\frac{3}{2}} = 1.7(5.77)^{\frac{3}{2}} = 23.56 \text{ cumecs / m}$$

U/s H.F.L = d/s H.F.L. + afflux = 193.0 + 1.0 = 194.0 m

U/s TEL = u/s H.F.L. + ha = 194.0 + 0.22 = 194.22 m

D/s TEL = d/s H.F.L. + ha (assumed same as of u/s) = 193.0 + 0.22 = 193.22 m

Head loss, H_L = 194.22 – 193.22 = 1.0 m

(*b*) *With 20% concentration and 0.5 m retrogression*

q = 1.2 × 23.56 = 28.27 cumecs/m

Head required, $$H = \left(\frac{28.27}{1.7}\right)^{\frac{2}{3}} = 6.52 \text{ m}$$

U/s TEL = U/s river bed level + H = 188.45 + 6.52 = 194.97 m

U/s H.F.L. = 194.97 – 0.22 = 194.75 m

D/s water level = 193.00 – 0.50 = 192.50 m

D/s TEL = 192.50 + 0.22 = 192.72 m

Head loss, HL = 194.97 – 192.72 = 2.25 m

Case 2. Pond level flow

(*a*) *Without concentration and retrogression*

Head over crest of — undersluices = 192.0 – 188.45 = 3.55 m

— river sluices = 192.0 – 189.45 = 2.55 m

Neglecting velocity head for this flow condition,

the total discharge passed through the barrage = 1.70 (3 × 10.0 – 0.1 × 6 × 3.55) $(3.55)^{3/2}$ + 1.84 (12 × 12.0 – 0.1 × 24 × 2.55) $(2.55)^{3/2}$ = 1349 cumecs

Average discharge intensity = 1349/208.0 = 6.49 cumecs/m

Normal scour depth, (R) $= 1.35\left(q^2/f\right)^{\frac{1}{3}} = 1.35\left[(6.49)^2/1.5\right]^{\frac{1}{3}} = 4.10 \text{ m}$

Velocity of approach Va = q/R = 6.49/4.10 = 1.58 m/s

Velocity head, ha $= V_a^2/2g = (1.58)^2/2 \times 9.81 = 0.13$ m

U/s TEL = 192.0 + 0.13 = 192.13 m

D/s flood level when a discharge of 1349 cumecs is passing through barrage = 191.75 m (read from Fig. 1)

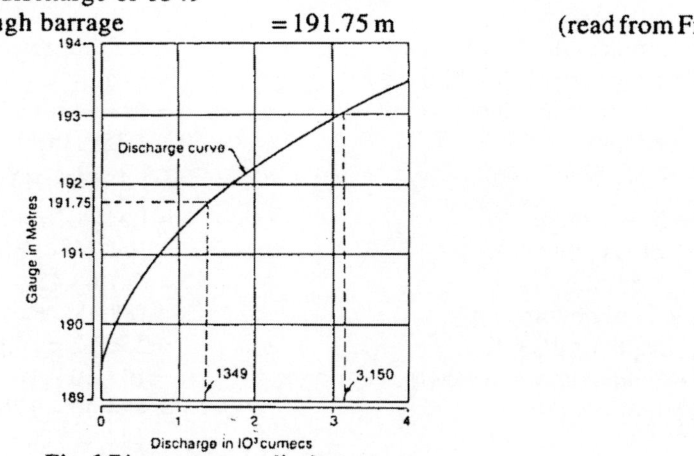

Fig. 1 River gauge — discharge curve

D/s TEL	= 191.75 + 0.13 = 191.88 m
Head loss, HL	= 192.13 – 191.88 = 0.25 m
Discharge intensity between piers of undersluices	= 1.7 (3.55 + 0.13)$^{3/2}$ = 12.00 cumecs/m

(b) With concentration and 0.5 m retrogression

Discharge intensity	= 12.00 × 1.20 = 14.40 cumec/m
Head	= (14.40/1.7)$^{2/3}$ = 4.16 m
U/s TEL	= 188.45 + 4.16 = 192.61 m
U/s water level	= 192.61 – 0.13 = 192.48 m
D/s water level with 0.5 m retrogression	= 191.75 – 0.5 = 191.25 m
D/s TEL	= 191.25 + 0.13 = 191.38 m
Head loss, HL	= 192.61 – 191.38 = 1.23 m

Hydraulic jump calculations are given in Table 1.1

Table 1.1 Hydraulic jump calculations

Design parameter	Unit	High flood condition		Pond level condition	
		case 1 (a)	case 1(b)	case 2(a)	case (b)
Discharge intensity	cumecs/m	23.56	28.27	12.00	14.40
U/s water level	El. (m)	194.0	194.75	192.0	192.48
D/s water level	El. (m)	193.0	192.50	191.75	191.25
U/s TEL	El. (m)	194.22	194.97	192.13	192.61
D/s TEL	El. (m)	193.22	192.72	191.88	191.38
Head loss (HL)	m	1.0	2.25	0.25	1.23
Ef_2 (from Fig.)	m	6.75	8.15	4.0	5.10
$Ef_1 = Ef_2 + Hl$	m	7.75	10.40	4.25	6.33
Level at which H.J. is formed	El. (m)	186.47	184.57	187.88	186.28
D_1 corresponding to Ef_1 (from Fig.)	m	2.25	2.10	1.80	1.40
D_2 corresponding to Ef_2 (from Fig.)	m	5.75	7.3	3.4	4.5
Length of floor 5 $(D_2 - D_1)$	m	17.5	26.0	8.0	15.5
Froud No. $F_1 = \frac{q}{\sqrt{g D_1^3}}$	—	2.23	2.97	1.59	2.78

D/s floor level	= Slightly lower than the level of formation of hydraulic jump, say 184.40 m
Floor length	= 26 m as given by worst condition of 20% concentration of flow

Depth of sheet piles

Undersluices discharge	= 625 cumecs
Overall waterway (3 × 10.0 + 2 × 2.0)	= 34 m
Average discharge intensity,	q = 625/34 = 18.38 cumecs/m
Scour depth,	R = 1.35 $(q^2/f)^{1/3}$ ∵ looseness factor is < 1
	$= 1.35 \left[(18.38)^2/1.5\right]^{1/3} = 8.2$ m
Depth of u/s pile	= 1.5 × 8.2 = 12 m
R.L. of u/s pile	= 194.75.– 12.0 = 182.75m
Depth of pile	= El. 188.45 – 182.75 = 6.0 m
∴ R.L. of bottom of pile	= 188.45 – 6.0 = 182.45 m.
∴ Depth of d/s pile	= 2 × 8.2 ≃ 16.0 m
(2 R below d/s water level)	= 192.50 – 16.50 = 176.00 m
Depth of d/s pile	= 184.40 – 176.00 ≈ 8.5 m
∴ R.L. of bottom of pile	= 184.40 – 8.50 = 175.90 m

Total floor length

Safe exit gradient, $G_E = H/d \dfrac{1}{\pi\sqrt{\lambda}}$

$$= \frac{\text{pond level} - \text{d/s floor level}}{\text{depth of d/s cutoff}} \frac{1}{\pi\sqrt{\lambda}}$$

$$\frac{1}{6} = \frac{(192.0 - 184.40)}{8.5} \times \frac{1}{\pi\sqrt{\lambda}}$$

$$\text{or } \frac{1}{\pi\sqrt{\lambda}} = 0.186$$

From Khosla exit gradient, curve (Fig.) for $\dfrac{1}{\pi\sqrt{\lambda}} = 0.186$, $\alpha = 5.0$

Also $\alpha = b/d$ or $b = 5.0 \times 8.5 \approx 43$ m.

The profile of floor by considering d/s floor length of 26 m, d/s glacis slope 3:1 and length of 12.15 m *i.e.*, 3 (188.45 – 184.40) and balance u/s floor length of 4.85 m *i.e.* (43.0 – 26.0 – 12.15).

Pressure calculations

Uplift pressures are determined by Khosla theory for which assume thickness of u/s & d/s cutoffs as 1.5 m.

Upstream pile line

$$H = \text{Pond level} - \text{d/s floor level (pumped dry)}$$
$$H = 192.0 - 184.40 = 7.6 \text{ m}$$
$$b = 43 \text{ m}, d = 6.0 \text{ m}$$
$$1/\alpha = d/b = 6/43 = 0.14$$

From Khosla curve (Fig.) $\Phi E_1 = 100\%$

$$100 - \Phi D = 100 - 21.5 = 78.5\%$$
$$\Phi D_1 = 78.5\%$$
$$\Phi C_1 = 66.5\%, \; 100 - \Phi E = 100 - 33.5 = 66.5\%$$
$$\Phi D_1 - \Phi C_1 = 12\%$$

Correction for ΦC_1

(*i*) Correction for interference of d/s pile $= 19\sqrt{D/b'}\,[(d+D)/b]$

where, D = (u/s floor level – floor thickness) – El. d/s pile

$$= (188.45 - 1.0) - 175.90 = 11.55 \text{ m}$$

$$b' = 43.0 - \left(\frac{1.5}{2} + \frac{1.5}{2}\right) = 41.50 \text{ m}$$

d = depth of u/s pile – floor thickness $= 6.0 - 1.0 = 5.0$ m

$b = 43.0$ m

Puting values, correction $= 19\sqrt{\dfrac{11.55}{41.50}}\left(\dfrac{5.0 + 11.55}{43.0}\right) = 3.86 \text{ m (+ve)}$

(*ii*) Correction for floor thickness $= \dfrac{(78.5 - 66.5)}{6.0} = 2.0\%$ (+ve)

$$\Phi C_1 \text{ corrected} = 66.50 + 3.86 + 2.0 = 72.36\%$$

Downstream pile

$$d = 8.5 \text{ m}, b = 43 \text{ m}$$

Table 1.2. Height/elevation of H.G. line above datum

Condition of flow	*U/S water level (m)*	*D/S water level (datum) (m)*	*Head (m)*	*Upstream pile line*			*Downstream pile line*		
				ϕE_1 *100%*	ϕD_1 *78.5%*	ϕC_1 *72.36%*	ϕE_2 *37.19%*	ϕD_2 *26%*	ϕC_2 *0%*
No flow (maximum static head)	192.00	184.40	7.6	7.6 192.0	5.97 190.37	5.49 189.89	2.83 187.23	1.89 186.38	0 184.40
High flood with concentration and retrogression	194.75	192.50	2.25	2.25 194.75	1.77 194.27	1.63 194.13	0.84 193.34	0.59 193.09	0 192.50
Flow at pond level with concentration & retrogression	192.48	191.25	1.23	1.23 192.48	0.97 192.22	0.89 192.14	0.46 191.71	0.32 191.57	0 191.25

$$1/\alpha = d/b = 8.5/43 = 0.198$$
$$\Phi\ E_2 = 40\%,\ \Phi\ D_2 = 26\%,\ \Phi\ C_2 = 0\%$$

Correction to $\Phi\ E_2$.

(*i*) Correction due to interference of u/s pile

$$= 19\sqrt{D/b'}\ \left[(d+D)/b\right]$$

$$D = (184.40 - 1.5) - 182.45 = 0.45 \text{ m}$$
$$d = (184.40 - 1.5) - 175.90 = 7.0 \text{ m}$$
$$b_1 = 41.50 \text{ m} \qquad b = 43 \text{ m}$$

Correction
$$= 19\sqrt{0.45/41.50}\left[(7.0+0.45)/43.0\right]$$
$$= 0.34\%\ (-\text{ve})$$

(*ii*) Correction due to floor thickness

$$= \frac{40-26}{184.40-175.90} \times 1.5 = 2.47\%\ (-\text{ve})$$

$\therefore$ ΦE_2 corrected $= 40 - 0.34 - 2.47 = 37.19\%$

Table 1.2 gives the calculations of levels of H.G.L.

The hydraulic jump profile under two conditions of flow viz, pre-jump and post-jump are computed in Table 1.3.

Pre-jump Profile

Table 1.3 Pre-jump profile

Distance from start of glacis (m)	*Glacis level (m)*	*High flood flow concentration and retrogression q = 28.27*		*Pond level flow with concentration and retrogression q = 14.4 cumecs/m*	
		Ef_1 = u/s TEL – glacis level (194.97 – Col. 2)	*D_1*	*Ef_1 = u/s TEL – glacis level (192.61 – Col. 2)*	*D_1*
1	2	3	4	5	6
0	188.45	6.52	4.0	4.16	2.5
3	187.45	7.52	3.0	5.16	1.8
6	186.45	8.52	2.5	6.16	1.5
6.51 (level at which H.J is formed at pond level)	186.28	8.69	2.45	6.33	1.4
9.0	185.45	9.52	2.3	—	—
11.64 (level at which H.J is formed at high flood level)	184.57	10.4	2.1	—	—

Post-jump profile

The post-jump profile calculation made from Fig. 2 are given in Table 1.4.

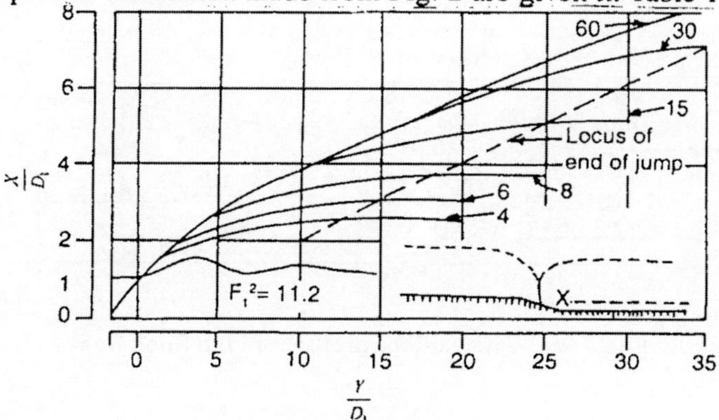

Fig. 2. Post-jump profile.

Table 1.4. Post-jump profile.

x/D (*on Fig. 2*)	*High flood condition* $F^2 = 8.82, D_1 = 2.1$			*Pond level flow condition* $F^2 = 7.73, D_1 = 1.4$		
	y/D₁ from Fig. 2	*y = (Col. 2 × 2.1)*	*x = (Col. 1 × 2.1)*	*y/D₁*	*y = (Col. 5 × 1.4)*	*x = (Col. 1 × 1.4)*
1	2	3	4	5	6	7
1.0	1.7	3.57	2.1	1.3	1.82	1.4
2.5	2.0	4.2	5.25	2.0	2.8	3.5
5.0	2.7	5.67	10.5	2.6	3.64	7.0
10.0	3.4	7.14	21.0	3.1	4.34	14.0
15.0	—	—	—	3.4	4.76	21.0
20.0	—	—	—	3.5	4.90	28.0

Bed protection

On upstream = Provide 4 rows of c.c. blocks 1.5 × 1.5 × 0.9 m in a length of 6.3 m (D = 6 m) over 0,6 m thick loose stones, including 75 mm jhurries between blocks filled with coarse sand and shingle.

On downstream = Provide 4 rows of c.c. blocks of 1.5 × 1.5 × 0.9 m over inverted filter 0.6 m in 6 m length of 0.75 D, 1.5 m wide curtain wall and 4 rows of c.c. blocks over 0.6 m stone spawls in 6.3 m length.

U/S and d/S aprons

Particular	*U/s apron*	*D/s apron*
Depth of scour below floor level (D)	6.0 m	8.5 m
Slope length $\sqrt{5}$ D	13.5 m	19 m
Quantity of stone, assuming 1.25 m thick loose stone = 13.5 × 1.25	17 m³	23.75 m³
1.5 m thick apron, length = 17/1.5 or 1.5 to 2.5 D *i.e.* 9 to 15 m	11 m	16 m
Adopt length	15 m	20 m

The hydraulic jump profiles for the (*i*) maximum flood condition with concentration and retrogression, and (*ii*) pond level flow with concentration and retrogression, determined above are plotted in Fig. 3*a* and Fig. 3*b* respectively. The maximum residual uplift pressure head in condition (*i*) is the maximum *i.e.* 7.16 m. The thickness of glacis is designed for 2/3 of this head *i.e.* 4.77 m (2/3 of 7.16) or for the static head condition (Fig. 3*c*) whichever is more. From Figs 3 (*a*) to (*c*) it is clear that dynamic flow condition governs the floor thickness to about 6 m beyond the hydraulic jump formation and thereafter static head governs the floor thickness. The floor thickness is thus as folows:

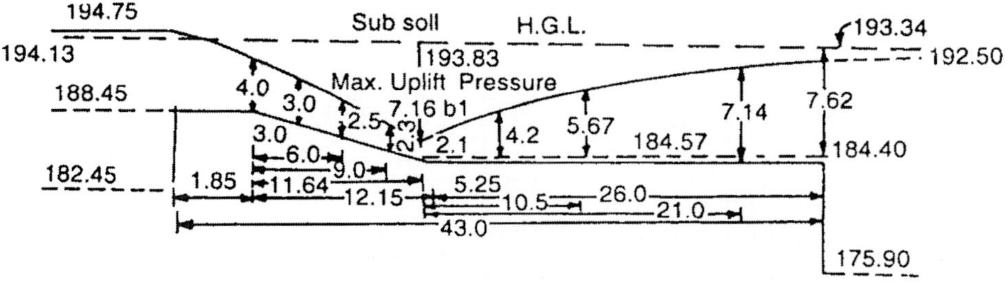

Fig. 3 (*a*). Water surface profile for high flood flow.

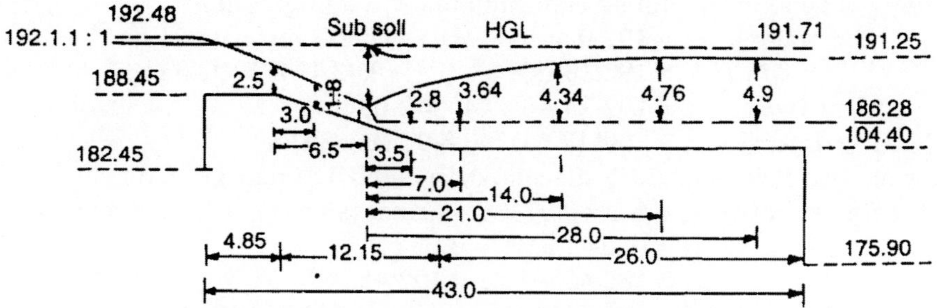

Fig. 3 (b). Water surface profile for pond level flow.

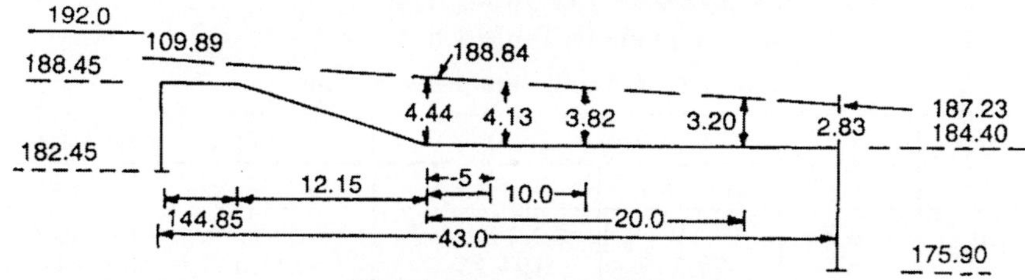

Fig. 3 (c). No. flow condition.

Location	Uplift pressure (U)	Floor thickness (U/1.24)
At H.J. formation	2/3 of 7.16 m = 4.77 m	3.85 say 4 m
5.25 m from H.J. formation	4.96 m	4 m
10 m from glacis toe	3.82 m	3.08 say 3.10 m
20 m from glacis toe	3.20 m	2.58 say 2.6 m
26 m from glacis toe	2.83 m	2.28 say 2.3 m

Upstream Floor

The subsoil hydraulic gradient line is below the upstream water level which means that all along the upbalanced head acting on the floor is counterbalanced by the self weight of water. Thus theoretically no floor thickness is required. However, provide a minimum floor thickness of 1.0 m as assumed and thickened near glacis as shown in Fig. 4.

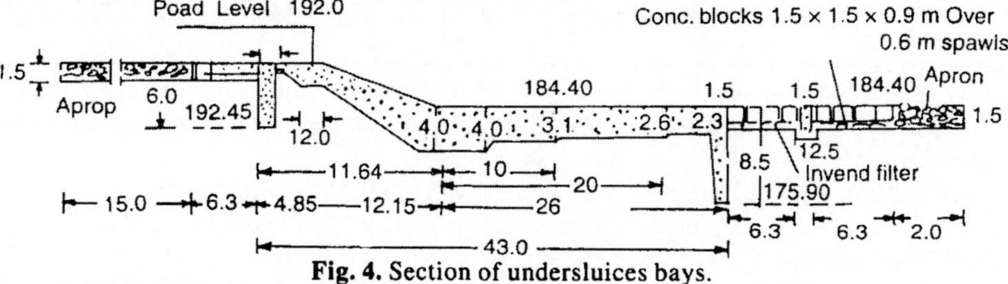

Fig. 4. Section of undersluices bays.

Design of River Bays

(a) High flood flow, with no concentration and retrogression

U/s water level = 194.0 m, D/s water level =193.0 m,

U/s TEL = 194.22 m, D/s TEL =193.22 m,

Head loss (HL) = 1.0 m, Head over crest =194.22 – 189.45 = 4.77 m,

Discharge intensity, $q = 1.84\ (4.77)^{3/2} = 19.17$ cumecs/m.

(b) High flood, with 20% concentration and 0.5 m retrogression

Discharge intensity, $q = 1.2 \times 19.17 = 23$ cumecs/m

Head over crest $= (23/1.84)^{2/3} = 5.39$ m

U/s TEL = 189.45 + 5.39 = 194.84 m

D/s TEL = 193.22 m

Head loss (HL) = 194.84 – 193.22 = 1.62 m

2. (*a*) Flow at pond level with no concentration and retrogression

U/S water level = 192.0 m, D/S water level =191.75 m

U/S TEL = 192.13 m, D/S TEL =191.88 m

Head over crest = 192.13 – 189.45 = 2.68 m

Discharge intensity, $q = 1.84\ (2.68)^{3/2} = 8.07$ cumecs/m

(*b*) Flow at pond level with 20% concentration and 0.5 m retrogression

Discharge intensity, $q = 1.2 \times 8.07 = 9.68$ cumecs/m

$H = (9.68/1.84)^{3/2} = 3.03$ m

U/S TEL = 189.45 + 3.03 = 192.48 m

D/S water level = 191.25 m, D/S TEL =191.38 m

HL = 192.48 – 191.38 = 1.1 m

Hydraulic jump calculations are made in Table 1.5.

Table 1.5. Hydraulic jump calculations

Item	*Unit*	*High flood condition*		*Pond level flow condition*	
		Case 1 (a)	*Case 1 (b)*	*Case 2 (a)*	*Case 2 (b)*
Discharge intensity	cumecs/m	19.17	23.0	8.07	9.68
U/S water level	m	194.0	194.75	192.0	192.48
D/S water level	m	193.0	192.50	191.75	191.25
U/S TEL	m	194.22	194.84	192.13	192.61
D/S TEL	m	193.22	193.22	191.88	191.38
Head loss, HL	m	1.0	1.62	0.25	1.23
Ef_2(from Fig.)	m	6.0	7.0	3.0	4.0
$Ef_1 = Ef_2 + HL$	m	7.0	8.62	3.25	5.23
Level at which hydraulic jump is formed (d/s TEL – Ef_2)	m	187.22	186.22	188.88	187.38
D_1 corresponding to Ef_1 (from Fig.)	m	1.9	2.0	1.4	0.95
D_2 corresponding to Ef_2 (from Fig.)	m	5.4	6.2	2.5	3.6
Floor length 5 (D_2–D_1)	m	17.5	21.0	5.5	13.25
Froude No. $F_1 = \frac{q}{\sqrt{gD_1^3}}$	m	2.34	2.60	1.56	3.34

D/s floor level: Slightly lower than the level of hydraulic jump formation say El. 186.0 m and floor length of 21.0 m as given by worst conditions of 20% concentration of flow and 0.5 m retrogression.

D/s glacis length with 3:1 slope = 3 (189.45 – 186.0) = 10.35 m

U/s glacis length with 2:1 slope = 2 (189.45 – 188.45) = 2.0 m

Depth of Sheet Piles

Average discharge intensity, q = 2541/174 = 14.60 cumecs/m

Normal scour depth, $R = \left(\frac{(14.60)^2}{1.5}\right)^{1/3} \approx 7.0$ m

R.L. of U/s pile line (1.5 R below U/s water level) = 194.75 – 7.0 × 1.5 = 184.25 m

Depth of U/s pile below floor level = 188.45 – 184.25 ≈ 4.5 m

R.L. of U/s pile line = 188.45 – 4.50 = 183.95 m

R.L. of d/s pile line (2R below d/s water level), although IS:6966 (Part I)-1989 provide for 1.25R = (192.50 – 7.0 × 2) = 178.50 m

Depth of d/s pile below floor level = 186.0 – 178.50 = 7.50 m

Total Floor length

Maximum head, H = 192.0 – 186.0 = 6 m

Depth of d/s pile, d = 7.50 m

$$G_E = \frac{H}{d} \cdot \frac{1}{\pi\sqrt{\lambda}}$$

$$\frac{1}{6} = \frac{6.0}{7.5} \cdot \frac{1}{\pi\sqrt{\lambda}}$$

$$\text{or } \frac{1}{\pi\sqrt{\lambda}} = 0.208$$

(or b/d) from Fig. 8.22 = 4.0

$\therefore$ $b = 4.0 \times 7.5 = 30$ m

Floor length on u/s side $= 30.0 - (2.0 + 2.0 + 10.35 + 21) = -5.35$ m.
Adopt overall length as 40 m so that u/s floor length = 4.65 m

Uplift Pressure

Assume floor thickness of 1.0 m at u/s pile line and 1.5 m at d/s pile line

U/s pile line, $b = 40$ m, $d = 4.5$

$$1/\alpha = d/b = \frac{4.5}{40.0} = 0.1125$$

From Fig. $\Phi E_1 = 0$

$\Phi D_1 = 80\%$ $100 - \Phi D = 100 - 20 = 80\%$

$\Phi C_1 = 70\%$ $100 - \Phi E = 100 - 30 = 70\%$

Correction for ΦC_1

(*i*) Correction due to interference of d/s pile

$$= 19\sqrt{\frac{D}{b'}}\,\frac{(d + D)}{b}$$

Where, D = (u/s floor level-floor thickness) – El. d/s pile

$D = (188.45 - 1.0) - 178.50 = 8.95$ m

$d = $ u/s pile depth $- 1.0 = 4.50 - 1.0 = 3.50$ m

$$b' = 40.0 - \left(\frac{1.5}{2} + \frac{1.5}{2}\right) = 38.50\,\text{m}$$

Putting values, correction $= 19\sqrt{\frac{8.95}{38.50}}\,\frac{(3.50 + 8.95)}{40.0} = 2.85$ m (+ve)

(*ii*) Correction due to thicknes of floor $= \frac{80 - 70}{4.5} \times 1.0 = 2.22$ m (+ve)

$\therefore$ ΦC_1 corrected $= 70 + 2.85 + 2.22 = 75.07\%$

Downstream pile line

$d = 7.5$ m, $b = 40$ m

$1/\alpha = d/b = 7.5/40 = 0.1875$

$\Phi E_2 = 38\%$, $\Phi D_2 = 25\%$, $\Phi C_2 = 0$

Correction for E_2

(*i*) Correction due to interference of u/s pile $= 19\sqrt{D/b'}\left(\frac{d + D}{b}\right)$

Where, $D = (186.0 - 1.5) - 183.95 = 0.55$ m

$d = (186.0 - 1.5) - 178.50 = 6.0$ m

$b' = 38.5$ m

$$\therefore \text{Correction} = 19\sqrt{0.55/38.50}\left[\frac{(6.0 + 0.55)}{40}\right] = 0.37\%\ (-\text{ve})$$

(*ii*) Correction due to thickness of floor

$$\frac{38 - 25}{186.0 - 178.50} \times 1.5 = 2.6\% \text{ (–ve)}$$

$\therefore \Phi E_2$ corrected $= 38.0 - 0.37 - 2.60 = 35.03\%$

Table 1.6 gives calculations of levels of H.G.L.

Table 1.6. Height/elevation of H.G. line above datum

Condition of flow	*U/s water level (m)*	*D/s water level (datum) (m)*	*Head (m)*	*Upstream pile line*			*Downstream pile line*		
				ΦE_1 100%	ΦD_1 80%	ΦC_1 75.07%	ΦE_2 35.03%	ΦD_2 25%	ΦC_2 0%
No flow (maximum static head)	192.00	186.0	6.0	6.0 192.0	4.80 190.80	4.50 190.50	2.10 188.10	1.5 187.50	0 186.0
High flood flow with concentration & retrogression	194.75	192.50	2.25	2.25 194.75	1.80 194.30	1.69 194.19	0.79 193.29	0.56 193.06	0 192.50
Flow at pond level with concentration & retrogression	192.48	191.25	1.23	1.23 192.48	0.98 192.23	0.92 192.17	0.43 191.68	0.31 191.56	0 191.25

Hydraulic jump profile

The hydraulic jump profile for the two conditions of flow *i.e.*, pre-jump and post-jump are computed in Table 1.7.

Table. 1.7. Pre-jump profile

Distance from start of glacis	*Glacis level (m)*	*High flood flow with D_1 concentration & retrogression q = 23 cumecs/m*		*Pond level flow with D_2 concentration & retrogression q = 9.68 cumecs/m*	
		Ef_1= u/s TEL — glacis level = 194.84 — Col.2		*Ef_1= u/s TEL — glacis level = 192.61 — Col.2*	
0	189.45	5.39	–	3.16	2.0
3	188.45	6.39	2.7	4.16	1.2
6	187.45	7.39	2.2	5.16	1.0
6.2 (Point at which HJ forms at Pond level)	187.38	7.41	2.18	5.18	0.95
9	186.45	8.39	2.05	–	–
9.69 (Point at which HJ forms at high flood)	186.22	8.62	2.0	–	–

Post-jump Profile

The post-jump profile calculation from Fig.2 is given in Table 1.8.

Table. 1.8. Post jump profile

x/D_1	*High flood flow*			*Pond level flow*		
	F_1^2 = 6.76	*D_1 = 2.0 m*		*F_1^2 = 11.15*	*D_1 = 0.95 m*	
	y/D_1	*y*	*x = col. 1 × 2.0*	*y/D_1*	*y*	*x = col 1 × 0.95*
1.0	1.3	2.6	2.0	1.3	1.24	0.95
2.5	2.0	4.0	5.0	2.0	1.9	2.38
5.0	2.5	5.0	10.0	2.8	2.66	4.75
10.0	3.0	6.0	20.0	3.4	3.23	9.5

The hydraulic jump profiles for the (*i*) maximum flood condition with concentration and retrogression and (*ii*) pond level flow with concentration and retrogression, determined above are plotted in Fig. 5*a* and Fig 5*b* respectively. The maximum residual uplift pressure head in condition (*i*) is the maximum *i.e.*, 5.56 m. The thickness of glacis is designed for 2/3 of this head 3.71 m *i.e.* (2/3 × 5.56) or for the static head condition (Fig. 5*c*) whichever is more. The maximum uplift pressure comes out to as under:

The Fig. 5(*a*) to (*c*) indicate that dynamic flow condition governs floor thickness for about 5 m beyond hydraulic jump formation and thereafter static head condition.

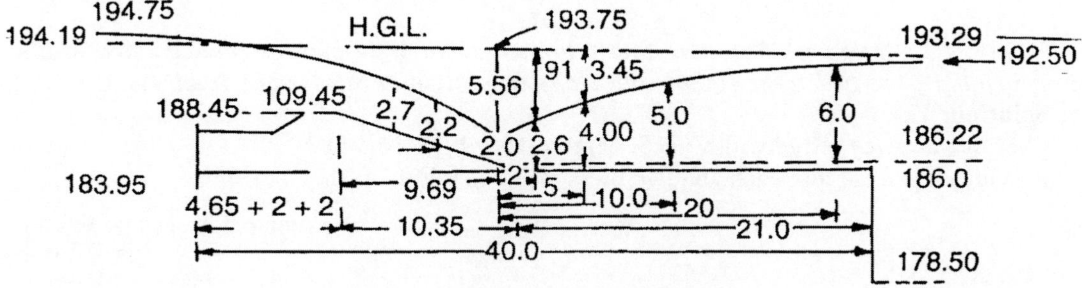

Fig. 5(*a*). High flood flow condition.

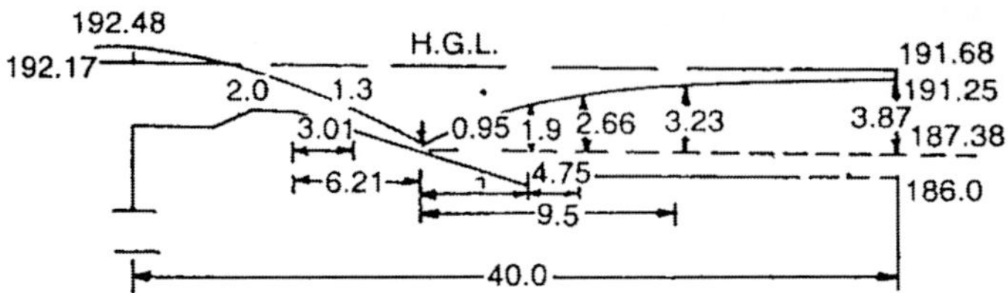

Fig. 5(*b*). Pond level flow condition.

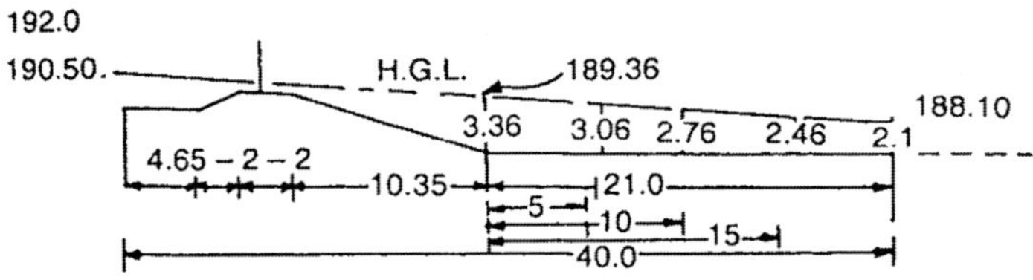

Fig. 5(*c*). No flow condition.

Location	*Uplift pressure (U)*	*Floor thickness (U/1.24)*
H.J. formation 2/3 × 5.56	3.71 m	2.99 adopt 4 m
2.0 m beyond	4.91 m	3.96
5.0 beyond	3.45 m	2.78 say 2.8 m
10 m beyond	2.76 m	2.23 say 2.5 m
15 m beyond	2.46 m	1.98 say 2.0 m
Bed protection	Same as for undersluices bays	
U/s & d/s aprons	u/s apron	d/s apron
Scour depth below floor level	4.5 m	7.5 m
Slope length $D\sqrt{5}$	10 m	17 m
Quantity of stone, assuming 1.25 m thick loose stones	12 m³	21 m³
1.5 m thick aprons,		
length = 12.0/1.5	8 m	14 m
or 1.5 to 2.5 D	7 to 11 m	11 to 19 m
Adopt	10 m	20 m

Fig. 6 shows the section of river bays.

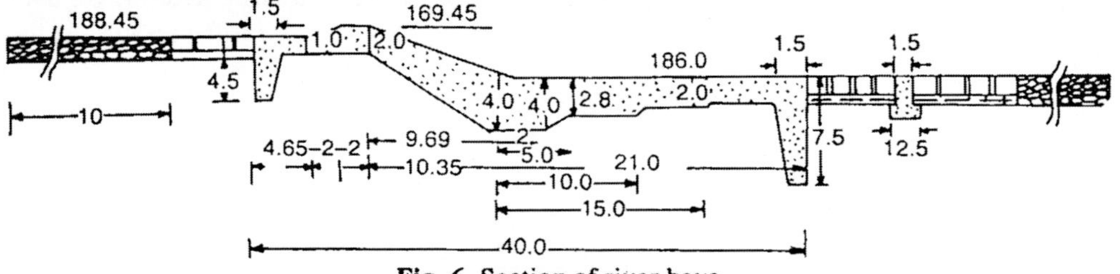

Fig. 6. Section of river bays.

Example 2. *Design a V.R. bridge over a branch canal of 12 cumecs capacity, bed width 12 m, depth 1.5 m. bed level 100.0, side slope 0.5 : 1, natural surface level 100.0 m.*

Solution:

Overall waterway = Bed width + F.S. depth = 15 + 1.5 = 16.5 m

Provide 3 spans of 5 m each and two piers of 0.7 m width* each.

$$\bullet \begin{pmatrix} \text{up to 3 m span} & = 0.6\text{m} \\ 3 - 4.5 & = 0.7\text{m} \\ 4.5 - 0.6 & = 0.8\text{m} \end{pmatrix}$$

Overall waterway provided = 3 × 5.0 + 2 × 0.7 = 16.4 m, hence O.K.

Assume slab thickness for IRC class A loading as 0.46 m

Road level	
Wearing coat	= 0.075 m say 0.08 m
Slab thickness	= 0.46 m
Free board	= 0.5 m
Scour depth below bed	= d/3 = 0.5 m
∴ Road level	= Bed level + depth + free board + slab thickness + wearing coat
	= 100.0 + 1.5 + 0.5 + 0.46 + 0.08 = 102.54 m

Wing wall

Length of return wall $= 1.5\,(\text{F.S.L.} + 0.15 - \text{N.S.L.}) + 0.5\text{ m}$

$= 1.5\,(101.50 + 0.15 - 99.5) + 0.5 = 3.75\text{ m}$

Length of straight portion $= \dfrac{1.0}{\sin 30^\circ} = 1.15\text{ m}$

Figs. 7 (*a*) & (*b*) show the drawing of the bridge.

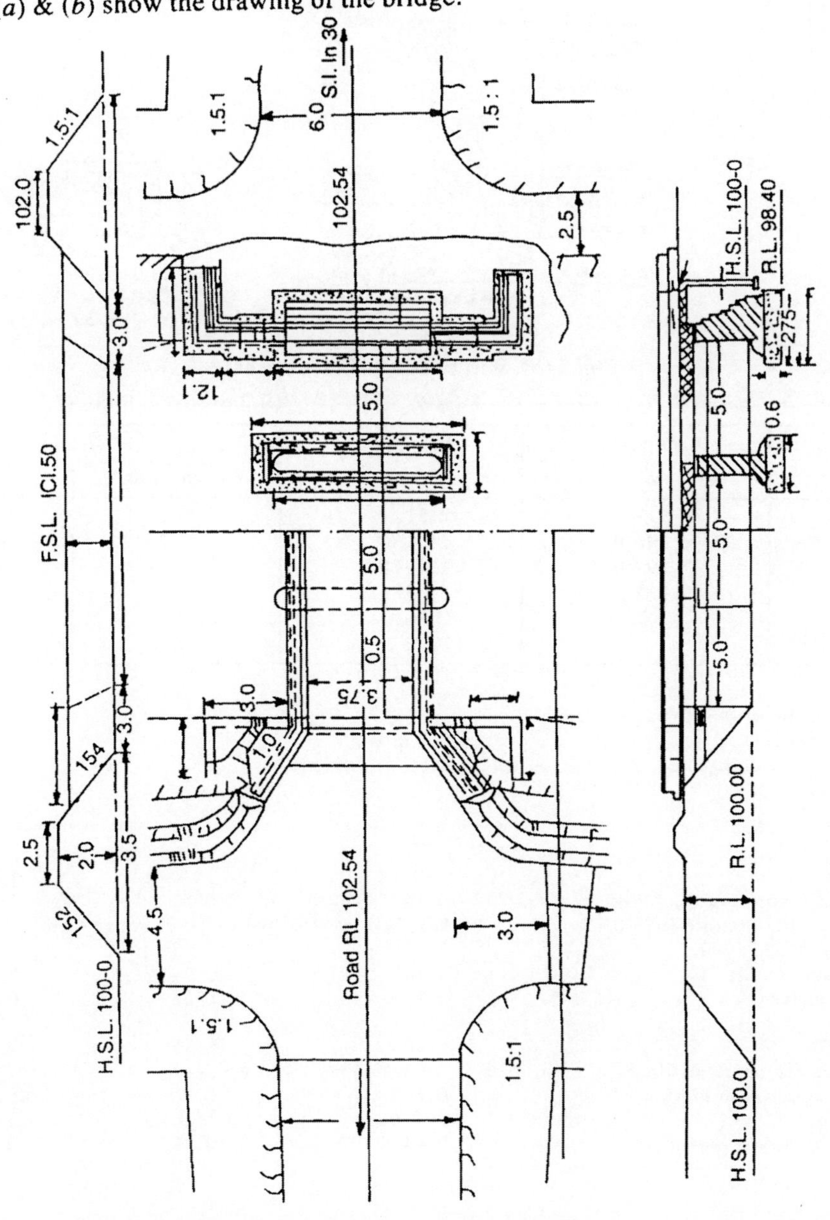

Fig. 7 (*a*). V.R. bridge on branch canal.

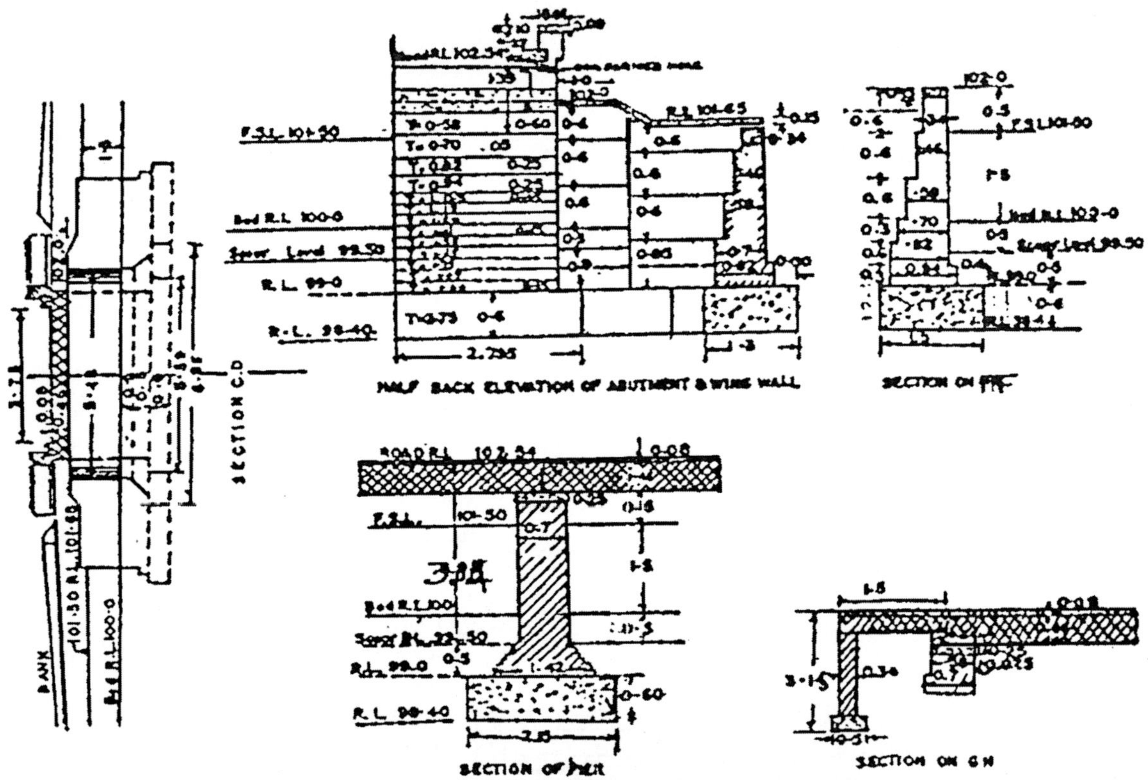

Fig. 7 (*b*). V.R. bridge on branch canal.

Example 3. *Design a proportional distributor for three distributaries with the following data:*

Particulars	*Parent*	*Distributary*	*Offtake*	
	u/s	*d/s*	*Right side*	*Left side*
Discharge	5.5	2.5	0.5	2.5 cumecs
F.S. level	200.0	199.50	199.75	199.70 m
Bed level	199.0	198.70	199.25 m	198.90 m
Bed width	9.0	5.75	2.5	5.75 m
Depth	1.0	0.80	0.50	0.80 m
Working head		0.5	0.25	0.30 m

Solution:

Throat width, Bt @ 0.6 B for parrent channel	3.5 m		
Coefficeint of discharge, C	1.69	1.62	1.63
Head over crest, H $(2.5/1.69 \times 3.5)^{\frac{2}{3}}$	0.56	0.56	0.56 m
Crest level	199.44	199.44	199.44 m
Height of crest above d/s bed level	0.74	0.19	0.54 m

being more than 0.15 m above highest bed level of the offtake, hence O.K.

Length of throat, Lt = 2H	1.12	1.12	1.12 m
Bt for offtake $(Q/CH^{3/2})$	3.50	0.74	3.66 m
Glacis and cistern q = Q/Bt	0.71	0.68	0.68 cumecs/m
Ef_2 (from Fig.)	0.70	0.55	0.56 m
Depth of cistern below d/s bed = $Ef_2 - D_n$ or 0.1 m minimum	0.1	0.1	0.1 m
Cistern level	198.60	199.15	198.80 m
Cistern depth below crest (d)	0.84	0.29	0.64 m

D/s glacis length (2.5d)	2.1	0.73	1.6 m
Cistern length = ($D_n + H_w$) or			
Le*– length of glacis, whichever is more	1.30	1.91	1.54 m
Downstream expansion	3.40	2.64	3.14 m
$Le^* = 3\left(\frac{B - Bt}{2}\right)$ or length of glacis + cistern, whichever is more	5.76	4.4	5.24 m
Radius of expansion $\left(\frac{Le^2}{B - Bt} + \frac{B - Bt}{4}\right)$	3.40	2.64	3.14
Downstream protection			
Side pitching length 3 ($D_n + H_w$)	4.0	2.0	3.0 m
Bed protection length ($D_n + H_w$)	1.3 say 1.25	0.75 say 1.0	1.1 say 1 m
Bed protection thickness	0.3	0.15	0.3 m
Upstream approach side			
Radius of common u/s wall	–	2.0	2.5 m
$La = \sqrt{3}\, H^{\frac{3}{2}}$	0.73	0.73	0.73 m
Height of crest above u/s bed, x	0.44	0.44	0.44 m
Radius for joining crest with u/s bed, $\frac{La^2 + x^2}{2x}$	0.83	0.83	0.83 m
Length of abutment for bridge	4.57	4.57	4.57 m

Fig. 8 shows the distributor designed as above.

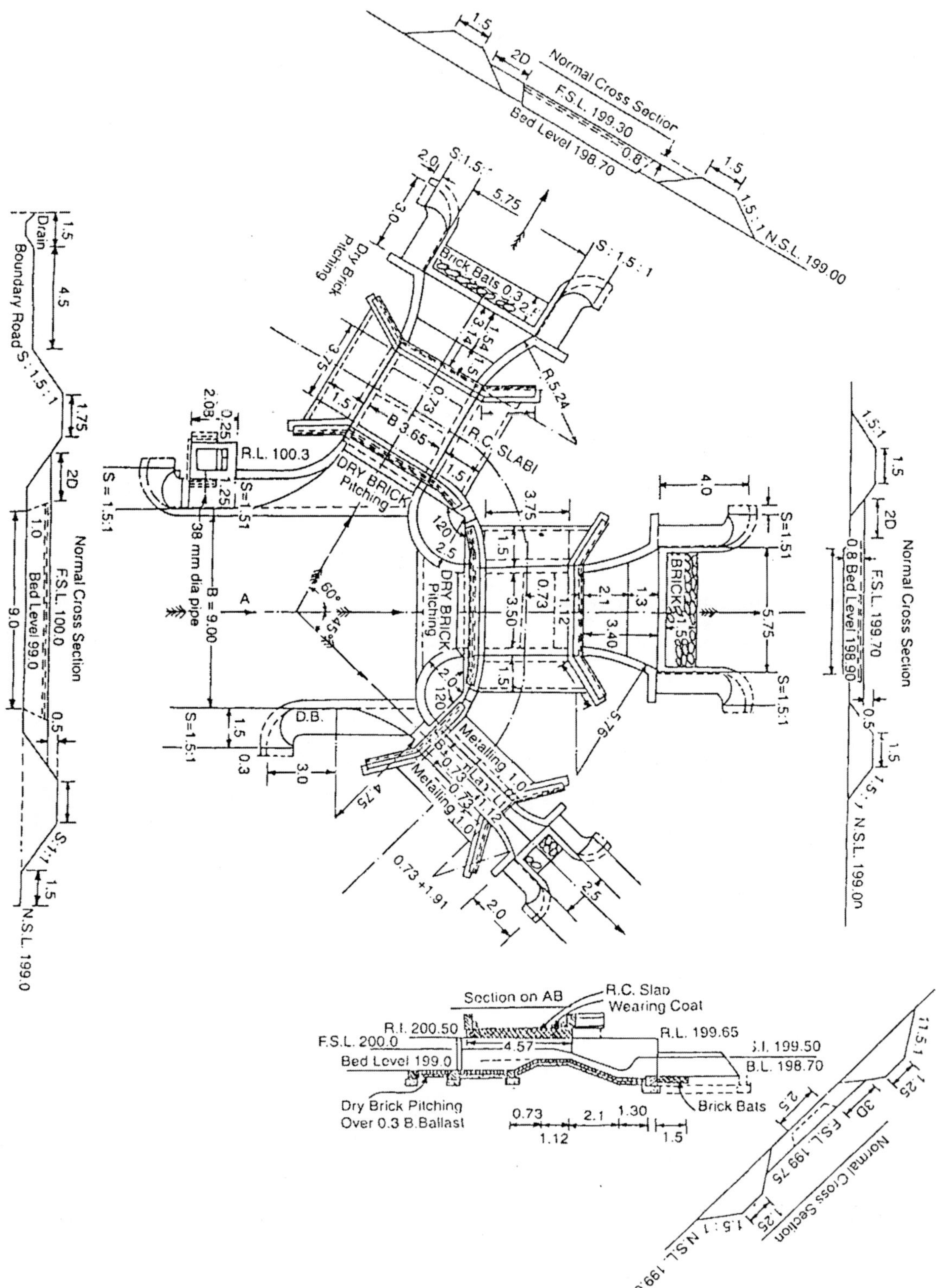

Fig. 8. Proportional distributor.

OBJECTIVE TYPE QUESTIONS

1. IRRIGATION

1. The object of irrigation is to
   (*a*) raise crops (*b*) provide water to fields
   (*c*) make water available to cultivators with respect to location, time and quantity as per the crop requirements
   (*d*) none of the above (*c*)
2. Irrigation water is essentially required in areas with annual rainfall
   (*a*) 100 cm (*b*) 50–100 cm (*c*) 50 – 24 cm (*d*) < 25 cm
3. Composed to conventional crops the (in all) high yielding varieties of crop have water requirements:
   (*a*) losser (*b*) same (*c*) higher (*d*) no relation (*c*)
4. Irrigation:
   (*a*) is eco-friendly with proper damage in command area
   (*b*) is eco-friendly even without adequate damage
   (*c*) has no relation with drainage (*a*)
5. Irrigation development
   (*a*) itself is not responsible for waterlogging rather lack of drainage results in development of waterlegging
   (*b*) itself is responsible for waterlogging
   (*c*) has no relation with the development of waterlogging (*a*)
6. Irrigation development is an instrument of:
   (*a*) national integration by bringing backward avid areas at par with developed areas
   (*b*) degrading areas into waterlogging (*c*) inflicint backwardness
   (*d*) none of the above (*a*)
7. Irrigated areas have per capita income:
   (*a*) 2-3 times unirrigated areas (*b*) less than rainfed areas cultivation
   (*c*) no relation to rainfed areas cultivation
8. Typical operation and maintenance under irrigation circle has command area of ha:
   (*a*) 1,00,000 – 2,00,000 (*b*) 1,00,000 – 3,00,000
   (*c*) 1,00,000 – 4,00,000 (*d*) 1,00,000 – 5,00,000 (*a*)
9. Typical divisional oppic responsible for operation and maintenance of the canal system in its charge has command area of has:
   (*a*) 30,000 – 40,000 (*b*) 40,000 – 50,000 (*c*) 40,000 – 60,000 (*d*) 40,000 – 70,000 (*c*)
10. The water availability of less than 1000 cubic metres per capita is considered as:
    (*a*) optimum (*b*) abundant (*c*). critical (*d*) none of the above
11. Present water use efficiency of 45% if increased to 60% means increasing the irrigated area without developing new water sources
    (*a*) 50% (*b*) 75% (*c*) 100% (*d*) 150% (*c*)

12. Water resources if properly managed can sustain the country's population of:
    (*a*) 1.5 billion (*b*) 2.0 billion (*c*) 2.3 billion (*d*) 2.5 6 billion (*d*)
13. Cultivated area in the region of moderate to low rainfall in India. Wherein water deficiencies lead to drought conditions is about:
    (*a*) 50% (*b*) 60% (*c*) 70% (*d*) 80%
14. Drought prone area of the country is about%
    (*a*) 20% (*b*) 25% (*c*) 30% (*d*) 33%
15. Irrigation potential of the country is about:
    (*a*) 87 M.ha (*b*) 100 M.ha (*c*) 113 M.ha (*d*) 125 M.ha (*c*)
16. Command area development envisages to:
    (*a*) identify command (*b*) demarcate command are
    (*c*) develop the command area (*d*) none of the above
17. Congenctive use of water for irrigation means:
    (*a*) use of rain water combined with full water
    (*b*) use of rain water and canal water
    (*c*) use of ground water and cannal water (*d*) all of the above (*c*)
18. The surface method of irrigation suitable for wheat, barsem and fodder is:
    (*a*) basin method (*b*) furrow method
    (*c*) border strip method (*d*) none of the above
19. Cannal water management envisages:
    (*a*) allocation and supply of irrigation water according to entitlement at the right time
    (*b*) equitable distribution of supplies to irrigators especially at the tril
    (*c*) optimum irrigation efficiencies commensurate with agricultural production
    (*d*) all of the above (*d*)
20. A major irrigation project has cultural command area of:
    (*a*) 5000 ha (*b*) 7000 ha (*c*) 10,000 ha (*d*) over 10,000 ha
21. Protective irrigation project is a project which
    (*a*) provides irrigation for protection against waterlogging
    (*b*) provides irrigation for protection against irrigation of villagers to cities
    (*c*) provides protection against famine (*d*) all of the above (*c*)
22. In areas with abürdance of both surface and ground waters, the latter should preferably be exploited for
    (*a*) low rainfall zones (*b*) high rainfall zones
    (*c*) local use (*d*) none of the above (*c*)
23. Canal automation is the arrangement of components:
    (*a*) electrical (*b*) electronic (*c*) mechanical (*d*) all of them (*d*)
24. Canal automation is the arrangement that commands or duicts:
    (*a*) regulation of canal system
    (*b*) collection of data regarding canal operation
    (*c*) transmission of data regarding canal operation to headquater
    (*d*) transmitting back, signals for canal operation at sensing location
    (*e*) all of them (*e*)

2. IRRIGATION WATER REQUIREMENT AND CROPS

1. Crop period is the time a crop takes:
    (*a*) from first watering to last watering (*b*) from sowing to its havesting
    (*c*) from sowing to last watering
    (*d*) from first watering at the time of sowing to its havesting (*b*)

2. Base period of a crop is the time between:
   (*a*) the instant of sowing to its havesting
   (*b*) first watering at the time of sowing to its last watering before havesting
   (*c*) sowing to last watering (*d*) none of the above (*b*)
3. Base period of a crop is:
   (*a*) the same as crop period (*b*) less than crop period
   (*c*) more than crop period (*d*) none of the above (*b*)
3. Paleo is the:
   (*a*) first watering before a crop is sown (*b*) first watering after a crop is sown
   (*c*) last watering (*d*) none of the above (*a*)
4. Kor watering is the:
   (*a*) first watering before a crop is sown (*b*) first watering after a crop is sown
   (*c*) first watering after a crop is grown (*d*) the watering of least depth (*b*)
5. Kor watering is the water application:
   (*a*) with maximum depth of water (*b*) with least depth of water
   (*c*) with same depth of water as the first watering
   (*d*) none of the above (*a*)
6. If duty for a crop with base period of 120 days is 1500 ha/cumec other its delta is (*AMIE w*'99)
7. The influtration capacity of a soil (decreases) with time. (*AMIE w*'99)
8. For row crops, e.g. potates, sugarbects etc., the most commonly adopted method of surface irrigation is: (*AMIE w*'96)
   (*a*) check floating (*b*) free flooding (*c*) basin flooding
   (*d*) furrow and corrugation irrigation (*d*)
9. The duty for a crop with a base period of 140 days is 3456 ha per cumec. The delta for the crop will be .... m. (0.35) (*AMIE w*'94)
10. The useful moisture of a soil is equal to its: (*AMIE w*'94)
    (*a*) field capacity (*b*) saturation capacity
    (*c*) moisture content at permanent wilting point
    (*d*) difference between field capacity and permanent wilting point within the root zone of plants. (*d*)
11. Permanent wilting point is: (*AMIE w*'94)
    (*a*) a characteristic of a plant (*b*) a soil characteristic
    (*c*) a soil characteristic modified by crop
    (*d*) dependent on soil water plant fertilizer interaction (*b*)
12. Conjunctive use of water in a basin means: (*AMIE w*'94)
    (*a*) combined use of water for irrigation and for hydro power generation
    (*b*) combined use of surface and ground water resources
    (*c*) use of irrigation water for both Rabi and Khanf crop
    (*d*) use of irrigation water by a cooperative of farmers (*b*)
    (*a*) Rabi season (*A*) last from 1st April to 30th September.
    (*b*) Kharif season (*B*) is the total quantity of water needed for growth of crop to maturely
    (*c*) Delta (*C*) is the relation between value of water and the
    ******
13. Rabi crops are sown and hairested in the month of: (*AMIE w*'97)
    (*a*) April — September (*A*) October — March
    (*b*) February and June (*B*) January and May (*b*)
14. Mixed cropping is defined as: (*AMIE w*'97)
    (*a*) two or more crops grown during any year

(*b*) two or more crops grown during the same crop season in different fields
(*c*) two or more crop seasons of the year
(*d*) growing of two or more crops together in the same field during the same crop season (*d*)

15. The internal of supply of irrigation water depends on the type of: (*AMIE w*'96)
(*a*) crop only (*b*) soil and crop
(*c*) crop and climate (*d*) soil, crop and climate (*d*)

16. Height of capillary fringe is: (*AMIE w*'99)
(*a*) more in fine grained soil (*b*) more in coarse graned soil
(*c*) does not depend upon soil characteristics
(*d*) depends on quantity of water in the soil (*b*)

17. Consumptive use is the sum of: (*AMIE w*'96)
(*a*) evapotranspiration and infiltration losses
(*b*) evapotranspriation plus application losses
(*c*) evapotranspiration plus amount of water used up in plant metorbolism
(*d*) evapotranspiration plus amount of water used up in plant meterbolism plus percolaton losses (*b*)

18. Consumptive use is: (*AMIE w*'94)
(*a*) water used up in plant metabolism
(*b*) sum of evapotranspiration and amount used up in plant metabolism
(*c*) sum of evapotranspiration and infiltration losses
(*d*) combined use of surface and ground water resources (*b*)

19. Blaneg-criddle formula is used to estimate: (*AMIE w*'97)
(*a*) open water evaporator (*b*) potential evaporation
(*c*) actual evapotranspiration (*d*) potential evapotranspiration (*d*)

20. For a crop the consumptive use of water is equal to the depth of water: (*AMIE s*'98)
(*a*) transpired (*b*) evaporated (*c*) transpired and evaporated
(*d*) used in transpiration, evaporation and evaporated from adjacent soil
(*c*)

Fill in the blanks:

(*i*) Field capacity is defined as
(*ii*) Delta for crop with duty 864 ha/cumec on the field and base period of 120 is .... cm
(*iii*) Infiltration capacity is more than infiltration rate when .... (*AMIE w*'97)
(*iv*) In a field measuring 30 ha, 40 cm of water was stored in the root zone when 6 cumecs of water was supplied for 8 hours. The field application efficiency is .... (*AMIE w*' 96)

21. Optimum depth of kor watering for rice crop is: (*AMIE w*' 94)
(*a*) 22 cm (*b*) 19 cm (*c*) 16 cm (*d*) 13 cm (*b*)

22. If the duty of water for a crop of base period of 120 days is 1 kg ha/cumec, then delta (Δ) of crop is (*AMIE w*' 93)
(*a*) 741 mm (*b*) 843 mm (*c*) 949 mm (*d*) 1056 mm (*a*)

23. An irrigation water has the ion concentration of Na^+, Ca^{++} and Mg^{++} as 20, 10 and 5 million equivalents per litre respectively. The sodium absorption ratio (5 AR) of this water is: (*AMIE w*' 01)
(*a*) 10×32 (*b*) 7×30 (*c*) 2×83 (*d*) 1×30

24. Consumptive use is defined as .... (*AMIE w*' 98)
25. The relationship between duty and delta is .... (*AMIE w*' 98)
26. Base period of an irrigated crop is the period from .... to .... (*AMIE w*' 98)
27. Water for irrigation can be used if it is: (*AMIE w*' 00)
(*a*) untreated effluent from a chemical industry

(*b*) surface water from a river carrying fines of high fertility
(*c*) ground water of high salinity
(*d*) surface water with fine rudiment and dissolved salts (*b*)

28. For a given soil: (*AMIE s*' 98)
(*a*) texture and structure represent same property
(*b*) texture implies presence of particles of different sizes
(*c*) framer can change the texture using cultivation method
(*d*) texture implies arrangements of a given size of particles (*d*)

29. For the application of Dercy's law for flow through porous media, Reynola s number should be: (*AMIE s*' 98)
(*a*) < 1.0 (*b*) < 500
(*c*) between 1 & 500 (*d*) a fixed value of 500
(*c*)

30. Soil in which ground water table rises as to affect the productivity of the soil due to air circulation to the plant root zone is called .... (waterlogged) (*AMIE s*' 93)

31. Salted soils are classified more rationally on the basis of (*AMIE s*' 98)
(*a*) concentration of dissolved salts (*b*) pH value
(*c*) pH value and concentration of dissolved salts
(*d*) concentration of dissolved salts, pH value and pressure of sodium ions (*d*

32. For plant's grow the useful soil moisture is: (*AMIE w*' 95)
(*a*) capillary water (*b*) gravity water (*c*) hygroscopic water (*d*) chemical water (*a*)

33. The volume of water that can be extracted by force of gravity from a unit volume of aquifer material is known as: (*AMIE w*' 95)
(*a*) specific yield (*b*) safe yield (*c*) specific retention (*d*) specific storage
(*a*)

34. Consumptive use is: (*AMIE w*' 95)
(*a*) water used up in plant metabolism
(*b*) sum of evapotionspiration and the amount used up in plant metabolism
(*c*) sum of evapotranspiration and infiltration losses
(*d*) combined use of surface and ground water resources (*b*)

35. The loss of water from plant surfaces and utilised in its growth is known as .... transpiration. (*AMIE s*' 93)

36. With the increase in quantity of water supplied, the yield of most crops: (*AMIE s*' 2000)
(*a*) continuously increases (*b*) continuously decreases
(*c*) increases up to certain limit and then decreases
(*d*) decreases up to certain limit and then increases

37. Khaif crops are sown and harvested in the months of:
(*a*) January and May (*b*) July and January
(*c*) April and September (*d*) October and March

38. Base period is .... than its crop period. (*AMIE s*' 2000)

39. Sugarcane according to croppendod is .... crop. (*AMIE s*' 2000)

40. In a zero of aeration, soil has pores filled with: (*AMIE s*' 2000)
(*a*) water (*b*) water and air (*c*) air (*d*) soluble salts

41. Monthly pan evaporation in a given area is 100 mm. For a given crop having crop coefficient as 0.70, the average daily consumptence use on the basis of pan evarporation data is: (*AMIE s*' 98)
(*a*) 2.33 mm (*b*) 70 mm (*c*) 700 mm (*d*) 35 mm (*a*)

42. For a given soil: (*AMIE s*' 94)
(*a*) texture and structure represent same property

(*b*) texture implies presence of particles of different sizes
(*c*) farmer can change the texture using cultivation method
(*d*) texture implies arrangements of a given size of particles

43. Salted soils are classified more rationally on the basis of: (*AMIE s*' 98)
(*a*) concentration of dissolved salts (*b*) pH value
(*c*) pH value and concentration of dissolved salts
(*d*) concentration of dissolved salts, pH value and presence of sodium ions

44. For the applicability of Darcy's law for flow through porous material, Reynolds number should be: (*AMIE s*' 98)
(*a*) < 1.0 (*b*) > 500
(*c*) 1 – 500 (*d*) a fixed value of 500

3. IRRIGATION SYSTEM

1. Inundation irrigation system requires construction of: (*AMIE w*' 96)
(*a*) only a diversion structure (*b*) only a canal regulation
(*c*) neither a diversion structure nor a canal regulation
(*d*) both a diversion structure and a canal regulator (*d*)

2. In an inundation irrigation system, there will be: (*AMIE s*' 94)
(*a*) neither a diversions work nor a regulator
(*b*) only a diversion structure across the river
(*c*) both a regulator on conel as well as a diversion structure across the river (*c*)

3. Canals taken directly from rivers with an without head regulators are called .... (Inundation canals) (*AMIE w*' 99)

4. Energy will be required to irrigate land (*AMIE w*' 2000)
(*a*) under gravity (*b*) much above river bed level
(*c*) in delta region (*d*) in all above cases (*b*)

5. Lift irrigation is flow: (*AMIE s*' 97)
(*a*) by gravity (*b*) from lower level to higher level
(*c*) by percolation (*d*) through sprinkler heads (*b*)

6. Where water must be raised by pumps or other means to irrigate crops is called .... irrigation. (lift) (*AMIE s*' 93)

7. Gravity irrigation is:
(*a*) the same a flow irrigation (*b*) the same as canal irrigation
(*c*) the same as well irrigation (*d*) the same as canal and well irrigation (*a*)

8. Perennial irrigation is:
(*a*) the supply of water for all the twelve months
(*b*) the supply of water throughout the crop season
(*c*) the supply of water throughout the crop period (*c*)

9. A well is best located in the command area:
(*a*) at lowest point to have minimum digging
(*b*) in the heart of the area at the highest point
(*c*) in the heart of the area at lowest point
(*d*) at any converiant location from accessibility
(*b*)

10. Tubewells located close by give:
(*a*) maximum discharge individually (*b*) minimum discharge individually
(*c*) same discharge as of individually (*d*) none of the above (*b*)

11. The most wasteful method of irrigation is:
(*a*) wild flooding (*b*) controlled flooding

(*c*) furrow irrigation (*d*) basin flooding (*a*)

12. The most efficient method of irrigation is:
(*a*) sprinkler irrigation (*b*) drip irrigation
(*c*) surface irrigation (*b*)

13. Transmissibility of an confined aquifer having its thickness 15 m and permeability 8×10^{-4} m/s is given by: (*AMIE s*' 98)
(*a*) 1.2×10^{-2} m²/s (*b*) 12×10^{-2} m²/s (*c*) 120×10^{-2} m²/s (*d*) none of the above (*a*)

14. The volume of water that can be extracted by force of gravity from a unit volume of aquifer material is known as: (*AMIE s*' 96)
(*a*) specific capacity (*b*) specific yield (*c*) specific retention (*d*) specific storage (*b*)

15. For unconfined aquifer the storage with unit: (*AMIE s*' 2000)
(*a*) is essentially same as specific yield (*b*) does not exist
(*c*) is essentially same as specific retention (*d*) is essentially same as possity

16. An outesian squifer is: (*AMIE s*' 2000)
(*a*) unconfired aquifer (*b*) a confined aquifer
(*c*) a water table aquifer (*d*) same as perched water table

17. Specific yield of an unconfired aquifer is: (*AMIE w*' 97)
(*a*) less than porosity (*b*) equal to porosity
(*c*) greater than porosity (*d*) unrelated to porosity (*a*)

18. If an unconfined aquifer rebound 5 m^2 of water for a water table drop of 2 m over horizontal area of 10 m^2, the specific yield is .... (*AMIE s*' 2000)

19. Soil with normal yield has pH value:
(*a*) below 5 (*b*) between 7.5 – 8.5
(*c*) between 8.5 – 9.0 (*d*) none (*c*)

20. The specific capacity of a well is defined as: (*AMIE w*' 98)
(*a*) drawdown per unit discharge (*b*) discharge per unit drawdown
(*c*) discharge per given drawdown (*d*) drawdown per given discharge

21. Ground water is free from: (*AMIE s*' 98)
(*a*) suspended solids but contains dissolved solids
(*b*) harmful bacteria but contain harmless bacteria
(*c*) floating impurities but have suspended imparities
(*d*) dissolved solids but have colloidal solids

22. A well of radius 100 mm penetrates fully confined aquifer of influence is 300 m in both cases, the percentage finalise in discharge is approximately: (*AMIE s*' 99)
(*a*) 25.5% (*b*) 20.5% (*c*) 15.5% (*d*) 9.5%

23. The usual capacity of an irrigation tubewell is .... cumec:
(*a*) 0.056 (*b*) 0.56 (*c*) 0.65 (*d*) none of them (*a*)

24. The optimum capacity of an irrigation tube well is .... cumec:
(*a*) 0.10 (*b*) 0.08 (*c*) 0.07 (*d*) 0.05 (*c*)

25. A good design of irrigation tubewell aims at:
(*a*) efficient utilisation of aquifer (*b*) long useful life
(*c*) low initial, operation and maintenance cost
(*d*) all of the above (*d*)

26. Blind pipe in a tubewell is:
(*a*) closed at both ends (*b*) has holes in its sides
(*c*) collects sand in it (*d*) non-water collection content (*d*)

27. Blind pipe in a tubewell is provided against in the hole:
(*a*) aquifer (*b*) aquifer (*c*) aquiclude (*d*) all of them (*a*)

28. Blind pipe in a tubewell is provided against .... station:
    (*a*) clayey (*b*) very fine sand
    (*c*) mixture of clay and very fine sand (*d*) none of them (*a*)
29. A confined aquifer has a longer of .... soil above it:
    (*a*) permeable (*b*) impermeable (*c*) semi-permeable (*d*) all of them (*b*)
30. An unconfined aquifer extends on top up to:
    (*a*) ground surface (*b*) water table (*c*) root zone of crops (*d*) none of them (*b*)
31. Aquichirde is a: (*AMIE s*' 97)
    (*a*) saturated bed through which appreciable leakage occurs
    (*b*) saturated bed through which no appreciable leakage occurs
    (*c*) bed that neither stores nor transmits appreciable quantity of water
    (*d*) none of them (*b*)
32. An artesian aquifer is: (*AMIE s*' 95)
    (*a*) an unconfined aquifer (*b*) leaky aquifer
    (*c*) perched aquifer (*d*) confined aquifer (*d*)
33. An artesian tubewell is eminently suitable for area:
    (*a*) plain (*b*) hilly (*c*) high water table (*d*) terai (*d*)
34. An artesian tubewell is run by:
    (*a*) electric pump (*b*) hand pump (*c*) automatically (*d*) all of them (*c*)
35. The yield of a tubewell can be increased in direct proportion to:
    (*a*) diameter (*b*) depression head (*c*) stronia length (*d*) all of them (*c*)
36. Development of a tubewell is necessitated to increase of tubewell:
    (*a*) depth (*b*) bore diameter (*c*) effective diameter (*d*) all of them (*c*)
37. Development of a tubewell is required in the case of tubewell:
    (*a*) new (*b*) sick (*c*) incrusted (*d*) all of them (*d*)
38. Development of a tubewell is done to:
    (*a*) remove coase material from the formation
    (*b*) remove fine material from the formation
    (*c*) obtain fine material living around the hole
    (*d*) all of the above (*b*)
39. Duty of an irrigation tubewell is .... than that of a canal irrigation:
    (*a*) higher (*b*) same (*c*) lower (*d*) none of them (*a*)
40. Duty of different irrigation tubewells is:
    (*a*) same (*b*) different (*c*) none of them (*b*)
41. The discharge per unit drawdown at a well is known as: (*AMIE w*' 97)
    (*a*) specific yield (*b*) specific discharge
    (*c*) specific capacity (*d*) specific transmissibility (*c*)
42. Specific yield of an unconfined aquifer is: (*AMIE w*' 98)
    (*a*) less than porosity (*b*) equal to porosity
    (*c*) greater than porosity (*d*) unrelated to porosity (*a*)
43. A 30 cm diameter well completely penetrates a confined aquifer. Under a steady state of pumping for a drawdown of 3m, the radius of influence is 300 m, the discharge observed was 1550 lpm. If the drawdown is increased to 4.5m, the discharge will be .... :
    (2325) (*AMIE s*' 98)
44. In the pumping well analysis of an unconfined aquifer, depicts assumptions fail: (*AMIE s*' 96)
    (*a*) in the imanediate neighbourhood of pumping well
    (*b*) for away from pumping well
    (*c*) every where rcif observation wells are present (*d*) nowhere (*a*)
45. In a strata as of gravel or coarse sand, the most suitable type of open well from the construction

point of view requires: (*AMIE s*' 94)

(*a*) an imperious living (*b*) a pervious living

(*c*) no living at all (*d*) partially impervious and balance no living

(*b*)

4. IRRIGATION METHODS

1. There is minimum wastage of water in: (*AMIE s*' 94)
   (*a*) furrow irrigation (*b*) check basin irrigation
   (*c*) sprinkler irrigation (*d*) border method of irrigation (*c*)
2. The loss of water during the growth of ceap is termed as .... (*AMIE s*' 99)
3. The irrigation method suitable for cotton, sugarcane, potatoes etc. is:
   (*a*) border strip method (*b*) furrow method
   (*c*) basin method (*d*) all of them (*b*)
4. The irrigation method suitable for wheat, leafy, vegetables, fodder etc. is:
   (*a*) border strip method (*b*) furrow method
   (*c*) basin method (*d*) all of them (*a*)
5. In sub-surface irrigation, water is applied beneath the ground surface at some depth usually .... cm
   (*a*) 200 – 100 (*b*) 100 – 75 (*c*) 75 – 30 (*d*) 30 (*c*)
6. In sub-surface irrigation, moisture move upwards through .... to meet irrigation requirements:
   (*a*) upward movement of water (*b*) capillary action
   (*c*) (*d*) all of them (*b*)
7. For successful sub-surface irrigation, a relatively imperious layer at .... m in substraction to prevent deep percolation:
   (*a*) 4–5 (*b*) 3–4 (*c*) 2–3 (*d*) 2–1 (*c*)
8. For successful sub-surface irrigation, subsoil water table is within .... in below ground surface:
   (*a*) 4 – 5 (*b*) 3 – 4 (*c*) 2 – 3 (*d*) 2 – 1 (*c*)
9. In a good sub-surface irrigation system, the efficiency is:
   (*a*) 45 – 50 (*b*) 55 – 60 (*c*) 60 – 65 (*d*) 70 – 75 (*d*)
10. The surface irrigation is essentially supplemented with efficient .... system. (water disposal)
11. The surface irrigation methods involve .... use of water compared to better irrigation methods:
    (*a*) efficient (*b*) better (*c*) wasteful (*d*) none of them
12. The commonly adopted method amongst various surface irrigation methods is:
    (*a*) wild flooding (*b*) controlled flooding
    (*c*) check flooding (*d*) centour furrow (*b*)
13. For irrigating orchards the commonly adopted surface irrigation method is:
    (*a*) controlled flooding (*b*) furrow method
    (*c*) basin method (*d*) contom furrow method (*c*)
14. The best method for irrigating undulating sandy land where uniform application is required with water of poor availability is .... irrigation:
    (*a*) sub surface (*b*) drip (*c*) sprinkler (*d*) surface (*c*)
15. For the conservation of water up to 50 percent, the best irrigation method is:
    (*a*) drip (*b*) sprinkler (*c*) furrow (*d*) basin (*b*)
16. The highest water saving method of irrigation is:

(a) sprinkler (b) drip (c) sub-surface (d) basin (b)

17. The drip irrigation method has water application efficiency of .... percent:
(a) 60 (b) 70 (c) 80 (d) 90 (d)
18. The sprinkler irrigation method has water application efficiency of .... percent:
(a) 60 (b) 70 (c) 80 (d) 90 (b)
19. Limited metting of soil surface to consence water is possible with .... irrigation method:
(a) basin (b) furrow (c) drip (d) sprinkler (c)
20. Localised application of fertilizer is possible with .... irrigation method:
(a) basin (b) furrow (c) drip (d) sprinkler (c)
21. The irrigation method which entails minimum evaporation losses is:
(a) basin (b) furrow (c) drip (d) sprinkler
22. The most economical but highly wasteful method of irrigation is:
(a) basin (b) furrow (c) wild flooding (d) sprinkler (c)
23. The irrigation method which merits in low head requirement than drip irrigation method is:
(a) sprinkler (b) sub surface (c) suction (d) none of them (c)
24. The automatic, uniform and controlled water supply to root zone of a plant is possible with .... irrigation method:
(a) drip (b) sprinkler (c) basin (d) suction (d)
25. Maximum water application efficiency is in: (*AMIE s'* 99)
(a) surface irrigation (b) lift irrigation (c) sprinkler irrigation (d) furrow irrigation (c)

5. UNLINED CANALS

1. Irrigation canals are generally aliquied along: (*AMIE s'* 98)
(a) ridge line (b) condom line (c) vollay line (d) straight line (a)
2. Value of n tends with water depths over z m because of .... (increase) (increasing turbulence).
3. Value of n tends to .... for very small channels .... depth because of .... (< 0.75 m) (relative increase in wall roughness).
4. For medium silt whose average gram size is 0.16 mm, lacey's silt factor is likely to be: (*Eng. Sec. Ex.* '97)
(a) 0.30 (b) 0.45 (c) 0.70 (d) 1.32 (c)
5. Canal design practices depend on:
(a) soil formation of bed and banks, (b) sediment transport characteristics
(c) operational needs (d) capacity
(e) economic considerations (f) all of them (f)
6. For 1 m depth of flow in a canal, with CVR equal to 0.90, the velocity of flow according to Kennedy's theory will be .... m/s. (*AMIE s'* 98)
7. Canal efficiency suffers from:
(a) silting and scouring (b) fanlty design slope/silt factor
(c) weed growth (d) all of these (d)
8. A stable canal section is that which neither silts nor scours with the given:
(a) discharge (b) water surface slope
(c) silt charge (d) all of the above (d)
9. According to Kennedy, non-silting and non-scouring velocity is called .... (critical velocity) (*AMIO w'* 01)
10. Hydranlic mean depth is the ratio between .... and .... (Area/wetted perimeter) (*AMIE w'* 98)
11. Kennedy mean velocity V_0 is given by:

(*a*) 0.55 m$D^{0.64}$ (*b*) 0.64 m $D^{0.55}$ (*c*) 0.55 m $D^{0.84}$ (*d*) 0.84 m $D^{0.64}$

12. Lacey's silt factor *f* is given by:
(*a*) $1.76\,(m_r)^{1/2}$ (*b*) $1.76\,(m_r)^{3/4}$ (*c*) $1.76\,(m_r)^{4/5}$ (*d*) none of these (*a*)

13. Rugosity coefficient *N* increases from minimum in:
(*a*) as the canal section gets old (*b*) poor maintained section
(*c*) Badly maintained sections (*d*) all of the above (*d*)

14. Wetted perimeter of an unlined canal as per Lacey is given by:
(*a*) $4.75\,Q^{0.5}$ (*b*) $5.75\,Q^{0.5}$ (*c*) $4.75\,Q^{0.75}$ (*d*) $4.75Q^{0.8}$

15. Critical velocity ratio, V/V_0, is not a dominant factor in the unlened canals taking off from:
(*a*) reservoirs (*b*) river (*c*) stream (*d*) all of the above
(*a*)

16. From head towards tail of a canal the kennedy CVR, V/V_0:
(*a*) increases (*b*) decreases (*c*) remains the same (*b*)

17. Kennedy CVR, V/V_0, value below critical implies canal:
(*a*) in require (*b*) silting tendency (*c*) scouring tendency (*b*)

18. Lacey's require theory was developed after extensive research in: (*AMIE* w' 00)
(*a*) Issael (*b*) Indonesia (*c*) India (*d*) Iraq (*c*)

19. Kennedy CRV, V/V_0, for fine sands compared to coaeser sand than the standard is:
(*a*) higher (*b*) the same (*c*) lower (*c*)

20. Kennedy CVR, V/V_0, ranges from head to tail as:
(*a*) 1.1 – 0.8 (*b*) 0.8 – 1.1 (*c*) 1.1 all through (*d*) 0.8 all through (*a*)

21. Compared to kennedy theory, a channel designed by Lacey's theory is:
(*a*) wides and shallow (*b*) same section (*c*) narrow and deep (*a*)

22. Kennedy adopted rugosity coefficient *N* as:
(*a*) 0.020 (*b*) 0.0225 (*c*) 0.025 (*d*) 0.0275 (*b*)

23. Compared to coarser silt, canal section for finer silt by lacey theory is:
(*a*) narrow and deeper (*b*) the same (*c*) shallow and wide (*a*)

24. Lacey considered channel section:
(*a*) rectangular (*b*) trapezoidal (*c*) aim-elliptical (*d*) elliptical (*c*)

25. Channels taking off from a reseveroir with low silt content and passing through hard soils or rocks are usually designed as per:
(*a*) Lacey theory (*b*) Kennedy theory
(*c*) Chezye–Kutter formulae (*d*) Garret diagrams
(*e*) Meanings formula (*c*)

26. Channels taking off from barrages or weirs carrying high sediment content are usually designed as per:
(*a*) Lacey theory (*b*) Kennedy theory
(*c*) Chezy-kutter formulae (*d*) Garret diagrams (*a*)

27. Incised unlined channels through non-alluvial soils carrying low silt content are usually designed by:
(*a*) Lacey theory (*b*) Kennedy theory
(*c*) Manning's formulae (*d*) Chezy-Kutter formulae (except *a*)

28. In Kennedy's equation for non-silting, non-scouring velocity, the critical velocity ratio was introduced to take into consideration the effect of canals taking off from a reservoir are usually designed for a discharge of .... percent dependability: (*AMIE* w' 97)
(*a*) 60% (*b*) 75% (*c*) 90% (*d*) 100% (*b*)

29. Hydel channels are usually designed for a discharge .... percent dependability.
(*a*) 60% (*b*) 75% (*c*) 90% (*d*) 100% (*c*)

30. A contom canal is generally aliquied:
(*a*) at right angle to contoms (*b*) conforming to contoms

(*c*) at watersheed line (*d*) none of the above (*b*)

31. A side slope channel is generally alquied:
(*a*) at right angle to contoms (*b*) conforming to contoms
(*c*) at watershed line (*d*) none of the above (*a*)

32. A protective canal is that which:
(*a*) has protected (from erosion) section
(*b*) provides protection against famine in command area
(*c*) yields protected (assumed) net revenue (*b*)

33. A productive canal is that which:
(*a*) is a means for enhancing production (agricultural) in the area
(*b*) is a means for hushing in economic production in the area
(*c*) earns net revenue to the state (*d*) none of the above (*b*)

34. Berm width in a canal in filling i.e., bed level higher than ground level is:
(*a*) 1D (*b*) 1.5 D (*c*) 2D (*d*) 3D
where *D* is designed water depth (*d*)

35. Berm width in a main/branch canal in cutting is:
(*a*) 1D (*b*) 1.5D (*c*) 2D (*d*) 3D (*b*)

36. Berm width in a distributing/minor in partial cutting is:
(*a*) 1D (*b*) 1.5D (*c*) 2D (*d*) none of them (*c*)

37. Lacey's theory is applicable to flow: (*AMIE s*' 97)
(*a*) through pipes (*b*) over spillways
(*c*) in allurial rivers and canal (*d*) in concrete lined canals (*c*)

38. Critical velocity ratio is used in connection with design of: (*AMIE s*' 97)
(*a*) allurial canals (*b*) earth dams (*c*) concrete lined only (*d*) barrages (*a*)

39. The average particle size of silt in mm is m, then Lacey''s silt factor *f* is proportional to: (*AMIE w*' 94)
(*a*) m^2 (*b*) $mm^{1/2}$ (*c*) $m^{1/2}$ (*d*) $m^{1/3}$ (*c*)

40. The coefficient of roughencess for an earthen canal in excellent condition is: (*AMIE s*' 94)
(*a*) 0.01 (*b*) 0.0225 (*c*) 0.025 (*d*) 0.7 (*b*)

41. The coefficient of roughness for an unlined canal:
(*a*) is directly proportional to discharge (*b*) is inversely proportional to discharge
(*c*) has no relation to discharge capacity (*d*) more of the above (*b*)

42. The scour depth for a channel in fine silt is .... than in coarse silt. (more) (*AMIE w*' 93)

43. The bed slope of a canal offtaking from a barrage is kept: (*AMIE w*' 93)
(*a*) steeper than the bed slope of the river (*b*) milder than bed slope of the river
(*c*) same as that of the river (*d*) none of the above

44. The bed slope of a canal:
(*a*) is directly proportional to discharge capacity
(*b*) is inversely proportional to discharge capacity
(*c*) has no relation with discharge capacity (*d*) none of the above (*b*)

45. According to Kennedy, non-silting and non-scouring velocity is called ....

46. Larcey's silt factor *f* for the average size of river bed material of 0.3 mm is: (*AMIE s*' 01)
(*a*) 0.96 (*b*) 1.0 (*c*) 1.2 (*d*) 0.3

47. Weed growth in unlined canal is .... if velocity of water is less than 60 cm/s. (possible) (*AMIE s*' 2000)

48. Percolation losses are more from a canal that is .... (*AMIE s*' 99)

49. Lacey's theory is applicable only to .... rivers and not .... (*AMIE s*' 99)

50. From canal water loss point of view, lining should preferably be done to:

(*a*) main canals and branches (*b*) distributions
(*c*) minors (*d*) depending on discharge capacity (*a*)

51. The usual water loss in an unlined canal systems is:
(*a*) 55% (*b*) 45% (*c*) 35% (*d*) 25% (*c*)

52. Canal should be .... in order to reduce the chances of waterlogging. (*AMIE* w' 01)

53. Seepage losses in a canal are more when it is: (*AMIE* s' 01)
(*a*) completely in filling (*b*) partly in filling and partly in cutting
(*c*) fully in cutting (*d*) lined

54. Current meters are used for measurement of: (*AMIE* s' 01)
(*a*) velocity (*b*) discharge (*c*) depth of flow (*d*) scour depth (*a*)

55. A canal is 80 km long and has an average surface width of 15 m. If evaporation measured in a class *A* pan is 0.5 cm/day, the volume of water evaporated in a month of 30 days is .... m^3. (1.25×10^5) (*AMIE* s' 96)

56. In order to keep saturation line in a canal bank well within the toe, it is necessary to provide: (*AMIE* w' 96)
(*a*) spoil bank (*b*) counter berm (*c*) free board (*d*) berm (*b*)

57. Hydranlic gradient line through a good clayey soil bank is considered as:
(*a*) 6 : 1 (*b*) 5 : 1 (*c*) 4 : 1 (*d*) 3 :1 (*c*)

58. Hydranlic gredrient line through a silty clay and clayey gravel soil bank is taken as:
(*a*) 6 : 1 (*b*) 5 : 1 (*c*) 4 : 1 (*d*) 3 : 1 (*a*)

59. Berm is a space provided at:
(*a*) F.S.L (*b*) bed level (*c*) natural ground (*d*) none of the above (*a*)

60. Counter berm or purshta is provided to:
(*a*) strengthen the bank in higher filling (*b*) prevent bank from slipping
(*c*) cover saturation line through to prevent seepage above
(*d*) to cater for all of the above (*c*)

61. Spoil bank is provided to:
(*a*) strengthen the bank (*b*) prevent bank from slopping
(*c*) disposal of surplus excanation from canal section
(*d*) to cater for all of the above (*c*)

62. From a canal section, losses due to seepage and evaporation:
(*a*) are the same (*b*) former are more than the latter
(*c*) former are less than the latter (*b*)

63. Discharge in a canal section is proportional to:
(*a*) $D^{1/2}$ (*b*) $D^{3/4}$ (*c*) $D^{5/4}$ (*d*) $D^{5/3}$ (*d*)

64. Mean velocity at a vertical in canal section is approximately at .... depth from the top:
(*a*) 0.5 (*b*) 0.6 (*c*) 0.75 (*d*) 0.8 (*b*)

65. Seepage losses in a canal are more when it is: (*AMIE* s' 01)
(*a*) completely in filling (*b*) partly in filling and partly in cutting
(*c*) fully in cutting (*d*) lined (*c*)

66. The chemical dilution method of stream ganging is ideally suited for measuring discharges in: (*AMIE* s' 94)
(*a*) a large allervial river
(*b*) steady flow in a small turbulent stream (*c*) a tidal estuary
(*d*) a stretch of river having heavy pollution
(*c*)

6. LINED CANALS

1. Canals are lined with the consideration of:
   (*a*) reduction in seeparge losses
   (*b*) improvement in discharging capacity of an existing canal
   (*c*) prevention of waterlogging
   (*d*) reduced section with consequent economy in excavation, land acquisition etc.,
   (*e*) all of the above (*e*)
2. Requirements of canal lining are:
   (*a*) highly impervious (*b*) hydranlically efficient
   (*c*) stable and drouable section (*d*) resistant to abrasion caused by sediments
   (*e*) all of the above (*e*)
3. Current concrete lining is .... impervious than tile lining.
   (*a*) more (*b*) equally (*c*) less (*a*)
4. The selection of section of line canal is governed by:
   (*a*) economy (*b*) higher velocity
   (*c*) higher sediment carrying capacity (*d*) discharge capacity
   (*e*) all of the above (*e*)
5. For a canal discharge of up to 85 cumecs, the lined section usually adopted is:
   (*a*) rectangular (*b*) trapezoidal (*c*) circular (*d*) horse shoe (*c*)
6. For a canal discharge above 85 cumecs, the lined section usually adopted is:
   (*a*) rectangular (*b*) trapezoidal (*c*) circular (*b*)
7. Lining of a canal is required when: (*AMIE w*' 93)
   (*a*) water table is above F.S.L. of canal (*b*) water table is below bed of canal
   (*c*) there is a natural impervious stration below the bed
   (*d*) none of the above (*d*)
8. For couraying the same discharge, a lined canal can have a bed slope .... then unlined canal: (*AMIE s*' 99)
   (*a*) steeper (*b*) milder (*c*) same (*d*) none of above (*b*)
9. For higher silt carrying capacity, the lined canal section should be:
   (*a*) trapezoidal (*b*) crip shaped (*c*) rectangular (*d*) horse shoe (*b*)
10. For achieving minimum cross sectional area and correspondingly cost of lining, the lined canal section should be:
    (*a*) trapezoidal (*b*) crip shaped (*c*) rectangular (*d*) horse shoe (*b*)
11. The structurally more stable lined canal section is:
    (*a*) trapezoidal (*b*) inverted arch section
    (*c*) rectangular (*d*) horse shoe (*b*)
12. In high cost land, a lined canal of limited capacity may have a section:
    (*a*) trapezoidal (*b*) crip shaped (*c*) rectangular (*d*) horse shoe (*b*)
13. The minimum free board recommended for lined canals carrying a discharge greater than 10 cumecs is: (*AMIE w*' 95)
    (*a*) 0.15 m (*b*) 0.60 m (*c*) 1.25 m (*d*) 0.75 m (*d*)
14. The side slope of lined section usually adopted is selected from the consideration of:
    (*a*) easy conduction (*b*) minimum
    (*c*) minimum evaporation losses
    (*d*) approximately the same as the angle of repose of the natural soil through which the canal passes (*d*)
15. Indicate type of lining for permissible maximum velocity of:
    (*a*) 2 m/s (*b*) 1/75 m/s (*c*) 1.5 m/s

(*a*) current concrete (*b*) tile linings (*c*) boulder lining

16. Drainage behind canal lining is required:
 (*a*) to eater for the drainage of canal water
 (*b*) to eater for excess hydranlic pressure behind the lining
 (*c*) when subgode is free draining (*d*) all of the above (*b*)
17. Pressure release valves are provided in canal lining:
 (*a*) to release excess pressure built up in canal section
 (*b*) to release excess hydrostatic pressure built behind the lining
 (*c*) when subgrade is free daining (*d*) for all the above (*b*)

7. DRAINAGE ENGINEERING

1. Indicate True or False: (*AMIE s' 98*)
 (*a*) Drainage of irrigated land is as essential as irrigation (True)
 There are areas where irrigation may not be required during monsoons but drainage is required (True)
 (*b*) Land drainage is an integral part of canal irrigation (True)
 (*c*) Land with subsoil water table within 1.5 m of natural ground surface during monsoon is called waterlogged
 (False)
 (*d*) Land with subsoil water table within 1.5 m of natural ground surface just prior to onset of monsoon is taken as waterlogged (True)
2. Leaching is a process by which: (*AMIE w' 98*)
 (*a*) water is excessively pumped from a waterlogged area
 (*b*) excess salt in the soil is pushed down by standing water above the soil
 (*c*) improving the soil structure and texture by adding suitable chemicals
 (*d*) sodium ions are removed from irrigation water
3. A land is sand to be waterlogged when:
 (*a*) water table rises to an extent that soil pores in root zone of crop becomes saturated
 (*b*) high water table restructs normal circulating of air
 (*c*) there is a decline in level of oxygen and increase in level of carbon dioxide
 (*d*) all of the above (*d*)
4. A land is considered prone to waterlogging when water table is:
 (*a*) within 1.5 m of ground surface (*b*) within 2 m of ground surface
 (*c*) within 3m of ground surface (*d*) all of the above (*b*)
5. Critical areas from waterlogging point of view one these where water table is:
 (*a*) within 1.5 m (*b*) within 2 m (*c*) within 2.3 m (*d*) all of above (*c*)
6. Land is sand to be waterlogged when: (*AMIE w' 98*)
 (*a*) gravity drainage is ceased (*b*) permanent wilting point is reached
 (*c*) salinity of soil increases (*d*) capillary fringe reaches roat zone of plants

 (*c*)
7. Mole drains are useful in equalising water levels between ditches far:
 (*a*) drainage (*b*) sub-irrigation
 (*c*) both drainage and sub-irrigation (*d*) none of them (*c*)
8. Canals meant for the purpose of draining off water from waterlogged areas are called: (*AMIE w' 99*)
 (*a*) seeparge canals (*b*) laterals (*c*) drains (*d*) watercomses (*c*)
9. Waterlogging in irrigated areas is due to:

(*a*) seepager from unlined canals and watercourses
(*b*) shift from protective to intensive irrigation
(*c*) inadequate drainage (*d*) all of the above (*d*)

10. Waterlogging seduces crop yield due to: (*AMIE s*' 98)
(*a*) lack of aeration in root zone of crop
(*b*) restriction in roots development due to low soil temperature
(*c*) creation of salinisation (*d*) all of above (*d*)

11. The discharge capacity of a drain is maximum at:
(*a*) outfall (*b*) head as in a canal
(*c*) any point along its length depending on topography
(*d*) all of above (*a*)

12. Rugosity coefficient, *n*, usually adopted in trapezoidal surface drains is:
(*a*) 0.018 (*b*) 0.020 (*c*) 0.025 (*d*) 0.030 (*d*)

13. Ideal minimum bed gradrint in an open drain is to have sufficient velocity at:
(*a*) low flows (*b*) mean flows (*c*) high flows (*d*) all of the above
(*a*)

14. For preventing silt deposition in a surface earthen drain. The minimum velocity should be .... m/s.
(*a*) 0.2 (*b*) 0.3 (*c*) 0.5 (*d*) 0.6 (*b*)

15. For preventing weeds and grass growth, the minimum velocity in an earthen drain should be .... m/s
(*a*) 0.3 (*b*) 0.6 (*c*) 0.8 (*d*) 1.0 (*b*)

16. For inhibiting growth of aquiatic plants, the minimum velocity in an earthen drain should not be less than .... m/s.
(*a*) 0.3 (*b*) 0.5 (*c*) 0.75 (*d*) 1 (*c*)

17. For lowering water table to greater depths, the effective drainage method is:
(*a*) open drain (*b*) mole drains (*c*) closed drain (*d*) all of them (*c*)

18. The alignment of surface drain should follow:
(*a*) sidge line (*b*) drainage line (*c*) shortest route (*d*) all of them (*b*)

19. Waterlogging takes place due to .... in ground water table.
(*AMIE s*' 2000)

20. Waterlogging is eliminated by (*AMIE s*' 97)
(*a*) deep ploughing (*b*) shallow ploughing
(*c*) irrigation (*d*) providing tile drains (*b*)

21. A land is known as waterlogged when: (*AMIE w*' 95 & *s*' 94)
(*a*) pH of the soil reaches a value of 8.5 (*b*) soil in root zone has high salinity
(*c*) soil pores in root zone are saturated with water
(*d*) there is water flowing on the land surface (*c*)

22. Alkali soils are redaimed by: (*AMIE s*' 94)
(*a*) leaching only (*b*) addition of gypsum and leaching
(*c*) addition of gypsum only (*d*) provision of drainage (*b*)

23. Canals should be .... in order to reduce the chances of waterlogging. (lined)
(*AMIE* 2001)

24. The land is said to be waterlogged, if the soil pores within: (*AMIE s*' 98)
(*a*) a depth of 40 cm is saturated (*b*) a depth of 60 cm is saturated
(*c*) root zone of crops is saturated (*d*) soil up to ground water table is saturated

25. Gypsum is added to improve drainage characteristics of: (*AMIE w*' 2000)
(*a*) saline soils (*b*) alkali soils
(*c*) soi!s not affected by sodium salts (*d*) none of the above (*a*)

26. The anti-waterlogging measure making use of deep rooted plants is called:
(*a*) agro-drainage (*b*) bio-drainage (*c*) deep drainage (*d*) none of them (*b*)
27. The most effective reclamation measure for areas with high salt content throughout the profile or saline underground water is .... drainage:
(*a*) surface (*b*) sub-surface (*c*) vertical (*d*) none of them (*b*)
28. Drainage is .... required when irrigation is extended to delta areas:
(*a*) not (*b*) essentially (*c*) sometimes (*c*)
29. Rice cultivation with abundant of water standing in the fields ....
(*a*) adds to waterlogging (*b*) is a land reclamation measure
(*c*) none of them (*d*) (*b*)
30. The drainage coefficient (D.C.) largely depends upon the rainfall. Of the effective rainfall of 24 h, the recommended value of D.C. is .... %: (*AMIE w' 94*)
(*a*) 0.5 (*b*) 1.0 (*c*) 1.5 (*d*) 2.0 (*b*)

8 & 9. MECHANICS & RIVER TRAINING

1. Rivers are classified on the basis of:
(*a*) discharging capacity (*b*) length
(*c*) waterway (*d*) location (*d*)
2. Increased river is a type of river:
(*a*) formed by the process of degradation (*b*) formed in the boulder stage of river
(*c*) formed by the process of aggradation (*d*) none of the above (*a*)
3. Himalayan rivers flow through:
(*a*) non-allurial soils (*b*) allurial soils (*c*) rocky soils (*d*) all of the above (*b*)
4. Non-Himalayan rivers are:
(*a*) snow-fed (*b*) perennial
(*c*) carry low sediment load (*d*) pass through non-friable rocks (*c*)
5. On a meandering river the current is strongest on: (*AMIE w' 00*)
(*a*) concave bend (*b*) concave band
(*c*) both banks (*d*) middle of the stream (*a*)
6. Toruosity of a meandering river is the ratio between .... and ....
(*AMIE w' 98*)
7. Meandering of a river is due to: (*AMIE s' 98*)
(*a*) sediment load of streams
(*b*) discharge and hydraulic properties of streams
(*c*) erodibility of the bed and banks of streams
(*d*) all the above (*c*)
8. High water training implies river training for:
(*a*) high flood discharge (*b*) high sediment transport
(*c*) high changes of river bed conditions (*d*) all of the above (*a*)
9. Low water training implies river training for:
(*a*) high flood (*b*) medium flood (*c*) low flood (*d*) project flood (*c*)
10. Longitudinal river training works are preferred to projecting works in rivers:
(*a*) carrying small bed loads (*b*) narrow river channel
(*c*) steep slopes and swift current (*d*) all of the above (*d*)
11. Projecting river training works are suitable for rivers:
(*a*) with stable bed (*b*) with unstable bed
(*c*) none of them (*b*)
12. Longitudinal river training works are provided for:

(a) guiding axis of flow (b) improving river alignment, trapping bed load
(c) protecting banks from erosion (d) all of them (d)

13. Guide banks are provided:
(a) in the direction of flow (b) all right angle to the flow
(c) at an angle to the flow (d) all of them (a)

14. Guide banks are provided to: (*AMIE w' 99*)
(a) decrease depth of river flow (b) corfine river width
(c) reduce flood peak (d) maintain minimum depth in river (b)

15. Guide banks are provided:
(a) in pairs (b) symmetrical in plan
(c) extending u/s and d/s of work (d) all of them (d)

16. Lengths of guide banks u/s and d/s of the work are:
(a) the same (b) more on u/s than on d/s
(c) less on u/s and more on d/s (d) as per site conditions (b)

17. The lengths of guide banks on upstream/downstream of the work of length L are:
(a) 1.5/0.5 (b) 1.0/1.0 (c) 1.5/0.25 (d) 1.25/0.5 (c)

18. Scour depth along guide banks is severer along:
(a) shank (b) nose
(c) transition from nose to shank (d) all of them (b)

19. A launching apron is a hydranlic structure provides safity against .... (scouring) (*AMIE s' 00*)

20. Scour depth is estimated by the formula:
(a) $0.473\,(q/f)^{1/3}$ (b) $0.473\,(q/f)^{1/3}$ (c) $0.473\,(q^2/f)^{1/3}$ (d) $0.473\,(q^{1/3}/f)$ (b)

21. Scour depth when looseness factor is < 1 is given by the relation
(a) $0.473\,(q/f)^{1/3}$ (b) $1.35\,(q^2/f)^{1/3}$ (c) $1.35\,(Q^2/f)^{1/3}$ (d) $1.35\,(q^{1/3}/f)$ (b)

22. Scour depth when looseness factor is > 1 is given by the relation
(a) $0.473\,(q/f)^{1/3}$ (b) $0.473\,(q/f)^{1/3}$ (c) $1.35\,(Q/f)^{1/3}$ (d) $1.35\,(q^2/f)^{1/3}$ (b)

23. The launched slope of apron is considered as:
(a) per site conditions (b) depending on flood discharge
(c) 2 : 1 (d) 3 : 1 (c)

24. Lacey's silt factor f for the average size of the river bed material of 0.3 mm is: (*AMIE s' 01*)
(a) 0.96 (b) 1.0 (c) 1.2 (d) 1.3

25. The scour depth is determined on the basis of diameter of silt encountered at:
(a) river bed (b) a certain depth below river bed
(c) at the depth of anticipated scour (d) at all of them (c)

26. The launched slope for apron of cement concrete blocks is assumed as:
(a) 2 : 1 (b) 1.5 : 1 (c) 3 : 1 (d) 1.25 : 1 (b)

27. The launched slope for apron of loose stones or boulders is assumed as:
(a) 2 : 1 (b) 1.5 : 1 (c) 1.25 : 1 (d) 3 :1 (a)

28. The width of apron usually adopted, with relation to scour depth D, is:
(a) $2D$ (b) $1.75\,D$ (c) $1.5\,D$ (d) $1.0\,D$ (c)

29. Spurs (groynes) are aligned on river bank in a direction
(a) normal to dominant flow discharge (b) at right angle pointing upstream
(c) at right angle pointing downstream (d) all of them (d)

30. Spurs serve to train a river in order to:
(a) reduce concentration of flow at the point of attack
(b) gruiding or deflecting axis of flow
(c) protecting on eroding bank tropping bed load

(*d*) all of them (*d*)

31. Spurs serve to:
(*a*) repel flow (*b*) attract flow (*c*) deflect flow (*d*) all of them (*d*)

32. Permeable spurs encourage:
(*a*) slow down of velocity for rapid sediment deposition
(*b*) water and sediment, transport capacity of the main river channel
(*c*) dropping of suspended and bed load deposition
(*d*) all of them (*d*)

33. Permeable spurs are suitable for:
(*a*) deep and narrow rivers (*b*) small and deep rivers
(*c*) light sediment carrying rivers (*d*) all of them (*a*)

34. Permeable spurs are suitable for:
(*a*) deep rivers carrying heavy sediment load
(*b*) deep rivers carrying light sediment load
(*c*) navigable rivers (*d*) small rivers with stop gradients (*a*)

35. Impermeable spurs:
(*a*) slow down velocity and promote sedimentation
(*b*) allow passage of flow through them
(*c*) attract, repel or deflect flow away from banks along the desired course
(*d*) effectively requlate the river course (*c*)

36. Permeable spurs are preferred than impermeansle spurs as they:
(*a*) cause less turbulence and eddy conditions
(*b*) effective in regulation of river courses
(*c*) make use of available construction materials (*d*) all of them
(*d*)

37. An artificial cut-off as a river training work:
(*a*) alters river regime violently
(*b*) aims at establishing ultimately hydranlic gradients which river can maintain
(*c*) improvement in flood carrying capacity (*d*) all of them (*d*)

38. Bandalling, as a river training work serves to:
(*a*) confine low river water flow in a single channel
(*b*) confine high river water flow in a single channel
(*c*) protect river bank (*d*) all of them (*a*)

39. Lerees are constructed .... to river flow: (*AMIE s*' 98)
(*a*) parallel (*b*) transverse (*c*) perpendicular (*d*) all of above (*c*)

40. Permeable spurs are used to: (*AMIE s*' 01)
(*a*) promote siltation along the banks of the river
(*b*) deflect the flow (*c*) attract the flow
(*d*) contract the width of the river for irrigation purposes (*c*)

41. River training works are seldom required in: (*AMIE s*' 2000)
(*a*) rock of stage of river (*b*) meandering river
(*c*) trough storge of river (*d*) deltaic stage of river (*c*)

42. Deflecting groynes are constructed on .... bank of a river. (*AMIE s*' 2000)

43. A groyne projecting from the river bank and with its nose pointing upstream is called a .... groyne (*AMIE w*' 00)

44. The depth of scour for a discharge intensity of 5 cumec/m and a silt factor of 1 is .... m (*AMIE s*' 97)

45. According to Gales the ratio of u/s length of guide bank to length of structure between abutments for flood dischanges more them 20,000 cumecs shall be:
(*a*) 1.3 (*b*) 1.4 (*c*) 1.5 (*d*) 1.6 (*c*)

46. A groyne pointing upstream in a river is known as: (*AMIE s*' 94)
(*a*) attracting groyne (*b*) repelling groyne (*c*) fending gropne (*d*) Denehy's gropne (*b*)
47. The scour depth as measured below highest flood level of a river will be .... in a bouldery river than an allervial river. (*AMIE s*' 99)
(*a*) (*b*) (*c*) same (*d*) none of them
48. The d/s expansion head of a guide bank is extended to subtend an angle at the centre between: (*AMIE w*' 94)
(*a*) 30° and 45° (*b*) 45° and 60° (*c*) 60° and 75° (*d*) 75° and 90° (*d*)
49. In a deflecting spur, the angle between the axis of the spur and the bank on u/s side is: (*AMIE w*' 93)
(*a*) acute (*b*) obture
(*c*) same as on d/s side (*d*) none of the above (*a*)
50. Repelling spur project from the bank and form an acute angle on .... the side (*AMIE s*' 99)
51. Spurs provided for the river training works are: (*AMIE w*' 01)
(*a*) parallel to the river banks (*b*) transverse to the river flow
(*c*) usually made of earth bends (*d*) provided to guide flow to a bridge (*b*)
52. A launching apron in hydranlic structure provides safety against .... (*AMIE s*' 01)
53. Equation given by Lacey is used for fixing the waterway of bridges and barrages in alluvial streams. ($4.83\,Q^{1/2}$) (*AMIE s*' 01)

10. HEADWORKS

1. Headworks is a hydranlic structure:
(*a*) appurtenant to a dam (*b*) to effect storage against raised crest
(*c*) to effect pondage with low set crest
(*d*) to serve as a safety valve to the canal head regulator (*c*)
2. Headworks on a river causes:
(*a*) aggradation on upstream (*b*) degradation in downstream
(*c*) feed offtaking canal (*s*) (*d*) pondage on upstream
(*e*) all of them (*e*)
3. A weir essentially comprises:
(*a*) high set crest (*b*) series of gates
(*c*) roadway across the river (*d*) all of them (*a*)
4. A barrage differs from a weir on account of:
(*a*) low set crest (*b*) effective control over river flow
(*c*) series of gates to effect pondage (*d*) all of them (*d*)
5. A weir essentially has:
(*a*) no control over high floods (*b*) more efflux
(*c*) shutters in part length only (*d*) all of them (*d*)
6. A barrage is a hydranlic structure provided on: (*AMIE w*' 00)
(*a*) deep narrow valleys (*b*) wide, shallow valley
(*c*) rock foundation (*d*) silt and clay and gravel (*d*)
7. The main function of diversion head works provided at the offtake of a canal from a river is: (*AMIE w*' 94)
(*a*) to raise water level in river (*b*) to control floods
(*c*) store water (*d*) control silt entry into canal (*c*)
8. Retrogression of d/s levels generally considered in the design of weirs or barrages is: (*AMIE w*' 94)

(*a*) higher at high river stage than at low water levels
(*b*) same at high stage and low water levels
(*c*) higher at low water levels stage than at high flood stage
(*d*) independent of the stage of flow (*c*)

(*a*)	Underslinces	(*A*)	regulates the supply of water in canal
(*b*)	canal head regulator	(*B*)	enables fish to migrate to their spawning grounds upstream of the barrage
(*c*)	Banage bays	(*C*)	preseric a clear and defined river channel approaching canal regulator
(*d*)	Fish ladder	(*D*)	are operated to release excess water during floods

(*a* – *C*), (*b* – *A*), (*c* – *D*), (*d* – *B*)

9. The design flood, for a permanent barrage commonly adopted in India is the: (*AMIE w*' 00)
(*a*) probable maximum flood (*b*) standard project flood
(*c*) flood with a return period of 1000 years (*d*) flood with a return period of 50 years
10. In a dirversion head works, the crest of underslince portion is kept: (*AMIE s*' 01)
(*a*) at minimum river bed level
(*b*) higher than crest level of canal head regulator
(*c*) at the same level as that of canal head regulator
(*d*) higher than the minimum bed level of the river
11. Fish laden is provided on the side of: (*AMIE w*' 96)
(*a*) divide wall (*b*) diaphragm wall (*c*) corewall (*d*) wingwall (*a*)
12. For a barrage, the exist gradient is independent of : (*AMIE w*' 96)
(*a*) applied water head (*b*) horizontal floor length
(*c*) u/s cutoff depth (*d*) d/s cutoff depth (*c*)
13. The term piping used in connection with diversion structures on permeable soil is associated with: (*AMIE w*' 96)
(*a*) drainage of seepage water (*b*) measurement of uplift pressure
(*c*) failure intionted by boiling (*d*) consolidation of foundation (*c*)
14. For consolidation grouting, depth of B-holes drilled through the foundation rock generally various between:
(*a*) 5 and 10 m (*b*) 10 and 15 m (*c*) 15 and 20 m (*d*) 20 and 25 m
15. According to Kherla, to keep the structure safe against piping, exit gradient to be provided lies between:
(*a*) 0.10 and 0.15 (*b*) 0.15 and 0.20 (*c*) 0.20 and 0.26 (*d*) 0.25 and 0.30 (*c*)
16. The characteristic feature of a barrage is: (*AMIE w*' 95)
(*a*) provision of raised crest (*b*) creation of storage reservoir on u/s
(*c*) provision of a series of gates across the river for flow regulation
(*d*) that is built in delta areas only (*c*)
17. Fish ladden is a component part in .... (headworks) (*AMIE s*' 2000)
18. In a hydranlic structure the position of hydranlic gradient line is 2 m from the top of impervious floor. The relative density of the material of impervious floor is 2.4. From uplift consideration floor at this location should be .... (0.93 m) (*AMIE s*' 01)
19. The nose of the divide wall of a barrage is founded on: (*AMIE s*' 99)
(*a*) firm rock which may be at any depth (*b*) well foundation
(*c*) pile foundation (*d*) none of the above (*b*)
20. By increasing length of floor on u/s of were axis, the uplift pressure will .... below the floor d/s of weir: (*AMIE s*' 99)
(*a*) increase (*b*) decrease (*c*) remain same (*d*) none of above
21. Tunnels of a sediment excluder are so designed that: (*AMIE s*' 99)

(*a*) longest turnnel is farthest from canal head regulation
(*b*) longest tunnel is nearest to canal head regulation
(*c*) all tunnels are of the same length (*d*) none of the above

22. Khesla's theory is applicable to structures founded on: (*AMIE s*' 99)
(*a*) rock (*b*) any type of material
(*c*) permeable material only (*d*) none of the above (*c*)

23. Khosla theory of independent variables is applicable to structure founded on: (*AMIE w*' 93)
(*a*) rock (*b*) clay (*c*) sands (*d*) concrete (*c*)

24. According to Khosla's theory the exit gradient is given by: (*AMIE s*' 93)
(*a*) $\frac{d}{H}\frac{1}{\pi\lambda}$ (*b*) $\frac{H}{d}\frac{1}{\pi\sqrt{\lambda}}$ (*c*) $\frac{H}{d}\pi\sqrt{\lambda}$ (*d*) $\frac{d}{H}\lambda$ (*b*)

25. White carrying weir designs on permeable soils, the correction for mutual interference of sheet pile is NOT applicable on intermediate pile if the outer pile:
(*a*) goes deeper than intermediate pile and is farther from intermediate pile by more than twice its own length
(*b*) goes only just as deep as the intermediate pile and is within a distance of one and half times its own length
(*c*) does not go as deep as intermediate pile, no matlen what the horizontal distance between them is
(*d*) is safe against dolecterions exit godral (*b*)

26. Depth-area-duration curve would seem to resemble: (*Engg. Service Exam.* '98)
(*a*) ares of a circle concave upwards with duration increasing outward
(*b*) first quadrant limbs of hyperbolae with duration increasing outward
(*c*) third quadrant limbs of hyperbolae with duration decreasing outward
(*d*) first quadrant limbs of hyperbolae with duration decreasing outward (*a*)

11. CONTROL STRUCTURES

1. Control structures in a canal system are required:
(*a*) to feed offtakes during low supplies
(*b*) localise malfunctioning of the head regulator
(*c*) to help attain designed conditions in the channel
(*d*) all of them (*d*)

2. Silt control measures are adopted:
(*a*) at headworks (*b*) in head reach of canal
(*c*) at offtake points (*d*) all of them (*d*)

3. Silt excluder to exclude coarser sediments is provided:
(*a*) in river in conjunction with canal head regulator
(*b*) in canal in head reach (*c*) at offtaking canal heads
(*d*) all of them (*d*)

4. Silt ejector is provided:
(*a*) in river in conjunction with canal head regulator
(*b*) in canal in head reach
(*c*) at offtaking canal head (*d*) all of them (*b*)

5. Escape provided across the channel at its fag end to maintain required full supply is called .... (tail cluster) (*AMIE s*' 01)

6. For a metering flume crest over which head is measured should be: (*AMIE s*' 99)
(*a*) sharp (*b*) broad
(*c*) either sharp or broad (*d*) none of the above (*b*)

7. To minimise the head losses at the entry of a branch canal taking off from main canal the centre line of the branch canal should make on angle with the centre line of main canal that is:
(*a*) obtuse in the direction of flow (*b*) acute in direction of flow
(*c*) obtuse or arcuate (*d*) none of the above (*c*)
8. In the design of canal, falls are provided: (*AMIE s*' 01)
(*a*) to dissipate energy
(*b*) when ground slope is steeper than the canal bed slope
(*c*) when canal bed slope is steeper than ground slope
(*d*) to reduce hydranlic losses
9. Hydraulic jump occurs when flow charges from: (*AMIE s*' 01)
(*a*) supercritical to subcritical (*b*) subartical to supercritical
(*c*) critical to turbulent (*d*) laminan to turbulent
10. A cistern in a canal fall serves as a .... (stilling basin) (*AMIE w*' 00)
11. The flow occurring in a turnnel with water level above the roof at the exit end is .... flow (*AMIE s*' 99)
12. A drop in canal bed is generally provided, if (*AMIE s*' 98)
(*a*) ground slope exceeds designed bed slope
(*b*) design bed slope exceeds ground slope
(*c*) ground slope is same as designed bed slope
(*d*) none of the above
13. Proportional distributor is a cross regulator to serve for:
(*a*) proportionate distribution of supplies among offtake without any regulation
(*b*) drawal of a constant proportion of supply by each offtake when supply fluctnates in the parent channel
(*c*) proportionate distribution of silt amongst offtakes and parent channels
(*d*) all of them (*d*)
14. Cross regulator is not essentially provided:
(*a*) when available working head of offtake is more than half the full supply depth in the parent channel
(*b*) when offtake can be fed with two-third supply in the parent channel
(*c*) offtake discharge is disproportionately less than that of parent channel
(*d*) all of them (*d*)
15. Canal escape provided to:
(*a*) escape supply into offtaking channel
(*b*) to escape surplus or excess water from canal
(*c*) to escape canal water for emptying it in the event of a breach
(*d*) periodical flushing o˜ silt prone head reach of a canal head: (except *a*)
16. The ideal location for a canal escape is below canal
(*a*) 2 km (*b*) 3 km (*c*) 5 km (*c*)
17. In a power channel escape is located:
(*a*) below head (*b*) near a drainage
(*c*) upstream of penstock (*d*) downstream of penstock
(*c*)
18. Uplift pressure on floor of syphon - aqueduct due to seepage is maximum when:
(*a*) canal is running full but drainage dry (*b*) drainage is running full but canal dry
(*c*) both canal and drainage are running full (*d*) both canal and drainage are dry (*a*)
19. Cross regulator in main canals are provided to: (*AMIE w*' 94)
(*a*) regulate water supply in offtaking channel

(*b*) regulate water supply in main canal (*c*) regulate excessive flood water
(*d*) head up water for adequate supply into offtaking channel (*d*)

20. A cross regulator is provided on a main canal: (*AMIE w*' 93)
(*a*) to minimise the amount of silt entering the branch canal
(*b*) to let maximum silt be carried into the branch canal
(*c*) for no specific purpose (*d*) to carry canal across drain (*a*)

21. Silt ejector is provided on: (*AMIE w*' 98)
(*a*) river just u/s of undersluices (*b*) canal just d/s of canal head regulator
(*c*) river, some distance d/s of underslinces
(*d*) canal, some distance d/s of canal head regulator (*d*)

22. Escape provided across the channel at its fag end to maintain required full supply level is called ....

23. Silt excluder is provided: (*AMIE s*' 93)
(*a*) in canal on d/s side of the head regulator
(*b*) in river on d/s side of the barrage
(*c*) in river far away from the weir on d/s side
(*d*) in river adjacent to the head regulator (*d*)

24. Assuming useful life of a hydraulic structure as 100 years, minimum risk would be involved by selecting a flood of return period .... years: (*AMIE s*' 98)
(*a*) 100 (*b*) 1000 (*c*) 50 (*d*) > 1000 (*d*)

12. CROSS DRAINAGE WORKS

1. When the canal with bed level higher than high flood level in drain cross each other, then the cross drainage structure is called .... (Aqueduct) (*AMIE w*' 99)
2. Cross drainage works are not needed when canal is aliqued as .... (watershed canal) (*AMIE w*' 99)
3. In a syphon aqueduct, the maximum uplift pressure on the floor occurs when: (*AMIE s*' 01)
(*a*) canal is dry and drain is running full (*b*) canal is running full and drain is dry
(*c*) both canal and drain are running (*d*) both are dry
4. A cross drainage work in which the canal water and the drainage meet at the same level is called a .... (*AMIE w*' 00)
5. A canal on a watershed requires a: (*AMIE s*' 2000)
(*a*) level crossing (*b*) siphon (*c*) superpassage (*d*) none of above
6. A superpassage is a cross drainage work provided when bed of natural drain is: (*AMIE s*' 2000)
(*a*) at same level as canal bed (*b*) well above F.S.L. in canal
(*c*) below FSL of canal (*d*) below bed level of canal
(*b*)
7. A level crossing is a cross drainage work where: (*AMIE s*' 97)
(*a*) canal flows under a specified level above the drain
(*b*) canal flows under pressure
(*c*) canal has a free flow
(*d*) none of above (*d*)
8. A syphon aqueduct is provided when:
(*a*) canal bed level is well above HFL of drain
(*b*) canal bed level lies between bed level & HFL of drain
(*c*) drain bed level is well above FSL of canal
(*d*) drain bed level lies between bed level & FSL of canal

9. The velocity of drainage water in the levels of syphon aqueduct is normally limited to: (*AMIE w'* 94)

(*a*) 1 to 2 m/s (*b*) 2 to 3 m/s (*c*) 3 to 4 m/s (*d*) 4 to 5 m/s (*b*)

(*a*) In an aqueduct	(*A*) drainage water is carried in a trough sup-posted on piers while irrigation canal passes below with its FSL above the slab of trough above
(*b*) In a level crossing	(*B*) drainage water is carried in a trough sup-posted on pier while irrigation canal passes below with its FSL above the slab of the trough above
(*c*) In a superpassage	(*C*) canal and drain approach each other practi-tically at the same level
(*d*) In a syphon	(*D*) irrigation canal is carried across the drain in a trough supported on piers with canal bed level well above HFL of the drain

$(a - D)$, $(b - C)$, $(c - A)$, $(d - A)$

(*a*) Cross drainage works include	(*A*) notches, escapes slinces and syphon well drops
(*b*) Works for regulation of water level include	(*B*) bridges and cuberts
(*c*) Communication works include	(*C*) spurs, groynes, bunds, mathresses and aprons
(*d*) River training works	(*D*) aqueducts, syphons, syphon aqueducts, superpasseges and level crossings

$(a - D)$, $(b - A)$, $(c - B)$, $(d - C)$

10. In a syphon-aqueduct, the most severe condition of uplift on the floor occurs when: (*AMIE s'* 93)
    (*a*) canal runs full drain is dry but water table is at stream bed
    (*b*) canal is dry and drain is a high flood level
    (*c*) canal runs dry and drain also runs dry
    (*d*) Both canal and drain run full (*b*)
11. Cross drainage work where the bed levels of canal and drain are equal is called (*AMIE w'* 99)
    (*a*) aqueduct (*b*) level crossing (*c*) superpassage (*d*) siphon-aqueduct (*d*)
12. A cala on a watershed requires: (*AMIE w'* 00)
    (*a*) level crossing (*b*) a siphon (*c*) a superpassage (*d*) none of the above (*d*)
13. In a level crossing type of cross drainage works: (*AMIE s'* 99)
    (*a*) canal is above drain (*b*) canal is below drain
    (*c*) canal and drain are at the same level (*d*) none of the above (*c*)
14. A canal syphon is a cross drainage work in which the cala is taken .... the river (below) (*AMIE s'* 99)
15. A drainage syphon is a cross drainage in which drainage is taken .... the canal (below)
16. Syphon-aqueduct is the cross drainage work provided to carry canal over drain when .... (*AMIE s'* 99)
17. The structure constructed to allow drainage water to flow under pressure through an inverted syphon below a canal is called: (*AMIE s'* 98)

(*a*) syphon (*b*) superpassage (*c*) syphon-aqueduct (*d*) super-aqueduct

18. In cross drainage works, the function of an inlet is to admit drain water into a: (*AMIE w*' 98)
(*a*) canal (*b*) field (*c*) drain (*d*) none of the above (*a*)

13. BRIDGES AND CULVERTS

1. A high level bridge is a bridge:
(*a*) on a national highway
(*b*) of unconventional height pass highest anticipated flood without submergence
(*c*) none of the above (*d*) all of the above (*b*)
2. A submersible bridge is designed to be overtopped in floods for not more than .... days at a time:
(*a*) 3 (*b*) 5 (*c*) 7 (*d*) 10 (*a*)
3. A submersible bridge is designed to be overtopped in floods for not more than .... times in a year:
(*a*) 3 (*b*) 5 (*c*) 6 (*d*) 10 (*c*)
4. A canseway is provided to:
(*a*) to provide caution to traffic flow
(*b*) with a dip to pass flood water over it being economical
(*c*) same as submersible bridge (*d*) none of the above (*b*)
5. A culvert is a small bridge with a span up to .... *m*
(*a*) 2.5 (*b*) 3 (*c*) 5 (*d*) 6 (*d*)
6. Minimum vent dimension of a culvert is .... *m*
(*a*) 0.5 (*b*) 0.75 (*c*) 1.0 (*d*) 1.5 (*d*)
7. The bassel length of a culvert is .... than that of a bridge:
(*a*) same (*b*) shorter
(*c*) longer (*d*) as per site condition (*c*)
8. The difference in water levels u/s and d/s of a bridge is known as .... (head loss) (*AMIE s*' 01)
9. If a river in alluvial plain has a dominant discharge of 1600 cumecs, the waterway for a bridge on the river should be approximately: (*AMIE w*' 97)
(*a*) 250 m (*b*) 240 m (*c*) 190 m (*d*) 180 m
10. The dominant discharge of a river in alluvial plain is 3600 cumecs. The waterway on the river, according to Lacey's formula, should be .... *m* (*AMIE w*' 96)
11. For a given discharge, efflux for a pipe culvert will be: (*AMIE w*' 93)
(*a*) same as for a large opening (*b*) independent of the size of the opening
(*c*) more than that for a large opening (*d*) none of the above (*c*)
12. If dominant discharge of an alluvial river is 1800 cumecs, the waterway for a bridge on the river should be: (*AMIE s*' 93)
(*a*) 101.5 m (*b*) 151.5 m (*c*) 201.5 m (*d*) 251.5 m (*c*)
13. If a river in alluvial plain has dominant discharge of 1600 cumecs, the waterway for a bridge on this river should be .... *m* (approximate) (*AMIE s*' 98)
14. The head due to sudden expansion is expressed by: (*AMIE s*' 93)
(*a*) $(V_1^2 - V_2^2)/2q$ (*b*) $(V_1 - V_2)^2/q$ (*c*) $(V_1 - V_2)^2/2q$ (*d*) $[(V_1 - V_2)/2q]^2$ (*a*)
15. A rectangular channel of 4 m width carries water at the rate of 8 cumecs under critical conditions. Specific energy for the flow is

$$D_c = \sqrt[3]{(8/4)^2/9.82} = 0.74 \text{ m}$$

$$E_{fc} = D_e\, 7.4 + [(8/4 \times 0.74)^2]\, 2 \times 9.81$$

16. If confugate depths for a horizontal rectangular channel 4 m wide are 0.2 m and 1 m, the discharge in the channel is .... cumecs (*AMIE s'* 93)

$$4.34 \text{ cumecs } 4 \left(1 = -\frac{0.2}{2} + \sqrt{\frac{0.2^2}{4} + \frac{2q^2}{9.81 \times 0.2}}\right)$$

17. If y_1 and y_2 are depths of flow before and after the hydranlic jump in a rectangular channel then the loss of energy is given by: (*AMIE s'* 93)

(*a*) $\frac{(y_2 - y_1)^3}{4y_1y_2}$ (*b*) $\frac{(y_2 - y_1)^2}{4y_1y_2}$ (*c*) $\frac{(y_2 - y_1)^3}{y_1y_2}$ (*d*) $\frac{(y_2 - y_1)^2}{y_1y_2}$ (*a*)

14. CANAL OUTLETS

1. Canal outlet is a device to:
   (*a*) pass out excess supply to waste (*b*) deliver supply to watercourse
   (*c*) escape canal supply to a drainage (*d*) all of above (*b*)
2. Canal outlets are generally not provided on:
   (*a*) main canals and branches (*b*) major distributions
   (*c*) distributions (*a*)
3. A direct outlets is that which:
   (*a*) provides direct canal supply to a watercourse
   (*b*) directly feeds a minor or sub minor
   (*c*) is rarely provided on main canal and branches
   (*d*) all of the above (*c*)
4. An outlet should essentially:
   (*a*) pass a known discharge (*b*) constant discharge
   (*c*) be a measure of discharge (*d*) all of them (*d*)
5. The area irrigated by an outlet is called .... : (chark or outlet area)
6. The upper limit of an outlet capacity is .... cumecs:
   (*a*) 0.012 (*b*) 0.05 (*c*) 0.06 (*d*) 0.03 (*b*)
7. The lower limit of an outlet capacity is .... cumecs:
   (*a*) 0.09 (*b*) 0.06 (*c*) 0.03 (*d*) 0.020 (*c*)
8. The maximum length of a watercourse is .... km:
   (*a*) 6 (*b*) 5 (*c*) 3 (*d*) 2 (*c*)
9. The size of outlet area or chark is determined by;
   (*a*) natural drainage lines (*b*) village boundary
   (*c*) railway or national highway
   (*d*) upper and lower limits of permissible outlet discharge
   (*e*) all of the above (*e*)
10. An outlet with 0.015 cumec discharge may irrigate only .... area irrigated by an outlet of 0.03 cumec:
    (*a*) 3/4 (*b*) 1/2 (*c*) 1/3 (*d*) 1/4 (*c*)
11. An outlet should be proportional in discharge distribution that is it:
    (*a*) drains same discharge inspective of varying supply in the channel
    (*b*) drains more or less discharge corresponding to the varying discharge in the channel
    (*c*) Drains less discharge (*d*) drains excess discharge (*b*)
12. Modular outlet is that the discharge of which is
    (*a*) independent of water levels in the distributary
    (*b*) independent of water levels in the watercourse

(*c*) independent of water levels both in the distributary or watercourse
(*d*) all of the above (*c*)

13. Sami-modular outlet is that the discharge of which is:
(*a*) independent of water level in the distributary/minor
(*b*) independent of water level in the watercourse
(*c*) independent of water levels in the distributary and watercourse
(*d*) all of above (*b*)
14. An APM outlet is:
(*a*) orifice (*b*) weir (*c*) meter (*d*) none of them (*a*)
15. An open flume outlet is:
(*a*) orifice (*b*) weir (*c*) meter (*d*) none of them (*b*)
16. A semi-modular outlet can distribute supply more or less:
(*a*) equitably in variations in levels in watercourse
(*b*) equitably in variations in supply in distributary/minor
(*c*) all of the above (*d*) none of the above (*b*)
17. A non-modular outlet supplies water depending on
(*a*) water level in the distributary/minor (*b*) water level in the water course
(*c*) water levels in the canal/watercourse (*d*) all of the above (*c*)
18. Adjustability of semi-modular outlets is most converient and economical in:
(*a*) A.P.M. (*b*) open flume (*c*) both of them (*d*) none of them
(*a*) by raising or lowering roof block.
19. In the ascending order of consumption of head, the outlets are:
(*a*) A.P.M., O.F., pipe (*b*) A.P.M., pipe, O.F.
(*c*) O.F., A.P.M. pipe (*d*) Pipe, O.F. A.P.M.
20. In the ascending order of silt conduction, the outlets are:
(*a*) A.P.M., O.F., pipe (*b*) A.P.M., pipe, O.F.
(*c*) pipe, O.F., A.P.M. (*d*) O.F., Pipe, A.P.M. (*a*)
21. In a new channel, the outlet generally provided is:
(*a*) A.P.M. (*b*) O.F. (*c*) pipe (*d*) all of them
(*c*) except at tail
22. The outlet type provided at end in immediate vicinity of a fall is:
(*a*) A.P.M. (*b*) O.F. (*c*) pipe (*d*) all of them (*b*)
23. The outlet type provided at tail in an old channel is:
(*a*) A.P.M. (*b*) O.F. (*c*) pipe (*d*) all of them (*b*)
24. The tail cluster is essentially a combination or singly of the type:
(*a*) A.P.M. (*b*) O.F. (*c*) pipe (*d*) all of them (*b*)
25. For more or less equitable distribution of supply, the outlet types prefered, from head reach to tail are:
(*a*) A.P.M., O.F., pipe (*b*) A.P.M., pipe, O.F.
(*c*) pipe, O.F., A.P.M. (*d*) O.F., A.P.M., pipe (*a*)
26. For proportionality, weir type outlets should be set at .... depth:
(*a*) 0.9 (*b*) 0.8 (*c*) 0.6 (*d*) 0.3 (*a*)
27. For proportionality, pipe outlet is set at .... depth
(*a*) 0.9 (*b*) 0.8 (*c*) 0.6 (*d*) 0.3 (*d*)
28. Sinsitivity index for a wide trapezoidal channel is .... *F*:
(*a*) 5/3 (*b*) 1 (*c*) 3/4 (*d*) 1/2 (*a*)
29. A device to release water from a canal to the water course is called an .... (outlet)
(*AMIE w' 00*)
30. Flexibility of an outlet is the ratio of: (*AMIE s' 00*)

(*a*) rate of change of discharge in an outlet to that in parent channel
(*b*) rate of change of discharge in an outlet to normal depth in the channel
(*c*) rate of depth of crest level of outlet to F.S.L. of channel to its F.S. depth
(*d*) head recovered to head put in

31. Biggs module is an example for .... outlet. (*AMIE s' 00*)
32. A module is said to be proportional when flexibility equals: (*AMIE s' 96*)
(*a*) 0.8 (*b*) 0.9 (*c*) 1.0 (*d*) 1.1 (*c*)
33. Distribution outlet is said to be proportional, if its flexibility is: (*AMIE w' 94*)
(*a*) equal to unity (*b*) less than unity
(*c*) slightly greater than unity (*d*) equal to two (*a*)
34. An irrigation outlet whose discharge is independent of water levels in the distributary and water course and which ensures constant supply within certain working limits is called a .... (modular) (*AMIE s' 93*)
35. By considering channel index as 1.8, the setting of an orifice type irrigation outlet to have proportionality is: (*AMIE w' 99*)
(*a*) 0.67 (*b*) 0.28 (*c*) 0.9 (*d*) 0.15 (*c*)
36. The adjustable proportional module is a: (*AMIE w' 99*)
(*a*) modular outlet (*b*) a submerged pipe outlet
(*c*) an open flume outlet (*d*) a semi-modular outlet (*d*)
37. An outlet which maintain a constant discharge, inspective of fluctuations in water levels of supplying canal and watercourse is known as .... outlet: (*AMIE s' 98*)
(*a*) non-modular (*b*) semi-modular (*c*) flexible modular (*d*) rigid modular (*d*)
38. A submerged pipe outlet from a minor feeds into a well chamber across the bank. Water from the well chamber overflows a weir-like face into the field channel. The flow into the field channel: (*Eng. Su. Exam. '98*)
(*a*) will increase with any drop in the water level in the minor
(*b*) will not be much affected by any change in the water level in the minor
(*c*) will decrease with any drop in the water level in the minor
(*d*) will be proportional to the flow in the minor when the minor flows between set ranges: (*c*)

15. HYDROLOGY

1. Hydrology is the science which deals with the water:
(*a*) in atmosphere (*b*) an earth surface
(*c*) below ground (*d*) of the earth in all states (*d*)
2. Watershed is the line joining the points having: (*AMIE w' 00*)
(*a*) maximum elevation across the different cross sections
(*b*) minimum elevation across the different cross sections
(*c*) average elevation (*d*) none of the above (*a*)
3. In India 85 – 90% of the annual rainfall is contributed by:
(*a*) South-West monsoon (*b*) North-East monsoon
(*c*) cyclonic rainfall (*d*) all of them (*a*)
4. About 90 of the rainfall in India is received during the months:
(*a*) May-October (*b*) June-September (*c*) July-August (*d*) October-February (*b*)
5. About 90 of the rainfall is received approximately in .... hours:
(*a*) 100 (*b*) 200 (*c*) 300 (*d*) 400 (*a*)
6. Cold front precipitation is: (*AMIE s' 96*)

(*a*) very intense of short duration (*b*) light and short duration
(*c*) very intense and considerable duration (*d*) showery and of considerable duration

7. Normal time of onset of monsoon in India is: (*AMIE s*' 98)
(*a*) Early June at Kerala (*b*) Early June at Bombay
(*c*) Late October at Kerala (*d*) Early June at Channai

8. Normal average annual rainfall over then whole of India is estimated as .... cm: (*AMIE w*' 95)
(*a*) 88 (*b*) 217 (*c*) 119 (*d*) 290

9. The average annual rainfall over whole of India is estimated as .... cm: (*AMIE s*' 94)
(*a*) 189 (*b*) 319 (*c*) 89 (*d*) 117

10. The average annual rainfall is secomputed after every .... years:
(*a*) 10 (*b*) 20 (*c*) 30 (*d*) 35

11. The rainfall cycle in India is considered to be of .... years:
(*a*) 20 (*b*) 25 (*c*) 30 (*d*) 35 (*d*)

12. The monthly rainfall at a place is generally indicated as above or below normal. The term normal means: (*AMIE w*' 94)
(*a*) rainfall in the same month of previous year
(*b*) rainfall expected based on previous month's data
(*c*) average rainfall of previous twelve months
(*d*) computed from 30 years of part record (*d*)

13. Rainfall of less than 0.1 mm is recorded as a:
(*a*) trace (*b*) select (*c*) light (*d*) hail (*a*)

14. Precipitation is the same as:
(*a*) rainfall (*b*) snowfall (*c*) hail, select (*d*) all of them (*d*)

15. A raing day is defined as the day when .... mm or more of rainfall is recorded:
(*a*) 2.5 (*b*) 5 (*c*) 7.5 (*d*) 10 (*a*)

16. A rainfall with an intensity of 5 mm/h is classified as: (*AMIE w*' 98)
(*a*) trace (*b*) light rain (*c*) moderate rain (*d*) heavy rain (*a*)

17. Effective precipitation is defined as: (*AMIE s*' 98)
(*a*) precipitation minus loss due to evaporation
(*b*) precipitation minus loss due to infiltration
(*c*) precipitation occurred during crop period
(*d*) available water stored within root zone (*b*)

18. Water year in India starts from the first day of: (*AMIE w*' 97)
(*a*) January (*b*) April (*c*) June (*d*) November (*c*)

19. Wet yeat is a year in which: (*AMIE s*' 97)
(*a*) a specified rainfall occurs
(*b*) rainfall is same as previous year rainfall
(*c*) rainfall is more than previous year's rainfall
(*d*) rainfall is more than average rainfall over the several consecutive previous years (*d*)

20. The commonly used raingange in India is .... (Symon) (*AMIE w*' 01)

21. The standard symon's raingange has a collecting area of diameter .... cm: (*AMIE s*' 94 & *w*'95)
(*a*) 5.08 (*b*) 10 (*c*) 12.7 (*d*) 25.4 (*c*)

22. Thiessen polygon is used in connection with: (*AMIE s*' 97)
(*a*) average rainfall determination (*b*) catchment area computation
(*c*) drawing hydrograph (*d*) estimating infiltration (*a*)

23. Thiesson polygon method is used to find .... (Average rainfall over a basin)

(*AMIE s'* 2000)

24. The method used for estimating missing rainfall data is: (*AMIE w'* 97)
    (*a*) normal ratio method (*b*) station year method
    (*c*) plotting positive method (*d*) rational method (*a*)
25. Self recording gange records: (*AMIE s'* 97)
    (*a*) stream discharge (*b*) stream velocity
    (*c*) variation of water surface level with time
    (*d*) velocity of cyclonic storm (*c*)
26. Isolytel method is used for determination of: (*AMIE s'* 97)
    (*a*) evapotranspiration (*b*) seepage loss
    (*c*) precipitation (*d*) intensity of flood
27. Which of the following pairs are not correctly matched: (*AMIE s'* 01)
    (*a*) Waterloss Evaporation
    (*b*) Percolation Soil erosion
    (*c*) Runoff Stream flow
    (*d*) Storm Precipitation
28. Match List-I with List-II and select the correct answer from the codes given below the lists: (*AMIE s'* 01)

List I	*List II*	*Codes*	*A*	*B*	*C*	*D*
(*a*) Isolyet	(1) Pressure	(a)	1	2	4	3
(*b*) Isobars	(2) Rainfall	(*b*)	2	1	3	4
(*c*) Isolves	(3) Time of travel	(*c*)	2	1	4	3
(*d*) Isochrones	(4) Velocity	(*d*)	3	4	2	1

29. Isolyetal method is employed for determination of: (*AMIE s'* 97)
    (*a*) evapotranspiration (*b*) seepage loss
    (*c*) precipitation (*d*) intensity of flood (*c*)
30. Transpiration is a process by which: (*AMIE s'* 97)
    (*a*) plant transfers water from one part to the other
    (*b*) plant absorbs water through its roots
    (*c*) plant excludes travel of particles through it
    (*d*) plant dissipates water vapour into atmosphere (*d*)
31. Infiltration process is:
    (*a*) movement of water over the land surface
    (*b*) movement of water through the soil
    (*c*) absorption of water by the surface of the soil
32. Mean annual rainfall at station X follow Gaussian distribution having its mean value as P mean and standard deviation σ. The maximum value of annual rainfall can be estimated as: (*AMIE w'* 00)
    (*a*) Pmean + 3σ (*b*) Pmean – 3σ (*c*) Pmean + σ (*d*) Pmean + 4σ
33. For a storm of 2-*h* duration, rainfall rates in cm/h at 20 minutes intervals are 2.5, 2.5, 10, 7.5, 5.1 and 1.25. If ϕ index is 3 cm/h, total surface runoff is:
    (*a*) 4.53 cm (*b*) 45.3 cm (*c*) 0.453 cm (*d*) none of the above
34. Infiltration capacity is: (*AMIE s'* 97)
    (*a*) capacity of a soil by which water infiltrates through it under specified head
    (*b*) capacity of a soil for the maximum rate of absorption of rain water at incidence on its surface
    (*c*) rate at which enters into the infiltration well
    (*d*) rate at which water enters through the unit length of an infiltration galley

35. Rational method conclates: (*AMIE s*' 97)
    (*a*) runoff coefficient with intensity of rainfall
    (*b*) runoff coefficient with drainage area
    (*c*) drainage area with intensity of rainfall
    (*d*) intensity of rainfall with runoff
36. Rational method gives discharge from a catchment as $Q = 2.78$ CIA in which:
    (*a*) Q is cumecs, I is in cm/h and A is in sq/km
    (*b*) Q is in the m/h, I is in min./h and A is sq./km.
    (*c*) Q is in cum/dry, I is in cm/day and A is in m^2
    (*d*) Q is in cumec/h, I is in mm/h and A is in sq. km.
37. If area of drainage basin is doubled while keeping the other factors same, the peak discharge according to Dicken's formula will increase by a factor of (*AMIE w*' 00)
    (*a*) 2 (*b*) 1.68 (*c*) 1.59 (*d*) 1.5
38. Double-mass curve technique is adopted to:
    (*a*) estimate the missing data
    (*b*) obtain intensities of rainfall at various durations
    (*c*) check the consistency of data
    (*d*) obtain the storage capacity needed for a reservoir
39. The drainage density is the:
    (*a*) average length of streams per unit drainage area within the basin
    (*b*) stream discharge per unit drainage area
    (*c*) drainage area per unit stream length
    (*d*) number of streams per unit drainage area
40. Streams that contribute to the ground water are called: (*AMIE w*' 98)
    (*a*) effluent streams (*b*) ground water stream
    (*c*) influent strems (*d*) perennial stream (*a*)
41. An ephemeral stream: (*AMIE w*' 96)
    (*a*) is one which always carries some flow
    (*b*) does not have any base flow contribution
    (*c*) is one which has limited contribution of ground water in the reason
    (*d*) is one which carries only snawnelt water
    (*b*)
42. The probable maximum flood discharge is:
    (*a*) an impossibly large flood discharge
    (*b*) a flood discharge that is maximum over the part years
    (*c*) an externally large but physically possible flood at the site
    (*d*) a flood with maximum probability of occurrence
43. Lysimeter method is used for direct measurement of .... (*AMIE w*' 01)
44. Water enters the soil at capacity rate when intensity of rainfall is: (*AMIE w*' 98)
    (*a*) more than infiltration capacity (*b*) less than infiltration capacity
    (*c*) just equal to infiltration capacity (*d*) non-uniform and of long duration
45. Rainfall hyotograph shows the variation of: (*AMIE w*' 98)
    (*a*) cumulative rainfall with time (*b*) rainfall intensity with time
    (*c*) rainfall depth over area (*d*) rainfall intensity with cumulative rainfall
46. Direct runoff is made up of: (*AMIE w*' 94)
    (*a*) surface runoff, prompt interflow and channel precipitation
    (*b*) surface runoff, infiltration and evapotranspiration
    (*c*) overland flow and quick return flow
    (*d*) infiltration, rainfall and evaporation annual (*c*)
47. A meanannual runoff of 1 cumec from a catchment of area 10 km^2 represients on effective

rainfall .... cm. (315) $\left(\dfrac{1\times 365\times 24\times 60\times 60}{10\times 10^6}\right)$

48. Wind velocity at a height 2 m above ground is 5 kph, its value at a height of 10 m above ground can be expected to be about .... k (6.29) $\left(\dfrac{5}{V}=\dfrac{2}{10}\right)^{1/7}$ (*AMIE w' 98*)

49. If wind velocity 1m above ground is 4 km/h, the value at a height 9 m above the ground is expected to be .... km/h. 5.4777 $\left[\dfrac{4}{V_2}=\left(\dfrac{1}{9}\right)^{1/7}\right]$ (*AMIE w' 97*)

(*a*) surface runoff is that form of — (*A*) precipitation in which drops of water are under 0.5 mm in diameter and intensity is usually less than 1 mm per hour.

(*b*) Infiltration in that form of — (*B*) hydrologic cycle in which water moves from the surface of the ground into the soil

(*c*) Drizzle is that form of — (*C*) precipitation in which water drops are greater than 0.5 mm diameter

(*d*) Rain is that form of — (*D*) precipitation which travels over land surface or just below it.

(*a* – *D*), (*b* – *B*), (*c* – *A*), (*d* – *C*) (*AMIE w' 93*)

50. Glaze is a term used in convection with: (*AMIE s' 97*)
   (*a*) classificatio of precipitation (*b*) classification of catchments
   (*c*) classification of sediment transport (*d*) none of the above (*a*)

51. Infiltration process is: (*AMIE w' 96*)
   (*a*) movement of water over the land surface
   (*b*) movement of water through the soil
   (*c*) absorption of water by the surface of soil
   (*d*) none of the above (*b*)

52. Infiltration capacity is: (*AMIE w' 96*)
   (*a*) capacity of a soil by which water infiltrates through it under a specified head
   (*b*) capacity of a soil for maximum rate of absorption of rain water at incidence on its surface
   (*c*) rate at which water enters into the infiltration well
   (*d*) rate at which water enters through the unit length of an infiltration gallery (*b*)

53. Transpiration is a process by which plant: (*AMIE s' 97*)
   (*a*) transfers water from one part to the other
   (*b*) absorbs water through its roots
   (*c*) excludes the travel of particles through it
   (*d*) dissipates water vapour into atmosphere (*d*)

54. A hyetograph is a plot of: (*AMIE w' 96*)
   (*a*) cumulative rainfall Vs time (*b*) rainfall intensity Vs time
   (*c*) rainfall depth Vs duration (*d*) discharge Vs time (*c*)

55. Hydrograph is a plot of: (*AMIE s' 98*)
   (*a*) rainfall intensity against time (*b*) discharge against time
   (*c*) cumulative rainfall against time (*d*) cumulative discharge against time

56. Rational method gives discharge from a catchment as $Q = 2.78$ CIA in which:
   (*a*) Q is cumec, I is in cm/h and A is in km^2
   (*b*) Q is in harm, I is in mm/h and A is in km^2
   (*c*) Q is in m^3/day, I is in cm/day and A is in km^2

(*d*) Q is in m³/h, I is in mm/h and A is in km²

57. If the area of drainage basin is doubled while keeping the other factors same, the peak discharge according to Dickan's formula will increase by a factor of: (*AMIE w*' 2000)
(*a*) 2 (*b*) 1.68 (*c*) 1.59 (*d*) 1.5
58. Rational method correlates: (*AMIE s*' 97)
(*a*) runoff coefficient with intensity of rainfall
(*b*) runoff coefficient with drainage area
(*c*) drainage area with intensity of rainfall (*d*) intensity of rainfall with runoff (*d*)
59. If the peak of a 3–h UH is 1000 m^3s^{-1}, the flood hydrograph peak for a net rainfall of 150 mm and a base flow of 200 m^3s^{-1} is: (*AMIE s*' 99)
(*a*) 14800 m^3s^{-1} (*b*) 12000 m^3s^{-1} (*c*) 15000 m^3s^{-1} (*d*) 18000 m^3s^{-1}
60. The area below a direct runoff hydrograph is:
(*a*) always > that below the corresponding rainfall histogram
(*b*) always < that below the corresponding rainfall histogram
(*c*) always equal to that below the corresponding rainfall histogram
(*d*) Sometimes > and sometimes < that below the corresponding rainfall histogram
61. The relation between probability and recurrence interval T is given by .... (*AMIE s*' 99)
62. Soure tropical cyclones in the bay of Bengal that cause intense rainfall are most likely to occur in the month of:
(*a*) March (*b*) June (*c*) November (*d*) January
63. An anticyclone is: (*AMIE s*' 99)
(*a*) a low pressure system with clockwise winds in the northern hemisphere
(*b*) a high pressure system with clockwise winds in northern hemisphere
(*c*) zone of low pressure with anticlockwise winds in northern hemisphere
(*d*) zone of high pressure with anticlockwise winds in northern hemisphere

Fill up the blanks:

1. The area between isohypts 450 mm and 550 mm is 100 sq km. The average depth of annual precipitation over the entire basin of 250 sq km is .... mm.
2. The relation between probability and recurrence interval T is given by ....
3. If rainfall excess of 1 cm occurs instalancously them the resulting hydrograph is called ....
4. There is a 5% probability that the design storm will occur in the next 4 years. The return period is .... years.
5. Phytometer is used for measuring ....
6. A one-day rainfall of 10 cm at a place was found to have return period of 100 years. The probability that a one day rainfall of this or larger magnitude will occur in the next year is .... (0.01) (*AMIE w*' 94)
7. The mass curve of rainfall of a storm is plot of: (*AMIE s*' 96)
(*a*) rainfall depths for various equal durations plotted in decreasing order
(*b*) rainfall depths for various equal durations plotted in increasing order
(*c*) rainfall intensity Vs time in chronological order
(*d*) accumulated precipitation Vs time in chronological order (*d*)
8. A line joining points having equal time of travel of surface runoff to the catchment outlet is called an: (*AMIE w*' 97)
(*a*) isochorone (*b*) isovel (*c*) isobar (*d*) isopture
9. A stream receiving contribution from ground water is called: (*AMIE s*' 97)
(*a*) influent (*b*) effluent (*c*) ephenseral (*d*) basal (*b*)

10. Using ideal conditions, the eraporation from sea water as compared to fresh water is: (*AMIE w'* 98)
    (*a*) < that of fresh water (*b*) = that of fresh water
    (*c*) > that of fresh water (*d*) no definite conclusion can be made (*a*)
11. The annual maximum values of flood for 22 years are arranged in decreasing order. Using weibull formula, the flood magnitude having a risk 5, has return period of about: (*AMIE w'* 97)
    (*a*) 5–0 yr. (*b*) 4–5 yr. (*c*) 4–0 yr. (*d*) 3.5 yr. (*b*)
12. According to Ryve's formula for estimating floods, the peak discharge is proportional to: (*AMIE s'* 98)
    (*a*) A (*b*) $A^{1/2}$ (*c*) $A^{2/3}$ (*d*) $A^{3/4}$
13. According to Dicken's formula for estimating floods, the peak discharge is proportional to: (*AMIE w'* 97)
    (*a*) A (*b*) $A^{1/2}$ (*c*) $A^{2/3}$ (*d*) $A^{3/4}$ (*d*)
14. The relation between probability *P* and recurrence interval *T* is given by:
    (*a*) $PT = 1$ (*b*) $PT^2 = 1$ (*c*) PT^{-1} (*d*) $PT^{-1} = 1$
15. Rainfall hyotograph shows the variation of:
    (*a*) cumulative rainfall with time (*b*) rainfall intensity with time
    (*c*) rainfall depth over area (*d*) rainfall intensity with cumulative rainfall
16. An siohyet is a line joining points having: (*AMIE s'* 94)
    (*a*) equal eraporation value (*b*) equal barometric pressure
    (*c*) equal height above MSL (*d*) equal rainfall depth in a given (*c*)
17. Pan evaporation is a hyotograph is a plot of: (*AMIE s'* 94)
    (*a*) cumulative rainfall Vs time (*b*) rainfall depth Vs duration
    (*c*) evaporation rate Vs time (*d*) rainfall intensity Vs time (*d*)
18. The total observed runoff volume during a 6-h storm with uniform intensity of 1.5 cm/h is 21.6×10^6 m^3, If the area of basm is 300 km^2, then average infiltration rate is .... (0.3 cm/h) (*AMIE s'* 95)
19. A 6-h storm had 6 cm of rainfall and resulting runoff was 3 cm. If index remains at the same value, runoff due to 12 cm of rainfall in 9h in the catchment is .... cm 7.5. (*AMIE w'* 94)
20. Envelope curve is a method to: (*AMIE w'* 99)
    (*a*) determine peak flood (*b*) determine infiltration capacity
    (*c*) estimate interception (*d*) route the flood through reservoir (*a*)
21. The probable maximum flood is: (*AMIE w'* 98)
    (*a*) an impossibly large flood discharge
    (*b*) a flood discharge that is maximum over the past year
    (*c*) an extremely large but physically possible flood
    (*d*) a flood with maximum probability of occurrence
22. The drainage density is the: (*AMIE w'* 98)
    (*a*) average length of streams per unit drainage area within the basim
    (*b*) stream discharge per unit drainage area
    (*c*) drainage area per unit stream length
    (*d*) number of streams per unit drainage area
23. Stream that contribute to the ground water are called: (*AMIE w'* 98)
    (*a*) effluent stream (*b*) ground water streams
    (*c*) influent streams (*d*) permannial streams

24. Double-mass curve technique is adopted to: (*AMIE w*' 98)
   (*a*) estimate missing data
   (*b*) obtain intensities of rainfall at various durations
   (*c*) check consistency of data
   (*d*) obtain storage capacity needed for a reservoir
25. A plot between rainfall intensity Vs time is called: (*AMIE w*' 95)
   (*a*) mass curve (*b*) hyetograph (*c*) isohyet (*d*) hydrograph (*b*)
26. Double mass analysis is adopted to: (*AMIE s*' 94)
   (*a*) estimate missing rainfall data
   (*b*) obtain intensities of rainfall at various durations
   (*c*) check consistency of data
   (*d*) obtain amount of storage needed to maintain (*c*)
27. Slope-area method is extensively used in: (*AMIE s*' 96 & *s*' 01)
   (*a*) development of rating curve
   (*b*) estimation of flood discharge based on high-water marks
   (*c*) cases where shifting control exists
   (*d*) cases where back-water effect is present (*b*)
28. Determine discharge of a stream from rating curve, at a section one needs: (*AMIE w*' 94)
   (*a*) current meter readings at various verticals at the section
   (*b*) slope of the water surface at the section
   (*c*) stage at the section
   (*d*) surface velocity at various sections (*c*)
29. Other conditions remaining the same runoff is maximum from precipitation occurring over (*AMIE s*' 97)
   (*a*) sandy strata (*b*) rocky area
   (*c*) alluvial treat after a day spell (*d*) boulders (*b*)
30. Severse tropical cyclones in the Bay of Bengal that cause intense rainfall are most likely to occur in the month of
   (*a*) March (*b*) June (*c*) November (*d*) January
31. An anticydone is:
   (*a*) a low pressure system with clockwise winds in northern hemisphere
   (*b*) a high pressure system with clockwise winds in northern harmsphere
   (*c*) zone of low pressure with anticlockwise winds in northern hemisphere
   (*d*) zone of high pressure with antichockwise winds in northern hamisphere
32. In a river carrying a discharge of 141-42 cumecs, slope was 1 in 6000. If during a flood, the stage at *A* was 3.6 m and water surface slope was 1 m 3000, the flood discharge (in cumecs) was approximately .... : 200 $\left[\dfrac{Q}{141.42} = \left(\dfrac{1/3000}{1/6000}\right)^{1/2}\right]$ (*AMIE s*' 96)
33. According to Dichen's formula for estimating floods, the peak discharge is proportional to:
   (*a*) A (*b*) $A^{1/2}$ (*c*) $A^{2/3}$ (*d*) $A^{3/4}$
   where, A is the catchment area (km^2)
34. The annual maximum values of flood for 22 years are arranged in decreasing order. Using weibull formula, the flood magnitude having a rank 5.
35. A hydrograph is a relation between time in hours and:
   (*a*) rainfall (mm) (*b*) peredation (cm)
   (*c*) surface river off (cumecs) (*d*) evaporation (cm)

36. Due to valley storage the peak discharge of an outflow hydrograph is:
(*a*) increased (*b*) reduced (*c*) not affected
(*d*) sometimes increased and sometimes reduced

37. If the peak of a 3 hour UH is 1000 cumecs, the flood hydrograph peak for a net rainfall of 150 mm and a base flow of 200 cumecs is:
(*a*) 14800 cumces (*b*) 12000 cumec (*c*) 15200 cumecs (*d*) 18000 cumecs

38. The area below a direct runoff hydrograph is:
(*a*) always greater them that below the corresponding rainfall histogram;
(*b*) always less them that below the corresponding rainfall histogram
(*c*) always equal to that below the corresponding rainfall histogram
(*d*) sometimes greater and sometimes less than that below the corresponding rainfall histogram

39. Unit duration of unit hydrograph is derived from effective rainfall of periods equal to about .... of the peaking time
(*a*) 3/4 (*b*) 1/2 (*c*) 1/4 (*d*) none of them

40. A tolerance of .... % is usually accepted in the variation of the unit duration from the desired unit duration derived from effective rainfalls of periods equal to about 1/4th of the peaking time:
(*a*) 45 (*b*) 35 (*c*) 25 (*d*) 15 (*c*)

41. Match the sentences of column A to that of column B

(*i*)

Column A	Column B
(*a*) Ryve's formula	(*A*) to determine average depth of rainfall
(*b*) Theissen's polygon	(*B*) for estimation of peak flood
(*c*) Lacey's formuale	(*C*) design of alluvial channels

(*a*– *B*), (*b*– *A*), (*c*– *C*)

(*ii*)

(*a*) Isohyet	(*A*) A line joining points on catchment having equal time of interval
(*b*) Isochrone	(*B*) lines having equal depth of evaporation
(*c*) Isopleths	(*C*) line joining points of equal rainfall magnitude

(*a*– *C*), (*b*– *A*), (*c*– *B*) (*AMIE w*'97)

(*iii*)

(*a*) Lysimeter	(*A*) measurement of stage
(*b*) Current meter	(*B*) measurement of evapotranspiration
(*c*) Bubble gange	(*C*) measurement of velocity of stream flow

(*a*– *B*), (*b*– *C*), (*c*– *A*)

(*iv*)

(*a*) Hydranlic jump	(*A*) Spillway
(*b*) Seepage	(*B*) Hydroelectric power project
(*c*) Penstock	(*C*) Earth dam

42. The area of a drainage basim is twice that of an adjoining are. Excepting area, other characteristics of the two basins are identical. Q_1 and Q_2 are the peak discharges, computed by Dickan's formula, for the smaller and lårger basin respectively. Then $Q_2 = Q_2 \times$ .... 1.68 (*AMIE w*' 96)

43. A rainfall of 1.2 cm occurred in a 6h storm. If ϕ index is 0.25 cm/h, rainfall excess of this storm was .... (*AMIE s*' 2000)

44. A one day rainfall of 1 cm at a place *x* was found to have a return period of 100 years. The probability that a one day rainfall of this or laryer morgnitude will not occur at *x* during the next 50 years is: 0.60 (*AMIE s*' 96)

45. Due to valley storage, the peak discharge of an outflow hydrograph is: (*AMIE w*' 97)

(*a*) increased (*b*) reduced (*c*) not affected
(*d*) sometimes increased and sometimes reduced (*b*)

46. According to Dicken's formula for estimating floods, the peak discharge is proportional to (*AMIE w*' 97)
(*a*) A (*b*) $A^{1/2}$ (*c*) $A^{2/3}$ (*d*) $A^{5/3}$ (*d*)

47. According to Dicken's formale, peak flood discharge is proportional to: (*AMIE s*' 93)
(*a*) area (*b*) $(\text{area})^{1/2}$ (*c*) $(\text{area})^{2/3}$ (*d*) $\text{area}^{3/4}$ (*b*)

48. The relation between probability P and recurrence interval T is (*AMIE w*' 98)
(*a*) $PT = 1$ (*b*) $PT^2 = 1$ (*c*) $PT^{-1} = 1$ (*d*) $P^{-1}T = 1$

49. The empirical formula for estimation of design flood of a catchment involving drainage area and its shape is .... formula: (*AMIE s*' 96)
(*a*) Dicken's (*b*) Inglis (*c*) Janis (*d*) Dredge or Burge (*d*)

50. Flow duration curve is a plot of: (*AMIE s*' 96)
(*a*) stream discharge against percentage of time the flow is equalled or exceeded
(*b*) stream discharge against time in chronological order
(*c*) accumulated stream flow against time
(*d*) base flow against percentage of time the flow is exceeded (*a*)

51. Virigin flow denotes flow in the river: (*AMIE s*' 96)
(*a*) domenstream of ganging station (*b*) upstream of a ganging station
(*c*) unaffected by works of man (*d*) at a ganging station
(*c*)

52. Evaporation depends on:
(*a*) temperature (*b*) humidity (*c*) wind velocity (*d*) all the above

53. Humidity is measured by a: (*AMIE s*' 00)
(*a*) hydrometer (*b*) hyetometer (*c*) hygrometer (*d*) pysheiometer

54. In a storm the average depth of rainfall: (*AMIE s*' 00)
(*a*) increases with area (*b*) decreases linearly with the area
(*c*) has no relationship with area (*d*) decreases exponentially with the area

55. Potential evapotramspiration is: (*AMIE s*' 00)
(*a*) evaporation when there is sufficient moisture available to a fully vegetated area
(*b*) evaporationspiration of a forest area
(*c*) actual evapotionspiration of a crop before application of irrigation water
(*d*) amount of water needed to bring the moisture content of a soil to its field capacity

56. In estimating the maximum floods for design purposes in the absence of any other data, the ϕ index is usually assumed as: (*AMIE s*' 00)
(*a*) 2 cm/h (*b*) 1 cm/h (*c*) 0.5 cm/h (*d*) 0.1 cm/h

57. A flood hydrograph due to an isolated storm consists of: (*AMIE s*' 00)
(*a*) surface runoff, interflow and base flow phases
(*b*) surface runoff only (*c*) base flow only
(*d*) surface runoff and base flow phases

58. The stream that provides water to water table is termed as (*AMIE s*' 00)
(*a*) affluent (*b*) influent (*c*) ephemeral (*d*) effluent

59. Base flow reparation is performed (*AMIE w*' 96)
(*a*) on a U.H to get direct runoff hydrograph
(*b*) on a flood hydrograph to obtain magnitude of effective rainfall
(*c*) on flood hydrographs to obtain magnitude of effective rainfall
(*d*) on hydrographs of effluent streams only (*b*)

60. A catchment with an area of 756 km² has a 6 h unit hydrogbraph which is triangular with a

base of 70 h. The peak discharge of DRH due to 5 cm rainfall excess in 6 h from the catchment is .... (*AMIE s'* 00)

61. A river has an average surface width of 20 m. If the evaporation measured and the vicinity of the river is 0.5 m m/day, the volume of water evaporated in 60 km stretch of the river in a month of 30 days (in m^3) is .... (*AMIE s'* 00)
62. The 4-h unit hydrograph of a catchment may be taken as triangular. If the base width is 72 h and peak discharge is 50 cumecs, the area of the catchment is .... km^2 (*AMIE w'* 97)
63. A catchment area of .... km^2 and perimeter 887.6 km has compactness coefficient 1.4. (32,000 km^2) .... m^3/day of flow enters into a drain from a C.A. of 14 ha with a drainage coefficient of 21.5 mm (3010) (*AMIE s'* 97)
64. Storm map method is a method for the determination of: (*AMIE s'* 97)
    (*a*) peak flood (*b*) mapping of storm
    (*c*) flood hydrograph (*d*) intensity of storm (*d*)
65. The points connecting equal evapotionspirtion is known as: (*AMIE w'* 98)
    (*a*) isobar (*b*) isopleth (*c*) isolyet (*d*) isoval
66. ϕ index is defined as .... (*AMIE w'* 97)
67. A hyotograph is the plot of .... (*AMIE w'* 97)
68. Water enters the soil of capacity rate when intensity of rainfall is: (*AMIE s'* 2000)
    (*a*) more than infiltiation capacity of the soil
    (*b*) less than infilration capacity of the soil
    (*c*) equal to infiltration capacity of the soil
    (*d*) slightly less than infiltration capacity of the soil
69. The ordinates of unit hydrograph may be obtained by dividing the ordiantes of the direct runoff hydrograph of a storm by: (*AMIE s'* 2000)
    (*a*) duration of unit hydrograph (*b*) storm duration
    (*c*) base flow volume (*d*) direct runoff volume
70. Rational method gives discharge from a catchment as $Q = 2.78$ CIA, in which:
    (*a*) Q is in m^2/s, I is in cm/h and A is in km^2
    (*b*) Q is in ha-m/h, I is in mm/h and $A$is in km^2
    (*c*) Q is in m^3/day, I is cm/day, and A is in m^2
    (*d*) Q is in m^3/h, I is in mm/h and A is m km^2
71. In a zone of aeration, soil has pores filled with
    (*a*) water (*b*) water and air (*c*) air (*d*) soluble salts
72. For a storm of 2-h duration, rainfall rates in cm/h at 20 minutes intervals are 2.5, 2.5, 10.0, 7.5, 5.1 and 1.25. If ϕ index is 3 cm/h, the total runoff is:
    (*a*) 4.53 cm (*b*) 45.3 cm (*c*) 0.453 cm (*d*) none ot the above
73. Water shed is a line joining the points having. The requent peak algorithem is a technique used in the estimation of .... (*AMIE s'* 98)
74. .... is used commonly measure actual evapotranspiration (*AMIE w'* 98)
75. Fan shaped catchment gives .... runoff them fan shaped leaf shaped catchment (*AMIE w'* 98)
76. Match the following:

Part A	Part B
(*i*) Aeration	(*I*) calloidal matter and incrorg anism
(*ii*) Screening	(*II*) living organism
(*iii*) Sedimentation aided with coagulation	(*III*) elements causing taste and odour
(*iv*) Filteration	(*IV*) creacking clogging filter beds
(*v*) Chemical treatment sedimentation and filtration	(*V*) dissolved matter

(*vi*) Disinfection (*VI*) floating matter

77. A major catchment is divided into 4 sub basins with areas 900, 700, 1000 and 1400 sq/km respectively. If average annual rainfall on the bains are 100, 80, 100 and 110 respectively, then average rainfall of major catchment as a whole is .... cm (*AMIE w' 98*)
78. A storm with uniform intensity occuring over a basin for a period of 6 h produced an effective rainfall of 15 cm. The peak discharge of its hydrograph is 930 cumecs. Over the same basin, if another storm of same duration but with effective rainfall of 7.5 cm occurred, the peak ordinate of its hydrograph is .... cumecs. (*AMIE w' 98*)
79. The area of a direct runoff hydrigraph is always: (*AMIE w' 98*)
    (*a*) greater than that below corresponding rainfall histogram
    (*b*) less than that below corresponding rainfall histogram
    (*c*) equal to that below corresponding rainfall histogram
    (*d*) independent of corresponding rainfall histogram
80. A 6-hr rainfall of 60 mm at a place A was found to have a return period of 40 years. The probability that at A 6-ha precipitation of this magnitude or larger amount will occur at least once in 20 successive years is:
    (*a*) 0.397 (*b*) 0.309 (*c*) 0.603 (*d*) 0.025 (*c*)
81. For the estimation of flood peak, the value of Q in Dicken's formula is given by: (*AMIE s' 01*)
    (*a*) $CM^{3/4}$ (*b*) $CM^{3/3}$ (*c*) $CM^{3/2}$ (*d*) $CM^{4/3}$ (*a*)
82. The ordinate of unit hydrograph may be obtained by dividing the ordinates of direct runoff hydrograph of a storm by (*AMIE s' 2000*)
    (*a*) duration of unit hydrograph (*b*) storm duration
    (*c*) base flow volume (*d*) direct runoff volume
83. Due to valley storage, the peak discharge of an outflow hydrograph (*AMIE s' 2000*)
    (*a*) increases (*b*) reduces
    (*c*) is not affected (*d*) not any of the above
84. A hyetograph is a plot against time of: (*AMIE s' 96*)
    (*a*) cumulative rainfall (*b*) rainfall intensity
    (*c*) discharge (*d*) stage
85. A hydrograph is a relation between time in hours and: (*AMIE s' 97*)
    (*a*) rainfall (mm) (*b*) perrolation (cm)
    (*c*) surface runoff (cumec) (*d*) evaporation (cm) (*c*)
86. Base flow reparation is used in connection with: (*AMIE w' 97*)
    (*a*) seepage flow (*b*) infiltration (*c*) evaporation (*d*) stream flow (*d*)
87. The time base of a hydrograph increases with: (*AMIE s' 97*)
    (*a*) increase in intensity of storm (*b*) decrease in storm duration
    (*c*) increase in time of concentration (*d*) decrease in influtration capacity of storm (*c*)
88. Increase in duration of unit hydrograph causes an increase in the: (*AMIE w' 98*)
    (*a*) peak dischange (*b*) rainfall intensity
    (*c*) equilibrium discharge (*d*) time base (*d*)
89. Term base flow denotes delayed flow reaching a stream from: (*AMIE s' 96*)
    (*a*) ground water only (*b*) ground water and snow melt
    (*c*) ground water and interflow (*d*) ground water, snow melt and interflow (*a*)
90. The use of rational formula for estimating floods is limited to catchments of sizes less than:
    (*AMIE w' 95*)
    (*a*) 5000 km^2 (*b*) 500 km^2 (*c*) 50 km^2 (*d*) 5 km^2 (*c*)

91. A 2-h unit hydrograph of a catchment is triangular in shape with a time base of 60 h and peak ordinate of 30 cumecs. The area of catchment is .... km^2 648 $\left(\frac{30\times60\times60\times60}{1cm\times10^{-2}\times10^{-6}}\right)$

(*AMIE w' 95*)

92. A triangular direct runoff hydragraph due to a storm has a time base of 80 h and a peak flow of 50 cumecs occurring at 20 h from the start. If the catchment area is 144 km^2, the rainfall excess in the storm was .... cm (5) $\left(\frac{1}{2}\times50\times60\times60\times80\times10^2/144\right)$

(*AMIE s' 96*)

93. The basic assumptions of unit hydrograph theory are: (*AMIE w' 94*)
   (*a*) linear response and non-linear time variance
   (*b*) linear response and time invariance
   (*c*) non-linear response and time invariance
   (*d*) linear response and linear time invariance (*b*)
94. The basic assumptions of unit hydrograph theory are: (*AMIE w' 98 & s' 01*)
   (*a*) linear response and time variance (*b*) linear response and time invariance
   (*c*) non-linear response and time variance (*d*) non-linear response and time invariance
95. Standard project flood as compared to probable maximum flood is: (*AMIE w' 97*)
   (*a*) larger (*b*) equal (*c*) smaller (*d*) uinrelated (*c*)
96. The annual maximum values of flood for 22 years are arranged in decreasing order. Using weibull formula. The flood magnitude having a rank of 5 has the return period of about: (*AMIE w' 97*)
   (*a*) 5 – 0 yrs (*b*) 4 – 5 yrs (*c*) 4 – 0 yrs (*d*) 3 – 5 yrs (*b*)
97. The method used for estimating missing rainfall data is: (*AMIE w' 97*)
   (*a*) normal ratio method (*b*) station year method
   (*c*) plotting positive method (*d*) natioanl method (*a*)
98. The empirical formula involving only drainage area for estimation of peak dischange for the catchment of North India is .... formula: (*AMIE w' 94*)
   (*a*) Ryve's (*b*) Inglis
   (*c*) Nawab Jung Bahadur (*d*) Dicken's (*d*)
99. Increase of depth increases the rate of evaporation in: (*AMIE w' 94*)
   (*a*) summer season only (*b*) winter season only
   (*c*) both summer and winter seasons (*d*) rany reason only (*b*)
100. In connection with stream flow measurement, slope area method is extensively used in: (*AMIE w' 94*)
   (*a*) development of rating curve
   (*b*) estimation of flood dischange based on high water marks
   (*c*) cases where shifting control exist (*d*) cases where backwater effect is present (*b*)
101. Orographic precipitation occurs due to air masses being lifted to higher altitudes by: (*AMIE w' 94*)
   (*a*) presence of mountain barriers (*b*)density difference of air masses
   (*c*) a prontal action (*d*) extra-trapical cyclones (*d*)
102. The flow mass curve is an integral cumec of: (*AMIE w' 94*)
   (*a*) hydrograph (*b*) hyetograph
   (*c*) flow duration curce (*d*) S-curce (*a*)
103. When a river starts meandering, the sediment carrying capacity: (*Engg. Sev. Exam.'98*)
   (*a*) first decreases and ultimately increases (*b*) first increased and ultimately decreases

(*c*) remains unaffected as the plan changes continue
(*d*) changes unaffected all the time leaving permanent braids (*c*)

104. A channel of bed slope 0.0009 carries a discharge of 30 cumcs when the depth of flow is 1 m. What is the discharge (in cumecs) carried by a similar channel at the same depth of flow if the slope is increased to 0.0001? (*Engg. Sev. Exam.*'97)
(*a*) 10 (*b*) 15 (*c*) 60 (*d*) 90 (*a*)

105. Mean precipitation over an area is best obtained from ganged amounts by .... method (*Engg. Sev. Exam.*'97)
(*a*) arithmetic mean (*b*) Thiessen
(*c*) linearly interpolated isohytel (*d*) orographically weighted isohyetal method

106. The following steps are involved in arriving at or unit hydrograph
(*a*) estimating surface runoff in depth (*b*) estimating surface runoff in volume
(*c*) reparation of base flow
(*d*) dividing surface runoff ordinate by depth of runoff

107. The correct requiems of these steps is: (*Engg. Sev. Exam.*'97)
(*a*) 3, 2, 1, 4 (*b*) 2, 3, 4, 1 (*c*) 3, 1, 2, 4 (*d*) 4, 3, 2, 1 (*a*)

108. Probability of a 10 yr flood to occur at least once in next 4 yrs is .... % (*Engg. Sev. Exam.*'97)
(*a*) 25 (*b*) 35 (*c*) 50 (*d*) 65 (*b*)

109. The standard project flood is: (*Engg. Sev. Exam.*'97)
(*a*) derived from probable maximum precipitation in the region
(*b*) derived from severemort meteorologrical conditions any where in the country
(*c*) flood with return period of 100 yrs
(*d*) the same as probable maximum flood (*b*)

16. RESERVOIRS AND DAMS PLANNING

1. A fresh water reservoir is the reservoir:
(*a*) which receives fresh water directly from the catchment
(*b*) in which inflow-outflow cycle is such that water is replenished at bast once, preferably twice in a year
(*c*) in which entire inflow is stood (*d*) all of the above (*b*)

2. A fresh water reservoir has capacity-inflow ratio:
(*a*) high (*b*) low (*c*) medium (*b*)

3. The storage capacity of a reservoir is determined by:
(*a*) mass curve (*b*) analytical method
(*c*) bar graph method (*d*) all of these (*d*)

4. Reservoir yield is the:
(*a*) amount of water stored
(*b*) amount of water that can be released in a certain interval of time
(*c*) amount of water available (in storage) during critical dry period
(*d*) all of the above (*b*)

5. Firm on safe yield of a reservoir is the:
(*a*) amount of storageat full reservoir level
(*b*) amount of storage at normal reservoir level
(*c*) maximum amount of supply or yield guaranteed during an critical dry period
(*d*) all of the above (*c*)

6. No storage is required if the slope of demand curve is .... then that of mass curve:
(*a*) always greater (*b*) always smaller
(*c*) sometimes equal and sometimes greater (*d*) sometimes equal and sometimes smaller (*a*)

7. Secondary yield of a reservoir is the:
   (*a*) quantity of water available in excess of safe yield during periods of high flows
   (*b*) quantity of water available between full reservoir level and normal reservoir level
   (*c*) quantity of water available between maximum reservoir level and full reservoir level
   (*d*) quantity of water available between maximum drawdown level and full reservoir level (*a*)
8. Trap efficiency of a reservoir is the ratio of:
   (*a*) sediment retained to the sediment inflow into the reservoir
   (*b*) sediment inflow to sediment outflow
   (*c*) sediment retained to sediment outflow (*d*) all of the above (*a*)
9. Trap efficiency of a reservoir is determined by:
   (*a*) Brume curve (*b*) strange curve
   (*c*) gange-discharge curve (*d*) none of these (*a*)
10. Rate of reservoir is considered to .... with the advance of years:
   (*a*) full off (*b*) remains the same
   (*c*) increases (*d*) no relation (*a*)
11. Sedimentation survey of Indian reservoirs indicate that the rate of sedimentation very between ha m annually/100 sq. km.
   (*a*) 3.75 to 9.5 (*b*) 4.75 to 12.0 (*c*) 4.75 to 14.25 (*d*) 5 – 10 (*c*)
12. ISI recommended provision for sedimentation at the rate of .... ha m/hr/sq. km:
   (*a*) 0.01 to 0.02 (*b*) 0.02 to 0.05 (*c*) 0.05 to 0.10 (*d*) 0.10 to 0.2 (*d*)
13. Rate of reservoir sedimentation is determined by:
   (*a*) inflow-outflow method (*b*) capacity survey method
   (*c*) both of them (*c*)
14. The sedimentation surveys of major reservoirs one made when they lose .... percent of their capacity:
   (*a*) 1 (*b*) 2 (*c*) 3 (*d*) 5 (*c*)
15. Minimum drawdown level of a reservoir in India is fixed as is likely to develop after .... years of its operation:
   (*a*) 50 (*b*) 100 (*c*) 200 (*d*) 500 (*b*)
16. Full service time of an irrigation project is considered to be to less than .... years after the start of operation:
   (*a*) 50 (*b*) 100 (*c*) 200 (*d*) 300 (*a*)
17. Retarding reservoir for flood control is provided with: (*AMIE w*' 98)
   (*a*) uncontrolled outlets and spillway (*b*) uncontrolled outlets and controlled spillway
   (*c*) controlled outlets and uncontrolled spillway
   (*d*) controlled outlets and spillway
18. Evaporation is maximum: (*AMIE s*' 97)
   (*a*) when temperature is very low (*b*) when temperature is high
   (*c*) when it is raining (*d*) during night (*b*)
19. Other factors remaining the same, open water evaporation rate increases as: (*AMIE w*' 96)
   (*a*) sunshine hours increase (*b*) altitude decreases
   (*c*) barometric pressure increases (*d*) water salt content decreases (*d*)
20. Evaporation is measured by using: (*AMIE s*' 94)
   (*a*) an open pan (*b*) lysimeter (*c*) infiltrometer (*d*) neutron tube (*c*)
21. The chemical suitable for use as a water evaporation retarding and reducing agent is: (*AMIE s*' 94)

(*a*) butyl alcohol (*b*) methyl alcohol (*c*) cetyl alcohol (*d*) ethyl alcohol (*b*)

22. The average pan coefficient for standard Us weather Burean class A Pan is : (*AMIE w*' 94)
(*a*) 0.55 (*b*) 0.70 (*c*) 0.85 (*d*) 0.95 (*b*)

23. The average pan coefficient for standard Us weather Burean class A Pan is: (*AMIE w*' 95 & *s*' 96)
(*a*) 0.60 (*b*) 0.65 (*c*) 0.70 (*d*) 0.75 (*c*)

24. Pan coefficient is always .... take evaporation:
(*a*) < (*b*) > (*c*) = (*d*) does not depend on

25. If Pan evaporation indicates a total yearly evaporation of 1.5 m, then annual water loss by evaporation from reservoir of water surface of 100 sq km will be: (*AMIE s*' 2000)
(*a*) 1.5×10^3 ha m (*b*) 1.5×10^4 ha m (*c*) 1.5×10^5 ha m (*d*) 1.5×10^6 ha m

26. Pan evaporation is always .... than lake evaporation (*AMIE w*' 98)
(*a*) less (*b*) more (*c*) equal (*d*) none of them (*b*)

27. Indicate type of dam suitable for the foundation comprising:
(*i*) Impervious solid rock
(*a*) earthen dam (*b*) rockfill dam (*c*) concrete dam (*d*) all of them (*d*)
(*ii*) Coarse sand and grand:
(*a*) earthen dam (*b*) rockfill dam (*c*) concrete dam (*d*) none of them (*b*)
(*iii*) Heavy overburden on rock larger:
(*a*) gravity dam (*b*) stone marony dam
(*c*) buttress dam (*d*) all of them (*c*)
(*iv*) Clay, plastic clay:
(*a*) earthen dam (*b*) low gravity dam (*c*) stone masony dam (*d*) none of them
(*a*) with very flat slopes

28. Indicate type of dam suitable for:
(*i*) V-shaped narrow gorge:
(*a*) earthen dam (*b*) gravity dam (*c*) arch dam (*d*) all of them (*c*)
(*ii*) U-shaped narrow gorge:
(*a*) earthen dam (*b*) overflow concrete dam
(*c*) arch dam (*d*) none of them (*b*)
(*iii*) Low-wide gorge:
(*a*) earthen dam (*b*) concrete gravity dam
(*c*) arch dam (*d*) all of them (*a*)
(*iv*) site where pervious and impervious soils and stores are available:
(*a*) earth dam (*b*) concrete dam (*c*) arch dam (*d*) all of them (*a*)
(*v*) a site where spillway location is available separate from dam:
(*v*) earthen dam (*b*) arch dam (*c*) concrete dam (*d*) all of them (*a*)
(*vi*) a site where there is a requirement of large capacity spillway:
(*a*) earthen dam (*b*) arch dam (*c*) overflow dam (*d*) none of them (*c*)
(*vii*) s site where excessive uplift pressure is approchended:
(*a*) earthen dam (*b*) arch dam (*c*) gravity dam (*d*) rockfill dam (*b*)

True of False

(*i*) In a zoned dam, the entire dam section except the impervious core forms part of drainage system. (True)

(*ii*) In a zoned dam, the permeability increases from the centre towards each slope. (True)

(*iii*) In a homogeneous dam, the finer the soil, the higher the dam and flatter are the slopes. (True) (*iv*) A purely homogenous dam is essentially of low height and of relatively bigger section with flatter slopes. (True)

(*v*) Rock toe is essentially provided in earthen dam, homogencous dam and zoned dam to take care of seepage. (True)

(*vi*) In a massive section of zoned earthen dam, the water barrier is provided by the imperious zone only (True)

1. Volume of water stored between normal reservoir level and maximum reservoir level is known as: (*AMIE w*' 96)
   (*a*) line storage (*b*) surcharge storage
   (*c*) valley storage (*d*) bank storage (*b*)
2. The difference in elevation between top of dam and full reservoir level is called: (*AMIE s*' 98)
   (*a*) berm (*b*) free board (*c*) height of bank (*d*) wave height (*b*)
3. Area increment method is studied in connection with: (*AMIE s*' 97)
   (*a*) sediment deposition (*b*) population estimate
   (*c*) discharge measurement (*d*) reservoir capacity estimate (*d*)
4. In a reservoir with steep slope, sediments deposit take place in .... reach of the reservoir
   (*a*) upper (*b*) middle (*c*) lower (*d*) any where (*c*)
5. Large sediments deposit takes place in .... reaches in reservoirs of short length.
   (*a*) upper (*b*) middle (*c*) lower (*d*) all of them (*c*)
6. Longe sediments deposit takes place in .... in reservoir of longer length:
   (*a*) upper (*b*) middle (*c*) lower (*d*) all of them (*a*)
7. In a river where in flood in early part of monsoon, more sediments are deposited in .... elevation
   (*a*) high (*b*) intermediate (*c*) lower (*d*) all of them (*c*)
8. In a river in which loads occur the end of monsoon, more sediments are deposited
   (*a*) upper (*b*) middle (*c*) lower (*d*) all of them (*a*)
9. In a reservoir, the line storage is defined as the volume of water between: (*AMIE s*' 98)
   (*a*) full reservoir level and minimum reservoir level
   (*b*) top of dam and full reservoir level
   (*c*) top of dam and minimum reservoir level
   (*d*) top of dam and base of the dam (*a*)
10. The dead storage capacity of a storage reservoir is usually decided by: (*AMIE w*' 99)
    (*a*) using mass curve of inflow and demand
    (*b*) using the request peak algoithen and optimisation technique
    (*c*) based on the expected life and sedimentation characteristics of the proposed reservoir
    (*d*) economic analysis to get the minimisation of the cost of the reservoir
11. Flood hydrograph routed through a reservoir shows: (*AMIE s*' 98)
    (*a*) reduction in peak discharge (*b*) only increase in basin lag
    (*c*) reduction in peak discharge, increased base width
    (*d*) reduction in peak discharge, increased base width and increase in basin lag (*d*)
12. The equation used in all types of flood routing is: (*AMIE w*' 97)
    (*a*) equation of motion (*b*) energy equation
    (*c*) continuity equation (*d*) momentum equation (*c*)
13. A linear reservoir is one in which linear relationship holds between: (*AMIE w*' 97)
    (*a*) storage and elevation (*b*) storage and tune
    (*c*) storage and inflow rate (*d*) storage and outflow rate (*d*)
14. Flood routing considers the following factors: (*AMIE s*' 97)

(*a*) inflow and outflow hydrographs (*b*) spillway discharge and reservoir storage
(*c*) inflow hydrograph and spillway discharge
(*d*) all of the above factors (*c*)

15. The requance of sediment deposition in a reservoir is:
(*a*) coarse material in bottom set being heavier and fine material
(*b*) bottom-set mixture of fine and coarse material, foreset beds of coarse material and top-set beds of very corse material
(*c*) just reverse of the above (*b*)

16. Density current any sediment of:
(*a*) coarse material (*b*) fine material
(*c*) mixture of coarse and fine material (*d*) very coarse material (*b*)

17. Multipurpose project is one which is constructed to: (*AMIE w*' 98)
(*a*) be used in combination with other reservoirs
(*b*) supply water for more than one year
(*c*) serve more than one purpose
(*d*) serve one main purpose always and other purposes occasionally (*c*)

18. In a multipurpose reservoir, the requirement for flood control is: (*AMIE w*' 00)
(*a*) most compatible with other requirements
(*b*) compatible with water power requirement
(*c*) compatible with requirements for irrigation and drinking water
(*d*) non-compatiable with other requirements

19. The sequent peak alongwith is used in the esternation of: (*AMIE w*' 97)
(*a*) surcharge storage capacity of a reservoir
(*b*) peak discharge at a channel section
(*c*) maximum alternation owing to storage effect
(*d*) minimum storage capacity of reservoir (*d*)

20. Using identical conditions, the evaporation from sea water as compared to fresh water is: (*AMIE w*' 97)
(*a*) less than that of fresh water (*b*) equal to that of fresh water
(*c*) more than that of fresh water (*d*) no definte conclusion can be made (*a*)

21. Surcharge storage of servoir to the volume of water stored between: (*AMIE w*' 94)
(*a*) normal pool level and maximum pool level
(*b*) maximum pool level and minimum pool level
(*c*) minimum pool level and normal pool level
(*d*) normal pool level and river bed level (*a*)

22. In a flow-mass curve study, the demand line drawn from a sidge in the curve did not intersect the mass curve again. This represents that: (*AMIE w*' 94)
(*a*) the reservoir was not full at the beginning
(*b*) storge was not adequate
(*c*) demand connot be met by the inflow as the reservoir will not refill
(*d*) reservoir if wasting water by spill (*c*)

17. EARTH AND ROCKFILL DAMS

1. Berms on downstream slope of the dam serve to:
(*a*) increase sleep stability (*b*) reduce surface erosion
(*c*) provide roadway (*d*) all of them (*d*)

2. Berm on upstream slope of the dam serves to:
(*a*) increase slope stability (*b*) reduce surface erosion
(*c*) support foundation of riprap (*d*) all of them (*c*)

3. Settlement allowance generally provided as a percent of dam embankment height above designed top level is:
(*a*) 1 – 2 (*b*) 2 – 3 (*c*) 3 – 4 (*d*) 4 – 5 (*a*)
4. Filter length usually adopted is .... of the distance of two of dam to centre of dam crest:
(*a*) 15% (*b*) 20% (*c*) 25% (*d*) 30% (*c*)
5. Flow net is employed to determine:
(*a*) seepage discharge (*b*) pore pressure
(*c*) both of them (*d*) none of them (*c*)
6. Dumped riprap is considered .... effective, them hand placed riprap, for wave protection:
(*a*) more (*b*) less (*c*) equally (*a*)
6. Dumped riprap is considered .... to hand placed riprap for wave protection:
(*a*) superior (*b*) inferior (*c*) similar (*a*)
7. Dam embankment slope has .... effect on wake side in dumped riprap compared to hand placed riprap:
(*a*) appreciable (*b*) no (*c*) equal (*b*)
8. Thickness of hand placed riprap is usually .... percent of dumped riprap:
(*a*) 25 (*b*) 40 (*c*) 50 (*d*) 60 (*c*)
9. The depth of cutoff in a zoned dam is about .... percent of stram height:
(*a*) 10% (*b*) 15% (*c*) 25% (*d*) 15–45% (*d*)
10. The width at bottom of cutoff in a zoned dam is about .... percent of dam height:
(*a*) 10 – 15 (*b*) 10 – 20 (*c*) 10 – 30 (*d*) 10 – 50 (*c*)
11. In a diaphragm dam, the central impervious core has a base width of .... times the water head against the dam:
(*a*) 0.10 – 0.20 (*b*) 0.15 – 0.25 (*c*) 0.20 – 0.30 (*d*) 0.3 – 0.5
12. The hydranlic fill dam is suitable where .... material is available:
(*a*) clayey (*b*) sandy (*c*) sandy and gravelly (*d*) all of them (*c*)
13. The top width of a high earthen dam varies from .... *m*:
(*a*) 2 – 5 (*b*) 5 – 6 (*c*) 6 – 8 (*d*) 6 – 12 (*c*)
14. The minimum top width of a low earthen dam is about .... *m*:
(*a*) 2 (*b*) 3 (*c*) 5 (*d*) 6 (*d*)
15. Minimum free board in the difference in elevation between:
(*a*) dam crest level and maximum reservoir level
(*b*) dam top level and normal reservoir level
(*c*) dam top level and full reservoir level (*a*)
16. Wave run up is maximum along:
(*a*) dumped riprap (*b*) hand placed riprap
(*c*) concrete pavement riprap (*c*)
17. A core in earthen dam is considered them if its width is less than .... percent of water head:
(*a*) 10 (*b*) 15 (*c*) 15 – 20 (*c*)
18. Core thickness of less than .... percent of water head is not usually used except where a large leakage through the core is not considered servoir consequence:
(*a*) 5 (*b*) 10 (*c*) 15 (*d*) 20 (*b*)
19. A core in earthen dam is considered thick if its width is .... percent of water head:
(*a*) 10 – 15 (*b*) 15 – 20 (*c*) 20 – 30 (*d*) 30 – 50 (*d*)
(*i*) The pervious and filter zones serve the same purpose *i.e.*, drainage system of the dam (True)

(*ii*) The pervious zone, in addition to drainage import stability and protect the impervious core (True)

(*iii*) Within the pervious zone, individual loads are directed so that the material is graded in coarseners towards the outer slopes. (True)

(*iv*) The inner pervious zone acts as a drain to control the line of seepage (False)

(*v*) The inner pervious zone provides stability against sudden drawdown and protects the impervious zoned (True)

(*vi*) The outer pervious zone acts as a drain to control the line of seepage. (True)

(*vii*) The outer pervious zone provides stability against sudden drawdown (False)

(*viii*) Free draining material is essentially required in upstream pervious zone. (True)

(*ix*) The upstream pervious zone may be of non-free draining material if it is not available (False)

(*x*) The downstream pervious zone may be of non-free draining material, if it is not available (True)

20. Phreatic line in seepage analysis is defined as the line on which pressure is: (*AMIE s*' 01)
    (*a*) equal to atmosphere (*b*) < atmosphere
    (*c*) < atmosphere (*d*) varying
21. From the upstream slope of an earthen dam the most critical condition is:
    (*a*) sudden drawdown condition (*b*) steady seepage condition
    (*c*) neither sudden drawdown nor steady seepage condition
    (*d*) during construction when reservoir is allowed to be filled
22. Seepage endangers the stability of an earthen dam built on pervious foundation because of piping which depends on: (*AMIE w*' 97)
    (*a*) height of dam (*b*) quantity of seepage flow
    (*c*) value of exit gradient (*d*) total storge capacity of reservoir
23. Piping under an earthen dam starts developing from the .... side of the structure.
24. Stability of the upstream slope of an earthen dam becomes critical for the condition of:
    (*a*) reservoir full (*b*) reservoir empty
    (*c*) sudden drawdown (*d*) slow filling of reservoir
25. The upstream face of an earthen dam is considered as:
    (*a*) equiptoential line (*b*) stream line
    (*c*) streak line (*d*) path line
26. The upstream face of an earth dam is: (*AMIE w*' 01)
    (*a*) a flow line (*b*) an equipotential line
    (*c*) a phreatic line (*d*) a stream line
27. In gorges with steep and strong side slopes, the most suitable type of dam is: (*AMIE w*' 01)
    (*a*) earth dam (*b*) gravity dam (*c*) buttress dam (*d*) arch dam
28. The commonly used earth dam is .... (*AMIE w*' 01)
29. Inverted filter for providing foundation diamage has:
    (*a*) multi-layers of soil particles of same permeability
    (*b*) multi-layers in which permeability increases from bottom to top
    (*c*) multi-layers in which permeability increases from top to bottom
    (*d*) only one layer or soil (*b*)
30. Failure of an earth dam due to flow of water under the foundation and emerging on the d/s side is termed as failure due to .... (piping) (*AMIE w*' 00)
31. A coffer dam is: (*AMIE s*' 94)
    (*a* kind of gravity dam
    (*b*) an earthen dam of small height built to protect important areas

(*c*) a temporary structure constructed to exclude water from the work area during construction

(*d*) an embankment built along a river to regulate the river (*c*)

32. Stability of upstream slope of an earth dam becomes critical for the condition of: (*AMIE s*' 99)

(*a*) reservoir full (*b*) reservoir empty

(*c*) sudden drawdown (*d*) slow filling of reservoir (*c*)

33. Seepage endangers the stability of an earthen dam built on pervious foundation become of piping which depends on: (*AMIE w*' 97)

(*a*) height of dam (*b*) quantity of seepage flow

(*c*) value of exit gradvit (*d*) total reservoir storage capacity (*c*)

34. For the same height and top width, the width at the bottom of the earth dam is: (*AMIE w*' 99)

(*a*) less than that of a concrete dam (*b*) same as that of concrete dam

(*c*) more them that of a concrete dam (*d*) none of the above (*c*)

(*a*) For an earthen dam (*A*) rock in abutments must be stronger than in foundation

(*b*) For an arch dam (*B*) foundation may be rock or soil

(*c*) For a solid gravity dam (*C*) foundation should consist of pervious material

(*d*) For a barrage (*D*) foundation rock should be stronger than in abutments

$(a-B)$, $(b-A)$, $(C-D)$, $(d-C)$ (*AMIE w*' 93)

35. Piping under an earthen dam starts developing from the .... side of the structure (*AMIE s*' 99)

36. Dam failures due to foundation problems are assessed as:

(*a*) 40% (*b*) 30% (*c*) 20% (*d*) 10% (*a*)

37. Dam failure due to inadequate spillway are assessed as:

(*a*) 40% (*b*) 33% (*c*) 23% (*d*) 15% (*c*)

38. Dam failures due to uneven settlement are assessed as:

(*a*) 40% (*b*) 30% (*c*) 20% (*d*) 10% (*d*)

39. A rockfill dam generally has rockfill to import .... to the dam:

(*a*) watetighness (*b*) stability (*c*) drainage system (*d*) all of the above (*b*)

40. The mass stability in rockfill dam is developed by:

(*a*) comenting agent binding the particles together

(*b*) friction and interaction fo one particle on another

(*c*) cementing agent as well as friction and interaction of one particle on another

(*d*) impervious membrane on upstream (*b*)

41. Compared to earthen dams the rockfill dams are considered as:

(*a*) more flexible (*b*) less flexible (*c*) equally flexible (*d*) none of them (*b*)

42. Compared to gravity concrete dams, rockfill dams are considered as .... flexible:

(*a*) more (*b*) less (*c*) equally (*d*) none (*a*)

43. The rockfill dams have foundation requirements .... than earthen dams:

(*a*) more flexible (*b*) less flexible (*c*) rigid (*d*) more rigid (*d*)

44. Compared to gravity dams rockfill dams have foundation requirements:

(*a*) strict and rigid (*b*) slightly strict and rigid

(*c*) not so struct and rigid (*c*)

45. The rockfull dam usually requires spillway of .... type:

(*a*) chute (*b*) shaft (*c*) sephon (*d*) side channel (*d*)

46. The upstream side slope of rockfill dam usually is:
(*a*) 3 : 1 (*b*) 2 : 1 (*c*) 1.5 : 1 (*d*) 1.3 : 1 & 1 : 1 (*d*)
47. The overall base width of a rockfill dam usually is .... times
(*a*) 4 to 5 (*b*) 3 to 4 (*c*) 2.5 to 3 (*d*) 1.5 to 2.5 (*c*)
48. Inverted filter for providing foundation drainage has: (*AMIE s*' 98)
(*a*) multilayers of soil particles of same permeability
(*b*) multilayers in which permeability increases from top to bottom
(*c*) multilayers in which permeability increases from bottom to top
(*d*) only one layer of soil (*c*)
49. The crest width of an earthen dam has .... bearning on the embankment or foundation stability: (*AMIE s*' 98)
(*a*) appreciable (*b*) no appreciable
(*c*) sometimes appreciable sometimes inappreciable
(*d*) none of the above (*b*)

18. GRAVITY DAMS

1. A gravity dam derives its stability from:
(*a*) reinforcement provided in the dam body
(*b*) weight of water resting against sloping upstream face of dam
(*c*) weight of dam itself
(*d*) batter provided on d/s face of the dam (*c*)
2. In a gravity, slightly curved in plan, the external forces are resisted by:
(*a*) reinforcement provided in the dam today
(*b*) weight of the dam (*c*) arch action
(*d*) none of them (*b*)
3. Gravity dams transfer load to foundation by:
(*a*) arch action (*b*) cantilever action
(*c*) arch and cantilever action both (*d*) none of them (*b*)
4. Gravity dams essentially require:
(*a*) strong and watertight abutments (*b*) strong foundation
(*c*) both of them (*d*) none of them (*c*)
5. Gravity dams are most suited across:
(*a*) high-wide gorges (*b*) narrow gorges with steep side slopes
(*c*) V-shaped gorge (*d*) low-wide gorge (*b*)
6. In a gravity dam, the quantity concrete is determined by:
(*a*) strength of concrete (*b*) dead weight of dam
(*c*) both strength and weight of concrete (*d*) none of them (*b*)
7. In a gravity dam:
(*a*) high strength concrete in central part of the dam is not optimally stressed
(*b*) lesser efficient use of concrete strength
(*c*) large uplift force on dam base (*d*) massive dam seciion for stability
(*e*) all of them (*f*) none of them (*e*)
8. The elementary profile of a gravity dam composed of a triangle with apex at the maximum water surface is .... are:
(*a*) ideal (*b*) practical (*c*) theoretical (*d*) none of them (*b*)
9. The economical top width of a gravity dam is usually .... percent of dam height:
(*a*) 22 (*b*) 18 (*c*) 14 (*d*) 10 (*c*)

10. Free board usually provided in a gravity dam usually is .... percent of dam height:
(*a*) 7 (*b*) 6 (*c*) 5 (*d*) 4 (*c*)
11. The base width of a gravity dam usually is:
(*a*) $H/\sqrt{\rho - c}$ (*b*) $H^{1/2}/\rho - c$ (*c*) $H/\rho - c$ (*d*) $H/(\rho - c)^2$
(H is dam height and c is coefficient of uplift pressure, ρ is unit weight of dam material) (*a*)
12. Galleries in gravity dams are provided for:
(*a*) drainage (*b*) inspection (*c*) drilling
(*d*) gates and control equipment operation (*e*) all of them (*e*)
13. The downstream slope is flattened and batter provided on upstreamside of a high dam to maintain:
(*a*) uplift within permissible
(*b*) resultant of all forces to pass through a lower middle-third point
(*c*) resultant nearer to the centre of base
(*d*) all of them (*c*)
14. Full use of structural strength of concrete is made in:
(*a*) gravity dam (*b*) arch dam (*c*) buttress dam (*d*) none of them (*c*)
15. Self weight of a gravity dam: (*AMIE s*' 97)
(*a*) favours stability (*b*) canses overturning
(*c*) causes sliding (*d*) causes uplift
16. If the eccentricity of the resultant falls outside the middle third, the ultimate failure of dam occurs by: (*AMIE w*' 01)
(*a*) tension (*b*) crushing (*c*) sliding (*d*) overturning
17. For triangular dam section of height H, for just no tension, under action of water pressure, self weight and uplift pressure, the minimum base width required with usual notion is .... (*AMIE s*' 2000)
(*a*) $H/S-1$ (*b*) $H/(S-1)^{-1}$ (*c*) $H/S^{1/2}$ (*d*) $H/(S-1)^{1/2}$
18. Drainage holes provided inside the body of a concrete gravity dam are located on the .... side of the grout curtain.
19. For triangular dam section of height H, for just no tension, under action of water pressure, self-weight and uplift pressure, the minimum base width with usual notation is
(*a*) $H/S-1$ (*b*) $H/(S-1)^{-1/2}$ (*c*) $H/S^{1/2}$ (*d*) $H/(S-1)^{1/2}$
20. A straight concrete gravity dam should have: (*AMIE w*' 00)
(*a*) strong abutments (*b*) clay foundations
(*c*) strong foundation (*d*) none of above (*c*)
21. The structure provided in a concrete dam to release excess water during floods is called a .... excess water during floods is called a .... (*AMIE w*' 00)
22. Passage provided in gravity dam to drain seepage water is called .... (*AMIE s*' 00)
23. For no tension at any point on a gravity dam, the criterion to be satisfied is: (*AMIE w*' 97)
(*a*) resultant of all the forces must always pass through the mid point of the base of the dam
(*b*) resultant force for all conditions of loading must pass through the middle third of the base
(*c*) resultant of all forces must pass through the upstream extensity of the middle third of the base
(*d*) resultant of all forces must always pass through the downstream extremity of the middle third of the base
24. A straight concrete gravity dam should have (*AMIE w*' 2000)
(*a*) strong abutments (*b*) clay foundations
(*c*) strong foundation (*d*) none of the above

25. The self weight of a gravity dam (*AMIE s*' 97)
    (*a*) favours stability (*b*) canses overturning
    (*c*) canses sliding (*d*) danses uplift (*a*)
26. In order to reduce uplift on a gravity dam, type of grouting done is: (*AMIE w*' 96)
    (*a*) curtain grouting near heel (*b*) curtain grouting near toe
    (*c*) consolidation grouting near heel (*d*) consolidation grouting near toe (*a*)

19. ARCH AND BUTTRESS DAMS

1. In an arch dam loads are resisted by:
   (*a*) weight of dam (*b*) arch action (*c*) both of them (*d*) none of them (*c*)
2. An arch dam depends principally on .... for its stability:
   (*a*) weight of dam (*b*) arch action (*c*) none of them (*b*)
3. The arch action in an arch dam results in:
   (*a*) massive dam section
   (*b*) economy in quantity of dam section material
   (*c*) wider base of dam (*d*) none of them (*b*)
4. An arch dam mostly transfer the water load :
   (*a*) vertically on foundation (*b*) horizontally to abutments
   (*c*) horizontally on foundation (*d*) vertically on abutments (*b*)
5. The usually recommended ratio of crest length to dam height of an arch dam is:
   (*a*) 10 (*b*) 9 (*c*) 7 (*d*) 5 (*d*)
6. Uplift on base of an arch dam is:
   (*a*) very important (*b*) important (*c*) not importance (*c*)
7. The suitable gorge for an arch dam is:
   (*a*) V-shaped wide (*b*) U-shaped wide
   (*c*) both U and V shaped narrow (*d*) both U and V shaped wide
   (*c*), particularly V-shaped
8. Constant radius arch dam is suitable for .... gorge:
   (*a*) C-shaped (*b*) D-shaped (*c*) V and U shaped (*d*) none of them
   (*c*) particularly U-shaped
9. Variable radius arch dam is suitable for .... gorge:
   (*a*) V-shaped wide (*b*) U-shaped wide
   (*c*) narrow V and U shaped (*d*) none of them (*c*)
10. Constant radius arch dam is .... in section:
    (*a*) trapezoidal (*b*) rectangular (*c*) triangular (*d*) none of them (*c*)
11. The top of dam preferred for a narrow gorge with strong abutments is: (Arch dam)
    (*AMIE w*' 98)

Choose the correct answer:

1. An arch dam is constructed with: (*AMIE s*' 97)
   (*a*) earth (*b*) concrete (*c*) boulders (*d*) none of above
2. A core wall is provided in:
   (*a*) earth dam (*b*) canal regulater (*c*) aqueduct (*d*) gravity dam (*a*)
3. The resultant of all forces must always pass through:
   (*a*) the mid point of the base of dam
   (*b*) resultant force for all conditions of all forces must always pass through the upstream extremity of the middle third of the base
   (*c*) the resultant of all the forces must always pass through the downstream extremity of the middle third of the base

(*a*) standard project flood (*b*) maximum known flood
(*c*) extraordinary flood (*d*) probable maximum precipitation flood

20. In the design of approach channel of a spillway the head loss through it is:
(*a*) no consequence (*b*) little consequence (*c*) important
(*c*) as it effects discharge capacity of the spillway

21. A high spillway is that the ratio of height of which measured from the river bed to the design head is greater than:
(*a*) 1 (*b*) 1.33 (*c*) 1.5 (*d*) 2.0 (*b*)

22. A low spillway is that the ratio of height of which measured from the river bed to the design head is less than:
(*a*) 1.0 (*b*) 1.33 (*c*) 1.5 (*d*) 2.0 (*b*)

23. In siphon spillway the operating head is .... than the head of water over the crest:
(*a*) same (*b*) higher (*c*) lower (*d*) all of them

24. The discharging capacity per unit length of siphon spillway is disproportionately .... than other types of spillways:
(*a*) lower (*b*) higher (*c*) same (*d*) none of them (*b*)

25. The dissipating arrangement below a spillway in sound rock foundation is usually:
(*a*) hydraulic jump stillway basin (*b*) ski-jump bucket
(*c*) well-stilling basin (*d*) all of them (*b*)

26. The dissipating arrangement below a spillway in alluvial soil foundation is usually:
(*a*) hydranlic jump stilling basin (*b*) upturned bucket
(*c*) ski-jump (*d*) all of them (*a*)

27. At a hydranlic jump, the depths at two sides are 0.4 m and 1.4 m. The head loss in jump is nearly .... m: (*Engg. Ser. Exam.* '97)
(*a*) 1.0 (*b*) 0.9 (*c*) 0.7 (*d*) 0.45 (*d*)

28. In a side channel spillway, the initial and final flows are approximately
(*a*) parallel (*b*) at right angle (*c*) straight (*d*) none of them (*b*)

29. In a side channel spillway, the flow after passing over ogce crest is carried away by a channel running essentially .... to the crest:
(*a*) at right angle (*b*) parallel (*c*) forward (*d*) backward (*b*)

30. The velocity distribution along the length of a side channel spillway is:
(*a*) regular (*b*) irregular
(*c*) sometimes regular sometimes irregular (*d*) none of them (*b*)

31. Tunnel spillay is eminently adaptable in .... carryons where there is danger to open channel from snow or rock slides:
(*a*) wide (*b*) narrow (*c*) steep (*d*) none of them (*b*)

32. Tunnel spillway may have .... discharge capacity:
(*a*) large (*b*) limited (*c*) intermediate (*d*) all of them (*b*)

33. Morning glory spillway is the other name of:
(*a*) tunnel spillway (*b*) shaft spillway (*c*) siphon spillway (*d*) none of them (*b*)

34. Shaft spillway can be adopted in .... carryons where the abutments rise steeply:
(*a*) wide (*b*) narrow
(*c*) wide as well as narrow (*d*) none of them (*b*)

35. A siphon spillway consists of two legged conduct formed in the shape of .... :
(*a*) L-tube (*b*) inverted U-tube
(*c*) inverted V-tube (*d*) all of them (*b*)

36. The two-legged content of a siphon spillway is essentially of:
(*a*) equal length (*b*) unequal length
(*c*) sometimes of equal and sometimes of inequal length (*d*) none of them (*b*)

37. Rordial gate is also known as: (*AMIE s*' 99)
(*a*) Taunter gate (*b*) stoney gate (*c*) drum cycrete (*d*) rolling gate (*a*)
38. An example of vertical lift gate is .... (slide gate) (*AMIE w*' 99)
39. River intake is usually situated on: (*AMIE s*' 99)
(*a*) convex side of curve (*b*) concave side of curve
(*c*) straight reach (*d*) d/s side of sever outfall (*a*)
40. A river intake may be situated on a: (*AMIE w*' 98)
(*a*) straight reach (*b*) convex side of a curve
(*c*) d/s side of outfall sever (*d*) proximity to industry (*b*)
41. Multiple-level port of intakes are usually provided in a reservoir to drain:
(*a*) sufficient quantity of water during all the seasons
(*b*) quanlity water obtainable from different levels during all the seasons
(*c*) comparatively warm and cold water during all the seasons
(*d*) none of the above (*b*)
42. In resverois with wide variations in water surface elevation, the intake is better located at .... stage so total one inlet is always submerged and operative and minimum head is always available:
(*a*) highest (*b*) intermediate (*c*) lowest
(*d*) between highest and intermediate (*c*)
43. For minimum silt entry from the reservoir, the intake port level is kept at .... :
(*a*) lowest points or pockets (*b*) highest points on pockets
(*c*) sufficiently above the reservoir bed (*d*) dead storage level (*c*)
44. The usual location of intake in an earthen dam is in .... water near upstream toe of dam:
(*a*) shallow (*b*) deep (*c*) neither shallow nor deep
(*d*) sometimes shallow sometimes deep (*b*)
45. In a reservoir with widey fluctuations in water level and/or to draw quality water, the intake is usually of .... type:
(*a*) submerged (*b*) tower (*c*) rock filled cribs
(*d*) none of the above (*b*)
46. For most design conditions usually .... percent of the transhracks are considered clogged:
(*a*) 70 (*b*) 60 (*c*) 50 (*d*) 30 (*c*)
47. The intake with ports submerged at all operating reservoir levels, there is .... risk of damage to trashracks due to clogging by debris or ice:
(*a*) always (*b*) frequently (*c*) rarely (*d*) little (*d*)
48. Large size port openings in reservoir intakes are provided to:
(*a*) draw maximum supply of water (*b*) drawn optimum supply of water
(*c*) minimise entrance velocity at trashrack as also to avoid difficiently with entrained air at low velocities
(*d*) all of them (*c*)
49. Small transhracks susceptible to clogging by ice formation in cold regions are provided with:
(*a*) electrical heating (*b*) compressed air (*c*) anti-freeze arrange (*d*) all of them (*a*)
50. Large transhracks susceptible to clogging by ice formation in cold regions are provided with:
(*a*) electrical heating (*b*) compressed air
(*c*) anti-freeze arrangment (*d*) all of them (*b*)
51. The required area of trashrack is governed by:
(*a*) withdrawal discharge through intake (*b*) limiting velocity through rock
(*c*) discharge as well as velocity (*d*) none of them (*b*)

21. WATER POWER ENGINEERING

1. The primary power at a hydro plant is determined considering:
   (*a*) minimum stream flow (*b*) available pondage
   (*c*) probable load curve (*d*) all of these (*d*)
2. Maximum of primary power is obtained by operation of storage by:
   (*a*) yearly-use method (*b*) safe yield method (*b*)
3. Indicate method of studying effect of storage on a power plant situated:
   (*a*) at reservoir (*b*) at some distance from reservoir.
   (*a*) mass curve method
   (*b*) hydrograph method with accompanying reservoir depletion curve.
4. Wheel capacity and available power is best determined by:
   (*a*) flow-duration curve (*b*) hydrograph
   (*c*) mass curve (*a*)
5. For economical location penstock is laid:
   (*a*) on grade varied as per topography
   (*b*) on as flat grade as possible consistent with keeping below hydraulic gradient for greater part of its length and sudden drop near power plant.
   (*c*) on as steep grade as possible consistent with keeping below hydraulic gradient and sudden drop near power plant (*b*)
6. The simplest method to eliminate positive water-hammer pressure is to provide–to take the rejected flow:
   (*a*) simple surge tank at upper end of penstock
   (*b*) pressure relief valve at turbine
   (*c*) none of them (*d*) both of them (*d*)
7. Pressure relief is not effective for:
   (*a*) positive water hammer (*b*) negative water hammer
   (*c*) none of these (*d*) both of these (*b*)
8. A high ground is available close to a power house for the location of surge tank. Indicate the type of tank you will prefer:
   (*a*) simple surge tank (*b*) tower type tank (*c*) restricted tank
   (*d*) tank resting on ground containing penstock and an orifice for inflow and outflow provided on top of conduit (*d*)
9. Substructure of a power house should provide space for:
   (*a*) turbine governors (*b*) generators for horizontal units
   (*c*) switch board (*d*) low tension switches and other auxiliaries
   (*e*) all (*e*)
10. In out door type power plant:
    (*a*) substructure is omitted (*b*) superstructure is omitted
    (*c*) intermediate structure is omitted (*d*) none is omitted (*b*)
11. Substructure for an impulse turbine is relatively:
    (*a*) simple (*b*) complicated
    (*a*) because turbines are small in size and draft tubes are not there to be provided.
12. Transmission line is a line meant for:
    (*a*) transmission of water from reservoir to power plant
    (*b*) transport of power from generation source to load centre
    (*c*) transmission of power from one load centre to another.
    (*b*) and (*c*)

13. For large blocks of power and long distances, the cost per kW transmitted is less at:
(*a*) lower voltage (*b*) higher voltage
(*c*) not related to voltage (*d*) maximum number of circuits (*b*)
14. High voltage for transmission is used to:
(*a*) transmit large amount of power over long distances
(*b*) transmit large amount of power over short distances
(*c*) transmit small amount of power over long distances
(*d*) large terminals on tap-off stations are involved (*a*)
15. Load on a turbine is increased suddenly or a great amount at a time:
(*a*) usually (*b*) seldom (*b*)
16. The function of surge tank is to:
(*a*) avoid flow in reverse direction (*b*) smoothen the flow
(*c*) act as a reservoir for emergency condition
(*d*) relieve the pipe line of excessive pressure transients. (*d*)
17. In the determination of area of surge tank factor of safety of 2.0 and 1.6 is applicable to:
(*a*) simple (*b*) restricted (*c*) differential 2.0 (*a*), 1.6 (*b*) & (*c*)
18. Classify reservoir type power intakes with respect to head :
(*a*) 50 m (*b*) 15-30 m (*c*) 15 (*a*) high (*b*) medium (*c*) low head
19. Indicate type of hydro plant:
(*a*) up to 100 kW (*b*) up to 2 MW
(*c*) up to 25 MW (*a*) micro (*b*) mini (*c*) small
20. A small hydro plant has firm discharge as average of the lean period inflows in a dependable year:
(*a*) 100% (*b*) 90% (*c*) 80% (*d*) 75% (*d*)
21. Tidal plant can give only ____________ supply of power. (*intermittent*)
22. Tidal plant is not suitable for ____________ area. (*plain*)
23. In a hydro power project, reservoir is allowed to be depleted up to:
(*a*) dead storage level (*b*) minimum drawdown level
(*c*) flood storage level (*d*) normal reservoir level (*b*)
24. Dependability of live storage in a power project is:
(*a*) 100% (*b*) 90% (*c*) 75% (*d*) 60% (*b*)
25. Dependability of a run-of-river power project is:
(*a*) 100% (*b*) 90% (*c*) 75% (*d*) 60% (*b*)
26. Storage capacity of a reservoir is determined by:
(*a*) mass curve method (*b*) ratio method
(*c*) calendar year method (*a*)
27. Trap efficiency of a reservoir indicates:
(*a*) water trapped (*b*) sediment retained
(*c*) sediment inflow (*d*) sediment outflow (*b*)
28. Trap efficiency is low for a reservoir with:
(*a*) small capacity-inflow ratio (*b*) large capacity-inflow ratio (*a*)
29. Provision for sedimentation (ha m/yr/sq km of C.A) made as per ISI recommendation is:
(*a*) 0.01 to 0.02 (*b*) 0.1 to 0.2
(*c*) 0.15 to 0.25 (*d*) 0.25 to 0.35 (*a*)
30. Evaporation from reservoir surface is usually the highest in the month of ----------, give reason.
(*a*) April (*b*) May
(*c*) June (*d*) July
(*b*) because in June water spread is the lowest, being the pre-filling month.

31. V-shaped narrow valley is suitable for:
(*a*) Rockfill dam (*b*) Overflow concrete dam
(*c*) Arch dam (d) Earthen dam (*c*)

32. Indicate classification of the dams:
(*a*) > 15 m high (*b*) 12 m high, crest length > 500 m.
(both high dams)

33. Surge tank is necessarily provided in:
(*a*) long penstocks (*b*) short length penstocks
(*c*) surface penstocks (*d*) embedded penstocks (*a*)

34. Surge tank provision is necessary in tailrace consisting of:
(*a*) open channel (*b*) long pressure tunnel
(*c*) short length pressure tunnel (*d*) long free flow tunnel (*b*)

35. Water hammer effect vis-a-vis turbine operation is headrace and tailrace tunnels is:
(*a*) indentical (*b*) reverse (*b*)

36. Pressure release valve can cater for:
(*a*) turbine gate closure and consequent rejecting surplus water
(*b*) gate opening and supplying additional water
(*c*) negative water pressure (*d*) positive water pressure (*a*) and (*d*)

37. Location of surge tank is the best where ground:
(*a*) drops rapidly towards power house (*b*) rises rapidly towards power house
(*c*) midway in rising and dropping ground (*a*)

38. Function of a surge tank is to:
(*a*) Store water on load rejection
(*b*) furnish additional water during increased load damand
(*c*) both to store and supply water (*d*) none (*c*)

39. Maximum pressure occurs in full flowing-power tunnel at the time of:
(*a*) load rejection (*b*) load acceptance (*a*)
(*a*) load rejection (*b*) load acceptance (*b*)

40. Minimum pressure occurs in full flowing power tunnel at the time of:
(*a*) load rejection (*b*) load acceptance (*b*)
(*c*) Headrace (*d*) Tailrace (*a*)

41. In hydraulic tunnels located near mountain slopes, the governing criteria is:
(*a*) lateral cover (*b*) vertical cover (*a*)

42. Indicate tunnel sections excavated through:
(*a*) hard rock (*b*) loose and weak rock
(*a*) arched roof
(*b*) circular or horse shoe section

43. Indicate tunnel shape from construction point of view for
(*a*) smaller tunnel (*b*) large tunnel
(*a*) D–shape
(*b*) Horse shoe

44. Maximum flow occurs in–when ratio of depth to diameter is:
Circular section (*a*) 0.90 (*b*) 0.92 (*c*) 0.93 (*d*) 0.94 (*c*)
Horse shoe section (*a*) 0.90 (*b*) 0.92 (*c*) 0.93 (*d*) 0.94 (*d*)

45. Usual velocity (m/s) in–power tunnels is:
(*a*) 3 (*b*) 4 (*c*) 5 (*d*) 6 (*d*)
(*a*) 6 (*b*) 7 (*c*) 8 (*d*) 9 (*d*)
(*a*) 5 (*b*) 6 (*c*) 7 (*d*) 8 (*d*)

46. The load factor for the peak day of the year determines the required:
(*a*) generating capacity (*b*) water storage (*c*) pondage (*a*)

47. The requirement of peak week or month determines the amount of:
(*a*) water storage (*b*) generating capacity (*c*) none of the two (*a*)
48. Load factor for a power system serving highly industrial area vis-a-vis residential area is:
(*a*) same (*b*) high (*c*) low (*b*)
49. Load factor for a power system serving residential area compared to highly industrial area is:
(*a*) high (*b*) same (*c*) low (*c*)
50. If load factor is high, the unit cost of energy is comparatively:
(*a*) high (*b*) low (*c*) same (*b*)
51. With high load factor the power system is operating at:
(*a*) near capacity (*b*) near best capacity (*c*) idle for large part of time
(*a*) & (*b*)
52. During high and low stream flow, the functions of thermal and hydro plants are sometimes:
(*a*) the same (*b*) reversed (*b*)
53. The optimum point of operation in a combined system usually occurs when water power capacity is somewhere between ________ of total capacity.
(*a*) one-half to one-third (*b*) one-third to two-third
(*c*) one-half to three-fourth (*b*)
54. A system supplying hydro power alone is at a disadvantage if its carry-over storage is _____ to meet demand of a severe drought
(*a*) adequate (*b*) inadequate (*c*) optimum (*d*) surplus (*b*)
55. The peak load of a 40 MW capacity power plant is 30MW. The loads having maximum demands of 25, 10, 5 and 7 MW are connected to the system. The diversity factor of the plant is __________ . (1.57)
56. In (3) if the load factor for the year is 50%, the energy supplied by the system is __________ kWh/year. (205.86×10^6
57. The fact that a conductor carries more current on the surface as compared to core is known as fault.
(*a*) skin effect (*b*) corona
(*c*) permeability (*d*) unsymmetrical fault (*a*)
58. Presence of ozone as a result of corona is harmful because:
(*a*) it gives bad odour (*b*) it corrodes the material
(*c*) it transfers energy to the ground (*d*) reduces power factor (*b*)
59. The function of steel wire in a ACSR conductor is:
(*a*) take care of surges (*b*) prevent corona
(*c*) reduce inductance (*d*) provide additional mechanical stength (*d*)
60. A 30 km transmission line carrying power at 33 kV is known as:
(*a*) short transmission line (*b*) medium transmission line
(*c*) high power line (*d*) ultra high voltage line (*b*)
61. The permissible voltage variation in a distribution system at consumer end is:
(*a*) ± 5% (*b*) ± 6%
(*c*) ± 7% (*d*) ± 10%
62. The sag of a transmission line is least affected by:
(*a*) self weight of conductors (*b*) temperature of surrounding air
(*c*) current through conductor (*d*) ice deposited on conductor (*c*)
63. A pumped storage hydroelectric plant is a:
(*a*) high head plant (*b*) run-off river plant
(*c*) base load plant (*d*) peak load plant (AMIES' 01) (*d*)

64. The correct sequence, in the direction of flow of water, of installations in a hydro power plant is:

(*a*) reservoir, penstock, turbine, surge tank

(*b*) reservoir, surge tank, penstock, turbine

(*c*) reservoir, penstock, surge tank, turbine

(*d*) reservoir, surge tank, turbine, penstock (AMIES '01) (*b*)

65. Surge tank in hydropower plants are provided for:

(*a*) controlling surge in river (*b*) controlling water hammer in penstock pipes

(*c*) constant generation of power (*d*) removal of silt from reservoir (AMIES '98) (*b*)

66. In hydel installation power generation is calculated on the basis of:

(*a*) gross head (*b*) rated head

(*c*) design head (*d*) net head (AMIES '97) (*d*)

67. In a hydro power scheme, the pipe between tank and power house is known as ....

(AMIES '01) (Penstock)

68. Run off river plant is one which:

(*a*) has a limited pondage (*b*) has no pondage at all

(*c*) is a high head plant (*d*) is a pumped storage plant (AMIES '99) (*a*)

69. A medium head hydroelectric scheme is one where:

(*a*) gross head is between 15 to 60 m

(*b*) net head operating on turbine is between 15 to 60 m

(*c*) total loss of head in penstock is between 15 to 60 m

(*d*) difference in level of tailwater and centre line of turbine is between 15 to 60 m. (*b*)

(AMIE W'95)

70. Match the following:

(*i*) spiral casing	(*a*) are employed to regulate flow into the runner and to direct the water onto the runner at an appropriate angle.
(*ii*) Guide vanes	(*b*) transforms the major part of the exit kinetic energy into pressure energy and thereby improves the turbine efficiency
(*iii*) Draft tubes	(*c*) transforms hydraulic energy into mechanical energy
(*iv*) Runner	(*d*) distributes the flow over the periphery of the runner so that the entrance velocity is same at inlet tips.
(*v*) Penstock	(*e*) the flow direction is changed from radial to axial direction.
(*vi*) Whirl chamber	(*f*) Water Conveying Conduit

(AMIE W' 01)

BIBLIOGRAPHY

A. ARTICLES AND BOOKS

Abbet, Robert W. (1956). American civil engineering practice, Vol. I to III. New York: John Willey and Sons.

Ahmed, Nazir (1951). Lining of channels, and short review of laboratory testscarried out on various types of lining material, Proc. 4th meeting IAHR.

Ahuja, D.R. and Mehndirata, K.R. Canal lining — a review, CBIP Publication No. 82.

Allende, J.L. (1978). Dam drainage techniques and skills, 10th Congress Irr. and Drg. trans. Vol. IV.

Arakeri, H.R., Donahue, Roy (1984). Principles of soil conservation and water management, M/s Oxford & IBH Co. New Delhi.

Arora K.R. (1996). Irrigation, water power and water resources engineering: Standard Publishers distributors, 1705 B, Nai Sarak, Delhi-110006.

Bharat singh. (1975). Irrigation Engineering. Roorkee: Nem Chand and Sons.

Biswas, K. (1980). Environmental assessment of hydel power projects. Bhagirath.

Brandley, J.N. and Peterker A.J. (1957). Hydraulic design of stilling basins. J. Hydraulic division, ASCE.

Chaturvedi, A.C. (1977). Water quality and availability in India. Third international hydrology symposiun, Fort collins, Colorado.

Chaudhari, N.M. (1980). Role of evaporation in consumptive use computation. Journal of Indian Institution of Hydrologists, IV, 3 and 4, p.9.

Chaudhury, T.K. (1979). Use of unconventional materials in the construction of Nurpur spurs, river Hooghly, Journal of I and P, 36, pp. 183, 186.

Chow, V.T. (1959). Open channel hydraulics. New York: McGraw-Hill Book Co.

Comrie, j. (1961). Civil engineering reference book, Vols. I, II, III and IV, London: Butterworths.

Dakshinamurthi, C. (1972). Soil management for efficient water use. Indian Farming.

Datar, S.V., Mohammed, P. and Mani, A. (1967). An automatic radio-reporting raingauge system. Journal of I and P, 24, 2, pp. 145-148.

Datta, O.P. (1976). Design features of cross drainage works on Sunder nagar hydel channel. Indian Journal of Power and River Valley Development, XXVI, 4, p. 96.

Davis, C.V. and Sorenson, K.E. (1969). A handbook of applied hydraulics. New York: McGraw-Hill Book Co.

Deshmukh, M.B. Punjabi R.M., Paseband, S.P. (1987). Structures irrigation and conservation of water resources. First National Water Conservation. Vol. III CBIP, New Delhi.

Dhillon, G.S. (1967). Estimation of seepage losses from channels. Indian Journal of Power and River Valley Development, p. 16.

Dhillon, G.S. and Gurdial Singh (1981). History of evolution of tubewell filters in Punjab region. Institution of Engineers (India) sponsored Seminar on Water Resources and Its Development and Management, Pre-session Proc. pp. 9-24.

Diwan, P.L.(2000). Sustenance of irrigated agriculture and drainage.

Drake, A.W. (1967). Fundamentals of applied probability theory, New York: McGraw-Hill Book Co.

Eakin, H.A. (1939) Silting of reservoirs. Tech. Publication No. 524, U.S. Deptt. of Agriculture.

Einstein, H.A. (1940). Suspended load in streams. Trans. American Geo-physical Union, Vol. 21.

Fair, G.M. and Greyer, J.C. (1961). Elements of water supply and waste-water disposal. New York: John Wiley and Sons.

Gales, R. (1938). The principles of river training for railway bridges and their application on the Hardinge Bridge over lower Ganga at Sara. Govt. of India Technical paper No. 5167.

Garg, S.K. (1977). Water resources and hydrology. Delhi: Khanna Publishers.

Garg, S.K. (1978). Irrigation engineering and hydraulic strucrtures. Delhi: Khanna Publishers.

Garg, S.P. (1977). Use of vegetation and bamboos in river training works. I and P. Journal, 34.

Garg, S.P. Asthana, B.N. and Jain, S.K. (1971). Design of guide banks for alluvial rivers, Journal of Institution of Engineers (India) 52, pp. 24-28.

Goel, R.S. (2000). Environment management in hydro power: M/s Oxford IBH Publishing CO, New Delhi.

Goel, R.S. (2000). Environmental impacts assessment of water resources projects: M/s Oxford & IBH Publishing CO, New Delhi.

Goel, R.S. (1999). Hydro power and river valley development: M/s Oxford & IBH Publishing CO. New Delhi.

Goldberg and Gornat, B. (1972). Trickle irrigation, a new approach to irrigation totally dependent on plastic polymer materials. 8th Congress. Trans. Vol. III. International Commission on Irrigation and Drainage.

Golze, A.R. (1977). Handbook of dam engineering, New York: Van Nostrand Rheinhold Co.

Govardhan Lal (1964). Design of small bridges and culverts, New Delhi: CWC Publication.

International Symposium on reclamation and management of salt affected soils (pp. 121 & 122).

Gulati, T.D. (1969). Training of rivers. Bhagirath, XV, pp. 88-91.

Inglis, C.C. (1949). The behaviour and control of rivers and canals. Research Publication No. 13, Part II. Central Water Irrigation and Navigation Research Station, Poona.

Jogleker, D.V. (1972). Manual on river behaviour, control and training. New Delhi: C.B.I. and P.

Jogleker, D.V. (1961). Density currents in reservoirs and the simulation in models. Journal of Institution of Engineers (India). 42, I part CII, p.21.

Kanwar Sain (1957). Canal lining in India. Trans. ICID, 3rd Congress on Irr. and Drg., New Delhi, II, Q.R. 11, pp. 145.

Kapoor, A.S. (2000). Bio-drainage-feasibility and planning aspects.

Khanna, P.N. (1971). Indian practical civil engineers handbook, New Delhi: Engineer's Publishers.

Khosla, A.N. (1939). Silting of reservoirs. New Delhi: C.B.I.P. Publication.

Khosla, A.N., Bose, N.K. and Taylor, E.M. (1962). Design of weirs on permeable foundations. New Delhi: C.B.I.P.

Khushalani, K.B. and Khushalani Manohar (1984) Irrigation practice and design, Vol I to III, New Delhi, Oxford & IBH Publishing Co.

Leliavsky, S. (1965). Design textbook in civil engineering, Vols. I, II, III, IV and V. New Delhi: Oxford & IBH Publishing Co.

Linsley, R.K. and Franzini, J.B. (1979). Water resources engineering. New York: McGraw-Hill book Co.

Linsley, R.K., Kholer, M.A. and Paulhus, J. (1975). Hydrology for Engineers, New York: McGraw-Hill Book Co.

Linsley, R.K., Kholer, M.A. and Paulhus, J. (1949). Applied hydrology, New York: McGraw-Hill Book Co.

Mehbub, S.I. and Gulati, N.D. (1951). Irrigation outlets. Delhi: Atma Ram and Sons.

Michael, A.M. (1978). Irrigation theory and practice. New Delhi: Vikas Publishing House.

Montague, A.M.R. (1953). Irrigation canal falls. C.B.I.P. Publication.

Moorthy, R.S.N. (1970). Type design of irrigation structures. Satyavelu, R. Raghunatha Swamy.

Murthy, Y.K. (1976). Environmental impacts of water development projects, Journal of I. and P., 33, 1, pp. 17-19.

Naegamvala, J.P. (1974). Water resources development, Bhagirath, XXI, pp.40-41.

Nag, B.S. and Kathpalia G.N. (1975). Water resources of India. Proc. of 2nd World Congress on Water Resources, Vol. II, International Water Resources Association, New Delhi.

Narasimham, R.K.V. (1955). Energy dissipation below spillways. Journal of Institution of Engineers (India), pp. 495-515.

Naidu, B.S.K. (1999): Strategy for growth and environmental sustainability of hydro power projects in India.

Mohile, A.D. (1999). River valley development – A comprehensive approach.

Olivier, H. (1972). Irrigation and Water Resources Engineering. London: Edward Arnold.

Padhye, M.G. (1981). Water Resources development — some aspects of research and development. Journal of Institution of Engineers (India), 62, Part C 13, pp. 169-170.

Patel, C.C. (1980). National perspective for water resources development. Water World, Vol. III. pp. 89.

Patel, C.C. (1980). Storage reservoirs in India, Bhagirath, XXVII, pp. 3-5.

Patel, C.C. (1977). Institutional arrangements for water planning and management in India. World Congress held in Argentina.

Priyani, V.B. (1967). The fundamental principles of irrigation and water power. Anand; Charotar Book Stall.

Punmia, B.C. and Pande, B.B.L. (1979). Irrigation and water power engineering. Delhi: Standard Publishers.

Raj, K.P. (1981). National perspective for water resources development — an agricultural engineers view. Water World, IV, I, pp. 32.

Rao, G.V.K. (1980). Irrigation development in India — Tasks for future. Bhagirath, July-Sept. 1980.

Rao.J. Viswesweria Rao; The theory and applications of flow nets.

Rao, K.L. (1981). National water plan for drought areas. Water World, III.

Rao, K.L. (1975). India's water wealth. New Delhi: Orient Longman.

Rolston, D.E. (1972). Soil-water-plant relationship in water management. Indian Farming, pp. 46-47.

Sahasrabudha, S.R. (1978). Irrigation engineering and hydraulic structures. Ludhiana: Katson Publishing House.

Sharma, H.D. (1995). Embankment dams: M/s Oxford & IBH Publishing Co. New Delhi.

Sharma, H.D. and Chawla, A.S. (1977). Manual on ground water and tubewells. Tech. Report No. 18. New Delhi. C.B.I. and P. Publication.

Sharma, K.R. (1959). Irrigation engineering, Vols. I, II, III. Jullundur: India Printers.

Sharma, R.K. (1979 through 2001). A textbook of hydrology and water resources. New Delhi: Dhanpat Rai Publication Pvt. Ltd., Daryaganj, Delhi.

Sharma, R.K. (1984). Text book of irrigation engineering and hydraulic structures: M/s Oxford & IBH Publishing Co., New Delhi.

Sharma, R.K. Text book of irrigation engineering Vol. I, II, III. (1990-1992): M/s Oxford & IBH Publishing Co., New Delhi.

Sharma, R.K. et.al. Socio-economic aspects of water management at Beas Dam — a case study, presented at 64th R&D Session CBIP, 1988.

Sharma, R.K. et.al. Accelerated hydropower development — international conference on hydro power (1998).

Sharma, R.K. and Gupta, G.D. Watershed management — international conference on watershed management (1998).

Sharma, R.K. Hydrological impacts of Pong reservoir — international conference on emerging trends in hydrology.

Sharma, R.K. Beas water and pisciculture development — H.P. Fishery Deptt. (1999) Roorkee (1997).

Sharma, R.K. Hydrometeorological instruments along Pong reservoir — CBIP, ICOLD, BBMB symposium (1997).

Sharma, R.K. et.al. Capacity evaluation of Pong reservoir — CBIP, ICOLD, BBMB seminar (1996).

Sharma, R.K. et.al. Sediment measurement at Pong reservoir — CBIP, R&D session, 1993.

Sharma, R.K. et.al. Dam safety at Pong — an appraisal — CBIP, R&D session, 1991.

Sharma, R.K. et.al. Monitoring environmental impacts of Pong reservoir — CBIP, R&D session, 1992.

Sharma, R.K. Water conservation — need of day — BBMB water day, 1991.

Sharma, R.K. et.al. Environment and reservoir silting — BBMB seminar, 1990.

Sharma, R.K. et.al. Sustainable economic development by Pong dam — a case study-Wapcos Institution of engineers, 1999.

Sharma, R.K. et.al. Water quality of Pong reservoir — a case study-National symposium, (1990).

Sharma, R.K. et.al. Sediment measurement & Control at Pong reservoir, a case study, presented in Indo-British Workshop on Sediment measurement and control, Chandigarh, India, 1988.

Sharma, R.K. et.al. Hydro power – its expeditious development, presented in First National Water Convention, New Delhi (1987).

Sharma, R.K. Lecture on environmental impacts of Bhakra-Beas system. Wapcos, Institution of engineers (India), New Delhi (2000).

Sharma, R.K. et.al. Sedimentation in Pong reservoir — a case study, presented in 53rd R&D Session of CBIP, 1986.

Shri Mohan (1972). Irrigation methods. Indian Farming, pp. 50-54.

Singal, A.M., Gupta, R.K. (1980). Canal lining in closure.

Thatte, C.D. (2000). Challenges in development and management of water resources.

Tanwar, B.S. (1980). Electric logging techniques applied to ground water exploration and water wells. I. and P. Journal.

Tideman, E.M. (1998). Watershed Management — Guidelines for Indian Conditions. New Delhi: Omega Scientific Publications.

Urquhart, L. (1959). Civil engineering handbook. New York: McGraw-Hill Book Co.

Varma, C.V.J. (1978). Efficiency in the use and reuse of water in India.

Varshney, R.S., Gupta, S.C. and Gupta, R.L. (1972). Theory and design of irrigation structures, Roorkee: Nem Chand and Bros.

Withers, B. and Stanley, (1974). Irrigation design and practice. London: B.T. Batsford Ltd.

Wu, I. Pai, and Giltru, H.M. (1974). Drip irrigation design, based on uniformity, Trans. ASCE, 17, pp. 157-158.

Yogindra Prasad (1999): Strategies and experiences of Hydro Power Development in India.

B. MANUALS AND REPORTS

P.W.D. Handbook Bombay (1960). Vols. I and II. Government of Maharashtra Publication.

Water (1955). United States Deptt. of Agriculture. New Delhi: Oxford & IBH Publishing Co.

Hydraulics of Alluvial Streams (1974). Status report, Govt. of India Publication. pp. 10, 11, 77.

Water Measurement manual. USBR Publication.

Symposium of canal lining (1968). Central Board of Irrigation and Power Publication No. 82.

Design practices of irrigation canals in the world (1972). New Delhi: International Commission on Irrigation and Drainage.

Design of small dams (1974). USBR Publication, New Delhi: Oxford & IBH Publishing Co.
Drainage Manual (1978). USBR Publication, New Delhi: Oxford & IBH Publishing Co.
Development of Irrigation, Drainage & flood control in India (1991). CWC Publication.
Symposium on waterlogging, causes and measures for its prevention (1972). Vols. I and II. C.B.I. P. Publication 46, 118. 5th Annual Board Session.
Ministry of Irrigation, Govt. of India (1980). National perspective for water resources development in India. Water World, Vol. III.
Manual of irrigation practice, Vol. II. Punjab Irrigation Publication.
Premier Irrigation Equipment Pvt. Ltd. Calcutta (1981). Publication Nos., S-12 Issue 2, No. S-9, Issue no. 5, No. S-19, Issue 1, S-20 Issue No. 2, S-10, Issue 4.
Sprinkler irrigation system, Jindal Aluminium Ltd. Bangalore.
Saline water conservation (1960). Collection of Papers, Symposium.
Embankment manual management by C.W. and P.C.
Total catchment management (1987). CBIP, New Delhi.
Development of economic alternatives to stone for river protection works (1970). C.B.I. and P. Tech. Report No. 5.
Earth Manual (1957). U.S.B.R. Publication.
Concrete Manual (1956). U.S.B.R. Publication.
A guide for estimating water requirement of crops — Water Management Division of Department of Agriculture, New Delhi.
Manual on barrages and weirs on permeable foundation, Vol. I and II. (1985). CBIP Publication no. 179.

C. BUREAU OF INDIAN STANDARDS SPECIFICATIONS

IS: 1893–1975 Criteria for earthquake resistant design of structures:
IS: 3872–1966 Code of practice for lining of canals with burnt clay tiles:
IS: 3873–1966 Code of practice for laying in situ cement concrete lining on canals:
IS: 4410 (Part III, VIII, IX, XVII): Glossary of terms relating to river valley projects, dams and spillway.
IS: 4558–1995 Under drainage for lined canals:
IS: 4701–1968 Code of practice for earthwork gauge on canals:
IS: 4987–1994: Establishing network of rainguage stations:
IS: 5477 (Part–I, II and III): Fixing capacity of reservoirs:
IS: 5968–1987: Planning and layout of canal system:
IS: 6004–1971: Criteria for hydraulic design of sediment ejector:
IS: 6531–1994: Canal head regulator.
IS: 8826–1978: Design of large earth and rock fill dams:
IS: 6939–1992: Evaporation from reservoirs:
IS: 6966–1973: Criteria for hydraulic design of barrages and weirs:
IS: 6986 (Part I) –1989: Hydraulic design of barrages and weirs:
IS: 7784 (Part–I) – 1995: Code of practice for designs of cross drainage works:
IS: 7894–1975: Code of practice for stability analysis of earth dams:
IS: 8389–1983: Installation and use of reservoir.
IS: 8408–1994: Criteria for river training works for barrages and weirs:
IS: 8444–1977: Design of under–seepage control measures for earth and rockfill dams:
IS: 4515–1993: Stone pitched lining for canals.
IS: 9452 (Part I, II and III) –1998: Measurement of seepage losses from canals.
IS: 71114–1973: Hydraulic design of cross regulator for canals:
IS: 10430–1982: Design fill dams type of lining:
IS: 9429–1980: Drainage system for earth and rock fill dam of:

IS: 12379–1988: Lining of water resources and field channels:
IS: 10907–1984: Design of farm drainage tile or pipe system:
IS: 1266–1993: Fixing intensity of irrigation:
IS: 7986–1993: Canal outlets:
IS: 6522–1972: Design of silt vanes and sediment control in offtaking canals:
IS: 10646–1991: Cement concrete lining:
IS: 8237–1985: Protection of slope for reservoir:
IS: 8835–1978: Planning and design of surface drains:
IS: 9447–1980: Assessments of seepage losses from canals:
IS: 10635–1993: Free board requirement in embankment dams:
IS: 7323–1994: Operation of reservoirs:
IS: 11130–1984: Structural design of barrages and weirs:
IS: 6572–1998: Design of solid gravity dams:
IS: 10799 (Part I) – 1984: Design and installation of trickle irrigation systems: